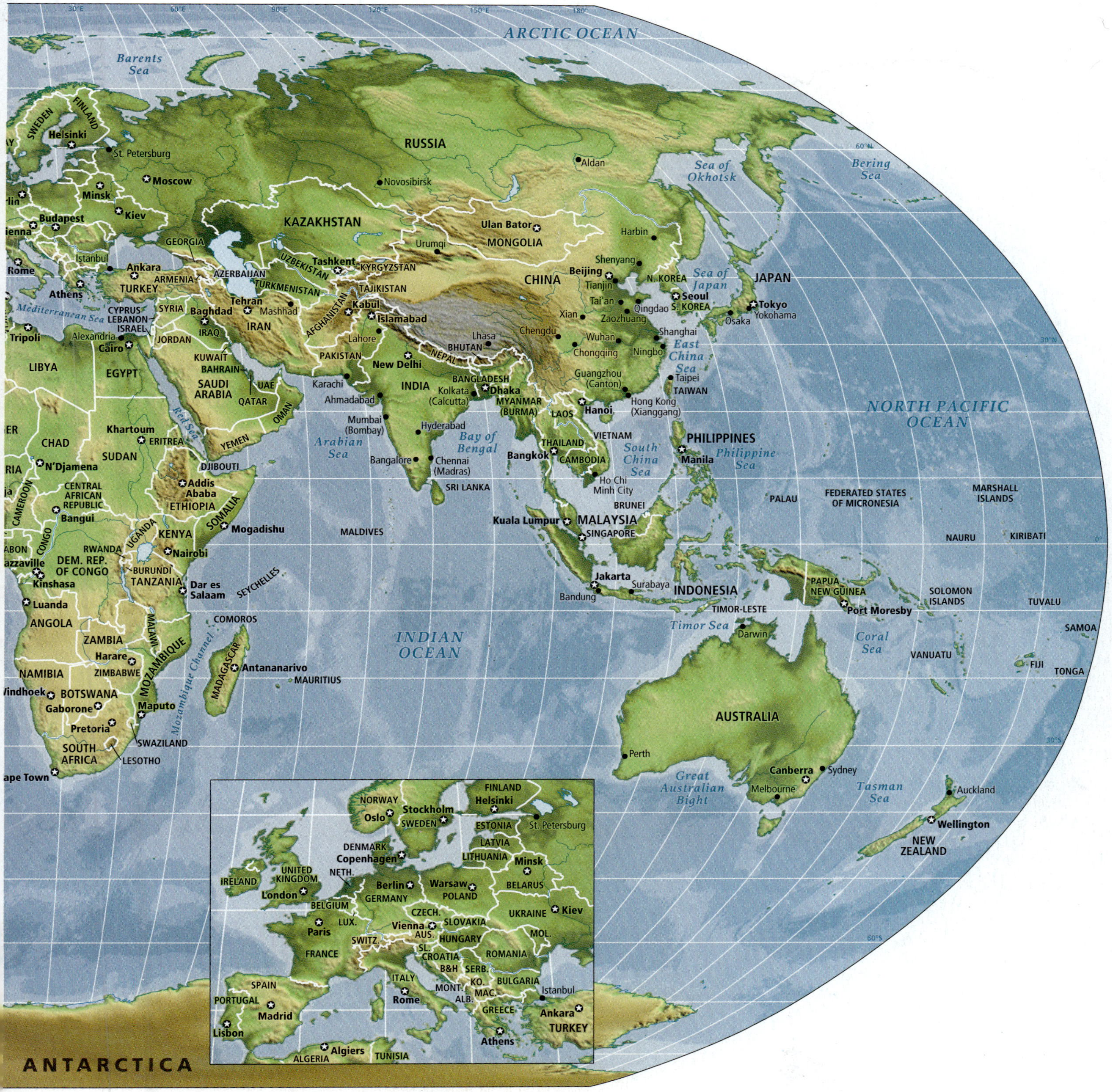

ARCTIC OCEAN
Barents Sea
RUSSIA
FINLAND
SWEDEN
Helsinki
St. Petersburg
Moscow
Minsk
Kiev
Budapest
Novosibirsk
Aldan
Sea of Okhotsk
Bering Sea
KAZAKHSTAN
GEORGIA
Istanbul
Ankara
ARMENIA
AZERBAIJAN
UZBEKISTAN
TURKMENISTAN
Tashkent
KYRGYZSTAN
TAJIKISTAN
Urumqi
Ulan Bator
MONGOLIA
Harbin
Shenyang
Beijing
Tianjin
Tai'an
CHINA
N. KOREA
Seoul
S. KOREA
Sea of Japan
JAPAN
Tokyo
Yokohama
Osaka
Qingdao
Zaozhuang
Xian
Chengdu
Wuhan
Chongqing
Shanghai
Ningbo
East China Sea
Taipei
TAIWAN
Guangzhou (Canton)
Hong Kong (Xianggang)
Lhasa
BHUTAN
NEPAL
Rome
Athens
TURKEY
Mediterranean Sea
CYPRUS
LEBANON
ISRAEL
SYRIA
Baghdad
IRAQ
Tehran
Mashhad
IRAN
AFGHANISTAN
Kabul
Islamabad
Lahore
PAKISTAN
New Delhi
INDIA
Karachi
Ahmadabad
Mumbai (Bombay)
Hyderabad
Bangalore
Chennai (Madras)
SRI LANKA
Arabian Sea
Bay of Bengal
BANGLADESH
Kolkata (Calcutta)
Dhaka
MYANMAR (BURMA)
LAOS
Hanoi
THAILAND
Bangkok
CAMBODIA
VIETNAM
Ho Chi Minh City
South China Sea
PHILIPPINES
Manila
Philippine Sea
NORTH PACIFIC OCEAN
PALAU
FEDERATED STATES OF MICRONESIA
MARSHALL ISLANDS
NAURU
KIRIBATI
BRUNEI
Kuala Lumpur
MALAYSIA
SINGAPORE
Jakarta
Bandung
Surabaya
INDONESIA
TIMOR-LESTE
PAPUA NEW GUINEA
Port Moresby
SOLOMON ISLANDS
TUVALU
SAMOA
VANUATU
FIJI
TONGA
Coral Sea
Timor Sea
Darwin
AUSTRALIA
Perth
Great Australian Bight
Canberra
Sydney
Melbourne
Tasman Sea
Auckland
Wellington
NEW ZEALAND
Tripoli
Alexandria
Cairo
JORDAN
KUWAIT
BAHRAIN
SAUDI ARABIA
QATAR
UAE
OMAN
YEMEN
LIBYA
EGYPT
Red Sea
Khartoum
ERITREA
CHAD
SUDAN
N'Djamena
DJIBOUTI
Addis Ababa
ETHIOPIA
SOMALIA
Mogadishu
CENTRAL AFRICAN REPUBLIC
Bangui
CAMEROON
CONGO
KENYA
UGANDA
RWANDA
Nairobi
DEM. REP. OF CONGO
Kinshasa
BURUNDI
TANZANIA
Dar es Salaam
SEYCHELLES
MALDIVES
INDIAN OCEAN
Luanda
ANGOLA
ZAMBIA
MALAWI
COMOROS
Harare
ZIMBABWE
MOZAMBIQUE
Mozambique Channel
MADAGASCAR
Antananarivo
MAURITIUS
NAMIBIA
Windhoek
BOTSWANA
Gaborone
Maputo
Pretoria
SWAZILAND
SOUTH AFRICA
LESOTHO
Cape Town
ANTARCTICA

FINLAND
Helsinki
NORWAY
Oslo
Stockholm
SWEDEN
St. Petersburg
ESTONIA
LATVIA
LITHUANIA
DENMARK
Copenhagen
Minsk
IRELAND
UNITED KINGDOM
NETH.
London
Berlin
Warsaw
POLAND
BELARUS
GERMANY
BELGIUM
LUX.
Paris
CZECH.
SLOVAKIA
UKRAINE
Kiev
Vienna
AUS.
HUNGARY
MOL.
SWITZ.
FRANCE
SL.
CROATIA
ROMANIA
B&H
SERB.
SPAIN
ITALY
MONT.
KO.
BULGARIA
Istanbul
PORTUGAL
Madrid
Rome
ALB.
MAC.
GREECE
Ankara
TURKEY
Lisbon
Algiers
ALGERIA
TUNISIA
Athens

WORLD REGIONAL GEOGRAPHY

WITHOUT SUBREGIONS

Global Patterns, Local Lives

Sixth Edition

LYDIA MIHELIČ PULSIPHER
Geography Professor Emeritus,
University of Tennessee

ALEX A. PULSIPHER
Geographer and Independent Scholar

with the assistance of

CONRAD "MAC" GOODWIN
Anthropologist/Archaeologist and
Independent Scholar

W. H. Freeman and Company
A Macmillan Higher Education Company
New York

To the youngest members of our family: Louis, Vincent, and Sam

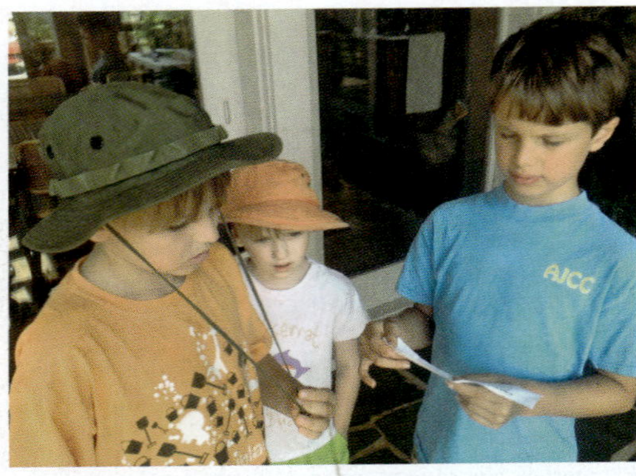

Publisher: **Steven Rigolosi**
Developmental Editor: **Elaine Epstein**
Senior Project Editor: **Vivien Weiss**
Marketing Manager: **Tom Digiano**
Cover and Text Designer: **Blake Logan**
Art Manager: **Matthew McAdams**
Assistant and Supplements Editor: **Stephanie Ellis**
Maps: **University of Tennessee, Cartographic Services Laboratory, Will Fontanez, Director; Maps.com**
Photo Editors: **Jennifer MacMillan, Hilary Newman, Nick Ciani**
Photo Researcher: **Alex Pulsipher**
Production Manager: **Susan Wein**
Composition: **MPS Limited**
Printing and Binding: **RR Donnelley**
Front cover & title page: **Niko Guido/Getty Images**
Back cover: **(left) Bartosz Hadyniak/Getty Images; (center) Richard Nowitz/Getty Images; (right) Tristan Savatier/Getty Images**

Library of Congress Control Number: 2013951040

ISBN-13: 978-1-4641-1069-6
ISBN-10: 1-4641-1070-0 (with subregions)
ISBN-10: 1-4641-1069-7 (without subregions)

© 2014, 2011, 2008, 2006, 2002 by W. H. Freeman and Company. All rights reserved.

Printed in the United States of America
First printing

W. H. Freeman and Company
41 Madison Avenue
New York, NY 10010
Houndmills, Basingstoke RG21 6XS, England
www.whfreeman.com/geography

ABOUT THE AUTHORS

Lydia Mihelič Pulsipher is a cultural-historical geographer who studies the landscapes of ordinary people through the lenses of archaeology, geography, and ethnography. She has contributed to several geography-related exhibits at the Smithsonian Museum of Natural History in Washington, D.C., including "Seeds of Change," which featured the research she and Conrad Goodwin did in the eastern Caribbean. Lydia Pulsipher has ongoing research projects in the eastern Caribbean (historical archaeology) and in central Europe, where she is interested in various aspects of the post-Communist transition. Her graduate students have studied human ecology issues in the Caribbean and border issues and issues of national identity and exclusion in several central European countries. She has taught cultural, gender, European, North American, and Mesoamerican geography at the University of Tennessee at Knoxville since 1980; through her research, she has given many students their first experience in fieldwork abroad. Previously she taught at Hunter College and Dartmouth College. She received her B.A. from Macalester College, her M.A. from Tulane University, and her Ph.D. from Southern Illinois University. For relaxation, she works in her gardens, makes jam, and bakes rhubarb pies.

Alex A. Pulsipher is an independent scholar in Knoxville, Tennessee, who has conducted research on vulnerability to climate change, sustainable communities, and the diffusion of green technologies in the United States. In the early 1990s, while a student at Wesleyan University in Connecticut, Alex spent time in South Asia working for a sustainable development research center. He then completed his B.A. at Wesleyan, writing his undergraduate thesis on the history of Hindu nationalism. Beginning in 1995, Alex contributed to the research and writing of the first edition of *World Regional Geography* with Lydia Pulsipher. In 1999 and 2000, he traveled to South America, Southeast Asia, and South Asia, where he collected information for the second edition of the text and for the Web site. In 2000 and 2001, he wrote and designed maps for the second edition. He participated in the writing of the fourth edition and in restructuring the content of, and creating photo essays and maps for, the fifth edition. Alex worked extensively on the first and second editions of *World Regional Geography Concepts*. He has a master's degree in geography from Clark University in Worcester, Massachusetts.

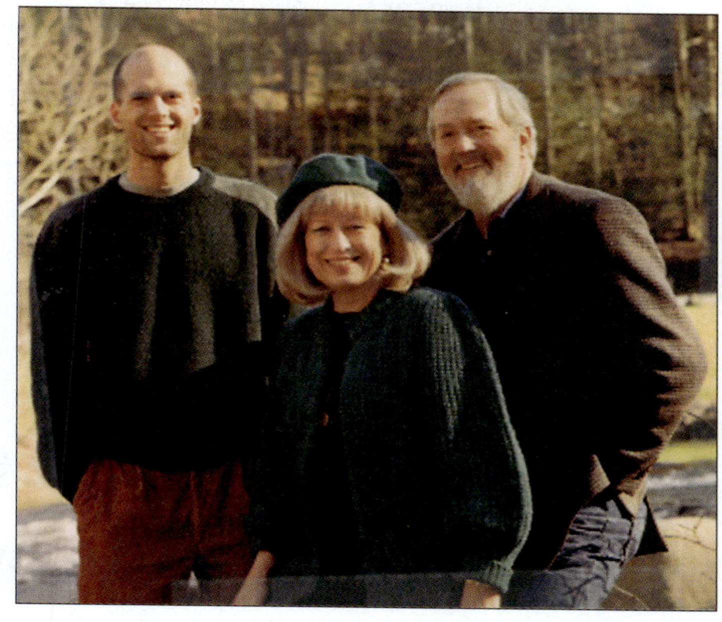

Conrad "Mac" Goodwin assisted in the writing of *World Regional Geography* and the *Concepts* version in many ways. Mac, Lydia's husband, is an anthropologist and historical archaeologist with a B.A. in anthropology from the University of California, Santa Barbara; an M.A. in historical archaeology from the College of William and Mary; and a Ph.D. in archaeology from Boston University. He specializes in sites created during the European colonial era in North America, the Caribbean, and the Pacific. He has particular expertise in the archaeology of agricultural systems, gardens, domestic landscapes, and urban spaces. In addition to work in archaeology and on the textbook, for the past 10 years he has been conducting research on wines and winemaking in Slovenia, and delivering papers on these topics at professional geography meetings. For relaxation, Mac works in his organic garden, builds stone walls (including a pizza oven), and is a slow-food chef.

BRIEF CONTENTS

	Preface	xii
CHAPTER 1	Geography: An Exploration of Connections	1
CHAPTER 2	North America	58
CHAPTER 3	Middle and South America	108
CHAPTER 4	Europe	152
CHAPTER 5	Russia and the Post-Soviet States	194
CHAPTER 6	North Africa and Southwest Asia	232
CHAPTER 7	Sub-Saharan Africa	280
CHAPTER 8	South Asia	326
CHAPTER 9	East Asia	372
CHAPTER 10	Southeast Asia	416
CHAPTER 11	Oceania: Australia, New Zealand, and the Pacific	458
	Epilogue: Antarctica	496
	Glossary	G-1
	Photo Credits	PC-1
	Figure Credits	FC-1
	Text Sources and Credits	TC-1
	Index	I-1

CONTENTS

CHAPTER 1
Geography: An Exploration of Connections 1

Where Is It? Why Is It There? Why Does It Matter? 2
What Is Geography? 3
Geographers' Visual Tools 4
Understanding Maps 4 • Legend and Scale 4 • Longitude and Latitude 4 • Map Projections 7 • Geographic Information Science (GISc) 7

The Detective Work of Photo Interpretation 8

The Region as a Concept 9
Thematic Concepts and Their Role in This Book 12

POPULATION 12
Global Patterns of Population Growth 12
Local Variations in Population Density and Growth 13
Age and Sex Structures 14
Population Growth Rates and Wealth 15

GENDER 17
Gender Roles 17
Gender Issues 17

DEVELOPMENT 19
Measuring Economic Development 19
Geographic Patterns of Human Well-Being 20
Sustainable Development and Political Ecology 22
Human Impact on the Biosphere 22

FOOD 23
Agriculture: Early Human Impacts on the Physical Environment 23
Agriculture and Its Consequences 23
Modern Food Production and Food Security 24

URBANIZATION 27
Why Are Cities Growing? 27
Patterns of Urban Growth 28

GLOBALIZATION 30
What Is the Global Economy? 30
Workers in the Global Economy 31
The Debate over Globalization and Free Trade 32

POWER AND POLITICS 33
The Expansion of Democracy 34
What Factors Encourage Successful Democratization? 36
Democratization and Geopolitics 36
International Cooperation 37

WATER 37
Calculating Water Use per Capita 38
Who Owns Water? Who Gets Access to It? 38
Water Quality 38 • Water and Urbanization 39

GLOBAL CLIMATE CHANGE 40
Drivers of Global Climate Change 42
Climate-Change Impacts 42
Vulnerability to Climate Change 42
Responding to Climate Change 43

PHYSICAL GEOGRAPHY PERSPECTIVES 43
Landforms: The Sculpting of the Earth 44
Plate Tectonics 44 • Landscape Processes 47
Climate 47
Temperature and Air Pressure 48 • Precipitation 48 • Climate Regions 49

HUMAN AND CULTURAL GEOGRAPHY PERSPECTIVES 52
Ethnicity and Culture: Slippery Concepts 52
Values 52
Religion and Belief Systems 54
Language 54
Race 54

PHOTO ESSAYS
1.3 Understanding Maps 5
1.4 The Detective Work of Photo Interpretation 8
1.13 Human Impacts on the Biosphere 24
1.15 Urbanization 28
1.18 Power and Politics 34
1.24 Vulnerability to Climate Change 44
1.29 Climate Regions of the World 50
1.30 Major Religions of the World 53

CHAPTER 2
North America 58

THE GEOGRAPHIC SETTING 61
Physical Patterns 61
Landforms 61 • Climate 61

Environmental Issues 62
Loss of Habitat for Plants and Animals 62 • Oil Drilling 62 • Logging 65 • Coal Mining and Use 65 • Urbanization and Habitat Loss 65 • Climate Change and Air Pollution 66 • Water Resource Depletion, Pollution, and Marketization 66

Human Patterns over Time 70
The Peopling of North America 70 • The European Transformation 71 • Expansion West of the Mississippi and Great Lakes 72 • European Settlement and Native Americans 74 • The Changing Regional Composition of North America 74

CURRENT GEOGRAPHIC ISSUES 75

Political Issues 75
Post-9/11 Geopolitical Relationships 75 • The United States and Canada Abroad 75 • Dependence on Oil Questioned 76

Relationships Between Canada and the United States 78
Asymmetries 78 • Similarities 79 • Interdependencies 79 • Democratic Systems of Government: Shared Ideals, Different Trajectories 79 • The Social Safety Net: Canadian and U.S. Approaches 80 • Gender in National Politics 81

Economic Issues 82
North America's Changing Food-Production Systems 82 • Changing Transportation Networks and the North American Economy 84 • The New Service and Technology Economy 85 • Globalization and the Economy 86 • Repercussions of the Global Economic Downturn Beginning in 2007 88 • Women in the Economy 89

Sociocultural Issues 89
Urbanization and Sprawl 89 • Immigration and Diversity 93 • Race and Ethnicity in North America 96 • Religion 98 • Gender and the American Family 99

General Population Patterns 101
The Geography of Population Change in North America 102 • Mobility in North America 102 • Aging in North America 102 • Geographic Patterns of Human Well-Being 104

PHOTO ESSAYS

2.4 Climates of North America 63
2.5 Human Impacts on the Biosphere in North America 64
2.6 Vulnerability to Climate Change in North America 67
2.9 A Visual History of North America 72
2.12 Power and Politics in North America 76
2.21 Urbanization in North America 90

CHAPTER 3
Middle and South America 108

THE GEOGRAPHIC SETTING 112

Physical Patterns 112
Landforms 112 • Climate 114

Environmental Issues 117
Tropical Forests, Climate Change, and Globalization 117 • Environmental Protection and Economic Development 119 • The Water Crisis 120

Human Patterns over Time 122
The Peopling of Middle and South America 122 • European Conquest 123 • A Global Exchange of Crops and Animals 124 • The Legacy of Underdevelopment 124

CURRENT GEOGRAPHIC ISSUES 126

Economic and Political Issues 126
Economic Inequality and Income Disparity 127 • Phases of Economic Development 128 • The Present Era of Foreign Direct Investment 131 • The Informal Economy 132 • Regional Trade and Trade Agreements 132

Food Production and Contested Space 133

Power, Politics, and the Move Toward Democracy 135
The Drug Trade, Conflict, and Democracy 137 • Foreign Involvement in the Region's Politics 138

Sociocultural Issues 138
Population Patterns 138 • Cultural Diversity 142 • Race and the Social Significance of Skin Color 142 • The Family and Gender Roles 142 • Migration and Urbanization 145 • Religion in Contemporary Life 148

PHOTO ESSAYS

3.5 Climates of Middle and South America 115
3.8 Human Impacts on the Biosphere in Middle and South America 118
3.10 Vulnerability to Climate Change in Middle and South America 121
3.11 A Visual History of Middle and South America 122
3.18 Power and Politics in Middle and South America 136
3.25 Urbanization in Middle and South America 146

CHAPTER 4
Europe 152

THE GEOGRAPHIC SETTING 155

Physical Patterns 156
Landforms 156 • Vegetation and Climate 157

Environmental Issues 159
Europe's Impact on the Biosphere 159 • Air Pollution 159 • Freshwater and Seawater Pollution 161 • Europe's Vulnerability to Climate Change 162

Human Patterns over Time 164
Sources of European Culture 165 • The Inequalities of Feudalism 166 • The Role of Urbanization in the Transformation of Europe 166 • European Colonialism: The Founding and Acceleration of Globalization 168 • Urban Revolutions in Industry and Democracy 169 • Urbanization and Democratization 170 • Two World Wars and Their Aftermath 171

CURRENT GEOGRAPHIC ISSUES 174

Economic and Political Issues 174
Steps in Creating the European Union 174 • The Euro and Debt Crises 176 • Food Production and the European Union 178 • Europe's Growing Service Economies 179

Sociocultural Issues 180
Population Patterns 181 • Immigration and Migration: Needs and Fears 184 • Changing Gender Roles 187 • Social Welfare Systems and Their Outcomes 189

PHOTO ESSAYS
4.4 Climates of Europe 158
4.5 Human Impacts on the Biosphere in Europe 160
4.7 Vulnerability to Climate Change in Europe 163
4.9 A Visual History of Europe 166
4.13 Power and Politics in Europe 172
4.20 Urbanization in Europe 183

CHAPTER 5
Russia and the Post-Soviet States 194

THE GEOGRAPHIC SETTING 198
Physical Patterns 198
Landforms 199 • Climate and Vegetation 199
Environmental Issues 201
Urban and Industrial Pollution 201 • The Globalization of Nuclear Pollution 201 • The Globalization of Resource Extraction and Environmental Degradation 203 • Water Issues 203 • Climate Change 204
Human Patterns over Time 205
The Rise of the Russian Empire 205 • The Communist Revolution and Its Aftermath 208 • World War II and the Cold War 210

CURRENT GEOGRAPHIC ISSUES 211
Economic and Political Issues 211
Oil and Gas Development: Fueling Globalization 212 • Economic Reforms in the Post-Soviet Era 212 • Food Production in the Post-Soviet Era 215 • Democratization in the Post-Soviet Years 217 • Corruption and Organized Crime 220
Sociocultural Issues 220
Population Patterns 221 • Gender: Challenges and Opportunities in the Post-Soviet Era 226 • Religious Revival in the Post-Soviet Era 229

PHOTO ESSAYS
5.5 Climates of Russia and the Post-Soviet States 200
5.6 Human Impacts on the Biosphere in Russia and the Post-Soviet States 202
5.8 Vulnerability to Climate Change in Russia and the Post-Soviet States 206
5.10 A Visual History of Russia and the Post-Soviet States 208
5.17 Power and Politics in Russia and the Post-Soviet States 218
5.22 Urbanization in Russia and the Post-Soviet States 224

CHAPTER 6
North Africa and Southwest Asia 232

THE GEOGRAPHIC SETTING 236
Physical Patterns 237
Climate 237 • Landforms and Vegetation 237
Environmental Issues 239
An Ancient Heritage of Water Conservation 240 • Could There Be New Sources of Water? 240 • Water and Food Production 240 • Vulnerability to Climate Change 242
Human Patterns over Time 246
Agriculture and the Development of Civilization 246 • Agriculture and Gender Roles 247 • The Coming of Monotheism: Judaism, Christianity, and Islam 248 • The Spread of Islam 248 • Western Domination, State Formation, and Antidemocratic Practices 250

CURRENT GEOGRAPHIC ISSUES 251
Sociocultural Issues 251
Religion in Daily Life 251 • Family Values and Gender 254 • Gender Roles and Gendered Spaces 254 • The Rights of Women in Islam 255 • The Lives of Children 256 • Changing Population Patterns 257 • Population Growth and Gender Status 258 • Urbanization, Globalization, and Migration 259 • Human Well-Being 262
Economic and Political Issues 262
Globalization, Development, and Fossil Fuel Exports 263 • Economic Diversification and Growth 265 • Power and Politics: The Arab Spring 267 • The Role of the Press, Media, and Internet in Political Change 269 • Democratization and Women 270
Three Worrisome Geopolitical Situations in the Region 272
Situation 1: Fifty Years of Trouble Between Iraq and the United States, 1963 to 2013 272 • Situation 2: The State of Israel and the "Question of Palestine" 272 • Situation 3: Failure of the Arab Spring in Syria 275

PHOTO ESSAYS
6.5 Climates of North Africa and Southwest Asia 238
6.7 Human Impacts on the Biosphere in North Africa and Southwest Asia 241
6.10 Vulnerability to Climate Change in North Africa and Southwest Asia 244
6.15 A Visual History of North Africa and Southwest Asia 250
6.23 Urbanization in North Africa and Southwest Asia 261
6.28 Power and Politics in North Africa and Southwest Asia 270

CHAPTER 7 Sub-Saharan Africa 280

THE GEOGRAPHIC SETTING 284

Physical Patterns 284
Landforms 284 • Climate and Vegetation 285

Environmental Issues 285
Deforestation and Climate Change 285 • Agricultural Systems, Food, Water, and Vulnerability to Climate Change 289 • Wildlife and Climate Change 293

Human Patterns over Time 294
The Peopling of Africa and Beyond 295 • Early Agriculture, Industry, and Trade in Africa 295 • The Scramble to Colonize Africa 297 • Power and Politics in the Aftermath of Independence 299

CURRENT GEOGRAPHIC ISSUES 300

Economic and Political Issues 300
Commodity-Based Economic Development and Globalization in Africa 300 • Successive Eras of Globalization 300 • Regional and Local Economic Development 303 • Power and Politics 306 • Gender and Democratization 310

Sociocultural Issues 310
Population Patterns: Growth, Density, and the Demographic Transition 311 • Gender Issues 317 • Religion 319

Geographic Patterns of Human Well-Being 322

PHOTO ESSAYS

7.5 Climates of Sub-Saharan Africa 286
7.6 Human Impacts on the Biosphere in Sub-Saharan Africa 288
7.8 Vulnerability to Climate Change in Sub-Saharan Africa 291
7.11 A Visual History of Sub-Saharan Africa 294
7.21 Power and Politics in Sub-Saharan Africa 308
7.24 Urbanization in Sub-Saharan Africa 314

CHAPTER 8 South Asia 326

THE GEOGRAPHIC SETTING 330

Physical Patterns 330
Landforms 330 • Climate and Vegetation 331

Environmental Issues 332
South Asia's Vulnerability to Climate Change 332 • Responses to Water Issues Related to Global Climate Change 337 • Deforestation 337 • Industrial Air Pollution 340

Human Patterns over Time 340
The Indus Valley Civilization 340 • A Series of Invasions 341 • Globalization and the Legacies of British Colonial Rule 342

CURRENT GEOGRAPHIC ISSUES 345

Sociocultural Issues 345
The Texture of Village Life 346 • The Texture of City Life 346 • Language and Ethnicity 349 • Religion, Caste, and Conflict 350 • Geographic and Social Patterns in the Status of Women 353 • Gender and Democratization 355 • Population Patterns 357

Economic Issues 361
Economic Trends 361 • Food Production and the Green Revolution 361 • Microcredit: A South Asian Innovation for the Poor 363 • Economic Reform: Globalization and Competitiveness 364

Political Issues 366
Religious Nationalism 366 • The Growing Influence of Women and Young Voters 368 • Power and Politics 368

PHOTO ESSAYS

8.5 Climates of South Asia 333
8.7 Vulnerability to Climate Change in South Asia 336
8.8 Human Impacts on the Biosphere in South Asia 338
8.13 A Visual History of South Asia 344
8.15 Urbanization in South Asia 348
8.31 Power and Politics in South Asia 367

CHAPTER 9 East Asia 372

THE GEOGRAPHIC SETTING 376

Physical Patterns 376
Landforms 376 • Climate 377

Environmental Issues 379
Climate Change: Emissions and Vulnerability in East Asia 379 • Food Security and Sustainability in East Asia 381 • Three Gorges Dam: The Power of Water 384 • Air Pollution: Choking on Success 385

Human Patterns over Time 387
Bureaucracy and Imperial China 387 • Confucianism Molds East Asia's Cultural Attitudes 388 • Why Did China Not Colonize an Overseas Empire? 389 • European and Japanese Imperialism 389 • China's Turbulent Twentieth Century 391 • Japan Becomes a World Leader 392 • Chinese and Japanese Influences on Korea, Taiwan, and Mongolia 392

CURRENT GEOGRAPHIC ISSUES 394

Economic and Political Issues 394
The Japanese Miracle 394 • Mainland Economies: Communists in Command 395 • Market Reforms in China 396 • Urbanization, Development, and Globalization in East Asia 397 • Power and Politics in East Asia 402

Sociocultural Issues 406
Population Patterns 406 • Responding to an Aging Population 406 • China's One-Child Policy 407 • Population Distribution 408 • Human Well-Being 409 • Cultural Diversity in East Asia 411 • East Asia's Most Influential Cultural Export: The Overseas Chinese 413

PHOTO ESSAYS

9.4 Climates of East Asia 378
9.5 Vulnerability to Climate Change in East Asia 380
9.9 Human Impacts on the Biosphere in East Asia 386
9.12 A Visual History of East Asia 390
9.17 Urbanization in East Asia 398
9.21 Power and Politics in East Asia 403

CHAPTER 10
Southeast Asia 416

THE GEOGRAPHIC SETTING 420
Physical Patterns 420
Landforms 420 • Climate and Vegetation 422
Environmental Issues 424
Climate Change and Deforestation 424 • Climate Change and Food Production 427 • Climate Change and Water 428 • Responses to Climate Change 429
Human Patterns over Time 431
The Peopling of Southeast Asia 431 • Diverse Cultural Influences 431 • European Colonization 431 • Struggles for Independence 433

CURRENT GEOGRAPHIC ISSUES 434
Economic and Political Issues 434
Strategic Globalization: State Aid and Export-Led Economic Development 434 • Economic Crisis and Recovery: The Perils of Globalization 436 • Regional Trade and ASEAN 437 • Pressures For and Against Democracy 440
Sociocultural Issues 443
Population Patterns 443 • Urbanization 448 • Emigration Related to Globalization 448 • Cultural and Religious Pluralism 451 • Gender Patterns in Southeast Asia 454 • Globalization and Gender: The Sex Industry 455

PHOTO ESSAYS

10.7 Climates of Southeast Asia 423
10.9 Human Impacts on the Biosphere in Southeast Asia 426
10.11 Vulnerability to Climate Change in Southeast Asia 430
10.13 A Visual History of Southeast Asia 434
10.18 Power and Politics in Southeast Asia 441
10.23 Urbanization in Southeast Asia 449

CHAPTER 11
Oceania: Australia, New Zealand, and the Pacific 458

THE GEOGRAPHIC SETTING 461
Physical Patterns 462
Continent Formation 462 • Island Formation 463 • Climate 463 • Fauna and Flora 465
Environmental Issues 466
Global Climate Change 467 • Invasive Species and Food Production 467 • Globalization and the Environment in the Pacific Islands 469
Human Patterns over Time 472
The Peopling of Oceania 473 • Arrival of the Europeans 474 • The Colonization of Australia and New Zealand 475 • Oceania's Shifting Global Relationships 475

CURRENT GEOGRAPHIC ISSUES 476
Economic and Political Issues 476
Globalization, Development, and Oceania's New Asian Orientation 476 • The Stresses of Asia's Economic Development "Miracle" on Australia and New Zealand 477 • The Advantages and Stresses of Tourism 479 • The Future: Diverse Global Orientations? 480
Population Patterns 481
Disparate Population Patterns in Oceania 481 • Urbanization in Oceania 482 • Human Well-Being in Oceania 482
Sociocultural Issues 485
Ethnic Roots Reexamined 485 • Power and Politics: Different Political Cultures 487 • Forging Unity in Oceania 489 • Gender Roles in Oceania 490

PHOTO ESSAYS

11.5 Climates of Oceania 464
11.8 Human Impacts on the Biosphere in Oceania 468
11.9 Vulnerability to Climate Change in Oceania 470
11.14 A Visual History of Oceania 476
11.18 Urbanization in Oceania 483
11.23 Power and Politics in Oceania 488

EPILOGUE: ANTARCTICA 496
Glossary G-1
Photo Credits PC-1
Figure Credits FC-1
Text Sources and Credits TC-1
Index I-1

PREFACE

In this text, we portray the rich diversity of human life across the world and humanize geographic issues by representing the daily lives of women, men, and children in the various regions of the globe. Our goal is to make global patterns of trade and consumption meaningful for students by showing how these patterns affect not only world regions but also ordinary people at the local level. In striving to reach this goal, we have made this sixth edition of *World Regional Geography* as current, instructive, and visually appealing as possible.

CONTINUING IN THE SIXTH EDITION

Thematic Concepts

Teaching world regional geography is never easy. Many instructors have found that focusing their courses on a few key ideas makes their teaching more effective and helps students retain information. With that goal in mind, we have identified nine thematic concepts that provide a few basic hooks on which students can hang their growing knowledge of the world and each of its regions. These thematic concepts are listed here in the order in which they are first covered in Chapter 1:

- Population: What are the major forces driving population growth or decline in a region? How have changes in gender roles influenced population growth? How are changes in life expectancy, family size, and the age of the population influencing population change?
- Gender: How do the lives and livelihoods of men and women differ, and how do gender roles influence societies in a region? To what extent do men and women differ in their contributions to family and community well-being? From what do persistent disparities in income, education, and rights between genders arise?
- Development: How do shifts in economic, social, and other dimensions of development affect human well-being? What paths have been charted by the so-called developed world, and how are they relevant, or irrelevant, to the rest of the world? What new "homegrown" solutions are emerging from the so-called less-developed countries?
- Food: How do food production systems impact environments and societies in a region? How has the use of new agricultural technologies impacted farmers? How have changes in food production created pressure to urbanize?
- Urbanization: Which forces are driving urbanization in a particular region? How have cities responded to growth? How is the region affected by the changes that accompany urbanization—for example, changes in employment, education, and access to health care?
- Globalization: How has a particular region been impacted by globalization, historically and currently? How are lives changing as flows of people, ideas, products, and resources become more global?
- Power and Politics: What are the main differences in the ways that power is wielded in societies? Which types of governance tend to arise from centralized power? Which tend to arise when individuals in a society have more say in the development of policies and the ways that governments are run?
- Water: How do issues of water scarcity, water pollution, and water management affect people and environments in a particular region? How might global climate change and changes in food production systems affect water resources?
- Climate Change: What are the indications that climate change is underway? How are places, people, and ecosystems in a particular region vulnerable to the shifts that climate change may bring? How are people and governments in the region responding to the threats posed by global warming? Which human activities contribute significant amounts of greenhouse gases?

Photos

Photos are a rich source of geographic information, and at the beginning of each regional chapter, a series of photos surrounding the regional map introduce the reader to landscapes within the region. **Photo Essay** figures illustrate particular thematic concepts. For example, a photo essay about urbanization might include a map of urban patterns in that region as well as photos that illustrate various aspects of current urban life in that part of the world. Photos also are central to new features on **Local Lives** and **Visual Histories**.

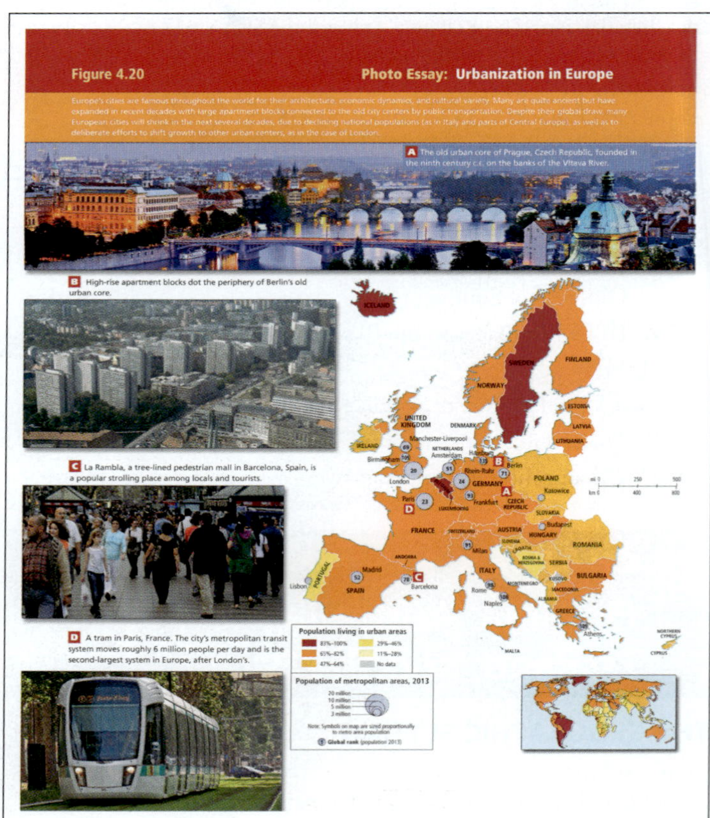

NEW TO THE SIXTH EDITION

New special features include **Geographic Insights, Visual Histories, On the Bright Side** commentaries, **Local Lives,** features, and **Thinking Geographically** questions.

The Thematic Concepts form the basis for this edition's new **Geographic Insights**. For each chapter, alone and in combination, the thematic concepts form the basis of five to six learning objectives that we call **Geographic Insights.** These insights are stated at the beginning of each chapter and discussed at the relevant point in the text. They also are reviewed in "Things to Remember" sections found throughout the chapter, as well as in new questions posed in the "Geographic Insights Review and Self-Test" section at the end of each chapter.

In this edition there are **Visual Histories** (with timelines) for each region, which use images to illustrate key points in the region's history. **On the Bright Side** commentaries explore some of the more hopeful patterns and opportunities emerging within each region. Three new **Local Lives** photo features in each region chapter add further human interest by showing regional customs related to foodways, people and animals, and festivals.

> **Environmental Issues**
>
> **Geographic Insight 1**
>
> **Climate Change, Food, and Water:** Sub-Saharan Africa is particularly vulnerable to climate change because subsistence occupations are sensitive to even slight variations in temperature, rainfall, and water availability. In large part because of poverty, political instability, and having little access to cash, the region does not have much resilience to the effects of climate change.

To help instructors make use of all these new photo features in their teaching, the text offers **Thinking Geographically** questions with many photo essays and photo figures. The answers can be found on this book's Web site, where they form the basis of computer-graded exercises that can be assigned and automatically graded and entered into each instructor's grade book.

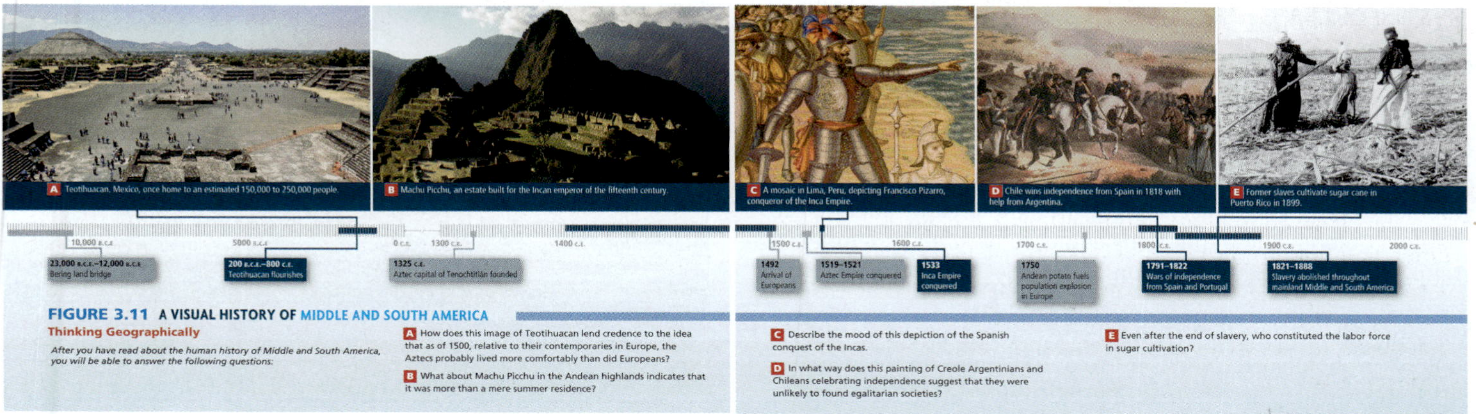

FIGURE 3.11 A VISUAL HISTORY OF MIDDLE AND SOUTH AMERICA

Thinking Geographically

After you have read about the human history of Middle and South America, you will be able to answer the following questions:

A How does this image of Teotihuacan lend credence to the idea that as of 1500, relative to their contemporaries in Europe, the Aztecs probably lived more comfortably than did Europeans?

B What about Machu Picchu in the Andean highlands indicates that it was more than a mere summer residence?

C Describe the mood of this depiction of the Spanish conquest of the Incas.

D In what way does this painting of Creole Argentinians and Chileans celebrating independence suggest that they were unlikely to found egalitarian societies?

E Even after the end of slavery, who constituted the labor force in sugar cultivation?

FIGURE 8.14 LOCAL LIVES **FESTIVALS IN SOUTH ASIA**

A A village festival in Pakistan features *kabaddi*, a popular South Asian sport in which teams take turns sending a "raider" across a field center line. That person must tag, or in some cases wrestle to the ground, members of the other team and then return to his or her own side without taking a breath. Kabaddi has been played at the Indian National Games since 1939 and at the Asian Games since 1991.

B Celebrants in Kolkata, India, during Holi—a festival celebrating the end of winter and beginning of spring. Holi evolved from temple worship practices involving the application of color to statues. In a riotous and celebratory atmosphere, people of different ages, genders, castes, and economic backgrounds temporarily disregard their differences and hurl the colors of the coming spring at each other.

C Pilgrims during the 2010 Kumbh Mela bathe in the Ganga River at Haridwar. During this event, which is held every 3 years, Hindus purify themselves by bathing in the sacred waters of the river. In 2013, the 45-day-long event attracted over 100 million participants.

Restructured Chapters

Each chapter includes a variety of features to support the teaching and learning of world regional geography.

Things to Remember At the close of every main section, a few concise statements review the important points in the section. The statements emphasize some key themes while encouraging students to think through the ways in which the material illustrates these points. They also review the Geographic Insights that begin each chapter.

> **THINGS TO REMEMBER**
>
> - Globalization encompasses many types of worldwide and interregional flows and linkages, especially the ways in which goods, capital, labor, and resources are exchanged among distant and very different places.
>
> Geographic Insight 5
>
> - **Globalization and Development** Throughout the world, globalization is transforming patterns of economic development as local self-sufficiency is giving way to global interdependence and international trade.
> - Under true free trade, all economic transactions are conducted without interference or regulation in an open marketplace. Under fair trade, an alternative to free trade, consumers are asked to pay a fair price to producers, and producers are expected to pay living wages and uphold environmental and safety standards in the workplace.

Geographic Insights Review and Self-Test At the end of each chapter, a series of questions, many tied to the chapter's Geographic Insights, encourage students to more broadly analyze the chapter content. These questions could be used for assignments, group projects, or class discussion.

Marginal Glossary of Key Terms Terms important to the chapter content are boldfaced on first usage and defined on the page on which they appear. The terms are listed at the end of the chapter, with the page numbers where they are defined. The key terms are also listed alphabetically and defined in the glossary at the end of the book.

> **The Social Safety Net: Canadian and U.S. Approaches**
>
> The Canadian and U.S. governments have responded differently to the displacement of workers by economic change. Ultimately, these differences derive from prevailing political positions and widely held notions each country has about what the government's role in society should be. In Canada there is broad political support for a robust **social safety net**, the services provided by the government—such as welfare, unemployment benefits, and health care—that prevent people from falling into extreme poverty. In the United States there is much less support for these programs and a great deal of contention over nearly all efforts to strengthen the U.S. social safety net.
>
> **social safety net** the services provided by the government—such as welfare, unemployment benefits, and health care—that prevent people from falling into extreme poverty

Consistent Base Maps This edition focuses on improving further what has often been cited as a principal strength of this text: high-quality, relevant, and consistent maps. To help students make conceptual connections and to compare regions, every chapter contains the following:

- Regional map with landscape photos at the beginning of each chapter
- Political map
- Climate map with photos of different climate zones

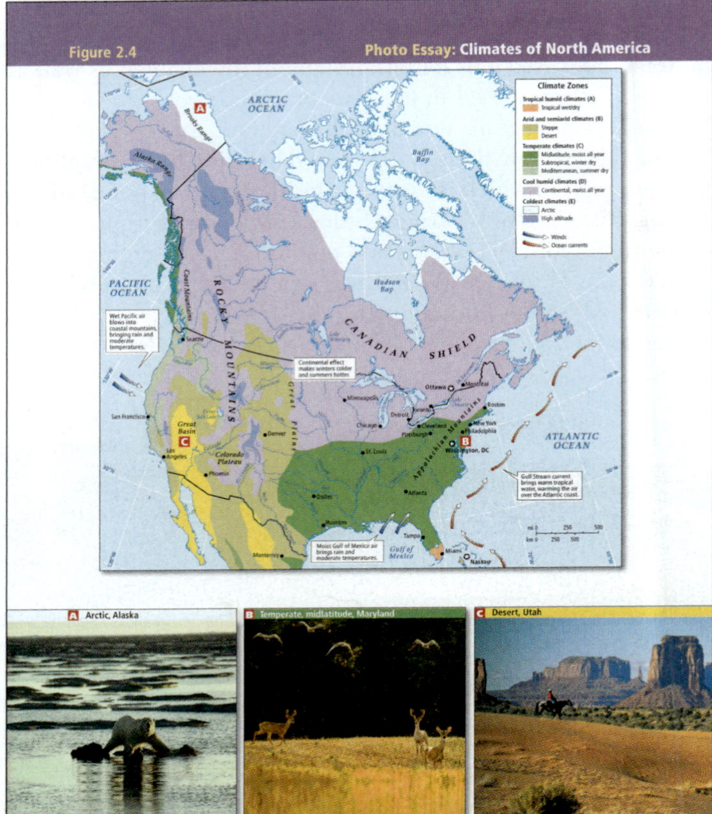

- Map of the human impacts on the biosphere, with photo essay
- Map of the region's vulnerability to climate change, with photo essay
- Urbanization map, with photo essay
- Map of regional power and politics, with photo essay
- Map of population density
- Maps of geographic patterns of human well-being

New Photos

An ongoing aim of this text has been to awaken students to the circumstances of people around the world, and photos are a powerful way to accomplish this objective. This edition continues our tradition of promoting careful attention to photos by including in Chapter 1 a short lesson on photo interpretation. Students are encouraged to use these skills as they look at every photo in the text, and instructors are encouraged to use the photos as lecture themes and to help generate analytical class discussions.

Each photo was chosen to complement a Thematic Concept or situation described in the text. All photos are numbered and referenced in the text, making it easier for students to integrate the text with the visuals as they read. Moreover, the photos—like all of the book's graphics, including the maps—have been given significant space and prominence in the page layout. The result is a visually engaging, dynamic, and instructive text. All photo credits are at the end of the book, listed according to figure numbers.

Figure 11.1D

Figure 11.8D

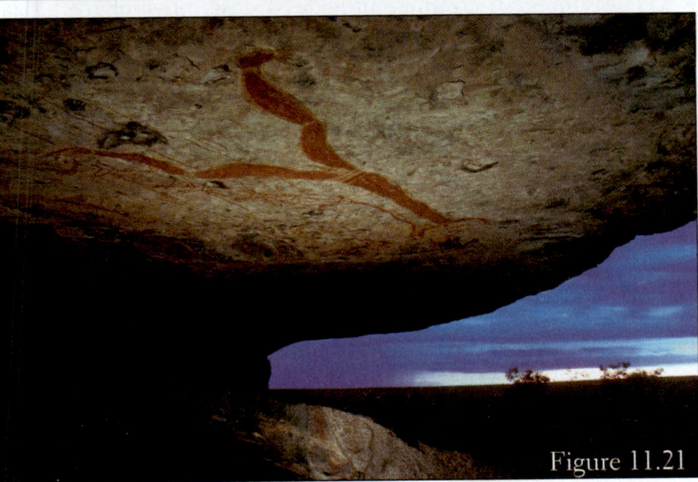
Figure 11.21

Videos

More than 300 videos clips (an average of 27 per chapter) are available with the sixth edition. Most videos are 2 to 6 minutes long and cover key issues discussed in the text. They can help instructors gain further expertise or can be used to generate class discussion. Each video is keyed to the text with an icon at the point in the discussion where it is most relevant. These videos, along with a related multiple-choice quiz, can be accessed at www.whfreeman.com/geographyvideos. Questions can be automatically graded and entered into a grade book. To access the videos, students need a password that can be bundled free with this textbook.

Up-to-Date Content

Because the world is constantly changing, it is essential that a world regional geography text be as current as possible. To that end, the sixth edition discusses the Arab Spring and its aftermath in North Africa and Southwest Asia; the varying effects of the global recession on regions, countries, and individuals; civil unrest in once-stable Thailand; the new influence of Arabic media outlets, such as Al Jazeera, which now affect thinking around the world; and the consequences of global climate change on Pacific Island nations. Some of the major content areas of the book that have been updated include:

- Political revolutions and conflict in North Africa and Southwest Asia (the Middle East)
- Maps reflecting the new country of South Sudan (some data remain based on Sudan as a whole because South Sudan has not yet begun reporting statistics)
- The global economic recession and its effect on migrants, labor outsourcing, and job security in importing and exporting countries
- Domestic and global implications of the U.S. political, economic, and military stances
- The role of terrorism in the realignment of power globally and locally
- Immigration and the ways it is changing countries economically and culturally
- Recent economic crises in the European Union which may bring about significant reorganization that has consequences for the original EU members, new and potential member states, and the global community
- Changing gender roles, particularly in developing countries
- The increasing role and influence of Islam around the world
- Climate change and its environmental, political, and economic implications

THE ENDURING VISION: GLOBAL AND LOCAL PERSPECTIVES

The Global View

In addition to the new features and enhancements to the text, we retain the hallmark features that have made the first five editions of this text successful for instructors and students. For the sixth edition, we continue to emphasize global trends and the interregional linkages that are changing lives throughout the world, including those trends related to changing gender roles. The following linkages are explored in every chapter, as appropriate:

- The **multifaceted economic linkages** among world regions. These include (1) the effects of colonialism; (2) trade; (3) the role in the world economy of transnational corporations such as Walmart, Norilsk Nickel, Nike, and Apple; (4) the influence of regional trade organizations such as ASEAN and NAFTA; and (5) the changing roles of the World Bank and the International Monetary Fund as the negative consequences of structural adjustment programs become better understood.
- **Migration.** Migrants are changing economic and social relationships in virtually every part of the globe. The societies they leave are changed radically by the migrants' absence, just as the host societies are changed by their presence. The text explores the local and global effects of foreign workers in places such as Japan, Europe, Africa, the Americas, and Southwest Asia, and the increasing number of refugees resulting from conflicts around the world. Also discussed are long-standing migrant groups, including the Overseas Chinese and the Indian diasporas.
- **Mass communications and marketing techniques** are promoting **world popular culture** across regions. The text integrates coverage of popular culture and its effects in discussions of topics such as tourism in the Caribbean and Southeast Asia; the wide-ranging impact of innovations originating in modernizing economies, such as Ushahidi in Kenya; and the blending of Western and traditional culture in places such as South Africa, Malawi, Japan, and China.
- **Gender issues** are covered in every chapter with the aim of covering more completely the lives of ordinary people. Gender is intimately connected to other patterns, including internal and global migration, and these connections and other region-wide gender patterns are illustrated in a variety of maps and photos and in vignettes that illustrate gender roles as played out in the lives of individuals. The lives of children, especially with regard to their roles in families, are also covered, often in concert with the treatment of gender issues.

The Local Level

Our approach pays special attention to the local scale—a town, a village, a household, an individual. Our hope is, first, that stories of individual people and families will make geography interesting and real to students; and second, that seeing the effects of abstract processes and trends on ordinary lives will dramatize the effects of these developments for students. Reviewers have mentioned that students particularly appreciate the personal vignettes, which are often stories of real people (with names disguised). For each region, we examine the following local phenomena:

- **Local lives:** We use photo essays to focus on particular regional customs and traditions as they relate to foodways, festivals, and the relationship of animals to the people of a region.
- **Cultural change:** We look closely at changes in the family, gender roles, and social organization in response to urbanization, modernization, and the global economy.
- **Impacts on well-being:** Ideas of what constitutes "well-being" differ from culture to culture, yet broadly speaking, people everywhere try to provide a healthful life for themselves in a community of their choosing. Their success in doing so is affected by local conditions, global forces, and their own ingenuity.
- **Issues of identity:** Paradoxically, as the world becomes more tightly knit through global communications and media, ethnic and regional identities often become stronger. The text examines how modern developments such as the Internet and related technologies are used to reinforce particular cultural identities, often bringing educated emigrants back to help with reforms or to facilitate rapid responses in crises.
- **Local attitudes toward globalization:** People often have ambivalent reactions to global forces. They are repelled by the seeming power of these forces, fearing effects on their own lives and livelihoods and on local traditional cultural values, but they are also attracted by the economic opportunities that may emerge from greater global integration. The text looks at how the people of a region react to cultural and economic globalization.

ONE VISION, TWO VERSIONS: WITH OR WITHOUT SUBREGIONAL COVERAGE

To better serve the different needs of diverse faculty and curricula, three versions of this textbook are available.

World Regional Geography with Subregions, Sixth Edition (1-4641-1070-0)
The sixth edition continues to employ a consistent structure for each chapter. Each chapter beyond the first is divided into three parts: **The Geographic Setting, Current Geographic Issues,** and **Subregions.**

The subregion coverage provides a descriptive characterization of particular countries and places within the region that expands on coverage in the main part of the chapter. For example, the sub-Saharan Africa chapter considers the West, Central, East, and Southern Africa subregions, providing additional insights into differences in well-being and into social and economic issues across the African continent.

World Regional Geography Without Subregions, Sixth Edition (1-4641-1069-7)
The briefer version provides essentially the same main text coverage as the version described above, omitting only the subregional sections. This version contains all the types of pedagogy found in the main version.

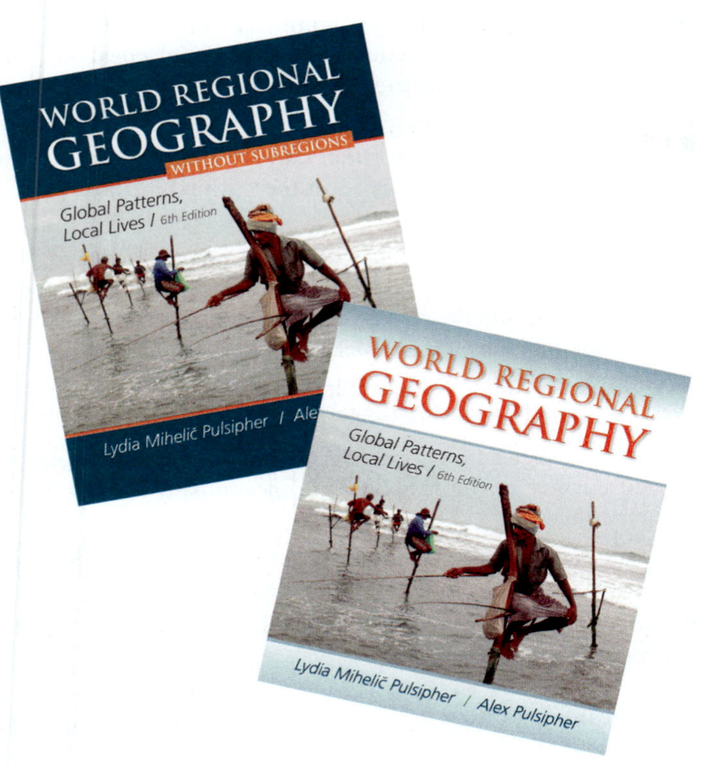

World Regional Geography Concepts, Third Edition
(1-4292-5366-5)
This more compact version is designed to allow instructors to cover all world regions in a single semester.

FOR THE INSTRUCTOR: A WEALTH OF RESOURCES ONLINE AT LAUNCHPAD FOR PULSIPHER, *WORLD REGIONAL GEOGRAPHY*

LaunchPad **www.whfreeman.com/pulsipher6e**
(before August 2014)
www.whfreeman.com/launchpad/pulsipher6ewithsubregions
(after August 2014)
www.whfreeman.com/launchpad/pulsipher6ewithoutsubregions
(after August 2014)

The authors have taught world regional geography many times and understand the need for quick, accessible aids to instruction. Many of the new features were designed to streamline the job of organizing the content of each class session, with the goal of increasing student involvement through interactive discussions. Ease of instruction and active student involvement were the principal motivations behind the book's key features—the thematic concepts and geographic insights, the photo essays, the content maps that facilitate region-to-region comparisons, the photo features on local lives and regional customs, the feature commentaries that highlight bright and emerging reasons for optimism, and the wide selection of videos.

All of the following are available on the book's companion Web site, at www.whfreeman.com/pulsipher6e. Many resources offer free and open access. Premium resources are available on LaunchPad, a complete course management system featuring full gradebook and reporting capacities. For a demo of LaunchPad, please email us at geography@whfreeman.com.

- **All text images** in PowerPoint and JPEG formats with enlarged labels for better projection quality.
- **PowerPoint lecture outlines** by Bharath Ganesh, University College London. The main themes of each chapter are outlined and enhanced with images from the book, providing a pedagogically sound foundation on which to build personalized lecture presentations.
- **Instructor's resource manual** by Jennifer Rogalsky, State University of New York, Geneseo, and Helen Ruth Aspaas, Virginia Commonwealth University, contains suggested lecture outlines, points to ponder for class discussion, and ideas for exercises and class projects. It is offered as chapter-by-chapter Word files to facilitate editing and printing.
- **Test Bank** by Rebecca Johns, University of South Florida, expanded from the original test bank created by Jason Dittmer, University College London, and Andy Walter, West Georgia University. The Test Bank is designed to match the pedagogical intent of the text and offers more than 2500 test questions (multiple choice, short answer, matching, true/false, and essay) in a Word format that makes it easy to edit, add, and resequence questions. A **computerized test bank** (powered by Diploma) with the same content is also available. Please use the following ISBNs to request your computerized test bank on disc: 1-4641-2119-2 (with subregions), 1-4641-2128-1 (without subregions).
- **Clicker questions** by Rebecca Johns, University of South Florida. Prepared in Word, clicker questions allow instructors to jump-start discussions, illuminate important points, and promote better conceptual understanding during lectures.
- **Syllabus posting** online
- An integrated **gradebook** that records students' performance on online and video quizzes

Course Management

All instructor and student resources are also available via **BlackBoard, WebCT, Canvas, Angel, Moodle, Sakai,** and **Desire2Learn**. W. H. Freeman offers a course cartridge that populates your site with content tied directly to the book.

W. H. Freeman *World Regional Geography* DVD

This DVD, available free to adopters of the sixth edition, builds on the book's purpose of putting a face on geography by giving students and instructors access to the fascinating personal stories of people from all over the world. The DVD contains 35 projection-quality video clips from 3 to 7 minutes in length, with over 300 videos also available online. An **instructor's video manual** is also included on the DVD.

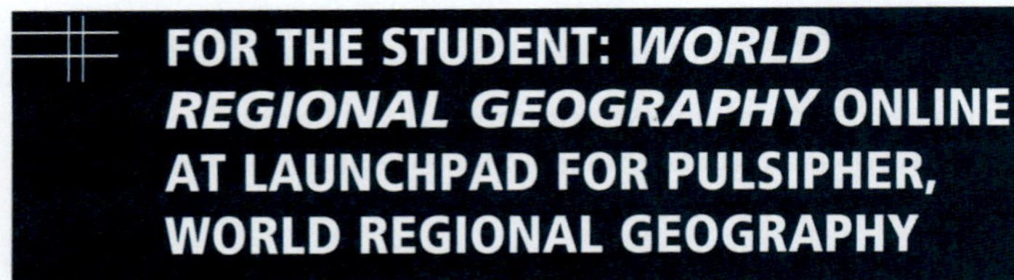

LaunchPad **www.whfreeman.com/pulsipher6e** (before August 2014)
www.whfreeman.com/launchpad/pulsipher6ewithsubregions (after August 2014)
www.whfreeman.com/launchpad/pulsipher6ewithoutsubregions (after August 2014)

A wealth of resources to support the textbook are available online, including the following free and open assets on the companion Web site:

- **Chapter quizzes:** These multiple-choice quizzes help students assess their mastery of each chapter.
- **Thinking Geographically questions:** These multiple-choice questions relate to select photos found throughout the book. The question sets form the basis of computer-graded exercises that can be assigned and automatically graded and entered into the instructor's online grade book.
- **Thinking Critically About Geography:** These activities, fully updated for the sixth edition, allow students to explore a set of current issues, such as deforestation, human rights, or free trade, and see how geography helps clarify our understanding of them. Linked Web sites are matched with a series of questions or with brief activities that help students think about the ways in which they themselves are connected to the places and people they read about in the text.
- **Map Builder software and Map Builder exercises:** The Map Builder program allows students to create layered thematic maps on their own, while Map Builder Exercises offer a specific activity for each chapter in the second edition.
- **Map learning exercises:** Students can use this interactive feature to identify and locate countries, cities, and the major geographic features of each region.
- **Blank outline maps:** Printable maps of the world, and of each region, are available for note taking, exam review, or both, as well as for preparing assigned exercises.
- **Flashcards:** Matching exercises teach vocabulary and definitions.
- **Audio pronunciation guide:** This spoken guide helps students learn to pronounce place names, regional terms, and names of historical figures.
- **World recipes and cuisines:** From *International Home Cooking*, the United Nations International School cookbook, this provides students with the opportunity to explore foods from around the world.

LaunchPad offers all the instructor and student resources listed above, as well as premium resources available only on the portal:

- An eBook of *World Regional Geography*, complete and customizable. Students can quickly search the text and personalize it just as they would the printed version; complete with highlighting, bookmarking, and note-taking features
- **A Guide to Using Google Earth** for the novice, plus step-by-step **Google Earth** exercises for each chapter
- Selected articles from *Focus on Geography* magazine (one for each chapter in the textbook) and accompanying quizzes for each article
- **Physical geography videos** for instructors who want to cover physical geography topics in more detail
- Online **news feeds** for highly respected magazines such as the *Economist*

For more information or to schedule a demo of LaunchPad, please contact your W. H. Freeman sales representative.

NEW! Learning Curve

LEARNINGCurve **Learning Curve** is an intuitive, fun, and highly effective formative assessment tool that is based on extensive educational research. It is a key asset of LaunchPad. Students can use Learning Curve to test their knowledge in a low-stakes environment that helps them improve their mastery of key concepts and prepare for lectures and exams. This adaptive quizzing engine moves students from basic knowledge through critical thinking and synthesis skills as they master content at each level. For a demo, visit www.learningcurveworks.com.

Rand McNally's *Atlas of World Geography*

This atlas, available at a greatly reduced price when bundled with the textbook, contains:

- Fifty-two physical, political, and thematic maps of the world and continents; 49 regional, physical, political, and thematic maps; and dozens of metro-area inset maps
- Geographic facts and comparisons, covering topics such as population, climate, and weather
- A section on common geographic questions, a glossary of terms, and a comprehensive 25-page index

ACKNOWLEDGMENTS

The authors wish to acknowledge the many geographers whose insights and suggestions have informed this book.

World Regional Geography, Sixth Edition

Victoria Alapo
Metropolitan Community College

Jeff Arnold
Southwestern Illinois College

Shaunna Barnhart
Pennsylvania State University

Dean Butzow
Lincoln Land Community College

Philip Chaney
Auburn University

Christine Hansell
Skyline College

Heidi LaMoreaux
Santa Rosa Junior College

Kent Mathewson
Louisiana State University

Julie Mura
Florida State University

Michael Noll
Valdosta State University

Tim Oakes
University of Colorado, Boulder

Kefa M. Otiso
Bowling Green State University

Sam Sweitz
Michigan Technological University

Jeff Ueland
Bemidji State University

Ben Wolfe
Metropolitan Community College, Blue River

World Regional Geography Concepts, First Edition

Gillian Acheson
Southern Illinois University, Edwardsville

Tanya Allison
Montgomery College

Keshav Bhattarai
Indiana University, Bloomington

Leonhard Blesius
San Francisco State University

Jeffrey Brauer
Keystone College

Donald Buckwalter
Indiana University of Pennsylvania

Craig Campbell
Youngstown State University

John Comer
Oklahoma State University

Kevin Curtin
George Mason University

Ron Davidson
California State University, Northridge

Tina Delahunty
Texas Tech University

Dean Fairbanks
California State University, Chico

Allison Feeney
Shippensburg University of Pennsylvania

Eric Fournier
Samford University

Qian Guo
San Francisco State University

Carole Huber
University of Colorado, Colorado Springs

Paul Hudak
University of North Texas

Christine Jocoy
California State University, Long Beach

Ron Kalafsky
University of Tennessee, Knoxville

David Keefe
University of the Pacific

Mary Klein
Saddleback College

Max Lu
Kansas State University

Donald Lyons
University of North Texas

Barbara McDade
University of Florida

Victor Mote
University of Houston

Darrell Norris
State University of New York, Geneseo

Gabriel Popescu
Indiana University, South Bend

Claudia Radel
Utah State University

Donald Rallis
University of Mary Washington

Pamela Riddick
University of Memphis

Jennifer Rogalsky
State University of New York, Geneseo

Tobie Saad
University of Toledo

Charles Schmitz
Towson University

Sindi Sheers
George Mason University

Ira Sheskin
University of Miami

Dmitri Siderov
California State University, Long Beach

Steven Silvern
Salem State College

Ray Sumner
Long Beach City College

Stan Toops
Miami University

Karen Trifonoff
Bloomsburg University of Pennsylvania

Jim Tyner
Kent State University

Michael Walegur
University of Delaware

Scott Walker
Northwest Vista College

Mark Welford
Georgia Southern University

World Regional Geography, Fifth Edition

Gillian Acheson
Southern Illinois University, Edwardsville

Greg Atkinson
Tarleton State University

Robert Begg
Indiana University of Pennsylvania

Richard Benfield
Central Connecticut State University

Fred Brumbaugh
University of Houston, Downtown

Deborah Corcoran
Missouri State University

Kevin Curtin
George Mason University

Lincoln DeBunce
Blue Mountain Community College

Scott Dobler
Western Kentucky University

Catherine Doenges
University of Connecticut, Stamford

Jean Eichhorst
University of Nebraska, Kearney

Brian Farmer
Amarillo College

Eveily Freeman
Ohio State University

Hari Garbharran
Middle Tennessee State University

Abe Goldman
University of Florida

Angela Gray
University of Wisconsin, Oshkosh

Ellen Hansen
Emporia State University

Nick Hill
Greenville Technical College

Johanna Hume
Alvin Community College

Edward Jackiewicz
California State University, Northridge

Rebecca Johns
University of South Florida, St. Petersburg

Suzanna Klaf
Ohio State University

Jeannine Koshear
Fresno City College

Brennan Kraxberger
Christopher Newport University

Heidi Lannon
Santa Fe College

Angelia Mance
Florida Community College, Jacksonville

Meredith Marsh
Lindenwood University

Linda Murphy
Blinn Community College

Monica Nyamwange
William Paterson University

Adam Pine
University of Minnesota, Duluth

Amanda Rees
Columbus State University

Benjamin Richason
St. Cloud State University

Amy Rock
Kent State University

Betty Shimshak
Towson University

Michael Siola
Chicago State University

Steve Smith
Missouri Southern State University

Jennifer Speights-Binet
Samford University

Emily Sturgess Cleek
Drury University

Gregory Taff
University of Memphis

Catherine Veninga
College of Charleston

Mark Welford
Georgia Southern University

Donald Williams
Western New England College

Peggy Robinson Wright
Arkansas State University, Jonesboro

World Regional Geography, Fourth Edition

Robert Acker
University of California, Berkeley

Joy Adams
Humboldt State University

John All
Western Kentucky University

Jeff Allender
University of Central Arkansas

David L. Anderson
Louisiana State University, Shreveport

Donna Arkowski
Pikes Peak Community College

Jeff Arnold
Southwestern Illinois College

Richard W. Benfield
Central Connecticut University

Sarah A. Blue
Northern Illinois University

Patricia Boudinot
George Mason University

Michael R. Busby
Murray State College

Norman Carter
California State University, Long Beach

Gabe Cherem
Eastern Michigan University

Brian L. Crawford
West Liberty State College

Phil Crossley
Western State College of Colorado

Gary Cummisk
Dickinson State University

Kevin M. Curtin
University of Texas, Dallas

Kenneth Dagel
Missouri Western State University

Jason Dittmer
University College London

Rupert Dobbin
University of West Georgia

James Doerner
University of Northern Colorado

Ralph Feese
Elmhurst College

Richard Grant
University of Miami

Ellen R. Hansen
Emporia State University

Holly Hapke
Eastern Carolina University

Mark L. Healy
Harper College

David Harms Holt
Miami University

Douglas A. Hurt
University of Central Oklahoma

Edward L. Jackiewicz
California State University, Northridge

Marti L. Klein
Saddleback College

Debra D. Kreitzer
Western Kentucky University

Jeff Lash
University of Houston, Clear Lake

Unna Lassiter
California State University, Long Beach

Max Lu
Kansas State University

Donald Lyons
University of North Texas

Shari L. MacLachlan
Palm Beach Community College

Chris Mayda
Eastern Michigan University

Armando V. Mendoza
Cypress College

Katherine Nashleanas
University of Nebraska, Lincoln

Joseph A. Naumann
University of Missouri, St. Louis

Jerry Nelson
Casper College

Michael G. Noll
Valdosta State University

Virginia Ochoa-Winemiller
Auburn University

Karl Offen
University of Oklahoma

Eileen O'Halloran
Foothill College

Ken Orvis
University of Tennessee

Manju Parikh
College of Saint Benedict and St. John's University

Mark W. Patterson
Kennesaw State University

Paul E. Phillips
Fort Hays State University

Rosann T. Poltrone
Arapahoe Community College, Littleton, Colorado

Waverly Ray
MiraCosta College

Jennifer Rogalsky
State University of New York, Geneseo

Gil Schmidt
University of Northern Colorado

Yda Schreuder
University of Delaware

Tim Schultz
Green River Community College, Auburn, Washington

Sinclair A. Sheers
George Mason University

D. James Siebert
North Harris Montgomery Community College, Kingwood

Dean Sinclair
Northwestern State University

Bonnie R. Sines
University of Northern Iowa

Vanessa Slinger-Friedman
Kennesaw State University

Andrew Sluyter
Louisiana State University

Kris Runberg Smith
Lindenwood University

Herschel Stern
MiraCosta College

William R. Strong
University of North Alabama

Ray Sumner
Long Beach City College

Rozemarijn Tarhule-Lips
University of Oklahoma

Alice L. Tym
University of Tennessee, Chattanooga

James A. Tyner
Kent State University

Robert Ulack
University of Kentucky

Jialing Wang
Slippery Rock University of Pennsylvania

Linda Q. Wang
University of South Carolina, Aiken

Keith Yearman
College of DuPage

Laura A. Zeeman
Red Rocks Community College

World Regional Geography, Third Edition

Kathryn Alftine
California State University, Monterey Bay

Donna Arkowski
Pikes Peak Community College

Tim Bailey
Pittsburg State University

Brad Baltensperger
Michigan Technological University

Michele Barnaby
Pittsburg State University

Daniel Bedford
Weber State University

Richard Benfield
Central Connecticut State University

Sarah Brooks
University of Illinois, Chicago

Jeffrey Bury
University of Colorado, Boulder

Michael Busby
Murray State University

Norman Carter
California State University, Long Beach

Gary Cummisk
Dickinson State University

Cyrus Dawsey
Auburn University

Elizabeth Dunn
University of Colorado, Boulder

Margaret Foraker
Salisbury University

Robert Goodrich
University of Idaho

Steve Graves
California State University, Northridge

Ellen Hansen
Emporia State University

Sophia Harmes
Towson University

Mary Hayden
Pikes Peak Community College

R. D. K. Herman
Towson University

Samantha Kadar
California State University, Northridge

James Keese
California Polytechnic State University

Phil Klein
University of Northern Colorado

Debra D. Kreitzer
Western Kentucky University

Soren Larsen
Georgia Southern University

Unna Lassiter
California State University, Long Beach

David Lee
Florida Atlantic University

Anthony Paul Mannion
Kansas State University

Leah Manos
Northwest Missouri State University

Susan Martin
Michigan Technological University

Luke Marzen
Auburn University

Chris Mayda
Eastern Michigan University

Michael Modica
San Jacinto College

Heather Nicol
State University of West Georgia

Ken Orvis
University of Tennessee

Thomas Paradis
Northern Arizona University

Amanda Rees
University of Wyoming

Arlene Rengert
West Chester University of Pennsylvania

B. F. Richason
St. Cloud State University

Deborah Salazar
Texas Tech University

Steven Schnell
Kutztown University

Kathleen Schroeder
Appalachian State University

Roger Selya
University of Cincinnati

Dean Sinclair
Northwestern State University

Garrett Smith
Kennesaw State University

Jeffrey Smith
Kansas State University

Dean Stone
Scott Community College

Selima Sultana
Auburn University

Ray Sumner
Long Beach City College

Christopher Sutton
Western Illinois University

Harry Trendell
Kennesaw State University

Karen Trifonoff
Bloomsburg University

David Truly
Central Connecticut State University

Kelly Victor
Eastern Michigan University

Mark Welford
Georgia Southern University

Wendy Wolford
University of North Carolina, Chapel Hill

Laura A. Zeeman
Red Rocks Community College

World Regional Geography, Second Edition

Helen Ruth Aspaas
Virginia Commonwealth University

Cynthia F. Atkins
Hopkinsville Community College

Timothy Bailey
Pittsburg State University

Robert Maxwell Beavers
University of Northern Colorado

James E. Bell
University of Colorado, Boulder

Richard W. Benfield
Central Connecticut State University

John T. Bowen Jr.
University of Wisconsin, Oshkosh

Stanley Brunn
University of Kentucky

Donald W. Buckwalter
Indiana University of Pennsylvania

Gary Cummisk
Dickinson State University

Roman Cybriwsky
Temple University

Cary W. de Wit
University of Alaska, Fairbanks

Ramesh Dhussa
Drake University

David M. Diggs
University of Northern Colorado

Jane H. Ehemann
Shippensburg University

Kim Elmore
University of North Carolina, Chapel Hill

Thomas Fogarty
University of Northern Iowa

James F. Fryman
University of Northern Iowa

Heidi Glaesel
Elon College

Ellen R. Hansen
Emporia State University

John E. Harmon
Central Connecticut State University

Michael Harrison
University of Southern Mississippi

Douglas Heffington
Middle Tennessee State University

Robert Hoffpauir
California State University, Northridge

Catherine Hooey
Pittsburg State University

Doc Horsley
Southern Illinois University, Carbondale

David J. Keeling
Western Kentucky University

James Keese
California Polytechnic State University

Debra D. Kreitzer
Western Kentucky University

Jim LeBeau
Southern Illinois University, Carbondale

Howell C. Lloyd
Miami University of Ohio

Judith L. Meyer
Southwest Missouri State University

Judith C. Mimbs
University of Tennessee, Chattanooga

Monica Nyamwange
William Paterson University

Thomas Paradis
Northern Arizona University

Firooza Pavri
Emporia State University

Timothy C. Pitts
Edinboro University of Pennsylvania

William Preston
California Polytechnic State University

Gordon M. Riedesel
Syracuse University

Joella Robinson
Houston Community College

Steven M. Schnell
Northwest Missouri State University

Kathleen Schroeder
Appalachian State University

Dean Sinclair
Northwestern State University

Robert A. Sirk
Austin Peay State University

William D. Solecki
Montclair State University

Wei Song
University of Wisconsin, Parkside

William Reese Strong
University of North Alabama

Selima Sultana
Auburn University

Suzanne Traub-Metlay
Front Range Community College

David J. Truly
Central Connecticut State University

Alice L. Tym
University of Tennessee, Chattanooga

World Regional Geography, First Edition

Helen Ruth Aspaas
Virginia Commonwealth University

Brad Bays
Oklahoma State University

Stanley Brunn
University of Kentucky

Altha Cravey
University of North Carolina, Chapel Hill

David Daniels
Central Missouri State University

Dydia DeLyser
Louisiana State University

James Doerner
University of Northern Colorado

Bryan Dorsey
Weber State University

Lorraine Dowler
Pennsylvania State University

Hari Garbharran
Middle Tennessee State University

Baher Ghosheh
Edinboro University of Pennsylvania

Janet Halpin
Chicago State University

Peter Halvorson
University of Connecticut

Michael Handley
Emporia State University

Robert Hoffpauir
California State University, Northridge

Glenn G. Hyman
International Center for Tropical Agriculture

David Keeling
Western Kentucky University

Thomas Klak
Miami University of Ohio

Darrell Kruger
Northeast Louisiana University

David Lanegran
Macalester College

David Lee
Florida Atlantic University

Calvin Masilela
West Virginia University

Janice Monk
University of Arizona

Heidi Nast
DePaul University

Katherine Nashleanas
University of Nebraska, Lincoln

Tim Oakes
University of Colorado, Boulder

Darren Purcell
Florida State University

Susan Roberts
University of Kentucky

Dennis Satterlee
Northeast Louisiana University

Kathleen Schroeder
Appalachian State University

Dona Stewart
Georgia State University

Ingolf Vogeler
University of Wisconsin, Eau Claire

Susan Walcott
Georgia State University

These world regional geography textbooks have been a family project many years in the making. Lydia Pulsipher came to the discipline of geography at the age of 5, when her immigrant father, Joe Mihelič, hung a world map over the breakfast table in their home in Coal City, Illinois, where he was pastor of the New Hope Presbyterian Church, and quizzed her on the location of such places as Istanbul. They soon moved to the Mississippi Valley of eastern Iowa, where Lydia's father, then a professor at the Presbyterian theological seminary in Dubuque, continued his geography lessons on the passing landscapes whenever Lydia accompanied him on Sunday trips to small country churches. Lydia's sons, Anthony and Alex, got their first doses of geography in the bedtime stories she told them. For plots and settings, she drew on Caribbean colonial documents she was then reading for her dissertation. They first traveled abroad and learned about the hard labor of field geography when, at age 12 and 8, respectively, they were expected to help with the archaeological and ethnographic research conducted by Lydia and her colleagues on the eastern Caribbean island of Montserrat. It was Lydia's brother John Mihelič who first suggested that Lydia, Alex, and Mac write a book like this one, after he too came to appreciate geography. He has been a loyal cheerleader during the process, as have family and friends in Knoxville, Montserrat, California, Slovenia, and beyond.

The author team was aided by Ola Johansson (University of Pittsburgh at Johntown), who revised Chapter 5 (Russia and the Post-Soviet States) and Chapter 9 (East Asia) for this new edition and improved the final work in innumerable ways. Graduate students and faculty colleagues in the geography department at the University of Tennessee have been generous in their support, serving as helpful impromptu sounding boards for ideas. Ken Orvis, especially, has advised us on the physical geography sections of all editions. Yingkui (Philippe) Li provided information on glaciers and climate change; Russell Kirby wrote one of the vignettes based on his research in Vietnam; Toby Applegate, Alex Pulsipher (in his capacity as an instructor), Michelle Brym, and Sara Beth Keough helped the authors understand how to better assist instructors; and Ron Kalafsky, Tom Bell, Margaret Gripshover, and Micheline Van Riemsdijk chatted with the authors many times on specific and broad issues related to this textbook.

Maps for this edition were conceived by Mac Goodwin and Alex Pulsipher and produced by Will Fontanez and the University of Tennessee cartography shop staff and by Maps.com under the direction of Mike Powers. Alex Pulsipher created and produced the photo essays and chose all the photos used in the book.

Liz Widdicombe and Sara Tenney at W. H. Freeman were the first to facilitate the idea that together we could develop a new direction for *World Regional Geography*, one that included the latest thinking in geography written in an accessible style and well illustrated with attractive, relevant maps and photos. In accomplishing this goal, we are especially indebted to our first developmental editor, Susan Moran, and to the W. H. Freeman staff for all they have done in the first years and since to ensure that this book is well written, beautifully designed, and well presented to the public.

We would also like to gratefully acknowledge the efforts of the following people at W. H. Freeman: Steven Rigolosi, publisher for this sixth edition, who has been extraordinarily supportive and resourceful; Elaine Epstein, developmental editor, who has remained calm and congenial under great pressure; Vivien Weiss, senior project editor; Tom Digiano, marketing manager; Anna Paganelli, copyeditor; Blake Logan, design manager; Matt McAdams, art manager; Susan Wein, production manager; and Stephanie Ellis, assistant editor.

Given our ambitious new photo program, we are especially grateful for Blake Logan's brilliant work and responsiveness as designer for the sixth edition, as well as for Hilary Newman and Jennifer MacMillan's guidance and direction as our photo editors for the sixth edition. We are also grateful to the supplements authors, who have created what we think are unusually useful, up-to-date, and labor-saving materials for instructors who use our book.

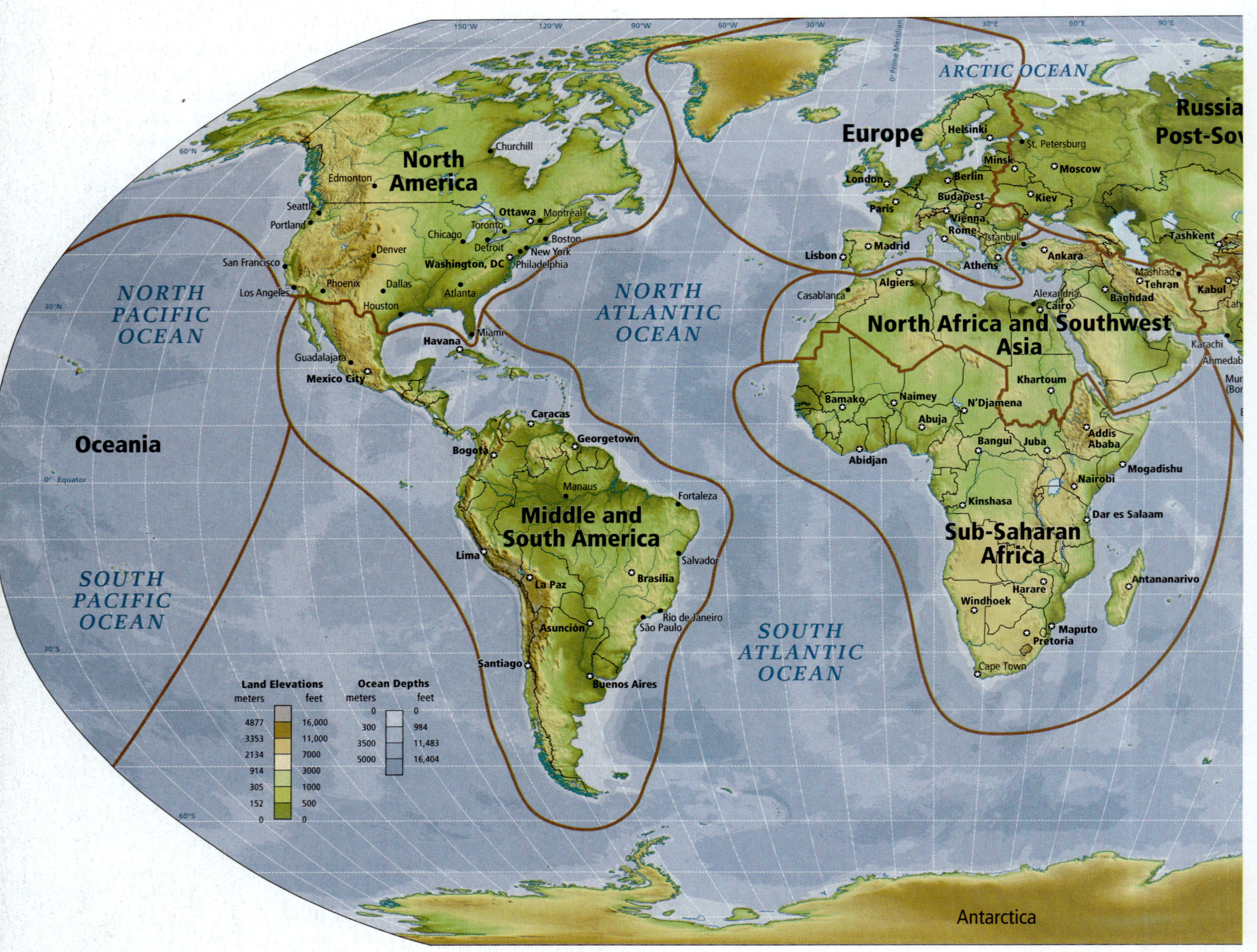

ARCTIC OCEAN
NORTH PACIFIC OCEAN
NORTH ATLANTIC OCEAN
SOUTH PACIFIC OCEAN
SOUTH ATLANTIC OCEAN
North America
Europe
Oceania
Middle and South America
North Africa and Southwest Asia
Sub-Saharan Africa
Antarctica
Churchill
Edmonton
Seattle
Portland
San Francisco
Los Angeles
Denver
Phoenix
Dallas
Houston
Chicago
Detroit
Toronto
Ottawa
Montréal
Boston
New York
Philadelphia
Washington, DC
Atlanta
Miami
Havana
Guadalajara
Mexico City
Caracas
Georgetown
Bogotá
Manaus
Fortaleza
Lima
Salvador
La Paz
Brasília
Rio de Janeiro
São Paulo
Asunción
Santiago
Buenos Aires
Helsinki
St. Petersburg
Minsk
Moscow
London
Berlin
Kiev
Paris
Budapest
Vienna
Rome
Istanbul
Madrid
Lisbon
Athens
Ankara
Tashkent
Mashhad
Tehran
Kabul
Baghdad
Casablanca
Algiers
Alexandria
Cairo
Karachi
Khartoum
Bamako
Niamey
N'Djamena
Abuja
Abidjan
Bangui
Juba
Addis Ababa
Mogadishu
Nairobi
Kinshasa
Dar es Salaam
Antananarivo
Harare
Windhoek
Maputo
Pretoria
Cape Town
Land Elevations
meters feet
4877 16,000
3353 11,000
2134 7000
914 3000
305 1000
152 500
0 0
Ocean Depths
meters feet
0 0
300 984
3500 11,483
5000 16,404

1 Geography: An Exploration of Connections

FIGURE 1.1 Regions of the world.

GEOGRAPHIC INSIGHTS

After you read this chapter, you will be able to discuss the following geographic insights as they relate to the nine thematic concepts, which are explained further on page 12:

1. Physical and Human Geographers: The primary concerns of both physical and human geographers are the study of the Earth's surface and the interactive physical and human processes that shape the surface.

2. Regions: The concept of *region* is useful to geographers because it allows them to break up the world into manageable units in order to analyze and compare spatial relationships. Nonetheless, regions do not have rigid definitions and their boundaries are fluid.

3. Gender and Population: The shift toward greater gender equality is having an influence on population growth patterns, patterns of economic development, and the distribution of power within families, communities, and countries.

4. Food and Urbanization: Modernization in food production is pushing agricultural workers out of rural areas toward urban areas where jobs are more plentiful but where food must be purchased. This circumstance often leads to dependency on imported food.

5. Globalization and Development: Increased global flows of information, goods, and people are transforming patterns of economic development.

6. Power and Politics: There are major differences across the globe in the ways that power is wielded in societies. Modes of governing that are more authoritarian are based on the power of the state or community (or tribal) leaders. Modes that are more democratic give the individual a greater say in how policies are developed and governments are run. There are also many other ways of managing political power.

7. Climate Change and Water: Water and other environmental factors often interact to influence the vulnerability of a location to the impacts of climate change. These vulnerabilities have a spatial pattern.

Where Is It? Why Is It There? Why Does It Matter?

Where are you? You may be in a house or a library or sitting under a tree on a fine fall afternoon. You are probably in a community (perhaps a college or university), and you are in a country (perhaps the United States) and a region of the world (perhaps North America, Southeast Asia, or the Pacific). Why are you where you are? Some answers are immediate, such as "I have an assignment to read." Other explanations are more complex, such as your belief in the value of an education, your career plans, and your or someone's willingness to sacrifice to pay your tuition. Even past social movements that opened up higher education to more than a fortunate few may help explain why you are where you are.

The questions *where* and *why* are central to geography. Think about a time you had to find the site of a party on a Saturday night, the location of the best grocery store, or the fastest and safest route home. You were interested in location, spatial relationships, and connections between the environment and people. Those are among the interests of geographers.

Geographers seek to understand why different places have different sights, sounds, smells, and arrangements of features. They study what has contributed to the look and feel of a place, to the standard of living and customs of the people, and to the way people in one place relate to people in other places. Furthermore, geographers often think on several scales, from the local to the global. For example, when choosing the best location for a new grocery store, a geographer might consider the physical characteristics of potential sites, the socioeconomic circumstances of the neighborhood, traffic patterns locally and in the city at large, as well as the store's location relative to the main population concentrations for the whole city. She would probably also consider national or even international transportation routes, possibly to determine cost-efficient connections to suppliers.

To make it easier to understand a geographer's many interests, try this exercise. Draw a map of your most familiar childhood landscape. Relax, and recall the objects and experiences that were most important to you there. If the place was your neighborhood, you might start by drawing and labeling your home. Then fill in other places you encountered regularly, such as your backyard, your best friend's home, or your school. Figure 1.2 shows the childhood landscape remembered by Julia Stump in Franklin, Tennessee.

Consider how your map reveals the ways in which your life was structured by space. What is the scale of your map? That is, how much space did you decide to illustrate on the map? The amount of space your map covers may represent the degree of freedom you had as a child, or how aware you were of the world around you. Were there places you were not supposed to go? Does your map reveal, perhaps subtly, such emotions as fear, pleasure, or longing? Does it indicate your sex, your ethnicity, or the makeup of your family? Did you use symbols to show certain features? In making your map and analyzing it, you have engaged in several aspects of geography:

- Landscape observation
- Descriptions of the Earth's surface and consideration of the natural environment
- Spatial analysis (the study of how people, objects, or ideas are related to one another across space)
- The use of different scales of analysis (your map probably shows the spatial features of your childhood at a detailed *local scale*)
- Cartography (the making of maps)

As you progress through this book and this course, you will acquire geographic information and skills. Perhaps you are planning to travel to other lands or are thinking about investing in East Asian timber stocks. Maybe you are searching for a good place to market an idea or are trying to understand current events in your town within the context of world events. Knowing how to practice geography will make your task easier and more engaging.

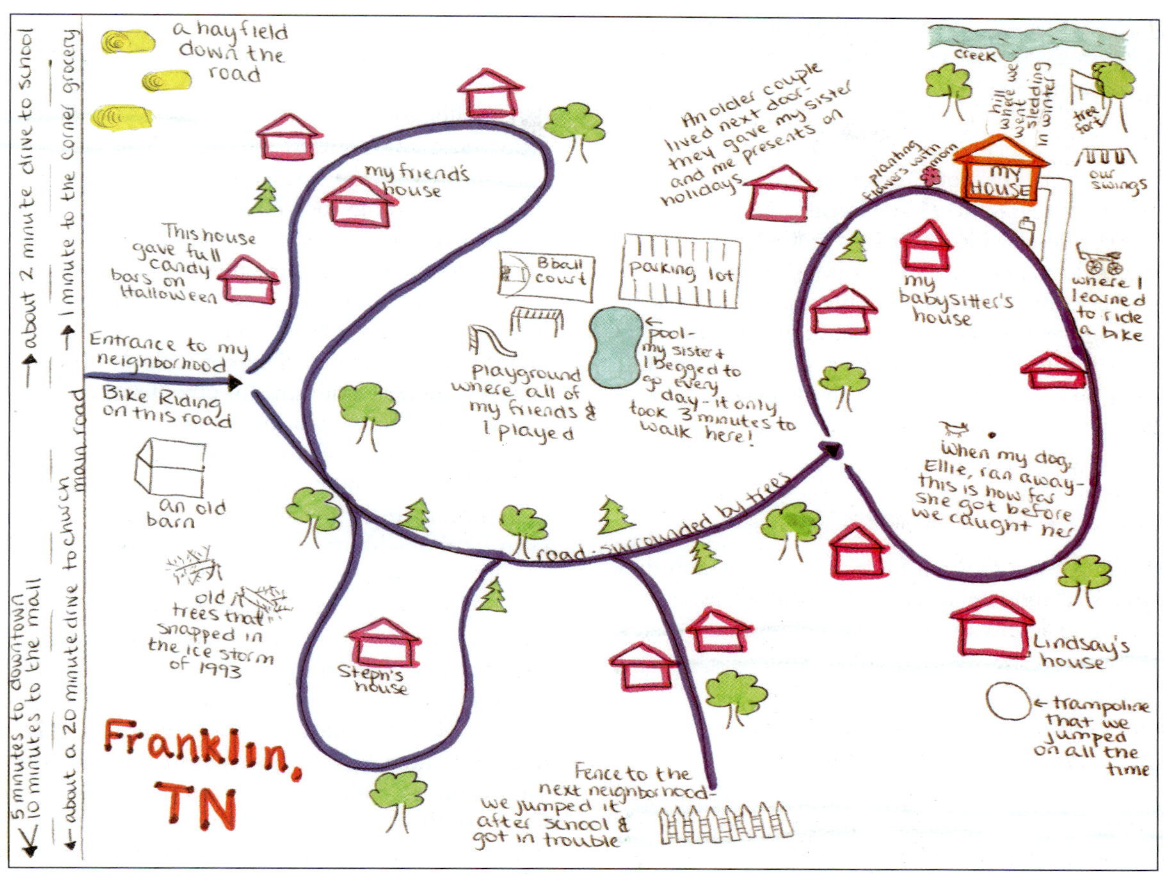

FIGURE 1.2 A childhood landscape map. Julia Stump drew this map of her childhood landscape in Franklin, Tennessee, as an exercise in Dr. Pulsipher's world geography class.

What Is Geography?

> **Geographic Insight 1**
>
> **Physical and Human Geography:** The primary concerns of both physical and human geographers are the study of the Earth's surface and the interactive physical and human processes that shape the surface.

Geography is the study of our planet's surface and the processes that shape it. Yet this definition does not begin to convey the fascinating interactions of human and environmental forces that have given the Earth its diverse landscapes and ways of life.

Geography, as an academic discipline, is unique in that it links the physical sciences—such as geology, physics, chemistry, biology, and botany—with the social sciences—such as anthropology, sociology, history, economics, and political science. **Physical geography** generally focuses on how the Earth's physical processes work independently of humans, but increasingly, physical geographers have become interested in how physical processes may affect humans and how humans affect these processes in return. **Human geography** is the study of the various aspects of human life that create the distinctive landscapes and regions of the world.

Physical and human geography are often tightly linked. For example, geographers might try to understand:

- How and why people came to occupy a particular place.
- How people use the physical aspects of that place (climate, landforms, and resources) and then modify them to suit their particular needs.
- How people may create environmental problems.
- How people interact with other places, far and near.

Geographers usually specialize in one or more fields of study, or subdisciplines. Some of these particular types of geography are mentioned over the course of the book. Despite their individual specialties, geographers often cooperate in studying **spatial interaction** between people and places and the **spatial distribution** of relevant phenomena. For example, in the face of increasing global warming, climatologists, cultural geographers, and economic geographers work together to understand the spatial distribution of carbon dioxide emissions, as well as the cultural and economic practices that

physical geography the study of the Earth's physical processes: how they work and interact, how they affect humans, and how they are affected by humans

human geography the study of patterns and processes that have shaped human understanding, use, and alteration of the Earth's surface

spatial interaction the flow of goods, people, services, or information across space and among places

spatial distribution the arrangement of a phenomenon across the Earth's surface

might be changed to limit such emissions. This could take the form of redesigning urban areas so that people can live closer to where they work, or encouraging food production in locations closer to where the food will be consumed.

Many geographers specialize in a particular region of the world, or even in one small part of a region. Regional geography is the analysis of the geographic characteristics of a particular place, the size and scale of which can vary radically. The study of a region can reveal connections among physical features and ways of life, as well as connections to other places. These links are key to understanding the present and the past, and are essential in planning for the future. This book follows a "world regional" approach, focusing on general knowledge about specific regions of the world. We will see just what geographers mean by *region* a little later in this chapter.

Geographers' Visual Tools

Among geographers' most important tools are maps, which they use to record, analyze, and explain spatial relationships, as you did on your childhood landscape map. Geographers who specialize in depicting geographic information on maps are called **cartographers**.

Understanding Maps

A map is a visual representation of the surface of the Earth used to record, display, analyze, and explain spatial relationships. Figure 1.3 on pages 5–6 explains the various features of maps.

Legend and Scale

The first thing to check on a map is the **legend**, which is usually a small box somewhere on the map that provides basic information about how to read the map, such as the meaning of the symbols and colors used (see parts A–C and the "Legend" box in Figure 1.3). Sometimes the scale of the map is also given in the legend.

In cartography, *scale* has a slightly different meaning than it does in general geographic analysis. **Scale** on a map refers to the relationship between the size of things on the map and the actual size they have on the surface of the Earth. It is usually represented with a scale bar (see Figure 1.3D–G) but is also sometimes represented by a ratio (for example 1:8000) or a fraction (1/8000), which indicates what one unit of measure on the map equals in the same units on the ground. For example, 1:8000 in. means that 1 inch on the map represents 8000 inches (about an eighth of a mile) on the surface of the Earth.

cartographer geographers who specialize in depicting geographic information on maps

legend a small box somewhere on a map that provides basic information about how to read the map, such as the meaning of the symbols and colors used

scale (of a map) the proportion that relates the dimensions of the map to the dimensions of the area it represents; also, variable-sized units of geographical analysis from the local scale to the regional scale to the global scale

longitude the distance in degrees east and west of Greenwich, England; lines of longitude, also called meridians, run from pole to pole (the line of longitude at Greenwich is 0° and is known as the prime meridian)

latitude the distance in degrees north or south of the equator; lines of latitude run parallel to the equator, and are also called parallels

A scale of 1/800 is considered to be larger than a scale of 1/8000 because the features on a 1/800 scale map are larger and can be shown in greater detail. The larger the scale of the map, the smaller the area it covers. A larger-scale map shows things larger; a smaller-scale map shows more things—with each thing smaller, less visible. You can remember this with the following statement: "Things look larger on a larger-scale map."

In parts A–C of Figure 1.3, different *scales of imagery* are demonstrated using maps, photographs, and a satellite image. Read the captions carefully to understand the scale being depicted in each image. Throughout this book, you will encounter different kinds of maps at different scales. Some will show physical features, such as landforms or climate patterns at the regional or global scale. Others will show aspects of human activities at these same regional or global scales—for example, the routes taken by drug traders. Yet other maps will show patterns of settlement or cultural features at the scale of countries or regions, or cities, or even local neighborhoods (Figure 1.3D–G).

It is important to keep the two types of scale used in geography—*map scale* and *scale of analysis*—distinct, because they have opposite meanings! In spatial analysis of a region such as Southwest Asia, scale refers to the spatial extent of the area that is being discussed. Thus a large-scale analysis means a large area is being explored. But in cartography, a large-scale map is one that shows a given area blown up so that fine detail is visible, while a small-scale map shows a larger area in much less detail. In this book, when we talk about scale we are referring to its meaning in spatial analysis (larger scale = larger area), unless we specifically indicate that we are talking about scale as used in cartography (larger scale = smaller area). In the Understanding Maps: Scale box in Figure 1.3, the largest-scale map is that on the left (D); the smallest is on the right (G).

Longitude and Latitude

Most maps contain lines of latitude and longitude, which enable a person to establish a position on the map relative to other points on the globe. Lines of **longitude** (also called *meridians*) run from pole to pole; lines of **latitude** (also called *parallels*) run around the Earth parallel to the equator (see Figure 1.3H).

Both latitude and longitude lines describe circles, so there are 360° (the symbol ° refers to degrees) in each circle of latitude and 180° in each pole-to-pole semicircle of longitude. Each degree spans 60 minutes (designated with the symbol ′), and each minute has 60 seconds (designated with the symbol ″). Keep in mind that these are measures of relative linear space on a circle, not measures of time. They do not even represent real distance because the circles of latitude get successively smaller to the north and south of the equator until they become virtual dots at the poles.

The globe is also divided into hemispheres. The Northern and Southern hemispheres are on either side of the equator. The Western and Eastern hemispheres are defined as follows. The prime meridian, 0° longitude, runs from the North Pole through Greenwich, England, to the South Pole. The half of the globe's surface west of the prime meridian is called the Western Hemisphere; the half to the east is called the Eastern Hemisphere.

FIGURE 1.3 Understanding Maps

The Legend

Being able to read a map legend is crucial to understanding the maps in this book. The colors in the legend convey information about different areas on the map. In the population density map below, the lowest density (0–3 persons per square mile), is colored light tan. A part of North America with this density is shown in the map inset to the right of the legend. On the far right is a picture of this area. Two other densities (27–260 and more than 2600 people per square mile or 1000 per square kilometer) are also shown in this manner.

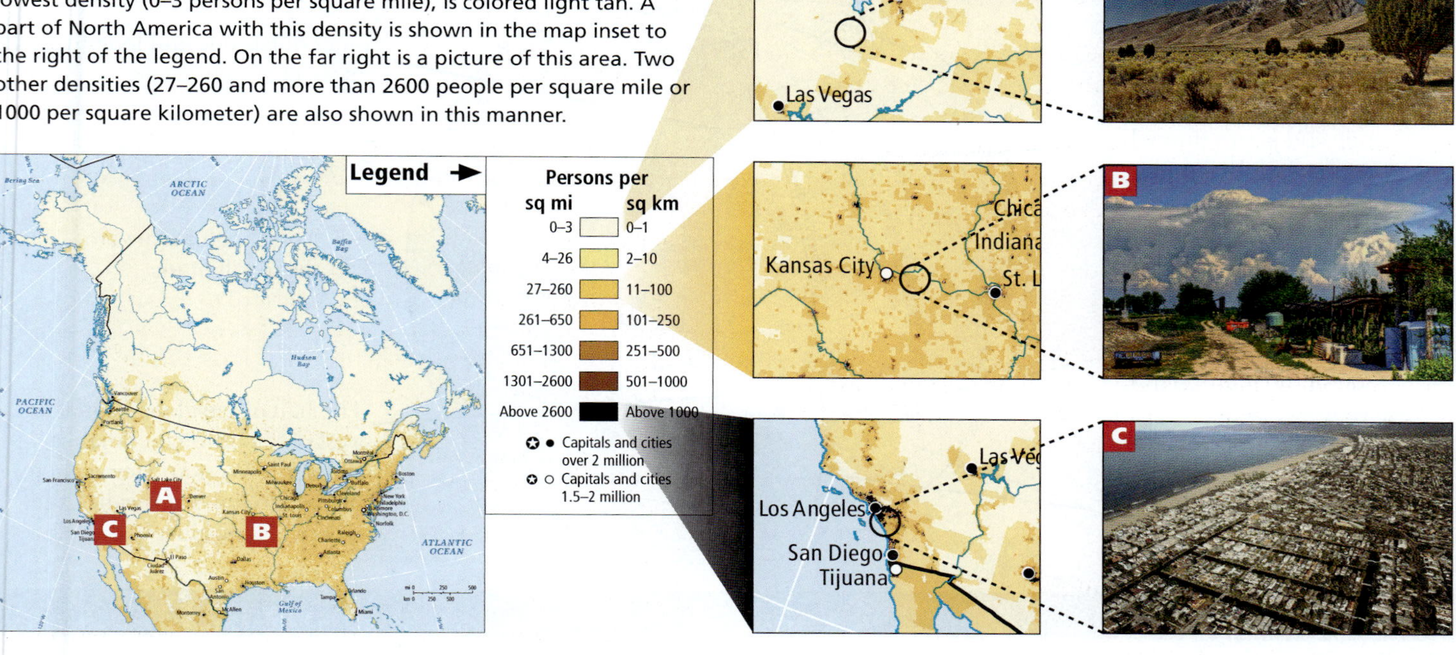

Scale

Maps often display information at different spatial scales, which means that lengths, areas, distances, and sizes can appear dramatically different on otherwise similar maps. This book often combines maps at several different scales with photographs taken by people at Earth's surface and photographs taken by satellites or astronauts in space. All of these visual tools convey information at a spatial scale. Here are some of the map scales you might encounter in this book. The scale is visible below each image.

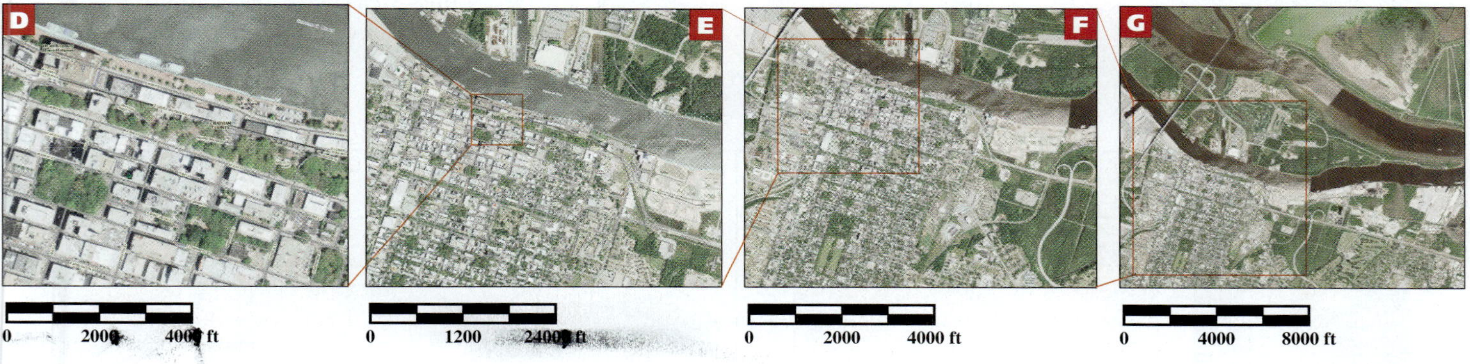

Representation of Scale

Here are some representations of map scale that you may encounter on maps in this book and elsewhere.

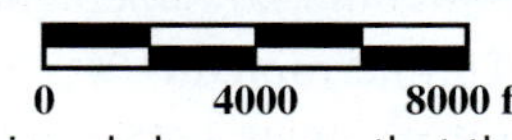

This scale bar means that the length of the entire box represents 8000 feet on the ground.

mi 0 1000 2000 3000 4000 5000

km 0 1000 2000 3000 4000 5000 6000 7000 8000

This scale bar works like the one on the left, but also gives lengths in miles and kilometers.

1:8000

This means that 1 unit of measure (an inch, or finger width) equals 8000 similar units of measure on the ground.

FIGURE 1.3 Understanding Maps *(continued)*

Latitude and Longitude

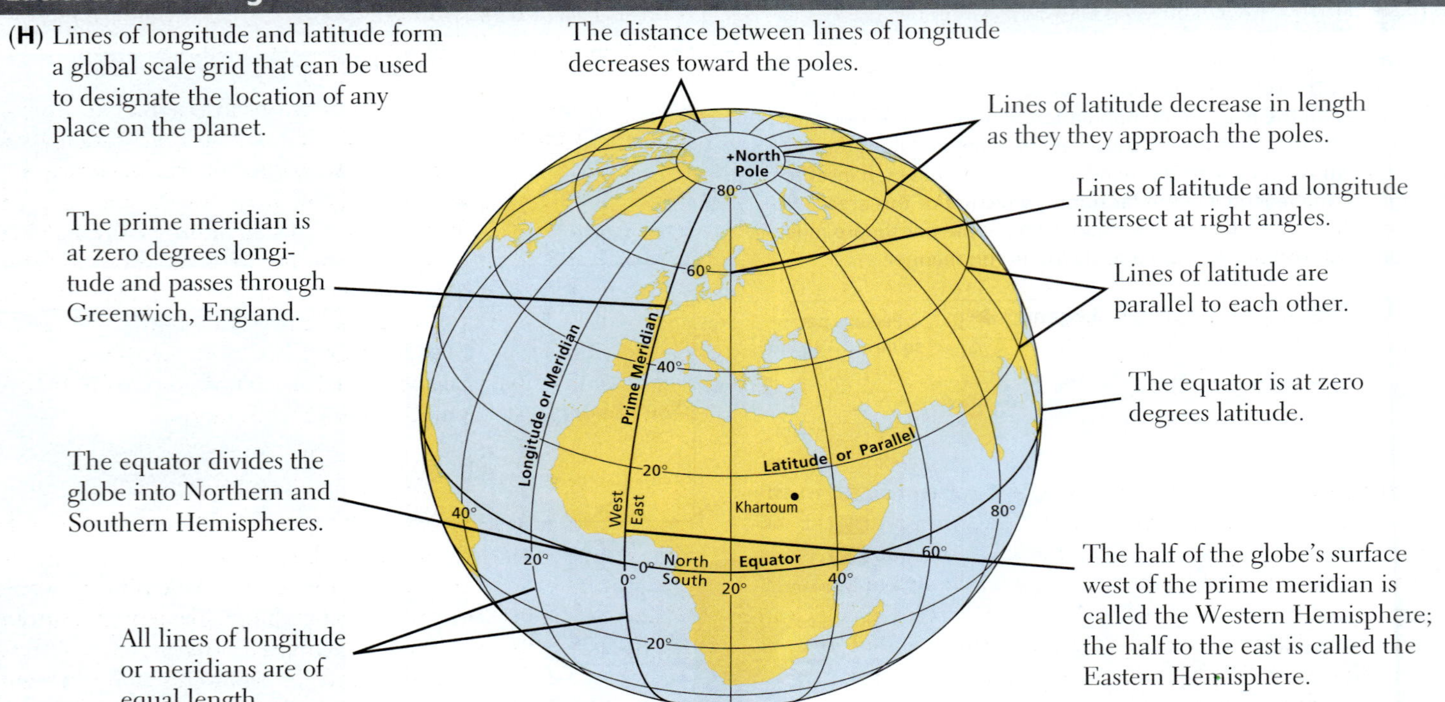

(H) Lines of longitude and latitude form a global scale grid that can be used to designate the location of any place on the planet.
The distance between lines of longitude decreases toward the poles.
Lines of latitude decrease in length as they they approach the poles.
Lines of latitude and longitude intersect at right angles.
Lines of latitude are parallel to each other.
The equator is at zero degrees latitude.
The prime meridian is at zero degrees longitude and passes through Greenwich, England.
The equator divides the globe into Northern and Southern Hemispheres.
All lines of longitude or meridians are of equal length.
The half of the globe's surface west of the prime meridian is called the Western Hemisphere; the half to the east is called the Eastern Hemisphere.
North Pole
80°
60°
40°
20°
0°
Prime Meridian
Longitude or Meridian
Latitude or Parallel
Khartoum
West
East
North
South
Equator
80°
60°
40°
20°
20°
40°

Projections

(I) Albers Projection

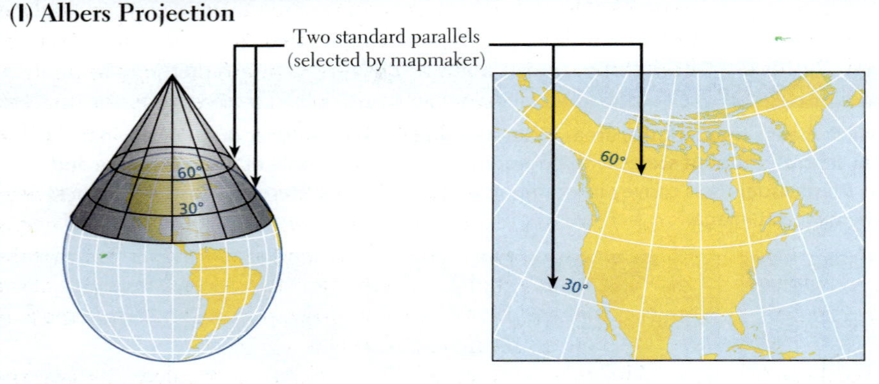

Pros: Minimal distortion near two parallels (lines of latitude)

Cons: Areas farther away from these lines will have distortion.

(J) Mercator Projection

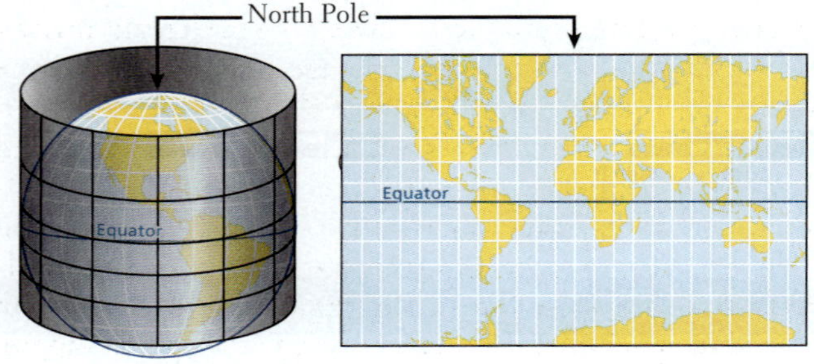

Pros: A straight line between two points on this map gives an accurate compass direction between them. Minimal distortion within 15 degrees of the equator.

Cons: Extreme distortion near the poles, especially above 60 degrees latitude.

(K) Robinson Projection

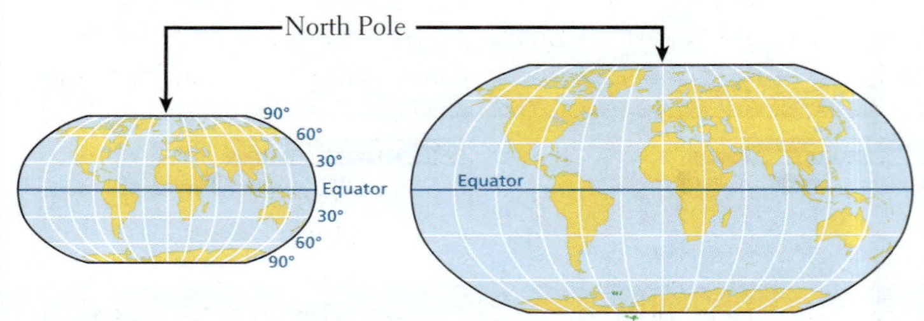

Pros: Uninterrupted view of land and ocean. Less distortion in high latitudes than in the Mercator projection.

Cons: The shapes of landmasses are slightly distorted due to the curvature of the longitude lines.

The longitude lines both east and west of the prime meridian are labeled from 1° to 180° by their direction and distance in degrees from the prime meridian. For example, 20 degrees east longitude would be written as 20° E. The longitude line at 180° runs through the Pacific Ocean and is used roughly as the international date line; the calendar day officially begins when midnight falls at this line.

The equator divides the globe into the Northern and Southern hemispheres. Latitude is measured from 0° at the equator to 90° at the North Pole or South Pole.

Lines of longitude and latitude form a grid that can be used to designate the location of a place. In Figure 1.3H, notice the dot that marks the location of Khartoum below the 20th parallel in eastern Africa. The position of Khartoum is 15° 35′ 17″ N latitude by 32° 32′ 3″ E longitude.

Map Projections

Printed maps must solve the problem of showing the spherical Earth on a flat piece of paper. Imagine drawing a map of the Earth on an orange, peeling the orange, and then trying to flatten out the orange-peel map and transferring it exactly to a flat piece of paper. The various ways of showing the spherical surface of the Earth on flat paper are called **map projections**. All projections create some distortion. For maps of small parts of the Earth's surface, the distortion is minimal. Developing a projection for the whole surface of the Earth that minimizes distortion is much more challenging.

For large midlatitude regions of the Earth that are mainly east/west in extent (North America, Europe, China, Russia), an *Albers projection* is often used. As you can see in Figure 1.3I, this is a conic, or cone-shaped, projection. The cartographer chooses two standard parallels (lines of latitude) on which to orient the map, and these parallels have no distortion. Areas along and between these parallels display minimal distortion. Areas farther to the north or south of the chosen parallels have more distortion. Although all areas on the map are proportional to areas on the ground, distortion of actual shape is inherent in the projection, because, as previously discussed, parts of the globe are being projected onto flat paper.

The *Mercator projection* (see Figure 1.3J) has long been used by the general public, but geographers rarely use this projection because of its gross distortion near the poles. To make his flat map, the Flemish cartographer Gerardus Mercator (1512–1594) stretched out the poles, depicting them as lines equal in length to the equator! As a result, Greenland, for example, appears about as large as Africa, even though it is only about one-fourteenth Africa's size. Nevertheless, the Mercator projection is still useful for navigation because it portrays the shapes of landmasses more or less accurately, and because a straight line between two points on this map gives the compass direction between them; but actual distance measurements are distorted.

map projections the various ways of showing the spherical Earth on a flat surface

Geographic Information Science (GISc) the body of science that underwrites multiple spatial analysis technologies and keeps them at the cutting edge

The *Robinson projection* (see Figure 1.3K) shows the longitude lines curving toward the poles to give an impression of the Earth's curvature, and it has the advantage of showing an uninterrupted view of land and ocean; however, as a result, the shapes of landmasses are slightly distorted. In this book we often use the Robinson projection for world maps.

Maps are not unbiased. Most currently popular world map projections reflect the European origins of modern cartography. For example, Europe or North America is often placed near the center of the map, where distortion is minimal; other population centers, such as East Asia, are placed at the highly distorted periphery. For a less-biased study of the modern world, we need world maps that center on different parts of the globe. Another source of bias in maps is the convention that north is always at the top of the map. Some cartographers think that this can lead to a subconscious assumption that the Northern Hemisphere is somehow superior to the Southern Hemisphere.

Geographic Information Science (GISc)

The acronym **GISc** is now widespread and usually refers to **Geographic Information Science**, the body of science that supports spatial analysis technologies. GISc is multidisciplinary, using techniques from cartography (mapmaking), geodesy (measuring the Earth's surface), and photogrammetry (the science of making reliable measurements, especially by using aerial photography). Other sciences, such as cognitive psychology and spatial statistics (geomatics or geoinformatics) are increasingly being used to give greater depth and breadth to three-dimensional spatial analysis. GISc, then, can be used in medicine to analyze the human body, in engineering to analyze mechanical devices, in architecture to analyze buildings, in archaeology to analyze sites above and below ground, and in geography to analyze the Earth's surface and the space above and below the Earth's surface.

GISc is a burgeoning field in geography, with wide practical applications in government and business and in efforts to assess and improve human and environmental conditions. GIS (without the *c*) is an older term that refers to *geographic information systems* and is applied to the computerized analytical systems that are the tools of this newest of spatial sciences.

The now widespread use of GISc, particularly by governments and corporations, has dramatically increased the amount of information that is collected and stored, and changed the way it is analyzed and distributed. These changes create many new opportunities for solving problems, for example, by increasing the ability of local governments to plan future urban growth. However, these technologies also raise serious ethical questions. What rights do people have over the storage, analysis, and distribution of information about their location and movements, which can now be gathered from their cell phones? Should this information reside in the public domain? Should individuals have the right to have their location-based information suppressed from public view? Should a government or corporation have the right to sell information to anyone, without special permission, about where people spend their time and how frequently they go to particular places? Progress on these societal questions has not kept pace with the technological advances in GISc.

THE DETECTIVE WORK OF PHOTO INTERPRETATION

Most geographers use photographs to help them understand or explain a geographic issue or depict the character of a place. Interpreting a photo to extract its geographical information can sometimes be like detective work. Below are some points to keep in mind as you look at the pictures throughout this book; try them out first with the photo on this page in Figure 1.4.

(A) Landforms: Notice the lay of the land and the landform features. Is there any indication of how the landforms and humans have influenced each other? Is environmental stress visible?

(B) Vegetation: Notice whether the vegetation indicates a wet or dry, or warm or cold environment. Can you recognize specific species? Does the vegetation appear to be natural or influenced by human use?

(C) Material culture: Are there buildings, tools, clothing, foods, plantings, or vehicles that give clues about the cultural background, wealth, values, or aesthetics of the people who live where the picture was taken?

(D) What do the people in the photo suggest about the situation pictured?

(E) Can you see evidence of the global economy, such as goods that probably were not produced locally?

(F) Location: From your observations, can you tell where the picture was taken or narrow down the possible locations?

You can use this system to analyze any of the photos in this book or elsewhere. Practice by analyzing the photos in this book before you read their captions. Here is an example of how you could do this with the photo below (Figure 1.4):

(A) Landforms:

1. The flat horizon suggests a plain or a river delta.

   Environmental stress is visible in several places.

2. This oily liquid doesn't look natural. Could it be crude oil?

   What would have caused the landscape transformation? Maybe an oil spill?

(B) Vegetation:

3. This looks like a palm tree. There are quite a few palm trees and other trees.

   This must be the tropics and be fairly wet and warm.

(C) Material culture:

There is not much that is obviously material culture here, just a single person. The whole area might be abandoned.

(D) People:

4. The clothing on this person doesn't look like he made it. It looks mass produced.

   This suggests that he has access to goods produced some distance away, maybe in a nearby city. Or possibly, he could buy things in a market where imported goods are sold.

(E) Global economy: See (D).

(F) Location: This could be somewhere tropical where there could have been an oil spill. Hint: Use this book! See Figure 6.26 for a figure showing OPEC members. This suggests that the photo could be of Venezuela, Ecuador, Nigeria, Angola, or Indonesia. Suggestion: Look for relevant information in Chapter 7.

FIGURE 1.4 Oil development and the environment. A man walks through swampy land in 2010. An international development company began extracting products from this area 50 years ago. A recent UN report stated that now these areas need one of the world's largest cleanups, which could take up to 30 years and cost over a billion dollars. The area has experienced around 300 incidences of pollution a year since the 1970s, causing an unknown number of deaths. More information about this situation can be found in Chapter 7.

The Region as a Concept

Geographic Insight 2

Regions: The concept of *region* is useful to geographers because it allows them to break up the world into manageable units in order to analyze and compare spatial relationships. Nonetheless, regions do not have rigid definitions and their boundaries are fluid.

A **region** is a unit of the Earth's surface that contains distinct patterns of physical features and/or distinct patterns of human development. It could be a desert region, a region that produces rice, or a region experiencing ethnic violence. Geographers rarely use the same set of attributes to describe any two regions. For example, the region of the southern United States might be defined by its distinctive vegetation, architecture, music, foods, and historical experience. Meanwhile Siberia, in eastern Russia, could be defined primarily by its climate, vegetation, remoteness, and sparse settlement.

region a unit of the Earth's surface that contains distinct patterns of physical features and/or distinct patterns of human development

Another issue in defining regions is that they may shift over time. The people and the land they occupy may change so drastically in character that they no longer can be thought of as belonging to a certain region, and become more closely aligned with another, perhaps adjacent, region. Examples of this are countries in Central Europe, such as Poland and Hungary, which, for more than 40 years, were closely aligned with Russia and the Soviet Union, a vast region that stretched across northern Eurasia to the Pacific (Figure 1.5A). Poland and Hungary's borders with western Europe were highly militarized and shut to travelers. With the demise of the Soviet Union in the early 1990s and the drastic political and economic changes that then came about, Poland and Hungary became members of the European Union (EU) in 2004 (Figure 1.5B). Their western borders are now open, while their eastern borders are now more heavily guarded in order to keep unwelcome immigrants and influences

FIGURE 1.5 Changing country alliances and relationships in Europe, pre-1989 (A) and 2013 (B).

(A) Pre-1989 alignment of countries in Europe and the Soviet Union.

FIGURE 1.5 (B) Post-2004 alignments of the European Union and of Russia and the Post-Soviet States.

out of the European Union. But through all this change—on the ground, in the border regions—people share cultural features (language, religion, historical connections) even as they may define each other as having very different regional allegiances. For this reason, we say regional borders can be *fuzzy*, meaning that they are hard to determine precisely.

A recurring problem in world regional textbooks is the changing nature of regional boundaries and the fact that on the ground they are not clear lines but linear zones of fuzziness. For example, when we first designed this textbook in the mid-1990s, the changes to Europe were just beginning; its eastern limits were under revision as the Soviet Union disintegrated. There were also hints that as the Soviet Union disappeared, countries of Central Asia, long within the Soviet sphere—places like Kazakhstan, Turkmenistan, Kyrgyzstan, Uzbekistan—should perhaps have been defined as constituting a new region of their own. Some suggested that these four, plus Turkey, Syria, Lebanon, Israel, Iraq, Iran, Saudi Arabia, the Emirates, as well as Afghanistan, Pakistan, and possibly even western China, should become a new post-Soviet world region of Central Asia (Figure 1.6). The suggestion was that this region would be defined by what was thought to be a common religious heritage (Islam), plus long-standing cultural and historical ties and a difficult environment marked by water scarcity but also by rich oil and gas resources.

In fact, such a facile lumping together of these very different countries into a region based on such criteria would misrepresent the current situation as well as the past. A world region known as Central Asia may emerge eventually (and there are groups of interested parties discussing that possibility right now), but such a region will be slow to take shape and will be based on criteria very different from common religion, historical experiences, and environmental features. First of all, these supposed uniting features are actually a fallacy. There are many different versions of Islam practiced from the Mediterranean to western China and from southern Russia to the Hindu Kush. Also, to the extent that there are common historical experiences, they are actually linked more to European and Russian colonial

FIGURE 1.6 Hypothetical map of a possible Central Asian world region. A hypothetical region called Central Asia could consist of Kazakhstan, Turkmenistan, Kyrgyzstan, and Uzbekistan, plus Turkey, Iraq, Iran, the Caucasus, and also Afghanistan and Pakistan. For cultural reasons, it might even include parts of western China, but that is not considered here. The Caucasus countries have had a long association with Central Asia; economically, they might find this a better association than Europe, which is halfheartedly courting them for oil reasons. Just which region Saudi Arabia plus Jordan, Syria, Kuwait, Yemen, and Israel would fall into is debatable—perhaps a region called the Eastern Mediterranean and North Africa. The Emirates, despite their location on the Arabian Peninsula, appear to be angling for a leading economic role in such a new Central Asian region. The reader is reminded to consider the commentary in the text in considering the viability of this possible region.

exploitation than to an ancient and deeply uniting Central Asian cultural heritage. Finally, while environments generally defined by water scarcity are common to all the countries listed, oil and gas resources are not uniformly distributed at all. It could be that a Central Asian identity will eventually develop, perhaps centered on the leadership of Turkey or Dubai, which is trying hard to define itself as the affluent capital of such a region; but thus far, the region has not coalesced. Therefore in this book, the countries listed above are to be found in the three different regions shown in Figure 1.6.

If regions are so difficult to define and describe, why do geographers use them? To discuss the whole world at once would be impossible, so geographers seek a reasonable way to divide the world into manageable parts. There is nothing sacred about the criteria or the boundaries for the world regions we use. They are just practical aids to learning. In defining each of the world regions for this book, we have considered such factors as physical features, political boundaries, cultural characteristics, history, how the places now define themselves, and what the future may hold. We are constantly reevaluating regional boundaries and, in this edition, have made some changes. For example, the troubled new country of South Sudan is no longer administered from Khartoum. Because its population is majority Black African, it is now considered part of sub-Saharan Africa, to which it is more culturally aligned (see Chapter 7).

global scale the level of geography that encompasses the entire world as a single unified area

world regional scale the regional scale of analysis that encompasses all the regions of the world

local scale the level of geography that describes the space where an individual lives or works; a city, town, or rural area

world region a part of the globe delineated according to criteria selected to facilitate the study of patterns particular to the area

This book organizes the material into three *regional scales of analysis*: the **global scale**, the **world regional scale**, and the **local scale**. The term *scale of analysis* refers to the relative size of the area under discussion. At the global scale, explored in this chapter, the entire world is treated as a single area—a unity that is more and more relevant as our planet operates as a global system. We use the term **world region** for the largest divisions of the globe, such as East Asia and North America (see Figure 1.1). We have defined ten world regions, each of which is covered in a separate chapter. Each regional chapter considers the interaction of human and physical geography in relation to cultural, social, economic, population, environmental, and political topics. Because it is at the local scale that we live our daily lives—in villages, towns, city neighborhoods—we show how global or regional patterns affect individuals where they live.

In summary, regions have the following traits:

- A region is a unit of the Earth's surface that contains distinct environmental or cultural patterns.
- No two regions are necessarily defined by the same set of attributes.
- Regional definitions and the territory included often change.
- The boundaries of regions are usually indistinct and hard to agree upon.
- Regions can vary greatly in size (scale).

Thematic Concepts and Their Role in This Book

Within the world regional framework, this book is also organized around nine thematic concepts of special significance in the modern world—concepts that operate in every world region (but with varying intensity) and interact in numerous ways.

The sections that follow explain each of nine thematic concepts, how they are related to one another, and how they tie in with the main concerns of geographers: ***Population*** (pages 12–17), ***Gender*** (pages 17–19), ***Development*** (pages 19–23), ***Food*** (pages 23–27), ***Urbanization*** (pages 27–30), ***Globalization*** (pages 30–33), ***Power and Politics*** (pages 33–37), ***Water*** (pages 37–40), and ***Climate Change*** (pages 40–43). While the concepts are relevant to every region and provide a useful framework for your rapidly accumulating knowledge of the world, the way they relate to current issues varies from region to region. For example, in some regions, water is chiefly a scarcity issue; in others, managing water resources equitably under conditions of climate change is the main concern; in still others, the use of newly discovered groundwater resources will be a focus.

The nine concepts are not presented in the same order in each chapter, but rather are discussed as they naturally come up in the coverage of issues.

THINGS TO REMEMBER

Geographic Insight 1

- **Physical and Human Geographers** These geographers study the Earth's surface and the processes that shape it. There is a continual interaction between physical and human processes.
- The careful analysis of photographs can lead to important geographic insights.

Geographic Insight 2

- **Regions** A region is a unit of the Earth's surface that has a combination of distinct physical and/or human features; the complex of features can vary from region to region, and regional boundaries are rarely clear or precise.

POPULATION

To study population is to study the growth and decline of numbers of people on Earth, their distribution across the Earth's surface in terms of age and sex, as well as the reasons that prompt them to move. Over the last several hundred years, the global human population has boomed, but growth rates are now slowing in most societies and, in a few, have even begun to decline (a process that is also called negative growth). Much of this pattern has to do with changes in economic development and gender roles that have reduced incentives for large families.

Global Patterns of Population Growth

It took between 1 million and 2 million years (at least 40,000 generations) for humans to evolve and to reach a global population of 2 billion, which happened around 1945. Then, remarkably, in just 66 years—by October of 2011—the world's population more than tripled to 7 billion (Figure 1.7). What happened to make the population grow so quickly in such a short time?

The explanation lies in changing relationships between humans and the environment. For most of human history, fluctuating food availability, natural hazards, and disease kept human death rates high, especially for infants. Out of many pregnancies, a couple might raise only one or two children. Also, pandemics, such as the Black Death in Europe and Asia in the 1300s, killed millions in a short time.

An astonishing upsurge in human population began about 1500, at a time when the technological, industrial, and scientific revolutions were beginning in some parts of the world. Human life expectancy increased dramatically, and more and more people lived long enough to reproduce successfully, often many times over. The result was an exponential pattern of growth (Figure 1.7).

This pattern is often called a *J curve* because the ever-shorter periods between doubling and redoubling of the population cause an abrupt upward swing in the growth line when depicted on a graph.

Today, the human population is growing in most regions of the world, more rapidly in some places than in others. The reason for this growth is that currently a very large group of young people has reached the age of reproduction, and more will shortly join them. Nevertheless, the *rate* of global population growth is slowing. Since 1993 it has dropped from 1.7 percent per year to roughly 1.1 percent per year, which is where it stands today. If present slower growth trends continue, the world population may level off at between 7.8 billion and 10 billion before 2050. However, this projection is contingent on couples in less-developed countries having the education and economic security to choose to have smaller families and the ability to practice birth control using the latest information and technology.

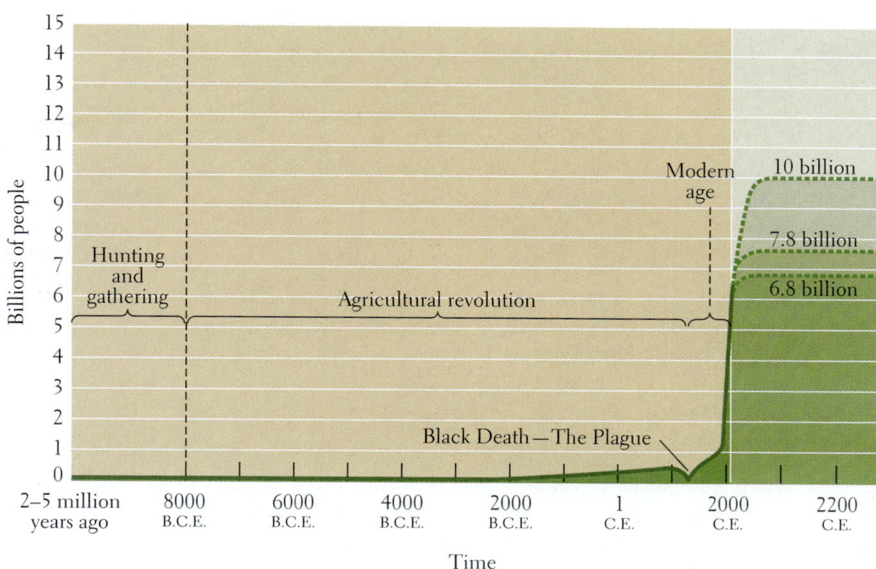

FIGURE 1.7 Exponential population growth: The J curve. The curve's J shape is a result of successive doublings of the population. It starts out nearly flat, but as doubling time shortens, the curve bends ever more sharply upward. Note that B.C.E. (before the common era) is equivalent to B.C. (before Christ); C.E. (common era) is equivalent to A.D. (anno domini).

In a few countries (especially Japan in Asia; and Russia and Ukraine plus Romania, Bulgaria, and Hungary in Central Europe), the population is actually declining and rapidly aging, due primarily to low birth and death rates. This situation could prove problematic, as those who are elderly and dependent become more numerous, posing a financial burden on a declining number of working-age people. HIV/AIDS is affecting population patterns to varying extents in all world regions. In Africa, the epidemic is severe; as a result, several African countries have sharply lowered life expectancies among young and middle-aged adults. **11. WORLD POPULATION TO BE CONCENTRATED IN DEVELOPING NATIONS, AS TOTAL EXPECTED TO REACH 9 BILLION BY 2050**

Local Variations in Population Density and Growth

If the more than 7 billion people on Earth today were evenly distributed across the land surface, they would produce an *average population density* of about 121 people per square mile (47 per square kilometer). But people are not evenly distributed (Figure 1.8). Nearly 90 percent of all people live north of the equator, and most of them live between 20° N and 60° N latitude. Even within that limited territory, people are concentrated on about 20 percent of the available land. They live mainly in zones that have climates warm and wet enough to support agriculture: along rivers, in lowland regions, or fairly close to the sea. In general, people are located where resources are available.

Usually, the variable that is most important for understanding population growth in a region is the rate of natural increase (often called the *growth rate*). The **rate of natural increase (RNI)** is the relationship in a given population between the number of people being born (the **birth rate**) and the number dying (the **death rate**), without regard to the effects of **migration** (the movement of people from one place or country to another).

The rate of natural increase is expressed as a percentage per year. For example, in 2011, the annual birth rate in Austria (in Europe) was 9 per 1000 people, and the death rate was 9 per 1000 people. Therefore, the annual rate of natural increase was 0 per 1000 (9 – 9 = 0), or 0 percent.

For comparison, consider Jordan (in Southwest Asia). In 2011, Jordan's birth rate was 31 per 1000, and the death rate was 4 per 1000. Thus the annual rate of natural increase was 27 per 1000 (31 – 4 = 27), or 2.7 percent per year. At this rate, Jordan's population will double in just 29 years.

Total fertility rate (TFR), another term used to indicate trends in population, is the average number of children a woman in a country is likely to have during her reproductive years (15–49). The TFR (2011) for Austrian women is 1.4; for Jordanian women, it is 3.8. As education rates for women increase, and as they postpone childbearing into their late 20s, total fertility rates tend to decline.

Another powerful contributor to population growth is **immigration** (in-migration). In Europe, for example, the rate of natural increase is quite low, but the region's economic power attracts immigrants from throughout the world. Austria was attracting immigrants at the rate of 0.03 percent of its population in 2011, while Jordan was losing population to **emigration** (out-migration) at the rate of 0.04 percent of the population per year. In 2010,

rate of natural increase (RNI) the rate of population growth measured as the excess of births over deaths per 1000 individuals per year without regard for the effects of migration

birth rate the number of births per 1000 people in a given population, per unit of time (usually per year)

death rate the ratio of total deaths to total population in a specified community, usually expressed in numbers per 1000 or in percentages

migration the movement of people from one place or country to another, often for safety or economic reasons

total fertility rate (TFR) the average number of children that women in a country are likely to have at the present rate of natural increase

immigration in-migration (see also *migration*)

emigration out-migration (see also *migration*)

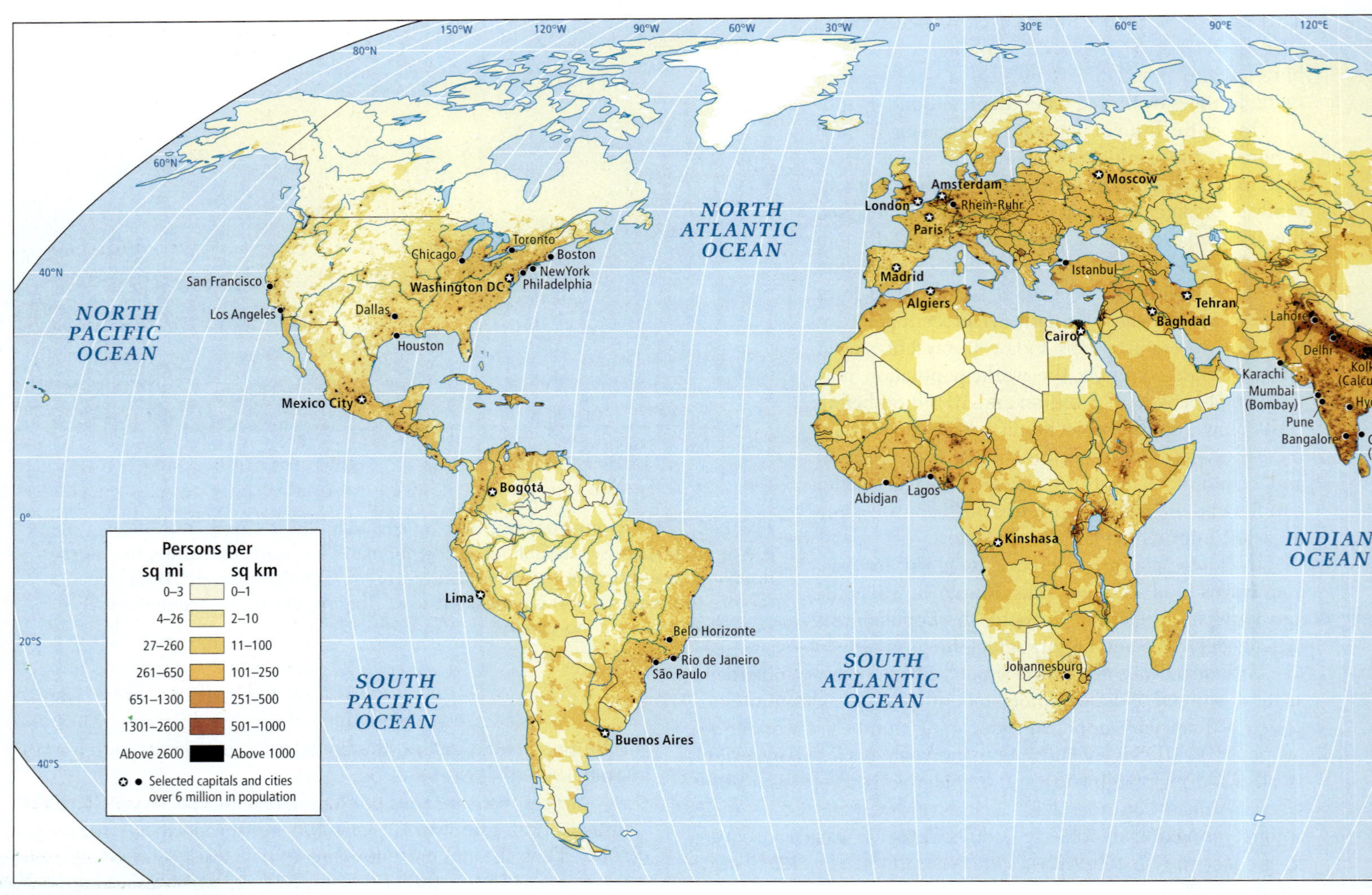

FIGURE 1.8 World population density.

international migrants accounted for about 85 percent of the European Union's population growth; they were important additions to the labor force in an era of declining births. All across the world people are on the move, seeking to improve their circumstances; often they are fleeing war, natural disasters, or economic recessions; understandably, developed countries are favored destinations.

Age and Sex Structures

Age and sex structures of a country's population reflect past and present social conditions, and can help predict future population trends. The age distribution, or age structure, of a population is the proportion of the total population in each age group. The sex structure is the proportion of males and females in each age group.

population pyramid a graph that depicts the age and gender structures of a political unit, usually a country

The **population pyramid** is a graph that depicts age and sex structures. Consider the population pyramids for Austria and Jordan (Figure 1.9). Notice that Jordan's is a true pyramid with a wide bottom, and that the largest groups are in the age categories 0 through 14.

In contrast, Austria's pyramid has an irregular vertical shape that tapers in toward the bottom. The base, which began to narrow rather relentlessly about 40 years ago, indicates that there are now fewer people in the youngest age categories than in the teen, young adulthood, or middle-age groups; those over 70 greatly outnumber the youngest (ages 0 to 4). This age distribution is due to two trends: many Austrians now live to an old age, and in the last several decades, Austrian couples have chosen to have only one child, or none. If these trends continue, Austrians will need to support and care for large numbers of elderly people, and those responsible will be an ever-declining group of working-age people.

Population pyramids also reveal *sex imbalance* within populations. Look closely at the right (female) and left (male) halves of the pyramids in Figure 1.9. In several age categories, the sexes are not evenly balanced on both sides of the line. In the Austria pyramid, there are more women than men near the top (especially in the age categories of 70 and older). In Jordan, there are more males than females near the bottom (especially ages 0 to 24).

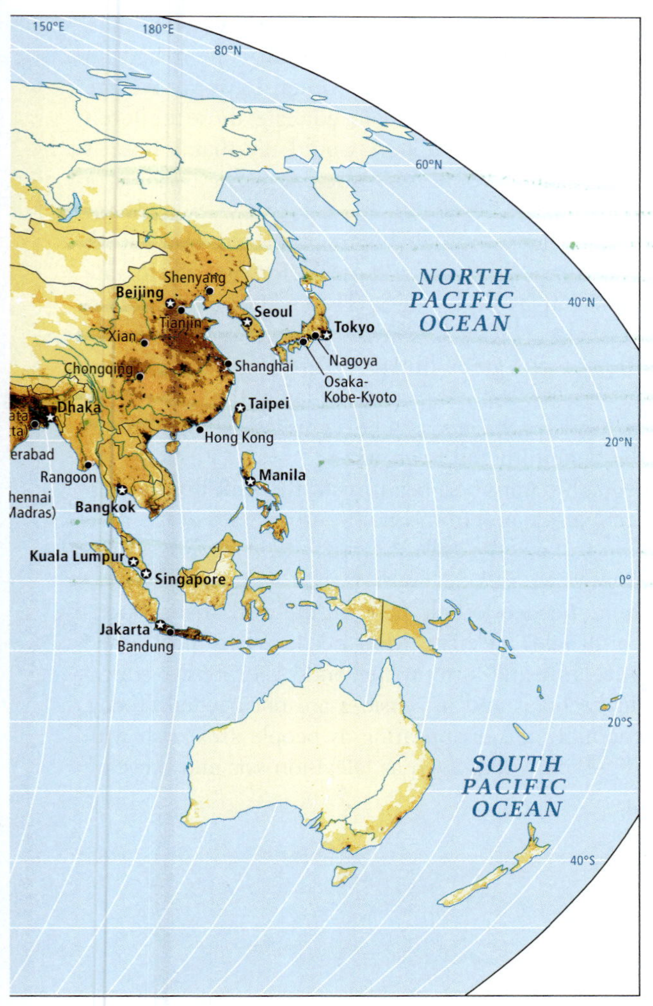

Demographic research on the reasons behind statistical sex imbalance is relatively new, and many different explanations are proposed. In Austria, the predominance of elderly women reflects the fact that in countries with long life expectancies, women live about 5 years longer than men (a trend that is still poorly understood). But the sex imbalance in younger populations has different explanations. The normal ratio worldwide is for about 95 females to be born for every 100 males. Because baby boys on average are somewhat weaker than girls, the ratio normally evens out naturally within the first 5 years. However, in many places the ratio is as low as 80 females to 100 males, and continues throughout the life cycle. The widespread cultural preference for boys over girls (discussed in later chapters) becomes more observable as couples choose to have fewer children. Some fetuses, if identified as female, are purposely aborted; also, girls and women are sometimes fed less well and receive less health care than males, especially in poverty-stricken areas. Because of this, females are more likely to die, especially in early childhood. The gender imbalance of more males than females can be especially troubling for young adults, because the absence of females will mean that young men will find it hard to find a marriage mate.

Population Growth Rates and Wealth

Although there is a wide range of variation, regions with slow population growth rates usually tend to be affluent, and regions with fast growth rates tend to have widespread poverty. The reasons for this difference are complicated; again, Austria and Jordan are useful examples.

In 2011, Austria had an annual **gross national income (GNI) per capita** of $38,400 (PPP).[1] This figure represents the total production of goods and services in a country in a given year, divided by the mid-year population; Austria's is among the highest on Earth (the comparable figure for the

gross national income (GNI) per capita the total production of goods and services in a country in a given year divided by the mid-year population

FIGURE 1.9 Population pyramids for Austria and Jordan, 2012.

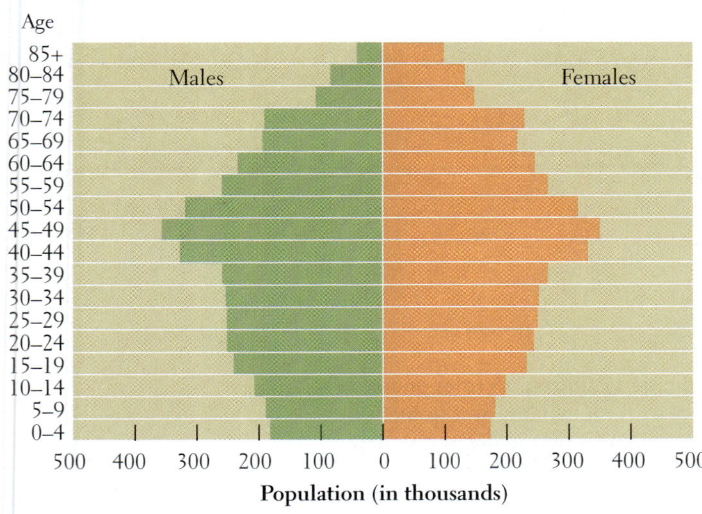

(A) Population pyramid for Austria.

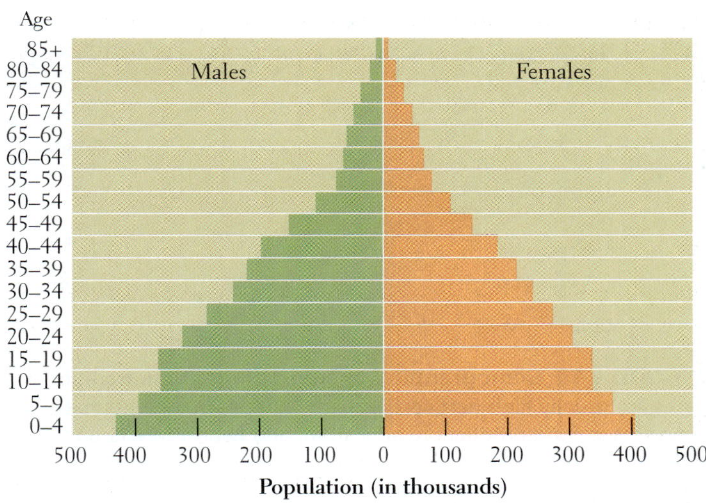

(B) Population pyramid for Jordan.

[1]All GNI per capita figures cited in this book are adjusted for purchasing power parity (PPP), so they represent comparable purchasing power for the year cited.

United States is $45,640). Austria has a very low infant mortality rate of 3.7 per 1000. Its highly educated population is 100 percent literate, employed largely in technologically sophisticated industries and services. Large amounts of time, effort, and money are required to educate a child to compete in this economy. Many Austrian couples choose to have only one or two children, probably because of the cost of raising a child and because it is highly likely that children will survive to adulthood and thus will be available as companions for aging parents. Furthermore, the social welfare system in Austria ensures that citizens will not be impoverished in old age.

By contrast, Jordan has a GNI (PPP) per capita of $5730 in addition to a high infant mortality rate of 23 per 1000. Much everyday work is still done by hand, so each new child is a potential contributor to the family income at a young age. There is little social welfare for elderly citizens and a much greater risk of children not surviving into adulthood. Having more children helps ensure that someone will be there to provide care for aging parents and grandparents.

demographic transition the change from high birth and death rates to low birth and death rates that usually accompanies a cluster of other changes, such as change from a subsistence to a cash economy, increased education rates, and urbanization

subsistence economy an economy in which families produce most of their own food, clothing, and shelter

Circumstances in Jordan, nonetheless, have changed rapidly over the last 25 years. In 1985, Jordan's GNI per capita was less than $1900, infant mortality was 77 per 1000, and women had an average of 8 children. Now Jordanian women have an average of only 3.8 children, with the need to educate those children increasingly recognized. Geographers would say that Jordan is going through a **demographic transition**, meaning that a period of high birth and death rates is giving way to a period of much lower birth and death rates. In the middle phases, however, when death rates decline more rapidly than birth rates (Jordan's death rate is just 4 per 1000 but its birth rate is 34 per 1000), population numbers can increase rapidly as attitudes toward optimal family size slowly adjust (see the green line in Figure 1.10). Populations may eventually stabilize but because of the lag in the lowering of birth rates, population numbers may be significantly higher than they were at the beginning of the transition.

The demographic transition begins with the shift from subsistence to cash economies and from rural to urban ways of life. In a **subsistence economy**, a family, usually in a rural setting, produces most of its own food, clothing, and shelter, so there is little need for cash. Many children are needed to help perform the work that supports the family, so birth rates are high. Most needed skills are learned around the home, farm, and surrounding lands, so expensive educations at technical schools and universities are not needed. Today, subsistence economies are disappearing as people seek cash with which to buy food and goods such as television sets and bicycles.

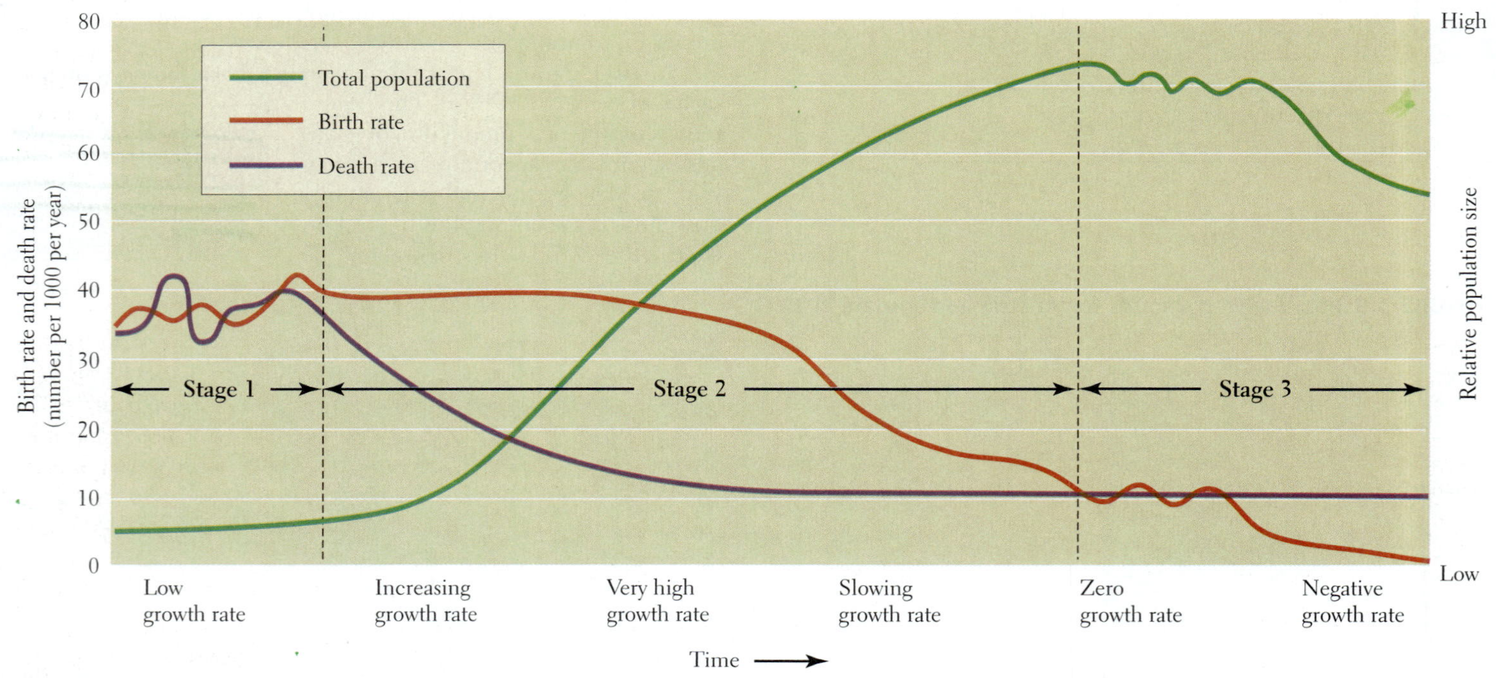

FIGURE 1.10 Demographic transition diagram. In traditional societies (Stage 1), both birth rates and death rates are usually high (left vertical axis), and population numbers (right vertical axis) remain low and stable. With advances in food production, education, and health care (Stage 2), death rates usually drop rapidly, but strong cultural values regarding reproduction remain, often for generations, so birth rates drop much more slowly, with the result that for decades or longer, the population continues to grow significantly. When changed social and economic circumstances enable most children to survive to adulthood and it is no longer necessary to produce a cadre of family labor (Stage 3), population growth rates slow and may eventually drift into negative growth. At this point, demographers say that the society has gone through the demographic transition.

In a **cash economy**, which tends to be urban but may be rural, skilled workers, well-trained specialists, and even farm laborers are paid in money. Each child needs years of education to qualify for a good cash-paying job and does not contribute to the family budget while in school. Having many children, therefore, is a drain on the family's resources. Perhaps most important, the higher development levels of cash economies mean those populations are more likely to have better health care, increasing the likelihood that each child will survive to adulthood.

cash economy an economic system that tends to be urban but may be rural, in which skilled workers, well-trained specialists, and even farm laborers are paid in money

THINGS TO REMEMBER

- Over the last several hundred years, global population growth has been rapid. Although growth will continue for many years, rates are now slowing in most places and in a few places growth has reversed and the population is even shrinking.
- The circumstances that lead to lower population growth rates also lead to the overall aging of populations; these two phenomena are found nearly everywhere on Earth today, but to varying degrees.

GENDER

Note the difference between the terms *gender* and *sex*. **Gender** indicates how a particular social group defines the differences between the sexes. **Sex** refers to the biological category of male or female but does not indicate how males or females may behave or identify themselves. Gender definitions and accepted behavior for the sexes can vary greatly from one social group to another. Here we consider gender.

For women, the historical and modern global gender picture is puzzlingly negative. In nearly every culture, in every region of the world, and for a great deal of recorded history, women have had (and still have) an inferior status. Exceptions are rare, although the intensity of this second-class designation varies considerably. On average, females have less access to education, medical care, and even food. They start work at a younger age and work longer hours than males. Around the world, people of both sexes still routinely accept the idea that males are more productive and intelligent than females. The puzzling question of how and why women became subordinate to men has not yet been well explored because, oddly enough, few thought the question significant until the last 50 years.

In nearly all cultures, families prefer boys over girls because, as adults, boys will have greater earning capacity (Table 1.1 on page 18) and more power in society. This preference for boys has some unexpected side effects. For example, the preference may lead to fewer girls being born; eventually there will be a shortage of marriageable women, leaving many men without the hope of forming a family. Currently, there is concern in several Asian societies that the scarcity of young women could lead to antisocial behavior on the part of discouraged young men. **6. WOMEN STILL LAG BEHIND MEN IN TOP BOARDROOM JOBS**

Gender Roles

Geographic Insight 3

Population and Gender: The shift toward greater gender equality is having an influence on population growth patterns, patterns of economic development, and the distribution of power within families, communities, and countries.

Geographers have begun to pay more attention to **gender roles**—the socially assigned roles for males and females—in different culture groups. In virtually all parts of the world, and for at least tens of thousands of years, the biological fact of maleness and femaleness has been translated into specific roles for each sex. The activities assigned to men and to women can vary greatly from culture to culture and from era to era, but they remain central to the ways societies function. Indeed, increasing attention to gender roles by geographers has been driven largely by interest in the shifts that occur when traditional ways are transformed by modernization.

There are some striking consistencies regarding traditional gender roles around the globe, and over time. Men are expected to fulfill public roles, while women fulfill private roles. Certainly there are exceptions in every culture, and customs are changing, especially in wealthier countries. But generally, men work outside the home in positions such as executives, animal herders, hunters, farmers, warriors, or government leaders. Women keep house, bear and rear children, care for the elderly, grow and preserve food, and prepare the meals, among many other tasks. In nearly all cultures, women are defined as dependent on men—their fathers, husbands, brothers, or adult sons—even when the women may produce most of the family sustenance.

Gender Issues

Because their activities are focused on the home, women tend to marry early. One quarter of the girls in developing countries are mothers before they are 18. This is crucial in that pregnancy is the leading cause of death among girls 15 to 19 worldwide, primarily because immature female bodies are not ready for the stress of pregnancy and birth. Globally, babies born to women under 18 have a 60 percent greater chance of dying in infancy than do those born to women over 18.

Typically, women also have less access to education than men (globally, 70 percent of youth who leave school early are girls). They are less likely to have access to information and paid employment, and so have less access to wealth and political power. When they do work outside the home (as is the case increasingly in every world region), women tend to fill lower-paid positions, such as laborers, service workers, or lower-level

gender the ways a particular social group defines the differences between the sexes

sex the biological category of male or female; does not indicate how males or females may behave or identify themselves

gender roles the socially assigned roles for males and females

TABLE 1.1 Comparisons of male and female income in selected countries, 2011

Country	HDI rank	Female income (PPP U.S.$)	Male income (PPP U.S.$)	Female income as percent of male income
Austria	19	22,528	55,934	40
Barbados	47	15,119	23,507	64
Botswana	118	8823	17,952	49
Canada	6	30,005	45,763	66
Japan	12	20,572	44,892	46
Jordan	95	2456	8581	29
Kuwait	63	24,531	65,010	38
Poland	39	13,886	24,292	57
Russia	66	15,191	23,284	65
Saudi Arabia	56	7157	36,727	19
Sweden	10	32,990	41,830	79
United Kingdom	28	28,354	42,217	67
United States	4	35,346	56,918	62

Sources: Income data from "Estimated earned income," *The Global Gender Gap Report 2011* (Geneva: World Economic Forum), 44, table D3, http://reports.weforum.org/global-gender-gap-2011/.

HDI data from *United Nations Human Development Report 2011*, 128, table 1, at http://hdr.undp.org/en/reports/global/hdr2011/40

professionals. And even when they work outside the home, most women retain their household duties, so they work a *double day*.

Gender and sex categories can be confusing in matters such as the extent to which physical differences between males and females may affect their social roles. On average, males and females are equally strong and durable, but women's physical capabilities are somewhat limited during pregnancy and nursing. However, from the age of about 45, women are no longer subject to the limits of pregnancy, and most contribute in some significant way to the well-being of their adult children and grandchildren, an important social role. A growing number of biologists suggest that the evolutionary advantage of menopause in midlife is that it gives women the time and energy and freedom to help succeeding generations thrive. This notion—sometimes labeled the *grandmother hypothesis*—seems to have worldwide validity (Figure 1.11). Certainly, grandfathers can play nurturing roles, but they tend to not be involved in intimate care; and women tend to live 5 years longer than men.

Perhaps more than for any other culturally defined human characteristic, significant agreement exists that gender is important, but just how gender roles are defined varies greatly across places and over time.

Traditional notions of gender roles are now being challenged everywhere. In many countries, including conservative Muslim countries, females are acquiring education at higher rates than males. Although it will take females a while to catch up, eventually, acquiring education should make women competitive with men for jobs and roles in public life as policy makers and government officials, not just as voters. Unless discrimination

FIGURE 1.11 The grandmother hypothesis. In every culture and community worldwide, grandmothers contribute to the care and education of their grandchildren. This quality has played an essential role in human evolution. A grandmother in Xian, China, helps her toddler granddaughter learn to walk.

Thinking Geographically

What are five things grandmothers in every world region are likely to worry about, regarding their grandchildren?

persists, women should also begin to earn pay equal to that of men (Table 1.1). Research data suggest that there is a ripple effect that benefits the whole group when developing countries pay attention to the needs of girls:

- Girls in developing countries, who get 7 or more years of education, marry 4 years later than average and have 2.2 fewer children.
- An extra year of secondary schooling over the average for her locale boosts a girl's lifetime income by 15 to 25 percent.
- The children of educated mothers are healthier and more likely to finish secondary school.
- When women and girls earn income, 90 percent of their earnings are invested in the family, compared to just 40 percent of males' earnings.

Considering only women's perspectives on gender, however, misses half the story. Men are also affected by strict gender expectations, often negatively. For most of human history, young men have borne a disproportionate share of burdensome physical tasks and dangerous undertakings. Until recently, mostly young men left home to migrate to distant, low-paying jobs. Overwhelmingly, it has been young men (the majority of soldiers) who die in wars or suffer physical and psychological injuries from combat.

This book will return repeatedly to the question of gender disparities because they play such a central role in the potential for a brighter future in every country on Earth.

THINGS TO REMEMBER

- Gender—the sexual category of a person—is both a biological and a cultural phenomenon. Gender indicates how a particular group defines the social differences between the sexes. Sex is the biological category of male or female.
- There are some global consistencies in gender disparities: typically, males have public roles and females have private roles; in every country, the average woman earns less than the average man.

Geographic Insight 3

- **Gender and Population** Generally, as modernization takes hold, there is a move toward greater equality between the sexes. This shift is influencing population growth rates because women, presented with opportunities, choose to have fewer children.

DEVELOPMENT

The economy is the forum in which people make their living, and resources are what they use to do so. *Extractive resources* are resources that must be mined from the Earth's surface (mineral ores) or grown from its soil (timber and plants). There are also *human resources*, such as skills and brainpower, which are used to transform extractive resources into useful products (such as refrigerators or bread) or bodies of knowledge (such as books or computer software). Economic activities are often divided into three *sectors of the economy:* the **primary sector** is based on **extraction** (mining, forestry, and agriculture); the **secondary sector** is **industrial production** (processing, manufacturing, and construction); and the **tertiary sector** is **services** (sales, entertainment, and financial). Of late, a fourth, or **quaternary sector** has been added to cover intellectual pursuits such as education, research, and IT (information technology) development. Generally speaking, as people in a society shift from extractive activities, such as farming and mining, to industrial, service, and intellectual activities, their material standards of living rise—a process typically labeled **development**.

The development process has several facets, one of which is a shift from economies based on extractive resources to those based on human resources. In many parts of the world, especially in poorer societies (often referred to as "underdeveloped" or "developing"), there are now shifts away from labor-intensive and low-wage, often agricultural, economies toward higher-wage but still labor-intensive manufacturing and service economies (including Internet-based economies). Meanwhile, the richest countries (often labeled "developed") are lessening their dependence on labor-intensive manufacturing and shifting toward more highly skilled mechanized production or knowledge-based service and technology (quaternary) industries. As these changes occur, societies must provide adequate education, health care, and other social services to help their people contribute to economic development.

Measuring Economic Development

The most long-standing measure of development has been **gross domestic product (GDP) per capita**. GDP is simply an economic measure that refers to the total market value of all goods and services produced in a country in a given year. A closely related index that is now used more often by international agencies is gross national income (GNI), the total value of income in a country. When GDP or GNI is divided by the number of people in the country, the result is per capita GDP or GNI. This book now uses primarily the GNI per capita statistics.

primary sector an economic sector of the economy that is based on extraction (see also *extraction*)

extraction mining, forestry, and agriculture

secondary sector an economic sector of the economy that is based on industrial production (see also *industrial production*)

industrial production processing, manufacturing, and construction

tertiary sector an economic sector of the economy that is based on services (see also *services*)

services sales, entertainment, and financial services

quaternary sector a sector of the economy that is based on intellectual pursuits such as education, research and IT (information technology) development

development a term usually used to describe economic changes such as the greater productivity of agriculture and industry that lead to better standards of living or simply to increased mass consumption

gross domestic product (GDP) per capita the total market value of all goods and services produced within a particular country's borders and within a given year, divided by the number of people in the country

Using GNI per capita as a measure of how well people are living has several disadvantages. First is the matter of wealth distribution. Because GNI per capita is an average, it can hide the fact that a country has a few fabulously rich people and a great mass of abjectly poor people. For example, a GNI per capita of U.S.$50,000 would be meaningless if a few lived on millions per year and most lived on less than $10,000 per year.

Second, the purchasing power of currency varies widely around the globe. A GNI of U.S.$18,000 per capita in Barbados might represent a middle-class standard of living, whereas that same amount in New York City could not buy even basic food and shelter. Because of these purchasing power variations, in this book GNI (and occasionally GDP) per capita figures have been adjusted for **purchasing power parity (PPP)**. PPP is the amount that the local currency equivalent of U.S.$ will purchase in a given country. For example, according to *The Economist*, on January 14, 2012, a Big Mac at McDonald's in the United States cost U.S.$4.20. In the **Euro zone** (those countries in the European Union that use the Euro currency), the very same Big Mac cost the equivalent of U.S.$4.43, while in India it cost U.S.$1.62. Of course, for the consumer in India, where annual per capita GNI PPP is $3280, this would be a rather expensive meal. On the other hand, at $4.20, the Big Mac would be an economy meal in the United States, where the GNI per capita PPP is $45,640, or in the Euro area, where $34,000 is the average GNI PPP per capita (see "Big Mac index," at http://www.economist.com/node/21542808).

A third disadvantage of using GDP PPP or GNI PPP per capita is that both measure only what goes on in the **formal economy**—all the activities that are officially recorded as part of a country's production. Many goods and services are produced outside formal markets, in the **informal economy**. Here, work is often traded for *in-kind* payments (food or housing, for example) or for cash payments that are not reported to the government as taxable income. It is estimated that one-third or more of the world's work takes place in the informal economy. Examples of workers in this category include anyone who contributes to her/his own or someone else's well-being through unpaid services such as housework, gardening, herding, animal care, or elder and child care. *Remittances*, or pay sent home by migrants, become part of the formal economy of the receiving society if they are sent through banks or similar financial institutions—because records are kept and taxes levied. If they are transmitted "off the books" (perhaps illegally), such as via mail or cash, they become part of the informal economy.

There is a gender aspect to informal economies. Researchers studying all types of societies and cultures have shown that, on average, women perform about 60 percent of all the work done, and that much of this work is unpaid and in the informal economy. Yet only the work women are paid for in the formal economy appears in the statistics, so economic figures per capita ignore much of the work women do. Statistics also neglect the contributions of millions of men and children who work in the informal economy as subsistence farmers, traders, service people, or seasonal laborers.

A fourth disadvantage of GDP PPP or GNI PPP per capita is that neither takes into consideration whether these levels of income are achieved at the expense of environmental sustainability, human well-being, or human rights.

purchasing power parity (PPP) the amount that the local currency equivalent of U.S.$1 will purchase in a given country

Euro zone those countries in the European Union that use the Euro currency

formal economy all aspects of the economy that take place in official channels

informal economy all aspects of the economy that take place outside official channels

human well-being various measures of the extent to which people are able to obtain a a healthy and socially rewarding standard of living in an environment that is safe and sustainable

United Nations Human Development Index (HDI) index that calculates a country's level of well-being, based on a formula of factors that considers income adjusted to PPP, data on life expectancy at birth, and data on educational attainment

United Nations Gender Equality Index (GEI) a composite measure reflecting the degree to which there is inequality in achievements between women and men in three dimensions: reproductive health, political and educational empowerment, and labor-force participation. A high rank indicates that the genders are tending toward equality.

Geographic Patterns of Human Well-Being

Some development experts, such as the Nobel Prize–winning economist Amartya Sen, advocate a broader definition of development that includes measures of **human well-being**. This term generally means a healthy and socially rewarding standard of living in an environment that is safe and sustainable. The following section explores the three measures of human well-being that are used in this book.

Global GNI per capita PPP is mapped in Figure 1.12A on page 21. Comparisons between regions and countries are possible, but as discussed above, GNI per capita figures ignore all aspects of development other than economic ones. For example, there is no way to tell from GNI per capita figures how quickly a country is consuming its natural resources, or how well it is educating its young, maintaining its environment, or seeking gender and racial equality. Therefore, along with the traditional GNI per capita figure, geographers increasingly use several other measures of development.

The second measure used in this book is the **United Nations Human Development Index (HDI)**, which calculates a country's level of well-being with a formula of factors that considers income adjusted to PPP, data on life expectancy at birth (an indicator of overall health care), and data on educational attainment (see Figure 1.12B).

The third measure of well-being used here, the **United Nations Gender Equality Index (GEI)** is a composite measure reflecting the degree to which there is inequality in achievements between women and men in three dimensions: reproductive health, empowerment, and the labor market. A high rank indicates that the genders are tending toward equality. Ranks are from most equal (1) to least equal (146) (see Figure 1.12C).

Together, these three measures reveal some of the subtleties and nuances of well-being and make comparisons between countries somewhat more valid. Because the more sensitive indices (HDI and GEI) are also more complex than the purely economic GNI PPP per capita, they are all still being refined by the United Nations.

FIGURE 1.12 Global maps of human well-being.

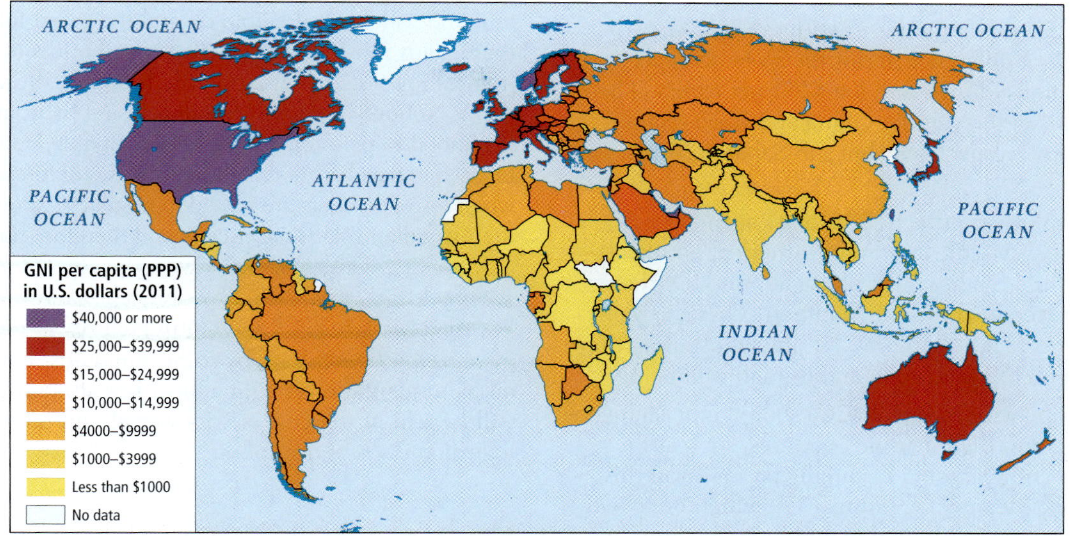

(A) Gross national income (GNI) per capita, adjusted for purchasing power parity (PPP).

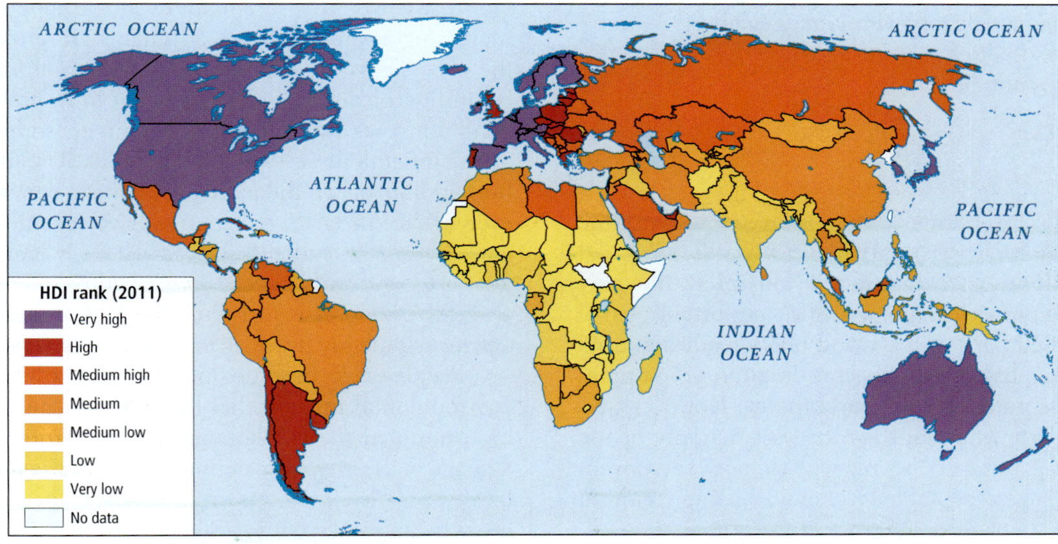

(B) Human Development Index (HDI).

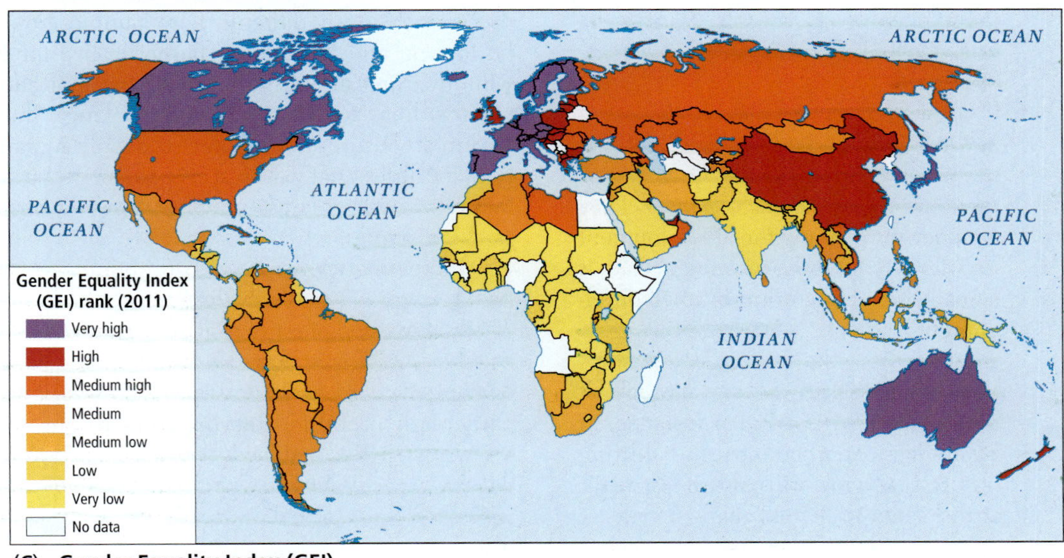

(C) Gender Equality Index (GEI).

A geographer looking at these maps might make the following observations:

- The map of GNI per capita figures (see Figure 1.12A) shows a wide range of difference across the globe with very obvious concentrations of high and low GNI per capita. The most populous parts of the world—China and India—rank medium-low and low, respectively, but sub-Saharan Africa ranks the lowest.
- The HDI rank map (see Figure 1.12B) shows a similar pattern, but look closely. Middle and South America, Southeast Asia, China, Italy, and Spain rank a bit higher on HDI than they do on GNI; several countries in southern Africa, as well as Iran, rank lower on HDI than on GNI, thus illustrating the disconnect between GNI per capita and human well-being.
- The map of GEI rank (see Figure 1.12C) shows some strange anomalies, with certain high-income countries, such as the United States (47) and Qatar (111) ranking far lower on this index than on HDI. Meanwhile, China (35), which once had a reputation for gender discrimination and has a medium-low GNI, ranks 12 points higher than the United States on the GEI scale, in the same category with New Zealand (32) and Ireland (33).

Sustainable Development and Political Ecology

The United Nations (UN) defines **sustainable development** as the effort to improve present living standards in ways that will not jeopardize those of future generations. Sustainability has only recently gained widespread recognition as an important goal—well after the developed parts of the world had already achieved high standards of living based on mass consumption of resources accompanied by mass pollution of environments. However, sustainability is particularly important for the vast majority of the Earth's people who do not yet enjoy an acceptable level of well-being. Without sustainable development strategies, efforts to improve living standards for those who need it most will increasingly be foiled by degraded or scarce resources.

sustainable development the effort to improve present standards of living in ways that will not jeopardize those of future generations

political ecologists geographers who study the interactions among development, politics, human well-being, and the environment

Anthropocene a geologic epoch during which humans have had an overwhelming impact on Earth's biosphere

biosphere the entirety of the Earth's integrated physical spheres, with humans and other impacts included as part of nature

ecological footprint the amount of biologically productive land and sea area needed to sustain a person at the current average standard of living for a given population

Geographers who study the interactions among development, politics, human well-being, and the environment are called **political ecologists**. They are known for asking the "Development for whom?" question, meaning, "Who is actually benefiting from so-called development projects?" Political ecologists examine how the power relationships in a society affect the ways in which development proceeds. For instance, in a Southeast Asian country, the clearing of forests to grow oil palm trees might at first seem to benefit many people. It would create some jobs, earn profits for the growers, and raise tax revenues for the government through the sale of palm oil, an important and widely used edible oil and industrial lubricant. However, these gains must be balanced against the loss of highly biodiverse tropical forest ecosystems and the human cultures that depend on them. Not only are forest dwellers losing their lands and means of livelihood to palm oil agribusiness, valuable knowledge that could be used to develop more sustainable uses of forest ecosystems is lost when forest dwellers are forced to migrate to crowded cities, where their woodland skills are useless and therefore soon forgotten.

Political ecologists are raising awareness that development should be measured by the improvements brought to overall human well-being and long-term environmental quality, not just by the income created. By these standards, converting forests to oil palm plantations might appear less attractive, since only a few will benefit at the cost of widespread and often irreversible ecological and social disruption.

Human Impact on the Biosphere

Concerns over the sustainability of development grow out of increasing awareness that we humans are profoundly impacting the ecosystems we depend on. This is nothing new. From the beginning of human life, in seeking to improve our own living conditions, we have overused resources, sometimes with disastrous consequences. What is new is our awareness of the scale of human impacts on the planet, which can now be found virtually everywhere. In fact, geoscientists have recently identified a new geologic epoch, called the **Anthropocene**, which is defined as the time during which humans have had an overwhelming impact on Earth's biosphere. Just when the Anthropocene began is under debate—some say it was as long as 10,000 years ago.

People have become increasingly aware of the significant environmental impacts that humans are causing; this awareness has prompted the development of numerous proposals to limit damage to the **biosphere**, defined here as the entirety of the Earth's integrated physical systems, with humans and their impacts included as part of nature. Societies have become so transformed by intensive use of Earth's resources that reversing this level of use is enormously difficult. For example, how possible would it be for you and your entire family to live for even just one day without using any fossil fuels for transportation, home heating and cooling, food buying and cooking? Intensive per capita resource consumption is now so deeply ingrained, especially in rich countries, that with just 20 percent of the world's population, the rich countries consume more than 80 percent of the available world resources.

Increasingly, human consumption of natural resources is being examined through the concept of the **ecological footprint**. This is a method of estimating the amount of biologically productive land and sea area needed to sustain a human at the average current standard of living for a given population (country). It is particularly useful for drawing comparisons. For example, about 4.5 acres is the worldwide average of the biologically productive area needed to support one person—this would be one individual's ecological footprint. Because of the lifestyle in the

United States, one person's ecological footprint averages about 24 acres, about 18 in Canada, and just 4 acres in China. You can calculate your own footprint using the Global Footprint Network's calculator, at http://tinyurl.com/6d2wyl4. A similar concept more closely related to global warming is the *carbon footprint*, which measures the greenhouse gas emissions a person's activities produce. To calculate your family's carbon footprint, use Carbon Footprint's calculator, at http://tinyurl.com/2el2j5.

Because the biosphere is a global ecological system that integrates all living things and their relationships, it is important to raise awareness that actions in widely separated parts of Earth have a cumulative effect on the whole. Figure 1.13 on pages 24–25 shows a global map of the relative intensity of human biosphere impacts. The map includes photo insets that show particular trouble spots in South America (Figure 1.13E, F), Europe (Figure 1.13A), South Asia (Figure 1.13B), and Southeast Asia (Figure 1.13C, D).

THINGS TO REMEMBER

- The term development has until recently referred to the rise in material standards of living that usually accompanies the shift from extractive economic activities, such as farming and mining, to industrial and service economic activities.
- Measures of development are being redefined to mean improvements in overall average well-being and progress in overall environmental sustainability.
- For development to happen, social services, such as education and health care, are necessary to enable people to contribute to economic growth.
- Human impact on the Earth's biosphere is now so profound that geoscientists have designated a new geologic epoch called the Anthropocene.

FOOD

Over time, food production has undergone many changes. It started with hunting and gathering and over the millennia evolved into more labor-intensive, small-scale, subsistence agriculture. For the vast majority of people, subsistence remained the mode of food production for many thousands of years until a series of innovations and ideas that changed the way goods were manufactured. The **Industrial Revolution**, broadly from 1750 to 1850, opened the door to the development of what has become modern and mechanized commercial agriculture, involving an intensive use of machinery, fuel, and chemicals. Modern processes of food production, distribution, and consumption have greatly increased the supply and, to some extent, the security of food systems. However, this has come at the cost of environmental pollution that may strain future food production and food security.

Agriculture: Early Human Impacts on the Physical Environment

Agriculture includes animal husbandry, or the raising of animals, as well as the cultivation of plants. The ability to produce food, as opposed to being dependent on hunting and gathering, led to a host of long-term changes. Human population began to grow more quickly, rates of natural resource use increased, permanent settlements eventually developed into towns and cities, and ultimately, human relationships became more formalized. Some would say these are the steps that led to civilization.

Very early humans hunted animals and gathered plants and plant products (seeds, fruits, roots, and fibers) for their food, shelter, and clothing. To successfully use these wild resources, humans developed an extensive folk knowledge of the needs of the plants and animals they favored. The transition from hunting in the wild to tending animals in pens and pastures and from gathering wild plant products to sowing seeds and tending plants in gardens, orchards, and fields probably took place gradually over thousands of years.

Where and when did plant cultivation and animal husbandry first develop? Genetic studies support the view that at varying times between 8000 and 20,000 years ago, people in many different places around the globe independently learned to develop especially useful plants and animals through selective breeding, a process known as **domestication**.

Why did agriculture and animal husbandry develop in the first place? Certainly the desire for more secure food resources played a role, but the opportunity to trade may have been just as important. It is probably not a coincidence that many of the known locations of agricultural innovation lie along early trade routes—for example, along the Silk Road that runs through Central Asia from the eastern Mediterranean to China. In such locales, people would have had access to new information and new plants and animals brought by traders.

Agriculture and Its Consequences

Agriculture made possible the amassing of surplus stores of food for lean times, and allowed some people to specialize in activities other than food procurement. It also may have led to several developments now regarded as problems: rapid population growth, concentrated settlements where diseases could easily spread, environmental degradation, and paradoxically, malnutrition or even famine.

Through the study of human remains, archaeologists have learned that it was not uncommon for the nutritional quality of human diets to decline as people stopped eating diverse wild food species and began to eat primarily

Industrial Revolution a series of innovations and ideas that occurred broadly between 1750 and 1850, which changed the way goods were manufactured

agriculture the practice of producing food through animal husbandry, or the raising of animals, and the cultivation of plants

domestication the process of developing plants and animals through selective breeding to live with and be of use to humans

one or two species of domesticated plants and animals. Evidence of nutritional stress (shorter stature, malnourished bones and teeth) has been found repeatedly in human skeletons excavated in sites around the world where agriculture was practiced.

Whereas agriculture could support more people on a given piece of land than hunting and gathering, as populations expanded and as more land was turned over to agriculture, natural habitats were destroyed, reducing opportunities for hunting and gathering. Furthermore, the storage of food surpluses not only made it possible to trade food, but also made it possible for people to live together in larger concentrations, which marked the beginning of urban societies and, coincidentally, facilitated the spread of disease. Moreover, land clearing increased vulnerability to drought and other natural disasters that could wipe out an entire harvest. Thus, as ever-larger populations depended solely on cultivated food crops, episodic famine may have become more common and affected more people.

Figure 1.13 Photo Essay: Human Impacts on the Biosphere

Thinking Geographically

After you have read about human impacts on the biosphere, you will be able to answer the following questions:

A In which sector of the economy is mining?

B What form of pollution does this photo show most directly?

C, **D** What is the evidence that shifting cultivation may be contributing to deforestation in the region depicted?

E, **F** How is logging in Brazil linked to rising CO_2 levels and the global economy?

Humans have had enormous impacts on the biosphere. The map and insets show varying levels of these impacts on the biosphere as of 2002. The depictions here are derived from a synthesis of hundreds of studies. High-impact areas are associated with intense urbanization. Medium-to-high-impact areas areas are associated with roads, railways, agriculture, or other intensive land uses. Low-to-medium-impact areas are experiencing biodiversity loss and other disturbances related to human activity. For more details, go to www.whfreeman.com/pulsipher.

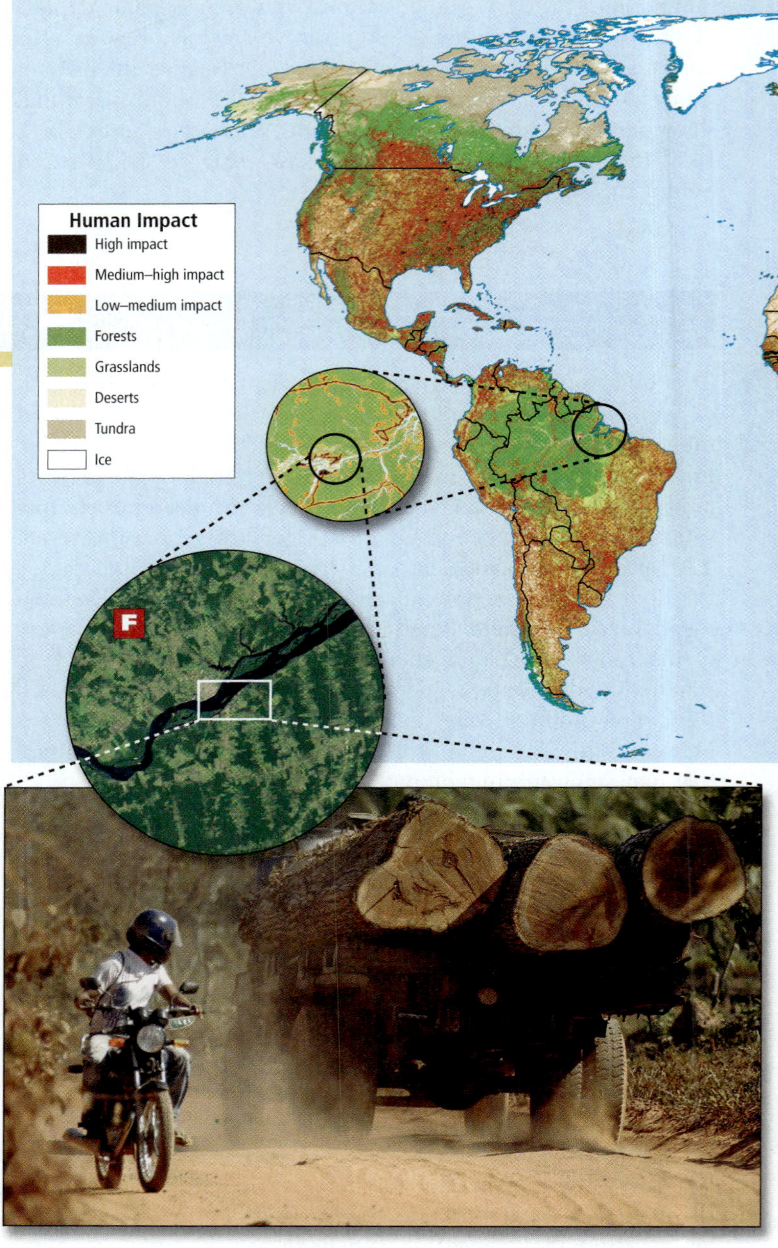

E **Development and Deforestation:** In the Brazilian Amazon, deforestation often occurs in regularized spatial patterns, such as the "fishbone" pattern (see satellite image inset **F**). This pattern results from regulations that determine the location of roads used for settlement and logging. Whole logs are brought by road to rivers, where they are put on barges and taken to a port for export.

Modern Food Production and Food Security

For most of human history, people lived in subsistence economies. However, over the past five centuries of increasing global interaction and trade, people have become ever more removed from their sources of food. Today, occupational specialization means that food is increasingly mass-produced. Far fewer people work in agriculture than in the past, and now most humans work for cash to buy food and other necessities.

A side effect of this dependence on money is that the **food security**—the ability of a state to consistently supply a sufficient amount of basic food to the entire population—of individuals and families can be threatened by economic disruptions, even in distant places. As countries become more involved with the global economy, either they may import more food or their own food production can become vulnerable to price swings. For example, a crisis in food security began to develop in 2007 when the world price of corn spiked. Speculators in alternative energy, thinking that corn would be an ideal raw material with which to make ethanol—a substitute for gasoline—invested heavily in this commodity, creating a shortage. As a result, global corn prices rose beyond the reach of those who depended on corn as a dietary staple. Then, between the sharp price rise in oil in 2008 and the recession of 2007–2009, the global cost of basic foods rose 17 percent. When oil prices rise, all foods produced and transported with machines get more expensive.

food security the ability of a state to consistently supply a sufficient amount of basic food to the entire population

A Development and Mining: Perhaps no other human activity has as striking an impact on the landscape as mining. This open-pit coal mine is located in one of the most industrialized and densely inhabited parts of Germany.

B War and Political Conflict: War can have a devastating effect on the environment. Women outside Kabul, Afghanistan, must now carry water by hand from distant sources that have not been polluted or damaged by war.

C and D Food and Deforestation: Farmers practicing shifting cultivation (see Chapter 3, page 122) plant "hill rice" in Burma. Shifting cultivation is an ancient technique that can be sustained indefinitely, given sufficient land and fallow periods long enough for forest to regrow (20 years or more). Today, more and more forest is being turned over to short-fallow cultivation—3 to 6 years of cultivation—resulting in a loss of habitat and biodiversity (see inset D).

There were other contributing causes to food insecurity. When the global recession—partially caused by rising oil prices—eliminated jobs in many world locations, migrant workers could no longer send remittances to their families who then no longer had money with which to buy food. These episodes called into question the sustainability of current food production and acquisition systems. In developing countries, household economies were so ruined that parents sold important assets; went without food, to the detriment of their long-term health; and stopped sending children to school. UN statistics show real reversals of progress in human well-being in 2007–2008. Figure 1.14 identifies countries in which undernourishment is an ongoing problem and periodic food insecurity is especially intense. The situation had improved somewhat by 2012 for a few countries. See www.fao.org/hunger/en/.

Another way that modernized agriculture impacts food security is through its reliance on machines, chemical fertilizers, and pesticides. Paradoxically, when the shift to this kind of agriculture was introduced in the 1970s into developing countries like India and Brazil, it was called the **green revolution**. In fact, the green revolution is not "green" in the modern sense of being environmentally savvy. When successfully implemented, the results of green revolution agriculture were at first spectacular: soaring production levels and high profits for those farmers who could afford the additional investment. In the early years of this movement, it seemed to scientists and developers that the world was literally getting greener. But often, poorer farmers couldn't afford the machinery and chemicals. Also, since greatly increased production leads to lower crop prices on the market, these poorer farmers lost money. To survive, they often were forced to sell their land and move to crowded cities. Here they joined masses of urban poor whose food security was chronically precarious.

green revolution increases in food production brought about through the use of new seeds, fertilizers, mechanized equipment, irrigation, pesticides, and herbicides

carrying capacity the maximum number of people that a given territory can support sustainably with food, water, and other essential resources

Green revolution agriculture can also impact food security by damaging the environment. As rains wash fertilizers and pesticides into streams, rivers, and lakes, these bodies of water become polluted. Over time, the pollution destroys fish and other aquatic animals vital to food security. Hormones fed to farm animals to hasten growth may enter the human food chain. Soil degradation can also increase as green revolution techniques (such as mechanical plowing, tilling, and harvesting) leave soils exposed to rains that wash away natural nutrients and the soil itself. Indeed, many of the most agriculturally productive parts of North America, Europe, and Asia have already suffered moderate to serious loss of soil through erosion. Globally, soil erosion and other problems related to food production affect about 7 million square miles (2000 million hectares), putting at risk the livelihoods of a billion people.

At least in the short term, green revolution agriculture raised the maximum number of people that could be supported on a given piece of land, or its **carrying capacity**. However, it is unclear how

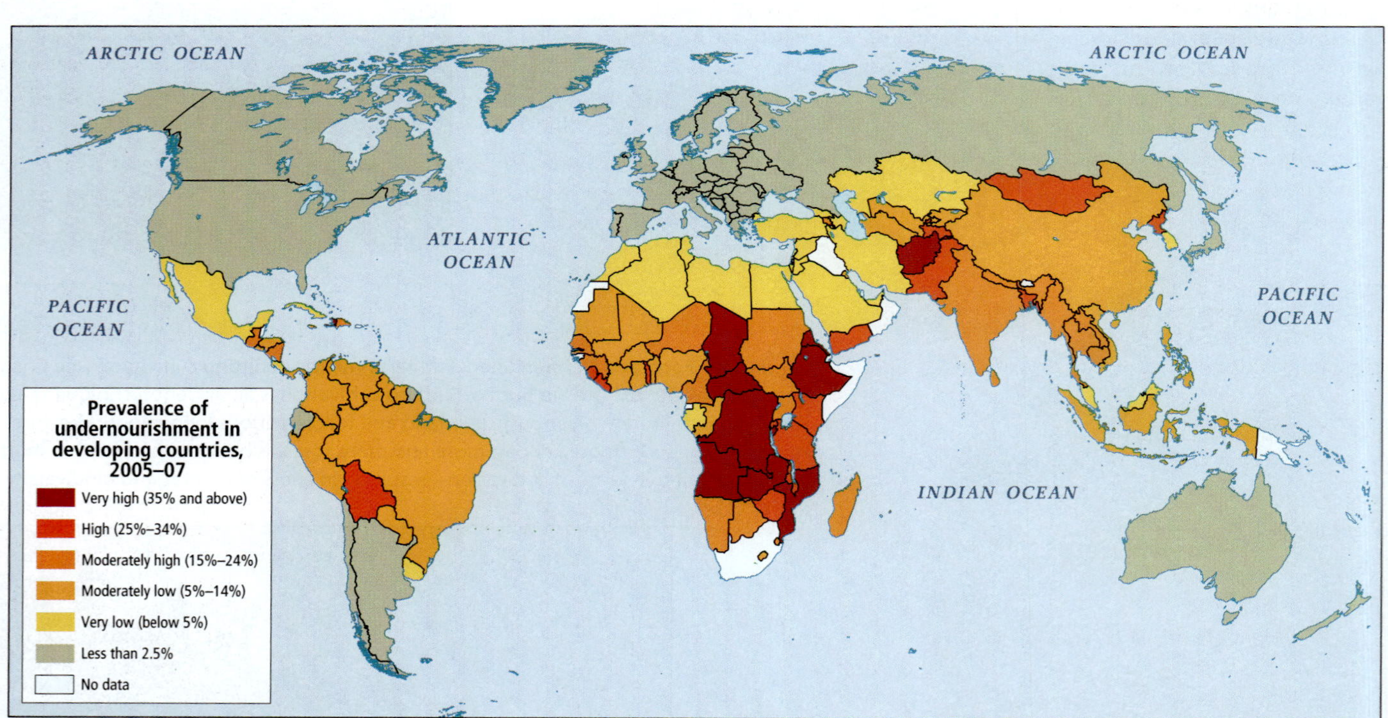

FIGURE 1.14 Global map of undernourishment, using data from 2007. This map is based on 2007 data (the only data presently available), before the global recession (not yet reflected in 2007 data) put many people back into a state of hunger. What shows here is that the proportion of people suffering from undernourishment—the lack of adequate nutrition to meet their daily needs—had declined in the developing world, most notably in India, over the past several years. However, hundreds of millions of people remain affected by chronic hunger. As you can see from the map, those who are the most affected are people in much of sub-Saharan Africa, parts of South Asia, Mongolia, North Korea, Guatemala, Nicaragua, and Bolivia, and Haiti and the Dominican Republic in the Caribbean.

sustainable these green revolution gains in food production are. In the 25 years between 1965 and 1990, total global food production rose between 70 and 135 percent (varying from region to region). In response, populations also rose quickly during this period. These successes in improving agricultural production and carrying capacities led the general public to assume that technological advances would perpetuate these increases; indeed, it is now estimated that to feed the population projected for 2050, global food output must increase by another 70 percent. Yet scientists from many disciplines estimate that within the next 50 years, environmental problems such as water scarcity and global climate change will limit, halt, or even reverse increases in food production. How can these discrepancies between expectations and realities be resolved?

The technological advances that could make present agricultural systems more productive are increasingly controversial. In North America, **genetic modification (GM)**, the practice of splicing together the genes from widely divergent species to achieve particular characteristics, is increasingly being used to boost productivity. However, outside of North America, many worry about the side effects of such agricultural manipulation. Europeans have tried (unsuccessfully) to keep GM food products entirely out of Europe, fearing that they could lead to unforeseen ecological consequences or catastrophic crop failures. They point out that the main advance in GM agriculture has been the production of seeds that can tolerate high levels of environmentally damaging herbicides, such as Roundup. The use of GM crops thus could lead to more, not less, environmental degradation. Even if GM crops prove safe, in developing countries where farmers' budgets are tiny, genetically modified seeds are much more expensive than traditional seeds. They must be purchased anew each year because GM plants do not produce viable seeds, as do plants from traditional seeds.

As a result of the uncertainties of GM crops and the potential negative side effects of new agricultural technologies, many are returning to the much older idea of **sustainable agriculture**—farming that meets human needs without poisoning the environment or using up water and soil resources. Often these systems avoid chemical inputs entirely, as in the case of increasingly popular methods of *organic agriculture*. However, while these systems can be productive, they are less so than conventional green revolution systems and often require significantly more human labor, resulting in higher food prices. Increased dependence on sustainable and organic systems could therefore lead to food insecurity for some poor people, especially in cities.

genetic modification (GM) in agriculture, the practice of splicing together the genes from widely divergent species to achieve particular desirable characteristics

sustainable agriculture farming that meets human needs without poisoning the environment or using up water and soil resources

According to the UN Food and Agriculture Organization, one-fifth of humanity subsists on a diet too low in total calories and vital nutrients to sustain adequate health and normal physical and mental development (see Figure 1.14). As we will see in later chapters, this massive hunger problem is partly due to political instability, corruption, and inadequate distribution systems. When, for whatever reason, food is scarce, it tends to go to those who have the money to pay for it. **3. DEFORESTATION: WORLDWIDE CONCERNS**

THINGS TO REMEMBER

- While modern processes of food production and distribution have greatly increased the supply of food, environmental damage and market disruptions could strain future food production and compromise food security.
- Many farmers are unable to afford the chemicals and machinery required for commercial agriculture or new, genetically modified seeds. Because their production is low, they cannot compete on price and may be forced to give up farming, often migrating to cities.
- Sustainable agriculture is farming that meets human needs without harming the environment or depleting water and soil resources.

URBANIZATION

In 1700, fewer than 7 million people, or just 10 percent of the world's total population, lived in cities, and only 5 cities had populations of several hundred thousand people or more. The world we live in today has been transformed by **urbanization**, the process whereby cities, towns, and suburbs grow as populations shift from rural to urban livelihoods. Now a little over half of the world's population lives in cities, and there are more than 400 cities of more than 1 million.

Why Are Cities Growing?

Geographic Insight 4

Food and Urbanization: Modernization in food production is pushing agricultural workers out of rural areas toward urban areas where jobs are more plentiful but where food must be purchased. This circumstance often leads to increased dependency on imported food.

For some time, economic changes, such as the mechanization of food production that has drastically reduced the need for labor while greatly increasing the food supply, have been pushing people out of rural areas, while the development of manufacturing and service economies and the possibility of earning cash incomes have been pulling them into cities. This process is called the **push/pull phenomenon of urbanization**. Numerous cities, especially in poorer parts of the world, have been unprepared for the massive inflow of rural migrants, many of whom now live in polluted **slum** areas plagued by natural and human-made hazards, poor housing, and inadequate access to food, clean water, education, and social services (see Figure 1.15C on page 29). Often a substantial portion of the migrants' cash income goes to support their still-rural families (see Figure 1.15B).

urbanization the process whereby cities, towns, and suburbs grow as populations shift from rural to urban livelihoods

push/pull phenomenon of urbanization conditions, such as political instability or economic changes, that encourage (push) people to leave rural areas, and urban factors, such as job opportunities, that encourage (pull) people to move to urban areas

slum densely populated area characterized by crowding, run-down housing, and inadequate access to food, clean water, education, and social services

Patterns of Urban Growth

The most rapidly growing cities are in developing countries in Asia, Africa, and Middle and South America. The settlement pattern of these cities bears witness to their rapid and often unplanned growth, fueled in part by the steady arrival of masses of poor rural people looking for work. Cities like Mumbai (in India), Cairo (in Egypt), Nairobi (in Kenya), and Rio de Janeiro (in Brazil) sprawl out from a small affluent core, often the oldest part, where there are upscale businesses, fine old buildings, banks, shopping centers, and residences for wealthy people. Surrounding these elite landscapes are sprawling mixed commercial, industrial, and middle-class residential areas, interspersed with pockets of extremely dense slums. Also known as *barrios*, *favelas*, *hutments*, *shantytowns*, *ghettos*, and *tent villages*, these settlements provide housing for the poorest of the poor, who provide low-wage labor for the city (see Figure 1.15B). Housing is often self-built out of any materials the residents can find: cardboard, corrugated metal, masonry, and scraps of wood and plastic. There are usually no building codes, no toilets with sewer connections, and no clean water; electricity is often obtained from illegal and dangerous connections to nearby power lines. Schools are few and overcrowded, and transportation is provided only by informal, nonscheduled van-based services. Because these slums can pop up on scraps of vacant land virtually overnight, an overflow of people, often unflatteringly called *squatters*, may soon be living in unhealthy conditions. The workers who labor to build soaring modern skyscrapers in one part of a city may find that they have to sleep on the street or in a hovel with no water or plumbing just a few blocks away.

The UN estimates that currently over a billion people live in urban slums, with that number to increase to 2 billion by 2030. Life in these areas can be insecure and chaotic as criminal gangs often assert control through violence and looting—all actions that

Figure 1.15 Photo Essay: Urbanization

Thinking Geographically

After you have read about urbanization, you will be able to answer the following questions:

A To what group of urban migrants does this skydiving man probably belong?

B In what kind of neighborhood of Dhaka would you guess this man lives?

C This photo exemplifies what problem commonly faced by rapidly growing cities?

Urbanization and urban areas. In the map below, the color of the country indicates the percentage of the population living in urban areas. The circles represent the populations of the world's largest urban areas in 2013 (blue circle) and 2020 (black circle).

A Cities have always been centers of innovation, entertainment, and culture, in large part because they attract both money and talented people. This resident of Shanghai is engaging in a current fad: parachuting off one of the new skyscrapers that now dominate the city's skyline.

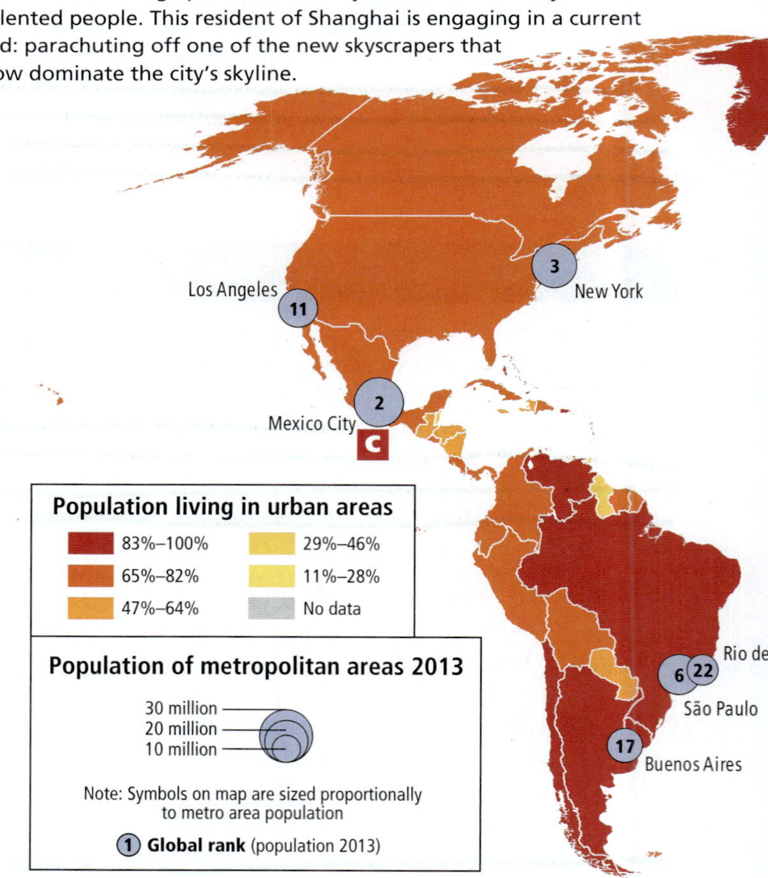

ON THE BRIGHT SIDE

Urban Migrant Success Stories

Slums are only part of the story of urbanization today. Those who are financially able to come to urban areas for education and complete their studies tend to find employment in modern industries and business services. They constitute the new middle class and leave their imprint on urban landscapes via the high-rise apartments they occupy and the shops and entertainment facilities they frequent (see Figure 1.15 map, A). Cities such as Mumbai in India, São Paulo in Brazil, Cape Town in South Africa, and Shanghai in China are now home to this more educated group of new urban residents, many of whom may have started life on farms and in villages.

In the past, most migrants in cities were young males, but increasingly they are young females. Cities offer women more than better-paying jobs. They also provide access to education, better health care, and more personal freedom.

B A recent migrant to Dhaka pulls a cart loaded with goods. Many of the world's fastest-growing cities are attracting more people than they can support with decent jobs, housing and infrastructure.

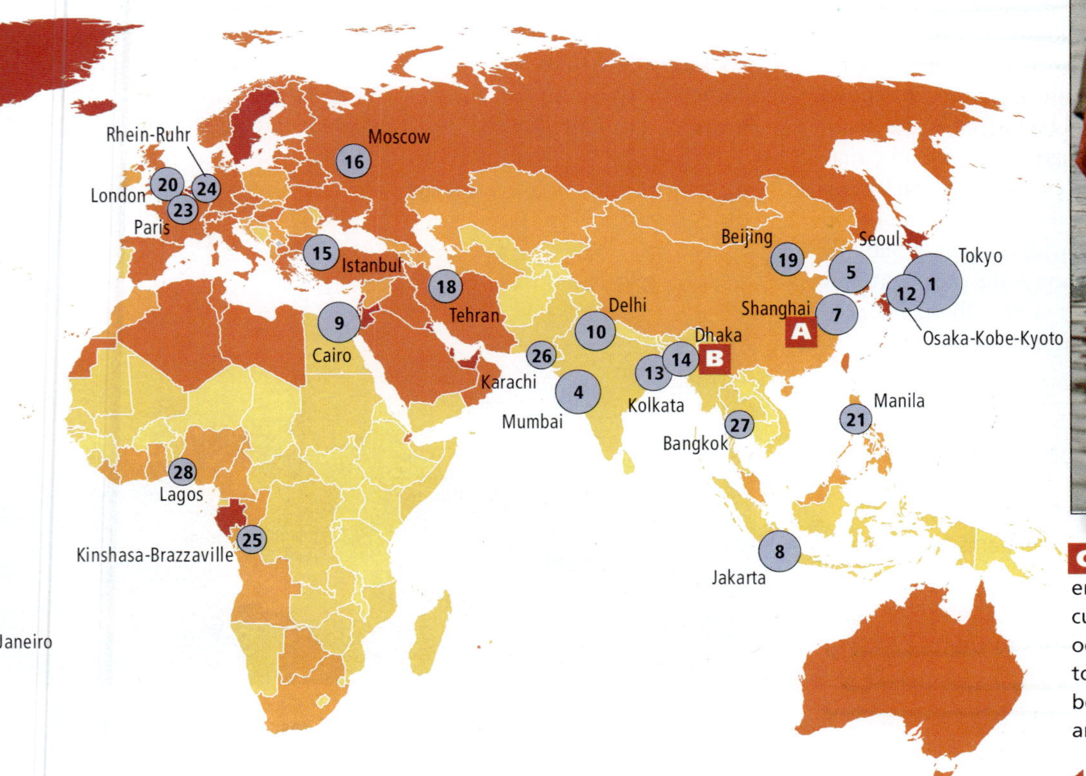

C Some cities struggle with major environmental problems. Mexico City, currently the world's second-largest city, occasionally suffers from severe flooding due to its location on an old, now-sinking, lake bed, and to its antiquated drainage and sewage infrastructure.

are especially likely during periods of economic recession and political instability. 331. UN HABITAT AGENCY SAYS HALF THE WORLD POPULATION LIVES IN CITIES

THINGS TO REMEMBER

- Today about half of the world's population live in cities; there are over 400 cities with more than 1 million people and about 28 cities of more than 10 million people.

Geographic Insight 4

- **Food and Urbanization** Changes in agriculture and food production are pushing people out of rural areas, while the development of manufacturing and service economies is pulling them into cities.
- For some, urbanization means improved living standards, while for others it means being forced into slums with inadequate food, water, and social services.

GLOBALIZATION

Throughout the world, globalization is transforming patterns of economic development, as local self-sufficiency is giving way to global interdependence and international trade. While for a time there was anticipation that globalization would lead to more prosperity for all, the economic recession that began in 2007 cast a spotlight on the unpredictable effects of global interdependence. In that year, economic disruptions in the United States and Europe resulted in powerful ripple effects that reached around the world. Foreclosures in the U.S. housing market meant that European banks that had invested in U.S. mortgages faltered and some failed; consumers in the United States and Europe shopped and vacationed less; some business borrowers as well as indebted country governments, such as Greece and Ireland, could not make loan payments and, as a result, many factory workers across the world lost their jobs. These connections between distant regions are known as **interregional linkages**. The term **globalization** refers to the worldwide changes brought about by many types of these interregional linkages and flows that reach well beyond economies. Attitudes and values are modified because of global connections and resultant culture change; ethnic identity may be reinforced or erased; personal or collegial relationships may be established between people who will never actually meet. Globalization is the most complex and far-reaching of the thematic concepts described in this book.

interregional linkages economic, political, or social connections between regions, whether contiguous or widely separated

globalization the growth of interregional and worldwide linkages and the changes these linkages are bringing about

global economy the worldwide system in which goods, services, and labor are exchanged

What Is the Global Economy?

Geographic Insight 5

Globalization and Development: Increased global flows of information, goods, and people are transforming patterns of economic development.

The **global economy** includes the parts of any country's economy that are involved in global flows of resources—mined minerals, agricultural commodities, manufactured products, money, and people and their ideas. Most of us participate in the global economy every day. For example, books like this can be made from trees cut down in Southeast Asia or Siberia and shipped to a paper mill in Oregon. Many books are now printed in Asia because labor costs are lower there. Globalization is not new. At least 2500 years ago, silk and other goods were traded along the Central Asian Silk Road that connected Greece and then Rome in the Mediterranean with distant China, an expanse of over 4000 miles (6437 kilometers).

European colonization was an early expansion of globalization. Starting in about 1500 C.E., European countries began extracting resources from distant parts of the world that they had conquered. The colonizers organized systems to process those resources into higher-value goods to be traded wherever there was a market. Sugarcane, for example, was grown on Caribbean and Brazilian plantations with slave labor from Africa (Figure 1.16) and made locally into crude sugar, molasses, and rum. It was then shipped to Europe and North America, where it was further refined and sold at considerable profit. The global economy grew as each region produced goods for export, rather than just for local consumption. At the same time, regions also became increasingly dependent on imported food, clothing, machinery, energy, and knowledge.

The new wealth derived from the colonies, and the ready access to global resources led to Europe's Industrial Revolution. No longer was one woman producing the cotton or wool for clothes by spinning thread, weaving the thread into cloth, and sewing a garment. Instead, these separate tasks were spread out

FIGURE 1.16 European use of colonial resources. Among the first global economic institutions were Caribbean plantations like the one shown here in a nineteenth-century painting of St. Croix Island. In the eighteenth century, thousands of sugar plantations in the British West Indies, subsidized by the labor of slaves, provided vast sums of money for England and helped fund the Industrial Revolution.

among many workers, often in distant places, with some people specializing in producing the fiber and others in spinning, weaving, or sewing. These innovations in efficiency were followed by labor-saving improvements such as mechanized reaping, spinning, weaving, and sewing.

This larger-scale mechanized production accelerated globalization, as it created a demand for raw materials and a need for markets in which to sell finished goods. European colonizers managed to integrate the production and consumption of their colonial possessions in the Americas, Africa, and Asia. For example, in the British Caribbean colonies, hundreds of thousands of enslaved Africans wore rough garments made of cheap cloth woven in England from cotton grown in British India. The sugar that slaves produced on British-owned plantations with iron equipment from British foundries was transported to European markets in ships made in the British Isles of trees and resources from the North American colonies and other parts of the world.

Until the early twentieth century, much of the activity of the global economy took place within the colonial empires of a few European nations (Britain, France, the Netherlands, Germany, Belgium, Spain, and Portugal). By the 1960s, global economic and political changes brought an end to these empires, and now almost all colonial territories are independent countries. Nevertheless, the global economy persists in the form of banks and **multinational corporations**, such as Shell, Chevron, IBM Canada, Walmart, Coca-Cola, Bechtel, Apple, British Petroleum, Toyota, Google, and Cisco, that operate across international borders. These corporations extract resources (including intellectual properties) from many places, make products in factories located where they can take advantage of cheap labor and transportation facilities, and market their products wherever they can make the most profit. Their global influence, wealth, and importance to local economies enable the multinationals to influence the economic and political affairs of the countries in which they operate.

multinational corporation a business organization that operates extraction, production, and/or distribution facilities in multiple countries

Workers in the Global Economy

VIGNETTE Sixty-year-old Olivia lives near Soufrière on St. Lucia, an island in the Caribbean (**Figure 1.17A** on page 32). She, her daughter Anna, and her three grandchildren live in a wooden house surrounded by a leafy green garden dotted with fruit trees. Anna has a tiny shop at the side of the house, from which she sells various small everyday items and preserves that she and her mother make from the garden fruits.

On days when the cruise ships dock, Olivia heads to the market shed on the beach with a basket of homegrown goods. She calls out to the passengers as they near the shore, offering her spices and snacks for sale. In a good week she makes U.S.$50. Her daughter makes about U.S.$100 per week in the shop and is constantly looking for other ways to earn a few dollars.

Olivia and Anna support their family of five on about U.S.$170 a week (U.S.$8840 per year). From this income they take care of their bills and other purchases, including school fees for the granddaughter who will go to high school in the capital next year and perhaps college if she succeeds. Their livelihood puts them at or above the standard of living of most of their neighbors.

In Malacca, Malaysia, 30-year-old Setiya, an illegal immigrant from Tegal, Indonesia, is boarding a bus that will take him to a jobsite where he is helping build a new tourist hotel (see Figure 1.17B). Like a million other Indonesians attracted by the booming economy, he snuck into Malaysia, risking arrest, because in Malaysia average wages are four times higher than at home.

This is Setiya's second trip to Malaysia. His first trip was to do work legally on a Malaysian oil palm plantation. Upon arrival, however, Setiya found that he would have to work for 3 months just to pay off his boat fare from Indonesia. Not one to give in easily, he quietly went to another city and found a construction job earning U.S.$10 a day (about U.S.$2600 a year), which allowed him to send money home to his family in Indonesia.

In 2008, Malaysia announced it was expelling 500,000 foreign workers; Setiya was one of them. The country was suffering from growing unemployment due to the global recession, and its leaders wanted to save more jobs for locals by expelling foreign workers. When the recession is alleviated, young fathers from Indonesia (like Setiya) may once again risk trips on leaky boats to illegally enter Malaysia and Singapore in order to support their families.

Fifty-year-old Tanya works at a fast-food restaurant outside of Charleston, South Carolina, making less than U.S.$7.50 an hour. She had been earning U.S.$8 an hour sewing shirts at a textile plant until it closed and moved to Indonesia. Her husband is a delivery truck driver for a snack-food company.

Between them, Tanya and her husband make $30,000 a year, but from this income they must cover all their expenses, including mortgage, gasoline, and car payments. In addition, they help their daughter, Rayna, who quit school after eleventh grade and married a man who is now out of work. They and their baby live at the back of the lot in an old mobile home (see Figure 1.17C).

With Tanya's now-lower wage (almost $1000 less a year), there will not be enough money to pay the college tuition for her son, who is in high school. He had hoped to become an engineer, and would have been the first in the family to go to college. For now, he is working at the local gas station.

These people, living worlds apart, are all part of the global economy. Workers around the world are paid startlingly different rates for jobs that require about the same skill level. Varying costs of living and varying local standards of wealth make a difference in how people live and regard their own situation. Though Tanya's family has the highest income by far, compared to their neighbors they live in poverty, and their hopes for the future are dim. Olivia's family, on the other hand, are not well off, but they do not think of themselves as poor because they have what they need, others around them live in similar circumstances, and their children seem to have a future. They can subsist on local resources, and the tourist trade promises continued cash income. But their subsistence depends on circumstances beyond their control; in an instant, the cruise-line companies can choose another port of call. Setiya, by far the poorest, seems trapped by his status as an illegal worker, which robs him of many of his rights. Still, the higher pay that he can earn in Malaysia offers him a possible way out of poverty. *[Source: From Lydia Pulsipher's and Alex Pulsipher's field notes. For detailed source information, see Text Credit pages.]* ■

The Debate over Globalization and Free Trade

The term **free trade** refers to the unrestricted international exchange of goods, services, and capital. Free trade is an ideal that has not been achieved and probably never will be. Currently, all governments impose some restrictions on trade to protect their own national economies from foreign competition, although such restrictions are far fewer than in the 1980s. Restrictions take two main forms: *tariffs* and *import quotas*. Tariffs are taxes imposed on imported goods that increase the cost of those goods to the consumer, thus giving price advantages to competing, locally made goods. Import quotas set limits on the amount of a given good that may be imported over a set period of time, thus curtailing supply and keeping prices high, again to protect local producers.

These and other forms of trade protection are subjects of contention. Proponents of free trade argue that the removal of all tariffs and quotas encourages efficiency, lowers prices, and gives consumers more choices. Companies can sell to larger markets and take advantage of mass-production systems that lower costs further. As a result, businesses can grow faster, thereby providing people with jobs and opportunities to raise their standard of living. These pro–free trade arguments have been quite successful, and in recent decades, restrictions on trade imposed by individual countries have been greatly reduced. Several *regional trade blocs* have been formed; these are associations of neighboring countries that agree to lower trade barriers for one another. The main ones are the North American Free Trade Agreement (NAFTA), the European Union (EU), the Southern Common Market in South America (Mercosur), and the Association of Southeast Asian Nations (ASEAN).

One of the main global institutions that supports the ideal of free trade is the **World Trade Organization (WTO)**, whose stated mission is to lower trade barriers and to establish ground rules for international trade. Related institutions, the *World Bank* (officially named the International Bank for Reconstruction and Development) and the *International Monetary Fund* (IMF), both make loans to countries that need money to pay for economic development or financial adjustment projects. Before approving a loan, the World Bank or the IMF may require a borrowing country to reduce and eventually remove tariffs and import quotas. These requirements are part of larger *structural adjustment policies* (SAPs)—belt-tightening measures that the IMF imposes on countries seeking loans, such as the requirement to close or privatize government enterprises and to reduce government services, mostly to the detriment of the poor.

Those opposed to free trade and the SAPs that have long promoted it argue that a less-regulated global economy can be chaotic, leading to rapid cycles of growth and decline that only increase global wealth disparity and can wreak havoc on smaller national economies. Labor unions point out that as corporations relocate factories and services to poorer countries where wages are lower, jobs are lost in richer countries, creating poverty. In the poorer countries, multinational corporations often work with governments to prevent workers from organizing labor unions that could

free trade the unrestricted international exchange of goods, services, and capital

World Trade Organization (WTO) a global institution made up of member countries whose stated mission is the lowering of trade barriers and the establishment of ground rules for international trade

FIGURE 1.17 Workers in the global economy.

A Soufrière, St. Lucia.

B Setiya, along with other migrant workers boarding a bus that will take them to the construction site in Malacca, Malaysia, before being expelled from the country, along with 500,000 other foreign workers, during the recession that began in 2008.

C The trailer at the back of Tanya's lot, where her daughter lives.

Thinking Geographically

A What about this photo might suggest that Olivia and her neighbors share a similar standard of living?

B What about this photo suggests that Setiya is a low-paid worker?

C How does this photo suggest that Tanya's low income is being perpetuated in the life of her daughter, Reyna?

ON THE BRIGHT SIDE

Reforms at the IMF and World Bank

In response to the now widely recognized failures of SAPs, and the overemphasis on the power of markets to guide development, the IMF and the World Bank have made some changes. SAPs have been replaced with "Poverty Reduction Strategy Papers," or PRSPs. Each country in need works with World Bank and IMF personnel to design a broad-based plan for both economic growth *and* poverty reduction. PRSPs still push market-based solutions and aim toward reducing the role of government in the economy. These programs remain highly bureaucratic, but they do focus on poverty reduction rather than just "development via structural adjustment" per se. They promote the maintenance of education, health, social services, and broader participation in civil society. PRSPs also include the possibility that all or some of a country's debt be "forgiven" (written off by the IMF and the World Bank), thus alleviating one of the worst side effects of SAPs—debt that bankrupts that country—that has stopped progress in poor countries and, by 2011, began to affect even developed countries in Europe such as Greece, Italy, Ireland, Spain, and Portugal.

bargain for **living wages**, minimum wages high enough to support a healthy life. Environmentalists argue that in newly industrializing countries, which often lack effective environmental protection laws, multinational corporations tend to use highly polluting and unsafe production methods to lower costs. Many fear that a "race to the bottom" in wages, working conditions, government services, and environmental quality is underway as countries compete for profits and potential investors. Multinational corporations that have recently agreed to address worker abuses include Nike, Walmart, and Apple.

In October 2011, activists around the world joined what was dubbed the *Occupy Wall Street* movement, protesting what they saw as the downside of free trade: the growing disparity of wealth worldwide and the unfettered role of global financial institutions, commodity speculators, and the superrich in molding the global economy to suit their ends. Labeling the superrich the "1 percent" and the rest of the world's population the "99 percent," the "Occupy" demonstrations spread throughout Europe and to hundreds of cities in Asia, Africa, Europe, Oceania, and the Americas.

Fair trade, proposed as an alternative to free trade, is intended to provide a fair price to producers and to uphold environmental and safety standards in the workplace. Economic relationships surrounding trade are rearranged in order to provide better prices for producers from developing countries. For example, "fair trade" coffee and chocolate are now sold widely in North America and Europe. Prices are somewhat higher for consumers, but the extreme profits of middlemen are eliminated; growers of coffee and cocoa beans, who produce for "fair trade" companies, receive living wages and improved working conditions.

In evaluating free trade, globalization, and fair trade, consider how many of the things you own or consume were produced in the global economy—your computer, clothes, furniture, appliances, car, and foods. These products are cheaper for you to buy, and your standard of living is higher as a result of lower production costs as well as competition among many global producers. However, you or someone you know may have lost a job because a company moved to another location where labor and resources are cheaper. Underpaid workers (even children) working under harsh conditions that possibly generate high levels of pollution may have made those cheap products. If workers can't earn living wages, often some family members end up migrating, perhaps without the proper papers, to earn a better wage. Given all these factors, consider the advantages and drawbacks of both free trade and fair trade.

living wages minimum wages high enough to support a healthy life

fair trade trade that values equity throughout the international trade system; now proposed as an alternative to free trade

THINGS TO REMEMBER

- Globalization encompasses many types of worldwide and interregional flows and linkages, especially the ways in which goods, capital, labor, and resources are exchanged among distant and very different places.
- Geographic Insight 5 — **Globalization and Development** Throughout the world, globalization is transforming patterns of economic development as local self-sufficiency is giving way to global interdependence and international trade.
- Under true free trade, all economic transactions are conducted without interference or regulation in an open marketplace. Under fair trade, an alternative to free trade, consumers are asked to pay a fair price to producers, and producers are expected to pay living wages and uphold environmental and safety standards in the workplace.

POWER AND POLITICS

For geographers studying globalization, the social and spatial distribution of political power is an area of increasing interest (see Figure 1.18 on pages 34–35). Recent years have produced what appears to be a transition toward political systems guided by competitive elections, a process often called **democratization**. This trend is of particular interest because it runs counter to **authoritarianism**, a form of government that subordinates individual freedom to the power of the state or elite regional

democratization the transition toward political systems guided by competitive elections

authoritarianism a political system that subordinates individual freedom to the power of the state or of elite regional and local leaders

and local leaders. In democratic systems of government, beyond the right to participate in free elections, average individuals have more *economic and political freedoms*, such as the right to be an entrepreneur, the right to peacefully protest government policies, the right to access information of all types, the right to marshal public support through the media (including print, broadcast, the Internet, and social media) for particular programs or candidates or election issues, and the right to take action against injustice, especially through legal systems.

Geographers do not necessarily conclude that democracy is the "best" system of government. Indeed, many geographers are critical of the imposition of democracy, often by foreign governments or organizations, in places where long-standing cultural traditions support other political arrangements. They say that a democratic form of government should be achieved democratically. Nevertheless, few would deny that the shift toward more democratic systems of government over the past century and into recent times has been extremely significant, if not always peaceful. In this regard, geographers and other scholars are particularly interested in the role of democracy at the state and local levels, in addition to the roles that social movements, international organizations, and a free media play in democratization.

23. PROMOTING DEMOCRACY: A CONTROVERSIAL THEME

The Expansion of Democracy

Geographic Insight 6

Power and Politics: There are major differences across the globe in the ways that power is wielded in societies. Modes of governing that are more authoritarian are based on the power of the state or community (or tribal) leaders. Modes that are more democratic give the individual a greater say in how policies are developed and governments are run. There are also many other ways of managing political power.

The twentieth century saw a steady expansion of democracy throughout the world, with more and more countries holding elections of their leaders—at least at the national level. While democracy has become an ideal to which most countries aspire, there is little agreement about just what constitutes democratic institutions. Can the principles of a specific religion be part of democratic constitutions? Is it possible to have democratic government at the national level but quite authoritarian rule at the local level? What about forms of democratic participation other than voting, such as the ability to speak openly to leaders, to protest, to lobby for or against particular laws (Figure 1.18A)?

Figure 1.18 Photo Essay: Power and Politics

Thinking Geographically

After you have read about power and politics, you will be able to answer the following questions:

A Why has Juan Manuel Santos' relationship with the Colombian military come under criticism?

B Why is the participation of women in the Arab Spring significant?

C, **D** Of the factors mentioned in the text as necessary for democracy to flourish, which are obviously present in **C** and missing in **D**?

The map and accompanying photo essay show two related trends in political power. Generally speaking, countries with fewer political freedoms and lower levels of democratization also suffer the most from violent conflict. Countries are colored on the map according to their score on a "democracy index" created by *The Economist* magazine, which uses a combination of statistical indicators to capture elements crucial to the process of democratization. These elements include the ability of a country to hold peaceful elections that are accepted as fair and legitimate, the ability of people to participate in elections and other democratic processes, the strength of political freedoms (such as a free media and the right to hold political gatherings and peaceful protests), and the ability of governments to enact the will of their citizens and be free of corruption. Also displayed on the map are major conflicts initiated or in progress since 1990 that have resulted in at least 10,000 casualties. Aspects of the connections between democratization and armed conflict are explored in the photo captions.

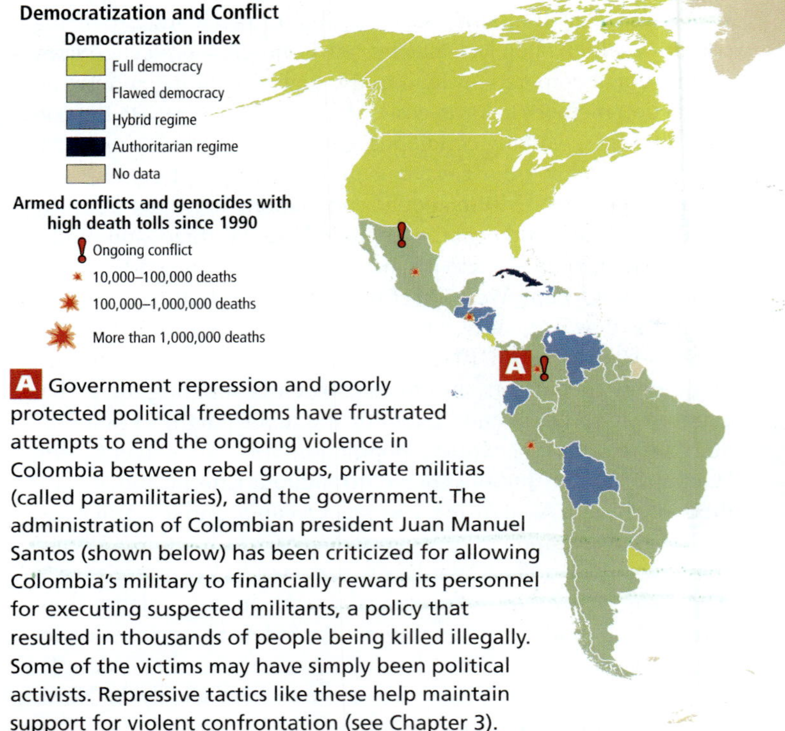

A Government repression and poorly protected political freedoms have frustrated attempts to end the ongoing violence in Colombia between rebel groups, private militias (called paramilitaries), and the government. The administration of Colombian president Juan Manuel Santos (shown below) has been criticized for allowing Colombia's military to financially reward its personnel for executing suspected militants, a policy that resulted in thousands of people being killed illegally. Some of the victims may have simply been political activists. Repressive tactics like these help maintain support for violent confrontation (see Chapter 3).

B A Syrian woman holds an artillery shell fired by the Syrian military on civilians during the country's civil war. Sparked by demonstrations for greater political freedoms in 2011, part of the larger wave of protests known as the Arab Spring, Syria's civil war developed when the government responded with a harsh military crackdown on the civilian population. The resulting civil war has led to nearly 100,000 deaths, over a million international refugees, and 2.5 to 3 million displaced people within Syria (see Chapter 6).

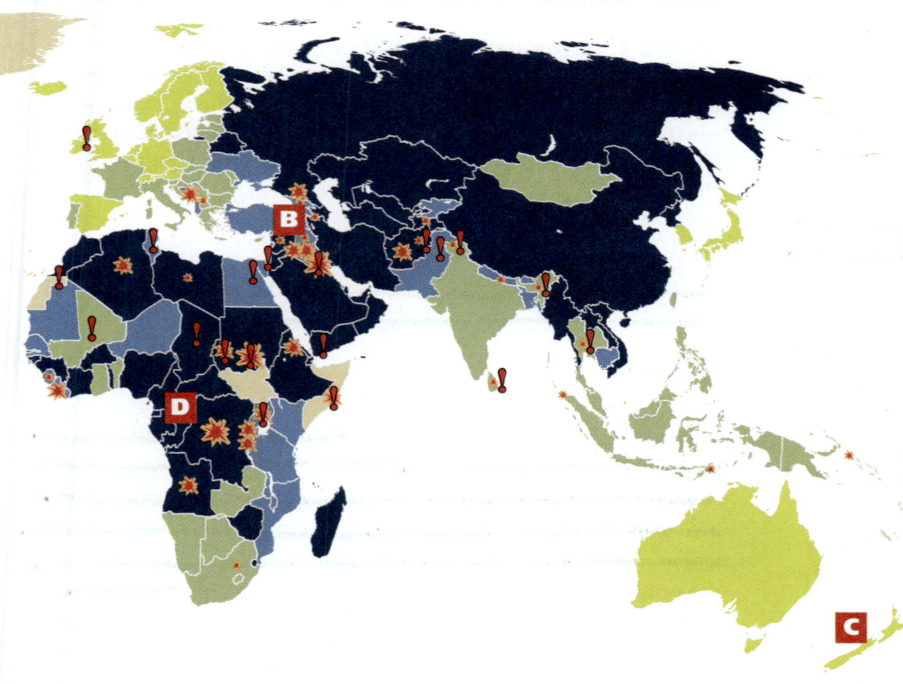

C New Zealand has well-protected political freedoms and is among the world's most democratized nations. Shown above is a member of the Service and Food Workers Union who is picketing and soliciting support from passing motorists for better wages and working conditions for its largely immigrant and female members (see Chapter 11).

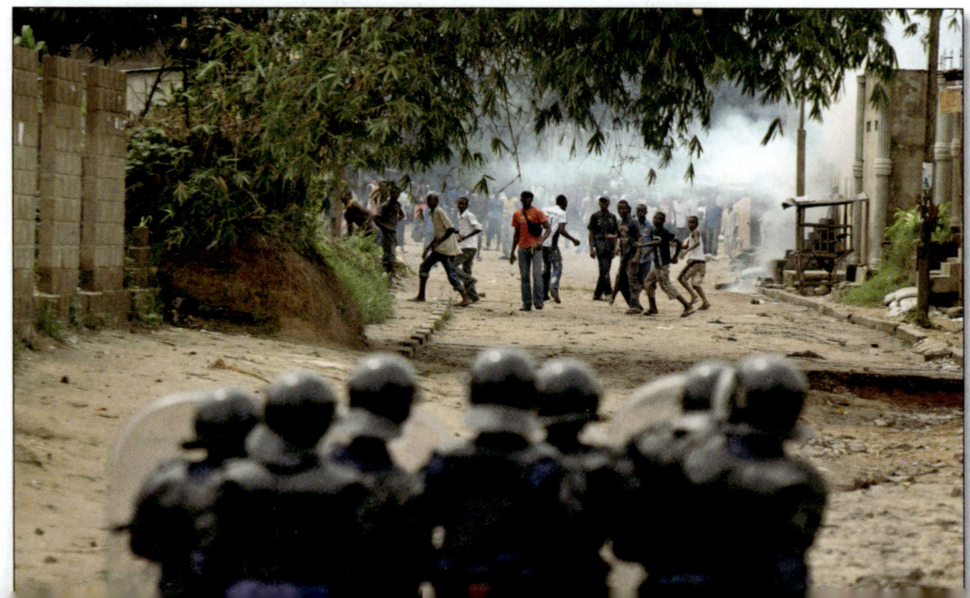

D Political campaigns often become compromised during times of war or conflict. Shown here are supporters of the political opposition in Kinshasa, Congo, being dispersed by police with tear gas during elections in 2011. Accusations that the vote counting was being rigged added to tensions related to the country's ongoing civil war that has resulted in more than 5.4 million deaths so far, mostly due to disease and malnutrition among people displaced by the violence (see Chapter 7).

Are not these also essential components of democracy? The Arab Spring movements, which commenced in early 2011 in a number of countries in North Africa and Southwest Asia, demonstrated that popular will can achieve amazing turnovers of power, as was the case in Tunisia, Egypt, and Libya. However, in Egypt in the 2012 elections, the Muslim Brotherhood—a male-dominated Islamist political party with authoritarian leanings—won an easy majority, causing concern that voters had quashed broader democratic reforms. Popular uprisings don't always work well, smoothly, or at all (see Figure 1.18B, D). In Syria, Bahrain, and Yemen, where opposition demonstrations against authoritarian regimes also began early in 2011, the bloodshed continued for more than a year—and who shall ultimately have power remains unresolved. In all six of these countries, only time will tell if activists, unfamiliar with marshalling broad-based political will, can design systems that satisfy enough people to create a calm interim during which these governing systems can be refined peacefully.

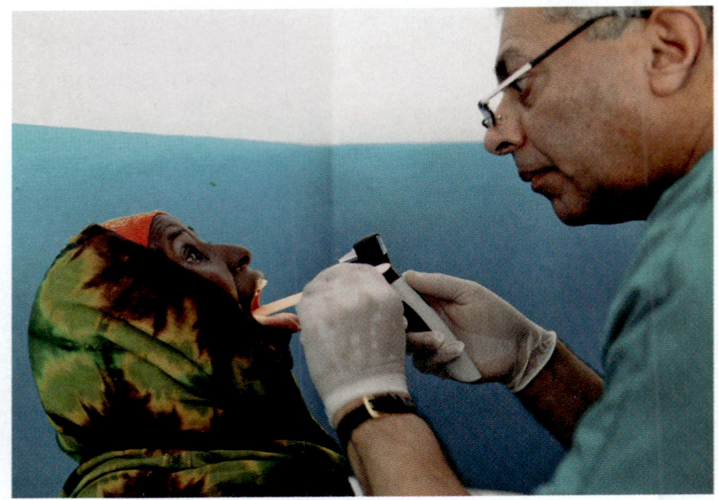

FIGURE 1.19 Nongovernmental organizations (NGOs) in action. A doctor examines a patient as part of an NGO-based effort to provide medical assistance and food aid to famine-stricken parts of Somalia. This doctor volunteers with South Africa–based "Gift of the Givers," which is the largest NGO started and staffed completely by Africans.

What Factors Encourage Successful Democratization?

Here are some of the most widely agreed-upon factors that support the idea that the public should have an active role in their own governance:

- **Peace:** Peace is essential to creating an environment in which "free and fair" elections can take place.
- **Broad prosperity:** As a broader segment of the population gains access to more than the bare essentials of life, there is generally a shift toward freer elections, which put more power in the hands of citizens. Whether general prosperity must be in place before truly stable democracy can be established is still widely debated, as is the question of whether prosperity necessarily leads to democracy.
- **Education:** Better-educated people tend to want a stronger voice in how they are governed. Although democracy has spread to countries with relatively undereducated populations, leaders in such places sometimes become more authoritarian once elected.
- **Civil society:** The social groups and traditions that function independently of the state and its institutions can foster a sense of unity and informed common purpose among the general population. **Civil society** institutions can include the media, *nongovernmental organizations* (NGOs, discussed below and in Figure 1.19), political parties, universities, unions, community service organizations such as Rotary and Lions clubs, and in some cases, religious organizations (see Figure 1.18C).

civil society the social groups and traditions that function independently of the state and its institutions to foster a sense of unity and informed common purpose among the general population

geopolitics the strategies that countries use to ensure that their own interests are served in relations with other countries

capitalism an economic system based on the private ownership of the means of production and distribution of goods, driven by the profit motive and characterized by a competitive marketplace

communism an ideology, based largely on the writings of the German revolutionary Karl Marx, that calls on workers to unite to overthrow capitalism and establish an egalitarian society in which workers share what they produce; as practiced, communism was actually a socialized system of public services and a centralized government and economy in which citizens participated only indirectly through Communist Party representatives

Democratization and Geopolitics

As the map in Figure 1.18 shows, democratization at the global level has not yet been achieved (in fact, it is not a goal shared by all). A possible explanation is that democratization is often at odds with **geopolitics**, the strategies that countries use to ensure that their own interests are served in relations with other countries. Geopolitics was perhaps most obvious during the *Cold War era*, the period from 1946 to the early 1990s when the United States and its allies in western Europe faced off against the Union of Soviet Socialist Republics (USSR) and its allies in eastern Europe and Central Asia. Ideologically, the United States promoted a version of free market **capitalism**—an economic system based on the private ownership of the means of production and distribution of goods, driven by the profit motive and characterized by a competitive marketplace. By contrast, the USSR and its allies favored what was called **communism**, but what was actually a state-controlled economy—a socialized system of public services and a centralized government in which citizens participated only indirectly through the Communist Party.

The Cold War became a race to attract the loyalties of unallied countries and to arm them. Sometimes the result was that unsavory dictators were embraced as allies by one side or the other. Eventually, the Cold War influenced the internal and external policies of virtually every country on Earth, often oversimplifying complex local issues into a contest of democracy versus communism.

In the post–Cold War period of the 1990s, geopolitics shifted. The Soviet Union dissolved, creating many independent states, nearly all of which began to implement some democratic and free market reforms. Globally, countries jockeyed for position in what looked like a possible new era of trade and amicable prosperity rather than war. But throughout the 1990s, while the developed countries enjoyed unprecedented prosperity, many unresolved political conflicts emerged in southeastern Europe, Central and South America, Africa, Southwest Asia, and South Asia. Too often these disputes erupted into bloodshed and the systematic attempt to remove (through **ethnic cleansing**) or exterminate (through **genocide**) all members of a particular ethnic or religious group.

The new geopolitical era ushered in by the terrorist attacks on the United States on September 11, 2001, is still evolving. Because of the size and the global power of the United States, the attacks and the U.S. reactions to them affected virtually every international relationship, public and private. The ensuing adjustments, which will continue for years, are directly or indirectly affecting the daily lives of billions of people around the world.

International Cooperation

So far there has been no serious effort to engage the principles and practices of democratization at the global scale. However, many aspects of globalization favor international cooperation over national self-interest.

The prime example today of international cooperation is the **United Nations (UN)**, an assembly of 193 member states. The member states sponsor programs and agencies that focus on economic development, general health and well-being, democratization, peacekeeping assistance in "hot spots" around the world, humanitarian aid, and scientific research. Thus far, various countries have been unwilling to relinquish *sovereignty*, the right of a country to conduct its internal affairs as it sees fit without interference from outside. Consequently, the United Nations has limited legal authority and often can enforce its rulings only through persuasion. Even in its peacekeeping missions, there are no true UN forces. Rather, troops from member states wear UN designations on their uniforms and take orders from temporary UN commanders.

Nongovernmental organizations (NGOs) are an increasingly important embodiment of globalization. In such associations, individuals, often from widely differing backgrounds and locations, agree on political, economic, social, or environmental goals. For example, some NGOs, such as the World Wildlife Fund, work to protect the environment. Others, such as Doctors Without Borders, provide medical care to those who need it most, especially in conflict zones. The Red Cross and Red Crescent provide emergency relief after disasters, as do Oxfam and Catholic Charities and Gift of Givers (Figure 1.19). The educational efforts of an NGO such as Rotary International can raise awareness among the global public about important issues such as childhood vaccinations.

NGOs can be an important component of civil society, yet there is some concern that the power of huge international NGOs might undermine democratic processes, especially in small, poor countries. Some critics feel that NGO officials are a powerful, do-gooder elite that does not interact sufficiently well with local people. A frequent target of such criticism is Oxfam International (a networked group of 17 NGOs), the world leader in emergency famine relief. Oxfam was a major provider of relief after the 2004 Indian Ocean tsunami and the 2010 Haiti earthquake. It has now expanded to cover long-term efforts to reduce poverty and injustice, which Oxfam sees as the root causes of famine. This more politically active role has brought Oxfam into conflict with local officials and with WTO policies on trade. The wisest NGOs solicit input from a wide range of individuals and organizations—a feature that political scientists consider essential to building civil society. **22. NGOs PLAY LARGER ROLE IN WORLD AFFAIRS**

ethnic cleansing the deliberate removal of an ethnic group from a particular area by forced migration

genocide the deliberate destruction of an ethnic, racial, or political group

United Nations (UN) an assembly of 193 member states that sponsors programs and agencies that focus on economic development, security, general health and well-being, democratization, peacekeeping assistance in "hot spots" around the world, humanitarian aid, and scientific research

nongovernmental organizations (NGOs) associations outside the formal institutions of government in which individuals, often from widely differing backgrounds and locations, share views and activism on political, social, economic, or environmental issues

THINGS TO REMEMBER

- Democratization is the transition toward political systems that are guided by competitive elections and other forums in which individuals can have a greater voice in policy formation and in how their governments are run.
- Democratic systems are gradually replacing many authoritarian regimes worldwide, but the result of this apparent trend is unknown.
- **Geographic Insight 6** **Power and Politics** Peace, broad prosperity, education, and civil society are among the most widely agreed-upon factors necessary for democratization to flourish.
- Increasingly, the type of government that a country decides upon is of geopolitical significance in that official international organizations, trade associations, and NGOs from other countries have a stake in the decision and may try to influence it.

WATER

Water is emerging as the major resource issue of the twenty-first century. Demand for clean water skyrockets as people move out of poverty. A huge increase in water consumption occurs with the modernization of the production of foods, the manufacture of goods, the creation of energy, and the development of services (like cleaning or even entertainment). Modernization also inevitably creates *water pollution*. With clean, fresh water becoming scarce in so many parts of the world, water disputes are proliferating. This is especially true where rivers cross international borders and upstream users use more than their perceived fair share or pollute water for downstream users. Controversy also surrounds the sale of clean water. When water becomes a commodity rather than a free good, as it was before modern times, water prices can increase so much that the poor lose access, which then affects health and sanitation.

Calculating Water Use per Capita

Humans require an average of 5 to 13 gallons (20 to 50 liters) of clean water per day for basic domestic needs: drinking, cooking, and bathing/cleaning. Per capita domestic water consumption tends to increase as incomes rise; the average person in a wealthy country consumes as much as 20 times the amount of water, per capita, as the average person in a very poor country. However, domestic water consumption is only a fraction of a person's actual water consumption. Per capita **virtual water** is the volume of water required to process all the goods and services a person consumes in a year. Every apple eaten or cup of coffee drunk requires many liters of water for production and distribution. When we add an individual's domestic water consumption to her virtual water consumption for an entire year, we have that person's total annual **water footprint**. The more one consumes, the larger one's virtual water footprint. Table 1.2 shows the amounts of water used to produce some commonly consumed products. (As you look at Table 1.2 and read further, note that there are 1000 liters, or 263 gallons, in a cubic meter (m^3).)

virtual water the volume of water used to produce all that a person consumes in a year.

water footprint the water used to meet a person's basic needs for a year, added to the person's annual virtual water consumption

TABLE 1.2 The global average virtual water content of everyday products*

Product†	Virtual water content (in liters)
1 potato	25
1 cup tea	35
1 slice of bread	40
1 apple	70
1 glass of beer	75
1 glass of wine	120
1 egg	135
1 cup of coffee	140
1 glass of orange juice	170
1 pound of chicken meat	2000
1 hamburger	2400
1 pound of cheese	2500
1 pair of bovine leather shoes	8000

*Virtual water is the volume of water used to produce a product.

†To see the virtual water content of additional products, go to www.waterfootprint.org/?page=files/productgallery.

Like domestic consumption, personal water footprints vary widely according to physical geography, standards of living, and rates of consumption (Figure 1.20). Moreover, the amount of virtual water used to produce 1 ton of a specific product varies widely from country to country due to climate conditions, along with agricultural and industrial technology and efficiency. For example, on average, to produce 1 ton of corn in the United States requires 489 m^3 of virtual water, whereas in India the same amount of corn requires 1935 m^3 of virtual water; in Mexico, 1744 m^3; and in the Netherlands, just 408 m^3. In the case of corn, water can be lost to *evapotranspiration* in the field, to the evaporation of standing irrigation water, and to evaporation as water flows to and from the field. A further component of virtual water is that the water that becomes polluted in the production process is also lost to further use.

For help calculating your individual water footprint, go to the Water Footprint Web site, at http://www.waterfootprint.org/?page=cal/waterfootprintcalculator_indv.

Who Owns Water? Who Gets Access to It?

Though many people consider water a human right that should not cost anything to access, water has become the third most valuable commodity, after oil and electricity. Water in wells and running in streams and rivers is increasingly being *privatized*. This means that its ownership is being transferred from governments—which can be held accountable for protecting the rights of all citizens to water access—to individuals, multinational corporations, and other private entities that manage the water primarily for profit. Governments often privatize water under the rationale that private enterprise will make needed investments that will boost the efficiency of water distribution systems and the overall quality of the water supply. Regardless of whether or not these potential gains are actually realized, privatization usually brings higher water costs to consumers. This can become quite controversial. For example, in 1999 in the city of Cochabamba, Bolivia, water costs rose beyond what the urban poor could afford, following the sale of the city public water agency to a group of multinational corporations led by Bechtel of San Francisco, California. The result was a nationwide series of riots that led finally to the abandonment of privatization.

Water Quality

About one-sixth of the world's population does not have access to clean drinking water, and dirty water kills over 6 million people each year. More people die this way than in all of the world's armed conflicts for an average year. In the poorer parts of cities in the developing world, many people draw water with a pail from a communal spigot, sometimes from shallow wells, or simply from holes dug in the ground (Figure 1.21). Usually this water should be boiled before use, even for bathing. These water quality and

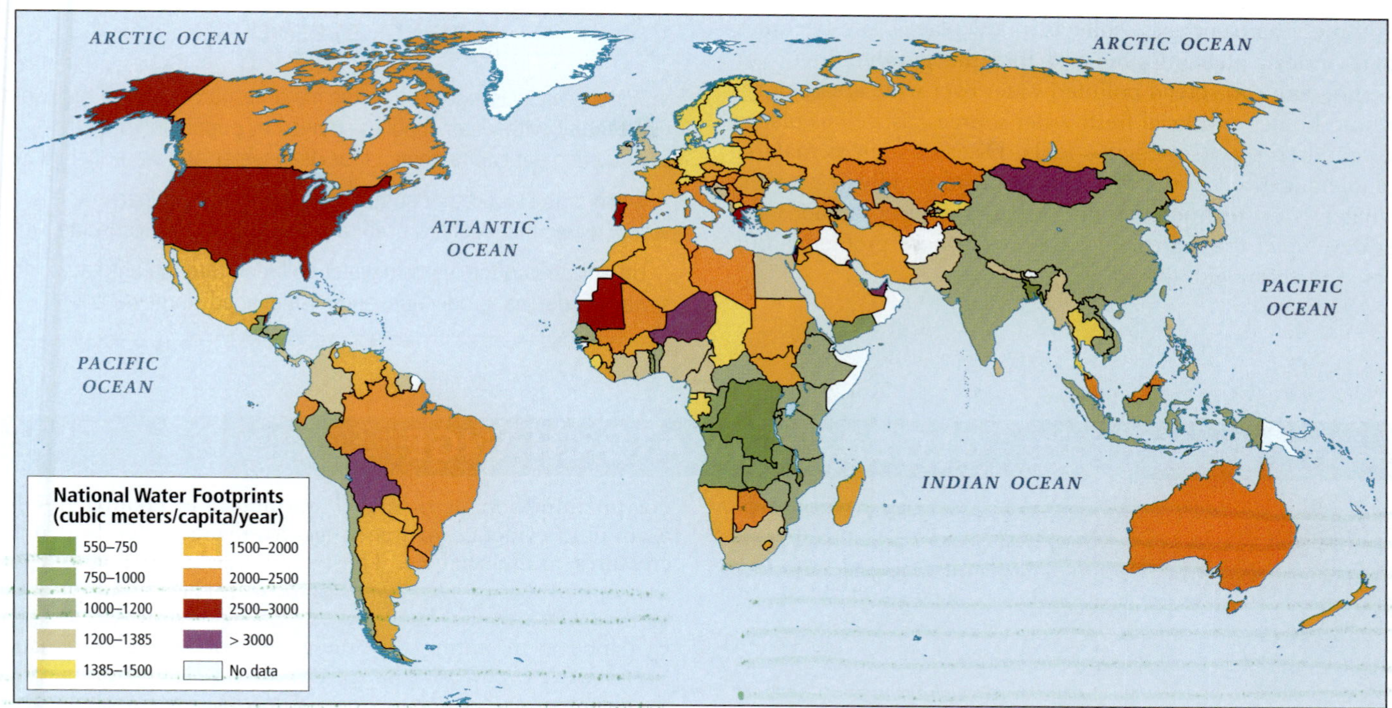

FIGURE 1.20 Map of national water footprints, 2005. Average national water footprint per capita (m3 per capita per year). The color green indicates that the nation's water footprint is equal to or smaller than the global average. Countries in red have a water footprint beyond the global average. (The latest year for which data are available is 2005.)

access problems help explain why so many people are chronically ill and why 24,000 children under the age of 5 die every day from waterborne diseases.

In Europe and the United States, demand for higher water quality has resulted in a $100 billion bottled water industry. However, there are few standards of quality for bottled water; and in addition to generating mountains of plastic bottle waste, the bottled water industry often acquires its water from sources (springs, ponds, deep wells) that are publicly owned. Hence, consumers are unwittingly paying for the same high-quality water they could consume free as a public commodity.

245. CLEAN WATER PROJECT IMPROVES LIVES IN SENEGAL

Water and Urbanization

Urban development patterns dramatically affect the management of water. In most urban slum areas in poor countries, and even in places like the United States and Canada, crucial water management technologies such as sewage treatment systems can be entirely absent. Germ-laden human waste from toilets and kitchens may be deposited in urban gutters that drain into creeks, rivers, and bays. And yet, retrofitting adequate wastewater collection and treatment systems, in cities already housing several million inhabitants, is often deemed prohibitively costly, especially in poor countries.

Independent of sewage, water inevitably becomes polluted in cities as parking lots and rooftops replace areas that were once covered with natural vegetation. Rainwater quickly

FIGURE 1.21 Access to water. A young girl gathers water from a shallow open well in a slum in Mumbai, India. Mumbai has vast low-lying slums that have no sanitation and whose sewage pollutes nearby waterways. Flooding of these waterways would spread deadly epidemics of waterborne illnesses via open wells such as the one shown here.

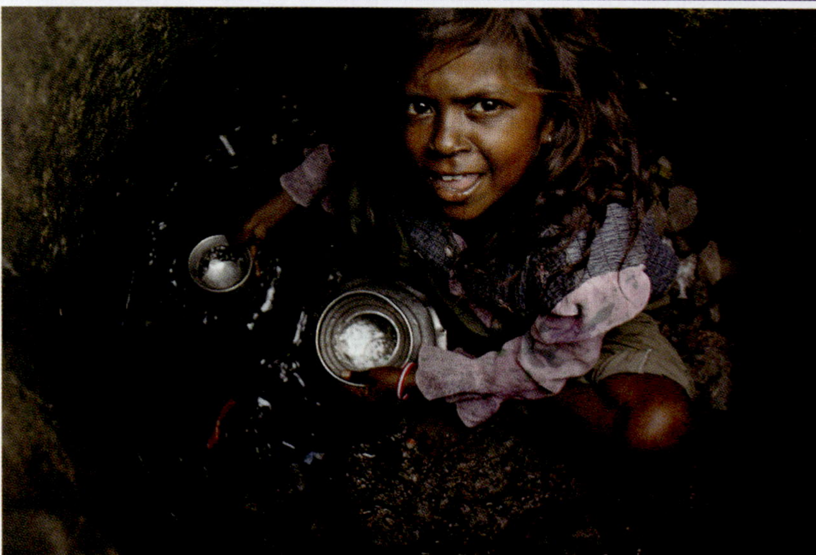

Thinking Geographically

Why might this child be endangering her health by drinking this water?

runs off these hard surfaces, collects in low places, and becomes stagnant instead of being absorbed into the ground. In urban slums, flooding can spread polluted water over wide areas, carrying it into homes and local fresh water sources, as well as places where children play (see Figure 1.21). Diseases such as malaria and cholera, carried in this water, can spread rapidly as a result. Fortunately, new technologies and urban planning methods are being developed that can help cities avoid these problems, but they are not yet in widespread use.

THINGS TO REMEMBER

- Water scarcity is emerging as the major resource issue of the twenty-first century, with water supplies strained by population growth, skyrocketing per capita demand for clean water, and water pollution.
- Much of the water that humans use they never see because it is *virtual water*: that which is used to produce what they consume.
- Urbanization often leads to water pollution through untreated sewage, industrial production, and mismanaged storm water.

GLOBAL CLIMATE CHANGE

Planet Earth is continually undergoing **climate change**, a slow shifting of climate patterns due to the general cooling or warming of the atmosphere. The present trend of **global warming**—which refers to the observed warming of Earth's climate as atmospheric levels of greenhouse gases increase—is extraordinary because it is happening more quickly than climate changes in the past and appears to be linked primarily to human agency. **Greenhouse gases** (carbon dioxide, methane, water vapor, and other gases) are essential to keeping Earth's incoming and outgoing radiation balanced in such a way that the Earth's surface is maintained in a temperature range hospitable to life. Similar to the glass panes of a greenhouse, these gases allow solar radiation to pass through the atmosphere and strike the Earth's surface; these gases also allow much of this radiation to bounce back into space as surface radiation. But some radiation is re-reflected back to the Earth, keeping its surface (like the interior of a greenhouse) warmer than it would be if incoming and outgoing radiation were in balance. Figure 1.22 shows this system of incoming and outgoing radiation and the role of greenhouse gases in trapping some of the outgoing radiation and sending it back to warm the Earth. Now, as greenhouse gases are being released at accelerating rates by the burning of fossil fuels and by the effects of deforestation, the evidence indicates that the Earth is warming at a critical rate.

climate change a slow shifting of climate patterns due to the general cooling or warming of the atmosphere

global warming the warming of the Earth's climate as atmospheric levels of greenhouse gases increase

greenhouse gases gases, such as carbon dioxide and methane, released into the atmosphere by human activities, which become harmful when released in excessive amounts

Most scientists now agree that there is an urgent need to reduce greenhouse gas emissions to avoid catastrophic climate change in coming years. Climatologists, biogeographers, and other scientists are documenting long-term global warming and cooling trends by examining evidence in tree rings, fossilized pollen and marine creatures, and glacial ice. These data indicate that the twentieth century was the warmest century in 600 years, with the decade of the 1990s as the hottest since the late nineteenth century. Evidence is mounting that these are not normal fluctuations. Very long-term climate-change patterns indicate that we should be heading into a cooling pattern, but instead it is estimated that, at present rates of warming, by 2100 average global temperatures could rise between 2.5°F and 10°F (about 2°C to 5°C).

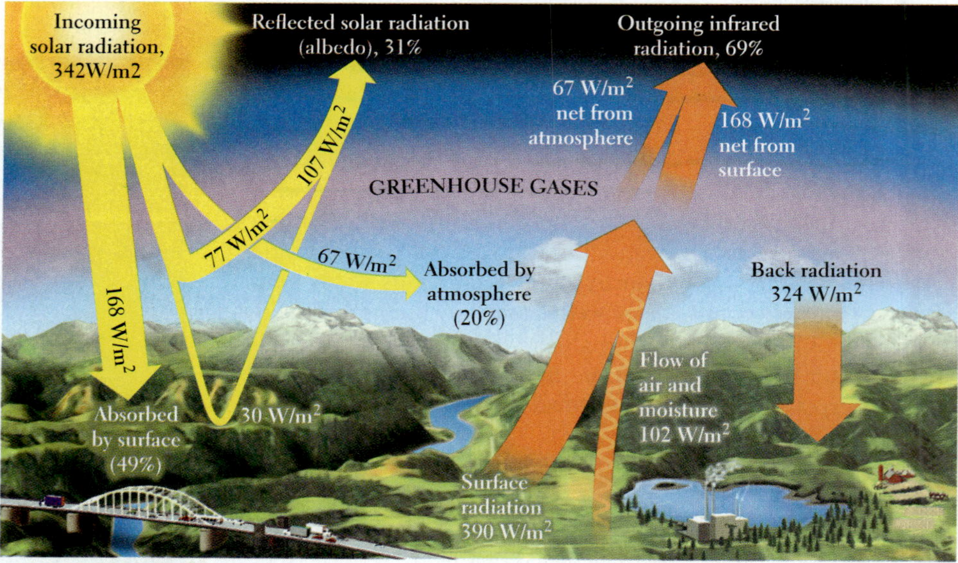

FIGURE 1.22 The balance of incoming and outgoing radiation and the greenhouse effect. To maintain an even temperature, Earth has to balance energy coming in with energy going out. Energy coming in is mostly sunshine, and energy going out is mostly radiant heat. Here the sunshine, or incoming solar radiation, is shown in yellow: some reflects right back into outer space, a little gets absorbed in the air, and about half warms the ground. The numbers represent averages—obviously there's usually more sunshine at noon than at midnight! Heat, mainly infrared radiation, is shown in orange: quite a lot bounces and flows near the surface in various forms. Clouds, dust, smoke, water vapor, and certain other gases tend to keep it there. But what finally reaches outer space almost exactly balances the amount of sunshine absorbed. These days, scientists find that extra greenhouse gases released by humans are causing Earth to retain extra energy—outgoing infrared radiation seems to average nearly one Watt per square meter (W/m²) less than incoming solar radiation, so average temperatures on Earth are rising.

However, a key problem is that those most responsible for global warming (the world's wealthiest and most industrialized countries) have the least incentive to reduce emissions because they are the least vulnerable to the changes global warming causes in the physical environment. Meanwhile, those most vulnerable to these changes (poor countries with low human development and with large slums in low-lying coastal wetlands) are the least responsible for the growth in greenhouse gases and have the least power in the global geopolitical sphere; hence they have the least ability to affect the level of emissions (Figure 1.23).

330. U.S. GOVERNMENT SCIENTISTS CALL FOR URGENT ACTION ON GLOBAL WARMING

FIGURE 1.23 Greenhouse emissions around the world in 2010, total per country and per capita.

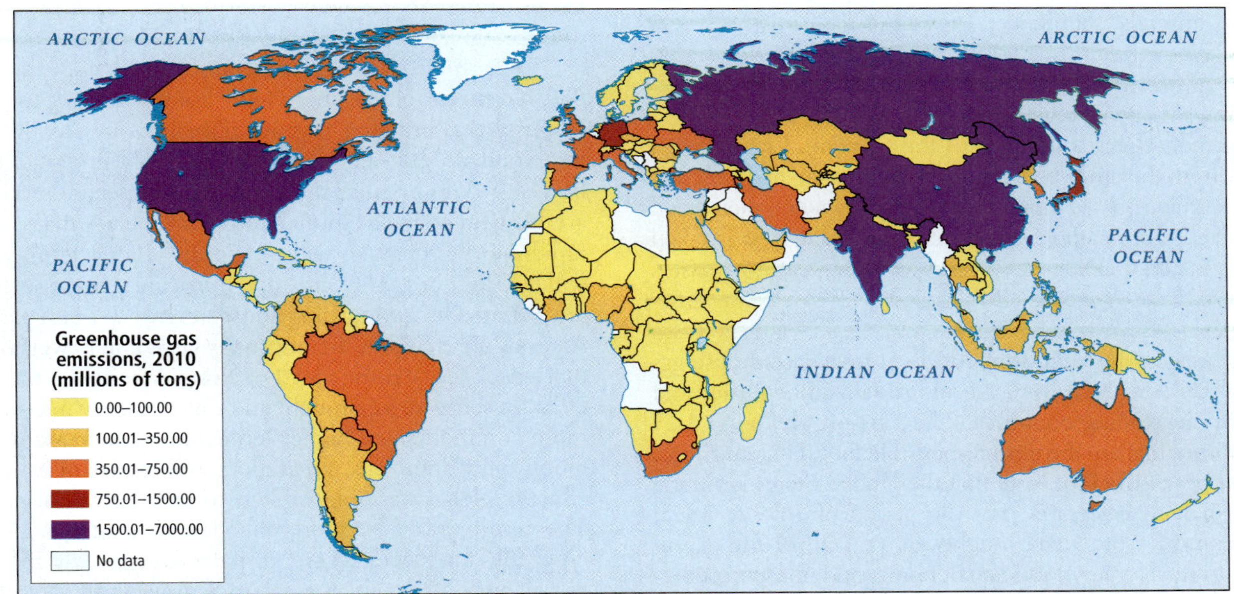

(A) Total emissions by millions of tons, 2010. China is now in first place as an emitter (23.3 percent), with emissions rising sharply since 2005; the United States is second (18.11 percent), achieving a small reduction since 2005; India increased emissions and is now third (5.78 percent); Russia is fourth (5.67 percent); and Japan fifth (4.01 percent). These top five countries contribute 56.9 percent of the world's greenhouse gas emissions.

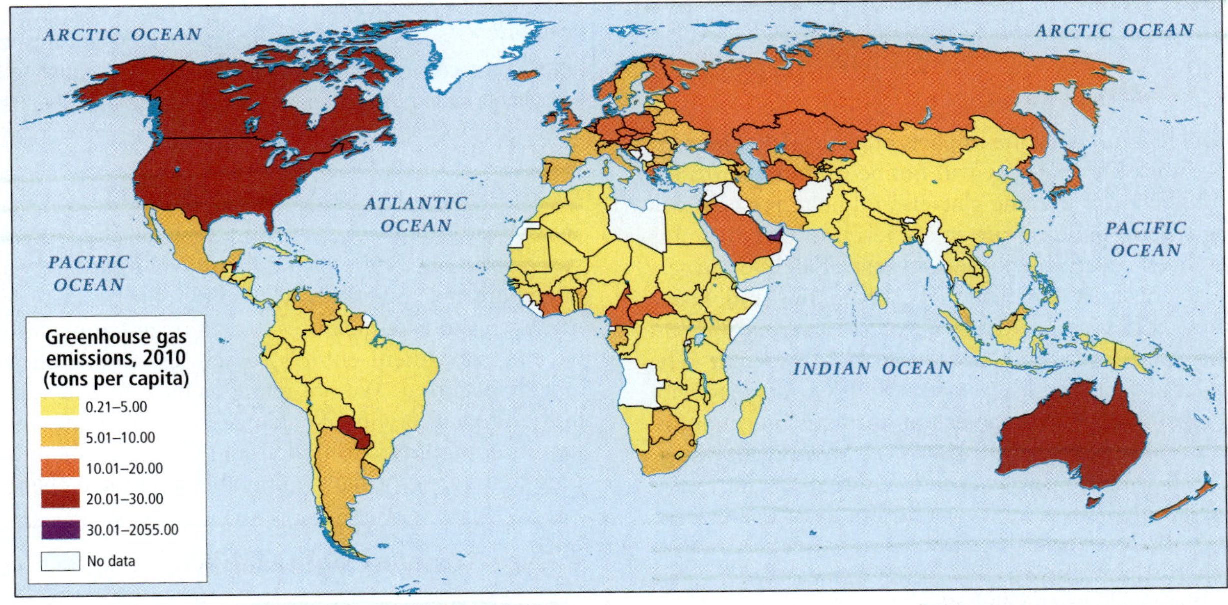

(B) Tons of emissions per capita. Jamaica leads with 47.58 tons of GHG per capita; the United Arab Emirates are second (40.10 tons of GHG per capita); followed by Bahrain (34.9 tons per capita) and Paraguay (29.9 tons per capita). The United States is eighth, at 22.22 tons per capita. China is far down the list, at 3.39 tons per capita.

Drivers of Global Climate Change

Greenhouse gases exist naturally in the atmosphere. It is their heat-trapping ability that makes the Earth warm enough for life to exist. Increase their levels, as humans are doing now, and the Earth becomes warmer still.

Electricity generation, vehicles, industrial processes, and the heating of homes and businesses all burn large amounts of CO_2-producing fossil fuels such as coal, natural gas, and oil. Even the large-scale raising of grazing animals contributes methane through the animals' flatulence. Unusually large quantities of greenhouse gases from these sources are accumulating in the Earth's atmosphere, and their presence has already led to significant warming of the planet's climate.

Widespread deforestation worsens the situation. Living forests take in CO_2 from the atmosphere, release the oxygen, and store the carbon in their biomass. As more trees are cut down and their wood is used for fuel, more carbon enters the atmosphere, less is taken out, and less is stored. The loss of trees and other forest organisms produces as much as 30 percent of the buildup of CO_2 in the atmosphere. The use of fossil fuels accounts for the remaining 70 percent.

In percentages, the largest producers of total greenhouse gas emissions (GHG) in 2010 were the industrialized countries and large, rapidly developing countries. The caption of Figure 1.23 lists the countries that are the most responsible for GHG emissions, total (A) and per capita (B). Note that the United States is among the leaders in both categories. For the period of 1859 to 1995, developed countries produced roughly 80 percent of the greenhouse gases from all types of industrial, home, and transport sources, and developing countries produced 20 percent. But by 2007, the developing countries were catching up, accounting for nearly 30 percent of total emissions. As developing countries industrialize over the next century and continue to cut down their forests, they will release more and more greenhouse gases every year. If present patterns hold, greenhouse gas contributions by the developing countries will exceed those of the developed world by 2040.

Climate-Change Impacts

While it is not clear just what the impacts of rising global temperatures will be, it is clear that they will not be uniform across the globe. One prediction is that the glaciers and polar ice caps will melt, causing a corresponding rise in sea level. In fact, this phenomenon has been observable for several years. Satellite imagery analyzed by scientists at the National Aeronautics and Space Administration (NASA) shows that between 1979 and 2005—just 26 years—the polar ice caps shrank by about 23 percent. The amount of polar ice cap shrinkage wavers from year to year, with the ice caps regaining ice to some extent during winter months, but the overall trend in recent years is 12 percent shrinkage per decade. The polar ice caps normally reflect solar heat back into the atmosphere, but as the ice melts, the dark, open ocean absorbs solar heat. This warming of the oceans not only hastens ice cap melting, it is changing ocean circulation, which is the engine of weather and climate globally.

The melting of the ice caps has several other effects. Already, trillions of gallons of meltwater have been released into the oceans. If this trend continues, at least 60 million people in coastal areas and on low-lying islands could be displaced by rising sea levels. Another issue is the melting of high mountain glaciers, which are a major source of water for many of the world's large rivers, such as the Ganga, the Indus, the Brahmaputra, the Huang He (Yellow), and the Chang Jiang (Yangtze). Similar melting effects on crucial rivers are expected in South America. Scientists have monitored mountain glaciers across the globe for more than 30 years; while some are growing, the majority are melting rapidly. Over the short term, melting mountain glaciers could result in flooding in many rivers, but eventually river flows will decrease as mountain glaciers shrink or disappear entirely.

Over time, higher temperatures will shift northward in the Northern Hemisphere and southward in the Southern Hemisphere, bringing warmer climate zones to these regions. Such climate shifts could lead to the displacement of large numbers of people because the zones where specific crops can grow are likely to change. Animal and plant species that cannot adapt rapidly to the changes will disappear. Higher temperatures also will lead to stronger tornados and hurricanes, because these storms are powered by warm, rising air (see Figure 1.24 on pages 44–45). An example is Sandy, the unusually large and powerful hurricane that struck the Atlantic Coast of North America in the autumn of 2012. In some areas, drought and water scarcity may also become more common since higher temperatures increase water evaporation rates from soils, vegetation, and bodies of water. Another effect of global warming is likely to be a shift in ocean currents. The result would be more chaotic and severe or dry weather, especially for places where climates are strongly influenced by ocean currents, such as western Europe (see "Vegetation and Climate" in Chapter 4 on pages 157–158).

Vulnerability to Climate Change

Geographic Insight 7

Climate Change and Water: Water and other environmental factors often interact to influence the vulnerability of a location to the impacts of climate change. These vulnerabilities have a spatial pattern.

A place's vulnerability to climate change can be thought of as its risk of suffering damage to human or natural systems as a result of such impacts as sea level rise, drought, flooding, or increased storm intensity. Some of these vulnerabilities are water related; others are not. Scientists who study climate change agree that while we can take measures to minimize temperature increases, we can't stop them entirely, much less reverse those that have already occurred. We are going to have to live with and adapt to the impacts of climate change for quite some time. The first step in doing this is to understand how and where humans and ecosystems are especially vulnerable to climate change.

A Case Study of Vulnerability: Mumbai, India

Three concepts are important in understanding a place's vulnerability to climate change: *exposure*, *sensitivity*, and *resilience*. Here we explore them in the context of the vulnerability to water-related climate-change impacts in Mumbai, India.

Exposure refers to the extent to which a place is exposed to climate-change impacts. For example, a low-lying coastal city like Mumbai, India (see Figure 8.1), is highly exposed to sea level rise. *Sensitivity* refers to how sensitive a place is to those impacts. For example, many of Mumbai's inhabitants are very sensitive to sea level rise because they are extremely poor and can only afford to live in low-lying slums that have no sanitation and thus whose nearby waterways are polluted. Flooding of these waterways would spread epidemics of waterborne illness that could kill many. *Resilience* refers to a place's ability to "bounce back" from the disturbances that climate-change impacts create. Mumbai's resilience to sea level rise is bolstered by the paradox that, despite its widespread poverty, it is the wealthiest city in South Asia. This wealth enables it to afford relief and recovery systems that could help it deal with sea level rise over the short and long term. Over the short term, Mumbai benefits from more and better hospitals and emergency response teams than any other city in South Asia. Over the long term, Mumbai's well-trained municipal planning staff can create and execute plans to help sensitive populations, like people living in slums, adapt to sea level rise. This could be achieved, for example, through planned relocation to higher ground or by building sea walls and dikes that could keep sea waters out of low-lying slum areas. Of course, Mumbai's overall vulnerability is more complicated than these examples suggest because the city faces many more climate-change impacts than just sea level rise. However, these examples help us understand vulnerability to climate change as a combination of exposure, sensitivity, and resilience.

One effort at understanding the global pattern of vulnerability to climate change can be seen in Figure 1.24, which features a map of vulnerability to climate change and photos of the types of problems that are already observable (see Figure 1.24A, C–E). The pattern of greatest vulnerability involves all of the world regions that have the lowest human development. This reflects the reality that both sensitivity and resilience are related to the many factors that influence human development. For example, many of the qualities that make Mumbai more sensitive to sea level rise are less present in urban areas located in highly developed countries. New York City has virtually none of the large, unplanned lowland slums found in Mumbai. In addition, numerous world-class hospitals, emergency response teams, plus large and well-trained municipal planning staffs boost New York's resilience. Nonetheless, Hurricane Sandy showed that even highly resilient areas, such as the New York City metropolitan area, can suffer overwhelming devastation. ■

Responding to Climate Change

In 1997, an agreement known as the **Kyoto Protocol** was adopted. The protocol called for scheduled reductions in CO_2 emissions by the highly industrialized countries of North America, Europe, East Asia, and Oceania. The agreement also encouraged, though it did not require, developing countries to curtail their emissions. One hundred eighty-three countries had signed the agreement by 2009. The only developed country that had not signed was the United States, then and still one of the world's largest per capita producers of CO_2 (Canada withdrew in 2011).

In Copenhagen in December 2009, 181 countries attempted to halt the rising concentrations of CO_2 in the atmosphere by 2020. However, wealthy nations did not offer to curb emissions sufficiently enough to make a difference. The only progress that came out of Copenhagen was an agreement to help developing countries achieve clean-energy economies and otherwise adapt to climate change; but even for this, adequate funds were not allocated.

In December 2011, the UN Framework Convention on Climate Change (UNFCCC) convened, this time in Durban, South Africa, again with the goal of significant emission control. After grueling negotiations, all 190 countries, including the top three emitters—China, the United States, and India—agreed to a plan to cut emissions significantly by 2020. But was this sufficient progress? According to climate scientists and those who are vulnerable to the effects of climate change, 2020 is too late to begin significant changes. Rather, to avoid a temperature rise above 2°C, carbon emissions would need to be in decline well before 2020 and the language of the agreements would need to be far more ironclad. As of 2012, a consortium of scientists, known as Yale Environment 360, acknowledged that there has been no appreciable improvement in CO_2 emission control (see "Thinking the Unthinkable: Engineering Earth's Climate," at http://tinyurl.com/73txae7).

Kyoto Protocol an amendment to a United Nations treaty on global warming, the Protocol is an international agreement, adopted in 1997 and in force in 2005, that sets binding targets for industrialized countries for the reduction of emissions of greenhouse gases

THINGS TO REMEMBER

- Planet Earth is continually undergoing climate change—a slow shifting of climate patterns that results from general cooling or warming of the atmosphere.
- Human activities that emit large amounts of carbon dioxide, methane, and other greenhouse gases are trapping heat in the atmosphere, causing what is now widely recognized as global warming.

Geographic Insight 7

- **Climate Change and Water** Many climate-change impacts are water related, such as glacial melting, flow levels in rivers, sea level rise, drought, and flooding. This makes it crucial to understand how places are vulnerable to climate impacts.
- There has been little progress in creating global agreement on controlling CO_2 emissions.

PHYSICAL GEOGRAPHY PERSPECTIVES

As discussed early in this chapter, physical geography is concerned with the processes that shape the Earth's landforms, climate, and vegetation. In this sense, physical geographers are similar to scientists from other disciplines who focus on these phenomena, though as geographers they often look at problems and depict the results of their analysis spatially. In this book, physical geography provides a backdrop for the many aspects of human geography we discuss. Physical geography is a fascinating, large, and growing field of study worth exploring in greater detail. What follows are just the basics of landforms and climate.

Landforms: The Sculpting of the Earth

The processes that create the world's varied **landforms**—mountain ranges, continents, and the deep ocean floor—are some of the most powerful and slow-moving forces on Earth. Originating deep beneath the Earth's surface, these *internal processes* can move entire continents, often taking hundreds of millions of years to do their work. However, it is *external processes* that form many of the Earth's landscape features, such as a beautiful waterfall or a rolling plain. These more rapid and delicate processes take place on the surface of the Earth. Geomorphologists study the processes that constantly shape and reshape the Earth's surface.

Plate Tectonics

Two key ideas related to internal processes in physical geography are the *Pangaea hypothesis* and *plate tectonics*. The geophysicist Alfred Wegener first suggested the Pangaea hypothesis in 1912. This hypothesis proposes that all the continents were once joined in a single vast continent called Pangaea (meaning "all lands"), which then fragmented over time into the continents we know today (see Figure 1.25 on page 46). As one piece of evidence for his theory, Wegener pointed to the neat fit between the west coast of Africa and the east coast of South America.

For decades, most scientists rejected Wegener's hypothesis. We now know, however, that the Earth's continents have been assembled into supercontinents a number of times, only to break apart again. All of this activity is made possible by plate tectonics, a process of continental motion discovered in the 1960s, long after Wegener's time.

landforms physical features of the Earth's surface, such as mountain ranges, river valleys, basins, and cliffs

plate tectonics the scientific theory that the Earth's surface is composed of large plates that float on top of an underlying layer of molten rock; the movement and interaction of the plates create many of the large features of the Earth's surface, particularly mountains

The study of **plate tectonics** has shown that the Earth's surface is composed of large plates that float on top of an underlying layer of molten rock. The plates are of two types. Oceanic plates are dense and relatively thin, and they form the floor beneath the oceans. Continental plates are thicker and less dense. Much of their surface rises above the oceans, forming continents. These massive plates drift slowly, driven by the circulation of the underlying

Figure 1.24 Photo Essay: Vulnerability to Climate Change

Thinking Geographically

After you have read about vulnerability to climate change, you will be able to answer the following questions:

A Does this photo relate most to short-term or long-term resilience?

B Of the four thumbnail maps in this graphic, which depicts the information that best explains why the Iberian Peninsula is so much less vulnerable to climate change than is North Africa?

C What sign of an orderly response to disaster is visible in this picture?

D How can you tell that food supplies are low in this camp?

This map shows overall human vulnerability to climate change based on a combination of human and environmental factors. Areas shown in darkest brown are vulnerable to floods, hurricanes, droughts, sea level rise, or other impacts related to climate change. When a population is exposed to an impact that it is sensitive to and has little resilience to, it becomes vulnerable. For example, many populations are exposed to drought, but generally speaking, the poorest populations are the most sensitive. However, sensitivity to drought can be compensated for if adequate relief and recovery systems, such as emergency water and food distribution systems, are in place. These systems lend an area a level of resilience that can reduce its overall vulnerability to climate change. A place's vulnerability to climate change can be thought of as a result of its sensitivity, exposure, and resilience in the face of multiple climate impacts.

E Climate change and hurricanes. Climatologists predict that hurricanes will increase in intensity as the planet warms. Indeed, there has been an increase in powerful storms in recent decades. Poverty (high sensitivity) and inadequate recovery systems (low resilience) make much of Central America particularly vulnerable to many hurricane-related impacts. Shown here is flood damage along the Aguan River, Honduras, caused by Hurricane Mitch in 1999. Over 9000 people died in the storm, making Mitch the second-most-deadly hurricane in history.

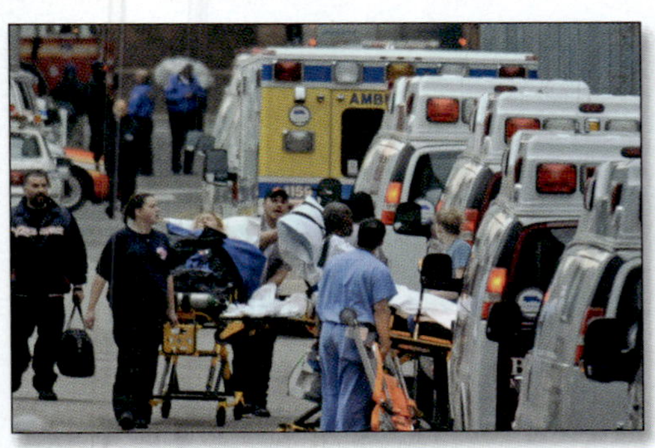

A United States: High Resilience, Low Vulnerability. Effective and well-funded recovery and relief systems give the United States high resilience to climate impacts. This contributes to generally low vulnerability. Shown here are ambulances in New York City responding to Hurricane Sandy in 2012.

B Spain and Morocco: The Multiple Dimensions of Vulnerability. A wide variety of information is used to make the global map of vulnerability shown below. For example, the contrast in vulnerability between Spain and Morocco relates to (among other things) differences in climate, population density and distribution, and human development (which is itself based on many factors).

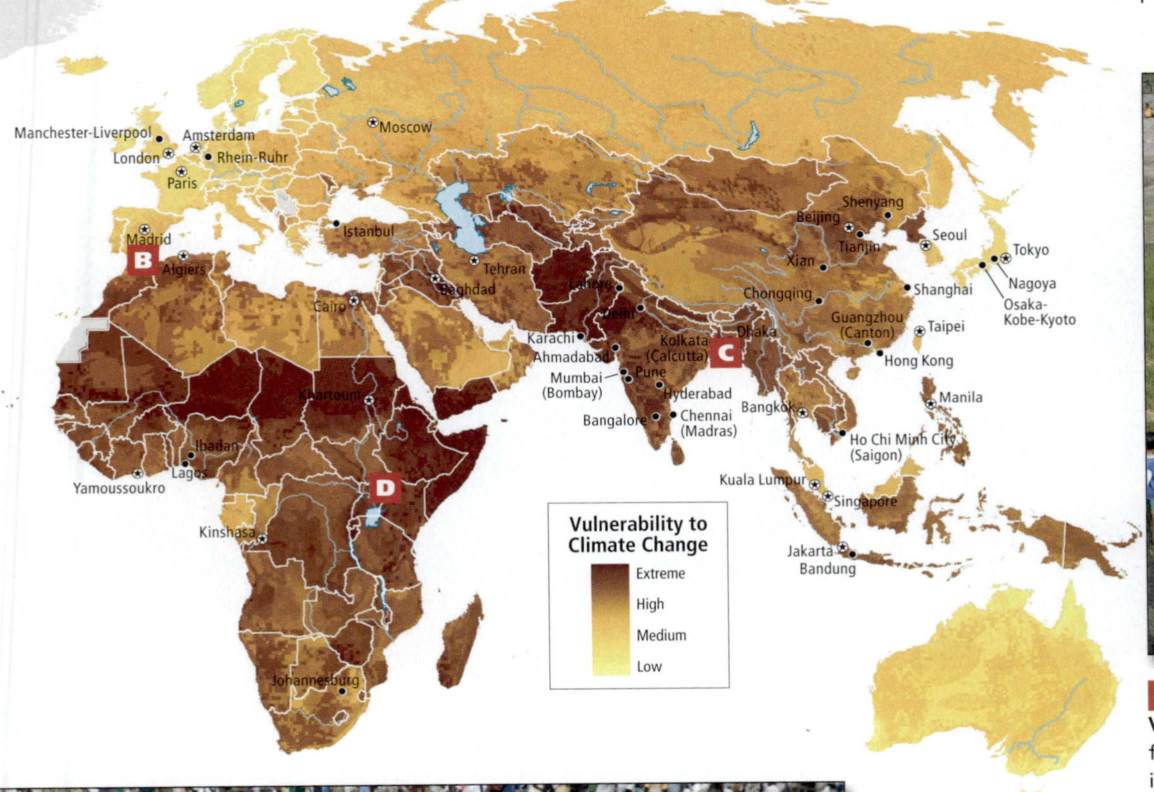

C India: Moderate Resilience, High Vulnerability. Rural Indians line up for food and water after Cyclone Aila in 2009. Advances in government-led disaster recovery have increased India's resilience, taking most areas out of the "extreme vulnerability" of nearby Pakistan and Afghanistan. Nevertheless, much of India remains in a state of "high vulnerability" with high exposure and sensitivity to sea level rise, flooding, hurricanes (cyclones), drought, and other disturbances that climate change can create or intensify.

D Northern Uganda and South Sudan: High to Extreme Vulnerabilty. The overall situation here is somewhat similar to India **(C)** but for different reasons. Sensitivity to drought is somewhat lower as access to water is better. However, armed conflict reduces resilience in much of Uganda, as many are forced to live in refugee camps dependent on food aid donated by foreigners. Shown here are refugees in northern Uganda, picking up bits of donated grain that has been dropped.

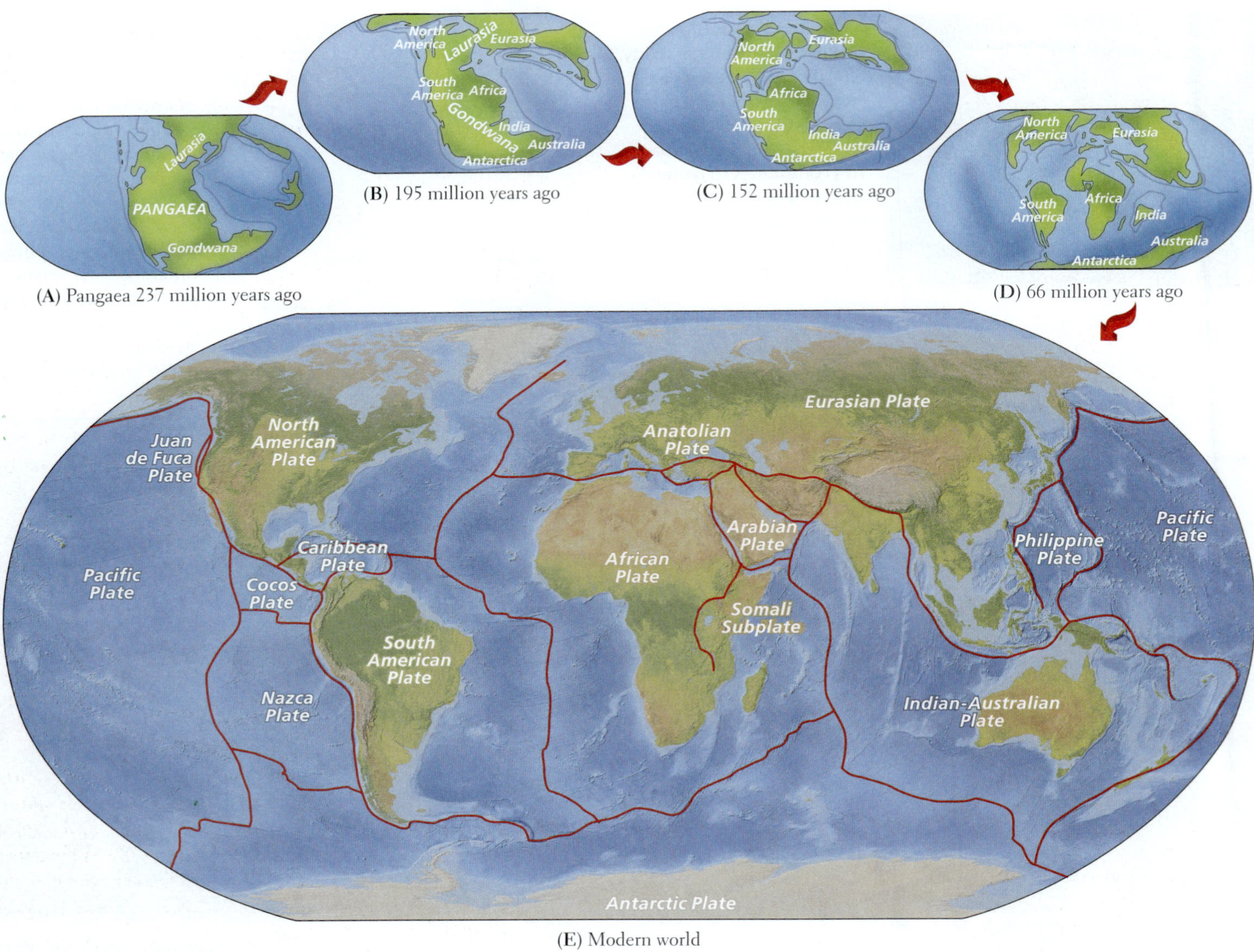

FIGURE 1.25 The breakup of Pangaea. The modern world map **(E)** depicts the current boundaries of the major tectonic plates. Pangaea is only the latest of several global configurations that have coalesced and then fragmented over the last billion years.

molten rock flowing from hot regions deep inside the Earth to cooler surface regions and back. The creeping movement of tectonic plates fragmented and separated Pangaea into pieces that are the continents we know today (Figure 1.25E).

Plate movements influence the shapes of major landforms, such as continental shorelines and mountain ranges. Huge mountains have piled up on the leading edges of the continents as the plates carrying them collide with other plates, folding and warping in the process. Plate tectonics accounts for the long, linear mountain ranges that extend from Alaska to Chile in the Western Hemisphere and from Southeast Asia to the European Alps in the Eastern Hemisphere. The highest mountain range in the world, the Himalayas of South Asia, was created when what is now India, situated at the northern end of the Indian-Australian Plate, ground into Eurasia. The only continent that lacks these long, linear mountain ranges is Africa. Often called the "plateau continent," Africa is believed to have been at the center of Pangaea and to have moved relatively little since the breakup. However, as Figure 1.25 shows, parts of eastern Africa—the Somali Subplate and the Arabian Plate—continue to separate from the continent (the African Plate).

Humans encounter tectonic forces most directly as earthquakes and volcanoes. Plates slipping past each other create the catastrophic shaking of the landscape we know as an earthquake. When plates collide and one slips under the other, this is known as *subduction*. Volcanoes arise at zones of subduction or sometimes in the middle of a plate, where gases and molten rock (called magma) can rise to the Earth's surface through fissures and holes in the plate. Volcanoes and earthquakes are particularly

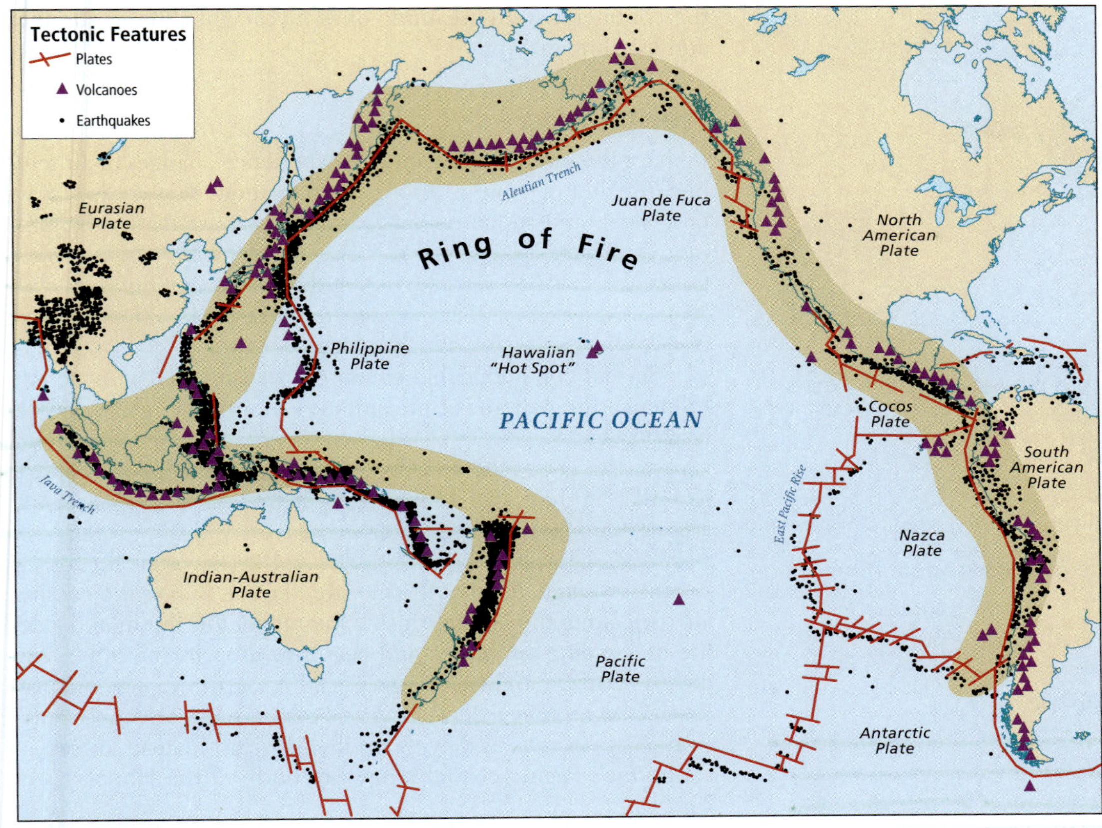

FIGURE 1.26 Ring of Fire. Volcanic formations encircling the Pacific Basin form the Ring of Fire, a zone of frequent earthquakes and volcanic eruptions.

common around the edges of the Pacific Ocean, an area known as the **Ring of Fire** (Figure 1.26).

Landscape Processes

The landforms created by plate tectonics have been further shaped by external processes, which are more familiar to us because we can observe them daily. One such process is **weathering**. Rock, exposed to the onslaught of sun, wind, rain, snow, ice, freezing and thawing, and the effects of life-forms (such as plant roots), fractures and decomposes into tiny pieces. These particles then become subject to another external process, **erosion**. During erosion, wind and water carry away rock particles and any associated decayed organic matter and deposit them in new locations. The deposition of eroded material can raise and flatten the land around a river, where periodic flooding spreads huge quantities of silt. As small valleys between hills are filled in by silt, a **floodplain** is created. Where rivers meet the sea, floodplains often fan out roughly in the shape of a triangle, creating a **delta**. External processes tend to smooth out the dramatic mountains and valleys created by internal processes.

Human activity often contributes to external landscape processes. By altering the vegetative cover, agriculture and forestry expose the Earth's surface to sunlight, wind, and rain. These agents in turn increase weathering and erosion. Flooding becomes more common because the removal of vegetation limits the ability of the Earth's surface to absorb rainwater. As erosion increases, rivers may fill with silt, and deltas may extend into the oceans.

Climate

The processes associated with climate are generally more rapid than those that shape landforms. **Weather**, the short-term and spatially limited expression of climate, can change in a matter of minutes. **Climate** is the long-term balance of temperature and moisture that keeps weather patterns fairly consistent from year to year. By this definition, the last major global climate change took place about 15,000 years ago, when the glaciers of the last ice age began to melt. As we have seen, a new global climate change is occurring in our own time, produced in part by human activity.

Solar energy is the engine of climate. The Earth's atmosphere, oceans, and land surfaces absorb huge amounts of solar energy, and the differences in the amounts they absorb account for part of the variations in climate we observe. The most intense direct solar energy strikes the Earth more or less head-on in a broad band stretching about 30° north and south of the equator. The fact that Earth's axis sits at a 23° angle as it orbits the sun—an angle that does not change—means that where the band of greatest solar intensity strikes the Earth varies in regular sequence over the course of a year, creating seasons. Just how this yearly seasonal pattern works is illustrated by Figure 1.27 on page 48. The highest average temperatures on the Earth's surface occur within this band of greatest solar intensity. Moving away from the equator, solar energy strikes the Earth's surface less directly—at more of an obtuse (wide) angle. This wide angle reduces its heating effect of the sun's rays and results in lower average annual temperatures.

Ring of Fire the tectonic plate junctures around the edges of the Pacific Ocean; characterized by volcanoes and earthquakes

weathering the physical or chemical decomposition of rocks by sun, rain, snow, ice, and the effects of life-forms

erosion the process by which fragmented rock and soil are moved over a distance, primarily by wind and water

floodplain the flat land around a river where sediment is deposited during flooding

delta the triangular-shaped plain of sediment that forms where a river meets the sea

weather the short-term and spatially limited expression of climate that can change in a matter of minutes

climate the long-term balance of temperature and precipitation that characteristically prevails in a particular region

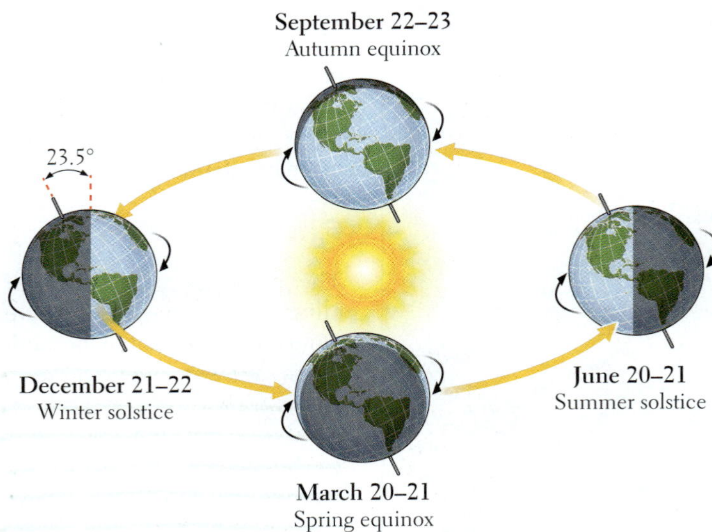

FIGURE 1.27 Diagram of the angle of Earth's orientation to the sun.

Temperature and Air Pressure

The wind and weather patterns we experience daily are largely a result of variations in solar energy absorption that create complex patterns of air temperature and *air pressure*. To understand air pressure, think of air as existing in a particular unit of space—for example, a column of air above a square foot of the Earth's surface. Air pressure is the amount of force (due to the pull of gravity) exerted by that column on that square foot of surface. Air pressure and temperature are related: The gas molecules in warm air are relatively far apart and are associated with low air pressure. In cool air, the gas molecules are relatively close together (dense) and are associated with high air pressure.

As the sun warms a unit of cool air, the molecules move farther apart. The air becomes less dense and exerts less pressure. Air tends to move from areas of higher pressure to areas of lower pressure, creating wind. If you have been to the beach on a hot day, you may have noticed a cool breeze blowing in off the water. This happens because land heats up (and cools down) faster than water, so on a hot day, the air over the land warms, rises, and becomes less dense than the air over the water. This causes the cooler, denser air to flow inland. At night the breeze often reverses direction, blowing from the now cooling land onto the now relatively warmer water.

These air movements have a continuous and important influence on global weather patterns and are closely associated with land and water masses. Because continents heat up and cool off much more rapidly than the oceans that surround them, the wind tends to blow from the ocean to the land during summer and from the land to the ocean during winter. It is almost as if the continents were breathing once a year, inhaling in summer and exhaling in winter.

precipitation dew, rain, sleet, and snow

orographic rainfall rainfall produced when a moving moist air mass encounters a mountain range, rises, cools, and releases condensed moisture that falls as rain

monsoon a wind pattern in which in summer months, warm, wet air coming from the ocean brings copious rainfall, and in winter, cool, dry air moves from the continental interior toward the ocean

Precipitation

Perhaps the most tangible way we experience changes in air temperature and pressure is through rain or snow. **Precipitation** (dew, rain, sleet, and snow) occurs primarily because warm air holds more moisture than cool air. When this warmer, moist air rises to a higher altitude, its temperature drops, which reduces its ability to hold moisture. The moisture condenses into drops to form clouds and may eventually fall as rain or snow.

Several conditions that encourage moisture-laden air to rise influence the patterns of precipitation around the globe. When moisture-bearing air is forced to rise as it passes over mountain ranges, the air cools, and the moisture condenses to produce rainfall (Figure 1.28). This process, known as **orographic rainfall**, is most common in coastal areas where wind blows moist air from above the ocean onto the land and up the side of a coastal mountain range. Most of the moisture falls as rain as the cooling air rises along the coastal side of the range. On the inland side, the descending air warms and ceases to drop its moisture. The drier side of a mountain range is said to be in the *rain shadow*. Rain shadows may extend for hundreds of miles across the interiors of continents, as they do on the Mexican Plateau, or east of California's Pacific coastal ranges, or north of the Himalayas of Eurasia.

A central aspect of Earth's climate is the "rain belt" that exists in equatorial areas. Near the equator, moisture-laden tropical air is heated by the strong sunlight and rises to the point where it releases its moisture as rain. Neighboring nonequatorial areas also receive some of this moisture when seasonally shifting winds move the rain belt north and south of the equator. The huge downpours of the Asian summer monsoon are an example.

In the summer **monsoon** season, the Eurasian continental landmass heats up, causing the overlying air to expand, become less dense, and rise. The somewhat cooler, yet moist, air of the Indian Ocean is drawn inland. The effect is so powerful that the equatorial rain belt is sucked onto the land (see Figure 8.4, Winter and Summer Monsoons). The result is tremendous, sometimes catastrophic, summer rains throughout virtually all of South and Southeast Asia and much of coastal and interior East Asia. The reverse happens in the winter as similar forces pull the equatorial rain belt south during the Southern Hemisphere's summer.

Much of the moisture that falls on North America and Eurasia is *frontal precipitation* caused by the interaction of large air masses of different temperatures and densities. These masses develop when air stays over a particular area long enough to take on the temperature of the land or sea beneath it. Often when we listen to a weather forecast, we hear about warm fronts or cold fronts. A *front* is the zone where warm and cold air masses come into contact, and it is always named after the air mass whose leading edge is moving into an area. At a front, the warm air tends to rise over the cold air, carrying warm clouds to a higher, cooler altitude. Rain or snow may follow. Much of the

FIGURE 1.28 Orographic rainfall.

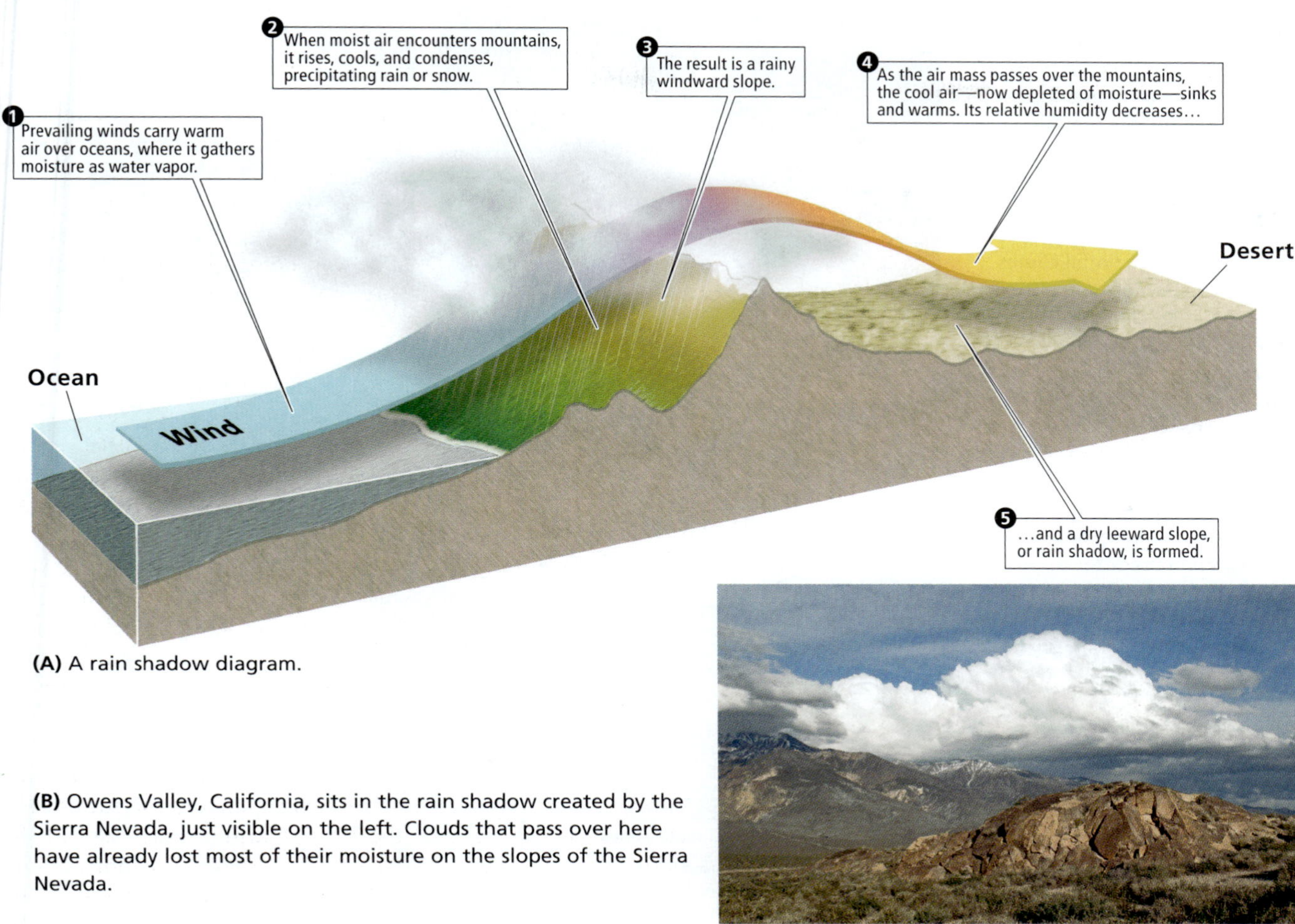
1 Prevailing winds carry warm air over oceans, where it gathers moisture as water vapor.
2 When moist air encounters mountains, it rises, cools, and condenses, precipitating rain or snow.
3 The result is a rainy windward slope.
4 As the air mass passes over the mountains, the cool air—now depleted of moisture—sinks and warms. Its relative humidity decreases…
5 …and a dry leeward slope, or rain shadow, is formed.
Ocean
Wind
Desert

(A) A rain shadow diagram.

(B) Owens Valley, California, sits in the rain shadow created by the Sierra Nevada, just visible on the left. Clouds that pass over here have already lost most of their moisture on the slopes of the Sierra Nevada.

rain that falls along the outer edges of a hurricane is the result of frontal precipitation.

Climate Regions

Geographers have several systems for classifying the world's climates. The systems are based on the patterns of temperature and precipitation just described. This book uses a modification of the widely known *Köppen classification system*, which divides the world into several types of climate regions, labeled A through E on the climate map in Figure 1.29 on pages 50–51. As you look at the regions on this map, examine the photos, and read the accompanying climate descriptions, the importance of climate to vegetation becomes evident. Each regional chapter includes a climate map; when reading these maps in any chapter, refer to the written descriptions in Figure 1.29 as necessary. Keep in mind that the sharp boundaries shown on climate maps are in reality much more gradual transitions.

THINGS TO REMEMBER

- Physical geography focuses on the processes that shape the Earth's landforms and its climate, and on how human practices interact with physical patterns.
- Landforms are shaped by internal processes, such as plate tectonics, and external processes, such as weathering.
- Seasons occur because of the fact that as the Earth orbits the sun over the course of a year, its axis is positioned at a consistent angle of 23°.
- Variations in air pressure are caused by heating and cooling and account for phenomena such as wind and precipitation patterns.
- Climate is the long-term balance of temperature and precipitation that keeps weather patterns fairly consistent from year to year. Weather is the short-term and spatially limited expression of climate that can change in minutes.

Figure 1.29 Photo Essay: Climate Regions of the World

A Tropical humid climates. In *tropical wet climates*, rain falls predictably every afternoon and usually just before dawn. The *tropical wet/dry climate*, also called a *tropical savanna*, has a wider range of temperatures and a wider range of rainfall fluctuation than the tropical wet climate.

B Arid and semiarid climates. *Deserts* generally receive very little rainfall (2 inches or less per year). Most of that rainfall comes in downpours that are extremely rare and unpredictable. *Steppes* have climates that are similar to those of deserts, but more moderate. They usually receive about 10 inches more rain per year than deserts and are covered with grass or scrub.

C Temperate climates. *Midlatitude temperate climates* are moist all year and have short, mild winters and long, hot summers. *Subtropical climates* differ from midlatitude climates in that subtropical winters are dry. *Mediterranean climates* have moderate temperatures but are dry in summer and wet in winter.

D Cool humid climates. Stretching across the broad interiors of Eurasia and North America are *continental climates*, which either have dry winters (northeastern Eurasia) or are moist all year (North America and north-central Eurasia). Summers in cool humid climates are short but can have very warm days.

E Coldest climates. *Arctic* and *high-altitude climates* are by far the coldest and are also among the driest. Although moisture is present, there is little evaporation because of the low temperatures. The Arctic climate is often called *tundra*, after the low-lying vegetation that covers the ground. The high-altitude version of this climate, which can be far from the Arctic, is more widespread and subject to greater daily fluctuations in temperature. High-altitude microclimates, such as those in the Andes and the Himalayas, can vary tremendously, depending on factors such as available moisture, orientation to the sun, and vegetation cover. As one ascends in altitude, the changes in climate loosely mimic those found as one moves from lower to higher latitudes. These changes are known as *temperature-altitude zones* (see Figure 3.6 on page 116).

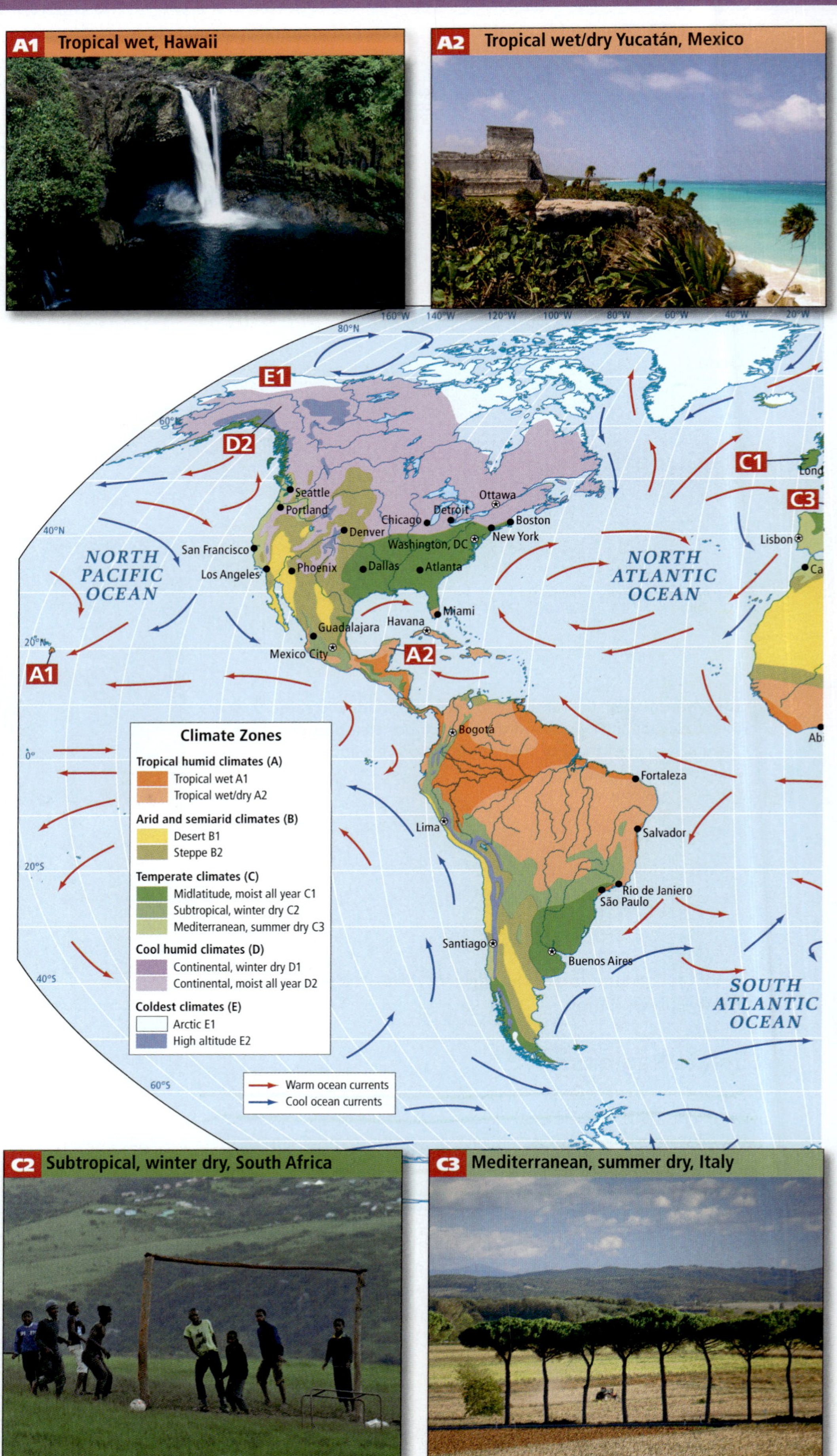

B1 Desert, Namibia

B2 Steppe, Mongolia

C1 Midlatitude, moist all year, United Kingdom

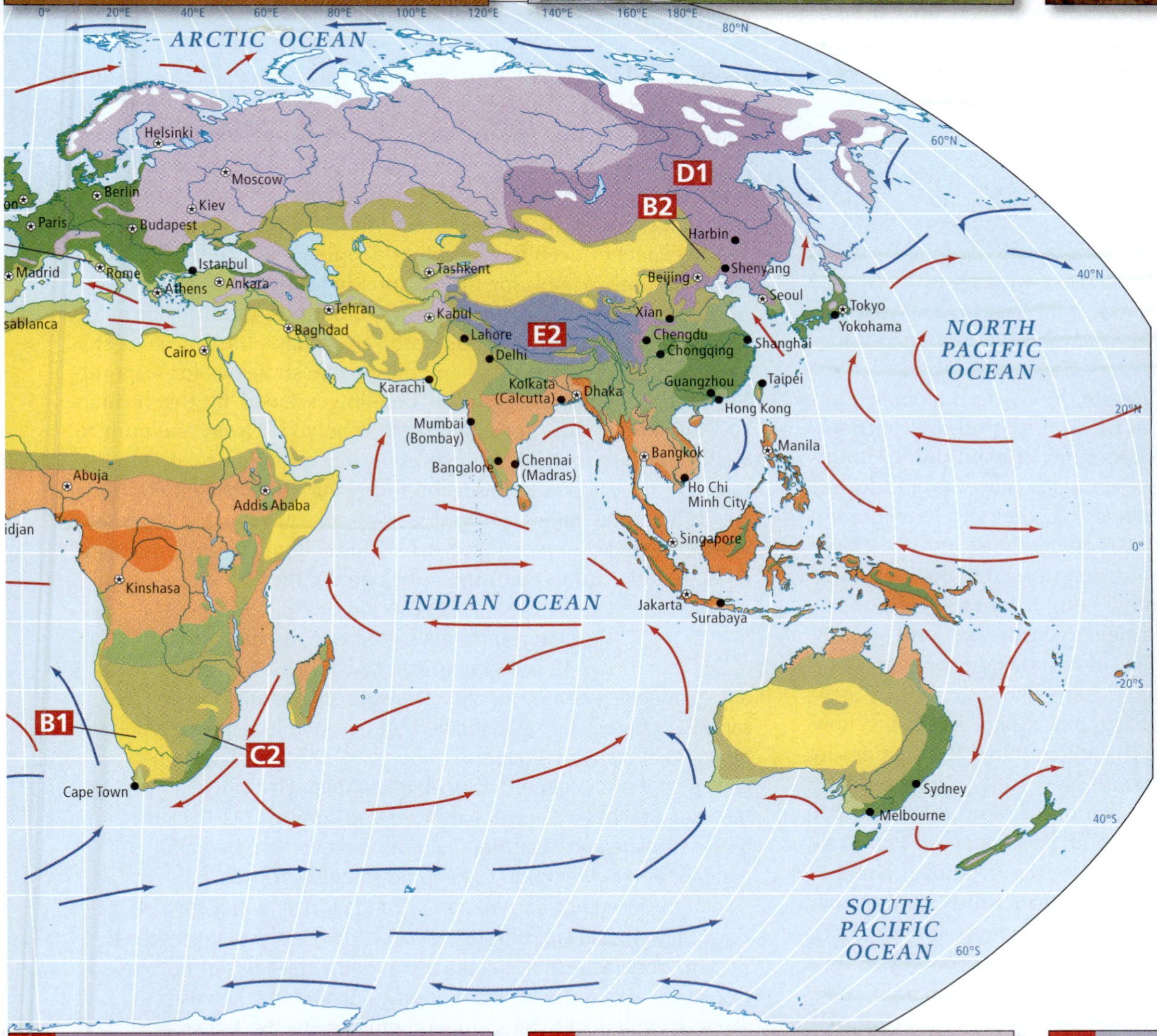

ARCTIC OCEAN

NORTH PACIFIC OCEAN

INDIAN OCEAN

SOUTH PACIFIC OCEAN

0° 20°E 40°E 60°E 80°E 100°E 120°E 140°E 160°E 180°E

80°N 60°N 40°N 20°N 0° 20°S 40°S 60°S

Helsinki, Moscow, Berlin, Kiev, Paris, Budapest, Madrid, Rome, Istanbul, Athens, Ankara, Casablanca, Tehran, Baghdad, Tashkent, Kabul, Lahore, Delhi, Karachi, Cairo, Kolkata (Calcutta), Dhaka, Mumbai (Bombay), Bangalore, Chennai (Madras), Abuja, Addis Ababa, Kinshasa, Cape Town, Harbin, Shenyang, Beijing, Seoul, Tokyo, Yokohama, Xian, Chengdu, Chongqing, Shanghai, Guangzhou, Taipei, Hong Kong, Manila, Bangkok, Ho Chi Minh City, Singapore, Jakarta, Surabaya, Sydney, Melbourne

D1 B2 E2 B1 C2

E2 High altitude, Tibet

D1 Continental, winter dry, Russia

D2 Continental, moist all year, Alaska

E1 Arctic tundra, Canada

HUMAN AND CULTURAL GEOGRAPHY PERSPECTIVES

Human geographers are interested in the economic, social, and cultural practices of a people, and in the spatial patterns these factors create. An important component of human geography is cultural geography, which focuses on culture as a complex of important distinguishing characteristics of human societies. **Culture** comprises everything people use to live on Earth that is not directly part of biological inheritance. Culture is represented by the ideas, materials, methods, and social arrangements that people have invented and passed on to subsequent generations, such as methods of producing food and shelter. Culture includes language, music, tools and technology, clothing, gender roles, belief systems, and moral codes, such as those prescribed in Confucianism, Islam, and Christianity.

Ethnicity and Culture: Slippery Concepts

A group of people who share a location, a set of beliefs, a way of life, a technology, and usually a common ancestry and sense of common history form an **ethnic group**. The term *culture group* is often used interchangeably with ethnic group. Both of the concepts of culture and ethnicity are imprecise, especially as they are popularly used. For instance, as part of the modern globalization process, migrating people often move well beyond their customary cultural or ethnic boundaries to cities or even distant countries. In these new places they take on many new ways of life and beliefs—their culture actually changes, yet they still may identify with their cultural or ethnic origins.

The Kurds in Southwest Asia are an example of the tenacity of this ethnic or cultural group identity. Long before the U.S. war in Iraq, the Kurds were asserting their right to create their own country in the territory where they have lived as nomadic herders since before the founding of Islam. Syria, Iraq, Iran, and Turkey now claim parts of the traditional Kurdish area. Many Kurds are now educated urban dwellers, living and working in modern settings in Turkey, Iraq, Iran, or even London and New York, yet they actively support the cause of establishing a Kurdish homeland. Although urban Kurds think of themselves as ethnic Kurds and are so regarded in the larger society, they do not follow the traditional Kurdish way of life. We could argue that these urban Kurds have a new identity within the Kurdish culture or ethnic group. Or they may be in a *transcultural* position, moving from one culture to another.

culture all the ideas, materials, and institutions that people have invented to use to live on Earth that are not directly part of our biological inheritance

ethnic group a group of people who share a common ancestry and sense of common history, a set of beliefs, a way of life, a technology, and usually a common geographic location of origin

multicultural society a society in which many culture groups live in close association

Another problem with the imprecision of the concept of culture is that it is often applied to a very large group that shares only the most general of characteristics. For example, one often hears the terms American culture, African American culture, or Asian culture. In each case, the group referred to is far too large to share more than a few broad characteristics.

It might fairly be said, for example, that U.S. culture is characterized by beliefs that promote individual rights, autonomy, and individual responsibility. But when we look at specifics, just what constitutes the rights and responsibilities of the individual are quite debatable. In fact, American culture encompasses many subcultures that share some of the core set of beliefs, but disagree over parts of the core and over a host of other matters. The same is true, in varying degrees, for all other regions of the world. When many culture groups live in close association, the society may be called **multicultural**.

Values

Occasionally you will hear someone say, "After all is said and done, people are all alike," or "People ultimately all want the same thing." It is a heartwarming sentiment, but an oversimplification. True, we all want food, shelter, health, love, and acceptance; but culturally, people are not all alike, and that is one of the qualities that makes the study of geography interesting. We would be wise not to expect or even to want other people to be like us. It is often more fruitful to look for the reasons behind differences among people than to search hungrily for similarities. Cultural diversity has helped humans to be successful and adaptable animals. The various cultures serve as a bank of possible strategies for responding to the social and physical challenges faced by the human species. The reasons for differences in behavior from one culture to the next are usually complex, but they are often related to differences in values. Consider the following vignette that contrasts the values and *norms* (accepted patterns of behavior based on values) held by modern and urban individualistic cultures with those held by rural, community-oriented cultures.

VIGNETTE One recent rainy afternoon, a beautiful 40-something Asian woman walked alone down a fashionable street in Honolulu, Hawaii. She wore high-heeled sandals, a flared skirt that showed off her long legs, and a cropped blouse that allowed a glimpse of her slim waistline. She carried a laptop case and a large fashionable handbag. Her long, shiny black hair was tied back. Everyone noticed and admired her because she exemplified an ideal Honolulu businesswoman: beautiful, self-assured, and rich enough to keep herself well-dressed.

In the village of this woman's grandmother—whether it be in Japan, Korea, Taiwan, or rural Hawaii—the dress that exposed her body to open assessment and admiration by strangers of both sexes would signal that she lacked modesty. The fact that she walked alone down a public street—unaccompanied by her father, husband, or female relatives—might even indicate that she was not a respectable woman. That she at the advanced age of 40 was investing in her own good looks might be assessed as pathetic self-absorption. Thus a particular behavior may be admired when judged by one set of values and norms but may be considered questionable or even disreputable when judged by another. *[Source: From Lydia Pulsipher's field notes. For detailed source information, see Text Credit pages.]* ■

If culture groups have different sets of values and standards, does that mean that there are no overarching human values or standards? This question increasingly worries geographers,

Figure 1.30 **Photo Essay:** Major Religions of the World

The small symbols on the map indicate a localized concentration of a particular religion within an area where another religion is predominant.

A Islam, Egypt

B Hinduism, India

C Indigenous religion, West Papua, Indonesia

ARCTIC OCEAN

ARCTIC OCEAN

ATLANTIC OCEAN

PACIFIC OCEAN

PACIFIC OCEAN

INDIAN OCEAN

A B C D E F

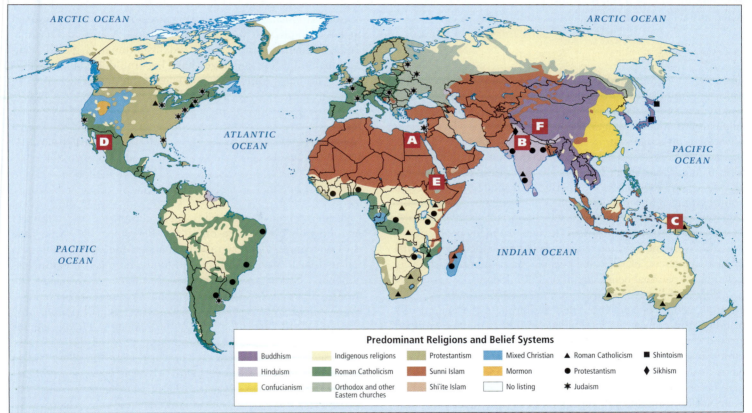

Predominant Religions and Belief Systems

Buddhism	Indigenous religions	Protestantism	Mixed Christian	▲ Roman Catholicism	■ Shintoism
Hinduism	Roman Catholicism	Sunni Islam	Mormon	● Protestantism	♦ Sikhism
Confucianism	Orthodox and other Eastern churches	Shi'ite Islam	No listing	✱ Judaism	

D Indigenous religion, Mexico

E Christianity, Ethiopia

F Buddhism, Tibet

ON THE BRIGHT SIDE

Altruism

Recognizing all the ills that have emerged from racism and similar prejudices, we need not infer that human history has been marked primarily by conflict and exploitation or that these conditions are inevitable. Actually, humans have probably been so successful as a species because of a strong inclination toward *altruism*, the willingness to sacrifice one's own well-being for the sake of others. It is probably our capacity for altruism that causes us such deep distress over the relatively infrequent occurrences of inhumane behavior.

who try to be sensitive both to the particularities of place and to larger issues of human rights. Those who lean too far toward appreciating difference could end up tacitly accepting inhumane behavior, such as the oppression of minorities or violence against women. Acceptance of difference does not preclude judgments of extreme customs or points of view. Nonetheless, although it is important to take a stand against cruelty of all sorts, deciding when and where to take that stand is rarely easy.

Religion and Belief Systems

The religions of the world are formal and informal institutions that embody value systems. Most have roots deep in history, and many include a spiritual belief in a higher power (such as God, Yahweh, or Allah) as the underpinning for their value systems. Today religions often focus on reinterpreting age-old values for the modern world. Some formal religious institutions—such as Islam, Buddhism, and Christianity—proselytize; that is, they try to extend their influence by seeking converts. Others, such as Judaism and Hinduism, accept converts only reluctantly. Informal religions, often called belief systems, have no formal central doctrine and no firm policy on who may or may not be a practitioner. The fact that many people across the world combine informal religious beliefs with their more formal religious practices, of whatever persuasion, accounts for the very rich array of personal beliefs found on Earth today.

Religious beliefs are often reflected in the landscape. For example, settlement patterns often demonstrate the central role of religion in community life: village buildings may be grouped around a mosque, a temple, a synagogue, or a church, and the same can be said for urban neighborhoods. In some places, religious rivalry is a major feature of the landscape. Certain spaces may be clearly delineated for the use of one group or another, as in Northern Ireland's Protestant and Catholic neighborhoods.

Religion has also been used to wield power. For example, during the era of European colonization, religion (Christianity) was a way to impose a change of attitude on conquered people. And the influence lingers. Figure 1.30A–F on page 53 shows the distribution of the major religious traditions on Earth today; it demonstrates some of the religious consequences of colonization. Note, for instance, the distribution of Roman Catholicism in the parts of the Americas, Africa, and Southeast Asia, all places colonized by European Catholic countries.

Religion can also spread through trade contacts. In the seventh and eighth centuries, Islamic people used a combination of trade and political power (and less often, actual conquest) to extend their influence across North Africa, throughout Central Asia, and eventually into South and Southeast Asia (see Figure 6.14 map).

Language

Language is one of the most important criteria used in delineating cultural regions. The modern global pattern of languages reflects the complexities of human interaction and isolation over several hundred thousand years. Between 2500 and 3500 languages are spoken on Earth today, some by only a few dozen people in isolated places. Many languages have several *dialects*—regional variations in grammar, pronunciation, and vocabulary.

The geographic pattern of languages has continually shifted over time as people have interacted through trade and migration. The pattern changed most dramatically around 1500, when the languages of European colonists began to replace the languages of the people they conquered. For this reason, English, Spanish, Portuguese, or French are spoken in large patches of the Americas, Africa, Asia, and Oceania. Today, with increasing trade and instantaneous global communication, a few languages have become dominant. English is now the most important language of international trade, but Arabic, Spanish, Chinese, Hindi, and French are also widely used (see Figure 1.31 on page 55). At the same time, other languages are becoming extinct because children no longer learn them within families. **250. NEW DOCUMENTARY FILM TRACKS LANGUAGES**

Race

Like ideas about gender roles, ideas about race affect human relationships everywhere on Earth. Paradoxically, while race is of enormous social significance across the world, biologists tell us that from a scientific standpoint race is a meaningless concept! The characteristics we popularly identify as **race** markers—skin color, hair texture, and face and body shape—have no significance as biological categories. All people now alive on Earth are members of one species, *Homo sapiens sapiens*. For any supposed *racial trait*, such as skin color or hair texture or facial features, there are wide variations within human groups. Meanwhile, many invisible biological characteristics, such as blood type and DNA patterns, cut across skin color distributions and other so-called *racial attributes* and are shared across what are commonly viewed as different races. In fact, over the last several thousand years there has been such massive gene flow among moving human populations that no modern group presents a discrete set of biological characteristics. Although any two of us may look quite different, from the biological point of view we are all simply *Homo sapiens sapiens* and closely related.

Some of the easily visible features of particular human groups evolved to help them adapt to environmental conditions. For example, biologists have shown that darker skin (containing a high proportion of protective melanin pigment) evolved in regions close to the equator, where

race a social or political construct that is based on apparent characteristics such as skin color, hair texture, and face and body shape, but that is of no biological significance

ARCTIC OCEAN
ARCTIC OCEAN
ATLANTIC OCEAN
PACIFIC OCEAN
PACIFIC OCEAN
INDIAN OCEAN

Predominant Languages

Indo-European family
Germanic group (incl. English, German)
Balto-Slavic group (incl. Russian, Ukrainian)
Celtic
Greek
Albanian
Iranian group
Armenian
Romance group (incl. Spanish, Portuguese, French, Italian)
Indo-Aryan group (incl. Hindi, Bengali, Urdu, Punjabi, Marathi)

Caucasian family

Afro-Asiatic family
Semitic group (incl. Arabic)
Kushitic group
Berber group

Khoisan family

Niger-Congo family

Nilo-Saharan family

Uralic family

Sino-Tibetan family
Sinitic (Chinese) languages
Tibetic-Burmic languages

Altaic family
Turkic group
Mongolian group
Tungus-Manchu group
Japanese and Korean

Tai family

Dravidian family
(incl. Telugu, Tamil)

Austro-Asiatic family
Mon-Khmer group
Munda group
Vietnamese

Austronesian family
(incl. Malay, Indonesian)

Indigenous languages

No listing

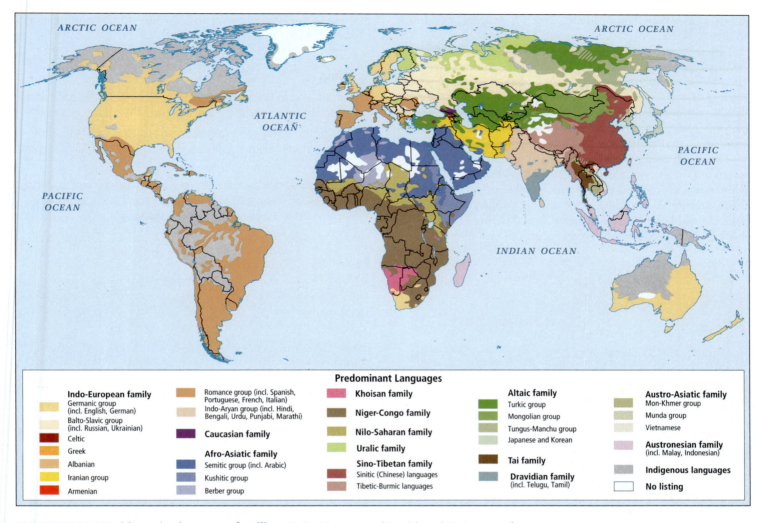

FIGURE 1.31 World's major language families. Distinct languages (Spanish and Portuguese, for example) are part of a larger group (Romance languages), which in turn is part of a language family (Indo-European).

sunlight is most intense (see Figure 1.32; see also Figure 1.27 on page 48). All humans need the nutrient vitamin D, and sunlight striking the skin helps the body absorb vitamin D. Too much of the vitamin, however, can result in improper kidney functioning. Dark skin absorbs less vitamin D than light skin and thus would be a protective adaptation in equatorial zones. In higher latitudes, where the sun's rays are more dispersed, light skin facilitates the sufficient absorption of vitamin D; darker-skinned people at these higher latitudes may need to supplement vitamin D to be sure they get enough, since D deficiencies can result in several health risks. Meanwhile, light-skinned people with little protective melanin in their skin—if they live in equatorial or high-intensity sunlit zones (parts of Australia, for example)—will need to protect against too much vitamin D, serious sunburn, and skin cancer. Similar correlations have been observed between skin color, sunlight, and

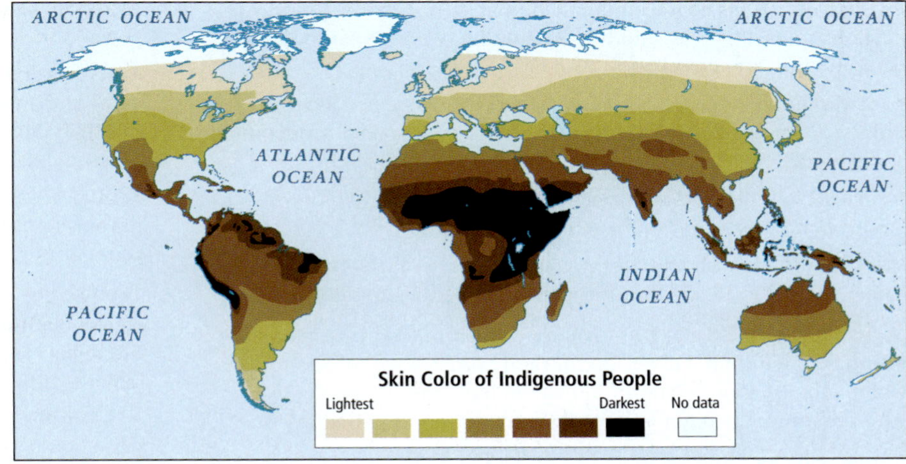

FIGURE 1.32 Skin color map for indigenous people as predicted from multiple environmental factors. Skin plays a twin role with respect to the sun: protection from excessive UV radiation and absorption of enough sunlight to trigger the production of vitamin D.

another essential vitamin, folate, which if deficient can result in birth defects.

Over time, race has acquired enormous social and political significance as humans from different parts of the world have encountered each other in situations of unequal power. *Racism*—the belief that genetic factors, usually visually apparent ones such as skin color, are a primary determinant of human abilities and even cultural traits—has often been invoked to justify the enslavement of particular groups, or confiscation of their land and resources. Racism is insidious and hard to stamp out, in part because in the eyes of many, of all skin tones, whiteness is considered the "normal condition," while all other colors are simply nonwhite. Race and its implications in North America will be covered in Chapter 2, and the topic will be discussed in several other world regions as well (see also a TED—Technology, Entertainment, and Design—talk by anthropologist Nina Jablonski, at http://tinyurl.com/l7wg7m).

THINGS TO REMEMBER

- Human geographers are interested in the economic, social, and cultural practices of a people and in the spatial patterns these practices create.
- Cultural geography seeks to understand human variability on Earth through a variety of lenses: culture and ethnicity, religion, language, and race.
- Race is biologically meaningless, yet it has acquired enormous social and political significance. Evolutionary science and the study of social history can aid in dispelling racist misconceptions.

Geographic Insights: Review and Self-Test

1. Physical and Human Geographers: The primary concerns of both physical and human geographers are the study of the Earth's surface and the interactive physical and human processes that shape the surface.

- What are some of the ways in which the Earth's physical and human processes interact?
- Name some circumstances that can benefit from collaboration between physical and human geographers.

2. Regions: The concept of *region* is useful to geographers because it allows them to break up the world into manageable units in order to analyze and compare spatial relationships. Nonetheless, regions do not have rigid definitions and their boundaries are fluid.

- How might you understand the place where you live as part of a region? What combination of distinct physical and/or human features describes this region?

3. Gender and Population: The shift toward greater gender equality is having an influence on population growth patterns, patterns of economic development, and the distribution of power within families, communities, and countries.

- How is the shift toward greater equality between the genders influencing population growth rates, patterns of economic development, and politics?
- How do your plans for a career and a family compare to those of your grandparents and great-grandparents?
- Ask your parents who the first woman in your family was who had a career. How, if at all, did this impact the number of children she had?

4. Food and Urbanization: Modernization in food production is pushing agricultural workers out of rural areas toward urban areas where jobs are more plentiful but where food must be purchased. This circumstance often leads to dependency on imported food.

- How is the food you eat more like or less like a product of green revolution agriculture or other food production systems, such as organic agriculture?
- How far back in your family history would you need to go to find ancestors who lived in a rural area and grew almost all their own food?
- Assuming you do not live in a rural area, how has urban life affected the diet and level of physical activity of you and/or your family?

5. Globalization and Development: Increased global flows of information, goods, and people are transforming patterns of economic development.

- How is globalization evident in your life—from the clothes you wear, to your favorite foods, to your career plans?
- To what extent does your circle of friends show the effects of globalization?

6. Power and Politics: There are major differences across the globe in the ways that power is wielded in societies. Modes of governing that are more authoritarian are based on the power of the state or community (or tribal) leaders. Modes that are more democratic give the individual a greater say in how policies are developed and governments are run. There are also many other ways of managing political power.

- Give several examples from recent world events of shifts from authoritarian modes of government to more democratic systems in which individuals have a greater say in the running of their governments.
- Are peace, broad prosperity, education, and civil society features of life in your home country? How do the political freedoms you enjoy shape your daily life?

7. Climate Change and Water: Water and other environmental factors often interact to influence the vulnerability of a location to the impacts of climate change. These vulnerabilities have a spatial pattern.

- How is the place where you live vulnerable to climate change?
- Assess your exposure, sensitivity, and resilience to water-related climate-change impacts.

Critical Thinking Questions

1. Some people argue that it is acceptable for people in the United States to consume at high levels because their consumerism keeps the world economy going. What are the weaknesses in this idea?

2. What are the causes of the huge increases in migration, legal and illegal, that have taken place over the last 25 years?

3. What would happen in the global marketplace if all people had a living wage?

4. As people live longer and decide not to raise large families, how can they beneficially spend the last 20 to 30 years of their lives?

5. Given the threats posed by global warming, what are the most important steps to take now?

6. As you read about differing ways of life, values, and perspectives on the world, reflect on the appropriateness of force as a way of resolving conflicts.

7. What are some possible careers that would address one or more of the issues raised in this chapter?

8. Reflect on the reasons why some people have much and others have little.

9. What would be some of the disadvantages and advantages of abolishing gender roles in any given culture?

10. Consider the ways that access to the Internet enhances prospects for world peace and the ways that it contributes to discord.

Chapter Key Terms

agriculture 23
Anthropocene 22
authoritarianism 33
biosphere 22
birth rate 13
capitalism 36
carrying capacity 26
cartographer 4
cash economy 17
civil society 36
climate 47
climate change 40
communism 36
culture 52
death rate 13
delta 47
democratization 33
demographic transition 16
development 19
domestication 23
ecological footprint 22
emigration 13
erosion 47
ethnic cleansing 37
ethnic group 52
Euro zone 20
extraction 19
fair trade 33
floodplain 47
food security 24
formal economy 20
free trade 32
gender 17
gender roles 17
genetic modification (GM) 27
genocide 37
Geographic Information Science (GISc) 7
geopolitics 36
global economy 30
global scale 12
global warming 40
globalization 30
green revolution 26
greenhouse gases 40
gross domestic product (GDP) per capita 19
gross national income (GNI) per capita 15
human geography 3
human well-being 20
immigration 13
industrial production 19
Industrial Revolution 23
informal economy 20
interregional linkages 30
Kyoto Protocol 43
landforms 44
latitude 4
legend 4
living wages 33
local scale 12
longitude 4
map projections 7
migration 13
monsoon 48
multicultural society 52
multinational corporation 31
nongovernmental organization (NGO) 37
orographic rainfall 48
physical geography 3
plate tectonics 44
political ecologists 22
population pyramid 14
precipitation 48
primary sector 19
purchasing power parity (PPP) 20
push/pull phenomenon of urbanization 27
quaternary sector 19
race 54
rate of natural increase (RNI) 13
region 9
Ring of Fire 47
scale (of a map) 4
secondary sector 19
services 19
sex 17
slum 27
spatial distribution 3
spatial interaction 3
subsistence economy 16
sustainable agriculture 27
sustainable development 22
tertiary sector 19
total fertility rate (TFR) 13
United Nations (UN) 37
United Nations Gender Equality Index (GEI) 20
United Nations Human Development Index (HDI) 20
urbanization 27
virtual water 38
water footprint 38
weather 47
weathering 47
world region 12
world regional scale 12
World Trade Organization (WTO) 32

Canadian Rockies, Alberta
Central Lowlands, Louisiana
ARCTIC OCEAN
PACIFIC OCEAN
CANADA
UNITED STATES
MEXICO
Hudson Bay
Gulf of Mexico
Bering Sea
Rocky Mountains
Great Plains
Land Elevations
meters
feet
1:24,000,000
Azimuthal Equidistant Projection

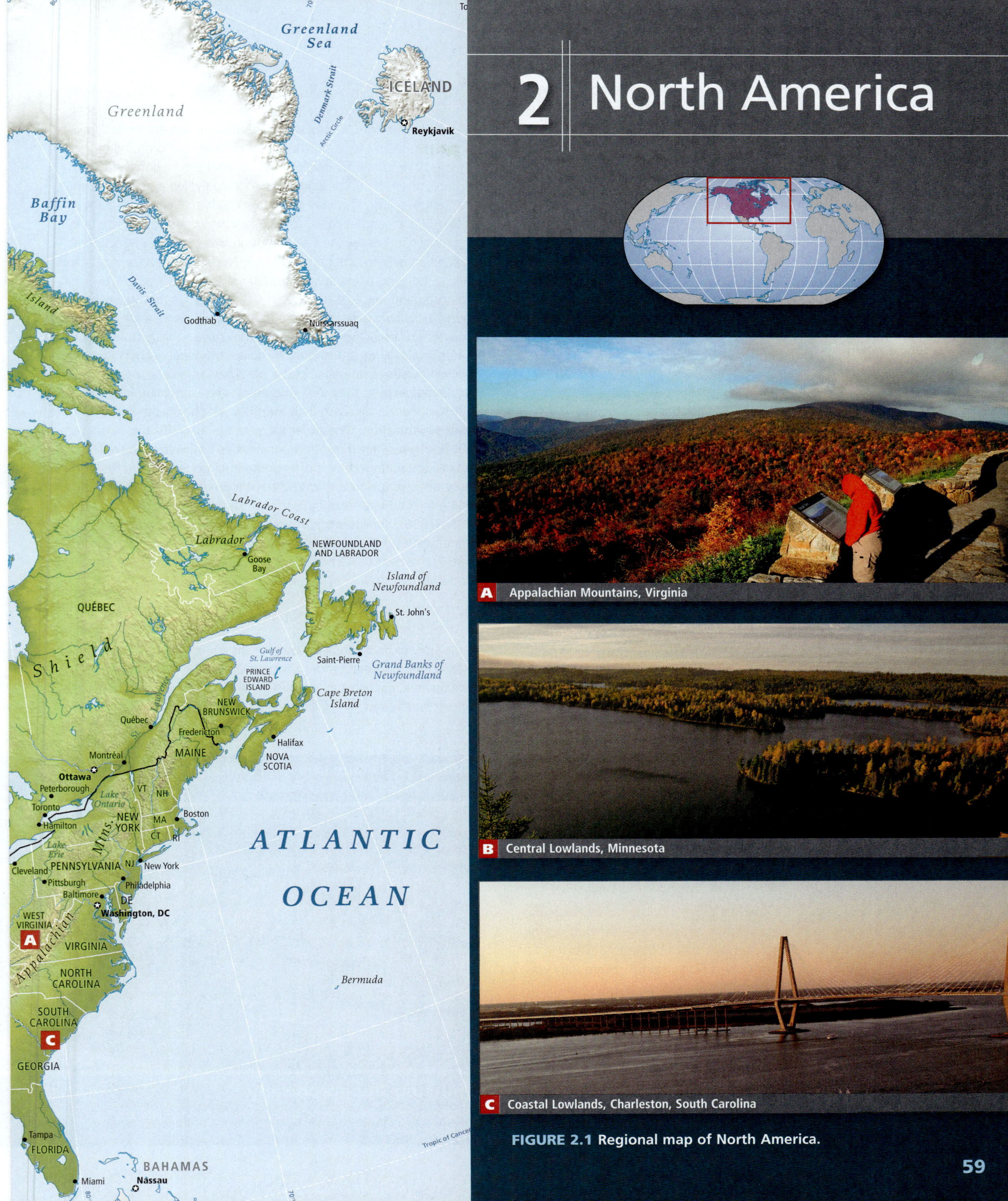

2 North America

FIGURE 2.1 Regional map of North America.

GEOGRAPHIC INSIGHTS: NORTH AMERICA

After you read this chapter, you will be able to discuss the following geographic insights as they relate to the nine thematic concepts:

1. **Climate Change and Urbanization:** North America's large per capita production of greenhouse gases is related to its standard of living and dominant pattern of urbanization.

2. **Water and Food:** North American food production systems contribute to water scarcity and water pollution.

3. **Power and Politics:** The interpretation of the ideal of democracy by Canada and the United States has influenced North America's involvement in the affairs of countries around the world.

4. **Globalization and Development:** Globalization has transformed economic development in North America, thereby changing trade relationships and the kinds of jobs available in this region.

5. **Population and Gender:** Side effects of changes in gender roles and in opportunities for women include some improvements in education and pay equity; smaller families; and the aging of North America's population.

The North American Region

North America (see **Figure 2.1** on pages 58–59) is one of the largest and wealthiest regions of the world. It encompasses many environments and a complex array of local cultures and economic activities that interact with each other across wide distances. The nine thematic concepts in this book are explored as they arise in the discussion of regional issues, with interactions between two or more themes featured, as in the geographic insights above. Vignettes illustrate one or more of the themes as they are experienced in individual lives.

GLOBAL PATTERNS, LOCAL LIVES

VIGNETTE Javier Aguilar, a 39-year-old father of three, has worked as an agricultural laborer in California's Central Valley for 20 years. He and hundreds of thousands like him tend the fields of crops (**Figure 2.2**) that feed the nation, especially during the winter months. Aguilar used to make $8.00 an hour, but now he is unemployed and standing in a church-sponsored food line. "If I don't work, [we] don't live. And here all the work is gone," he says, grimly.

In Mendota, also in California's Central Valley, young Hispanic men wait on street corners to catch a van to the fields. None come. By March of 2009, the unemployment rate in Mendota was 41 percent and rising, and Mayor Robert Silva said his community was dying on the vine. He saw the trouble spreading. Many small businesses were closing. Silva worried about drug use, alcohol abuse, family violence, and malnutrition that can accompany severe unemployment in any community.

This level of unemployment in the heart of the nation's biggest producer of fruit, nuts, and vegetables was partly due to drought and partly to a global economic recession. The drought is related to natural dry cycles as well as global climate change. Water is increasingly scarce in the Central Valley, and access to irrigation water has been cut to force conservation. Meanwhile, the global economic recession, commencing in 2007 and continuing into 2012, reduced overall demand for California's fresh fruits and vegetables as families turned to cheaper foods or those grown closer to home. All of these stresses have forced farmers to remove from production as much as 1 million of the 4.7 million acres once cultivated and irrigated in the Central Valley. This may eventually result in a loss of as many as 80,000 jobs and as much as $2.2 billion in California agriculture and related industries. And even though by early 2012, agricultural production in Mendota had partially revived, unemployment remained over 16 percent, nearly twice the national average.

The confluence of troubles in California's Central Valley has been particularly devastating for low-wage Hispanic male agricultural workers like Javier Aguilar, because they have so little to fall back on in terms of savings, education, or skills. Low-wage–earning Hispanic women may be somewhat better off. They are likely to remain employed because they are in domestic and caregiving jobs for which demand is likely to increase; they also have somewhat higher levels of education. One potential silver lining in this tale is the fact that economic hardship often encourages unskilled workers like Aguilar to take advantage of government-sponsored adult education programs. His children may also have a greater opportunity than did he to graduate from high school. During the Great Depression of the 1930s, similar programs encouraged people to go back to school, causing the nation's high school graduation rate to jump from 20 percent to 60 percent. [*Source: New York Times; PBS NewsHour. For detailed source information, see Text Credit pages.*] ■

THINGS TO REMEMBER

- The story of Javier Aguilar illustrates the complexity of the relationships between various thematic concepts—water, food, development, gender, and global climate change. Reviewing how these concepts are linked here will help you throughout the book.

FIGURE 2.2 Lettuce harvest in California. Agricultural laborers pick lettuce in California's Central Valley.

Thinking Geographically

What factors might cause lettuce production to decline during a recession?

THE GEOGRAPHIC SETTING

What Makes North America a Region?
This world region is relatively easy to define. It consists of Canada and the United States, and they are linked because of their geographic proximity, similar history, and many common cultural, economic, and political features. The nature of the relationship between these two countries will be discussed in numerous places throughout the chapter.

Terms in This Chapter
The term *North America* is used to refer to both countries. Even though it is common on both sides of the border to call the people of Canada "Canadians" and people in the United States "Americans," this text will use the term United States, or U.S., rather than "America," for the United States. Other terms relate to the growing cultural diversity in this region. The text uses the term **Hispanic** to refer to all Spanish-speaking people from Middle and South America, although their ancestors may have been African, Asian, or Native American. When writing about the Southwest, where **Latino** is the preferred term, it is used instead of Hispanic.

Physical Patterns

The continent of North America is a huge expanse of mountain peaks, ridges and valleys, expansive plains, long winding rivers, myriad lakes, and extraordinarily long coastlines. Here the focus is on a few of the most significant landforms.

Landforms

A wide mass of mountains and basins, known as the Rocky Mountain zone, dominates western North America (see Figure 2.1D). It stretches down from the Bering Strait in the far north, through Alaska, and into Mexico. This zone formed about 200 million years ago when, as part of the breakup of the supercontinent Pangaea (see Figure 1.25 on page 46), the Pacific Plate pushed against the North American Plate, thrusting up mountains. These plates still rub against each other, causing earthquakes along the Pacific coast of North America.

The much older, and hence more eroded, Appalachian Mountains stretch along the eastern edge of North America from New Brunswick and Maine to Georgia. This range resulted from very ancient collisions between the North American Plate and the African Plate.

Between these two mountain ranges lies the huge central lowland of undulating plains that stretches from the Arctic to the Gulf of Mexico. This landform was created by the deposition of deep layers of material eroded from the mountains and carried to this central North American region by wind and rain and by the rivers flowing east and west into what is now the Mississippi drainage basin.

During periodic ice ages over the last 2 million years, glaciers have covered the northern portion of North America. In the most recent ice age (between 25,000 and 10,000 years ago), the glaciers, sometimes as much as 2 miles (about 3 kilometers) thick, moved south from the Arctic, picking up rocks and soil and scouring depressions in the land surface. When the glaciers later melted, these depressions filled with water, forming the Great Lakes. Thousands of smaller lakes, ponds, and wetlands that stretch from Minnesota and Manitoba to the Atlantic were formed in the same way (see Figure 2.1B). Melting glaciers also dumped huge quantities of soil throughout the central United States. This soil, often many meters deep, provides the basis for large-scale agriculture but remains susceptible to wind and water erosion.

East of the Appalachians, the Atlantic coastal lowland stretches from New Brunswick to Florida. This lowland then sweeps west to the southern reaches of the central lowland along the Gulf of Mexico. In Louisiana and Mississippi, much of this lowland is filled in by the Mississippi River delta—a low, flat, swampy transition zone between land and sea. The delta was formed by massive loads of silt deposited during floods over the past 150-plus million years by the Mississippi, North America's largest river system. The delta deposit originally began at what is now the junction of the Mississippi and Ohio rivers at Cairo, Illinois; slowly, as ever more sediment was deposited, the delta advanced 1000 miles (1600 kilometers) into the Gulf of Mexico.

The construction of levees along the riverbanks of the Mississippi River has drastically reduced flooding. Because of this flood control, much of the silt that used to be spread widely across the lowlands during floods is being carried to the extreme southern part of the Mississippi delta, where it now drops off the continental shelf and falls into the deep waters of the Gulf of Mexico. In the absence of this silt, the rest of the southern Mississippi River delta is now sinking, a process known as subsidence (Figure 2.3). As the land sinks, salt water from the Gulf intrudes inland, killing fresh-water–adapted plants and destroying swamps and wetlands.

Climate

The landforms across this continental expanse influence the movement and interaction of air masses and contribute to its enormous climate variety (Figure 2.4 on page 63). Along the southern west coast of North America, the climate is generally mild (Mediterranean)—dry and warm in summer, cool and moist in winter. North of San Francisco, the coast receives moderate to heavy rainfall. East of the Pacific coastal mountains, climates are much drier because as the moist air sinks into the warmer interior lowlands, it tends to hold its moisture (see Figure 1.29A on page 50). This interior region becomes increasingly arid as it stretches east across the Great Basin (see Figure 2.4C) and Rocky Mountains. Many dams and reservoirs for irrigation projects have been built to make agriculture and urbanization possible. Because of the low level of rainfall, however, efforts to extract water for agriculture and urban settlements are exceeding the capacity of ancient underground water basins (**aquifers**) to replenish themselves.

On the eastern side of the Rocky Mountains, the main source of moisture is the Gulf of Mexico. When

Hispanic (Latino) a term used to refer to all Spanish-speaking people from Middle and South America, although their ancestors may have been black, white, Asian, or Native American

aquifers ancient natural underground reservoirs of water

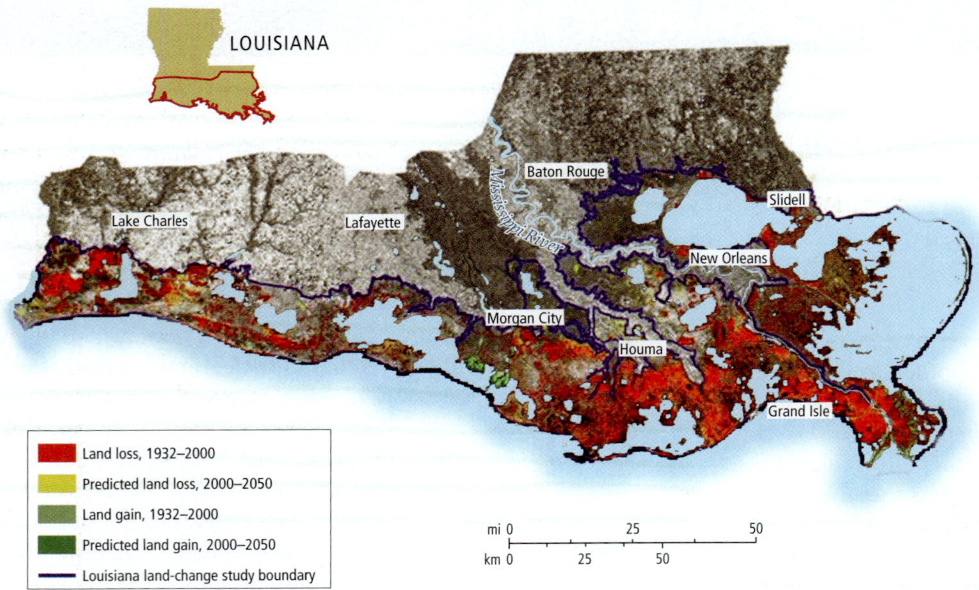

FIGURE 2.3 Wetland loss in Louisiana. The Louisiana coastline and the lower Mississippi River basin are vital to the nation's interests. They are the end point for the vast Mississippi drainage basin, and provide coastal wildlife habitats, recreational opportunities, and transportation lanes that connect the vast interior of the country to the ocean and to offshore oil and gas. Most important, the wetlands provide a buffer against damage from hurricanes. Unfortunately, Louisiana has lost one-quarter of its total wetlands over the last century, largely due to human impacts on natural systems. The remaining 3.67 million acres constitute 14 percent of the total wetland area in the lower 48 states.

the continent is warming in the spring and summer, the air masses above it rise, sucking in warm, moist, buoyant air from the Gulf. This air interacts with cooler, drier, heavier air masses moving into the central lowland from the north (see Figure 2.4A) and west, often creating violent thunderstorms and tornadoes. Generally, central North America is wettest in the eastern (see Figure 2.4B) and southern parts and driest in the north and west (see Figure 2.4C). Along the Atlantic coast, moisture is supplied by warm, wet air above the Gulf Stream—a warm ocean current that flows north from the eastern Caribbean and Florida, which follows the coastline of the eastern United States and Canada before crossing the north Atlantic Ocean.

The large size of the North American continent creates wide temperature variations. Because land heats up and cools off more rapidly than water, temperatures in the interior of the continent are hotter in the summer and colder in the winter than in coastal areas, where temperatures are moderated by the oceans.

THINGS TO REMEMBER

- North America has two main mountain ranges, the Rockies and the Appalachians, separated by expansive plains through which run long, winding rivers.
- The size and variety of landforms influence the movement and interaction of air masses, creating enormous climatic variation. Because land heats up and cools off more rapidly than water, temperatures in the interior of the continent are higher in the summer and colder in the winter than in coastal areas.

Environmental Issues

North America's wide range of resources, plus seemingly limitless stretches of forest and grasslands, long diverted attention from the environmental impacts of settlement and development. Increasingly, however, it is impossible to ignore the many environmental consequences of the North American lifestyle. This section focuses on a few of those consequences: habitat loss, climate change and air pollution, depletion and pollution of water resources and fisheries, and the accumulation of hazardous waste.

Loss of Habitat for Plants and Animals

Before the European colonization of North America, which began soon after 1500, the environmental impact of humans in the region was relatively low. Though North America was by no means a pristine paradise when Europeans arrived, subsequently millions of acres of forests and grasslands that had served as habitats for native plants and animals were cleared to make way for European-style farms, cities, and industries (Figure 2.5 on page 64). This was particularly true in the area that became the United States.

Oil Drilling

In many coastal and interior areas of North America, oil extraction is a large and potentially environmentally devastating industry. This often-overlooked reality was made clear in the spring and summer of 2010, when U.S. waters in the Gulf of Mexico became the site of the largest accidental marine oil spill in world history. In April of 2010, an explosion aboard the Deepwater Horizon, an offshore oil-drilling rig run by British Petroleum (BP), caused it to sink. One result was a massive leak from the rig's wellhead. Due to its location more than a mile beneath the sea surface, the wellhead could not be capped for almost 4 months, during which time it spewed out at least 200 million gallons of oil. While some of this oil made its way to the surface, where it damaged shorelines in all the Gulf Coast states, most of the oil remains beneath the surface due to the use of chemical dispersants by BP. There is mounting concern that this oil poses an ongoing threat to the Gulf's many

Photo Essay: Climates of North America

Climate Zones

Tropical humid climates (A)
- Tropical wet/dry

Arid and semiarid climates (B)
- Steppe
- Desert

Temperate climates (C)
- Midlatitude, moist all year
- Subtropical, winter dry
- Mediterranean, summer dry

Cool humid climates (D)
- Continental, moist all year

Coldest climates (E)
- Arctic
- High altitude

- Winds
- Ocean currents

Wet Pacific air blows into coastal mountains, bringing rain and moderate temperatures.

Continental effect makes winters colder and summers hotter.

Gulf Stream current brings warm tropical water, warming the air over the Atlantic coast.

Moist Gulf of Mexico air brings rain and moderate temperatures.

ARCTIC OCEAN
PACIFIC OCEAN
ATLANTIC OCEAN
Gulf of Mexico
Hudson Bay
Baffin Bay
Brooks Range
Alaska Range
Coast Mountains
ROCKY MOUNTAINS
CANADIAN SHIELD
Great Plains
Great Basin
Colorado Plateau
Appalachian Mountains
Yukon
Mackenzie
Peace
Athabasca
Saskatchewan
Lake Winnipeg
Lake Superior
Lake Michigan
Lake Huron
Lake Erie
Lake Ontario
St. Lawrence
Columbia
Snake
Missouri
Yellowstone
Platte
Great Salt Lake
Colorado
Rio Grande
Pecos
Red
Brazos
Arkansas
Mississippi
Ohio
Tennessee
Seattle
San Francisco
Los Angeles
Phoenix
Denver
Dallas
Houston
Monterrey
Minneapolis
Chicago
St. Louis
Detroit
Toronto
Cleveland
Pittsburgh
Ottawa
Montréal
Boston
New York
Philadelphia
Washington, DC
Atlanta
Tampa
Miami
Nassau
A
B
C
170°W
150°W
140°W
130°W
120°W
70°W
60°W
70°N
80°N
50°N
40°N
30°N
mi 0 250 500
km 0 250 500

A Arctic, Alaska

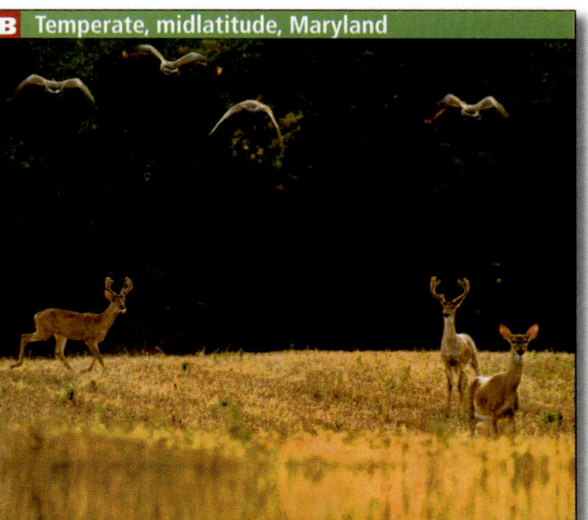

B Temperate, midlatitude, Maryland

C Desert, Utah

Figure 2.5 Photo Essay: Human Impacts on the Biosphere in North America

While parts of North America remain relatively unimpacted by humans, much of this region has seen low-to-medium impacts, and the parts where most people live are highly impacted.

E In remote areas such as Alaska, mining and oil industries have significant landscape impacts. This photo shows the Trans-Alaska Pipeline, which runs for 800 miles.

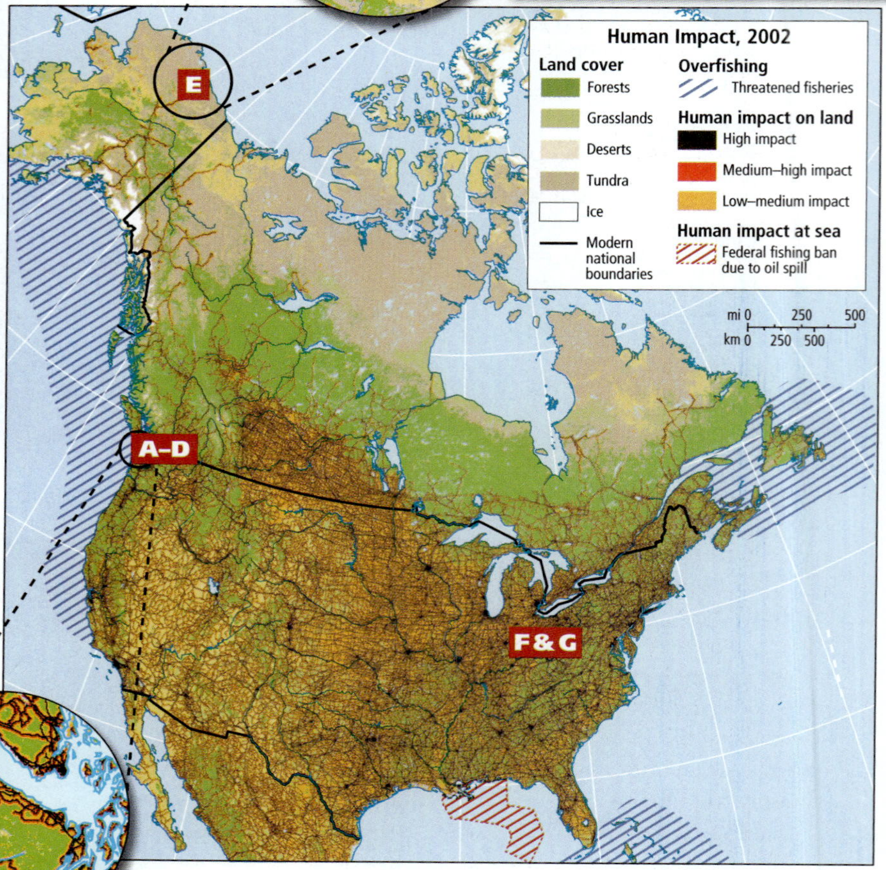

Human Impact, 2002

Land cover: Forests, Grasslands, Deserts, Tundra, Ice, Modern national boundaries

Overfishing: Threatened fisheries

Human impact on land: High impact, Medium–high impact, Low–medium impact

Human impact at sea: Federal fishing ban due to oil spill

Photo **A** shows a logger just outside the Olympic National Park in Washington State making the first cut in the process of felling an 800-year-old, 120-foot-tall cedar. Part of an irreplaceable old-growth forest, the tree is worth about $10,000 at the saw mill. Clear-cutting, in which all trees on a plot of land are cut down **B**, has become an increasingly controversial method of logging. Since 1971, over 30 percent of the forests on Washington State's Olympic Peninsula **C** and **D** have been clear-cut. The clear-cutting of old-growth forests that have never been cut is a practice that has been declining in recent years, but still continues in some places.

F Mountaintop removal coal mining in West Virginia. Shown here is Kayford Mountain, the top of which has been extensively mined, with the tailings pushed into the valley in the foreground.

G Polluted water seeps out of the Kayford Mountain mine site.

Thinking Geographically

After you have read about the human impacts on the biosphere in North America, you will be able to answer the following questions:

A Why might it make economic and environmental sense to keep old-growth trees standing, such as the one shown here?

E What are some threats to the environment posed by the Trans-Alaska Pipeline?

F What kind of mining is mountaintop removal?

G What clues are there that this water may be toxic?

fish, shrimp, and other aquatic species. Moreover, the dispersants used by BP are a health concern because of their high toxicity, which has already led to their banning in Europe. Controversy surrounded the U.S. government's reopening of many Gulf fisheries in July of 2010, and independent studies found that both oil and dispersants in Gulf seafood did not agree with more optimistic government reports. Clearly, the effects of the Deepwater Horizon spill will be felt for years to come.

In other places too, oil extraction has a dramatic effect on the environment. Along the northern coast of Alaska, the Trans-Alaska Pipeline runs southward for 800 miles to the Port of Valdez (see Figure 2.5E). Often running above ground to avoid shifting as the earth freezes and thaws, the pipeline poses a constant risk of rupture, which potentially could result in devastating oil spills. Moreover, the pipeline interferes with migrations of caribou and other animals that Alaska's indigenous people have depended on for food in the past. Protests about threats to the environment from oil extraction in Alaska tend to be quieted by the yearly rebate of several thousand dollars from oil revenues received by each Alaskan.

Canada, with one-tenth the population of the United States, has the largest proven oil reserves in the world after Saudi Arabia, sufficient to meet its needs plus provide export capacity. It is the largest foreign supplier of oil to the United States, but over 90 percent of this oil is in hard-to-access oil sands in western Canada. The costs of extracting this oil in a process called *fracking* are high in terms of environmental impacts, energy expended, and water resources used. Transport of the extracted oil across the North American continent through the Keystone Pipeline system (part of it already in use, part yet to be built) also poses environmental and aesthetic threats.

Logging

Though widespread forest clearing for agriculture is now rare, logging is common throughout North America, where it remains especially important along the northern Pacific coast and in the southeastern United States. Logging in these areas provides most of the construction lumber and much of the paper used in Canada, the United States, and increasingly, in parts of Asia.

Although the logging industry provides jobs and an exportable commodity, it has been depleting the continent's forests. Environmentalists have focused on the damage created by the logging industry, especially on **clear-cutting**, the dominant logging method used throughout North America. In this method, all trees on a given plot of land are cut down, regardless of age, health, or species (see Figure 2.5A–D). Clear-cutting destroys wild animal and plant habitats, thereby reducing species diversity; it leaves forest soils uncovered and highly susceptible to erosion. Concerns about the environmental impacts of logging are boosted by the dominance of service-sector jobs in the major logging states and provinces. For example, in the Pacific Northwest, even in many remote areas where logging was once the backbone of the economy, people are now dependent on tourism and other occupations that rely on the beauty of intact forest ecosystems.

Coal Mining and Use

In many remote interior areas of North America, coal mining is also a large and environmentally damaging industry. Strip mining, where vast quantities of earth and rock are removed in order to extract underlying coal, can result in visual wastelands and in huge piles of mining waste called "tailings" that pollute waterways and threaten communities that depend on well water. Particularly damaging is a form of strip mining known as *mountaintop removal*, in which the entire top of a mountain may be leveled and the tailings pushed into surrounding valleys, resulting in the pollution of whole watersheds (see Figure 2.5F and G). When coal is burned, it is difficult to find a safe place to store the huge volumes of toxic ash that result. Left to dry out, the material becomes airborne and contributes to air pollution, so it is stored wet. In December 2008, a large earthen dike of wet coal ash burst after a heavy rainstorm, spilling 1.5 billion gallons of toxic sludge (the largest industrial spill in U.S. history) over 300 acres of beautiful lakeshore and forestland in rural Tennessee. The sludge, containing heavy metals and harmful chemicals, ruined the ecology of the immediate area and threatened hundreds of square miles with water and air pollution. Cleanup was continuing in 2013 at the cost of well over $1 billion.

Urbanization and Habitat Loss

An important aspect of urbanization is **urban sprawl** (see the "Urbanization and Sprawl" section on pages 89–93). In many areas, and for several decades, middle- and upper-income urbanites have sought lower-density suburban neighborhoods. Farms that were once highly productive have given way to expansive, low-density urban and suburban residential developments where pavement, golf courses, office complexes, and shopping centers cover the landscape. In the process, natural habitats are being degraded even more intensely than they were by farming. The loss of farmland and natural habitat in the urban fringe affects recreational land and the ability to produce local, affordable food for urban populations.

As North American native plants and animals have been forced into ever smaller territories, many have died out entirely and have been replaced by nonnative species (European and African grasses and the domestic cat, for example) brought in by humans either purposely or inadvertently (see Figure 2.23 on page 92). Estimates vary, but at least 4000 nonnative species have invaded North America. An example is the Asian snakehead fish, which is rapidly invading the Potomac River, where it eats baby bass and fiercely competes with native fish for food.

29. SNAKEHEAD REPORT

clear-cutting the cutting down of all trees on a given plot of land, regardless of age, health, or species

urban sprawl the encroachment of suburbs on agricultural land

Climate Change and Air Pollution

Geographic Insight 1

Climate Change and Urbanization: North America's large per capita production of greenhouse gases is related to its standard of living and dominant pattern of urbanization.

On a per capita basis, North Americans are among the highest contributors of greenhouse gases to the Earth's atmosphere. A very few small countries have higher per capita emissions. With only 5 percent of the world's population, North America produces 26 percent of the greenhouse gases released globally by human activity. This large share can be traced to North America's high consumption of fossil fuels, which in turn is related to several factors. One of these is North America's dominant pattern of urbanization, characterized by vast and still-growing suburbs where people depend on their automobiles for almost all of their transportation needs. Freestanding dwellings and businesses spread out across the land also require more energy to heat and cool than do the densely packed, high-rise buildings typical of cities in most other world regions. North American industrial and agricultural production is also highly dependent on fossil fuels.

Canada's government was one of the first in the world to commit to reducing the consumption of fossil fuels. Until recently, the United States resisted such moves, fearing damage to its economy. Both countries are now exploring alternative sources of energy, such as solar, wind, geothermal, and nuclear power. So far, neither country has been able to reduce its levels of greenhouse gas emissions, or even the rate at which these emissions are growing. However, both Canada and the United States possess the technological capabilities needed to lead the world in shifting over to cleaner sources of energy.

31. ENERGY REPORT

32. GREEN BUILDING

Vulnerability to Climate Change Both Canada and the United States are vulnerable to the effects of climate change. Dense population centers on the Gulf of Mexico and on the Atlantic coast are highly exposed to hurricanes, which may become more violent as oceans warm (Figure 2.6B, D). Sea level rise and coastal erosion, due to the thermal expansion of the oceans, is already affecting many coastal areas along the Arctic coast of North America (see Figure 2.6A). Here, at least 26 coastal villages are being forced to relocate inland, at an estimated cost of $130 million per village. Meanwhile, many arid farming zones will dry further as higher temperatures reduce soil moisture, making irrigation crucial (see Figure 2.6C). In all of these areas, resilience to climate change is bolstered over the short term by excellent emergency response and recovery systems. Long-term resilience is boosted by careful planning as well as by North America's large and diverse economy, which can provide alternative livelihoods to people highly exposed to climate impacts.

30. GLACIER NATIONAL PARK

smog a combination of industrial emissions, car exhaust, and water vapor that frequently hovers as a yellow-brown haze over many cities, causing a variety of health problems.

acid rain falling precipitation that has formed through the interaction of rainwater or moisture in the air with sulfur dioxide and nitrogen oxides emitted during the burning of fossil fuels, making it acidic

Air Pollution In addition to climate change, most greenhouse gases contribute to various forms of air pollution, such as smog and acid rain. **Smog** is a combination of industrial emissions, car exhaust, and water vapor that frequently hovers as a yellow-brown haze over many North American cities, causing a variety of health problems. These same emissions also result in **acid rain**, which is created when pollutants dissolve in falling precipitation and make the rain acidic. Acid rain can kill trees and, when concentrated in lakes and streams, poisons fish and wildlife.

The United States, with its large population and extensive industry, is responsible for the vast majority of acid rain in North America. Due to continental weather and wind patterns, however, the area most affected by acid rain encompasses a wide swath on both sides of the eastern U.S.–Canada border (Figure 2.7 on page 68). The eastern half of the continent, which includes the entire eastern seaboard from the Gulf Coast to Newfoundland, is significantly affected by acid rain.

Thinking Geographically

After you have read about the vulnerability to climate change in North America, you will be able to answer the following questions:

A Describe the coastal erosion illustrated by this photo.

B What elements of this community's emergency response system may be compromised by flooding?

C How might irrigation reduce vulnerability to climate change, but also result in higher greenhouse gas emissions?

D Why are stronger hurricanes more likely as the climate warms?

Water Resource Depletion, Pollution, and Marketization

Geographic Insight 2

Water and Food: North American food production systems contribute to water scarcity and water pollution.

People who live in the humid eastern part of North America find it difficult to believe that water is becoming scarce even there. Consider the case of Ipswich, Massachusetts, where the watershed is drying up as a result of overuse. There, innovators are saving precious water through conservation strategies in their homes and businesses (see the *Rainwater Cash* video). Elsewhere, as populations grow and per capita water usage increases, conflicts over water are more and more common. **34. RAINWATER CASH**

Water Depletion In North America, water becomes increasingly precious the farther west one goes. The Great Lakes Agreement (formally known as the Great Lakes–St. Lawrence River Basin Sustainable Water Resources Agreement) is an agreement between the eight states and two Canadian provinces that border this largest of the Earth's freshwater bodies (see the map in Figure 2.1) to manage the Great Lakes waters more wisely than they have been in the past. Cities as far away as those in Alabama have eyed the lakes as a potential water source for their growing needs. Long before people were aware of the ramifications of making drastic changes to

Figure 2.6 Photo Essay: Vulnerability to Climate Change in North America

North America's wealth and its well-developed emergency response systems give it high resilience that reduce its overall vulnerability. However, certain regions are highly exposed to temperature increases, drought, hurricanes, and sea level rise.

A The location of Shishmaref on the Arctic Sea leaves it exposed to coastal erosion, which may be increasing here due to higher air and sea temperatures.

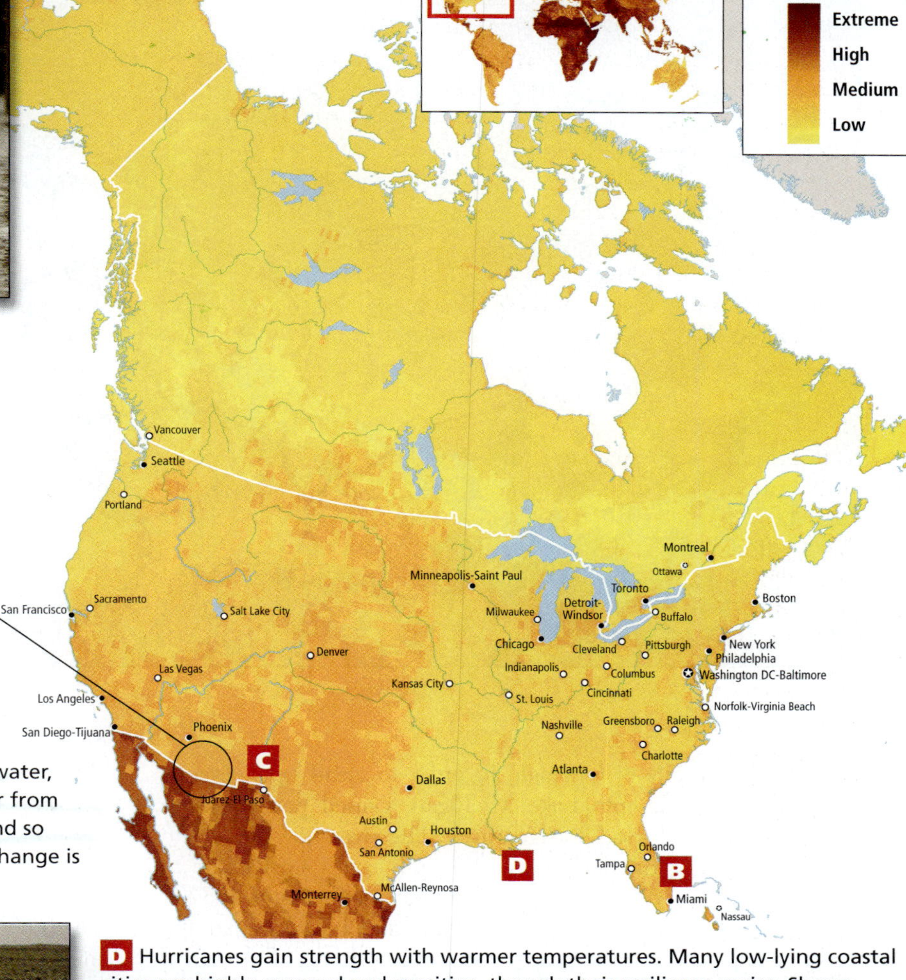

B Florida's low elevations leave it exposed to sea level rise and flooding during hurricanes (shown here).

The stark contrast in vulnerability between the United States and Mexico results from the countries' very different sensitivity and resilience to water scarcity. As temperatures rise, this already dry borderland area will have less water. The United States has a much better water infrastructure, reducing its sensitivity to drought. Meanwhile, more developed emergency response systems in the United States increase its resilience to water shortages. With such drastic differences, the U.S.–Mexico border could become an even more contentious zone as the climate changes.

C Higher temperatures raise the rate at which plants lose water, increasing the need for irrigation. Farms that get their water from shrinking aquifers are thus less able to increase irrigation, and so are highly sensitive to the rising temperatures that climate change is bringing.

D Hurricanes gain strength with warmer temperatures. Many low-lying coastal cities are highly exposed and sensitive, though their resilience varies. Shown here is a fire that destroyed several homes in New Orleans after flooding caused by Hurricane Katrina, which made it impossible for fire trucks to respond.

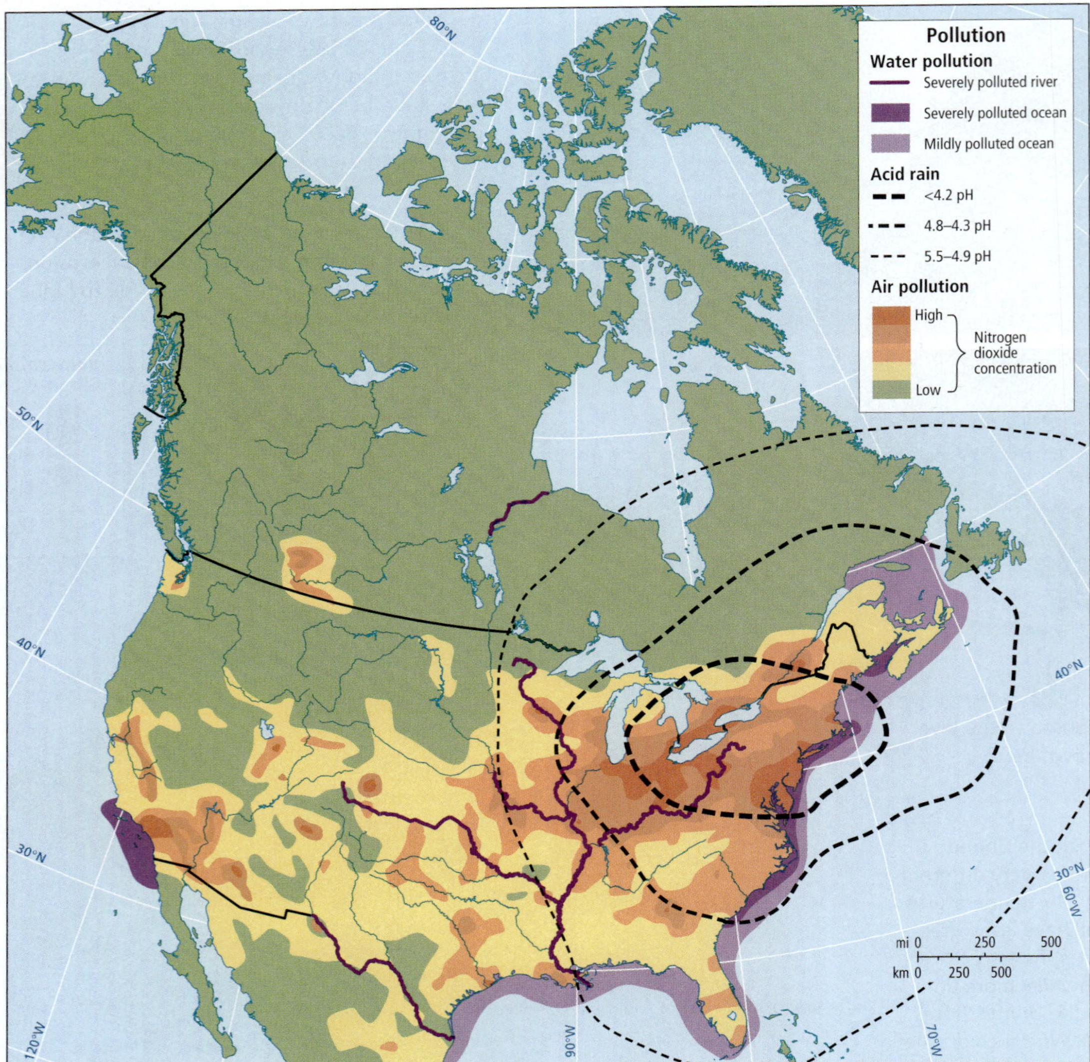

FIGURE 2.7 Air and water pollution in North America. This map shows two aspects of air pollution, as well as polluted rivers and coastal areas. Red and yellow indicate concentrations of nitrogen dioxide (NO_2), a toxic gas that comes primarily from the combustion of fossil fuels by motor vehicles and power plants. This gas interacts with rain to produce nitric acid, a major component of acid rain, as well as toxic organic nitrates that contribute to urban smog. The map also shows polluted coastlines (including all of the coastline from Texas to New Brunswick) as well as severely polluted rivers, which include much of the Mississippi River and its tributaries.

ecosystems, the city of Chicago, in order to clean up sewage it had been dumping into Lake Michigan, gained permission to reverse the flow of the Chicago River. The river now flows into the Mississippi, ultimately transferring Chicago's wastewater to the Gulf of Mexico. This diversion, which moves water at 3200 cubic feet per second from the Great Lakes, has opened the doors to requests from cities, not even on the shores of the Great Lakes, to gain access to the lake water. The transnational Great Lakes Agreement is an attempt to rationally manage this vulnerable inland freshwater sea. (See a clip from NPR's *Science Friday* of the *Inland Seas* video at http://www.sciencefriday.com/videos/watch/10113.)

On the North American Great Plains, rainfall is highly variable from year to year. To make farming more secure and predictable, taxpayers across the continent have subsidized the building of pumps and stock tanks for farm animals, and aqueducts and reservoirs for crop irrigation. However, irrigation is increasingly based on fossil water that has been stored over the millennia in *aquifers*. The *Ogallala aquifer* (Figure 2.8) underlying the Great Plains is the largest in North America. In parts of the Ogallala, water is being pumped out at rates that exceed natural replenishment by 10 to 40 times.

As mentioned in the opening vignette, fruit and vegetable crops in California are routinely irrigated with water from surrounding states. Such irrigation, involving expensive and massive interstate engineering projects, accounts for some of the water that goes into the *virtual water footprint* of U.S. consumers, as discussed in Chapter 1 (page 38; see also Table 1.2 on page 38). This water also supplies the cities of Southern California, which are built on land that was once desert. Water is pumped from hundreds of miles away and over mountain ranges. California uses more energy to move water than some states use for all purposes. Moreover, irrigation in Southern California, especially that which is drawn from the Colorado River, deprives Mexico of this much-needed resource. The mouth of the Colorado (which is in Mexico) used to be navigable; now, because of massive diversions, it is dry and sandy—only a mere trickle of water gets to Mexico.

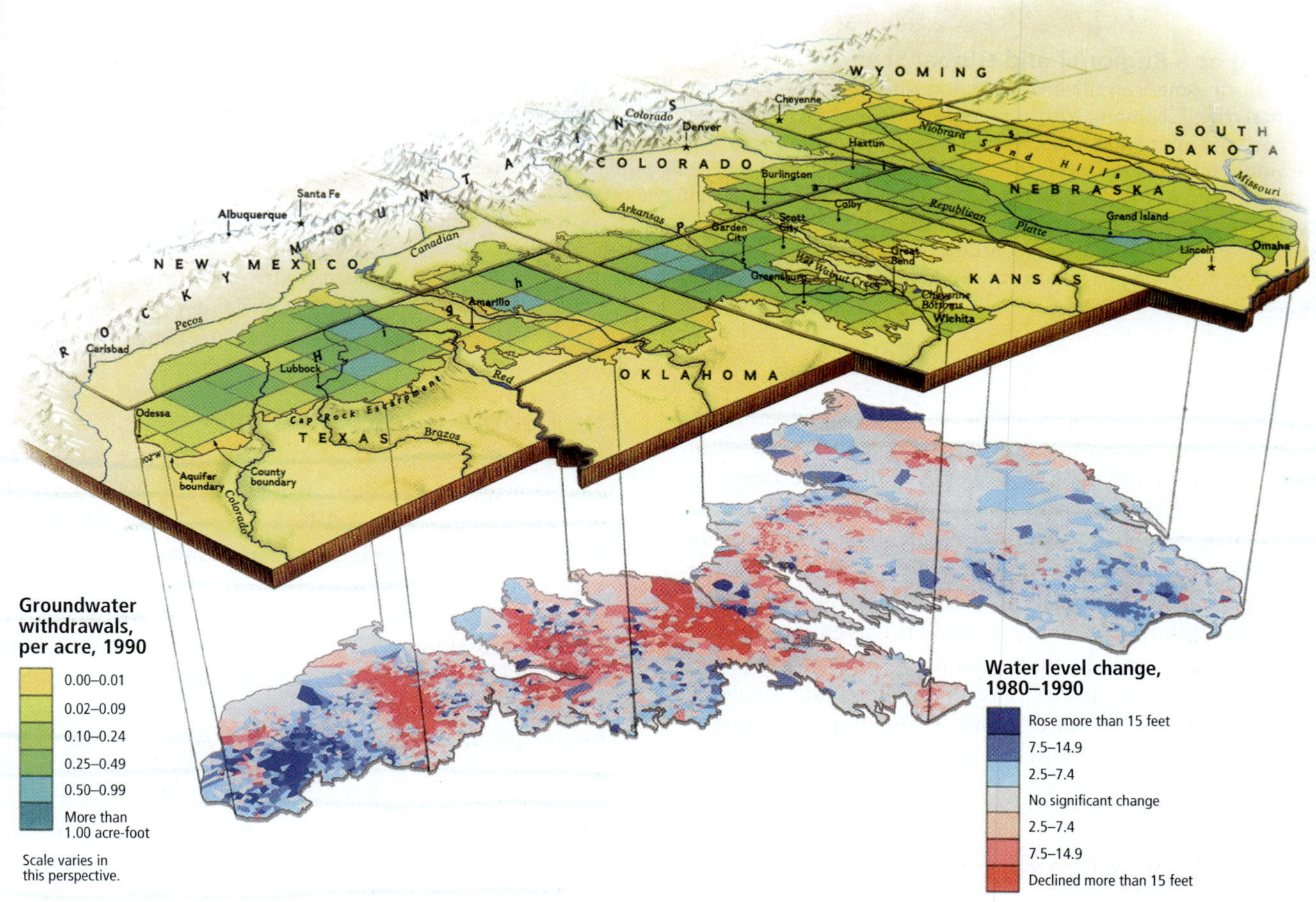

FIGURE 2.8 The Ogallala aquifer. Between the 1940s and the 1980s, the aquifer lost an average of 10 feet (3 meters) of water overall, and more than 100 feet (30 meters) of water in some parts of Texas. Then, during the 1980s, abundant rain and snow meant less water decline in the aquifer. However, in the Ogallala area the climate fluctuates from moderately moist to very dry, but dry periods are lengthening. A drought began in mid-1992 and has returned every few years, causing large agribusiness firms to pump Ogallala water to supplement scarce precipitation. Since 1992, water levels in the aquifer declined an average of 1.35 feet per year and now exceed replenishment rates many times over.

Increasingly, citizens in western North America are recognizing that the use of scarce water for irrigating agriculture, raising livestock, and keeping lawns and golf courses green in desert environments is unsustainable. Conflicts over transporting water from wet regions to dry ones, or from sparsely inhabited to urban areas, are ongoing and have halted some new water projects. However, government subsidies have kept water artificially cheap, and in the past, new water supplies have always been found and harnessed, creating little incentive to change.

Water Pollution In the United States, 40 percent of rivers are too polluted for fishing and swimming, and over 90 percent of *riparian areas* (the interface between land and flowing surface water) have been lost or degraded. Pollution in the rivers of North America comes mainly as storm-water runoff from agricultural areas, urban and suburban developments, and industrial sites. However, a recently discovered type of water pollution is that of trace elements of pharmaceuticals, such as male and female hormones, that are excreted by humans and not removed during water purification processes. These chemicals then make their way into rivers and lakes, where they enter the food system in drinking water or through fish. **35. DRUGS AND WATER SUPPLY**

In the 1970s, scientists studying coastal areas began noticing *dead zones* where water is so polluted that it supports almost no life. Dead zones occur near the mouths of major river systems that have been polluted by fertilizers and pesticides washed from farms and lawns when it rains. A large dead zone is in the Gulf of Mexico near the mouth of the Mississippi, and similar zones have been found in all U.S. coastal areas. Even Canada, where much lower population density means that rivers are generally cleaner, has dead zones on its western coast.

Water Marketization The answer to the question, "Who owns our water?" is usually answered with "We do," but to what extent is water really public property? North Americans are used to paying for water, but the cost has usually been just high enough to

ON THE BRIGHT SIDE

Green Living at a Regional and Global Scale

Increasingly, North American citizens are asking what they can do in daily life to ameliorate looming environmental crises. Solutions, such as *greener living*—which involves recycling, driving less, growing a food garden, and improving home energy efficiency—can collectively make an important impact.

Environmental disasters can inspire green policies: After the BP spill, public pressure resulted in a tenfold increase in federal support for environmental protection in the region. Numerous locally focused and federally funded projects have been helping farmers to both control soil erosion and to stop the pollution of aquifers. Cattle in coastal zones are now fenced in small herds so they don't overgraze; they are provided with drinking troughs so they don't stand in streams to drink and defecate.

cover extraction, purification, and delivery in pipes. Now, threats of polluted drinking water are beginning to change the way water is viewed. For example, in the last few years, in response to aggressive advertising, the public has been buying bottled water even when tap water is perfectly safe. A number of North American communities with abundant fresh water have agreed to sell water to beverage companies for bottling, without understanding that massive water withdrawals from local aquifers can cause geologic subsidence, loss of aquatic habitats, and the depletion and pollution of natural wells and springs. Eventually, realizing belatedly that their own low-cost access to water is threatened, expensive litigation against the water bottling companies is often the only recourse.

THINGS TO REMEMBER

Geographic Insight 1

- **Climate Change and Urbanization** North America produces 26 percent of the world's greenhouse gases, even though it has only 5 percent of the world's population. This is related to the region's pattern of urban development and its fossil fuel–dependent industrial and agricultural production methods.
- While many parts of North America are exposed to multiple climate-change impacts, North America is also a highly resilient region. Its overall vulnerability to climate change is generally low compared to other world regions.

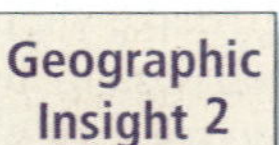

- **Water and Food** North American food production systems have contributed to water pollution in part because large sections of the countryside have been cleared for cultivation, a practice that results in eroded soil entering streams and rivers; and also because the chemicals used in large-scale agriculture are washed by rain and irrigation runoff into the nation's waterways and aquifers.
- Water pollution is now a major problem, especially in the United States, where 40 percent of the rivers are too polluted for fishing or swimming.

Human Patterns over Time

In prehistoric times, humans came from Eurasia via Alaska, dispersing to the south and east. Beginning in the 1600s, waves of European immigrants, enslaved Africans, and their descendants, spread over the continent, primarily from east to west. Today, immigrants are coming mostly from all of Asia and from Middle and South America. They are arriving mainly in the Southwest and West, where they remain concentrated. In addition, internal migration is still a defining characteristic of life for most North Americans, who are among the world's most mobile people. On average they move nearly 12 times in a lifetime.

The Peopling of North America

Recent evidence suggests that humans first came to North America from northeastern Asia at least 25,000 years ago or perhaps earlier, most arriving during an ice age. At that time, the global climate was cooler, polar ice caps were thicker, and sea levels were lower. The Bering land bridge, a huge, low landmass wider than 1000 miles (1600 kilometers), connected Siberia to Alaska. Bands of hunters crossed by foot or small boats into Alaska and traveled down the west coast of North America.

The Original Settling of North America By 15,000 years ago, humans had reached nearly to the tip of South America and had spread deep into that continent. By 10,000 years ago, global temperatures began to rise. As the ice caps melted, sea levels rose and the Bering land bridge was submerged beneath the sea.

Over thousands of years, the people settling in the Americas domesticated plants, created paths and roads, cleared forests, built permanent shelters, and sometimes created elaborate social systems. About 3000 years ago, corn was introduced from Mexico (into what is now the U.S. southwestern desert) along with other Mexican domesticated crops, particularly squash and beans. These food crops are thought to be closely linked to settled life, resulting in North America's prehistoric population growth.

These foods provided surpluses that allowed some community members to engage in activities other than agriculture, hunting, and gathering, making possible large, city-like regional settlements. For example, by 1000 years ago, the urban settlement of Cahokia, including suburban settlements (in what is now central Illinois across the Mississippi from St. Louis), covered 5 square miles (12 square kilometers) and was home to an estimated 30,000 people (see Figure 2.9A on page 72). Here people could specialize in crafts, trade, or other activities beyond the production of basic necessities.

The Arrival of the Europeans North America was completely transformed by the sweeping occupation of the continent by Europeans. In the sixteenth century, Italian, Portuguese, and English explorers came ashore along the eastern seaboard of North America, and the Spanish explorer Hernando De Soto, made his way from Florida deep into the heartland of the continent in the 1540s. In the early seventeenth century, the British established colonies along the Atlantic coast in what is now

Virginia (1607) and Massachusetts (1620). The Dutch explored the Atlantic seaboard looking for trading opportunities, and the French explored the northern interior of the continent, entering via the St. Lawrence River. Over the next two centuries, colonists and settlers from northern Europe, assisted by enslaved Africans, built villages, towns, port cities, and plantations along the eastern coast. By the mid-1800s, they had occupied most Native American lands into the central part of the continent.

Disease, Technology, and Native Americans The rapid expansion of European settlement was facilitated by the vulnerability of Native American populations to European diseases. Having long been isolated from the rest of the world, Native Americans had no immunity to diseases such as measles and smallpox. Transmitted by Europeans and Africans who had built up immunity to them, these diseases killed up to 90 percent of Native Americans within the first 100 years of contact. It is now thought that diseases spread by early expeditions, such as De Soto's into Florida, Georgia, Tennessee, and Arkansas, so decimated populations in the North American interior that fields and villages were abandoned, the forest grew back, and later explorers erroneously assumed the land had never been occupied.

Technologically advanced European weapons, trained dogs, and horses also took a large toll. Often the Native Americans had only bows and arrows. Some Native Americans in the Southwest acquired horses from the Spanish and learned to use them in warfare against the Europeans, but their other technologies could not compete. Numbers reveal the devastating effect of European settlement on Native American populations. Roughly 18 million Native Americans lived in North America in 1492. By 1542, after just a few Spanish expeditions, only half that number survived. By 1907, slightly more than 400,000, or a mere 2 percent, remained.

The European Transformation

European settlement erased many of the landscapes familiar to Native Americans and imposed new ones that fit the varied physical and cultural desires of the new occupants.

The Southern Settlements European settlement of eastern North America began with the Spanish in Florida in the mid-1500s and the establishment of the British colony of Jamestown in Virginia in 1607. By the late 1600s, large plantations in the colonies of Virginia, the Carolinas, and Georgia were cultivating cash crops such as tobacco and rice.

To secure a large, stable labor force, Europeans brought enslaved African workers into North America, beginning in 1619. Within 50 years, enslaved Africans were the dominant labor force on some of the larger Southern plantations (see Figure 2.9B). By the start of the Civil War in 1861, slaves made up about one-third of the population in the Southern states and were often a majority in the plantation regions. North America's largest concentrations of African Americans are still in the southeastern states (Figure 2.10).

The plantation system concentrated wealth in the hands of a small class of landowners, who made up just 12 percent of Southerners in 1860. Planter elites kept taxes low and invested their money in Europe or the more prosperous northern colonies, instead of in **infrastructure** at home. As a result, the road, rail, communication networks, and other facilities necessary for economic growth were not built.

More than half of Southerners were poor white farmers. Both they and the general slave population lived simply, so their meager consumption did not provide a demand for goods. Hence, there were few market towns and almost no industries. Plantations tended to be self-sufficient and generated little *multiplier effect*. Enterprises like small shops, garment making, small restaurants and bars, small manufacturing, and transportation and repair services that normally "spin off" from, or serve, main industries developed only minimally in the South. Antagonism between the weak southern economy and the stronger, more diversified northern economies was a main cause of the Civil War (1861–1865), perhaps equal to the abolition movement to free enslaved Africans and their descendents. After the war, while the victorious North returned to promoting its own industrial development, the plantation economy declined, and the South sank deeply into poverty. The South remained economically and socially underdeveloped well into the 1970s.

The Northern Settlements Throughout the seventeenth century, relatively poor subsistence farming communities dominated the colonies of New England and southeastern Canada. There were no plantations and few slaves, and not many cash crops were exported. What exports there were consisted of raw materials like timber, animal pelts, and fish from the Grand Banks off Newfoundland and the coast of Maine. Generally, farmers lived in interdependent communities that prized education, ingenuity, self-sufficiency, and thrift.

By the late 1600s, New England was implementing ideas and technology from Europe that led to the first industries. By the 1700s, diverse industries were supplying markets in North America and the Caribbean with metal products, pottery, glass, and textiles. By the early 1800s, southern New England, especially the region around Boston, became the center of manufacturing in North America. It drew largely on young male and female immigrant labor from French Canada and Europe.

The Mid-Atlantic Economic Core The colonies of New York, New Jersey, Pennsylvania, and Maryland eventually surpassed in population and in wealth New England and southeastern Canada. This mid-Atlantic region benefited from more fertile soils, a slightly warmer climate, multiple deepwater harbors, and better access to the resources of the interior. By the end of the Revolutionary War in 1783, the mid-Atlantic region was on its way to becoming the **economic core**, or the dominant economic region, of North America. Port cities such as New York, Philadelphia, and Baltimore prospered as the intermediary for trade between Europe and the vast American Continental Interior.

In the early nineteenth century, both agriculture and manufacturing grew and diversified, drawing immigrants from much of northwestern Europe. As farmers became more successful, they bought mechanized

infrastructure road, rail, and communication networks and other facilities necessary for economic activity

economic core the dominant economic region within a larger region

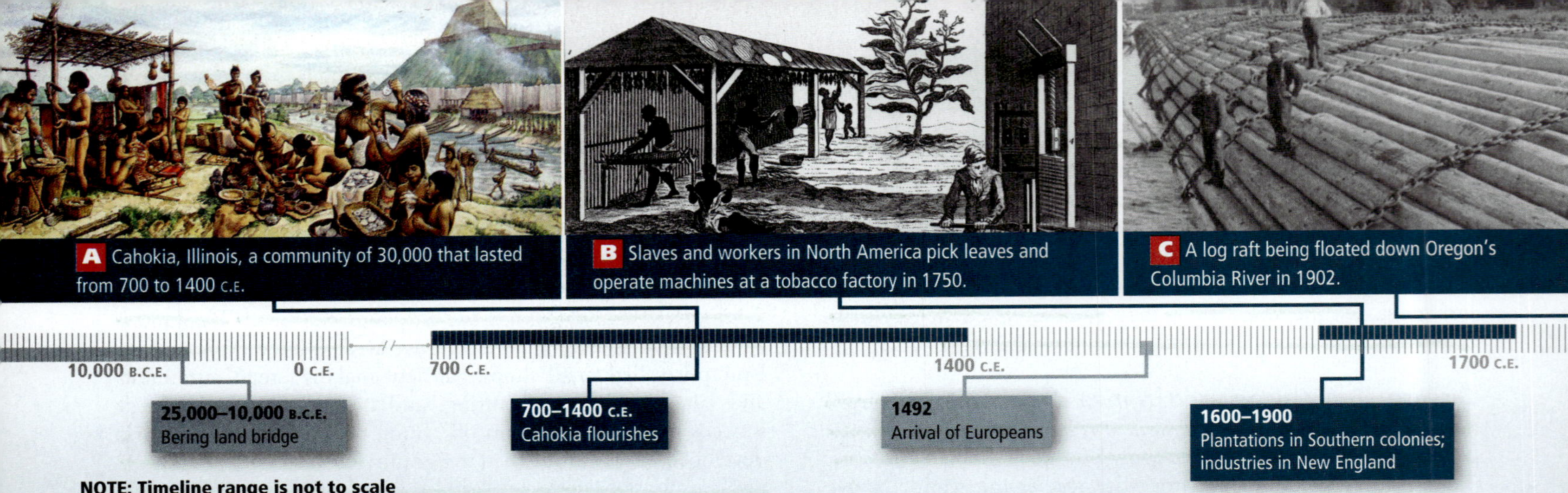

FIGURE 2.9 A VISUAL HISTORY OF NORTH AMERICA

Thinking Geographically

After you have read about the human history of North America, you will be able to answer the following questions.

A How did food crops such as corn, beans, and squash influence the development of North American settlements like Cahokia?

B Why did the plantation system inhibit the establishment of roads, communication networks, and small enterprises that could have boosted other forms of economic activity in the Southern colonies?

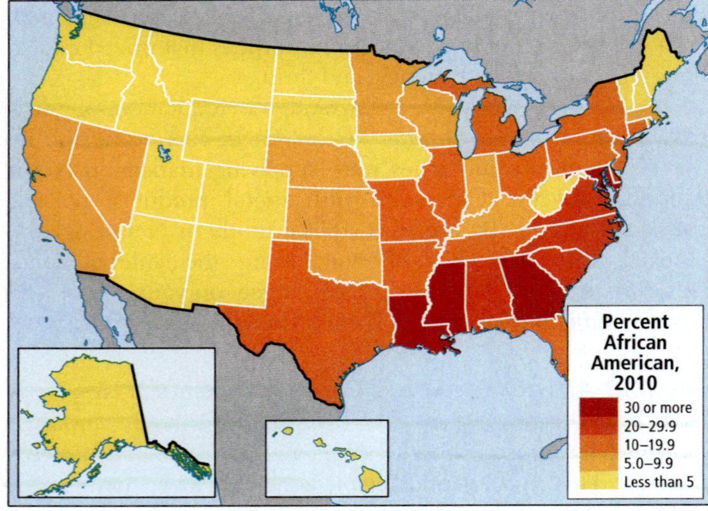

FIGURE 2.10 Percent of African American population in each state, 2010. The percent refers only to those persons who selected "Black, African Am., Negro" as their only race in the 2010 census. It does not include those who selected more than one race that included black.

equipment, appliances, and consumer goods made in nearby cities. By the mid-nineteenth century, the economy of the core was increasingly based on the steel industry, which diffused westward to Pittsburgh and the Great Lakes industrial cities of Cleveland, Detroit, and Chicago. The steel industry further stimulated the mining of deposits of coal and iron ore throughout the region and beyond. Steel became the basis for mechanization, and the region was soon producing heavy farm and railroad equipment, including steam engines.

By the early twentieth century, the economic core stretched from the Atlantic to St. Louis on the Mississippi (including many small industrial cities along the river, plus Chicago and Milwaukee), and from Ottawa to Washington, DC. It dominated North America economically and politically well into the mid-twentieth century. Most other areas produced food and raw materials for the core's markets and depended on the core's factories for manufactured goods (see Figure 2.9D).

Expansion West of the Mississippi and Great Lakes

The east-to-west trend of settlement continued as land in the densely settled eastern parts of the continent became too expensive for new immigrants. By the 1840s, immigrant farmers from central and northern Europe, as well as European descendants born in eastern North America, were pushing their way beyond the Great Lakes and across the Mississippi River, north and west into the Great Plains of Canada and the United States (Figure 2.11).

The Great Plains Much of the land west of the Great Lakes and the Mississippi River was dry grassland or prairie. The soil usually proved very productive in wet years, and the area became known as North America's breadbasket. But the naturally arid character of this land eventually created an ecological disaster for Great Plains farmers. In the 1930s, after 10 especially dry years, a series

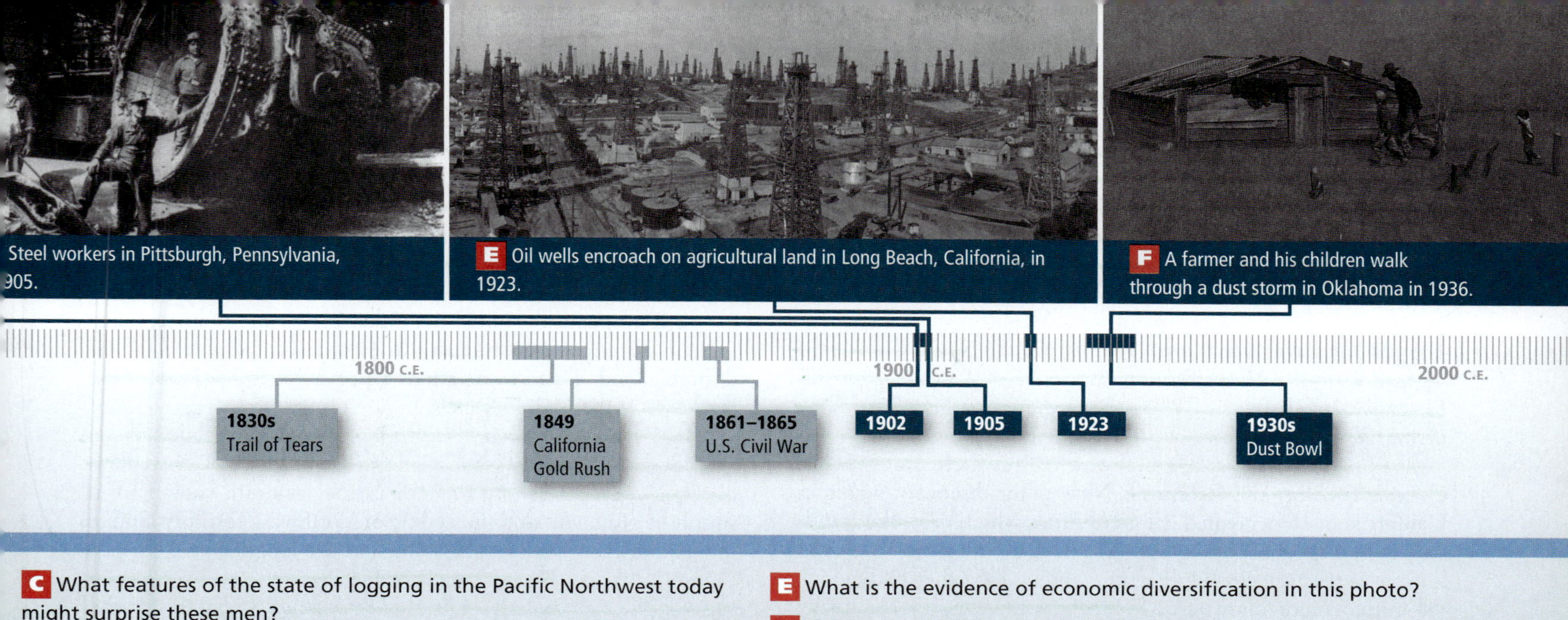

Steel workers in Pittsburgh, Pennsylvania, 1905.

E Oil wells encroach on agricultural land in Long Beach, California, in 1923.

F A farmer and his children walk through a dust storm in Oklahoma in 1936.

C What features of the state of logging in the Pacific Northwest today might surprise these men?

D How did the steel industry stimulate development in the "economic core" of North America?

E What is the evidence of economic diversification in this photo?

F What were the causes of the Dust Bowl?

FIGURE 2.11 Nineteenth-century transportation.

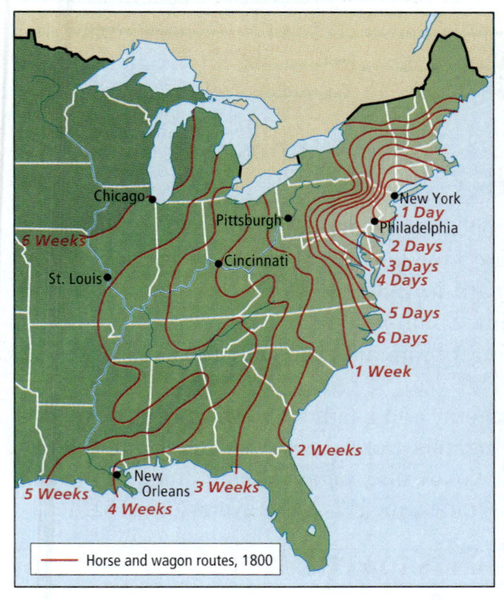

(A) Travel times from New York City, 1800. It took a day to travel by wagon from New York City to Philadelphia and a week to go to Pittsburgh.

(B) Travel times from New York City, 1857. The travel time from New York to Philadelphia was now only 2 or 3 hours and to Pittsburgh less than a day because people could go part of the way via canals (dark blue). Via the canals, the Great Lakes, and rivers, they could easily reach principal cities along the Mississippi River, and travel was less expensive and onerous.

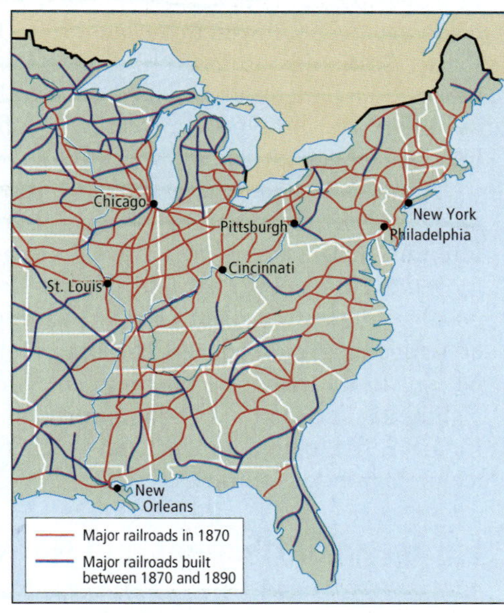

(C) Railroad expansion by 1890. With the building of railroads, which began in the decade before the Civil War, the mobility of people and goods increased dramatically. By 1890, railroads crossed the continent, though the network was most dense in the eastern half.

of devastating dust storms blew away topsoil by the ton. This hardship was made worse by the widespread economic depression of the 1930s. Many Great Plains farm families packed up what they could and left what became known as the Dust Bowl (see Figure 2.9F), heading west to California and other states on the Pacific Coast.

The Mountain West and Pacific Coast Some Europeans skipped over the Great Plains entirely, alerted to the possibilities farther west. By the 1840s, they were coming to the valleys of the Rocky Mountains, to the Great Basin, and to the well-watered and fertile coastal zones of what was then known as the Oregon Territory and California. News of the discovery of gold in California in 1849 created the *Gold Rush*, which drew thousands with the prospect of getting rich quickly. The vast majority of gold seekers were unsuccessful, however, and by 1852 they had to look for employment elsewhere. Farther north, logging eventually became a major industry (see Figure 2.9C).

The extension of railroads across the continent in the nineteenth century facilitated the transportation of manufactured goods to the West and raw materials and eventually fresh produce to the East. Today, the coastal areas of this region, often called the Pacific Northwest, have booming, diverse, high-tech economies and growing populations. Perhaps in response to their history, residents of the Pacific Northwest are on the forefront of so many efforts to reduce human impacts on the environment that the region has been nicknamed "Ecotopia."

The Southwest People from the Spanish colony of Mexico first colonized the Southwest in the late 1500s. Their settlements were sparse. As immigrants from the United States expanded into the region, many interested in what became a booming cattle-raising industry, Mexico found it increasingly difficult to maintain control. By 1850, nearly the entire Southwest was under U.S. control.

By the twentieth century, a vibrant agricultural economy had developed in central and southern California; irrigated agriculture was made possible by massive government-sponsored water-movement projects. The mild Mediterranean climate made it possible to grow vegetables almost year-round. With the advent of refrigerated railroad cars, fresh California vegetables could be sent to the major population centers of the East. Southern California's economy rapidly diversified to include oil (see Figure 2.9E), entertainment, and a variety of engineering- and technology-based industries.

European Settlement and Native Americans

As settlement relentlessly expanded west, Native Americans (called First Nations people in Canada) living in the eastern part of the continent, who had survived early encounters with Europeans, were occupying land that European newcomers wished to use. During the 1800s, almost all the surviving Native Americans were either killed in one of the innumerable skirmishes with European newcomers, absorbed into the societies of the Europeans (through intermarriage and acculturation), or forcibly relocated west to relatively small reservations with few resources. The largest relocation, in the 1830s, involved the Choctaw, Seminole, Creek, Chickasaw, and Cherokee of the southeastern states. These people had already adopted many European methods of farming, building, education, government, and religion. Nevertheless, they were rounded up by the U.S. Army and marched to Oklahoma, along a route that became known as the Trail of Tears because of the more than 4000 who died along the way.

As Europeans occupied the Great Plains and prairies, many of the reservations were further shrunk or relocated onto even less desirable land. Today, reservations cover just over 2 percent of the land area of the United States.

In Canada the picture is somewhat different. Reservations now cover 20 percent of Canada, mostly due to the creation of the Nunavut Territory in 1999 in the far north (now known simply as Nunavut) and the ceding of Northwest Territory land to the Tłı̨chǫ First Nation (also known as the Dogrib) in 2003 (see Figure 2.1 map). These Canadian First Nations stand out as having won the right to legal control of their lands. In contrast to the United States, it had been unusual for native groups in Canada to have legal control of their territories.

After centuries of mistreatment, many Native American and First Nations people still live in poverty, and as in all communities under severe stress, rates of addiction and violence are high, especially in the United States. However, in recent decades some tribes have found avenues to greater affluence on the reservations and territories by establishing manufacturing industries; developing fossil fuel, uranium, and other mineral deposits under their lands; or opening gambling casinos. One measure of this economic resurgence is population growth. Expanding from a low of 400,000 in 1907, the Native American population now stands at approximately 6 million since 2010 in the United States, less than 2 percent of the population. In Canada, First Nations people account for 1,172,785 people, or 3.8 percent of the population.

The Changing Regional Composition of North America

The regions of European-led settlement still remain in North America, but they are now less distinctive. The economic core region is less dominant in industry, which has spread to other parts of the continent. Some regions that were once dependent on agriculture, logging, or mineral extraction now have high-tech industries as well. The West Coast, in particular, has boomed with a high-tech economy and a rapidly growing population that includes many immigrants from Asia and Middle and South America. The West Coast also benefits from trade with Asia, which now surpasses trade with Europe in volume and value.

THINGS TO REMEMBER

- By the late 1600s, New England was implementing ideas and technology from Europe, which led to some of the first industries in North America.
- The push to settle the Great Plains, the Mountain West, and the Pacific Coast attracted many primarily European immigrants.
- By the early twentieth century, North America's economic core was well developed. Stretching from the Atlantic Ocean to St. Louis, Chicago, and Milwaukee, and from Ottawa to Washington, DC, it dominated

North America economically and politically well into the mid-twentieth century.

- In 1492, roughly 18 million Native Americans lived in North America. By 1542, after only a few European expeditions, there were half that many. By 1907, only about 2 percent of the original population remained; however, by 2010, the Native American population had partially rebounded, to approximately 6 million.

CURRENT GEOGRAPHIC ISSUES

A huge regional economy and still-plentiful resources privilege North America, but the region faces complex challenges posed by globalization, an increasingly diverse population, and rising energy and environmental concerns. These long-standing issues must be understood now in the context of a globalized world in which economic downturns impact life everywhere. Meanwhile, both Canada and the United States face mounting pressure to help resolve a wide range of global conflicts and disaffections arising from widening disparities of wealth and opportunity at home and abroad. The two countries address these responsibilities in different ways. At the global level, Canada, with a smaller population yet large resource base, exercises primarily persuasive and diplomatic skills; it dispenses more nonmilitary foreign aid per citizen than does the United States. The United States, because of its huge economy and leadership in geopolitical affairs, tends to take a more proactive global role that often leads to joint military actions with allies against perceived mutual threats. Canada provides more social services for its citizens than does the United States. It also is more energy independent.

Political Issues

Many political issues in this region are linked to national identity and the image North Americans would like their countries to have. Although North Americans do not participate particularly well in voting (an estimated 57.5 percent of voting age U.S. citizens turned out for the presidential election of 2012, and 61.1 percent of voting age Canadians voted in 2011), citizens of both countries are quite active in less formal democratic institutions, such as community service groups. Here we will look at just a few defining political issues of the present day and especially at the ways in which the two countries are similar and different in terms of political culture.

Post-9/11 Geopolitical Relationships

Immediately after the attacks in the United States on September 11, 2001 (9/11), the international community extended warm sympathy to the country and generally supported the strategies of then President George W. Bush. In the fall of 2001, his administration, with the advice and consent of Congress, launched what was called the War on Terror, defined as a defense of the American way of life and of democratic principles. The first target was Afghanistan, which was then thought to be host to Osama bin Laden and the elusive Al Qaeda network that openly claimed credit for masterminding the 9/11 attacks. The aim was to capture bin Laden—accomplished in 2011, under President Barack Obama, with bin Laden killed by U.S. special forces in Abbottabad, Pakistan—and to remake Afghanistan into a stable ally, an aim still unaccomplished by 2013. Although NATO (North Atlantic Treaty Organization) forces (including Canadian troops) joined U.S. forces in Afghanistan, the war proved difficult to resolve because of heavy resistance from tribal leaders within the country and from militant Muslim insurgents in adjacent Pakistan. Some of the loss of momentum in Afghanistan can be attributed to the fact that in the spring of 2003, President Bush brought the War on Terror to Iraq, and U.S. strategic attention and troop support was diverted.

The war in Iraq persisted from 2003 to 2011, with no real resolution. As of December 2011, U.S. troops had withdrawn from Iraq. By 2012, the United States was following a plan of withdrawing troops and operating instead in an advisory role in Afghanistan. Troops are scheduled to be out in 2014, but the advisory role is to continue in some as yet undefined way. Most analysts agree that despite official withdrawals, a large U.S. military presence is likely to remain in both Iraq and Afghanistan for years to come because of the U.S. long-term strategic interests in the larger Central and Southwest Asian regions (**Figure 2.12A, B, and map**).

The United States and Canada Abroad

> **Geographic Insight 3**
> **Power and Politics:** The interpretation of the ideal of democracy by Canada and the United States has influenced North America's involvement in the affairs of countries around the world.

While it is an official policy of both the U.S. and Canadian governments to promote democracy abroad, the two countries approach the project from different perspectives. The United States tends to see itself as the protector of democracy on a global scale, often using an approach that involves the military. Canada, seeking assiduously to distinguish itself from the United States, takes a more "live and let live" approach. Canada's foreign policies and foreign aid projects tend to be geared toward enhancing civil society by making grants for museums, cultural events, and social services that strengthen local identity and citizen participation. U.S. policies are often correlated to the global distribution of its military bases. U.S. foreign aid sometimes is given to projects meant to enhance human well-being, but it can also be in the form of military assistance. Three main concentrations of military bases and spending, where the United States has for years had strong strategic and economic interests, can be seen on the map in Figure 2.12 in Europe; North Africa and Southwest Asia (Egypt, Israel, Iraq, Jordan), and Afghanistan and Pakistan in South Asia; and island and peninsular Southeast Asia. The first and third concentrations of bases and spending (those in Europe and Southeast Asia) relate to strategic and economic interests dating from World War II that have remained relevant due to the Cold War (see Chapter 1, page 36), the subsequent collapse of the Soviet Union (see Chapter 5, pages 211–213), and the opening up of China (see Chapter 9, pages 401–403; see also Figure 2.12C). The second concentration relates directly to U.S. interests in Iraq, Iran, Afghanistan, Pakistan, Israel, and Egypt and the oil and mineral resources of South and Southwest Asia in general.

The map in Figure 2.12 also shows that there are many places where the United States does not have significant bases and where spending on foreign aid is at low or moderate levels. These are generally places where the United States has fewer strategic and economic interests. For example, sub-Saharan Africa's poverty and ongoing political instability has kept U.S. economic interests few, relative to other, wealthier regions (see Figure 2.12D). (The same could be said of Haiti in the Caribbean; see Figure 2.12E). If U.S. policies to support democracy abroad were a strong factor in the allocation of assistance, one might expect that the focus of U.S. spending on foreign aid and military assistance would be in the parts of sub-Saharan Africa and elsewhere that have low levels of democratization. And yet these parts of the world have almost no U.S. bases and only low levels of U.S. spending on foreign aid and military assistance.

39. CONSEQUENCES REPORT

40. SECURITY AND PERSONAL LIBERTY REPORT

41. U.S. IMAGE REPORT

Dependence on Oil Questioned

The 9/11 attacks in 2001 alerted the U.S. public to the fact that the United States was largely dependent on foreign oil. At that time, about 25 percent of imported oil came from the Organization of Petroleum Exporting Countries (OPEC), dominated by countries along the Persian Gulf, where the terrorists had originated. Roughly 46 percent of oil imported into the United States comes from countries that are not part of OPEC (Figure 2.13A on page 78). Of this 46 percent, Canada supplies about 21 percent and Mexico about 12 percent (see Figure 2.13A), with 20 other countries supplying only tiny percentages. Meanwhile, Canada, which has the potential to be petroleum independent, actually imports nearly 44 percent of its oil. International oil corporations dominate the oil industry in both countries. Most of Canada's oil fields are in the west, and at least half of that oil is exported to the United States, where it is refined. Some stays in the United States, and some is sold back to the more heavily populated eastern provinces. The rest of oil

Figure 2.12 Photo Essay: Power and Politics in North America

Thinking Geographically

After you have read about power and politics in North America, you will be able to answer the following questions:

A What economic interests does the United States have in Europe?

B What strategic and economic interests make it unlikely that the United States will withdraw completely from Afghanistan and Iraq in the near future?

D What has kept U.S. economic interests in Africa at a low level relative to those in other regions?

E What has kept U.S. economic interests in Haiti at a low level relative to those in other countries?

Political freedoms are well protected and democratization is at a relatively high level in North America. On a global scale, it is often argued that the United States uses its power to promote political freedoms and democratization. There is some truth to this. On the other hand, and as just one example, some U.S. officials who planned the Iraq War point out that U.S. strategic and economic interests influenced U.S. foreign aid more than the promotion of democracy did. Nongovernment analysts point especially to the desire to control Iraq's oil resources and those of its neighbors. If the main U.S. interest abroad were the promotion of democracy, then sub-Saharan Africa would be a major focus of U.S. foreign aid and military installations, especially given the many violent conflicts that plague this region. As the map below shows, sub-Saharan Africa receives relatively little aid and has few U.S. military bases. Nevertheless, the United States has increased its presence there in recent years.

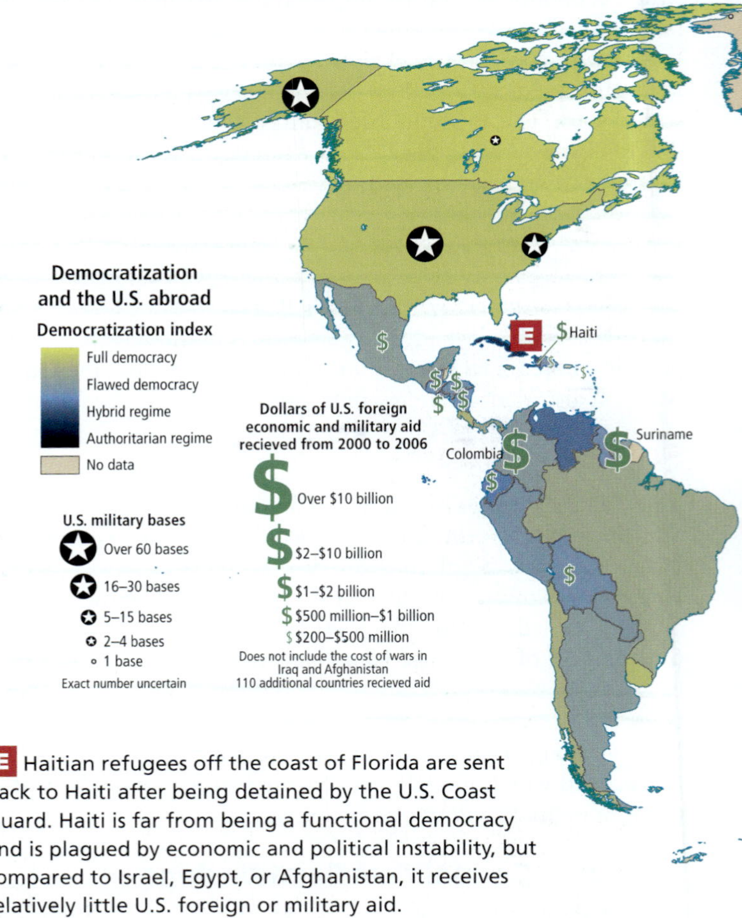

E Haitian refugees off the coast of Florida are sent back to Haiti after being detained by the U.S. Coast Guard. Haiti is far from being a functional democracy and is plagued by economic and political instability, but compared to Israel, Egypt, or Afghanistan, it receives relatively little U.S. foreign or military aid.

A A Canadian soldier wounded in Afghanistan is lifted off a plane at Ramstein Air Base, one of 260 U.S. bases in Germany. The large U.S. military presence in Europe is a legacy of World War II and the Cold War era that followed. The purpose of the military presence was to discourage potential aggression from the (now-defunct) Soviet Union against U.S. allies in Western Europe.

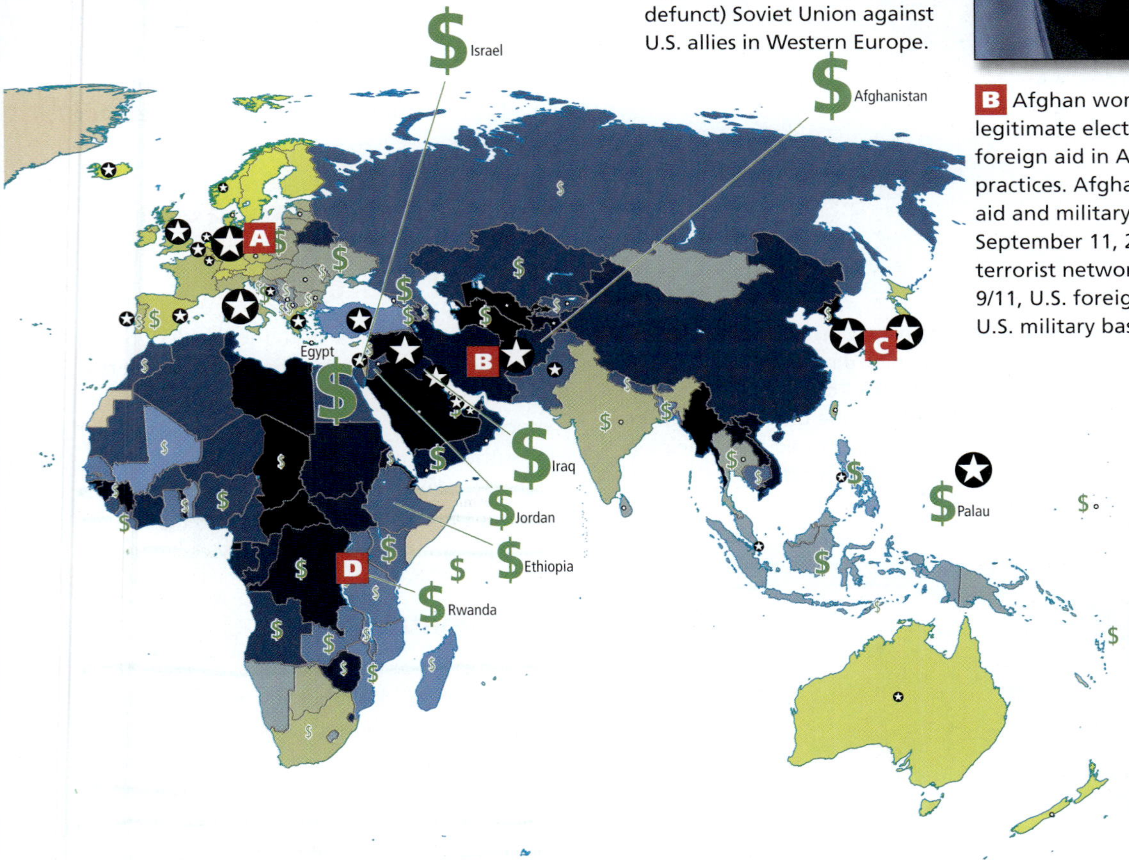

B Afghan women line up to vote in the country's first legitimate elections in decades. A major part of U.S. foreign aid in Afghanistan is used to promote democratic practices. Afghanistan became a major recipient of U.S. aid and military intervention only after the attacks of September 11, 2001 (masterminded by the Al Qaeda terrorist network, based, in part, in Afghanistan). Before 9/11, U.S. foreign aid to Afghanistan was much lower and U.S. military bases there were nonexistent.

C A U.S. military base in Japan, one of 130 in the entire country, is surrounded by dense urban development. These bases are first and foremost a projection of U.S. power designed to counter any future aggression by China, North Korea, the former Soviet Union, or by Japan and South Korea themselves.

D A Kenyan police officer investigates a burning road block set up in the aftermath of disputed presidential elections in 2013. Democracy is fragile throughout much of sub-Saharan Africa, and elections are often plagued by violence. While the United States is giving increasing support to democracy in sub-Saharan Africa, U.S. foreign aid to the area is relatively small and historically has done little to promote democracy in the region. U.S. military bases are few.

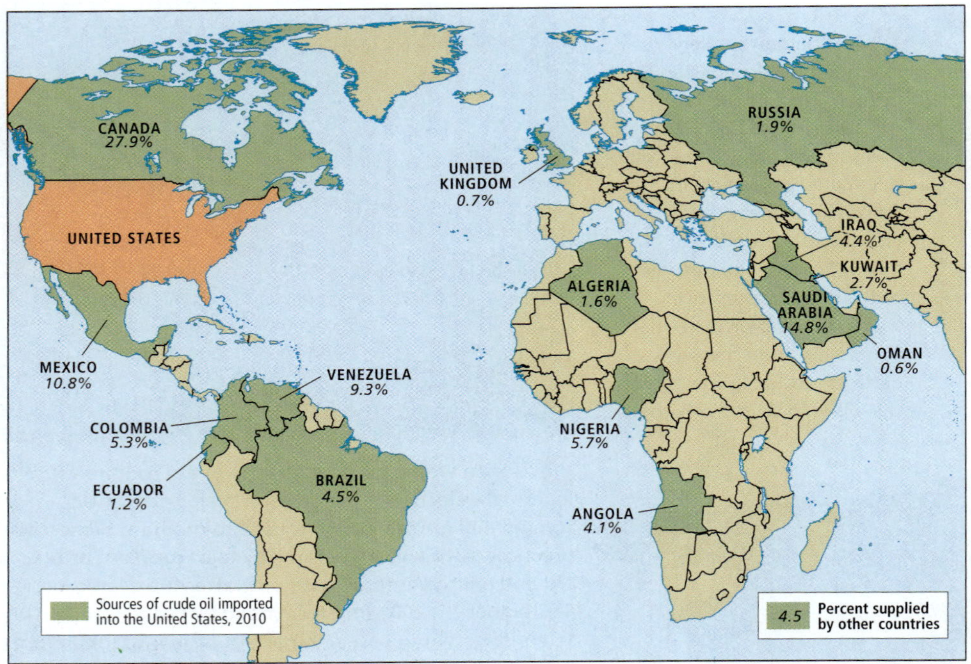

FIGURE 2.13 (A) Sources of average daily crude oil imported into the United States in September 2012. The United States is dependent on crude oil from many locations around the world and imports nearly 60 percent of the crude it uses. This map shows the percentage of average daily crude (millions of barrels) imported into the United States from the top 15 exporting nations in September 2012. The United States produced the remaining 40 percent from its own sources. Canada's exports into the United States rose from 21 percent in September 2009 to nearly 27 percent by September 2012.

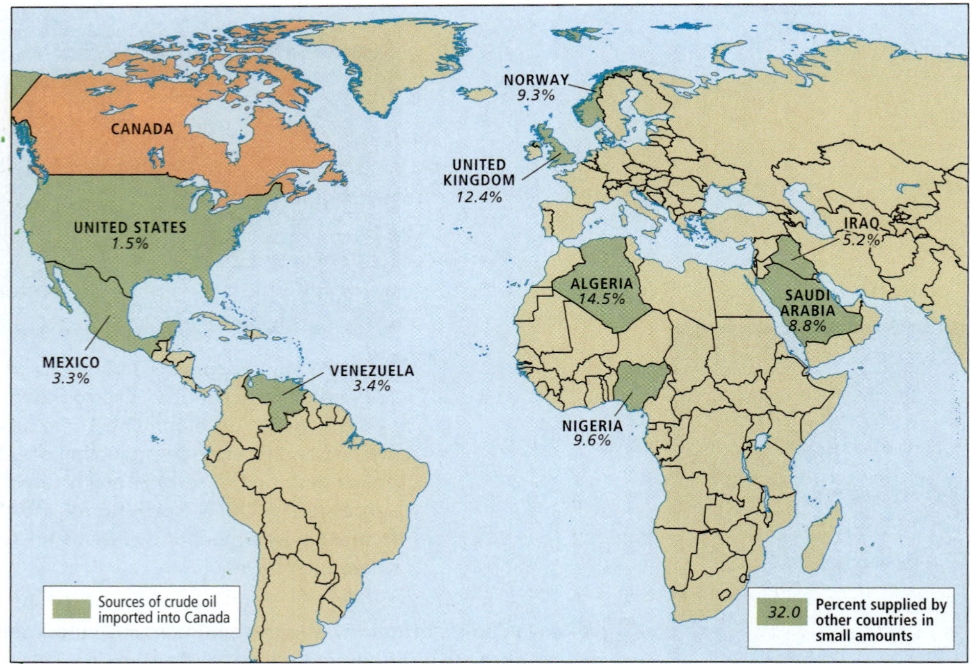

FIGURE 2.13 (B) Sources of average daily crude oil imported into Canada in 2010. In 2010, Canada produced 56.1 percent of the crude oil it used from its own resources, with most coming from western Canada. More than half of the crude oil from western Canada was shipped to the United States, with western Canada receiving most of the rest. However, Canada also imported 43.9 percent of its crude oil, with most of the imported crude going to the eastern provinces.

imported into eastern Canada comes mostly from OPEC sources (see Chapter 1, page 8). This rate of dependency makes many in both Canada and the United States uncomfortable, especially given the ongoing potentially antagonistic relationships with the Persian Gulf states. Finding more domestic sources of energy has become a major issue for the United States. This has boosted interest in renewable energy, as well as increased pressure in the United States to allow extraction of oil and gas from shale deposits and to drill for oil in coastal waters—though this last strategy was dealt a blow by the massive spill in 2010 involving the Deepwater Horizon, the British Petroleum installation in the Gulf of Mexico.

Relationships Between Canada and the United States

Citizens of Canada and the United States share many characteristics and concerns. Indeed, in the minds of many people—especially those in the United States—the two countries are one. Yet that is hardly the case. Three key factors characterize the interaction between Canada and the United States: *asymmetries*, *similarities*, and *interdependencies*.

Asymmetries

Asymmetry means "lack of balance." Although the United States and Canada occupy about the same amount of space (Figure 2.14), much of Canada's territory is cold and sparsely inhabited. The U.S. population is about ten times the Canadian population. While Canada's economy is one of the largest and most productive in the world, producing U.S.$1.4 trillion (PPP) in goods and services in 2011, it is dwarfed by the U.S. economy, which is more than ten times larger, at $15.04 trillion (PPP) in 2011.

In international affairs, Canada quietly supports civil society efforts abroad. The United States is an economic, military, and political superpower preoccupied with maintaining a world leadership role. In framing foreign affairs policy, the United States references Canada's position only as an afterthought, in part because the country is so secure an ally. But for Canada, managing its relationship with the United States is a top foreign policy priority. As former Canadian Prime Minister Pierre Trudeau once told the

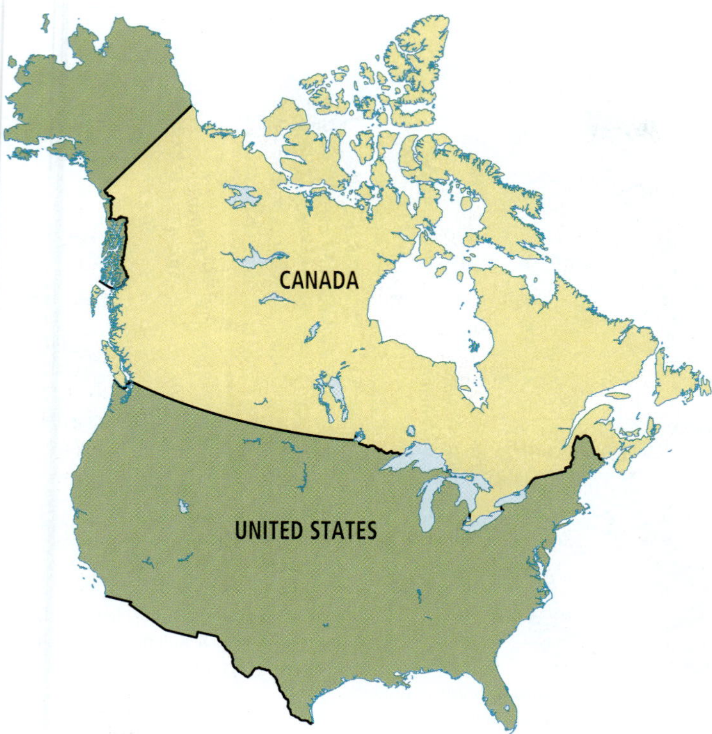

FIGURE 2.14 Political map of North America.

U.S. Congress, "Living next to you is in some ways like sleeping with an elephant: No matter how friendly and even-tempered the beast, one is affected by every twitch and grunt." In 2008, for example, the robust, debt-free Canadian economy slowed abruptly, due in large part to the slump in U.S. consumer spending, which hurt Canadian exports such as oil, gas, and cars.

Similarities

Notwithstanding the asymmetries, the United States and Canada have much in common. Both are former British colonies and have retained English as the dominant language. Both also experienced settlement and exploration by the French. From their common British colonial experience, they developed comparable democratic political traditions. Both are federations (of states or provinces), and both are representative democracies, with similar legal systems.

Not the least of the features they share is a 4200-mile (6720-kilometer) border, which until 2009 contained the longest sections of unfortified border in the world. For years, the Canadian border had just 1000 U.S. border guards, while the Mexican border, which is half as long, had nearly 10,000 agents. In 2009, the Obama administration decided to equalize surveillance of the two borders for national security reasons. Where some rural residents used to pass unobserved in and out of Canada and the United States many times in the course of a routine day, now there are drone aircraft with night-vision cameras and cloud-piercing radar scanning the landscape for smugglers, illegal immigrants, and terrorists. Canadians and Americans, who formerly thought of their common border as nonexistent, now endure intrusive technology, barricades, and crossing delays. Tourism and commerce are also negatively affected by these efforts to keep both countries safer. **42. U.S. BORDER SECURITY REPORT**

Well beyond the border, Canada and the United States share many other landscape similarities. Their cities and suburbs look much the same. The billboards that line their highways and freeways advertise the same brand names. Shopping malls and satellite business districts have followed suburbia into the countryside, encouraging similar patterns of mass consumption and urban sprawl. The two countries also share similar patterns of ethnic diversity that developed in nearly identical stages of immigration from abroad.

Interdependencies

Canada and the United States are perhaps most intimately connected by their long-standing economic relationship. The two countries engage in mutual tourism, direct investment, migration, and most of all, trade. The equivalent of nearly U.S.$1.5 billion is traded daily between the two countries. Canada is a larger market for U.S. goods than are all 27 countries in the European Union. By 2005, that trade relationship had evolved into a two-way flow of U.S.$1 trillion annually (Figure 2.15 on page 80). Each country is the other's largest trading partner. In 2010, seventy-five percent of Canada's imports came from the United States and 75 percent of its exports went to the United States. The United States, in turn, sells 18.97 percent of its exports to Canada and buys 14.34 percent of its imports from Canada.

Notice, however, that there is asymmetry even in the realm of interdependencies: Canada's smaller economy is much more dependent on the United States than the reverse. Nonetheless, as many as 1 million U.S. jobs are dependent on the relationship with Canada.

Democratic Systems of Government: Shared Ideals, Different Trajectories

Canada and the United States have similar democratic systems of government, but there are differences in the way power is divided between the federal government and provincial or state governments. There are also differences in the way the division of power has changed since each country achieved independence.

Both countries have a federal government, in which a union of states (provinces, in Canada) recognizes the sovereignty of a central authority, while many governing powers are retained by state/provincial or local governments. In both Canada and the United States, the federal government has an elected executive branch, elected legislatures, and an appointed judiciary. In Canada, the executive branch is more closely bound to follow the will of the legislature. At the same time, the Canadian federal government has more and stronger powers (at least constitutionally) than does the U.S. federal government.

Over the years, both the Canadian and U.S. federal governments have moved away from the original intentions of their constitutions. Canada's originally strong federal government has become somewhat weaker. This change is largely in response to demands by provinces, such as the French-speaking province of Québec (see the map in Figure 2.1), for greater autonomy over local affairs.

Meanwhile, the initially more limited federal government in the United States has expanded its powers. The U.S. federal

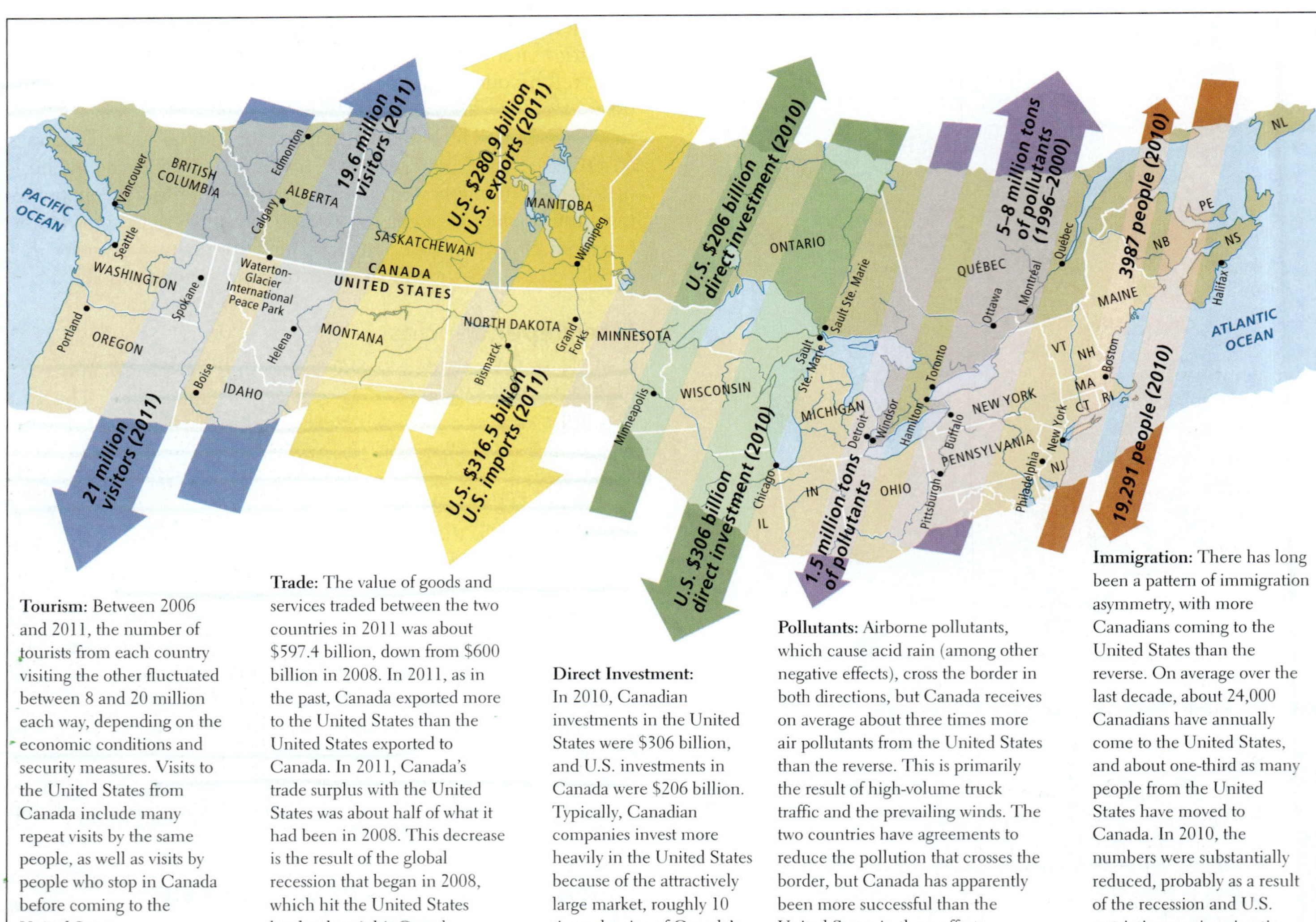

FIGURE 2.15 Transfers of tourists, goods, investment, pollution, and immigrants between the United States and Canada. Canada and the United States have the world's largest trading relationship. The flows of goods, money, and people across the long Canada–U.S. border are essential to both countries. However, because of its relatively small population and economy, Canada is more reliant on the United States than the United States is on Canada. All amounts shown are in U.S. dollars.

government's original source of power was its mandate to regulate trade between states. Over time, this mandate has been interpreted ever more broadly. Now the U.S. federal government powerfully affects life even at the local level, primarily through its ability to dispense federal tax monies via such means as grants for school systems, federally assisted housing, military bases, drug and food regulation and enforcement, urban renewal, and the rebuilding of interstate highways. Money for these programs is withheld if state and local governments do not conform to federal standards. This practice has made some poorer states dependent on the federal government. However, it has also encouraged some state and local governments to enact more enlightened laws than they might have done otherwise. For example, in the 1960s the federal government promoted civil rights for African American citizens by requiring states to end racial segregation in schools in order to receive federal support for their school systems.

The Social Safety Net: Canadian and U.S. Approaches

The Canadian and U.S. governments have responded differently to the displacement of workers by economic change. Ultimately, these differences derive from prevailing political positions and widely held notions each country has about what the government's role in society should be. In Canada there is broad political support for a robust **social safety net**, the services provided by the government—such as welfare, unemployment benefits, and health care—that prevent people from falling into extreme poverty. In the United States there is much less support for these programs and a great deal of contention over nearly all efforts to strengthen the U.S. social safety net.

social safety net the services provided by the government—such as welfare, unemployment benefits, and health care—that prevent people from falling into extreme poverty

For many decades, Canada has spent more per capita than the United States on social programs. These programs, especially unemployment benefits, have generally made the financial lives of working Canadians more secure. These policies also reflect the workings of Canada's democracy; voters have supported tax hikes to fund several major expansions of the social safety net during the twentieth century.

Compared to Canada, the United States provides much less of a social safety net. For years the prevailing political argument against expansion of the U.S. social safety net has been that the system offers lower taxes for businesses, corporations, and the wealthy, which in turn invest their surpluses in expansions or in new businesses, thus creating jobs and new taxpayers. These benefits of low taxes are assumed to *trickle down* to those most in need. The fit of this position with the reality of what actually happens has long been challenged, especially during times of economic recession.

Two Health-Care Systems The contrasts between health-care systems in the two countries are indeed striking. The Canadian system, reformed in the 1970s, is heavily subsidized and covers 100 percent of the population. In the United States, health care has been largely private, relatively expensive, and tied to employment, which made changing jobs difficult for those with preexisting conditions. Also, many small businesses did not provide health insurance coverage for employees, who had to pay for it themselves or go without. A recent reform, dubbed "Obamacare" after President Obama, who pushed for the change, will cover most of the 47 million people who previously had no coverage; and government subsidies for care for the elderly, the disabled, children, veterans, and the poor will continue.

The financial viability and medical outcomes of Obamacare are not yet known because the plan will not be fully implemented until 2014, but the most recent statistics show why change was needed. In 2011, the United States spent more per capita on health care than any other industrialized country—$8680 per capita, for a total of 17.9 percent of GDP. Canada spent $5948 per capita, or just 11.6 percent of its GDP (GNI figures are not available)—yet Canada had better health outcomes, outranking the United States on most indicators of overall health, such as infant mortality, maternal mortality, and deaths per 1000 (Table 2.1). **48. SICKO REPORT**

Gender in National Politics

North America has some powerful political contradictions with regard to gender. While women voters are an ever more potent political force, successful female politicians are not as numerous as one might expect. Women cast the deciding votes in the U.S. presidential election of 1996, voting overwhelmingly for Bill Clinton. In the 2006 interim elections, 55 percent of women voted for the Democratic Party candidates, registering strong anti-war sentiment and making it possible for the Democratic Party to gain many seats in Congress. In 2008 and 2012, Barack Obama won 55 percent of the female vote. Deciding factors in the election included issues that are often priorities for women, such as reproductive rights, health care, family leave, equal pay, day care, and ending the wars in Iraq and Afghanistan. **49. WOMEN VOTING REPORT**

And yet, as elected political leaders, women have not had the success warranted by their powerful role as voters. Of the 535 people in the U.S. Congress as of 2013, only 98 members, or 18.3 percent, were women. Gender equity was a bit closer at the state level, where 24.1 percent of legislators were women as of the 2012 elections. In Canada, women have had somewhat more success. As of 2012, Canadian women constituted 25 percent of the House of Commons in Parliament and held 24 percent of provincial legislative seats. Neither country does well by comparison to the world at large in that, as Figure 2.16 (see page 82) shows, by 2012 many countries have added substantial female representation to legislatures.

Things are somewhat more equitable for Canadian women at the executive level. In 2011, Canada elected four women as provincial governors and another as executive of Nunavut. (There are ten provinces and three territories.) After the 2012 election, the United States had 7 out of a possible 50 female governors. Canada briefly had a female prime minister in 1993. While the

TABLE 2.1 Health-related indexes for Canada and the United States

Country	Health care cost as a percentage of GDP	Percentage of population with no insurance	Deaths per 1000[†]	Infant mortality per 1000 live births in 2009[*]	Maternal mortality per 100,000 live births in 2008[‡]	Life expectancy at birth (years) in 2011[*]	Annual health expenditures per capita (PPP U.S.$) in 2009
Canada	10.9	0	7	6	7	78.5	$5948[§]
United States	16.2	15.5	8	8	17	81	$8680[**]

Sources: [*]United Nations Development Program, *Human Development Report 2011* (New York: United Nations Development Programme), Tables 1, 9, and 10, at http://hdr.undp.org/en/reports/global/hdr2011/. Retrieved January 2012.

[†]Population Reference Bureau, *2011 World Population Data Sheet*, at http://www.prb.org/Publications/Datasheets/2011/world-population-data-sheet/data-sheet.aspx. Retrieved January 2012.

[‡]Medical News Today, "Maternal Mortality Rises in the USA, Canada and Denmark, and Falls in China, Egypt, Ecuador and Bolivia," April 13, 2010, at http://www.medicalnewstoday.com/articles/185154.php.

[§]News Medical, "Health Care Spending in Canada Expected to Reach $183.1 Billion in 2009: CIHI," November 19, 2009, at http://www.news-medical.net/news/20091119/Health-care-spending-in-Canada-expected-to-reach-241831-billion-in-2009-CIHI.aspx.

[**]Centers for Medicare & Medicaid Services, "National Health Expenditures," at https://www.cms.gov/NationalHealthExpendData/25_NHE_Fact_Sheet.asp#TopOfPage. Retrieved July 2012.

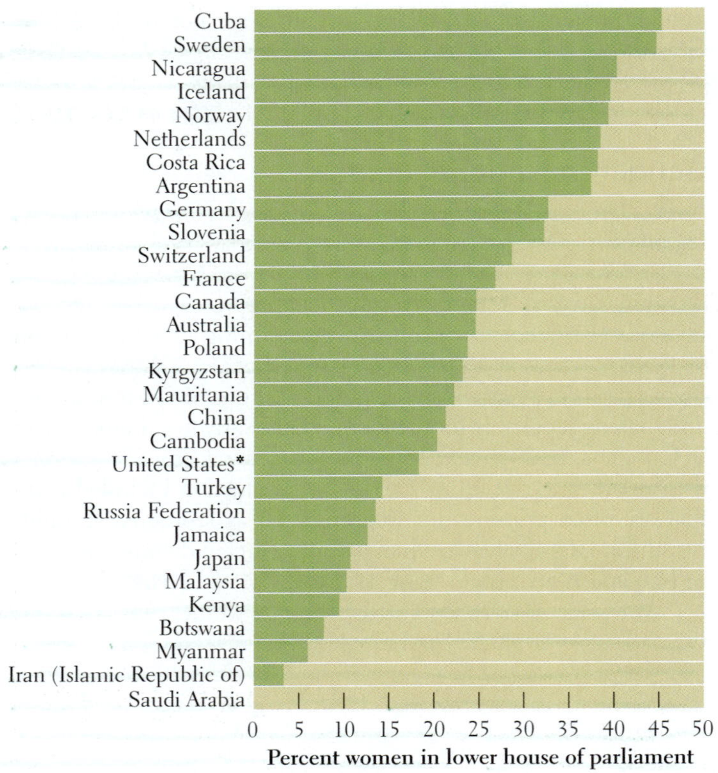

* Percentage for the entire U.S. Congress is used, which has 100 senators (20 are women) and 435 congresspeople (77 are women)

FIGURE 2.16 Percentage of women in the lower house of Parliament (legislature) in selected countries in October 2012.

United States has never had a female president, in 2008 Hillary Clinton was the first woman to have a serious chance at becoming the U.S. president when she competed with Barack Obama to become the Democratic Party candidate.

THINGS TO REMEMBER

Geographic Insight 3

- **Power and Politics** Canada and the United States are vocally proud of their democratic traditions. While promoting democracy abroad is an oft-stated ideal of U.S. official foreign policy (it is less of one for Canada), in actuality the global distribution of U.S. spending on foreign aid and military assistance indicates that U.S. strategic and economic interests take precedence over promoting democracy.
- Canada continues to have a more robust social safety net, including health care, than does the United States.
- While women voters are an ever more potent political force in the United States and Canada, the number of successful female politicians has not been commensurate with the percentage of women in the general population of each country.

Economic Issues

The economic systems of Canada and the United States, like their political systems, have much in common. Both countries evolved from societies based mainly on family farms. Then came an era of industrialization, followed by a move to primarily service-based economies. Both countries have important technology sectors and their economic influence reaches worldwide.

North America's Changing Food-Production Systems

Agriculture remains the spatially dominant feature of North American landscapes, yet less than 1 percent of North Americans are engaged directly in agriculture. North America benefits from an abundant supply of food; it produces food for foreign as well as domestic consumers (Figure 2.17). At one time, exports of agricultural products were the backbone of the North American economy. However, because of growth in other sectors, agriculture now accounts for less than 1.2 percent of the United States's GDP and less than 2 percent of Canada's. Moreover, because both countries are involved in the global economy, food in both countries is increasingly being imported. 43. FOOD GLOBALIZATION

The shift to mechanized agriculture in North America brought about sweeping changes in employment and farm management. In 1790, agriculture employed 90 percent of the American workforce; in 1890, it employed 50 percent. Until 1910, thousands of highly productive family-owned farms, located over much of the United States and southern Canada, provided for most domestic consumption and the majority of all exports. Today, many family farms are being replaced by corporation-owned farms.

Family Farms Give Way to Agribusiness When family farms began to mechanize in the late nineteenth century with mechanical corn-seed planters and steam-powered threshing machines, such inventions reduced the need for labor to just family members and expanded the amount of land that a family farm could manage. To remain profitable over time, ever larger investments were required in land, sophisticated machinery, fertilizers, pesticides, and farm buildings. By 2000, only younger, wealthier farmers were able to invest in these "green revolution" methods (see Chapter 1, page 26). Some prospered (see the vignette on page 83), while some with fewer resources chose to sell their land, hoping to make a profit sufficient for retirement. Indebtedness and bankruptcies are increasingly common experiences for North America's small farmers. Very recently, some family farmers have been able to earn added income by leasing the mineral rights of their land for natural gas production.

A major part of the transformation of North American agriculture has been the growth of large **agribusiness** corporations that sell machinery, seeds, and chemicals. These corporations may themselves produce crops on land they bought from retiring farmers, or they may contract with independent farms to produce crops for the corporation. Agribusiness has facilitated the transition to green revolution methods. The corporations' financial resources have enabled them to invest in research, develop new products, and provide loans and

agribusiness the business of farming conducted by large-scale operations that purchase, produce, finance, package, and distribute agricultural products

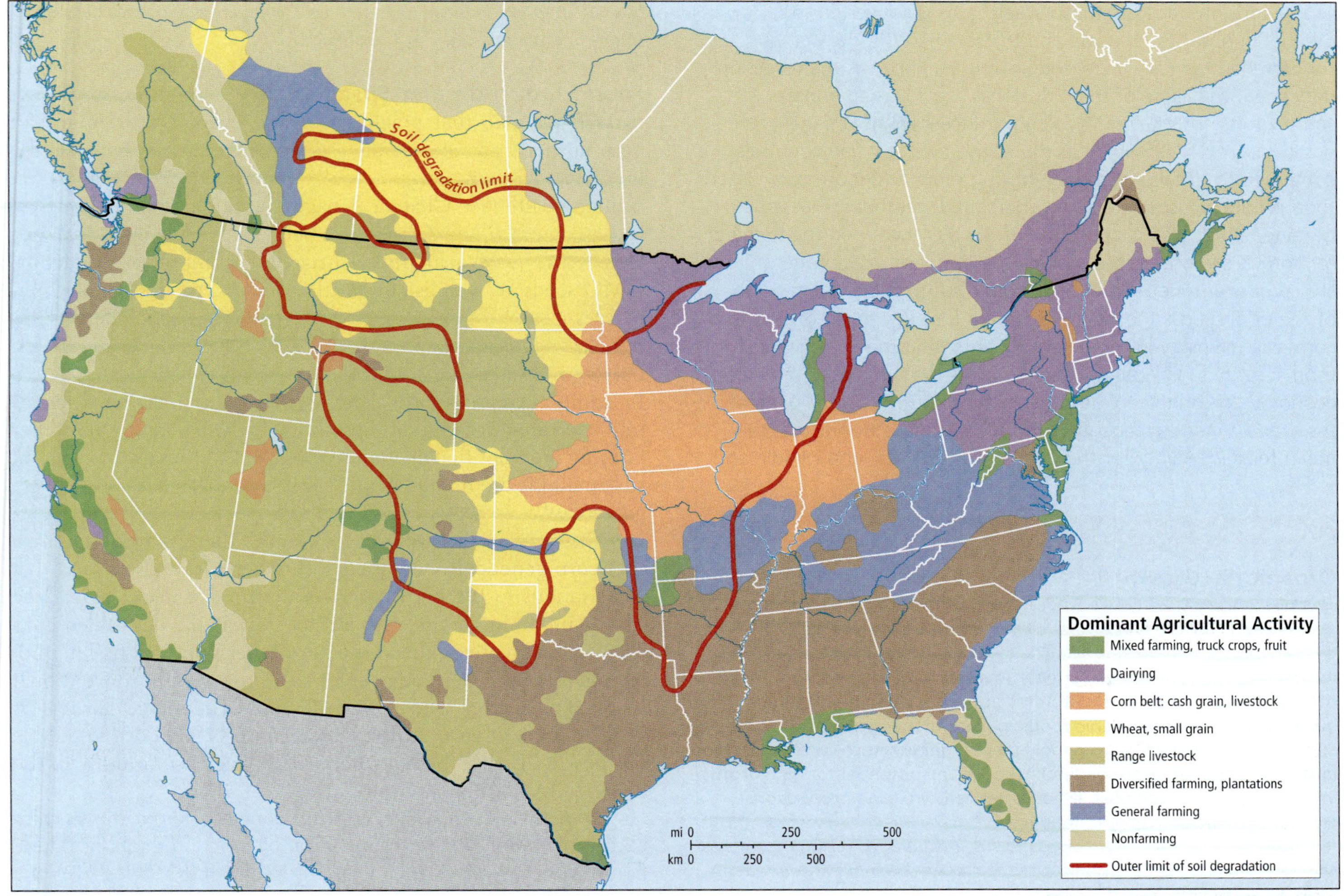

FIGURE 2.17 Agriculture in North America. Throughout much of North America, some type of agriculture is possible. The major exceptions are the northern reaches of Canada and Alaska and the dry mountain and basin region (the Continental Interior) that lies between the Great Plains and the Pacific coastal zone. However, in some marginal areas, such as Southern California, southern Arizona, and the Utah Valley, irrigation is needed for cultivation. (Hawaii is not included here because it is covered in Oceania, Chapter 11.)

cash to individual farmers. Over time, many farmers have been pressured to enter into contracts and other relationships with agribusiness corporations, leaving these farmers with little actual control over which crops are grown and what farming methods are used.

While green revolution agriculture provides a wide variety of food at low prices for North Americans, the shift to these production methods has depressed local economies and created social problems in many rural areas. Communities in places such as rural Alberta and Saskatchewan in Canada, and Iowa, Nebraska, the Dakotas, and Kansas in the United States were once made up of farming families with similar middle-class incomes, social standing, and commitment to the region. Today, farm communities are increasingly composed of a few wealthy farmer-managers amid a majority of poor, often migrant Hispanic or Asian laborers who work on large farms and in food-processing plants for wages that are too low to provide a decent standard of living. They also struggle to be accepted into the communities where they live.

VIGNETTE The flip side of this story of the decline in North America of the family farm way of life is represented by the burgeoning prosperity of farmers like 37-year-old Twitter user Brandon Hunnicutt, who runs his family's 3600 acre (1457 hectare) farm above the vast Ogallala Aquifer in Nebraska. He uses the aquifer water to irrigate in dry times and his computer to garner the latest information on nearly every aspect of his operation—from the diagnosis of a crop disease to the right moment for marketing his crops. In the midst of the global recession, he was selling corn at twice the price he received a few years before.

How can this be? In 2007, the rising price of petroleum products spurred the U.S. government to require that ethanol made from corn be added to gasoline in order to lower dependency on foreign energy sources. Now, more than one-third of the corn produced in the United States is converted to ethanol in order to fuel cars and trucks. By 2008, the demand for corn had radically increased, as had its price. As the recession deepened, the value of the U.S. dollar was lowered in order to spur exports. This gave U.S. corn producers an advantage in the world market. Meanwhile, as the economies of Asian countries continued to grow, Asia began buying more and more corn-based food products, especially sweeteners. Bad weather in Russia and Ukraine, which also produce grains, further increased the demand for corn. Corn prices rose globally, to the great advantage of large U.S. producers like Brandon Hunnicutt but to the great anguish of the poor across the world, who had formerly relied on cheap corn as a mainstay in their diets. As shown in the discussion of food security in Chapter 1 (page 24), the vicissitudes of the global economy can bring very different outcomes to different parts of the world. Financial security for farmers like the Hunnicutts can mean *food insecurity* for the poor in Africa, Asia, and Middle and South America. *[Source: NPR staff, All Things Considered. For detailed source information, see Text Credit pages.]*

ON THE BRIGHT SIDE

Organically Grown

Throughout North America, there is a modest but growing revival of small family farms that supply **organically grown** (produced without chemical fertilizers and pesticides) vegetables and fruits and grass-fed meat directly to consumers. Farmers' markets have popped up across the country (Figure 2.18), and customers in ever greater numbers are now buying locally grown organic foods, paying higher prices than those in traditional grocery stores. This movement is gaining such wide favor that large corporate farms are beginning to see the potential for high profits in sustainable (i.e., organic), if not locally grown, food production.

Greater interest in sustainable food production may reduce harmful impacts on the environment as farms shift over to organic methods and consumers learn the advantages of paying more for higher quality, toxin-free vegetables and fruits.

FIGURE 2.18 A Place of the Heart. A Tennessee wife-and-husband team sell produce from their farm, A Place of the Heart, at a local farmers' market. They also deliver a basketful of vegetables each week to a set of customers who pay a seasonal fee of $700 for the produce.

Food Production and Sustainability Can green revolution agriculture, like that practiced by successful North American corn farmers, persist over time? First, there is concern over the long-term impacts of genetically modified (GMO) crops on human health (see Chapter 1, page 27). Such crops are too new to have been adequately tested. Second, modernized strategies to increase yields—including irrigation, computerized tractors, GMO seeds, as well as chemical fertilizers, pesticides, and herbicides—can have negative environmental impacts. These methods can contaminate food, use scarce water resources, pollute nearby streams and lakes, and even affect distant coastal areas. In addition, many North American farming areas have lost as much as one-third of their topsoil due to deep plowing and other farming methods that create soil erosion. But these worries about sustainability are not the whole story. Scientists have shown that, given a little time, they can adjust green revolution technologies to enhance sustainability. For example, there is evidence now that corn genetically modified with a gene from a bacteria that is poisonous to the corn borer—a destructive insect that came from Europe in about 1917 and has repeatedly devastated corn crops—has greatly reduced the corn borer population. The protection has spread to adjacent fields grown with conventional seeds. The GMO seed companies now encourage farmers to grow traditional corn in fields adjacent to GMO corn to reduce the possibility that the corn borer will develop immunity to the genetically modified poison.

Nonetheless, there are growing concerns about the sustainability of rapidly changing patterns of food production. These concerns, expressed in a variety of books and articles over the last two decades, point out that federal subsidies to American agriculture have encouraged the drift to factory farms where animals are bred, tended to in crowded conditions, and fed chemicals to enhance weight gain. Vast quantities of animal waste produce environmental hazards. Furthermore, the current subsidized system has encouraged the monoculture of corn, soybeans, wheat, and rice, leading to the production of high-sugar, high-carbohydrate foods that contribute to obesity and diabetes. The critiques coalesce around the idea of changing the subsidy structure to favor small farmers, encouraging them to go back to more traditional methods, at least in food agriculture.

Changing Transportation Networks and the North American Economy

An extensive network of road and air transportation, enabling high-speed delivery of people and goods, is central to the productivity of North America's economy. The development of this system began as early as 1800 (see Figure 2.11) and edged toward a nationwide network with the development of inexpensive mass-produced automobiles in the 1920s. Where railroads and waterborne shipping had once dominated transportation, trucks could now deliver cargo more quickly and

organically grown products produced without chemical fertilizers and pesticides

conveniently. Beginning in the 1950s, automobile- and truck-based transport was boosted by the Interstate Highway System—a 45,000-mile (72,000-kilometer) network of high-speed, multilane roads. Because this network is connected to the vast system of local roads, it can be used to deliver manufactured products faster and with more flexibility than can be done using the rail system. Thus the highways have made possible the dispersal of industry and related services into suburban and rural locales across the country, where labor, land, and living costs are lower.

After World War II, air transportation also enabled economic growth in North America. The primary niche of air transportation is business travel, because face-to-face contact remains essential to American business culture despite the growth of telecommunications and the Internet. Because many industries are widely dispersed in numerous medium-size cities, air service is organized as a *hub-and-spoke network*. Hubs are strategically located airports, such as those in Atlanta, Chicago, Dallas, and Los Angeles. These airports are used as collection and transfer points for passengers and cargo continuing on to smaller locales. Most airports are also located near major highways, which provide an essential link for high-speed travel and cargo shipping.

The New Service and Technology Economy

> **Geographic Insight 4**
> **Globalization and Development:** Globalization has transformed economic development in North America, thereby changing trade relationships and the kinds of jobs available in this region.

As the trend toward global economic interdependency has taken hold, the North American job market has become more oriented toward knowledge-intensive jobs that require education and specialized professional training, often in high-tech fields or management. Meanwhile, the relatively low-skill, mass-production industrial jobs upon which the region's middle class was built have increasingly moved abroad.

46. U.S. GEOGRAPHY REPORT

The Decline in Manufacturing Employment By the 1960s, the geography of manufacturing was changing. In the old economic core, higher pay and benefits and better working conditions won by labor unions led to increased production costs. Many companies began moving their factories to the southeastern United States, where wages were lower due to the absence of labor unions. **44. U.S. LABOR TRANSITION REPORT**

In 1994, the **North American Free Trade Agreement (NAFTA)** was passed. In response, many manufacturing industries, such as clothing, electronic assembly, and auto parts manufacturing, began moving farther south to Mexico or overseas. In these locales, labor was vastly cheaper. Furthermore, employers saved on production costs because laws mandating environmental protection as well as safe and healthy workplaces were absent or less strictly enforced.

Another factor in the decline of manufacturing employment is automation. The steel industry provides an illustration. In 1980, huge steel plants, most of them in the economic core, employed more than 500,000 workers. At that time, it took about 10 person-hours and cost about $1000 to produce 1 ton of steel. Spurred by more efficient foreign competitors in the 1980s, 1990s, and 2000s, the North American steel industry applied new technology to lower production costs, improve efficiency, and increase production. By 2006, steel was being produced at the rate of 0.44 person-hours per ton and at a cost of about $165 per ton. As a result, the steel industry in the United States reorganized with much steel produced in small, highly efficient mini-mills. In total, the steel industry now employs fewer than half the workers it did in 1980. Throughout North America, this efficiency trend has resulted in far fewer people producing more of a given product at a far lower cost than was the case 30 years ago. Remarkably, even as employment in manufacturing has declined over the last three decades, the actual amount of industrial production has steadily increased.

Growth of the Service Sector The economic base of North America is now a broad *tertiary sector* in which people are engaged in various services such as transportation, utilities, wholesale and retail trade, health, leisure, maintenance, finance, government, information, and education.

As of 2012, in both Canada and the United States, about 75 percent of jobs and a similar percentage of the GNI were in the tertiary, or service, sector. There are high-paying jobs in all the service categories, but low-paying jobs are far more common. The largest private employer in the United States is the discount retail chain Walmart (1.4 million employees in 2011), where the average wage is $12 an hour, or $24,000 a year, full time. This is just barely above the poverty level for a family of four in the United States. Because many Walmart employees are part time, a large percentage do not receive benefits, such as health care. Walmart creates primarily retail jobs because, for the most part, its wares are manufactured abroad.

Service jobs are often connected in some way to international trade. They involve the processing, transport, and trading of agricultural and manufactured products and related information that are either imported to or exported from North America. Hence, international events can shrink or expand the numbers of these jobs rather precipitously.

The Knowledge Economy An important subcategory of the service sector involves the creation, processing, and communication of information—what is often labeled the *knowledge economy*, or the *quaternary sector*. The knowledge economy includes workers who manage information, such as those employed in finance, journalism, higher education, research and development, and many aspects of health care.

Industries that rely on the use of computers and the Internet to process and transport information are freer to locate where they wish than were the manufacturing industries of the old economic core, which

> **North American Free Trade Agreement (NAFTA)** a free trade agreement made in 1994 that added Mexico to the 1989 economic arrangement between the United States and Canada

depended on locally available steel and energy, especially coal. These newer industries are more dependent on highly skilled managers, communicators, thinkers, and technicians, and are often located near major universities and research institutions. They may also be located in clusters of companies involved with the subset of the knowledge economy, known as the *information technology* (or *IT*) sector, which deals with computer software, hardware, and the management of digital data.

Crucial to the knowledge economy is the Internet, which was first widely available in North America and has emerged as an economic force more rapidly there than in any other region in the world. China surpassed North America in actual numbers of Internet users in about 2005. It now has 23 percent of the world's users; however, only about 36 percent of China's population has Internet access, the Internet is tightly controlled in China, and purchasing via the Internet has not yet taken off there. North America, with only 5 percent of the world's population, accounted in 2011 for 11.6 percent of the world's Internet users. Roughly 78 percent of the population of the United States uses the Internet, as does 85 percent of the Canadian population, compared to 67 percent in the European Union and 30.2 percent for the world as a whole. The total economic impact of the Internet in North America is hard to assess, but retail Internet sales increase every year. Indeed, during the recession of 2008 to 2011, while overall purchases were down, online purchases in the United States and Canada steadily increased.

Internet-based social networking (Facebook, MySpace, Twitter, YouTube, and others) has now moved well beyond mere personal communication to play a rapidly expanding role in marketing for large retail firms and especially for small businesses, which use these facilities to keep almost constant contact with their customers.

The Internet has also entered the political sphere, with social networking having played a large role in recruiting volunteers and eliciting cash contributions, especially during recent U.S. election cycles and in all subsequent elections. Beginning in 2009, social networking through Facebook, Twitter, and YouTube became an integral part of public communication, often referred to on daily TV news programming; it played a particularly active role in the Occupy Wall Street protests and is now a major source of information for North Americans about international events, such as the Arab Spring demonstrations around the Mediterranean.

digital divide the discrepancy in access to information technology between small, rural, and poor areas and large, wealthy cities that contain major governmental research laboratories and universities

The growth of Internet-based activity in North America makes access to the Internet increasingly crucial. Unfortunately, a **digital divide** has developed, because nearly a fifth of the North American population is not yet able to afford computers and Internet connections.

Globalization and the Economy

Today, North America remains the wealthiest and most technologically advanced presence in the global economy. The United States and Canada impact the world economy because of the size of their economies—together, they are almost as large as the economy of the entire European Union—and their affluent consumers. North America's advantageous position in the global economy is also a reflection of its geopolitical influence—its ability to mold the pro-globalization free trade policies that suit the major corporations and the governments of Canada and the United States.

Free trade has not always been emphasized the way it is now. Before North America's rise to prosperity and global dominance, trade barriers were important aids to its development. For example, when it became independent of Britain in 1776, the new U.S. government imposed tariffs and quotas on imports and gave subsidies to domestic producers. This protected fledgling domestic industries and commercial agriculture, allowing its economic core region to flourish.

Now, because both are wealthy and globally competitive exporters, Canada and the United States see tariffs and quotas in other countries as obstacles to North America's economic expansion abroad. Thus, they usually advocate heavily for the reduction of trade barriers worldwide. Critics of these free trade policies point out a number of fallacies in the present North American position on free trade. First, North America once needed tariffs and quotas to protect its fledgling firms, much like many currently poorer countries still need to do. Furthermore, both the United States and Canada, contrary to their own free trade precepts, still give significant subsidies to their farmers. These subsidies make it possible for North American farmers to sell their crops on the world market at such low prices that farmers elsewhere are hurt or even driven out of business (see the vignette on page 60). For example, many Mexican farmers have lost their small farms because of competition from large U.S. corporate farms, which receive subsidies from the U.S. government. U.S. corporate farms can now sell their produce in Mexico or even relocate there under NAFTA agreements. The critics add that beyond agriculture, the benefits of free trade in North America go mostly to large manufacturers and businesses and their managers, while many workers end up losing their jobs to cheaper labor overseas, or see their incomes stagnate.

NAFTA Trade between the United States and Canada has been relatively unrestricted for many years. The process of trade barrier reduction formally began with the Canada–U.S. Free Trade Agreement of 1989. With the creation of NAFTA in 1994, Mexico has now been included. The major long-term goal of NAFTA is to increase the amount of trade between Canada, the United States, and Mexico. Today, it is the world's largest trading bloc in terms of the GDP of its member states.

The extent of the impacts of NAFTA are hard to assess because it is difficult to tell whether the many observable changes are due to the actual agreement or to other changes in regional and global economies. However, a few things are clear. NAFTA has increased trade, and many companies are making higher profits because they now have larger markets. Since 1990, exports among the three countries have increased in value by more than 300 percent. NAFTA's exports to the world economy, by value, have increased by about 300 percent for the United States and Canada and by 600 percent for Mexico. Some U.S. companies, such as Walmart, expanded aggressively into Mexico after NAFTA was passed. Mexico now

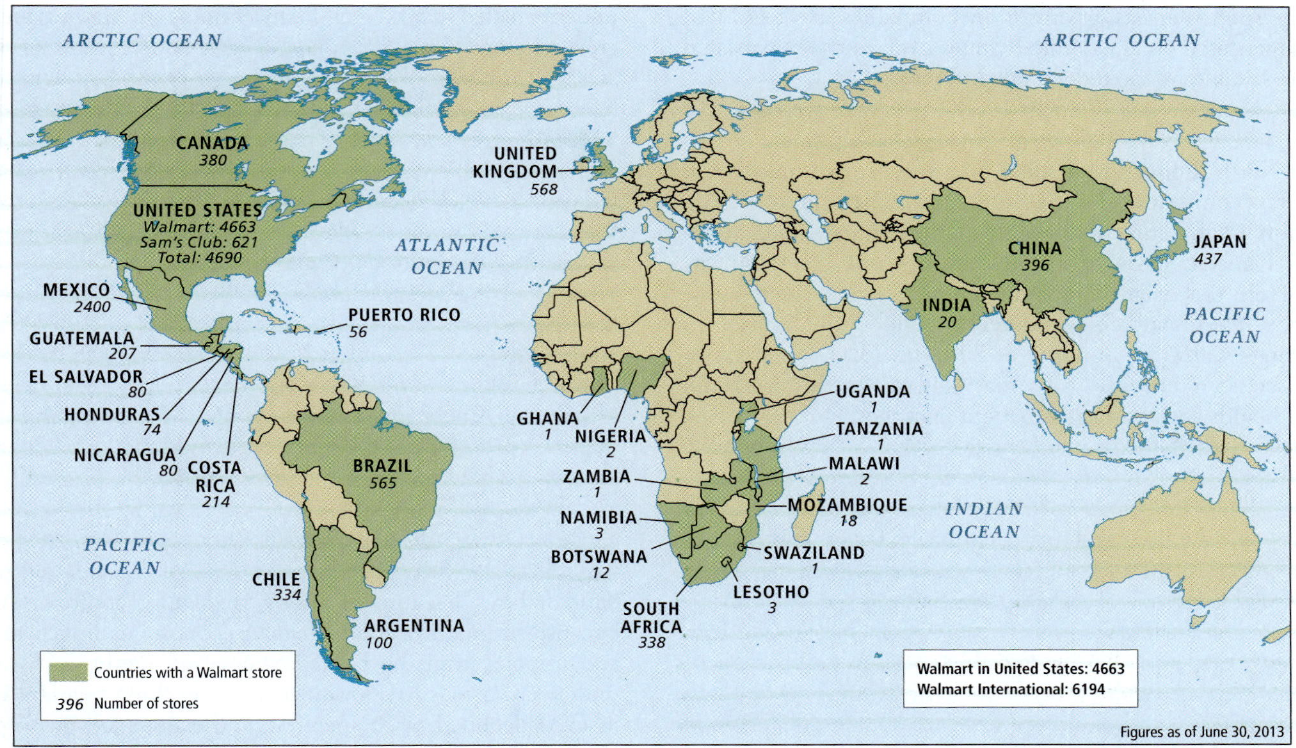

FIGURE 2.19 Walmart on the global scale. As of April 2013, Walmart had 4663 store operations in the United States and 5561 in 26 other countries. Walmart draws it products from over 70 countries, deals with over 61,000 U.S. businesses, and is instrumental in generating over 3 million U.S. jobs. Walmart itself employs over 1.3 million workers in the United States and 900,000 workers in other countries.

has more Walmarts (2088 retail stores) than any country except the United States (Figure 2.19). Canada has 333 Walmarts; the United States has 3868.

NAFTA seems to have worsened the long-standing tendency for the United States to spend more money on imports than it earns from exports. This imbalance is called a **trade deficit**. Before NAFTA, the United States had much smaller trade deficits with Mexico and Canada. After the agreement was signed, these deficits rose dramatically, especially with Mexico. For example, between 1994 and 2009, the value of U.S. exports to Mexico increased by about 153 percent, while the value of imports increased 265 percent.

NAFTA has also resulted in a net loss of around 1 million jobs in the United States. Increased imports from Mexico and Canada have displaced about 2 million U.S. jobs, while increased exports to these countries have created only about 1 million jobs. Some new NAFTA-related jobs do pay up to 18 percent more than the average North American wage. However, those jobs are usually in different locations than the ones that were lost, and the people who take them tend to be younger and more highly skilled than those who lost jobs. Former factory workers often end up with short-term contract jobs or low-skill jobs that pay the minimum wage and carry no benefits.

trade deficit the extent to which the money earned by exports is exceeded by the money spent on imports

As the drawbacks and benefits of NAFTA are being assessed, talk of extending it to the entire Western Hemisphere has stalled. Such an agreement, which would be called the Free Trade Area of the Americas (FTAA), would have its own drawbacks and benefits. A number of countries, such as Brazil, Bolivia, Ecuador, and Venezuela, are wary of being overwhelmed by the U.S. economy. Even in the absence of such an agreement, trade between North America and Middle and South America is growing faster than trade with Asia and Europe. This emerging trade is discussed in Chapter 3.

The Asian Link to Globalization Another way in which the North American economy is becoming globalized is through the lowering of trade barriers with Asia. One huge category of trade with Asia is the seemingly endless variety of goods imported from China—everything from underwear to the chemicals used to make prescription drugs. China's lower wages make its goods cheaper than similar products imported from Mexico, despite Mexico's proximity and membership in NAFTA. Indeed, many factories that first relocated to Mexico from the southern United States have now moved to China to take advantage of its enormous supply of cheap labor. Despite adjustments in the Chinese economy that may raise prices, trade with China promises to remain quite robust for some time.

U.S. and Canadian companies also want to take advantage of China's vast domestic markets. For example, the U.S. fast-food chain Kentucky Fried Chicken (KFC) now has more than 2100 locations in 450 cities in China, and its business is growing rapidly there. The U.S. government has a powerful incentive to encourage such overseas expansion by companies like KFC that are headquartered in the United States and whose repatriated profits are taxable by the federal government.

Asian investment in North America is also growing. For example, Japanese and Korean automotive companies located plants in North America to be near their most important pool of car buyers—commuting North Americans. They often located their plants in the rural mid-South of the United States or in southern Canada, close to arteries of the Interstate Highway System. Here the Asian companies found a ready, inexpensive labor force. These workers could access high-quality housing in rural settings within a commute of 20 miles (32 kilometers) or so from secure automotive jobs that paid reasonably well and included health and retirement benefit packages.

Japanese and Korean carmakers succeeded in North America even as U.S. auto manufacturers such as General Motors were struggling. This was primarily because more-advanced Japanese automated production systems—requiring fewer but better-educated workers—produce higher-quality cars, which sell better both in North America and globally. Some analysts predict that foreign carmakers will eventually take over the entire North American market, while others predict that the old American car companies, several of which are in the process of restructuring, will turn out more competitive, smaller, better-built, and more fuel-efficient cars. Note that General Motors saw its highest profit in its history in 2011 ($7.6 billion), paid back its bailout loan, and is producing autos that are increasingly popular with North American consumers (see the *New York Times* article, "G.M. Reports Big Profit; Europe Lags," at http://tinyurl.com/8yf8fk7, for more on this).

IT Jobs Face New Competition from Developing Countries By the early 2000s, globalization was resulting in the *offshore outsourcing* of information technology (IT) jobs. A range of jobs—from software programming to telephone-based, customer-support services—shifted to lower-cost areas outside North America. By the middle of 2003, an estimated 500,000 IT jobs had been outsourced, and forecasts are that 3.3 million more will follow by 2020. During the recent recession, IT jobs were still being created at a faster rate than they were being ended by economic contraction, yet the overall trend pointed to fewer and fewer of the world's total IT jobs staying in North America. New IT centers are now located in India, China, Southeast Asia, the Baltic states in North Europe, Central Europe, and Russia. In these areas, large pools of highly trained, English-speaking young people work for wages that are 20 to 40 percent of their American counterparts' pay. Some argue that rather than depleting jobs, outsourcing will actually help job creation in North America by saving corporations money, which will then be reinvested in new ventures. The viability of this pro-outsourcing argument remains unproven. **45. U.S. COMPETITIVENESS**

Repercussions of the Global Economic Downturn Beginning in 2007

The severe worldwide economic downturn that began in 2007 came on the heels of a long global upward trajectory of economic expansion. A booming housing industry and related growth in the underregulated banks that finance home mortgages fueled this growth in the United States. In 2007, the housing industry collapsed as it became clear that much of the growth in previous years had been based on banks allowing millions of buyers to purchase homes with mortgages that were well beyond their means. Thus the boom in the construction industry, based on erroneous assumptions, came to a sudden halt. When too many homebuyers could no longer afford their mortgage payments, the banks that had lent them money, and/or the institutions to which the banks had sold these mortgages, started to fail. The bank failures produced worldwide ripple effects because so many foreign banks were involved in the U.S. housing market. Between September 2008 and March 2009, the U.S. stock market fell by nearly half, wiping out the savings and pensions for millions of Americans. Similar plunges followed in foreign stock markets, ultimately resulting in a worldwide economic downturn as businesses could no longer find money to fund expansion.

As the recession intensified, job losses in the United States caused a sharp drop in consumption, which further affected world markets, including those in Canada. Canada did not have bank failures because of strong regulatory controls. However, because so much of the Canadian economy is linked to exports and imports from the United States (see Figure 2.15), Canada underwent a massive slowdown. During three quarters in 2009, its GNI declined by 3.3 percent, and exports fell by 16 percent. Canada also suffered a high rate of job losses; but the recovery was quicker than in the United States, possibly because household consumption was buttressed by Canada's strong social safety net, plus some Canadian incomes were stabilized by global demand for Canadian energy resources. However, Canadian household indebtedness increased sharply into 2011. A Bank of Canada official concluded in 2011 that while Canada weathered the recession fairly well, the recession brought changes that may be lasting.

Growing concern with the U.S. national debt (Canada's national debt is roughly comparable in size), which consists of the money borrowed by issuing treasury securities to cover expenses that exceed income from taxes and other revenues, exacerbated the perception of financial disaster in the United States. There are two components to the U.S. debt: debt held by the public, including that held by investors, the Federal Reserve System, and foreign, state, and local governments ($10.5 trillion as of January 2012); and debt held in accounts administered by the federal government, such as that borrowed from the Social Security Trust Fund ($4.7 trillion as of January 2012).

To whom does the United States owe its debt? About 70 percent of the total debt is owed to Americans who own Treasury securities or to various federal government entities, like the Social Security Trust Fund. About 29 percent is owed to foreign governments (8 percent to China, 5 percent to Japan, 2 percent to the United Kingdom, and smaller amounts to Brazil, Taiwan, and Hong Kong).

The debt constituted a major issue during the U.S. 2012 election cycle. Republicans held that the debt was dangerous and needed to be reduced immediately by drastically reducing government services. Democrats held that some services should be cut and that revenues should be raised by taxing the very rich (who, at the time, were paying taxes at significantly lower rates than middle-income people); also, creating jobs through government programs would alleviate the recession.

Women in the Economy

> **Geographic Insight 5**
> **Population and Gender:** Side effects of changes in gender roles and in opportunities for women include some improvements in education and pay equity; smaller families; and the aging of North America's population.

While women have made steady gains in achieving equal pay and overall participation in the labor force, there are still important ways in which they lag behind their male counterparts. On average, U.S. and Canadian female workers earn about 80 cents for every dollar that male workers earn for doing the same job (Figure 2.20). For example, a female architect earns approximately 80 percent of what a male architect earns for performing comparable work. This situation is actually an improvement over previous decades. During World War II, when large numbers of women first started working in male-dominated jobs, North American female workers earned, on average, only 57 percent of what male workers earned. The advances made by this older generation of women and the ones that followed have transformed North American workplaces. For the first time in history, women now represent more than half of the North American labor force, though most still work for male managers.

ON THE BRIGHT SIDE

Women in Business and Education

Throughout North America, women entrepreneurs are increasingly active, starting nearly half of all new businesses. While women-owned businesses do tend to be small and less financially secure than those in which men have dominant control, credit opportunities for businesswomen are improving as women go into upper-level jobs in banking. Change is also coming to the private sector, where a 2012 study showed that male executives now prefer to hire qualified women because they are particularly ambitious and willing to gain advanced qualifications.

In secondary and higher education, North American women have equaled or exceeded the level of men in most categories. In 2008 in the United States, 33 percent of women age 25 and over held an undergraduate degree, compared to just 26 percent of men. This imbalance is likely to increase because in 2010, U.S. women age 25 to 29 were receiving 7 percent more undergraduate degrees than were men.

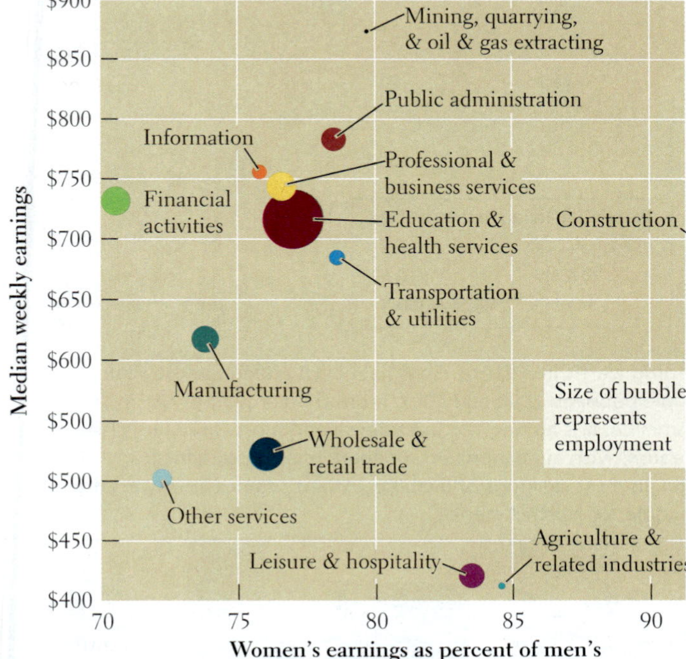

FIGURE 2.20 Women's earnings and employment by industry, 2009. On average, women earned 80 percent of men's median income in 2009 if they worked full time in a wage or salary job: $657 a week compared to men's $819.

THINGS TO REMEMBER

- North American farms have become highly mechanized operations that need few workers. To be profitable, however, they require huge investments in land and machinery, as well as the use of fertilizers and pesticides that contribute to water pollution.
- In the twentieth century, the mass production of inexpensive automobiles and trucks, as well as development of the Interstate Highway System, fundamentally changed how people and goods move across the continent.
- **Geographic Insight 4** **Globalization and Development** Global economic interdependency has meant that certain kinds of jobs in North America are vulnerable to outsourcing. Employment is increasingly in knowledge-intensive jobs that require higher education for professions or specialized technical training. Meanwhile, low-skill, mass-production industrial jobs are increasingly being moved abroad. Also, NAFTA has affected both job loss and job creation.
- Flows of trade and investment between North America and Asia have increased dramatically in recent decades.
- **Geographic Insight 5** **Population and Gender** The participation by women in the economy is beginning to rival that of men, yet women's average pay remains low. Younger women are focusing on educational achievement.

Sociocultural Issues

North American attitudes about urbanization, immigration, race, ethnicity, and religion are changing rapidly, with some embracing change and others resisting it. Shifting gender roles are redefining the North American family, and an aging population is raising new concerns about the future.

Urbanization and Sprawl

A dramatic change in urban spatial patterns has transformed the way most North Americans live. Since World War II,

Figure 2.21 Photo Essay: Urbanization in North America

North America is highly urbanized, with 79 percent of the population living in cities. Some of the wealthiest cities in the world are located here, though only a few have achieved high levels of "livability." Many cities, especially those in the United States, are characterized by sprawling development patterns that require high levels of dependence on the automobile.

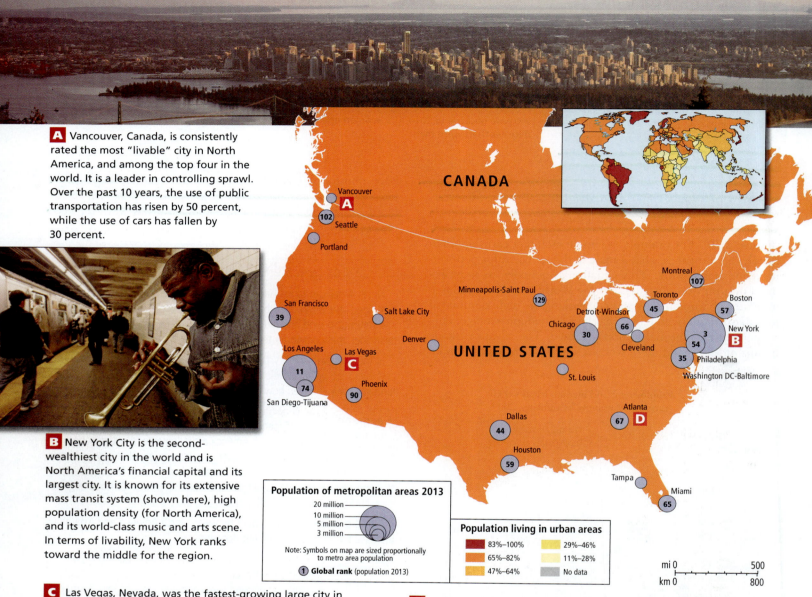

A Vancouver, Canada, is consistently rated the most "livable" city in North America, and among the top four in the world. It is a leader in controlling sprawl. Over the past 10 years, the use of public transportation has risen by 50 percent, while the use of cars has fallen by 30 percent.

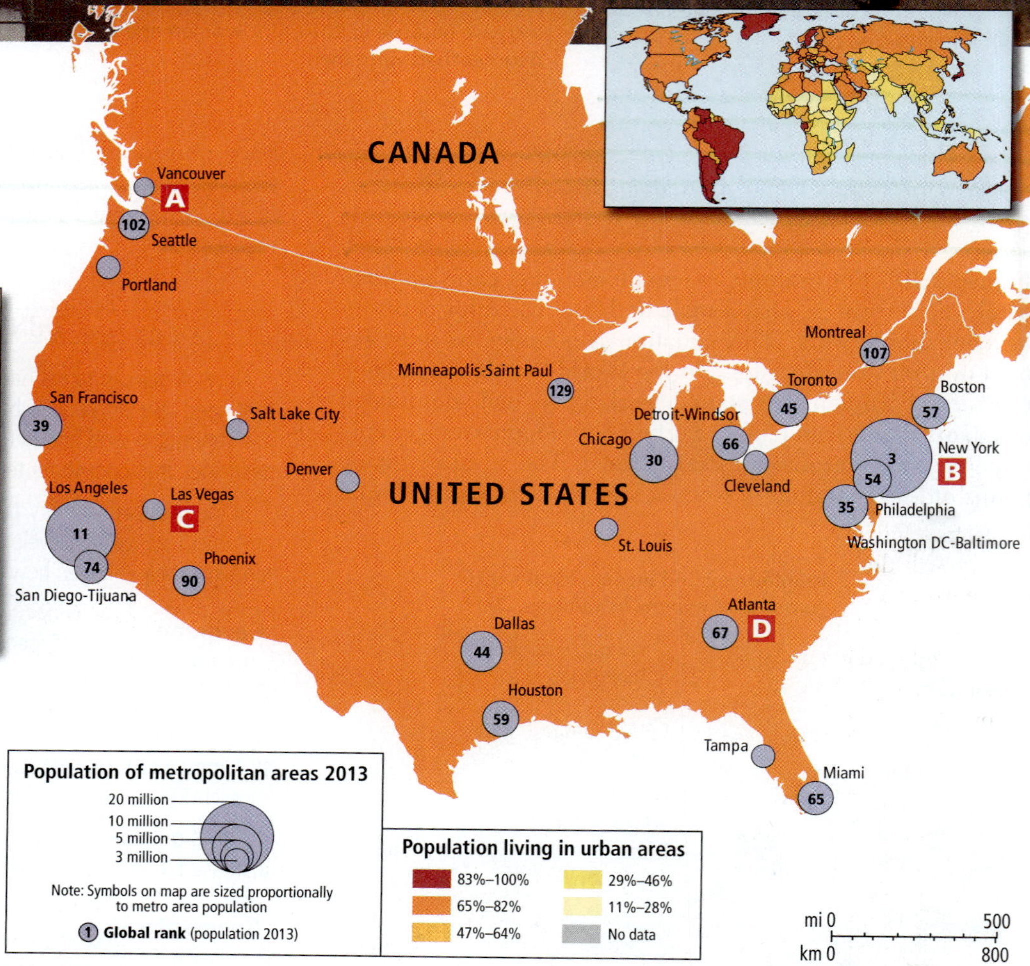

B New York City is the second-wealthiest city in the world and is North America's financial capital and its largest city. It is known for its extensive mass transit system (shown here), high population density (for North America), and its world-class music and arts scene. In terms of livability, New York ranks toward the middle for the region.

CANADA
UNITED STATES
Vancouver A
102 Seattle
Portland
San Francisco 39
Salt Lake City
Los Angeles 11
Las Vegas C
74 San Diego-Tijuana
90 Phoenix
Denver
Minneapolis-Saint Paul 129
Chicago 30
Detroit-Windsor 66
Cleveland
St. Louis
Dallas 44
Houston 59
Atlanta 67 D
Tampa
Miami 65
Montreal 107
Toronto 45
Boston 57
New York 3 B
54 Philadelphia
35 Washington DC-Baltimore

Population of metropolitan areas 2013
20 million
10 million
5 million
3 million
Note: Symbols on map are sized proportionally to metro area population
① Global rank (population 2013)

Population living in urban areas
83%–100%
65%–82%
47%–64%
29%–46%
11%–28%
No data

mi 0 500
km 0 800

C Las Vegas, Nevada, was the fastest-growing large city in North America between 2000 and 2007, with 31.8 percent growth. However, the recession that began in 2007 led to population decline by 2010 as jobs in tourism shrank. For many, leaving was complicated by the inability to sell homes that had lost value. In most rankings of livability, Las Vegas comes in toward the lower end for North America.

D Morning traffic in Atlanta, Georgia, which added almost a million people between 2000 and 2007, more than any other city in North America. It is among the least dense and most sprawling cities in the region, with almost universal dependence on cars for transportation. In most rankings of livability, Atlanta comes in toward the lower middle for North America.

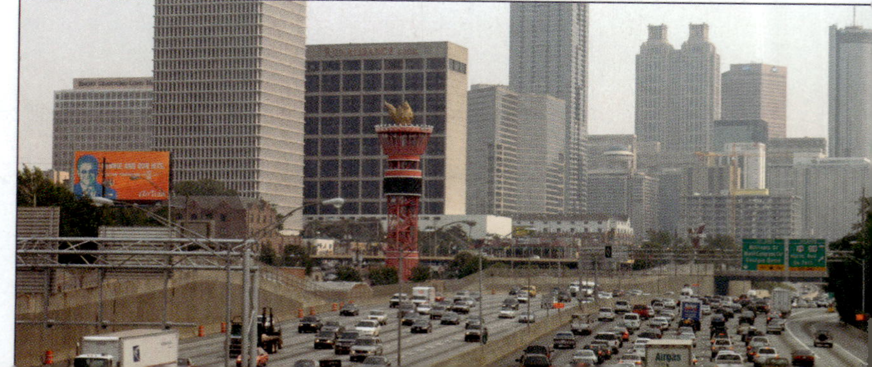

Thinking Geographically

After you have read about urbanization in North America, you will be able to answer the following questions:

A Is Vancouver's growth pattern typical for North America?

B What factors reduce livability in New York?

C What factors reduce livability in Las Vegas?

D What factors reduce livability in Atlanta?

North America's urban populations have increased by about 150 percent, but the amount of land they occupy has increased almost 300 percent (Figure 2.21). This is primarily because of suburbanization and urban sprawl, companion processes to urbanization.

Today, close to 80 percent of North Americans live in **metropolitan areas**—cities of 50,000 or more, plus their surrounding suburbs and towns. Most of these people live in car-dependent suburbs built since World War II, where urban life bears little resemblance to that in the central cities of the past.

In the nineteenth and early twentieth centuries, cities in Canada and the United States consisted of dense inner cores and less-dense urban peripheries that graded quickly into farmland. Starting in the early 1900s, central cities began losing population and investment, while urban peripheries—the **suburbs**—began growing. Workers were drawn by the opportunity to raise their families in single-family homes in secure and pleasant surroundings on lots large enough for vegetable gardens. Some continued to work in the city, traveling to and from on streetcars. After World War II, suburban growth dramatically accelerated as cars became affordable to more workers (Figure 2.22). With the growing popularity of the family car, interest in publicly funded transport for spreading urban areas was reduced to city bus systems. Retrofitting highly developed urban areas such as Atlanta or San Francisco with light rail systems or subways has proven to be very expensive.

As North American suburbs grew and spread out, nearby cities often coalesced into a single urban mass. The term **megalopolis** was originally coined to describe the 500-mile (800-kilometer) band of urbanization stretching from Boston through New York City, Philadelphia, and Baltimore, to south of Washington, DC. Other megalopolis formations in North America include the San Francisco Bay Area, Los Angeles and its environs, the region around Chicago, and the stretch of urban development from Eugene, Oregon, to Vancouver, British Columbia.

metropolitan areas cities of 50,000 or more and their surrounding suburbs and towns

suburbs populated areas along the peripheries of cities

megalopolis an area formed when several cities expand so that their edges meet and coalesce

This pattern of urban sprawl requires residents to drive automobiles to complete most daily activities such as grocery shopping or commuting to work. Increasing air pollution and emission of greenhouse gases that contribute to climate change are two major side effects of the dependence on vehicles that comes with urban sprawl. Another important environmental consequence is habitat loss brought about by suburban development expanding into farmland, forests, grasslands, and deserts.

Farmland and Urban Sprawl Urban sprawl drives farmers from land that is located close to urban areas, because farmland on the urban fringe is very attractive to real estate developers. The land is cheap compared to urban land, and it is easy to build roads and houses on since farms are generally flat and already cleared. As farmland is turned into suburban housing, property taxes go up and surrounding farmers who can no longer afford to keep their land sell it to housing developers. In North America each year, 2 million acres of agricultural and forest lands make way for urban sprawl.

Advocates of farmland preservation argue that beyond food and fiber, farms also provide economic diversity, soul-soothing scenery, and habitat for some wildlife. For example, the town of

FIGURE 2.22 Urban sprawl in Phoenix, Arizona.

(A) Phoenix grew rapidly between 1950 and 2000, resulting in a demand for housing. The photo is of Sun City, a new, car-oriented suburb designed for senior citizens located about 30 miles from Phoenix's center.

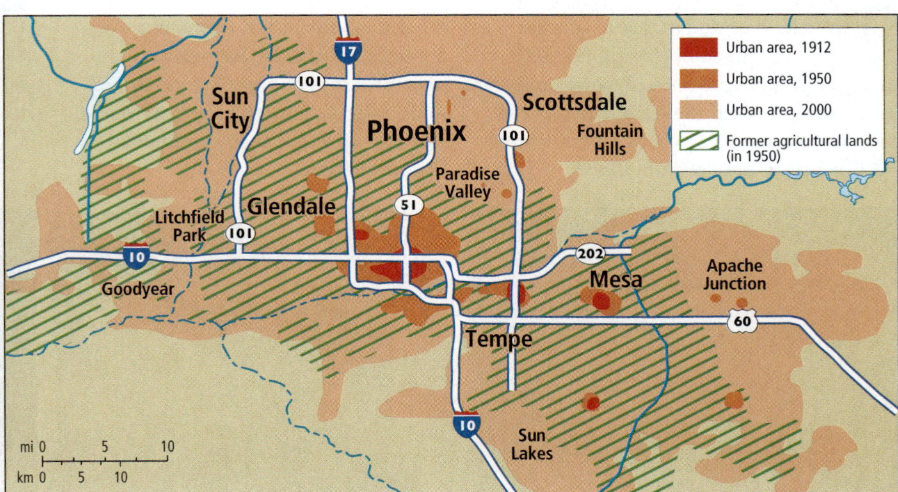

(B) This current map shows the original urban settlement and how it has grown from 1912 to the present. A preference for single-family homes means that the city is sprawling into the surrounding desert rather than expanding vertically into high-rise apartments.

Pittsford, New York (a suburb of Rochester), decided that farms were a positive influence on the community. Mark Greene's 400-acre, 200-year-old farm lay at the edge of town. As the population grew, the chances of the farm remaining in business for another generation looked dim. Land prices and property taxes rose, and the Greene family could not meet their tax payments. Residents in the new suburban homes, sprouting up on what had been neighboring farms, pushed local officials to halt normal farm practices, such as noisy nighttime harvesting or planting, spreading smelly manure, or importing bees to pollinate fruit trees. Pittsford, however, decided to stand by the farmers by issuing $410 million in bonds so that it could pay Greene and six other farmers for promises that they would not sell their 1200 acres to developers, but would instead continue to farm them.

VIGNETTE For Canada's *First Nations people* (the preferred term in Canada for Native Americans), the chance to indulge in the extravagance of urban sprawl is viewed positively. The Tsawwassen First Nation is one of 630 native groups that are negotiating with the Canadian government to gain full control over their ancestral territories. The land they recently received by a treaty lies 20 miles south of Vancouver, near the Strait of St. George. Eager to finally enjoy some of the fruits of development, they are planning a 2-million-square-foot "big box" and indoor shopping mall in the midst of what has been rural farmland. The surrounding communities, which had long shielded their village-like settlements against urban sprawl, feel that there are enough shopping opportunities already. The Tsawwassen see the mall as a long-overdue chance for the prosperity that their neighbors already enjoy. *[Source: Adapted from The Atlantic Cities. For detailed source information, see Text Credit pages.]* ■

Smart Growth and Livability The term *smart growth* has been coined for a range of policies aimed at stopping sprawl by making existing urban areas more "*livable.*" *Livability* relates to factors that lead to a higher quality of life in urban settings, such as safety, good schools, affordable housing, quality health care, numerous and well-maintained parks that offer recreational opportunities for people and their pets (Figure 2.23), and well-developed public transportation systems. Some indexes of livability for major cities worldwide are put out every year by *The Economist* magazine and the Mercer Quality of Living Survey.

In smart growth planning, environmental benefits are envisioned as well. The focus on life lived locally, where mass transit and walking replace the family car, and where gardening for food is a community entertainment, will result in lower energy use, air pollution, and CO_2 emissions. While smart growth and the shift toward livability are just starting to take hold throughout North America, there are a few cities, such as Vancouver, British Columbia (see Figure 2.21A), and Portland, Oregon, where these principles are much further along in implementation.

Some older cities are enjoying a renaissance as people move back to them in search of the greater livability. Places like New York City are arguably the furthest along in crucial aspects of smart growth, given their high density and widespread use of mass transit (see Figure 2.21B). However, the high cost of living in these places can reduce their livability.

It may be a while before smart growth and livability make much of an impact in North America's fastest-growing cities, such as Atlanta, Dallas, and Las Vegas. Here, in recent years, vast car-dependent suburbs have spread over enormous areas. Not surprisingly, none of these cities ranks high on livability indexes (see Figure 2.21C). Car dependence and rapid growth have brought

FIGURE 2.23 LOCAL LIVES **PEOPLE AND ANIMALS IN NORTH AMERICA**

A Labrador retrievers are the most popular breed of dog in North America. Fishers on the island of Newfoundland, part of Canada's province of Newfoundland and Labrador, bred these dogs to retrieve fishing nets, which accounts for the breed's webbed paws and love of water.

B The ancestors of the Maine coon cat, one of the largest of the popular domestic cat breeds, were developed to control rodent populations aboard ships trading along the shore of Maine and other parts of the Atlantic coast of North America.

C Alaskan Malamutes pull a sled, which is the job that they were bred for by the Kuuvangmiut (formerly known as the Mahlemut) Inuit tribe of northwestern Alaska. Their large size and heavy build makes them ideal for pulling heavily loaded sleds.

Atlanta massive traffic jams and some of the worst air quality in the United States (see Figure 2.21D). Meanwhile, livability in the Las Vegas metro area is reduced by poor air and water quality as well as high crime rates.

Inner-city decay is another impact of sprawl on livability. This is especially true in the United States, where many inner cities are dotted with large tracts of abandoned former industrial land and neighborhoods debilitated by persistent poverty and loss of jobs. Old industrial sites that once held factories or rail yards are called **brownfields**. Because they are often contaminated with chemicals and covered with obsolete structures, blacktop, and concrete, they can be very expensive to redevelop for other uses. Also left behind in the inner cities are the least-skilled and least-educated citizens, many of whom were drawn in generations ago by the promise of jobs that have long since moved out to the suburbs or overseas. Often the majority population in these inner cities is a mixture of African Americans, Asians, Hispanics, and new immigrants. Some are relatively affluent and are leading efforts to renew old city centers. Others are in great need of the very services—health care, schools, and social support (including churches, synagogues, and mosques)—that have moved to the suburbs.

The new emphasis on smart growth and livability has also reinforced the **gentrification** of old, urban residential districts. As affluent people invest substantial sums of money in renovating old houses and apartments, poor inner-city residents can be displaced in the process. The effect of gentrification on the displaced poor appears to be somewhat less harsh in Canada than in the United States, primarily because Canada's stronger social safety net better ensures social services, housing, and help with housing maintenance. Some U.S. cities (for example, Knoxville, Tennessee; Portland, Oregon; and Charlotte, North Carolina) have initiated *New Urbanism* projects for inner-city residents displaced by gentrification. These projects aim to rehouse people in pleasant, newly built, walkable urban neighborhoods with conveniently located services (Figure 2.24).

ON THE BRIGHT SIDE

The Greening of Detroit

Detroit, once the center of American auto industries, began to decline in the 1970s when a series of faulty autos besmirched the reputation of the "Motor City." Over the decades, thousands of jobs were lost and the city's population shrank as global competition led to factory closures. As of 2010, the depressing decay of a now much smaller Detroit began to sprout with greenery and flowers as inner-city Detroiters discovered the joys of gardening on urban wastelands. These green oases draw neighbors together and reconnect people with natural cycles. Children learn that food is something one can grow, not only buy. As the auto industry revives in Detroit, and some workers return to what are now very different automated factories, urban gardening is likely to retain its charm as a *green* solution.

FIGURE 2.24 New Urbanism Celebration, Florida, built in the 1990s, was designed to resemble older cities such as Charleston, South Carolina; New Orleans, Louisiana; Alexandria, Virginia; and Vicksburg, Mississippi.

Immigration and Diversity

Immigration has played a central role in populating both the United States and Canada. Most people in North America descend from European immigrants; but there are many who have roots in Africa, Asia, Middle and South America, and Oceania. New waves of migration from Middle and South America and parts of Asia promise to make North America a region where people of non-European or mixed descent will become the majority (Figure 2.25). Most major North American cities are already characterized by ethnic diversity, and in some, recent immigration has led to near majorities of foreign-born residents. Houston, Texas, for example, was the most ethnically diverse metropolitan area in the United States in 2010, with 40 percent of the population Anglo, 36.7 percent Latino, 16.8 percent African American, and 6.5 percent Asian. **50. IMMIGRATION AND POPULATION REPORT**

In the United States, the spatial pattern of immigration is also changing. For decades, immigrants settled mainly in coastal or border states such as New York, Florida, Texas, or California. However, since about 1990, immigrants are increasingly settling in interior states such as Illinois, Colorado, Nevada, and Utah (see Figure 2.25). The influence of immigrants can also be seen in many aspects of life in North America, including the popularity of ethnic cuisines (Figure 2.26 on page 94).

Why Do People Decide to Immigrate to North America?

The decision to leave one's homeland is usually not an easy one. There are almost always **push factors** that cause people to consider the drastic move of leaving family, friends, and familiar places to strike out into the unknown, with what are usually limited resources. These push factors can be civil or political unrest, or some kind of discrimination, but most often it is lack of economic opportunity and curbs on upward mobility that overcome the natural resistance to leaving home. Recently, the forces of globalization have often been the root cause of these push factors. For example, when NAFTA encouraged large U.S. agribusinesses to relocate to Mexico to grow corn on a massive scale, small Mexican corn farmers were dislodged from land they had

brownfields old industrial sites whose degraded conditions pose obstacles to redevelopment

gentrification the renovation of old urban districts by affluent investors, a process that often displaces poorer residents

push factors factors that get people to consider the drastic move of leaving family, friends, and a familiar place to strike out into the unknown, with what are usually unknown resources

FIGURE 2.25 Percent of total foreign-born people within each state in (A) 2000 and (B) 2010.

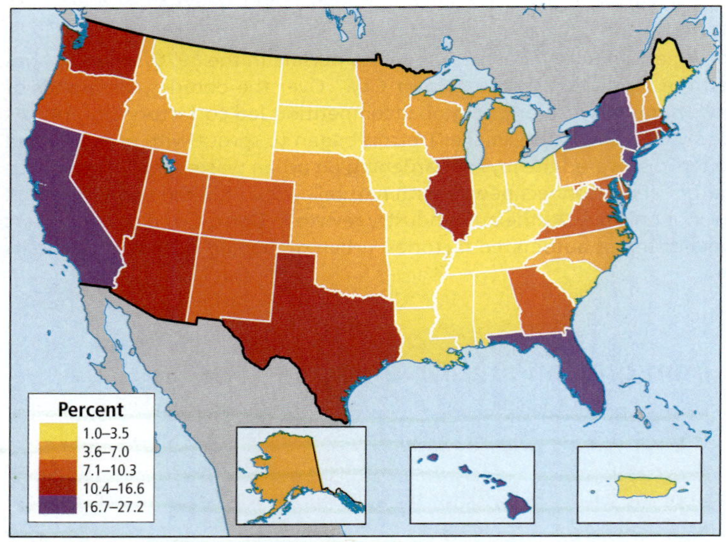

(A) Percent of foreign-born of total population within each state in 2000.

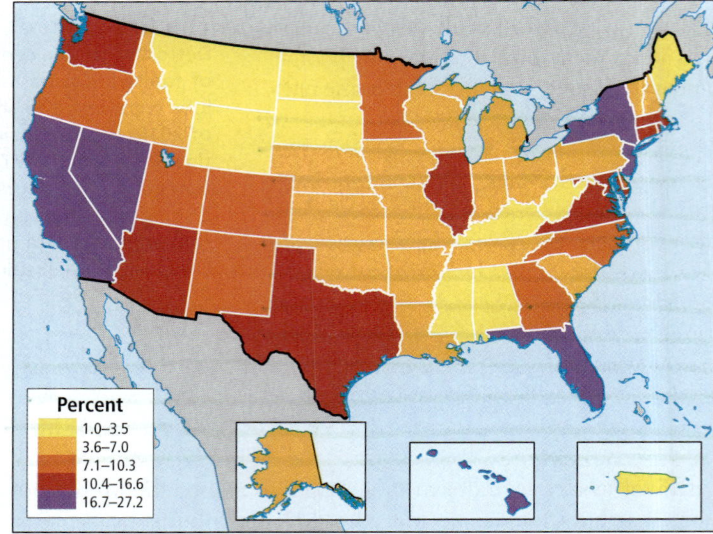

(B) Percent of foreign-born of total population within each state, as of 2010.

traditionally worked; others found they could not compete on price and had to give up farming. Yet all these farmers had families to feed and educate, so often fathers and their brothers and sons migrated north in search of work. In Texas, California, or Tennessee, they worked in construction or as field laborers, lived frugally, and sent most of their earnings home. Some entered the United States legally, but many—because of the complexities and costs of getting papers to immigrate—entered illegally. Legal U.S. immigration visas are available in far fewer numbers than both the potential immigrants and the jobs available for them to fill.

Do New Immigrants Cost U.S. Taxpayers Too Much Money? Many North Americans are concerned that immigrants burden schools, hospitals, and government services. And yet

FIGURE 2.26 LOCAL LIVES FOODWAYS IN NORTH AMERICA

A Pork ribs are smoked and slow cooked in the "pit" at Spoon's Barbecue in Charlotte, North Carolina. Originally, Native Americans would first dig a pit or trench and fill it with burning coals or hot rocks. They would place meat on top and either slow cook it in the open at low temperature for several hours or cover the entire pit and let it cook for a day or more. Most barbeque is now cooked in a metal smoker, still called a "pit," which is often attached to a trailer so that it can be brought to special events.

B Poutine, a French Canadian dish, now hugely popular throughout Canada, consists of French fries served with cheese curds (a tasty by-product of cheese making), covered in gravy. One story holds that a take-out customer of Fernand LaChance's restaurant in Warwick, Québec, in the 1950s, asked for the combination, to which the owner responded "Ça va faire une maudite poutine!" ("That's going to make a cursed mess!"), thus giving the dish its name. The gravy was later added to keep the fries warm.

C New England–style clam chowder served in a sourdough bread bowl is a classic San Francisco dish that springs from two American traditions. New England clam chowder was developed by fishers and consists of clams and other seafood, potatoes, and seasonings, cooked in a broth to which hard "sea biscuits" were mixed to thicken the chowder. In San Francisco, sourdough bread came via pioneer families and was later popularized by French bakers. It has a particularly sour, tangy flavor that may be due to the climate of the San Francisco Bay Area.

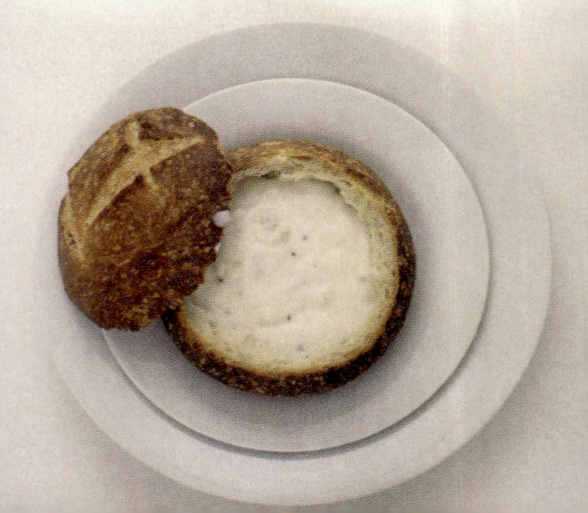

numerous studies have shown that, over the long run, immigrants contribute more to the U.S. economy than they cost. Legal immigrants have passed an exhaustive screening process that assures they will be self-supporting. As a result, most start to work and pay taxes within a week or two of their arrival in the country. Those immigrants, who draw on taxpayer-funded services such as welfare, tend to be legal refugees fleeing a major crisis in their homeland; they are dependent only in the first few years after they arrive. More than one-third of immigrant families are firmly within the middle class, with incomes of $45,000 or more. Even undocumented (illegal) immigrants play important roles as payers of payroll taxes, sales taxes, and indirect property taxes through rent. Because they fear deportation, illegal immigrants are also the least likely to take advantage of taxpayer-funded social services, such as food stamps or subsidized housing.

On average, immigrants are healthier and live longer than native U.S. residents, according to a 2004 study by the National Institutes of Health (NIH). Thus they represent less drain on the health care and social service systems than do native residents. The NIH attributes this difference to a stronger work ethic, a healthier lifestyle that includes more daily physical activity, and the more nutritious eating patterns of new residents compared to those of U.S. society at large. Unfortunately, these healthy practices tend to diminish the longer immigrants are in the country, and the more healthy status does not carry over to immigrants' children, who are nearly as likely to suffer from obesity as native-born children.

Do Immigrants Take Jobs Away from U.S. Citizens? The least-educated, least-skilled American workers are the most likely to end up competing with immigrants for jobs. In a local area, a large pool of immigrant labor can drive down wages in fields like roofing, landscaping, and general construction. Immigrants with little education now fill many of the very-lowest-paid service, construction, and agricultural jobs.

It is often argued that U.S. citizens have rejected these jobs because of their low pay, which results in immigrants being needed to fill the jobs. Others argue that these jobs might pay more and thus be more attractive to U.S. citizens if there were not a large pool of immigrants ready to do the work for less pay. Research has failed to clarify the issue. Some studies show that immigrants have driven down wages by 7.4 percent for U.S. natives without a high school diploma. However, other studies show no drop in wages at all. Employers across the country speak of the superior work ethic of immigrants, especially in the construction industry, which makes them happy to pay these workers decent wages.

Professionals in the United States occasionally compete with highly trained immigrants for jobs, but such competition usually occurs in occupations where there is a scarcity of native-born people who are trained to fill these positions (Figure 2.27). The computer engineering industry, for example, regularly recruits abroad in such places as India, where there is a surplus of highly trained workers. In this case, it is unclear whether these skilled immigrants are driving down wages. Until the recession caused jobs to be eliminated, it was quite clear that there were not enough sufficiently trained Americans to fill the available positions. In

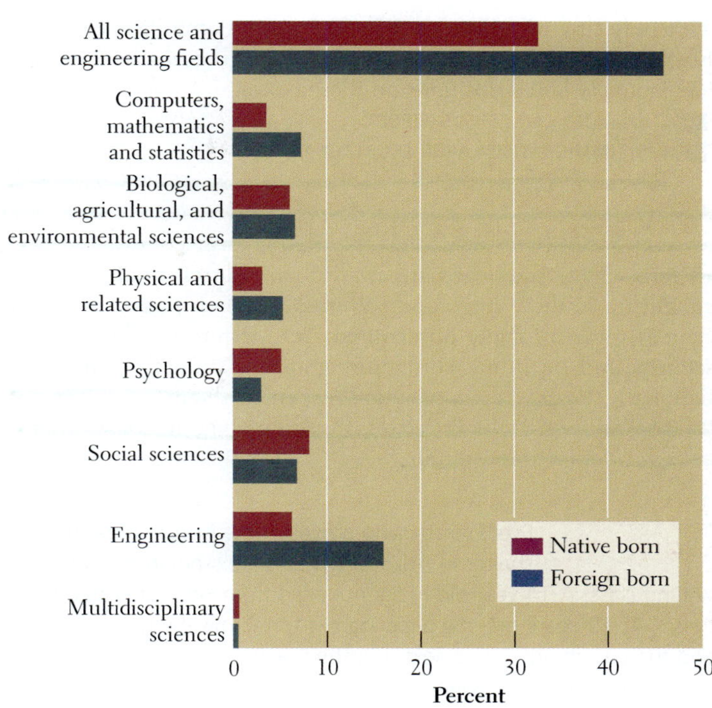

FIGURE 2.27 Percent of the population aged 25 and older with a bachelor's degree or higher by nativity and field of degree, 2010.

fact, India's growth in IT jobs and related services has been so impressive that many who came to North America to work in these jobs have elected to go back to India to work in the newly burgeoning IT industry there, a development that has not been welcomed by their American employers. **253. SKILLED FOREIGN WORKERS IN U.S. MAY HAVE TO LEAVE**

Are Too Many Immigrants Being Admitted to the United States? Many people are concerned that immigrants are coming in such large numbers that they will strain resources here. This point of view has some credibility. Immigrants and their children accounted for 78 percent of the U.S. population growth in the 1990s. At this rate of entry, by the year 2050, the U.S. population is projected to reach 422.6 million (down 20 million from estimates in 2008), with immigration accounting for the majority of the increase. But one of the problems with making these projections is that no one really knows how much undocumented (illegal) immigration there really is. Most research indicates that it has reached unprecedented levels over the past 30 years, possibly exceeding legal immigration rates since the mid-1990s. While estimates of the illegal immigrant population currently in the United States range from 7 to 20 million, reports in 2012 indicated that migration—legal and illegal—was down to less than half of previous years.

Undocumented immigrants tend to lack skills, and they are not screened for criminal background, as are all legal immigrants. However, research also shows that undocumented migrants are not only less likely to partake of social services, but also less likely to participate in criminal behavior than the general population, with only tiny percentages of them having committed offenses. Further analysis of the 2010 census will help researchers formulate factually based positions on these issues.

47. AMERICAN CENSUS REPORT

51. IMMIGRATION LABOR SHORTAGE REPORT

Research shows that the decision to migrate without legal documentation is a very difficult one, undertaken because of severely limited economic, educational, and social opportunities at home. Few Americans are aware that children of all ages get caught up in these migration patterns, and their migratory experiences remain largely unrecorded. Yet, despite tougher border security and declining adult immigration nationally and internationally, the numbers of unaccompanied children stopped at the border continually increases, illustrating yet one more of the complexities of immigration.

VIGNETTE Every youth who arrives at the border detention shelter in Arizona writes a life story in a creative writing class. A 14-year-old indigenous girl from Honduras started hers: "At 10-years-old, my papa started to tell me the good things and the bad things." The next day, the teacher asked what this meant. After sitting in silence for 30 minutes, the girl wrote: "While my father is a good man, he did not always do good things. Some of these things created a bad situation for me." Afraid to tell her mom, she called her brothers to help her escape.

Three years before, at the peak of the global recession, her then-teenage brothers, unable to find work in their rural area, decided to take the month-long journey to the United States by jumping trains through Mexico and then walking for days across the mountainous and arid terrain of Texas. Once out of Texas, they found migrant farm work in Southern states. The brothers were working in Florida when they spoke to their sister on the phone.

Within days, the brothers sent their sister all of their savings—$8000—and arranged for a *coyote* to accompany the girl to the United States. As an unaccompanied and attractive female, she was more vulnerable to exploitation, abuse, and sex trafficking. Like most detained in the border shelter, she is unwilling to discuss what happened on her journey, describing it as "the necessary suffering to become someone." Luckily, she made it without falling victim to drug trafficking and fatal gang violence along the border.

In the 2 months it took her to arrive, she lost contact with her brothers. Lacking family contacts and money, her deportation is likely. Her biggest desire is "to work hard and give the money back to my brothers." If she is allowed to stay, her brothers will expect her to work to help pay the bills. *[Source: Material for this vignette was pieced together by Lydia Pulsipher in April 2012 from personal correspondence with Elizabeth Kennedy and Stuart Aitken, geographers who participate in the ISYS Unaccompanied Minors project.]* ■

Race and Ethnicity in North America

Despite strong scientific evidence to the contrary, people across the world still perceive skin color and other visible anatomical features to be significant markers of intelligence and ability. This racialized view of skin color is partially a remnant of European colonialism, which was based on assigning some to a perpetually low status, while others were given privilege. As discussed previously (see Chapter 1, pages 54–56), the science of biology tells us there is no scientific validity to such assumptions and practices. The same is true for **ethnicity**, which is the cultural counterpart to race, in that people may ascribe overwhelming (and unwarranted) significance to cultural characteristics, such as religion, family structure, or gender customs. Thus, race and ethnicity are very important sociocultural factors not because they *have* to be, but because people *make* them so.

Extending Equal Opportunity Numerous surveys show that a large majority of Americans of all backgrounds favor equal opportunities for minority groups. Nonetheless, in both the United States and Canada, many middle-class African Americans, Native Americans, and Hispanics (and to a lesser extent, Asian Americans) report experiencing both overt and covert discrimination that affects them economically as well as socially and psychologically. And indeed, even a cursory examination of statistics on access to health care, education, and financial services shows that, on average, Americans have quite unequal experiences based on their racial and ethnic characteristics.

Diversity is increasing across North America, and acceptance of diversity as the norm is increasing as well; take, for example, the enthusiasm with which various cultural festivals are attended by Americans regardless of their ethnic heritage (Figure 2.28). Nonetheless, the issue of race remains important in both the United States and Canada, where whiteness is considered the norm and African Americans, Native Americans (First Nations people in Canada), and Hispanics often find themselves defined as exceptions to the norm. Prejudice has clearly hampered the ability of these three groups to reach social and economic equality with Americans of other ethnic backgrounds. In the United States, and to a considerably lesser degree in Canada, despite the removal of legal barriers to equality, these groups (with the notable exception of Asians) still have higher poverty rates and thus lower life expectancies, higher infant mortality rates, lower levels of academic achievement, and more unemployment than other groups.

Figure 2.29 shows the changes in the ethnic composition of the North American population from 1950 to 2010, and the projected changes for 2050. In 2001, Hispanics overtook African Americans as the largest minority group in the United States. Because of a higher birth rate and a high immigration rate, the Hispanic population increased by 58 percent in the 1990s, to 14.7 percent of the total population by 2010. By 2010, Asian Americans made up just 5 percent of the U.S. population, but their numbers had increased by 80 percent between 1990 and 2010. Of the foreign-born population in Canada in 1981, Asians made up 14.1 percent and those of European birth, 66.7 percent. Asians appear to be the fastest-growing nonnative-born group in both countries. It is estimated that by 2031, if present trends continue, 55 percent of North Americans who are born abroad will be Asians.

Income Discrepancies Over the past few decades, many non–Euro-Americans and those of mixed heritage have joined the middle class, achieving success in the highest ranks of government and business. In particular, African Americans have completed

ethnicity the quality of belonging to a particular culture group

FIGURE 2.28 LOCAL LIVES **FESTIVALS IN NORTH AMERICA**

A Mardi Gras Indian Chief Golden Comanche leads a procession through a neighborhood in New Orleans, Louisiana. *Mardi Gras* stands for "Fat Tuesday." In Christianity, it marks the last day before the start of Lent, a 40-day period of fasting and abstinence that precedes Easter. Mardi Gras Indians originated as a tribute to Native American communities that once harbored runaway African American slaves.

B Revelers play with fire at Burning Man in Black Rock Desert, Nevada. This gathering is a weeklong "art event and transitory community based on radical self-expression and self-reliance." Begun in 1986 as a bonfire on the beach in San Francisco, Burning Man is guided by principles of inclusive participation, anticommercialism, and the goal of not leaving any physical trace of the festival after it happens each year.

C A bull rider is thrown at the Calgary Stampede, a 10-day rodeo, exhibition, and festival in Alberta, Canada. Drawing over a million people, it is billed as the "Greatest Outdoor Show on Earth." Started in 1886 as an effort to draw settlers from the East, the highlight of today's Stampede is its rodeo, which offers more prize money than any other.

advanced degrees in large numbers, and more than one-third now live in the suburbs. Yet, as overall groups, African Americans, Hispanics, and Native Americans remain the country's poorest and least educated people (Figure 2.30 on page 98).

The Culture of Poverty Is anything other than prejudice holding back some in these ethnic minority groups? Some social scientists suggest that persistently disadvantaged Americans of all ethnic backgrounds may suffer from a *culture of poverty*, meaning that poverty itself forces coping strategies that are counterproductive to social advancement. Plans to get higher education, for example, must be abandoned because of lack of funds. This and low social status have bred the perception among those trapped in the culture of poverty that there is no hope and therefore no point in trying to succeed. This perception is perpetuated and even exaggerated by a daily existence on the fringes of American mainstream life. The reality is that these poorest people rarely encounter examples of success, and they have few opportunities to complete a decent education, be hired into jobs with livable wages, or find a decent dwelling.

One component in the culture of poverty is the growing numbers of low-income, single-parent families. In 2005, only 23 percent of Euro-American children and 17 percent of Asian American children lived in single-parent families, while 65 percent of African Americans, 49 percent of Native Americans, and 36 percent of Hispanics did. The reasons for these stark

U.S. Population by Race and Ethnicity, 1950, 2010, and 2050 (projected)

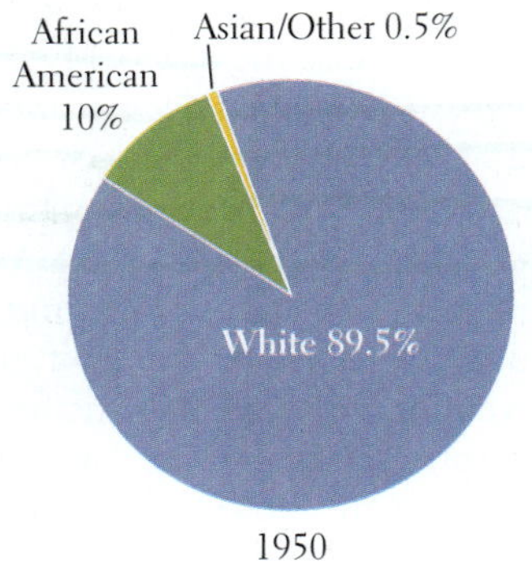

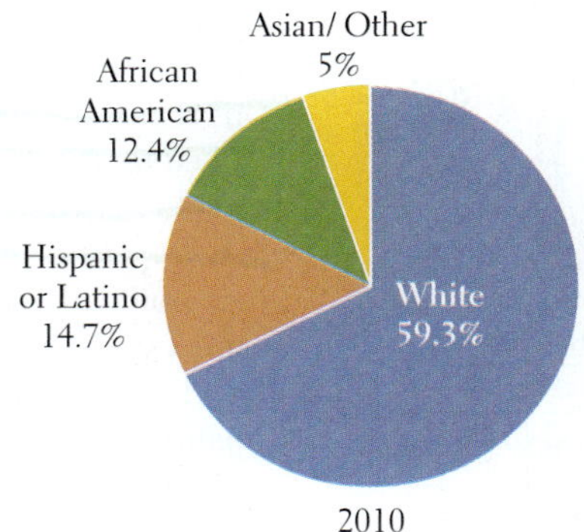

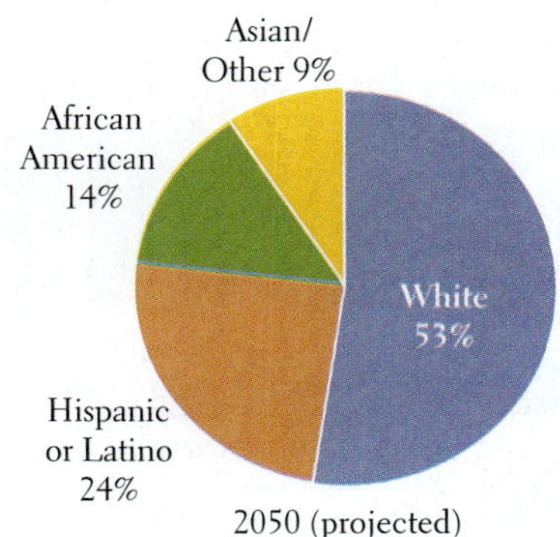

FIGURE 2.29 The changing ethnic composition in the United States, 1950, 2010, and 2050. As the percentage of ethnic minorities increases in the United States, the percentage of whites decreases. By 2050, if present reproductive trends continue, whites will constitute only slightly over half of the population.

ON THE BRIGHT SIDE

What Class Can Do

Curing poverty may be the answer to lowering racial and ethnic tensions. Evidence for a diminishing of racial and ethnic prejudice is to be found in situations where privileged Americans of any racial background share middle- and upper-class neighborhoods, workplaces, sports teams, places of worship, and marriages. In these cases, the attention paid to skin color and ethnicity decreases markedly.

differences are often systemic. For example, many states prohibit or severely limit the amount of income support allowed to intact families who fall on hard times. In these states, it is easier to qualify for welfare if the family is missing one parent. Fathers who lose their jobs may simply absent themselves so the mothers can qualify for assistance. In these situations, children usually stay with their mothers, and fathers often are no longer active in their support and upbringing. This perpetuates the idea that males are not active, participating family members. The assistance packages always fall far short of meeting actual needs, and the enormous responsibilities of both child rearing and breadwinning are often left in the hands of undereducated young mothers who, being in need themselves, are unable to help their children advance.

Another particularly geographic aspect of the *culture of poverty* is that it is part of a larger problem of economic and social spatial segregation based on class. In both the United States and Canada, the increasingly prosperous middle class, of whatever race or ethnicity, has moved to the suburbs—even to gated residential enclaves. Their children seldom encounter children from poorer families. Those who are successful have difficulty developing empathy for the poor because they know no poor; in turn, the very poor rarely have the chance to associate with models of economic success.

Religion

Because so many early immigrants to North America were Christian in their home countries, Christianity is currently the predominant religious affiliation claimed in North America. In the United States, 76 percent identified themselves as Christian in 2008; in Canada, 70 percent did. In the United States, most Christians are Protestants; in Canada, most are Catholic. Despite the numerical dominance of Christianity, about 15 percent of people in both countries reported no religious affiliation. This group includes atheists, agnostics, and the majority who simply claim no religion by choice. Minority religions are growing in both countries. Virtually every medium-sized city across the region has at least one synagogue, mosque, and Buddhist or Hindu temple. In some localities, adherents of Judaism, Islam, Hinduism, or Buddhism are numerous enough to constitute a prominent cultural influence (see, for example, California and Ontario in Figure 2.31).

Because of the constitutional provisions against the establishment of religion, the U.S. Census Bureau is legally prohibited from collecting data on religious affiliation, so this information is available only from private agencies, many of which have a vested interest in what they report. The Pew Forum on Religion and Public Life (see http://www.pewforum.org) is a reliable source of information and conducts many polls on religious issues.

The Geography of Christian Subgroups There are many versions of Christianity in North America, and their geographic distributions are closely linked to the settlement patterns of the immigrants who brought them here (see Figure 2.31). Roman Catholicism dominates in regions where Hispanic, French, Irish, and Italian people have settled—in southern Louisiana, the Southwest, and the far Northeast in the United States, and in Québec and other parts of Canada. Lutheranism is dominant where Scandinavian people have settled, primarily in Minnesota and the eastern Dakotas. Mormons dominate in Utah.

Baptists, particularly Southern Baptists and other evangelical Christians, are prominent in the Bible Belt, which stretches across the Southeast from Texas and Oklahoma to the Eastern Seaboard. Evangelical Christianity is such an important part of community life in the South that frequently the first question newcomers to the region are asked is what church they attend.

The Relationship of Religion and Politics Just how interactive religion and politics should be in North American life has long been a controversial issue in the United States, more so than in Canada. This is true in large part because the framers of the U.S. Constitution, in an effort to ensure religious freedom, supported the idea that church and state should remain separate. In the past three decades, however, many conservative Christians have successfully pushed for a closer integration of religion and public life. Their political goals include banning abortion, promoting prayer in the public schools, teaching the biblical version of creation instead of evolution, and preventing gays and lesbians from participating openly in the public sphere and marrying each other.

New immigrants have brought their own faiths and belief systems, and they are contributing to the debate about religion and public life. Muslims, as yet a tiny minority, have had trouble achieving the same rights as other religious minorities, such as Buddhists or Jews. Some immigrants leave their traditional faith and adopt another. Such is the case with some 15 percent

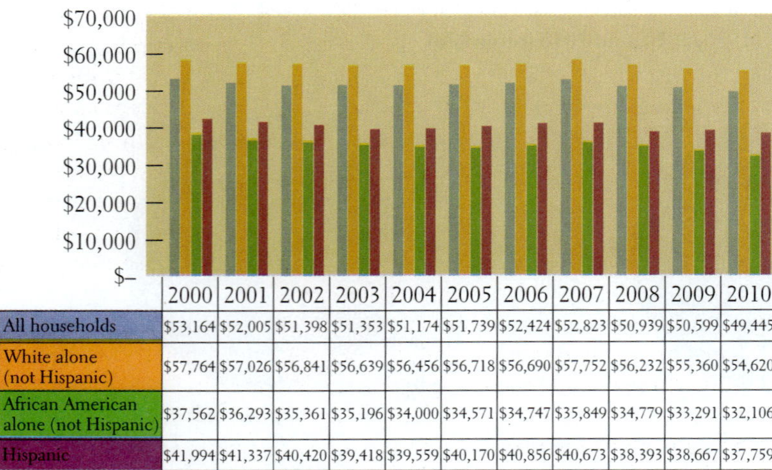

	2000	2001	2002	2003	2004	2005	2006	2007	2008	2009	2010
All households	$53,164	$52,005	$51,398	$51,353	$51,174	$51,739	$52,424	$52,823	$50,939	$50,599	$49,445
White alone (not Hispanic)	$57,764	$57,026	$56,841	$56,639	$56,456	$56,718	$56,690	$57,752	$56,232	$55,360	$54,620
African American alone (not Hispanic)	$37,562	$36,293	$35,361	$35,196	$34,000	$34,571	$34,747	$35,849	$34,779	$33,291	$32,106
Hispanic	$41,994	$41,337	$40,420	$39,418	$39,559	$40,170	$40,856	$40,673	$38,393	$38,667	$37,759

FIGURE 2.30 Median household income by race and ethnicity, 2010.

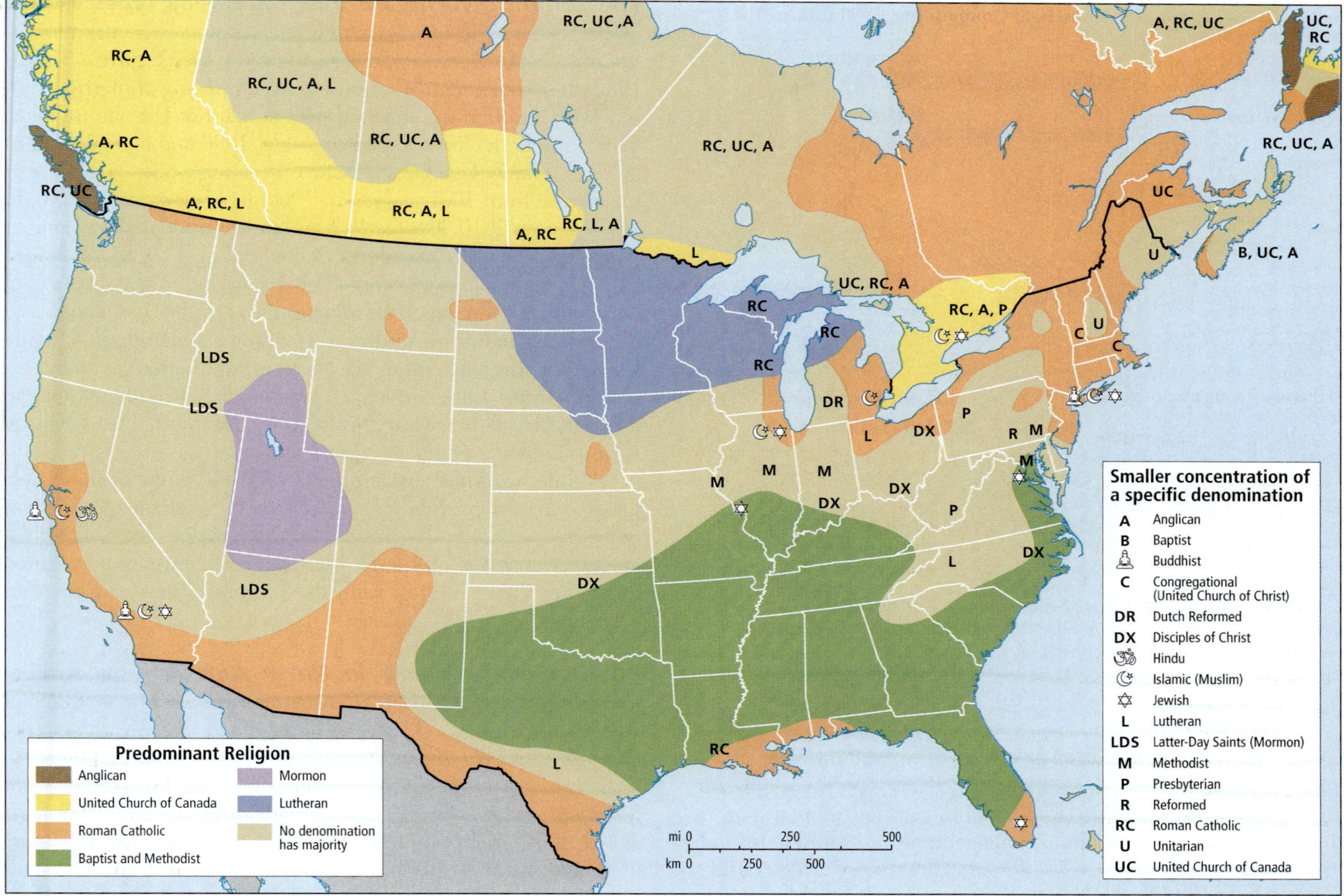

FIGURE 2.31 Religious affiliations across North America.

of Hispanic immigrants to the United States who have left the Roman Catholic faith, either while still in their home country or after immigrating, and are now evangelical Christians. In Canada, Catholicism is growing.

National surveys have consistently indicated that a substantial majority of North Americans favor the separation of church and state and support personal choice in belief and behavior. However, strident voices for particular points of view are increasingly heard, and the outcome of contentions in North American religious and political life is not yet apparent.

Gender and the American Family

The family has repeatedly been identified as the institution most in need of support in today's fast-changing and ever more impersonal North America. A century ago, most North Americans lived in large extended families of several generations. Families pooled their incomes and shared chores. Aunts, uncles, cousins, siblings, and grandparents were almost as likely to provide daily care for a child as were the mother and father. Though it has long been present in society, the **nuclear family**, consisting of a married father and mother and their children, became especially widespread in the post-1900 industrial age.

The Nuclear Family Becomes a Shaky Norm Beginning after World War I, and especially after World War II, many young people left their large kin groups on the farm and migrated to distant cities, where they established new nuclear families. Soon suburbia, with its many similar single-family homes, seemed to provide the perfect domestic space for the emerging nuclear family.

This compact family type suited industry and business because it had no firm ties to other relatives and so was portable. Many North Americans born since 1950 moved as many as ten times before reaching adulthood. The grandparents, aunts, and uncles who were left behind missed helping raise the younger generation and had no one to look after them in old age; this separation of older from younger generations is one reason nursing homes for the elderly proliferated.

In the 1970s, the whole system began to come apart. It was a hardship to move so often. Suburban sprawl meant onerous commutes to jobs for men and long, lonely days at home for women. Women began to want their own careers, and rising consumption patterns made their incomes increasingly useful to family economies. By the 1980s, seventy percent of the women born between 1947 and 1964

nuclear family a family consisting of married or common-law parents and their children

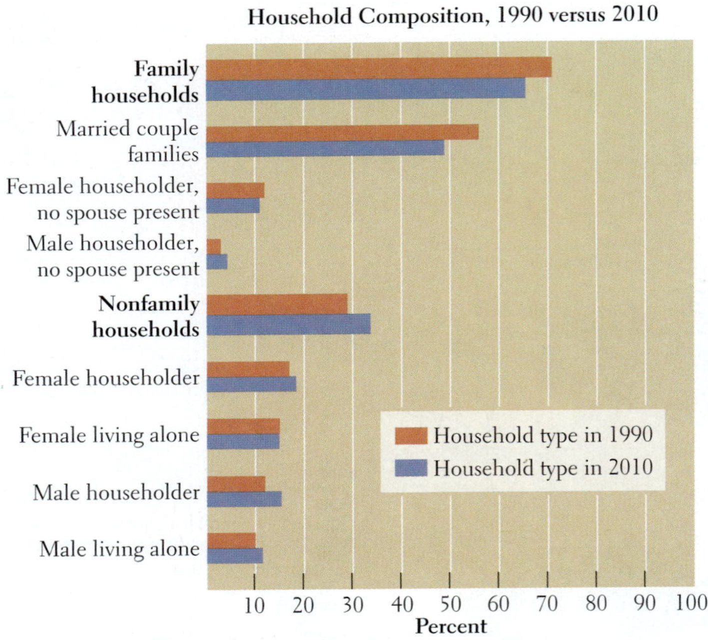

FIGURE 2.32 U.S. households by type, 1990–2010.

were in the paid workforce, compared with 30 percent of their mothers' generation.

Once employed, however, a woman could not easily move to a new job location if she had an upwardly mobile husband. Also, working women could not manage all of the family's housework and child care, as well as a job. Some married men began to handle part of the household management and child care, but the demand for commercial child care grew sharply. With family no longer around to strengthen the marital bond and help with child care, and with the new possibility of self-support available to women in unhappy marriages, divorce rates rose into the mid-1970s, while birth rates fell. Those who married after 1975, however, have had a slowly declining rate of divorce. This may be related to the growing number of couples with college educations, as this group is less likely to divorce.

The Current Diversity of American Family Types There is no longer a typical American household, only an increasing diversity of household forms and ways of family life (Figure 2.32). In 1960, the nuclear family—households consisting of married couples with children—comprised 74.3 percent of U.S. households. This number declined to 56 percent by 1990 and to 49.3 percent by 2010. Family households headed by a single person (female or male) rose from 10.7 percent in 1960, to 15 percent in 1990, to 18 percent in 2010. Nonfamily households (unrelated by blood or marriage) rose from 15 percent of the total in 1960, to 29 percent in 1990, to nearly 34 percent in 2010. Added to the diversity of family forms is the fact that, increasingly, families in Canada and the United States are multiracial and/or multicultural. Usually this family diversity comes via cross-cultural marriage, and often children from these marriages are biracial. President Obama, for example, with a white North American mother and a black African father, is a biracial American.

While most of these families work well, a few of these new forms may not provide adequately for the welfare of children. In 2006, more than 29 percent of U.S. children lived in single-parent households. Although most single parents are committed to rearing their children well, the responsibilities can be overwhelming.

Children and Poverty in North America Single-parent families tend to be hampered by economic hardship and lack of education (Figure 2.33). Young women head the vast majority of single-parent households, and their incomes, on average, are less than two-thirds those of single male heads of household. Low income and low levels of education are closely linked, as the graph in Figure 2.33B shows. One result of the pattern of single-parent households having lower education levels and incomes is that children in the United States are disproportionately poor. In 2010 in the United States, 22 percent of children aged 0 to 17 (16.4 million) lived in poverty, whereas only 11.4 percent of adults did. In Canada, 14.2 percent of children were poor. By comparison, in Sweden, just 2.4 percent of children lived in poverty; in Ireland, 12.4 percent; and in Poland, 12.7 percent.

(A) Median U.S. Income by Type of Household, 2007

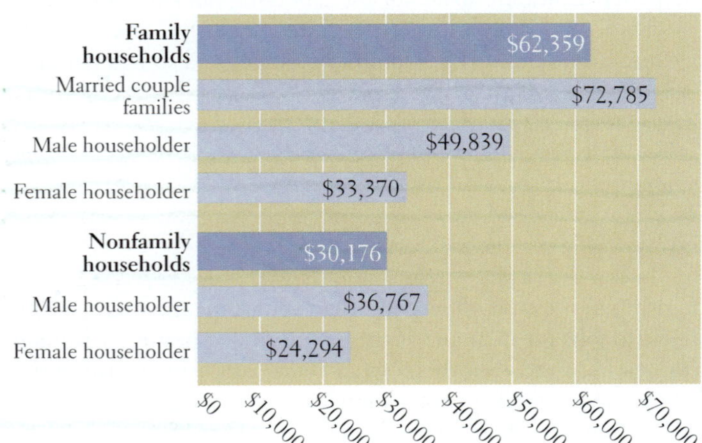

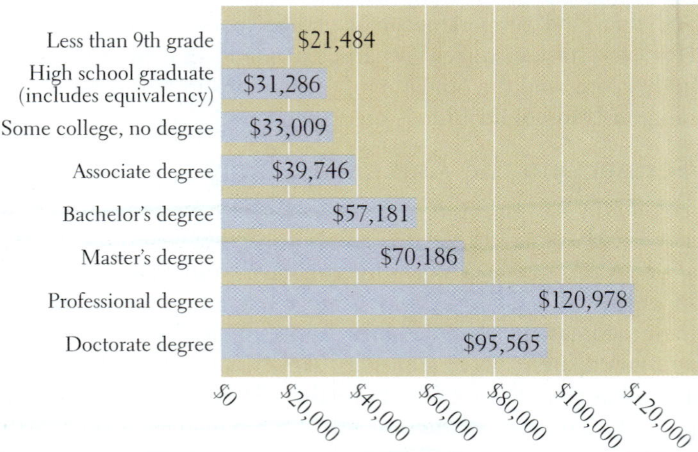

FIGURE 2.33 Median U.S. income by (A) type of household and (B) level of education, 2007.

THINGS TO REMEMBER

- North America has some of the world's wealthiest cities, most of which are in the eastern United States. Nearly 80 percent of North Americans now live in metropolitan areas.
- Since World War II, North America's urban populations have increased by about 150 percent, but the amount of land they occupy has increased almost 300 percent, a phenomenon known as urban sprawl.
- A steady increase in migration from Middle and South America and parts of Asia promises to make North America a region where most people are of non-European descent.
- Statistics on access to health care, education, and financial services by race and ethnicity show that, on average, Americans have quite uneven experiences based on their racial and ethnic characteristics.
- There is no longer a typical American household, only an increasing diversity of household forms.
- A large percentage of children in the United States live in poverty, compared to other developed countries.

General Population Patterns

The population map of North America (Figure 2.34) shows the uneven distribution of the more than 346 million people who

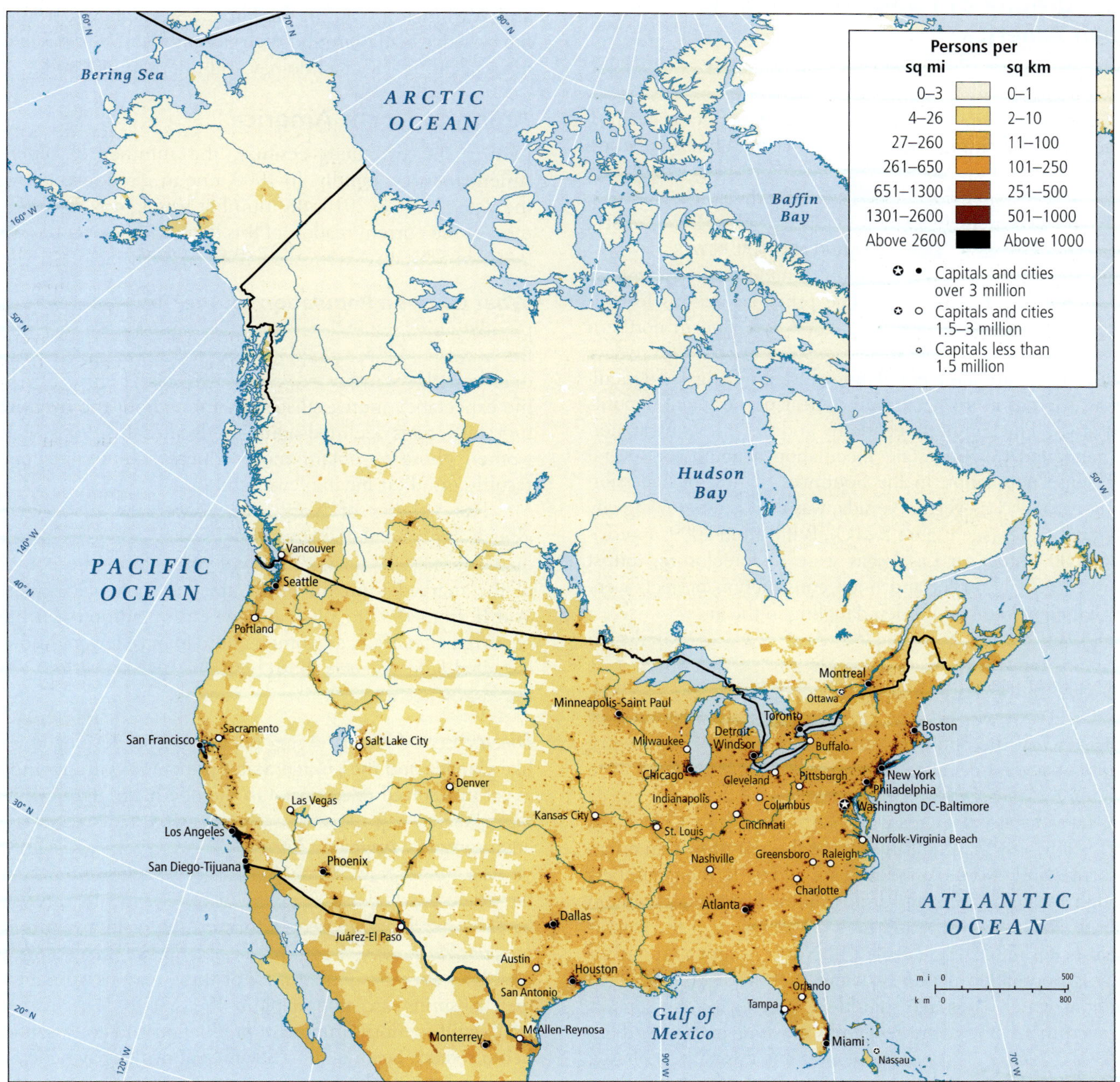

FIGURE 2.34 Population density in North America.

live on the continent. Canadians make up just under one-tenth (34.5 million) of North America's population. They live primarily in southeastern Canada, close to the border with the United States. The population of the United States is 311.7 million, with the greatest concentration of people shifting away from the old economic core into other regions of the country that are now growing much faster. Just how this trend will play out, however, is uncertain. In 2004, the U.S. Census Bureau projected that between 2000 and 2030, the Northeast and Middle West would grow by less than 10 percent, while the South and West would grow by more than 40 percent. This general trend may persist, but the 2008 economic recession, water shortages, and the energy crisis have limited expansion, especially in the West but also in the South.

The Geography of Population Change in North America

In many farm towns and rural areas in the *Middle West* (the large central farming region of North America), populations are shrinking. As family farms are consolidated under corporate ownership, labor needs are decreasing and young people are choosing better-paying careers in cities. Middle Western cities are growing only modestly yet are becoming more ethnically diverse, with rising populations of Hispanics and Asians in places such as Indianapolis, St. Louis, and Chicago. **52. LATINO POLITICAL POWER**

In the western mountainous interior, traditionally settlement has been light (see Figure 2.34). The principal reasons for this low density are rugged topography, lack of rain, and in northern or high altitude zones, a growing season that is too short to sustain agriculture. There are some population clusters in irrigated agricultural areas, such as in the Utah Valley, near rich mineral deposits or resort areas. The gambling economy and frenetic construction activity generated by real estate speculation account for several knots of dense population in the Southwest. Until the recession beginning in 2007, Las Vegas, Nevada, was the fastest-growing city in the United States (see Figure 2.21C). But by mid-2009, approximately 67,000 homes in Las Vegas were in foreclosure, tourist arrivals were sharply down, and hotel construction projects were abruptly halted with crane jibs left dangling in the air.

Along the Pacific Coast, a band of growing population centers stretches north from San Diego to Vancouver (see Figure 2.21A) and includes Los Angeles, San Francisco, Portland, and Seattle. These are all port cities engaged in trade around the **Pacific Rim** (all the countries that border the Pacific Ocean). Over the past several decades, these North American cities have become centers of technological innovation.

The rate of natural increase in North America (0.5 percent per year) is low, less than half the rate of the rest of the Americas (1.2 percent). Still, North Americans are adding to their numbers fast enough through births and immigration that the population could reach more than 422 million by 2050. Many of the important social issues now being debated in North America are linked to changing population patterns—issues such as legal and illegal immigration; which language should be used in public schools and in government offices; the geography of voting patterns; urbanization; cultural diversity; mobility; and the social effects of an aging population.

Pacific Rim a term that refers to all the countries that border the Pacific Ocean

Mobility in North America

Every year, almost one-fifth of the U.S. population and two-fifths of Canada's population relocate. Some people are changing jobs and moving to cities; others are attending school, retiring to a warmer climate or a smaller city or town, or merely moving across town, to the suburbs, or to the countryside. Still other people are arriving from outside the region as immigrants.

Urbanization remains a powerful force behind this mobility. Dynamic economies and the search for lower production and living costs are drawing employers, employees, and retirees to urban areas. In Canada, people are going to cities in the southeast and on the West Coast. In the United States, people are moving to the South, Southwest, and Pacific Northwest. Cities in these areas, such as Vancouver and Toronto in Canada, and Atlanta, Georgia, and Washington, DC, in the United States, have sprouted satellite or "edge" cities around their peripheries, often based on businesses dealing with technology and international trade.

Aging in North America

During the twentieth century, the number of older North Americans grew rapidly. In 1900, one in 25 individuals was over the age of 65; by 2010, the number was 1 in 8. By 2050, when most of the current readers of this book will be over 65, it is likely that 1 in 5 North Americans will be elderly.

What Causes a Population to Age? This aging of the region's population relates both to longer life expectancy, which increases the number of older people, and to a declining birth rate, which decreases the number of younger people (Figure 2.35). While life expectancy increased steadily throughout the twentieth century, birth rates declined significantly after the 1960s when more women chose to obtain more education and pursue careers, resulting in their having fewer children.

The number of North Americans over the age of 65 is already high by global standards and will increase dramatically over the next 20 years. This will result from the combination of the marked jump in birth rate that took place after World War II, from 1947 to 1964, and the ensuing drop in birthrate that occurred after 1965, as women chose to have fewer children. The so-called *baby boomers* (born between 1947 and 1965) constitute the largest age group in North America.

Dilemmas of Aging Aging populations in developed countries like Canada and the United States present us with unfamiliar and as-yet unresolved dilemmas. On the one hand, it is widely agreed that globally, population growth should be reduced to lessen the environmental impact of human life on Earth, especially in those societies (such as in North America) that consume the most. On the other hand, slower population growth means that there will be fewer working-age people to keep the economy going and to provide the financial and physical help the increasing number of elderly people will require. For these reasons, immigrants and their offspring are important to the system. To reiterate: first, immigrants perform important production and consumption roles in our society. Second, their Social Security tax payments are helping to support the large group of retiring boomers; while they were working, the boomers' contributions were paid into trust funds that

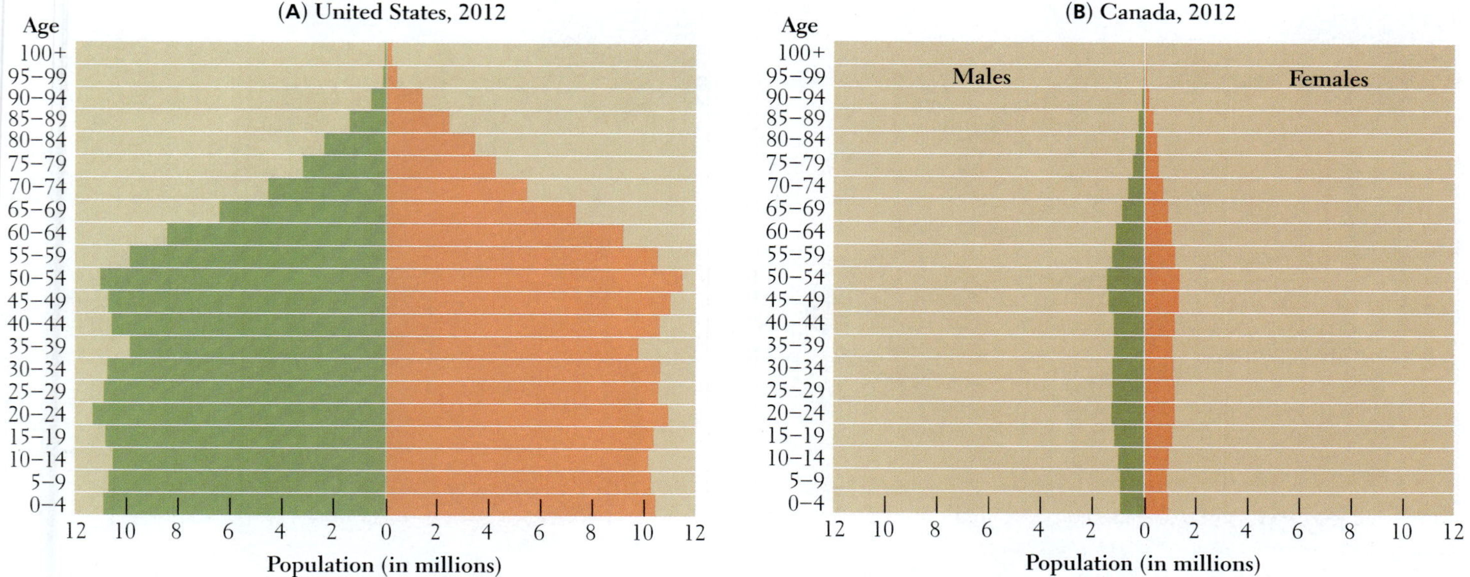

FIGURE 2.35 Population pyramids for the United States and Canada, 2012. The "baby boomers," born between 1947 and 1964, constitute the largest age group in North America, as indicated by the wider middle portion of these population pyramids.

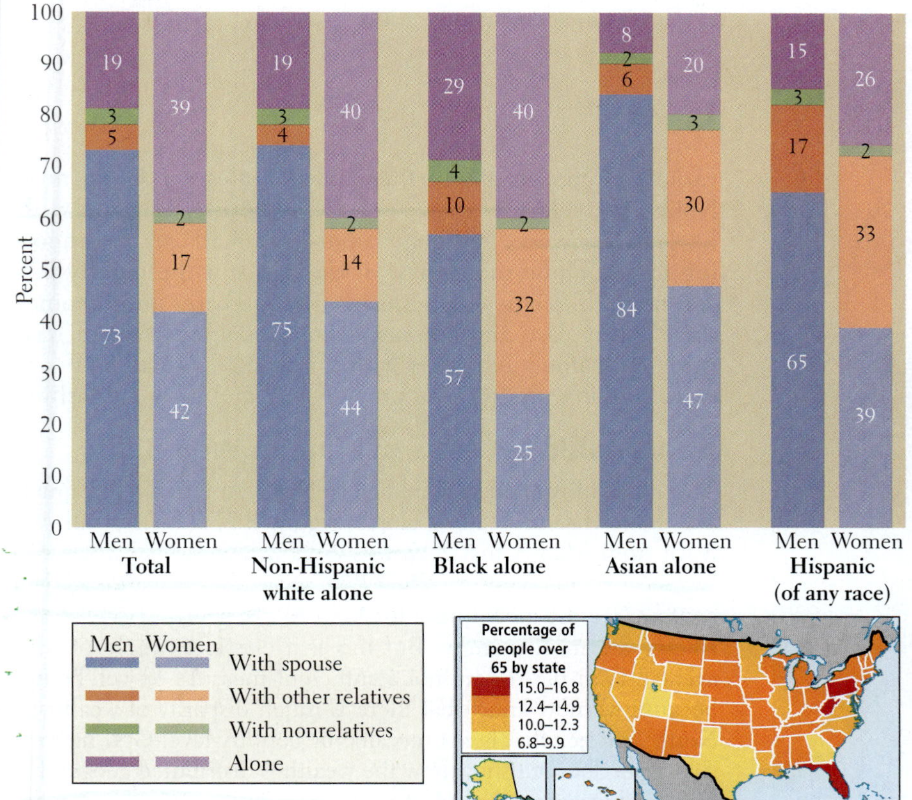

FIGURE 2.36 Living arrangements of U.S. people age 65 and over by sex, race, and Hispanic origin, 2007.

supported then-current Social Security recipients. Finally, because of the increasing demand for caregivers for the aged, some immigrants will fill in for the diminished supply of boomer kinfolk as paid caregivers and companions.

Given that the population is aging, should retirement ages be extended from 65 to perhaps 75? Might this deprive young people of access to jobs? These worries have greatly preoccupied Europeans, for whom retirement ages can be as low as 58 and whose populations are aging faster than North Americans. Economists who have studied this issue say that it clearly makes sense to extend the retirement age. First, this means those over 65 will remain self-supporting longer; second, the talents and experience of these people will remain in circulation; third, statistics across the developed countries attest to the fact that high elderly employment rates are associated with high youth *employment*, not unemployment. Working people of any age spend their incomes and create jobs by so doing. A society that pays an increasing proportion of its workers to not work cannot prosper.

Eventually, most elderly people will need special attention of some sort. To avoid the social isolation that so many elderly are already encountering, it will be necessary to develop affordable alternative living arrangements. Figure 2.36 shows how the elderly of

FIGURE 2.37 Maps of human well-being.

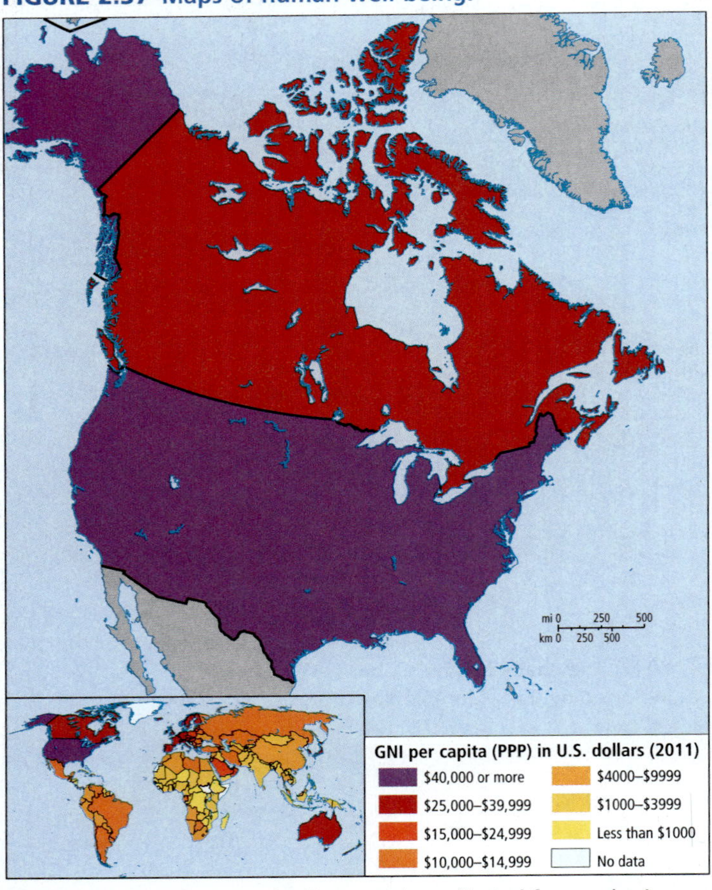

(A) Gross national income (GNI) per capita, adjusted for purchasing power parity (PPP).

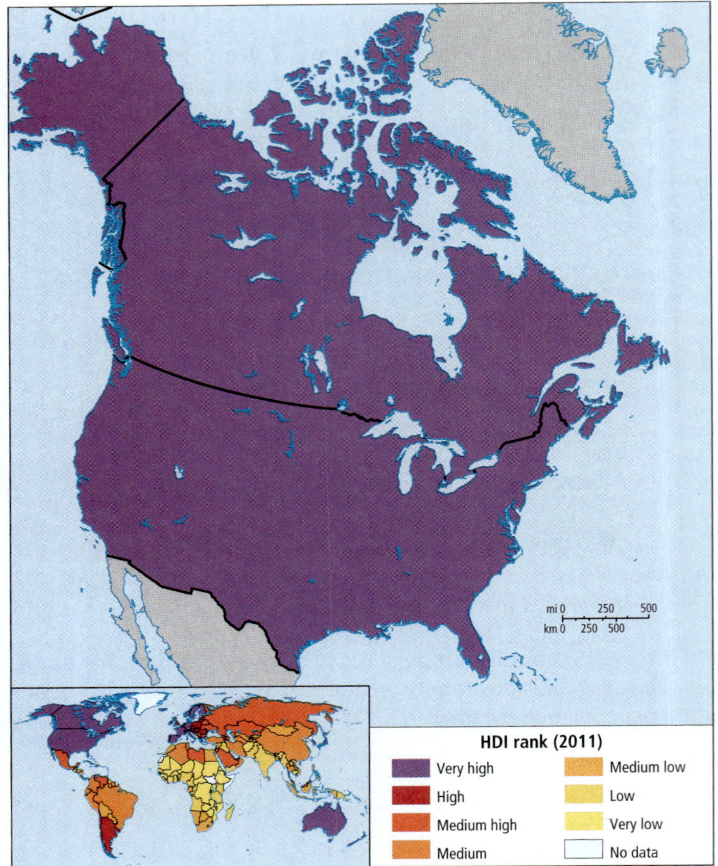

(B) Human Development Index (HDI).

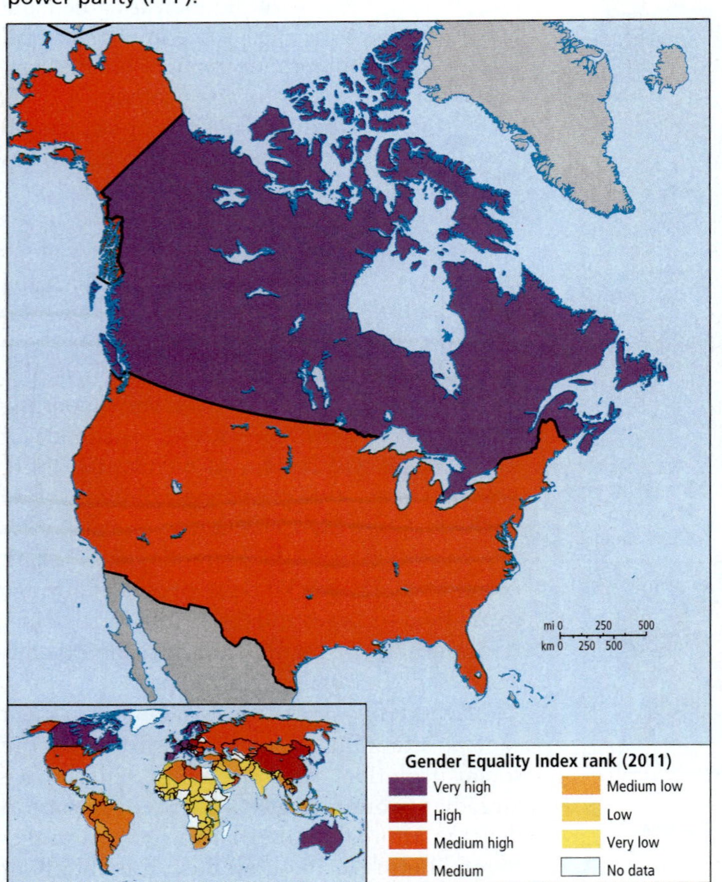

(C) Gender Equality Index (GEI) rank.

various ethnic groups in the United States are living now. Notice how elderly women in all groups are the ones most likely to live alone. Although Americans of all ages are choosing to live alone more than in the past, for the elderly this can be problematic as physical frailty increases and incomes shrink. Day care and cohousing for the elderly, where residents look after each other with the aide of a small staff, are two affordable strategies to provide care and companionship.

Geographic Patterns of Human Well-Being

The maps of human well-being for North America tell a quick and somewhat misleading story (Figure 2.37). First of all, we learn from Map A that both countries in this region have very high average per capita GNI (PPP) figures and that in the global context (see the global inset for Map A), these two countries are among the most wealthy. But this is a classic example of why average GNI figures are not a sufficient measure of well-being, because we now know that there is much disparity of wealth in North America that is not revealed by country-level GNI figures. This map does not show how the wealth is distributed geographically or between social classes. It gives no hint that 22 percent of children in the United States and 14 percent in Canada are classed as poor despite having at least one working parent, that the situation has deteriorated over the last decade, or that in these countries the richest 10 percent have more than 14 times the wealth of the poorest 10 percent. The truth is that at the local scale the poor are geographically separated from the middle class so completely that poor children often experience life only in very deprived communities.

Virtually the same problem is presented by Map B, which shows the ranks of these two countries on the United Nations Human Development Index (HDI). Because the data are presented for the entire countries, not at the province or state level, the fact that many parts of both countries provide very well for their citizens in terms of income, health care, and education boosts the average figures for all. Glossed over is the reality that there are significant pockets of poverty, ill health, and illiteracy in virtually every state and province (they are less devastating in Canada because of its more robust safety net).

Map C is a composite measure reflecting the degree to which there is inequality in achievements between women and men in three dimensions: reproductive health, empowerment, and the labor market. A high rank indicates that the genders are tending toward equality. But notice that disparity in pay is not used, probably because statistics on pay by gender are not being submitted by many governments. Map C reveals that the United States lags behind Canada and other places on Earth, in gender equality. Of those with comparable per capita GNI figures, Australia and North and West Europe are doing better at equalizing pay and opportunity for female and male citizens. In fact, nowhere on Earth has actual gender equity been achieved.

THINGS TO REMEMBER

- The population of North America is changing in distribution and becoming more diverse.
- North Americans are highly mobile. Every year, for a variety of reasons, almost one-fifth of the U.S. population and two-fifths of the Canadian population move to a different location.
- North America's population is rapidly aging, and by 2050, one in five people will be over the age of 65. There will be fewer young people to work, pay taxes, and take care of the elderly.
- Birth rates in North America declined significantly after the 1960s as women chose to obtain more education and pursue careers, resulting in their having fewer children.

Geographic Insights: North America Review and Self-Test

1. Climate Change and Urbanization: North America's large per capita production of greenhouse gases is related to its standard of living and dominant pattern of urbanization.

• What aspects of North American lifestyles have increased greenhouse gases? What efforts have been made to control these emissions—and are there any you would participate in?

• How does urban sprawl influence the average North American's consumption of gasoline?

• How might the livable city movement and smart growth affect greenhouse gas emissions? What are some responses to reducing greenhouse gases that you would be willing to take part in yourself?

2. Water and Food: North American food production systems contribute to water scarcity and water pollution.

• What activities in North America are major contributors to water pollution and water scarcity, and what are the likely resulting impacts for future water use?

• What effect do green revolution agriculture techniques have on aquatic ecosystems?

3. Power and Politics: The interpretation of the ideal of democracy by Canada and the United States has influenced North America's involvement in the affairs of countries around the world.

• What role does the urge to spread democracy play in the foreign policy of North America?

• How does Canada's approach to foreign relations differ from that of the United States?

• What role did the ideal of democracy play in justifying the Iraq and Afghanistan wars?

• To what extent is the allocation of U.S. foreign aid linked to U.S. strategic military interests?

4. Globalization and Development: Globalization has transformed economic development in North America, thereby changing trade relationships and the kinds of jobs available in this region.

• How has globalization transformed the North American job market, including the steel industry?

• How did fair trade principles lead to NAFTA goals, and what has been the effect of NAFTA on manufacturing jobs and immigration in North America?

• What are some exceptions to U.S. support of free trade principles?

5. Population and Gender: Side effects of changes in gender roles and in opportunities for women include some improvements in education and pay equity; smaller families; and the aging of North America's population.

• How do education and work opportunities for contemporary women affect the lives and geographical mobility of the North American population—and in what ways does this affect you?

• What are factors that have created an aging North American population, and how do those factors connect with immigration and household types in Canada and the United States?

• What are three ways in which you personally are likely to be affected by the aging of North America's population?

• Why are immigrants crucial to North America's aging population?

• What is the connection between single-parent households and the fact that 22 percent of U.S. children (14 percent of Canadian) live in poverty?

Critical Thinking Questions

1. Discuss the ways in which North American culture is adapted to the high rate of spatial (geographic) mobility engaged in by Americans.

2. North American family types are changing. Explain the general patterns and discuss whether or not the nuclear family is or should be the sought-after norm.

3. The influence of globalization is now felt in small, even isolated, places in North America. Pick an example from the text or from your own experience and explain at least four ways in which this place is now connected to the global economy or global political patterns.

4. Some say that North Americans profit from having undocumented workers produce goods and services. Explain why this may be the case and discuss if and how the situation should be changed.

5. People like to say that the attacks on 9/11 changed everything for North Americans (especially for those living in the United States). What do they mean by this? How were U.S. attitudes toward the rest of the world modified? Will these modifications be useful or destructive in the long run?

6. When you compare the economies of the United States and Canada and the ways they are related, what are the most important factors to mention?

7. When U.S. farmers or producers get subsidies from the federal government, what is intended? Why do farmers and other food producers in poor countries say that this practice hurts them?

8. The United States and Canada have quite different approaches in their treatment of citizens who experience difficulties in life. Explain those different approaches and the apparent philosophies behind them.

Chapter Key Terms

acid rain 66
agribusiness 82
aquifers 61
brownfields 93
clear-cutting 65
digital divide 86
economic core 71
ethnicity 96
gentrification 93
Hispanic (Latino) 61
infrastructure 71
megalopolis 91
metropolitan areas 91
North American Free Trade Agreement (NAFTA) 85
nuclear family 99
organically grown 84
Pacific Rim 102
push factors 93
smog 66
social safety net 80
suburbs 91
trade deficit 87
urban sprawl 65

A Sierra Madre, Mexico

B Andes, Chile

C Tierra del Fuego, Argentina

3 Middle and South America

D Soufrière Hills Volcano, Montserrat

G Rio de Janeiro, Brazil

H Yucatán Lowlands, Mexico (with Mayan temple)

F Amazonian Basin, Brazil (from the air)

E Amazonian Basin, Brazil (from space)

I Pampas, Argentina

FIGURE 3.1 Regional map of Middle and South America.

GEOGRAPHIC INSIGHTS: MIDDLE AND SOUTH AMERICA

After you read this chapter, you will be able to discuss the following geographic insights as they relate to the nine thematic concepts:

1. Climate Change and Deforestation: Deforestation in this region contributes to climate change by removing trees, which as living plants naturally absorb carbon dioxide. As the cut vegetation decays, large amounts of carbon dioxide are released into the atmosphere. Sustainable alternatives to deforestation are being tried.

2. Water: Despite the region's abundant water resources, parts of the region are experiencing a water crisis related to unplanned urbanization; marketization; corruption; inadequate water infrastructure; and the use of water to irrigate commercial agriculture.

3. Globalization, Development, Power, and Politics: The integration of this region with the world economy has led to growing disparities of wealth and opportunity, forcing millions to migrate in search of work. A political backlash against these patterns of development has brought to power some leaders who question the benefits of globalization as well as others who seek to help their countries compete globally.

4. Food and Urbanization: The shift from small-scale subsistence food production to mechanized agriculture for export has fueled migration to cities. Many places previously were self-sufficient with regard to food; they now depend on imported food, making them susceptible to food insecurity.

5. Population: A population explosion in the twentieth century has given this region about 596 million people, more than 10 times the population of the region in 1492. Now social and economic changes, including urbanization and the availability of contraception, are giving women options other than raising a family.

6. Gender: Traditional gender roles are changing as women work outside the home and manage families while many men are migrating to other places for work. Also, urbanization is reducing the influence of extended families, opening the door for many types of social change.

The Middle and South American Region

Middle and South America, like North America, is a region of great physical and cultural diversity (see Figure 3.1 on pages 108–109); but here the cultural mix is, if anything, richer and there are wider disparities of wealth than in any other world region. In spite of many similarities, modern history in Middle and South America has been very different from that of North America. The nine thematic concepts in this book are explored as they arise in the discussion of regional issues, with interactions between two or more themes featured, as in the geographic insights above. Vignettes—like the one that follows about the Secoya—illustrate one or more of the themes as they are experienced in individual lives.

GLOBAL PATTERNS, LOCAL LIVES

VIGNETTE The boat trip down the Aguarico River in Ecuador took me into a world of magnificent trees, river canoes, and houses built high up on stilts to avoid floods. I was there to visit the Secoya, a group of 350 indigenous people locked in negotiations with the U.S. oil company Occidental Petroleum over its plans to drill for oil on Secoya lands. Oil revenues supply 40 percent of the Ecuadorian government's budget and are essential to paying off its national debt. The government had threatened to use military force to compel the Secoya to allow drilling.

The Secoya wanted to protect themselves from pollution and cultural disruption. As Colon Piaguaje, chief of the Secoya, put it to me, "A slow death will occur. Water will be poorer. Trees will be cut. We will lose our culture and our language, alcoholism will increase, as will marriages to outsiders, and eventually we will disperse to other areas." Given all the impending changes, Chief Piaguaje asked Occidental to use the highest environmental standards in the industry. He also asked the company to establish a fund to pay for the educational and health needs of the Secoya people.

Like the Secoya, indigenous peoples around the world are facing environmental and cultural disruption arising from economic development efforts. Chief Piaguaje based his predictions for the future on what has happened in other parts of the Ecuadorian Amazon that have already had several decades of oil development.

The U.S. company Texaco was the first major oil developer to establish operations in Ecuador. From 1964 to 1992, its pipelines and waste ponds leaked almost 17 million gallons of oil into the Amazon Basin, enough to fill about 1900 fully loaded oil tanker trucks, or 35 Olympic-size swimming pools. Although Texaco sold its operations to the government and left Ecuador in 1992, its oil wastes continue to leak into the environment from hundreds of open pits (**Figure 3.2**).

In 1993, some 30,000 people sued Texaco for damages in New York State, where the company (now owned by and called Chevron) is headquartered, for damages from the pollution. Those suing were both indigenous people and settlers who had established farms along Texaco/Chevron's service roads. Several epidemiological studies concluded that oil contamination has contributed to higher rates of childhood leukemia, cancer, and spontaneous abortions among people who live near the pollution created by Texaco/Chevron. Also, oil development has had many negative effects on the environment. Air and water pollution have increased rates of illness. The wildlife that the Secoya used to depend on, such as tapirs, have disappeared almost entirely because of overhunting by new settlers from the highlands who are working in the oil industry.

In 2002, the Ecuadorian suit against Chevron was dismissed by the U.S. Court of Appeals, but it was refiled in Ecuador in 2003. In 2011 the Ecuadorian court ruled against Chevron, assessing damages of U.S.$18 billion. Chevron appealed to the Ecuadorian Supreme Court. This decision could go against Chevron; however, because Chevron no longer has any assets in Ecuador, the villagers may never collect. *[Sources: Alex Pulsipher's field notes; Amazon Watch, 2006; Oxfam America, 2005; National Public Radio. For detailed source information, see Text Credit pages.]* ■

FIGURE 3.2 Pollution from oil development in Ecuador. A local resident samples one of the several hundred open waste pits that Texaco left behind in the Ecuadorian Amazon. Wildlife and livestock trying to drink from these pits are often poisoned or drowned. After heavy rains, the pits overflow, polluting nearby streams and wells.

The rich resources of Middle and South America have attracted outsiders since the first voyage of Christopher Columbus in 1492. Europe's encounter with this region marked a major expansion of the global economy. However, during most of the period of expansion, Middle and South America occupied a disadvantaged position in global trade, supplying cheap raw materials that aided the Industrial Revolution in Europe and then North America but reaping few of the profits. These extractive industries did little to advance economic development within the region, as most profits went to foreign investors, and the negative environmental effects were largely ignored. In recent years, countries such as Ecuador, Mexico, Bolivia, Brazil, and Venezuela have worked to control their own resources, develop local industries, keep profits at home, and limit environmental pollution. Meanwhile, trade blocs within the region are creating conditions in which these countries can prosper from trade with each other.

INDIGENOUS PEOPLES (4&5)
DOGRIB (2)
NUNAVUT (2)
CREE (2)
CHOCTAW (2)
LOCAL POPULATION KAZAKH HERDERS (5)
INDIGENOUS HUNTERS & FISHERS (5)
AINU (9)
KURDS (6)
INDIGENOUS PEOPLES M/F (9)
INDIGENOUS PEOPLES (9)
ETHNIC MINORITIES XXX (10)
INDIGENOUS PEOPLES (3)
INDIGENOUS PEOPLES (3)
INDIGENOUS PEOPLES (3)
INDIGENOUS HERDERS (7)
DINKA, NUER, & PEOPLES of DARFUR (6)
BADAGA, TODA, KOTA, KURUMBA PEOPLES (8)
INDIGENOUS PEOPLES (8)
KOMORO (10)
INDIGENOUS PEOPLES (11)
INDIGENOUS PEOPLES (7)
FISHING PEOPLES (8)
FOREST PEOPLE (10)
INDIGENOUS PEOPLES (10)
ABORIGINES (11)
NATIVE FIJIANS (11)
MAORI (11)
COLOMBIA
SECOYA
Puerto Misahualli
Aguarico River
ECUADOR
PERU
(3)

Issues Facing Indigenous Peoples

Tourism
M/F Gender customs
Assimilation
Religous/ethnic violence
Petroleum pollution
Mining pollution
Deforestation
Water depletion
Nuclear pollution
Hydroelectric projects
Land and resource rights
XXX Sex trade
Drug wars
War/genocide
Fish depletion
Wild animal hunting
(11) Indicates chapter in text where issue is discussed

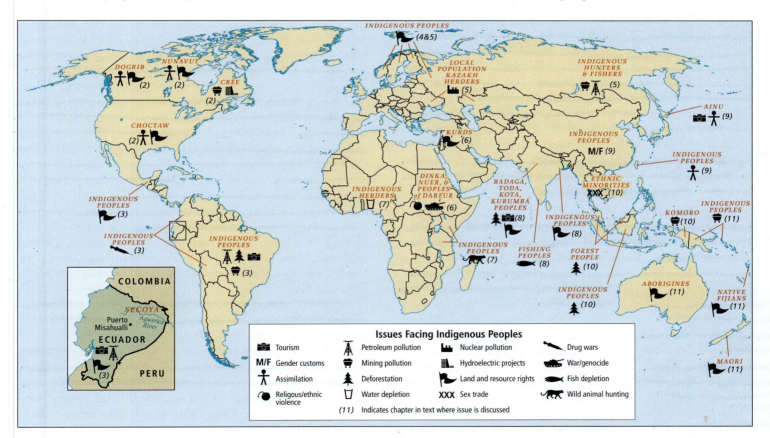

FIGURE 3.3 Indigenous peoples and environmental issues in this text. Issues relating to indigenous peoples are mentioned in many places in this book. In all cases, the issues are in some way related to interactions with the outside world and usually to uses of indigenous peoples' resources by the outside world.

Secoya villagers' efforts to secure a damage settlement against a powerful multinational corporation is indicative of changing attitudes in this region and others (Figure 3.3 on page 111) toward outside developers. Governments are now somewhat more transparent and savvy when they lure investors to develop new extractive, manufacturing, and service industries. Local people are more aware that they must be vigilant to ensure that development serves their interests.

THINGS TO REMEMBER

- The region of Middle and South America, heavily impacted by European colonialism, is making a shift away from raw materials–based industries to more profitable manufacturing and service-based industries.
- As people in this region gain more control over their own resources, some are searching for more sustainable ways to develop.

THE GEOGRAPHIC SETTING

What Makes Middle and South America a Region?

Physically, the Middle and South America region consists of Mexico, which geologically is part of the North American continent, the **isthmus** (land bridge) of Central America, and the continent of South America. For the last 500 years, the region of Middle and South America has been defined by a colonial past very different from that of Canada and the United States. Most of the countries were at one time colonies of Spain. The exceptions are Brazil, which was a colony of Portugal, and a few small countries that were possessions of Britain, France, the Netherlands, Germany, or Denmark (see Figure 3.12 on page 124).

Today, Middle and South America is a region of contrasts and disparities. Culturally, this region has large **indigenous** populations that have contributed to every aspect of life, blending with and changing the European, African, and Asian cultures introduced by the colonists. Social stratification based on class, race, and gender is notable; economically, the gap between rich and poor is the widest of any world region. Politically, the region's more than three dozen countries exhibit a range of governing ideologies, from the socialism of Cuba to the capitalism of Chile. Yet despite these contrasts and disparities, there are significant commonalities across the region, such as the Spanish language, Catholicism, and development trajectories increasingly connected to the global economy.

Terms in This Chapter

In this book, **Middle America** refers to Mexico, Central America (the narrow ribbon of land, or isthmus, that extends south of Mexico to South America), and the islands of the Caribbean (Figure 3.4). **South America** refers to the continent south of Central America. The term *Latin America* is not used in this book because it describes the region only in terms of the Roman (Latin-speaking) origins of the former colonial powers of Spain and Portugal. It ignores the region's large indigenous groups, its African, Asian, and Northern European populations, as well as the many mixed cultures, often called *mestizo* cultures, that have emerged. In this chapter, we use the term *indigenous groups* or *peoples* rather than *Native Americans* to refer to the native inhabitants of the region.

isthmus a narrow strip of land that joins two larger land areas

indigenous native to a particular place or region

Middle America in this book, a region that includes Mexico, Central America, and the islands of the Caribbean

South America the continent south of Central America

subduction zone a zone where one tectonic plate slides under another

Physical Patterns

Middle and South America extend south from the midlatitudes of the Northern Hemisphere across the equator through the Southern Hemisphere, nearly to Antarctica (see Figure 3.1 on pages 108–109). This vast north–south expanse combines with variations in altitude to create the wide range of climates in the region. Tectonic forces have shaped the primary landforms of this huge territory to form an overall pattern of highlands to the west and lowlands to the east.

Landforms

There are a wide variety of landforms in Middle and South America, and this variety contributes to the many different climatic zones in the region. But for ease in learning, landforms are here divided into just two categories: highlands and lowlands.

Highlands A nearly continuous chain of mountains stretches along the western edge of the American continents for more than 10,000 miles (16,000 kilometers) from Alaska to Tierra del Fuego at the southern tip of South America (see Figure 3.1C). The middle part of this long mountain chain is known as the Sierra Madre in Mexico (see Figure 3.1A), by various names in Central America, and as the Andes in South America (see Figure 3.1B, C). It was formed by a lengthy **subduction zone**, which runs thousands of miles along the western coast of the continents (see Figure 1.25 on page 46). Here, two oceanic plates—the Cocos Plate and the Nazca Plate—plunge beneath three continental plates—the North American Plate, the Caribbean Plate, and the South American Plate.

In a process that continues today, the leading edge of the overriding plates crumples to create mountain chains. In addition, molten rock from beneath the Earth's crust ascends to the surface through fissures in the overriding plate to form volcanoes. Such volcanoes are the backbone of the highlands that run through Middle America and the Andes of South America (see Figure 3.1). Although these volcanic and earthquake-prone highlands have been a major barrier to transportation, communication, and settlement, people now live close to quiescent volcanoes and in earthquake-prone zones, which can pose deadly hazards (for example, the 2010 earthquakes in Peru and Chile).

The chain of high and low mountainous islands in the eastern Caribbean is also volcanic in origin, created as the Atlantic

FIGURE 3.4 Political map of Middle and South America.

Plate thrusts under the eastern edge of the Caribbean Plate. It is not unusual for volcanoes to erupt in this active tectonic zone. On the island of Montserrat, for example, people have been living with an active, and sometimes deadly, volcano for more than a decade (see Figure 3.1D). Eruptions have taken the form of violent blasts of superheated rock, ash, and gas (known as *pyroclastic flows*) that move down the volcano's slopes with great speed and force. The unusually strong earthquake in Haiti in January 2010 was also the result of plate tectonics.

Lowlands Vast lowlands extend over most of the land to the east of the western mountains. In Mexico, east of the Sierra Madre, a coastal plain borders the Gulf of Mexico (see Figure 3.1H). Farther south, in Central America, wide aprons of sloping land descend to the Caribbean coast. In South America, a huge wedge of lowlands, widest in the north, stretches from the Andes east to the Atlantic Ocean. These South American lowlands are interrupted in the northeast and the southeast by two modest highland zones: the Guiana Highlands and the Brazilian Highlands (see Figure 3.1G). Elsewhere in the lowlands, grasslands cover huge, flat expanses, including the llanos of Venezuela, Colombia, and Brazil, and the pampas of Argentina (see Figure 3.1I).

The largest feature of the South American lowlands is the Amazon Basin, drained by the Amazon River and its tributaries (see Figure 3.1E, F). This basin lies within Brazil and the neighboring countries to its west. The Earth's largest remaining expanse of tropical rain forest gives the Amazon Basin global significance as a reservoir of **biodiversity**. Hundreds of thousands of plant and animal species live here.

biodiversity the variety of life forms to be found in a given area

The basin's water resources are also astounding. Twenty percent of the Earth's flowing surface waters exist here, running in rivers so deep that ocean liners can steam 2300 miles (3700 kilometers) upriver from the Atlantic Ocean all the way to Iquitos, jokingly referred to as Peru's "Atlantic seaport." The vast Amazon River system starts as streams high in the Andes. These streams eventually unite as rivers that flow eastward toward the Atlantic. Once they reach the flat land of the Amazon Plain, their velocity slows abruptly; fine soil particles, or **silt**, then sink to the riverbed. When the rivers flood, silt and organic material transported by the floodwaters renew the soil of the surrounding areas, nourishing millions of acres of tropical forest. Not all of the Amazon Basin is rain forest, however. Variations in weather and soil types, as well as human activity, have created grasslands and seasonally dry deciduous tropical forests in some areas.

Climate

From the jungles of the Caribbean and the Amazon to the high, glacier-capped peaks of the Andes to the parched moonscape of the Atacama Desert and the frigid fjords of Tierra del Fuego, the climate variety of Middle and South America is enormous (Figure 3.5). Climates are essentially the result of interactions between temperature and moisture. In this region, the wide range of temperatures reflects both the great distance the landmass spans on either side of the equator and the tremendous variations in altitude across the region's landmass (the highest point in the Americas is Aconcagua in Argentina, at 22,841 feet [6962 meters]). Patterns of precipitation are affected both by the local shape of the land and by global patterns of wind and ocean currents that bring moisture in varying amounts.

Temperature-Altitude Zones Four main **temperature-altitude zones**, shown in Figure 3.6 on page 116, are commonly recognized in the region. As altitude increases, the temperature of the air decreases by about 1°F per 300 feet (1°C per 165 meters) of elevation. Thus temperatures are highest in the lowlands, which are known in Spanish as the *tierra caliente*, or "hot land." The *tierra caliente* extends up to about 3000 feet (1000 meters), and in some parts of the region these lowlands cover wide expanses. Where moisture is adequate, tropical rain forests thrive, as does a wide range of tropical crops, such as bananas, sugarcane, cacao, and pineapples. Many coastal areas of the *tierra caliente*, such as northeastern Brazil, have become zones of plantation agriculture that support populations of considerable size.

Between 3000 and 6500 feet (1000 to 2000 meters) is the cooler *tierra templada* ("temperate land"). The year-round, spring-like climate of this zone drew large numbers of indigenous people in the distant past and, more recently, has drawn Europeans. Here, such crops as corn, beans, squash, various green vegetables, wheat, and coffee are grown.

Between 6500 and 12,000 feet (2000 to 3600 meters) is the *tierra fria* ("cool land"). A variety of crops, such as wheat, fruit trees, potatoes, and cool-weather vegetables—cabbage and broccoli, for example—do very well at this altitude. Many animals—such as dogs, llamas, sheep, and guinea pigs (Figure 3.7 on page 117)—are raised for pets, food, and fiber. Several modern population centers are in this zone, including Mexico City, Mexico, and Quito, Ecuador.

Above 12,000 feet (3600 meters) is the *tierra helada* ("frozen land"). In the highest reaches of this zone, vegetation is almost absent, and mountaintops emerge from under snow and glaciers. A remarkable feature of such tropical mountain zones is that in a single day of strenuous hiking, one can encounter many of the climate types found on earth.

Precipitation The pattern of precipitation throughout the region is influenced by the interaction of global wind patterns with mountains and ocean currents (see Figure 1.29, pages 50–51). The **trade winds** sweep off the Atlantic, bringing heavy seasonal rains to places roughly 23° north and south of the equator (see the map in Figure 3.5). Winds from the Pacific bring seasonal rain to the west coast of Central America, but mountains block that rain from reaching the Caribbean side, which receives heavy rainfall from the northeast trade winds.

The Andes are a major influence on precipitation in South America. They block the rains borne by the trade winds off the Atlantic into the Amazon Basin and farther south, creating a rain shadow on the western side of the Andes in northern Chile and southwestern Peru (see Figure 3.4). Southern Chile is in the path of eastward-trending winds that sweep off of the southern Pacific Ocean, bringing steady, cold rains that support forests similar to those of the Pacific Northwest of North America. The Andes block this flow of wet, cool air and divert it to the north. They thereby create another extensive rain shadow on the eastern side of the mountains along the southeastern coast of Argentina (Patagonia).

Adjacent oceans and their currents also influence the pattern of precipitation. Along the west coasts of Peru and Chile, the cold surface waters of the Peru Current bring cold air that cannot carry much moisture. The combined effects of the Peru Current and the central Andes rain shadow have created what is possibly the world's driest desert, the Atacama of northern Chile (see Figure 3.5B).

El Niño One aspect of the Peru Current that is only partly understood is its tendency to change direction every few years (on an irregular cycle, possibly linked to sunspot activity). When this happens, warm water flows eastward from the western Pacific, bringing warm water and torrential rains, instead of cold water and dry weather, to parts of the west coast of South America. The phenomenon was named **El Niño**, or "the Christ Child," by Peruvian fishermen, who noticed that when it does occur, it reaches its peak around Christmastime.

El Niño also has global effects, bringing cold air and drought to normally warm and humid western Oceania and unpredictable weather patterns to Mexico and the southwestern United States. The El Niño phenomenon in the western Pacific is discussed further in Chapter 11, where Figure 11.7 (page 466) illustrates its trans-Pacific effects.

silt fine soil particles

temperature-altitude zones regions of the same latitude that vary in climate according to altitude

trade winds winds that blow from the northeast and the southeast toward the equator

El Niño periodic climate-altering changes, especially in the circulation of the Pacific Ocean, now understood to operate on a global scale

Figure 3.5 Photo Essay: Climates of Middle and South America

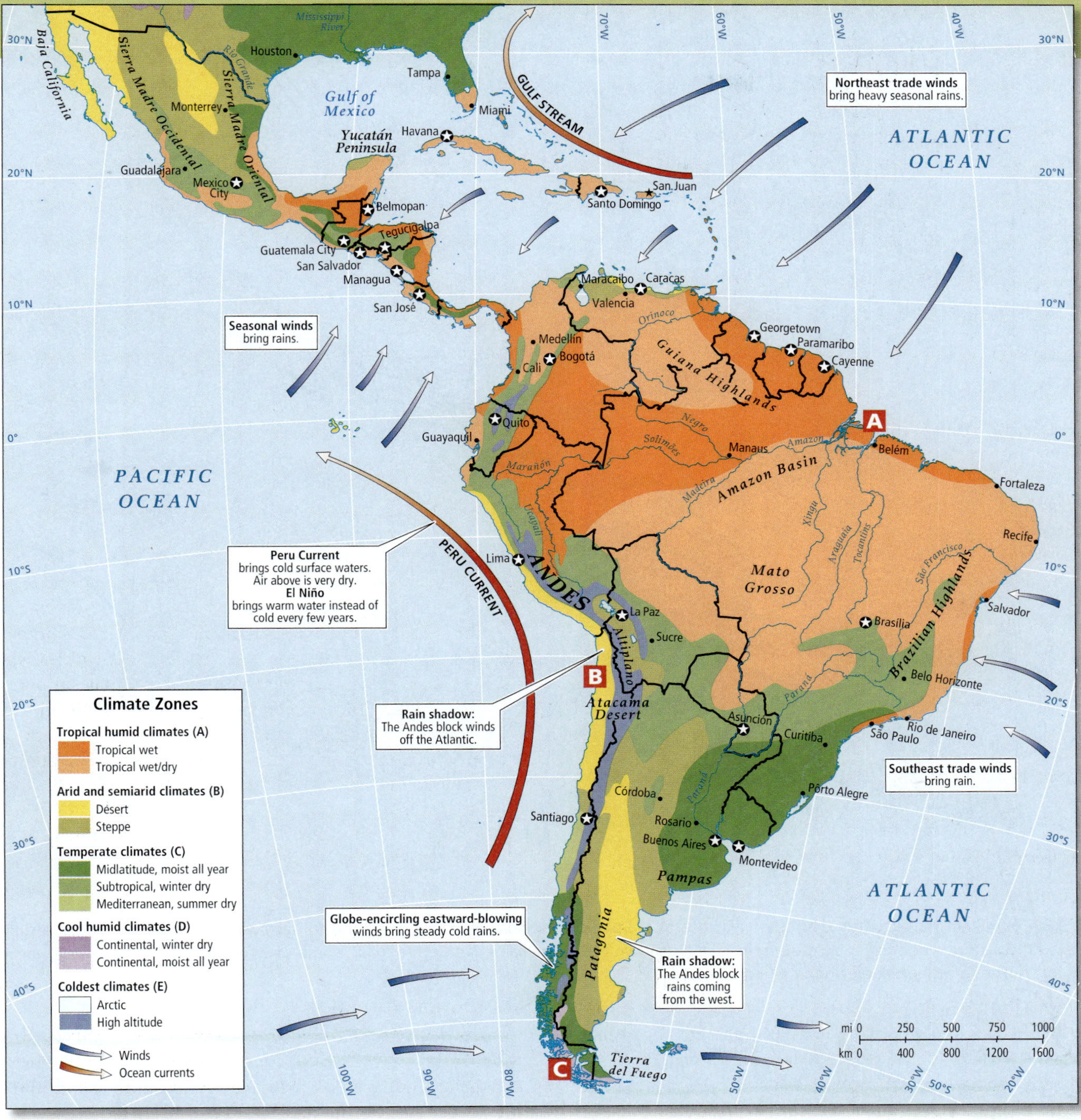

Northeast trade winds bring heavy seasonal rains.
Seasonal winds bring rains.
Peru Current brings cold surface waters. Air above is very dry.
El Niño brings warm water instead of cold every few years.
Rain shadow: The Andes block winds off the Atlantic.
Southeast trade winds bring rain.
Globe-encircling eastward-blowing winds bring steady cold rains.
Rain shadow: The Andes block rains coming from the west.
GULF STREAM
PERU CURRENT
ATLANTIC OCEAN
PACIFIC OCEAN
Gulf of Mexico
Yucatán Peninsula
Baja California
Sierra Madre Occidental
Sierra Madre Oriental
Guiana Highlands
Amazon Basin
Mato Grosso
Brazilian Highlands
ANDES
Altiplano
Atacama Desert
Pampas
Patagonia
Tierra del Fuego
Houston
Tampa
Miami
Havana
Monterrey
Guadalajara
Mexico City
Belmopan
Tegucigalpa
Guatemala City
San Salvador
Managua
San José
Santo Domingo
San Juan
Maracaibo
Caracas
Valencia
Medellín
Bogotá
Cali
Georgetown
Paramaribo
Cayenne
Quito
Guayaquil
Manaus
Belém
Fortaleza
Recife
Lima
La Paz
Sucre
Salvador
Brasília
Belo Horizonte
Asunción
Rio de Janeiro
São Paulo
Curitiba
Porto Alegre
Córdoba
Rosario
Santiago
Buenos Aires
Montevideo
A
B
C
Climate Zones
Tropical humid climates (A)
Tropical wet
Tropical wet/dry
Arid and semiarid climates (B)
Desert
Steppe
Temperate climates (C)
Midlatitude, moist all year
Subtropical, winter dry
Mediterranean, summer dry
Cool humid climates (D)
Continental, winter dry
Continental, moist all year
Coldest climates (E)
Arctic
High altitude
Winds
Ocean currents
mi 0 250 500 750 1000
km 0 400 800 1200 1600

A Tropical wet, Belém, Brazil

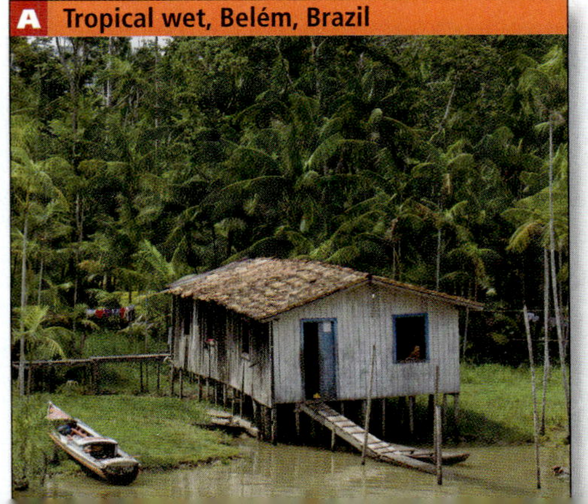

B Desert, Atacama, Chile

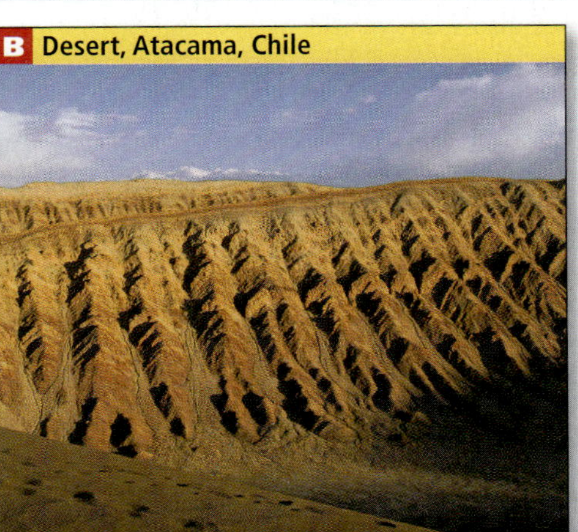

C Continental, moist all year, Tierra del Fuego

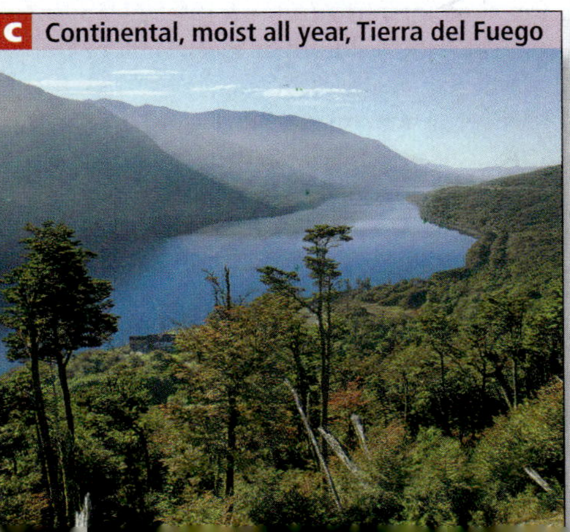

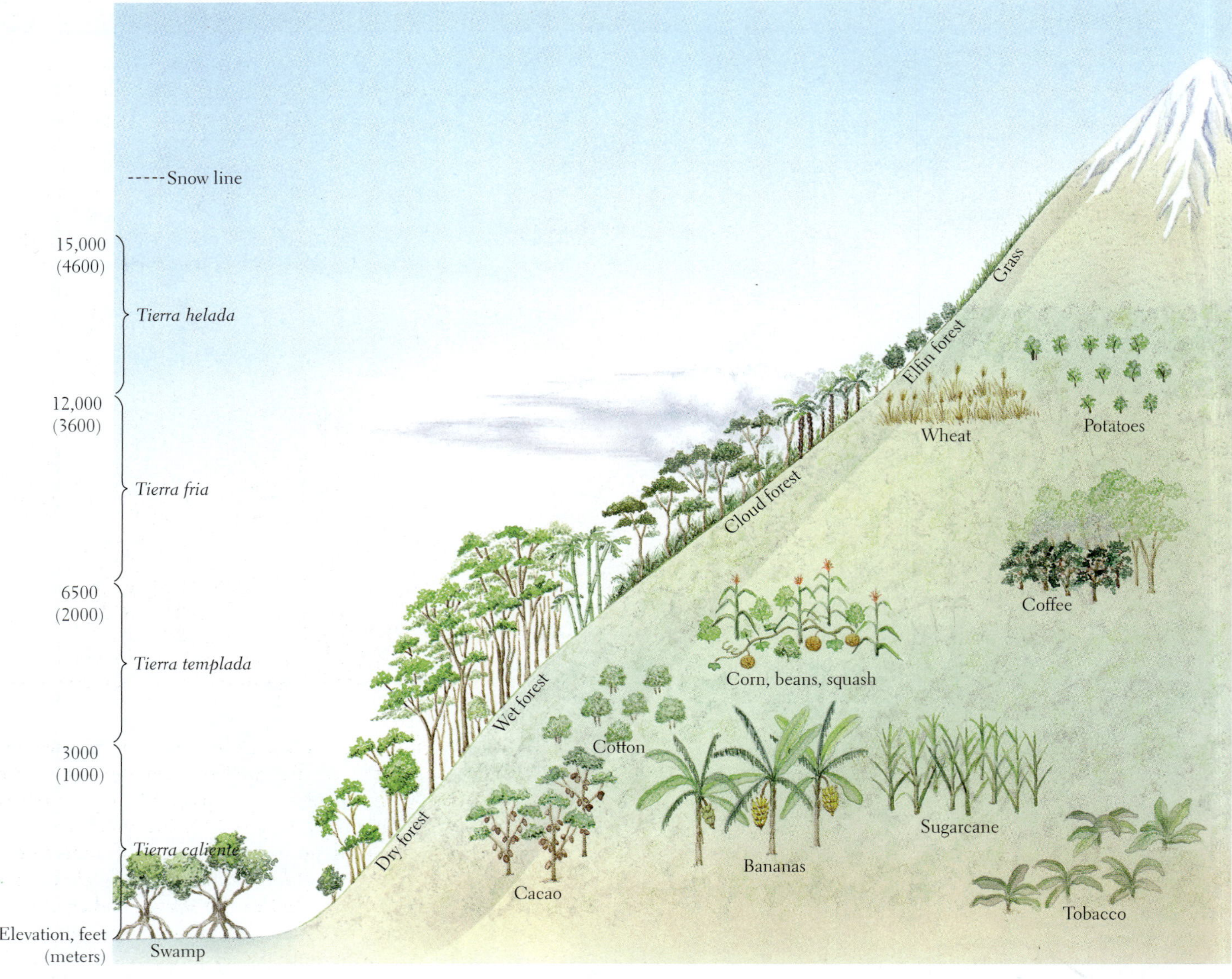

FIGURE 3.6 Temperature-altitude zones of Middle and South America. Temperatures tend to decrease as altitude increases, resulting in changes to the natural vegetation on mountainsides, as shown here. The same is true for crops, some of which are suited to lower, warmer elevations and some to higher, cooler ones.

Hurricanes In this region, many coastal areas are threatened by powerful storms that can create extensive damage and loss of life. These form annually, primarily in the Atlantic Ocean north of the equator and close to Africa. A tropical storm begins as a group of thunderstorms. A few hurricanes also form in the southeastern Pacific and can affect the western coasts of Middle America before turning west toward Hawaii. When enough warming wet air comes together, the individual storms organize themselves into a swirling spiral of wind that moves across the Earth's surface. The highest wind speeds are found at the edge of the eye, or center, of the storm. Once wind speeds reach 75 miles (120.7 kilometers) per hour, such a storm is officially called a *hurricane*. Hurricanes usually last about 1 week; because they draw their energy from warm surface waters, they slow down and eventually dissipate as they move over cooler water or land. As the populations of coastal areas grow, more people are being exposed to the effects of hurricanes. Some scientists also think that climate change is leading to an increase in the number and intensity of hurricanes (see Figure 3.10B on page 121).

THINGS TO REMEMBER

- The rain forests of the Amazon Basin are planetary treasures of biodiversity that also play a key role in regulating the Earth's climate.
- There are four temperature-altitude zones in the region that influence where and how people live and the crops they can grow.
- Significant environmental hazards include earthquakes, volcanic eruptions, and hurricanes.

FIGURE 3.7 LOCAL LIVES PEOPLE AND ANIMALS IN MIDDLE AND SOUTH AMERICA

A The Mexican hairless dog, or *xolo*, was bred some 3000 years ago as a hunting dog and companion, and for food. The xolo was considered sacred by the Aztecs, Mayans, and other indigenous groups, all of whom believed the dogs helped their masters' souls pass safely through the dangers of the underworld realm of Mictlan.

B *Cuy*, or guinea pigs, have been raised throughout the Andes region for at least 5000 years. Primarily a source of meat, they are also considered spiritual mediums by traditional Andean healers who use them to diagnose illnesses in people.

C A llama being led by a girl in traditional Quechua dress in Cusco, Peru. Domesticated by pre-Incan peoples thousands of years ago, llamas were raised for their fur, their meat, and their labor as pack animals, and later started to be used as guard animals for sheep, which were introduced by the Spanish.

Environmental Issues

> **Geographic Insight 1**
> **Climate Change and Deforestation:** Deforestation in this region contributes to climate change by removing trees, which as living plants naturally absorb carbon dioxide. As the cut vegetation decays, large amounts of carbon dioxide are released into the atmosphere. Sustainable alternatives to deforestation are being tried.

Environments in Middle and South America have long inspired concern about the use and misuse of the Earth's resources. Millennia before Europeans arrived in this region, human settlements in the Americas had major environmental impacts. However, today's impacts are particularly severe because both population density and per capita consumption have increased so dramatically. Moreover, local environments now supply global demands.

Tropical Forests, Climate Change, and Globalization

A period of rapid deforestation that has had global repercussions began in the 1970s in Brazil with the construction of the Trans-Amazon Highway. Migrant farmers followed the new road into the rain forest. They began clearing terrain to grow crops, prompting a few observers to warn of an impending crisis of deforestation. Initially, concern focused on the loss of plant and animal species, as the rain forests of the Amazon Basin are some of the most biodiverse on the planet. (Please note that the Amazon includes parts of Colombia, Peru, Bolivia, and Brazil; see Figure 3.1 on pages 108–109.) However, climate change now dominates concerns about deforestation in the Amazon and the rest of this region. As explained in Chapter 1 (see page 42), forests release oxygen and absorb carbon dioxide (CO_2), the greenhouse gas most responsible for global warming. The loss of large forests, such as those in the Amazon Basin, contributes to global warming by releasing CO_2 once locked in the woody bodies of trees. This CO_2 release happens as trees are burned to make way for crops and roads. When there are fewer trees, less CO_2 can be absorbed from the atmosphere. Together, all of the Earth's tropical rain forests absorb about 18 percent of the CO_2 added to the atmosphere yearly by human activity. Because 50 percent of the Earth's remaining tropical rain forests are in South America, keeping these forests intact is crucial to controlling climate change.

Middle and South American rain forests are being diminished by multiple human impacts (Figure 3.8 on page 118). One of the biggest impacts is caused by the clearing of land to raise cattle and grow crops such as soybeans (for animal feed), sugarcane (for ethanol), and African oil palm (for cooking oil) (see Figures 3.8C and 3.8E). Brazil now ranks as the world's fourth-largest emitter of greenhouse gases, after the United States, China, and Indonesia. In the Amazon, Middle America, and elsewhere in the region, forests are cleared to create pastures for beef cattle, many of which are destined for the U.S. fast-food industry. If deforestation continues in Middle America at the present rates, the natural forest cover will be entirely gone in 20 years.

Hardwood logging and the extraction of underlying minerals, including oil, gas, and precious stones, are also contributing to deforestation. Investment capital comes from Asian multinational

Figure 3.8 Photo Essay: Human Impacts on the Biosphere in Middle and South America

While there are a wide variety of human impacts on the environments of this region, here we focus on the processes of land cover and land use change that lead to the conversion of forests to grazing land or farmland. The forces guiding this process are complex, driven by poor people's need for livelihoods, governments' desire to assert control over lightly populated areas, and the demands for wood, meat, and food that arise in distant urban centers and the global market. Together these forces have led to a rapid loss of forest cover throughout much of this region. Brazil loses more forest cover each year than does any other country on the planet.

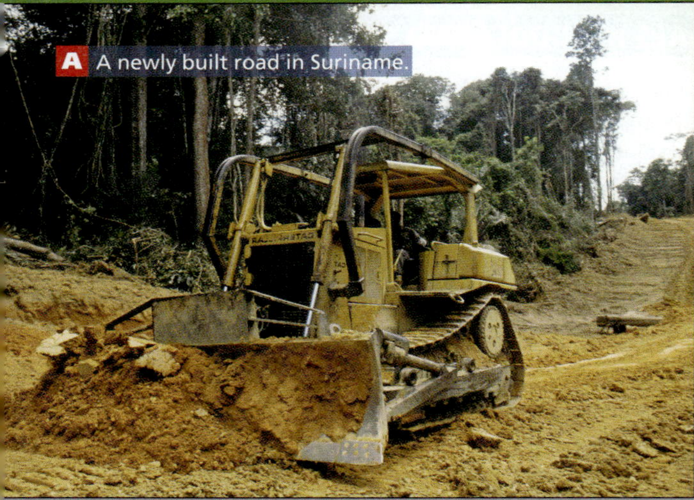

A A newly built road in Suriname.

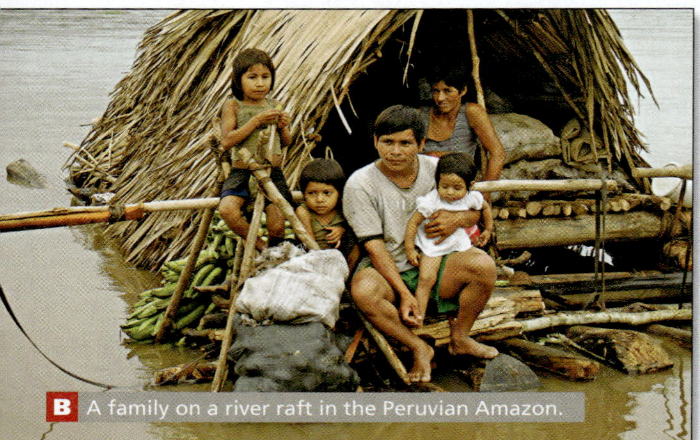

B A family on a river raft in the Peruvian Amazon.

C African oil palm is planted on land recently cleared for agriculture.

D Cattle on recently burned land in Rondonia, Brazil.

E Soy fields recently cleared of forest in Minas Gerais, Brazil.

Thinking Geographically

After you have read about human impacts on the biosphere in Middle and South America, you will be able to answer the following questions:

A What environmental impact is clearly visible in this photo?

D What land use likely preceded cattle grazing on the land shown in this photograph?

E Where are the soybeans grown on this land likely to end up?

companies that have turned to the Amazon forests after having logged up to 50 percent of the tropical forests in Southeast Asia. Moreover, the construction of access roads to support these activities continues to accelerate deforestation by opening new forest areas to migrants. The governments of Peru, Ecuador, and Brazil encourage impoverished urban people to occupy cheap land along the newly built roads (see Figure 3.8A) and encourage deforestation by supplying chainsaws to the settlers. However, the settlers have a difficult time learning to cultivate the poor soils of the Amazon. After a few years of farming, they often abandon the land, now eroded and depleted of nutrients, and move on to new plots. Ranchers may then buy the worn-out land from these failed small farmers to use as cattle pastures (see Figure 3.8D).

While Middle and South America are significantly contributing to global warming by adding CO_2 to the Earth's atmosphere, many people in this region are particularly vulnerable to the increasing threats of climate change. Figure 3.10 on page 121 illustrates several such cases: shortages of clean water brought on by intensifying droughts and by glacial melting; vulnerability to rising sea levels; and the effects of increasingly violent storms.

Major international efforts are under way to combat climate change by preserving the world's remaining forests, especially the crucial tropical rain forests found in this region. Integrating sustainable yet profitable agroforestry (tree cropping) into existing forests has been found to be among the cheapest ways to reduce greenhouse gas emissions. When well-funded and carefully executed, such efforts have the potential to save forests throughout the region. However, implementing complex agroforestry programs, which will require cooperation between diverse groups, will be an enormous challenge (see page 431 in Chapter 10).

Environmental Protection and Economic Development

In the past, governments in the region argued that economic development was so desperately needed that environmental regulations were an unaffordable luxury. Now, there are increasing attempts to embrace both economic development and environmental protection. One example of this new approach is ecotourism.

Ecotourism Many countries are now trying to earn money from the beauty of still-intact natural environments through **ecotourism**. This form of tourism encourages visitors from wealthier areas to appreciate ecosystems and wildlife that do not exist where they live (Figure 3.9; see also On the Bright Side on page 120).

Ecotourism has its downsides, however. Mismanaged, it can be similar to other kinds of tourism that damage the environment and return little to the surrounding community. While the profits of ecotourism can potentially be used to benefit local communities and environments, the profit margins may be small.

ecotourism nature-oriented vacations, often taken in endangered and remote landscapes, usually by travelers from affluent nations

VIGNETTE Puerto Misahualli, a small river boomtown in the Ecuadorian Amazon, is currently enjoying significant economic growth. Its prosperity is due to the many European, North American, and other foreign travelers who come for experiences that will bring them closer to the now-legendary rain forests of the Amazon.

The array of ecotourism offerings can be perplexing. One indigenous man offers to be a visitor's guide for as long as desired, traveling by boat and on foot, camping out in "untouched forest teeming with wildlife." His guarantee that they will eat monkeys and birds does not seem to promise the nonintrusive, sustainable experience the visitor might be seeking. At a well-known "eco-lodge," visitors are offered a plush room with a river view, a chlorinated swimming pool, and a fancy restaurant serving "international cuisine."

FIGURE 3.9 Ecotourism in the Amazon.

(A) A tourist poses at the bottom of a giant ceiba tree in the Ecuadorian Amazon.

(B) An Amazon River dolphin being fed by an ecotourism guide.

(C) A tourist explores a walkway suspended in the canopy of tall rainforest trees in the Ecuadorian Amazon.

ON THE BRIGHT SIDE

Alternatives to Deforestation

Ecotourism is now the most rapidly growing segment of the global tourism and travel industry, which by some measures is the world's largest industry, with $3.5 trillion spent annually. Many Middle and South American nations also have spectacular national parks that can provide a basis for ecotourism. As an alternative to deforestation, development based on ecotourism has the potential to preserve this region's biodiversity, reduce emissions of greenhouse gases, and provide less affluent and indigenous people with a chance to use their skills to teach tourists.

All of this is on a private 740-acre nature reserve separated from the surrounding community by a wall topped with broken glass. It seems more like a fortified resort than an eco-lodge.

By contrast, the solar-powered Yachana Lodge has simple rooms and local cuisine. Its knowledgeable resident naturalist is a veteran of many campaigns to preserve Ecuador's wilderness. Profits from the lodge fund a local clinic and various programs that teach sustainable agricultural methods that protect the fragile Amazon soils while increasing farmers' earnings from surplus produce. The nonprofit group running the Yachana Lodge—the Foundation for Integrated Education and Development—earns just barely enough to sustain the clinic and agricultural programs. *[Source: Alex Pulsipher's field notes in Ecuador.]*

The Water Crisis

Geographic Insight 2

Water: Despite the region's abundant water resources, parts of the region are experiencing a water crisis related to unplanned urbanization; marketization; corruption; inadequate water infrastructure; and the use of water to irrigate commercial agriculture.

Although Middle and South America receive more rainfall than any other world region and have three of the world's six largest rivers (in volume), parts of the region are experiencing water crises (Figure 3.10). Most of the factors causing the water crises are induced by humans, who in turn are now exposed to a variety of water-related stresses. For example, the air pollution that hangs over so many cities in Middle and South America contributes to global warming, which in turn accelerates glacial melting in the Andes. The glaciers feed rivers that are the main source of water for millions (see Figure 3.10D). Should the glaciers actually disappear, the rivers will at best run only seasonally, devastating communities and industries that depend on them.

Lax environmental policies allow industries to pollute both air and water with few restraints. Waterways along the Mexican border with the United States are polluted by numerous factories set up to take advantage of NAFTA-related trade with the United States and Canada (see discussion of maquiladoras on page 131). Only a small percentage of the often highly toxic water discharges are treated and disposed of properly. And in Mexico City, it is estimated that as much as 90 percent of urban wastewater goes untreated.

Rapid urbanization, combined with corruption, inadequate investment, and misguided policies, has left many people without access to clean water or sanitation. In some of the largest cities, the water infrastructure is so inadequate that as much as 50 percent of the fresh water is lost because of leaky pipes (see Figure 3.10A). Often as much as 80 percent of the population has no access to decent sanitation; toilets are sometimes entirely lacking, leaving people to relieve themselves on the city streets. This poses a major health hazard. Uncontrolled dumping of wastewater so pollutes local water resources that cities must bring in drinking water from distant areas.

Behind these immediate problems lie systemic policy failures and inadequate planning, such as that in Cochabamba, Bolivia (discussed in Chapter 1 on page 38). There, efforts to improve the city's inadequate water supply system focused on "marketizing" the water system. The water supply, long thought of as a public resource, was sold to a group of foreign corporations led by Bechtel of San Francisco, California. The hope was that Bechtel, in return for profits, would make investments in infrastructure that Cochabamba's notoriously corrupt water utility would not make. Unfortunately, Bechtel, unfamiliar with the actual living conditions of the majority of citizens, immediately increased water prices to levels that few urban residents could afford, while doing little to improve water supply or delivery systems. At one point Bechtel even charged urban residents for water taken from their own wells and rainwater harvested off their roofs! Popular protests forced Bechtel to abandon Cochabamba's water utility, which remains plagued by corruption and an inadequate infrastructure.

Thinking Geographically

After you have read about vulnerability to climate change in Middle and South America, you will be able to answer the following questions:

A What clues can be seen in this photo that the neighborhood lacks a centralized water distribution system?

B From which direction are hurricanes most likely to hit Honduras?

C In addition to exposure to tropical storms and hurricanes, what else contributes to Haiti's vulnerability to climate change?

D Why is glacial melting of particular concern to cities in Bolivia?

THINGS TO REMEMBER

Geographic Insight 1

- **Climate Change and Deforestation** The rapid loss of forests in this region is a major contributor to climate change. Trees absorb huge amounts of carbon dioxide in their bodies; the carbon dioxide is released when forests are cleared through burning.

Geographic Insight 2

- **Water** Despite considerable water resources, parts of this region are experiencing a water crisis related to rapid urbanization, corruption, inadequate investment in infrastructure, misguided policies, and the use of water to irrigate commercial agriculture. Many people have inadequate access to sanitation, and the dumping of waste in rivers is widespread.

Figure 3.10 Photo Essay: Vulnerability to Climate Change in Middle and South America

Adapting to the multiple stresses that climate change is bringing to this region is proving to be quite a challenge. Water-related troubles, such as drought, hurricanes, flooding, and glacial melting are combining with growing populations and persistent poverty to create a complex landscape of vulnerability to climate change.

A A poor neighborhood in Nogales, Mexico, where there is no centralized water infrastructure and where people depend on water gathered off of the roofs of their homes, or else brought in by truck. The higher temperatures that climate change is bringing could make this area, where water is already scarce, even drier. The millions of Mexicans who have moved to work in factories along the U.S.–Mexico border are thus highly vulnerable to climate change.

B A major bridge in Honduras washed out by Hurricane Mitch, which killed 18,000 people. Hurricanes are likely to intensify as temperatures rise with climate change. Poor countries like Honduras are particularly vulnerable to the damage these storms bring.

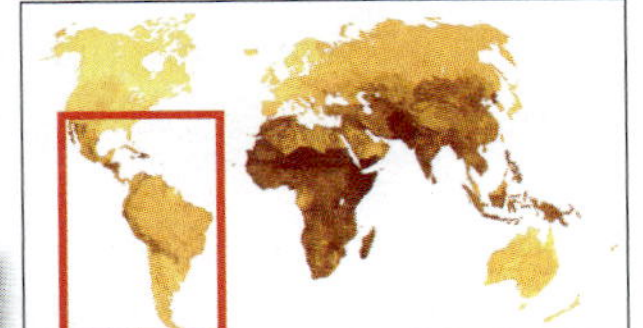

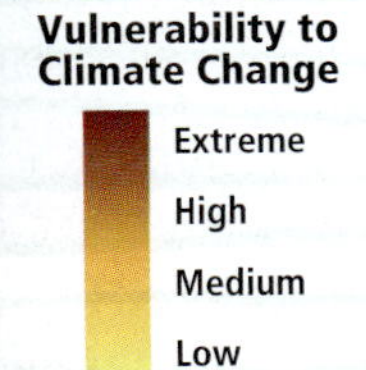

C Haitians examine damage to crops from rain and flooding associated with Hurricane Sandy. So much of Haiti's forest cover has been removed that even mild tropical storms can cause catastrophic flooding. Much of Haiti's impoverished population is already dependent on foreign aid, a situation likely to worsen with climate change.

D La Paz and many smaller cities and towns in Bolivia's drought-prone highlands receive much of their drinking water from glaciers in the Andes. Higher temperatures are causing these glaciers to melt rapidly. Many have already disappeared and the rest could be gone in 15 years.

A Teotihuacan, Mexico, once home to an estimated 150,000 to 250,000 people.

B Machu Picchu, an estate built for the Incan emperor of the fifteenth century.

FIGURE 3.11 A VISUAL HISTORY OF MIDDLE AND SOUTH AMERICA

Thinking Geographically

After you have read about the human history of Middle and South America, you will be able to answer the following questions:

A How does this image of Teotihuacan lend credence to the idea that as of 1500, relative to their contemporaries in Europe, the Aztecs probably lived more comfortably than did Europeans?

B What about Machu Picchu in the Andean highlands indicates that it was more than a mere summer residence?

Human Patterns over Time

The conquest of Middle and South America by Europeans set in motion a series of changes that helped create the ways of life found in this region today. The conquest wiped out much of the indigenous civilizations and set up new societies in their place. Many cultural features from the time of European colonialism endure to this day, as do some vestiges of the precolonial era.

The Peopling of Middle and South America

Recent evidence suggests that between 25,000 and 14,000 years ago, groups of hunters and gatherers from northeastern Asia spread throughout North America after crossing the Bering land bridge on foot, or moving along shorelines in small boats, or both. Some of these groups ventured south across the Central American isthmus, reaching the tip of South America by about 13,000 years ago.

By 1492, there were 50 to 100 million indigenous people in Middle and South America. In some places, population densities were high enough to threaten sustainability. People altered the landscape in many ways. They modified drainage to irrigate crops, they terraced hillsides, and they built paved walkways across swamps and mountains. They constructed cities with sewer systems and freshwater aqueducts, raised huge earthen and stone ceremonial structures that rivaled the pyramids of Egypt (Figure 3.11A).

The indigenous people also practiced the system of **shifting cultivation** that is still common in wet, hot regions in Central America and the Amazon Basin. In this system, small plots are cleared in forestlands, the brush is dried and burned to release nutrients in the soil, and the clearings are planted with multiple crop species. Each plot is used for only 2 or 3 years and then abandoned for several decades—long enough to allow the forest to regrow. If there is sufficient land, this system is highly productive per unit of land and labor, and is sustainable for long periods of time. However, if population pressure increases to the point that a plot must be used before it has fully regrown and its fertility has been restored, yields decrease drastically.

The **Aztecs** of the high central valley of Mexico had some technologies and social systems that rivaled or surpassed those of Asian and European civilizations of the time. Particularly well developed were urban water supplies, sewage systems, and elaborate marketing systems. Historians have concluded that, on the whole, by 1500 C.E., Aztecs probably lived more comfortably than their contemporaries in Europe.

In 1492, the largest state in the region was that of the **Incas**, stretching from what is today southern

shifting cultivation a productive system of agriculture in which small plots are cleared in forestlands, the dried brush is burned to release nutrients, and the clearings are planted with multiple species; each plot is used for only 2 or 3 years and then abandoned for many years of regrowth

Aztecs indigenous people of high central Mexico noted for their advanced civilization before the Spanish conquest

Incas indigenous people who ruled the largest pre-Columbian state in the Americas, with a domain stretching from southern Colombia to northern Chile and Argentina

C A mosaic in Lima, Peru, depicting Francisco Pizarro, conqueror of the Inca Empire.

D Chile wins independence from Spain in 1818 with help from Argentina.

E Former slaves cultivate sugar cane in Puerto Rico in 1899.

C Describe the mood of this depiction of the Spanish conquest of the Incas.

D In what way does this painting of Creole Argentinians and Chileans celebrating independence suggest that they were unlikely to found egalitarian societies?

E Even after the end of slavery, who constituted the labor force in sugar cultivation?

Colombia to northern Chile and Argentina. The main population clusters were in the Andes highlands, where the cooler temperatures at these high altitudes eliminated the diseases of the tropical lowlands, while proximity to the equator guaranteed mild winters and long growing seasons. For several hundred years, the Inca Empire was one of the most efficiently managed empires in the history of the world. Highly organized systems of labor were used to construct paved road systems, elaborate terraces, irrigation systems, and great stone cities in the Andean highlands (see Figure 3.11B). Incan agriculture was advanced, particularly in the development of crops, which included numerous varieties of potatoes and grains.

European Conquest

The European conquest of Middle and South America was one of the most significant events in human history (see Figure 3.11C). It rapidly altered landscapes and cultures and, through disease and slavery, ended the lives of millions of indigenous people.

Columbus established the first Spanish colony in 1492 on the Caribbean island of Hispaniola (presently occupied by Haiti and the Dominican Republic). After learning of Columbus's exploits, other Europeans, mainly from Spain and Portugal on Europe's Iberian Peninsula, conquered the rest of Middle and South America.

The first part of the mainland to be invaded was Mexico, home to several advanced indigenous civilizations, most notably the Aztecs. The Spanish were unsuccessful in their first attempt to capture the Aztec capital of Tenochtitlán, but they succeeded a few months later after a smallpox epidemic decimated the native population. The Spanish demolished the grand Aztec capital in 1521 and built Mexico City on its ruins (Figure 3.11A).

A tiny band of Spaniards, again aided by a smallpox epidemic, conquered the Incas in South America. Out of the ruins of the Inca Empire, the Spanish created the Viceroyalty of Peru, which originally encompassed all of South America except Portuguese Brazil. The newly constructed capital of Lima flourished, in large part as a transshipment point for enormous quantities of silver extracted from mines in the highlands of what is now Bolivia.

Diplomacy by the Roman Catholic Church prevented conflict between Spain and Portugal over the lands of the region. The Treaty of Tordesillas of 1494 divided Middle and South America at approximately 46° W longitude (**Figure 3.12**). Portugal took all lands to the east and eventually acquired much of what is today Brazil; Spain took all lands to the west.

The superior military technology of the Spanish and Portuguese sped the conquest of Middle and South America. A larger factor, however, was the vulnerability of the indigenous people to diseases carried by the Europeans. In the 150 years following 1492, the total population of Middle and South America was reduced by more than 90 percent to just 5.6 million. To obtain a new supply of labor to replace the dying indigenous people, the Spanish initiated the first shipments of enslaved Africans to the region in the early 1500s.

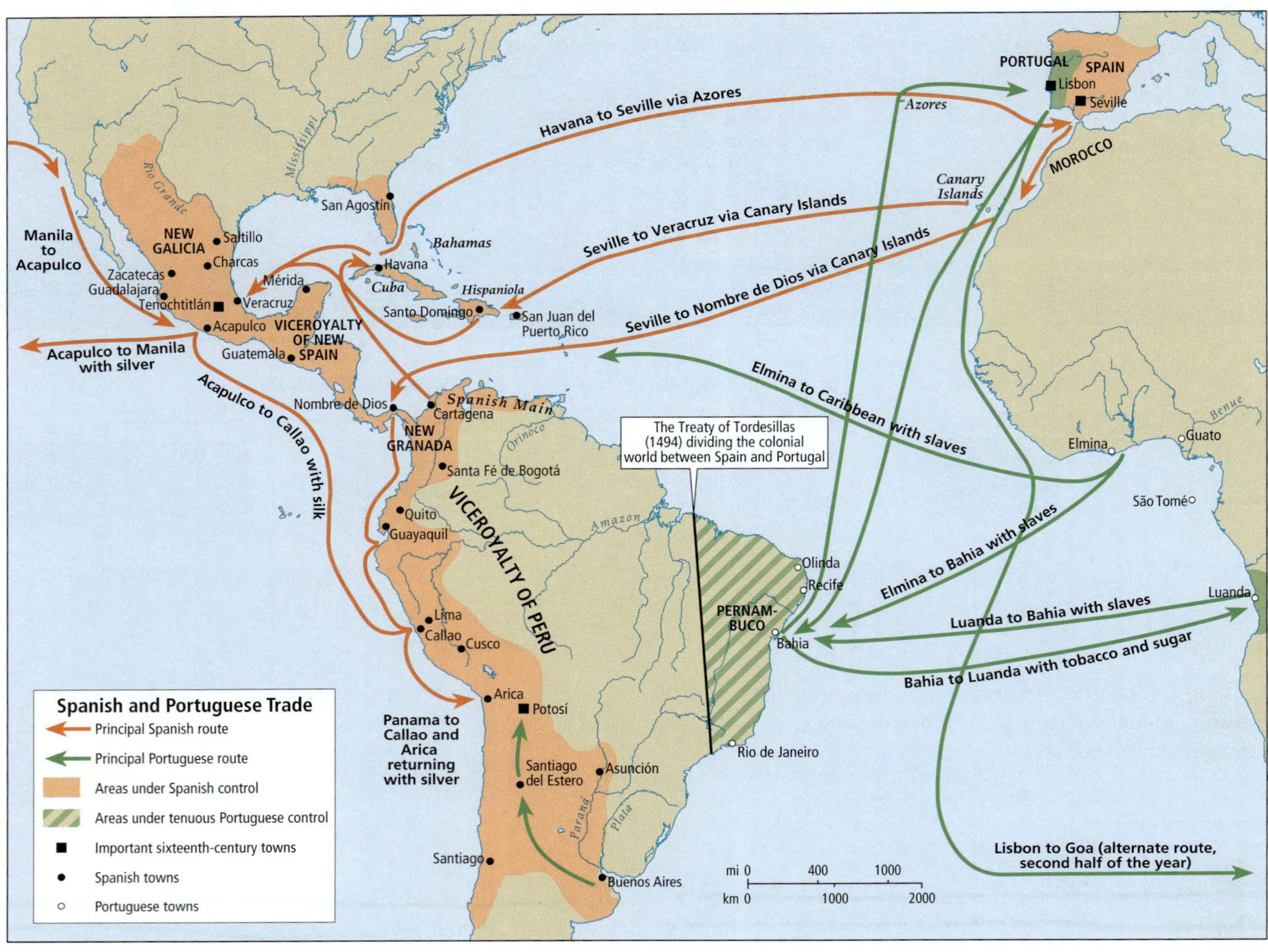

FIGURE 3.12 Spanish and Portuguese trade routes and territories in the Americas, circa 1600. The major trade routes from Spain to its colonies led to the two major centers of its empire, Mexico and Peru. The Spanish colonies could trade only with Spain, not directly with one another. By contrast, there were direct trade routes from Portuguese colonies in Brazil to Portuguese outposts in Africa. Many millions of Africans were enslaved and traded to Brazilian plantation and mine owners (as well as to Spanish, British, French, and Dutch colonies in Middle and South America).

By the 1530s, a mere 40 years after Columbus's arrival, all major population centers of Middle and South America had been conquered and transformed by Iberian colonial policies. Figure 3.12 shows how the colonies soon became part of extensive regional and global trade networks, the latter with Europe, Africa, and Asia.

A Global Exchange of Crops and Animals

From the earliest days of the conquest, plants and animals were exchanged between Middle and South America, Europe, Africa, and Asia via the trade routes illustrated in Figure 3.12. Many plants essential to agriculture in Middle and South America today—rice, sugarcane, bananas, citrus, melons, onions, apples, wheat, barley, and oats, for example—were all originally imports from Europe, Africa, or Asia. When disease decimated the native populations of the region, the colonists turned much of the abandoned land into pasture for herd animals imported from Europe, including sheep, goats, oxen, cattle, donkeys, horses, and mules.

Just as consequential were plants first domesticated by indigenous people of Middle and South America. These plants have changed diets everywhere and have become essential components of agricultural economies around the globe. The potato, for example, had so improved the diet of the European poor by 1750 that it fueled a population explosion. Manioc (cassava) played a similar role in West Africa. Corn, peanuts, vanilla, and cacao (the source of chocolate) are globally important crops to this day, as are peppers, pineapples, and tomatoes (Table 3.1).

The Legacy of Underdevelopment

In the early nineteenth century, wars of independence left Spain with only a few colonies in the Caribbean (see Figure 3.11D).

TABLE 3.1 Globally important domesticated plants that originated in the Americas

Type	Names and places of origin
Seeds (Quinoa)	Amaranth—*Amaranthus cruentus*, S. Mexico, Guatemala Beans—*Phaseolus* (4 species), S. Mexico Maize (corn)—*Zea mays*, valleys of Mexico Peanut—*Arachis hypogaea*, central lowlands of S. America Quinoa—*Chenopodium quinoa*, Andes of Chile and Peru Sunflower—*Helianthus annuus*, southwest and southeast N. America
Tubers (Potato)	Manioc (cassava)—*Manihot esculenta*, lowlands of Middle and S. America Potato (numerous varieties)—*Solanum tuberosum*, Lake Titicaca region of Andes Sweet potato—*Ipomoea batatas*, S. America Tannia—*Xanthosoma sagittifolium*, lowland tropical America
Vegetables (Tomato)	Chayote (christophene)—*Sechium edule*, S. Mexico, Guatemala Peppers (sweet and hot)—*Capsicum* (various species), many parts of Middle and S. America Squash (including pumpkin)—*Cucurbita* (4 species), tropical and subtropical America Tomatillo (husk tomato)—*Physalis ixocarpa, Mexico*, Guatemala Tomato (numerous varieties)—*Lycopersicon esculentum*, highland S. America
Fruit (Pineapple)	Avocado—*Persea americana*, S. Mexico, Guatemala Cacao (chocolate)—*Theobroma cacao*, S. Mexico, Guatemala Papaya—*Carica papaya*, S. Mexico, Guatemala Passion fruit—*Passiflora edulis*, central S. America Pineapple—*Ananas comosus*, central S. America Prickly pear cactus (tuna)—*Opuntia* (several species), tropical and subtropical America Strawberry (commercial berry)—*Fragaria* (various species), genetic cross of Chilean berry and wild berry from N. America Vanilla—*Vanilla planifolia*, S. Mexico, Guatemala, perhaps Caribbean
Ceremonial and drug plants (Coca)	Coca (cocaine)—*Erythroxylon coca*, Eastern Andes of Ecuador, Peru, and Bolivia Tobacco—*Nicotiana tabacum*, tropical America

Figure 3.13 on page 126 shows the European colonizing countries and the dates of independence for the various Middle and South American countries. The supporters of the nineteenth-century revolutions were primarily **Creoles** (people of mostly European descent born in the Americas) and relatively wealthy **mestizos** (people of mixed European, African, and indigenous descent). The Creoles' access to the profits of the colonial system had been restricted by **mercantilism**, and the mestizos were excluded by racist colonial policies. Once these groups gained power, however, they became a new elite that controlled the state and monopolized economic opportunity. Because they did little to expand economic development or access to political power for the majority

Creoles people mostly of European descent born in the Caribbean

mestizos people of mixed European, African, and indigenous descent

mercantilism the policy by which European rulers sought to increase the power and wealth of their realms by managing all aspects of production, transport, and commerce in their colonies

FIGURE 3.13 The colonial heritage of Middle and South America. Most of Middle and South America was colonized by Spain and Portugal, but important and influential small colonies were held by Britain, France, and the Netherlands. Nearly all the colonies had achieved independence by the late twentieth century. Those for which no date appears on the map are still linked in some way to the colonizing country.

of their populations (see Figure 3.11E), the revolutions were incomplete.

Today, the economies of Middle and South America are much more complex and technologically sophisticated than they once were. Nevertheless, disparities persist: About 30 percent of the population is poor and lacks access to land, adequate food, shelter, water and sanitation, and basic education. Meanwhile, a small elite class enjoys levels of affluence equivalent to those of the very wealthy in the United States. These conditions are in part the lingering result of colonial economic policies that favored the export of raw materials and fostered undemocratic privileges for elites and outside investors who often spent their profits elsewhere rather than reinvesting them within the region. These policies will be further discussed in the "Economic and Political Issues" section below.

THINGS TO REMEMBER

- Shifting cultivation of small plots is a traditional and potentially sustainable agricultural strategy still used by some throughout the region.
- Spain and Portugal were the primary colonizing countries of Middle and South America, with smaller colonies held by Britain, France, Denmark, and the Netherlands. By the mid-nineteenth century, most of the larger colonies had gained independence.
- Exploitative development, incomplete revolutions, a dependence on raw materials exports, and the failure of elites to reinvest profits all led to underdevelopment in this region.

CURRENT GEOGRAPHIC ISSUES

Historically, wealth and political and economic power in the countries of Middle and South America have been concentrated in the hands of a few. Globalization, economic modernization, and the transformation from rural to urban societies have not changed that reality, although recent shifts toward democratization are beginning to give more political power to the majority. Culturally, this region retains diverse influences from Europe, Africa, and indigenous peoples. Meanwhile, shifting gender roles and new religious movements are powerful agents of change.

Economic and Political Issues

Geographic Insight 3

Globalization, Development, Power, and Politics: The integration of this region with the world economy has led to growing disparities of wealth and opportunity, forcing millions to migrate in search of work. A political backlash against these patterns of development has brought to power some leaders who question the benefits of globalization as well as others who seek to help their countries compete globally.

For decades, the region of Middle and South America has been one of the poorer world regions—not as poor, on average, as sub-Saharan Africa, South Asia, or Southeast Asia—and has had serious economic challenges. The unique colonial and postcolonial history of this region has resulted in wide economic inequality and income disparity, as well as persistent political disempowerment of poor people. Because these issues are so connected, overall discussions of economic and political issues are combined in this chapter. As the discussions will show, however, in recent years the region has emerged from a long series of efforts to change its economic trajectory and includes several countries that are now being mentioned as the economic powerhouses of the future, perhaps even positioned to help Europeans resolve their issues with economic security.

Economic Inequality and Income Disparity

With the exception of a few small countries, the **income disparity**—the gap between rich and poor—in Middle America is one of the biggest in the world. According to the most recent United Nations (UN) figures, the richest 10 percent of the population was between 19 and 94 times richer than the poorest 10 percent, depending on which country was being analyzed (Table 3.2). In recent years, disparities have been shrinking in Brazil, Chile, Mexico, and Venezuela, primarily because of new government economic and social policies aimed at reducing wide disparities. However, in Bolivia, Guatemala,

income disparity the gap in income between rich and poor

TABLE 3.2 Income disparities in selected countries[a]

Country	HDI rank, 2001	1987–1995[c] Richest 10% to poorest 10%	HDI rank, 2003	1998–2000[c] Richest 10% to poorest 10%	HDI rank, 2009	2004–2007[c,d] Richest 10% to poorest 10%
		Ratio of wealth of richest 10% to poorest 10% of the population[b]				
Middle and South America						
Bolivia	104	91:1	114	25:1	113	94:1
Brazil	69	49:1	65	66:1	75	41:1
Chile	39	34:1	43	43:1	44	26:1
Colombia	62	43:1	64	43:1	77	60:1
Guatemala	108	29:1	119	29:1	122	34:1
Mexico	51	26:1	55	35:1	53	21:1
Peru	73	23:1	82	22:1	78	26:1
Venezuela	61	24:1	69	44:1	58	19:1
Other selected countries						
China	87	13:1	104	13:1	92	13:1
France	13	9:1	17	9:1	8	9:1
Jordan	88	9:1	90	9:1	96	10:1
Philippines	70	16:1	85	17:1	105	14:1
South Africa	94	33:1	111	65:1	129	35:1
Thailand	66	12:1	64	13:1	87	13:1
Turkey	82	14:1	96	13:1	79	17:1
United States	6	17:1	7	17:1	13	16:1

[a] The UN used data from 1987–1995, 1998–2000, and 2004–2007 on either income or consumption to calculate an approximate representation of how much richer the wealthiest 10 percent of the population is than the poorest 10 percent. The lower the ratio, the more equitable the distribution of wealth in the country.

[b] Decimals rounded up or down.

[c] Survey years fall within this range.

[d] Ratios are from the *United Nations Development Report* 2009.

Sources: United Nations Human Development Report 2001, Table 12; UNHDR 2003, Table 13; UNHDR 2009, Table M.

Colombia, and Peru, disparities have recently increased. The poverty rate for the region as a whole is around 33 percent, but in the poorer countries (Haiti, Guatemala, Honduras, Nicaragua, Peru, Bolivia, and Paraguay), more than half the population now lives in poverty.

Phases of Economic Development

The current economic and political situation in Middle and South America derives from the region's history, which can be divided into three major phases: the **early extractive phase**, the import substitution industrialization (ISI) phase, and the current structural adjustment and marketization phase, which includes sharply rising investment from abroad. All three phases have helped entrench wide income disparities despite a consensus that more egalitarian development is desirable.

Early Extractive Phase From the time European conquerors first arrived, economic development was guided by a policy of mercantilism in which Europeans extracted resources and controlled much of the economic activity in the American colonies in order to increase the power and wealth of their "mother" country. Even after independence was granted, many extractive enterprises were owned by foreign investors who had few incentives to build stable local economies.

A small flow of foreign investment and manufactured goods entered the region, while a vast flow of raw materials left for Europe and beyond. The money to fund the farms, plantations, mines, and transportation systems that enabled the extraction of resources for export came from abroad, first from Europeans and later from North Americans and other international sources. One example is the still highly lucrative Panama Canal. The French first attempted to build the canal in 1880. It was later completed and run by the United States, and finally turned over to Panamanian control in 1999. The profits from these ventures were usually banked abroad, depriving the region of investment funds and tax revenues that could have made it more economically independent. Industries were slow to develop in the region, so even essential items, such as farm tools and household utensils, had to be purchased from Europe and North America at relatively high prices. Many people simply did without.

early extractive phase a phase in Central and South American history, beginning with the Spanish conquest and lasting until the early twentieth century, characterized by a dependence on the export of raw materials

hacienda historically, a large agricultural estate in Middle or South America, not specialized by crop and not focused on market production

plantation a large factory farm that grows and partially processes a single cash crop

A number of economic institutions arose in Middle and South America to supply food and raw materials to Europe and North America. Large rural estates called **haciendas** were granted to colonists as a reward for conquering territory and people for Spain. For generations these essentially feudal estates were then passed down through the families of those colonists. Over time, the owners, who often lived in a distant city or in Europe, lost interest in the day-to-day operations of the haciendas. Productivity on these lands was generally low and hacienda laborers remained extremely poor. Nevertheless, haciendas produced a diverse array of products (cattle, cotton, rum, sugar) for local consumption and export.

Plantations were large factory farms, meaning that in addition to growing crops such as sugar, coffee, cotton, or (more recently) bananas, some processing for shipment was done on site. Plantation owners made larger investments in equipment, and these farms were more efficient and profitable than haciendas. However, plantations had relatively little local economic impact. Instead of employing local populations, plantation owners imported slave labor from Africa. The equipment that the plantations used was usually imported from Europe, which was also where plantation owners preferred to invest their profits. As a result, little money was available to support the development of local industries that could have grown up around the plantations.

First developed by the European colonizers of the Caribbean and northeastern Brazil in the 1600s, plantations became more common throughout Middle and South America by the late nineteenth century. Unlike haciendas, which were often established in the continental interior in a variety of climates, plantations were for the most part situated in tropical coastal areas with year-round growing seasons. Their coastal and island locations gave them easier access to global markets via ocean transport.

As markets for meat, hides, and wool grew in Europe and North America, the livestock ranch emerged, specializing in raising cattle and sheep. Today, commercial ranches serving such global markets as the fast-food industry are found in the drier grasslands and savannas of South America, Central America, northern Mexico, and even in the wet tropics on freshly cleared rain forest lands.

Mining was another early extractive industry (Figure 3.14). Important mines (primarily gold and silver at first) were located on the island of Hispaniola and in north-central Mexico, the Andes, the Brazilian Highlands, and many other locations. Extremely inhumane labor practices were common in all of these mines. Today, oil and gas have been added to the mineral extraction industry, but rich mines throughout the region continue to produce gold, silver, copper, tin, precious gems, titanium, bauxite, and tungsten.

Profits from the region's mines, ranches, plantations, and haciendas continued to leave the region, even after the countries gained independence in the nineteenth century. One of the main reasons for this was that wealthy foreign investors retained control of many of the extractive enterprises.

Import Substitution Industrialization Phase In the 1950s, there were waves of protests against the continuing domination of the economy and society by local elites and a few foreign investors. Many governments—Mexico and Argentina most prominent among them—proclaimed themselves socialist democracies. To boost their economic independence, these governments tried to

FIGURE 3.14 A copper mine in northern Chile, where some of the largest copper mines in the world are found. Copper accounts for 13 percent of Chile's GDP, and Chile produces one-third of the world's copper, more than any other country. Chile's copper mines are highly dependent on imported machinery, such as large dump trucks that are made in the United States.

keep money and resources within the region through policies of **import substitution industrialization (ISI)**. Subsidies, among other measures, encouraged the local production of machinery and other commonly imported items.

To encourage local people to buy manufactured goods from local suppliers, governments placed high tariffs on imported, usually higher-quality manufactured goods. The money and resources kept within each country were expected to provide the basis for further industrial development. These policies were intended to create well-paying jobs, raise living standards for the majority of people, and ultimately replace the extractive industries as the backbone of country economies.

The state-owned manufacturing sectors on which the success of ISI depended were never able to produce goods of a high-enough quality to compete with those produced in Asia, Europe, and North America. This was largely because the design, as well as the technological and managerial skills needed to run globally competitive factories, were lacking. Moreover, local consumers were not numerous or prosperous enough to support these industries. With demand for ISI products weak among domestic and foreign consumers alike, employment in these industries stagnated, tax revenues remained low, and social programs could not be adequately funded.

Not all state-owned corporations were losing propositions. ISI still survives in some countries, and periodically ISI schemes are considered for reimplementation in such places as Bolivia and Venezuela. Brazil—with its aircraft, armament, oil exports, and auto industries—and Mexico—with its oil and gas industries—both had success with some ISI programs. Interest in state-supported manufacturing industries continues as a way to keep profits in-country; however, the general trend is now toward more market-oriented management and global competitiveness. Most countries still depend on the export of raw materials.

Beginning in the early 1970s, sharp increases in oil prices and decreases in global prices of raw materials ended a period of relative prosperity that had begun in the early 1950s. Reluctant to let go of the dream of rapid development, governments and private interests across the region continued to pursue ambitious plans to modernize and industrialize their national economies. They did this even as prices for raw material exports—their main source of income—fell. With false optimism, they paid for dam, road, and factory projects by borrowing millions of dollars from major international banks, most of which were in North America or Europe. When, in 1980, a global economic recession hit, not only were development plans halted, but a number of governments in the region also found that they were unable to repay their loans (Figure 3.15 on page 130).

Structural Adjustment and the Marketization Phase

Alarmed over the mounting debt of their clients, foreign banks that had made loans to governments in the region took action. The International Monetary Fund (IMF) developed and enforced policies that mandated profound changes in the organization of national economies. To ensure that sufficient money would be available to repay loans taken out from foreign banks to finance the now-discredited ISI phase, the IMF required **structural adjustment policies (SAPs)**—belt-tightening measures. These SAPs were based on concepts of **privatization** (the selling of formerly government-owned industries and firms to private investors), **marketization** (the development of a *free market* economy in support of *free trade*), and *globalization* (the opening of national economies to global investors; see Chapter 1, pages 30–33). At the time, these concepts were considered the soundest ways for countries to achieve economic expansion and thus to repay the debts to banks in North America, Europe, and Asia.

In the case of privatization, the investors to whom government firms were sold were multinational corporations located in North America, Europe, and Asia. Thus the SAP era resulted in industries across the region being returned to foreign ownership. The other main SAP policy, marketization, required that governments remove tariffs on imported goods of all types in order to obtain further loans. The removal of tariffs was meant to make the market more competitive, but many local industries failed as a result.

Crucially, SAPs also reversed the ISI-era trend of expanding government social programs and building infrastructure. To free up funds for debt

import substitution industrialization (ISI) policies that encourage local production of machinery and other items that previously had been imported at great expense from abroad

structural adjustment policies (SAPs) policies that require economic reorganization toward less government involvement in industry, agriculture, and social services; sometimes imposed by the World Bank and the International Monetary Fund as conditions for receiving loans

privatization the selling of formerly government-owned industries and firms to private companies or individuals

marketization the development of a free market economy in support of free trade

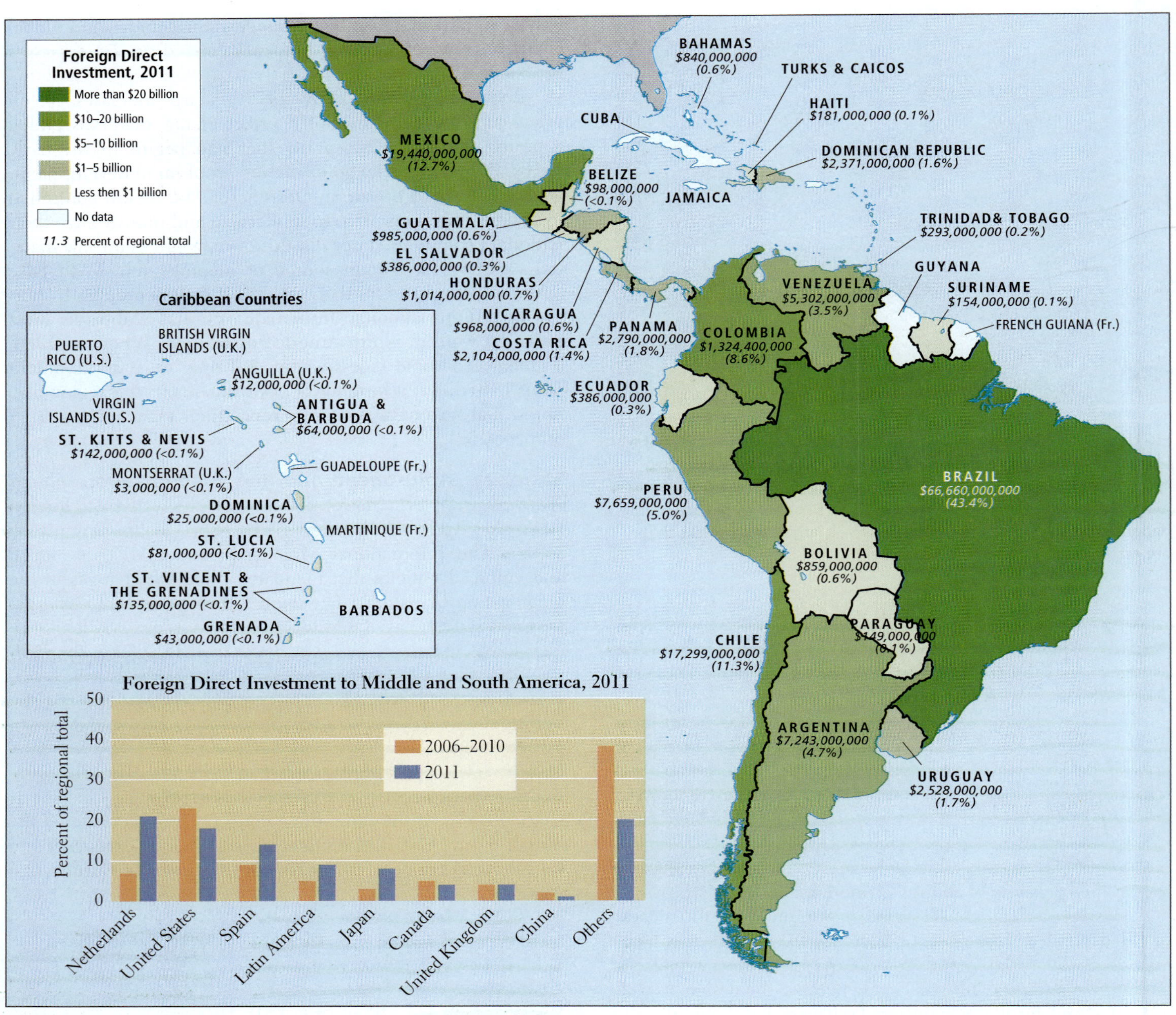

FIGURE 3.15 Origins of Foreign Direct Investment in Latin America and the Caribbean, 2006–2011. FDI flows into the region reached U.S.$153.448 billion in 2011, the highest total ever—and only a few years after a global financial crisis hit in 2008. The Netherlands was the leading investor in 2011, in large part because it is a conduit for investments from third-party countries. The United States was the second-leading investor, with 18% of the total; Spain was third; and Latin American countries were fourth in investing in other countries within the region. Brazil was the largest recipient of FDI in 2011, getting nearly half of the total, followed by Mexico and Chile.

repayment, governments were required to fire many civil servants and drastically reduce spending on public health, education, job training, day care, water systems, sanitation, and infrastructure building and maintenance. In other words, SAPs took government services away from the poor and middle-class workers. While severely cutting these badly needed government programs, SAPs encouraged the expansion of industries that were already earning profits by lowering taxes on their activities, thus further shrinking government revenues to pay for services. In the countries of Middle and South America, as in most developing countries, the most profitable industries remained those based on the extraction of raw materials for export. **61. PERUVIANS STRUGGLE TO GAIN HEALTH CARE ACCESS**

Export Processing Zones A major component of SAPs was the expansion of manufacturing industries in **Export Processing Zones (EPZs)**, also known as *free trade zones*—specially created areas within a country where, in order to attract foreign-owned factories, taxes on imports and exports are not charged. The main benefit to the host country is the employment of local labor, which eases unemployment and brings money into the economy. Products are often assembled strictly for export to foreign markets.

There are EPZs in nearly all countries on the Middle and South American mainland and on some Caribbean islands (in many other world regions also). However, the largest of the EPZs is the conglomeration of assembly factories, called **maquiladoras**, that are located along the Mexican side of the U.S.–Mexico border. Although these factories do provide employment, living conditions are difficult and exactly the role they will ultimately play in Mexican economic development is unclear, as the following vignette illustrates.

VIGNETTE Orbalin Hernandez has just returned to his self-built shelter in the town of Mexicali on the border between Mexico and the United States. Recently fired for taking off his safety goggles while loading TV screens onto trucks at Thompson Electronics, he had just gone to the personnel office to ask for his job back. Because there were no previous problems with him, he was rehired at his old salary of $300 per month ($1.88 per hour). Thompson is a French-owned electronics firm that took advantage of NAFTA (see page 132) when it moved to Mexicali from Scranton, Pennsylvania, in 2001. There, its 1100 workers had been paid an average of $20 per hour. Now 20 percent of the Scranton workers are unemployed, and many feel bitter toward the Mexicali workers.

Nonetheless, in 2006, Orbalin and his fellow Mexicali workers were being told that their wages were too high to allow their employers to compete with companies located in China, where in 2005, workers with the same skills as Orbalin earned just $0.35 an hour. Indeed, that year, 14 Mexicali plants closed and moved to Asia. Those firms remaining in Mexicali cut wages and reduced benefits.

Though Orbalin's salary at Thompson-Mexicali is barely enough to support himself, his wife, Mariestelle, and four children, they are grateful for the job. They are originally from a farming community in the Mexican state of Tabasco, where, for people with no high school education, wages averaged $60 per month.

By 2009, it appeared that the global recession was affecting Chinese–Mexican relations in new ways. Rising costs for fuel and storage, and some questions about the quality of Chinese goods, meant that it made less sense to move a Mexican factory to China. In fact, in 2012, China began to invest in Mexico in order to produce such things as motorcycles and trucks for the Mexican market. It is not yet clear whether this trend will be short-lived or the wave of the future. *[Source: NPR, ZNet, Bloomberg Businessweek, Maquiladora Portal. For detailed source information, see Text Credit pages.]* ■

Voter Backlash Against SAPs The SAP era did not produce the sustained economic growth that was expected to relieve the debt crisis and create broader prosperity. SAPs failed to stimulate economic growth in large part because they encouraged greater dependence on exports of raw materials just when prices for these items were falling in the global market.

Beginning in the late 1990s, millions of voters within this region have registered their opposition to SAPs. Since 1999, presidents who explicitly opposed SAPs have been elected in eight countries in the region: Argentina, Bolivia, Brazil, Chile, Ecuador, Nicaragua, Uruguay, and Venezuela. Both Brazil and Argentina made considerable sacrifices to pay off their debts and liberate themselves from the restrictions of SAPs. Venezuela, under President Hugo Chávez, emerged as a leader of the SAP backlash. In 2005, the Chávez administration **nationalized** foreign oil companies operating in Venezuela. Since then it has also used its own considerable oil wealth to help other countries, such as Argentina, Bolivia, and Nicaragua, pay off their debts. Bolivia, led by Evo Morales (often an ally of Chávez), reversed the SAP policy of privatization by nationalizing that country's natural gas industry. Loans taken from the IMF dropped from $48 billion in 2003 to just $1 billion in 2008. During this same period, IMF policy reversed itself from promoting SAPs to promoting Poverty Reduction Strategies (PRSs) that attempt to combine marketization with strong investment in social programs. This new IMF position has been maintained across the globe now for several years, suggesting that voters in Middle and South America have had an important impact on a major global financial institution (see page 32 in Chapter 1).

63. NEW BOLIVIAN ENERGY POLICY CAUSES CONCERN

67. U.S.–VENEZUELA ENERGY TIES ENDURE DESPITE DETERIORATING POLITICAL RELATIONS

The Present Era of Foreign Direct Investment

The most recent development in this region is the emergence of several countries—most notably Brazil, Mexico, and Chile—that are not only improving in economic health but are stepping into global economic leadership roles. These three have received notably increased flows of **foreign direct investment (FDI)**. FDI is defined as investment funds coming in to enterprises from outside the country (Figure 3.15). In fact, FDI flows to Middle and South America in 2011 were larger than to any other world region; in Brazil, specifically, FDI increased by more than 80 percent in just 9 months of 2011, with the money flowing primarily into the telecommunications, food, metal works, and petroleum industries. Furthermore, Brazil is often mentioned as part of a global foursome known as the BRIC countries: Brazil, Russia, India, and China. These are emerging economic powers, rich in resources and poised for transformation into highly developed, highly

Export Processing Zones (EPZs) specially created legal spaces or industrial parks within a country where, to attract foreign-owned factories, duties and taxes are not charged

maquiladoras foreign-owned, tax-exempt factories, often located in Mexican towns just across the U.S. border from U.S. towns, that hire workers at low wages to assemble manufactured goods which are then exported for sale

nationalize to seize private property and place under government ownership, with some compensation

foreign direct investment (FDI) investment funds that come in to enterprises from outside the country

ON THE BRIGHT SIDE

Contributions from the Informal Economy

Throughout this region, nearly every citizen depends on the informal economy in some way as either a buyer or a seller. Critics argue that informal workers are just treading water, making too little to ever expand their businesses significantly. Moreover, the bribes they have to pay to avoid arrest or fines are much less beneficial to the economy as a whole than the taxes paid by legitimate businesses. Work in the informal economy is also risky because there is neither protection of workers' health and safety nor sick leave, retirement, and disability benefits (also often lacking in the formal economy). Nevertheless, the informal economy can be a lifesaver during times of economic recession. For example, after a recession hit Peru in 2000, sixty-eight percent of urban workers were in the informal sector. They generated 42 percent of the country's total GDP, mostly through street-vending work.

In some instances, the informal economy can serve as an incubator for new businesses that may expand, eventually providing legitimate jobs for family and friends. Those who work in the informal economy as well as those who migrate and send remittances are often unfairly overlooked as major contributors to the economies of the region.

productive powerhouses. One measure of the region's rising reputation as financially stable is the fact that in 2012, Christine Lagarde, director of the IMF, asked for help from Brazil, Mexico, and Peru in designing and funding a bailout package for overly indebted countries in the European Union—the first time countries in Middle and South America have played such a role.

Interest in investing in the region has risen significantly and shifted, as shown in Figure 3.15. The Netherlands now plays the biggest role, primarily because it acts on behalf of a consortium of investor countries. Spain has played a major FDI role for a number of years. The absolute amounts invested by the United States are rising, but because of the strong interest of others, the U.S. proportion has decreased. China quietly invested in the region in relatively small ways for several decades, but recently it has increased its investments and appears to be most interested in buying or leasing (for an extended period) large swatches of agricultural land in Brazil, Argentina, and Chile. In the Bahamas, China has a major upscale casino/resort tourism investment called Baha Mar. Six hotels with a total of 2250 rooms are planned, with construction to be done by the Chinese State Construction Engineering Corporation.

The Role of Remittances Migration for the purpose of supporting family members back home with remittances is an extremely important economic activity throughout this region (see further discussion on page 145). Each year, remittances amount to roughly double the U.S. foreign aid to the region. Most such remittance migrations are within countries. While the amounts of cash sent home are small, these remittances are crucial for families that may not have much other income.

One report (by geographer Dennis Conway and anthropologist Jeffrey H. Cohen) suggests that couples that migrate to the United States from indigenous villages in Mexico typically work at menial jobs and live frugally in order to save a substantial nest egg. Then they may return home for several years to build a house (usually a family self-help project) and buy furnishings. When the money runs out, the couple, or just the husband or wife, may migrate again to save up another nest egg.

The Informal Economy

For centuries, low-profile businesspeople throughout Middle and South America have operated in the informal economy, meaning that they support their families through inventive entrepreneurship but are not officially recognized in the statistics and do not pay business, sales, or income taxes. Most are small-scale operators involved with street vending or recycling used items such as clothing, glass, or waste materials.

Regional Trade and Trade Agreements

The growth in international trade and foreign investment in the region, encouraged in part by SAPs, has been joined by growth in regional free trade agreements. Such agreements reduce tariffs and other barriers to trade among a group of neighboring countries. The two largest free trade agreements within the region are NAFTA and UNASUR.

The **North American Free Trade Agreement (NAFTA)** established a free trade bloc in 1994, consisting of the United States, Mexico, and Canada and containing more than 450 million people. The main goal of NAFTA is to reduce barriers to trade, thereby creating expanded markets for the goods and services produced in the three countries. Since 1994, the value of the economies of these three countries has grown steadily; by 2011, it was worth at least $18 trillion. A subsequent effort by the United States to create a NAFTA-like trade bloc for all of the Americas (FTAA) has stalled in recent years, largely because of dissatisfaction with the effects of globalization and fear of U.S. domination.

UNASUR, a union of South American nations, was organized in May of 2008; it supersedes **Mercosur** and the Andean Community of Nations, two previous customs unions. The 12 member countries (Argentina, Bolivia, Brazil, Chile, Colombia, Ecuador, Guyana, Paraguay, Peru, Suriname, Uruguay, and Venezuela) contain nearly 400 million people, with a combined economy worth nearly $8 trillion per year. UNASUR appears to want to emulate the European Union in that it has adopted resolutions on a multitude of political, social, and trade issues, not the least of which is to reduce disparities of wealth and opportunity. **62. MANY VENEZUELANS UNCERTAIN ABOUT CHÁVEZ'S TWENTY-FIRST-CENTURY SOCIALISM**

The overall record of regional free trade agreements so far is mixed. While they have increased trade, the benefits of that trade usually are not spread evenly among regions or among all sectors of society. In Mexico, for instance, the benefits of NAFTA have

North American Free Trade Agreement (NAFTA) a free trade agreement made in 1994 that added Mexico to the 1989 economic arrangement between the United States and Canada

UNASUR a union of South American nations that was organized in May of 2008; it supersedes Mercosur and the Andean Community of Nations, two previous customs unions

Mercosur a free trade zone created in 1991 that links the economies of Brazil, Argentina, Uruguay, and Paraguay to create a common market

gone mainly to wealthy investors, concentrated in the northern states that border the United States. There, the agreement facilitated the growth of maquiladoras by easing cross-border finances and transit. But significantly, as many as one-third of small-scale farmers throughout Mexico have lost their jobs as a result of increased competition from U.S. corporate agriculture, which now, with NAFTA, has unrestricted access to Mexican markets. In particular, corn imports from the United States have driven down the price of corn in local markets, driving small farmers out of business.

The record of UNASUR is too short to be evaluated; but the shift in the global economy since UNASUR's creation in 2008 may be creating conditions that will place this trade bloc in an advantageous position vis-à-vis markets far outside the region: China, India, Europe, and even North America.

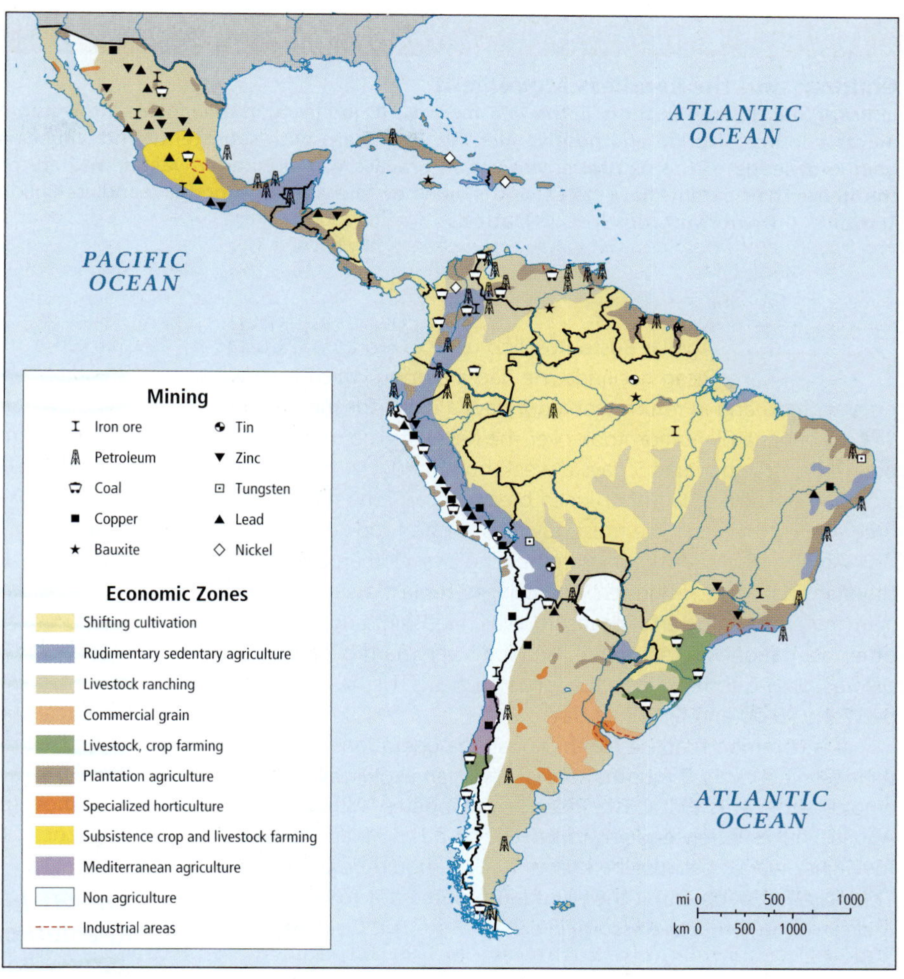

FIGURE 3.16 Agricultural and mineral zones in Middle and South America.

THINGS TO REMEMBER

- Three phases of economic development—extractive, import substitution industrialization (ISI), and structural adjustment policy (SAP)—dominated the region into the twenty-first century.
- Geographic Insight 3 **Globalization, Development, Power, and Politics** Development strategies of the past have left this region with the widest gap between rich and poor in the world; more recently, leaders in the region have challenged the benefits of globalization. In some countries, efforts to develop and integrate the region with the global economy are succeeding, but whether this trend will lead to an easing of disparities of wealth and opportunity remains to be seen.
- Since the 1990s, free trade policies have been adopted in some industries, and regional trading blocs have been formed and re-formed, with mixed results.

Food Production and Contested Space

Geographic Insight 4

Food and Urbanization: The shift from small-scale subsistence food production to mechanized agriculture for export has fueled migration to cities. Many places previously were self-sufficient with regard to food; they now depend on imported food, making them susceptible to food insecurity.

The agricultural lands of Middle and South America (Figure 3.16) are examples of what geographers call **contested space**, where various groups are in conflict over the right to use a specific territory as each sees fit. The conflicts take place at various scales, as the following four case studies demonstrate, and all are in one way or another related to inequities that have been in place at least since colonial times. They also relate to a region-wide shift toward mechanized agriculture for export (see Chapter 1, page 24).

Local Scale: A Banana Worker in Costa Rica For centuries, while people labored for very low wages on haciendas, the hacienda owners would at least allow them a bit of land to grow a small garden or some cash crops. In recent years, however, SAPs encouraged a shift to large-scale, mechanized, export-oriented agriculture, also known as green revolution agriculture. The rationale was that these operations could earn larger profits that could help countries pay off debts faster. SAPs made it easier for foreign multinational corporations, such as Del Monte, to use their large financial resources to buy up many haciendas and other farms, converting them to plantations. However, this also forced many rural people off lands they once cultivated, and onto plantations, where they now work as migrant laborers for low wages, as the following story illustrates.

contested space any area that several groups claim or want to use in different and often conflicting ways, such as the Amazon or Palestine

ON THE BRIGHT SIDE

Children and the Landless Movement

In Brazil, the children of those in the MST movement find living in the landless protestor encampments a source of pride and positive identity. They learn both subsistence and leadership skills, gain knowledge of the natural environment, practice communitarian values, and are proud of the stance their parents have taken, as the video by Michalis Kontopodis, "Landless Children/Sem Terrinha" (http://tinyurl.com/9pxeclw), attests.

VIGNETTE Aguilar Busto Rosalino used to work on a Costa Rican hacienda. He had a plot on which to grow his own food and in return worked 3 days a week for the hacienda. Since a banana plantation took over the hacienda, he rises well before dawn and works 5 days a week from 5:00 A.M. to 6:00 P.M., stopping only for a half-hour lunch break. Because he now lacks the time and land to farm, he must buy most of his food.

Aguilar places plastic bags containing pesticide around bunches of young bananas. He prefers this work to his last assignment of spraying a more powerful pesticide, which left him and 10,000 other plantation workers sterile. He works very hard because he is paid according to how many bananas he treats. Usually he earns between $5.00 and $14.50 a day.

It is common practice for these banana operations to fire their workers every 3 months so that they can avoid paying the employee benefits that Costa Rican law mandates. Although Aguilar makes barely enough to live on, he has no plans to press for higher wages because he knows that he would be put on a "blacklist" of people that the plantations agree not to hire. *[Source: Andrew Wheat. For detailed source information, see Text Credit pages.]*

Provincial Scale: The Zapatista Rebellion In the southern Mexican state of Chiapas, agricultural activists and indigenous leaders have mobilized armed opposition to the economic and political systems that have left them poor and powerless. The Mexican government redistributed some hacienda lands to poor farmers early in the twentieth century, but most fertile land in Chiapas is still held by a wealthy few who use green revolution agriculture to grow cash crops for export. The poor majority farm tiny plots on infertile hillsides. In 2000, about three-fourths of the rural population was malnourished, and one-third of the children did not attend school.

The Zapatista rebellion (named for the hero of the 1910 Mexican revolution, Emiliano Zapata) began on the day the North American Free Trade Agreement took effect in 1994. The Zapatistas view NAFTA as a threat because it diverts the support of the Mexican federal government from land reform to large-scale, export-oriented green revolution agriculture. The Mexican government used the army to suppress the rebellion.

In 2003, after 9 years of armed resistance, the Zapatista movement redirected part of its energies. It began a nonviolent political campaign to democratize local communities by setting up people's governing bodies parallel to local official governments. Before the national elections in 2006, the Zapatistas toured the country in an effort to turn the political climate against globalization and toward greater support for indigenous people and the poor. The election ended in a near tie, but in the end, the Zapatista-favored candidate lost. The Zapatistas continue to be active in efforts to reform Mexican politics to include more grassroots participation.

National Scale: Brazil's Landless Movement The trends in agriculture described above for Middle America have sparked rural resistance in Brazil as well. Sixty-five percent of Brazil's arable and pasture land is owned by wealthy farmers who make up just 2 percent of the population. Since 1985, more than 2 million small-scale farmers have been forced to sell their land to larger farms that practice green revolution agriculture. Because the larger farms specialize in major export items, such as cattle and soybeans, they have been favored by governments wishing to increase exports. As a result, many poor farmers have been forced to migrate.

To help these farmers, organizations such as the Movement of Landless Rural Workers (MST) began taking over unused portions of some large farms. Since the mid-1980s, the MST has coordinated the occupation of more than 51 million acres of Brazilian land (an area about the size of Kansas). Some 250,000 families have gained land titles, while the elite owners have been paid off by the Brazilian government and have moved elsewhere. There are now movements with goals similar to those of MST in Ecuador, Venezuela, Colombia, Peru, Paraguay, Mexico, and Bolivia.

World Regional Scale: The Persistence of Green Revolution Agriculture Instances of workers contesting the use of agricultural land are exceptions to overall trends that favor the growth of large-scale green revolution agriculture (using chemicals, mechanization, and irrigation). Because of the moneymaking potential and political power of large-scale agriculture, conflicts are only rarely resolved in favor of small farmers and agricultural workers. Moreover, in rapidly urbanizing countries, green revolution agriculture is seen as the only way to supply cheap food for the millions of city dwellers. Ironically, many of these new urbanites were once farmers capable of feeding themselves; they moved to the cities because they were unable to compete with the new green revolution systems.

Many areas have had large farms for centuries—the haciendas mentioned on page 128—that are now being converted to green revolution agriculture, similar to farms that can be found in the midwestern United States. A wide belt of large-scale farming/ranching stretches through the Argentine pampas and into Brazil (see Figure 3.16). And, as we saw in the earlier discussion of environmental issues (see page 117), rice, corn, soybeans, and meat animals, which are often produced by green revolution farms on once-forested lands in and around the Amazon Basin, are all major exports for countries like Brazil and Argentina. Despite the recent political tensions involving small farmers and landless agricultural workers, large-scale agriculture appears to have a solid place in the future of this region.

At a medium scale are zones of modern mixed farming (those that produce meat, vegetables, and specialty foods for sale

FIGURE 3.17 LOCAL LIVES **FOODWAYS IN MIDDLE AND SOUTH AMERICA**

A Corn tortillas are prepared in Mexico City. Corn has been a pillar of diets in Mexico and the countries of Central America for thousands of years. Now cultivated throughout the world, corn was first domesticated from a wild grass in southern Mexico.

B An Argentine *asado*, or barbeque. Here beef is roasted over an open bed of coals, using a special *parilla*, or grill. This method of cooking is a variant of methods first developed by the indigenous peoples of the grasslands of Argentina. Asado is the national dish of Argentina; it is prepared widely throughout Uruguay, Paraguay, and southern Brazil.

C *Lomo Saltado*, a Peruvian dish with Asian influences. First developed in restaurants started by Chinese immigrants, it is a mix of Chinese and Peruvian ingredients and culinary traditions. It consists of strips of beef or pork marinated in vinegar and stir-fried with onions, parsley, tomatoes, and other vegetables, then served over rice and french fries.

in urban centers), located on large and small plots around most major urban centers. Figure 3.17 illustrates how the foodways of the area are based largely on indigenous plants (corn, potatoes, tomatoes) and methods of cooking (baking, grilling), and involve beef, pork, rice, seasonings, and cooking methods (frying) imported from Europe, Asia, and Africa. 57. HAITI'S RISING COST OF FOOD WORRIES AID GROUPS

THINGS TO REMEMBER

- Food production is undergoing a shift away from small-scale, often subsistence-level production toward large-scale, green revolution agriculture that is aimed at earning cash through growing food for export.
- Geographic Insight 4 **Food and Urbanization** The shift toward large-scale green revolution agriculture has forced many small-scale farmers, who cannot afford the investment in machinery or chemicals, off their land and into the cities. In significant numbers, ordinary people have begun to effectively protest policies that have left them landless and poor.
- Contentions, some violent and some peaceful, have developed in widespread parts of the region in response to the shift toward large-scale agriculture. Nonetheless, many governments now see large-scale agriculture as the only way to feed the region's large urban populations.

Power, Politics, and the Move Toward Democracy

After decades of elite and military rule, almost all countries in the region now have multiparty political systems and democratically elected governments. In the last 30 years, there have been repeated peaceful and democratic transfers of power in countries once dominated by rulers who seized power by force. These **dictators** often claimed absolute authority, governing with little respect for the law or the rights of their citizens. Their authority was based on alliances between the military, wealthy rural landowners, wealthy urban entrepreneurs, foreign corporations, and even foreign governments such as the United States.

Although democracy has transformed the politics of the region, problems remain (as seen in Figure 3.18 on page 136). Elections are sometimes poorly or unfairly run and their results are frequently contested. Elected governments are sometimes challenged by citizen protests or threatened with a **coup d'état**, in which the military takes control of the government by force. Such coups are usually a response to policies that are unpopular with large segments of the population, powerful elites, the military, or the United States. In the last decade, coups threatened Honduras, Venezuela, Colombia, Ecuador, and Bolivia; but in the more recent past, peaceful democratic elections have become the norm in this region.

Hugo Chávez and Venezuela's Fragile Democracy In Venezuela, the landslide election of Hugo Chávez as president in 1998 appeared for a time to advance the cause of democracy there. Elites had long dominated Venezuelan politics, and profits from the country's rich oil deposits had not trickled down to the poor. One-third of the population lived on $2 (PPP) a day, and per capita GNI (PPP) was lower in 1997 than it had been in 1977. Chávez campaigned as a champion of the poor, calling for the government to subsidize job creation, community health care, and food for the

dictator a ruler who claims absolute authority, governing with little respect for the law or the rights of citizens

coup d'état a military- or civilian-led forceful takeover of a government

Figure 3.18 Photo Essay: Power and Politics in Middle and South America

An expansion of political freedoms is well under way in Middle and South America. Decades of elite and military rule, combined with rampant corruption, have led to popular revolutionary movements that have used both force and elections to take control of the government. In the worst cases, wars and brutal repression by governments and by militants have cost citizens many political freedoms. The drug trade has also emerged as a major obstacle, with huge amounts of illicit cash being used to pay off elected officials, civil servants, police forces, and the military. Nevertheless, in the region as a whole, there has been a dramatic expansion of political freedom and a decline in violence in recent decades.

A Federal agents search cars at a checkpoint in Juarez, Mexico. Mexico's ongoing drug war has led to between 60,000 and 100,000 deaths over the past several years and corruption that has eroded local democratic institutions, especially near the U.S. border. Most of Mexico's drug production supplies the United States.

B Colombian police officers, many crippled in combat with a rebel group, undertake a protest march destined for the capital in Bogotá. Popular revolutionary movements and a drug war drive this complex conflict, the worst in the region. Democratic institutions and the rule of law have been badly damaged, with many high-ranking officials linked to extreme human rights and civil liberties abuses.

C Cuban revolutionaries Che Guevara (left), Camilo Cienfuegos (center), and Juan Antonio Mella are immortalized in an aging mural in Havana. Since coming to power in 1959, Cuba's communist government has been criticized for jailing and sometimes executing leaders of the political opposition.

Thinking Geographically

After you have read about power and politics in Middle and South America, you will able to answer the following questions:

A What evidence can you find in the photo of a violent situation in Mexico?

B What are the indications that these men were once involved in violent combat?

C Why would the Cuban government lionize these men in a mural?

large underclass. He also limited the profits of the largely foreign-managed oil industry and redirected the profits to support government social programs. These policies, all of which were enacted soon after he was elected, eroded Chávez's original support from the small middle class and elites, who began to fear a turn toward broad government control. When the U.S. government expressed alarm, Chávez became a major critic of U.S. and other foreign influence in the region. Chávez helped lead the region-wide backlash against SAPs and condemned the long history of U.S. intervention in the internal affairs of countries across the region.

66. SOCIAL PROGRAMS AT ROOT OF CHÁVEZ'S POPULARITY

Chávez was reelected in 2000, and then briefly deposed in a coup d'état in 2001, in which the United States participated covertly. He was quickly reinstated, however, by a groundswell of popular support among the large underclass, who stood to benefit from his policies. His presidency was sustained in a referendum in 2004 and again in the landslide election of 2006. In 2009, voters approved a constitutional amendment removing presidential term limits, thereby allowing Chávez to run for president indefinitely. Some saw this as a triumph of the *populist movement* (see definition on page 148), others as a loss for democracy and a drift back toward dictatorship (see the Figure 3.18 map). Hugo Chavez died in March of 2013 from cancer, after having been treated numerous times by specialists in Cuba.

68. WILL CHÁVEZ INHERIT CASTRO'S REVOLUTIONARY MANTLE?

69. VENEZUELANS REJECT CONSTITUTIONAL CHANGES

The Drug Trade, Conflict, and Democracy

The international illegal drug trade is a major contributor of corruption, violence, and subversion of democracy throughout the region. Cocaine and marijuana are the primary drugs of trade, with most drugs produced in or passing through northwestern South America, Central America, and Mexico. Figure 3.19 illustrates the geographic distribution of cocaine seizures in 2008. (Note: Such seizures may not reveal activity in areas where law enforcement is lax or entirely co-opted by drug traders.)

Production of addictive drugs is illegal in all of Middle and South America. However, public figures, from the local police on up to high officials, are paid to turn a blind eye to the industry. Most coca and marijuana growers are small-scale farmers of indigenous or mestizo origin, working in remote locations where they can make a better income for their families from these plants than from other cash crops. Many rent land from the *drug cartels* which get the drugs to global markets. The cartels are monopolistic and violent, and those who oppose them—whether farmers, journalists, or law enforcement officials—often end up kidnapped, tortured, and brutally murdered.

In Colombia, the illegal drug trade has financed all sides of a continuing civil war that has threatened the country's democratic traditions and displaced more than 1.5 million people over the past several decades (see Figure 3.18B). Mexico also has for several years faced increasing threats to its political stability (see Figure 3.18A) from drug cartels that control a hugely profitable U.S.-oriented drug trade. **72. THOUSANDS KIDNAPPED IN COLOMBIA IN LAST DECADE**

U.S. policy has emphasized stopping the production of illegal drugs in Middle and South America and interrupting trade flows rather than curtailing the demand for drugs in the United States. As a result, the U.S. *war on drugs* has led to a major U.S. presence

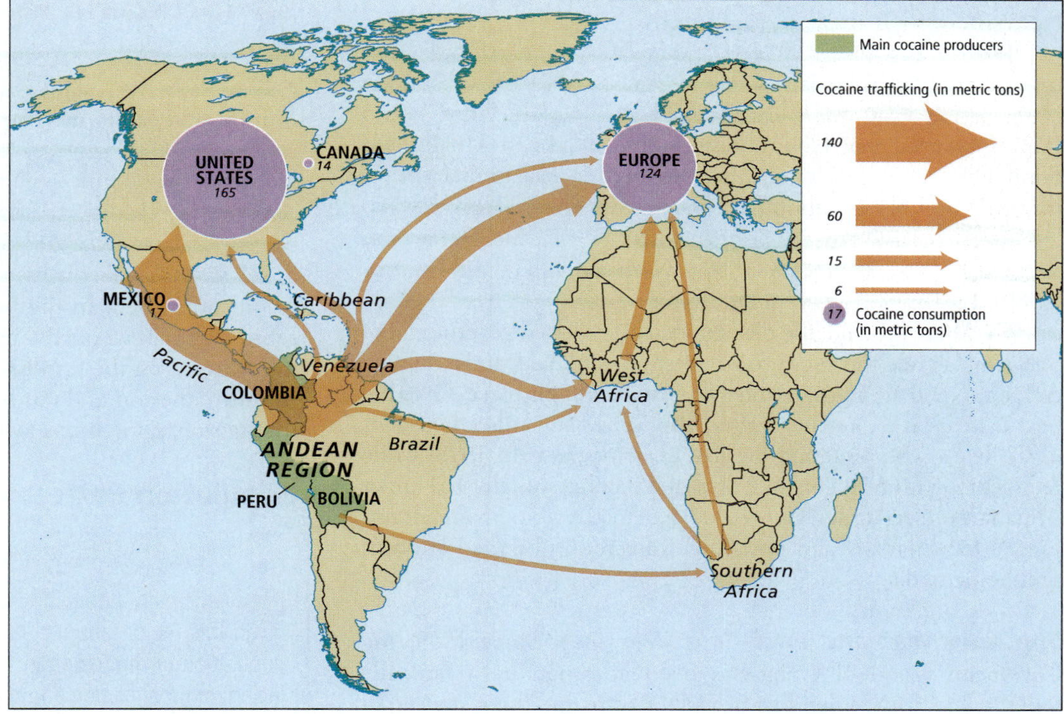

FIGURE 3.19 Linkages: Cocaine sources, trafficking routes, and seizures worldwide. Colombia, Peru, and Bolivia are the most important sources of cocaine cultivation and production in the world. The big cocaine markets are in the Americas; use has declined by 36 percent in the United States since 1998 and increased in South America. Cocaine use in Western Europe and West Africa is on the rise, due in part to new trade routes through West Africa to Europe.

in the region that is focused on supplying intelligence, eradication chemicals, military equipment, and training to military forces in the region. One consequence of this is that U.S. military aid to Middle and South America is now about equal to U.S. aid for education and other social programs in the region. Meanwhile, drug production in Middle and South America is now greater than ever, exceeding demand in the U.S. market, where street prices for many drugs have fallen in recent years. Concerned and informed citizens of Middle and South America have increasingly been calling for radical changes in international drug policies, including the legalization of drug use in the main (U.S.) market, which they assert would cause criminals to lose interest in the drug trade.

Foreign Involvement in the Region's Politics

Interventions in the region's politics by outside powers have frequently compromised democracy and human rights. Although the former Soviet Union, Britain, France, and other European countries have wielded much influence, by far the most active foreign power in this region has been the United States.

In 1823, the United States proclaimed the Monroe Doctrine to warn Europeans that no further colonization would be tolerated in the Americas. Subsequent U.S. administrations interpreted this policy more broadly to mean that the United States itself had the sole right to intervene in the affairs of the countries of Middle and South America, and it has done so many times. The official goal for such interventions was usually to make countries safe for democracy, but in most cases the driving motive was to protect U.S. political and economic interests.

During the past 150 years, at various times, U.S.-backed unelected political leaders, many of them military dictators, have been installed in many countries in the region. After World War II, worried that Communism would diffuse to Middle and South America, the United States focused special military scrutiny on a number of countries. Since the 1960s, the United States has funded armed interventions in Cuba (1961), the Dominican Republic (1965), Nicaragua (1980s), Grenada (1983), and Panama (1989). Perhaps the most infamous intervention occurred in Chile in 1973. With U.S. aid, the elected socialist-oriented government of Salvador Allende was overthrown. Allende was killed, and a military dictator, General Augusto Pinochet, installed in his place. Over the next 17 years, the Pinochet regime imprisoned and killed thousands of Chileans who protested the loss of democracy. In 1998, while recuperating from surgery in England, Pinochet was arrested under a Spanish judicial order. He would have stood trial in Spain but was judged too mentally and physically impaired to be tried and was sent home to die.

The Cold War and Post–Cold War Eras Since 1959, the Caribbean island of Cuba has been governed by a radically socialist government led first by Fidel Castro and then, starting in 2006, by Fidel's brother Raúl. The revolution that brought Fidel Castro to power transformed a plantation and tourist economy once known for its extreme income disparities into one of the most egalitarian in the region. However, because Castro adopted socialism and allied Cuba with the Soviet Union, relations with the United States became extremely hostile. The United States has funded many efforts to destabilize Cuba's government, and it actively discourages other countries from trading with Cuba. The United States maintains an ongoing embargo against Cuba, and so far international efforts to lift the embargo have failed.

73. EU SPLIT OVER DEVELOPMENT COMMISSIONER'S PROPOSAL TO NORMALIZE RELATIONS WITH CUBA

With help from the Soviet Union, Castro managed to dramatically improve the country's life expectancy, literacy, and infant mortality rates. Unfortunately, he also imprisoned or executed thousands of Cubans who disagreed with his policies, many of whom fled to southern Florida. Following the demise of the Soviet Union, Castro opened the country to foreign investment. Many countries responded, and Cuba is now a major European tourist destination. However, the Castro regime is intact, political repression in Cuba persists, and relations with the United States remain distant.

THINGS TO REMEMBER

- Democracy has transformed the politics of the region, yet problems remain. Elected governments are at times challenged by citizen protests or with a coup d'état because policies are unpopular with large segments of the population, powerful elites, or the military.
- The drug trade has emerged as a major obstacle to democracy, with huge amounts of illicit cash being used to pay off elected officials, civil servants, police forces, and the military.

Sociocultural Issues

Under colonialism, a series of social structures evolved that guided daily life—standard ways of organizing the family, community, and economy. They included rules for gender roles, race relations, and religious observance. These social structures, combined with economic systems, influenced population distribution and growth. Traditional social structures, still widely accepted in the region, are nonetheless changing in response to economic development, urbanization, migration, and globalization. The results are varied. In the best cases, change is leading to a new sense of initiative on the part of women, men, and the poor. In the worst cases, the result is a loss of economic viability and community cohesion and the breakdown of family life, with some of the most extreme impacts felt by indigenous communities.

Population Patterns

> **Geographic Insight 5**
>
> **Population:** A population explosion in the twentieth century has given this region about 596 million people, more than 10 times the population of the region in 1492. Now social and economic changes, including urbanization and the availability of contraception, are giving women options other than raising a family.

Today, populations in Middle and South America continue to grow but at an ever-slower pace than in the past because of lower birth rates. At the same time, a major migration from rural to urban areas is redistributing people and transforming traditional ways of life. A second trend, international migration, is also growing as many people leave their home countries, temporarily or permanently, to seek opportunities in the United States and Europe, and also in neighboring countries. As of 2011, about 596 million people were living in Middle and South America, close to 10 times the highest estimated population of the region in 1492.

Population Distribution The population density map of this region reveals a very unequal distribution of people. Comparing the population map (Figure 3.20) with the landforms map (see Figure 3.1), you will find areas of high population density in a variety of environments. Some of the places with the highest

FIGURE 3.20 Population density in Middle and South America.

sities, such as those around Mexico City and in Colombia and Ecuador, are in highland areas. But high concentrations are also found in lowland zones along the Pacific coast of Central America and especially along the Atlantic coast of South America. The cool uplands (*tierra templada*) were densely occupied even before the European conquest. Most coastal lowland concentrations, in the *tierra caliente*, are near seaports with vibrant globalizing economies and a cosmopolitan social life that attracts people.

Population Growth During the twentieth century, cultural and economic factors combined with improvements in health care to create a population explosion. High birth rates were sustained in part because the Roman Catholic Church discourages systematic family planning, and cultural mores encouraged men and women to reproduce prolifically. In agricultural areas, children were seen as sources of wealth because they could do useful farm and household work at a young age and eventually would care for their aging elders. Worry over infant death rates persisted, so some parents had four or more children to be sure of raising at least a few to adulthood. But, by 1975, child death rates were one-third what they had been in 1900. Most of the recent population growth has been related to longer life expectancy, not to high birth rates.

By 1990, improved living conditions (food, shelter, sanitation habits) and improved education and medical care, plus urbanization, all contributed to declining rates of natural increase, as shown in Figure 3.21; however, because 28 percent of the region's population is under age 15, even if couples choose to have only one or two children, the population will continue to grow as this large group reaches the age of reproduction. Nonetheless, population projections for the year 2050 were recently revised downward from 778 million to 746 million. Though significantly lower than previous forecasts, for a region struggling to increase living standards, supplying 150 million more people with food, water, homes, schools, and hospitals will be a challenge.

By the 1980s, migration out of the region and lower birth rates were further curbs to population growth. Now the region is undergoing a demographic transition (see Figure 1.10 on page 16). Between 1975 and 2011, the annual rate of natural increase for the entire region fell from about 1.9 percent to 1.2 percent—a rate of growth now equal to the world average. However, the low death rate continues to contribute to population growth in most of the region.

HIV-AIDS The global epidemic of HIV-AIDS is now taking a significant toll on populations throughout this region. In 2009, about 2.9 million people were HIV-positive. Of these, at least 60,000 were children. Some Caribbean islands have the highest HIV-AIDS infection rates outside sub-Saharan Africa, where the rate is 5 percent of the population. On the impoverished island of Haiti, 2.2 percent of the age 15–49 population has HIV-AIDS; in the Bahamas, 3 percent does. In most of the Caribbean, however, aggressive education programs about HIV, along with high literacy rates and the relatively high status of women, are limiting the number of infections.

The Venezuelan journalist Silvana Paternostro, in her book *In the Land of God and Man: Confronting Our Sexual Culture* (1998), argues that across Middle and South America, cultural practices discourage the open discussion of sex. Men are rarely expected to be monogamous, and many visit prostitutes, exposing their wives and any children they might bear to infection. A wife would be loath to ask her husband to use a condom. Drug use is also a significant factor in the spread of HIV.

While this general inclination to keep all kinds of sexual relationships in the closet persists, over the last decade attitudes have begun to change, due in part to television novellas that have explored many formerly taboo sexual issues, including HIV and especially homosexuality. In 2010, Argentina legalized gay marriage, as did Mexico City.

Human Well-Being The wide disparities in wealth that are a feature of life in Middle and South America have been discussed above. Too often, development efforts linked to urbanization and globalization have only increased the gap between rich and poor. Each of the three maps in Figure 3.22 shows one of the various indicators of well-being used by the United Nations to show how countries are doing in providing a decent life for their citizens.

Map A in Figure 3.22 is of average GNI per capita (PPP). It is immediately apparent that no country in this region falls in the

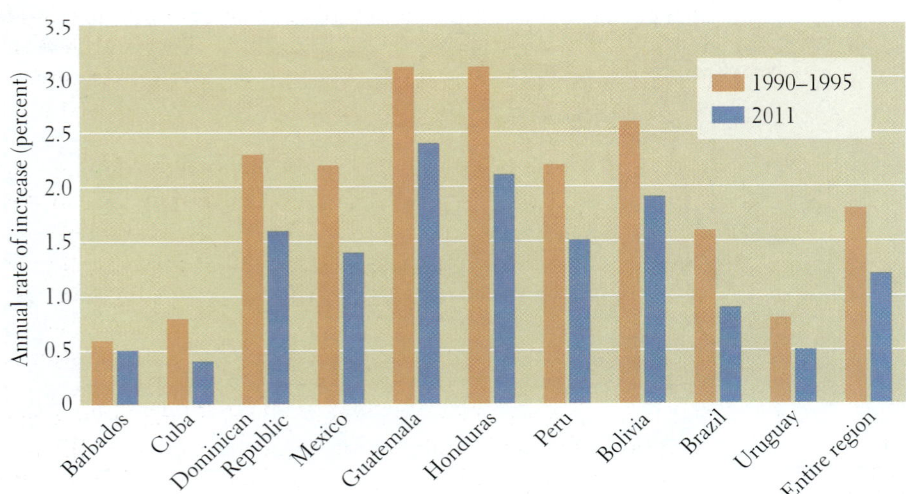

FIGURE 3.21 Trends in average rates of natural population increases, 1990–1995 and 2011. The orange and blue columns show that rates of natural increase have declined throughout the region and are projected to continue to do so into the future. Nevertheless, in many countries, natural population increase remains high enough to outstrip efforts to improve standards of living. Note that the rates of natural increase are for just a few selected countries and for the entire region.

FIGURE 3.22 Maps of human well-being.

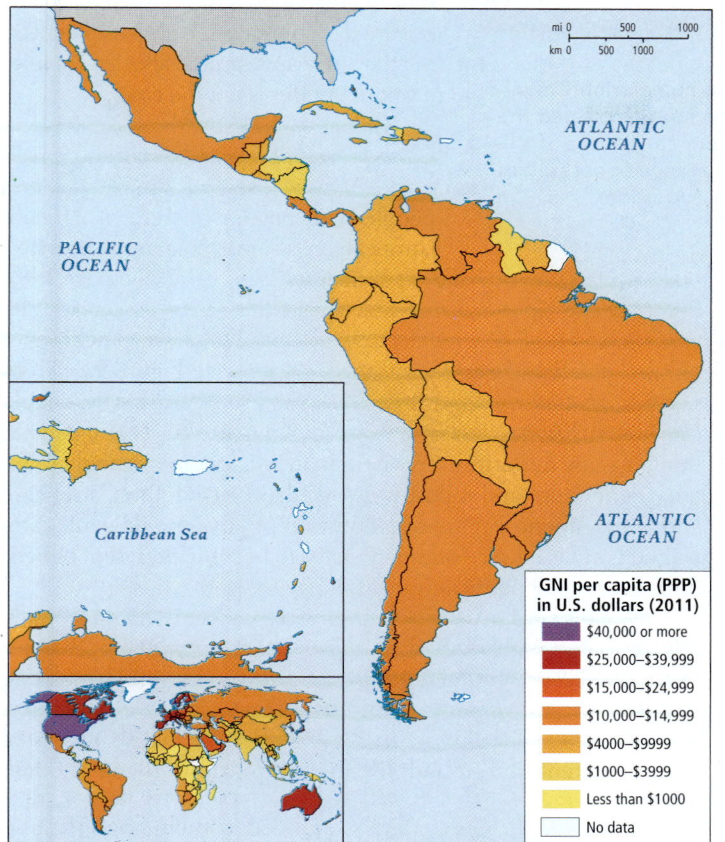

(A) Gross national income (GNI) per capita, adjusted for purchasing power parity (PPP).

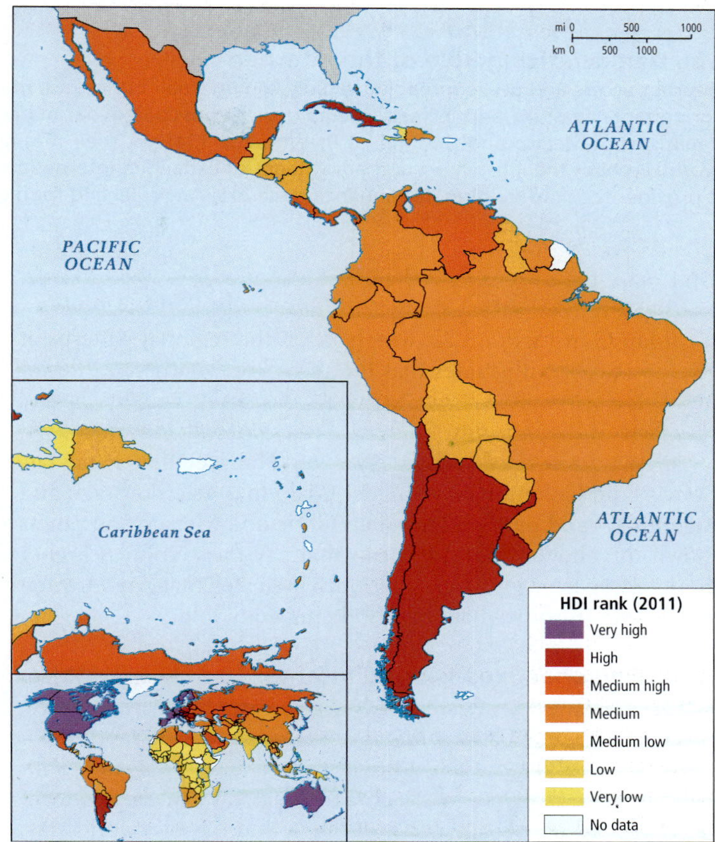

(B) Human Development Index (HDI).

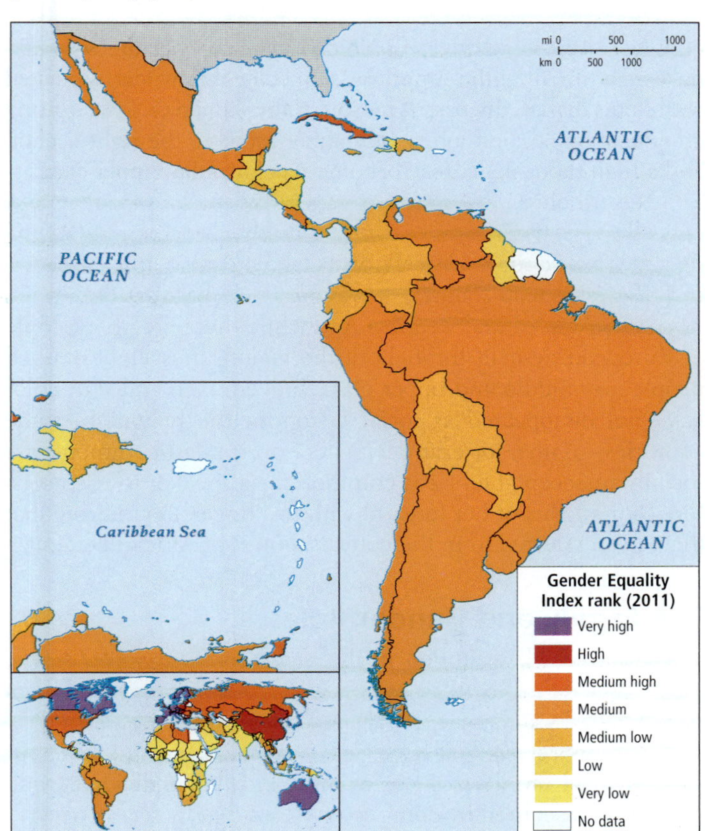

(C) Gender Equality Index (GEI).

two highest categories, and just a few islands in the Caribbean (Trinidad and Tobago, Barbados, and Antigua and Barbuda) fall into the third-highest category. Most of the Caribbean is in the global middle class, with one major exception, Haiti. Across the mainland, only Mexico, Panama, Venezuela, Uruguay, Argentina, Brazil, and Chile have an average per capita GNI between U.S.$10,000 and U.S.$14,999. All other countries in the region have average per capita GNIs that are less than $9999, many of them significantly less. Also, most of these countries have rather wide disparities in wealth (see Table 3.2 on page 127) that are masked on a map like this. In Brazil, for example, the most affluent 10 percent have incomes that are 41 times greater than the bottom 10 percent, and in actuality, some Brazilians survive on less than $1000 a year, while others have incomes well over $100,000 per year.

Map B in Figure 3.22, which charts the HDI rank for all countries in the region, shows that most countries rank in the medium and high ranges in providing the basics (education, health care, and income) for their citizens. The global inset map confirms this mid-range status for Middle and South America. Most of Africa, Central Asia, and South Asia rank lower, while North America, Europe, Australia, Japan, and South Korea rank higher. Much of East Asia, Russia, and some of the post-independent states of the former Soviet Union rank in the same range as Middle and South America. Of course, the UN HDI is a countrywide index and does not give any indication of how well-being is distributed within a given country. On this map and the

ON THE BRIGHT SIDE

The Democratizing Role of the Internet

Internet access and use is spreading rapidly, aiding small businesses, making human rights monitoring more feasible, and increasing employment possibilities for many. The Internet helped the Zapatistas of Mexico and the MST of Brazil (page 134) get their message out to the rest of the world. Perhaps the most important advantages of expanded Internet access are the opportunities it provides for skills training, for enhancing schools' curricula, and for higher education.

GNI map, the countries of the Central American isthmus stand out as poorer than Mexico, the Caribbean, and South America.

Map C in Figure 3.22, which shows the regional patterns of gender equality, illustrates that the region of Middle and South America does not yet begin to approach gender equality in pay or access to opportunity; it ranks even lower than some parts of Africa in terms of gender equality. The fact that Paraguay, Bolivia, and Colombia have higher GEI ranks than do Brazil and Argentina—and much higher than Chile and Mexico and others shown in yellow—does not mean that pay for women is higher there (compare with Map A); it only indicates that pay for men and women is somewhat closer to being equitable.

Rapid Regional Expansion of Internet Use Overall, Middle and South America are more advanced in terms of information technology than are many other developing regions of the world—a fact that could open up many new economic opportunities for the region in the near future. Brazil ranks fifth in the world in terms of total number of Internet users, and has two world-class technology hubs in the environs of São Paulo. Mexico, with 36.9 percent of its population connected, has recently launched a program to give its citizens access to training and higher education via the Internet (Figure 3.23). Similar efforts throughout the region are part of a long-term shift, especially in urban areas, toward more technologically sophisticated and better-paid service sector employment. In 2000, only 3.2 percent of the region's population used the Internet, but that figure had risen to nearly 40 percent by 2010, an increase of more than 1000 percent.

Cultural Diversity

The region of Middle and South America is culturally complex because many distinct indigenous groups were already present when the Europeans arrived, and then many cultures were introduced during and after the colonial period. From 1500 to the early 1800s, some 10 million people from many different parts of Africa were brought to plantations on the islands and in the coastal zones of Middle and South America. After the emancipation of African slaves in the British-controlled Caribbean islands in the 1830s, more than half a million Asians were brought there from India, Pakistan, and China as indentured agricultural workers. Their cultural impact remains most visible in Trinidad and Tobago, Jamaica, and the Guianas. In some parts of Mexico, Central America, the Amazon Basin, and the Andean Highlands, indigenous people have remained numerous. To the unpracticed eye, they may appear little affected by colonization, but this is not the case.

acculturation adaptation of a minority culture to the host culture enough to function effectively and be self-supporting

assimilation the loss of old ways of life and the adoption of the lifestyle of another culture

extended family a family that consists of related individuals beyond the nuclear family of parents and children

Today in the Caribbean and along the east coast of Central America as well as the Atlantic coast of Brazil, mestizos (people who have a mixture of African, European, and some indigenous ancestry) make up the majority of the population (Figure 3.24 on page 144). Some of the most colorful festivals in the Americas—especially versions of Mardi Gras and Carnival—are based on a melding of cultural practices from these three heritages (see Figure 3.24A). In some areas, such as Argentina, Chile, and southern Brazil, people of Central European descent are also numerous. The Japanese, though a tiny minority everywhere in the region, increasingly influence agriculture and industry, especially in Brazil, the Caribbean, and Peru. Alberto Fujimori, a Peruvian of Japanese descent, campaigned on his ethnic "outsider" status to become president of Peru during a time of political upheaval.

In some ways, diversity is increasing as the media and trade introduce new influences from abroad. At the same time, the processes of **acculturation** and **assimilation** are also accelerating, meaning that diversity is being erased to some extent as people adopt new ways. This is especially true in the biggest cities, where people of widely different backgrounds live in close proximity to one another.

Race and the Social Significance of Skin Color

People from Middle and South America, especially those from Brazil, often proudly claim that race and color are of less consequence in this region than in North America. They are only partially right. In all of the Americas, skin color is now less associated with status than in the past. A person of any skin color, by acquiring an education, a good job, a substantial income, the right accent, and a high-status mate, may become recognized as upper class.

Nevertheless, the ability to erase the significance of skin color through one's actions is not quite the same as race having no significance at all. Overall, those who are poor, less educated, and of lower social standing tend to have darker skin than those who are educated and wealthy. And while there are poor people of European descent throughout the region, most light-skinned people are middle and upper class. Indeed, race and skin color have not disappeared as social factors in the region. In some countries—Cuba, for example, where overt racist comments are socially unacceptable—it is common for a speaker to use a gesture (tapping his or her forearm with two fingers) to indicate that the person referred to in the conversation is of African descent.

The Family and Gender Roles

The basic social institution in the region is the **extended family**, which may include cousins, aunts, uncles, grandparents, and more distant relatives. It is generally accepted that the individual should sacrifice many of his or her personal interests to those of the extended family and community and that individual well-being is best secured by doing so.

FIGURE 3.23 Internet use in Middle and South America, December 2011. The numbers on the map indicate the number of Internet users in each country; percents indicate the percentage of the country's population that uses the Internet. By December 2011, over 235 million people (39.5 percent of the population) in Middle and South America were using the Internet; the number of users rose slightly more than 1200 percent since 2000. Argentina has the highest percentage of users (67) in South America; St. Lucia (88.5 percent) the highest in the Caribbean; and Costa Rica (43.7 percent) the highest in Middle America.

The arrangement of domestic spaces and patterns of socializing illustrate these strong family ties. Families of adult siblings, their mates and children, and their elderly parents frequently live together in domestic compounds of several houses surrounded by walls or hedges. Social groups in public spaces are most likely to be family members of several generations rather than unrelated groups of single young adults or married couples, as would be the case in Europe or the United States. A woman's best friends are likely to be her female relatives. A man's social or business circles will include male family members or long-standing family friends.

Gender roles in the region have been strongly influenced by the Roman Catholic Church. The Virgin Mary is held up as the model for women to follow through a set of values known as ***marianismo***, which emphasizes chastity, motherhood, and service to the family. In this

marianismo a set of values based on the life of the Virgin Mary, the mother of Jesus, that defines the proper social roles for women in Middle and South America

still-widespread tradition, the ideal woman is the day-to-day manager of the house and of the family's well-being. She trains her sons to enter the wider world and her daughters to serve within the home. Over the course of her life, a woman's power increases as her skills and sacrifices for the good of all are recognized and enshrined in family lore.

Her husband, the official head of the family, is expected to work and to give most of his income to his family. Still, men have much more autonomy and freedom to shape their lives than women because they are expected to move about the larger community and establish relationships, both economic and personal. A man's social network is considered just as essential to the family's prosperity and status in the community as is his work.

In addition, there is an overt double sexual standard for males and females. While all expect strict fidelity from a wife to a husband in mind and body, a man is much freer to associate with the opposite sex. Males measure themselves by the model of ***machismo***, which considers manliness to consist of honor, respectability, fatherhood, household leadership, attractiveness to women, and the ability to be a charming storyteller. Traditionally, the ability to acquire money was secondary to other symbols of maleness. Increasingly, however, a new, market-oriented culture prizes visible affluence as a desirable male attribute.

machismo a set of values that defines manliness in Middle and South America

Many factors are transforming these traditional family and gender roles. With infant mortality declining steeply, couples are now having only two or three children instead of five or more. Moreover, because most people still marry young, parents are generally free of child-raising responsibilities by the time they are 40. Left with 30 or more years of active life to fill in other ways (life expectancies in the region average in the mid-70s), middle-aged mothers and grandmothers are increasingly working in urban factory or office jobs that put to use the organizational and problem-solving skills they developed while managing a family. Employment outside the home allows women greater independence and helps them contribute to the needs of the extended family. Some women are even moving into high-level jobs, such as management positions, professorships, and positions that traditionally went to men—for example, the presidencies of Chile, Argentina, and Brazil. Many of the traditional cultural restrictions on the mobility of women have been relaxed, with the result that women are better able to use their talents and abilities to contribute economically and socially to family well-being.

FIGURE 3.24 LOCAL LIVES FESTIVALS IN MIDDLE AND SOUTH AMERICA

A A member of a "bateria," or percussion band, is parading during Rio de Janeiro's *Carnaval*, the largest in the world. Carnaval, which is celebrated in much of the United States as Mardi Gras, takes place in the days and weeks before Lent, the 40-day period of fasting and abstinence preceding Easter. Elaborate celebrations take place in Baranquilla in Colombia, Port-au-Prince in Haiti, Port of Spain in Trinidad, and Montevideo in Uruguay.

B A procession commemorating Jesus walks on an *alfombra*, or carpet, made of colored sawdust, painstakingly drawn by hand on the streets of Antigua, Guatemala, during *Semana Santa*, or Holy Week, the last week of Lent. The processions will destroy the alfombras. This tradition is an adaptation of the Mayan practice of making elaborate ground designs with flowers and feathers for Mayan royalty to walk on as they made their way to important ceremonies.

C A richly decorated grave in Tijuana, Mexico, during *Dia de los Muertos*, or Day of the Dead, a holiday celebrated throughout Mexico on November 1. Families gather in cemeteries, where they build private altars and other decorations dedicated to deceased loved ones. The holiday is a "Christianized" version of an Aztec festival dedicated to the goddess Mictecacihuatl, Queen of the Underworld, whose job is to watch over the bones of the dead.

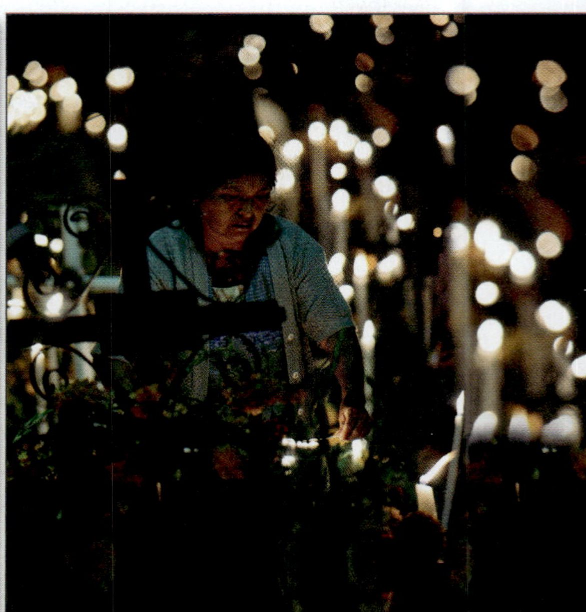

Migration and Urbanization

Geographic Insight 6

Gender: Traditional gender roles are changing as women work outside the home and manage families while many men are migrating to other places for work. Also, urbanization is reducing the influence of extended families, opening the door for many types of social change.

Since the early 1970s, Middle and South America have seen very rapid urban growth as rural people migrate to cities and towns. More than 70 percent of the people in the region live in settlements of at least 2000 people. Increasingly, one city, known as a **primate city**, is vastly larger than all the others, accounting for a large percentage of the country's total population (Figure 3.25). Examples in the region are Mexico City (with 23 million people, a bit more than 21 percent of Mexico's total population); Managua, Nicaragua (33 percent of that country's population); Lima, Peru (29 percent); Santiago, Chile (34 percent); and Buenos Aires, Argentina (36 percent).

The concentration of people into just one or two large cities in a country leads to uneven spatial development and to government policies and social values that favor the largest urban areas. Wealth and power are concentrated in one place, while distant rural areas, and even other towns and cities, have difficulty competing for talent, investment, industries, and government services. Many provincial cities languish as their most educated youth leave for the primate city.

Brain Drain It always takes some resourcefulness to move from one place to another. Those people who already have some years of education and strong ambition are the ones who migrate to cities. The loss of these resourceful young adults in which the community has invested years of nurturing and education is referred to as **brain drain**. Brain drain happens at several scales in Middle and South America: through rural-to-urban migration from villages to regional towns, and from towns and small cities to primate cities. There is also international brain drain when migrants move to North America and Europe; for example, one in four U.S. doctors are foreign born. Families often encourage their children to migrate so that they can benefit from the remittances, goods, and services that migrants send back to their home communities.

Urban Landscapes A lack of planning to accommodate the massive rush to the cities has created urban landscapes that are very different from the common U.S. pattern. In the United States, a poor, older inner city is usually surrounded by affluent suburbs, with clear and planned spatial separation of residences by income as well as separation of residential, industrial, and commercial areas. In Middle and South America, by contrast, both affluent and working-class areas have become unwilling neighbors to unplanned slums filled with poor migrants. Too destitute even to rent housing, these "squatters" occupy parks and small patches of vacant urban land wherever they can be found, as depicted in the diagram in Figure 3.26 on page 147.

Unplanned Neighborhoods The best known of these unplanned communities are Brazil's **favelas** (Figure 3.27 on page 147). In other countries they are known as *slums, shantytowns, colonias, barrios,* or *barriadas*. The settlements often spring up overnight, without city-supplied water or electricity and with housing made out of whatever is available. Once the settlements are established, efforts to eject squatters usually fail, though police are often called in to make an attempt. The impoverished are such a huge portion of the urban population that even those in positions of power will not challenge them directly. Nearby wealthy neighborhoods simply barricade themselves with walls and security guards.

The squatters frequently are enterprising people who work hard to improve their communities. They often organize to press governments for social services. Some cities, such as Fortaleza in northeastern Brazil, even contribute building materials so that favela residents can build more permanent structures with basic indoor plumbing. Over time, as shacks and lean-tos are transformed through self-help initiatives into crude but livable suburbs, the economy of favelas can become quite vibrant, with much activity in the informal sector. Housing may be intermingled with shops, factories, warehouses, and other commercial enterprises. Favelas and their counterparts can become centers of pride and support for their residents, where community work, folk belief systems, crafts, and music (for example, Favela Funk) flourish. Many of the best steel bands of Port of Spain, Trinidad, have their homes in the city's shantytowns.

primate city a city, plus its suburbs, that is vastly larger than all others in a country and in which economic and political activity is centered

brain drain the migration of educated and ambitious young adults to cities or foreign countries, depriving the communities from which the young people come of talented youth in whom they have invested years of nurturing and education

favelas Brazilian urban slums and shantytowns built by the poor; called *colonias, barrios,* or *barriadas* in other countries

VIGNETTE Favelas are everywhere in Fortaleza, Brazil. The city grew from 30,000 to 300,000 people during the 1980s. By 2006, there were more than 3 million residents, most of whom had fled drought and rural poverty in the interior of the country. City parks of just a square block or two in middle-class residential areas were suddenly invaded by squatters. Within a year, 10,000 or more people were occupying crude, stacked concrete dwellings in a single park, completely changing the ambience of the surrounding upscale neighborhood. In the early days of the migration, lack of water and sanitation often forced the migrants to relieve themselves on the street.

One day, while strolling on the Fortaleza waterfront, Lydia Pulsipher chanced to meet a resident of a beachfront favela who invited her to join him on his porch. There, he explained how he and his wife had come to the city 5 years before, after being forced to leave the drought-plagued interior when a newly built irrigation reservoir flooded the rented land their families had cultivated for generations. With no way to make a living, they set out on foot for the city. In Fortaleza, they constructed the building they used for home and work from objects they collected along the beach. Eventually they were able to purchase roofing tiles, which gave the building an air of permanency. At the time of the visit, he maintained a small refreshment stand and his wife a beauty parlor that catered to women from the beach settlements. *[Source: Lydia Pulsipher's field notes. For detailed source information, see Text Credit pages.]* ■

Figure 3.25 Photo Essay: Urbanization in Middle and South America

Urbanization is at the heart of many transformations in this region. The last several decades have seen cities grow extremely fast, and 77 percent of the region now lives in cities. New opportunities have opened up for migrants from rural areas, but new stresses have emerged as well. The low-skilled jobs they can get often don't pay well, and many people are forced to live in unplanned neighborhoods on the outskirts of huge cities.

A Santiago is one of the region's primate cities, home to 34 percent of Chile's population. It dominates Chile's economy, generating 40 percent of the country's GDP.

B A child works as a street vendor in Salvador, Brazil. Many new urban migrants work in similar informal sector jobs that generally pay less and are much less secure than formal sector jobs.

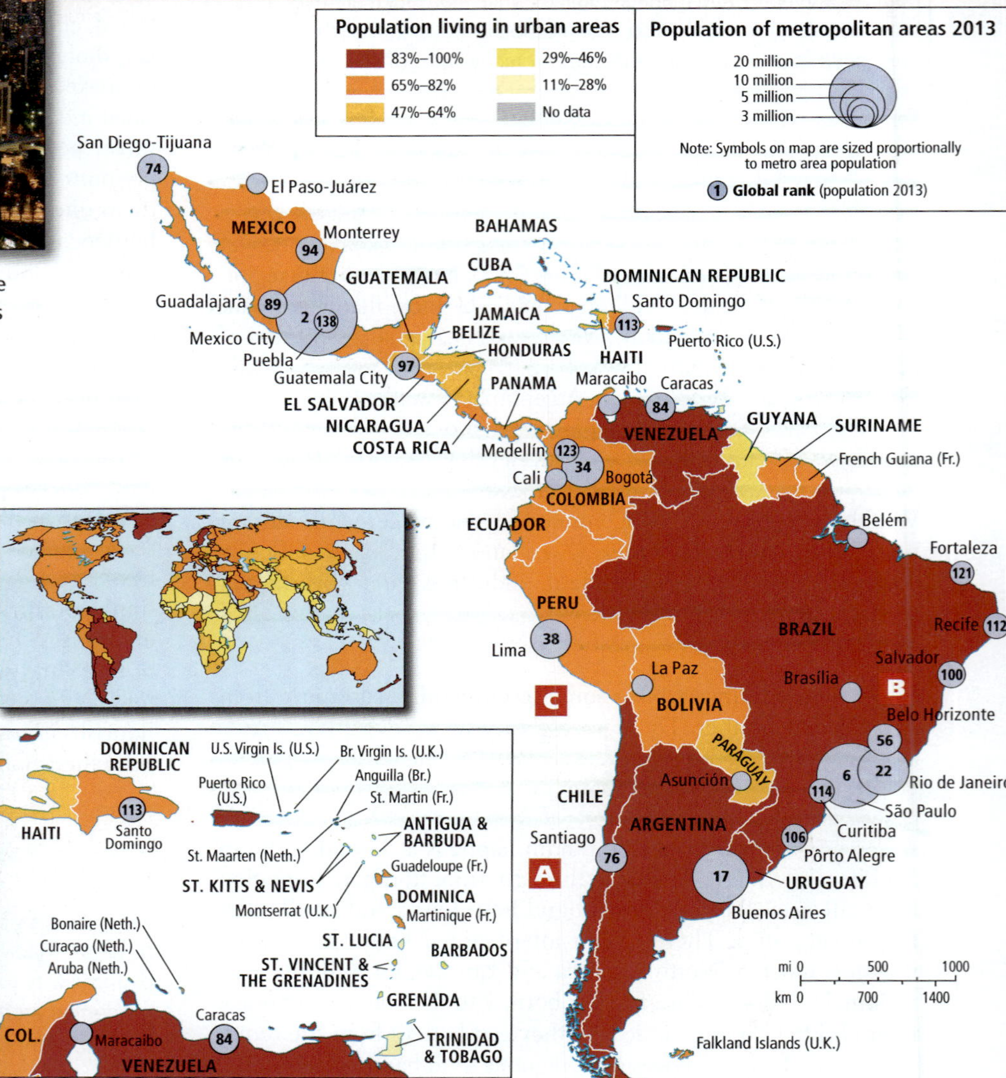

C Police evict 2500 people from an informal squatter settlement recently formed outside of Lima, Peru. Structures made out of woven grass mats are the first to go up, followed by wooden shacks and eventually cement or brick houses. Water and electricity may eventually be extended to such settlements, but only after years or decades of occupation.

Thinking Geographically

After you have read about urbanization in Middle and South America, you will be able to answer the following questions:

A What about this picture suggests that there may be considerable wealth in Santiago?

B Why might work in the informal economy be considered insecure?

C What are some defining characteristics of "informal" squatter settlements?

Planned Elite Landscapes

In contrast to colorful but problematic favelas are districts in nearly every city that are modern, planned, and similar to elite urban districts across the world. Examples of such urban landscapes can be found in Rio de Janeiro, São Paulo, and other cities in Brazil's southeastern industrial heartland. These cities have districts that are elegant and futuristic showplaces, resplendent with the very latest technology and graced with buildings and high-end shops that rival those in New York, Singapore, Kuala Lumpur, and even Tokyo. Brazilian planners now acknowledge, however, that in the rush to develop these modern urban landscapes, developers neglected to underwrite the parallel development of a sufficient urban infrastructure (for an exception, see the discussion of the southern city of Curitiba on page 148). In short supply are sanitation and water systems, an up-to-date electrical grid, transportation facilities, schools, housing, and medical facilities—all necessary to sustain modern business, industry, and a socially healthy urban population. But the Brazilians should not be criticized too severely for what in hindsight seems like an obvious error. During the 1970s and 1980s, development theory set forth by the World Bank and other financial institutions held that social transformation and infrastructure development would just naturally evolve. They would be normally occurring side effects of investment in what were called **urban growth poles**. The debt crisis of the 1980s, which ate up investment profits with skyrocketing *inflation* (a

urban growth poles locations within cities that are attractive to investment, innovative immigrants, and trade, and thus attract economic development like a magnet

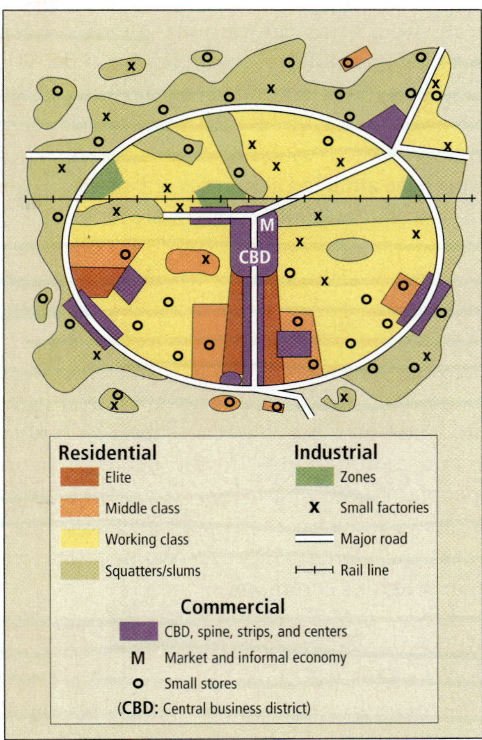

FIGURE 3.26 Crowley's model of urban land use in mainland Middle and South America. William Crowley, an urban geographer who specializes in Middle and South America, developed this model to depict how residential, industrial, and commercial uses are mixed together, with people of widely varying incomes living in close proximity to one another and to industries. Squatters and slumdwellers ring the city in an irregular pattern.

FIGURE 3.27 A favela in Rio de Janeiro. In Rio de Janeiro, a favela clings to a once-vacant hillside that was considered too steep for apartment buildings. Now home to thousands of people, it is still served mainly by footpaths and narrow alleyways. Water is only available sporadically, so many residents store water in tanks on their roof.

Thinking Geographically

Why are favelas often located on steep hills, and why do residents often store water on their roofs?

rapid rise in consumer prices caused by the falling value of the country's currency on the international market), put an end to such optimistic forecasts. Only very recently has there been global recognition that planned development of human capital through education, health services, and community development, whether privately or publicly funded, is a primary part of building strong urban economies.

Urban Transport Issues Transportation in large, rapidly expanding urban areas with millions of poor migrants can be a special challenge. Favelas are often on the urban fringes, far from available low-skill jobs and ill-served by roads and public transport (see Crowley's model, Figure 3.26). Workers must make lengthy, time-consuming, and expensive commutes to jobs that pay very little. The entire urban population gets caught up in endless traffic jams that cripple the economy and pollute the environment. Rapid growth seems to have outstripped the capacities of even the most enlightened urban planning.

Yet the southern Brazilian city of Curitiba has carefully oriented its expansion around a master plan (dating from 1968) that includes an integrated transportation system funded by a public-private collaboration. Eleven hundred minibuses making 12,000 trips a day bring 1.3 million passengers from their remote neighborhoods to terminals, where they meet express buses to all parts of the city. The large pedestrian-only inner-city area is a boon for merchants because minibus commuters spend time shopping rather than looking for parking. Being able to get to work quickly and cheaply has helped the poor find and keep jobs. The reduced use of cars means that emissions and congestion are lowered and urban living is more pleasant.

Gender and Urbanization Interestingly, rural women are just as likely as rural men to migrate to the city. This is especially true when employment is available in foreign-owned factories that produce goods for export. Companies prefer women for such jobs because they are a low-cost and usually passive labor force. Factors that push people out of rural areas—for instance, the shift toward green revolution agriculture—often affect women intensely because women are rarely considered for jobs such as farm-equipment operators or mechanics. In urban areas, unskilled migrant women and children usually find work as street vendors (see Figure 3.25B) or domestic servants.

Male urban migrants tend to depend on short-term, low-skill day work in construction, maintenance, small-scale manufacturing, and petty commerce. Many work in the informal economy as street vendors, errand runners, car washers, and trash recyclers—and some turn to crime. In an urban context, both men and women sorely feel the loss of family ties and village life; the chances for recreating the family life they once knew are extremely low.

Once relocated to a city, migrant families may disintegrate because of extreme poverty and malnutrition, long commutes for both working parents, and poor quality of day care for children. At too young an age, children are left alone or sent into the streets to scavenge for food or to earn money for the family.

populist movements popularly based efforts, often seeking relief for the poor

liberation theology a movement within the Roman Catholic Church that uses the teachings of Jesus to encourage the poor to organize to change their own lives and to encourage the rich to promote social and economic equity

Religion in Contemporary Life

Judaism, Islam, Hinduism, and indigenous beliefs are found across the region, but the majority of the people are at least nominal Christians, most of them Roman Catholic. While the Roman Catholic Church remains highly influential and relevant to the lives of believers, it has had to contend both with popular efforts to reform it and with increasing competition from other religious movements. From the beginning of the colonial era, the church was the major partner of the Spanish and Portuguese colonial governments. It received extensive lands and resources from colonial governments, and in return built massive cathedrals and churches throughout the region, sending thousands of missionary priests to convert indigenous people. The Roman Catholic Church encouraged the poor converts to accept their low status, obey authority, and postpone rewards for their hard work until heaven.

Poor people throughout the region embraced the faith, and they still make up the majority of the church members. Many are indigenous people who put their own spin on Catholicism, creating multiple folk versions of the Mass with folk music, greater participation of women in worship services, and interpretations of Scripture that vary greatly from European versions. A range of African-based belief systems (Candomblé, Umbanda, Santería, Obeah, and Voodoo) combined with Catholic beliefs is found in Brazil, northern South America, and Middle America and the Caribbean—wherever the descendants of Africans have settled. These African-based religions have attracted adherents of European or indigenous backgrounds as well, especially in urban areas.

The power of the Roman Catholic Church began to erode in the nineteenth century in places such as Mexico. **Populist movements**, aimed at addressing the needs of the poor masses, seized and redistributed some church lands. They also canceled the high fees the clergy had been charging for simple rites of passage such as baptisms, weddings, and funerals. Over the years, the Catholic Church became less obviously connected to the elite and more attentive to the needs of poor and non-European people. By the mid-twentieth century, the church was abandoning many previous racist policies and ordaining indigenous and clergy of African descent. Women were also given a greater role in religious ceremonies.

In the 1970s, a Catholic movement known as **liberation theology** was begun by a small group of priests and activists. They sought to reform the church into an institution that could combat the extreme inequalities in wealth and power common in the region. The movement portrayed Jesus Christ as a social revolutionary who symbolically spoke out for the redistribution of wealth when he divided the loaves and fish among the multitude. The perpetuation of gross economic inequality and political repression was viewed as sinful, and social reform as liberation from evil.

The legacy of liberation theology is now fading. At its height in the 1970s and early 1980s, liberation theology was

evangelical Protestantism a Christian movement that focuses on personal salvation and empowerment of the individual through miraculous healing and transformation; some practitioners preach to the poor the "gospel of success"—that a life dedicated to Christ will result in prosperity for the believer

the most articulate movement for region-wide social change. It had more than 3 million adherents in Brazil alone. But the Vatican then objected to this popularized version of Catholicism, and today its influence is diminished. The first pope from the Americas, Francis of Argentina, while critical of capitalism and an advocate for the poor, eschews connections with liberation theology, which now must compete with newly emerging evangelical Protestant movements. **79. POPE OPENS TOUR OF LATIN AMERICA IN BRAZIL**

Evangelical Protestantism has diffused from North America into Middle and South America and is now the region's fastest-growing religious movement. About 10 percent of the population, or at least 50 million people, are adherents. The movement is growing rapidly in Brazil, Chile, the Caribbean, Mexico, and Middle America, especially among poor and middle classes in both rural and urban settings. It does not, however, share liberation theology's emphasis on combating the region's extreme inequalities in wealth and power. Some evangelical Protestants teach a "gospel of success," stressing that those true believers who give themselves to a new life of hard work and clean living will experience prosperity of the body (wealth) as well as of the soul.

The movement is *charismatic*, meaning that it focuses on personal salvation and empowerment of the individual through miraculous healing and psychological transformation. Evangelical Protestantism is not hierarchical in the same way as the Roman Catholic Church, and there is usually no central authority; rather, there are a host of small, independent congregations led by entrepreneurial individuals who may be either male or female.

Perhaps two of the most important contributions of evangelical Protestantism are the focus (as in Umbanda and other African-based religions) on helping people, especially urban migrants, cope with modern life; and the emphasis placed on involving women in leadership roles in church activities and services.

THINGS TO REMEMBER

Geographic Insight 5

- **Population** A population explosion occurred during the early twentieth century as improved health care lowered death rates.
- Birth rates started to decline by the mid-1980s as more women began to delay childbearing in order to pursue work outside the home. At the same time, populations became more urbanized, which reduced the need for large families to operate farms.
- Rapid growth of Internet use is part of a long-term shift toward more technologically sophisticated service sector employment.
- Gender roles in the region have been strongly influenced by the Catholic Church through the ideals of *marianismo* and *machismo*.

Geographic Insight 6

- **Gender** Middle-aged women are increasingly working in factory or office jobs once dominated by men; they put to use the organizational and problem-solving skills developed while supervising a family.
- Urbanization is putting new pressures on the extended family, opening up new opportunities for women and fostering new belief systems.
- Evangelical Protestantism came to the region from North America and is now the fastest-growing religious movement in the region.

Geographic Insights: Middle and South America Review and Self-Test

1. Climate Change and Deforestation: Deforestation in this region contributes to climate change by removing trees, which as living plants naturally absorb carbon dioxide. As the cut vegetation decays, large amounts of carbon dioxide are released into the atmosphere. Sustainable alternatives to deforestation are being tried.

• What factors are driving the cycle of deforestation in this region? How is the market economy a factor?

• How are the processes of global warming and deforestation linked?

• What are some possibly sustainable alternatives to deforestation?

2. Water: Despite the region's abundant water resources, parts of the region are experiencing a water crisis related to unplanned urbanization; marketization; corruption; inadequate water infrastructure; and the use of water to irrigate commercial agriculture.

• What is driving increases in overall water consumption?

• Why is per capita water use going up?

3. Globalization, Development, and Democratization: The integration of this region with the world economy has led to growing disparities of wealth and opportunity, forcing millions to migrate in search of work. A political backlash against these patterns of development has brought to power some leaders who question the benefits of globalization, while others seek to help their countries compete globally.

• What are some factors that are hampering the equitable distribution of wealth?

• Why did import substitution fail as a strategy to develop manufacturing and service-based industries?

• Why has there been so much dissatisfaction with SAPs?

4. Food and Urbanization: The shift from small-scale, subsistence food production to mechanized agriculture for export has fueled migration to cities. Many places previously were self-sufficient with regard to food; they now depend on imported food, making them susceptible to food insecurity.

• How have new food production techniques changed the demand for agricultural labor?

• Why has food insecurity increased with the advent of modern, mechanized agriculture?

• How have cities responded to the waves of migrants from rural areas?

5. Population: A population explosion in the twentieth century has given this region about 596 million people, more than 10 times the population of the region in 1492. Now social and economic changes, including urbanization and the availability of contraception, are giving women options other than raising a family.

• What accounts for the population explosion of the last century?

• What are the urbanization factors that have resulted in smaller family size and more options for women?

6. Gender: Traditional gender roles are changing as women work outside the home and manage families in the absence of migrating males. Also, urbanization is reducing the influence of extended families, opening the door for many types of social change.

• How does migration to the city influence extended family relationships?

• Why does the opportunity to gain an education so strongly affect a woman's fertility?

Critical Thinking Questions

1. If the European colonists had come to Middle and South America in a different frame of mind—if, say, they were simply looking for a new place to settle and live quietly—how do you think the human and physical geography of the region would be different today?

2. Explain two main ways in which tectonic processes account for the formation of mountains in Middle and South America.

3. Reflecting on the whole chapter, pick some locations that impressed you with regard to the ways in which people are dealing with either environmental issues or with issues of income/wealth disparity or human well-being. Explain your selections.

4. Discuss the ways in which you see the historical circumstances of colonization affecting modern approaches to economic problems in Mexico, Bolivia, Brazil, Venezuela, or Cuba.

5. Describe the main patterns of migration in this region and discuss the effects of migration on both the sending and receiving societies.

6. Name three factors you see as important in increasing democratic participation in this region. In which countries would you say these factors are making the biggest difference?

7. Explain how the Amazon Basin and its resources constitute an example of contested space.

8. Argue for or against the proposition that free trade blocs such as NAFTA and UNASUR (previously Mercosur) help the lowest-paid workers have some upward mobility.

9. How would you respond to someone who suggested that Middle and South America were helped toward development and modernization by the experience of European colonization?

Chapter Key Terms

acculturation 142
assimilation 142
Aztecs 122
biodiversity 113
brain drain 145
contested space 133
coup d'état 135
Creoles 125
dictator 135
early extractive phase 128
ecotourism 119
El Niño 114
evangelical Protestantism 149
Export Processing Zones (EPZs) 131
extended family 142
favelas 145
foreign direct investment (FDI) 131
hacienda 128
import substitution industrialization (ISI) 129
Incas 122
income disparity 127
indigenous 112
isthmus 112
liberation theology 148
machismo 144
maquiladoras 131
marianismo 143
marketization 129
mercantilism 125
Mercosur 132
mestizos 125
Middle America 112
nationalize 131
North American Free Trade Agreement (NAFTA) 132
plantation 128
populist movements 148
primate city 145
privatization 129
shifting cultivation 122
silt 114
South America 112
structural adjustment policies (SAPs) 129
subduction zone 112
temperature-altitude zones 114
trade winds 114
UNASUR 132
urban growth poles 147

ARCTIC OCEAN
ATLANTIC OCEAN
North Sea
Baltic Sea
Mediterranean Sea
Arctic Circle
Prime Meridian
Reykjavik
ICELAND
Faroe Islands
Torshavn
NORWAY
SWEDEN
FINLAND
Hammerfest
Vadsø
Tromsø
Inari
Ivalo
Kola Peninsula
White Sea
Narvik
Kiruna
Lapland
Bodo
Umea
Gulf of Bothnia
Lieksa
Kuopio
Vaasa
Trondheim
Alesund
Lake Onega
Lake Ladoga
Turku
Helsinki
St. Petersburg
Gulf of Finland
Bergen
Oslo
Lake Vanern
Stockholm
Tallinn
ESTONIA
Gulf of Riga
Aluksne
LATVIA
Riga
Goteborg
Lake Vattern
Alborg
DENMARK
Copenhagen
Lund
Malmo
LITHUANIA
Vilnius
Kaunas
RUSSIA
Kaliningrad
Minsk
European
BELARUS
Kirkwall
SCOTLAND
Aberdeen
Glasgow
Edinburgh
UNITED KINGDOM
NORTHERN IRELAND
Belfast
Irish Sea
Douglas
REPUBLIC of IRELAND
Dublin
Liverpool
Leeds
Manchester
Sheffield
Cork
St. George's Channel
Newcastle
WALES
ENGLAND
Norwich
Birmingham
Cambridge
Oxford
Southampton
London
Plymouth
English Channel
SCHLESWIG-HOLSTEIN
Kolobrzeg
Gdansk
POLAND
North
Hamburg
NETHERLANDS
Amsterdam
Bremen
Hannover
Berlin
Poznan
Warsaw
Brest
The Hague
Rotterdam
Essen
Duisburg
Dortmund
Brugge
Antwerp
Brussels
Dusseldorf
Cologne
Bonn
Leipzig
Dresden
Wroclaw
Lodz
Calais
Lille
BELGIUM
Liege
GERMANY
Katowice
Krakow
Land Elevations
meters
feet
4877
16,000
3353
11,000
2134
7000
914
3000
305
1000
152
500
0
0
1:14,800,000
Albers Equal Area Conic Projection
Le Havre
LUXEMBOURG
Frankfurt
Prague
CZECH REPUBLIC
Brno
UKR
Luxembourg
Saarschleife
Nuremberg
Mannheim
SLOVAKIA
Presov
Zvolen
Paris
Metz
Stuttgart
Bratislava
Miskolc
Carpathian Mts.
Iasi
Le Mans
Strasbourg
Danube
Munich
Vienna
Budapest
Salzburg
Loire
Heldsburg
Nantes
FRANCE
Dijon
Bern
Zurich
Vaduz
LIECHTENSTEIN
AUSTRIA
HUNGARY
Kecskemet
Kuhtai
Graz
Szeged
ROMANIA
SWITZERLAND
Moulins
Geneva
Lyon
Ljubljana
SLOVENIA
Pecs
Arad
Zagreb
Bay of Biscay
Trieste
Bucharest
Bordeaux
Grenoble
Alps
Milan
Venice
Craiova
Turin
BOSNIA & HERZEGOVINA
Belgrade
La Coruña
Gijon
Bilbao
Toulouse
Rhone
SAN MARINO
CROATIA
SERBIA
Pleven
Vigo
Genoa
ITALY
Sarajevo
BULGARIA
Braga
Pyrenees
Nice
MONACO
Viareggio
Florence
Nis
Sofia
Porto
Marseille
Adriatic Sea
KOSOVO
Duero
ANDORRA
Bastia
Podgorica
MONTENEGRO
Skopje
MACEDONIA
PORTUGAL
Zaragoza
Corsica
VATICAN CITY
Apennines
Madrid
Rome
Barcelona
Bari
Tirane
ALBANIA
Thessaloniki
Lisbon
SPAIN
Naples
Korce
GREECE
Aegean
Iberian Peninsula
Valencia
Palma
Islas Baleares
Sardinia
Kerkira
Volos
Balkans
Cordoba
Tyrrhenian Sea
Ionian Sea
Faro
Seville
Guadalquivir
Granada
Cagliari
Athens
Messina
Palermo
Pirgos
Strait of Gibraltar
Málaga
Gibraltar
Algiers
Sicily
Tunis
Rabat
Casablanca
MOROCCO
Atlas Mountains
ALGERIA
TUNISIA
MALTA
Valletta

4 Europe

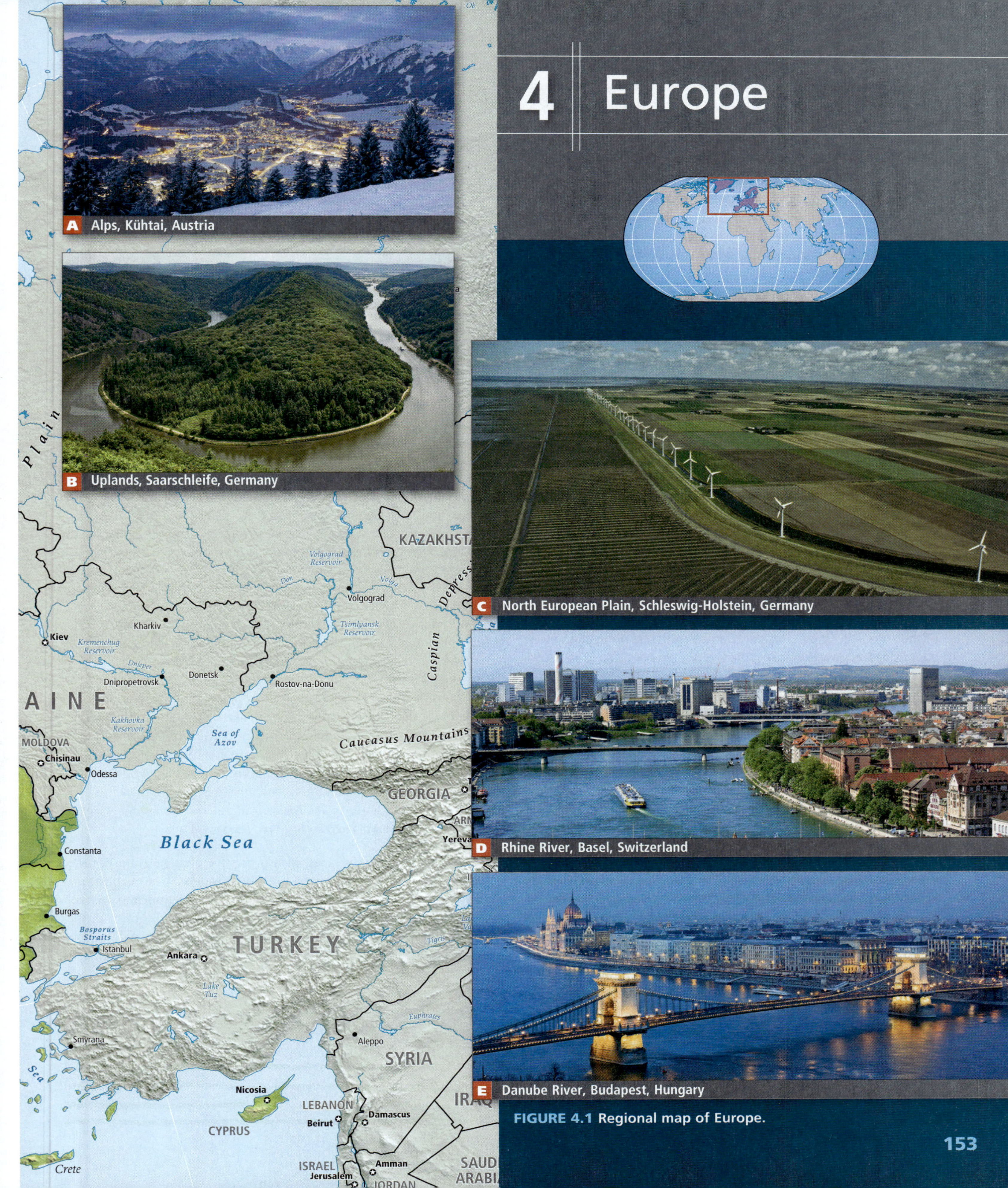

A Alps, Kühtai, Austria

B Uplands, Saarschleife, Germany

C North European Plain, Schleswig-Holstein, Germany

D Rhine River, Basel, Switzerland

E Danube River, Budapest, Hungary

FIGURE 4.1 Regional map of Europe.

GEOGRAPHIC INSIGHTS: EUROPE

After you read this chapter, you will be able to discuss the following geographic insights as they relate to the nine thematic concepts:

1. Water: Water is an issue in multiple ways. More droughts and floods are likely with climate change. Transportation via water is widespread and energy efficient, but the pollution of rivers and seas, especially the Mediterranean Sea, is increasing. European consumption patterns use *virtual water* from parts of the world that have deficits of water.

2. Climate Change: The European Union (EU) is a leader in the global response to climate change. Goals to cut greenhouse gas emissions are complemented by many other strategies to save energy and resources.

3. Globalization and Development: As European powers like Spain, Portugal, the United Kingdom, and the Netherlands conquered vast overseas territories, they created trade relationships that transformed Europe economically and culturally and laid the foundation for the modern global economy. Today, Europeans live well, but must struggle to remain globally competitive.

4. Urbanization, Power, and Politics: Rates of urbanization and democratization in Europe are linked, but both rates vary across the region. Northern Europe is the most urbanized (close to 80 percent) and has the most democratic participation. Levels of both urbanization and democratization fall to the south and east.

5. Food: Concerns about food security in Europe have led to heavily subsidized and regulated agricultural systems. Subsidies to agricultural enterprises account for nearly 40 percent of the budget of the European Union. Although large-scale food production by agribusiness is now dominant, there is a revival of organic and small-scale sustainable techniques.

6. Population and Gender: Europe's population is aging, as working women are choosing to have only one or two children. Immigration by young workers is partially countering this trend.

The European Region

The region of Europe (see Figure 4.1 on pages 152–153) has played a central role in world history and geography over the last 500 years as a colonizer of distant territories and as a major player in the development of capitalism. Then, in the twentieth century, Europe brought the world to the edge of disaster with two massive wars. Since then it has miraculously revived, prospered, and designed a model for regional trade integration that has been copied in other regions. Europe is now at another important juncture: Can it perfect its union and extend it, or will the union dissolve in the face of growing subregional financial and social disparities? The nine thematic concepts in this book are explored as they arise in the discussion of regional issues, with interactions between two or more themes featured, as in the geographic insights above. Scattered through the chapter, vignettes illustrate one or more of the themes as they are experienced in individual lives.

European Union (EU) a supranational organization that unites most of the countries of West, South, North, and Central Europe

GLOBAL PATTERNS, LOCAL LIVES Grigore Chivu (a pseudonym) wanders through the empty pig pens on his farm near Lugoj in western Romania. For generations, his family has made a meager but rewarding living by raising hogs and then processing and selling the meat. A few years ago, just before Romania joined the European Union (EU), he had more than 250 pigs. At Christmas, he and thousands of pig farmers across the country would slaughter a certain number of their pigs and preserve the meat, using time-honored methods. Pig slaughtering was a time of high spirits, community cooperation, and celebration, as the farmers contemplated the coming feast and their profits. Flavorful sausages, which they smoked and hung high in the rafters of their kitchens for further drying, were slowly parceled out to select customers, providing a steady income well into summer.

When Romania entered the European Union in 2007, all farmers were required to conform to EU standards for processing meat. The old methods were no longer allowed, and for some the new standards were prohibitively expensive.

At first the farmers were uncertain just how to respond. Before they could organize a butchering cooperative that conformed to EU standards, and for which development funds were available, an American company stepped into the breach. Smithfield, a U.S. Fortune 500 meat company based in Virginia, was expanding operations into Central Europe, where it planned to produce pork and pork products on an industrial scale and market them globally, especially in China. Eager to enter the EU market, Smithfield enlisted the help of Romanian politicians and got permission (and even EU subsidies) to establish a conglomerate that included feed production, pig breeding, modernized sanitary barns for fattening thousands of hogs in small cages, and slaughterhouses.

As the old, picturesque Romanian agricultural landscape is transformed into one of huge factory farms, the number of independent pig farmers has been reduced by more than 90 percent (Figure 4.2). Unable to compete with the lower prices Smithfield can charge, Grigore Chivu, like thousands of his fellow pig farmers, was considering migrating to Western Europe where stronger economies meant better-paying jobs. However, because of his age and traditional farming background, Chivu would be eligible for only menial labor. Meanwhile in Romania, he is too young for a pension. *[Source: Margareta Lelea; New York Times. For detailed source information see Text Credit pages.]* ■

Grigore Chivu is faced with disruptive change as his country adjusts to new circumstances in the European Union, but he is also part of a global revolution in food production and marketing that is changing food patterns across the world. For example, Smithfield, recently purchased by a Chinese firm, will market pork products, produced and processed at low cost in Romania, to West Africa, where the low prices are putting African pig farmers out of business. A video, *Pig Business*, explores the many sides of the change from small farm production of hogs to large, corporate farm production. The section on Romania begins at 42 minutes (see http://tinyurl.com/96vh9ja).

The **European Union (EU)** is a supranational organization that unites most of the countries of West, South, North, and Central Europe (Figure 4.3 on page 156). In principle, throughout the European Union, people, goods, and money can move freely.

FIGURE 4.2 Pig farms in Romania and the United States.

(A) Pigs on a farm in Romania that raises small animals.

(B) An industrial hog farm in Michigan, U.S., capable of raising thousands of pigs at a time.

Thinking Geographically

What about **(A)** suggests small-scale rearing of pigs?

Many people, like Grigore Chivu, decide to migrate only reluctantly, and do so only because poorer regions do not have the political power to push for useful reforms.

There are now 28 countries in the European Union, Croatia being the most recent addition in 2013, with a few more hoping to join in the next several years. The newest members (with the exceptions of Malta and Cyprus) are formerly Communist countries in Central Europe, with lower standards of living and, until the recent EU economic crisis, usually higher unemployment rates than countries in western Europe. As Central European economies faltered, hundreds of thousands of workers from Central Europe took advantage of the EU principle that (theoretically) citizens of any member state may move to any other EU state. More recently, with rising job losses in western Europe, many have returned home.

Since 2008, the European Union has been under increasing financial stress as a number of less industrialized and less prosperous countries (Greece, for example) have found it so hard to make loan payments that for them national bankruptcy is a possibility. Residents of original and more prosperous EU members—such as France and especially Germany—resent not only having to support these failing economies, but also fear that the migration of workers from the European Union's south and east is bringing very different people into close contact with each other, resulting in political tensions and increased costs for social and educational services. Others, especially the migrants themselves, lament the **cultural homogenization** that is wiping out traditional ways. Meanwhile, economic planners and employers across Europe argue that allowing workers to migrate is crucial to Europe's economic growth and competitiveness on a global scale.

cultural homogenization the tendency toward uniformity of ideas, values, technologies, and institutions among associated culture groups

THINGS TO REMEMBER

- Throughout the European Union, smaller family-run farms are giving way to larger farms run by corporations. This change is strongest in those Central Europe countries that have been recently admitted to the European Union, and has resulted in many farm family members migrating to other EU countries in search of work.

THE GEOGRAPHIC SETTING

What Makes Europe a Region?

Physically, Europe is a peninsula extending off the western end of the huge Eurasian continent. On this giant peninsula of Europe are many peninsular appendages, large and small. Norway and Sweden share one of the larger appendages. Other large peninsulas are the Iberian Peninsula (occupied by Portugal and Spain), as well as those of Italy and Greece. All extend into either the Atlantic Ocean or into the various seas that are adjacent to the Atlantic (see the Figure 4.1 map). The perimeter of the region of Europe to the north, west, and south is primarily oceanic coastline; this unique access to the world ocean facilitated European exploration and colonization of distant territories. The eastern physical and cultural border of Europe has long been harder to define. Just where Europe ends and Asia begins has been a

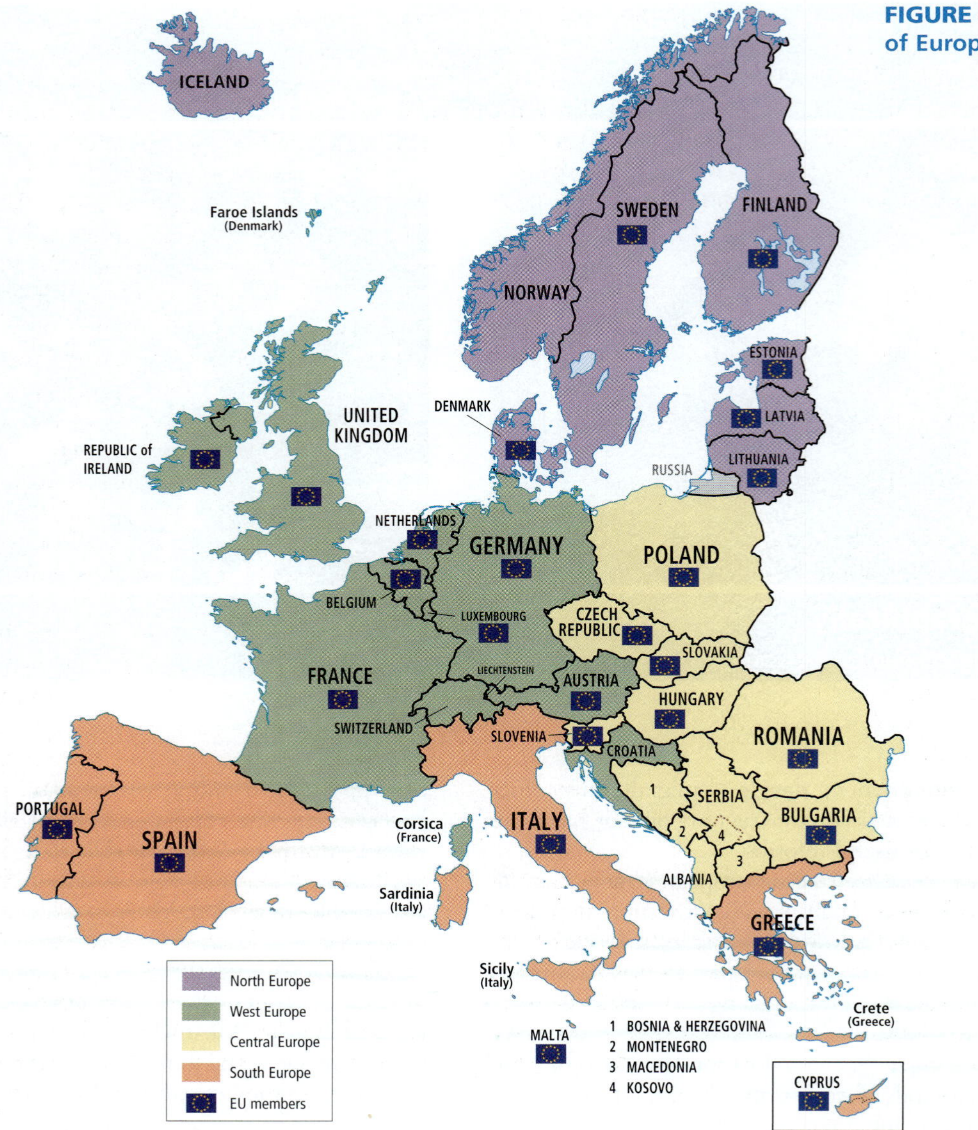

FIGURE 4.3 Political map of Europe.

source of contention for millennia. In this book, at this time, the eastern limit of the European region is taken to be the border of the European Union. Other potential parts of Europe—Ukraine, Belarus, western Russia, Moldova, and the Caucasus—are covered in Chapter 5, Turkey in Chapter 6.

Terms in This Chapter

This book divides Europe into four subregions (Figure 4.3)—*North, West, South,* and *Central Europe. Central Europe* refers to all those countries formerly in the Soviet sphere that are now in the European Union, as well as the countries that were formerly in Yugoslavia, plus Albania.

For convenience, we occasionally use the term *western Europe* to refer to all the countries that were *not* part of the experiment with communism in the Soviet sphere and in Yugoslavia. That is, *western Europe* comprises the combined subregions of *North Europe* (except Estonia, Latvia, and Lithuania), *West Europe* (except the former East Germany), and *South Europe*. When we refer to the countries that were part of the Soviet sphere up to 1989, we use the pre-1989 label *eastern Europe*. *EU-28* is the term for just the 28 member countries of the European Union. When we refer to the group of countries collectively known to some as the *Balkans* (Albania, Bosnia and Herzegovina, Bulgaria, Croatia, Macedonia, Montenegro, Romania, Serbia, and Slovenia), we use the term *southeastern Europe*.

Physical Patterns

Europe's physical geography is shaped by its many peninsulas that reach into surrounding oceans and seas, and by its variable landforms, all of which affect the region's climates and vegetation.

Landforms

Although European landforms are fairly complex, the basic pattern is mountains, uplands, and lowlands, all stretching roughly west to east in wide bands. As you can see in Figure 4.1 (in the map and in photo A), Europe's largest and highest mountain chain, the *Alps*, stretches west to east through the middle of the continent, from southern France through Switzerland and Austria. It extends into the Czech Republic and Slovakia, and curves southeast as the Carpathian Mountains into Romania. This network of mountains is mainly the result of pressure from the collision of the northward-moving African Plate with the southeasterly moving Eurasian Plate (see Figure 1.25 on page 46). Europe lies on the westernmost extension of the Eurasian Plate.

South of the main Alps formation, lower mountains extend into the peninsulas of Iberia and Italy, and along the Adriatic Sea through Greece to the southeast. The northernmost mountainous formation is shared by Scotland, Norway, and Sweden. These northern mountains are old (about the age of the Appalachians in North America) and have been worn down by glaciers and millions of years of erosion.

Extending northward from the central Alpine zone is a band of low-lying hills and plateaus curving from Dijon (France) through Frankfurt (Germany) to Krakow (Poland). These uplands (see Figure 4.1B) form a transitional zone between the high mountains and lowlands of the *North European Plain*, the most extensive landform in Europe. (see Figure 4.1C). The plain begins along the Atlantic coast in western France and stretches in a wide band around the northern flank of the main European peninsula, reaching across the English Channel and the North Sea to take in southern England, southern Sweden, and most of Finland. The plain continues east through Poland, then broadens to the south and north to include all the land east to the Ural Mountains in Russia.

Crossed by many rivers and holding considerable mineral deposits, the coastal lowland of the North European Plain is an area of large industrial cities and densely occupied rural areas. Over the past thousand years, people have transformed the natural seaside marshes and vast river deltas into farmland, pastures, and urban areas by building dikes and draining the land with wind-powered pumps. This is especially true in the low-lying Netherlands, where concern over climate change and sea level rise is considerable.

The rivers of Europe link its interior to the surrounding seas. Several of these rivers are navigable well into the upland zone, and Europeans have built large industrial cities on their banks. The Rhine carries more traffic than any other European river, and the course it has cut through the Alps and uplands to the North Sea also serves as a route for railways and motorways (see Figure 4.1D). The area where the Rhine flows into the North Sea is considered the economic core of Europe. Here Rotterdam, Europe's largest port, is located. The Danube River, larger and much longer than the Rhine, flows southeast from Germany, connecting the center of Europe with the Black Sea. As the European Union expands to the east, the economic and environmental roles of the Danube River basin, including the Black Sea, are getting increased attention (see Figure 4.1E).

Vegetation and Climate

Nearly all of Europe's original forests are gone, some for more than a thousand years. Today, forests with very large and old trees exist only in scattered areas, especially on the more rugged mountain slopes (see Figure 4.1A, B) and in the northernmost parts of Norway, Sweden, and Finland. Today forests are intensively managed and are regenerating on abandoned farmland. Although forests now cover about one-third of Europe, the dominant vegetation is crops and pasture grass.

Europe has three main climate types: temperate midlatitude, Mediterranean, and humid continental (**Figure 4.4** on page 158). The **temperate midlatitude climate** dominates in western Europe, where the influence of the Atlantic Ocean is very strong (see Figure 4.4A). A broad, warm-water ocean current called the **North Atlantic Drift** brings large amounts of warm water to the coasts of Europe. It is really just the easternmost end of the Gulf Stream, which carries water from the Gulf of Mexico north along the eastern coast of North America and across the North Atlantic to Europe (see Figure 2.4 on page 63).

The air above the North Atlantic Drift is relatively warm and wet. Eastward-blowing winds push it over North and West Europe and the North European Plain, bringing moderate temperatures and rain deep into the Eurasian continent. These factors create a climate that, although still fairly cool, is much warmer than elsewhere in the world at similar latitudes. To minimize the effects of heavy precipitation runoff, people in these areas have developed elaborate drainage systems for their houses and communities. Forests are both evergreen and deciduous. There is some concern that global climate change could eventually weaken the North Atlantic Drift, leading to a significantly cooler Europe.

Farther to the south, the **Mediterranean climate** prevails—warm, dry summers and mild, rainy winters (see Figure 4.4B). In the summer, warm, dry air from North Africa shifts north over the Mediterranean Sea as far north as the Alps, bringing high temperatures and clear skies. Crops grown in this climate, such as olives and grapes, citrus, apple and other fruits, and wheat, must be drought-resistant or irrigated. In the fall, this warm, dry air shifts to the south and is replaced by cooler temperatures and rainstorms sweeping in off the Atlantic. Overall, the climate here is mild, and houses along the Mediterranean coast are often open and airy to afford comfort in the hot, sunny summers. During the short, mild winter, life moves indoors, where wood fires heat one or two small rooms.

In Central Europe, without the moderating influences of the Atlantic Ocean and the Mediterranean Sea, the climate is more extreme. In this region of **humid continental climate**, summers are fairly hot, and the winters become longer and colder the farther north or deeper into the interior of the continent one goes (see Figure 4.4C). Here, houses tend to be well insulated, with small windows, low ceilings, and steep roofs that can shed snow. Crops must be adapted for much shorter growing seasons, and include corn and other grains plus fruit trees and a wide variety of vegetables adapted to the cold, especially root crops and cabbages.

temperate midlatitude climate as in south-central North America, China, and Europe, a climate that is moist all year with relatively mild winters and long, mild to hot summers

North Atlantic Drift the easternmost end of the Gulf Stream, a broad warm-water current that brings large amounts of warm water to the coasts of Europe

Mediterranean climate a climate pattern of warm, dry summers and mild, rainy winters

humid continental climate a midlatitude climate pattern in which summers are fairly hot and moist, and winters become longer and colder the deeper into the interior of the continent one goes

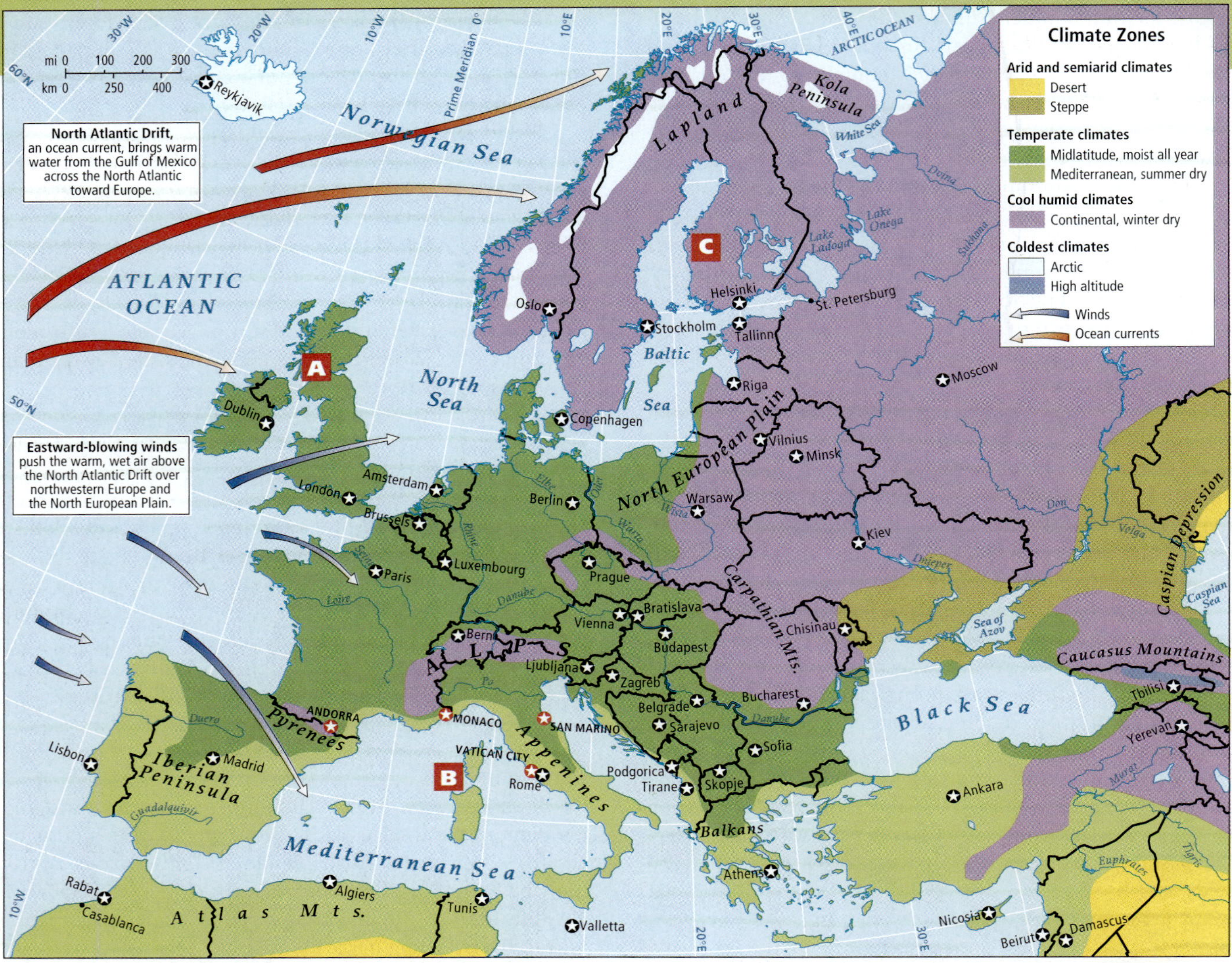

Climate Zones

Arid and semiarid climates
- Desert
- Steppe

Temperate climates
- Midlatitude, moist all year
- Mediterranean, summer dry

Cool humid climates
- Continental, winter dry

Coldest climates
- Arctic
- High altitude

- Winds
- Ocean currents

North Atlantic Drift, an ocean current, brings warm water from the Gulf of Mexico across the North Atlantic toward Europe.

Eastward-blowing winds push the warm, wet air above the North Atlantic Drift over northwestern Europe and the North European Plain.

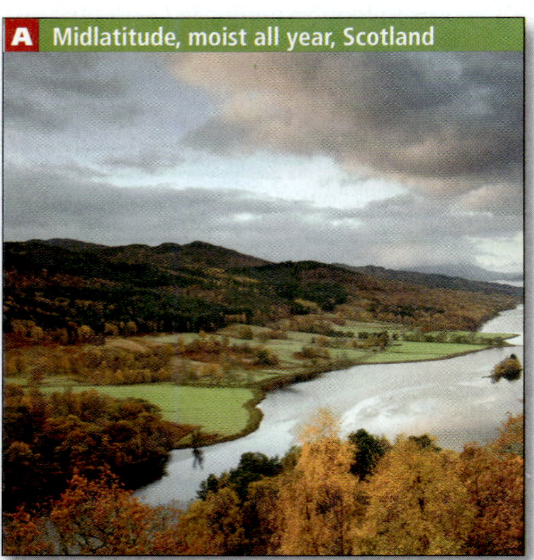

A Midlatitude, moist all year, Scotland

B Mediterranean, summer dry, Corsica

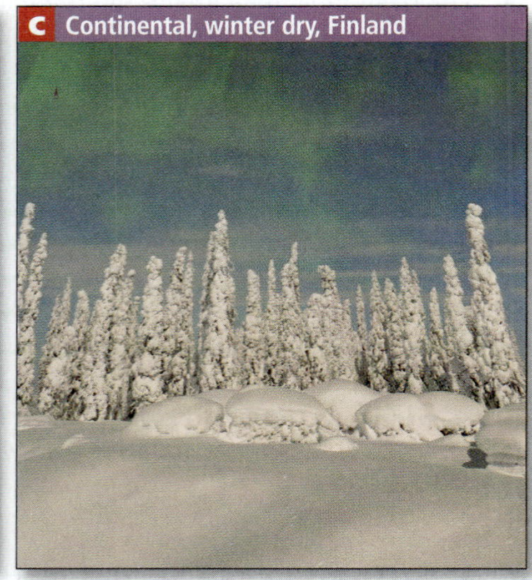

C Continental, winter dry, Finland

THINGS TO REMEMBER

- Europe is a region of peninsulas upon peninsulas.
- Europe has three main landforms—mountain chains, uplands, and the lowlands of the vast North European Plain.
- Europe has three principal climates—the temperate midlatitude, the Mediterranean climate, and the humid continental climate.

ON THE BRIGHT SIDE

Energy Solutions

The European Union wants to increase its use of renewable energy in order to reduce fuel imports and thereby increase energy security, stimulate the economy with new energy-related jobs, and combat climate change. The EU's goals are to reduce CO_2 emissions by 20 percent, increase use of renewable energy by 20 percent, and power 60 percent of EU homes with renewably generated electricity, all by 2020. A wide array of alternative energy projects is now attracting significant investment. While wind power is generally the favored technology, solar power technology is surging. Denmark, Germany, and Spain are leading the way; Germany, for instance, has deemphasized nuclear power and has become a leader in the production and use of wind and solar power. Germany recently opened a large wind power plant in the North Sea; in 2012, it set a record by producing 12 percent of its daily electricity through solar cells.

Environmental Issues

Europeans are increasingly aware of just how dramatically they have transformed their environments over the past several thousand years, and are now taking action on environmental issues at the local and global scales. Nevertheless, Europe's air, seas, and rivers remain some of the most polluted in the world, and the European Union still has a long way to go to meet its stated environmental goals of clean air and water; sustainable development in agriculture, industry, and energy use; and maintenance of biodiversity (Figure 4.5).

Europe's Impact on the Biosphere

There is a geographic pattern to the ways in which human activities over time have transformed Europe's landscapes. Western Europe shows the effects of dense population and heavy industrialization, and Central Europe reveals the results of long decades of willful disregard for the environment. Central Europe continues to have a major impact on the biosphere through the air and water pollution it generates (see Figure 4.5B). Furthermore, all of Europe is especially vulnerable to a number of the potential effects of climate change: changes in the North Atlantic Drift, which may lead to both cooler climates and drier, hotter climates in various parts of Europe and more climate variability from year to year.

Europe's Energy Resources Europe's main energy sources have shifted over the years from wood to coal and more recently to petroleum and natural gas, and in some countries to nuclear power. However, Europe is leading the move to alternative energy sources in response to rising energy costs and in an effort to cut greenhouse gas emissions.

The 28 members of the European Union (the EU-28) get a large portion of their fuel supplies from Russia—30 percent of their crude oil and 34 percent of their natural gas, as of 2011. Most of the gas now comes via pipelines through Belarus and Ukraine, but Turkey supplies Russian gas to Europe via the Black Sea and hosts a pipeline that carries gas from the Caspian Sea to Europe. Russia is negotiating for another trans-Turkey pipeline to carry oil and gas to the Mediterranean. While Europeans fear that this dependency will be used against them by Russia, which periodically withholds and releases flows to Europe through Ukraine and Belarus, Russia is actually also dependent on the EU fuel trade because 80 percent of Russia's oil exports and 70 percent of its natural gas exports go to the European Union. Another 30 percent of the gas and oil the European Union consumes comes from various Middle Eastern producers. Large oil and gas deposits in the North Sea, most controlled by Norway (not an EU member), have alleviated Europe's dependence on "foreign" sources of energy, but the production of oil from the North Sea has already peaked and is expected to run out by 2018.

82. WINTER ON THE WAY

The use of nuclear power to generate electricity has been more common in Europe than in North America. The EU-28 depends on nuclear power for 30 percent of its total needs. In France, 78 percent of the electricity is generated by nuclear power (compared with only 20 percent in the United States). However, support for nuclear power has declined, partly in response to the disastrous nuclear accidents in 1986 in Chernobyl, Ukraine, and in 2011 in Fukushima, Japan. The safe disposal of nuclear waste products is also a concern.

Air Pollution

At present, there is significant air pollution in much of Europe, but it is particularly heavy over the North European Plain. This is a region of heavy industry, dense transportation routes, and large and affluent populations. The intense fossil fuel use associated with such lifestyles results not only in the usual air pollution but also in *acid rain*, which can fall far from where it was generated (see the Figure 4.5 map).

There are also high levels of air pollution in the former communist states of Central and North Europe. Mines in Central Europe produce highly polluting soft coal (see Figure 4.5B) that is burned in out-of-date factories. Central Europe produces high per capita emissions from inefficient burning of oil and gas; it also receives air pollution blown eastward from western Europe. In Upper Silesia, Poland's leading coal-producing area, acid rain has destroyed forests, contaminated soils and the crops grown on them, and raised water pollution to deadly levels. Residents have higher rates of birth defects, high rates of cancer, and lowered life expectancies than comparable populations in Europe. Industrial pollution was one of Poland's greatest obstacles to entry into the European Union; Poland barely met the EU's requirements.

Central Europe's severe environmental problems developed in part because the Marxist theories and policies promoted by the Soviet Union portrayed nature as existing only to serve human needs. During the Soviet era, little pollution data was collected and public protest against pollution was prohibited, but the recent shift toward democracy has enabled greater action in places like

Figure 4.5 Photo Essay: Human Impacts on the Biosphere in Europe

Most of Europe has been transformed by human activity. Western Europe has some of the most heavily impacted landscapes and ecosystems, but the formerly communist countries of Central Europe also have severe environmental problems. Meanwhile, agricultural intensification is creating new problems in South Europe. The map shows some of the impacts on Europe's land, sea, and air.

A The Garzweiler open-pit coal mine in Germany, one of the largest in the world, covers 25.3 square miles (66 square kilometers). Its operations forced the abandonment of 12 villages and towns, and threaten local groundwater resources. The coal this mine produces is a highly polluting variety called lignite. Its high sulfur content is responsible for much of Europe's acid rain problem.

B A hotel in Lodz, Poland, in the shadow of a Soviet-era power plant. Central Europe has thousands of similar facilities, many of which are highly polluting and in close proximity to human settlements. Most were built during or before the era of Soviet domination, when environmental safeguards were rare.

C A sea of plastic greenhouses covers most of Campo de Dalía, a coastal plain in Spain's Almería province that has seen intense growth in year-round export-oriented vegetable production over the past several decades. Pesticides and fertilizers are heavily used, and health effects are beginning to be seen among the mainly Moroccan migrant workers and their families who work and live among the greenhouses. Meanwhile, water shortages are so severe that a desalination plant is being built nearby.

Thinking Geographically

After you have read about the human impacts on the biosphere in Europe, you will be able to answer the following questions:

A How might this large, open pit threaten groundwater resources?

B What clue(s) in the picture shows that large numbers of people may be exposed to pollution from the power plant?

C How might the large areas covered by plastic contribute to a water shortage?

Hungary and Bulgaria, where popular protest has resulted in reductions in air pollution.

The new market economies in the former Soviet bloc countries are improving energy efficiency and reducing emissions. Power plants, factories, and agriculture are polluting less, and the countries with the worst emissions records, such as Poland, have been making the most progress.

Freshwater and Seawater Pollution

Sources of water pollution in Europe include insufficiently treated sewage, chemicals and silt in the runoff from agricultural plots and residential units, consumer packaging litter, petroleum residues, and industrial effluent. Most inland waters contain a variety of such pollutants. Any pollutants that enter Europe's inland wetlands, rivers, streams, and canals eventually reach Europe's surrounding coastal waters. Thus far, the Atlantic Ocean, the Arctic Ocean, and the northern reaches of the North Sea are able to disperse most chemical pollutants dumped into them because they are part of, or closely connected to, the circulating flow of the world ocean. In contrast, the Baltic Sea, southern North Sea, Mediterranean Sea, and Black Sea are nearly landlocked bodies of water that do not have the capacity to flush themselves out quickly; thus, all are prone to accumulating chemical pollution such as chlorophyll (see the red areas in Figure 4.6) and plastic litter.

> **Geographic Insight 1**
>
> **Water:** Water is an issue in multiple ways. More droughts and floods are likely with climate change. Transportation via water is widespread and energy efficient, but the pollution of rivers and seas, especially the Mediterranean Sea, is increasing. European consumption patterns use *virtual water* from parts of the world that have deficits of water.

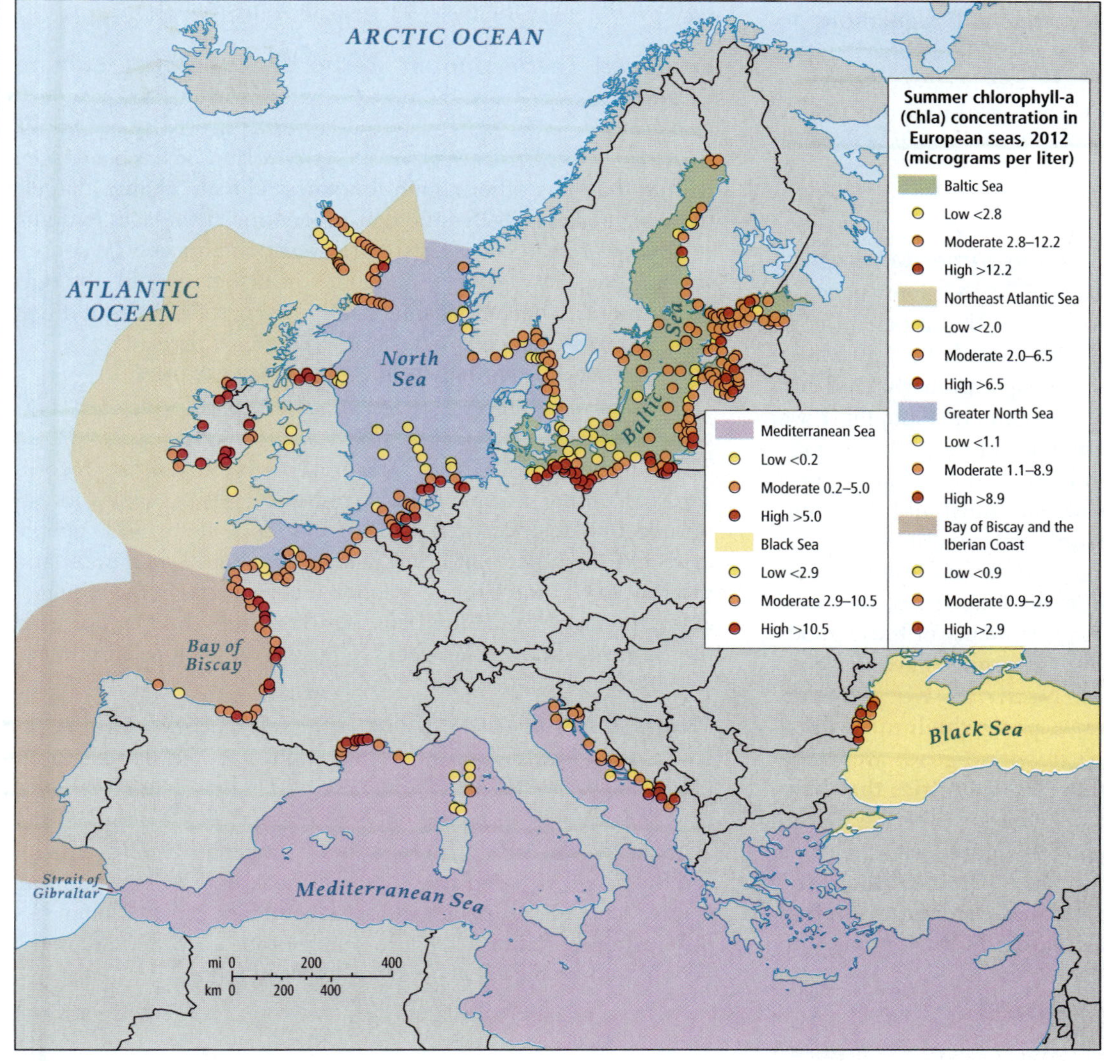

FIGURE 4.6 Pollution of the seas. Chlorophyll concentration is one measure of pollution levels in large bodies of water. Europe's exceptionally long and convoluted coast affords easy access to the world's oceans. However, pollution of the nearly landlocked Baltic, Black, and Mediterranean seas (red and yellow dots) is causing increasing concern. Lesser concentrations of chlorophyll are significant in the Atlantic and Arctic oceans and in the Mediterranean and Black seas. For this map, chlorophyll was used to measure the amount of algae in bodies of water. Excessive algae growth is an indicator of pollution from fertilizers and sewage that fuel the growth of those organisms.

In the Mediterranean, the effect of the pollution that pours in from rivers, adjacent cities, industries, hotel resorts, and farms is exacerbated by the fact that there is just one tiny opening to the world ocean (see Figure 4.6). At the surface, seawater flows in from the Atlantic through the narrow Strait of Gibraltar and moves eastward. At the bottom of the sea, water exits through the same narrow opening, but only after it has been in the Mediterranean for 80 years! The natural ecology of the Mediterranean is attuned to this lengthy cycle, but the nearly 460 million people now living in the countries surrounding the sea have upset the balance. Their pollution stays in the Mediterranean for decades. As a result, fish catches have declined, beloved seaside resorts have become unsafe for swimmers, and agricultural workers have become sick.

There are 34 countries with coastlines on Europe's many seas, all with different economies, politics, and cultural traditions. Such diversity makes it difficult to cooperate to minimize pollution or even reduce the risk of severe pollution. Although most of the Mediterranean's pollution is generated by Europe and the European Union has been working since 1995 to lessen pollution in the Mediterranean Basin, rapidly growing populations and development on the North African and eastern Mediterranean coasts of the sea also pose environmental threats because these countries still lack adequate urban sewage treatment and environmental regulations to control agricultural and industrial wastes. Modernizing the ports along the southern and eastern Mediterranean shores—several large containership ports are planned or under construction—could make conforming to pollution standards more attractive.

The Wide Reach of Europe's Environmental Impact

The term *virtual water*, introduced in Chapter 1 on page 38 as the volume of water used to produce all that a person consumes in a year, has come into use lately largely because water is now recognized as being in short supply globally, and not endlessly renewable, as previously thought. In Chapter 1 we discussed average per capita water needs per year (the average amount is 18.25 cubic meters) and noted that to arrive at a person's annual total water footprint, one must add the water required for basic needs to a person's virtual water footprint (see Table 1.2 on page 38). Although Europeans do not import as much of their consumer goods as do Americans, they still consume one-fifth of the world's imports; many of these goods have a high *virtual water component* (meaning the water consumed in the production process). Nearly all EU countries import more virtual water than they export, which means that the countries that produce Europe's imported goods are losing in the water exchange. Many of these are countries that have very little water to begin with. As yet, the costs of the loss of the virtual water are not being adequately figured into the price of the products exported to Europe. In all fairness then, the environmental impacts of Europe's virtual water consumption globally should be counted against Europe's total impact on the biosphere.

Europe's Vulnerability to Climate Change

Geographic Insight 2

Climate Change: The EU is a leader in the global response to climate change. Goals to cut greenhouse gas emissions are complemented by many other strategies to save energy and resources.

Europe's wealth, technological sophistication, and well-developed emergency response systems make it more resilient to the consequences of climate change than most regions of the world. Nevertheless, some areas are much more vulnerable than others, due to their location, dwindling water resources (Figure 4.7A), rising sea levels (see Figure 4.7B), and the effects of poverty. Figure 4.7 illustrates some of the consequences that have been predicted.

European Leadership in Response to Global Climate Change

Europe leads the world in responsiveness to global climate change, with EU governments having agreed to cut greenhouse gas emissions by 20 percent by 2020. Europe has been more willing than any other region to address climate change, largely because it calculates that there are economic advantages to doing so. Recent research suggests that investments in energy conservation, alternative energy, and other measures would cost EU governments 1 percent of their GDP. By contrast, doing nothing about climate change could *shrink* GDP by 20 percent.

83. GERMAN INVESTMENTS IN CLEAN ENERGY PAY OFF

Europe's increasing concern about global warming may also be influenced by public alarm at recent abnormal weather. The summers of 2003 and 2012 broke high-temperature records across Europe. Crops failed, freshwater levels sank, forests burned, and deaths soared. In 2003 in France alone, 3000 people died. On the other hand, in 2002, 2006, and 2012, rainfall and snowfall in Central Europe reached record levels. In the spring of 2006, the rivers of Central Europe—the Elbe, the Danube, and the Morava—flooded for weeks. Unusual vacillations continue.

Progress in Green Behavior

By global standards, Europeans use large amounts of resources and contribute about one-quarter of the world's greenhouse gas emissions. However, one European resident averages only one-half the energy consumption of the average North American resident. Europeans live in smaller dwellings, which need less energy to heat; and air conditioning is rare. They drive smaller, more fuel-efficient cars and often use public transportation. Because communities are denser, many people walk or bicycle wherever they need to go.

Thinking Geographically

After you have read about vulnerability to climate change in Europe, you will be able to answer the following questions:

A What clues, other than the caption, can you see in this photo to indicate that the reservoir is at a very low level?

B Judging from the way these structures are arranged, how would you guess they work?

C How might owning a car decrease a family's sensitivity to flooding?

Figure 4.7 Photo Essay: Vulnerability to Climate Change in Europe

Despite Europe's wealth, technological sophistication, and well-developed emergency response systems, parts of this region are particularly vulnerable to climate change effects due to their location and social conditions. There may be too much or too little fresh water, sea levels may be rising, or people may be too poor to afford appropriate precautions.

A As the level of water in a reservoir in Catalonia, Spain, sinks, a once-submerged eleventh-century church reemerges. As temperatures rise, Spain's climate will become drier, evaporation will increase, and scarce water resources will shrink further, threatening agriculture and drinking water, especially along the Mediterranean coast.

B Part of the massive Delta Works that protect the Netherlands from rising sea levels during storms. A response to massive flooding in 1953, the Delta Works significantly reduces the Netherlands' sensitivity to sea level rise, making the Netherlands only slightly more vulnerable to climate change than the rest of Europe, even though 60 percent of its population lives below sea level.

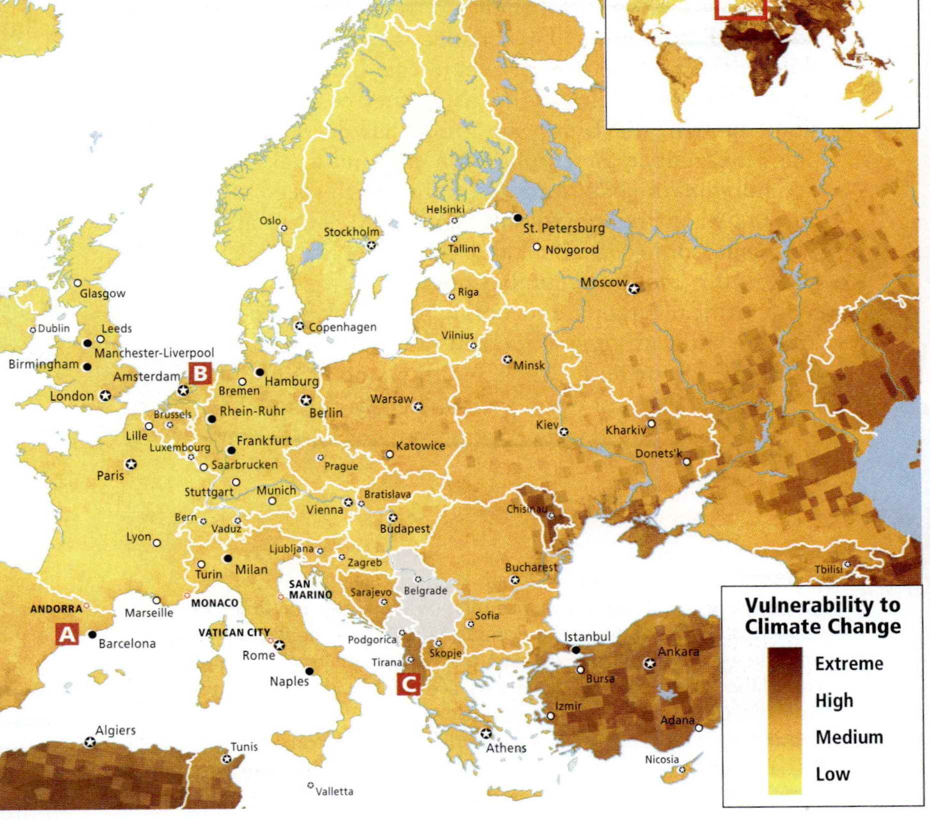

C Albania's high vulnerability to climate change relates to low incomes and inadequate infrastructure of many kinds, including transportation. Albania's sensitivity to many potential disturbances is thus high and its resilience is low. Many problems stem from misguided government policies. For example, until 1992 it was illegal for Albanians to own private automobiles. Even today, many poorer Albanians depend on horses and small, slow horse-drawn carts as their main mode of transportation.

Green environmentally conscious

These energy-saving practices are related in part to the high population densities and social customs of the region, and also to widespread explicit popular support for ecological principles. **Green** political parties influence national policies in all European countries, and Green policies are central to the agenda of the European Union. Results include strong regional advocacy for emission controls, well-entrenched community recycling programs, and grassroots work on local environmental concerns.

84. ICELAND: ENERGY TO SPARE
85. BRITAIN REQUIRES ENERGY INSPECTIONS FOR HOME SALES
86. MCD'S GARBAGE HEATS AND LIGHTS BRITISH CITY
87. LONDON LEADS BY EXAMPLE TO CURB POLLUTION, CLIMATE CHANGE

Changes in Transportation Although Europeans have for many years favored fast rail networks for both passengers and cargo (see Figure 4.20D on page 183) rather than private cars, trucks, and multilane highways, they have recently been drifting closer to the American model of private cars and trucks driven on sweeping freeways. This change is now happening in even the poorest parts of Europe where cars have been scarce (see Figure 4.7C). In response to rising fuel costs and CO_2 emissions, though, the European Union has developed long-term plans that reduce the emphasis on cars and trucks and involve designing *multimodal transport* to link high-speed rail to road, air, and water transportation (**Figure 4.8**). There are issues to overcome; for instance, not all EU railways use rails of a similar gauge, thus requiring cumbersome transfers of loads; and too many critical airports and ocean ports lack rail connections. Since 1 kilogram of gasoline can move 50 tons of cargo a distance of 1 kilometer by truck, but the same amount of gasoline can move 90 tons of cargo 1 kilometer by rail and 127 tons by waterway, highway transport is far from energy efficient. An EU report shows that private cars used for passenger transportation produce three times more CO_2 emissions than does rail-based public transportation.

Europe's long irregular coastline and the low cost of water transportation have been a boon for the development of links to global trade. Europe has numerous modern ocean ports that cater to container ships: Helsinki, Riga, Hamburg, Copenhagen, Antwerp, Rotterdam, Plymouth, Southampton, Le Havre, Marseille, Barcelona, and Koper. Soon, newly built Mediterranean ports in Morocco, Algeria, Tunisia, Malta, and Egypt, all catering at least in part to European markets, will join these ports. One-third of the world's container traffic now goes through the Mediterranean. The container ship industry is keenly attuned to the concerns about CO_2 and climate change, and is now using optimal (often slower) speeds that are carefully calculated to minimize fuel consumption and emissions while maximizing profits.

ON THE BRIGHT SIDE

Guerrilla Gardeners

Under cover of darkness, gardeners—who in the daytime are bureaucrats, stock traders, and computer jockeys—sneak into Central London to plant colorful flowers and foliage in traffic islands and roundabouts. These Green activists are part of a movement called Guerrilla Gardeners, which has quickly grown to more than 500 activists and a Web site, http://www.guerrillagardening.org.

Bypassing the town councils (which tend to impose crippling rules), the Guerrilla Gardeners make quick assaults late at night, armed with trowels, spades, mulch, and watering cans. Authorities seem unable to stop the guerrillas from covering neglected urban land with blooming hyacinths, tulips, marigolds, shrubs, and even trees. Yet more radical are the Seed Bombers, a group that packs flower seeds, soil, and water into compact parcels and tosses them into derelict patches of public land, where they shatter on impact, spewing forth seeds that produce plants capable of outcompeting the weeds. *[Source: NPR. For detailed source information, see Text Credit pages; see also Philip Booth's "Seed Bombers in Stroud," at http://www.youtube.com/watch?v=vY02FKd1Uco.]*

THINGS TO REMEMBER

Geographic Insight 1

• **Water** Europe currently has sufficient water overall, but climate change is likely to cause problems, especially in those areas that are already prone either to aridity or to flooding in low-lying areas. Europe's access to inland waters and the world ocean are a boon to cheap transport. Europeans, while generally conservation minded, nonetheless consume more than their share of the world's fresh water via *virtual water* in products imported from parts of the world that have deficient supplies of water.

Geographic Insight 2

• **Climate Change** Europe has emerged as the world's leader in combating global climate change. The European Union is using multiple strategies to reach its goal of cutting greenhouse gas emissions by 20 percent by 2020.

• European residents average only one-half the energy consumption of the average North American. Fuel costs and CO_2 emissions are increasingly being considered in the design of multimodal transport that links high-speed rail to road, air, and water transportation.

Human Patterns over Time

Over the last 500 years, Europe has profoundly influenced how the world trades, fights, thinks, and governs itself. The range of attempts to explain this influence varies widely. One argument is that Europeans are somehow a superior "breed" of humans. Another is that Europe's many bays, peninsulas, and navigable rivers have promoted commerce to a greater extent there than elsewhere. But, in fact, much of Europe's success is based on technologies and ideas it borrowed from elsewhere. For example, the concept of the peace treaty, so vital to current European and global stability, was first documented not in Europe but in ancient Egypt. In order to try to understand how Europe gained the leading role around the globe that it continues to have to this day, it is helpful to look at the broad history of this area.

FIGURE 4.8 The Trans-European transport network.

Sources of European Culture

Starting about 10,000 years ago, the practice of agriculture and animal husbandry gradually spread into Europe from the uplands and plains associated with the Tigris and Euphrates rivers in Southwest Asia and from farther east in Central Asia and beyond. Mining, metalworking, and mathematics also came to Europe from these places and from various parts of Africa. All of these borrowed innovations increased the possibilities for trade and economic development in Europe.

The first European civilizations were ancient Greece (800 to 86 B.C.E.) and Rome (753 B.C.E. to 476 C.E.). Located in southern Europe, both Greece and Rome initially interacted more with the Mediterranean rim, Southwest Asia, and North Africa than with the rest of Europe, which then had only a small and relatively poor, rural population. Later European traditions of science, art, and literature were heavily based on Greek ideas, which were themselves derived from yet earlier Egyptian and Southwest Asian (Arab and Persian) sources.

The Romans, after first borrowing heavily from Greek culture, also left important legacies in Europe. Many Europeans today speak *Romance* languages, such as Spanish, Portuguese, Italian, French, and Romanian, all of which are largely derived from Latin,

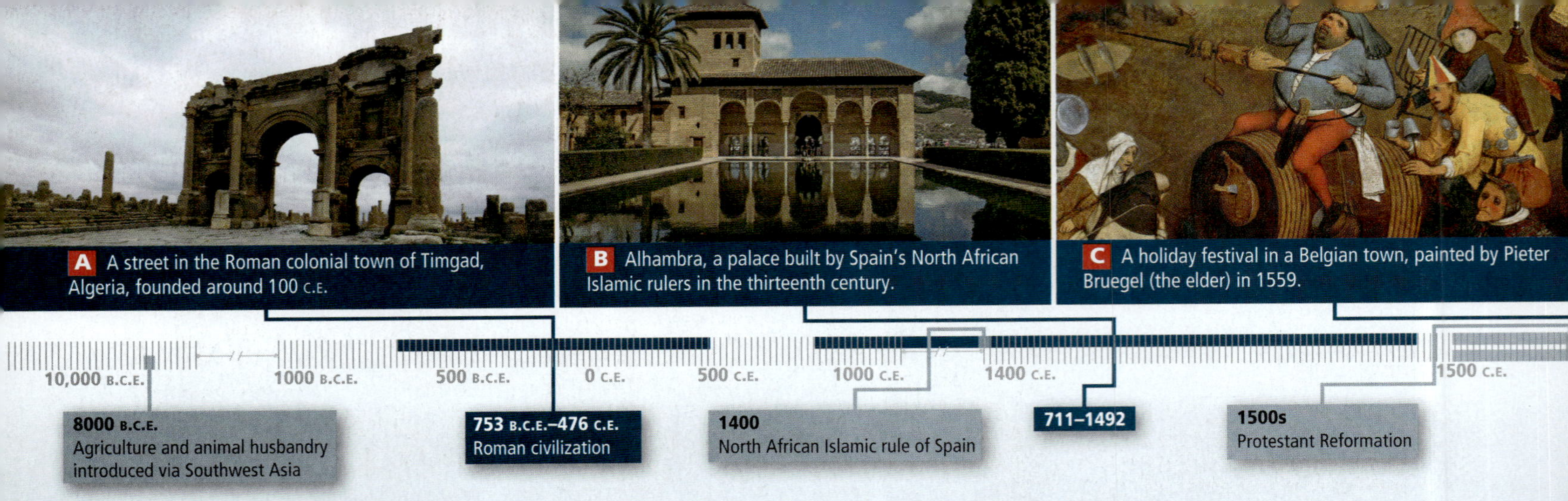

A A street in the Roman colonial town of Timgad, Algeria, founded around 100 C.E.

B Alhambra, a palace built by Spain's North African Islamic rulers in the thirteenth century.

C A holiday festival in a Belgian town, painted by Pieter Bruegel (the elder) in 1559.

FIGURE 4.9 A VISUAL HISTORY OF EUROPE

Thinking Geographically

After you have read about the human history of Europe, you will be able to answer the following questions:

A What about the photo of the ruins of Timgad reflects a particular Roman strategy for organizing space?

B What about the architecture of Alhambra suggests technological sophistication and an astute understanding of climate?

the language of the Roman Empire. European laws that determine how individuals own, buy, and sell land originated in Rome. Europeans then spread these legal customs throughout the world.

The practices that Romans used when colonizing new lands also shaped much of Europe. After a military conquest, the Romans secured control in rural areas by establishing large, plantation-like farms. Politics and trade were centered on new Roman towns built on a grid pattern that facilitated both commercial activity and the military repression of rebellions, as can be seen in Figure 4.9A. These same systems for taking and holding territory were later used when Europeans colonized the Americas, Asia, and Africa.

The influence of Islamic civilization on Europe is often overlooked. After the fall of Rome, while Europe was in a time known as the *early medieval period* (roughly 450 to 1300 C.E.; sometimes referred to as the Dark Ages), pre-Muslim (Arab and Persian) and then Muslim scholars preserved learning from Rome and Greece in large libraries, such as that in Alexandria, Egypt. Muslim Arabs originally from North Africa ruled Spain from 711 to 1492 (see Figure 4.9B). From the 1400s through the early 1900s, the Ottoman Empire (based in what is now Turkey) dominated much of southern Central Europe and Greece. The Arabs, Persians, and Turks all brought new technologies, food crops, architectural principles, and textiles to Europe from Arabia, Persia, Anatolia, China, India, and Africa. Arabs also brought Europe its numbering system, mathematics, and significant advances in medicine and engineering, building on ideas they picked up in South Asia.

Beginning 500 years ago, Europe began also to draw on a host of cultural features from the various colonies that were established in the Americas, Asia, and Africa. For example, many food crops now popular in Europe came mostly from the Americas (potatoes, corn, peppers, tomatoes, beans, and squash; see Table 3.1 on page 125) and Southwest and Central Asia (wheat, leafy greens, garlic, onions, and apples).

The Inequalities of Feudalism

As the Roman Empire declined, a social system known as *feudalism* evolved during the *medieval period* (450–1500 C.E.). This system originated from the need to defend rural areas against local bandits and raiders from Scandinavia and the Eurasian interior. The objective of feudalism was to have a sufficient number of heavily armed, professional fighting men, or knights, to defend a much larger group of *serfs*, who were legally bound to live on and cultivate plots of land for the knights. Over time, some of these knights became a wealthy class of warrior-aristocrats, called the *nobility*, who controlled certain territories. Some nobles gained power over other nobles, amassing vast kingdoms.

The often-lavish lifestyles and elaborate castles of the wealthier nobility were supported by the labors of the serfs (Figure 4.10). Most serfs lived in poverty outside castle walls and, much like slaves, were legally barred from leaving the lands they cultivated for their protectors.

The Role of Urbanization in the Transformation of Europe

While rural life followed established feudal patterns, new political and economic institutions were developing in Europe's towns and cities. Here, thick walls provided defense against raiders, and commerce and crafts supplied livelihoods, allowing the people to be more independent from feudal knights and kings. And Europeans, like people in other world regions, continued to develop distinctive cultures, including preferred pets (see Figure 4.11 on page 168).

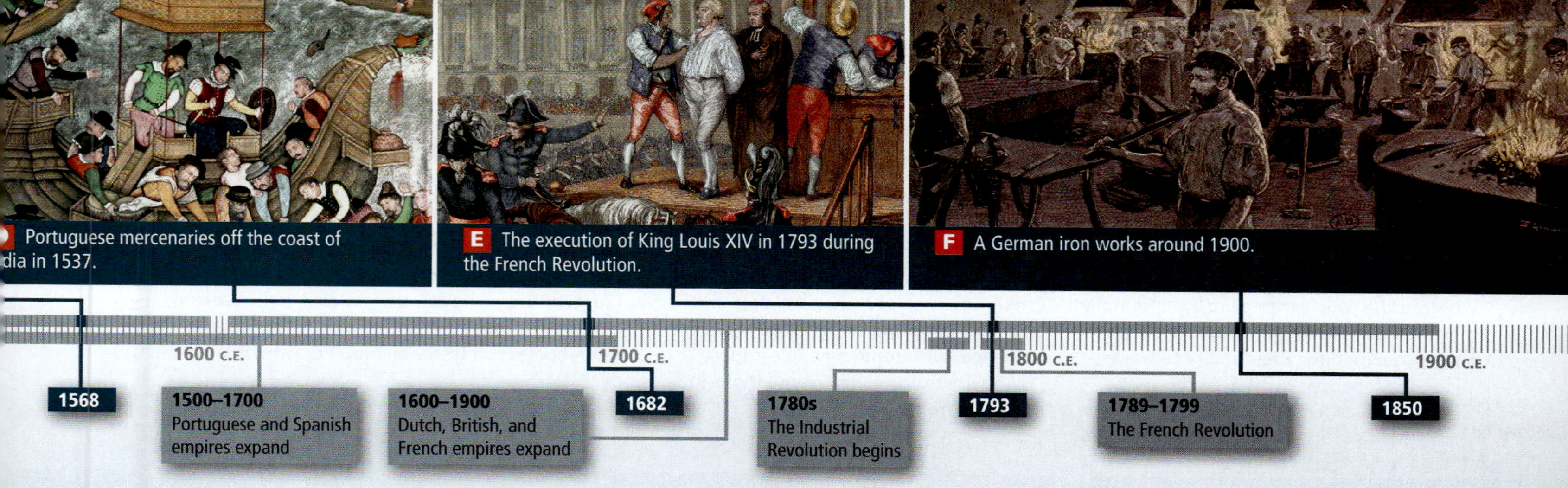

C What about this picture suggests relative prosperity in sixteenth-century Belgium?

D What about this picture suggests how the Portuguese felt about the value of trade with India?

E How is the execution of Louis XIV related to the evolution of democracy in Europe?

F What does this photo indicate about the impact of the Industrial Revolution on employment in Europe?

Located along trade routes, people in Europe's urban areas were exposed to new ideas, technologies, and institutions from Southwest Asia, India, and China. Some of these innovations, such as banks, insurance companies, and corporations, provided the foundations for Europe's modern economy. Over time, citizens of Europe's urban areas established a pace of social and technological change that left the feudal rural areas far behind.

Urban Europe flourished in part because of laws that granted basic rights to urban residents. With adequate knowledge of these laws, set forth in legal documents called *town charters*, people with few resources could protect their rights even if challenged by those who were more wealthy and powerful. Town charters provided a basis for European notions of *civil rights*, which have proved hugely influential throughout the world. With strong protections for their civil rights, some of Europe's townsfolk grew into a small middle class whose prosperity moderated the feudal system's extreme divisions of status and wealth (see Figure 4.9C). A related outgrowth of urban Europe was a philosophy known as **humanism**, which emphasizes the dignity and worth of the individual, regardless of wealth or social status.

The liberating influences of European urban life transformed the practice of religion. Since late Roman times, the Catholic Church had dominated not just religion but also politics and daily life throughout much of Europe. In the 1500s, however, a movement known as the *Protestant Reformation* arose in the urban centers of the North European Plain. Reformers, among them Martin Luther, challenged Catholic practices—such as holding church services in Latin, which only a tiny educated minority understood—that stifled public

humanism a philosophy and value system that emphasizes the dignity and worth of the individual, regardless of wealth or social status

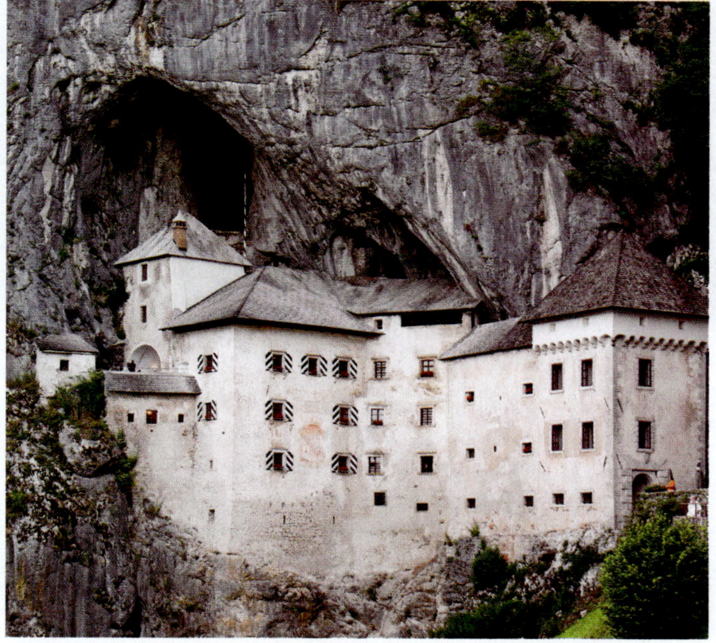

FIGURE 4.10 Predjama Castle in Slovenia. Predjama Castle, perched in the mouth of a cave in southwestern Slovenia, exemplifies the feudalism from which modern Europe eventually emerged. First mentioned in the historical record in the thirteenth century, Predjama became legendary as an impregnable fortress in the fifteenth century, when the Austrian Imperial Army laid siege to it for over a year, ignorant of a passage through the cave that kept the castle supplied. The Austrians finally succeeded in taking the castle and then the region when Predjama's owner, Erazem Leuger, was betrayed by one of his own men, who had informed the Austrians of the castle's greatest vulnerability. A cannonball killed Erazem in Predjama's unprotected outhouse.

FIGURE 4.11 LOCAL LIVES **PEOPLE AND ANIMALS IN EUROPE**

A The bloodhound, a scent-tracking dog, may have been developed by the Romans in the third century C.E. and later brought to England from France before the fourteenth century. Originally bred to track wild game, bloodhounds are renowned for their ability to track individual people by their body scent. Sometimes called a "sleuth hound," they are still used today by law enforcement.

B A rose-ringed parakeet, of the type that has been kept as a pet in Europe also since the days of the Romans. Prized for their sociable nature and ability to mimic human speech, parakeets and parrots are popular pets throughout Europe, with some having escaped captivity to form wild populations in many European cities.

C Originally bred by nomads along the Dalmatian (Croatian) coast as a guard dog, dalmatians became popular throughout Europe and North America during the nineteenth century. In the days of horse-drawn fire carriages, dalmatian "firehouse dogs" cleared the path in front of the carriages and made the horses run faster by nipping at their legs.

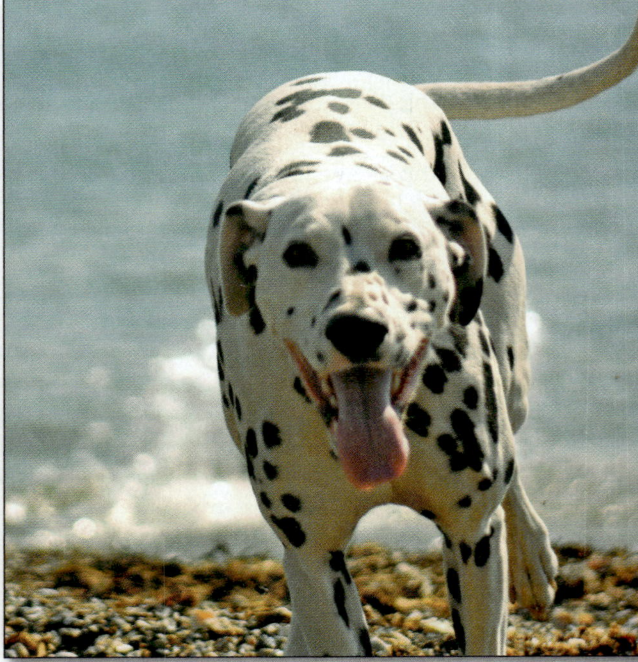

participation in religious discussions. Protestants also promoted individual responsibility and more open public debate of social issues, altering the perception of the relationship between the individual and society. These ideas spread faster with the invention of the European version of the printing press (1450 C.E.), which enabled widespread literacy.

European Colonialism: The Founding and Acceleration of Globalization

> **Geographic Insight 3**
>
> **Globalization and Development:** As European powers like Spain, Portugal, England, and the Netherlands conquered vast overseas territories, they created trade relationships that transformed Europe economically and culturally and laid the foundation for the modern global economy. Today, Europe is struggling to remain globally competitive.

A direct outgrowth of the greater openness and connectivity of urban Europe was the exploration and subsequent colonization of much of the world by Europeans. The increased commerce and cultural exchange began a period of accelerated *globalization* that persists today (see the discussion in Chapter 1 on pages 30–33).

In the fifteenth and sixteenth centuries, Portugal took advantage of advances in navigation, shipbuilding, and commerce to set up a trading empire in Asia and a colony in Brazil (see Figure 4.9D). Spain soon followed, founding a vast and profitable empire in the Americas and the Philippines. By the seventeenth century, however, England, the Netherlands, and France had seized the initiative from Spain and Portugal. All European powers implemented **mercantilism**, a strategy for increasing a country's power and wealth by acquiring colonies and managing all aspects of their production, transport, and trade for the colonizer's benefit (Figure 4.12).

> **mercantilism** a strategy for increasing a country's power and wealth by acquiring colonies and managing all aspects of their production, transport, and trade for the colonizer's benefit

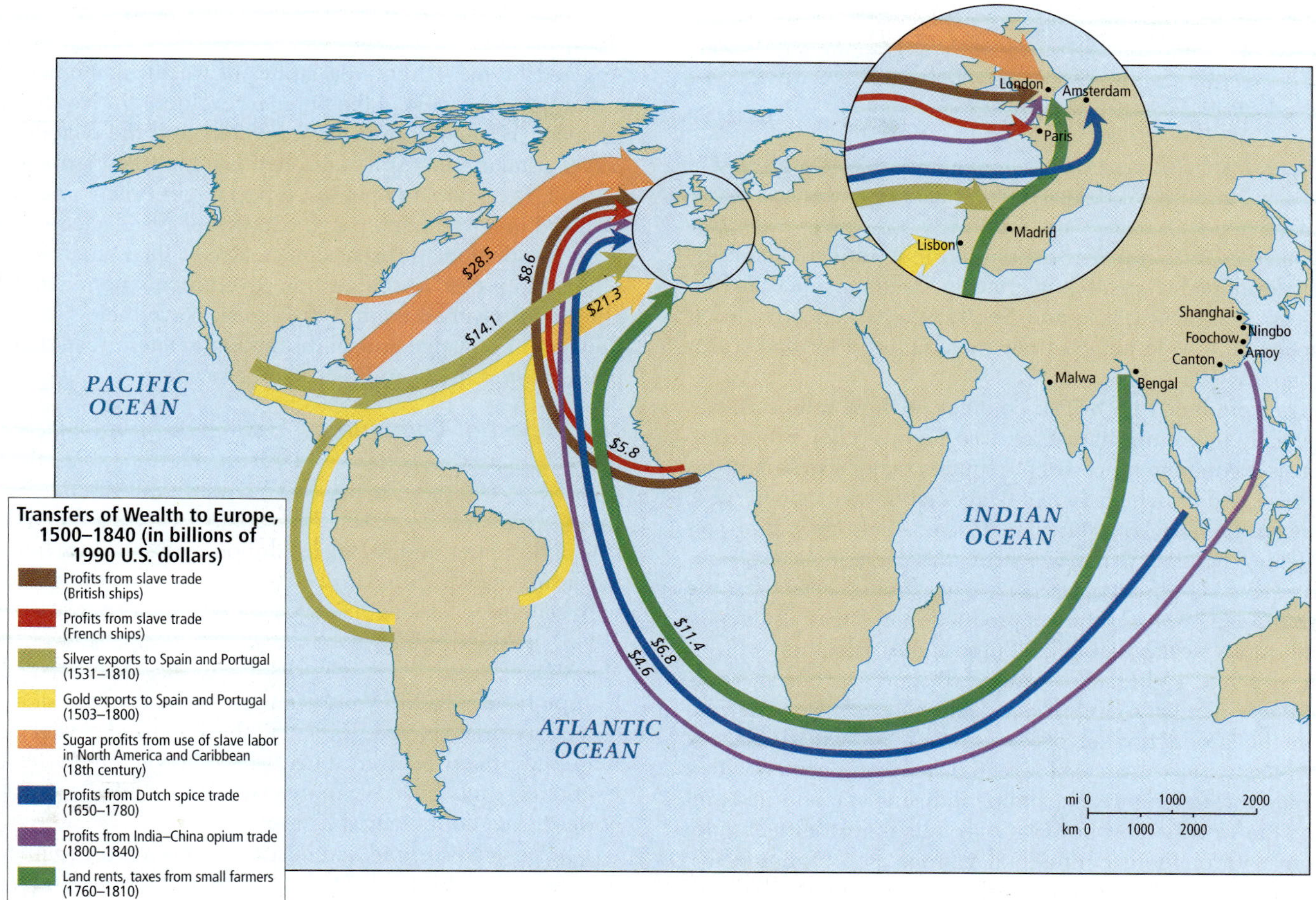

FIGURE 4.12 Transfers of wealth from the colonies to Europe. During the period of mercantilism, Europe received billions of dollars of income from its overseas colonies.

Mercantilism supported the Industrial Revolution in Europe (page 170) by supplying cheap resources from around the globe for Europe's new factories. The colonies also became markets for European-manufactured goods. **88. ART EXHIBIT ENCOMPASSES THE WORLD OF PORTUGUESE EXPLORATIONS**

By the mid-1700s, wealthy merchants in London, Amsterdam, Paris, Berlin, and other western European cities were investing in new industries. Workers from rural areas poured into urban centers in England, the Netherlands, Belgium, France, and Germany to work in new manufacturing industries and mining. Wealth and raw materials flowed in from colonial ports in the Americas, Asia, and Africa. Some cities, such as Paris and London, were elaborately rebuilt in the 1800s to reflect their roles as centers of global empires.

By 1800, London and Paris, each of which had a million inhabitants, were Europe's largest cities, a status that eventually brought them to their present standing as *world cities* (cities of worldwide economic or cultural influence). London is a global center of finance, and Paris is a cultural center that has influence over global consumption patterns, from food to fashion to tourism.

Also by the nineteenth century, the English, French, and Dutch (the people who live in the Netherlands, or Holland) overshadowed the Spanish and Portuguese colonial empires and extended their influence into Asia and Africa. By the twentieth century, European colonial systems had strongly influenced nearly every part of the world.

The overseas empires of England, the Netherlands, and eventually France were the beginnings of the modern global economy. The riches they provided shifted wealth, investment, and general economic development away from southern Europe and the Mediterranean and toward western Europe.

Urban Revolutions in Industry and Democracy

Geographic Insight 4

Urbanization, Power, and Politics: Rates of urbanization and democratization in Europe are linked, but both rates vary across the region. Northern Europe is the most urbanized (close to 80 percent) and has the most democratic participation. Levels of both urbanization and democratization fall to the south and east.

The wealth derived from Europe's colonialism helped fund two of the most dramatic transformations in a region already characterized by rebirth and innovation: the industrial and democratic revolutions. Both first took place in urban Europe.

The Industrial Revolution Europe's Industrial Revolution—particularly Britain's ascendancy as the leading industrial power of the nineteenth century—was intimately connected with colonial expansion and sugar production. In the seventeenth century, Britain developed a small but growing trading empire in the Caribbean, North America, and South Asia, which provided it with access to a wide range of raw materials and to markets for British goods.

Sugar, produced by British colonies in the Caribbean, was an especially important trade crop (see Figure 4.12). Sugar production is a complex process that requires major investments in equipment, and for which a great deal of labor was once needed. Slaves were forcibly brought from Africa to help grow and process sugar. The skilled management and large-scale organization needed for sugar later provided a production model for the Industrial Revolution. The mass production of sugar also generated enormous wealth that helped fund industrialization.

By the late eighteenth century, Britain was introducing mechanization into its industries, first in textile weaving and then in the production of coal and steel. By the nineteenth century, Britain was the world's greatest economic power, with a huge and growing empire, expanding industrial capabilities, and sporting the world's most powerful navy. Its industrial technologies spread throughout continental Europe (see Figure 4.9F), North America, and elsewhere, thereby transforming millions of lives in the process.

Urbanization and Democratization

Industrialization led to massive growth in urban areas in the eighteenth and nineteenth centuries. Extremely low living standards in Europe's cities created tremendous pressures for change in the political order, ultimately leading to democratization. Most early industrial jobs were dangerous and unhealthy, demanding long hours and offering little pay. Poor people were packed into tiny, airless spaces that they shared with many others. Children were often weakened and ill because of poor nutrition, industrial pollution, inadequate sanitation and health care, and being used as laborers. Water was often contaminated, and most sewage ended up in the streets. Opportunities for advancement through education were restricted to a tiny, wealthy elite.

The ideas and information gained by the few key people who did learn to read gave some the incentive to organize and protest for change. After lengthy struggles, democracy was expanded to Europe's huge and growing working class, and power, wealth, and opportunity were distributed more evenly throughout society. However, the road to democracy in Europe was rocky and violent, just as it is now in many parts of the world.

nationalism devotion to the interests or culture of a particular country, nation, or cultural group; the idea that a group of people living in a specific territory and sharing cultural traits should be united in a single country to which they are loyal and obedient

In 1789, the French Revolution led to the first major inclusion of common people in the political process in Europe. Angered by the extreme disparities of wealth in French society, and inspired by news of the popular revolution in North America, the poor rebelled against the monarchy and the elite-dominated power structure that still controlled Europe (see Figure 4.9E). As a result, the general populace, especially in urban areas, became involved in governing through democratically elected representatives. The democratic expansion created by the French Revolution ultimately proved short-lived as governments dominated by the elite soon regained control in France. Nevertheless, the French Revolution provided crucial inspiration to later urban democratic political movements in France and elsewhere.

The Impact of Communism During the struggles that resulted ultimately in the expansion of democracy, popular discontent erupted periodically in the form of new revolutionary political movements that threatened the established civic order. The political philosopher and social revolutionary Karl Marx framed the mounting social unrest in Europe's cities as a struggle between socioeconomic classes. His treatise *The Communist Manifesto* (1848) helped social reformers across Europe articulate ideas about how wealth could be more equitably distributed. East of Europe in Russia, Marx's ideas inspired the creation of a revolutionary communist state in 1917, the Union of Soviet Socialist Republics, often referred to as the USSR or the Soviet Union. Eventually, the Soviet Union extended its ideology and state power throughout Central Europe.

In West, North, and South Europe, communism gained some popularity, but this tended to lead to the formation of Communist political parties that operated peacefully within the context of democratizing political systems. In countries like France, Italy, and Spain, Communist political parties are still influential and often form governing coalitions with other parties.

Popular Democracy and Nationalism Throughout Europe, innumerable struggles between urban working-class people and the authorities continued during the nineteenth and early twentieth centuries, yielding two main results: workers throughout Europe gained the right to form unions that could bargain with employers and the government for higher wages and better working conditions; and there was an increase in democratization as most national governments eventually began to derive their authority from competitive elections in which all adult citizens could vote.

During the nineteenth and twentieth centuries, the development of democracy was also linked to the idea of **nationalism**, or allegiance to the state. The notion spread that all the people who lived in a certain area formed a nation, and that loyalty to that nation should supersede loyalties to family, clan, or individual monarchs. Eventually, the whole map of Europe was reconfigured, and the mosaic of kingdoms gave way to a collection of *nation-states*. All of these new nations were, at varying paces, transformed into democracies by the political movements arising in Europe's industrial cities. Nationalism was a major component of both World War I and II, so in the period after the wars—called the *post-war era*—and as the European Union was constructed, supernationalism was recognized as a problem and deemphasized.

Democracy and the Welfare State Channeled through the democratic process, public pressure for improved living standards moved most European governments toward becoming **welfare states** by the mid-twentieth century. Such governments accept responsibility for the well-being of their people, guaranteeing basic necessities such as education, employment, affordable food, and health care for all citizens. In time, government regulations on wages, hours, safety, and vacations established more harmonious relations between workers and employers. The gap between rich and poor declined, and overall civic peace increased. And although welfare states were funded primarily through taxes, industrial productivity and overall economic activity did not decline, but rather increased, as did general prosperity.

Modern Europe's welfare states have yielded generally adequate to high levels of well-being for all citizens. However, just how much support the welfare state should provide is still a subject of hot debate—one that is currently being addressed in different ways across Europe (see page 189).

Two World Wars and Their Aftermath

Despite Europe's many advances in industry and politics, by the beginning of the twentieth century, the region still lacked a system of collective security that could prevent war among its rival nations. Between 1914 and 1945, two horribly destructive world wars left Europe in ruins, no longer the dominant region of the world. At least 20 million people died in World War I (1914–1918) and 70 million in World War II (1939–1945). During World War II, Germany's Nazi government killed 15 million civilians during its failed attempt to conquer the Soviet Union. Eleven million civilians died at the hands of the Nazis during the **Holocaust**, a massive execution of 6 million Jews and 5 million gentiles (non-Jews), including ethnic Poles and other Slavs, **Roma** (Gypsies), disabled and mentally ill people, gays, lesbians, transgendered people, and political dissidents. **89. THE WORLD REMEMBERS VICTIMS OF HOLOCAUST**

During World War II, Russia joined the Allies (Great Britain and the British Commonwealth; France, except during the German occupation of 1940–1944; Poland; the United States; Belgium; Brazil; Czechoslovakia; Ethiopia; Greece; India; Mexico; the Netherlands; Norway; and Yugoslavia) in defeating Germany, the country seen as the instigator of both world wars. After World War II ended in 1945, finding a lasting peace became a primary goal that eventually led to the formation of the European Union. But first, a number of enduring changes took place. Germany was divided into two parts. West Germany became an independent democracy allied with the rest of western Europe—especially Britain and France—and the United States. Since Russia controlled East Germany and the rest of Central Europe (Latvia, Lithuania, Estonia, Poland, Czechoslovakia, Hungary, Romania, Bulgaria, Ukraine, Moldova, and Belarus) after the war, under Russia's dominant leadership these war-devastated countries solidified into an ideologically Communist, or Soviet, bloc. The line between East and West Germany was part of what was called the **Iron Curtain**, a long, fortified border zone that separated western Europe from Central Europe. Yugoslavia did not become part of the Soviet bloc but also kept itself insulated from the West, partly to protect its socialized economy.

The Cold War The division of Europe created a period of conflict, tension, and competition between the United States and the Soviet Union known as the **Cold War**, which lasted from 1945 to 1991. During this time, once-dominant Europe, and indeed the entire world, became a stage on which the United States and the Soviet Union competed for dominance. The central issue was the competition between **capitalism**—characterized by privately owned businesses and industrial firms that adjusted prices and output to match the demands of the market—and **communism** (actually a version of socialism), in which the state owned all farms, industry, land, and buildings. After 1945, in most of what we now call Central Europe, the Soviet Union forcibly implemented a communist-inspired economic model known as **central planning**, in which a central bureaucracy dictated prices and output, with the stated aim of allocating goods and services equitably across society according to need. This system was successful in alleviating long-standing poverty and illiteracy among the working classes and for women, but it was rife with waste, corruption, and bureaucratic bungling. It ultimately collapsed in the 1990s due to inefficiency, insolvency, high levels of environmental pollution, and public demands for more democratic participation.

The rest of post-war Europe, especially western Europe, successfully developed capitalist economies with financial support from the United States under the *Marshall Plan* and from the governments of the involved countries. Basic facilities, such as roads, housing, and schools, were rebuilt. Economic reconstruction proceeded rapidly in the decades after World War II and included special attention to the social welfare needs of the general public. **91. MARSHALL PLAN'S 61ST ANNIVERSARY**

Decolonization, Power and Politics, and Conflict in Modern Europe Europe's decline during the two world wars also led to the loss of its colonial empires. Many European colonies had participated in the wars, with some (India, the Dutch East Indies, Burma, and Algeria) suffering extensive casualties, and almost all emerging economically devastated. After

welfare state a government that accepts responsibility for the well-being of its people, guaranteeing basic necessities such as education, affordable food, employment, and health care for all citizens

Holocaust during World War II, a massive execution by the Nazis of 6 million Jews and 5 million gentiles (non-Jews), including ethnic Poles and other Slavs, Roma (Gypsies), disabled and mentally ill people, gays, lesbians, transgendered people, and political dissidents

Roma the now-preferred term in Europe for Gypsies

Iron Curtain a long, fortified border zone that separated western Europe from (then) eastern Europe during the Cold War

Cold War a period of conflict, tension, and competition between the United States and the Soviet Union that lasted from 1945 to 1991

capitalism an economic system characterized by privately owned businesses and industrial firms that adjust prices and output to match the demands of the market

communism an ideology and economic system, based largely on the writings of the German revolutionary Karl Marx, in which, on behalf of the people, the state owns all farms, industry, land, and buildings (a version of socialism)

central planning a communist economic model in which a central bureaucracy dictates prices and output, with the stated aim of allocating goods equitably across society according to need

World War II, Europe was less able to assert control over its colonies as their demands for independence grew. By the 1960s, most former European colonies had gained independence, often after bloody wars fought against European powers and their local allies.

Democracy's long history in Europe (if not in Europe's colonies) entered a new era after World War II, when most of West, South, and North Europe reorganized more strongly around democratic principles and humanitarian ideals. This process was supported by the United States, though several authoritarian European regimes, such as that of Spanish dictator Francisco Franco, who ruled from 1936–1975, also received U.S. support. Portugal, also an authoritarian European country after the war, remained a dictatorship and colonial power in Africa until 1974 (Figure 4.13D). Elsewhere in Europe, political freedoms were compromised by violence, as in the case of Northern Ireland, which until the late 1990s had sectarian violence between Catholics and Protestants so severe that free and fair democratic elections were not possible (see Figure 4.13C). Central Europe did not begin to democratize until the Soviet Union dissolved in 1991 as its former satellites, one by one, declared independence. Yugoslavia, a large communist republic in southern Central Europe, always firmly outside the Soviet bloc, also dissolved beginning in the early 1990s. There democratization was hampered by a powerful wave of ethnic xenophobia and violence (see Figure 4.13A, B).

THINGS TO REMEMBER

- During the medieval period, political and economic transformations in Europe's towns and cities challenged the feudal system that dominated rural areas.

Geographic Insight 3

- **Globalization and Development** The vast overseas colonies of various European states created trade relationships that persist in the modern global economy. Europe was transformed and continues to benefit.
- Most of the countries in the world have been ruled by a European colonial power at some point in their history.

Geographic Insight 4

- **Urbanization, Power, and Politics** The growth of cities set the stage for increasing the political power of ordinary people, but democratization in Europe has proceeded at different rates across the region and was interrupted by two world wars.
- The road to democracy in Europe was rocky and at times violent.

Figure 4.13 Photo Essay: Power and Politics in Europe

Thinking Geographically

After you have read about power and politics in Europe, you will be able to answer the following questions:

A What clues in the photo suggest war and not some other calamity?

C What are the political implications of the sign announcing entry into Free Derry? What might be the reasons for showing a child wearing a gas mask?

D Why could it be said that independence movements in Portugal's African colonies brought democracy to Portugal?

Patterns of political power and conflict in Europe and in former European colonies generally suggest that countries with fewer political freedoms and lower levels of democratization also suffer the most from violent conflict (see Figure 1.18 on pages 34–35 for more). Since World War II, the most violent conflicts in Europe have happened in southern Central Europe, where repressive governments frequently denied their populations basic political freedoms. Outside Europe, many long and brutal wars were fought to gain independence from European powers. In part, these wars stemmed from the inequitable extraction of resources from the "colonies" by Europeans, but the denial of basic political freedoms to the local non-European population also was a major impetus for independence. The map shows how low levels of democratization persist in the countries that fought wars of independence, decades after the formal end of colonial rule. Other factors also worked against democratization, but European imperialism helped set many of these countries on an authoritarian trajectory that has only recently started to change.

D A group of captured Angolan rebels during that country's war of independence from Portugal, which fought three major wars in Africa during the 1960s and 1970s to hold on to its colonies there. The rising human and financial costs of these wars eventually brought about a revolution in Portugal itself in 1974, which at the time was an undemocratic authoritarian dictatorship. The revolution resulted in independence for the colonies and the eventual establishment of democracy in Portugal. Today, none of Portugal's colonies have well-protected political freedoms, in part because of the authoritarian governing structures put in place by the Portuguese and because of the lengthy civil wars that followed decolonization.

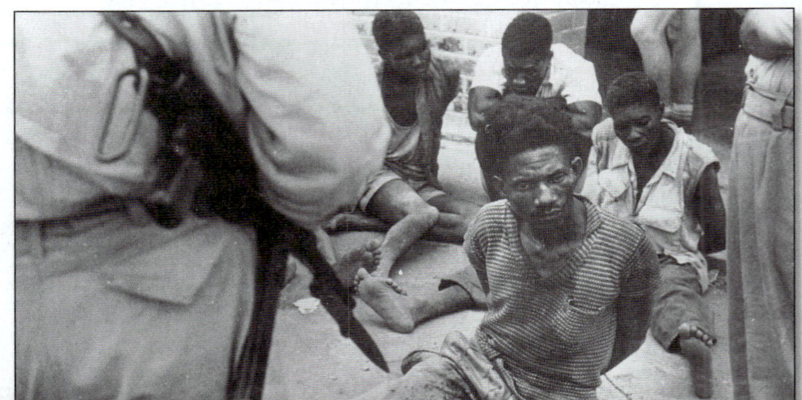

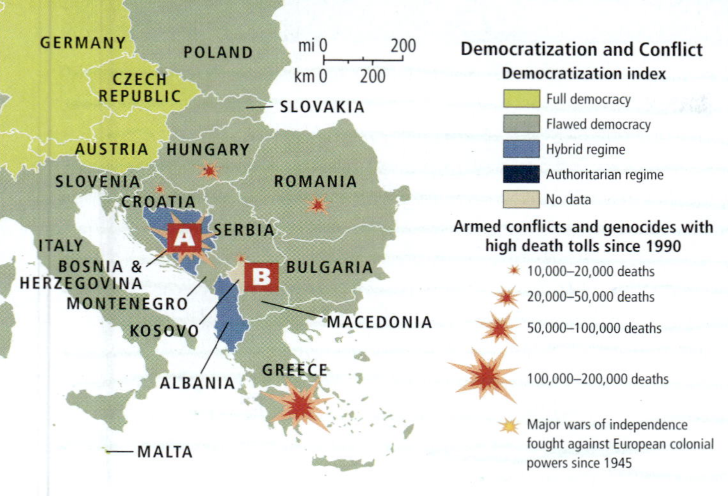

A The aftermath of Bosnia's civil war in a suburb of Sarajevo in 1996. Attempts by Serbia to limit political freedoms within the former Yugoslavia resulted in several wars during the 1990s. In Bosnia, a civil war that pitted different ethnicities against each other resulted in 175,000 deaths.

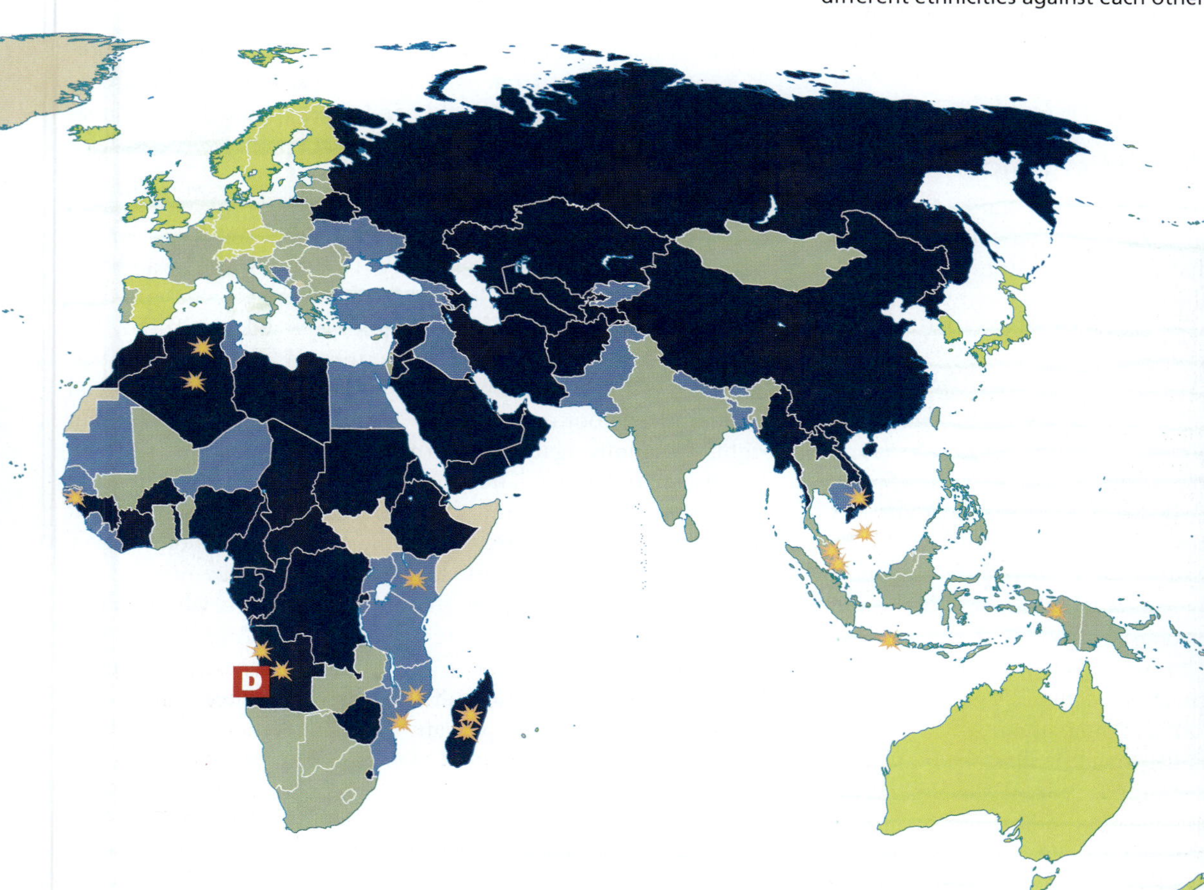

B A young member of the Kosovo Liberation Army in 1999. Like other parts of the former Yugoslavia, Kosovo fought for independence after Serbia's authoritarian policies denied people basic political freedoms.

C Two murals in Derry, Northern Ireland, mark an entrance to a "nationalist" neighborhood. The mural on the right depicts a boy wearing a gas mask and holding a homemade petrol bomb. Both murals were created during "the Troubles," a time of violent conflict from 1969 to 1998. One aspect of the still-sensitive conflict is competing claims about Northern Ireland's democratic legitimacy. The "nationalists" (who are usually Catholic) claim that Northern Ireland was undemocratically separated from the rest of Ireland in 1920 by the government of the United Kingdom during Ireland's war of independence from the U.K. The "loyalists" (usually Protestant) see Northern Ireland as an expression of the will of the voters in the state, most of whom are Protestant and want to remain part of the U.K.

CURRENT GEOGRAPHIC ISSUES

Europe today is in a state of transition as a result of two major changes that occurred during the 1990s: the demise of the Soviet Union (discussed on page 172 and throughout Chapter 5) and the rise of the European Union. These developments have already brought greater peace, prosperity, and world leadership to Europe. However, after years of increasing success and expansion, the European Union and its common currency are threatened by the grave financial difficulties of several member states, primarily in South Europe.

Economic and Political Issues

At the end of World War II, European leaders concluded that the best way to prevent the kind of hostilities that had led to two world wars would be to forge closer economic ties among the European nation-states. Over the next 50 years, a series of steps were taken that gradually united Europe economically and socially. More recently, tentative attempts at political unification have also been made, primarily for foreign affairs purposes.

Steps in Creating the European Union

The first step toward developing an economic union in Europe began in 1951 with the creation of a common market for coal and steel (two resources essential for war industries) that, in the rebuilding of Europe, enabled the monitoring of Germany's industrial rehabilitation. The next major step took place in 1958, when Belgium, Luxembourg, the Netherlands, France, Italy, and West Germany formed the *European Economic Community (EEC)*. The members of the EEC agreed to eliminate certain tariffs against one another and to promote mutual trade and cooperation. Denmark, Ireland, and the United Kingdom joined in 1973; Greece, Spain, and Portugal in the 1980s; and Austria, Finland, and Sweden in 1995, bringing the total number of member countries to 15. In 1992, the EEC became the European Union, with the intent to work toward a level of economic and social integration that would make possible the free flow of goods and people across national borders. The goal of eventually having sufficient social integration to make the borders truly open clarified the EU's ambitions as broader than merely achieving an economic union.

After the demise and fall of the Soviet Union, the European Union expanded into Central Europe. As early as the 1980s, Soviet control over Central Europe began to falter in the face of a workers' rebellion in Poland, known as *Solidarity*. By 1990, East Germany was reuniting with West Germany. The dissolution of the Soviet Union in 1991 brought on the collapse of many economic and political relationships in Central Europe. Thousands of workers lost their jobs. In some of the poorest countries, such as Romania, Bulgaria, and Serbia, social turmoil and organized crime threatened stability. Membership in the European Union became especially attractive to Central European political leaders and citizens who thought it would spur economic development for them that would be a result of investment by wealthier EU member countries from West and North Europe. **90. POLES CELEBRATE THE 25TH ANNIVERSARY OF SOLIDARITY**

Standards for EU membership, however, are rather demanding and specific. A country must have both political stability and a democratically elected government. Each country has to adjust its constitution to EU standards that guarantee the rule of law, human rights, and respect for minorities. Each must also have a functioning market economy that is open to investment by foreign-owned companies and that has well-controlled banks. Finally, farms and industries must comply with strict regulations governing the finest details of their products and the health of environments. Meeting these requirements has been a challenge for members such as Romania, Bulgaria, and Hungary.

Two very wealthy countries have chosen not to join the European Union: Switzerland and Norway. They, plus Iceland, have long treasured their neutral role in world politics. Moreover, they have been concerned about losing control over their domestic economic affairs. Iceland has now applied to join the European Union, primarily for financial security. During the recession that in Europe began in 2008–2009, Iceland lost enormous wealth due in large part to banking speculation in foreign markets; it also became highly indebted to the International Monetary Fund. Iceland is currently what is called a "candidate" EU country (Figure 4.14).

Croatia, in southeastern Europe, is likely to be the next country to join the European Union. Several countries on the perimeter of Europe are candidate EU countries. In addition to Iceland, they include Turkey, Macedonia, Montenegro, and Serbia. The likelihood of Turkey joining vacillates. Turkey has strained relations with the island country of Cyprus (which was admitted to the European Union in 2004, but due to financial problems may not remain), and Turkey has a history of human rights violations against minorities (especially against its large Kurdish population). There are also some issues regarding the separation of religion and the state; Turkey would be the first majority Muslim country to join the European Union. Turkey itself has some reservations. Its political advantage may lie not in Europe but as a leader in the Middle East, where, since the revolutions collectively referred to as the *Arab Spring* and during the crisis in Syria (see Chapter 6, page 234), Turkey has been a voice of calm and reason. A few other countries—Ukraine, Moldova, and perhaps even the Caucasian republics (Armenia, Azerbaijan, and Georgia)—may at some time be invited to join. However, there is strong opposition to this within Europe, and Europe's huge and potentially powerful neighbor, Russia, opposes the expansion of the European Union into what it regards as its sphere of influence. **93. EU TELLS TURKEY TO DEEPEN REFORMS**

EU Governing Institutions Somewhat similar to the United States, the European Union has one executive branch and two legislative bodies. The *European Commission* acts like an executive branch of government, proposing new laws and implementing decisions. Each of the 28 member states gets one commissioner, who is appointed for a 5-year term, subject to the approval of the European Parliament; but after 2014, the size of the Commission will be reduced to 18, with the right to appoint rotating commissioners. Commissioners are expected to uphold common interests and not those of their own countries. The entire Commission must

FIGURE 4.14 The current (2013) members of the EU and their dates of joining. The European Union, formed as the EEC in 1958 with the initial goal of economic integration, became the EU in 1992, and has led the global movement toward greater regional cooperation. It is older and more deeply integrated than its closest competitor, the North American Free Trade Agreement (NAFTA).

resign if censured by Parliament. The European Commission also includes about 25,000 civil servants who work in Brussels to administer the European Union on a day-to-day basis.

EU citizens directly elect the *European Parliament*, with each country electing a proportion of seats based on its population, much like the U.S. House of Representatives. The Parliament elects the president of the European Commission, who serves for 2½ years as a head of state and head of foreign policy. Laws must be passed in Parliament by 55 percent of the member states, which must contain 65 percent of the EU total population. In other words, a simple majority does not rule. The *Council of the European Union* is similar to the U.S. Senate in that it is the more powerful of the two legislative bodies. However, its members are not elected but consist of one minister of government from each EU country.

Economic Integration and a Common Currency Economic integration has solved a number of problems in Europe. Individual European countries have far smaller populations than their competitors in North America and Asia. This means that they have smaller internal markets for their products, so their companies earn lower profits than those in large countries. The European Union solved this problem by joining European national economies into a common market. Companies in any EU country now have access to a much larger market and can potentially make larger profits through **economies of scale** (reductions in the unit costs of production that occur when goods or services are efficiently mass produced, resulting in increased profits per

economies of scale reductions in the unit cost of production that occur when goods or services are efficiently mass produced, resulting in increased profits per unit

ON THE BRIGHT SIDE

The Nobel Peace Prize

Europe's long history of democracy has facilitated the healing of wounds. After both world wars, high levels of social and economic development have helped the countries of the region work cooperatively with each other to create and maintain the European Union. This accomplishment was recognized in October of 2012 with the awarding of the Nobel Peace Prize to the whole of the European Union, a prize that normally goes to an individual.

unit). Before the European Union, when businesses sold their products to neighboring countries, their earnings were diminished by tariffs and other regulations, as well as by fees for currency exchanges.

The official currency of the European Union is the **euro** (€). Seventeen EU countries now use the euro: Austria, Belgium, Cyprus, Estonia, Finland, France, Germany, Greece, Ireland, Italy, Luxembourg, Malta, the Netherlands, Portugal, Slovakia, Slovenia, and Spain. Countries that use the euro have a greater voice in the creation of EU economic policies, and the use of a common currency greatly facilitates trade, travel, and migration within the European Union. All non-euro member states, except Sweden and the United Kingdom, have currencies whose value is determined by that of the euro. Depending on global financial conditions, either the euro or the U.S. dollar is the preferred currency of international trade and finance.

euro the official (but not required) currency of the European Union as of January 1999

Economic Comparisons of the European Union and the United States The EU economy now encompasses over 500 million people (out of a total of 540 million in the whole of Europe)—roughly 200 million more than live in the United States. Collectively, the EU countries are wealthy. In 2011, their joint economy was \$15.65 trillion (PPP), about 2 percent larger than that of the United States, making the European Union the largest economy in the world. Furthermore, trade within the European Union amounts to about twice the monetary value of trade with the outside world.

In contrast to the United States, which usually imports far more than it exports—resulting in a trade deficit—the European Union often has maintained a trade balance in which the values of imports and exports are roughly equal. In terms of income, according to the World Bank, the average 2011 GNI per capita (PPP) for the European Union (\$33,982) was significantly less than that of the United States (\$48,450), although the disparity of wealth in the European Union was also lower by about one-third than in the United States, resulting in lower levels of poverty in the EU.

Careful regulation of EU economies results in generally slower economic growth than in the United States. Until the recent debt crises, these regulations made the European Union more resilient to global financial crises, such as the crisis beginning in 2008, largely due to stronger control (than in the United States) of the banking industry and financial markets.

Some Europeans believe that the European Union, which is already a global economic power, should become a global counterforce to the United States in political and military affairs. For this to happen, European countries would have to give up some political and financial sovereignty so that foreign policy could be initiated at the supranational or federal level.

The Euro and Debt Crises

In recent years, there have been major challenges to the euro, most significantly in the form of a debt crisis that has tested the traditional mechanisms that governments use to maintain economic stability. The debt crisis first emerged in Greece, where the fact that government spending was greatly outpacing tax revenues was disguised in official reports to the European Union. Also, Greece's trade deficit grew to unsustainable levels in part because of the global economic slowdown starting (in Europe) in 2008, which affected tourism and reduced demand for Greece's exports. Normally, governments respond to high levels of debt by borrowing from other countries and then reducing the value of their currency, relative to that of the lenders, which makes these debts easier to pay. However, this was not an option for any member of the Euro zone because no country could, by itself, manipulate the value of the euro. The EU's main lenders included Germany, France, banks across Europe and Asia, and the International Money Fund (IMF), as well as many Greek banks. Hindsight shows that none of these banks was thorough enough in researching the Greek financial situation. The fact that there is no European central bank with regulatory authority made it difficult to control the crisis. When Greece was revealed to be unable to repay its loans, the economies of other Euro zone countries in similar debt circumstances (such as Ireland, Italy, Portugal, Spain, and Slovenia) also began to falter as international investors started to pull their funding out.

In response to these crises, the European Union developed mechanisms to limit government debt in all countries that use the euro. However, as of 2013, the potential for currency and debt crises remains. Among the possible solutions are increasing the political and financial ties between EU countries so that economic policies could be hammered out and enforced at the federal level, and the creation of a central bank. However, many Euro zone countries, even those in financial difficulty—such as Greece, Cyprus, and Slovenia—strenuously object to giving EU institutions more control over their financial affairs.

The European Union, Globalization, and Development The European Union pursues a number of economic development strategies designed to ensure its ability to compete with the United States, Japan, and the developing economies of Asia, Africa, and South America. To increase Europe's export potential, the EU has focused on lowering the cost of producing goods in Europe. One strategy is to shift labor-intensive industries from the wealthiest EU countries in western Europe, where wages are high, to the relatively poorer, lower-wage member states of Central Europe (see Figure 4.14). This strategy has generally worked well, helping poorer European countries prosper while keeping the costs of doing business low enough to restrain European companies from moving away from Europe to places where costs are lower still. However, while the economies of Central Europe have grown, in Europe's wealthiest countries the

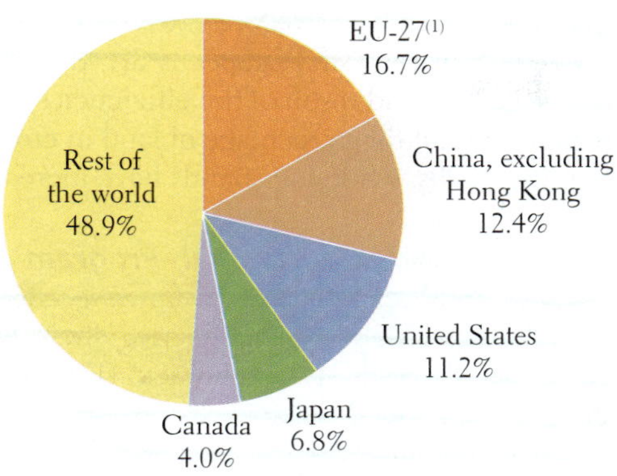

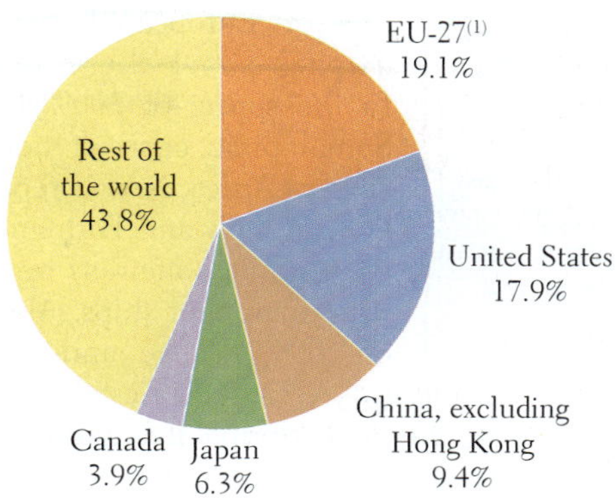

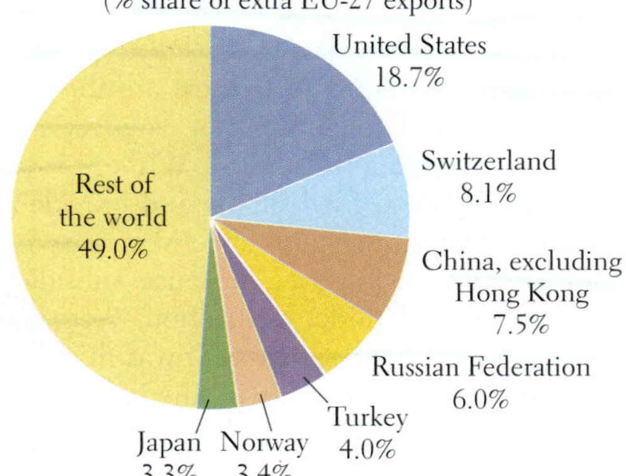

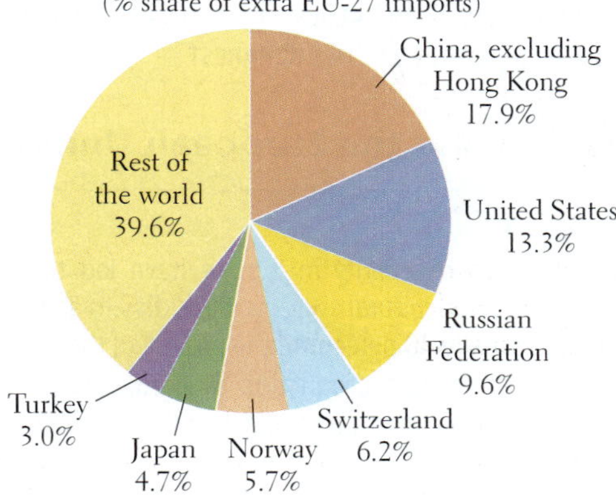

FIGURE 4.15 The EU's main trading partners and shares in world trade: 2008 and 2009.

resultant reduction of industrial capacity (*deindustrialization*) has led to higher unemployment. And despite these efforts, some EU firms have moved abroad to cut costs.

Like the United States, the European Union currently has a large share of world trade (Figure 4.15); it therefore exerts a powerful influence on the global trading system. The European Union often negotiates privileged access to world markets using *preferential trade agreements*, or *PTAs*, for European firms and farmers. The European Union also employs protectionist measures (tariffs) that favor European producers by making goods from outside the European Union more expensive. These higher-priced goods create added expense for European consumers but help ensure EU jobs and control over supplies. Of course, non-European producers shut out of EU markets are unhappy, and increasingly they have united to protest the EU's failure to open its economies to foreign competition. So far, such protests have met with little success.

NATO and the Rise of the European Union as a Global Peacemaker A new role for the European Union as a global peacemaker and peacekeeper is developing through the **North Atlantic Treaty Organization (NATO)**, which is based in Europe. During the Cold War, European and North American countries cooperated militarily through NATO to counter the influence of the Soviet Union. NATO originally included the United States, Canada, the countries of western Europe, and Turkey; it now includes almost all the EU countries as well.

Following the breakup of the Soviet Union, other nations came to assist the United States in the difficult task of addressing global security issues only after a major failure to avert a bloody ethnic conflict during the breakup of Yugoslavia in 1991. NATO has since focused mainly on providing the international security and cooperation needed to expand the European Union. When the United States invaded Iraq in 1993, most EU members opposed the war. As worldwide opposition to the United States built, the global status of the European Union rose. With the United States preoccupied in Iraq, NATO assumed more of a role as a global peacekeeper. It now provides a majority of the troops in Afghanistan, but enthusiasm for this war has waned sharply in Europe.

North Atlantic Treaty Organization (NATO) a military alliance between European and North American countries that was developed during the Cold War to counter the influence of the Soviet Union; since the breakup of the Soviet Union, NATO has expanded membership to include much of eastern Europe and Turkey, and is now focused mainly on providing the international security and cooperation needed to expand the European Union

Common Agricultural Program (CAP) an EU program, meant to guarantee secure and safe food supplies at affordable prices, that places tariffs on imported agricultural goods and gives subsidies to EU farmers

subsidies monetary assistance granted by a government to an individual or group in support of an activity, such as farming, that is viewed as being in the public interest

In addition to the major role now played in Afghanistan, NATO and the European Union are also helping patrol the world ocean. In the spring of 2009 during the Somali pirate crisis off the northeast coast of Africa, NATO reported that both French and Portuguese naval vessels foiled attempts by pirates to seize merchant ships. Also in May of 2009, France announced that in accordance with its role in NATO, it had established a base in Abu Dhabi. During the Arab Spring that started in 2011, NATO undertook a number of noncombat operations in the Mediterranean, aimed at monitoring developing situations in Algeria, Tunisia, Libya, and Mali.

94. NATO'S FUTURE ROLE DEBATED
95. NATO TO PROJECT DIFFERENT PHILOSOPHY
96. CONCERN OVER COMMON VALUES AT THE U.S.–EU SUMMIT
97. NATO LEADERS, PUTIN MEET IN BUCHAREST

Food Production and the European Union

> **Geographic Insight 5**
> **Food:** Concerns about food security in Europe have led to heavily subsidized and regulated agricultural systems. Subsidies to agricultural enterprises account for nearly 40 percent of the budget of the European Union. Although large-scale food production by agribusiness is now dominant, there is a revival of organic and small-scale sustainable techniques.

Food security is especially important to Europeans, probably because stories of post-World War II food shortages are still so vivid. Most food is now produced on large, mechanized farms that are efficient, require less labor, and are more productive per acre than were farms before the 1970s. One result is that only about 2.3 percent of Europeans are now engaged in full-time farming. A second result of the efficiencies of agricultural mechanization is that the percentage of land in crops has declined since the mid-1990s, while forestlands have increased.

The Common Agricultural Program (CAP) The drastic decline of labor and land in farming affects Europeans emotionally, who see it as endangering their cultural heritage and the goal of food self-sufficiency (Figure 4.16). To address these worries, the European Union established its wide-ranging **Common Agricultural Program (CAP)**, meant to guarantee secure and safe food supplies at affordable prices, provide a secure living for farmers, and preserve the quaint, rural landscapes nearly everyone associates with Europe.

The CAP helps farmers by placing tariffs on imported agricultural goods and by giving **subsidies** (in this case, payments to farmers) to underwrite the costs of production. Subsidies are expensive—payments to farmers are the largest category in the EU budget, still accounting for 38 percent of expenditures after negotiated reductions in 2013. While these policies do ensure a safe and sufficient food supply and provide a decent living standard for farmers, they also effectively raise food costs for millions of consumers. Moreover, because subsidies are based on the amount of land under cultivation, these payments favor large, often corporate-owned farms. It was this aspect of the CAP that the Smithfield company (with the help of its European partners) took advantage of in setting up its giant pig farms in Romania (recall the opening vignette).

Protective agricultural policies like tariffs and subsidies—also found in the United States, Canada, and Japan—are unpopular in the developing world. Tariffs lock farmers in poorer countries

FIGURE 4.16 LOCAL LIVES **FOODWAYS IN EUROPE**

A Tortellini, stuffed ring- or "navel"-shaped pasta from the region around Modena and Bologna, Italy. One legend claims that the gods Venus and Jupiter arrived at a tavern in Bologna one night, weary from an ongoing battle between Bologna and Modena. When they retired to their room, the innkeeper peeked through the keyhole and saw only Venus's navel. Inspired, he rushed off to the kitchen to create the first tortellini.

B Swedish "gravlax," raw salmon cured in salt and dill, is cut into thin slices. Gravlax was invented by fishermen who preserved salmon by salting it and then burying it in the sand on a beach above the high tide line. This process gave the dish its name. *Grav* means "grave" and *lax* means "salmon."

C In French *confit* of duck legs, the meat is salted, seasoned, and slowly cooked in its own fat. The meat is then stored in the fat, which can preserve it for several weeks. Confit is considered a regional specialty of southwestern France.

out of major markets. And subsidies encourage overproduction in rich countries (in order to collect more payments). The result is occasional gluts of farm products that are then sold cheaply on the world market, as is the case with Smithfield pork. This practice, called *dumping*, lowers global prices and thus hurts farmers in developing countries while it aids those in developed countries by reducing their supplies, which keeps their prices high.

Growth of Corporate Agriculture and Food Marketing

As small family farms disappear in the European Union—just as they did several decades ago in the United States—smaller farms are being consolidated into larger, more profitable operations run by European and foreign corporations (see Figure 4.5C). These farms tend to employ very few laborers and use more machinery and chemical inputs.

The move toward corporate agriculture is strongest in Central Europe. When communist governments gained power in the mid-twentieth century, they consolidated many small, privately owned farms into large collectives. After the breakup of the Soviet Union, these farms were rented to large corporations, which in turn further mechanized the farms and laid off all but a few laborers. Rural poverty rose. Small towns shrank as farmworkers and young people left for the cities. With EU expansion, the CAP has provided further incentives for large-scale mechanized agriculture, which is now found in Spain, Italy, and France.

A Case Study: Green Food Production in Slovenia

During the Communist era, Slovenia was unlike most of the rest of Central Europe in that the farms were not collectivized. As a result, the average farm size is just 8.75 acres (3.5 hectares). Although Slovenia has plenty of rich farmland, as standards of living have risen, it has become a net importer of food. Nonetheless, Slovenia's new emphasis on private entrepreneurship, combined with a growing demand throughout Europe for organic foods, has encouraged some Slovene farmers to carve out an organic niche for themselves, first in local markets and eventually as exporters. The case of Vera Kuzmic is illustrative (Figure 4.17). ■

VIGNETTE Vera Kuzmic (a pseudonym) lives 2 hours by car south of Ljubljana, Slovenia's capital. For generations, her family has farmed 12.5 acres (5 hectares) of fruit trees near the Croatian border. In the economic restructuring that took place after Slovenia became independent in 1991, Vera and her husband lost their government jobs. The Kuzmic family decided to try earning its living in vegetable market gardening because vegetable farming could be more responsive to market changes than fruit tree cultivation. By 2000, the adult children and Mr. Kuzmic were working on the land, and Vera was in charge of marketing their produce and that of neighbors whom she had also convinced to grow vegetables.

Vera secured market space in a suburban shopping center in Ljubljana, where she and one employee maintained a small vegetable and fruit stall (Figure 4.17). Her produce had to compete with less expensive, Italian-grown produce sold in the same shopping center—all of it produced on large corporate farms in northeastern Italy and trucked in daily. But Vera gained market share by bringing her customers special orders and by guaranteeing that only animal manure, and no pesticides or herbicides, were used on the fields. However, when Slovenia joined the European Union in 2004, she had to do more to compete with produce growers and marketers from across Europe who now had access to Slovene customers.

Anticipating the challenges to come, the Kuzmics' daughter Lili completed a marketing degree and the family incorporated their business. Lili is now its Ljubljana-based director, while Vera manages the farm. Lili's market research showed the wisdom of diversification. The Kuzmics still focus on Ljubljana's expanding professional population, who will pay extra for fine organic produce. But now, in a banquet facility on the farm built with CAP funds, Vera also prepares special dinners for bus-excursion groups from across Europe interested in witnessing traditional farm life and in tasting Slovene ethnic dishes made from homegrown organic crops. *[Source: Lydia Pulsipher. For detailed source information, see Text Credit pages.]* ■

FIGURE 4.17 Vera Kuzmic in her market stall in Ljubljana.

Thinking Geographically

How can you tell that Vera is involved in small-scale agriculture?

Europe's Growing Service Economies

As industrial jobs have declined across the region, most Europeans (about 70 percent) have found jobs in the service economy. Services, such as the provision of health care, education, finance, tourism, and information technology, are now the engine of Europe's integrated economy, drawing hundreds of thousands of new employees to the main European cities. For example, financial corporations that are located in London and serve the entire world play a huge role in the British economy, with many multinational companies headquartered in London.

A major component of Europe's service economy is *tourism*. Europe is the most popular tourist destination in the world, and one job in eight in the European Union is related to tourism. Tourism generates 13.5 percent of the EU's gross domestic product and 15 percent of its taxes, although this varies with global economic conditions. Europeans are themselves enthusiastic travelers, frequently visiting one another's countries to attend local festivals (Figure 4.18) as well as travelling to many distant locations throughout the world. This travel is made possible by the long paid vacations—usually 4 to 6 weeks—that Europeans are granted by employers. Vacation days can be taken a few at a time so that people can take numerous short trips. The most popular holiday destinations among EU members in 2011 were France, Spain, Italy, the United Kingdom, Germany, and Austria.

Service occupations increasingly involve the use of technology. While Europe has lagged behind North America in the development and use of personal computers and the Internet, it leads the world in cell phone use. The information economy is especially advanced in West Europe, though in South Europe and Central Europe, where personal computer ownership is lowest, public computer facilities in cafés and libraries are common.

THINGS TO REMEMBER

- The EU's original plan was to reach a level of economic and social integration that would make possible the free flow of goods and people across national borders; for the most part, those goals have been reached among the current 28 members.
- The European Union has one executive branch—the European Commission—and two legislative branches—the European Parliament, directly elected by EU citizens, and the Council of the European Union, whose members consist of one minister from each EU country.
- The European Union joined the members' national economies into a common market. By 2011, the EU's economy was almost $15.65 trillion (PPP), about 2 percent larger than that of the United States, making the European Union the largest economy in the world.
- A new role for the European Union as a global peacemaker and peacekeeper is developing through the North Atlantic Treaty Organization (NATO), which is based in Europe.

Geographic Insight 5

- **Food** Concerns about food security in Europe have caused EU members to invest tax money to ensure the survival of traditional farming and small-scale sustainable techniques, including the organic production of fruits, vegetables, and meats. The EU also subsidizes agricultural production on large corporate farms where agribusiness is now dominant.
- As industrial jobs have declined across the region, most Europeans (about 70 percent) have found jobs in the service economy.

Sociocultural Issues

The European Union was conceived primarily to promote economic cooperation and free trade, but its programs have social implications as well. As population patterns change across Europe, attitudes toward immigration, gender roles, and social welfare programs are also evolving. Christian religious factions and language rivalry, once divisive issues in the region, have now almost disappeared as a focus of disputes. Meanwhile,

FIGURE 4.18 LOCAL LIVES **FESTIVALS IN EUROPE**

A Every year in Buñol, Spain, participants in "La Tomatina" throw tomatoes at each other. Supposedly, this festival began in 1945 when a group of young men, not allowed in a parade, grabbed tomatoes from a nearby vegetable stand and hurled them. The young men returned the next year and repeated the brawl, though they brought their own tomatoes. Today, truckloads of tomatoes are brought in, such as shown here.

B Riders in "Il Palio," a horse race held each year in Siena, Italy. Jockeys from each of the city's 17 major neighborhoods circle the main plaza; they ride bareback and frequently get thrown from their horses. A colorful pageant precedes the race, which has been run since 1656, though similar races were run earlier.

C Glastonbury Festival of Contemporary Performing Arts (known as "Glasto") is held each year in Somerset, England. First organized in 1970 by a farmer who was inspired by a Led Zeppelin performance, Glastonbury has grown to become one of Europe's largest music festivals.

immigration, especially by people of the Muslim faith, is a source of apprehension, perhaps because these days few Europeans identify themselves as belonging to any religious faith and they are left uncomfortable by shows of religious fervor.

Population Patterns

> **Geographic Insight 6**
>
> **Population and Gender:** Europe's population is aging, as working women are choosing to have only one or two children. Immigration by young workers is only partially countering this trend.

Population patterns in Europe portended processes that are emerging around the world. Europe's high population density and urbanization trends are long-standing. A newer but prominent phenomenon, the aging of the populace and lower birth rates affect all manner of European social policies.

Population Distribution and Urbanization There are currently about 540 million Europeans. Of these, 500 million live within the European Union. The highest population densities stretch in a discontinuous band from the United Kingdom south and east through the Netherlands and central Germany into northern Switzerland (Figure 4.19). Northern Italy is another zone of high density, along with pockets in many countries along

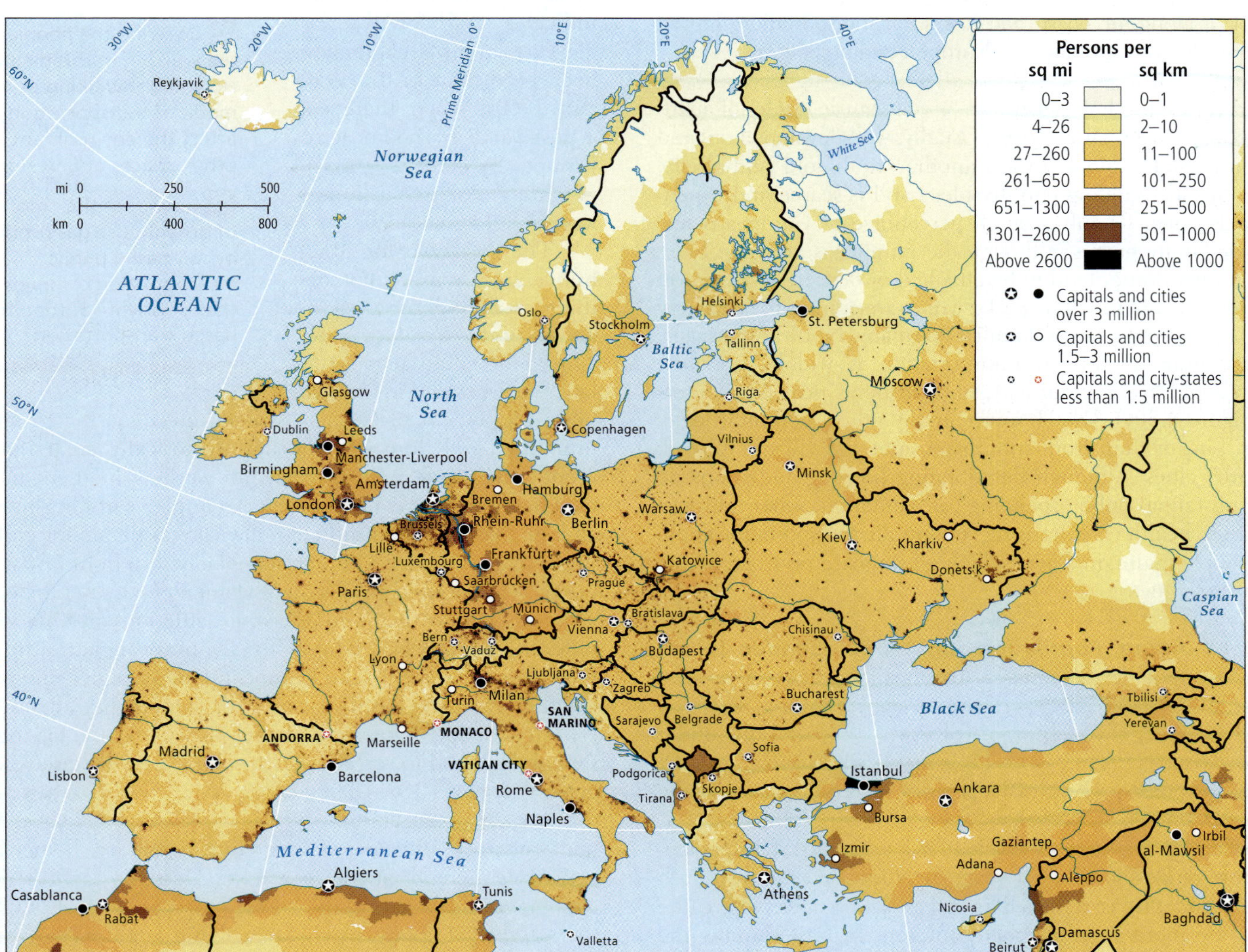

FIGURE 4.19 Population density in Europe. Europe's population is not growing, a fact that raises concerns about future economic conditions as the population ages. Many governments now encourage large families, with generous maternity and paternity leave (up to 10 months with full pay), free day care, and other incentives. However, few countries have seen much change in their population growth rate. A major reason for this is that more women now either work or want to work than ever before, so they are delaying childbearing and families remain small. As this trend shows no signs of decreasing, some officials are looking toward increased immigration as a possible solution. On the other hand, Europe has been attracting large numbers of immigrants for decades, and Europeans are increasingly wary of hosting large populations of foreigners who might not share their values.

the Mediterranean coast. Overall, Europe is one of the more densely occupied regions on Earth, as shown on the world population density map in Figure 1.8 on page 14.

Europe is a region of cities surrounded by well-developed rural hinterlands where urban dwellers often go to enjoy leisure hours. These cities are the focus of the modern European economy, which, though long based on agriculture, trade, and manufacturing, is now primarily service oriented. In West and North Europe, more than 75 percent of the population lives in urban areas. In South and Central Europe, urbanization rates are slightly lower, but village residents often commute to urban jobs. As noted earlier, many European cities began as trading centers more than a thousand years ago and still bear the architectural marks of medieval life in their historic centers (Figure 4.20A). These old cities are located either on navigable rivers in the interior or along the coasts because water transportation figured prominently (as it still does) in Europe's trading patterns.

Since World War II, nearly all the cities in Europe have expanded around their perimeters in concentric circles of apartment blocks (see Figure 4.20B). Usually, well-developed rail and bus lines link the blocks to one another, to the old central city, and to the surrounding countryside. Land is scarce and expensive in Europe, so only a small percentage of Europeans live in single-family homes, although the number is growing. Even single-family homes tend to be attached or densely arranged on small lots. Except in public parks, which are common, one rarely sees the sweeping lawns familiar to many North Americans. Because publicly funded transportation is widely available, many people live without cars, in apartments near city centers (see Figure 4.20C, D). However, many others commute daily by car, bus, or train from suburbs or ancestral villages to work in nearby cities. Although deteriorating housing and slums do exist, substantial public spending (on sanitation, water, utilities, education, health care, housing, and public transportation) help most people maintain a generally high standard of urban living.

Europe's Aging Population Linked to Gender Issues

Europe's population is aging as families are choosing to have fewer children and life expectancies are increasing. Between 1960 and 2012, the proportion of those 14 years and under declined from 27 percent to 16 percent, while those over 65 increased from 9 percent to 16 percent. By comparison, young people comprised 26 percent of the global population in 2012, while older generations accounted for 8 percent. Life expectancies now range close to 80 years in North, West, and South Europe, and this is reflected in urban landscapes, where the elderly are more common than children. In East Europe, life expectancies are notably lower: 66 years for men, 76 years for women.

Overall, Europe is now close to a negative rate of natural increase (<0.0), the lowest in the world. Birth rates are low, are shrinking across Europe, and are lowest in the wealthiest country—Germany. Increasingly, the one-child family is common throughout Europe, except among immigrants from outside the region, who are the major source of population growth. However, once they have assimilated into European life, most immigrants, too, choose to have small families.

The declining birth rate is illustrated in the population pyramids of European countries (Figure 4.21 on page 184), which look more like lumpy towers than pyramids. Examples are the population pyramid of Sweden and of the whole European Union (see Figure 4.21B, C). The pyramids' narrowing base indicates that for the last 35 years, far fewer babies have been born than in the 1950s and 1960s, when there was a post-war baby boom across Europe. By 2000, 25 percent of adult Europeans were having no children at all.

The reasons for these trends are complex. For one thing, more and more women want professional careers. This alone could account for late marriages and lower birth rates; 25 percent of German women are choosing to remain unmarried well into their thirties. Historically, many governments have made few provisions for working mothers, beyond paid maternity leave. Germany, for example, is just beginning to address that there is insufficient day care available for children under age 3. Because German school days run from 7:30 A.M. to noon or 1:00 P.M., many German women choose not to become mothers because they would have to settle for part-time jobs in order to be home by 1:00 P.M. To encourage higher birth rates, there is a move within the European Union to give one parent (mother or father) a full year off with reduced pay after a child is born or adopted; to provide better preschool care; to lengthen school days; and to serve lunch at school (see the discussion of gender on pages 187–188).

A stable population with a low birth rate has several consequences. Because fewer consumers are being born, economies may contract over time. Demand for new workers, especially highly skilled ones, may go unmet. Further, the number of younger people available to provide expensive and time-consuming health care for the elderly, either personally or through tax payments, will decline. Currently, for example, there are just two German workers for every retiree. Immigration provides one solution to the dwindling number of young people. However, Europeans are reluctant to absorb large numbers of immigrants, especially from distant parts of the world where cultural values are very different from those in Europe.

Thinking Geographically

After you have read about urbanization in Europe, you will be able to answer the following questions:

A What are the likely reasons why Prague is located along a river?

B What can you see in this picture that illustrates what we know about urban life in Europe?

C Describe the people you see walking in this picture. How might walking as a mode of transportation affect the environment and urban spaces of Barcelona?

D How does the amount of greenhouse gas produced by rail-based public transportation compare to that of private automobile–based transportation?

Figure 4.20

Photo Essay: Urbanization in Europe

Europe's cities are famous throughout the world for their architecture, economic dynamics, and cultural variety. Many are quite ancient but have expanded in recent decades with large apartment blocks connected to the old city centers by public transportation. Despite their global draw, many European cities will shrink in the next several decades, due to declining national populations (as in Italy and parts of Central Europe), as well as to deliberate efforts to shift growth to other urban centers, as in the case of London.

A The old urban core of Prague, Czech Republic, founded in the ninth century C.E. on the banks of the Vltava River.

B High-rise apartment blocks dot the periphery of Berlin's old urban core.

C La Rambla, a tree-lined pedestrian mall in Barcelona, Spain, is a popular strolling place among locals and tourists.

D A tram in Paris, France. The city's metropolitan transit system moves roughly 6 million people per day and is the second-largest system in Europe, after London's.

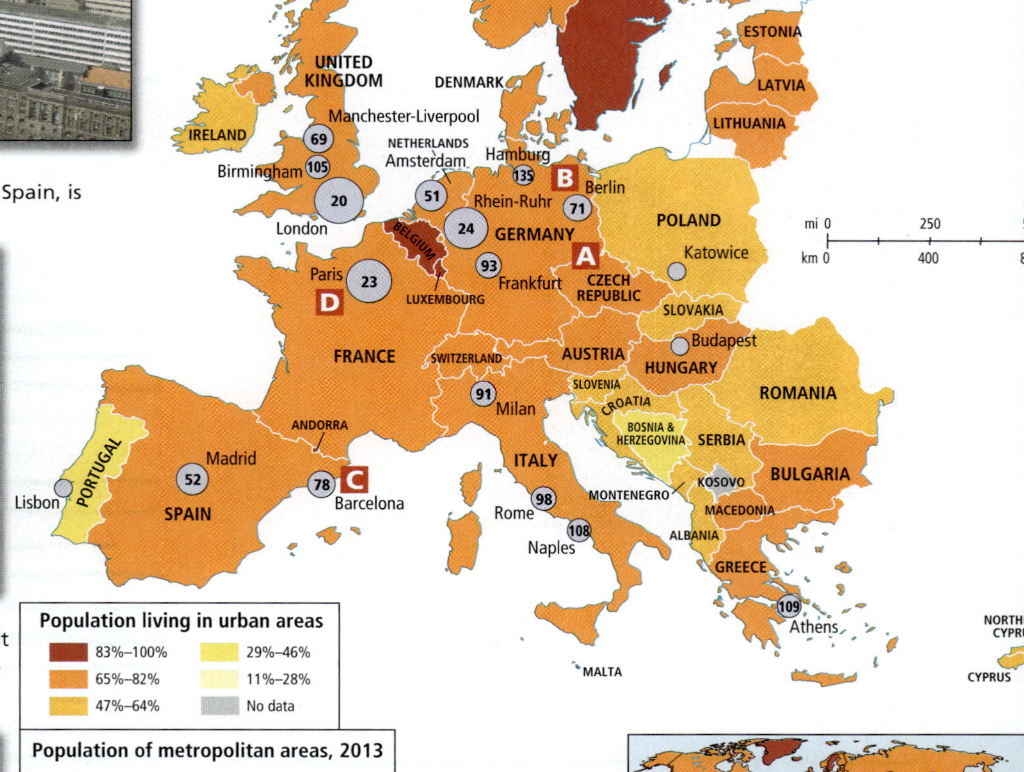

FIGURE 4.21 Population pyramids for Germany, Sweden, and the EU in 2012. These population pyramids have quite different shapes, but all exhibit a narrow base, indicating low birth rates. The EU pyramid (C) is at a different scale due to the much larger population.

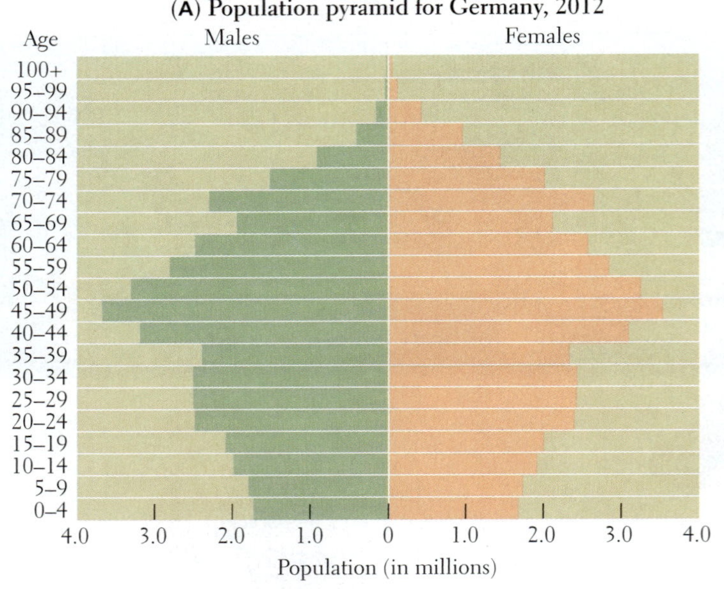

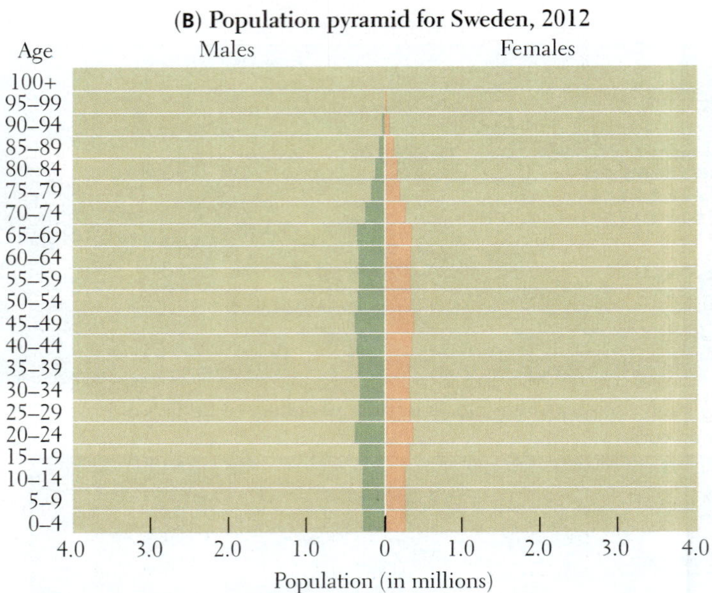

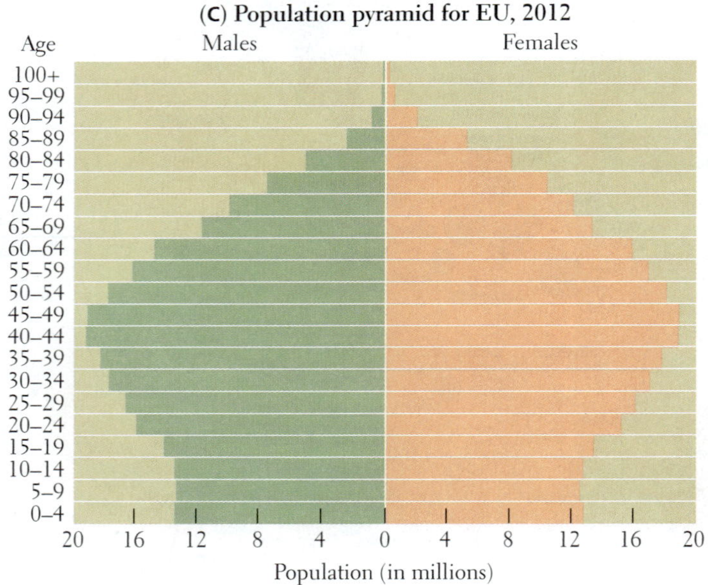

Immigration and Migration: Needs and Fears

Until the mid-1950s, the net flow of migrants was out of Europe, to the Americas, Australia, and elsewhere. By the 1990s, the net flow was into Europe. In the 1990s, most of the European Union (plus Iceland, Norway, and Switzerland) implemented the **Schengen Accord**, an agreement that allows free movement of people and goods across common borders. The accord has facilitated trade, employment, tourism, and most controversially, migration within the European Union. The Schengen Accord has also indirectly increased both the demand for immigrants from outside the European Union and their mobility once they are in the European Union (Figure 4.22). **92. AFTER 50 YEARS, EUROPE STILL COMING TOGETHER**

Attitudes Toward Internal and International Migrants and Citizenship Like citizens of the United States, Europeans have ambivalent attitudes toward migrants. The internal flow of migration is mostly from Central Europe into North, West, and South Europe. These Central European migrants are mostly treated fairly, although prejudices against the supposed backwardness of Central Europe are still evident. Immigrants from outside Europe, so-called *international immigrants*, meet with varying levels of acceptance.

International immigrants often come both legally and illegally from Europe's former colonies and protectorates across the globe. Many Turks and North Africans come legally as **guest workers** who are expected to stay for only a few years, fulfilling Europe's need for temporary workers in certain sectors. Other immigrants are refugees from the world's trouble spots, such as Afghanistan, Iraq, Haiti, and Sudan. Many also come illegally from all of these areas.

While some Europeans see international immigrants as important contributors to their

Schengen Accord an agreement signed in the 1990s by the European Union and many of its neighbors that allows for free movement across common borders

guest workers legal workers from outside a country who help fulfill the need for temporary workers but who are expected to return home when they are no longer needed

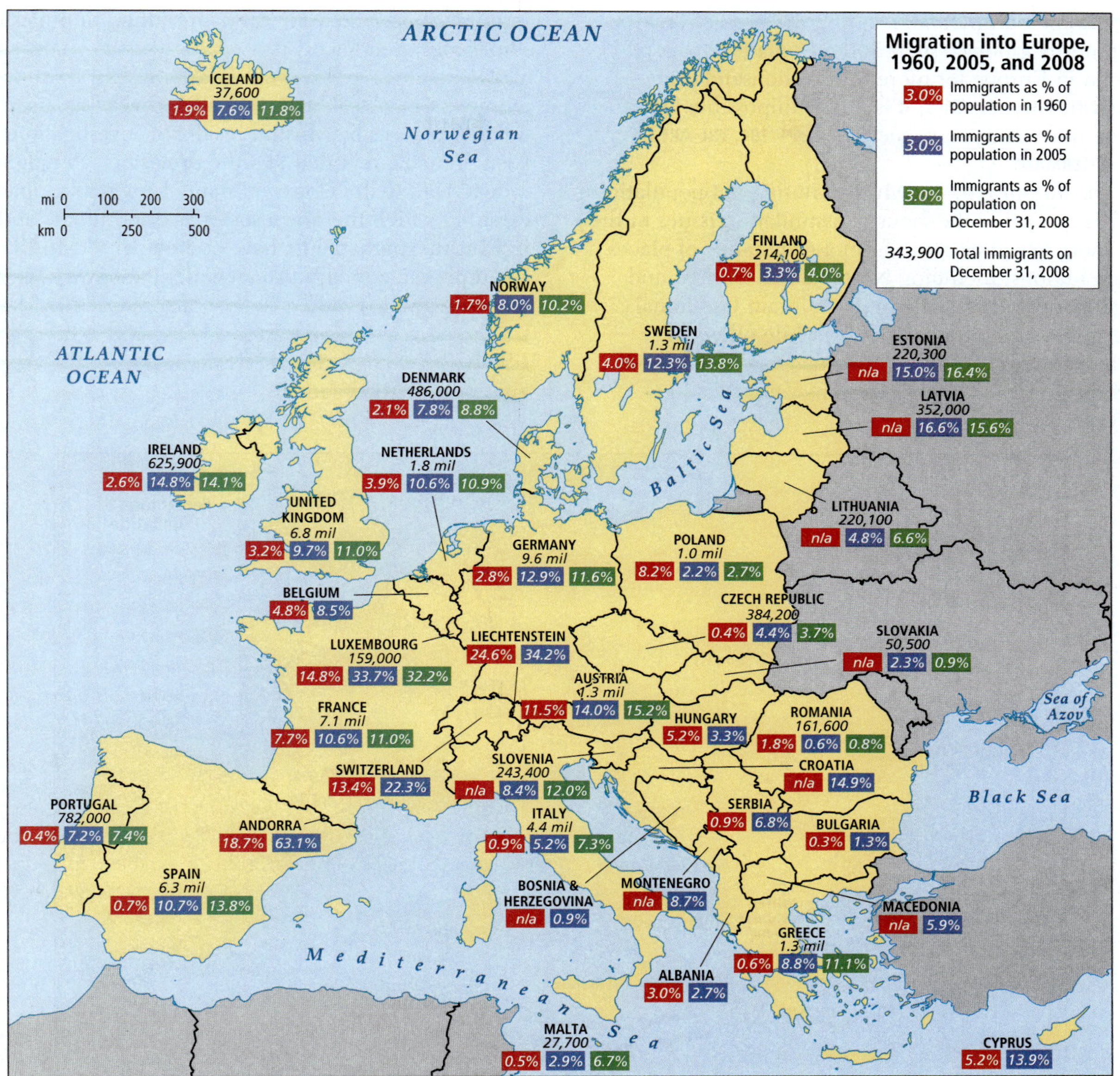

FIGURE 4.22 Migration into Europe, 1960, 2005, and 2008. Migration into Europe increased from 1960 through the 1990s and continued to increase into the twenty-first century for most countries. Migration continues to be a crucial issue in EU debates. The percentages of total population are shown in the three colored boxes for each country: the red box for 1960, the blue box for 2005, and the green box for 2008. The number of migrants living in each country at the end of 2008 is above the boxes.

economies—providing needed skills and labor and making up for the low birth rates—many oppose recent increases in immigration. Studies of public attitudes in Europe show that immigration is least tolerated in areas where incomes and education are low. Central and South Europe are the least tolerant of new immigrants, possibly because of fears that immigrants may drive down wages that are already relatively low. North and West Europe, with higher incomes and generally more stable economies, are the most tolerant. In response, the European Union is increasing its efforts to curb illegal immigration from outside Europe while encouraging EU citizens to be more tolerant of legal migrants.

assimilation the loss of old ways of life and the adoption of the lifestyle of another country

A Case Study: The Rules for Assimilation: Muslims in Europe

In Europe, culture plays as much of a role in defining differences between people as race and skin color. An immigrant from Asia or Africa may be accepted into the community if he or she has gone through a comprehensive change of lifestyle. **Assimilation** in Europe usually means giving up the home culture and adopting the ways of the new country. If minority groups—such as the Roma (see page 171), who have been in Europe for more than a thousand years—maintain their traditional ways, it is nearly impossible for them to blend into mainstream society. **101. THE ART OF INTEGRATION IN GERMANY**

Europe's small but growing Muslim immigrant population (Figure 4.23) is presently the focus of assimilation issues in the European Union. Muslims come from a wide range of places and cultural traditions, including North Africa, Turkey, and South Asia. Some of these immigrants maintain traditional dress, gender roles, and religious values, while others have assimilated into European culture. The deepening alienation that has boiled over in recent years among some Muslim immigrants and their children—resulting in protests, riots, and sometimes even terrorism—relates primarily to the systematic exclusion of these less-assimilated Muslims from meaningful employment, social services, and higher education. In the wake of protests by Muslim residents, investigations by the French media revealed that the protestors' complaints were indeed legitimate. However, many Europeans, unaware of the extent to which their own societies discriminate and engage in Muslim stigmatizing, have come to view Muslim protesters as simple malcontents. Meanwhile, it is young Muslims born in Europe and schooled in the lofty ideals of the European Union, who have never known life in any other place, who harbor the greatest resentment against the constricted opportunities they face. ■

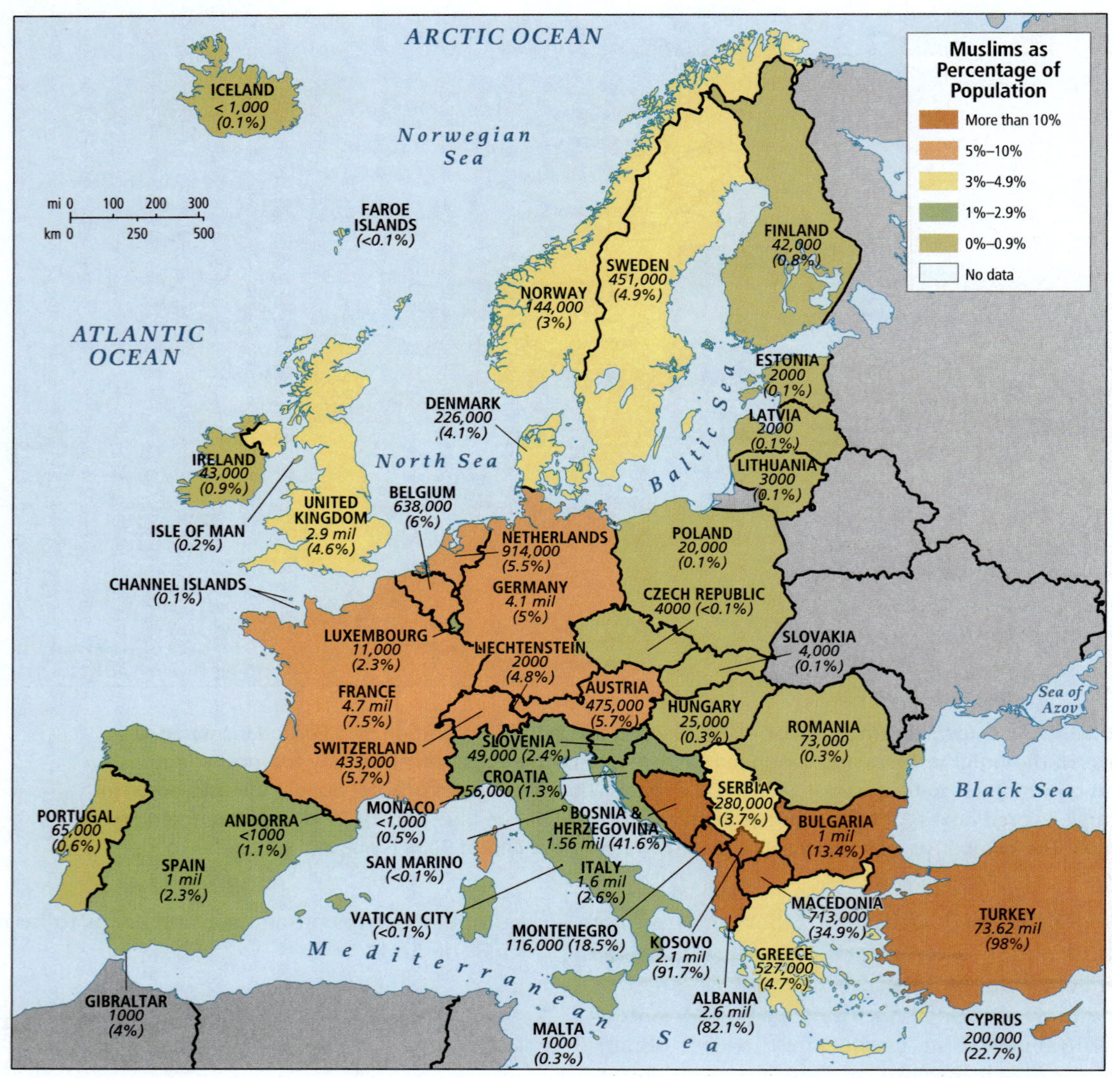

FIGURE 4.23 Muslims in Europe. Muslims are a small minority in most European countries, but their population has grown from a total of 29.6 million in 1990 to 44.1 million in 2010—amounting to about 6 percent of the total population—and is expected to rise to more than 58 million by 2030. Many intend to make the region their permanent home. Just how, or if, the assimilation of Muslims into largely secular Europe will proceed is a major topic of public debate.

In some cases, conflicts have also arisen over European perceptions of cultural aspects of Islam, such as the *hijab* (one of several traditional coverings for women). For example, in France in 2004, wearing of the hijab by observant Muslim schoolgirls became the center of a national debate about civil liberties, religious freedom, and national identity. French authorities wanted to ban the hijab but were wary of charges of discrimination. Eventually the French declared all symbols of religious affiliation illegal in French schools (including crosses and yarmulkes, the Jewish head covering for men).

102. RELIGIOUS TOLERANCE FACING TEST IN BRITAIN

103. LONDON'S MEGA MOSQUE STIRS CONTROVERSY

Changing Gender Roles

Gender roles in Europe have changed significantly from the days when most women married young and worked in the home or on the family farm. Increasing numbers of European women are working outside the home, and the percentage of women in professional and technical fields is growing rapidly (Figure 4.24). Nevertheless, European public opinion among both women and men largely holds that women are less able than men to perform the types of work typically done by men, and that men are less skilled at domestic, caregiving, and nurturing duties. In most places, men have greater social status, hold more managerial positions, earn on average about 15 percent more pay for doing the same work, and have greater autonomy in daily life than women—more freedom of movement, for example. These male advantages retain a stronger hold in Central and South Europe today than they do in West and North Europe.

Certainly, younger men now assume more domestic duties than did their fathers, but women who work outside the home usually still face what is called a **double day** in that they are expected to do most of the domestic work in the evening in addition to their job outside the home during the day. UN research shows that in most of Europe, women's workdays, including time spent in housework and child care, are 3 to 5 hours longer than men's. (Iceland and Sweden reported that women and men there share housework equally.) Women burdened by the double day generally operate with somewhat less efficiency in a paying job than do men. They also tend to choose employment that is closer to home and that offers more flexibility in the hours and skills required. These more flexible jobs (often erroneously classified as part-time) almost always offer lower pay and less opportunity for advancement, though not necessarily fewer working hours.

Many EU policies encourage gender equality. Managerial posts in the EU bureaucracy are increasingly held by women, and well over half the university graduates in Europe are now women. Despite this, the political influence and economic well-being of European women lag behind those of European men. In most European national parliaments, women make up less than a third of elected representatives.

double day the longer workday of women with jobs outside the home who also work as caretakers, housekeepers, and/or cooks for their families

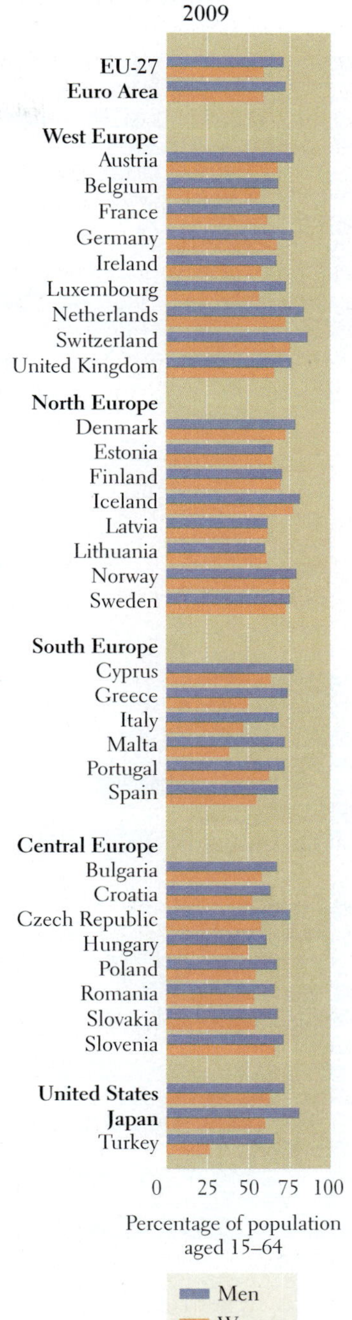

FIGURE 4.24 Women at work in Europe (2009). A majority of women in Europe work outside the home, and their numbers have increased from 52 percent in 1998 to 58.6 percent in 2009. The exceptions are in Greece, Hungary, Italy, and Malta, where less than 50 percent of women work outside the home. Included for comparison is Turkey, where slightly more than 25 percent of women work outside the home. These statistics, however, may not include some of the many women in Europe who work in the informal economy. In countries where the rate of women in the paid labor market is increasing, there is usually a concurrent decline in birth rates.

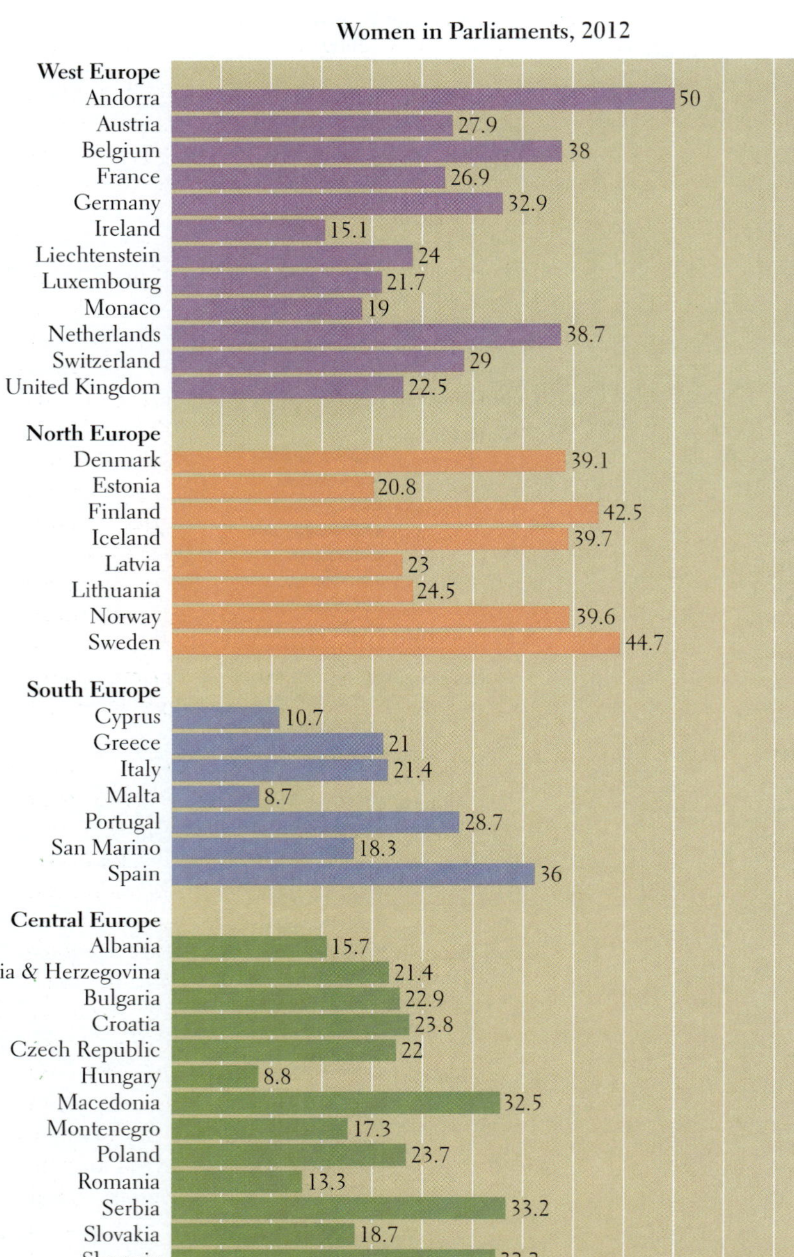

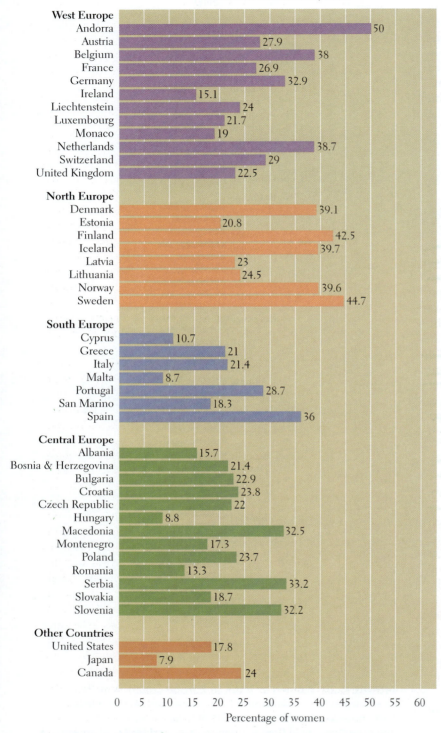

FIGURE 4.25 Women in national parliaments of Europe, 2013. Women constitute about half the adult population of the EU, but they have nowhere near their fair share of representation in European legislatures; hence, their influence on legislation is seriously restricted. Parliamentary elections in 2013 resulted in an increase in female representation in most countries. Notice that some regions are closer to equity than others, and also notice how the United States compares.

Only in North Europe, where several women have served as heads of government, do women come anywhere close to filling 50 percent of the seats in the legislature (Figure 4.25). In West Europe, the United Kingdom has had several female officials in high office, but elsewhere in the region this trend is only beginning. In 2009, Germany reelected Angela Merkel as its first woman chancellor (prime minister); in France, in 2007, Nicolas Sarkozy defeated female opponent Ségolène Royal, but then appointed women to a number of French cabinet-level positions. In 2012, fully one half of newly elected President François Hollande's cabinet was female. In 2013, Slovenia chose Alenka Bratušek as prime minister and for the first time elected a parliament that is nearly one-third women. Still, across Europe, women generally serve only in the lower ranks of government bureaucracies, where they implement policy but have limited power to formulate policy.

Although change is clearly underway in the European Union, economic empowerment for women has been slow on many fronts. For example, in 2006, unemployment was higher among women than among men in all but a few countries (the United Kingdom, Germany, the Baltic Republics, Ireland, Norway, and Romania), where the differences were slight—and 32 percent of women's jobs were part time, as opposed to only 7 percent of men's jobs. Throughout the European Union, women are paid less than men for equal work, despite the fact that young women tend to be more highly educated than young men.

Norway leads Europe in redefining gender in society. It does this by directing much of its most innovative work on gender equality toward advancing men's rights in traditional women's arenas. For example, men now have the right to at least a month of paternity leave with a newborn or adopted child. This policy recognizes a father's responsibility in child rearing and provides a chance for father and child to bond early in life.

ON THE BRIGHT SIDE

European Women in Leadership Positions

In 1986, Gro Harlem Brundtland, a physician, became Norway's first female prime minister; she appointed women to 44 percent of all cabinet posts, thus making Norway the first country in modern times to have such a high proportion of women in important government policy-making positions. By 2013, women made up 39.6 percent of Norway's parliament (Sweden had 44.7 percent; Finland 42.5 percent; and Iceland 39.7 percent). Norway also requires a minimum of 40 percent representation by each sex on all public boards and committees. Even though the rule has not yet been achieved in all cases, female representation averages over 35 percent for state and municipal agencies. In addition, employment ads are required to be gender neutral, and all advertising must be nondiscriminatory.

Social Welfare Systems and Their Outcomes

In nearly all European countries, tax-supported systems of **social welfare** or **social protection** (the EU term) provide all citizens with basic health care; free or low-cost higher education; affordable housing; old age, survivor, and disability benefits; and generous unemployment and pension benefits. Europeans generally pay much higher taxes than North Americans (the rate for EU countries is about 40 percent of GDP; for the United States, 27 percent; and for Canada, 30 percent); in return, they expect more in services. In some cases, European governments are able to deliver services more cheaply than the "free market" does elsewhere. For example, the European Union spends on average about $3000 per person for health care, while the United States spends more than $7000. Even at this much lower cost, the European Union has more doctors, more acute-care hospital beds per citizen, and better outcomes than the United States in terms of life expectancy and infant mortality.

social welfare (in the European Union, **social protection**) in Europe, tax-supported systems that provide citizens with benefits such as health care, affordable higher education and housing, pensions, and child care

Europeans do not agree on the goals of these welfare systems, or on just how generous they should be. Some argue that Europe can no longer afford high taxes if it is to remain competitive in the global market. Others say that Europe's economic success and high standards of living are the direct result of the social contract to take care of basic human needs for all. The debate has been resolved differently in different parts of Europe, and the resulting regional differences have become a source of concern in the European Union. With open borders, unequal benefits can encourage those in need to flock to a country with a generous welfare system and overburden the taxpayers there.

European welfare systems can be classified into four basic categories (Figure 4.26). *Social democratic welfare systems*, common in Scandinavia, are the most generous systems. They attempt to achieve equality across gender and class lines by providing extensive health care, education, housing, and child and elder care benefits to all citizens from cradle to grave. Child care is widely available, in part to help women enter the labor market. But early childhood training, a key feature of this system, is also meant to ensure that in adulthood every citizen will be able to contribute to the best of his or her highest capability, and that citizens will not develop criminal behavior or drug abuse. While finding comparable data is very difficult, surveys of crime victims in Scandinavia and the European Union show that generally the rate of crime in Scandinavia is lower than in other parts of Europe.

FIGURE 4.26 European social welfare/protection systems. The basic categories of social welfare systems shown here and described in the text should be taken as only an informed approximation of the existing patterns. In 2007 (the latest data available from the EU), the average expenditures for social protection in the EU-28 were about 26.7% of GDP. The map shows the percent of GDP spent on social protection by the EU-28 countries in 2007. The 2007 data do not include Croatia.

The goal of *conservative* and *modest welfare systems* is to provide a minimum standard of living for all citizens. These systems are common in the countries of West Europe. The state assists those in need but does not try to assist upward mobility. For example, college education is free or heavily subsidized for all, but strict entrance requirements in some disciplines can be hard for the poor to meet. State-supported health care and retirement pensions are available to all, but although there are movements to change these systems, they still reinforce the traditional "housewife contract" by assuming that women will stay home and take care of children, the sick, and the elderly. The "modest" system in the United Kingdom is considered slightly less generous than the "conservative" systems found elsewhere in West Europe. The two are combined here but not in Figure 4.26.

In countries with *rudimentary welfare systems*, citizens are not considered to inherently have the right to government-sponsored support. They are found primarily in South Europe and in Ireland. Here, local governments provide some services or income for those in need, but the availability of such services varies widely, even within a country. The state assumes that when people are in need, their relatives and friends will provide the necessary support. The state also assumes that women work only part time, and thus are available to provide child care and other social services for free. Such ideas reinforce the custom of paying women lower wages than men.

Post-Communist welfare systems prevail in the countries of Central Europe. During the Communist era, these systems were comprehensive, resembling the cradle-to-grave social democratic system in Scandinavia, except that women were pressured to work outside the home. Benefits often extended to nearly free apartments, health care, state-supported pensions, subsidized food and fuel, and early retirement. However, in the post-Communist era, state funding has collapsed, forcing many to do without basic necessities. In many post-Communist countries, welfare systems are now being revised, but usually with an eye to reducing benefits and extending work lives past age 65.

Geographic Patterns of Human Well-Being To a significant extent, the social welfare policies described above affect patterns of well-being, which, not surprisingly, have a geographical pattern. As we have observed, although Europe is one of the richest regions on Earth, there is still considerable disparity in wealth and well-being. GNI per capita is one measure of well-being because it shows in a general way how much people in a country have to spend on necessities, but it is an average figure and does not indicate how income is distributed. As a result, GNI per capita provides no way to determine if a few are rich and most poor, or if income is more equally apportioned.

Figure 4.27A, which shows GNI per capita (PPP) for all of Europe, illustrates two points. Most of the region has a GNI of at least $10,000 per capita, and only in three small countries in southeastern Europe—Albania, Bosnia and Herzegovina, and Macedonia—is the GNI less than $10,000 per capita. West and North Europe (except for Estonia, Latvia, and Lithuania) are the wealthiest regions. It is worth noting that Greece, which in the 1970s was still quite poor, joined the European Union in 1981 and began to realign its economy (industrializing, reducing its debt and rate of inflation) in order to meet EU specifications. While it is remarkable that in fewer than 30 years Greece appeared to have joined the ranks of Europe's wealthiest countries, in 2010, the extent to which Greece's prosperity was financed with expensive borrowing was revealed. These huge debt obligations have lowered the standard of living in Greece, at least for a while. The world map inset shows how Europe ranks in comparison to other parts of the world.

Figure 4.27B depicts the ranks of European countries in the United Nations Human Development Index (UN HDI), which is a calculation (based on adjusted real income, life expectancy, and educational attainment) of how well a country provides for the well-being of its citizens. It is possible to have a high GNI rank and a lower HDI rank if social services are inadequately distributed or if there is a wide disparity of wealth. Although the Figure 4.27A and B maps (of GNI per capita and HDI, respectively) look very similar, there are only a few countries in Europe that have both a high GNI and a high HDI—Norway, Lichtenstein, and Luxembourg. Most countries have a GNI ranking that is lower than their HDI rank, which indicates that all of these countries, regardless of their actual wealth, provide for their people relatively well, and wealth disparity is under control. In many cases this is because the countries have strong social safety nets or were until recently in communist societies that emphasized meeting basic needs for all.

The United Nations is not currently (2013) making data available on male and female earned income, but the reader should be aware that data from many agencies show that nowhere on Earth are the average wages of women and men equal. The UN data now uses the Gender Equality Index (GEI), which is based on reproductive health, political and educational empowerment, and labor-force participation (Figure 4.27C). European countries generally rank high, but some countries (Ireland, Slovenia, the United Kingdom, Malta, all of Central Europe, plus Greece), have a GEI ranking that is markedly lower than the rest of Europe. This raises concerns about equality between genders in these countries. On the other hand, many countries rank higher on the GEI scale than they do on the HDI. There is still cause for concern even in these countries, given what we know about ubiquitous disparities between female and male incomes, but in access to health care, education, and empowerment, they may be closer to equality between males and females.

THINGS TO REMEMBER

- Europe's cities are world famous for their architecture, economic dynamism, and cultural vitality. Many are quite ancient but have expanded in recent decades, with large apartment blocks connected by public transport to the old city centers. In West, North, and South Europe, about 80 percent of the population lives in urban areas.

FIGURE 4.27 Maps of human well-being.

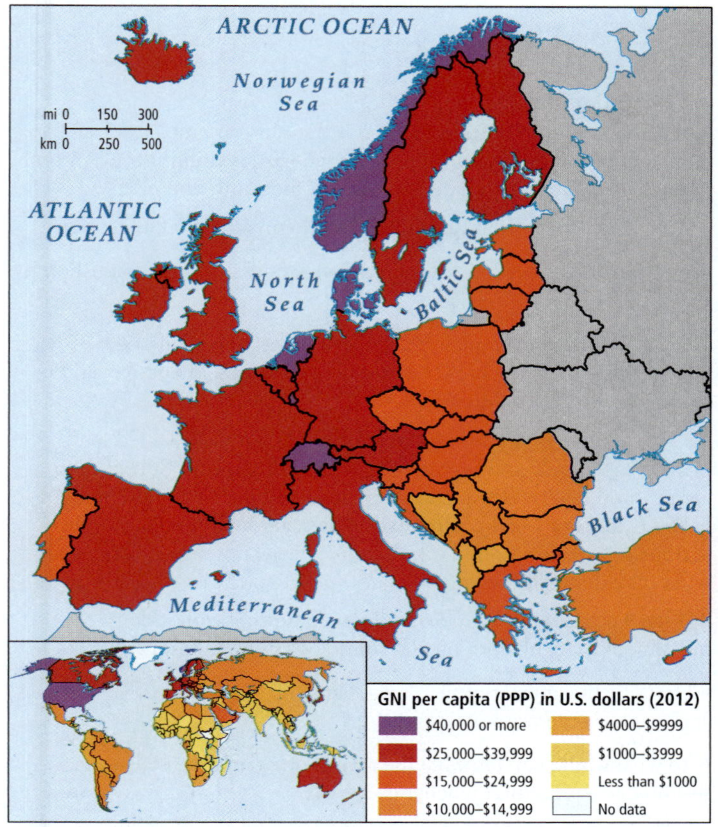

(A) Gross national income (GNI) per capita, adjusted for purchasing power parity (PPP).

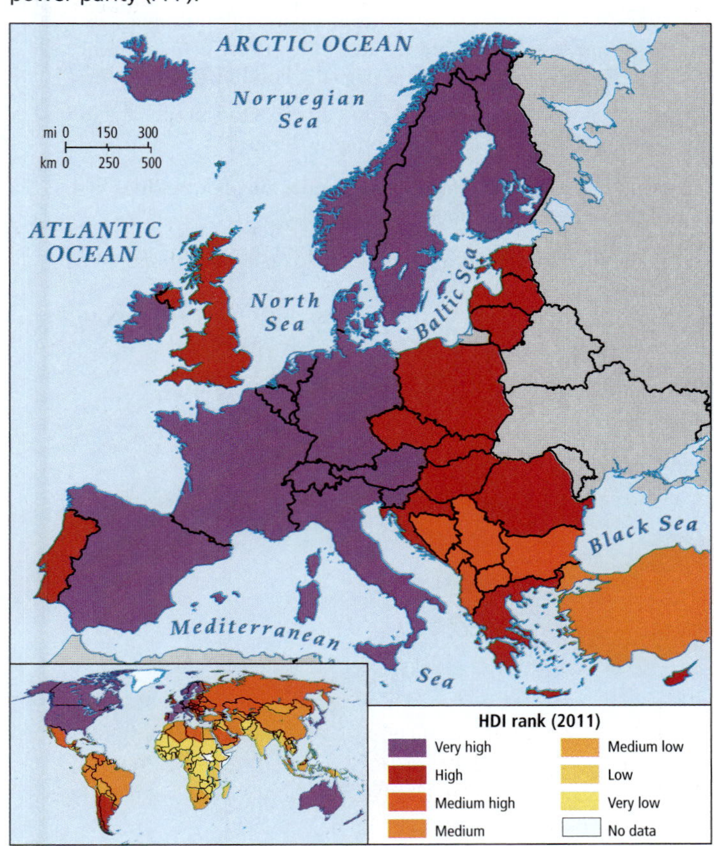

(B) Human Development Index (HDI).

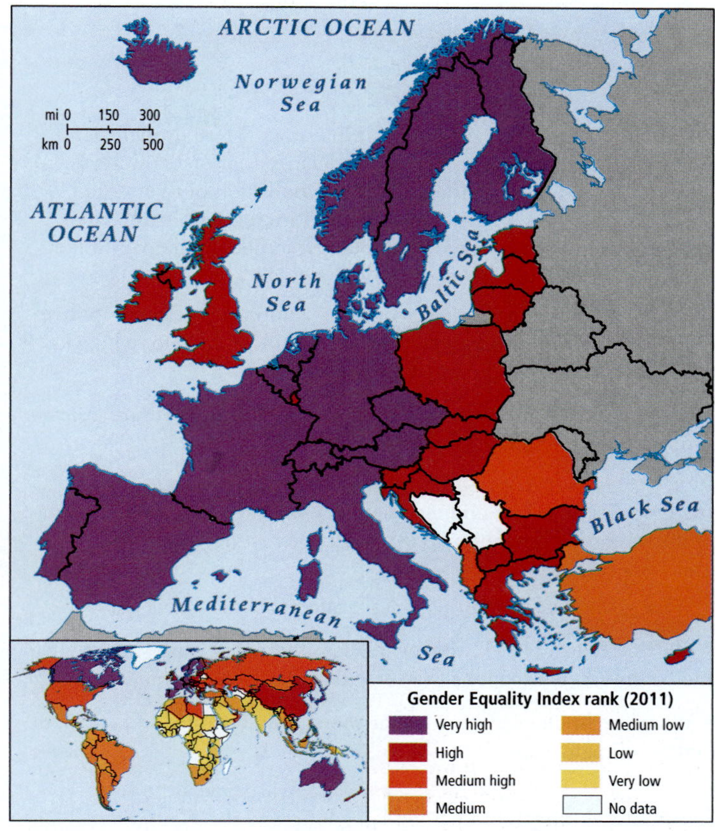

(C) Gender Equality Index (GEI).

THINGS TO REMEMBER *(continued)*

Geographic Insight 6

- **Population and Gender** Europeans are choosing to have fewer children; as a result, the population as a whole is aging. Small families are in part a result of women pursuing careers that require post-secondary education.
- Because Europe's population is not growing and the number of consumers is staying the same or decreasing, economies may contract over time. Some governments now encourage larger families, often to no avail, through generous maternity and paternity leave (up to 10 months with full pay), free day care, and other incentives.
- It is widely recognized that immigrants are necessary to Europe's future, as workers and as a way to counter aging populations. Yet Europeans are ambivalent about accepting people of other cultures into their midst. They are more willing to accept outsiders if they assimilate in all ways to European culture.
- Despite high levels of education, the political influence and economic well-being of European women lag behind those of European men; only a small percentage of women are in elected positions, and women have less access to jobs with high salaries.
- Europeans generally value social safety nets for all citizens, and standards of human well-being are generally high across Europe. Because Europeans vary in the types of systems they are willing to pay for, levels of well-being vary. All safety nets are currently being reevaluated.

Geographic Insights: Europe Review and Self-Test

1. Water: Water is an issue in multiple ways. More droughts and floods are likely with climate change. Transportation via water is widespread and energy efficient, but the pollution of rivers and seas, especially the Mediterranean Sea, is increasing. European consumption patterns use *virtual water* from parts of the world that have deficits of water.

• How does Europe's physical geography complicate problems of river and sea pollution? What could the European Union do to reduce its contribution to the pollution of Europe's seas?

• What products consumed in Europe entail the use of *virtual water* from places with deficits of water?

2. Climate Change: The European Union (EU) is a leader in the global response to climate change. Goals to cut greenhouse gas emissions are complemented by many other strategies to save energy and resources.

• What goals have governments set to address climate change?

• On a more local level, what evidence is there that European ways of life contribute less to global warming than North American ways?

• Which technological advances are being implemented in Europe to address climate change?

3. Globalization and Development: As European powers like Spain, Portugal, the United Kingdom, and the Netherlands conquered vast overseas territories, they created trade relationships that transformed Europe economically and culturally and laid the foundation for the modern global economy. Today, Europe is struggling to remain globally competitive.

• What was the original economic goal of colonialism?

• Describe in a general way the global map of European colonies.

• What are some of the ways that the colonial empire created by European countries affects life in Europe today?

• How do colonial relationships still affect modern trading patterns?

• Why is Europe now struggling to be competitive in global markets?

4. Urbanization, Power, and Politics: Rates of urbanization and democratization in Europe are linked, but both rates vary across the region. Northern Europe is the most urbanized (close to 80 percent) and has the most democratic participation. Levels of both urbanization and democratization fall to the south and east.

• How did the Industrial Revolution lead to urbanization in Europe?

• What is the connection between urbanization and democratization in Europe?

• What factors might explain the patterns of urbanization and democratization being higher in the north and west of Europe, and lower in the south and east?

5. Food: Concerns about food security in Europe have led to heavily subsidized and regulated agricultural systems. Subsidies to agricultural enterprises account for nearly 40 percent of the budget of the European Union. Although large-scale food production by agribusiness is now dominant, there is a revival of organic and small-scale sustainable techniques.

• What worries is the European Union addressing with its large investment in regulations of, and subsidies to, agricultural enterprises?

• Describe the characteristics and geographic distribution of agribusiness enterprises in Europe.

• To what would you attribute the revival of interest in small-scale sustainable agricultural practices?

• Why are EU subsidies to EU agriculture criticized by developing countries?

6. Population and Gender: Europe's population is aging, as working women are choosing to have only one or two children. Immigration by young workers is partially countering this trend.

• What are some of the potential consequences to Europe of aging populations and negative population growth rates?

• How do improvements in education and work opportunities for European women affect population growth?

• How do liberal immigration policies help an aging Europe? What about these policies worries Europeans?

Critical Thinking Questions

1. How do the issues of immigration in Europe compare with those in the United States? If you were a poor, undocumented immigrant searching for a way to support your family, would you choose the United States or Europe as a possible destination? Why?

2. What are three or four ways in which Europe is still linked to its former colonies?

3. How is the organization of urban space (housing, transportation, shopping) better suited to energy efficiency in Europe than in the United States?

4. What are some of the potential societal consequences of low birth rates in Europe?

5. How might the EU's status as the world's largest regional economy, and the rising importance of NATO, lead to changes in global power relationships?

6. What does the evolution of democracy in Europe suggest about current efforts to establish democracy in Iraq or Egypt?

7. Why might a pig farmer in West Africa criticize the EU's generous support for European agribusiness farmers?

8. Which European state-supported systems have offered women the greatest opportunities for employment outside the home?

9. Cite the evidence that Europe is seeking ways of life that contribute less to global warming.

10. Where is Europe's contribution to water-body pollution most severe? What could the European Union do to reduce its contribution to the pollution of Europe's oceans and seas?

Chapter Key Terms

assimilation 185
capitalism 171
central planning 171
Cold War 171
Common Agricultural Program (CAP) 178
communism 171
cultural homogenization 155
double day 187
economies of scale 175
euro 176
European Union (EU) 154
Green 164
guest workers 184
Holocaust 171
humanism 167
humid continental climate 157
Iron Curtain 171
Mediterranean climate 157
mercantilism 168
nationalism 170
North Atlantic Drift 157
North Atlantic Treaty Organization (NATO) 177
Roma 171
Schengen Accord 184
social welfare (in the European Union, **social protection**) 189
subsidies 178
temperate midlatitude climate 157
welfare state 171

A North European Plain, Belarus

B Caucasus Mountains, Georgia

C Volga River, Russia

D Urals, Russia

5 Russia and the Post-Soviet States

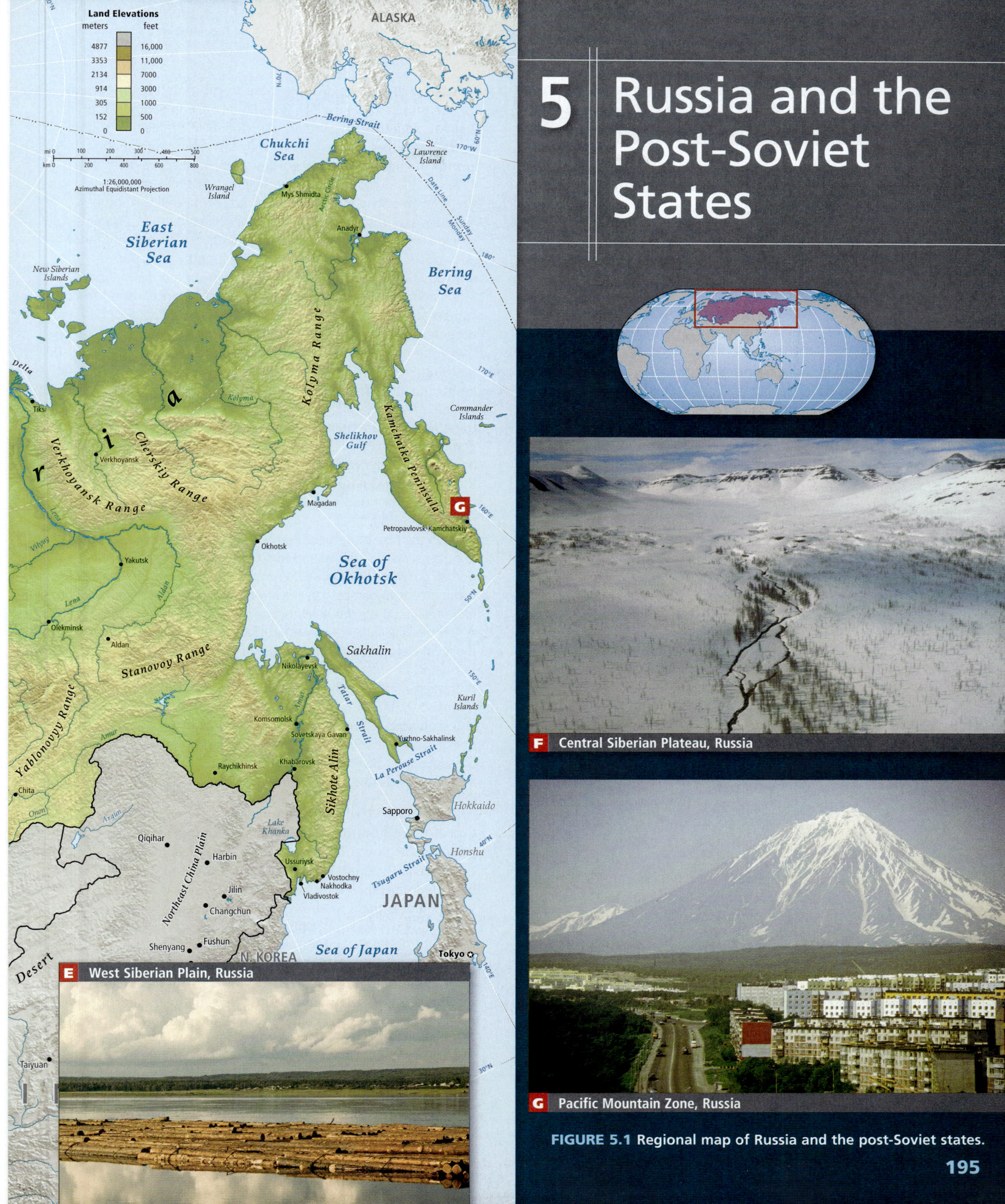

E West Siberian Plain, Russia

F Central Siberian Plateau, Russia

G Pacific Mountain Zone, Russia

FIGURE 5.1 Regional map of Russia and the post-Soviet states.

GEOGRAPHIC INSIGHTS: RUSSIA AND THE POST-SOVIET STATES

After you read this chapter, you will be able to discuss the following geographic insights as they relate to the nine thematic concepts:

1. Environment, Development, and Urbanization: Development and urbanization in this region have long taken precedence over environmental concerns. The region has severe environmental problems, especially in urban areas.

2. Climate Change, Food, and Water: Food production systems and water resources are particularly threatened by climate change in parts of this region, yet the region remains the third-largest global producer of greenhouse gases.

3. Globalization and Development: Since the fall of the Soviet Union, economic reforms and globalization have changed patterns of development in this region. Many Communist-era industries closed, and as jobs were lost and public assets sold to wealthy oligarchs, wealth disparity increased. The region is now dependent on its role as a leading global exporter of energy resources.

4. Power and Politics: Obstacles to democratization include a long history of authoritarian regimes and the popular desire to implement change rapidly, combined with the lack of mechanisms for public participation in the political process.

5. Population: Populations are shrinking in many parts of this region. This is a function of a continuously high level of participation by women in the workforce and overall poor living and housing standards.

6. Gender: Men and women have been differently affected by the massive economic and political changes in this region during the twentieth century. Women have slowly gained in economic power, while, on average, men have been adversely affected by economic instability, loss of jobs, and alcoholism.

The Russian and Post-Soviet States Region

Russia and the post-Soviet states (see Figure 5.1 on pages 194–195) make up a large portion of Eurasia, the world's largest continent. Despite its vastness, the region is not highly populated and has a climate that in many places is unforgiving. The nine thematic concepts are explored as they arise in the discussion of regional issues, with interactions between two or more themes featured, as in the geographic insights above. Vignettes illustrate one or more of the themes as they are experienced in individual lives.

GLOBAL PATTERNS, LOCAL LIVES It's a cold, snowy day in Moscow, the capital of the Russian Federation, in February 2012. Tens of thousands of protesters hold hands to form an enormous circle around the city, along one of the traffic-clogged ring roads that the capital is known for. National elections are coming up soon, and the demonstrators are trying to defeat the favorite for the presidential post, Vladimir Putin. Marina Segupova, a 28-year-old interior decorator, participates in the demonstration, which she hopes will spur further protest. Using a metaphor apt for the moment, she describes the protest movement: "We are a snowball and we are rolling." At the same time she acknowledges that change will not come easily. Another middle-class professional, the 56-year-old accountant Galina Venediktova, came to voice her opinion about corruption and poor public services. During that same month, other forms of protest were less conventional. The feminist punk band Pussy Riot staged an unauthorized "guerilla performance" at a Moscow cathedral, in which they decried the policies of Vladimir Putin and became a worldwide media sensation.

Union of Soviet Socialist Republics (USSR) the multinational union formed from the Russian empire in 1922 and dissolved in 1991; commonly known as the *Soviet Union*

Soviet Union see **Union of Soviet Socialist Republics**

Putin had already served 8 years as president, from 2000 to 2008. Prevented by the constitution from running again, he had an ally become president for 4 years, after which Putin staged a "comeback" that passed constitutional muster. Despite the efforts of Segupova, Venediktova, Pussy Riot, and other demonstrators, Putin is again serving as president of Russia after having decisively won the March 2012 election. The outcome of the election was hardly in doubt, due to the lack of other serious contenders for the presidency and Putin's dominance in state media reporting.

On the face of it, these events suggest that the political status quo has been maintained and that power is still concentrated within a small group of politically elite people who have a questionable commitment to democracy. However, the outburst of demonstrations indicates that the political landscape is slowly changing in Russia. A few months earlier, a contested election to Russia's parliament, the Duma, also resulted in widespread protests, as many felt that the victory of Putin's party was questionable. After Putin's election, protests continued in the streets and squares of central Moscow (**Figure 5.2**). In 2012, three members of Pussy Riot were put on trial and sentenced to prison for "hooliganism," which has reinforced the suspicion among many that political dissent in Russia will have consequences. *[Sources: Christian Science Monitor, New York Times, and Sky News. For detailed source information, see Text Credit pages.]* ■

As Russia slowly becomes wealthier, its people increasingly expect more from its democratic system, especially the growing middle class, the urbanites in large cities—such as Moscow and St. Petersburg—and the country's youth. But so far, politicians like Vladimir Putin have the support of the majority of Russians, who perceive them as strong leaders. The elderly and people in hardscrabble industrial cities, small towns, and rural areas view Putin as a force for stability, which is favored by many, considering the political turbulence Russia has had during the last two decades. Putin being reelected also has larger global geopolitical ramifications, as the relationship between Russia and the West sometimes involves suspicion on each side and sometimes, when it suits both regions, is one of cooperation.

Russia is the largest country in a region that has changed its political and economic systems entirely in a short period of time. Barely two decades ago, the **Union of Soviet Socialist Republics (USSR)**, more commonly known as the **Soviet Union**, was the largest political

FIGURE 5.2 Political demonstration in Moscow. Riot police in Moscow detain journalist and civil rights activist Alexandr Podrabinek during a rally protesting the inauguration of Vladimir Putin as president of Russia in 2012. This region is transitioning from a political culture in which authority was rarely questioned to one in which protests are becoming increasingly common. Nevertheless, many leaders remain highly authoritarian in their leadership style.

Thinking Geographically

Where in Russia are people more likely to support the policies of Vladimir Putin?

FIGURE 5.3 Political map of Russia and the post-Soviet states.

unit in the world, stretching from Central Europe to the Pacific Ocean. It covered one-sixth of Earth's land surface. In 1991, the Soviet Union broke apart, ending a 70-year era of nearly complete governmental control of the economy, society, and politics. Over the course of a few years, attempts were made to substitute the Communist Party's economic control with capitalist systems similar to those of Western countries, based on competition among private businesses. The transition has proven difficult.

Politically, the Soviet Union has been replaced by a loose alliance of Russia and 11 independent post-Soviet states—the European states of Ukraine, Belarus, and Moldova; the Caucasian states of Georgia, Armenia, and Azerbaijan; and the Central Asian states of Kazakhstan, Kyrgyzstan, Tajikistan, Turkmenistan, and Uzbekistan (Figure 5.3). Three former Soviet republics, the Baltic states of Lithuania, Latvia, and Estonia, are now part of the European Union (see Chapter 4). Russia, which was always the core of the Soviet Union, remains predominant in this alliance (which is formalized in an organization called the Commonwealth of Independent States) and is influential in the world because of its size, population, military, and huge oil and gas reserves.

Geopolitically, this region is still sorting out its relationships. The Cold War between the Soviet Union and the United States and its allies is over. Most former Soviet allies in Central Europe have already joined the European Union, and some other countries in this region may eventually do the same. Trade with Europe, especially in oil and gas, is booming, if fraught with conflict. In the future, the Central Asian states allied with Russia may align themselves with neighbors in Southwest Asia or South Asia (see Chapter 1, pages 9–12; see also Figure 1.6 on page 11). The far eastern parts of Russia—and Russia as a whole—are already finding common trading ground with East Asia and Oceania (Figure 5.4 on page 198).

After 70 years of authoritarian rule, elections are becoming the norm throughout this region. Whether these are actually free and fair elections is debatable, as opposition candidates are increasingly marginalized by lack of access to both print and broadcast media. Economic and political instability have led many voters to support strong leaders who tolerate little criticism. An explosion of crime and corruption has many wondering if the new democracies in this region are strong enough to endure.

THINGS TO REMEMBER

- The former Soviet Union has been replaced by a loose alliance of Russia and 11 post-Soviet states.
- Elections have become common in the region, although whether they are free and fair is uncertain. Democracies in the region are also threatened by crime and corruption.
- A process of geopolitical realignment is taking place, in which some post-Soviet states seek closer relations with Europe, while others are developing ties with nearby Asian countries.

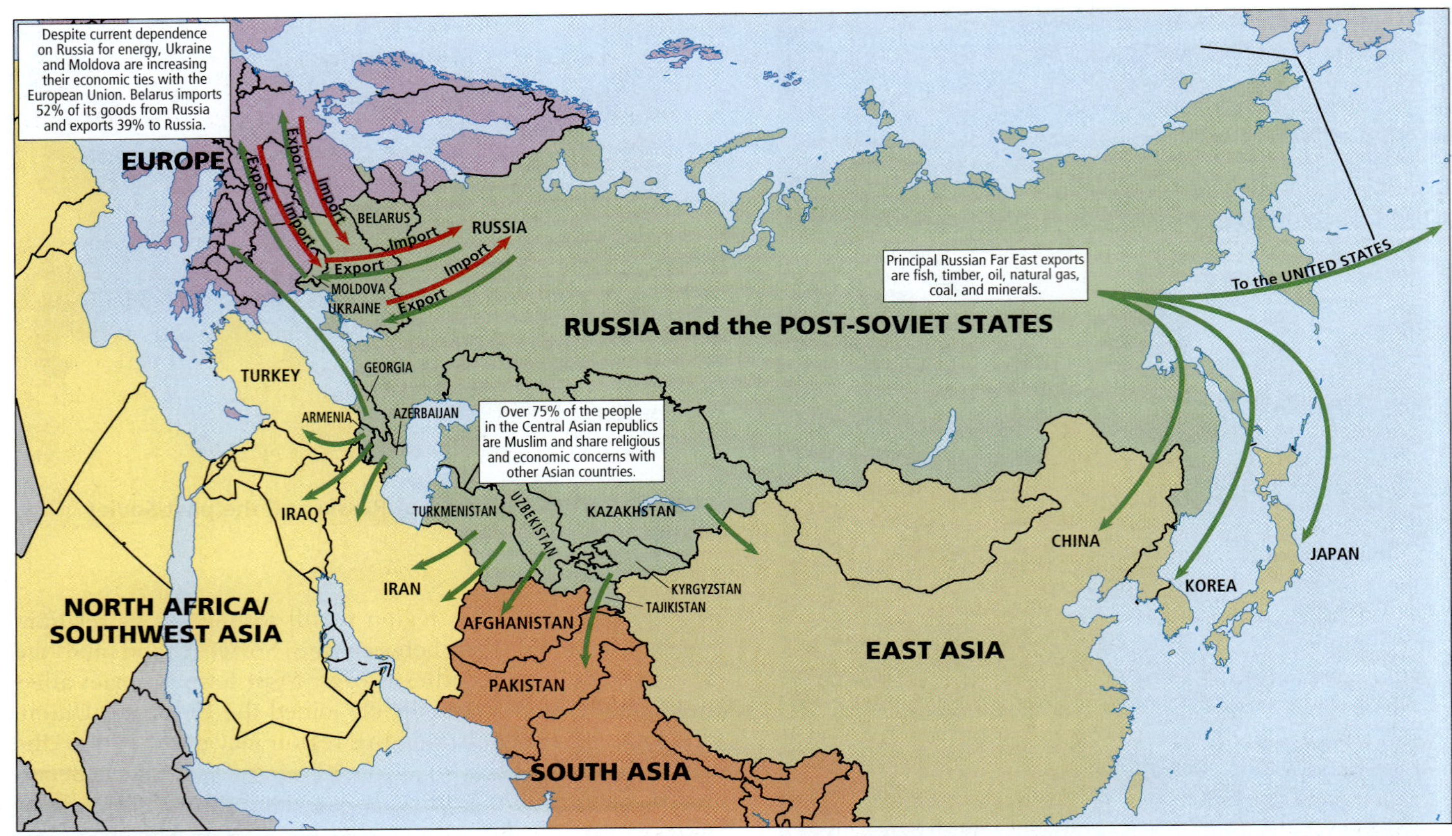

FIGURE 5.4 Russia and the post-Soviet states: Contacts with other world regions. The region of Russia and the post-Soviet states is in flux as all of the component countries rethink their geopolitical positions relative to one another and to adjacent regions. The arrows indicate various types of contact, ranging from economic to religious and cultural.

THE GEOGRAPHIC SETTING

What Makes Russia and the Post-Soviet States a Region?

Russia and the post-Soviet states are today associated as a region primarily because of their history from the nineteenth century onward. Imperial Russia underwent a convulsive revolution in 1917 that resulted in a new political and economic experiment with socialism that lasted until 1991. The leaders of this revolution continued with the imperial practice of territorial expansion that ultimately enveloped countries from the Baltic Sea to the Pacific and from the Arctic Sea to the mountains of Caucasia and southern Central Asia (see the Figure 5.1 map). Control of the economic and political systems of what eventually became known as the Soviet Union was maintained through authoritarian control reinforced with militarism. After 1991, central control was relaxed and the region is presently undergoing realignment. The Baltic republics have joined the European Union, and the Central Asian republics are seeking a redefinition of their association with Russia (see Figure 5.4).

Russian Federation Russia and its political subunits, which include 21 internal republics

Terms in This Chapter

There is no entirely satisfactory new name for the former Soviet Union. In this chapter we use *Russia and the post-Soviet states*. One reason is that economic, political, and social developments from the Soviet days still shape these countries today. Russia continues to be closely associated with these states economically, but they are independent countries, and their governments are legally separate from Russia's. Russia itself is formally known as the **Russian Federation** because it includes more than 21 (mostly) ethnic *internal republics* (also called *semiautonomous regions*)—such places as Chechnya, Ossetia, Tatarstan, and Buryatya—which all together constitute about one-tenth of the Federation's territory and one-sixth of its population. *Federation* should not be taken to mean that the internal republics share power equally with the central Russian government in Moscow.

Physical Patterns

The physical features of Russia and the post-Soviet states vary greatly over the huge territory they encompass. The region bears some resemblance to North America in size, topography, climate, and vegetation. Even if Russia is only about three-quarters the size of the Soviet Union, it is still the largest country in the world, nearly twice the size of the second-largest country, Canada.

Landforms

Because the region is relatively complex physically, a brief summary of its landforms is useful (see the Figure 5.1 map). Moving west to east, there is first the eastern extension of the North European Plain, including the plain of the Volga River (see Figure 5.1A, C), followed by the Ural Mountains (see Figure 5.1D). Russia's territories east of the Ural Mountains are usually called **Siberia**, and include the West Siberian Plain (see Figure 5.1E), followed by an upland zone called the Central Siberian Plateau (see Figure 5.1F), and finally, in the Far East, a series of mountain ranges bordering the Pacific (see Figure 5.1G). To the south there are mountains and uplands (the Caucasus, depicted in Figure 5.1B, and Central Asia, not pictured) as well as semiarid grasslands, or **steppes** (in western Central Asia, not pictured).

The eastern extension of the North European Plain rolls low and flat from the Carpathian Mountains in Ukraine and Romania, 1200 miles (about 2000 kilometers) east to the Ural Mountains (see Figure 5.1A, C, D). The part of Russia west of the Urals is often called *European Russia* because the Ural Mountains are traditionally considered part of the indistinct border between Europe and Asia. European Russia is the most densely settled part of the entire region (see Figure 5.20 on page 222) and its agricultural and industrial core. Its most important river is the Volga, which flows into the Caspian Sea. The Volga River and its tributaries form a major transportation route that connects many parts of the Russian North European Plain, including Moscow, to St. Petersburg and the Baltic and White seas in the north and to the Black and Caspian seas in the south (see Figure 5.1C).

The Ural Mountains extend in a fairly straight line south from the Arctic Ocean into Kazakhstan (see the Figure 5.1 map). The Urals, a low-lying range similar in elevation to the Appalachians, are not much of a barrier to humans and are only partially a barrier to nature (some European tree species do not extend east of the Urals). There are several easy passes across the mountains, and winds carry moisture all the way from the Atlantic into Siberia. Much of the Urals' once-dense forest has been felled to build and fuel industrial cities.

The West Siberian Plain, east of the Urals, is the largest wetlands plain in the world (see Figure 5.1E). A vast, mostly marshy lowland, about the size of the eastern United States, drains toward the north into the Arctic Ocean. Long, bitter winters mean that in the northern half of this area, a layer of permanently frozen soil (**permafrost**) lies just a few feet beneath the surface. Permafrost is formed when the ground warms up during the short summer, but the top layer of material insulates against this warming effect, leaving the subsurface always frozen. In the far north, the permafrost comes to within a few inches of the surface. Because water does not percolate down through this frozen layer, surface water accumulates into wetlands that form above the permafrost layer in the summer months, providing habitats for many migratory birds. In the far north lies a treeless area called the **tundra**, where only mosses and lichens can grow because of the extreme cold, the shallow soils, and the permafrost. The West Siberian Plain is one of the world's largest oil and natural gas producers, although its harsh climate and permafrost make extraction difficult.

The Central Siberian Plateau and the Pacific Mountain Zone, farther to the east, together equal the size of the United States (see Figure 5.1F, G). Permafrost prevails except along the Pacific coast. There, the ocean moderates temperatures; the many active volcanoes were and are created as the Pacific Plate sinks under the Eurasian Plate. Lightly populated places like the Kamchatka Peninsula, Sakhalin Island, and Sikhote-Alin on Russia's southeastern Pacific coast are havens for wildlife.

To the south of the West Siberian Plain, steppes and deserts stretch from the Caspian Sea to the Chinese border. To the west of these grasslands are the Caucasus Mountains (see Figure 5.1B) and to the southeast facing China and Mongolia are a series of other mountain chains, including Pamir and Tien Shan. The rugged terrain has not deterred people from crossing these mountains. From the Caucasus to the Pamir and Tien Shan, for tens of thousands of years people have exchanged plants (apples, onions, citrus, rhubarb, wheat), animals (horses, sheep, goats, cattle), technologies (cultivation, animal breeding, portable shelter construction, rug and tapestry weaving), and religious belief systems (principally Islam and Buddhism, but also Christianity, Hinduism, and Judaism).

Climate and Vegetation

The climates and associated vegetation in this large region are varied but are less so than in other regions because so much territory here is taken up by expanses of midlatitude grasslands and by northern forest and tundra that are cold much of the year. No inhabited place on Earth has as harsh a climate as the northern part of the Eurasian landmass occupied by Russia and, particularly, Siberia (shown in the map in Figure 5.5 on page 200). Winters are long and cold, with only brief hours of daylight. Summers are short and cool to hot, with long days. Precipitation is moderate, coming primarily from the west. In the northernmost areas, the natural vegetation is tundra grasslands. The major economic activities here are the extraction of oil, gas, and some minerals, as well as reindeer herding by the local indigenous population. Just south of the tundra lays a vast, cold-adapted coniferous forest known as **taiga** that stretches from northern European Russia to the Pacific (and from a global perspective, this coniferous belt also includes much of Alaska, Canada, and Scandinavia). The largest portion of taiga lies east of the Urals; and here forestry—often unrestrained by ecological concerns—is a dominant economic activity. The short growing season and the large areas of permafrost generally limit crop agriculture, except in the southern West Siberian Plain, where people grow grain.

Because massive mountain ranges to the south block access to warm, wet air from the Indian Ocean, most rainfall in the entire region comes from storms that blow in from the Atlantic Ocean far to the west (see the Figure 5.5 map). But by the time these air masses arrive, most of their moisture has been squeezed out over Europe. A fair amount of rain does reach Ukraine, Belarus, European Russia (see Figure 5.5A), and the Caucasian republics; these regions are especially important food production (vegetables, fruits, and grain) areas. The natural vegetation

Siberia a region of Russia that is located east of the Ural Mountains

steppes semiarid, grass-covered plains

permafrost permanently frozen soil that lies just a few feet beneath the surface

tundra a treeless area, between the ice cap and the tree line of arctic regions, where the subsoil is permanently frozen

taiga subarctic coniferous forests

Figure 5.5 Photo Essay: Climates of Russia and the Post-Soviet States

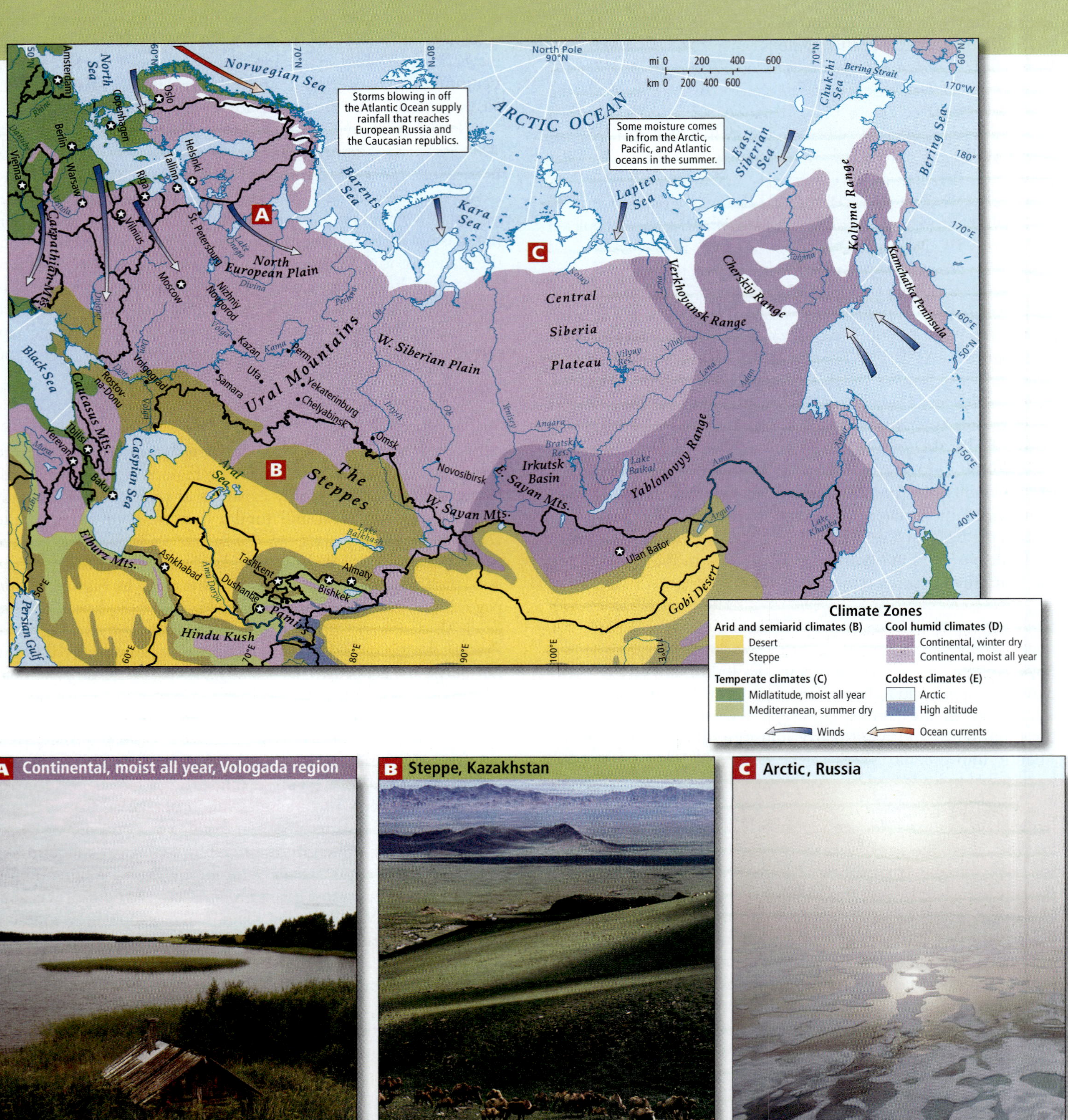

A Continental, moist all year, Vologada region

B Steppe, Kazakhstan

C Arctic, Russia

in these western zones is open woodlands and grassland, though in ancient times, forests were common.

East of the Caucasus Mountains, the lands of Central Asia have semiarid to arid climates (see Figure 5.5B) influenced by their location in the middle of a very large continent. The summers are scorching and short, the winters intense. In the desert zones, daytime-to-nighttime temperatures can vary by 50°F (28°C) or more. Northern Kazakhstan produces grain and grazing animals. The more southern areas (southern Kazakhstan, Uzbekistan, and Turkmenistan) have grasslands (steppes), which are also used for herding. In modern times, these grasslands have been used for irrigated commercial agriculture that is dependent upon glacially fed rivers, but most land is not useful for farming. The climates in the more mountainous area where Kazakhstan, Uzbekistan, Tajikistan, and Kyrgyzstan meet are varied and support a number of small-scale agricultural activities, some of them commercial.

The construction of land transportation systems has long been held back by the climate of the region. In the north especially, it is difficult to build during the long, harsh winters, which eventually give way to a spring period called the *rasputitsa*, or the "quagmire season," when melting permafrost (or, to a lesser extent, autumn rains) turns many roads and construction sites into impassable mud pits. Huge distances between populated places as well as the complex topography, especially in Siberia, add to the problem. As a result, few roads or railroads have ever been built outside western Russia (see Figure 5.15 on page 215).

Environmental Issues

Soviet ideology held that nature was the servant of industrial and agricultural progress, and that humans should dominate nature on a grand scale. While this sentiment was common throughout much of the world at the time of the formation of the USSR, it does seem to have been taken further here than elsewhere. Joseph Stalin, leader of the USSR from 1922 to 1953 and a major architect of Soviet policy, is famous for having said, "We cannot expect charity from Nature. We must tear it from her." During the Soviet years, huge dams, factories, and other industrial facilities were built without regard for their effect on the environment or on public health. Russia and the post-Soviet states now have some of the worst environmental problems in the world. By 2000, more than 35 million people in the region (15 percent of the population) were living in areas where the soil was poisoned and the air was dangerous to breathe (Figure 5.6A–C on page 202).

Since the collapse of the Soviet Union, the region's governments, beset with myriad problems, have been both reluctant to address environmental issues and incapable of doing so. As one Russian environmentalist put it, "When people become more involved with their stomachs, they forget about ecology." Pollution controls are complicated by a lack of funds and by an official unwillingness to correct past environmental abuses. Figure 5.6 shows just a few examples of human impacts on the region's environment, most of which are related to ongoing industrial pollution (as in Norilsk, in Figure 5.6A, B) or failure to properly dismantle contaminated industrial sites made obsolete when a more market-based economy was introduced (as in Pripyat, Ukraine, in Figure 5.6C). The extraction and sale of oil and gas on the global market is now a major source of income for Russia and several of the post-Soviet states, yet very little attention is being paid even today to the environmental impact of this activity (see Figure 5.6D).

Urban and Industrial Pollution

Geographic Insight 1

Environment, Development, and Urbanization: Development and urbanization in this region have long taken precedence over environmental concerns. The region has severe environmental problems, especially in urban areas.

Urban and industrial pollution was ignored during Soviet times as cities expanded quickly—with workers flooding in from the countryside—to accommodate the new industries that often generated lethal levels of pollutants. During the authoritarian Soviet regime, citizens' concerns about their living environment were suppressed, and this political environment did not allow for protests or an environmental movement as in North America and western Europe.

It is often difficult to link urban pollution directly to health problems because the sources of contamination are multiple and difficult to trace. Such **nonpoint sources of pollution** include untreated automobile exhaust, raw sewage, and agricultural chemicals that drain from fields into water supplies. In all urban areas of the region, air pollution resulting from the burning of fossil fuels is skyrocketing as more people purchase cars and as the industrial and transport sectors of the economy continue to grow.

Some cities were built around industries that produce harmful by-products. The former chemical weapons–manufacturing center of Dzerzhinsk has been listed by the nonprofit Blacksmith Institute as one of the ten most polluted cities in the world. It is competing with the city of Norilsk, where much of the vegetation has been killed off around the city's metal smelting complex—the largest facility of its kind in the world (see Figure 5.6B). Norilsk was the most polluted city in Russia as recently as 2011.

The Globalization of Nuclear Pollution

Russia and the post-Soviet states are also home to extensive nuclear pollution, and its effects have spread globally. The world's worst nuclear disaster occurred in Ukraine in 1986, when the Chernobyl nuclear power plant exploded. The explosion severely contaminated a vast area in northern Ukraine, southern Belarus, and Russia. It spread a cloud of radiation over much of Central Europe, Scandinavia, and eventually the entire planet. In the area surrounding Chernobyl, more than 300,000 people were evacuated from their homes, and the town of Pripyat, which used to be home to 50,000 people, has been an abandoned ghost town since the disaster (see Figure 5.6C). The ultimate health effects are impossible to assess exactly. The United Nations estimates that 6000 people developed cancer due to the accident, although many believe the number to be much higher.

278. SAFETY AT THE CENTER OF NUCLEAR POWER OPERATIONS WORLDWIDE

nonpoint sources of pollution diffuse sources of environmental contamination, such as untreated automobile exhaust, raw sewage, and agricultural chemicals, that drain from fields into water supplies

Figure 5.6 Photo Essay: Human Impacts on the Biosphere in Russia and the Post-Soviet States

The region is home to some of the worst pollution on the planet. More than 70 years of industrial development with few environmental safeguards have wreaked havoc on ecosystems. Industries are attempting to become cleaner, but decades may pass before significant improvements are realized.

A A cloud of pollution from a nearby nickel smelter hangs over the city of Norilsk. The smelter emits about 1 percent of all global emissions of sulfur dioxide, which helps form acid rain. Enough heavy metals have accumulated in the soils around Norilsk that they can now be mined commercially. The company that runs the smelter has promised improvements but estimates that there will not be any significant changes in emissions until between 2015 and 2020.

B This false-color satellite image shows Norilsk. Purple indicates areas where vegetation has been killed by pollution. The city is located just below the main lake seen in the upper part of the image.

mi 0 250 500 750 1000
km 0 250 500 750 1000

Human Impact, 2002

Land cover: Forests; Grasslands; Deserts; Tundra; Ice

Human impact on land: High impact; Medium–high impact; Low–medium impact

C The abandoned city of Pripyat in Ukraine, once home to 50,000 people, was evacuated after the explosion of the Chernobyl nuclear power plant.

Chernobyl nuclear power plant

D Oil production in the Russian Arctic. With such a short growing season, many arctic environments are extremely fragile. Some of the largest oil spills ever have happened here, and have caused severe damage to ecosystems and to the indigenous people who depend on them.

Thinking Geographically

After you have read about the human impact on the biosphere in Russia and the post-Soviet states, you will be able to answer the following questions:

C Why was Pripyat evacuated?

D In addition to oil pollution, from what other form of pollution has the arctic portion of this region suffered?

When the Soviet Union collapsed, many of the post-Soviet states inherited nuclear facilities and even nuclear weapons. One such country is Kazakhstan (see Figure 5.3). Remote areas of eastern Kazakhstan used to serve as a testing ground for Soviet nuclear devices. The residents were sparsely distributed and no one took the trouble to protect them from nuclear radiation. A museum in Kazakhstan now displays the preserved remains of hundreds of deformed human fetuses and newborns. On the positive side, Kazakhstan has disarmed the nuclear warheads that it inherited from the Soviet Union and has worked with U.S. authorities to safeguard remaining nuclear weapons material. Kazakhstan plans to use its nuclear technology to become a global player in the nuclear energy sector. The country is already the single biggest producer of uranium in the world, generating 35 percent of the global production total. In addition to mining and exporting the raw material, Kazakhstan also wants to produce nuclear fuel both for global markets and domestic nuclear power plants, and develop commercial repositories for radioactive waste from other countries. However, the global appetite for nuclear expansion after the 2011 Fukushima accident in Japan remains uncertain, and no reliably safe system has yet been found for storing nuclear waste until it is no longer radioactive. Environmental and human rights NGOs have raised strong opposition, pointing to the region's poor environmental track record. Kazakhstan's population may also be suspicious of nuclear power because they have lived for a long time with land poisoned by radioactive waste.

The Globalization of Resource Extraction and Environmental Degradation

Russia and the post-Soviet states have considerable natural and mineral resources (see Figure 5.14 on page 213). Russia itself has the world's largest natural gas reserves, major oil deposits, and forests that stretch across the northern reaches of the continent. Russia also has major deposits of coal and industrial minerals such as iron ore and nickel. The Central Asian states share substantial deposits of oil and gas, which are centered on the Caspian Sea and extend east toward China. The value of all these resources is determined in the global marketplace, and their extraction affects the global environment.

Despite obvious environmental problems, cities like Norilsk (see Figure 5.6A), which sits on huge mineral deposits, continue to attract investment crucial to the new Russian economy. The area is rich in nickel—used in steel and other industrial products—and other minerals, such as copper. Norilsk Nickel, a company that was privatized at the end of communism and is now the largest producer of nickel in the world, dominates the city. Because the Russian government considers many minerals to be a strategic resource for the country, it limits foreign ownership in the industry. Partly because of that, and because of its extensive and profitable home base, Russian-controlled Norilsk Nickel has globalized, despite its poor environmental record, by purchasing mining operations in Australia, Botswana, Finland, the United States, and South Africa. The company employs over 80,000 people worldwide.

ON THE BRIGHT SIDE

Controlling Nuclear Material

In the post-9/11 world, concern spread that military corruption in this region could put nuclear weapons from the former Soviet arsenal into the hands of terrorists. After huge military-funding cuts, weapons, uniforms, and even military rations were routinely sold on the black market. In 2007, Russian smugglers were caught trying to sell nuclear materials on the black market. Recent developments are more encouraging. In an effort to fight corruption and the temptation to sell nuclear materials in the black market, the Russian government has given impoverished military personnel long-delayed pay raises or termination compensation. In addition, all countries in the region are now cooperating with the International Atomic Energy Agency in controlling nuclear material (see the discussion on page 201).

Water Issues

Water was once assumed to be inexhaustible and usable for endless purposes, while needing little maintenance. Now most countries in this region recognize it as a vital resource.

Rivers, Irrigation, and the Loss of the Aral Sea To the south of Russia in the Central Asian states, glacially fed rivers have long served an irrigation role, supporting commercial cotton agriculture (see the discussion of the effects of global climate change on glaciers in Chapter 1, page 42). For millions of years, the landlocked Aral Sea, once the fourth-largest lake in the world, was fed by the Syr Darya and Amu Darya rivers, which bring snowmelt and glacial melt from the mountains in Kyrgyzstan and other nearby countries. Water was first diverted from the two rivers in the 1960s when the Soviet leadership ordered its use to irrigate millions of acres of cotton in naturally arid Kazakhstan and Uzbekistan. So much water was consumed by these projects that within a few years, the Aral Sea had shrunk measurably (Figure 5.7A on page 204). By the early 1980s, no water at all from the two rivers was reaching the Aral Sea, and by 2001, the sea had lost 75 percent of its volume and had shrunk into three smaller lakes. The region's huge fishing industry, on which the entire USSR population depended for some 20 percent of their protein, died out due to increasing water salinity. Once-active port cities were marooned many kilometers from the water. The decline and disappearance of the Aral Sea has been described as the largest human-made ecological disaster in history.

The shrinkage of the Aral Sea has caused many human health problems. Winds that sweep across the newly exposed seabed pick up salt and chemical residues, creating poisonous dust storms. At the southern end of the Aral Sea in Uzbekistan, 69 percent of the people report chronic illnesses caused by air pollution, a lack of clean water, and underdeveloped sanitation.

Efforts to increase water flows to the Aral Sea have been hampered by politics because the now-independent countries of Uzbekistan and Kazakhstan share the sea. Uzbekistan, in particular,

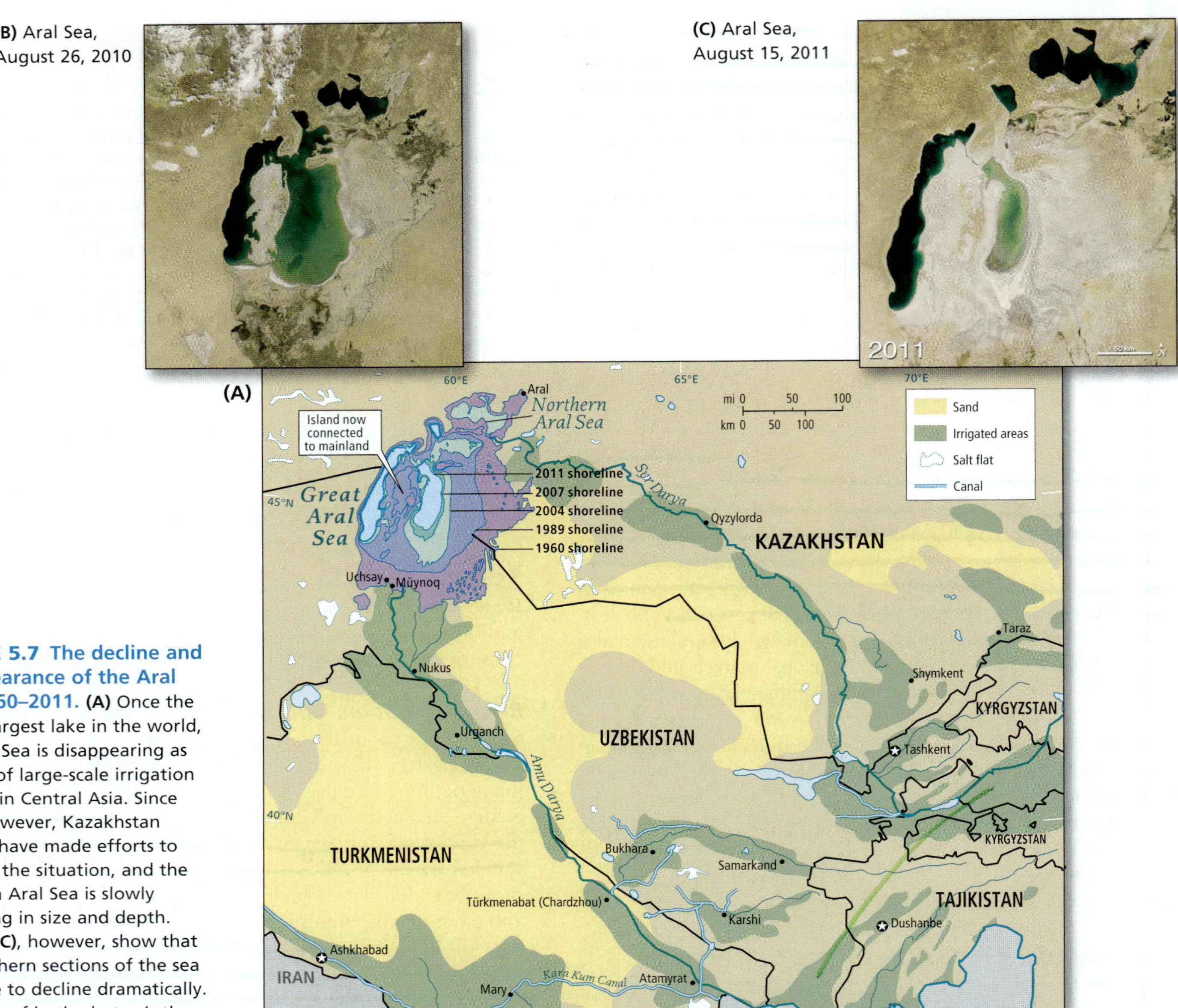

FIGURE 5.7 The decline and disappearance of the Aral Sea, 1960–2011. **(A)** Once the fourth-largest lake in the world, the Aral Sea is disappearing as a result of large-scale irrigation projects in Central Asia. Since 2005, however, Kazakhstan officials have made efforts to improve the situation, and the northern Aral Sea is slowly increasing in size and depth. **(B)** and **(C)**, however, show that the southern sections of the sea continue to decline dramatically. The scale of both photos is the same.

wants to continue massive irrigation programs, which have made it the world's seventh-largest cotton grower. Nevertheless, some restorative actions have been taken. Kazakhstan, which relies on oil more than cotton for income, built a dam in 2005 to keep fresh water in the northern Aral Sea. By the following year, the water had risen 10 feet, and the fish catch was improving. Kazakh fishers note that there is now more open public debate about what to do next regarding the Aral Sea. In the southern Aral Sea, however, water levels remain very low (see Figure 5.7B, C). In Uzbekistan, for instance, where 44 percent of the workforce is employed in agriculture, the limitation of irrigation has forced economic diversification, and cotton production now makes up just 11 percent of the country's exports, far less than the 45 percent it represented in 1990. **107. DYING SEA MAKES COMEBACK**

Climate Change

Geographic Insight 2

Climate Change, Food, and Water: Food production systems and water resources are particularly threatened by climate change in parts of this region, yet the region remains the third-largest global producer of greenhouse gases.

Climate change is an issue in this region for many reasons. While high levels of CO_2 and other greenhouse gases are produced here, Russia's forests also play a major role in absorbing global CO_2. Also, portions of the region have a medium level of vulnerability to the negative effects of a changing climate (shown on the map in Figure 5.8 on page 206), but Central Asia, which is prone to drought and water scarcity, has high to extreme levels of vulnerability to climate change effects.

Although Russia has reduced its CO_2 and other greenhouse gas emissions in recent years, it is still a major contributor of such emissions (see Figure 1.23 on page 41). In fact, Russia has the third-highest rate of greenhouse gas emissions in the world. It is lagging in the use of energy-efficient technologies and many wasteful practices are common, such as burning off or "flaring" of natural gas that comes to the surface when oil wells are drilled. The results are higher greenhouse gas emissions per capita when compared to European countries; however, they are not as high as those in the United States. Russia has shown some willingness to limit its own emissions by, for example, signing international treaties, such as the Kyoto Protocol and the Copenhagen Accord, to reduce greenhouse gas emissions. Such treaties, however, use 1990 as a baseline for emissions. Russia's compliance with international targets to reduce greenhouse emissions is largely an outcome of the country's post-1990 economic decline after the end of communism. Lower levels of industrial output have meant lower emissions.

The dependence in Central Asia on irrigated agriculture using water from rivers that are fed by glaciers in the Pamirs and other mountains in Kyrgyzstan and Tajikistan (see Figure 5.8), leaves the region vulnerable to global climate change. In recent decades, the glaciers have shrunk because of more extreme temperature ranges and decreases in rain and snowfall. If the glaciers have a decreased capacity to act as water-storage systems that supply melted ice flow to the rivers, the rivers may run dry during the summer, when irrigation is most needed for growing crops. Because most rain falls in the mountains in the winter and spring, Central Asia's agricultural systems would either have to adapt to a spring growing season (winters are too cold), or farmers would have to store water for use in the summer. Either proposition demands complex and expensive changes on a massive scale.

ON THE BRIGHT SIDE

Better Times Ahead?

While environmental pollution remains serious throughout this region, there may be reasons for optimism in the post-Soviet era. New patterns of economic development could bring better living standards, especially for urban residents who could pressure governments for higher environmental standards. Moreover, while freedom of expression and the press are still limited throughout the region, they are at least better than they were in Soviet times. This opens up the opportunity to make use of the media to expose the worst cases of pollution and environmental negligence. For example, in Sochi, Russia, a Black Sea resort town long cherished for its natural amenities, numerous environmental protests about the high levels of urban pollution have gained global attention in recent years. Such attention will probably increase, as Sochi will host the 2014 Winter Olympics. Similarly, much of the information presented above has come to light as a result of greater press freedom since the collapse of the Soviet Union.

THINGS TO REMEMBER

- The largest country in the world, Russia is nearly twice the size of the second-largest country, Canada. The region encompasses numerous landforms and associated climates.

Geographic Insight 1

- **Environment, Development, and Urbanization** Urban and industrial pollution were ignored during Soviet times as cities expanded quickly to accommodate new industries. In some places, nuclear contamination is also a serious concern.
- Industries are attempting to become cleaner, but decades may pass before significant improvements are realized.
- Global climate change is also a rising concern, with Russia a major contributor of greenhouse gases.

Geographic Insight 2

- **Climate Change, Food, and Water** Central Asia is especially vulnerable to climate change, given its dependence on irrigated agriculture that uses water from rivers fed by glacial melt.

Human Patterns over Time

The core of the entire region has long been European Russia, the large and densely populated homeland of the ethnic Russians. Expanding gradually from this center, the Russians conquered a large area inhabited by a variety of other ethnic groups. These conquered territories remained under Russian control as part of the Soviet Union (1917–1991), which attempted to create an integrated social and economic unit out of the disparate territories. The breakup of the Soviet Union has diminished Russian domination throughout this region, though Russia's influence is still considerable.

The Rise of the Russian Empire

For thousands of years, the militarily and politically dominant people in the region were **nomadic pastoralists** who lived on the meat and milk provided by their herds of sheep, horses, and other grazing animals, and used animal fiber to make yurts, rugs, and clothing. As possibly the first people to domesticate horses, their movements followed the changing seasons across the wide grasslands that stretch from the Black Sea to the Central Siberian Plateau. The nomads would often take advantage of their superior horsemanship and hunting skills to plunder settled communities. To defend themselves, permanently settled peoples gathered in fortified towns.

Towns arose in two main areas: the dry lands of greater Central Asia and the forests of Caucasia, Ukraine, and Russia. As early as 5000 years ago, Central Asia had settled communities that were supported by irrigated croplands (see Figure 5.10A on page 208). These communities were enriched by their central location for trade along what

nomadic pastoralists people whose way of life and economy are centered on tending grazing animals that are moved seasonally to gain access to the best pasture

Figure 5.8 Photo Essay: Vulnerability to Climate Change in Russia and the Post-Soviet States

Vulnerability to climate change is at medium levels throughout much of this region, but at high and extreme levels in Central Asia, which is highly exposed to drought and water scarcity. Both problems will worsen as temperatures rise and glaciers in the Pamirs and other mountains melt. The dependence of so many Central Asians on agriculture makes them very sensitive to these problems, and widespread poverty, combined with poorly developed disaster response systems, reduces their overall resilience to these and other disturbances.

A A river fed by glacial melting in Tajikistan. Most of Central Asia depends on rivers for drinking and irrigation water. If these rivers receive less water from glaciers, there will not be enough water to supply people and their crops.

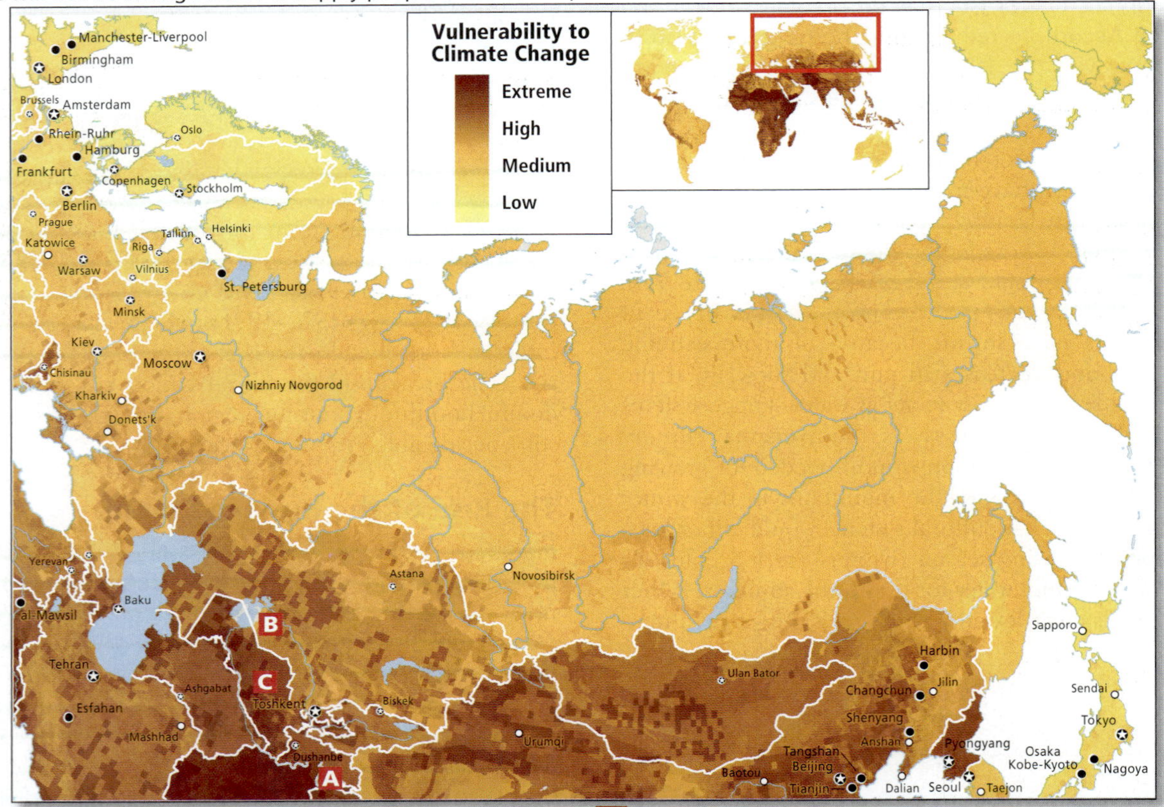

B A fishing boat stranded in what used to be part of the Aral Sea, which has shrunk dramatically because the river waters that feed the sea have been diverted to irrigate cotton and food crops and are now being overused. Even less water will reach the Aral Sea as glaciers continue melting.

C An irrigated cotton field in Uzbekistan. Forty-four percent of Uzbeks work in agriculture, and cotton is the country's leading export. If irrigation water were to become less available, the livelihoods of millions of Uzbeks, 33 percent of whom live below the poverty line, would be threatened.

Thinking Geographically

After you have read about the vulnerability to climate change in Russia and the post-Soviet states, you will be able to answer the following questions:

A Why would rivers like the one in this photo receive less water over the long term due to climate change?

B What major resource did the Aral Sea once provide in abundance?

became known as the Silk Road, that vast, ancient, interwoven ribbon of major and minor trading routes between China and the Mediterranean with lesser connections to other places (Figure 5.9). Many commodities were traded along the route, but the eastern tradition of silk fabric became highly valued in the west, which is why the term Silk Road was applied by later historians.

About 1500 years ago, the **Slavs**, a group of farmers including those known as the Rus (possibly of Scandinavian origin), emerged in what is now Poland, Ukraine, and Belarus. They moved east, founding numerous settlements, including the towns of Kiev in about 480 C.E. and Moscow in 1100. By 600, Slavic trading towns were located along all the rivers west of the Ural Mountains. The Slavs prospered from a lucrative trade route, over land, along the Volga River, and across the Black and Caspian Seas, that connected Scandinavia (North Europe) and Central and Southwest Asia (via Constantinople, modern-day Istanbul). Powerful kingdoms developed in Ukraine and European Russia. In the mid-800s, Greek missionaries introduced both Christianity (now known as "Orthodox Christianity"; see also Figure 5.10B) and the Cyrillic alphabet that we now associate with the Russian language. The Cyrillic alphabet is still used in most of the region's countries.

In the twelfth century, the Mongol armies of Genghis Khan conquered the forested lands of Ukraine and Russia. The **Mongols** were a loose confederation of nomadic pastoral people centered in East and Central Asia. Moscow's rulers became tax gatherers for the Mongols, dominating neighboring kingdoms and eventually growing powerful enough to challenge local Mongol rule. In 1552, the Slavic ruler Ivan IV ("Ivan the Terrible") conquered the Mongols, marking the beginning of the Russian empire. St. Basil's Cathedral, a major landmark in Moscow, commemorates the victory (Figure 5.11 on page 208).

By 1600, Russians centered in Moscow had conquered many former Mongol territories, integrating them into the growing empire and extending it to the east as shown in Figure 5.12 on page 209. The first major non-Russian area to be annexed was western Siberia (1598–1689). Russian expansion into Siberia (and even into North America, along its Pacific coast) resembled the spread of European colonial powers throughout Asia and the Americas. Russian colonists took land and resources from the Siberian populations and treated the people of those cultures as inferiors. Moreover, migrations of laborers from Russia to Siberia meant that indigenous Siberians were vastly outnumbered by the eighteenth century. By the mid-nineteenth century, Russia had also expanded its control to the south

Slavs a group of people who originated between the Dnieper and Vistula Rivers in modern-day Poland, Ukraine, and Belarus

Mongols a loose confederation of nomadic pastoral people centered in East and Central Asia, who by the thirteenth century had established by conquest an empire that stretched from Europe to the Pacific

FIGURE 5.9 The ancient Silk Road and related trade routes. Merchants who worked the Silk Road rarely traveled the entire distance. Instead, they moved back and forth along part of the road, trading with merchants to the east and west.

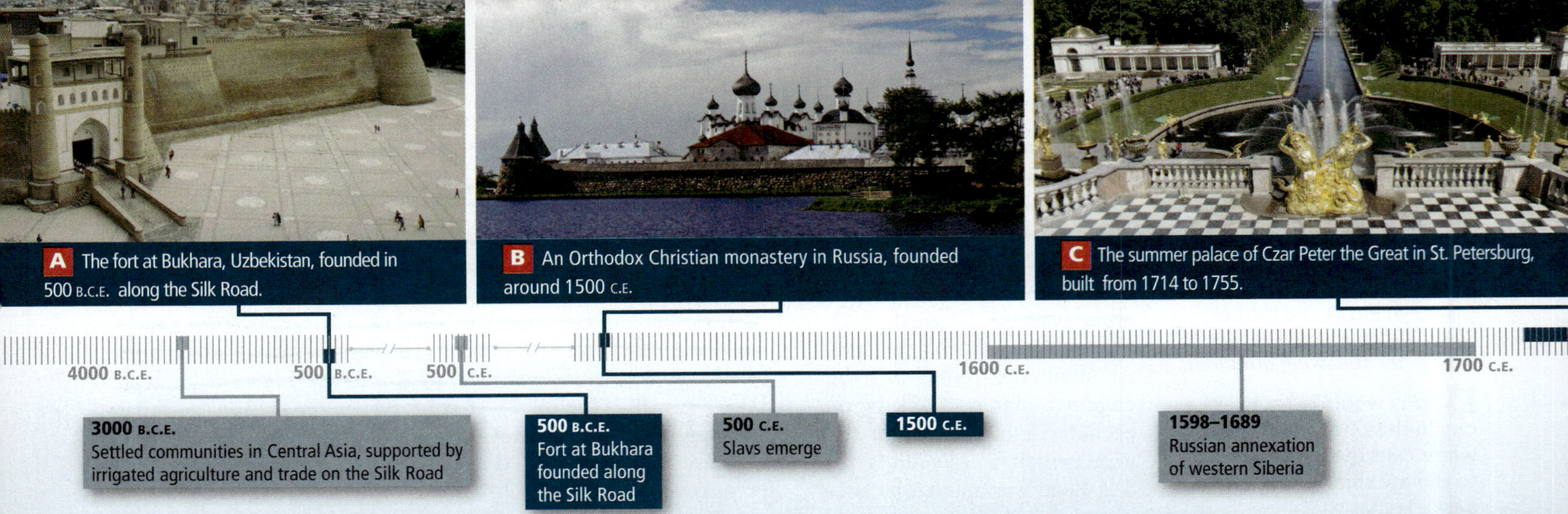

A The fort at Bukhara, Uzbekistan, founded in 500 B.C.E. along the Silk Road.
B An Orthodox Christian monastery in Russia, founded around 1500 C.E.
C The summer palace of Czar Peter the Great in St. Petersburg, built from 1714 to 1755.
4000 B.C.E.
500 B.C.E.
500 C.E.
1600 C.E.
1700 C.E.
3000 B.C.E. Settled communities in Central Asia, supported by irrigated agriculture and trade on the Silk Road
500 B.C.E. Fort at Bukhara founded along the Silk Road
500 C.E. Slavs emerge
1500 C.E.
1598–1689 Russian annexation of western Siberia

FIGURE 5.10 A VISUAL HISTORY OF RUSSIA AND THE POST-SOVIET STATES

Thinking Geographically

After you have read about the human history of Russia and the post-Soviet states, you will be able to answer the following questions:

A In addition to trade along the Silk Road, what supported Central Asian cities?

B From where was Orthodox Christianity introduced into this region?

czar title of the ruler of the Russian empire

in order to gain control of Central Asia's cotton crop, its major export, used in Russian industry.

The Russian empire was ruled by a powerful monarch, the **czar**, who lived in splendor (along with a tiny aristocracy) while the vast majority of the people lived short brutal lives in poverty (see Figure 5.10C, D). Many Russians were *serfs* who were legally bound to live on and farm land owned by an aristocrat. If the land was sold, the serfs were transferred with it. Serfdom was legally ended in the mid-nineteenth century. However, the inequalities of Russian society persisted into the twentieth century, fueling opposition to the czar. By the early twentieth century, a number of violent uprisings were underway.

The Communist Revolution and Its Aftermath

Periodic rebellions against the czars and elite classes took place over the centuries. By the mid-nineteenth century, these rebellions were being led not only by serfs, but also by members of the tiny educated and urban middle class. In 1917, at the height of Russian suffering during World War I, Czar Nicholas II was overthrown in a revolution. What followed was a civil war between rival factions with different ideas about how the revolution should proceed. Eventually the revolution brought a complete restructuring of the Russian economy and society, according to what at first promised to be a more egalitarian model—a model that was extended to adjacent countries.

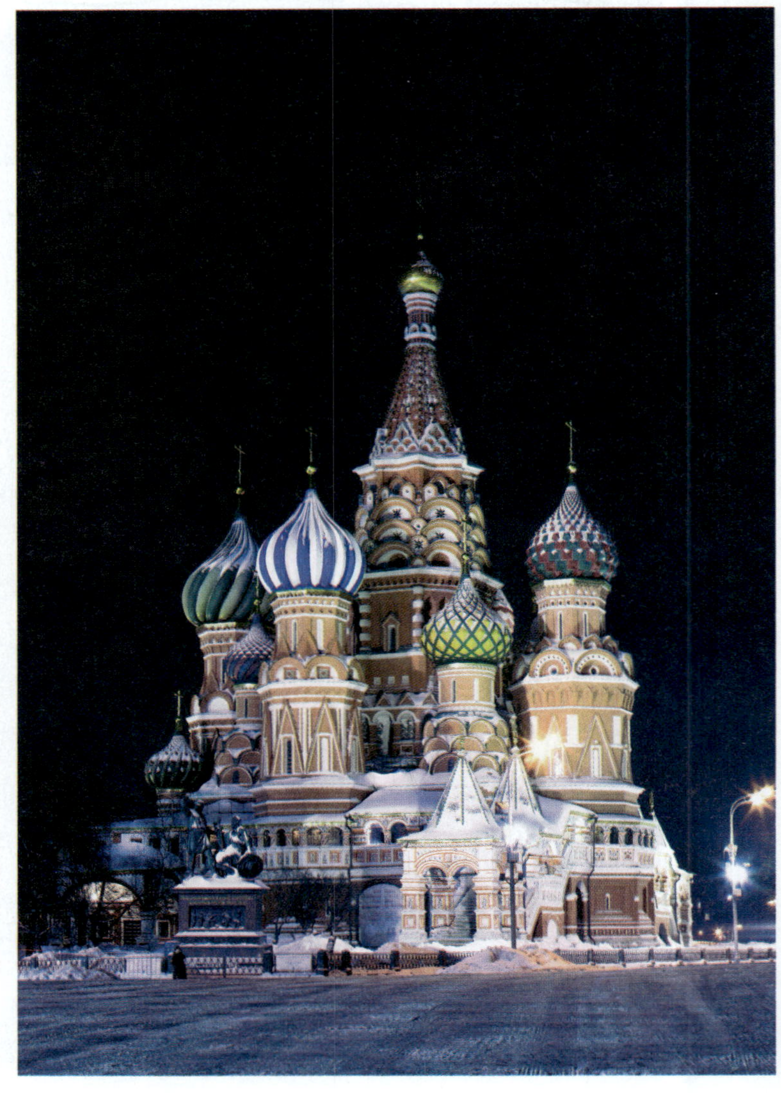

FIGURE 5.11 St. Basil's Cathedral, Moscow. Now the most recognized building in Russia, St. Basil's Cathedral was built between 1555 and 1561 to commemorate the defeat of the Mongols by Ivan the Terrible, the first Russian czar. A central chapel, capped by a pyramidal tower, stands amid eight smaller chapels, each with a colorful "onion dome." Each chapel commemorates a saint on whose feast day Czar Ivan had won a battle.

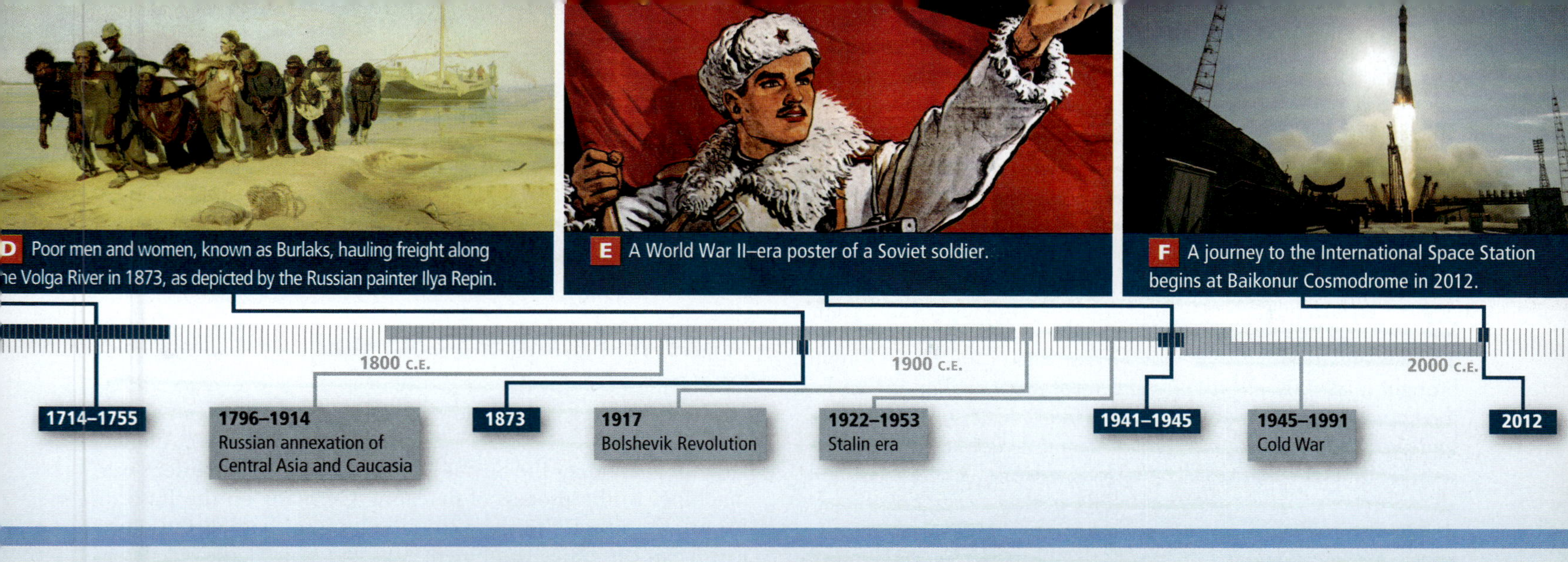

D Poor men and women, known as Burlaks, hauling freight along the Volga River in 1873, as depicted by the Russian painter Ilya Repin.

E A World War II–era poster of a Soviet soldier.

F A journey to the International Space Station begins at Baikonur Cosmodrome in 2012.

D How does this painting convey the poverty of the Burlaks?

E How many Soviets were killed during World War II?

Of the disparate groups that coalesced to launch the Russian revolution, the **Bolsheviks** succeeded in gaining control during the post-revolution civil war. The Bolsheviks were inspired by the principles of **communism** as explained by the German revolutionary philosopher Karl Marx. Marx criticized the societies of Europe as inherently flawed because they were dominated by **capitalists**—the wealthy minority who owned the factories, farms, businesses, and other means of production. Marx pointed out that the wealth of capitalists was actually created in large part by workers, who nonetheless remained poor because the capitalists undervalued their work. Under communism, workers were called upon to unite to overthrow governments that supported the capitalist system, take over the means of production (farms, factories, services), and establish a completely egalitarian society without government or currency. The philosophy held that people would work out of a commitment to the common good, sharing whatever they produced so that each had their basic needs met.

One Bolshevik leader, Vladimir Lenin, argued that the people of the former Russian empire needed a transition period in which to realize the ideals of communism.

Bolsheviks a faction of Communists who came to power during the Russian Revolution

communism an ideology, based largely on the writings of the German revolutionary Karl Marx, that calls on workers to unite to overthrow capitalists and establish an egalitarian society where workers share what they produce

capitalists usually a wealthy minority that owns the majority of factories, farms, businesses, and other means of production

FIGURE 5.12 Russian imperial expansion, 1300–1945. A long series of powerful rulers expanded Russia's holdings across Eurasia to the west and east. Expansion was particularly vigorous after 1700, when Russia acquired Siberia. Note that the centuries-long imperial expansion was reversed in 1991, when the Soviet Union was disbanded and the new post-Soviet states were created.

Accordingly, Lenin's Bolsheviks formed the **Communist Party**, which set up a powerful government based in Moscow. Lenin believed that the government should run the economy, taking land and resources from the wealthy and using them to benefit the poor majority.

The Stalin Era After Lenin suffered a series of strokes in 1922, Joseph Stalin stepped in to reorganize the society and economy in a highly authoritarian style. His 31-year rule of the Soviet Union brought a mixture of brutality and revolutionary change that set the course for the rest of the Soviet Union's history. He sought to increase general prosperity through rapid industrialization made possible by a **centrally planned**, or **socialist, economy**. The state owned all real estate and means of production, while government officials in Moscow managed all economic activity. This system became known as the *command economy*. In this economy, central control was used to determine where all factories, apartment blocks, and transportation infrastructure should be located, and to manage all production, distribution, and pricing of products. The idea was that under a socialist system, the economy would grow more quickly, which would hasten the transition to the idealized communist state in which everyone shared equally. This notion was reflected in the new name chosen for the country: the Union of Soviet Socialist Republics ("Soviet" means, roughly, "council").

The centerpiece of Stalin's vision was massive government investment in gigantic development projects, such as factories, dams, and chemical plants, some of which are still the largest of their kind in the world. To increase agricultural production, he forced farmers to join large government-run collectives.

Stalin's strategy of government control met many of his expectations. It brought rapid economic development and higher standards of living. Schools were provided for previously uneducated poor and rural children, and contributed to major technological and social advances. During the Great Depression of the 1930s, the Soviet Union's industrial productivity grew steadily even while the economies of other countries stagnated, allowing for the development of a large military.

However, Stalin's economic development model also had deep flaws. With production geared largely toward heavy industry (the manufacture of machines, transportation equipment, military vehicles, and armaments), less attention was paid to the demand for consumer goods and services that could have further improved living standards. Ultimately, communism also proved to be less innovative than capitalism and unable to develop advanced technologies to sustain economic growth over time.

The most destructive aspect of Stalin's rule, however, was his ruthless use of the secret police and mass executions to silence anyone who opposed him. Those who resisted were forced to work in prison labor camps, the so-called *gulags*, many of which were located in Siberia. Here prisoners mined for minerals and also built and worked in newly constructed industrial cities. Millions were executed or died from overwork under harsh conditions. Estimates vary greatly, but between 20 and 60 million people were killed as a result of Stalin's policies.

Communist Party the political organization that ruled the USSR from 1917 to 1991; other communist countries, such as China, Mongolia, North Korea, and Cuba, also have ruling Communist parties

centrally planned, or **socialist, economy** an economic system in which the state owns all land and means of production, while government officials direct all economic activity, including the locating of factories, residences, and transportation infrastructure

Cold War the contest that pitted the United States and western Europe, espousing free market capitalism and democracy, against the USSR and its allies, who were promoting a centrally planned economy and a socialist state

World War II and the Cold War

During its involvement in World War II in the years 1941–1945, the Soviet Union did more than any other country to defeat the armies of Nazi Germany (see Figure 5.10E). A failed attempt to conquer the Soviet Union had exhausted Hitler's war machine. In the process of defeating Germany on the Eastern Front, 23 million Soviets were killed, more than all the other European casualties combined. After the war, Stalin was determined to erect a buffer of allied communist states in Central Europe that would be the battleground of any future war with Europe. Because they were unwilling to risk another war, the leaders of the United States and the United Kingdom ceded control of much of Central Europe to the Soviet Union while they busied themselves with reconstructing western Europe. However, when Stalin's intention to utterly dominate Central Europe became clear, the United States and its allies organized to stop the Soviets from extending their power even further into Europe and elsewhere. The result was a politically and economically divided Europe separated by an "Iron Curtain" and the **Cold War**, a nearly 50-year-long global geopolitical rivalry that pitted the Soviet Union and its allies against the United States and its allies around the world (Figure 5.13).

A wide variety of economic and political problems led to the eventual collapse of the Soviet Union. Fundamental flaws with Soviet central planning led to innumerable inefficiencies and chronic mismanagement. Many problems also were related to the diversion of scarce resources to the military, including the Soviet space program, and away from much-needed economic and social development (see Figure 5.10F). Soviet efforts to promote communism in less developed countries were also expensive, as was extensive political, economic, and military coercion in Central Europe and Central Asia. As problems multiplied, so did resistance to Soviet control, especially in Central Europe in the late 1980s. Because the last Soviet leader, Mikhail Gorbachev, did not resort to violence to control the countries of Central Europe, they were able to abandon communism and instead establish democracy. Eventually, this dissension spread to the republics that made up the Soviet Union itself, and by 1991, the USSR had formally dissolved.

A war in Afghanistan also helped to stretch the USSR past the breaking point (see Figure 5.17A on page 218). In the 1970s in Afghanistan, which lies just to the south of the Central Asian states, there was contention among a small group who favored a Western-style democratic government, those who favored communism, and Muslims who favored a theocratic state based on Islamic law. The Soviets feared that the strongly anti-communist Muslim movement in Afghanistan would influence nearby Central Asian Soviet republics to unite under Islam and rebel against the USSR. In response, the USSR backed an unpopular

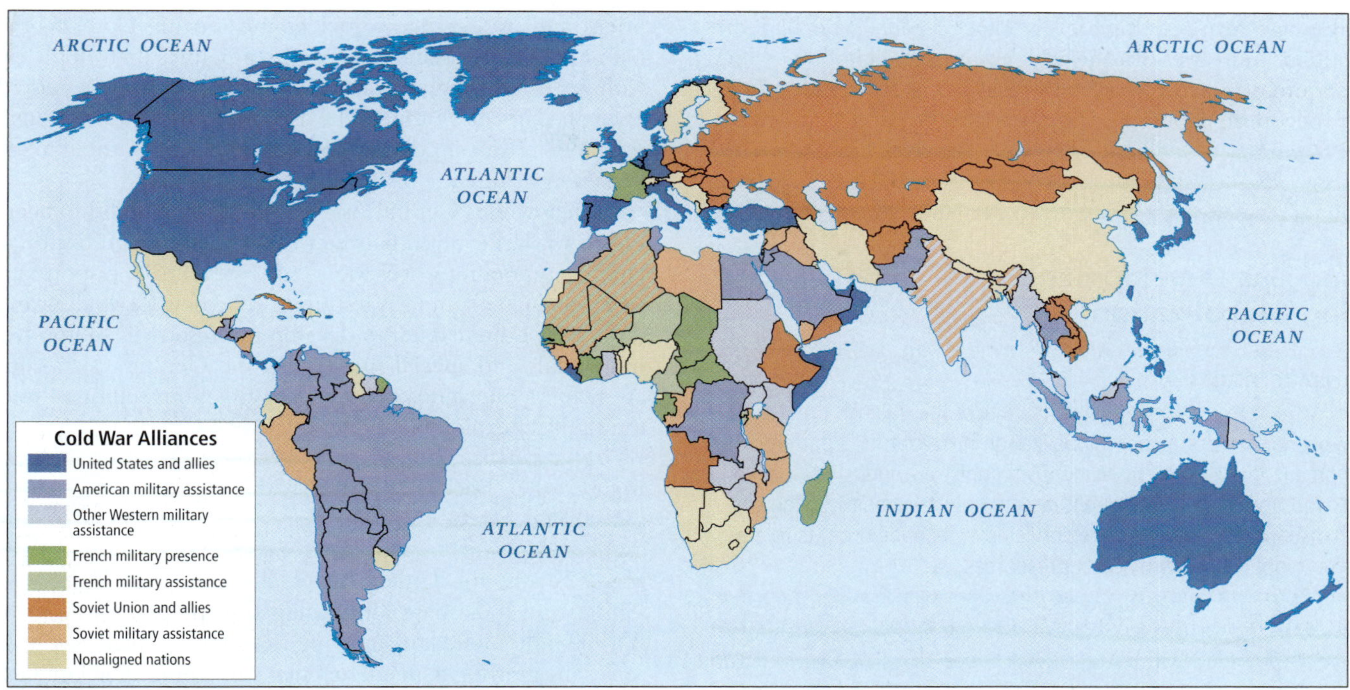

FIGURE 5.13 The Cold War in 1980. In the post–World War II contest between the Soviet Union and the United States, both sides enticed allies through economic and military aid. The group of countries militarily allied with the Soviet Union was known as the Warsaw Pact. Some countries remained nonaligned.

communist government that allowed Soviet military bases in Afghanistan. This brought strong resistance from the Muslim fundamentalist Mujahedeen movement, and in 1979, Russia launched a war to support the Afghan government. A bloody, decade-long conflict ensued, concluding with the Soviets being defeated by Afghan guerillas who were financed, armed, and trained by the United States. This is an example of the geopolitical struggles that took place during the Cold War. Within 2 years of the Soviet withdrawal from Afghanistan, the USSR disintegrated. The Soviet–Afghan conflict and the aftermath are discussed further in Chapter 8.

THINGS TO REMEMBER

- As early as 5000 years ago, Central Asia had settled communities that were supported by irrigated croplands and enriched by trade along what became known as the Silk Road.
- The Bolsheviks, a group inspired by the principles of communism, were the dominant leaders of the Russian Revolution of 1917. Their goal, never achieved, was an egalitarian society where people would work out of a commitment to the common good, sharing whatever they produced.
- Taking control in 1922, Joseph Stalin brought to the country a mixture of brutality and revolutionary change that set the course for the rest of the Soviet Union's history. He established a centrally planned, or socialist, economy in which the state owned all property and means of production, while government officials in Moscow directed all economic activity. This system became known as a command economy.
- After World War II, a nearly 50-year-long global geopolitical rivalry known as the Cold War pitted the Soviet Union and its allies against the United States and its allies.

CURRENT GEOGRAPHIC ISSUES

The goals of the Soviet experiment begun in 1917 were unique in human history: to reform, quickly and totally, an entire human society. The Soviet Union's collapse brought an even more rapid shift from centrally planned economies to market economies. The transition remains incomplete, and for many, life is proceeding amid economic, political, and social uncertainty.

Economic and Political Issues

When the Soviet Union disbanded in 1991 and the post-Soviet states embraced a market economy, everyone had to learn how capitalism works. The first decade of this new era was characterized by a disorderly tumble into a market economy that gave the advantage to a few high-level Soviet leaders who had run government industries in the old command economy. They often acquired previously publicly held assets at fire-sale prices, thereby becoming fabulously rich very quickly, and assumed the role of **oligarchs**—individuals who are so wealthy that they wield enormous, often clandestine, political control. These oligarchs continue to exercise power in government and private enterprise.

Oil and Gas Development: Fueling Globalization

> **Geographic Insight 3**
>
> **Globalization and Development:** Since the fall of the Soviet Union, economic reforms and globalization have changed patterns of development in this region. Many Communist-era industries closed, and as jobs were lost and public assets sold to wealthy oligarchs, wealth disparity increased. The region is now dependent on its role as a leading global exporter of energy resources.

Natural gas and crude oil have emerged as the region's most lucrative exports, introducing a new wave of globalization and fossil fuel–dependent economic development. While control of Russia's oil and gas resources still lies largely in the hands of the Russian government, the struggle to control Central Asia's oil and gas resources involves many multinational corporations and countries and could be a source of conflict for years to come (Figure 5.14).

By the year 2000, after an initial decade of economic instability, rising revenues from oil and gas began to finance Russia's economic recovery. In response, Russia's central government tightened control over the entire oil and gas sector. Today, Russia is the world's largest exporter of natural gas, and taxes on energy companies fund more than half of the federal budget. While this does guarantee that at least some of the revenues from the new fossil fuel–based industries benefit Russia's population, the remaining profits are concentrated in the hands of a few oligarchs.

Russia's relationship with Europe is intertwined with the lucrative oil and gas trade. The majority state-owned **Gazprom** is Russia's largest company as well as the world's largest gas company, and because Gazprom is state controlled, the Russian government can use the company to promote the country's geopolitical interests. Russia occasionally tries to use its gas exports to manipulate the politics and economies of Ukraine and Belarus and of countries in Central Europe. These countries are all much more reliant on gas from Gazprom than is the rest of Europe. EU member states do receive about 25 percent of their natural gas from Gazprom, but Russia must be careful because about 60 percent of Gazprom's profits are from sales to the European Union and the Russian economy is dependent upon these sales. In 2009, Russia curtailed Ukraine's access to Gazprom's gas in a dispute over prices and as a response to Ukraine moving closer to the EU rather being a Russian ally. Because pipelines that supply Europe with gas run through Ukrainian territory, gas shortages also developed in parts of Europe. The EU became concerned about its reliance on Russian energy and is now rapidly turning to alternative energy sources.

> **oligarchs** in Russia, those who acquired great wealth during the privatization of Russia's resources and who use that wealth to exercise power
>
> **Gazprom** in Russia, the state-owned energy company; it is the largest gas entity in the world
>
> **Group of Eight (G8)** an organization of eight countries with large economies: France, the United States, Britain, Germany, Japan, Italy, Canada, and Russia

Meanwhile, Central Asia has yet to find stability because a tug-of-war has evolved between Russia and foreign multinational energy corporations over the right to develop, transport, and sell Central Asian oil and gas resources. The new Central Asian states do not have the capital to develop the resources themselves, yet individuals and special interests in the Central Asian countries are reaping enormous financial rewards from selling to outsiders the rights to exploit oil and gas.

The pipeline routes are the key source of contention (see Figure 5.14). Because the region is landlocked, the only way to get oil and gas to world markets is via pipelines. The capacity and geography of the pipeline network determines how that takes place. Russia, the United States, the European Union, Turkey, China, and India have all developed or proposed various routes. A 1900-mile (3000-km) pipeline from the Caspian Sea to China serves the growing demand for energy in China. The United States maintains a military presence to protect its interests in countries such as Georgia and Uzbekistan, both of which, along with Turkey, have pipelines to Europe. Russia has built pipelines that traverse its own territory and reach into Europe. Each of these parties has made numerous efforts to encourage Central Asian countries to use their pipelines instead of those of their competitors.

108. ENERGY REVENUES AND CORRUPTION INCREASE IN RUSSIA

Russia's Relations with the Global Economy Both the European Union and the United States want to ensure that Russia's oil and gas wealth does not finance a return to the hostile relations of the Cold War. For this reason, Russia was invited to join the World Trade Organization (WTO) and the **Group of Eight (G8)**, an organization of eight affluent countries with large economies.

Looking to form an economic counterweight to the United States and the European Union, Russia has recently been meeting with Brazil, India, and China to form a trade consortium known as BRIC. All four countries have had spectacular economic growth over the last decade, though Russia's economy is arguably the slowest growing and least diverse in the group due to its dependence on fossil fuel exports.

Economic Reforms in the Post-Soviet Era

Transitioning from one economic system to another is a major undertaking for any society. Not only do such changes take a long time to implement fully, they also disrupt existing geographic patterns of production and affect the livelihood of many people, both in positive and negative ways. The transition from communism to a market economy in the post-Soviet states turned out to be more complex and problematic than many had anticipated.

Reforming the Command Economy The Soviet centrally planned command economy was notably less efficient than

Oil in Russia (2011–2012)	
Oil reserves	**60 billion barrels**
Oil reserves as percentage of world	4.07 percent
Saudi Arabian reserves	267.02 billion barrels
U.S. reserves	22.3 billion barrels (proved 2009)
Oil production	**10,228 million barrels per day**
Oil production as percentage of world	11.75 percent
U.S. oil production	5512 million barrels per day (2010)
Oil exports	**7.083 million barrels per day**
Oil exporter, rank	8
Oil exports to U.S.	580,000 barrels per day (2009)

Natural Gas in Russia (2011–2012)	
Gas reserves	**1680 trillion cubic feet**
Gas reserves as percentage of world	25.05 percent
Iranian reserves	1046 trillion cubic feet
U.S. reserves	272.5 trillion cubic feet
Gas production	**23,647 billion cubic feet**
Gas production as percentage of world	21.14 percent
U.S. gas production	1103.2 billion cubic feet (2009)
Gas exports	**5753 billion cubic feet**
Gas exporter, rank	90
Gas exports, to Europe	**4948 billion cubic feet**

FIGURE 5.14 Oil and natural gas: Russia and the post-Soviet states' resources and pipelines. For additional information and maps, see http://www.eia.gov/countries.

market economies at allocating resources. This was recognized during Mikhail Gorbachev's tenure as president of the USSR (1985–1991), and a private market economy began to be permitted. Under an overall policy of **perestroika**, or "restructuring," government decision making was decentralized and private entrepreneurs were allowed to legally sell goods and services. Gorbachev also introduced policies collectively known as **glasnost**, or "openness," which encouraged greater transparency, openness, and publicity in the workings of all levels of Soviet government. However, these reforms were short-lived, as the USSR ceased to exist shortly thereafter. **104. FORMER SOVIET UNION LAUNCHED SPACE AGE 50 YEARS AGO**

Privatization and the Lifting of Price Controls The Soviet economy consisted almost entirely of industries owned and operated by the government. These have now been sold to private companies or individuals through the process of **privatization**. The hope was that they would operate more efficiently in a competitive free market setting. The transition to a market economy was very rapid. During the 1990s,

perestroika literally, "restructuring"; the restructuring of the Soviet economic system that was done in the late 1980s in an attempt to revitalize the economy

glasnost literally, "openness"; the policies instituted in the late 1980s under Mikhail Gorbachev that encouraged greater transparency and openness in the workings of all levels of Soviet governments

privatization the sale of industries that were formerly owned and operated by the government to private companies or individuals

the Russian government pursued an economic strategy known as "shock therapy," which was an attempt to revitalize the sluggish economy by moving as quickly away from the old command economy as possible. Today, approximately 65 percent of Russia's economy is in private hands, a dramatic change from the virtually 100 percent state-owned economy of 1991.

Unfortunately, shock therapy came at a steep cost for many Russians. A major part of the reforms was the abandonment of government price controls that once kept goods affordable to all. Instead, private business owners now determine prices. The lifting of price controls led to inflation and skyrocketing prices in the 1990s for the many goods that were in high demand but also in short supply. While a few people became rich, many people lost much of their savings because of inflation and could barely pay for basic necessities such as food. Eventually, as opportunities opened and competition developed, the supply of goods increased and prices fell. But in the interim, many people suffered. The Russian state also lost tax revenue and only a bailout from the International Monetary Fund saved the country from bankruptcy.

Unemployment and the Loss of Social Services Since becoming privatized, most formerly state-owned industries have cut many jobs in an attempt to compete against more-efficient foreign companies. Losing a job is especially devastating in a former Soviet country because one also loses the subsidized housing, health care, and other social services that were often provided along with the job. The new companies that have emerged rarely offer full benefits to employees. There is little job security because most small private firms appear quickly and often fail. Discrimination is also a problem, given the absence of equal opportunity laws. Job ads often contain wording such as "only attractive, skilled people under 30 need apply."

Unemployment figures for the region vary widely. By 2010, official unemployment rates were 6.8 percent in Russia, but often much higher in Caucasia and Central Asia; for example, in some recent years, Turkmenistan has reported unemployment as high as 60 percent. In other countries, such as Belarus, unemployment is officially only about 1 percent, but actual unemployment may be higher because many remaining state-owned firms cannot pay employees still listed as workers. Many people do not bother to register as unemployed because of the absence of unemployment benefits. The rate of **underemployment**, which measures the number of people who are working too few hours to make a decent living or who are highly trained yet working at menial jobs, is even higher in all of these countries.

underemployment the condition in which people are working too few hours to make a decent living or are highly trained but working at menial jobs

The Difficult Legacies of Soviet Regional Development Schemes One of the reasons that so many state-owned firms have had to cut jobs is that they were components of largely failed regional development schemes. Numerous huge industrial areas were located in far-flung eastern portions of the Soviet Union, in large part to facilitate the political control of these areas. These projects were always held back by the vast distances and challenging physical geography that separated them from the main population centers in the west of the region. Because the region's rivers run mainly north to south, there was a need for land transportation systems, such as railroads and highways, that ran east to west. These proved exceedingly expensive to build and maintain, and many have fallen into disrepair in the post-Soviet era. Only one poorly paved road, and only one main rail line—the Trans-Siberian Railroad (Figure 5.15)—runs the full east–west length, connecting Moscow with Vladivostok, the main port city on the Russian Pacific coast.

By contrast, the road and rail network is fairly dense within western Russia, Belarus, and Ukraine, with Moscow as the main hub. This has had the effect of concentrating new economic development in these western areas. Hence, the industrial cities of eastern Russia have lost many jobs and their populations are shrinking because the Russian government is no longer interested in subsidizing regional economic development in the east.

Institutionalized Corruption The cost of doing business in Russia is also affected by a high level of corruption. In 2010, the organization Transparency International rated Russia the most corrupt of all major countries in the world. While Russia today is vastly different from the old Soviet Union, many officials have not made the transition to a more democratic mindset. Civil servants and the police are prone to being corrupted. They often expect businesses to offer bribes to receive licenses and to avoid other regulatory hurdles. This affects large corporations and small entrepreneurs alike. Even those who make it through the labyrinth of setting up a business must then face protection racketeers, as organized crime has become a fact of life during the post-Communist era. These crime syndicates have also developed international networks through which they control illicit activities abroad, especially in Europe.

The Growing Informal Economy A large share of the economic activity in the region takes place in the informal sector. To some extent, the new informal economy is an extension of the old one that flourished under communism. The black market of that time was based on currency exchange and the sale of hard-to-find luxuries. In the 1970s, for example, savvy Western tourists could enjoy a vacation on the Black Sea paid for by a pair or two of smuggled Levi's blue jeans and some Swiss chocolate bars. Today, many people who have lost stable jobs due to privatization now depend on the informal economy for their livelihood.

Workers in the informal economy tend to be very young unskilled adults, retirees on miniscule pensions, and those with only a low level of education. The majority of these workers operate out of their homes by selling cooked foods, vodka made in their bathtubs, clothing, or electronics that have been smuggled into the country. The World Bank has estimated that the informal sector makes up about 40 percent of the Russian economy. In the Caucasian countries, the number is as high as 50 to 60 percent. Such large percentages indicate that people may actually be better off financially than official GDP per capita figures suggest.

Despite the fact that the informal economy helps people survive, it is not popular with governments. Unregistered enterprises do not pay business and sales taxes and usually are so

FIGURE 5.15 Principal industrial areas and land transport routes of Russia and the post-Soviet states. The industrial, mining, and transportation infrastructure is concentrated in European Russia and adjacent areas. The main trunk of the Trans-Siberian Railroad and its spurs links industrial and mining centers all the way to the Pacific, but the frequency of these centers decreases with distance from the borders of European Russia.

underfinanced that they tend not to grow into job-creating formal sector businesses. In many cases, informal businesspeople must pay protection money to local gangsters (the so-called *Mafia tax*) to keep from being reported to the authorities.

VIGNETTE Natasha is an engineer in Moscow. She has managed to keep her job and the benefits it carries, but in order to better provide for her family, she sells secondhand clothes in a street bazaar on the weekends. "Everyone is learning the ropes of this capitalism business," she laughs. "But it can get to be a heavy load. I've never worked so hard before!" Asked about her customers, Natasha says, "Many are former officials and high-level bureaucrats who just can't afford the basics for their families any longer." Some are older retired people whose pensions are so low that they resort to begging on Moscow's elegant shopping streets close to the parked Rolls-Royces of Moscow's rich. These often highly educated and only recently poor people buy used sweaters from Natasha and eat in nearby soup kitchens. *[Source: This composite story is based on work by Alessandra Stanley, David Remnick, David Lempert, and Gregory Feifer.]*

Food Production in the Post-Soviet Era

Across most of the region, agriculture is precarious at best, either hampered by a short growing season and boggy soils or requiring expensive inputs of labor, water, and fertilizer. Because of Russia's harsh climates and rugged landforms, only 10 percent of its vast expanse is suitable for agriculture. The Caucasus mountain zones are some of the only areas in the region where rainfall adequate for agriculture coincides with a relatively warm climate and long growing seasons. Together with Ukraine and European Russia, this area is the agricultural backbone of the region (Figure 5.16 on page 216). The best soils are in an area stretching from Moscow south toward the Black and Caspian seas, and extending west to include much of Ukraine and Moldova. In Central Asia, irrigated agriculture is extensive, especially where long growing seasons support cotton, fruit, and vegetables.

FIGURE 5.16 Agriculture in Russia and the post-Soviet states. Agriculture in this part of the world has always been a difficult proposition, partly because of the cold climate and short growing seasons, and partly because soil fertility or lack of rainfall are problems in all but a few places (Ukraine, Moldova, and Caucasia).

Changing Agricultural Production During the 1990s, agriculture went into a general decline across the entire region. In most of the former Soviet Union, yields dropped by 30 to 40 percent compared to previous production levels. This was due mostly to the collapse of the subsidies and internal trade arrangements of the Soviet Union. Many large and highly mechanized collective farms suddenly found themselves without access to equipment, fuel, or fertilizers. Since that time, agricultural output has rebounded but even today remains below Soviet levels. Russia's commercial agricultural sector is still characterized by many large and inefficient farms. However, if global food prices are high and agricultural productivity increases, the most important farming areas might be able to expand grain production and turn the region into an viable food exporter. Meanwhile, Russians are heavily dependent on "homegrown" food—small household plots and gardens that produce 47 percent of total agricultural output.

In Central Asia, agriculture has been reorganized with less emphasis on collective farming of export crops and more on smaller, food-producing family farms. Grain and livestock farming (for meat, eggs, and wool) are increasing. Production levels per acre and per worker also are increasing. China, through the Asian Development Bank, has provided assistance in reducing the use of agricultural chemicals, which were used extensively during Soviet times.

Farmers in Georgia, blessed with warm temperatures and abundant moisture from the Black and Caspian seas, can grow citrus fruits and even bananas; they do so primarily on family, not collective, farms. Before 1991, most of the Soviet Union's citrus and tea came from Georgia, as did most of its grapes and wine. Georgia still exports some food to Russia, but because of political tensions between the two countries, Georgia is increasing the amount of food and wine it sells to Europe and other markets outside this region.

THINGS TO REMEMBER

Geographic Insight 3

• **Globalization and Development** Patterns of development changed after 1991 with the switch to a market economy. As state-owned firms closed or were sold to private parties, jobs were lost and wealth disparities increased drastically. Crude oil and natural gas emerged as the region's most lucrative exports, introducing a new wave of globalization and fossil fuel–dependent economic development.

- The Soviet centrally planned economy was less efficient than market economies in allocating resources. A legacy of Soviet central planning is the location of huge industrial projects in the farthest reaches of Soviet territory. Many of these projects are now in decline.
- The benefits of economic reforms in the post-Soviet era have been dampened by widespread corruption, organized crime, lower incomes for many, and the loss of social services.
- Across most of the region, agriculture is precarious, either hampered by a short growing season and boggy soils or requiring expensive inputs of labor, water, and fertilizer. European Russia and Ukraine is the region's main breadbasket.

Democratization in the Post-Soviet Years

Geographic Insight 4

Power and Politics: Obstacles to democratization include a long history of authoritarian regimes and the popular desire to implement change rapidly, combined with the lack of mechanisms for public participation in the political process.

Democratization in this region has not proceeded as quickly as has the introduction of a market economy. While several countries have held elections, many forms of authoritarian control remain. Elected representative bodies often act as rubber stamps for very strong presidents and exercise only limited influence on policy.

In Russia, Vladimir Putin, a former high official in the KGB (Russia's intelligence agency during the Cold War, now known as the Federal Security Service, or FSB), was elected president in 2000 and has been the country's de facto leader as either president or prime minister ever since. During this time, he has exercised tight control over political and economic policy and consolidated political power in Moscow. He has been popular for bringing relative peace and prosperity and for restoring Russia's image of itself as a world power, despite disallowing meaningful democratic reforms at the local, state, and national levels, and extending government control over the press and the media (see page 220). Ostensibly, criticism of the government is now allowed; but behind the scenes, critics fear retribution from the *securocrats* or *siloviki* (former FSB functionaries loyal to Putin) who control many governing institutions. Political analyst Olga Kryshtanovskaya found that in 2006, 77 percent of the people in top government positions had backgrounds in security agencies.

117. PUTIN CONFIRMATION

123. NEW RUSSIAN LEADER

Elsewhere in the region, progress toward true democracy is similarly limited. In Belarus, Ukraine, Moldova, the Caucasian republics (including Chechnya, discussed on pages 219–220), and Central Asia, authoritarian leaders are still the most common and are unaccustomed to criticism or to sharing power (see the map in Figure 5.17). Elections often involve systematically intimidating voters and arresting the political opposition.

The Color Revolutions and Democracy In a few of the post-Soviet states, there have been a number of relatively peaceful transitions to somewhat greater democracy—the so-called *color revolutions*, including the Rose Revolution in Georgia (2003–2004), the Orange Revolution in Ukraine (January 2005; see Figure 5.17B), and the Tulip Revolution in Kyrgyzstan (April 2005).

The democracy movements were named the "color revolutions" because they were led by coalitions of educated young adults who rallied under various symbolic colors meant to unite them despite differences. Funded largely by foreign NGOs and governments, including the United States, that wanted to hasten democratization, it appeared for a time that these movements would result in significant reforms. However, while elections have taken place, and some freedom of expression is tolerated, many forms of authoritarian control and practices that disrupt democratic procedures remain.

In Ukraine, which is now one of the more democratized countries of this region, the Orange Revolution occurred during the presidential elections of 2004. A candidate who favored closer ties with Russia, Viktor Yanukovych, won the election, but the results were so obviously rigged by the government that widespread protests went on for months. The protests, combined with international pressure, resulted in a new vote in which Viktor Yushchenko, who favored closer ties with the European Union and Ukrainian membership in NATO, won. In subsequent elections, however, the pro-Russian Yanukovych has returned to power.

Ukraine today is divided in terms of its geographic orientation. Areas to the west, which border Central Europe, tend to favor closer ties with Europe, while areas to the east, where many of Ukraine's Russian-speaking minority reside, lean toward Russia. The 2011 jailing and sentencing to 7 years of the Orange Revolution leader Yulia Tymoshenko, who is one of the region's few prominent female politicians, indicates that Ukraine's progress toward democracy is in jeopardy.

Cultural Diversity and Democracy Russia's long history of expansion into neighboring lands has left it with an exceptionally complex internal political geography. As the Russian czars and then the Soviets pushed the borders of Russia eastward toward the Pacific Ocean over the past 500 years, they conquered a number of non-Russian areas. Russians are now the main inhabitants of many of those areas. Other territories have been designated semiautonomous internal republics (Figure 5.18 on page 219), and they have significant ethnic minority populations that are descended from indigenous people or trace their origins to long-ago migrations from Germany, Turkey, or Persia.

Both the czars and the USSR had a policy of **Russification**, whereby large numbers of ethnic Russian migrants were settled in non-Russian

Russification the czarist and Soviet policy of encouraging ethnic Russians to settle in non-Russian areas as a way to assert political control

Since World War II, most conflicts in this region have been in Caucasia and Central Asia. Most occurred just before or shortly after the breakup of the Soviet Union. Since then, pro-democracy movements have emerged in several countries, where they hold the promise of bringing political change without violent conflict. However, their long-term effectiveness remains to be seen.

A Children in Afghanistan play on a tank dating from the Soviet Union's 10-year war in that country. The Soviets intervened in 1979 to prop up an authoritarian and unpopular Afghan government that had initiated radical reforms and murdered much of the political opposition. The war resulted in at least 1.5 million deaths and a humiliating defeat for the USSR. Fourteen thousand Soviet soldiers died, billions of dollars were spent, and practically nothing was gained. Public outrage over the Afghan war played a major role in the breakup of the Soviet Union. Since the fall of the USSR, democratization has proceeded at a much more rapid pace throughout the region.

B Ukrainian protesters pushing for a transition from an authoritarian to a more democratic political system during the Orange Revolution of 2004–2005. The movement followed a model pioneered in Serbia and Georgia, where student groups, funded by NGOs and governments in Europe and the United States, spearheaded grassroots political campaigns that featured massive nonviolent demonstrations. A similar "color revolution" occurred in Kyrgyzstan.

C Chechen police in a show of force in Grozny, Chechnya, where rebels have fought two unsuccessful wars for independence from Russia and have conducted ongoing terror campaigns against Russia since the fall of the Soviet Union. Political instability in Caucasia is one of the greatest forces working against democratization in the region.

Thinking Geographically

After you have read about power and politics in Russia and the post-Soviet states, you will be able to answer the following questions:

A How did the USSR's war in Afghanistan ultimately contribute to democratization in the Soviet Union?

B Who funded the color revolutions?

C How does Russia's unwillingness to let Chechnya secede relate to fossil fuel resources?

ethnic areas and given the best jobs and most powerful positions in regional governments. The goal was to force potentially rebellious regional minorities to conform to the state's goals. However, even during the Soviet period, but especially after, minorities organized to resist Russification and to enhance local ethnic identities.

Shortly after the breakup of the Soviet Union, several internal republics demanded greater autonomy, and two of them, Tatarstan and Chechnya, declared outright independence. Tatarstan has since been placated with greater economic and political autonomy. However, Chechnya's stronger resistance to Moscow's authority has led to the worst bloodshed of the post-Soviet era and raised many doubts about Russia's commitment to human rights.

Chechnya, located on the fertile northern flanks of the Caucasus Mountains, is home to 800,000 people (see the inset map in Figure 5.18). Partially in response to Russian oppression, the Chechens converted from Orthodox Christianity to the Sunni branch of Islam in the 1700s. Since then, Islam has served as an important symbol of Chechen identity and as an emblem of resistance to the Orthodox Christian Russians, who annexed Chechnya in the nineteenth century.

In 1942, during World War II, as the Germans continued to fight in Russia, a group of Chechen rebels simultaneously waged a guerilla war against the Soviets. Near the end of the war, Stalin exacted his revenge by deporting the majority of the Chechen population (as many as 500,000 people) to Kazakhstan and Siberia. Here they were held in concentration camps, and many died of starvation. The Chechens were finally allowed to return to their villages in 1957, but a heavy propaganda campaign portrayed them as traitors to Russia, a designation that greatly affected daily life.

FIGURE 5.18 Ethnic character of Russia and percentage of Russians in the post-Soviet states. Russia, with all of its internal republics and administrative regions, is formally called the Russian Federation. The ethnic character of many internal republics was changed by the policy of central planning, so Russians now form significant minorities in the republics. The pie charts for each country indicate the percentages of Russians in the post-Soviet states. In 2002, the ethnic makeup of Russia was 79.8 percent Russian, 3.8 percent Tatar, 2 percent Ukrainian, 1.2 percent Bashkir, 1.1 percent Chuvash, and 12.1 percent other.

In 1991, as the Soviet Union was dissolving, Chechnya declared itself an independent state. Russia saw this as a dangerous precedent that could spark similar demands by other cultural enclaves throughout its territory. Russia also wished to retain the agricultural and oil resources of the Caucasus; it had planned to build pipelines across Chechnya to move oil and gas to Europe from Central Asia. Russia responded to acts of terrorism by Chechen guerrillas with bombing raids and other military operations that killed tens of thousands and created 250,000 refugees. Presently, most Chechen guerillas have given up, most Russian combat troops have been pulled out of Chechnya, and Russia has begun making substantial investments in rebuilding the capital city of Grozny. While this shift is a welcome change, some Chechen rebels have continued their struggle by carrying out brutal terrorist attacks, many of which now take place in Moscow and other areas outside Chechnya, such as in Boston during the 2013 marathon. Chechnya remains highly militarized (see Figure 5.17C), and the ongoing conflict there continues to raise doubts about Russia's ability to peacefully address internal political dissent. **116. CHECHNYA HANGS ON TO UNEASY PEACE**

Just south of Chechnya, conflicts between Russia and Georgia over the ethnic republics of South Ossetia and Abkhazia have worsened in the post-Soviet era. Both republics became part of Georgia at the behest of Joseph Stalin (a native Georgian). He then moved Georgians and Russians into Ossetia and Abkhazia so that the native people became a minority in their own place. More recently, Russia has strategically supported the ethnic Ossetian and Abkhazian populations' agitation for secession from Georgia, even granting Russian citizenship to between 60 to 70 percent of the non-Georgian population. Russia may have done this in retaliation for Georgia's increasingly close relations with the United States and the European Union, as evidenced by its candidacy for NATO membership. A pipeline that links the oil fields of Azerbaijan with the Black Sea via Georgia is another source of contention (see Figure 5.14). The Black Sea route is important because it is an expedient way to ship oil to world markets, as the Black Sea connects to the Mediterranean Sea and, ultimately, to the Atlantic Ocean. Russia would like to enhance its control over the oil resources of the Caspian Sea region by routing pipelines through Russian-controlled territory.

Although in 2008, Georgia's military, attempting to gain control over rebelling parts of South Ossetia, engaged with the Russian army on Russian territory, that conflict ended without clear resolution. It is not clear at this time whether Abkhazia and South Ossetia will break away and become independent countries, become provinces within the Russian Federation, or be satisfied by offers of greater autonomy within Georgia's federal structure.

The Media and Political Reform During the Soviet era, all communication media were under government control. There was no free press, and public criticism of the government was a punishable offense. However, toward the end of the Soviet era, many journalists risked retribution for criticizing public officials and policies. Between 1991 and the early 2000s, the communications industry was a center of privatization, and several media tycoons emerged to challenge the authorities. Privately owned newspapers and television stations regularly criticized the policies of various leaders of Russia and the other states. It appeared that a free press was developing.

In Vladimir Putin's rise to power, however, the most outspoken newspapers and TV stations in Russia were shut down. Since then, critical analysis of the government has become rare throughout Russia. Journalists openly critical of government policies have been treated to various forms of punishment: censorship, exile, or violence. Since 2000, approximately 200 journalists have been killed under violent or unclear circumstances.

Corruption and Organized Crime

Any potential benefits of political and economic reforms in the post-Soviet era have been dampened by widespread corruption and the growth of organized crime. Many oligarchs became closely connected to the *Russian Mafia*, a highly organized criminal network dominated by former KGB and military personnel who control the thugs on the streets. The Russian Mafia extended its influence into nearly every corner of the post-Soviet economy, especially in illegal activities and the arms trade.

THINGS TO REMEMBER

Geographic Insight 4

- **Power and Politics** While several countries have held elections, many forms of authoritarian control remain. Elected representative bodies often act as rubber stamps for very strong presidents and exercise only limited influence on policy.
- Since World War II, most conflicts in this region have been in Caucasia and Central Asia, mainly just before or shortly after the breakup of the Soviet Union. In the years after the breakup, pro-democracy movements have emerged in several countries, where they hold the promise of bringing political change without violent conflict. However, their long-term effectiveness remains unclear.
- There continue to be limitations on the free press and related media across the region.
- The potential benefits of political and economic reforms in the post-Soviet era have been dampened by official corruption and the growth of organized crime.

Sociocultural Issues

When the winds of change began to blow through the Soviet Union in the 1980s, few anticipated the rapidity and depth of the transformations or the social instability that resulted. On the one hand, new freedoms have encouraged self-expression, individual initiative, and cultural and religious revival (Figure 5.19); on the other hand, the post-Soviet era has brought very hard times to many as their jobs and the social safety net have been obliterated.

Population Patterns

Geographic Insight 5

Population: Populations are shrinking in many parts of this region. This is a function of a continuously high level of participation by women in the workforce and overall poor living and housing standards.

This region shares some of the same population characteristics as the United States and Europe—low birth rates and an aging population—both usual features of wealthy and developed societies. But in Russia and the post-Soviet states, low birth rates are more extreme and are coupled with high mortality rates for middle-aged males, leading to a relatively low life expectancy. Several circumstances contribute to this situation.

Population Distribution and Urbanization The region of Russia and the post-Soviet states is one of the largest on Earth but is the least densely populated, with only about 282 million people. A broad area of moderately dense population forms an irregular triangle that stretches from Ukraine on the Black Sea north to St. Petersburg on the Baltic Sea and east to Novosibirsk, the largest city in Siberia. In this triangle, settlement is highly urbanized, but the cities are widely dispersed. The capital city of Moscow, with 11.5 million people, is a primate city in Russia, as are all other capital cities of the nations in this region (Figure 5.20).

Beyond Novosibirsk on the West Siberian Plain, settlement follows an irregular and sparse pattern of industrial and mining development in widely spaced cities, stretching east across Siberia. These economic activities and the cities they support are linked primarily to the route of the Trans-Siberian Railroad (see Figure 5.15). Nearly 90 percent of Siberia's people are concentrated in a few relatively large urban areas. The costs of maintaining these settlements in this remote and harsh environment are considerable. The low concentration of people is affected by the great distance from the main areas of economic activity and markets in European Russia. One telling statistic is that the population density in Russia's Far East is 140 times lower than it is just across the border in northeastern China, which has a similar physical environment.

Caucasia the mountainous region between the Black Sea and the Caspian Sea

A secondary spur of dense settlement extends south from Russia into **Caucasia**, the mountainous region between the Black Sea and the Caspian Sea, where there are several primate cities of well over 1 million people each. In Central Asia, another patch of relatively dense settlement is centered on the cities of Tashkent, Almaty (both over 1 million in population), and Astana, the new capital of Kazakhstan, which has 750,000 people. Along major Central Asian rivers during the Soviet period, the development of irrigated cotton farming and mineral extraction resulted in patches of high rural density, fueled partly by ethnic Russian immigration.

FIGURE 5.19 LOCAL LIVES FESTIVALS IN RUSSIA AND THE POST-SOVIET STATES

A A man performs stunts on horseback during the festival of Nauryz in Kyrgyzstan. Nauryz is an ancient Zoroastrian festival, originating in Iran, that marks the New Year and the first day of spring on March 21. Forbidden during the Soviet era, Nauryz celebrations are now a source of pride throughout Central Asia.

B Revelers at KaZantip, a 6-week-long electronic dance music festival held on Ukraine's Crimean Peninsula during the summer. More than 300 DJs entertain as many as 150,000 revelers each year at 14 festival venues, with electronic music playing 21 hours a day. A celebrated tradition is the notion that KaZantip is an independent republic, complete with its own constitution prohibiting a wide variety of "chauvinistic" behavior and promoting love and fun.

C Women in Richov, Belarus, eat pancakes during Maslenitsa, this region's answer to Mardi Gras. Celebrated 7 weeks before the Eastern Orthodox Christian Easter, Maslenitsa is a week-long festival that celebrates the end of winter and the last chance to revel before the onset of Lent (see page 97 in Chapter 2). During Maslenitsa, eating meat is forbidden, but dairy products and pancakes are consumed in large quantities.

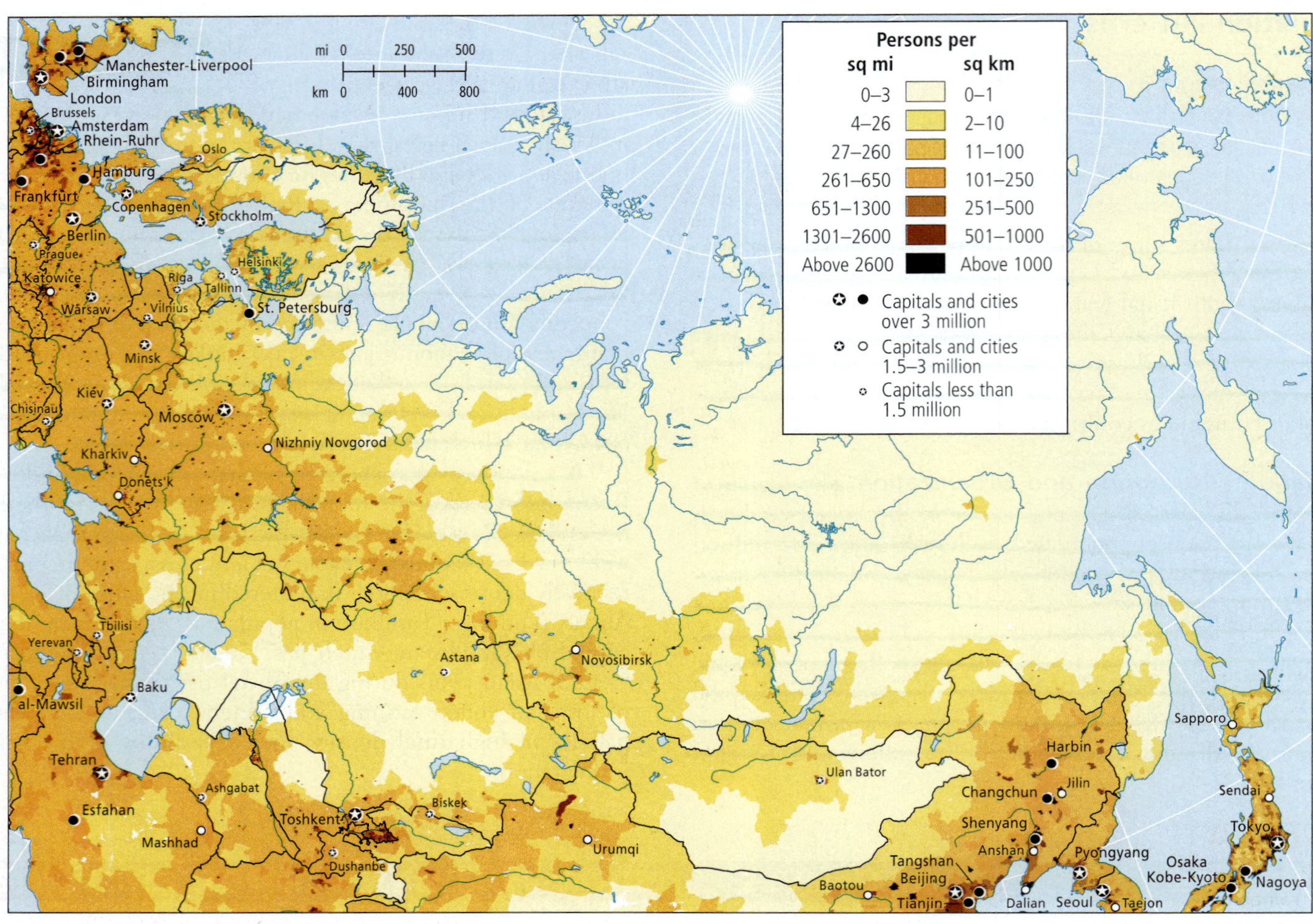

FIGURE 5.20 Population density in Russia and the post-Soviet states. Population trends in this region are highly uneven. While Central Asia and some countries in Caucasia are growing, populations in the rest of the region are shrinking. Some of this may be related to cultural differences or to varying levels of dependence on social welfare institutions that collapsed with the end of the Soviet Union.

VIGNETTE Yernar Zharkeshov is a 24-year-old university-educated young man who is moving back home to Kazakhstan after having spent time in Britain and Singapore. He is seeking new opportunities in his homeland, which are increasing because of extensive oil exports. The natural choice for him is Astana, a city that was designated the capital of Kazakhstan in 1997. Yernar quickly found a job as a government economist—and he is just one of thousands of young people who have migrated to the growing capital.

The idea of Astana came from Nursultan Nazarbayev, who has been president of Kazakhstan since its independence from the Soviet Union in 1991. Without much public debate, he moved the capital from Almaty in the south to the remote and sparsely populated steppe in the north, which is known for its harsh climate. Here he realized his vision for a grand and impressive new capital built more or less from scratch. Astana is meant to symbolize a new, forward-looking Kazakhstan. Such energy-driven urbanization makes Astana similar to Persian Gulf cities such as Dubai and Doha. Another appropriate comparison is to the decision by the Russian czar Peter the Great to build a new imposing imperial capital—St. Petersburg—during the eighteenth century. Nazarbayev spared no expenses when building Astana; the city is full of stunning buildings, either designed by world-renowned architects or by Nazarbayev himself (see Figure 5.22 on page 224).

Like many other migrants to Astana, Yernar Zharkeshov is an ethnic Kazakh. President Nazarbayev's unspoken motive for moving the capital is to consolidate the country's northern territories, which are currently dominated by ethnic Russians. The presence of these Russians is the result of the practice of Russification during the days of the Soviet Union. Astana has been designed to promote the opposite, or "Kazakhification" of the country's north. *[Source: National Geographic. For detailed source information, see Text Credit pages.]* ■

Post-Communist lifestyles in urban centers of Russia and the post-Soviet states are very different from the traditional lifestyles of the people of this region, who used to be more dependent on what nature could provide (Figure 5.21). Urban life for most remains shaped by the Communist-era legacies of central planning. Giant apartment blocks, designed by bureaucrats and built for industrial workers, dominate most cities, especially on the fringes of pre-Soviet urban cores. In the growing cities, such as Moscow and St. Petersburg, the housing shortages of the Soviet era have not abated since 1991. Cramped, drab apartments badly in need of repair, with shared kitchens and bathrooms, are common. At the community scale, inadequate sewage, garbage, and industrial waste management pose serious long-term health and environmental threats.

In Moscow, housing shortages are irregular but particularly severe because the city, as the focus of new investment in the region, has grown rapidly since 1991. Due to high demand, the cost of housing has increased so dramatically that Moscow has one of the highest costs of living in the world. Other cities face housing shortages, even as their populations shrink, because of general deterioration and the absence of funds for maintenance (Figure 5.22B). Shortages are also exacerbated by the ability of wealthy people to buy up multiple apartments and refurbish them into one luxurious dwelling. A few cities that are still growing have attracted both government and private investment to increase the housing stock.

Shrinking Populations With high death rates and low birth rates, this region is undergoing a unique, late-stage variant of the demographic transition (see Chapter 1, page 16). In all but the Caucasian and Central Asian states, populations are shrinking faster than in any other world region. During the Soviet era, the increase in women's opportunities to become educated and work outside the home curtailed population growth. Severe housing shortages were an additional disincentive to having children, and many families chose to have only one or two children. Free health care and adequate retirement pensions also helped lower incentives for large families. During the last years of the Soviet Union, the population was already shrinking, but this trend accelerated during the economic crisis brought on by the breakup of the USSR. Since then, Russia's population has shrunk more than 5 percent, to 143 million. The United Nations predicts that Russia will drop to just 126 million people by 2050. Populations are also shrinking in Belarus, Moldova, and Ukraine. Rural areas have been particularly affected. The Russian geographer Tatyana Nefedova has calculated that areas located more than 20 to 25 miles (35 to 40 kilometers) away from large cities, areas that she calls "Russia's black holes," are increasingly more deserted, with only the elderly remaining. Even in Central Asia and the Caucasian states, where higher birth rates should be resulting in robust growth, countries are still losing population

FIGURE 5.21 LOCAL LIVES — PEOPLE AND ANIMALS IN RUSSIA AND THE POST-SOVIET STATES

A A trained golden eagle used for hunting is released by its master in Kazakhstan. Training an eagle requires great devotion over a long period of time. Raised by hand by their master, eagles are kept blindfolded from birth to make them depend on and trust their human companions. A Kazakh proverb says that "as the man trains his eagle, so does the eagle train his man." A golden eagle is on the flag of Kazakhstan.

B Siberian huskies take a break from training for a long-distance sled race. One of the oldest dog breeds, huskies were bred several thousand years ago by the Chukchi people in far eastern Siberia, near Alaska (see Figure 5.1 on pages 194–195) to pull heavy loads over long distances. With their strength and stamina, huskies helped the Chukchi and other peoples explore and populate new territories. They were also central to U.S. penetration into Alaska in the early twentieth century.

C Reindeer are tended by a group of the indigenous Nenets people in western Siberia. Reindeer have been raised throughout northern Russia for thousands of years for their meat, hides, antlers, milk, and to pull sleds. Having weathered the collapse of Soviet state-run herding and meat-production economies, reindeer herders in some areas now find themselves in conflict with Russia's oil and gas producers, whose facilities often block important migration routes.

Figure 5.22 Photo Essay: Urbanization in Russia and the Post-Soviet States

An uneven pattern of urbanization is developing in this region. In several countries, the largest cities are growing rapidly, as they are centers of new development and globalization. Meanwhile, in Russia and elsewhere, many cities are shrinking as the overall population declines.

A Moscow is the region's primate city—its population of 11.5 million people is larger than the next three largest cities combined. Moscow already produces more than 20 percent of Russia's GDP, a percentage that is continuing to grow because of the city's role as the headquarters of multinational corporations operating in Russia. Population is likely to increase further in Moscow, and there may also be many undocumented migrants in the city who are undercounted by the Russian census.

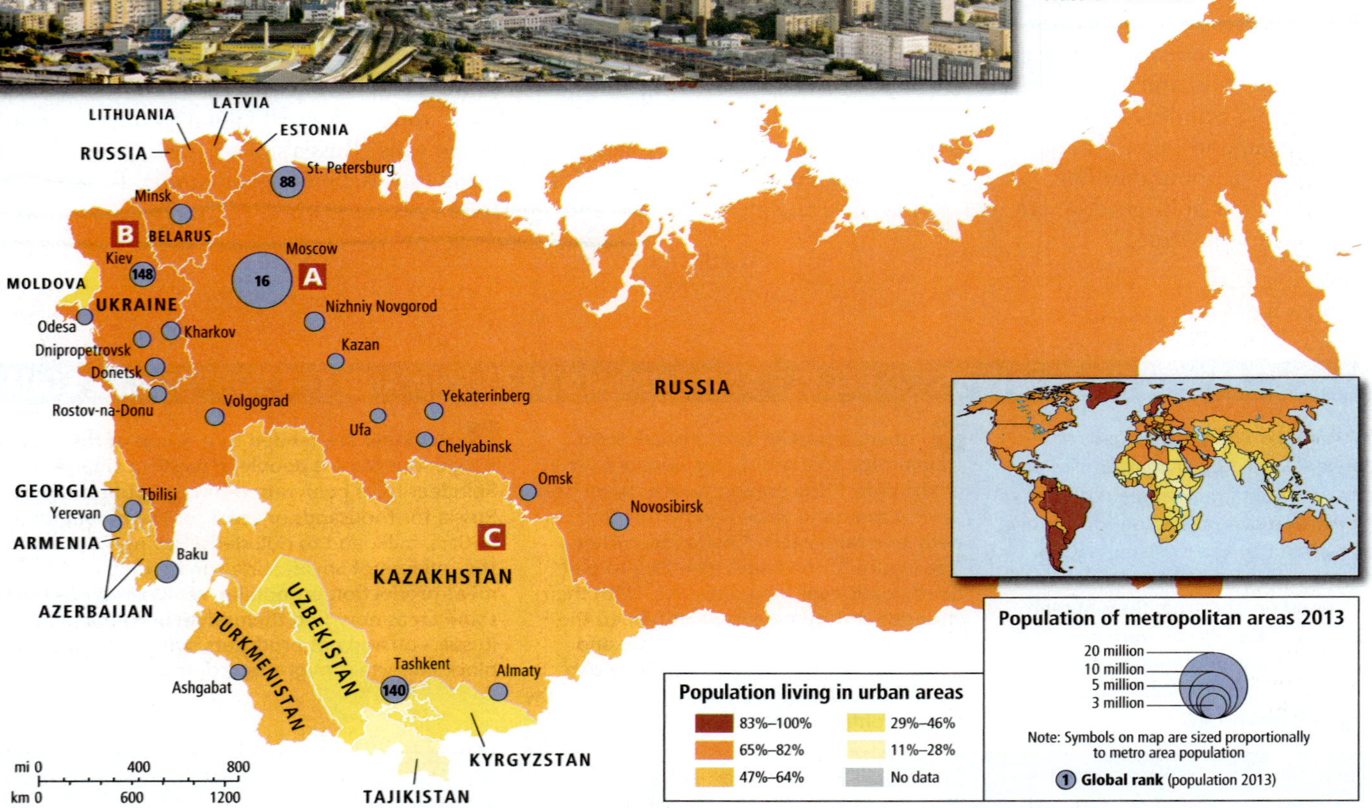

LITHUANIA
LATVIA
ESTONIA
RUSSIA
St. Petersburg 88
Minsk
BELARUS
Kiev 148
Moscow 16
MOLDOVA
UKRAINE
Nizhniy Novgorod
Odesa
Kharkov
Dnipropetrovsk
Kazan
Donetsk
Rostov-na-Donu
Volgograd
Yekaterinberg
Ufa
Chelyabinsk
RUSSIA
GEORGIA
Tbilisi
Yerevan
ARMENIA
Omsk
Novosibirsk
Baku
AZERBAIJAN
KAZAKHSTAN
UZBEKISTAN
TURKMENISTAN
Ashgabat
Tashkent 140
Almaty
KYRGYZSTAN
TAJIKISTAN
mi 0 400 800
km 0 600 1200

Population living in urban areas

- 83%–100%
- 65%–82%
- 47%–64%
- 29%–46%
- 11%–28%
- No data

Population of metropolitan areas 2013

20 million
10 million
5 million
3 million

Note: Symbols on map are sized proportionally to metro area population

① **Global rank** (population 2013)

B Abandoned buildings in Kiev, Ukraine, where the population is projected to decline by 0.02 percent by 2020, which is much less than for Ukraine as a whole, where the population is expected to decline by 5 percent by 2020.

C Astana, Kazakhstan's new capital, has undergone massive new development that has largely been supported by the increased exploitation of nearby oil resources by local and international companies. Astana doubled its population in the 10 years after it was designated the new capital.

Thinking Geographically

After you have read about urbanization in Russia and the post-Soviet states, you will be able to answer the following questions:

A Why is Moscow growing so rapidly?

B What larger-scale population trend does Ukraine's decline relate to?

C What is responsible for the growth of Astana?

because of emigration. Here, traditional rural lifestyles are often not economically viable anymore (see Figure 5.20). Many people migrate to Russia to find work. Russia's population decline would actually be larger than it already is if not for this stream of immigrants.

Surveys suggest that in the parts of this region that have low birth rates, people are now choosing to have fewer children primarily out of concern over gloomy economic prospects for the near future and also out of the desire to make money and have fun. Russia is attempting to reverse population loss by paying couples to raise more children and by attracting back Russians and their dependents who live abroad. In 2007 and 2008, Russia spent $300 million to send emissaries to the far corners of the Earth (Brazil, Egypt, Germany, and all the post-Soviet states) to lure "returnees." Only 10,300 were recruited.

In addition to negative birth and migration rates, much of the reason for the population decline is the declining life expectancy. In Russia, for example, between 1987 and 1994, male life expectancy dropped from 64.9 to 57.6 years. Since then, male life expectancy has recovered somewhat, but at 63 years, it is still the shortest in any developed country. Female life expectancy has not dropped as much and is now 75 years. Major causes of low life expectancies in the region are the loss of health care, which was usually tied to employment, and the physical and mental distress caused by lost jobs and social disruption. The higher male death rate is explained in large part by alcohol abuse and related suicides (women are less prone to both but tend to smoke in excess). More than half of all deaths among the working-age population are alcohol-related. On the positive side, the increases in life expectancy over the last few years indicate a modest improvement in health, which mirrors economic stabilization in Russia.

Population pyramids for several countries (Figure 5.23 on page 226) show the overall population trends in the region and reflect geographic differences in patterns of family structure and fertility. The pyramids for Belarus and Russia resemble those of European countries (for example, Germany, as shown in Figure 4.21A on page 184), but there are also differences. First, they are significantly narrower at the bottom, indicating that birth rates in the last several decades have declined sharply. Because so few children are born, in 20 years there will be few prospective parents, which means that low birth rates and population decline will continue into the foreseeable future. Also, the narrower point at the top for males shows their much shorter average life span compared to that of European men. The narrow top is also evident in Kazakhstan and Kyrgyzstan. In all countries, there has been a recent rebound in birth rates in the youngest age group, ranging from a slight increase in Russia to a significant rebound in Kazakhstan. Kyrgyzstan's pyramid indicates a younger population structure, which is common in less affluent economies. In the Central Asian countries, if emigration is controlled, population may expand during the next decades.

Geographic Patterns of Human Well-Being The recent increases in life expectancy and birth rates in some countries may indicate improvements in the general well-being of the population after the difficult period immediately following the Communist era. In Figure 5.24 (page 227), the color maps show that Russia and the post-Soviet states mostly fall in the middle range of the three well-being indicators that are presented. Only in three of the Central Asian states and Moldova is GNI per capita (PPP) low (less than U.S.$4000 per year). All countries in the region rank from medium high to medium low in the United Nations Human Development Index (HDI) and Gender Equality Index (GEI), although data on gender inequality are not available for all countries.

As we have observed elsewhere, GNI per capita can be used as a general measure of well-being because it shows in a broad way how much people in a country have to spend on necessities (see Figure 5.24A). However, because GNI per capita is averaged it does not indicate how income is distributed or how accessible health care, education, and other social services are to most people. In this particular region, the intermediate HDI ranks are in large part a holdover from the Soviet era, when socialism kept wages relatively equal (though they were usually lower on average outside of European Russia) and the strong social safety net provided basic health care, food, and housing for all. Since 1991, disparities in wealth have steadily increased within the Russian Federation and each of the post-Soviet states. We have observed that in Russia, a few people have become fabulously wealthy; their wealth has undoubtedly skewed the average figure. For many Russians, income is actually well below the country's $14,500 GNI per capita, as Russia has changed from a society based on economic equality to one where the gap between poverty and wealth is similar to that of the United States. The post-Soviet era has also revealed significant differences in income among the new states. The GNI per capita in Russia, Belarus, and Kazakhstan is more than $10,000, while in Kyrgyzstan and Tajikistan it is only about $2000.

Figure 5.24B depicts the ranks on the HDI, which is a calculation (based on adjusted real income, life expectancy, and educational attainment) of how adequately a country provides for the well-being of its citizens. It is possible to have a low GNI rank and a higher HDI rank if social services are sufficient or if there are no wide disparities of wealth. Georgia and Ukraine are two countries that stand out as having a good HDI (medium high) compared to their modest incomes (medium low). From a global perspective, Russia and the post-Soviet states rank slightly higher on HDI relative to their GDP, which may be a leftover result of the Soviet emphasis on broad education and basic social services that improved human well-being during most of the twentieth century. The lower HDI rankings are in the same Central Asian states and Moldova that also have low GNI per capita.

Figure 5.24C shows how countries rank in gender equality. First, as is true across the globe, nowhere in the region are women close to being equal to men. However, Russia and the post-Soviet states perform a little bit better with regard to gender equality than they do in terms of their GNI and HDI. All countries are in the

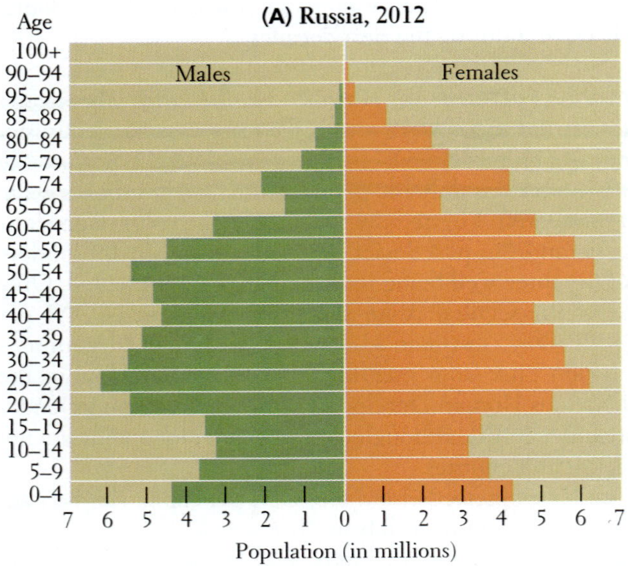

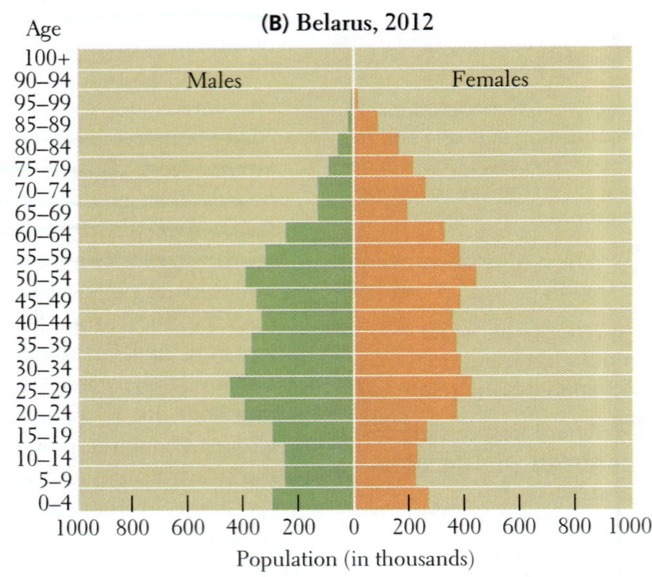

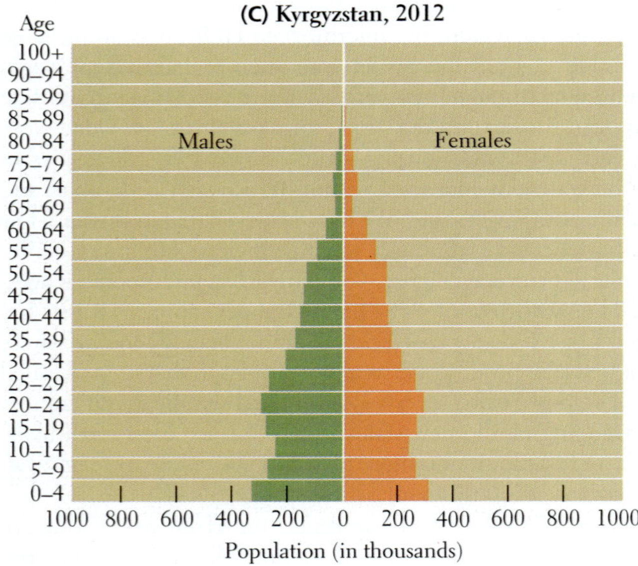

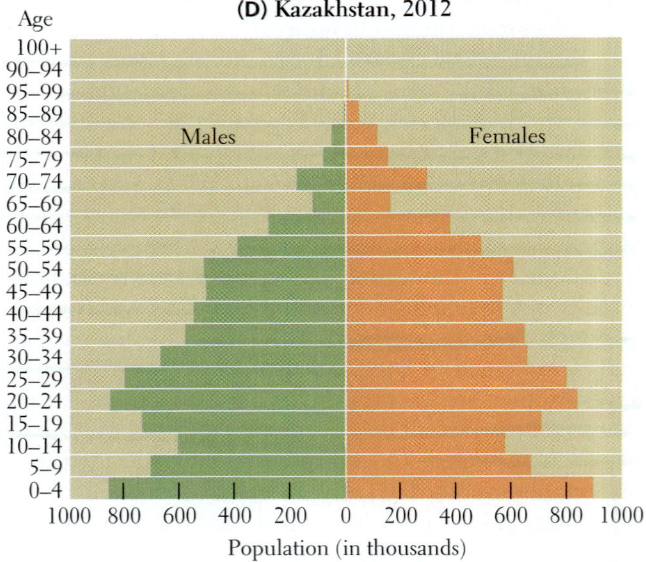

FIGURE 5.23 Population pyramids for Russia, Belarus, Kyrgyzstan, and Kazakhstan. Note that the pyramid for Russia is at a different scale (millions) than the other three pyramids (thousands) because of Russia's larger total population. This difference does not significantly affect the pyramid's shape.

medium or medium-high categories. This also means that, unlike for GNI and HDI, there is little variation across the region. Again, the Soviet system, which encouraged education and careers for women, is partly responsible for this record. The country with the lowest gender equality (Georgia) is not much more unequal than the country with the highest gender equality rank (Moldova). Unlike for HDI, the geographic gender equality differences that do exist among the countries in the region seem unrelated to economic differences; that is, some countries that have less gender equality have a low GNI, while other such countries have a medium GNI.

Gender: Challenges and Opportunities in the Post-Soviet Era

> **Geographic Insight 6**
>
> **Gender:** Men and women have been differently affected by the massive economic and political changes in this region during the twentieth century. Women have slowly gained economic power, while, on average, men have been adversely affected by economic instability, loss of jobs, and alcoholism.

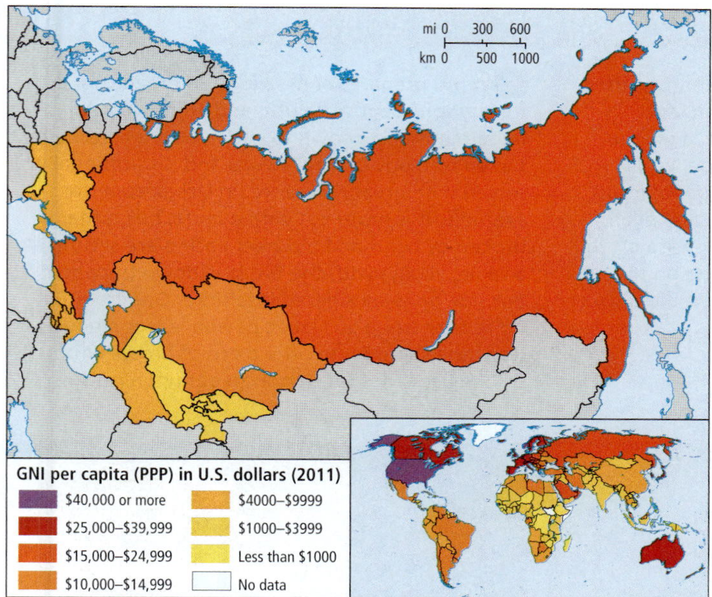

(A) GNI per capita, adjusted for purchasing power parity (PPP).

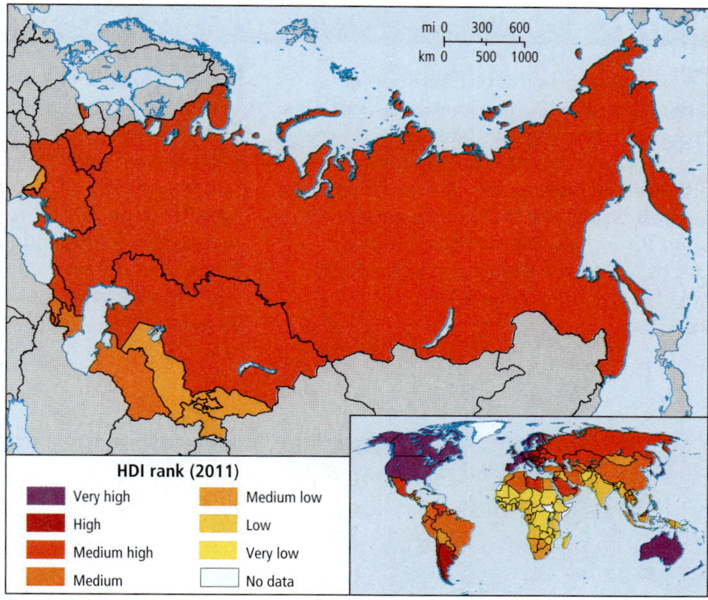

(B) HDI.

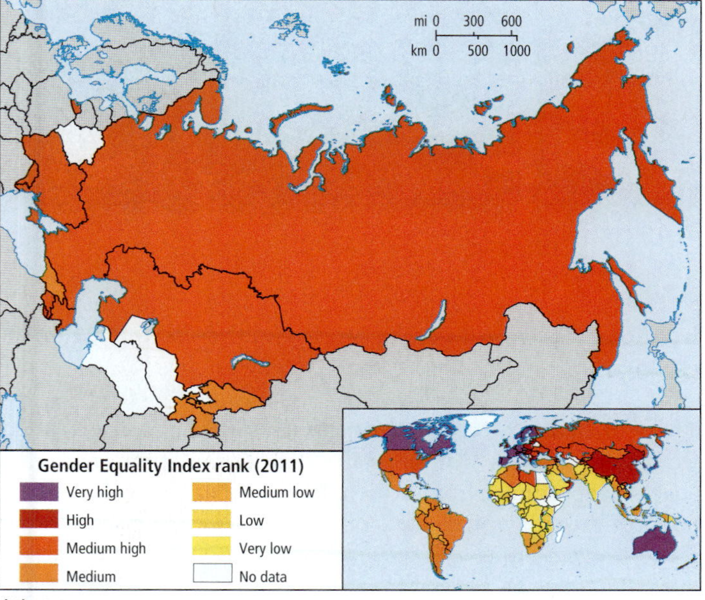

(C) GEI.

FIGURE 5.24 Maps of human well-being.

Soviet policies that encouraged all women to work for wages outside the home have transformed this region. By the 1970s, ninety percent of able-bodied women were working full time, giving the Soviet Union the highest rate of female paid employment in the world. However, the traditional attitude that women are the keepers of the home persisted. The result was the *double day* for women. Unlike men, most women worked in a factory or office or on a farm for 8 or more hours and then returned home to cook, care for children, shop daily for food, and do the housework (without the aid of household appliances). Because of shortages (the result of central-planning miscalculations), they also had to stand in long lines to procure necessities for their families.

These special pressures on working women affected diets because there was so little time or space for cooking and there were so few helpful appliances. The everyday cuisine in most of the region was a bit limited during the Communist era; only recently have market forces made a wider range of food available. Traditional recipes are now being revived. Figure 5.25 (page 228) shows several distinct dishes of the region that are prepared at home or in restaurants. The popularity of hearty soups and stews and the common use of root vegetables and grains (bread is an essential part of every meal) reflect the region's climate, soil, and agricultural systems.

When the first market reforms in the 1980s reduced the number of jobs available to all citizens, President Gorbachev encouraged women to go home and leave the increasingly scarce jobs to men. Many women lost their jobs involuntarily. By the late 1990s, women made up 70 percent of the registered unemployed, despite the fact that because of illness, death, or divorce, many if not most were the sole support of their families. Consequently, many had to find new jobs.

On average, the female labor force in Russia is now more educated than the male labor force. The same pattern is emerging in Belarus, Ukraine, Moldova, and parts of Muslim Central Asia. In Russia, the best-educated women commonly hold professional jobs, but they are unlikely to hold senior supervisory positions. They also are paid less than their male counterparts. In 2007, the wages of women professional workers averaged 36 percent less than those of men. Still, this region ranks higher in gender income equity than many others.

The Trade in Women During the economic boom stimulated by marketization and oil and gas wealth, the "marketing" of

FIGURE 5.25 LOCAL LIVES FOODWAYS IN RUSSIA AND THE POST-SOVIET STATES

A A bowl of borscht, or beet soup, a Ukrainian dish with many variants, is popular throughout this region. Beets are extremely nutritious root vegetables that can tolerate cold weather well, enabling farmers to make the most out of the short growing season found throughout much of this region. Borscht is served warm in the winter and chilled in the summertime.

B Pirozhki are baked yeast buns stuffed with meat, vegetables, or fruit, and glazed with egg to give them a golden color. Shown here are cabbage, egg, and dill pirozhki. They are a popular fast food throughout this region, and even in neighboring Mongolia. They are often made at home and, depending on the stuffing, can be an appetizer, main dish, or dessert.

C A girl prepares *non*—a kind of bread cooked throughout Central, South, and Southwest Asia (where it is also known as *naan*). Non is considered sacred by many Central Asians and is traditionally served with great reverence. In Uzbekistan, for example, it is never cut with a knife, but rather is broken by hand and put near each place setting at a table, always with the flat side down (serving it flat side up is considered insulting). One tradition holds that when an Uzbek person leaves a house, he or she should bite off a piece of bread, which will be kept for that person to eat upon return.

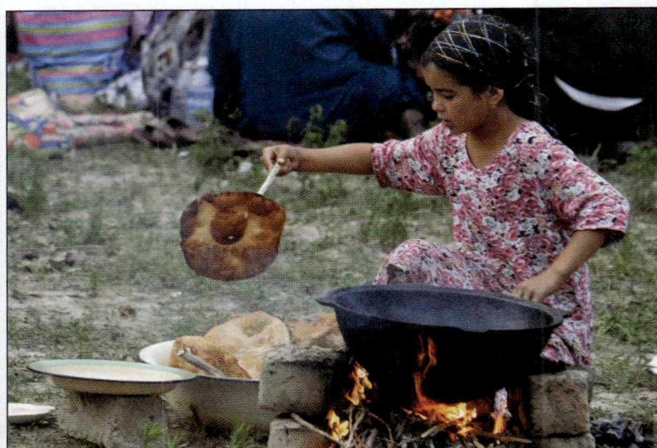

women became one of the less savory entrepreneurial activities. One part of this market is the Internet-based mail-order bride services aimed at men in Western countries. A woman in her late teens or early twenties, usually seeking to escape economic hardship, pays about $20 to be included in an agency's catalog of pictures and descriptions. (One Internet agency advertises 30,000 such women.) She is then assessed via e-mail or Facebook by the prospective groom, who then travels—usually to Russia or Ukraine—to meet women he has selected from the catalog and to potentially have one of them accompany him to the West to marry.

In recent years, there has also been a large increase in sex work. Precise numbers are hard to come by; however, a 2010 UN report estimates that 140,000 women have been smuggled into Europe, where they become part of a sex trade that generates $3 billion annually for racketeers. Women come from different parts of the world, but the post-Soviet states constitute the primary source region. The business of supplying sex workers is dominated by members of the Russian Mafia, who have been known to kidnap schoolgirls or deceive women who are looking for jobs as domestic servants or waitresses in Europe, and then force them to work as prostitutes. This is a form of **trafficking**, which is defined by the United Nations as the recruiting, transporting, and harboring of people through coercion for the purpose of exploiting them.

trafficking the recruiting, transporting, and harboring of people through coercion for the purpose of exploiting them

The Political Status of Women The most effective way for women to address institutionalized discrimination is to gain access to positions from which they can affect wide-ranging policies—usually as elected officials. Although women were granted equal rights in the Soviet constitution, they have never held much power. In 1990, women accounted for 30 percent of Communist Party membership but just 6 percent of the governing Central Committee. The very few in party leadership often held these positions at the behest of male relatives.

Ironically, since the fall of the Soviet Union, the political empowerment of women has advanced the least where democracy has developed the most, perhaps because of long-standing cultural biases against women in positions of power. Where governments are less democratic and more authoritarian, as in Belarus and Kyrgyzstan, women actually hold more legislative positions (Figure 5.26). However, in those countries, leaders who wanted to appear more progressive in the eyes of international donors may have promoted women undemocratically, choosing those who were least likely to work for change. Ironically, support among women for women's political movements is not widespread, as many fear being seen as anti-male or against traditional feminine roles.

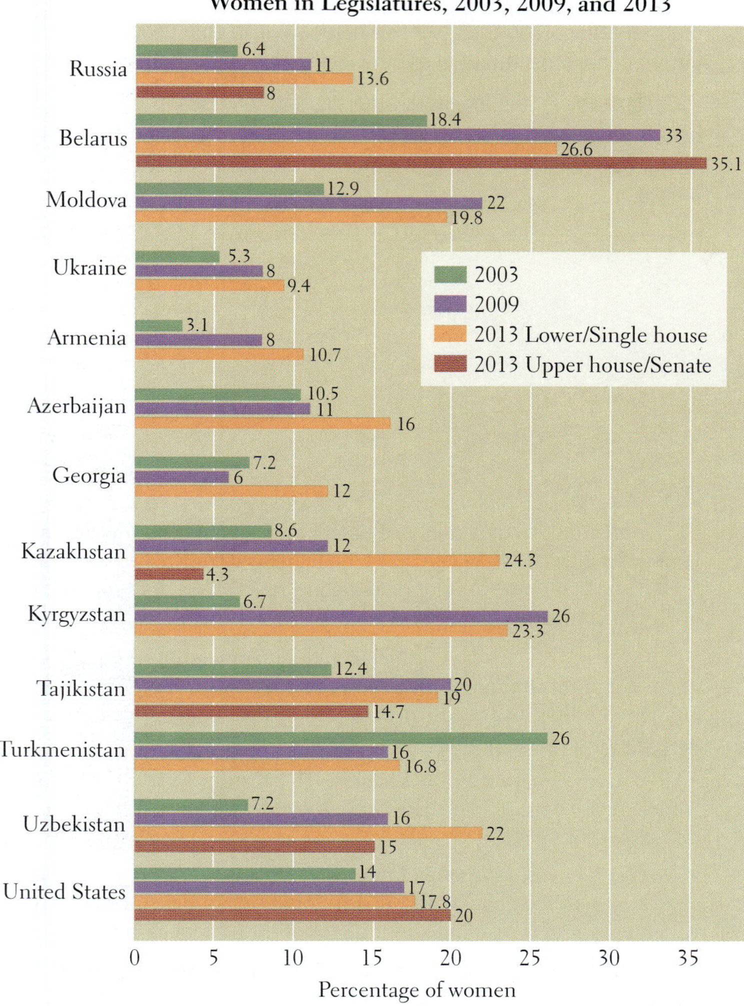

FIGURE 5.26 Women legislators, 2003 and 2013. This graph shows the percentage of legislators in Russia and the post-Soviet states (with the United States for comparison) who are women. Notice that over the last decade in some countries, the number of women in parliament has expanded and then contracted. Also, the countries with the largest percentage of female legislators are not necessarily those with the most open democratic participation.

Religious Revival in the Post-Soviet Era

The official Soviet ideology was atheism, and religious practice and beliefs were seen as obstacles to revolutionary change. The Orthodox Church was tolerated, but few went to church, in part because the open practice of religion could be harmful to one's career. Now, religion is a major component of the general cultural revival across the former Soviet Union. This may be happening because people turn to traditional practices during uncertain times, or as a way to assert their ethnic identity. The overall level of religious practice is moderate in Russia by global comparisons, but a number of people, especially those from indigenous ethnic minority groups, are turning back to ancient religious traditions. For example, the Buryats, from east of Lake Baikal in Siberia, who are related to the Mongols, are relearning the prayers and healing ceremonies of the Buryat version of Tibetan Buddhism, which they adopted in the eighteenth century. The shamans who lead them have organized into a guild to give official legitimacy to their spiritual work. They now pay taxes on their clergy income.

In Russia, Ukraine, Moldova, Belarus, Georgia, and Armenia, most people have some ancestral connection to Orthodox Christianity. Those with Jewish heritage form an ancient minority, mostly in the western parts of Russia and Caucasia, where they trace their heritage back to 600 B.C.E. Religious observance by both groups increased markedly in the 1990s, and many sanctuaries that had been destroyed or used for nonreligious purposes by the Soviets were rebuilt and restored.

A countertrend to the robust revival of Orthodox Christianity is the spread of evangelical Christian sects from the United States (Southern Baptists, Adventists, and Pentecostals). Evangelical Christianity first came to Russia in the eighteenth century, but after 1991, American missionary activity increased markedly. The notion often promoted by this movement—that with faith comes economic success—may be particularly comforting both to those struggling with hardship and to those adjusting to new prosperity.

In Central Asia, Islam was long repressed by the Soviets, who feared Islamic fundamentalist movements would cause rebellion against the dominance of Russia. Today, Islam is openly practiced and increasingly important politically across Central Asia, Azerbaijan, and some of the Russian Federation's internal ethnic republics, such as Chechnya and Tatarstan. Especially in the Central Asian states, however, the return to religious practices is often a subject of contention. Some local leaders still view traditional Muslim religious practices as obstacles to social and economic reform.

Both devout Muslims and more secular reformists are wary of religious extremists from Iran and Saudi Arabia who may pose a security risk. During the 1990s, Tajikistan fought a civil war against Islamic insurgents with links to the Taliban in Afghanistan—and the conflict in Chechnya involves combatants with similar links to Saudi militants. In 2000, Uzbekistan and Kyrgyzstan joined forces to eliminate an extremist Islamic movement. But many religious leaders and some human rights groups say that the fervor to eliminate radical insurgents has resulted in the persecution of ordinary devout Muslims, especially men, and that this persecution is recruiting anti-Western enthusiasts. Human Rights Watch, an organization that monitors human rights abuses worldwide, reports that Uzbekistan's government jails and tortures believers who worship independently outside the strict supervision of the state. The authorities are concerned that the Arab Spring (see Chapter 6, pages 267–269) will spread and lead to demands for more human rights and religious freedom in Uzbekistan too.

THINGS TO REMEMBER

Geographic Insight 5

• **Population** Population trends in this region are highly uneven. Russia and the countries bordering Europe have the most rapidly declining populations on Earth. Meanwhile, the Central Asian and Caucasian countries have higher birth rates, though they are still losing people to emigration.

- An uneven pattern of urbanization is developing in this region. In several countries, the largest cities are growing rapidly, as they are centers of new economic development and globalization. Meanwhile, in Russia and elsewhere, many cities and rural regions are shrinking as the overall population declines.

Geographic Insight 6

- **Gender** The Soviet Union gave strong incentives for women to work and become involved in politics, but in the post-Soviet era, gaps have been increasing between men and women in employment rates and political representation. New threats to women have also emerged, such as international prostitution rings, now active in some countries.

- Official Soviet ideology was atheism, and religious practice and beliefs were seen as obstacles to revolutionary change. In the post-Soviet era, religion is a major component of the general cultural revival.

Geographic Insights: Russia and the Post-Soviet States Review and Self-Test

1. Environment, Development, and Urbanization: Development and urbanization in this region have long taken precedence over environmental concerns. The region has severe environmental problems, especially in urban areas.

- Describe the challenges posed by the physical environments (landforms and climate) of the various parts of this region: the North European Plain, the Caucasus, the Central Asian steppes, Siberia, and the various mountain ranges (e.g., Urals, Pamirs, Tien Shan).
- How did Soviet attitudes contribute to environmental problems? What, if any, solutions have emerged? How have patterns of economic development affected the environment during the last two decades?
- The diffusion of Western economic practices has affected Russia and the post-Soviet states. Give examples of how these ideas and methods are changing urban landscapes in the region.

2. Climate Change, Food, and Water: Food production systems and water resources are particularly threatened by climate change in parts of this region, yet the region remains the third-largest global producer of greenhouse gases.

- Where in this region are food production systems and water resources most threatened by climate change?
- Where are irrigated agriculture systems particularly vulnerable to rising temperatures? How can they adapt?
- Which parts of this region produce the most greenhouse gases?

3. Globalization and Development: Since the fall of the Soviet Union, economic reforms and globalization have changed patterns of development in this region. Many Communist-era industries closed, and as jobs were lost and public assets sold to wealthy oligarchs, wealth disparity increased. The region is now dependent on its role as a leading global exporter of energy resources.

- Discuss how the transition from a communist system to a market economy was difficult for Russia and the post-Soviet states. Which areas have benefited and which ones have suffered from recent economic reforms?
- Discuss the evidence that Russia's natural resources are becoming the foundation of a new role for Russia in the global economy and in regional politics. How widely have the benefits from Russia's new fossil fuel–based economic development been shared? How does the situation in Russia compare with that in Central Asia?

4. Power and Politics: Obstacles to democratization include a long history of authoritarian regimes and the popular desire to implement change rapidly, combined with the lack of mechanisms for public participation in the political process.

- How have economic and political instability influenced the kind of leaders that have emerged? How have the color revolutions influenced democratization?
- Discuss the reasons for the diminishing of Russia's sphere of influence after 1991 and the changes in Russia's domestic politics since 2000, which appear to be leading a resurgence of its influence in global politics, particularly in Eurasia.

5. Population: Populations are shrinking in many parts of this region. This is a function of a continuously high level of participation by women in the workforce and overall poor living and housing standards.

- What Soviet-era developments helped curtail population growth? What is being done in the post-Soviet era to encourage population growth?
- How and why has access to health care and other social services changed for so many in this region since the fall of the USSR?

6. Gender: Men and women have been differently affected by the massive economic and political changes in this region during the twentieth century. Women have slowly gained economic power, while, on average, men have been adversely affected by economic instability, loss of jobs, and alcoholism.

- What is the social standing of women in Russia and the post-Soviet states compared to their standing in other world regions?
- How has the gap between men and women changed in terms of economic and political empowerment during the post-Soviet era? What new threats to women have emerged?

Critical Thinking Questions

1. What were the social circumstances that gave rise to the Russian Revolution? Explain why the revolutionaries adopted a form of communism grounded in rapid industrialization and central planning.

2. In general terms, discuss the ways in which the change from a centrally planned economy to a more market-based economy has affected career options and standards of living for the elderly, young professionals, unskilled laborers, members of the military, women, and former government bureaucrats.

3. Discuss the declining birth rate in Russia and other countries in the region. What are its major causes? What do you think the impact of this decline can have on neighboring countries and regions? Are similar declines happening elsewhere?

4. Given Russia's crucial role in World War II and the defeat of Germany, discuss what led to the Cold War between the Soviet Union and the West.

5. Given the changing options for women in the region, what do you anticipate will be the trend in population growth in the future? Will birth rates remain low or will they increase? Why?

6. Ethnic and national identities are affecting how the internal republics of Russia envision their futures. How might the geography of Russia change if these feelings were to intensify? In what ways might the global economy be affected?

Chapter Key Terms

Bolsheviks 209
capitalists 209
Caucasia 221
centrally planned, or **socialist, economy** 210
Cold War 210
communism 209
Communist Party 210
czar 208
Gazprom 212
glasnost 213
Group of Eight (G8) 212
Mongols 207
nomadic pastoralists 205
nonpoint sources of pollution 201
oligarchs 212
perestroika 213
permafrost 199
privatization 213
Russian Federation 198
Russification 217
Siberia 199
Slavs 207
Soviet Union 196
steppes 199
taiga 199
trafficking 228
tundra 199
underemployment 214
Union of Soviet Socialist Republics (USSR) 196

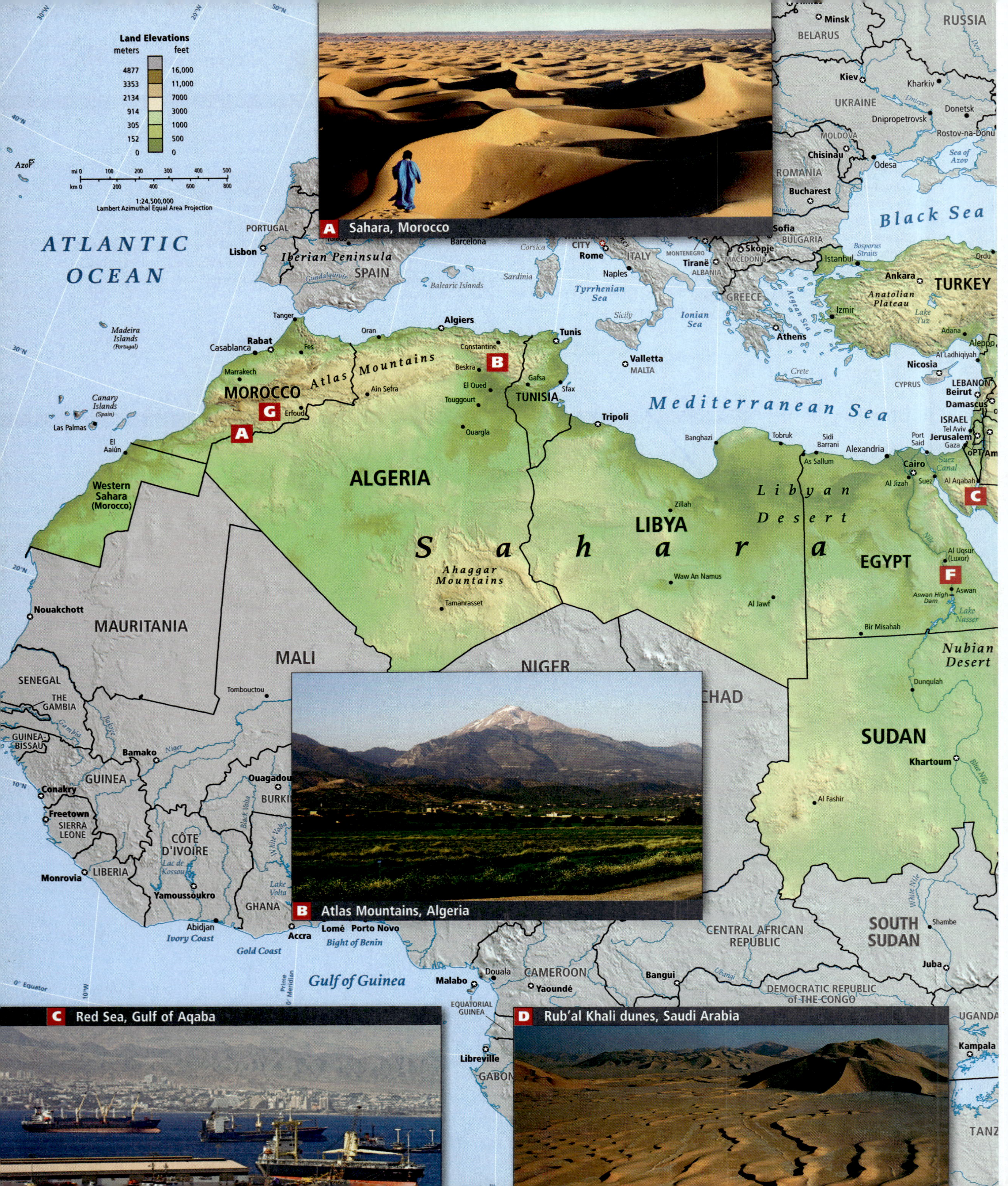

A Sahara, Morocco

B Atlas Mountains, Algeria

C Red Sea, Gulf of Aqaba

D Rub'al Khali dunes, Saudi Arabia

6 North Africa and Southwest Asia

E Mt. Ararat, Turkey

Volgograd
Caucasus
GEORGIA
Tbilisi
AZERBAIJAN
Baku
ARMENIA
Yerevan
TURKMENISTAN
Ashgabat
Mt. Ararat elev. 16,946
Tabriz
Van
Lake Van
Lake Urmia
Caspian Sea
Elburz Mts.
Tehran
Mashhad
AFGHANISTAN
Tabas
Mosul
Kermanshah
SYRIA
IRAN
Esfahan
Euphrates
Tigris
Baghdad
IRAQ
Kerman
PAKISTAN
Zagros Mts.
JORDAN
Abadan
Al Basrah
Shiraz
Kuwait
KUWAIT
Bushehr
Rafha
Bandar-e Abbas
Persian Gulf
Strait of Hormuz
Bandar Beheshti
Tabuk
Ad Dammam
BAHRAIN
Manamah
QATAR
Doha
Abu Dhabi
Dubai
Gulf of Oman
Muscat
Arabian Peninsula
Riyadh
UNITED ARAB EMIRATES
OMAN
Al Madinah (Medina)
SAUDI ARABIA
Red Sea
Jiddah
Makkah (Mecca)
D
Rub'al Khali
Al Jawarah
Port Sudan
Abha
Arabian Sea
YEMEN
ERITREA
Asmera
Sana'a
Taizz
Aden
Gulf of Aden
DJIBOUTI
Djibouti
Lake Tana
Somali Peninsula
Horn of Africa
SOMALIA
Addis Ababa
ETHIOPIA
Ethiopian Highlands
Lake Turkana
Mogadishu
KENYA
Nairobi
INDIAN OCEAN
Victoria
SEYCHELLES
Dar es Salaam
20°N
10°N
0° Equator
10°S
60°E

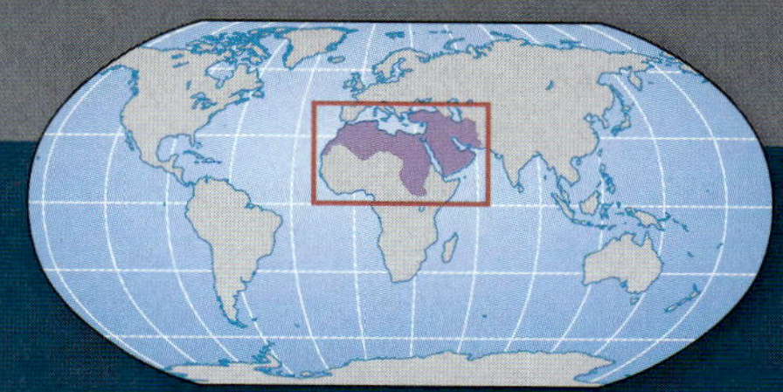

F Nile River, Egypt

G Wadi el Mellah, Morocco

FIGURE 6.1 Regional map of North Africa and Southwest Asia.

GEOGRAPHIC INSIGHTS: NORTH AFRICA AND SOUTHWEST ASIA

After you read this chapter, you will be able to discuss the following geographic insights as they relate to the nine thematic concepts:

1. Water, Food, and Climate Change: The population of this predominantly dry region is increasing faster than is access to cultivable land and water for agriculture and other human uses. Many countries here are dependent on imported food, and because global food prices are unstable, the region often has food insecurity. Climate change could reduce food output both locally and globally and make water, already a scarce commodity, a source of conflict. New technologies offer solutions to water scarcity but are expensive and unsustainable.

2. Gender and Population: Gender disparities are becoming more noticeable in this region as political change begins to occur. Women are dependent because they are generally less educated than men, have few options for working outside the home, and have little access to political power. Women's roles are usually confined to the home and to having and raising children. As a result, this region has the second-highest population growth rate in the world, after sub-Saharan Africa.

3. Urbanization and Globalization: Two patterns of urbanization have emerged in the region, both tied to the global economy. In the oil-rich countries, the development of spectacular new luxury-oriented urban areas is based on flows of money, credit, goods, and skilled people. But most cities are old, with little capacity to handle the mass of poor, rural migrants attracted by export industries aimed at earning foreign currency. In these older cities, many people live in overcrowded slums with few services.

4. Development and Globalization: The vast fossil fuel resources of a few countries have transformed economic development and driven globalization in this region. In these countries, economies have become powerfully linked to global flows of money, resources, and people. Politics have also become globalized, with Europe and the United States strongly influencing many governments.

5. Power, Politics, and Gender: Despite the predominance of elected bodies of government, and the increasing participation of women as voters, wealthy, politically connected males, clerics, and often the military retain the real power. Waves of protest, which included some women participants, swept this region, beginning with the Arab Spring of 2010, and resulted in the overthrow of several authoritarian governments. Within the region, official response to the Arab Spring has included both repression and reforms, but only a few of the positive changes hold promise for women.

The North African and Southwest Asian Region

The region of North Africa and Southwest Asia contains 21 countries plus the occupied Palestinian Territories (see Figure 6.1 on pages 232–233; see also Figure 6.2). Physically, the region is part of two continents: Africa and Eurasia. The North African countries of this region stretch from Western Sahara and Morocco on the Atlantic Ocean, through Algeria, Tunis, Libya, and Egypt along the Mediterranean Sea, plus Sudan, lying south of Egypt. Eurasia begins on the eastern side of the Red Sea with the Arabian Peninsula, which consists of Saudi Arabia, Yemen, Oman, the United Arab Emirates, Qatar, Bahrain, and Kuwait (see Figure 6.1C); the countries of the eastern Mediterranean *littoral* (shoreline), including Jordan, Israel, the occupied Palestinian Territories, Lebanon, Syria, and Turkey. Iraq and Iran lie inland. In political discussions, this region is often referred to as the *Middle East*, a term we do not use in this book because it arose during the colonial era and describes the region from a limited Euro-American geographic perspective. To someone in China or Japan, the region lies to the far west, and to someone in Russia, it lies to the south.

The nine thematic concepts in this book are explored as they arise in the discussion of regional issues, with interactions between two or more themes featured, as in the geographic insights on this page. Vignettes, like the one that follows about the rising role of women in this region, illustrate one or more of the themes as they are experienced in individual lives.

GLOBAL PATTERNS, LOCAL LIVES When in the spring of 2012, Mona Eltahawy, a Muslim Egyptian-American journalist, wrote an essay for *Foreign Policy* magazine entitled "Why Do They Hate Us?" she was angry. It is easy to see why. The *Arab Spring*—that seemingly merry embrace of democracy that washed across North Africa and into Egypt, beginning in 2010—held the promise of real change in her native land. For a time, the tens of thousands of largely peaceful demonstrators, male and female, in Cairo's Tahrir Square shared a common purpose: the resignation of the autocratic president, Hosni Mubarak. He had remained in power for nearly 30 years, aided by a powerful and oppressive military, and supported by internal corruption and large foreign aid packages from the United States.

For the many thousands of women—students and young professionals—who participated in the Cairo demonstrations, it was exhilarating to be welcomed by male compatriots, many of whom were members of the Muslim Brotherhood, a large and amorphous organization of men opposed to the corrupt Mubarak regime and its affiliations with the West. The Muslim Brotherhood, however, is also known for its stance against liberalizing women's rights. For a time, the common purpose of ousting Mubarak united everyone. Eltahawy, who grew up in Saudi Arabia, the United Kingdom, and Egypt, and who has focused her career on social justice and women's rights in Egypt, joined the throng of demonstrators. As the protests went on, not only did retaliation by the military and police toward male and female demonstrators become more and more brutal, but antagonism toward women by the Muslim Brotherhood demonstrators and even from young, more liberal male demonstrators grew, leading to random violence by men

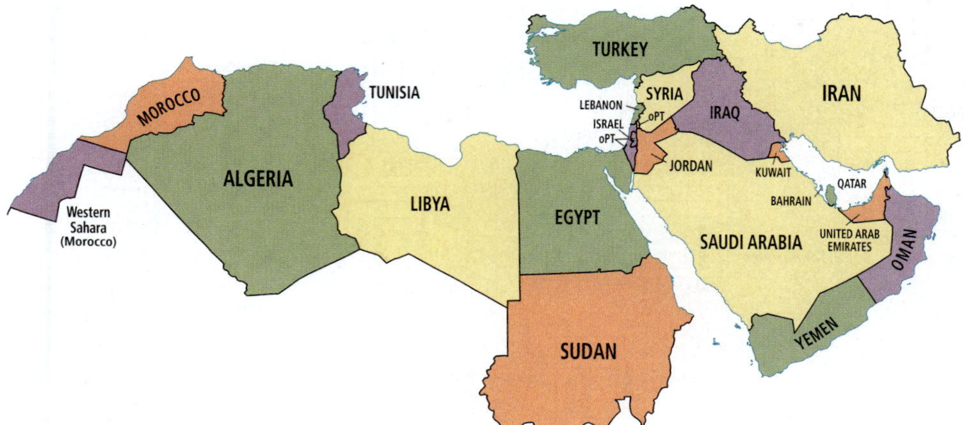

FIGURE 6.2 Political map of North Africa and Southwest Asia.

against women (**Figure 6.3**). Mubarak's resignation in February 2011 left undetermined the way in which a new government would be established; so as the military asserted control, the demonstrations continued. One day in the fall of 2011, Eltahawy, who continued to report on the increasingly frustrated demonstrators in Tahrir Square, was arrested along with some others. While in custody, both her arms were broken in a beating, and she was sexually assaulted—a type of treatment that she observed was especially aimed at women demonstrators.

FIGURE 6.3 Women and the Arab Spring. An Egyptian man holds up a widely circulated photo that shows riot police stripping and beating a woman during an Arab Spring protest in Cairo in 2011. Women have been very active in protests in Egypt and elsewhere in the region in recent years, and some groups, including parts of Egypt's government, have felt threatened by this. Some female protestors have become a target for sexual assaults designed to make others perceive them as dangerous, "dirty," or immoral.

In her essay, which she wrote some months after her release, Eltahawy raged against the repressive political authorities, whom she held in contempt, and against Muslim men as a whole, who she had hoped would have exhibited solidarity with women demonstrators. Her anticipation that gender equity would be one of the end products of the Arab Spring was dashed by what she saw and experienced, so she lashed out with a diatribe against the many large and small ways in which she perceives women are belittled—not by the Qur'an (see page 240), but by religious social practice in Egypt and elsewhere (**Figure 6.4**). In so expressing her rage, Eltahawy seems to have invigorated a cross-regional discussion on gender equity that is attracting serious male and female participants who are expressing a surprising level of agreement that the Arab Spring must include the human rights of women. *[Source: Foreign Policy, Al Jazeera. For more detailed information, see Text Credits page.]* ■

The Arab Spring, a movement that will likely continue evolving for decades, is just one more revolution brought on by the post-colonial awareness that all societies have inequities of some sort that must be addressed. The issues raised are deep and structural and political; that is, they have to do with realigning power so that whole segments of society—the poor, the uneducated, the educated unemployed, minorities, and women—will no longer be subjugated by systems that favor privileged elites.

The issue of gender equity in the region of North Africa and Southwest Asia is controversial because many people are convinced that only Westerners raise this issue, and that they do so in order to browbeat Muslim culture for having a set of values that places females in a special category. But Mona Eltahawy's article and, more importantly, the discussions that have erupted about it across the region indicate that gender equity is actually a central theme in the Arab Spring movement. The taboo against talking about gender is fading slightly, as demonstrated in broadcasts by the global Al Jazeera network based in Qatar (see pages 269–270). Some women in the region are beginning to say that it is they who will lead the transformation.

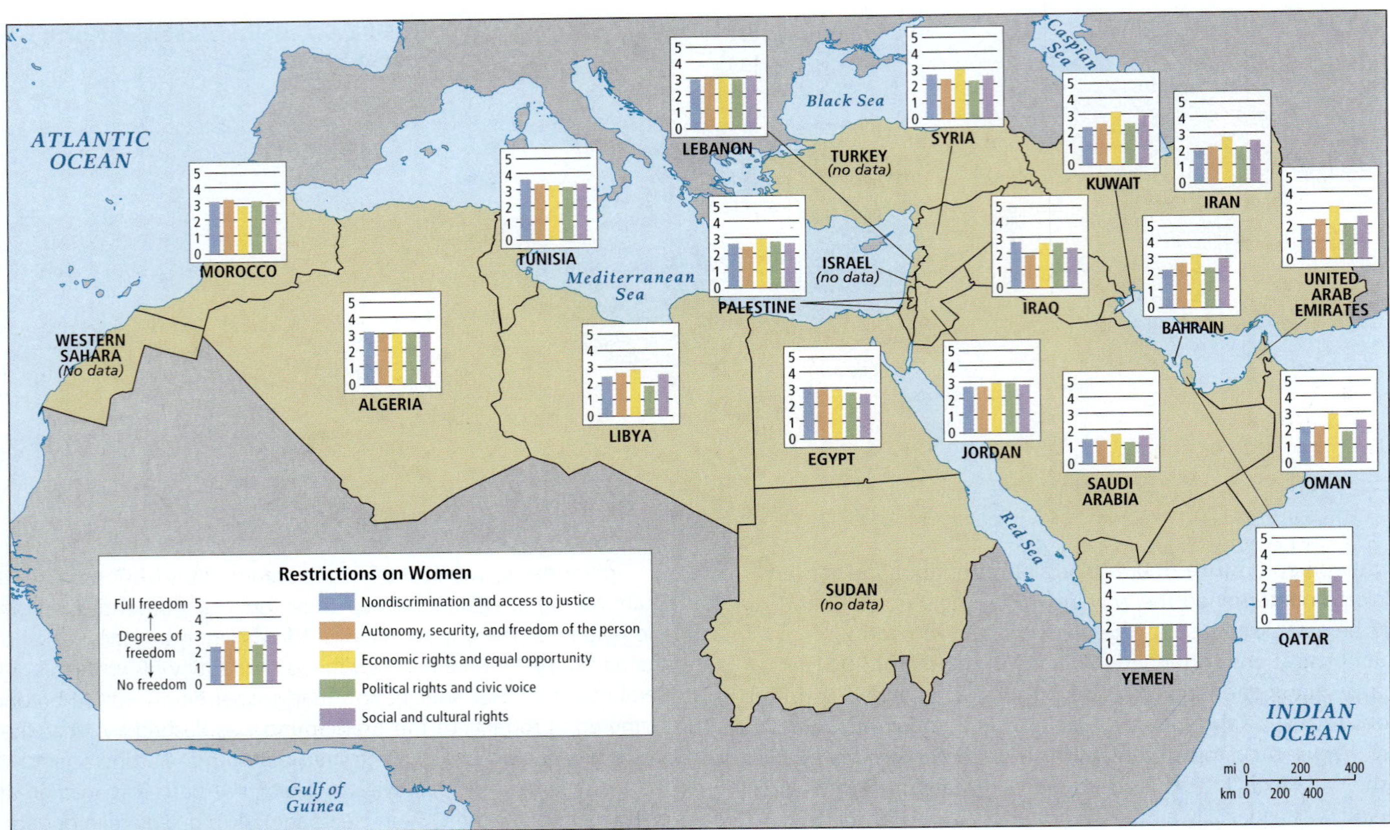

FIGURE 6.4 Variations in women's freedom to exercise their human rights in five categories (2009). Each country was evaluated on a scale of 1 to 5, with 1 being the lowest degree of freedom and 5 being the highest. A variety of factors was assessed for each nation and compiled into the five categories listed above. For reasons unknown, three countries—Turkey, Israel, and Sudan—were identified as not supplying data for this study. However, Freedom House reports elsewhere that in Israel, although women have achieved substantial parity at almost all levels of Israeli society, Arab communities in Israel and the occupied Palestinian Territories and ultra-Orthodox Jewish communities in Israel impose discrimination on women. In Turkey, the constitution grants women full equality before the law, but the World Economic Forum ranked Turkey at 124 out of 135 countries surveyed in its 2012 Global Gender Gap Index. Only about one-third of Turkish women participate in the labor force, and women hold just 14 percent of seats in Parliament. In Sudan, discrimination against women is severe.

THINGS TO REMEMBER

- The Arab Spring, a broad-based but not highly organized movement against corruption and toward more democratic institutions, has spread across the region since 2010.
- Women are increasingly participating in political movements but continue to face difficulties in doing so.
- Issues of gender equity are increasingly becoming central to the public dialogue.

THE GEOGRAPHIC SETTING

What Makes North Africa and Southwest Asia a Region?

To most outsiders, North Africa and Southwest Asia is a region characterized by five qualities: (1) It is the center of the religion of Islam; (2) it is the locus of a great deal of Earth's petroleum resources; (3) it has an overall scarcity of water; (4) it is predominantly Arab in ethnicity; and (5) it is a region where women are discriminated against. While these five features are all present, they alone are a far too simplistic picture of the region.

The vast majority of people practice **Islam**, a monotheistic religion that emerged between 601 and 632 C.E., a faith with multiple interpretations. Most Muslims are moderate in their thinking and accept the validity of other beliefs, especially that of Christianity, because Mary and Jesus both play important roles in Islam. Only a

Islam a monotheistic religion that emerged in the seventh century C.E. when, according to tradition, the archangel Gabriel revealed the tenets of the religion to the Prophet Muhammad

vocal minority of Muslims is drawn to ultra-fundamentalist versions of Islam known as *Islamism*. The term **Islamism,** which gained currency in the late twentieth century, refers to Muslim religious activist movement that seeks political power to curb what is seen as dangerous secular influences that are spreading as a result of globalization. Some Islamist movements characterize Western influence as corrupt and destructively self-serving. But Islamist movements also vary greatly, are stronger in some parts of the region than in others, and have waxed with economic recessions and waned with economic booms.

Fossil fuel reserves—made up of oil and natural gas formed over millions of years from the fossilized remains of dead plants and animals—are highly uneven in their distribution (see Figure 6.25 on page 264) and are found mainly around the Persian Gulf. These fossil fuels are extracted and exported throughout the world at tremendous profit, but the profits are not equitably distributed to the people in the countries from which they are generated. For example, in the oil-rich Gulf states (see the discussion of wealth disparities), profits go primarily to the members of a few large, privileged families, and only modest amounts are spread to the rest of the citizens.

Regarding water resources, the region is correctly thought of as being generally arid, as will be discussed in the next section on physical geography. The degree of aridity and its impact on development varies widely. Newly revealed underground water resources may change development options.

While Arabic culture and language is widespread, many people in the region are not Arab; the second and third most populous countries in the region, Turkey and Iran, are of non-Arab ethnicities, as are many minority populations, such as the Kurds, Berbers, Christians, and Jews.

Finally, the role and status of women, long a point of contention, is in transition, as the opening vignette makes clear. The experiences of women vary widely from country to country and from rural to urban areas. Even within a forward-looking country like Turkey, where urban women may be highly educated and outspoken and active in commerce, public life, and government, in rural areas women may lead secluded domestic lives with few educational opportunities. Women are beginning to lead the fight for gender equity in the region, and their life options are expanding.

Terms in This Chapter

In this book, as mentioned previously, we choose to not use the common term *Middle East*. The *Arab world* is used only where it applies, since many people in the region are not of Arab ethnicity.

We use the term **occupied Palestinian Territories (oPT)** to refer to Gaza and the West Bank, those areas where Israel still exerts control despite treaty agreements; the word "occupied" is not capitalized to show the supposed temporary quality of the occupation. The U.S. Department of State uses the term *Palestinian Territories*. The United Nations, upon recently agreeing to give the territory observer status, now refers to it as the *State of Palestine*.

Physical Patterns

Landforms and climates are particularly closely related in this region. The climate is dry and hot in the vast stretches of the relatively low, flat land; it is somewhat more moist where mountains capture orographic rainfall. The lack of vegetation perpetuates aridity. Without plants to absorb and hold moisture, the rare but occasionally copious rainfall simply runs off, evaporates, or sinks rapidly into underground aquifers, of which there are many across the region.

Islamism a grassroots religious revival in Islam that seeks political power to curb what are seen as dangerous secular influences; also seeks to replace secular governments and civil laws with governments and laws guided by Islamic principles

fossil fuel a source of energy formed from the remains of dead plants and animals

occupied Palestinian Territories (oPT) Palestinian lands occupied by Israel in since 1967

Climate

No other region in the world is as dry as North Africa and Southwest Asia (see Figure 6.5A). A belt of dry air that circles the planet between roughly 20° N and 30° N creates desert climates in the Sahara of North Africa, the eastern Mediterranean, the Arabian Peninsula, Kuwait, Iraq, and Iran.

The Sahara's size and location under this high-pressure belt of dry air make it a particularly hot desert region. In some places, temperatures can reach 130°F (54°C) in the shade at midday (see Figure 6.1A). With little water or moisture-holding vegetation to retain heat, nighttime temperatures can drop quickly to below freezing. Nevertheless, in even the driest zones, humans survive as traders and nomadic herders at scattered oases, where they maintain groves of drought-resistant plants such as date palms. Desert inhabitants often wear light-colored, loose, flowing robes that reflect the sunlight, retain body moisture during the day, and provide warmth at night.

In the uplands and at the desert margins, enough rain falls to nurture grass, some trees, and limited agriculture. Such is the case in northwestern Morocco; along the Mediterranean coast in Algeria, Tunisia, and Libya; in the highlands of Sudan, Yemen, and Turkey; and in the northern parts of Iraq and Iran (see Figure 6.5B–D). The rest of the region, generally too dry for cultivation, has for generations been the prime herding lands for nomads, such as the Kurds of Southwest Asia, the Berbers and Tuareg (a branch of Berbers) in North Africa, and the Bedouin of the steppes and deserts on the Arabian Peninsula. Recently, most nomads have been required to settle down; some of their lands are now irrigated for commercial agriculture, but the general aridity of the region means that sources of irrigation water (rivers and aquifers) are scarce.

Landforms and Vegetation

The rolling landscapes of rocky and gravelly deserts and steppes cover most of North Africa and Southwest Asia (see the Figure 6.1 map and Figure 6.5). In a few places, mountains capture sufficient moisture to allow plants, animals, and humans to flourish. In northwestern Africa, the Atlas Mountains, which stretch from Morocco on the Atlantic coast to Tunisia on the Mediterranean coast, lift damp winds that originate in the Atlantic Ocean, creating orographic rainfall of more than 50 inches (127 centimeters) per year on windward slopes. In some Atlas Mountain locations, there is enough snowfall to support a ski industry (see Figure 6.1B).

Figure 6.5 Photo Essay: Climates of North Africa and Southwest Asia

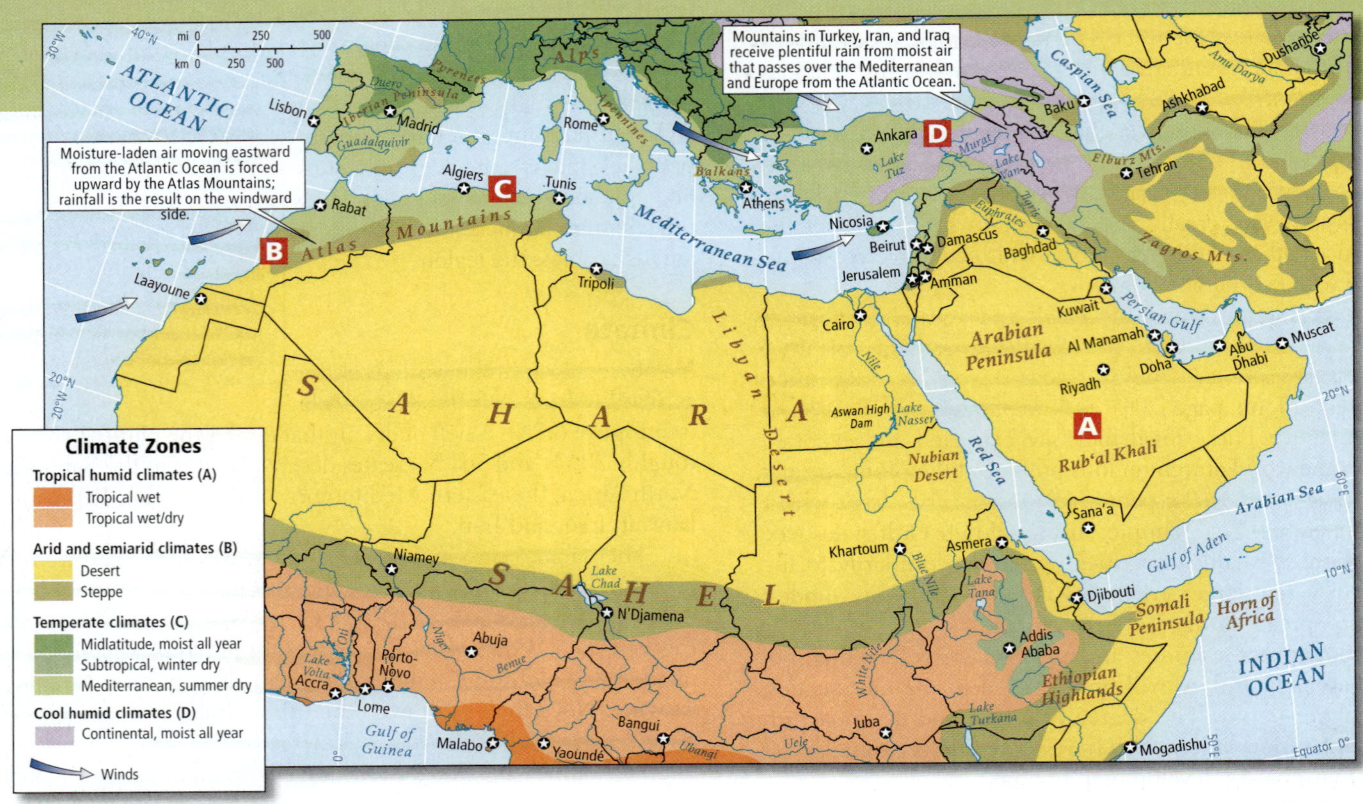

A Desert, Saudi Arabia

B Steppe, Morocco

C Mediterranean, summer dry, Algeria

D Continental, moist all year, Turkey

A rift that formed between two tectonic plates—the African Plate and the Arabian Plate—separates Africa from Southwest Asia (see Figure 1.25 on page 46). The rift, which began to form about 12 million years ago, is now filled by the Red Sea. The Arabian Peninsula lies to the east of this rift. There, mountains bordering the rift in the southwestern corner rise to 12,000 feet (3658 meters). They capture enough rainfall to sustain local agriculture, which has been practiced there for thousands of years. Irrigation has been used for millennia to increase yields.

Behind these mountains, to the northeast of the Arabian Peninsula, lies the great desert region of the Rub'al Khali. Like the Sahara, it has virtually no vegetation. The sand dunes of the Rub'al Khali, which are constantly moved by strong winds, are among the world's largest, some as high as 2000 feet (610 meters; see Figure 6.1D).

The landforms of Southwest Asia are more complex than those of North Africa. The Arabian Plate is colliding with the Eurasian Plate and pushing up the mountains and plateaus of Turkey and Iran (see Figure 1.25). Turkey's mountains lift damp air that passes over Europe and the Mediterranean from the Atlantic, resulting in considerable rainfall (see Figure 6.1E). Only a little rain makes it over the mountains to the interior of Iran, which is very dry. The tectonic movements that create mountains also create earthquakes, a common hazard in Southwest Asia.

There are only three major river systems in the entire region, and all have attracted human settlement for thousands of years. The Nile flows north from the moist central East African highlands (see Figure 6.1F). It crosses arid Sudan and desert Egypt and forms a large delta on the Mediterranean. The Euphrates and Tigris rivers both begin with the rain that falls in the mountains of Turkey; the rivers flow southeast to the Persian Gulf. Another much smaller river, the Jordan, starts as snowmelt in the uplands of southern Lebanon and flows through the Sea of Galilee to the Dead Sea. Most other streams are for most of the year dry riverbeds, or *wadis*, carrying water only after the generally light rains that fall between November and April (Figure 6.1G shows a wadi in Morocco).

North Africa and Southwest Asia were home to some of the very earliest agricultural societies. Today, rain-fed agriculture is practiced primarily in the highlands and along parts of the Mediterranean coast, where there is enough precipitation to grow citrus fruits, grapes, olives, and many vegetables, though supplemental irrigation is often needed. While irrigated agriculture is an age-old type of farming, it has become increasingly significant in modern times, both in the valleys of the major rivers, where seasonal flooding fills irrigation channels, and where aquifers are tapped by wells to provide water for cotton, wheat, barley, vegetables, and fruit trees.

Environmental Issues

Environmental concerns are only beginning to be an overt focus in this region. This is in part because for thousands of years people here have confronted the challenges of a naturally arid environment and have been quite attuned to its scarcities, which exceed those in most other places on Earth (Figure 6.6).

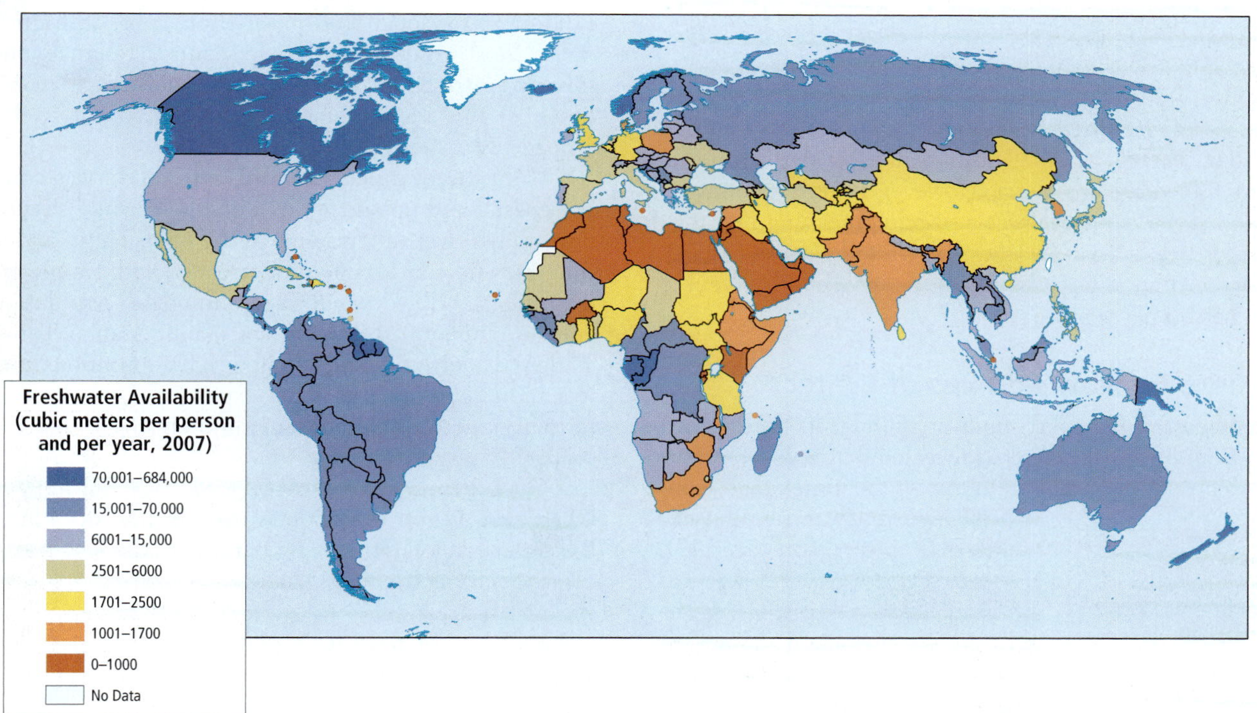

FIGURE 6.6 Global freshwater availability and stress. Freshwater stress and scarcity occur when so much water is withdrawn from rivers, lakes, and aquifers that not enough water remains to meet human and ecosystem requirements for sustainability.

An Ancient Heritage of Water Conservation

Geographic Insight 1

Water, Food, and Climate Change: The population of this predominantly dry region is increasing faster than is access to cultivable land and water for agriculture and other human uses. Many countries here are dependent on imported food, and because global food prices are unstable, the region often has food insecurity. Climate change could reduce food output both locally and globally and make water, already a scarce commodity, a source of conflict. New technologies offer solutions to water scarcity but are expensive and unsustainable.

The **Qur'an** (or **Koran**), the holy book of Islam, guides believers to avoid spoiling or degrading human and natural environments and to share resources, especially water, with all forms of life. In actual practice, the residents of this region conserve water better than most people in the world. Daily bathing is a religious requirement, and it often takes place in public baths where water use is minimized. For millennia, mountain snowmelt has been captured and moved to dry fields and villages via constructed underground water conduits called *qanats* that minimize evaporation. Likewise, traditional architectural designs are used to create buildings that stay cool by maximizing shade and airflow.

Despite their long history of water-conserving technologies and practices, however, this region's 477 million residents now have such limited water resources that even clever combinations of ancient and modern measures are no longer sufficient to ensure an adequate supply of water. Growing populations, water pollution, and unwise modern usages of water virtually guarantee that water shortages will be more extreme in the future (Figure 6.7A–D), especially with the likelihood of climate change–related drier conditions. As shown in the map in Figure 6.6, Turkey has the most plentiful water resources, followed by Iran, Iraq, and Syria. However, all other countries in the region suffer from *freshwater scarcity*, meaning that they are close to or below the minimum the United Nations considers necessary to support basic human development—1000 cubic meters per person per year.

Could There Be New Sources of Water?

A very recent reassessment of groundwater storage in North Africa by a consortium of British scientists revealed that water stored by nature in deep aquifers may amount to 100 times that available through renewable freshwater resources (Figure 6.8 on page 242). However, this *fossil water* that was deposited thousands of years ago is unlikely to be replenished and is unevenly distributed and difficult to access. Large quantities of fossil water are in sedimentary aquifers beneath Algeria, Libya, Egypt, and Sudan, but the water lies too deep to be brought up with simple pumps and is unlikely to be sufficient to meet the expected increases in demand for agricultural irrigation and for consumption by rapidly growing urban populations. Nonetheless, quantifying and mapping the stored groundwater provides a base for determining how to accommodate scarcities caused by climate variability.

Qur'an (or Koran) the holy book of Islam, believed by Muslims to contain the words Allah revealed to Muhammad through the archangel Gabriel

salinization a process that occurs when large quantities of water are used to irrigate areas where evaporation rates are high, leaving behind salts and other minerals

Water and Food Production

The predominant use of water in North Africa and Southwest Asia is for irrigated agriculture, even though agriculture does not contribute significantly to national economies. In Tunisia, for example, agriculture accounts for just 10 percent of GDP but 86 percent of all the water used. Only 13 percent of water is used in homes, and only 1 percent by industry. Some justification for the large amount of water used by agriculture in this water-stressed region may be found in the value that irrigated agriculture has in rural areas for local food production, jobs, and family economies. Most countries, to help control reliance on imports, subsidize irrigation for food and fiber crops in some way. Some of this region's food specialties appear in Figure 6.9 on page 242.

Until the twentieth century, agriculture was confined to a few coastal and upland zones where rain could support cultivation, and to river valleys (such as the Nile, Tigris, and Euphrates valleys), where farms could be irrigated with simple gravity-flow technology. However, to accommodate population growth and development, Libya, Egypt, Saudi Arabia, Tunisia, Syria, Turkey, Israel, and Iraq all now have ambitious mechanized irrigation schemes that have expanded agriculture deep into formerly uncultivable arid environments (see Figures 6.5B and 6.7B, C).

Over time, irrigation projects damage soil fertility through **salinization**. When irrigation is used in hot, dry environments, the water evaporates, leaving behind a salty residue of minerals or other contaminants. When too much residue accumulates, the plants are unable to grow or even survive. This human-induced loss of fertility is one of the largest environmental water issues that the world faces today in arid and semiarid regions.

Israel has developed relatively efficient techniques of *drip irrigation* that dramatically reduce the amount of water used, thereby limiting salinization and freeing up water for other uses.

Thinking Geographically

After you have read about the human impacts on the biosphere in North Africa and Southwest Asia, you will be able to answer the following questions:

B What about a dam creates soil moisture problems for downstream users of river water once the dam is operational?

C How might a dam like the Aswan on the upper part of the Nile River in Egypt contribute to soil salinity and lowered fertility in lowland regions once naturally flooded but now irrigated by pumped water?

Figure 6.7 Photo Essay: Human Impacts on the Biosphere in North Africa and Southwest Asia

War and water scarcity have resulted in major impacts on the biosphere in this region. The growth of population promises to increase these stressors on ecosystems.

A An old village is slowly swallowed by the desert near Douz, Tunisia. Many human activities have increased aridity, resulting in expanding deserts in this region.

Human Impact, 2002

Land cover

- Forests
- Grasslands
- Deserts
- Tundra
- Ice
- Modern national boundaries

Overfishing

- Threatened fisheries

Human impact on land

- High impact
- Medium–high impact
- Low–medium impact

Acid rain

- <4.2 pH
- 4.8–4.3 pH
- 5.5–4.9 pH

B Haditha Dam in Iraq. Dams make more water available for irrigation but reduce the amount of water available to downstream users.

C Outside Alexandria, Egypt, an automated irrigation system moves through a potato plantation located on what would otherwise be desert land.

D Smoke billows from the site of an attack on a pipeline in Iraq. War and insurgency have been major sources of environmental degradation in this region. The largest oil spill in history occurred during the first Gulf War (1991) when the Iraqi government spilled 300 million gallons of oil into the Persian Gulf in order to thwart a land invasion by the United States. Lebanon suffered a smaller, though similarly devastating spill in 2006 as a result of Israeli bombing raids.

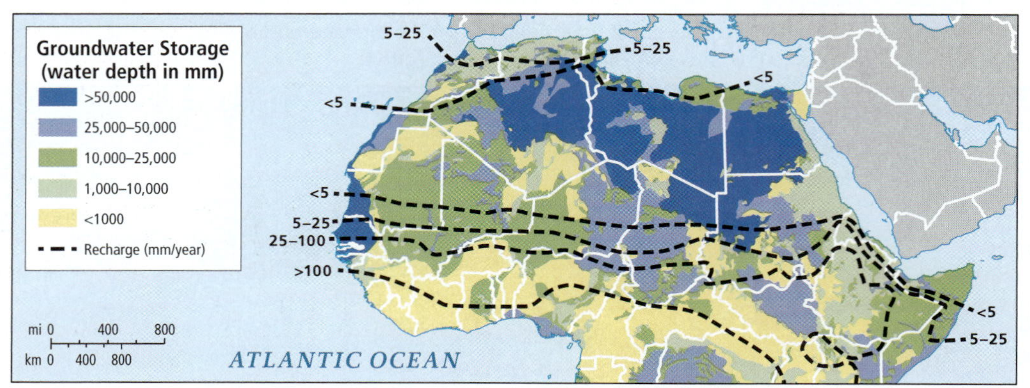

FIGURE 6.8 Water for the future? A map of newly discovered fossil groundwater deep below the deserts in North Africa. This ancient water deposit, already tapped in Libya, is expensive to extract and is unlikely to meet future needs because it is not being replenished.

Until very recently, however, poorer states have been unable to afford this somewhat complex technology, which requires an extensive network of hoses and pipes to deliver water to each plant. Arab countries resist depending on a technology developed by Israel—a country they deeply distrust.

Vulnerability to Climate Change

North Africa and Southwest Asia are especially vulnerable to the multiple effects of climate change, both in the region and beyond. Global warming is already changing rainfall patterns and increasing evaporation rates where water scarcity has long required accommodation (see Figure 6.7A; see also Figure 6.10B, C on page 244). These climatic changes, along with people's intensified use of water, are transforming more nondesert lands into deserts. Also, as the climate warms, a sea level rise of a few feet could severely impact the Mediterranean coast, especially the Nile Delta, one of the poorest and most densely populated lowland areas in the world (see Figure 6.10A), and the Persian Gulf coasts, where vast new

FIGURE 6.9 LOCAL LIVES FOODWAYS IN NORTH AFRICAN AND SOUTHWEST ASIA

A A stew is prepared in a Moroccan tagine, a heavy clay pot with two parts: a round bottom and a cone-shaped lid with a hole in the top. The bottom holds the food; the hole in the lid allows excess steam to escape, preventing the food from overcooking. The cone shape of the lid encourages condensation of moisture, thus reducing water loss and keeping flavors in the stew. Tagines are usually placed directly on hot coals or a grill and left for 3 or more hours.

B Falafel is fried in Gaza, Palestine. Either fava beans or chickpeas are mixed with parsley, scallions, garlic, and spices and then formed into a ball or patty that is deep fried. Falafel is often made into a flatbread sandwich and served with tomatoes, cucumbers, lettuce, and sauces based on tahini, a paste made from ground sesame seeds. The origin of this popular vegetarian street food, which has spread throughout this region and much of the world, is unclear. Egyptian Coptic Christians claim to have invented it as a substitute for meat during Lent. Others claim it originated in ancient Egypt, or even in South Asia.

C Khubz, or flatbread, is made in a tannur, a wood- or charcoal-fired oven. Possibly originating in ancient Sumer, khubz consists of a dough of flour, water, and salt that is rolled flat and then pressed onto the interior side walls of the tannur, where the radiant heat from the fire rapidly bakes it. Khubz is so central to diets in this region that in recent decades most governments have subsidized it, keeping prices artificially low to avoid political uprisings among the urban poor.

high-tech cities that lie at sea level would likely be flooded (see Figure 6.23A on page 261). Conversely, under some climate-change scenarios, periods of unusually intense rainfall could increase flooding.

Independent of any environmental effects of climate change, global efforts to reduce dependencies on fossil fuel consumption could devastate oil- and gas-based economies and transform the region's geopolitics, leaving countries vulnerable to losing some of their global power and to having lower overall income levels. Despite the enormous wealth from fossil fuel sales that has flowed into some of these countries over the past 40 years, few countries have undertaken the significant economic diversification needed to prepare for the reduced global consumption of the fossil fuels they sell.

Climate Change and Desertification Climate change could accelerate **desertification**, the conversion of nondesert lands into deserts (see Figure 6.7A). As temperatures increase, soil moisture decreases because of evaporation. As a result, plant cover is reduced, which causes less protection for arid soil. Bare patches of soil can become badly eroded by wind, and eventually sand dunes can blow onto formerly vegetated land.

In the grasslands (steppes) that often border deserts, a wide array of land use changes contribute to the general drying. For example, as groundwater levels fall because water is being pumped out for cities and irrigated agriculture, some plant roots no longer reach sources of moisture, causing the plants to die. In other cases, international development agencies have inadvertently contributed to desertification by encouraging nomadic herders to take up settled cattle ranching of the type practiced in western North America. Agencies encourage settlement because the mobility of nomads is viewed as complicating management for modern state bureaucracies where records must be kept, taxes collected, and children sent to school with regularity. However, ranching on fragile grasslands can lead to overgrazing, which can kill the grasses and increase evaporation. If irrigation is used to support ranching, such projects can also deplete groundwater resources, resulting in long-term desertification. Incidentally, encouraging nomads to settle may actually increase their vulnerability to climate change, because it is their very mobility and skills at cycling seasonally through multiple environments that have long helped nomads adapt to climate variability.

Imported Food, Virtual Water, and Global Climate Change Because agriculture is so difficult due to the aridity of this region, the diets of nearly all people here include imported food. To gain an understanding of total per capita water use, add the water used to produce this imported food, wherever it originates, to the overall water consumption by the citizens of this region. *Virtual water*, now a widely accepted term in water scarcity discussions, is the volume of water used to produce all that a person consumes in a year (see Chapter 1, page 38). For example, 1 kilogram (2.2 pounds) of beef requires 15,500 liters (4094 gallons) of water to produce, while 1 kilogram of goat meat requires just 4000 liters (1056 gallons). Goat meat, which is popular in this region and produced locally, has a much less significant water component than beef, but beef consumption is on the rise. One kilogram of corn requires 900 liters (238 gallons) of water, and 1 kilogram of wheat, 1350 liters (357 gallons). Beef, corn, and wheat are common imports; in fact, North Africa and Southwest Asia is the region with the world's largest consumption of imported wheat. The most disturbing aspect of dependence on imported food and virtual water is that climate change is reducing water availability in China, Russia, Ukraine, and the Americas—all places that supply wheat and other food to the countries of this region.

Strategies for Increasing Access to Water Some strategies have been developed for increasing supplies of fresh water, but each presents a set of difficulties. All of them are expensive, some have enormous potential to cause increased wasting of water, and most do not guarantee equal access to the increased supply.

Seawater desalination The fossil fuel–rich countries of the Persian Gulf have invested heavily in **seawater desalination** technologies that remove the salt from seawater, making it suitable for drinking or irrigating. Desalination plants supply 70 percent of Saudi Arabia's drinking water and some of its wheat field irrigation. However, the process of desalination uses huge amounts of energy, and the CO_2 emissions contribute to global warming. Furthermore, if all the costs of producing food with desalinated water were counted—the energy to desalinate, the cost of irrigation equipment, and the fact that irrigated soil inevitably loses productivity due to soil salinization—the wheat so produced would be too expensive to buy. Government subsidies for water desalination are now being reconsidered.

Groundwater pumping Many countries pump groundwater from underground aquifers to the surface for irrigation or drinking water. Under Muammar el-Qaddafi, Libya had invested some of its earnings from fossil fuels into one of the world's largest groundwater pumping projects, known as the *Great Man-Made River*. This project draws on part of the ancient fossil water deposit mentioned above to supply almost 2 billion gallons of water per day to Libya's coastal cities and to 600 square miles of agricultural fields. With this irrigated agriculture, Libya had hoped to grow enough food to end its current food imports and even supply food to the EU. However, the rate of natural replenishment of the aquifer is far slower than the rate of extraction, making this use of groundwater unsustainable. Hydrologists report fissures in the land surface above the aquifer that they associate with land sinking, or *subsidence*, as the water is withdrawn. Seawater intrusion is also a problem; as

desertification a set of ecological changes that converts arid lands into deserts

seawater desalination the removal of salt from seawater—usually accomplished through the use of expensive and energy-intensive technologies—to make the water suitable for drinking or irrigating

Figure 6.10 Photo Essay: Vulnerability to Climate Change in North Africa and Southwest Asia

Water scarcity, sea level rise, desertification, food scarcity, and political instability are some of the factors that make parts of this region highly vulnerable to climate change.

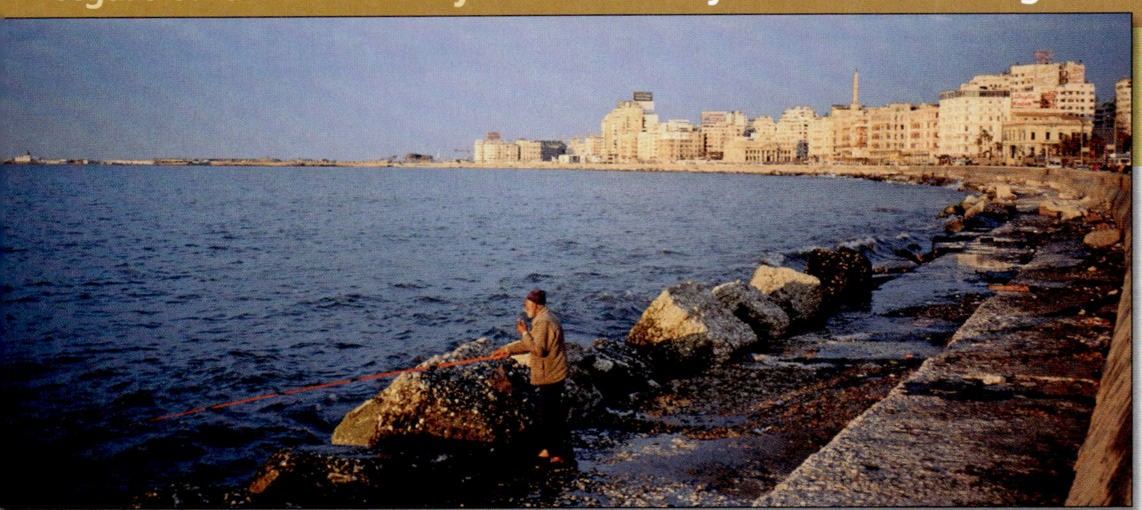

A Alexandria, Egypt, located on the Mediterranean coast of the Nile Delta region, is a low-lying city that is highly exposed to rising sea levels. Efforts to improve the seawall around the city, visible above, may be outpaced by rising sea levels. The entire Nile Delta is struggling to adapt to saltwater that is moving up rivers and permeating soils, making agriculture extremely difficult and polluting freshwater resources that cities depend on. Half of Egypt's population lives in the Nile Delta, and 80 percent of the country's imports and exports run through Alexandria.

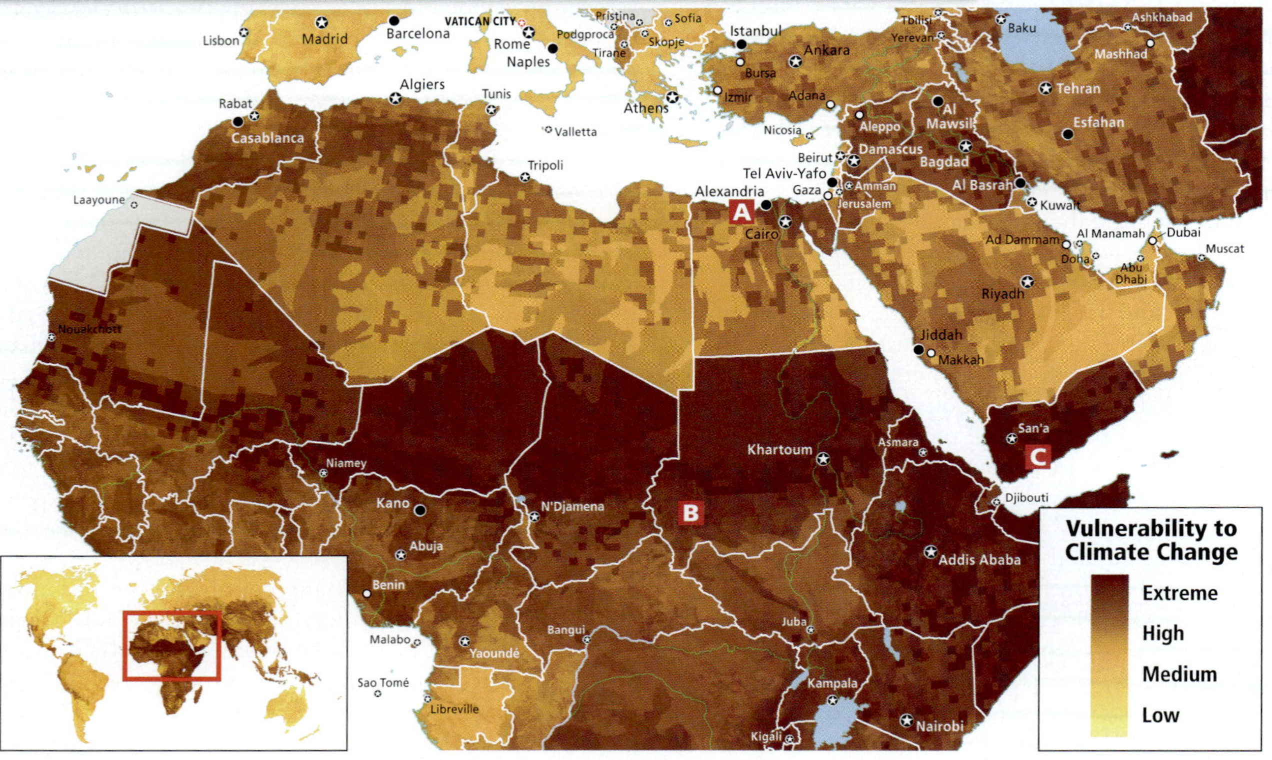

B Refugees from Darfur, Sudan, line up for food and water in neighboring Chad. Sudan has a high risk of drought, and widespread poverty has left the population very sensitive to any disruption in food or water supplies. Meanwhile, political instability has uprooted people from their homes and complicated emergency response and longer-term planning efforts that might create greater resilience.

C Residents of Wadi Mur, Yemen, ride out a sandstorm. The same factors that make Darfur in Sudan so vulnerable to climate change also apply to Yemen. However, Yemen also receives large numbers of refugees from Somalia and other parts of Africa, which further stretches resources and frustrates planning efforts.

Thinking Geographically

After you have read about the vulnerability to climate change in North Africa and Southwest Asia, you will be able to answer the following questions:

A What about Alexandria's location makes it particularly vulnerable to sea level rise?

B What are some of the factors in the region that complicate emergency responses to climate change, thus limiting resilience?

C How does Yemen's situation as a haven for refugees from nearby countries make it more vulnerable to the impacts of climate change?

fresh water is withdrawn, seawater flows into aquifers and wells in coastal zones, rendering the water brackish.

Dams and reservoirs Dams and reservoirs built on the regions' major river systems to increase water supplies have created new problems. In Egypt, for example, the natural cycles of the Nile River have been altered by the construction of the Aswan dams. Downstream of these two dams, water flows have been radically reduced, making it necessary to use supplemental irrigation, which leads to soil salinization (see Figure 6.7C), and the lack of floods means that fertility-enhancing silt is no longer deposited on the land. As a result, expensive fertilizers are now necessary. Moreover, with less water and silt coming downstream, parts of the Nile Delta are sinking into the sea. Upstream, the artificial reservoir created by the dams has flooded villages, fields, wildlife habitats, and historic sites and created still-water pools that harbor parasites. Dams can also cause hardship across borders. The Aswan High Dam sits on Egypt's border with Sudan, long an area of contention.

Dams as water-acquisition strategies strain cross-border relations The need to share the water of rivers that flow across or close to national boundaries often impedes national development efforts. Turkey's Southeastern Anatolia Project, which involves the construction of several large dams on the upper reaches of the Euphrates River for hydropower and irrigation purposes, has reduced the flow of water to the downstream countries of Syria and Iraq (Figure 6.11). In negotiations over who should get Euphrates water, for example, Turkey argues that it should be allowed to keep more water behind its dams because the river starts in Turkey and most of its water originates there as mountain rainfall. Meanwhile, Iraq's claim to the water is linked to the fact that the Euphrates travels the longest distance in Iraq. **285. THE JORDAN RIVER IS DYING**

THINGS TO REMEMBER

- Landforms and climates are particularly closely related in this region, and it is the driest region in the world. Rolling deserts and steppes cover most of the land. In a few places, mountains capture moisture, allowing plants, animals, and humans to flourish, but sustainability is becoming more difficult.

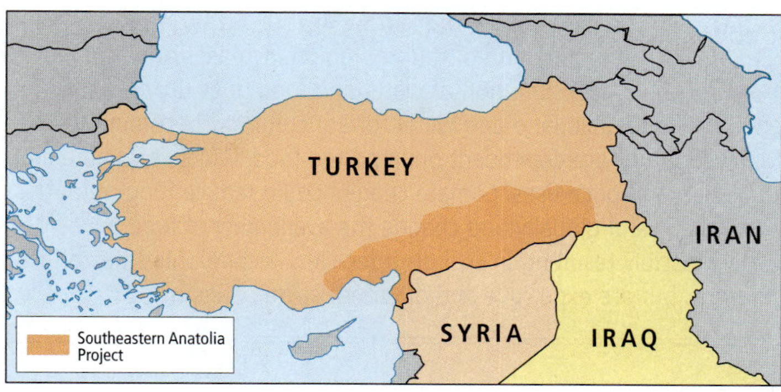

FIGURE 6.11 Dams on the Tigris and Euphrates drainage basins. Turkey's projects to manage the Tigris and Euphrates river basins through dam construction have international implications. Water that is retained in Turkey will not reach its neighbors. The main map shows Turkey's dams on the headwaters of the two rivers, as well as dams built in Syria and Iraq, all of which will have environmental effects, especially on the lower reaches of both rivers in Iraq. The smaller map shows the full extent of Turkey's Southeastern Anatolia Project.

Geographic Insight 1

• **Water, Food, and Climate Change** As the population of this dry region increases and the consumption of water and food goes up, obtaining enough cultivatable land and water for agriculture is becoming harder. Many countries are dependent on imported food, which is vulnerable to global food price rises. Climate change could reduce food output both locally and globally and change the availability of food and water, possibly resulting in civil disorder. New technologies may offer solutions but are expensive and often not sustainable.

Human Patterns over Time

Important developments in agriculture, societal organization, and urbanization took place long ago in this part of the world. Also, three of the world's great religions were born here: Judaism, Christianity, and Islam.

Agriculture and the Development of Civilization

Between 10,000 and 8000 years ago, formerly nomadic peoples founded some of the earliest known agricultural communities in the world. These communities were located in an arc formed by the uplands of the Tigris and Euphrates river systems (in modern Turkey and Iraq) and the Zagros Mountains of modern Iran (see the timeline and photo A in Figure 6.15 on page 250). This zone is often called the **Fertile Crescent** because of its once-plentiful fresh water, its fertile soil that is seasonally replenished by flooding, its open forests and grasslands, abundant wild grains, and its fish, goats, sheep, wild cattle, camels, horses, and other large animals (Figure 6.12 and Figure 6.13).

Fertile Crescent an arc of lush, fertile land formed by the uplands of the Tigris and Euphrates river systems and the Zagros Mountains, where nomadic peoples began the earliest known agricultural communities

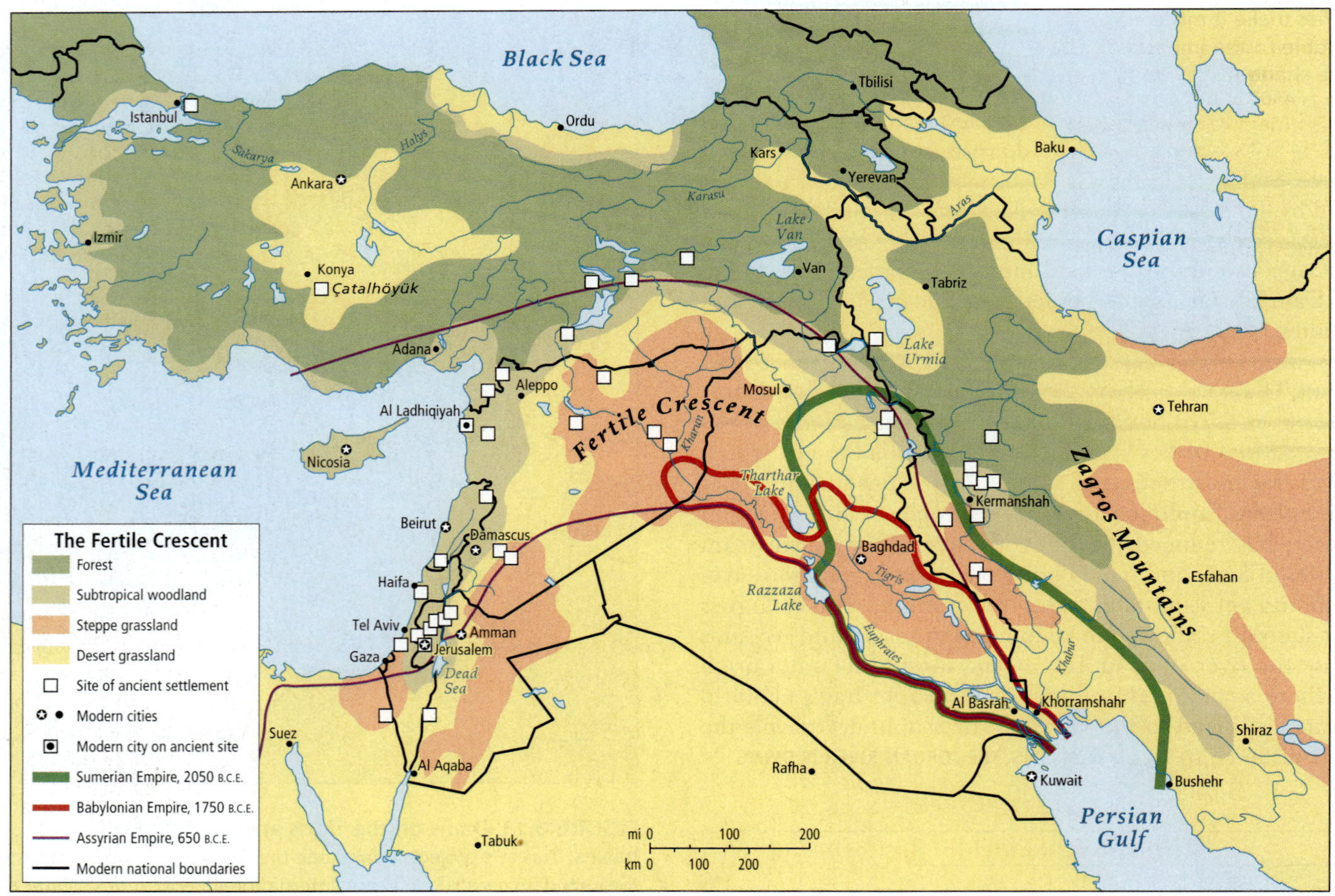

FIGURE 6.12 The Fertile Crescent, one of the earliest known agricultural sites. About 10,000 years ago, people in the Fertile Crescent began domesticating cereal grains, legumes, and animals, especially sheep and goats. The uses of domesticated animals spread into Europe and Africa as agricultural peoples traded their surpluses for other goods or moved into other regions. Three major empires developed successively in the eastern part of the Fertile Crescent: the Sumerian, the Babylonian, and the Assyrian.

The skills of these early people in domesticating plants and animals allowed them to build ever more elaborate settlements. The settlements eventually grew into societies based on widespread irrigated agriculture along the base of the mountains and in river basins, especially along the Tigris and Euphrates. Nomadic herders living in adjacent grasslands traded animal products for the grain and other goods produced in the settled areas.

Over several thousand years, agriculture spread to the Nile Valley, west across North Africa, north and west into Europe, and east to the mountains of Persia (modern Iran). Ultimately, other cultivation systems across the world were influenced by developments in the Fertile Crescent. Eventually, small settlements associated with agriculture took on urban qualities: dense populations, specialized occupations, concentrations of wealth, and centralized government and bureaucracies. For example, the agricultural villages of Sumer (in modern southern Iraq), which existed 5000 years ago, gradually turned into city-states that extended their influence over the surrounding territory. The Sumerians developed wheeled vehicles, oar-driven ships, and irrigation technology.

From time to time, nomadic tribes who had adopted the horse as a means of conquest banded together and, with devastating cavalry raids, swept over agricultural settlements. They then set themselves up as a ruling class, but soon these former nomads adopted the settled ways and cultures of the peoples they conquered and thus themselves became vulnerable to attack.

Agriculture and Gender Roles

Increasing research evidence suggests that the dawning of agriculture may have marked the transition to markedly distinct roles for men and women. Archaeologist Ian Hodder reports

FIGURE 6.13 LOCAL LIVES PEOPLE AND ANIMALS IN NORTH AFRICA AND SOUTHWEST ASIA

A A camel race in Jordan. Camels were probably domesticated between 4500 and 5000 years ago in the southern Arabian Peninsula. Extensively adapted to arid climates, camels have long been used to carry people and heavy loads across deserts. They are also prized for their lean meat and highly nutritious milk.

B A mummified cat from ancient Egypt, where cats were revered and sometimes worshiped as symbols of the cat goddess Bastet. Part of a wider tradition of animal worship in Egypt, cats were often treated with the same respect accorded humans, including mummification before burial. Though cat worship was officially banned in 390 B.C.E., cats have long been kept as pets and for pest control.

C An Arabian horse and his trainer at the stud farm of a Saudi prince in Saudi Arabia. Bred for war and raids by the nomadic, desert-dwelling Bedouin people of Southwest Asia, the Arabian horse is known for its endurance, intelligence, speed, stealth, and ability to survive on relatively little food and water. Arabian horses are one of the oldest horse breeds, with evidence of their domestication going back 4500 years. Now they are a part of most major horse breeds throughout the world.

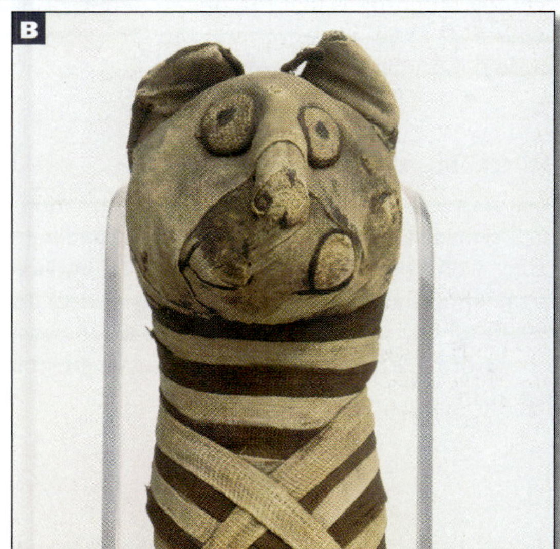

that at the 9000-year-old site of Çatalhöyük (see Figure 6.12), near modern Konya in south-central Turkey, where the economy was primarily hunting and gathering, there is little evidence of gender differences. Families were small and men and women performed similar chores in daily life. Both had comparable status and power, and both played key roles in social and religious life. Archaeological evidence elsewhere indicates that gender roles were also egalitarian in other pre-agricultural societies.

Scholars think that after the development of agriculture, as the accumulation of wealth and property became more important in human society, concerns about family lines of descent and inheritance emerged. This led in turn to the idea that women's bodies needed to be controlled so that a woman could not become pregnant by a man other than her mate and thus confuse lines of inheritance. From this core concern about secure lines of descent grew many practices aimed at reinforcing the idea that the mating of women had to be controlled, and by extension, that women's daily spatial freedom, interaction with men, and sexuality needed to be curtailed.

The Coming of Monotheism: Judaism, Christianity, and Islam

The very early religions of this region were based on a belief in many gods who controlled natural phenomena; such was the case through the Greek era and into the Roman period. Then, several thousand years ago, **monotheism**—the belief system based on the idea that there is only one god—began to emerge. The three major monotheistic world religions—Judaism, Christianity, and Islam—all have connections to the eastern Mediterranean; there the city of Jerusalem is sacred to all three. (Muslims also revere Makkah (Mecca) and Al Madinah (Medina) in Saudi Arabia.) All three religions have a connection to a sacred text: the Torah (the first five books in what Christians refer to as the Old Testament) for Jews; the Old and New Testaments of the Bible for Christians; and both the Bible and the Qu'ran for Muslims.

monotheism the belief that there is only one god

Judaism a monotheistic religion characterized by the belief in one god (Yahweh), a strong ethical code summarized in the Ten Commandments, and an enduring ethnic identity

diaspora the dispersion of Jews around the globe after they were expelled from the eastern Mediterranean by the Roman Empire, beginning in 73 C.E.; the term can now refer to other dispersed culture groups

Christianity a monotheistic religion based on the belief in the teachings of Jesus of Nazareth, a Jew, who described God's relationship to humans as primarily one of love and support, as exemplified by the Ten Commandments

Muslims followers of Islam

Judaism was founded approximately 4000 years ago. According to tradition, the patriarch Abraham led his followers from Mesopotamia (modern Iraq) to the shores of the eastern Mediterranean (modern Israel and the occupied Palestinian Territories), where he founded Judaism. Jewish religious history is recorded in the Torah. Judaism is characterized by the belief in one god, a strong ethical code summarized in the Ten Commandments, and an enduring ethnic identity reinforced by dietary and religious laws.

After the Jews rebelled against the Roman Empire, which culminated in their expulsion in 73 C.E. from the eastern Mediterranean, some were enslaved by the Romans and most migrated to other lands in a movement known as the **diaspora** (the dispersion of an originally localized people). Many Jews migrated across North Africa and Europe, and others went to various parts of Asia. After 1500, Jews were among the earliest European settlers in all parts of the Americas.

Christianity is based on the teachings of Jesus of Nazareth, a Jew who, claiming to be the son of God, gathered followers in the area of Palestine about 2000 years ago. Jesus, who became known as Christ (meaning *anointed one* or *Messiah*), taught that there is one God, who primarily loves and supports humans but who will judge those who do evil. This philosophy grew popular, and both Jewish religious authorities and Roman imperial authorities of the time saw Jesus as a dangerous challenge to their power.

After Jesus's execution in Jerusalem in about 32 C.E., his teachings were written down (the Gospels) by those who followed him, and his ideas as interpreted by these writers spread and became known as Christianity. Centuries of persecution ensued, but by 400 C.E., Christianity had become the official religion of the Roman Empire. However, following the spread of Islam after 622 C.E., only remnants of Christianity remained in Southwest Asia and North Africa.

Islam, now the overwhelmingly dominant religion in the region, emerged in the seventh century C.E., after the Prophet Muhammad transmitted the Qur'an to his followers by writing down what was conveyed to him by Allah ("God" in Arabic). Born in about 570 C.E., Muhammad was a merchant and caravan manager in the small trading town of Makkah (Mecca) on the Arabian Peninsula near the Red Sea (see Figure 6.15B). Followers of Islam, called **Muslims**, believe that Muhammad was the final and most important in a long series of revered prophets, which includes Abraham, Moses, and Jesus.

Unlike the many versions of Christianity, Islam has virtually no central administration and only an informal religious hierarchy (this is somewhat less true of the Shi'ite version of Islam; see the discussion on page 254). The world's 1 billion Muslims may communicate directly with God (Allah). A clerical intermediary is not necessary, though there are numerous clerical leaders who help their followers interpret the Qur'an. An important result of the lack of a central authority is that the interpretation of Islam varies widely within and among countries, from group to group, and from individual to individual.

The Spread of Islam

Among the first converts to Islam were the Bedouin—nomads of the Arabian Peninsula. By the time of Muhammad's death in 632 C.E., they were already spreading the faith and creating a vast Islamic sphere of influence. Over the next century, Muslim armies built an Arab–Islamic empire over most of Southwest Asia, North Africa, and the Iberian Peninsula of Europe (Figure 6.14).

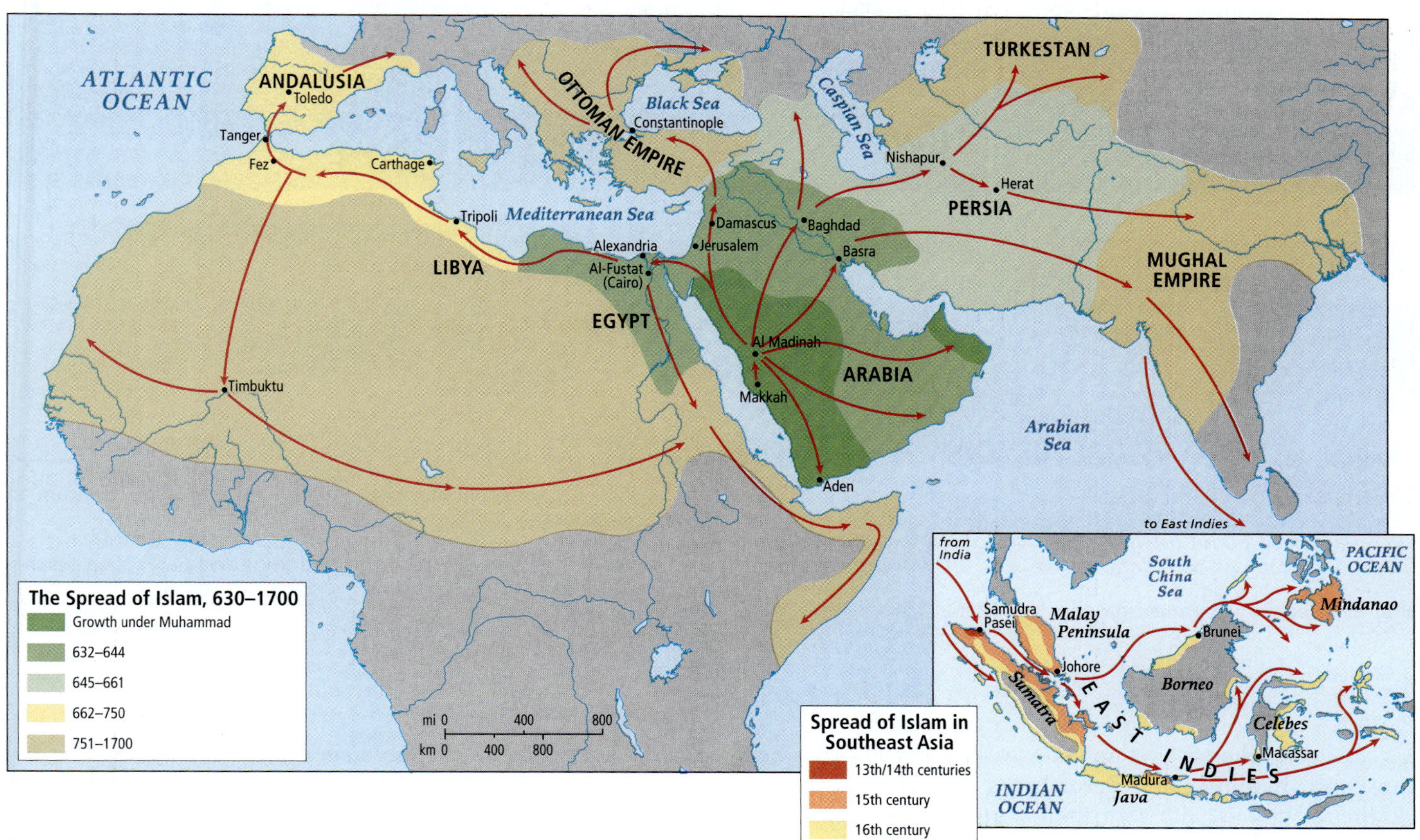

FIGURE 6.14 The spread of Islam, 630–1700. In the first 120 years following the death of the Prophet Muhammad in 632, Islam spread primarily by conquest. Over the next several centuries, Islam was carried to distant lands by both traders and armies.

While most of Europe was stagnating during the medieval period (450–1500 C.E.), the Arab–Islamic empire nurtured learning and economic development. Muslim scholars traveled throughout Asia and Africa, advancing the fields of architecture, history, mathematics, geography, and medicine. Centers of learning flourished from Baghdad (Iraq) to Toledo (Spain). During the early Arab–Islamic era, the development of banks, trusts, checks, receipts, and bookkeeping fostered vibrant economies and wide-ranging trade. The traders founded settlements and introduced new forms of living spaces. The architectural legacy of Arabs and Muslims lives on in Spain, India, Central Asia, the Americas, and in countless buildings across the world (see Figure 6.18 on page 255).

By the end of the tenth century, the Arab–Islamic empire had begun to break apart. From the eleventh to the fifteenth centuries, Mongols from eastern Central Asia (eventually converting to Islam by 1330) conquered parts of the Arab-controlled territory, forming the Muslim Mughal Empire, centered in what is now north India. Meanwhile, beginning in the 1200s, nomadic Turkic herders from Central Asia began to converge in western Anatolia (Turkey) where they eventually forged the **Ottoman Empire**, which became the most influential Islamic empire the world has ever known.

By the 1300s, the Ottomans had become Muslim, and by the 1400s they had defeated the Christian Byzantine Empire, centered in Constantinople, which was the successor to the Roman Empire. The Ottomans took over Constantinople, renamed it Istanbul, and soon controlled most of the eastern Mediterranean, Egypt, and Mesopotamia. By the late 1400s, they also controlled much of southeastern and central Europe and by the 1600s had taken over parts of coastal North Africa from the Arabs. Previously, in the 1490s, the Arab Muslims had lost their control of the Iberian Peninsula to Christian kingdoms. Today, Islam still dominates in a huge area that stretches from Morocco to western China and includes northern South Asia, as well as Malaysia, Brunei, and Indonesia in Southeast Asia (see the inset of Figure 6.14).

Ottoman Empire the most influential Islamic empire the world has ever known; begun in the 1200s when nomadic Turkic herders from Central Asia converged in western Anatolia (Turkey)

Once a location was completely conquered, the Ottoman Empire, like the Arab–Islamic empire before it, encouraged religious tolerance toward the conquered peoples so

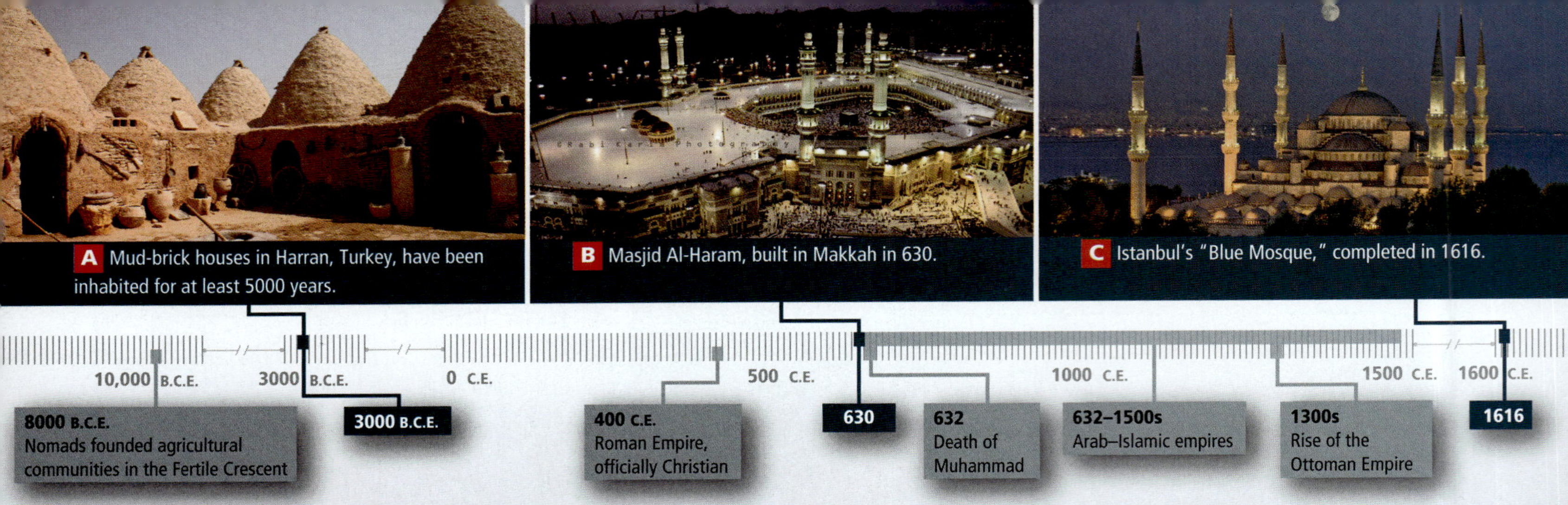

A Mud-brick houses in Harran, Turkey, have been inhabited for at least 5000 years.

B Masjid Al-Haram, built in Makkah in 630.

C Istanbul's "Blue Mosque," completed in 1616.

FIGURE 6.15 A VISUAL HISTORY OF NORTH AFRICA AND SOUTHWEST ASIA

Thinking Geographically

After you have read about the human history of North Africa and Southwest Asia, you will be able to answer the following questions:

A What allowed for urban settlements like Harran to develop?

B What ceremonial purpose is filled by the Masjid Al-Haram and the Kaaba in Makkah?

C What Turkish city had lavish buildings and parks that outshined anything in Europe until the nineteenth century?

long as they adhered to a religion with a sacred text. Jews, Christians, Buddhists, and Zoroastrians were allowed to practice their religions, although there were attractive economic and social advantages to converting to Islam. Multicultural urban life in Ottoman cities facilitated vast trading networks spanning the known world, and Istanbul became a cosmopolitan capital with elaborate buildings and lavish public parks that outshined anything in Europe until the nineteenth century (see Figure 6.15C).

Western Domination, State Formation, and Antidemocratic Practices

The Ottoman Empire ultimately withered in the face of a Europe made powerful by colonialism and the Industrial Revolution. Throughout the nineteenth century, North Africa provided raw materials for Europe in a trading relationship dominated by European merchants. By 1830, France was exercising direct control over parts of the North African territory of Algeria (see Figure 6.15D). France took control of Tunisia in 1881 and Morocco in 1912 (Spain controlled a bit of the Mediterranean coast and what is now called Western Sahara); Britain gained control of Egypt in 1882 and Sudan in 1898; and Italy took control of Libya in 1912 (Figure 6.16A on page 252).

World War I (1914–1918) brought the fall of the Ottoman Empire, which had allied itself with Germany. At the end of the war, the victorious powers dismantled the Ottoman Empire and all of the former Ottoman territories; only Turkey was recognized as an independent country. The rest of the formerly Ottoman-controlled territory was allotted to France and Britain as protectorates (see Figure 6.16B). On the Arabian Peninsula, Bedouin tribes were consolidated under Sheikh Ibn Saud in 1932, and Saudi Arabia began to emerge as an independent country.

World War II (1939–1945) further affected the political development of North Africa and Southwest Asia. Most significantly, in the aftermath of the Holocaust in Europe, the Jewish state of Israel was created in the eastern Mediterranean on land inhabited by Arab farmers and nomadic herders as well as by some Jews (see the discussion on page 273; see also Figure 6.15E).

By the 1950s, European and U.S. energy companies played a key role in influencing who ruled Iran and Saudi Arabia—countries where vast oil deposits were to become especially lucrative. In Egypt, a major cotton producer, European textile companies played a similar role. Officials in the governments of these countries received financial benefits from foreign companies and showed their loyalty to these companies with low taxes on oil and cotton exports and easy access to land. While a tiny ruling elite grew fabulously wealthy, even the modest oil and other tax revenues were not invested in creating opportunities for the vast majority of poor people. Over time, ever more political power accrued, especially to foreign energy companies. The United States and Western Europe supported those autocratic local leaders who were most sympathetic to the interests of their companies and their Cold War strategies vis-à-vis the Soviet Union (see Figure 6.15F). As a result of these concerns, both the Europeans and the Americans supported undemocratic governments and stood in the way of reforms that would have resulted in a more educated populace able to participate in the full range of democratic institutions.

D France's wars in Algeria in the 1830s.

E Jewish, Arab, and UN representatives negotiate a cease-fire during the 1948 war.

F The king of Saudi Arabia meets with the U.S. secretary of defense in 2011.

D When did France first assert control over the territory in what is now Algeria?

E The state of Israel was formally created in the aftermath of what event in Europe?

F Why did European and U.S. companies become involved in deciding who ruled Iran and Saudi Arabia?

THINGS TO REMEMBER

- About 10,000 years ago in the Fertile Crescent, formerly nomadic peoples founded some of the world's earliest known agricultural communities. The domestication of plants and animals allowed them to build ever more elaborate settlements that eventually grew into societies based on widespread irrigated agriculture.
- The three major monotheistic world religions—Judaism, Christianity, and Islam—all have their origins in this region in the eastern Mediterranean. In the region, Islam is by far the largest in numbers of adherents, and it is the principal faith in all of the region's countries except Israel.
- Beginning in the nineteenth century and continuing through the end of World War II, European colonial powers ruled or controlled most countries in the region.
- Following World War II, the state of Israel was created, and in countries with oil deposits and other useful assets, the United States and Western Europe supported those autocratic local leaders most likely to maintain a friendly attitude toward U.S. and European business and geopolitical strategic interests.

CURRENT GEOGRAPHIC ISSUES

For decades, social and political change in this region lagged behind economic development. Why did the wealth generated by oil not result in a spreading of opportunities, such as broad public education, an opening up of public discourse, and most importantly, an expansion of educational and political opportunities for women? As indicated in the opening vignette, change is now underway across this region and is gaining momentum, but there are countervailing conservative forces that could inhibit political reform, social change, and economic development.

Sociocultural Issues

This section explores the basics of the pervasive influence of Islam on the culture of the region and examines the broad social changes occurring with regard to family values, gender roles and gendered spaces, demographic change, urbanization and migration, and patterns of human well-being.

Religion in Daily Life

Ninety-three percent of the people in the region are followers of Islam; for them, the Five Pillars of Islamic Practice embody the central teachings of Islam. Not all Muslims are fully observant, but the Pillars have an impact on daily life, including festivals and religious holidays (Figure 6.17 on page 253).

The Pillars of Muslim Practice

1. A testimony of belief in Allah as the only God and in Muhammad as his messenger (prophet).

FIGURE 6.16 Colonial regimes in North Africa and Southwest Asia.

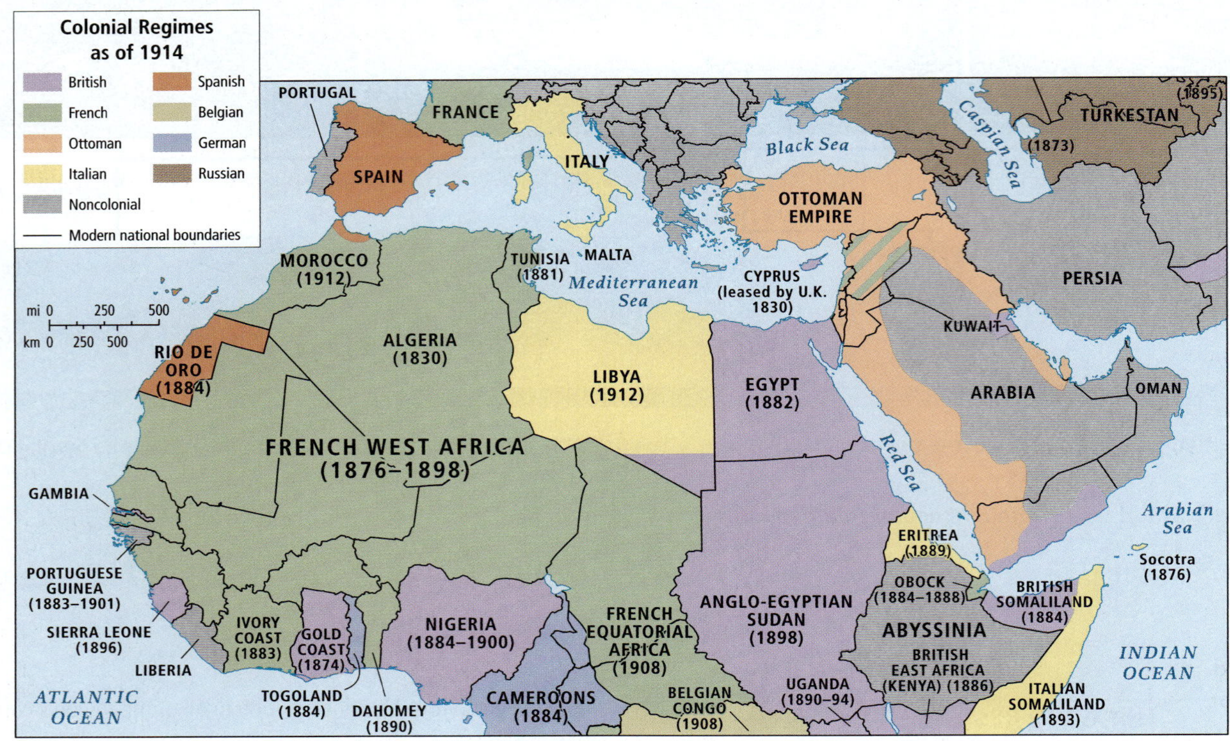

(A) European powers began influencing the affairs of the region in the nineteenth century and expanded their control by 1914 at the beginning of World War I. The dates on the map indicate when the Europeans took control of each country.

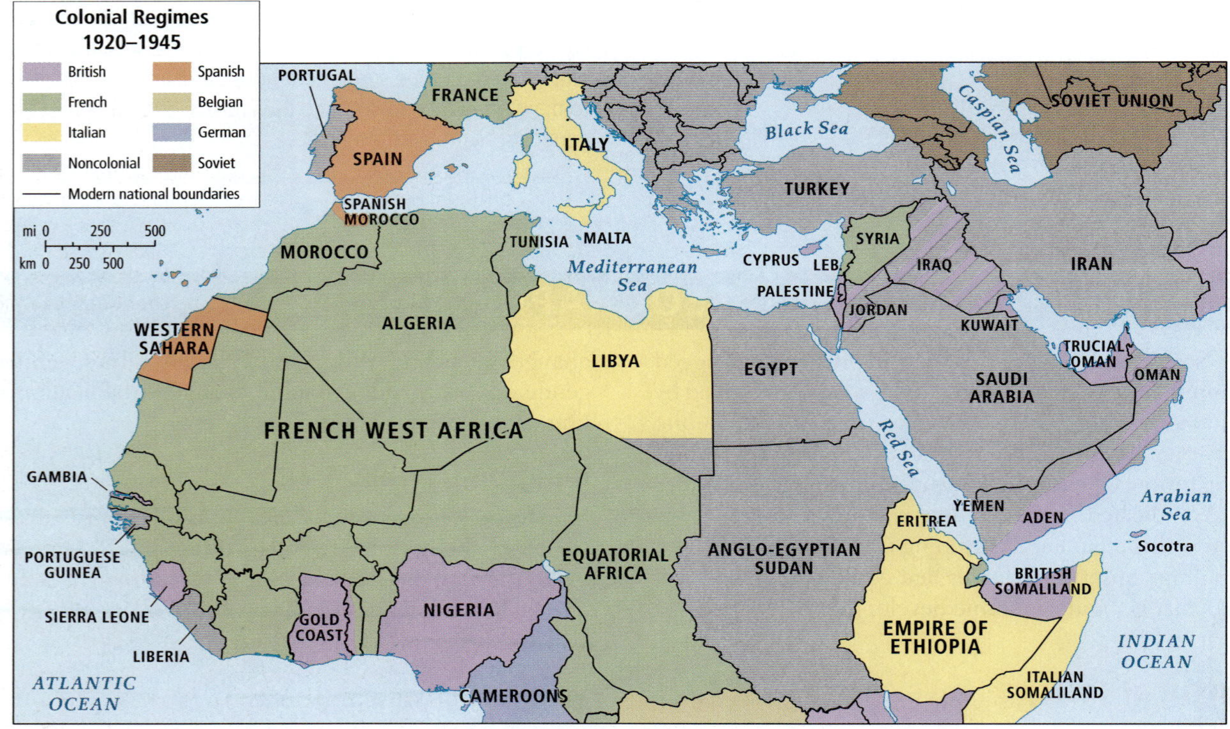

(B) Between 1920 and 1945, what was left of the Ottoman Empire in the eastern Mediterranean became protectorates administered by the British and French. The striped areas reflect the British colonial practice of allowing local rulers to govern while controlling many of their policies and actions.

2. Daily prayer at five designated times (daybreak, noon, midafternoon, sunset, and evening). Although prayer is an individual activity, Muslims are encouraged to pray in groups and in mosques. The call to prayer, broadcast five times a day in all parts of the region, is a constant reminder to all people to reflect on their beliefs.

3. Obligatory fasting (no food, drink, or smoking) during the daylight hours of the month of Ramadan, followed by a light celebratory family meal after sundown. Ramadan falls in the ninth month of the Islamic (lunar) calendar.

4. Obligatory almsgiving (*zakat*) in the form of a "tax" of at least 2.5 percent. The alms are given to Muslims in need. *Zakat* is based on the recognition of the injustice of economic inequity. Although it is usually an individual act, the practice of government-enforced *zakat* is returning in certain Islamic republics.

5. Pilgrimage (**hajj**) at least once in a lifetime to the Islamic holy places, especially the Masjid Al-Haram and the Kaaba in Makkah (Mecca), during the twelfth month of the Islamic calendar.

Saudi Arabia occupies a prestigious position in Islam, as it is the site of two of Islam's three holy shrines: Makkah, the birthplace of the Prophet Muhammad and of Islam; and Al Madinah (Medina), the site of the Prophet's mosque and his burial place. (The third holy shrine is in Jerusalem.) The fifth pillar of Islam has placed Makkah and Al Madinah at the heart of Muslim religious geography. Each year, a large private sector service industry, owned and managed by members of the huge Saud family, organizes and oversees the 5- to 7-day hajj for more than 2.5 million devout foreign visitors.

Islamic Religious Law and Variable Interpretations

Beyond the Five Pillars, Islamic religious law, called **shari'a**, or "the correct path," guides daily life according to the principles of the Qur'an. But there are many interpretations of the Qur'an, several renderings of shari'a, and a wide variety of versions of the observant Muslim life. Some Muslims believe that no other legal code is necessary in an Islamic society, because shari'a provides guidance in all matters of life, including worship, finance, politics, marriage, sex, diet, hygiene, war, and crime. Other Muslims think that secular law is more useful in modern societies that are increasingly multicultural because secular law makes allowances for different religious sensibilities.

In the region of North Africa and Southwest Asia, the debate about whether shari'a or secular law is best for the modern era gained scrutiny during and after the so-called Arab Spring beginning in 2010, when several countries started to frame new constitutions. Insofar as the interpretation of shari'a is concerned, the Muslim community is split into two major groups: **Sunni** Muslims, who today account for

hajj the pilgrimage to the city of Makkah (Mecca) that all Muslims are encouraged to undertake at least once in a lifetime

shari'a literally, "the correct path"; Islamic religious law that guides daily life according to the interpretations of the Qur'an

Sunni the larger of two major groups of Muslims who have different interpretations of shari'a

FIGURE 6.17 LOCAL LIVES **FESTIVALS IN NORTH AFRICA AND SOUTHWEST ASIA**

A Colombian singer Shakira performs at Mawazine, a world music festival held each year in Rabat, the capital of Morocco. Part of many conscious efforts to portray Morocco and Rabat as "open to the world," the festival has featured great traditional and international artists such as Cheb Khaled and Stevie Wonder. It has also come under criticism from Islamist politicians who see it as encouraging immoral behavior.

B A banquet is set in Gaza, Palestine, for Iftar, the sunset meal that occurs each night during the Muslim holy month of Ramadan, during which observant Muslims do not eat, drink, quarrel, deliberately mislead people with language, or have sex between sunrise and sunset. Extra prayers are made each day as a demonstration of one's submission to God. Wealthy Muslims often sponsor public banquets, such as the one shown below.

C An Iranian woman celebrates Newroz, an ancient New Year holiday of Zoroastrian origin enjoyed throughout Iran, parts of Turkey, and much of Central Asia (see Figure 5.19A on page 221). People jump over bonfires while singing a verse of purification that is meant to remove sickness and problems, replacing them with warmth and energy.

85 percent of the world community of Islam, and **Shi'ite** (or **Shi'a**) Muslims, who live primarily in Iran but also in southern Iraq, Syria, and southern Lebanon. Shi'ites recognize an authoritative priestly class whom they call *mullahs,* who have been known to mete out extreme punishments for supposed breaches of the rules of the Qur'an. These incidents that occasionally make the news are best defined as vigilante justice and are not condoned by observant Muslims, either Sunnis or Shi'ites.

The Sunni–Shi'ite split dates from shortly after the death of Muhammad, when divisions arose over who should succeed the Prophet and have the right to interpret the Qur'an for all Muslims. This division continues today. The original disagreements have been exacerbated by countless local disputes over land, resources, and philosophies. In Iraq, for example, long-standing conflict between Sunnis and Shi'ites intensified following the U.S. invasion in 2003, as rivalries arose over which group should control political power and fossil fuel resources.

Family Values and Gender

In part because Islam has so little religious hierarchy, the family is the most important institution in this region. Although the role of the family is changing, a person's family is still such a large component of personal identity that the idea of individuality is almost a foreign concept. Each person is first and foremost part of a family, and the defining parameter of one's role in the family is gender identity.

The head of the family is nearly always a man; even when a woman is widowed or divorced, she is under the tacit supervision of a male, perhaps her father or her adult son. An educated unmarried woman with a career outside the home is likely to live in the home of her parents or a brother and to defer to them out of respect in decision making. Traditionally, men are considered more stable and capable of making decisions, and thus, it is thought, they should be in charge. Women usually agree. These **patriarchal** ideas are beginning to change as the result of modernization. During the Arab Spring, a few brave women have challenged convention and taken a political stance (see the opening vignette; see also the discussion below).

Shi'ite (or Shi'a) the smaller of two major groups of Muslims who have different interpretations of shari'a; Shi'ites are found primarily in Iran and southern Iraq

patriarchal relating to a social organization in which the father is supreme in the clan or family

female seclusion the requirement that women stay out of public view

Gulf states Saudi Arabia, Kuwait, Bahrain, Oman, Qatar, and the United Arab Emirates

jihadists especially militant Islamists

Gender Roles and Gendered Spaces

Carefully specified gender roles are common in many cultures, and there is often a spatial component to these roles. In the region of North Africa and Southwest Asia, in both rural and urban settings, the ideal is for men and boys to go forth into *public spaces*—the town square, shops, the market. Women are expected to inhabit primarily *private spaces*. But there are many possible exceptions to those ideals.

To facilitate this ideal of public/private spaces for the sexes, traditional family compounds included a courtyard that was usually a private, female space within the home (Figure 6.18A, C); the only men who could enter it were relatives. For the urban upper classes, female space was an upstairs set of rooms with latticework or shutters at the windows, which increased the interior ventilation and from which it was possible to look out at street life without being seen (see Figure 6.18B). Today, the majority of people in the region live in urban apartments, yet even in these there is a demarcation of public and private space. One or two formally furnished reception areas are reserved for nonfamily visitors, and rooms deeper into the dwelling are for family-only activities. When guests are present, women in the family are usually absent or present only briefly. Customs vary not only from country to country but also from rural to urban settings, by social class, and by personal preference. Today, many women as well as men go out into public spaces, but the conditions under which women enter these spaces remain an issue and a woman who challenges convention does so at her own peril and that of her family's reputation.

The requirement that women stay out of public view (also known as **female seclusion**) is most strictly enforced in the more conservative Muslim countries of the **Gulf states** (Saudi Arabia, Kuwait, Bahrain, Oman, Qatar, and the United Arab Emirates). Women in these countries are generally expected to remain in private spaces except when on important business; then they are to be accompanied by a male relative. In the more secular and urbanized Islamic countries—Morocco, Jordan, Tunisia, Libya, Egypt, Turkey, Lebanon, Iran, and Iraq—women regularly engage in activities that place them in public spaces. In Jordan, for example, a group of women may go out together for a high-spirited evening dinner. Some women wear conservative religious clothing; others dress in Western styles. Increasingly, female doctors, lawyers, teachers, and businesspeople are found in even the most conservative societies. The map in Figure 6.4, accompanying the opening vignette, compares the various levels of restrictions on women across the region.

Affluent urban women may observe seclusion either very little (especially if they are highly educated) or, if they are relatively affluent but less educated, even more strictly than do rural women. Although rural women are often more traditional in their outlook, they have many tasks that they must perform outside the home: agricultural work, carrying water, gathering firewood, and buying or selling food in markets. Meanwhile upper-class women who can afford servants to perform daily tasks in public spaces can more easily stay secluded and afford the amenities (TV, DVDs, cell phones) that relieve the boredom and isolation of seclusion. It should also be noted that seclusion customs are in flux, being relaxed in some countries, such as Morocco, Turkey, and Tunisia, but becoming more strictly enforced in other countries, such as post–Arab Spring Egypt, Iran, and parts of North Africa where **jihadists** (especially militant Islamists) are newly active.

Many women in this region use clothing as a way to create private space (Figure 6.19 on page 256). This is done with the many

FIGURE 6.18 Domestic spaces. Many of the older cities and buildings in this region reflect the division between public space and private or domestic space.

(A) As seen from above, nearly every house in this old part of Baghdad has a courtyard with a garden in it.

(B) Projecting windows covered with lattice or louvers, known as a *mashrabiya*, in Jiddah, Saudi Arabia. *Mashrabiyas* allow women to look out on the street below and catch breezes without being seen.

(C) An interior view of a courtyard in Cairo. Courtyards are places where women can do chores and manage children while remaining secluded from the outside world.

varieties of the **veil**, which may be a garment that totally covers the person's body and face or just a scarf that covers the person's hair. In some cultures, even prepubescent girls wear the veil; in others, they go unveiled until their transition into adulthood is observed. The veil allows a devout Muslim woman to preserve a measure of seclusion when she enters a public space, thus increasing the territory she may occupy with her honor preserved. A modern young woman may choose to wear a headscarf with jeans and a t-shirt in order to signal to the public that she is both an up-to-date woman and an observant Muslim.

There is considerable debate about the origin and validity of female seclusion and whether veiling and seclusion are specifically Muslim customs. Scholars say that these ideas, as well as the custom of having more than one wife, actually predate Islam by thousands of years and do not derive from the teachings of the Prophet Muhammad. In fact, Muhammad may have been reacting against such customs when he advocated equal treatment of males and females. Muhammad's first wife, Khadija, did not practice seclusion and worked as an independent businesswoman whose counsel Muhammad often sought.

146. EDUCATION, ECONOMIC EMPOWERMENT ARE KEYS TO BETTER LIFE FOR MUSLIM WOMEN

The Rights of Women in Islam

This region has notably more restrictive customary and legal limits on women than any other. In some Gulf states, women cannot travel independently or even drive a car or shop without male supervision. Yet even here changes have recently been made. In the decade of the 2000s, some women in the Gulf states became more active in public life, education, and business. In Qatar, Sheikha Moza, the second of the three wives of the emir (Muslim ruler) of Qatar, Sheikh Hamad bin Khalifa Al-Thani, has implemented remarkable changes for women. They can now drive, attend university, be elected to political office, and work side by side with men; the sheikha has founded a battered women's shelter and serves as a UNESCO envoy. In the United Arab Emirates (UAE), a few female activists, along with other female lawyers and journalists, have begun to question persistent patriarchal attitudes. Saudi Arabia remains the most restrictive country, but even there it is now possible for a woman to register a business without first proving that she has hired a male manager. A few young Saudi women even staged mini-demonstrations by posting on YouTube videos of themselves driving. Although they were eventually punished by the state for this behavior, they inspired further challenges to convention.

The Arab Spring in North Africa demonstrated just how elusive political equality for women can be. In Tunisia, Libya, and Egypt, some women were active as public protesters against undemocratic regimes, but military and civilian supporters of the various regimes singled them out for particularly

veil the custom of covering the body with a loose dress and/or of covering the head—and in some places the face—with a scarf

FIGURE 6.19 Variations on the veil as portable seclusion. There is an almost infinite variety of interpretations of the veil.

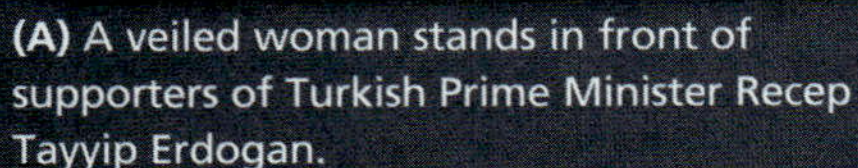
(A) A veiled woman stands in front of supporters of Turkish Prime Minister Recep Tayyip Erdogan.

(B) Schoolchildren in Iran wear a uniform that covers much of their hair and a suit that covers most of their body.

(C) In Tunisia, some women wear garments that cover all except their eyes and hands.

Thinking Geographically Why do these women choose to wear a veil (regardless of the style of veil they wear)?

harsh retribution. Among their fellow male demonstrators, it was usually the younger men who supported the women's rights and physically defended them when supporters of the various regimes attacked.

A practice that is a source of contention within this region and abroad is *polygyny* (see Chapter 7, page 317)—the taking by a man of more than one wife at a time. Although the Qur'an allows a man up to four wives, it generally does not encourage this and imposes financial limits on the practice by requiring that each wife be given separate and equal living quarters and support. While legal in most of the region (not Tunisia), polygyny is relatively rare, with less than 4 percent of males in North Africa practicing it. In Southwest Asia, polygyny—not covered in any reliable statistical surveys, but only in secondary reports—may be somewhat more common. About 5 percent in Jordan and 8 to 12 percent of marriages in Kuwait are polygamous. The social justifications given for polygyny further illustrate traditional ideas about women's rights in Islam. In common legal practice in many countries, an unmarried woman under 40 must be regarded as a minor requiring protection. Therefore, if a man takes her as a second or third wife, she gains support and safety. Similarly, a widow is saved from disgrace and poverty if her dead husband's brother takes her as an additional wife.

Female genital mutilation (FGM, defined and discussed in Chapter 7 on page 318) is not widely practiced in this region except in Egypt, where although it is illegal, as of 2013 according to the World Health Organization, more than 90 percent of women have undergone the practice, mostly before age 15.

Political and social equality with men can only improve with more economic independence for women (Figure 6.20). With the sole exception of Israel, women in this region generally do not work outside the home; when they do, they are paid, on average, only about 40 percent of what men earn for comparable work (see Figure 6.24C on page 263). Only India, Pakistan, and other South Asian countries have a similarly large wage gap based on gender.

The Lives of Children

Three sweeping statements can be made about the lives of children in the Islamic cultures of North Africa and Southwest Asia. First, in most families children contribute to the welfare of the family starting at a very young age. In cities, they run errands, clean the family compound, and care for younger siblings. In rural areas, they do all these chores and also tend grazing animals, fetch water, and tend gardens. Second, their daily lives take place overwhelmingly within the family circle. Both girls and boys spend their time within the family compound or in urban areas in adjacent family apartments. Their companions are adult female relatives and siblings and cousins of both sexes. Even teenage boys in most parts of the region identify more with family than with age peers.

In rural areas, prepubescent girls can move about in public space as they go about their chores in the village. After puberty they may be restricted to the family compound and required to wear the veil. The U.S. geographer Cindi Katz found that until puberty, rural Sudanese Muslim girls have considerably

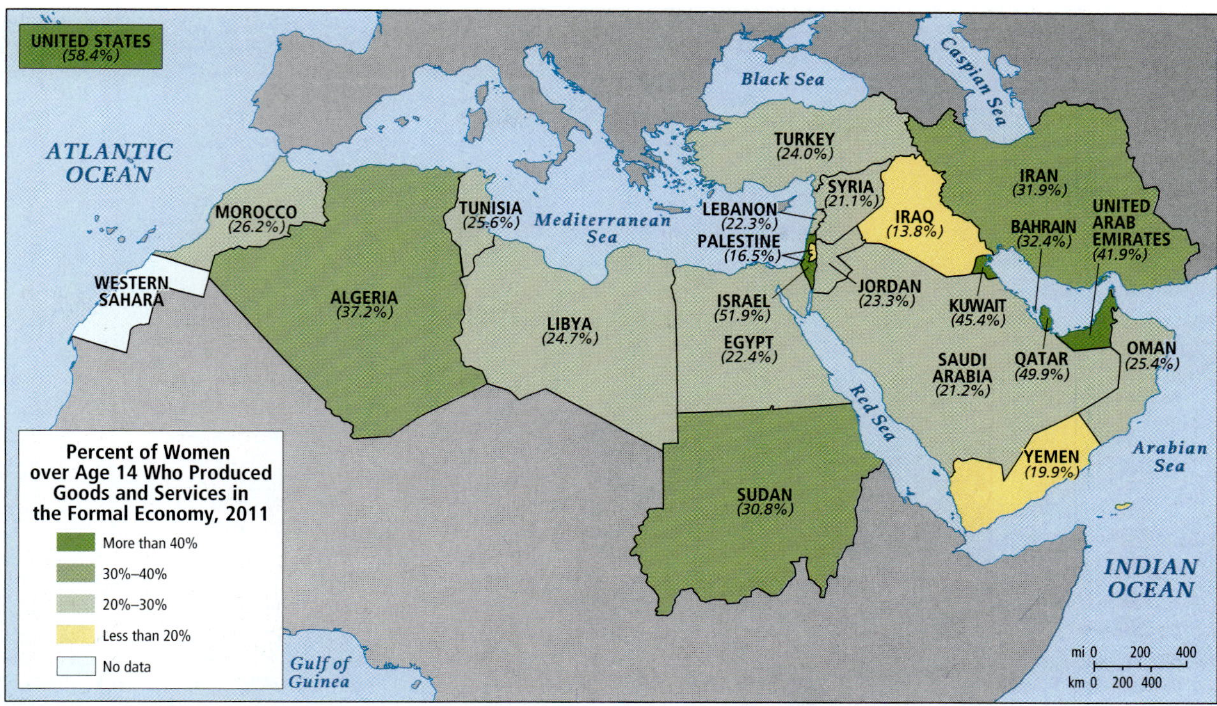

FIGURE 6.20 Percentage of the region's women who are in the labor force (2011). Between 2005 and 2011, the percentage of women who were earning wages in the labor force increased in 10 of the 22 countries and territories in the region, with the largest increase, more than 13 percentage points, in Qatar. The largest decreases (more than 10%) of women working were in Lebanon, Syria, and Yemen, and may be a result of major unrest in those countries.

more spatial freedom than do girls of similar ages in the United States, who are rarely allowed to range alone through their own neighborhoods.

The third observation is that school, television, and the Internet increasingly influence the lives of children and introduce them to a wider world. Most children go to school; many boys go for a decade or more, and increasingly girls go for more than a few years. In some countries (Algeria, Israel, Jordan, Lebanon, Libya, the occupied Palestinian Territories, Oman, Qatar, Tunisia, and the UAE), a larger percentage of girls than boys go to school. Like educated women everywhere, these girls will make life choices different from those of their mothers.

Even in rural areas, it is fairly common for the poorest families to have access to a television, which is often on all day, in part because it provides a window to the world for secluded women. Television can serve either to reinforce traditional cultural values or as a vehicle for secular perspectives, depending on which channels are watched.

THINGS TO REMEMBER

- For Muslims (93 percent of the population in this region), the Five Pillars of Islamic Practice embody the central teachings of Islam. Some Muslims are fully observant, some are not; but the Pillars have an impact on daily life for all. The call to prayer, broadcast five times a day in all parts of the region, is a reminder to all people to reflect on their beliefs.
- Given the wide diversity of thought, beliefs, and practices of Islam, the family is probably the most important societal institution for Muslims in the region.
- Most Muslim families in this region are patriarchal in structure; and the role and status of women, the spaces they occupy, and sometimes even the clothes they wear are carefully defined, circumscribed, and in some cases rigidly imposed.

Changing Population Patterns

Although the region as a whole is nearly twice as large as the United States, most of the population is concentrated in the few areas that are useful for agriculture. Vast tracts of desert are virtually uninhabited, while the region's 477 million people are packed into coastal zones, river valleys, and mountainous areas that capture orographic rainfall (compare Figure 6.21 with Figure 6.5). Population densities in these areas can be quite high. For example, some of Egypt's urban neighborhoods have over 260,000 people per square mile (100,000 people per square kilometer), a density 4 times higher than that of New York City, the densest city in the United States.

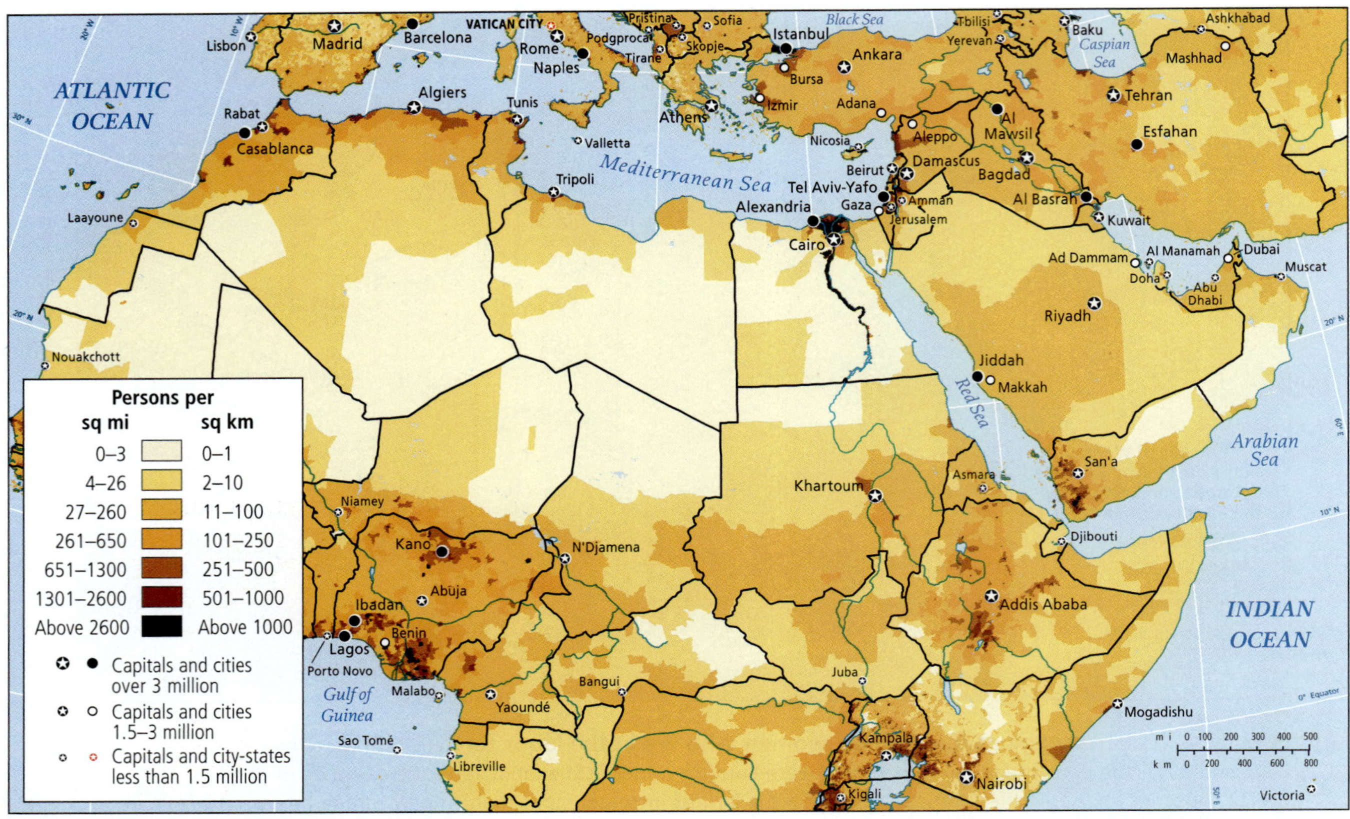

FIGURE 6.21 Population density in North Africa and Southwest Asia.

Population Growth and Gender Status

> **Geographic Insight 2**
>
> **Gender and Population:** Gender disparities are increasingly prominent in this region as political change begins to occur. Women are dependent because they are generally less educated than men, have few options for working outside the home, and have little access to political power. Women's roles are usually confined to the home and to having and raising children. As a result, this region has the second-highest population growth rate in the world, after sub-Saharan Africa.

Although fertility rates have dropped significantly since the 1960s, to 3.1 children per adult woman in 2012, they are still higher than the world average of 2.6. Only sub-Saharan Africa, at 5.1 children per woman, is growing faster. At current growth rates, the population of the region will be about 540 million by 2025. This growth will severely strain supplies of fresh water and food and will worsen shortages of housing, jobs, medical care, and education.

As noted in many world regions, population growth rates are higher in societies where women are not accorded basic human rights, are less educated, and work primarily inside the home. In places where women have opportunities to work or study outside the home, they usually choose to have fewer children. Figure 6.20 shows that as of 2011, considerably less than 50 percent of women across the region (except in Israel, Kuwait, and Qatar) worked outside the home at jobs other than farming. Moreover, on average only 74.8 percent of adult females can read—a fact that limits their employability—whereas 88 percent of adult males can read.

For uneducated women who work only at home or in family agricultural plots, children remain the most important source of personal fulfillment, family involvement, and power. This may partially explain why in 2011 a good deal less than half of women in this region were using modern methods of contraception, and only 54 percent were using any method of contraception at all. Both of these numbers are well below the world average of 55 and 62 percent, respectively. Other factors resulting in low use of contraception are male dominance over reproductive decisions and the unavailability or high cost of effective birth control products.

The deeply entrenched cultural preference for sons in this region is both a cause and a result of women's lower social and economic standing. It also contributes to population growth, as families sometimes continue having children until they have a desired number of sons. Moreover, some young females may not survive because of malnutrition and associated illnesses or because female fetuses are sometimes aborted. The result is that males slightly outnumber females in several age cohorts of the population pyramids in Figure 6.22 (see also the discussion in Chapter 1 on pages 14–15). In Qatar and the UAE, gender imbalance is extreme (see Figure 6.22C); in these countries, the unusually large numbers of males over the age of 15 is the result of the presence of numerous male guest workers. Opportunities for women are opening and this change is likely to reduce women's interest in having large families. Slower population growth will mean that fewer resources will have to be invested in housing, schools, and services (see "On the Bright Side" on page 269).

Urbanization, Globalization, and Migration

Geographic Insight 3

Urbanization and Globalization: Two patterns of urbanization have emerged in the region, both tied to the global economy. In the oil-rich countries, the development of spectacular new luxury-oriented urban areas is based on flows of money, credit, goods, and skilled people. But most cities are old, with little capacity to handle the mass of poor, rural migrants attracted by export industries aimed at earning foreign currency. In these older cities, many people live in overcrowded slums with few services.

(A) Iran, 2012

Age
100+
95–99
90–94
85–89
80–84
75–79
70–74
65–69
60–64
55–59
50–54
45–49
40–44
35–39
30–34
25–29
20–24
15–19
10–14
5–9
0–4
Males
Females
5 4 3 2 1 0 1 2 3 4 5
Population (in millions)

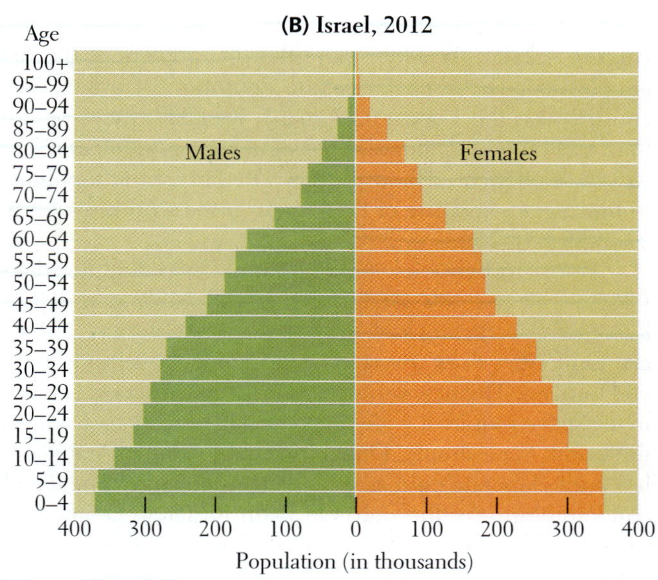

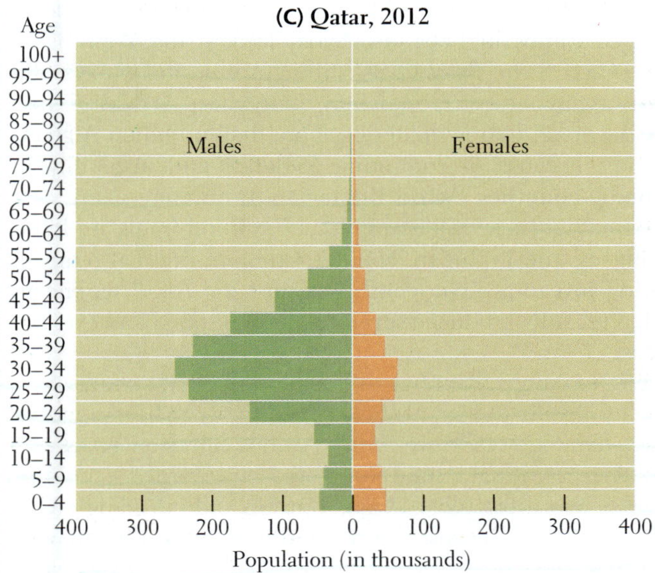

FIGURE 6.22 Population pyramids for Iran, Qatar, and Israel. The population pyramid for Iran **(A)** is at a different scale (millions) from those for Israel **(B)** and Qatar **(C)** (thousands). The imbalance of Qatar's pyramid in the 25–54 age groups is caused by the presence of numerous male guest workers. Note, too, that pyramids A and C show missing females in the younger age groups. (This is most easily observed by drawing lines from the ends of the male and female age bars to the scale at the bottom of the pyramid and comparing the numbers.)

Urbanization is transforming this region. Two highly globalized patterns of urbanization with distinct geographic signatures are now apparent: one pattern in the newly oil-rich countries, and another in those countries where development has primarily resulted in rural-to-urban migration with associated crowded slum housing.

Until the 1970s, most people lived in small rural settlements. Then, however, significant migration from villages to urban areas began to occur in response to economic forces driven by oil wealth and globalization, and by agricultural modernization that displaced small farmers. In locations other than the Gulf states, the shift toward green revolution–style export-oriented agriculture reduced the labor needed to produce crops headed for the world market. Even though the green revolution promoted needed export income for the countries, people who became part of the excess labor force had to leave the countryside for the cities. In North Africa, drought also instigated rural-to-urban migration (Figure 6.23B). By 2008, more than 70 percent of the region's people lived in urban areas (the definition of urban varies by country; see the Figure 6.23 map). By 2009, there were more than 434 cities with populations of at least 100,000 and 37 cities with more than 1 million people. The largest metropolitan area in the region, and one of the largest in the world, is Cairo, with 17 million residents in 2012.

The petroleum-rich Gulf states are now highly urbanized. Between 70 and 100 percent of the (still small) population now lives in urban areas, which are extravagant in design. These modern Gulf cities draw investment in high-tech ventures and high-end tourism. The new ventures and the construction of office and living space require a wide range of skilled workers from all over the world. For example, Dubai (one of the United Arab Emirates) has built elaborate (and as yet largely unoccupied) new high-rise condominiums for the rich. On Palm Jumeirah, a fanciful palm tree–shaped island and peninsula off the Gulf coastline, huge homes await buyers (see Figure 6.23A). The unexpectedly slow "take off" for the new Gulf cities is in large part the result of the global recession that began in 2008 and continues. The wealthier emirates subsidize the living standards of their citizens with oil profits, and for the most part only foreign contract laborers have low standards of living.

Outside the Gulf states, urban growth has been planned and financed far less well. For example, in 1950, Cairo had about 2.4 million residents, while today it is home to over 17 million people. These millions of new residents live in huge makeshift slums. Meanwhile, much of Cairo's middle class occupies the medieval interiors of the old city, where streets are narrow pedestrian pathways and plumbing and other services are chronically dysfunctional (see Figure 6.23C).

Internal and International Migration The prospect of better education opportunities, jobs, and living conditions pull rural internal migrants into Cairo and other cities outside the Gulf states. However, because stable well-paying jobs are scarce, many migrants end up working in the informal economy (see pages 132–133 in Chapter 3), sometimes as street vendors, as casual laborers, or as menial service providers. Unfortunately, education also does little to ensure employment; for decades this region has been noted for the large numbers of unemployed and underemployed university graduates.

In the Gulf states, on the other hand, there has been a deficit of trained native young people willing and able to work in white-collar jobs, yet surplus trained workers from neighboring countries have not been welcomed. Instead, immigrants come from all over the world to be temporary guest workers. Some work as laborers on construction sites and as low-wage workers throughout the service economy, but many work in shops and professions. In some countries, such as the UAE, guest workers make up 85 percent of the labor force. Most employers prefer Muslim guest workers, and over the last two decades, several hundred thousand Muslim workers have arrived from Palestinian refugee camps in Lebanon and Syria, as well as from Egypt, Pakistan, and India. Some female domestic and clerical workers come from Muslim countries in Southeast Asia, including the Philippines. Overwhelmingly, these immigrant workers are temporary residents with no job security and no right to become citizens. They remit most of their income to their families at home and often live in stark conditions alongside the opulence of those enriched by oil and gas.

Refugees comprise another category of migrants. This region has the largest number of refugees in the world. Usually they are escaping human conflict, but environmental disasters such as earthquakes or long-term drought also displace many people. When Israel was created in 1949, many Palestinians were placed in refugee camps in Lebanon, Syria, Jordan, the West Bank, and the Gaza Strip. Palestinians still constitute the world's oldest and largest refugee population, numbering at least 5 million. Elsewhere, Iran is sheltering more than a million Afghans and Iraqis because of continuing violence and instability in their home countries. Across the region, even more people are refugees within their own countries: 1.7 million Iraqis are internal refugees, and in Sudan, about 2.3 million internal refugees are living in camps. The conflict against the Assad regime in Syria has recently produced several million internal and external refugees.

Refugee camps often become semipermanent communities of stateless people in which whole generations are born, mature, and die. Although residents of these camps can show enormous ingenuity in creating a community and an informal economy, the cost in social disorder is high. Tension and crowding create health problems. Because birth control is generally unavailable, refugee women have an average of 5.78 children each.

Thinking Geographically

After you have read about urbanization in North Africa and Southwest Asia, you will be able to answer the following questions:

A What circumstances dealt a blow to high-end technology and tourism development in the Gulf states, such as Dubai?

B By 2008, what percent of the population in this region lived in urban areas?

C What has driven the migration of rural people to the region's old established cities, such as Cairo?

Figure 6.23 Photo Essay: Urbanization in North Africa and Southwest Asia

Globalization has brought two distinct patterns of urbanization to this region. In fossil fuel–rich countries, populations are more urbanized and cities have undertaken lavish building booms, drawing in laborers and highly skilled workers from across the globe. In countries without fossil fuel wealth, cities are receiving massive flows of poor rural migrants. This is partially a result of economic reforms (emphasizing export agriculture and industrialization) aimed at making rural and urban economies more globally competitive.

A Palm Jumeirah is an artificial, palm-shaped island in Dubai, UAE. Built by more than 40,000 workers, mostly low-wage migrants from South Asia, Palm Jumeirah will eventually house 65,000 people, mostly in villas and condos that cost millions of dollars each. Palm Jumeirah and several similar developments are central to Dubai's efforts to build a globalized tourism-based economy that will prosper long after the region's fossil fuel resources are exhausted.

Istanbul 15, Ankara 105, Izmir 146, Mashhad, Tehran 18, Esfahan 132, Algiers 42, Tunis, Casablanca 133, al-Mawsil 142, Aleppo 150, Baghdad 32, Beirut, Damascus 144, Tel Aviv-Yafo 117, Alexandria 65, Cairo 9, al-Basrah 76, Kuwait City, Dubai, Riyadh 78, Jiddah 128, Khartoum 97, San'a

TURKEY, MOROCCO, TUNISIA, ALGERIA, WESTERN SAHARA, LIBYA, EGYPT, SYRIA, LEBANON, ISRAEL, JORDAN, IRAQ, IRAN, KUWAIT, BAHRAIN, QATAR, UAE, OMAN, SAUDI ARABIA, YEMEN, SUDAN

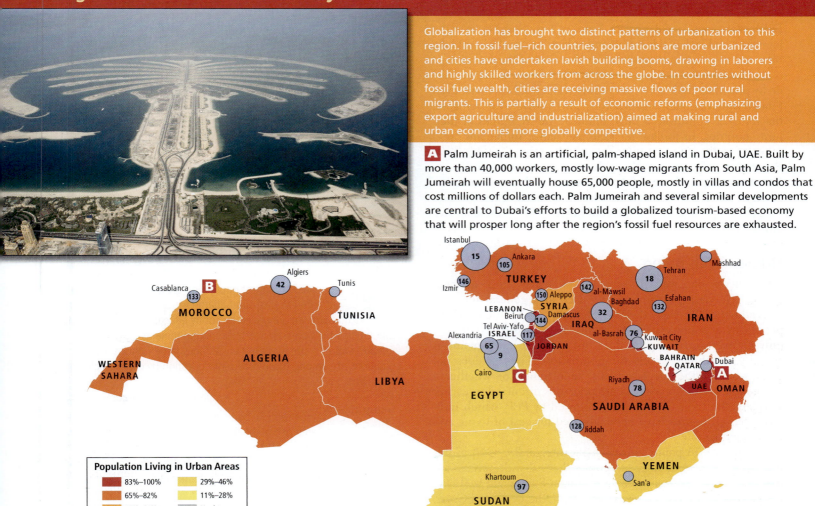

Population Living in Urban Areas

- 83%–100%
- 65%–82%
- 47%–64%
- 29%–46%
- 11%–28%
- No data

Population of Metropolitan Areas 2013

20 million, 10 million, 5 million, 3 million

Note: Symbols on map are sized proportionally to metro area population

① Global rank (population 2013)

mi 0 250 500
km 0 250 500

B A Moroccan man searches for valuables in a garbage dump outside Casablanca. The city's population is growing quickly, largely because people have been migrating from Morocco's drought-stricken interior, but a lack of jobs is resulting in escalating unemployment and crime.

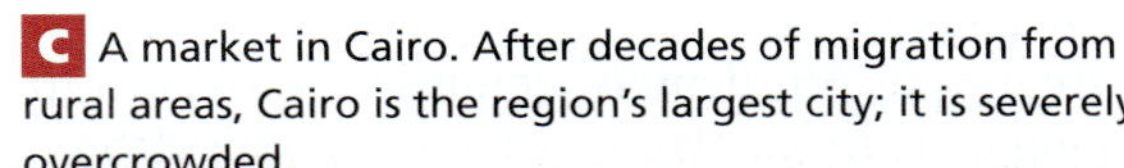

C A market in Cairo. After decades of migration from rural areas, Cairo is the region's largest city; it is severely overcrowded.

Disillusionment is widespread. Years of hopelessness, extreme hardship, and lack of education and rewarding employment take their toll on youth and adults alike. Because of these factors, it is easy for Islamists and jihadists to find new recruits, which they do by providing basic services to camp residents. Moreover, even though international organizations also provide for refugees, the refugees constitute a huge drain on the resources of their host countries. In Jordan, for example, native Jordanians are a minority in their own country because the more than 3 million Palestinian refugees and their children account for well over half the total population of the country, and their presence has changed life for all Jordanians. They have been joined by over 750,000 Iraqis, and as of 2013, more than 250,000 Syrians have crossed into Jordan.

Human Well-Being

As we have observed in previous chapters, gross national income (GNI) per capita can be used as a general measure of well-being because it shows broadly how much money people in a country have to spend on necessities, but it is only an average figure and does not indicate how income is actually distributed across the population. Thus, it does not show whether income is equally apportioned or whether a few people are rich and most are poor. (As discussed in earlier chapters, the term GNI is often followed by *PPP*, which indicates that all figures have been adjusted for cost of living and so are comparable.)

Figure 6.24A points to two situations regarding GNI in this region. First, from country to country there is wide variation—less than U.S.$2000 per person per year in Sudan and well over U.S.$47,000 per person per year in Kuwait, the UAE, and Qatar. Second, oil and gas wealth does not necessarily translate into high GNI per capita (compare Figure 6.24A with Figure 6.25 on page 264)—Iraq has significant oil wealth but only very modest GNI per capita. Libya, with more modest oil resources, has a much higher GNI than Iraq. Saudi Arabia, for all its oil wealth, does not have the top GNI per capita. Turkey, which has a diversified economy and very little oil, nonetheless has a GNI rank equal to that of Iran (in the middle range), which has significant oil and gas resources.

Figure 6.24B depicts countries' rank on the Human Development Index (HDI), which is a calculation (based on adjusted real income, life expectancy, and educational attainment) of how adequately a country provides for the well-being of its citizens. In this region, HDI ranks stretch from the medium-low to the high ranges, with Israel ranking the highest (17). If on the HDI scale a country drops below its approximate position on the GNI scale, one might suspect that the lower HDI ranking is due to poor distribution of wealth and poor social services: A critical mass of people are simply not getting a sufficient share of the national income to receive good health care and an education. Kuwait and Oman rank lower on HDI than they do on the GNI scale, indicating that they may not be managing their wealth well enough to ensure the well-being of all—thus, disparities are high. Libya, on the other hand, ranks medium-high on the HDI scale and only in the mid-range on GNI per capita. This probably signifies that at least prior to the revolution of 2011, Libya was investing in the social services that enhance human well-being.

Figure 6.24C shows the regional gender equality (GEI) patterns. Countries are ranked according to the degree to which women and men have equal access to reproductive health, political and educational empowerment, and the labor market. Of the 21 countries, only Israel, Kuwait, the UAE, Oman, Turkey, Morocco, Tunisia, and Libya have midlevel rankings. As shown on the world inset map in Figure 6.24C, this region ranks lower on gender equity than do most other parts of the world, including several parts of sub-Saharan Africa.

THINGS TO REMEMBER

Geographic Insight 2

- **Gender and Population** This region has the second-highest population growth rate in the world, after that of sub-Saharan Africa. Part of the reason for this is that women are generally poorly educated and tend not to work outside the home. Hence, childbearing remains crucial to a woman's status, a situation that encourages large families. These conditions are among those called into question by current political movements.

Geographic Insight 3

- **Urbanization and Globalization** Two patterns of urbanization, both stimulated by globalization, have emerged in the region. In the oil-rich countries, new luxury-oriented cities are tied to flows of money, goods, and people. Elsewhere, in older cities, economic reforms aimed at improving global competitiveness have brought massive migration by the rural poor into cities, creating crowding and slums.

- In terms of human well-being, this region is beginning to make improvements, but its overall rankings remain low. Recent political turmoil has reduced human well-being.

Economic and Political Issues

There are major economic and political barriers to peace and prosperity within North Africa and Southwest Asia. Wealth from fossil fuel exports remains in the hands of a few elite families. Most people are low-wage urban workers or relatively poor farmers or herders. The economic base is unstable because the main resources are fossil fuels and agricultural commodities, both of which are subject to wide price fluctuations on world markets. Meanwhile, in many poorer countries, large national debts and the need to stay competitive in the global market are forcing governments to streamline production and cut jobs and social services.

Political and economic cooperation in the region has been thwarted by a complex tangle of hostilities between neighboring countries. Many of these hostilities are the legacy of outside interference by Europe and the United States in regional politics, including colonial intrusions in earlier times and, more recently, the activities by global oil, gas, industrial,

FIGURE 6.24 Global maps of human well-being.

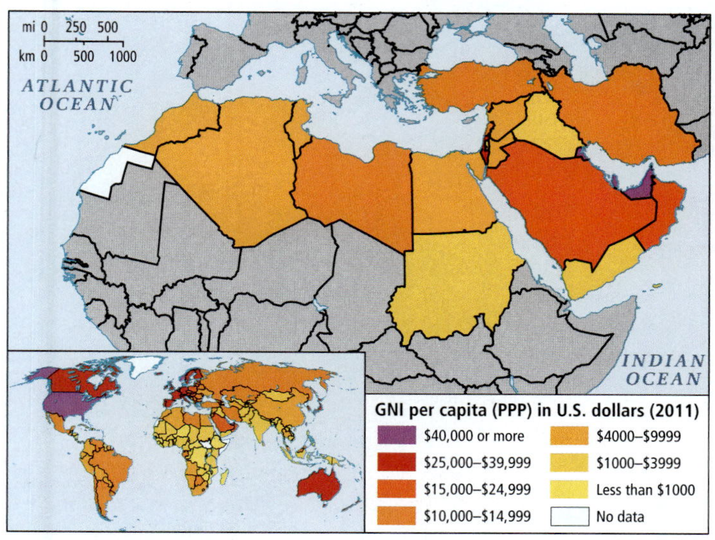

(A) Gross national income (GNI) per capita, adjusted for purchasing power parity (PPP).

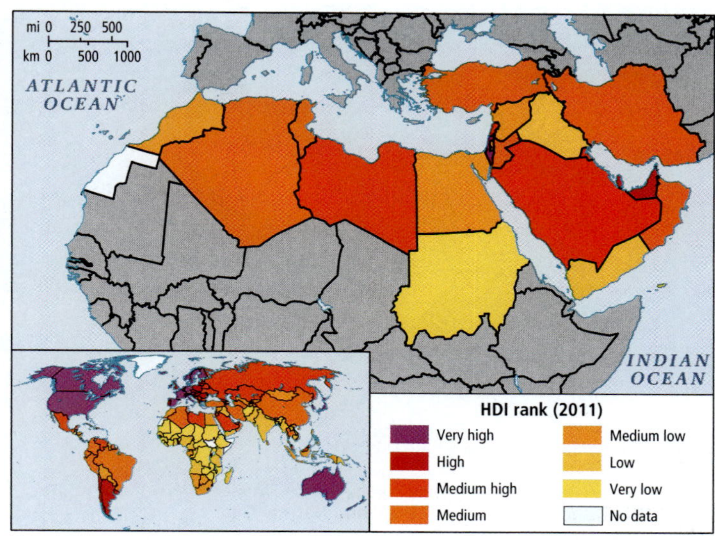

(B) Human Development Index (HDI).

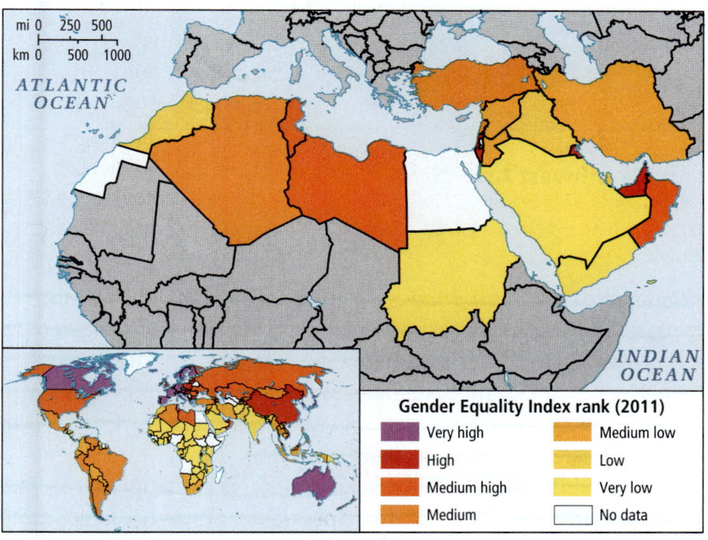

(C) Gender Equality Index (GEI).

and agricultural corporations. The Israeli–Palestinian conflict, which has profoundly affected politics throughout the region, is rooted in the persecution of the Jews in nineteenth-century Europe, culminating in the World War II Holocaust and its aftermath. The Iran–Iraq war of 1980–1988 and the Gulf War of 1990–1991 were instigated in part by pressures on petroleum resources, again from Europe and America. The long siege of violence in Iraq, though begun by a homegrown dictator, came to a head when the United States invaded and occupied Iraq, beginning in March of 2003 and formally ending on December 15, 2011.

Globalization, Development, and Fossil Fuel Exports

> **Geographic Insight 4**
>
> **Development and Globalization:** The vast fossil fuel resources of a few countries have transformed economic development in them and driven globalization in this region. In these countries, economies have become powerfully linked to global flows of money, resources, and people. Politics have also become globalized, with Europe and the United States strongly influencing many governments.

With two-thirds of the world's known reserves of oil and natural gas, this region is the major fossil fuel supplier to the world. Most oil and gas reserves are located around the Persian Gulf (see Figure 6.25) in the countries of Saudi Arabia, Kuwait, Iran, Iraq, Oman, Qatar, and the UAE. The North African countries of Algeria, Libya, and Sudan also have oil and gas reserves. It is important to note, however, that several countries in this region have minimal oil and gas resources or are net importers of fossil fuels: Morocco, Tunisia, Syria, Lebanon, Jordan, Israel, and Turkey. Meanwhile, Egypt and Sudan produce enough to supply most of their own needs, but export little. But regardless of their own petroleum resources, all countries in the region are affected by the geopolitics of oil and gas.

European and North American companies were the first to exploit the region's fossil fuel reserves early in the twentieth century. These companies paid governments a small royalty for the right to extract oil (natural gas was not widely exploited in this region until the 1960s, though gas extraction has grown rapidly since then). Oil was processed at on-site refineries owned by

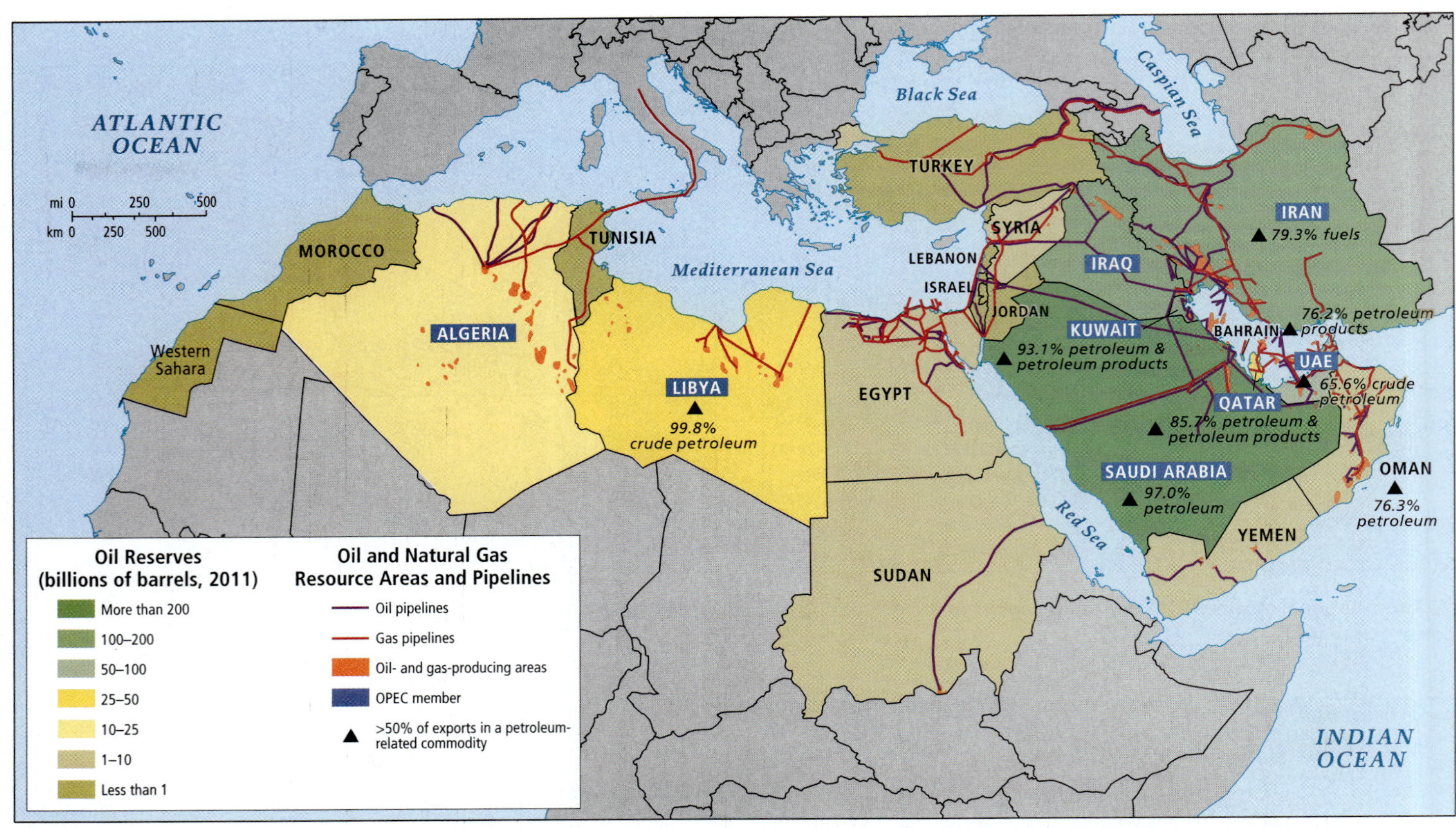

FIGURE 6.25 Economic issues: Oil and gas resources in North Africa and Southwest Asia.
(A) This map shows oil reserves, oil and gas resource areas, and pipelines.

the foreign companies and sold at very low prices, primarily to Europe and the United States and eventually to other countries, such as Japan.

Global oil and gas prices have risen and fallen dramatically since 1973, when governments in the Gulf states raised the price of oil and gas for several reasons. On the one hand, the Gulf states were imposing a kind of penalty in response to U.S. support for Israel in the 1973 Yom Kippur War between Israel and neighboring Arab countries (discussed further below). The oil-producing states were also reacting to a long-term trend of rising prices on products exported from European countries and the United States to the Gulf states. The oil price rise in 1973 was made possible by the founding back in 1960 of a **cartel**—a group of producers strong enough to control production and set prices for products—known as **OPEC (Organization of the Petroleum Exporting Countries)**. For a few months after the Yom Kippur War, there was a total halt of oil shipments from the Gulf states to the United States and to a few other countries that supported Israel. When the Gulf states resumed their shipments, they raised the price of oil. OPEC remains the main organization of oil-producing states; membership, which changes from time to time, now includes all of the states indicated in Figure 6.26. OPEC members cooperate to periodically restrict or increase oil production, thereby significantly influencing the price of oil on world markets. A move to create an OPEC-like cartel for natural gas is currently underway.

cartel a group of producers strong enough to control production and set prices for products

OPEC (Organization of the Petroleum Exporting Countries) a cartel of oil-producing countries—including Algeria, Angola, Iran, Iraq, Kuwait, Libya, Nigeria, Qatar, Saudi Arabia, the United Arab Emirates, Ecuador, and Venezuela—that was established to regulate the production and price of oil and natural gas

Many OPEC countries, which were exceedingly poor just 40 years ago, have become much wealthier since 1973, but they have also become more vulnerable to economic downturns. For example, oil income in Saudi Arabia shot up from U.S.$2.7 billion in 1971 to U.S.$110 billion in 1981, to U.S.$592.4 billion in 2012. But this wealth has not been widely shared within OPEC countries or within the region. The Gulf states were slow to invest their fossil fuel earnings at home in basic human resources, and they were slow to explore other economic strategies in case oil and gas ran out. Like their poorer neighbors, OPEC countries remain highly dependent on the industrialized world for their technology, manufactured goods, skilled labor, and expertise. Recently, the Gulf states have begun to invest heavily in roads, airports, new cities, irrigated agriculture, and petrochemical industries. Saudi Arabia is investing in six new cities in different parts of the country to spur investment and advance economic and social development. A good example of the scale of projects in the Gulf states is Palm Jumeirah (Palm Island) in Dubai, UAE (see Figure 6.23A), which is intended to be the

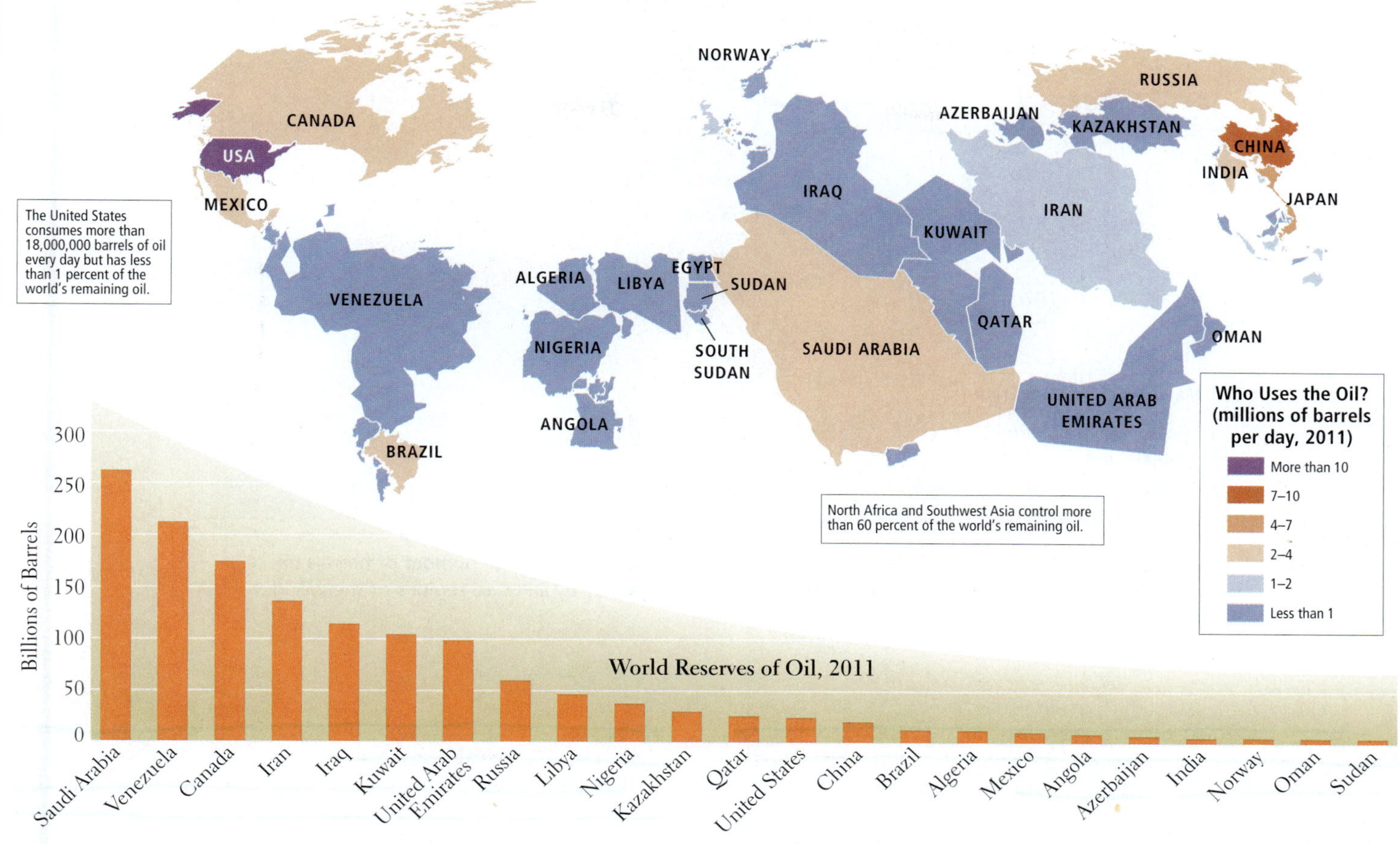

FIGURE 6.25 **(B)** This map contains a bar graph of oil reserves by country and a map in which the size of the countries corresponds to the amount of oil reserves held, and the color corresponds to the amount used.

centerpiece of Dubai's planned tourism economy—insurance against the day when the regional oil economy fails for whatever reason. The recession that started in 2008 badly damaged the economies of the Gulf states. Most of the massive building projects underway across the Gulf states screeched to a halt, including Palm Island and similar projects in the UAE. The recession demonstrated that Dubai's tourism economy is highly vulnerable to economic downturns, and as mentioned, with drastically fewer jobs available, countries that have many temporary workers, as in the Gulf states, have seen a steep decline in remittances sent home.

Resources have been poured primarily into visible physical development; much less oil wealth has been invested in education, social services, public housing, and health care. Libya, before the Arab Spring, and Kuwait are exceptions to this pattern, having developed ambitious plans for addressing all social service needs, with especially heavy investment in higher education. Iran is another possible exception, though because the Iranian government shares very little of its data with the rest of the world, it is hard to be certain how much it is investing in social services. Meanwhile, the non-OPEC countries (those that do not produce fossil fuels) do not share directly in the wealth of the Persian Gulf states, and for the most part, the oil-rich countries have not helped their poorer neighbors develop.

Many factors beyond OPEC's control strongly influence world oil prices. Recent rapid industrialization in China and India has increased their demand for fuel and other petroleum products, creating long-term pressures that are raising the price of oil. Employing a theory known as "peak oil," some experts argue that in an era of diminished oil reserves—which these experts say has already begun—prices will be driven higher. On the other hand, efforts to combat climate change with renewable energy sources have been increasingly successful and could dramatically reduce demand for oil. Current global oil flows are summarized in Figure 6.26.

Economic Diversification and Growth

Greater **economic diversification**—expansion into a wider range of economic development strategies—could have a significant impact on the region, bringing economic growth and broader prosperity and thereby limiting the damage caused by a drop in the prices of oil, gas, or other commodities on the world market.

economic diversification the expansion of an economy to include a wider array of activities

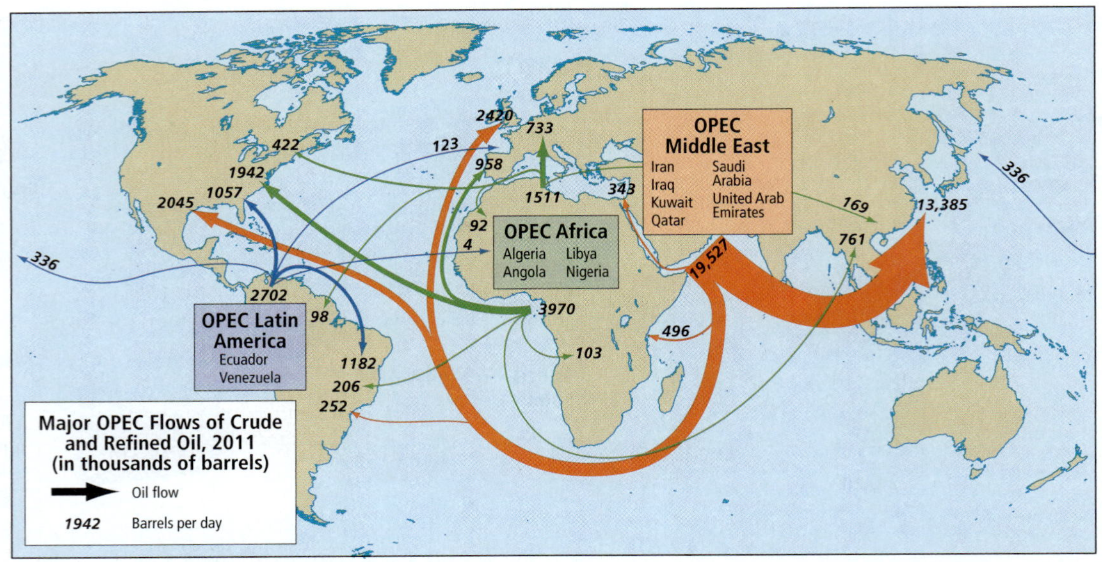

FIGURE 6.26 Major OPEC oil flows in 2011. The map shows the average number of barrels (in thousands) of crude oil and petroleum products distributed per day by OPEC to regions of the world.

By far the most diverse economy in the region is that of Israel, which has a large knowledge-based service economy and a particularly solid manufacturing base. Israel's goods and services and the products of its modern agricultural sector are exported worldwide. Turkey is the next most diversified, in part because—like Israel—it has never had oil income to fall back on. Egypt, Morocco, Libya, and Tunisia are also starting to move into many new economic activities. Some fossil fuel–rich countries tried to diversify into other industries. This was true in Syria until the dictator Bashar al-Assad refused to respond to civil protests from an economically stressed citizenry, which precipitated a revolution (see the next section). A somewhat more responsive government in the UAE developed an economy where only 25 percent of GDP is based directly on fossil fuels, and where trade, tourism, manufacturing, and financial services are now dominant. Even so, most Gulf states are still highly dependent on fossil fuel exports and the spin-off industries they generate.

Diversification has also been limited by economic development policies that favored *import substitution* (see pages 128–129 in Chapter 3). Beginning in the 1950s, many governments, such as those of Turkey, Egypt, Iraq, Israel, Syria, Jordan, Tunisia, and Libya, established state-owned enterprises to produce goods for local consumption. Among the major products were machinery and metal items, textiles, cement, processed food, paper, and printing. These enterprises were protected from foreign competition by tariffs and other trade barriers. With only small local markets to cater to, profitability was low and the goods were relatively expensive. Without competitors, the products tended to be shoddy and unsuitable for sale in the global marketplace. Meanwhile, the extension of government control into so many parts of the economy nurtured corruption and bribery and squelched entrepreneurialism. Many of the state-owned import substitution enterprises have subsequently either gone bankrupt or been sold to private investors.

Economic diversification and export growth were also limited by a lack of financing, private and public, from within the region. For example, rather than invest locally in mundane but needed consumer products, members of the Saudi royal family generally invest their wealth in more profitable private firms in Europe, North America, and Southeast Asia. Even when private and public investment has stayed in the region, it has gone into lavish projects, such as those in Dubai that give jobs to skilled Asians and may not necessarily prove to be economically profitable. Only recently have governments recognized that they need to invest locally in order to plan wisely for a time when oil and gas run out.

Finally, both international and domestic military conflicts and the ensuing political tensions have stymied economic diversification because they have resulted in some of the highest levels (proportionate to GDP) of military spending in the world. Military spending diverts funds from other types of development, such as health care and education (Figure 6.27). The top four spenders—Oman, Israel, Saudi Arabia, and Iraq—lead the world in military expenditures as a percentage of GDP. And all countries in the region, except Tunisia, are above the global average of 2 percent.

THINGS TO REMEMBER

Geographic Insight 4

- **Development and Globalization** The entire region has been influenced by the huge profits earned by a relatively few countries from petroleum resources. The economies of all countries in the region are linked to global flows of money, resources, and people. For many decades, Europe and the

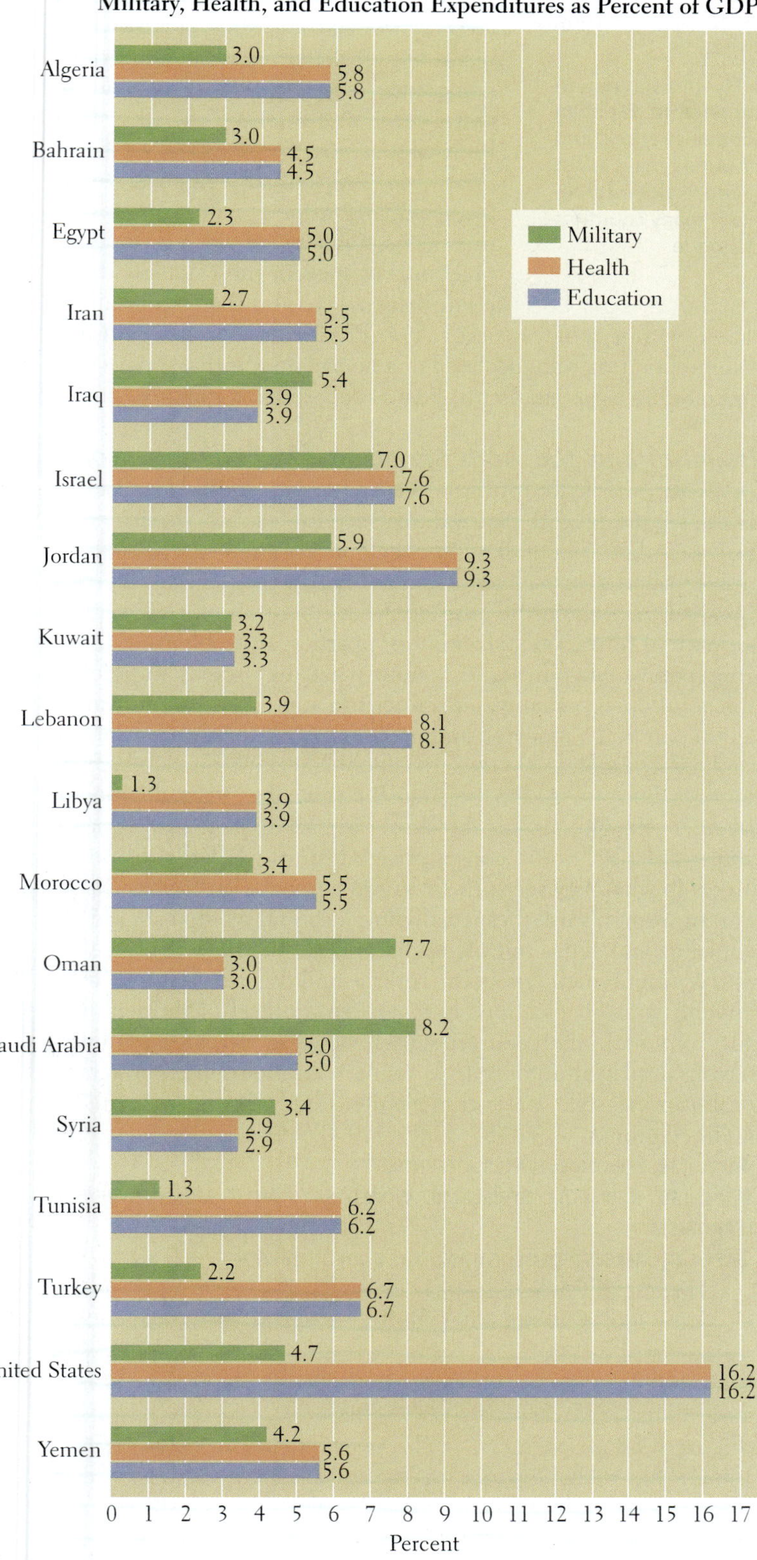

FIGURE 6.27 Military, health, and education expenditures as a percentage of GDP, 2008–2009. These graphs show those countries in the region that have data for each variable. Note that the data are from 2008–2009 and that expenditures may have changed due to the Arab Spring. The United States is included for comparative purposes.

United States have strongly influenced the governments of most of these countries.

- Profits in the oil- and gas-producing countries were low until OPEC initiated price increases in 1973. Since then, wealth has accrued mainly to a privileged few.
- Until very recently, countries in the region, with the exception of Turkey and Israel, had not diversified their economies. In the Gulf states, huge profits have not been invested at home, but rather abroad, thus limiting diversification. When import substitution policies have been tried, they have failed to lift countries out of poverty.

Power and Politics: The Arab Spring

Geographic Insight 5

Power, Politics, and Gender: Despite the predominance of elected bodies of government, and the increasing participation of women as voters, wealthy, politically connected males, clerics, and often the military retain the real power. Waves of protest, which included some women participants, swept this region, beginning with the Arab Spring of 2010, and resulted in the overthrow of several authoritarian governments. Within the region, official response to the Arab Spring has included both repression and reforms, but only a few of the positive changes hold promise for women.

The political protests that swept across North Africa in 2010 and into Syria by 2011 eventually resulted in the toppling of long-standing dictators in Tunisia, Egypt, and then Libya. The protest movement also presented an increasingly serious threat to President Assad in Syria. The early demonstrations were precipitated by high unemployment and rising food prices caused by the global recession that began in 2008; chronic poor living conditions; government corruption; and the absence of freedom of speech and other political freedoms. An undercurrent of dissent among women was also palpable in every country; a side effect was that movements for women's rights and human rights generally were rejuvenated (see the opening vignette).

The political disquiet reflected the fact that, for much of their history, most governments in this region gave citizens very little ability to influence how decisions were made. Constitutions were not constructed to facilitate widespread participation in public discourse or to protect women's and minorities' points of view. Laws were simply interpreted to suit the factions that held power. Elections were either nonexistent or were rigged to reelect those already in office or their chosen successors. Meanwhile, freedom of speech and of the press was strictly limited, especially if it involved criticism of the government. As a result, people across the region saw their governments as unresponsive and corrupt, and themselves as powerless to influence government in any way other than by massive protests.

Certainly, the role that Islam should play in society has been the most contentious issue of long-term significance, and it relates directly to determining which system of laws should be adopted as well as which protections women should be afforded.

ON THE BRIGHT SIDE

The Arab Spring as a Conduit for New Ideas

The potential for the Arab Spring to introduce political and social change quickly into this region was at first overstated. Certainly the trajectory of change is different and the pace slower than anticipated by optimists. As of this writing (August 2013), it is clear—as the demonstrations and debates have shown—that things will not return to the former status quo of purely male elitist control. The public shift toward support for democracy has the potential to reduce violence because the possibility of having a voice tends to curb anger. Furthermore, officials in many countries have now seen the power of public protests and are motivated to engage in dialogue.

Which First: Elections or Constitutions? The Arab Spring movement began in Tunisia and spread to Egypt, then to Libya, and eventually to every country in the region—including Iran—with the exceptions of Israel, Qatar, Turkey, and the UAE. The size and focus of the protests varied, but they all raised hopes for democratic reforms. Unfortunately, there were no viable reform models extant within the region to follow. Should constitutions be revised first to guarantee freedoms and equal participation for all, or should elections come first with the winning majority framing the new constitutions? Tunisia and Egypt chose to have elections first, but turnout was so low that not even one-third of the eligible voters participated. As a result in both cases, Islamist-oriented governments were elected. These governments then proceeded to adopt constitutions that did not adequately provide for the rights of more secular political groups, minorities, or women.

An American commentator on international politics, Fareed Zakaria, observed in 2013 that the thoughtful and cautious approaches to reform chosen by the kings of Jordan and Morocco—before full-blown Arab Spring movements developed—seem to have better served the evolution of democratic institutions in this region. In both cases, the respective kings negotiated with reformists to keep some of their royal powers while agreeing to give up others. They both supported the idea that lasting political reform should begin with constitutional reforms hammered out not by elected bodies, but by constitutional councils made up of a broad range of people. These councils were charged with making the political systems more democratic and inclusive, and in both cases women were significantly represented, as were religious minorities and those with secular points of view. Both kings were soundly criticized during the transition, but they accepted this as part of the process toward eventual constitutional monarchies. When elections were held, candidates from a wide range of positions won. In Morocco, for example, an Islamist party won a plurality of the seats in parliament (107 out of 395 seats) and formed the government; but with only a little over a quarter of the total seats, the Islamists had to accept cooperation and compromise as a necessity, an important precedent for the future.

Salafism an extreme, purist Qur'an-based version of Islam that has little room for adaptation to modern times

theocratic states countries that require all government leaders to subscribe to a state religion and all citizens to follow rules decreed by that religion

secular states countries that have no official state religion and in which religion has no direct influence on affairs of state or civil law

In Egypt, by contrast, where elections were the first priority, the highly organized Muslim Brotherhood mobilized their supporters and swept to power with only a minority of the electorate voting. Having won, the Muslim Brotherhood then controlled the writing of a constitution that gave the government authoritarian powers, rejected the already weak protections of women's rights that had been in place by, for example, reaffirming the custom that all women required the guardianship of a male, and undermined the rights of religious minorities. Freedoms of speech, most notably those of journalists, were also sharply curtailed. Then within a few weeks, after President Morsi, a Muslim Brotherhood stalwart, declared that he had powers that superseded those of the Egyptian courts, he was deposed in a military coup.

Islamism and the Arab Spring in a Globalizing World Ironically, the Arab Spring, with its seeming emphasis on political freedoms, actually opened up space for Islamist groups, especially Salafists, who took momentary advantage of the new openness to assert a very closed agenda. **Salafism** is an extreme, purist Qur'an-based version of Islam that has little room for adaptation to modern times. Globalization is a particular problem for these conservative Muslims who lament what they see as the global spread of open sexuality, consumerism, and hedonism, transmitted in part by TV, movies, and popular music. Salafists used to be explicitly peaceful, focusing mostly on the much-needed social services they provided to the poor. But since the 1990s, the idea of *jihad* (a struggle) against the West has gained popularity among young Salafists. Some conservative Muslims—Salafists or not—object to what they see as the Western emphasis on the liberalization of women's roles, especially the idea that women should be educated and active outside the home (see Figure 6.18).

As mentioned, in both Tunisia and Egypt, Islamists won enough votes in elections to give them control of the national government. In both countries, these newly empowered Islamists, having displayed authoritarian tendencies themselves, now face Arab Spring–style protest movements led by more secularly inclined groups.

139. WHAT MOTIVATES A TERRORIST?

140. NEW POLL OF ISLAMIC WORLD SAYS MOST MUSLIMS REJECT TERRORISM

144. JIHADIST IDEOLOGY AND THE WAR ON TERROR

Historically, Islamism has been rooted in the strong social, religious, and governmental authoritarianism common in this region, and in the lack of acceptance that people should have the right to organize popular political movements. Such movements are acceptable only if they are religious in nature (Figure 6.28 on pages 270–271). Governments in Saudi Arabia, Yemen, the UAE, Oman, and Iran are **theocratic states** in which Islam is the official religion and political leaders are considered to be divinely guided by both Allah and the teachings of the Qur'an. Elsewhere, such as in Turkey as well as in Tunisia, Libya, and even Syria (at least before the 2011 rebellion began), governments were officially **secular states**, where religious parties were not allowed and the law was neutral on matters of religion. In practice, however, even in some secular states, Islamic ideas influence government policies; authoritarian leanings mean that free speech and the right to hold public meetings have often been so severely limited that the only public spaces in which people have been allowed to gather are mosques, and the only public discussions free of censorship by the government have been religious discourses.

The militancy often associated with Islamism is characteristic of many popular political movements in this region, where challenges to the authority of governments are frequently met with violent repression (see the Figure 6.28 map). In both secular and theocratic governments, political freedoms are sometimes so weak that minorities have been denied the right to engage in their own cultural practices and to speak their own languages. Journalists and private citizens have been harassed, jailed, or even killed for criticizing governments or exposing corruption. While the Arab Spring protests that swept through this region were in part a response to these conditions, the extent to which the new governments and political reforms will protect political freedoms remains to be seen. **153. TURKEY VOTES FOR STABILITY**

ON THE BRIGHT SIDE

Reforms for Women Could Solve Labor Problems

An important impetus for change in the region is the increasing number of women who are becoming educated and employed outside the home. In a number of countries, women outnumber men in universities; most notably, in Saudi Arabia women make up 70 percent of university students (but only 5 percent of the workforce). While Saudi women activists have characterized their country as practicing a sort of apartheid with its own women, circumscribing or forbidding all manner of public activities, and while laws there still require a male guardian to accompany an adult woman when she leaves the house, change is in the wind. Recently, a Saudi prince, known for his interests in reform, suggested that one way to reduce dependence on foreign workers is to allow Saudi women to become drivers of taxis and trucks, as is now allowed in Jordan. [*Source:* http://www.trust.org/trustlaw/news/saudi-prince-questions-ban-on-women-driving]

Will There Be an Iranian Spring? When the political protests known as the Arab Spring erupted across North Africa, many wondered if Iran (which is not Arab) would join in the protests. A seeming prelude occurred in June 2009, when many Iranians suspected the presidential elections had been rigged. Tens of thousands of people in Tehran took to the streets in protest and on certain days their numbers exceeded 100,000. The main opposition force was the Green Party, which felt its candidate had been cheated by election fraud. A number of prominent Iranian private citizens joined with the Green Party protestors. The demonstrations first focused on election fraud, but when the Iranian police killed several unarmed demonstrators, the center of protest attention became the brutality of the autocratic, theocratic state and the need for democratic reforms as well as improvements in women's rights. But while the street protests became more and more radical, calling for an end to the Islamic state, no symbolic leader or rallying cause emerged around which all could unite.

The Iranian street protests persisted into 2010, when the activities of the Arab Spring in North Africa began. By February of 2013, protestors in Iran had a variety of topics on their minds; the need for democratic reforms was still central, as was Iran's international status, which was of concern for two reasons. Iran's ongoing uranium enrichment program and the threat that it would develop an atomic bomb had resulted in Western sanctions against Iran's oil trade (Iran's oil exports declined by as much as half in 2012 and the value of the country's currency plummeted; see also Figure 6.26 for patterns of oil distribution). Also, Iranian protestors objected to their country's support for Syrian president Bashar al-Assad as he repressed the increasingly strong rebellions against his regime. In the Iranian presidential elections of June 2013, with 72 percent of eligible voters participating, a moderate, Hassan Rouhani, was elected, winning three times as many votes as his nearest competitor. He called for improved international relations and a stronger economy less dependent on imports. Despite this apparent shift in focus for Iran's leadership, dissent continues to be expressed in Iran, so the durability of the political awakening remains in question.

The Role of the Press, Media, and Internet in Political Change

In some countries—Turkey, Israel, and Morocco, for example—the press is reasonably free and opposition newspapers are aggressive in their criticism of the government. But in much of the region, journalism can be a risky career, often leading to imprisonment. Egypt has a checkered history where the press is concerned. For years, the Mubarek government harassed Hisham Kassem, editor of the independent English-language weekly *Cairo Times*. When he became too critical of the government, he lost his license to publish in Egypt. For a time he took great pains to write and print his paper outside the country and smuggle it into Egypt, always risking arrest. Business leaders in Cairo began to provide backing for Kassem, enabling him to publish *Al-Masry Al-Youm*, which specialized in domestic issues of corruption, election fraud, and the need for an independent judiciary. Kassem and many independent journalists like him participated actively in the 2011 protests in Egypt that brought an end to the 30-year regime of Hosni Mubarak. In this he shared a common cause with Egypt's formerly repressed Islamist opposition group, the Muslim Brotherhood, which won control of the national government in elections in 2012. However, Kassem and many in the media returned to harsh criticism of the Muslim Brotherhood–led government for censoring print and broadcast journalists who were critical of the government.

In Saudi Arabia, a dozen newspapers are on the newsstands every morning, but all are owned or controlled by the royal Saud family, and all journalists are constrained by the fact that they may not print anything critical of Islam or of the Saud family, which numbers in the tens of thousands. When accidents happen or some malfeasance by a public official is revealed, the story is blandly reported, with little effort to explore the causes of events or their effects or to hold responsible officials accountable.

A beacon of journalistic change is the broadcasting network Al Jazeera, founded by the emir of Qatar, Sheikh Hamad bin Khalifa al-Thani, but now privately owned and independent of the Qatar government. Al Jazeera is credited with changing the climate for public discourse across the region and even with changing public opinion outside of the region through Al Jazeera English, the source of some of the information used in the opening vignette. The various versions of Al Jazeera now openly cover controversy, including reforms proposed by the most radical Iranian and Arab Spring protesters, with no apparent censorship.

147. AL JAZEERA LAUNCHES GLOBAL BROADCAST OPERATION

The role of the Internet and cell phone technology as a force for political change emerged first during the Iranian protests beginning in 2009, when the fact that so many Iranians had video-capable cell phones meant that the world instantly saw via the Internet the brutality of the Iranian police and army. Use of the Internet (e-mail, Facebook, crowdsourcing, and Twitter) became commonplace during the Arab Spring, when ordinary Egyptians and Libyans with nothing but an inexpensive cell phone were able to give real-time reports about activities in the streets of Cairo, Tripoli, and Banghazi. Twitter became the chief medium because it is free, mobile, and quick. The data from many sources could then be compiled via software, such as *Ushahidi*, invented and made popular by several Kenyan engineers (discussed in the opening vignette of Chapter 7).

Democratization and Women

Most countries in this region now allow women to vote, and two countries—Israel and Turkey—have elected female heads of state (prime ministers) in the past. Nevertheless, women who want to actively participate in politics still face many barriers. In the Gulf states, where women's political status is the lowest, circumstances vary from country to country. Oman gave all women the right to vote in 2003. In Kuwait in 2009, four highly educated women were elected to parliament; they quickly energized the pace of law making, often publicly criticizing their male colleagues, who were frequently absent for crucial votes. In Saudi Arabia, by contrast, women will be given the right to vote only in local elections in 2015, and only one Saudi woman serves as a public official. Across the region, women average less than 12 percent of national legislatures, the lowest of any world region and half the world average. By 2013, Saudi Arabia had appointed, not elected, women to 20 percent of Parliament seats, and only in Saudi Arabia, Iraq, Israel, Libya, Morocco, and Sudan did women make up more than

Figure 6.28 Photo Essay: Power and Politics in North Africa and Southwest Asia

Thinking Geographically

After you have read about democratization and conflict in North Africa and Southwest Asia, you will be able to answer the following questions:

A What circumstances made many suspicious of the U.S. claim that the 2003 war with Iraq was undertaken to bring democracy to that country?

B Why is the 1973 war between Israel and Egypt cited as one of the causes of political repression within countries of the region?

C What were the issues that motivated Syrians to take to the streets in prolonged demonstrations in 2011?

D What has changed for Kurds in Iraq that has lessened their use of violent protests?

Political freedoms are often poorly protected in this region, which is one of the least democratized in the world. It is no coincidence that this region also suffers disproportionately from violent conflict. Wars and terrorism here have limited the expansion of political freedoms. Authoritarian regimes often forcibly repress legitimate political opposition groups, and conflicts that might have been resolved peacefully through the political process drag on for years or even decades.

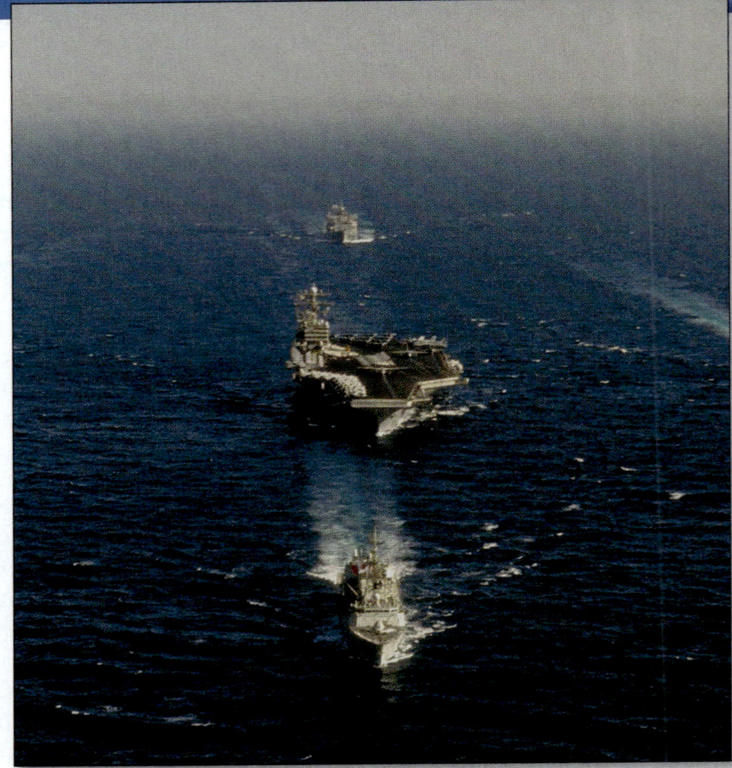

A U.S. Navy war ships patrol the Persian Gulf. U.S. and European militaries have long had a strong presence in this region because of special interest in the region's oil exports to the global economy. Part of the justification given for the U.S.-led invasion of Iraq in 2003 was that it was part of an effort to ensure greater political freedoms for the people of Iraq. However, many U.S. officials have openly admitted that safeguarding foreign access to the region's oil reserves was a more powerful motivator.

E Thousands celebrate in Tripoli, Libya, on the second anniversary of the end of Libya's civil war of 2011. Like the war in Syria, this conflict, in which 25,000 to 30,000 people were killed, was ignited by the government's brutal repression of Arab Spring protests.

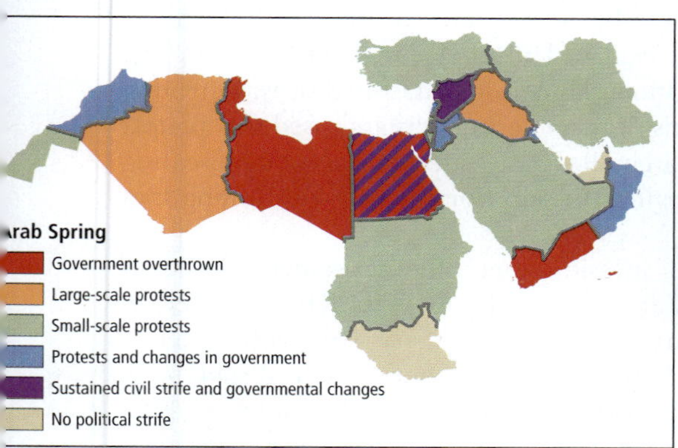

B An Egyptian man displays a Star of David painted on the bottom of his shoe—an act of symbolic desecration—during an anti-Israel rally. Ongoing conflict between Israel and its Arab neighbors has worked against political freedom throughout the region. Many Arab governments use the conflict with Israel to justify repressing legitimate political parties. Meanwhile, Israel recently banned anyone who has visited "enemy countries" (Lebanon, Syria, Iraq, Iran, Saudi Arabia, Sudan, Libya, and Yemen) from running for national office for 7 years.

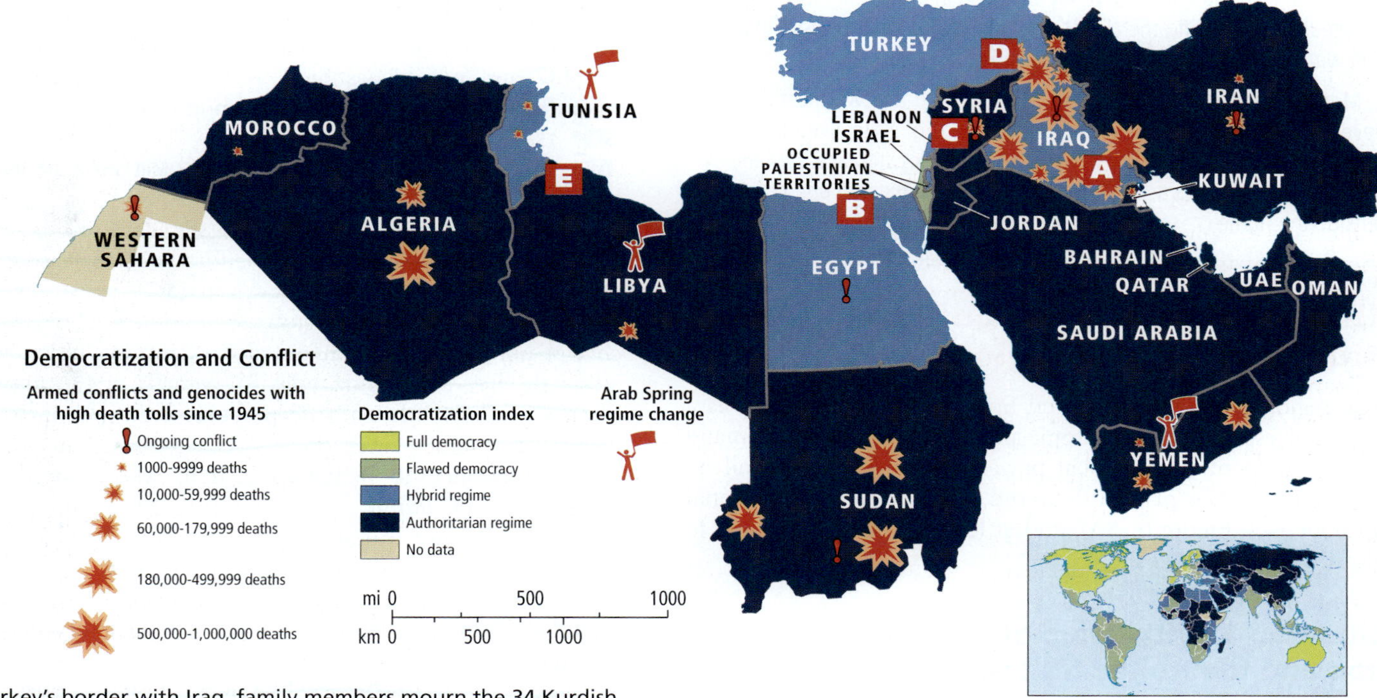

D Near Turkey's border with Iraq, family members mourn the 34 Kurdish civilians killed by the Turkish air force, who mistook them for rebels. Kurds form a majority of the population in southeastern Turkey and in neighboring parts of Iraq and Iran. Kurds have long been denied political freedoms and in some cases have been barred from even speaking their language. Though Kurdish militias still exist, democratization has recently reduced public support for violence, especially in northern Iraq.

C A rebel sniper in the city of Aleppo, Syria. The Syrian government's harsh crackdown on Arab Spring protests, beginning in 2011, was the spark that started the civil war. Over 100,000 people have been killed and millions of others have been displaced by the conflict.

15 percent of parliaments. Women lost ground in Egypt when the post–Arab Spring elections in 2012 brought fewer women to office than there were before. Moreover, there is a tendency—even among women—to not support female candidates.

THINGS TO REMEMBER

Geographic Insight 5

- **Power and Politics** While the immediate causes of the Arab Spring demonstrations may be economic, activists who have long promoted increased political freedoms as a path to peace in this region have been energized by the protests and continue their work to end authoritarian rule. Women participants have thus far been both encouraged and disappointed by the results.
- Demand for greater political freedoms is increasing, and public spaces for debate, other than mosques, are emerging. The press is reasonably free in some countries and severely curtailed in others. Al Jazeera is credited with changing the climate for public discourse across the region and with modifying public opinion outside the region.
- Most countries now allow women to vote, and the two most developed countries have elected female heads of state. Even though women who want to actively participate in politics still face barriers, patterns are changing as more women become educated and employed outside the home.

Three Worrisome Geopolitical Situations in the Region

The region of North Africa and Southwest Asia is known as a center of especially troublesome political issues that command the attention of major global powers. These issues can all be related in one way or another to the politics of the international oil trade (see Figure 6.28A), and at least one may be related to global climate change.

Situation 1: Fifty Years of Trouble Between Iraq and the United States, 1963 to 2013

The origins of the U.S. war with Iraq, beginning in 2003, lie in 1963, when the United States backed a coup that installed a pro-U.S. government that evolved into the regime of Saddam Hussein. For decades, the United States had an amicable relationship with the Iraqi government, driven in large part by Iraq's considerable oil reserves, the fourth largest in the world after those of Saudi Arabia, Canada, and Iran. The United States publicly supported Iraq in the 1980–1988 war between Iraq and Iran, but secretly supplied Iran with weapons during the Reagan administration (1981–1989) when it appeared that Iraq might become more troublesome if it won the war. Relations with Iraq took a dramatic turn for the worse in 1990 when Saddam Hussein, Iraq's dictator, invaded Kuwait. The United States, under President George H. Bush, forced Iraq's military out of Kuwait in the Gulf War of 1990–1991, and afterward placed Iraq under crippling economic sanctions, but they failed to tame Saddam Hussein (see Figure 6.7D).

After the terrorist attacks of September 11, 2001, the U.S. administration of George W. Bush (2001–2009) first launched a war on Afghanistan, but by 2003, shifted its focus to Iraq and its autocratic president, Saddam Hussein (see Chapter 2, page 75). The Bush administration claimed that Iraq had or was creating an arsenal of weapons of mass destruction and used this pretext to declare war on March 20, 2003, with the goals of confiscating Iraq's weapons, removing Hussein from power, and turning Iraq into a democracy.

After the initial invasion met little resistance, President Bush declared the war won on May 1, 2003. However, terrorist bombs and insurgent attacks soon erupted, with violence peaking between 2006 and 2007, after which it gradually decreased but did not cease. By March 2013, a total of 4422 U.S. troops had been killed and 31,926 had been wounded. Furthermore, over 150,000 veterans of the conflict had been diagnosed with some form of serious mental disorder, including posttraumatic stress (PTSD). The death toll for Iraqis, including civilians, was much higher, estimated as at least 111,407 and possibly as high as 1.3 million when counting Iraqi deaths caused by the harsh conditions created by the war.

136. IRAQ WAR ENTERS SIXTH YEAR

152. FIVE YEARS AFTER 'MISSION ACCOMPLISHED' IN IRAQ, WAR CONTINUES

The Bush administration's main stated justification for the war was democratization, but the failure to understand the long-simmering tensions between Iraq's major religious and ethnic groups crippled the democratization process. Sunnis in the central northwest had dominated the country under Saddam, although they constituted only 32 percent of the population. Shi'ites in the south, with 60 percent of the population, have dominated politics since the fall of Saddam. This has given greater influence to neighboring Iran, whose mostly Shi'ite population feels affinity for Iraqi Shi'as. Meanwhile, Kurds in the northeast, once brutally suppressed under Saddam Hussein, are allied with Kurdish populations in Turkey, Iran, and Syria, and resist cooperating with the Iraqi national government in Baghdad. **134. KURDISH NATIONALISTS IN IRAQ, TURKEY SEEK LAND OF THEIR OWN**

Recent studies of Iraqi public opinion indicate that a majority of Iraqis want a strong central government that can protect them from violent insurgents and that can maintain control of the country's large fossil fuel reserves. Polls also show that the vast majority of Iraqis want all fighting to stop and all U.S. and allied military forces to leave. All U.S. "combat" forces left Iraq in 2012, though thousands of personnel will remain for years to come as trainers and technical support for the Iraqi military.

Situation 2: The State of Israel and the "Question of Palestine"

The Israeli–Palestinian conflict has lasted more than 60 years and has included several major wars and innumerable skirmishes. It is a persistent obstacle to political and economic cooperation within the region, and it complicates the relations between many countries in this region and the rest of the world (see Figure 6.28B).

TABLE 6.1 Circumstances and human well-being among Palestinians and Israelis, 2012

Statistics show some stark differences in well-being between Palestinians and Israelis, including differences in overall satisfaction with life.

	Population (in millions) (mid-2012)	GNI PPP in U.S.$ per capita (2012)	Life expectancy at birth (2012)	Infant mortality rate (2012)	HDI ranks (2012)*	Overall life satisfaction (2011)**
Palestinians	4.3	3359	73	21	110	4.7
Israelis	7.9	27,660	82	3.4	16	7.4

* See http://hdrstats.undp.org/en/countries/profiles/PSE.html

** Ranked on a scale of 1 to 10, with 0 representing least satisfied and 10 representing most satisfied

Source: Population Reference Bureau 2012 Data Sheet

Human rights report: http://www.hrw.org/middle-eastn-africa/israel-palestine

Israel's excellent technical and educational infrastructure, its diverse and prospering economy, and the large aid contributions (public and private) it receives from the United States and elsewhere have made it one of the region's wealthiest, most technologically advanced and militarily powerful countries.

The Palestinian people, by contrast, are severely impoverished and undereducated after years of conflict, inadequate government, and meager living (see Table 6.1), often in refugee camps. Through a series of events over the past 60 years, Palestinians have lost most of the lands on which they used to live. They now live in two main areas—the West Bank (home to 2 million Palestinians) and the Gaza Strip (1.1 million), with another 2 million living as refugees in Jordan, where they outnumber ethnic Jordanians. The West Bank and Gaza are both highly dependent on Israel's economy. Israel often takes military action in these two zones in retaliation for Palestinian suicide bombings and rocket fire launched primarily from Gaza. The West Bank Palestinian territories continue to be encircled by security walls built by Israel, partly to defend against violence but also to curtail Palestinian access.

154. 60 YEARS AFTER ISRAEL'S FOUNDING, PALESTINIANS ARE STILL REFUGEES

131. ISSUES FROM 1967 ARAB–ISRAELI WAR REMAIN UNRESOLVED

The Creation of the State of Israel In response to centuries of discrimination and persecution in Europe and Russia, in the late nineteenth century, a small group of European Jews, known as **Zionists**, began to purchase land in a part of the Ottoman Empire known at the time as Palestine. Most sellers were wealthy non-Palestinian Arabs and Ottoman Turks living outside of Palestine. These absentee landowners had leased their lands for many years to Palestinian tenant farmers and herders or granted them the right to use the land freely. Such rights were negated by the sales to Zionists. In 1917, the British government adopted the Balfour Declaration, which favored the establishment of a national home for Jews in the Palestine territory, and which explicitly stated that "nothing shall be done which may prejudice the civil and religious rights of existing non-Jewish communities in Palestine, or the rights and political status enjoyed by Jews in any other country." It was at this time that the word *Palestine*, with roots far back in history, began to be used officially. In 1923 after World War I (see page 250), the United Kingdom received what is called a "mandate" from the League of Nations to administer the territory of Palestine (see Figure 6.29A on page 274).

On their newly purchased lands, the Zionists established communal settlements called *kibbutzim*, into which a small flow of Jews came from Europe and Russia. While Jewish and Palestinian populations intermingled in the early years, tensions emerged as Zionist land purchases displaced more and more Palestinians, seemingly in breach of the Balfour Declaration.

By 1946, following the death of 6 million Jews in Nazi Germany's death camps during World War II, strong sentiment had grown throughout the world in favor of a Jewish homeland in the space in the eastern Mediterranean that Jews had shared in ancient times with Palestinians and other Arab groups. Hundreds of thousands of Jews began migrating to Palestine, and against British policy, many took up arms to support their goal of a Jewish state.

Sixty Years of Conflict and the Two-State Solution In November 1947, after an intense debate in the UN General Assembly, the UN adopted the "Plan of Partition with an Economic Union" that ended the earlier mandate and called for the creation of both Arab and Jewish states (the *two-state solution* still under discussion today), with special international status for the city of Jerusalem (see Figure 6.29B).

The Palestinians and neighboring Arab countries fiercely objected to the establishment of a Jewish state and feared that they would continue to lose land and other resources. Then as now, the conflict between Jews and Palestinian Arabs was less about religion than control of land, settlements, and access to water. The sequence of changing allotments of land over time to Israel and to the Palestinians can be followed in Figure 6.29A–F.

On the same day that the British reluctantly ended the mandate and withdrew their forces, May 14, 1948, the Jewish Agency for Palestine unilaterally declared the creation of the state of Israel on the land designated to them by the Plan of Partition. Warfare between the Jews and Palestinians began immediately. Neighboring Arab countries—Lebanon, Syria, Iraq, Egypt, and Jordan—supporting the Palestinian Arabs, invaded Israel the next day. Fighting continued for several months, during which Israel prevailed militarily. An armistice was reached in 1949, and as a result,

Zionists those who have worked, and continue to work, to create a Jewish homeland (Zion) in Palestine

FIGURE 6.29 Israel and Palestine, 1923–1949; and after 1949, Israel and the Palestinian Territory.

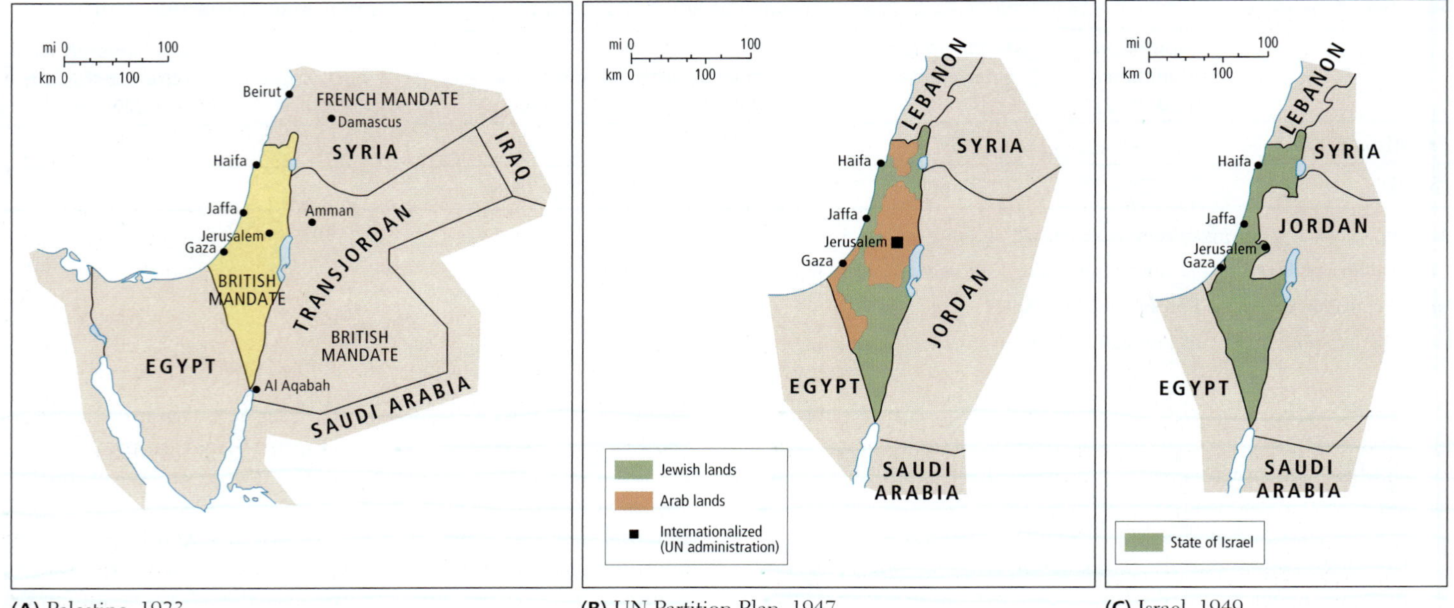

(A) Palestine, 1923 **(B)** UN Partition Plan, 1947 **(C)** Israel, 1949

Israel and Palestine, 1923–1949. (A) Palestine, 1923. Following World War I, Britain controlled what was called Palestine and is now Israel and Jordan (Transjordan was the precursor to Jordan). **(B)** The UN Partition Plan, 1947. After World War II, the United Nations developed a plan for separate Jewish and Palestinian (Arab) states. **(C)** Israel, 1949. The Jewish settlers did not agree to the partition plan; instead, they fought and won a war, creating the state of Israel.

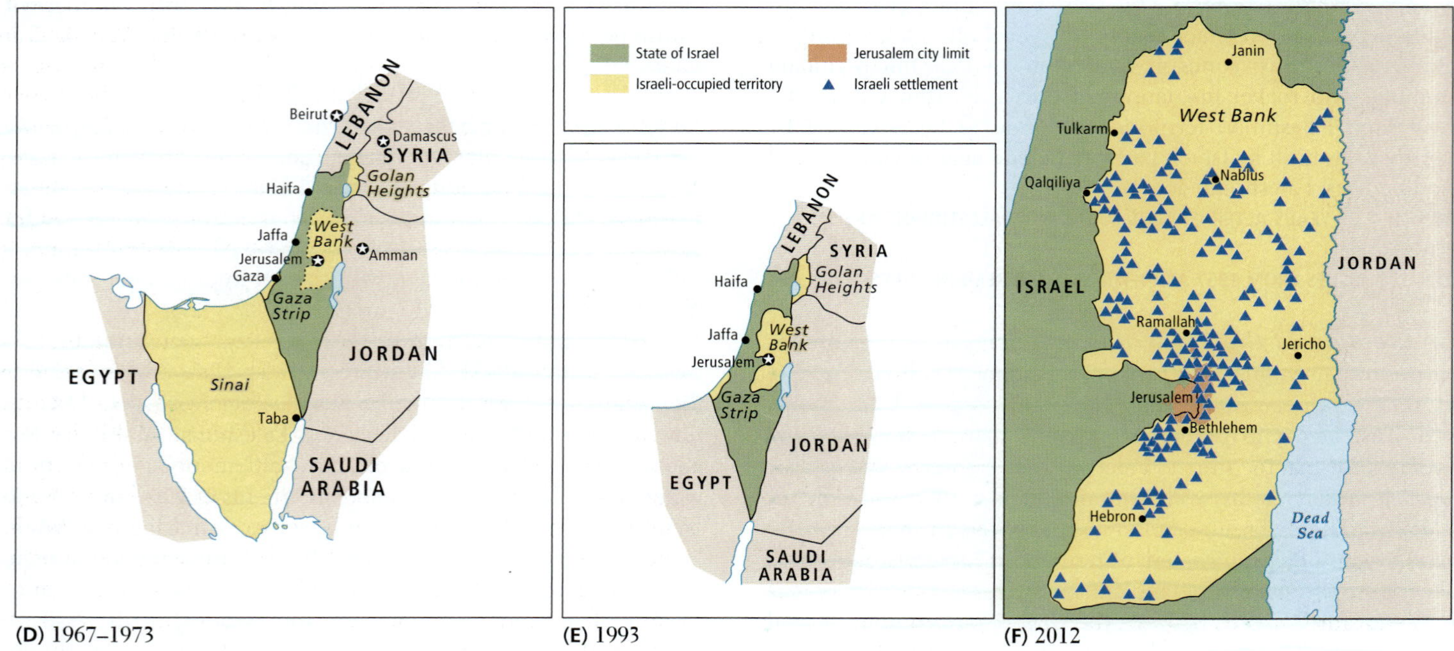

(D) 1967–1973 **(E)** 1993 **(F)** 2012

Israel and the Palestinian Territory after 1949. When the state of Israel was created in 1949, its Arab neighbors were opposed to a Jewish state. **(D)** In 1967, Israel soundly defeated combined Arab forces and took control of Sinai, the Gaza Strip, the Golan Heights, and the West Bank. **(E)** In subsequent peace accords, Sinai was returned to Egypt, but Israel maintained control over the Golan Heights and the West Bank, claiming that they were essential to Israeli security. **(F)** Although the Palestinians were granted some autonomy in the Gaza Strip and the West Bank, during the 1990s the Israelis, contrary to verbal agreements in the Oslo Accords, continued to build Jewish settlements in the West Bank and Golan Heights. By 2012, there were hundreds of extralegal Israeli settlements in these areas. This map shows only West Bank settlements.

the Palestinians' land shrank still further, with the remnants incorporated into Jordan and Egypt (see Figure 6.29C).

In the repeated conflicts over the next decades—such as the Six-Day War in 1967 and the Yom Kippur War in 1973—Israel again defeated its larger Arab neighbors, expanding into territories formerly controlled by Egypt (Sinai), Syria (Golan Heights), and Jordan (West Bank of the Jordan River; see also Figure 6.29D). Since 1948, hundreds of thousands of Palestinians have fled the war zones, with many forcibly removed to refugee camps in nearby countries. Some Palestinians stayed inside Israel and became Israeli citizens, but they have not been treated by the state as equal to Jewish Israelis.

128. HEZBOLLAH—SERVING MUSLIMS WITH GOD AND GUNS

133. LEBANESE OIL SPILL—COLLATERAL DAMAGE OF THE BOMBINGS

In 1987, the Palestinians mounted the first of two prolonged uprisings, known as the **intifada**, characterized by escalating violence. The first ran until 1993, when the Oslo Peace Accords provided that Israel withdraw from parts of the Gaza Strip and the West Bank, and that the Palestinian Authority be the entity that would enable Palestinians to govern themselves in their own state. The second intifada began in 2000 and continues into the present, primarily fueled by the expansion of Israeli settlements—in breach of the Oslo Accords—into Palestinian territories in Gaza, the West Bank, and the Golan Heights (see Figure 6.29F). Both sides have suffered substantial trauma and casualties. According to B'Tselem, an Israeli human rights group, far more Palestinians have died in this violence (since 2000, there have been 6630 Palestinian deaths versus 1097 Israeli deaths).

Over the years since 1947, the vision of a two-state solution has not died. In 2012 the Palestinians, led by their president, Mahmoud Abbas, officially petitioned the UN General Assembly for status as a nonmember observer state. The UN General Assembly voted 138 to 9 to approve this petition (41 nations abstained), and thus the Palestinians gained official recognition that they had never before achieved.

Territorial Disputes When Israel occupied Palestinian lands in 1967, the UN Security Council passed a resolution requiring Israel to return those lands, known as the occupied Palestinian Territories (oPT), in exchange for peaceful relations between Israel and neighboring Arab states. This *land-for-peace* formula, which set the stage for an independent Palestinian state, has been only partially fulfilled.

Despite the land-for-peace agreement, between 1967 and 2013, Israel secured ever more control over the land and water resources of the occupied territories. Israel took what appeared to be a major step toward peace in 2005 when it removed all Jewish settlements from the Gaza Strip, but this progress was negated by the blockade of Gaza's economy and the significantly stepped-up settlement in the West Bank. Between 2005 and mid-2013, some 95,000 new Israeli settlers were added and thousands of Palestinians were displaced.

The **West Bank barrier**, a high containment wall (Figure 6.30A) built beginning in 2003, encircles Jewish settlements on the West Bank and separates approximately 30,000 Palestinian farmers from their fields. It also blocks roads that once were busy with small businesses, effectively annexes 6 to 8 percent of the West Bank to Israel, and severely limits Palestinian access to much of the city of Jerusalem, most of which is now on the Israeli side of the barrier. The barrier, declared illegal by the World Court and the United Nations, and opposed by the United States, is nonetheless very popular among Israelis because it has reduced the number of Palestinian suicide bombings. In July 2013, new peace negotiations began.

132. WEST BANK BARRIER, NEW DIVIDE IN PALESTINIAN–ISRAELI CONFLICT

intifada a prolonged Palestinian uprising against Israel

West Bank barrier a 25-foot-high concrete wall in some places and a fence in others that now surrounds much of the West Bank and encompasses many of the Jewish settlements there

ON THE BRIGHT SIDE

Economic Interdependence as a Peace Dividend

Though rarely covered in the media, economic cooperation is a fact of life for Israelis and Palestinians and is a crucial component for any two-state solution. Economic ties between Palestine and Israel have long been essential to both. Israel is the largest trading partner of the West Bank and the Gaza Strip, and provides Palestinians with a currency (the Israeli shekel), electricity, and most imports. Israel needs the labor of tens of thousands of Palestinian workers in fields, factories, and homes (see Figure 6.30B).

Currently, there are several cooperative industrial parks that have been established jointly by the Palestinian Authority and the Israeli government in the West Bank and the Gaza Strip. Dubbed "peace parks," the goal of these parks is to use economic development to overcome conflict. Some people are critical of them because Israel tightly controls Palestinian access to the world market. For the peace parks to earn the trust of Palestinians may take some time, given Israel's history of discouraging industrial development in the occupied territories. Nevertheless, neighboring Israeli and Palestinian cities, such as Gilboa in Israel and Jenin in the West Bank, are pursuing the peace park idea. In 2012, Israelis and Palestinians quietly reached new agreements to increase Palestinian access to jobs in, and trade with, Israel. The agreements address tax procedures, increase the number of entrance permits, and help ease the movement of people through roadblocks.

Situation 3: Failure of the Arab Spring in Syria

The worries raised by the rebellion in Syria (beginning in 2011 and continuing through the present) are basically the same as those raised by all the Arab Spring movements. Can autocratic governments continue to block democratic reforms? Will Islamist factions succeed in taking control from more secular and/or moderate rebels? Will women be able to change patriarchal customs? Will gains in economic development be overwhelmed by the material destruction caused by civil war? In Syria, yet another concern has been raised that carries implications for the entire region: Climate change and the water scarcity that accompanies it (discussed later in this section) are now thought to be among the stressors that helped to bring on the Syrian rebellion.

The political causes of rebellion in Syria are many. Syria used to be one of the more developed countries in the region. Living together peacefully were several factions of Muslims, as well as Christians and even a few Jews; markets were full of life; women could have careers and go out alone past midnight; old men played board games while children played along

FIGURE 6.30 Conflict and cooperation between Israelis and Palestinians. Conflict and cooperation between Israelis and Palestinians have both increased in recent years.

(A) Palestinian women on their way to Jerusalem to celebrate Ramadan pass through a heavily guarded checkpoint at the West Bank Barrier in Bethlehem, Palestine. The barrier has hindered the flow of people and goods between Israel and the West Bank, severely damaging the latter's economy.

(B) Palestinian women working in a factory in Jerusalem. Israeli companies have access to the expertise, equipment, and investment capital that make the growth of manufacturing industries possible, while Palestinians supply cheap labor. Although economic partnerships are developing between Israelis and Palestinians, these efforts suffer when violence forces border closures.

community streets; tourists flocked to ancient historic sites; and Syrian TV dramas disseminated the Syrian hospitable way of life across the region. But after the authoritarian Assad family swept to power in a bloodless coup in 1970, Syrians had less and less to be proud of. Even before Bashar al-Assad took over what amounted to a police state from his father Hafez in 2000, Syria had gained a reputation for backing assassinations in Lebanon and bombings in Iraq, for building nuclear weapons factories, and for supporting *Hezbollah*—a Lebanese-based radical Shi'ite group that has close links to Iran and that opposes Israel and uses terrorist tactics.

Although for a time Syrians enjoyed the way the Assads thumbed their noses at the West (the United States and Europe), more educated and secular Syrians could see that their access to decent lives and participation in civic life was being ever more curtailed by the Assad regime. The trappings of a free society, such as shopping malls, Coca-Cola, Internet cafés, and Facebook, no longer pacified when debate, elections, and creativity were quelled by both a police state and a society rife with corruption even in the civil courts.

The Assad family belongs to the minority Alawite sect of Shi'ites, while the majority of Syrians are Sunni Muslims. To keep political equilibrium, the Assads favored a secular state, something like Turkey's; but Sunnis were kept out of governing circles. Sunni antagonism against Alawites grew because Alawites led particularly privileged lives, with many advantages coming to them from the Assad family.

Several pro-democracy protestors in Damascus who demanded the release of political prisoners were shot dead during March 2011. The protestors persisted, and from the spring of 2011 well into 2013, the Syrian government, backed by the army and aided by weapons and food assistance from Russia and Iran, escalated its attacks on the dissidents. In time, it became apparent that the rebel factions came from opposing points of view—some favoring pro-democratic secular changes, others wanting more Islamist or Salafist conservative reforms. The dissenters were unable to coalesce around common goals for Syria. Although for the most part the Syrian army remained loyal to Assad, some soldiers from Sunni parts of the country defected to the rebels; yet, despite all the destruction and loss of life, at no time did the Assad administration agree to peace-making concessions.

By early 2013, more than 100,000 people had died from violence and related causes, with children making up at least one-third of the dead. Aid agencies estimated that by March of 2013, a million Syrians lacked adequate food, 2.5 million were internally displaced, and 1 million had fled across borders to Turkey, Iraq, Lebanon, and Jordan. Western governments have been reluctant to intervene.

While the political causes of the Syrian civil rebellion are numerous, there is mounting evidence that drought and inappropriate human responses to it have played a role in fueling already well-stoked civil discontent. The Assad government denies that there is a drought, but 70 percent of the ancient *qanats* (water conduits) have already dried up. Drought is not even acknowledged in the government-controlled press, probably because the government does not want to admit that because of bad management, irrigation projects have failed, crops and herds are dying, and hundreds of thousands of agricultural families have had to move to cities, which are ill-equipped to house and feed them. Syria, the original home of wheat cultivation, must now buy wheat and other grains on a world market in which prices have doubled due to climate change crises in the global centers of wheat production: China, Russia, Ukraine, Australia, and the Americas.

THINGS TO REMEMBER

- The United States has had a long and meddlesome relationship with Iraq that dates back to the 1960s, when the United States tried to influence Iraq's internal affairs.
- While the modern conflict between Israel and the Palestinians is decades old, the situation remains dynamic and complex and continues to be a persistent obstacle to widespread political and economic cooperation in the region.
- Economic cooperation is already a fact of life for Israelis and Palestinians, and more such interaction is a crucial component of any solution to the current conflict. The entire region would benefit from the peace dividend of greater economic cooperation.
- Political protests against the autocratic Assad regime in Syria have become complicated by the involvement of competing points of view among Islamist, Salafist, secularist, and pro-democracy dissident factions.
- A further and as yet little-assessed factor is climate change and the drought Syria has suffered for more than a decade.

Geographic Insights: North Africa and Southwest Asia Review and Self-Test

1. Water, Food, and Climate Change: The population of this predominantly dry region is increasing faster than is access to cultivable land and water for agriculture and other human uses. Many countries here are dependent on imported food, and because global food prices are unstable, the region often has food insecurity. Climate change could reduce food output both locally and globally and make water, already a scarce commodity, a source of conflict. New technologies offer solutions to water scarcity but are expensive and unsustainable.

- How will people get enough water to grow food in the future, especially if climate change makes this dry region even drier?
- What factors could make it difficult to obtain enough water for agriculture? What technologies could provide potential solutions? How crucial is imported food for this region?

2. Gender and Population: Gender disparities are increasingly prominent in this region as political change begins to occur. Women are dependent because they are generally less educated than men, have few options for working outside the home, and have little access to political power. Women's roles are usually confined to the home and to having and raising children. As a result, this region has the second-highest population growth rate in the world, after sub-Saharan Africa.

- How does the low status of women contribute to this region's high population growth?
- Relative to other regions, how high is this region's population growth rate? How do women's education levels influence their decisions about how many children to have?

3. Urbanization and Globalization: Two patterns of urbanization have emerged in the region, both tied to the global economy. In the oil-rich countries, the development of spectacular new luxury-oriented urban areas is based on flows of money, credit, goods, and skilled people. But most cities are old, with little capacity to handle the mass of poor, rural migrants attracted by industries aimed at earning foreign currency. In these older cities, many people live in overcrowded slums with few services.

- How has globalization shaped different patterns of urbanization throughout the region?
- Why is present urbanization so different in the Gulf states than in the rest of the region? How have those economic reforms that are aimed at improving global competitiveness, especially in export-oriented agriculture, influenced urbanization outside the Gulf states?

4. Development and Globalization: The vast fossil fuel resources of a few countries have transformed economic development and driven globalization in this region. In these countries, economies have become powerfully linked to global flows of money, resources, and people. Politics have also become globalized, with Europe and the United States strongly influencing many governments.

• How have the huge fossil fuel reserves of some countries transformed economic development and driven globalization in this region?

• Characterize the flows of money, resources, and people that are features of globalization in the Gulf states.

• How might changes in job opportunities for women in this region relieve the dependency on imported labor? Where is this likely to be most important?

5. Power, Politics, and Gender: Despite the predominance of elected bodies of government, and the increasing participation of women as voters, wealthy, politically connected males, clerics, and often the military retain the real power. Waves of protest, which included some women participants, swept this region, beginning with the Arab Spring of 2010, and resulted in the overthrow of several authoritarian governments. Within the region, official response to the Arab Spring has included both repression and reforms, but only a few of the positive changes hold promise for women.

• What would you say are the chief signs that this region may be democratizing?

• How have recent events changed the political landscape of this region?

• What are some consequences of repression of the political opposition? What has been the effect of integrating political opposition movements into the democratic process?

Critical Thinking Questions

1. Which social forces in North Africa and Southwest Asia modify the power of religion?

2. Discuss how people in this region have affected world diets (including especially the Americas), first through the domestication of plants and animals and then through trade.

3. To what extent is the present-day map of North Africa and Southwest Asia related to the dismantling of the Ottoman Empire after World War I?

4. Considering the various factors that encourage relatively high fertility in this region, design themes for a public education program that would effectively encourage lower birth rates. Which population groups would you target? How would you incorporate cultural sensitivity into your project?

5. Consider the new forces that are affecting urban landscapes: immigration and globalization. Identify some of the expected effects on ordinary people of these abrupt changes to traditional living spaces.

6. Compare and contrast the public debate over the proper role of religion in public life in your country and in one country in this region (for example, Turkey, Morocco, Egypt, or Saudi Arabia). Contrast the roles of religious fundamentalists in the debates in your country and in the country chosen.

7. Gender is a complex subject in this region. Choose a rural location and an urban location and make a list of the forces in each that would affect the future of a 20-year-old woman. Describe those hypothetical futures objectively, that is, without using any judgmental terminology.

8. Why is it important to know that in some countries of this region agriculture may produce only a small amount of the GDP yet employ 40 percent or more of the people? What are some of the things such a relationship would indicate about the state of development in that country? What public policies would be appropriate in these circumstances—for example, should agriculture be deemphasized? How might this deemphasis affect food security?

9. Describe the circumstances that led to support in Europe and the United States for the formation of the state of Israel. Why did the West overlook the Palestinian people in this political undertaking?

10. Discuss the possibilities that scarcity of water is or will become a cause of violence in the region. What is the evidence against this happening?

Chapter Key Terms

cartel 264
Christianity 248
desertification 243
diaspora 248
economic diversification 265
female seclusion 254
Fertile Crescent 246
fossil fuel 237
Gulf states 254
hajj 253
intifada 275
Islam 236
Islamism 237
jihadists 254
Judaism 248
monotheism 248
Muslims 248
occupied Palestinian Territories (oPT) 237
OPEC (Organization of the Petroleum Exporting Countries) 264
Ottoman Empire 249
patriarchal 254
Qur'an (or Koran) 240
Salafism 268
salinization 240
seawater desalination 243
secular states 268
shari'a 253
Shi'ite (or Shi'a) 254
Sunni 253
theocratic states 268
veil 255
West Bank barrier 275
Zionists 273

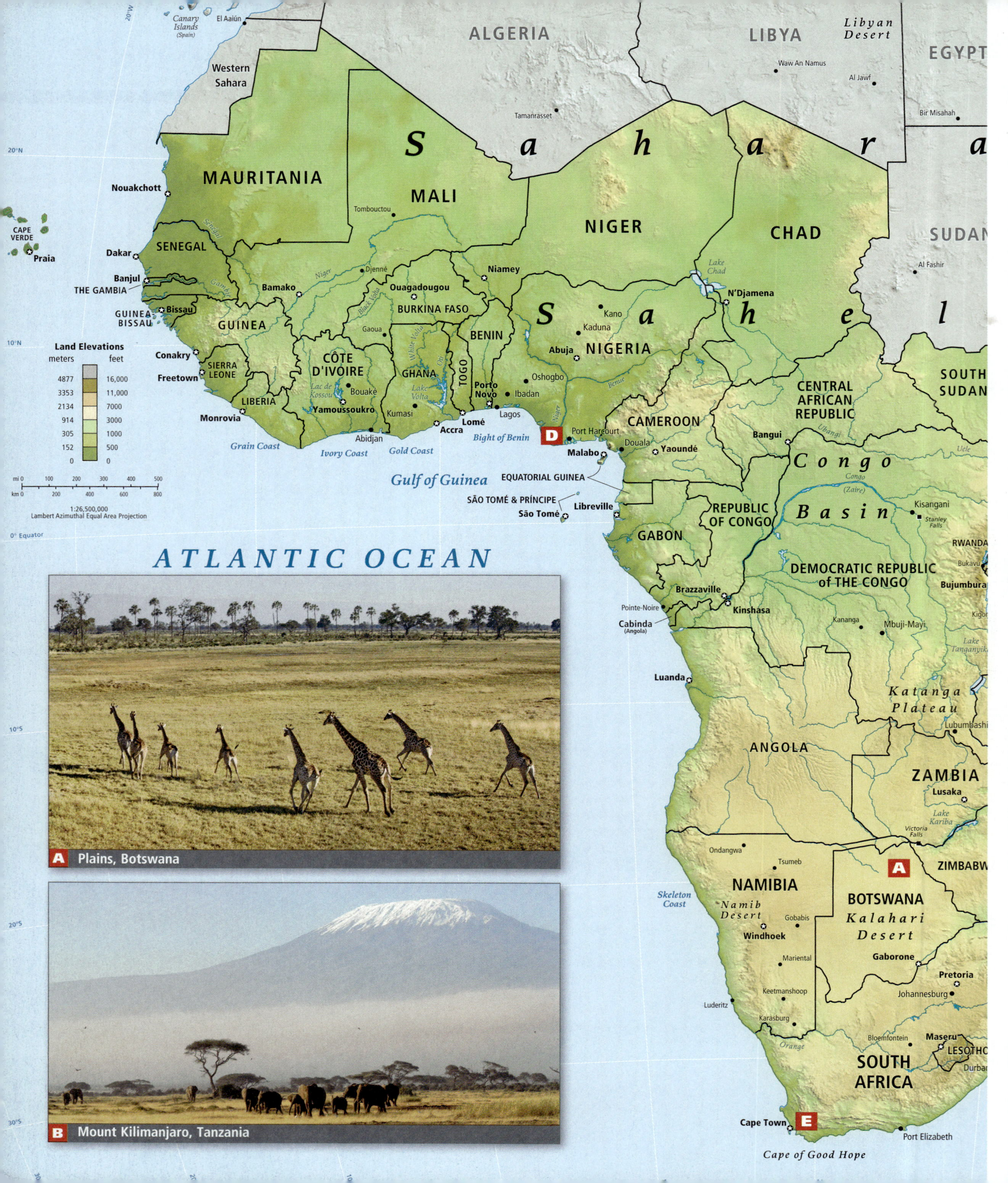
ALGERIA
LIBYA
Libyan Desert
EGYPT
Canary Islands (Spain)
El Aaiún
Western Sahara
Waw An Namus
Al Jawf
Tamanrasset
Bir Misahah
S a h a r a
MAURITANIA
MALI
NIGER
CHAD
SUDAN
Nouakchott
Tombouctou
CAPE VERDE
Praia
Dakar
SENEGAL
Banjul
THE GAMBIA
Bamako
Djenné
Niamey
Lake Chad
Al Fashir
GUINEA-BISSAU
Bissau
GUINEA
Ouagadougou
BURKINA FASO
N'Djamena
S a h e l
Kano
Kaduna
BENIN
NIGERIA
Abuja
Conakry
SIERRA LEONE
Freetown
LIBERIA
Monrovia
CÔTE D'IVOIRE
Bouaké
Yamoussoukro
GHANA
Lake Volta
Gaoua
TOGO
Porto Novo
Oshogbo
Ibadan
Lagos
Lomé
Accra
Kumasi
Abidjan
Grain Coast
Ivory Coast
Gold Coast
Bight of Benin
Gulf of Guinea
Port Harcourt
Malabo
Douala
CAMEROON
Yaoundé
CENTRAL AFRICAN REPUBLIC
Bangui
SOUTH SUDAN
EQUATORIAL GUINEA
SÃO TOMÉ & PRÍNCIPE
São Tomé
Libreville
GABON
REPUBLIC OF CONGO
C o n g o B a s i n
Kisangani
Stanley Falls
RWANDA
DEMOCRATIC REPUBLIC of THE CONGO
Bujumbura
Brazzaville
Kinshasa
Pointe-Noire
Cabinda (Angola)
Kananga
Mbuji-Mayi
Lake Tanganyika
Luanda
Katanga Plateau
Lubumbashi
ANGOLA
ZAMBIA
Lusaka
Lake Kariba
Victoria Falls
Ondangwa
Tsumeb
NAMIBIA
Skeleton Coast
Namib Desert
Windhoek
Gobabis
BOTSWANA
Kalahari Desert
Gaborone
ZIMBABWE
Mariental
Pretoria
Johannesburg
Keetmanshoop
Luderitz
Karasburg
Orange
Bloemfontein
Maseru
LESOTHO
SOUTH AFRICA
Durban
Cape Town
Port Elizabeth
Cape of Good Hope
ATLANTIC OCEAN
Land Elevations
meters feet
4877 16,000
3353 11,000
2134 7000
914 3000
305 1000
152 500
0 0
1:26,500,000
Lambert Azimuthal Equal Area Projection
0° Equator
A
D
E

A Plains, Botswana

B Mount Kilimanjaro, Tanzania

7 Sub-Saharan Africa

C Great Rift Valley, Kenya

D Niger River Delta, Nigeria

E Escarpment at Cape Town, South Africa

FIGURE 7.1 Regional map of sub-Saharan Africa.

GEOGRAPHIC INSIGHTS: SUB-SAHARAN AFRICA

After you read this chapter, you will be able to discuss the following geographic insights as they relate to the nine thematic concepts:

1. Climate Change, Food, and Water: Sub-Saharan Africa is particularly vulnerable to climate change because subsistence occupations are sensitive to even slight variations in temperature, rainfall, and water availability. In large part because of poverty, political instability, and having little access to cash, the region does not have much resilience to the effects of climate change.

2. Globalization and Development: During the era of colonialism, Europeans set up the pattern that continues to this day of basing the economies of sub-Saharan Africa on the export of raw materials. This leaves countries vulnerable in the global economy, where prices for raw materials can vary widely year to year. While a few countries are diversifying and industrializing, most still sell raw materials and all rely on expensive imports of food, consumer products, and construction and transportation equipment.

3. Power and Politics: Countries in the region are shifting from authoritarianism to elections and other democratic institutions, but the power structures in the region are resistant to democratization. While free and fair elections have brought about dramatic changes, distrusted elections have often been followed by surges of violence.

4. Gender, Power, and Politics: The education of African women and their economic and political empowerment are now recognized as essential to development. As girls' education levels rise, population growth rates fall; as the number of female political leaders increases, so does their influence on policies designed to reduce poverty.

5. Population, Urbanization, Food, and Water: In part because of lower infant mortality rates and the higher cost of rearing children in cities, urbanization has the positive effect of encouraging families to limit births. Nevertheless, urbanization often happens in an uncontrolled fashion, leaving many people to reside in impoverished slums without adequate services, sufficient food, and clean water.

The Sub-Saharan Region

Sub-Saharan Africa (see Figure 7.1 on pages 280–281) contains 48 countries and occupies a space bigger than North America and Europe combined. Contrary to popular opinion, it is not as yet particularly densely populated. This region is the ancient home of modern humans; it has megafauna that most of us will see only in zoos and an amazing variety of landscapes. But it is not a region that is well known to most and is often erroneously thought of as desperately poverty stricken and barely capable of helping itself. This chapter is intended to show a more objective, accurate picture of sub-Saharan Africa.

The nine thematic concepts in this book are explored as they arise in the discussion of regional issues, with interactions between two or more themes featured, as in the geographical insights on this page. Vignettes, like the one that follows, illustrate one or more of the themes as they are experienced in individual lives.

GLOBAL PATTERNS, LOCAL LIVES Few people in the East African country of Kenya have computers, but nine of ten Kenyans have mobile phones! They use these simple and inexpensive devices (not smart phones), with texting (SMS) capabilities and perhaps a camera for a multitude of tasks that in Europe or North America are done on computers. One of the most consequential tasks is mobile banking—sending and receiving money via mobile phone. More than 20 percent of the Kenyan gross domestic product (GDP) flows through the mobile banking system. This means that the smallest businessperson can order materials and pay bills and employees without having to pay for expensive equipment and accounting systems or making an expensive time-consuming trip to a bank in a distant city. Small farmers with crops to sell can use SMS to find out where the best market is at the moment and can maximize their profits by bypassing the middlemen who charge high fees.

Juliana Rotich (Figure 7.2) grew up in a small town in Kenya, just as Africa was first experiencing the mobile phone and the Internet revolutions in communication. She won admittance to a U.S. university, where she became an IT major. She then collaborated with friends in the online community to found and run Ushahidi, a web-based, open source reporting system with interactive mapping capabilities. Now widely used throughout the world, *Ushahidi,* which means "testimony" in Swahili, was originally designed for quickly sharing information about the violence that broke out in Kenya during the 2008 elections. A user can connect to Ushahidi through a phone or a Web site to instantly learn specific details about a situation. Information gathered from hundreds or even thousands of users (called *crowdsourcing*) can be verified, mapped, and disseminated in real time.

FIGURE 7.2 Juliana Rotich. Julia Rotich (standing) is a young Kenyan telecommunications specialist with a computer science degree from the University of Missouri. She founded and now directs the company that created *Ushahidi*, a crowdsourcing platform that enables mobile phone users to map crisis information in order to be able to quickly respond to the crises. Ushahidi is now used around the world. Rotich, who also writes a respected blog, *Afromusing*, is part of a growing number of educated Africans who focus on applying technology to African development needs.

Who benefits most when barriers to the use of technology are lowered and suddenly virtually everyone has a mobile phone? The businesspeople and farmers of Kenya have clearly been beneficiaries; but with its mapping capabilities, the Ushahidi application has also been used to help rescuers locate people buried by earthquakes in Haiti, Chile, and Japan. Ushahidi enabled crucial communication between dissidents during and after the Arab Spring revolutions in Libya and Egypt; it was used by Syrians protesting against the Assad government, and then during the Syrian civil war (**Figure 7.3**).

Some of the most exciting innovations that Ushahidi makes possible for Africans are those in disease control, health care, political reform, wildlife inventorying, and the control of corruption. Ushahidi increases transparency and helps people share their stories. *[Source: "99 Faces." For detailed source information, see Text Credit pages.]*

Modern sub-Saharan Africa (see Figure 7.1) has many success stories to tell. The tendency outsiders have of perpetuating the image of Africa as a place of disease, corruption, conflict, and poverty is now aggressively counteracted by assertive, highly competent women and men, such as Juliana Rotich, who use the powerful language of opportunity, optimism, and innovation to characterize the Africa they know today. In 2012, six of the ten fastest-growing economies in the world were in sub-Saharan Africa. World Bank studies (2010) indicate that some countries, such as Congo (Brazzaville), with some of the world's richest deposits of oil, gold, platinum, copper, and other strategic minerals, are taking control of mineral exploitation and using the profits to reduce poverty.

During the era of European colonialism (1850s–1950s), sub-Saharan Africa's massive wealth of human talent and natural resources flowed out of Africa. Even after African countries became politically independent in the 1950s, 1960s, and 1970s, wealth continued to flow out of Africa. Now, increasingly, wealth is being created in Africa for Africans. While much of the region is impoverished and parts are in armed conflict, there are many hopeful signs that corruption and discord will be alleviated and well-being significantly enhanced.

THINGS TO REMEMBER

- The many stereotypes of sub-Saharan Africa—that it is a region of conflict, disease, corruption, and poverty—are being countered by young, educated Africans who are constructing new public agencies and private businesses in their home countries, often grounded in modern technology.
- Africans are not only affecting their own countries, they are revising Africa's position vis-à-vis the rest of the world by devising innovative solutions that can be used in many situations.
- Some sub-Saharan countries now have economies that are among the fastest-growing in the world.

FIGURE 7.3 Ushahidi platform users worldwide. The map shows a selection of the 132 countries in which Ushahidi has been deployed.

THE GEOGRAPHIC SETTING

What Makes Sub-Saharan Africa a Region?

Sub-Saharan Africa is a region separate from North Africa for physical, cultural, and historical reasons. The Sahara Desert and the **Sahel**—that grassy transition zone between desert and wetter climes to the south—present major physical obstacles to human habitation (though trading caravans have traversed the Sahara and Sahel for thousands of years). Because of this, sub-Saharan Africa developed largely separate from North Africa. For millennia, sub-Saharan Africa was known to the outside world only by the accounts of a few travelers, such as those of the Arab explorer Ibn Battuta, who traveled widely throughout the region in the 1300s. In the mid-1400s, though, the Portuguese sent their exploratory fleets down the African west coast, and they began to realize that the zone of Africa south of the Sahara was quite intriguing.

The Europeans noted what to them were exotic qualities: dark-skinned people, with unique and varied ways of life and cultures, who possessed valuable trade items, such as precious minerals, exotic plants and animals, and fine textiles. But, for centuries, Europeans discounted the sophistication and complexity of sub-Saharan Africa's cultural and historical heritage of horticulture, weaving, mining, and metalwork. Everything south of the Sahara was lumped into "Black Africa," and for some this term is still used to distinguish the region from Africa as a whole. Now the reality is that sub-Saharan Africa is coming into an era of growth.

Sahel a band of arid grassland, where steppe and savanna grasses grow, that runs east–west along the southern edge of the Sahara

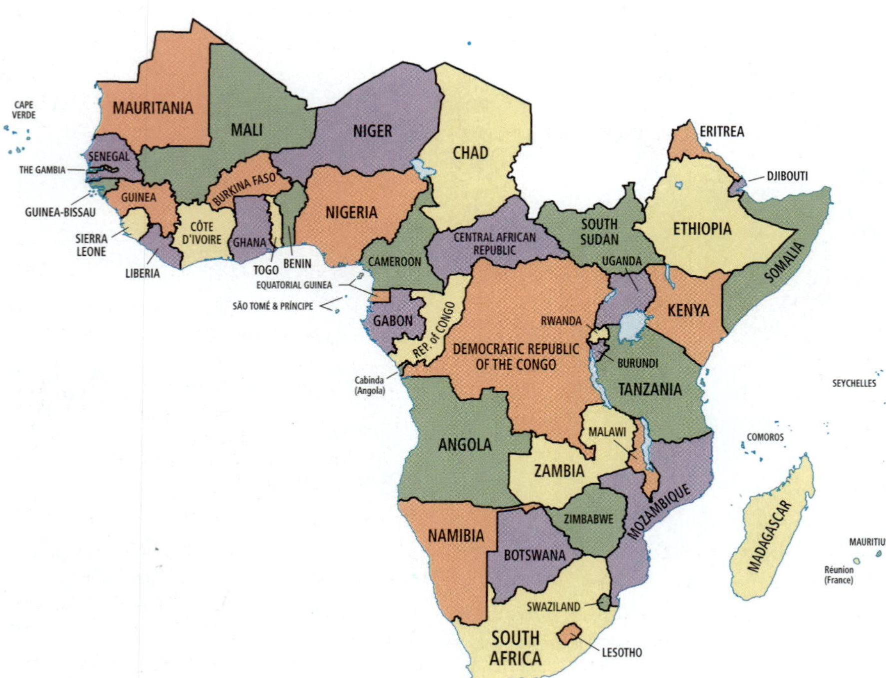

FIGURE 7.4 Political map of sub-Saharan Africa.

Terms in This Chapter

In this chapter, only occasionally do we refer to the whole continent, and then we refer to it simply as Africa. Sub-Saharan Africa is made up of the 48 countries shown in Figure 7.4. Notice that while the new country of South Sudan is included in this region, for cultural and historical reasons, the country of Sudan is not because its location on the Nile River and the strong Arab influence in the capital of Khartoum have brought Sudan into closer association with North Africa and Southwest Asia. Sudan is instead discussed in Chapter 6.

The names of African countries can often be confusing. For example, there are two neighboring countries called Congo—the Democratic Republic of the Congo and the Republic of Congo. Because these designations are both lengthy and easily confused, we abbreviate them in this text. The Democratic Republic of the Congo (formerly Zaire) carries the name of its capital in parentheses: Congo (Kinshasa). The Republic of Congo is called Congo (Brazzaville). Check the regional map in Figure 7.1 to see the locations of these countries and capitals.

Physical Patterns

The African continent is big—the second largest after Asia and about 2 million square miles (5.2 million square kilometers) bigger than North America. But Africa's great size is not matched by its surface complexity. Africa has no major mountain ranges, but it does have several high volcanic peaks, including Mount Kilimanjaro (19,324 feet [5890 meters] high; see Figure 7.1B) and Mount Kenya (17,057 feet [5199 meters] high). At their peaks, both have permanent snow and ice, though these features have shrunk dramatically due to global climate change and associated local environmental changes. More than one-fourth of the African continent is covered by the Sahara Desert, which reaches to the edge of the Mediterranean Sea.

Landforms

The surface of the continent of Africa can be envisioned as a raised platform, or plateau, bordered by fairly narrow and uniform coastal lowlands. The platform slopes downward to the north; it has an upland region with several high peaks in the southeast, and lower uplands in the northwest. The steep escarpments (long, high cliffs) between plateau and coast have obstructed transportation and hindered connections to the outside world. Africa's lengthy, uniform coastlines provide few natural harbors, Cape Town in South Africa being an exception (see Figure 7.1E).

Geologists usually place Africa at the center of the ancient supercontinent Pangaea (see Figure 1.25 on page 46). As landmasses broke off from Africa and moved away—North America to the northwest, South America to the west, and India to the northeast—Africa readjusted its position only slightly. Because its shift was so small, it did not pile up long, linear mountain ranges, as did the other continents when their plates collided with one another (see page 44).

Africa continues to break apart along its eastern flank. There the Arabian Plate has already split away and drifted to the northeast, leaving the Red Sea, which separates Africa and Asia. Another split, known as the Great Rift Valley, extends south from the Red Sea more than 2000 miles (3200 kilometers) (see Figure 7.1C). In the future, Africa is expected to split again along these rifts.

Climate and Vegetation

Most of sub-Saharan Africa has a tropical climate (see Figure 7.5 map). Average temperatures generally stay above 64°F (18°C) year-round everywhere except at the more temperate southern tip of the continent and in the cooler upland zones (hills, mountains, high plateaus). Seasonal climates in Africa differ more by the amount of rainfall than by temperature.

Most rainfall comes to Africa by way of the **intertropical convergence zone (ITCZ)**, a band of atmospheric currents that circle the globe roughly around the equator (see the inset map in Figure 7.5). At the ITCZ, warm winds converge from both the north and the south and push against each other. This causes the air to rise, cool, and release moisture in the form of rain. The rainfall produced by the ITCZ is most abundant in Africa near the equator, where dense tropical rainforests flourish in places such as the Congo Basin (see Figure 7.5A, D).

The ITCZ shifts north and south seasonally, generally following the area of Earth's surface that has the highest average temperature at any given time. Thus, during the height of summer in the Southern Hemisphere in January, the ITCZ might bring rain far enough south to water the dry grasslands, or steppes, of Botswana. During the height of summer in the Northern Hemisphere in August, the ITCZ brings rain as far north as the southern fringes of the Sahara, to the area called the Sahel, a band of arid grassland 200 to 400 miles (320 to 640 kilometers) wide that runs east-west along the southern edge of the Sahara (see Figure 7.5). At roughly 30° N latitude (the Sahara) and 30° S latitude (the Namib and Kalahari deserts) the air, which has dried out during its passage, descends, forming a subtropical high-pressure zone that shuts out lighter, warmer, moister air. As a result of this system (which is in no way precise), deserts tend to be found in Africa (and on other continents) in these zones about 30 degrees north and south of the equator.

The tropical wet climates that support equatorial rain forests are bordered on the north, east, and south by seasonally wet/dry subtropical woodlands (see Figure 7.5B). These give way to moist tropical savannas or steppes, where tall grasses and trees intermingle in a semiarid environment. These tropical wet, wet/dry, and steppe climates have provided suitable land for different types of agriculture for thousands of years, but the amount of moisture available in each of these climates can vary greatly, in cycles that are decades long. Farther to the north and south lie the true desert zones of the Sahara and the Namib and Kalahari (see Figure 7.5C). This banded pattern of African ecosystems is modified in many areas by elevation and wind patterns.

Without mountain ranges to block them, wind patterns can have a strong effect on climate in Africa. Winds blowing north along the east coast keep ITCZ-related rainfall away from the **Horn of Africa**, the triangular peninsula that juts out from northeastern Africa below the Red Sea. As a consequence, the Horn of Africa is one of the driest parts of the continent. Along the west coast of the Namib Desert, moist air from the Atlantic is blocked from moving over the desert by cold air above the northward-flowing water of the Benguela Current. Like the cool Peru Current off South America, the Benguela is an oceanic current that is chilled by its passage past Antarctica. Rich in nutrients, it supports a major fishery along the west coast of Africa (see the Figure 7.5 map).

Environmental Issues

> **Geographic Insight 1**
>
> **Climate Change, Food, and Water:** Sub-Saharan Africa is particularly vulnerable to climate change because subsistence occupations are sensitive to even slight variations in temperature, rainfall, and water availability. In large part because of poverty, political instability, and having little access to cash, the region does not have much resilience to the effects of climate change.

While Africans have generally contributed very little to the build-up of greenhouse gases in the atmosphere, deforestation in Africa by ordinary people and by commercial entities that operate globally is amplifying global climate change. Because of Africa's poverty, its people are much less able to adapt to climate change than those who generate most greenhouse gases. Even so, many Africans are developing strategies to cope with uncertainties related to the region's changing climate. **158. DISAPPEARING GLACIERS ON MT. KILIMANJARO RAISE ENVIRONMENTAL CONCERNS**

Deforestation and Climate Change

Sub-Saharan Africa contributes to CO_2 emissions and potential climate change primarily through deforestation. Trees absorb CO_2 as they photosynthesize, thus removing carbon from the air and storing it as biomass, a process known as **carbon sequestration**. When trees are burned or when they decompose after they die, they release the stored CO_2 into the atmosphere. Thus, deforestation is a major contributor to carbon buildup in the atmosphere and ultimately to climate change.

African countries lead the world in the *rate* of deforestation, the percentage of total forest area lost. Of the eight countries that had the world's highest rates of deforestation between 1990 and 2005,

intertropical convergence zone (ITCZ) a band of atmospheric currents that circle the globe roughly around the equator; warm winds from both north and south converge at the ITCZ, pushing air upward and causing copious rainfall

Horn of Africa the triangular peninsula that juts out from northeastern Africa below the Red Sea and wraps around the Arabian Peninsula

carbon sequestration the removal and storage of carbon taken from the atmosphere

Figure 7.5 **Photo Essay: Climates of Sub-Saharan Africa**

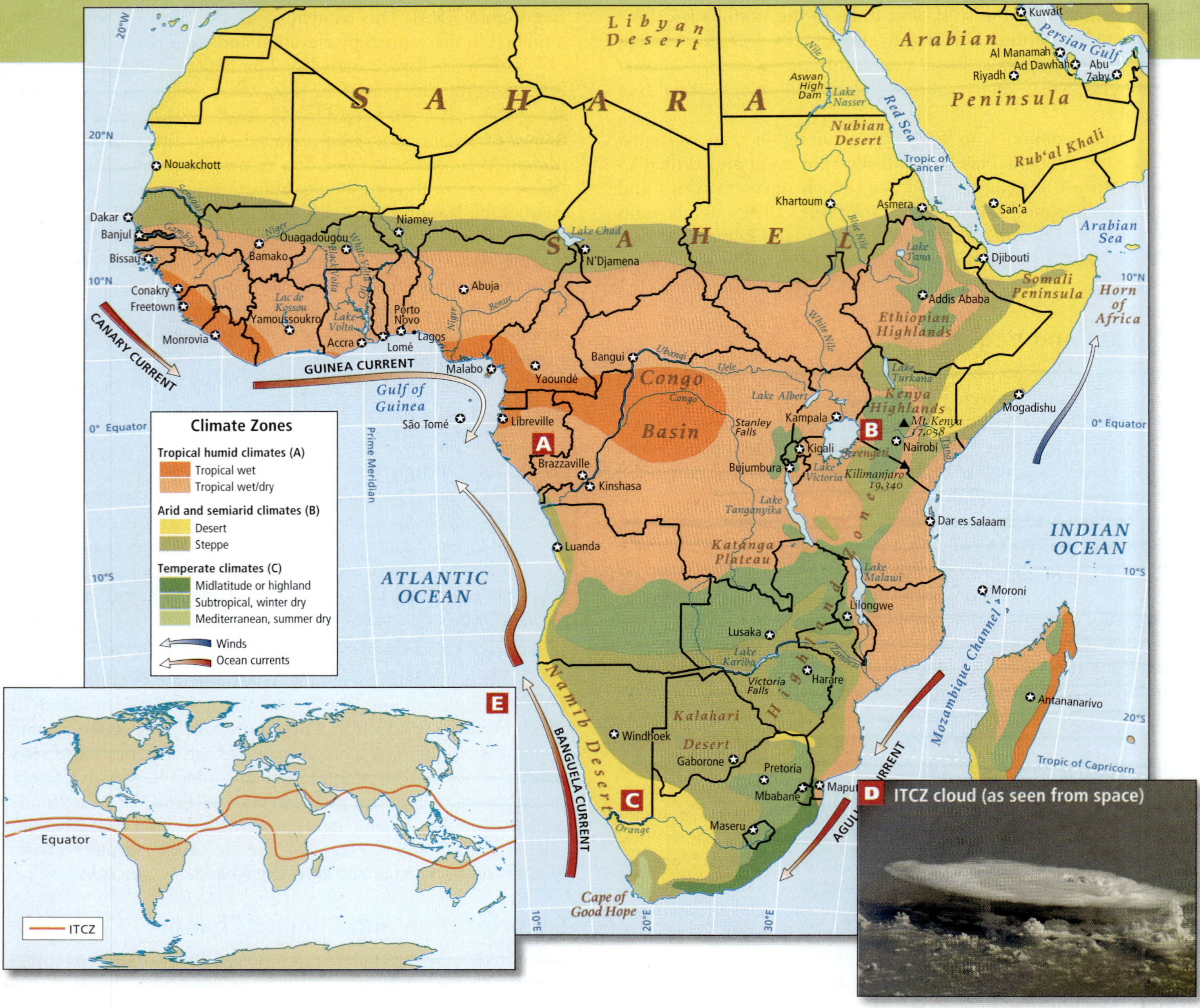

Climate Zones

Tropical humid climates (A)
- Tropical wet
- Tropical wet/dry

Arid and semiarid climates (B)
- Desert
- Steppe

Temperate climates (C)
- Midlatitude or highland
- Subtropical, winter dry
- Mediterranean, summer dry

Winds

Ocean currents

D ITCZ cloud (as seen from space)

E ITCZ

A Tropical wet, Gabon

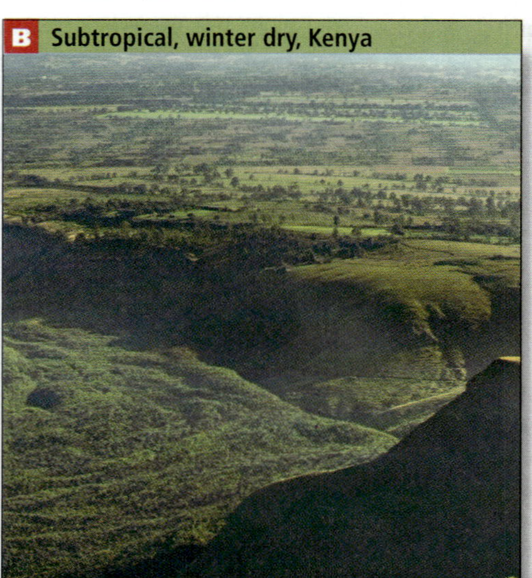

B Subtropical, winter dry, Kenya

C Desert in Namibia

six are in sub-Saharan Africa (Burundi, Togo, Nigeria, Benin, Uganda, and Ghana). This deforestation constitutes a major human environmental impact across the region (Figure 7.6 on page 288). The countries that have the most emissions from deforestation are Brazil and Indonesia, but Nigeria and Congo (Kinshasa) have the third and fourth most, respectively. **168. PLAN TO CLEAR-CUT UGANDAN FOREST RESERVE FOR GROWING SUGAR CANE SPARKS CONTROVERSY**

ON THE BRIGHT SIDE

The Value of Agroforestry

A strong promoter of agroforestry is the Green Belt Movement, founded by Dr. Wangari Maathai of Kenya. The Green Belt Movement helps rural women plant trees for use as fuelwood and to reduce soil erosion. The movement grew from Maathai's belief that a healthy environment is essential for democracy to flourish. Thirty years and 30 million trees after she began, Maathai was awarded the 2004 Nobel Peace Prize for her contributions to sustainable development, democracy, and peace.

VIGNETTE Liberian environmental activist Silas Siakor is an affable and unassuming fellow. But his casual style conceals a fierce dedication to his homeland and a remarkable sleuthing ability. At great personal risk, Siakor uncovered evidence that 17 international logging companies were bribing Liberia's then-president, Charles Taylor, with cash and guns. Taylor allowed the companies to illegally log Liberia's forests, which are home to many endangered species including forest elephants and chimpanzees. In return, the companies paid cash and provided weapons that Taylor used to equip his personal armies. Made up largely of kidnapped and enslaved children, Taylor's armies fought those of other Liberian warlords in a 14-year civil war that took the lives of 150,000 civilians. The war also spilled over into neighboring Sierra Leone, where another 75,000 people died. Meanwhile, the logging companies—based in Europe, China, and Southeast Asia—reaped huge fortunes from the tropical forests. **175. 'EZRA,' TRAGIC TALE OF CHILD SOLDIERS IN AFRICA**

Silas Siakor pulled together publicly available information that had been previously ignored by the international community and information provided by informed ordinary citizens in ports, villages, and lumber companies. He prepared a clear, well-documented report substantiating the massive logging fraud. In response to Siakor's report, the UN Security Council voted to impose sanctions to stop the timber trade and prosecute some of the people involved. Charles Taylor fled to Nigeria, but in 2006 he was turned over to face a war crimes tribunal in The Hague, Netherlands. In April of 2012 Taylor received a 50-year sentence.

Democratic elections followed in Liberia in 2006, after which Ellen Johnson-Sirleaf took office as Africa's first elected woman president. In a move that was bold—given the poverty and political instability of her country—she cancelled all contracts with timber companies, pending a revision of Liberian forestry law. By 2012, President Johnson-Sirleaf was herself being audited by the global Extractive Industries Transparency Initiative (EITI) for $8 billion worth of questionable resource contracts. Liberia, now calculated to be resource rich, is the first African country to submit to the EITI anticorruption audits. *[Sources: Goldman Environmental Prize, National Public Radio, and Silas Siakor. For detailed source information, see Text Credit pages.]* ■

The story of Silas Siakor and Ellen Johnson-Sirleaf illustrates how Liberia, like much of this region, is still in the process of developing strong democratic institutions that can sustainably develop the country's resources and use them to the benefit of its citizens.

Most of Africa's deforestation is driven by the growing demand for farmland and fuelwood, although logging by international timber companies is also increasing. Africans use wood (or charcoal made from wood) to supply nearly all their domestic energy (see Figure 7.6A, B). Wood remains the cheapest fuel available, in part because of African traditions that consider forests to be a free resource held in common. Even in Nigeria, a major oil producer, most people use fuelwood because they cannot afford petroleum products. Across the region, not only are charcoal prices rising as the forests disappear, but the smoke from the charcoal is causing asthma and other respiratory problems, as well as creating greenhouse gases.

In Maputo, the capital of Mozambique, women are engaged in a for-profit project that combines new technology with traditional crops and crop rotation techniques to create a sustainable fuel and food preparation system. The aim is to shift the 1.2 million inhabitants of Maputo from relying on charcoal—made from the disappearing old-growth forests in the north of the country—to using ethanol made from the roots of the cassava plant grown by the women on their own land. Newly developed small metal cookstoves run on the cassava-based ethanol. The project also encourages farmers to rotate their crops to preserve soil fertility. In any given year, a third of the the land is planted in a cash crop of cassava used to make ethanol and the rest of the land is used to plant multiple-species food gardens.

The multifaceted stove/ethanol/cassava pilot project in Maputo is funded by four American firms that hope to replicate it elsewhere around the world. They want to save the forests, relieve fuel shortages, halt rising fuelwood prices, and alleviate air pollution, while simultaneously improving food security and creating local enterprises: those that can build the innovative cookstoves and those that can produce cassava-based ethanol.

Elsewhere, African governments that are trying to slow deforestation are encouraging **agroforestry**—the farming of economically useful trees, along with the usual subsistence and cash crops. Using the farmed trees for fuel and construction helps reduce dependence on old-growth forests and can provide income through the sale of the farmed wood. By practicing agroforestry on the fringes of the Sahel, a family in Mali, for example, can produce fuelwood, fencing and building materials, medicinal products, and food—all on the same piece of land. This would double what could be earned from a single crop or animal herding. However, critics caution that agroforestry sometimes introduces trees that are *invasive species*, nonnative plants that threaten many

agroforestry growing economically useful crops of trees on farms, along with the usual plants and crops, to reduce dependence on trees from nonfarmed forests and to provide income to the farmer

Figure 7.6 Photo Essay: Human Impacts on the Biosphere in Sub-Saharan Africa

Deforestation, desertification, and increasing water use have had major impacts on sub-Saharan Africa's ecosystems.

Erosion caused by deforestation

Once-forested hills

Sisal cultivation

Rice cultivation

A Deforestation and several types of cultivation in a central Madagascar landscape. So much forest has been cleared for farming that hills are bare and eroded, and now have deep gullies.

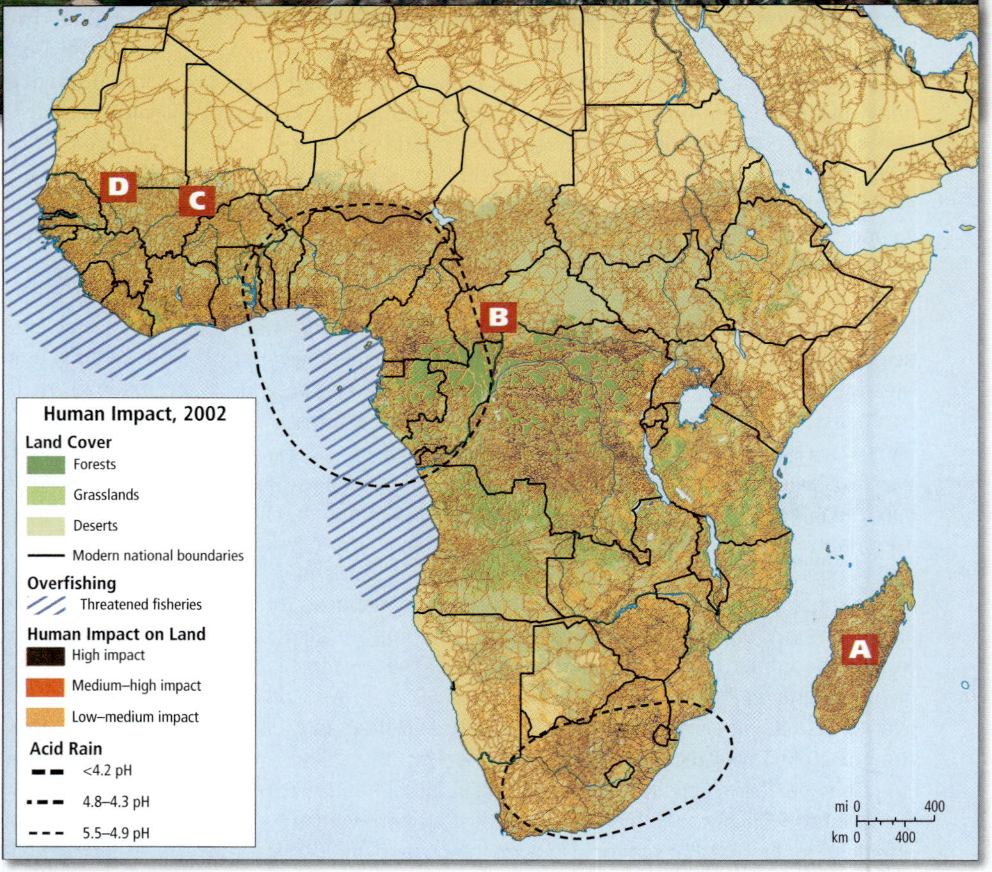

B A logger cuts a giant tree in the Central African Republic (CAR). Logging has degraded many forests; roads created by loggers have made areas that once were wild accessible to poor, subsistence farmers.

Human Impact, 2002

Land Cover
- Forests
- Grasslands
- Deserts
- Modern national boundaries

Overfishing
- Threatened fisheries

Human Impact on Land
- High impact
- Medium–high impact
- Low–medium impact

Acid Rain
- <4.2 pH
- 4.8–4.3 pH
- 5.5–4.9 pH

mi 0 400
km 0 400

C A fisher casts his net in the Niger River wetlands, where carefully synchronized resource-use patterns have been developed over millennia.

D A dam at Manantan in western Mali. A similar dam planned for upstream of the Niger River wetlands would alter the river's flow in order to supply farmers with irrigation water. The project would increase food security for farmers elsewhere in Mali but threaten the livelihoods of fishers, farmers, and pastoralists in the Niger River wetlands.

Thinking Geographically

After you have read about human impact on the biosphere in sub-Saharan Africa, you will be able to answer the following questions:

A Other than the caption, what clues are there to suggest this landscape has been cleared of forest?

B What about this photo suggests that the trees being logged are from oldgrowth forests?

C What about the fishing methods depicted in the photo suggests that catches are fairly small in size?

African ecosystems by outcompeting indigenous species.

Agricultural Systems, Food, Water, and Vulnerability to Climate Change

Agricultural systems in Africa are undergoing a rapid transition from being oriented around subsistence farming to being centered on commercial farming. Unfortunately, this change is happening in an era when the advantages of traditional agriculture are only beginning to be appreciated. Thus the social and environmental effects of turning rapidly to commercial production are not fully understood, especially given the as yet little-known ecological impacts of climate change.

Traditional Agriculture Most sub-Saharan Africans practice **subsistence agriculture**, usually done on small farms of about 2 to 10 acres (1 to 4 hectares) to provide food and other products for only the farmer's family. Most subsistence farmers also practice **mixed agriculture**, raising a diverse array of crops and a few animals as livestock. Many also fish, hunt, herd, and gather some of their food from forest or grassland areas. The wide variety of food produced means that the cuisine of Africa is also varied (Figure 7.7).

To maintain soil quality in a tropical wet or wet/dry climate, subsistence farmers have long used **shifting cultivation**, in which small patches of forest are cleared, with the detritus either burned or left to decay, and the clearings are cultivated with a wide variety of plants for 2 or 3 years and then are abandoned and left to regrow. After a few decades, the soil naturally replenishes its organic matter and nutrients and is ready to be cultivated again. However, when fallow periods are reduced, as is happening in many rural areas that have increasingly high population densities, the soil can become degraded.

These traditional, subsistence, mixed, and shifting cultivation food-acquisition techniques have advantages and disadvantages in a world of climate change. All of these closely related forms of agriculture

subsistence agriculture farming that provides food for only the farmer's family and is usually done on small farms

mixed agriculture farming that involves raising a variety of crops and animals on a single farm, often to take advantage of several environmental riches

shifting cultivation a productive system of agriculture in which small plots are cleared in forestlands, the dried brush is burned to release nutrients, and the clearings are planted with multiple species; each plot is used for only 2 or 3 years and then abandoned for many years of regrowth

FIGURE 7.7 LOCAL LIVES **FOODWAYS IN SUB-SAHARAN AFRICA**

A Ethiopian and Eritrean cuisine features spicy vegetable and meat stews and salads served with *injera*, a type of savory pancake made from fermented grains. Pieces of the injera are torn off and used to scoop up the rest of the meal. Injera is traditionally cooked on a clay plate over an open fire. Because cooking injera on an open fire requires a lot of wood, people also use more efficient stoves, including electric models that are popular in cities, to cook injera.

B Recently harvested cassava root is transported in Nigeria. Originally domesticated in Brazil, cassava was brought to Africa in the seventeenth century as part of the slave trade. Because it grows abundantly in a wide variety of soils and environments and is easy to cultivate, it has been a popular crop, despite its relatively low nutritional value. Sub-Saharan Africa produces more cassava than any world region, with Nigeria as the world's largest producer.

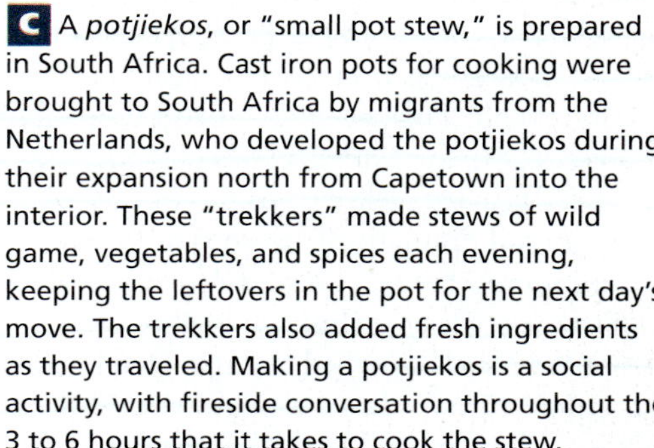

C A *potjiekos*, or "small pot stew," is prepared in South Africa. Cast iron pots for cooking were brought to South Africa by migrants from the Netherlands, who developed the potjiekos during their expansion north from Capetown into the interior. These "trekkers" made stews of wild game, vegetables, and spices each evening, keeping the leftovers in the pot for the next day's move. The trekkers also added fresh ingredients as they traveled. Making a potjiekos is a social activity, with fireside conversation throughout the 3 to 6 hours that it takes to cook the stew.

provide a diverse array of strategies for coping with the changes in temperature and rainfall that climate change may bring. As in much of sub-Saharan Africa, traditional Nigerian farmers (usually women) grow complex tropical gardens, often with 50 or more species of plants at one time. Some of the plants can handle drought, while others can withstand intense rain or heat.

However, the subsistence nature of most African farming can also leave families without much cash. If harvests are too low to provide surplus that can be sold, and hunting and gathering fail to provide supplementary food for the family, there may be insufficient money to buy food. While such situations can lead to famine, it is important to note that the most serious famines in Africa have occurred not because of low harvests but rather because of political instability that disrupts economies and food growing and distribution systems. **162. FELLOWSHIP PROGRAM AIMS TO OPEN DOORS FOR AFRICAN WOMEN IN AGRICULTURAL RESEARCH**

Commercial Agriculture Much of Africa is now shifting over to **commercial agriculture**, in which crops are grown deliberately for cash rather than solely as food for the farm family. This type of production also has advantages and disadvantages with respect to climate change. If harvests are good and prices for crops are adequate, farmers can earn enough cash to get them through a year or two of poor harvests. Having some cash income can also allow poor families to invest in other means of earning a living, such as opening a grinding mill to help process other farmers' harvests or building a stone oven to bake neighbors' bread for a fee.

However, some of the most common commercial crops, such as peanuts, cacao beans (used for making chocolate), rice, tea, and coffee, are often less adapted to environments outside their native range (Figure 7.8D). They are more likely to produce smaller yields if temperatures increase or water becomes scarce. Moreover, to maximize profits, these crops are often grown in large fields of only a single plant species. This can leave crops vulnerable to pests; if crops fail, farmers have no other garden foods to rely on. The potential for commercial agriculture to help farmers adapt to the uncertainties of global climate change is also limited by the instability of prices for commercial crops. Prices can rise or fall dramatically from year to year because of overproduction or crop failures both in Africa and abroad.

Commercial agriculture—whether small scale, locally managed, or large scale—owned and operated by international corporations (agribusiness), demands permanently cleared fields. But in the tropics, because soil fertility declines rapidly once the trees are removed, commercial crops are more likely to fail even under ideal climatic conditions. Chemical fertilizers can compensate for this loss, but are often too expensive for ordinary farmers, and when used repeatedly by agribusiness, may lose their effectiveness. Moreover, rains almost always wash much of the fertilizer into nearby waterways, thus polluting them. This may ultimately hurt farmers and fishers by spoiling drinking water and reducing the quantity of available fish, which are important sources of protein for rural people.

commercial agriculture farming in which crops are grown deliberately for cash rather than solely as food for the farm family

These aspects of commercial agricultural systems make them poorly adapted even to the most ideal climatic conditions, let alone the hotter and more drought-prone environments that climate change may bring. Foreign agricultural "experts" who often push commercial systems can be woefully ignorant of local conditions and the value of local cultivation techniques. For example, in Nigeria, it is women who grow most of the food for family consumption and who are guardians of the knowledge that makes Nigeria's traditional farming systems work. However, women are rarely included or even consulted by the foreign experts who promote commercial agricultural development projects. At best, women are employed as field laborers. Consequently, diverse subsistence agricultural systems based on numerous plant species and deep knowledge about the environment have been replaced by less stable commercial systems that are based on a single plant species, and that do not take into account the skills and knowledge base of women.

Agricultural scientists recently began to recognize past mistakes and have been trying to incorporate the traditional knowledge of African farmers, female and male, into more diverse commercial agriculture systems. For example, scientists at Nigeria's International Institute for Tropical Agriculture are developing cultivation systems that, like traditional systems, use many species of plants that help each other cope with varying climatic conditions. Most of these systems are designed for both subsistence and commercial agriculture, and thus can give families both a stable food supply and cash to help them ride out crop failures and pay school fees for their children (see Figure 7.8C).

Water Resources, Irrigation, and Water Management Alternatives Although sub-Saharan Africa has a large wet tropical zone, much of the region is seasonally dry (see the Figure 7.5 map); and because climate change is likely to result in changing rainfall patterns, the areas that have freshwater deficits are expected to increase, with a related increase in refugees (see Figure 7.8A). In water-deficit conditions, many people look to irrigation to provide more stability to agricultural systems (see Figure 7.6D). But does Africa have sufficient groundwater resources to supply irrigation systems plus fill all the other demands for fresh water by its modernizing societies? Recall that per capita water use expands exponentially with development (see Chapter 1, page 38).

New Groundwater Discoveries Is there sufficient groundwater in Africa? Until recently, the answer was a resounding no! However, a recent comprehensive review and quantitative

Thinking Geographically

After you have read about vulnerability to climate change in sub-Saharan Africa, you will be able to answer the following questions:

A How is poverty evident in this photo?

B What caused the death of many wildebeests in 2007?

C How is agricultural diversity illustrated in this photo?

D What kind of climate change in Africa most often causes a reduced yield for a crop like tea?

Figure 7.8 Photo Essay: Vulnerability to Climate Change in Sub-Saharan Africa

Much of this region is highly exposed to drought, flooding, and other climatic disturbances. Poverty and low access to cash increase sensitivity to climate change, while political instability makes it harder for governments to effectively use disaster management to help boost resilience.

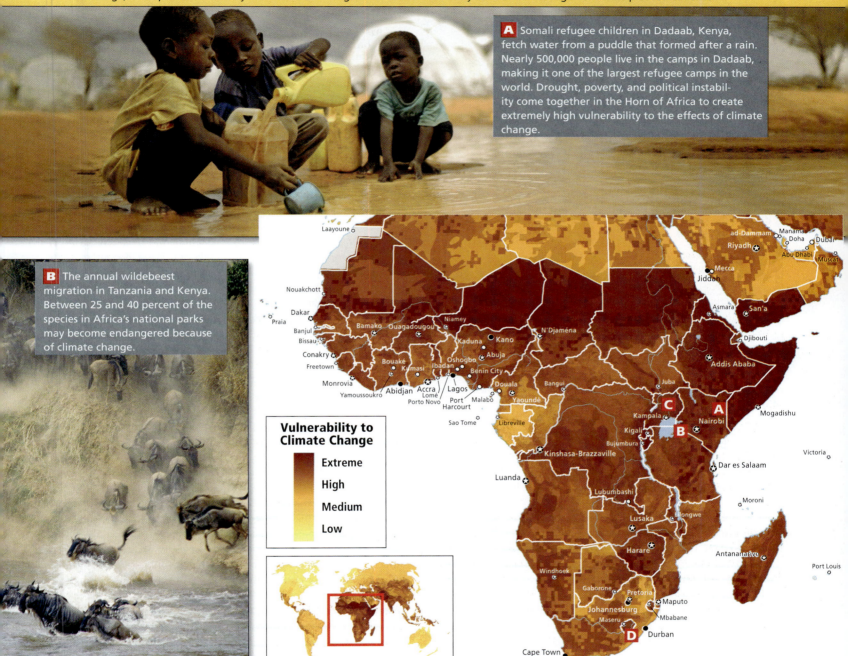

A Somali refugee children in Dadaab, Kenya, fetch water from a puddle that formed after a rain. Nearly 500,000 people live in the camps in Dadaab, making it one of the largest refugee camps in the world. Drought, poverty, and political instability come together in the Horn of Africa to create extremely high vulnerability to the effects of climate change.

B The annual wildebeest migration in Tanzania and Kenya. Between 25 and 40 percent of the species in Africa's national parks may become endangered because of climate change.

Vulnerability to Climate Change

Extreme

High

Medium

Low

C A landscape of subsistence and mixed agriculture in Uganda. These diversified systems lend farmers some resilience to climatic disturbances, but they provide little cash to help farmers buy other food they might need.

D Tea harvested for export in South Africa. Commercial crops like tea can increase the farmers' resilience to climate change by providing cash to buy any needed food. But tea, like many commercial crops, requires expensive and polluting fertilizers, and prices for tea are unstable.

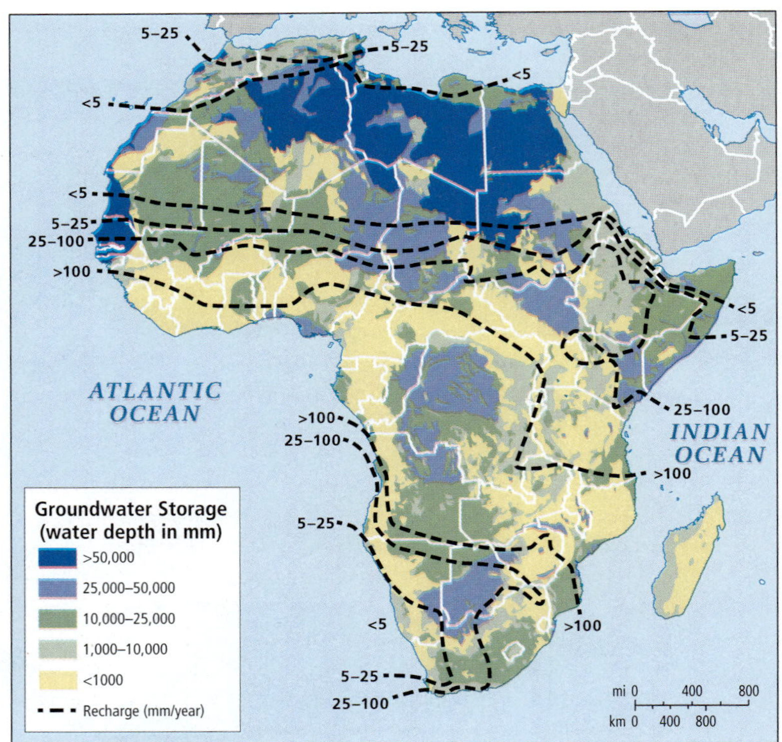

FIGURE 7.9 Newly exposed water resources. It was long thought that Africa had only shallow and sparse deposits of groundwater. But recent research has revealed that for the African continent, the quantity of water stored in aquifers is nearly 100 times that of previous estimates. Some is near the surface and relatively accessible, but because much is fossil water deposited as many as 5000 years ago, it cannot be *recharged* (replenished) quickly under present drier climate conditions. Also, many densely populated areas have only small deposits. Compare this map of the depth of groundwater deposits and number of years needed to recharge with the population map (see Figure 7.22 on page 311).

mapping of all available data on **groundwater** for the African continent (Figure 7.9) revealed that the quantity of groundwater (water naturally stored in aquifers as many as 5000 years ago, during wetter climate conditions) was nearly 100 times that of previous estimates. In many cases, the water was close enough to the surface to be accessed via inexpensive bore well pumps. This does not mean that Africa suddenly has an inexhaustible supply of water; in fact, the largest reservoirs of groundwater storage are in lightly populated areas (such as North Africa and Botswana; see Figure 7.1A). The most heavily populated country, Nigeria, has only a relatively small supply, and those aquifers are in many cases no longer being recharged by nature. Nonetheless, this news on groundwater resources is welcome because climate change is likely to reduce precipitation in irregular patterns across the continent.

groundwater water naturally stored in aquifers as long as 5000 years ago during wetter climate conditions

pastoralism a way of life based on herding; practiced primarily on savannas, on desert margins, or in the mixture of grass and shrubs called *open bush*

New Large- and Small-Scale Irrigation Projects

Government officials continue to favor large-scale irrigation projects to address the need for increased food supplies in the face of present and future water scarcity. Africa has a number of major rivers—the Nile, the Congo, the Zambezi, and the Limpopo—and climate change will affect all of them. Here we focus on the Niger, one of Africa's most important rivers, which flows through several very different ecological zones. The Niger rises in the tropical wet Guinea Highlands and carries summer floodwaters northeast into the normally arid lowlands of Mali and Niger (see the Figure 7.5 map and Figure 7.6C, D). There the waters spread out into lakes and streams that nourish wetlands. For a few months of the year (June through September), the wetlands ensure the livelihoods of millions of fishers, farmers, and pastoralists. These people share the territory in carefully synchronized patterns of land use that have survived for millennia. Wetlands along the Niger produce 8 times more plant matter per acre than the average wheat field. They provide seasonal pasture for millions of domesticated animals, and serve as an important habitat for wildlife.

The governments of Mali and Niger now want to dam the Niger River and channel its water into irrigated agribusiness projects that they hope will help feed the more than 26 million people in both countries. However, the dams will forever change the seasonal rise and fall of the river. The irrigation systems may also pose a threat to human health. Systems that rely on surface storage not only lose a great deal of water through evaporation and leakage, the standing pools of water they create often breed mosquitoes that spread tropical diseases, such as malaria, and harbor the snails that host schistosomiasis, a debilitating parasite that enters the skin of humans who spend time standing in still water to fish or do other chores.

Many smaller-scale alternatives are available. In some parts of Senegal, for example, farmers are using hand- or foot-powered pumps to bring water from rivers or ponds directly to the individual plants that need it. This is in some ways a more modern version of traditional African irrigation practices whereby water is delivered directly to the roots of the plants by human, often female, water brigades. Smaller-scale projects provide the same protection against drought that larger systems offer, but are much cheaper and simpler to operate for small farmers. They also avoid the social dislocation and ecological disruption of larger projects. Already in successful use for 15 years, these low-tech pumps will help farmers adjust to the drier conditions that may come with climate change.

Herding and Desertification

Herding, or **pastoralism**, is practiced by millions of Africans, primarily in savannas, on desert margins, or in the mixture of grass and shrubs called *open bush*. Herders live off the milk, meat, and hides of their animals. They typically circulate seasonally through wide areas, taking their animals to available pasturelands and trading with settled farmers for grain, vegetables, and other necessities.

Many traditional herding areas in Africa are now undergoing *desertification*, the process by which arid conditions spread to areas that were previously moist (see Chapter 6, page 243). The drying out is often the result of the loss of native vegetation; traditional herding may be partially to blame, but economic development schemes that encourage cattle raising are also at fault. Cattle need more water and forage than do traditional herding animals and so can place greater stress on native grasslands than goats or camels do. Agricultural intensification in the Sahel also contributes to desertification, as scarce water resources are diverted to irrigation, leaving dry, vegetation-less soils exposed to wind erosion.

Over the last century, desertification has shifted the Sahel to the south. For example, the *World Geographic Atlas* in 1953 showed Lake Chad situated in a forest well south of the southern edge of the Sahel. By 1998, the Sahara itself was encroaching on Lake Chad (see the Figure 7.5 map) and the lake had shrunk to a tenth of the area it occupied in 1953.

Wildlife and Climate Change

Africa's world-renowned wildlife faces multiple threats from both human and natural forces, all of which could become more severe with increases in global climate change. The Intergovernmental Panel on Climate Change, a body of scientists tasked by the United Nations with assessing scientific information relevant to understanding climate change, estimates that 25 to 40 percent of the species in Africa's national parks may become endangered as a result of climate change.

Wildlife managers are developing new management techniques to help animals survive. For example, one of the greatest wildlife spectacles on the planet took a tragic turn in 2007. The annual 1800-mile-long (2900-kilometer-long) natural migration of more than a million wildebeest, zebras, and gazelles in Kenya's Maasai Mara game reserve requires animals to traverse the Mara River. In the best of times, this is a difficult migration that usually results in a thousand or so animals drowning. In the past, the park's managers have taken a hands-off approach to the migration, considering the losses normal. However, in 2007, extremely heavy rains, possibly related to global climate change, swelled the Mara River to record levels. When the animals tried to cross, 15,000 drowned (see Figure 7.8B). Park managers are now taking a more active role in helping the migrating animals cope with unusual climatic conditions that may worsen with climate change. This may involve stopping animals from attempting a river crossing or directing them to a safer crossing.

169. PROTECTING NATURE IN GUINEA COLLIDES WITH HUMAN NEEDS

Farmers' dependence on hunting wild game (*bushmeat*) for part of their food and income is already a major threat to wildlife in much of Africa. For example, farmers who need food or extra income are killing endangered species, such as gorillas, chimpanzees, and especially elephants, in record numbers (Figure 7.10C). If crop harvests are diminished by global climate change, many farmers will become even more dependent on bushmeat and on income from selling contraband, such as ivory. The threat to wild populations of various species has led to calls to expand and establish new protected areas for wildlife.

Africa's national parks constitute one-third of the world's preserved national parkland. The parks are struggling to deal with *poaching* (illegal hunting) within the park boundaries by members of surrounding communities. Poaching is often fueled by demand outside Africa for exotic animal parts (tusks, hooves, penises) as medicines and aphrodisiacs, especially in Asia, where efforts to educate consumers about the damage done by their purchases are only beginning.

Some parks are now using profits from ecotourism to fund development in nearby communities that previously depended in part on poaching. For example, in 1985, wildlife poaching

FIGURE 7.10 **LOCAL LIVES** **PEOPLE AND ANIMALS IN SUB-SAHARAN AFRICA**

A The Rhodesian ridgeback, or the African lion hound, helps lion hunters by distracting the lions. Highly intelligent and able hunters themselves (they can kill animals as large as baboons), ridgebacks are now used mostly as guard dogs. Ridgebacks were developed by interbreeding the indigenous dogs of southern Africa's nomadic KhoiKhoi people with a variety of European dogs.

B African grey parrots, highly intelligent and able to mimic human speech, are popular pets worldwide. One bird famously greeted Dr. Jane Goodall, known for her research with chimpanzees, by saying, "Got a chimp?" (The bird had seen a photo of Goodall with a chimpanzee.) As many as 21 percent of the wild African grey parrots are caught each year, and the International Union for the Conservation of Nature (IUCN) lists the parrots as a vulnerable species.

C Elephants are a major attraction at Africa's many national parks and nature reserves, but park space cannot sustain large numbers of elephants whose natural impulse is to pull down trees to eat leaves. In response, some parks have reduced their elephant populations. The parks, though, may be the only hope for the survival of elephants who are hunted for their ivory tusks. Hunters and poachers have for years tended to take large-tusked elephants, leading to natural selection for elephants with smaller tusks or with no tusks at all—a once-rare trait that is now found in about 30 percent of the population.

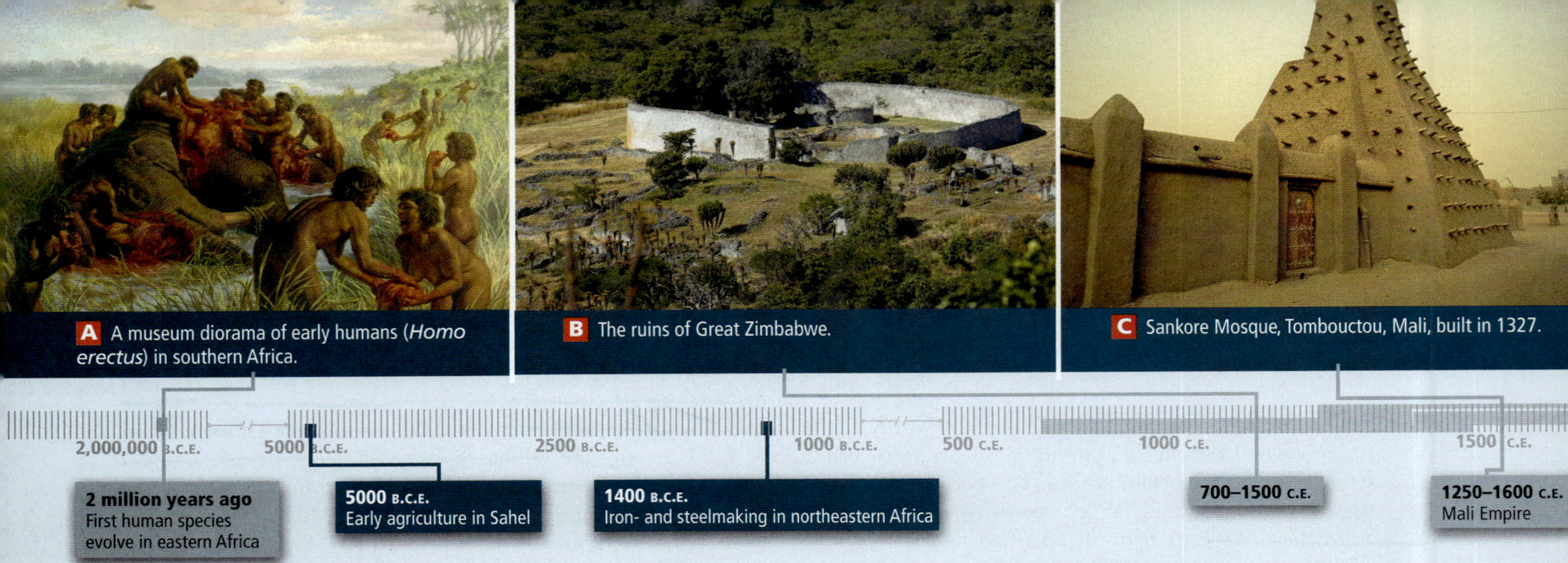

A A museum diorama of early humans (*Homo erectus*) in southern Africa.

B The ruins of Great Zimbabwe.

C Sankore Mosque, Tombouctou, Mali, built in 1327.

FIGURE 7.11 A VISUAL HISTORY OF SUB-SAHARAN AFRICA

Thinking Geographically

After you have read about the human history of sub-Saharan Africa, you will be able to answer the following questions:

A What are the clues in the picture that food may have been a source of conflict between early humans and between humans and other animals?

B Beyond the central fortlike structure, describe any further evidence of human habitation.

C What does the existence of a mosque in Mali in 1327 indicate about the geography of trading patterns in this part of Africa?

threatened the animal population in Zambia's Kasanka National Park. Park managers decided to generate employment for the villagers through tourism-related activities. They built tourist lodges and wildlife-viewing infrastructure and started cottage industries to make products to sell to the tourists. Funding also goes to local clinics and schools, and students are included in research projects within the park. Local farmers have expanded into alternative livelihoods, such as beekeeping and agroforestry. Today, poaching in Kasanka is very low, its wildlife populations are booming, tourism is growing, and local communities have an ongoing stake in the park's success. Sub-Saharan Africa's rich array of animals has long played an economic role in daily life. Three species are discussed in Figure 7.10.

THINGS TO REMEMBER

- The surface of the continent of Africa can be envisioned as a raised platform, or plateau, bordered by fairly narrow and uniform coastal lowlands.
- Most of sub-Saharan Africa has a tropical wet or wet/dry climate, except at the southern tip and in the cooler uplands. Seasonal climates in Africa differ more by the amount of rainfall than by temperature.
- Most rainfall comes to Africa by way of the intertropical convergence zone (ITCZ), a band of atmospheric currents that circle the globe roughly around the equator. The effects of desertification are most dramatic in the region called the Sahel.

Geographic Insight 1

- **Climate Change, Food, and Water** Africa's food production systems, which are still largely subsistence-based but include some mixed agriculture and hunting, are threatened by climate change. The effects of climate change are made worse by poverty and political instability.
- The primary way in which sub-Saharan Africa contributes to CO_2 emissions and potential climate change is through deforestation. Much of Africa's deforestation is driven by the growing demand for farmland and fuelwood, but corruption on the part of local authorities and international timber companies also plays a role.
- Long-standing physical challenges, such as deforestation, desertification, and increasing water scarcity, have had a major impact on sub-Saharan Africa's ecosystems. Commercial food production for Africa's cities has many ecological drawbacks. These challenges are likely to increase with climate change.
- Newly discovered groundwater resources hold promise for alleviating some of Africa's water scarcities.

Human Patterns over Time

Africa's rich past has often been misunderstood and dismissed by people from outside the region. European slave traders and colonizers called Africa the *Dark Continent* and assumed it was a place where little of significance in human history had occurred. The substantial and elegantly planned cities of Benin

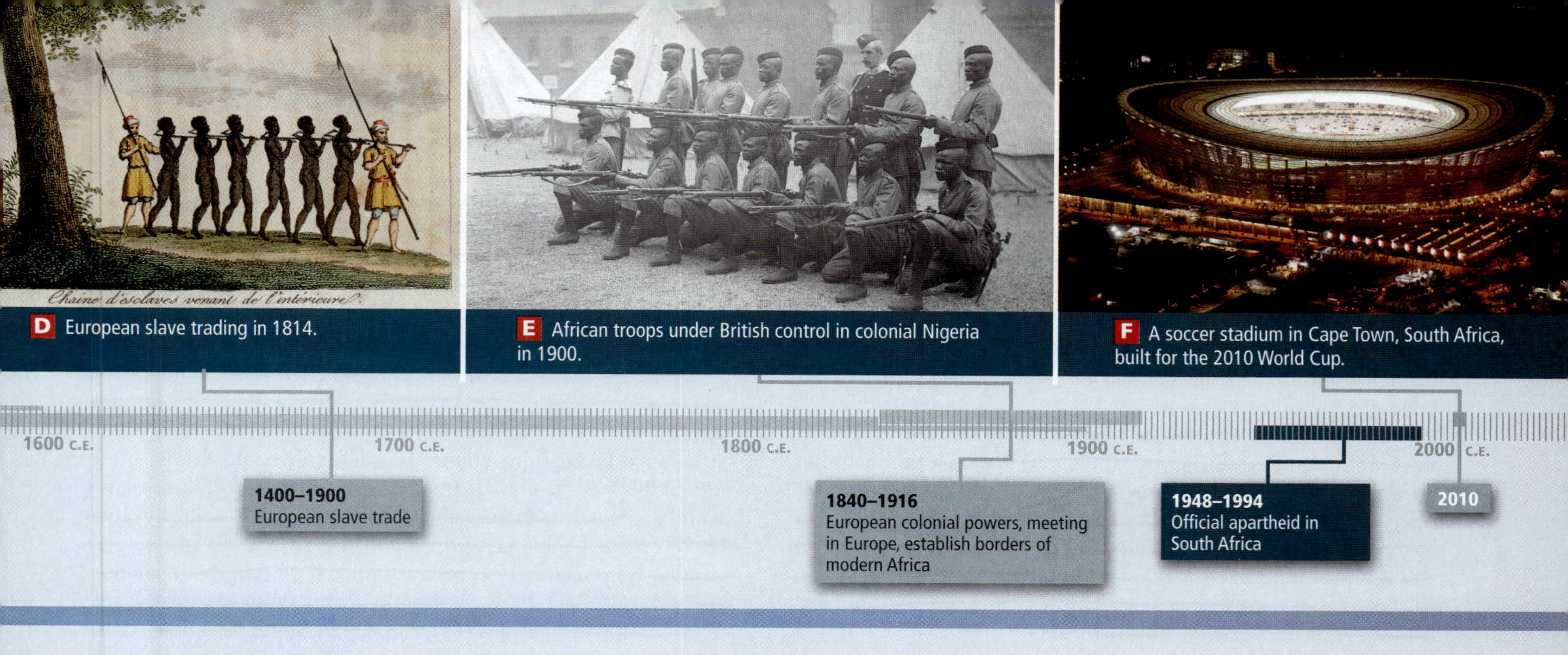

D European slave trading in 1814.

E African troops under British control in colonial Nigeria in 1900.

F A soccer stadium in Cape Town, South Africa, built for the 2010 World Cup.

D What is the evidence of European involvement in the slave trade in this image?

E What about this photo suggests hierarchies of status in British colonial Africa?

F What about this stadium conveys the wealth and Westernization of South Africa?

in western Africa, Djenné in the Niger River basin, and Loango in the Congo Basin, which European explorers encountered in the 1500s, never became part of Europe's image of Africa. Even today, most people outside the continent are unaware of Africa's internal history or its contributions to world civilization, let alone its role in the very emergence of humankind.

The Peopling of Africa and Beyond

Africa is the original home of humans (**Figure 7.11A**). It was probably in eastern Africa (in what are today the highlands of Ethiopia, Kenya, and Uganda) that the first human species (*Homo erectus*) evolved more than 2 million years ago. *Homo erectus* differed anatomically from humans today. These early, tool-making humans ventured out of Africa, reaching north of the Caspian Sea and beyond as early as 1.8 million years ago. Anatomically modern humans (*Homo sapiens*) evolved about 250,000 years ago from earlier hominoids (probably *Homo erectus*) in eastern Africa. Like the migrations of earlier human species, those of modern humans radiated out of Africa, first toward the eastern Mediterranean (*Homo sapiens* reached the eastern Mediterranean about 90,000 years ago), then spreading across mainland and island Asia and only later turning west into Europe. Eventually, after coexisting in Eurasia and Europe with earlier dispersions of *Homo erectus* and other *Homo* species (all originating in Africa), *Homo sapiens* outcompeted them all.

Early Agriculture, Industry, and Trade in Africa

In Africa, people began to cultivate plants as far back as 7000 years ago in the Sahel and the highlands of present-day Sudan and Ethiopia. Agriculture was brought south to equatorial Africa 2500 years ago and to southern Africa about 1500 years ago. Trade routes spanned the African continent, extending north to Egypt and Rome and east to India and China. Gold, elephant tusks, and timber from tropical Africa were exchanged for salt, textiles, beads, and a wide variety of other goods.

About 3400 years ago, people in the vicinity of Lake Victoria learned how to smelt iron and make steel. By 700 C.E., when Europe was still recovering from the collapse of the Roman Empire, a remarkable civilization with advanced agriculture, iron production, and gold-mining technology had developed in the highlands of southeastern Africa in what is now Zimbabwe. This empire, now known as the Great Zimbabwe Empire, traded with merchants from Arabia, India, Southeast Asia, and China, exchanging the products of its mines and foundries for silk, fine porcelain, and exotic jewelry. The Great Zimbabwe Empire collapsed around 1500 for reasons not yet understood (see Figure 7.11B).

Complex and varied social and economic systems existed in many parts of Africa well before the modern era. Several influential centers made up of dozens of linked communities developed in the forest and the savanna of the western Sahel.

There, powerful kingdoms and empires rose and fell, such as Djenné; that of Ghana (700–1000 C.E.), centered in what is now Mauritania and southwestern Mali; and the Mali Empire (1250–1600 C.E.), centered a bit to the east on the Niger River in what became the famous Muslim trading and religious center of Tombouctou (Timbuktu), originally founded about 1100 C.E. by Tuareg nomads as a seasonal camp (see Figure 7.11C). Some rulers periodically sent large trade caravans carrying gold and salt through Tombouctou on to Makkah (Mecca), where their opulence was a source of wonder.

Africans also traded slaves. Long-standing customs of enslaving people captured during war fueled this trade. The treatment of slaves within Africa was sometimes brutal and sometimes relatively humane. Long before the beginning of Islam, a slave trade developed with Arab and Asian lands to the east. After the spread of Islam began around 700 C.E., the slave trade continued, and Muslim traders exported close to 9 million African slaves to parts of Southwest, South, and Southeast Asia (Figure 7.12). When slaves were traded to non-Africans, indigenous checks on brutality were abandoned. For example, to ensure sterility and to help promote passivity, Muslim traders often preferred to buy castrated male slaves.

The course of African history shifted dramatically in the mid-1400s, when Portuguese sailing ships began to appear off Africa's west coast. The names given to stretches of this coast by the Portuguese and other early European maritime powers reflected their interest in Africa's resources: the Gold Coast, the Ivory Coast, the Pepper Coast, and the Slave Coast.

By the 1530s, the Portuguese had organized a slave trade with the Americas. The trading of slaves by the Portuguese, and then by the British, Dutch, and French, was more widespread and brutal than any trade of African slaves that preceded it. African slaves became part of the elaborate production systems supplying the raw materials and money that fueled Europe's Industrial Revolution.

To acquire slaves, the Europeans established forts on Africa's west coast and paid nearby African kingdoms with weapons, trade goods, and money to make slave raids into the interior. Some slaves were taken from enemy kingdoms in battle. Many more were kidnapped from their homes and villages in the forests and savannas. Most slaves traded to Europeans were male because they brought the highest prices and the raiding kingdoms preferred to keep captured women for their reproductive capacities. Between 1600 and 1865, about 12 million captives were packed aboard cramped and filthy ships and sent to the Americas. One-quarter or more of them died at sea. Of those who arrived in the Americas, about 90 percent went to plantations in South America and the Caribbean. Between 6 and 10 percent were sent to North America (see Figure 7.12).

The European slave trade severely drained parts of Africa of human resources and set in motion a host of damaging social responses within Africa that are not completely understood even today (see Figure 7.11D). The trade enriched those African

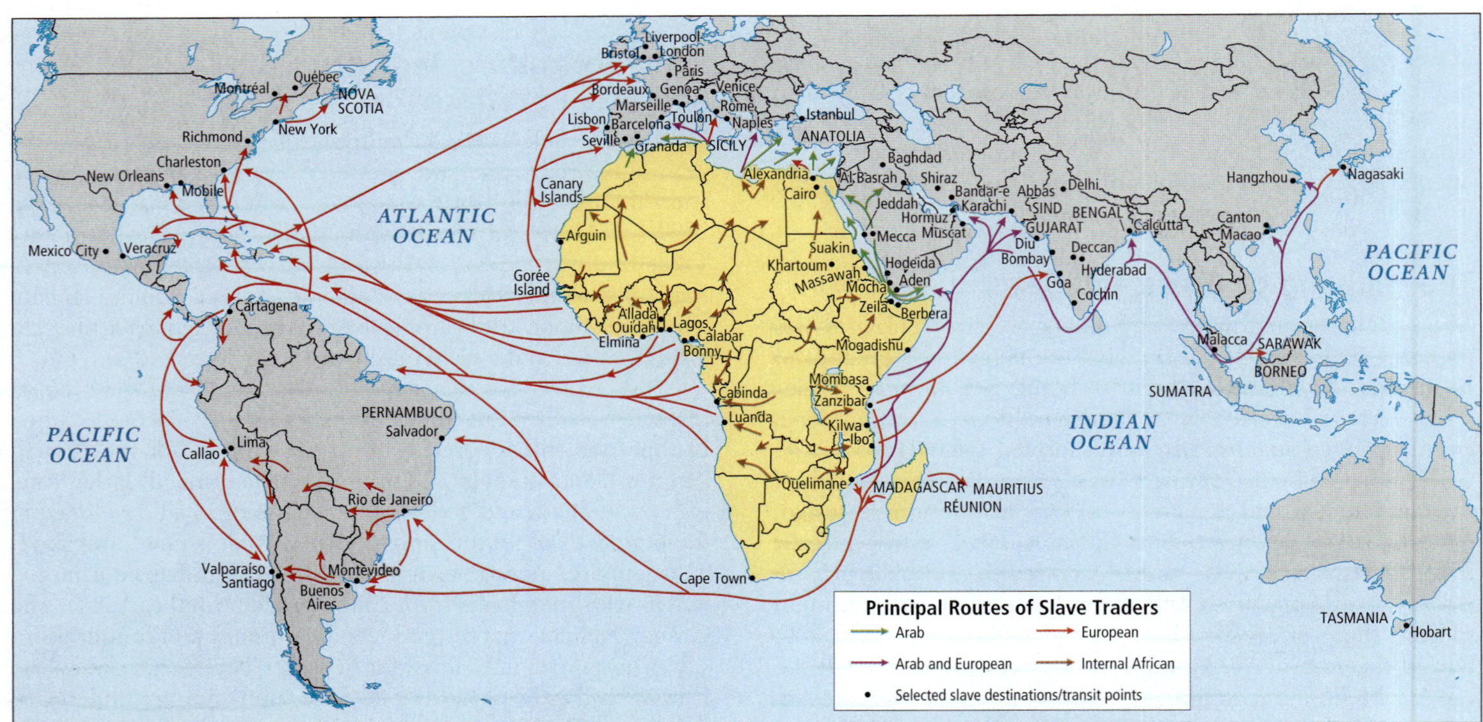

FIGURE 7.12 Map of the African slave trade. The origin of the African slave trade is indigenous and predates Islam. Outside Africa, many slave markets were established around the Mediterranean and the Indian Ocean and eventually, after 1500 C.E., millions of slaves were brought to the Americas.

FIGURE 7.13 Modern slavery in Niger. A man born as a slave in Niger is presented with a document that declares his freedom in 2008. Though slavery was "abolished" in Niger in 1960, and made a criminal offense in 2003, it remains deeply embedded in society. At least 43,000 people are enslaved in Niger, most of them born into the status. Slaves have no rights, little access to education, and will spend most of their lives herding cattle, tending fields, or working in their "master's" house.

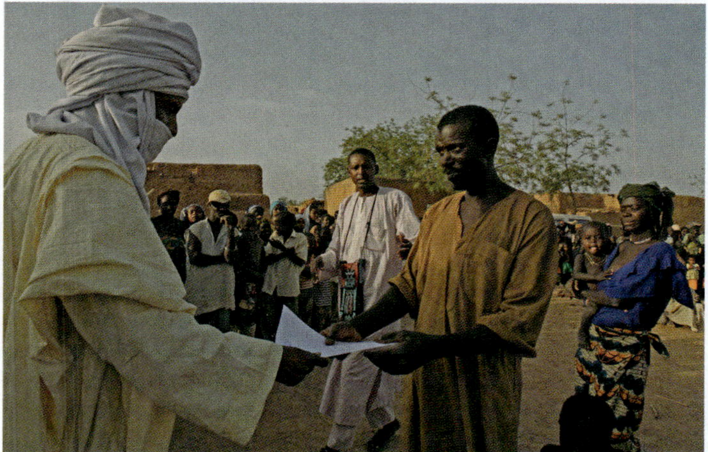

Thinking Geographically

What does this photo suggest about the social relationships of the man being freed from slavery in Niger?

kingdoms that could successfully conquer their neighbors and impoverished or enslaved more peaceful or less powerful kingdoms. It also encouraged the slave-trading kingdoms to be dependent on European trade goods and technologies, especially guns.

Slavery persists in modern Africa and is a growing problem that some argue exceeds the transatlantic slave trade of the past. Today, slavery is most common in the Sahel region, where several countries have made the practice officially illegal only in the past few years (Figure 7.13). People may become enslaved during war. They may be sold by their parents or relatives to pay off debts or forced into slavery when migrating to find a job in a city. Some are even enslaved by gangsters in Europe, where they become street vendors, domestic servants, or prostitutes. In Côte d'Ivoire, Burkina Faso, and Nigeria, forced child labor is used increasingly in commercial agriculture, such as cacao (chocolate) production. In Liberia, Sierra Leone, Cameroon, and Congo, young boys have been enslaved as miners and warriors. It is hard to know exactly how many Africans are currently enslaved, but estimates range from several million to more than 10 million.

The Scramble to Colonize Africa

The European slave trade wound down by about the mid-nineteenth century, as Europeans found it more profitable to use African labor within Africa to extract raw materials for Europe's growing industries.

European colonial powers competed avidly for territory and resources, and by World War I, only two areas in Africa were still independent (Figure 7.14 on page 298): Liberia on the west coast (populated by former slaves from the United States) and Ethiopia (then called Abyssinia) in East Africa. Ethiopia managed to defeat early Italian attempts to colonize it. Otto von Bismarck, the German chancellor who convened the 1884 Berlin Conference at which the competing European powers formalized the partitioning of Africa, revealed the common arrogant European attitude when he declared, "My map of Africa lies in Europe." With some notable exceptions, the boundaries of most African countries today derive from the colonial boundaries set up between 1840 and 1916 by European administrators and diplomats to suit their purposes (see Figure 7.11E). These territorial divisions lie at the root of many of Africa's current problems.

In some cases, the boundaries were purposely drawn to divide tribal groups and thus weaken them. In other cases, groups who used a wide range of environments over the course of a year were forced into smaller, less diverse lands or forced to settle down entirely, thus losing their traditional livelihoods. As access to resources shrank, hostilities between competing groups developed. Colonial officials often encouraged these hostilities, purposely replacing strong leaders with leaders who could be manipulated. Food production, which was the mainstay of most African economies until the colonial period, was discouraged in favor of activities that would support European industries: cash-crop production (cotton, rubber, palm oil) and mineral and wood extraction. Eventually, many formerly prosperous Africans were hungry and poor.

One of the main objectives of European colonial administrations in Africa, in addition to extracting as many raw materials as possible, was to create markets in Africa for European manufactured goods. The case of South Africa provides insights into European expropriation of African lands and the subjugation of African peoples. In this case, these aims ultimately led to the infamous system of racial segregation known as *apartheid* (see below).

A Case Study: The Colonization of South Africa

The Cape of Good Hope is a rocky peninsula that shelters a harbor on the southwestern coast of South Africa. Portuguese navigators seeking a sea route to Asia first rounded the Cape in 1488. The Portuguese remained in nominal control of the Cape until the 1650s, when the Dutch took possession with the intention of establishing settlements. Dutch farmers, called *Boers*, expanded into the interior, bringing with them herding and farming techniques that used large tracts of land and depended on the labor of enslaved Africans. The British were also interested in the wealth of South Africa, and in 1795, they seized control of areas around the Cape of Good Hope. When slavery was outlawed throughout the British Empire in 1834, large numbers of slave-owning Boers migrated to the northeast in order to keep their slaves. There, in what became known as the Orange Free State and the Transvaal, the Boers often came into violent conflict with African inhabitants.

In the 1860s, extremely rich deposits of diamonds and gold were unearthed in these areas, securing the Boers' economic

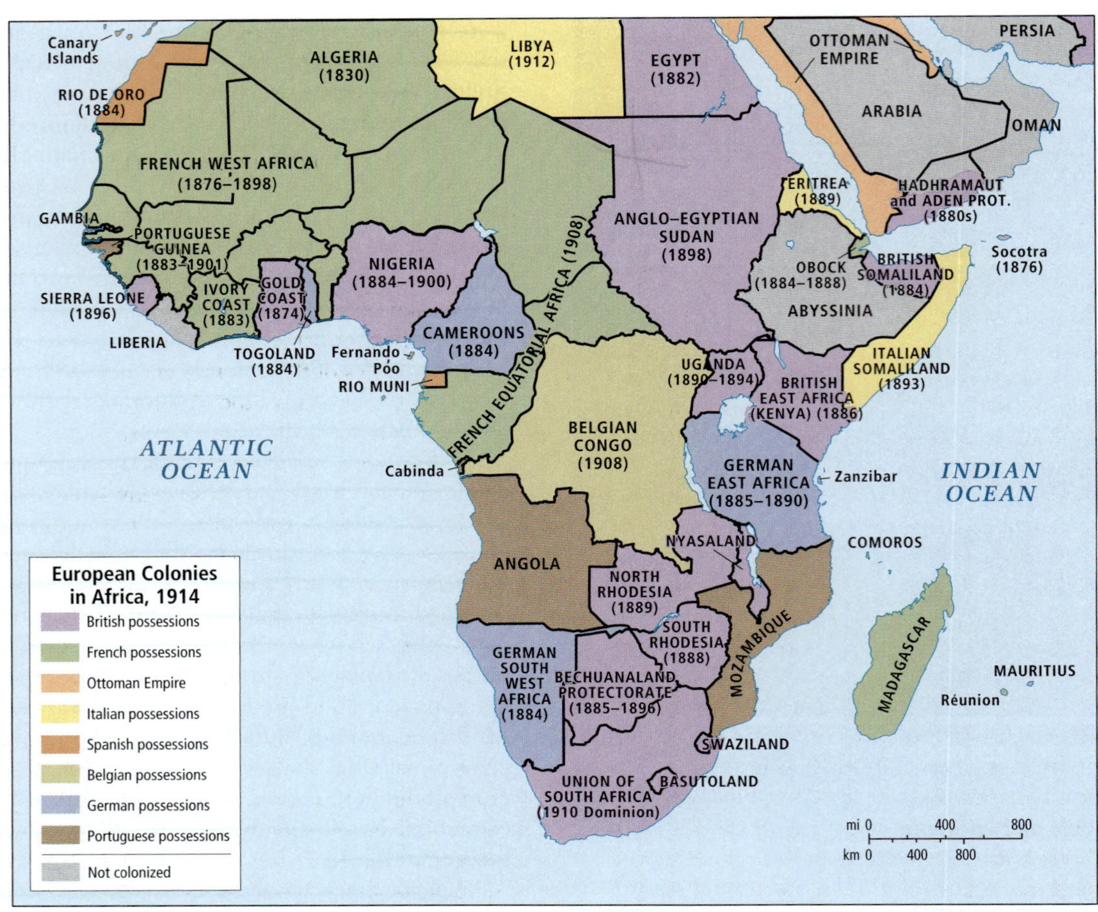

FIGURE 7.14 The European colonies in Africa in 1914. The dates on the map indicate the beginning of officially recognized control by the European colonizing powers. Countries without dates were informally occupied by colonial powers for a few centuries.

future. African men were forced to work in the diamond and gold mines under extreme hardship and unsafe conditions for minimal wages. They lived in unsanitary compounds that travelers of the time compared to large cages.

Britain, eager to claim the wealth of the mines, invaded the Orange Free State and the Transvaal in 1899, waging the Boer War. This brutal war gave the British control of the mines briefly, until resistance by Boer nationalists forced the British to grant independence to South Africa in 1910. This independence, however, applied to only a small minority of whites: the Boers (who have since been known as Afrikaners) and some British people who chose to remain. Not until 1994 would full political rights (voting, freedom of expression, freedom of assembly, freedom to live where one chose) be extended to black South Africans—who made up more than 80 percent of the population—and to Asians and mixed race or "coloured" people (2 percent and 8 percent of the population, respectively).

apartheid a system of laws mandating racial segregation. South Africa was an apartheid state from 1948 until 1994

In 1948, the long-standing segregation of South African society was reinforced by **apartheid**, a system of laws that required everyone except whites to carry identification papers at all times, to live in racially segregated areas, and to use segregated transportation and other infrastructure (Figure 7.15). Eighty percent of the land was reserved for the use of white South Africans, who at that time made up just 10 percent of the population. Black, Asian, and "coloured" people were assigned to ethnically based "homelands." The South African government considered the homelands to be independent enclaves within the borders of, but not legally part of, South Africa. Nevertheless, the South African government exerted strong influence in them. Democracy theoretically existed throughout South Africa, but nonwhites were allowed to vote only in the homelands.

The African National Congress (ANC) was the first and most important organization that fought to end racial discrimination in South Africa. Formed in 1912 to work nonviolently for civil rights for all South Africans, the ANC grew into a movement with millions of supporters, most of them black but some of them white. Its members endured decades of brutal repression by the white minority. One of the most famous members to be imprisoned was Nelson Mandela, a prominent ANC leader, who was jailed for 27 years and finally released in 1990.

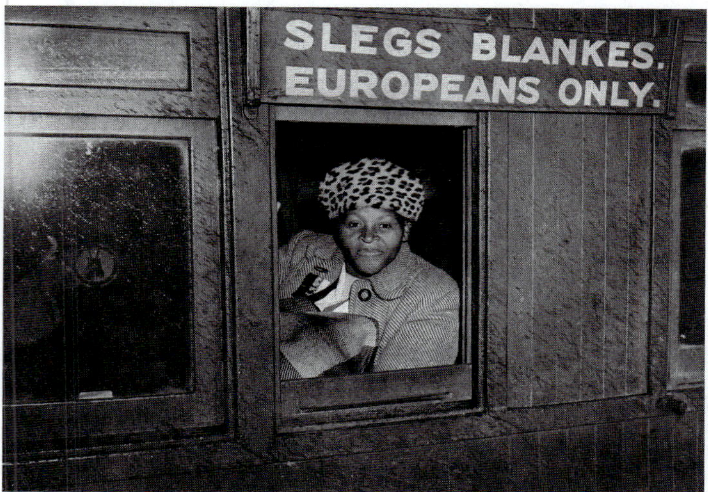

FIGURE 7.15 Apartheid in South Africa. A South African woman protests apartheid by using a "whites only" car on a train in 1952. Africa's long history of European domination officially ended only in 1994 with the first national elections in which black South Africans were allowed to vote.

Violence increased throughout South Africa until the late 1980s, when it threatened to engulf the country in civil war. The difficulties of maintaining order, combined with international political and economic pressure, forced the white-dominated South African government to initiate reforms that would end apartheid. A key reform was the dismantling of the homelands. Finally, in 1994, the first national elections in which black South Africans could participate took place. Nelson Mandela, the long-jailed ANC leader, was elected the country's president and proved to be an extraordinary leader. He was awarded the Nobel Peace Prize in 1993. Today in South Africa, finding new leaders with the unselfish, clear-eyed vision Mandela had has proven difficult. The thorny process of dismantling systems of racial discrimination and corruption continues, with varying amounts of success. ■

Power and Politics in the Aftermath of Independence

The era of formal European colonialism in Africa was relatively short. In most places, it lasted for about 80 years, from roughly the 1880s to the 1960s. In 1957, Ghana became the first sub-Saharan African colonial state to become independent. The last sub-Saharan African country to gain independence was South Sudan in 2011, not from a European power, but from its neighbor Sudan, after more than 40 years of civil war.

Africa entered the twenty-first century with a complex mixture of enduring legacies from the past and looming challenges for the future. Although it has been liberated from colonial domination, most old colonial borders remain intact (compare the colonial borders in Figure 7.14 on page 298 with the modern country borders in Figure 7.4 on page 284). Often these borders exacerbate conflicts between incompatible groups by joining them into one resource-poor political entity; other borders divide potentially powerful ethnic groups, thus diminishing their influence, or cut off nomadic people from resources formerly used on a seasonal basis.

For many years and with few exceptions, governments continued to mimic the colonial bureaucratic structures and policies that distanced them from their citizens. Corruption and abuse of power by bureaucrats, politicians, and wealthy elites stifled individual initiative, civil society, and entrepreneurialism, creating instead frustration and suspicion. Too often, coups d'état were used to change governments. Democracy, where it existed, was often weakly connected only to voting and not to true participation in policy making at the community, regional, and national levels.

Other relics of the colonial era are the dependence of African economies on relatively expensive imported food and manufactured goods as well as the production of agricultural and mineral raw materials for which profit margins are low and prices on the global market highly unstable. Thus, many sub-Saharan African countries often compete against each other and remain economically entwined in trade relationships that rarely work to their advantage.

The wealthiest country in the region remains South Africa, where the economy has highly profitable manufacturing and service sectors as well as extractive sectors. However, South Africa is no longer assured of remaining the economic leader of the region. Too often, politicians are insensitive to the widespread poverty and hardship around them, content to live well themselves. This was demonstrated during the summer of 2010 when South Africa hosted the 2010 World Cup for soccer, the world's largest sporting event and a first for this region, which up to that point had never hosted an Olympics or an event of similar scale. The preparations involved demolishing the homes of tens of thousands of poor urban residents who were relocated to make way for the elegant new stadiums (see Figure 7.11F). More recently, in an August 2012 crackdown reminiscent of those under apartheid, police shot dead 34 miners during a strike at a platinum mine in Marikana in Northwest Province. Other strikes followed.

THINGS TO REMEMBER

- Anatomically modern humans (*Homo sapiens*) evolved from earlier hominoids in eastern Africa about 250,000 years ago. By about 90,000 years ago, *Homo sapiens* had reached the eastern Mediterranean. They joined and eventually replaced other hominid species from Africa that had migrated throughout Eurasia starting 1.8 million years ago.
- Agriculture, industry, and international trade have an early history in Africa; agriculture began as early as 7000 years ago in northeast sub-Sahara; iron smelting 3400 years ago; and trade extended across the Indian Ocean to Asia long before Europeans came to Africa.
- Slavery was practiced in Africa within several powerful kingdoms, and then pre-Islam Arab traders extended the custom around the Mediterranean and to Asia. Muslims continued the practice. The European slave trade to the Americas severely drained the African interior of human resources and set in motion a host of damaging social responses within Africa. Slavery persists in parts of modern Africa.
- The main objectives of European colonial administrations in Africa were to extract as many raw materials as possible; to create markets

in Africa for European manufactured goods and food exports; and to divide and conquer by splitting powerful ethnic groups or by placing groups hostile to one another under the same jurisdiction, governed by complicit indigenous leaders.

- During the era of European colonialism, human talent and natural resources flowed out of Africa on a massive scale. Even after the countries of the region became politically independent, wealth continued to flow out of Africa because of fraud and unfavorable terms of trade.
- Into the twenty-first century, governments have continued to mimic colonial policies, including corruption, in ways that have kept sub-Saharan African countries economically dependent. This has stifled individual initiative, civil society, and entrepreneurialism. Democracy, where it exists, is often weak.
- South Africa trod a rocky road through colonial oppression and apartheid to its present relatively precarious position as the wealthiest country in sub-Saharan Africa.

CURRENT GEOGRAPHIC ISSUES

Most countries of sub-Saharan Africa have been independent of colonial rule for about 50 years. While many countries are still struggling with the lingering effects of the colonial era, others have moved forward.

Economic and Political Issues

Sub-Saharan Africa emerged from the exploitation and dependence of the colonial era just in time to get sucked into the vortex of globalization, where small or weak countries are often left with little control over their fates.

Commodity-Based Economic Development and Globalization in Africa

> **Geographic Insight 2**
> **Globalization and Development:** During the era of colonialism, Europeans set up the pattern that continues to this day of basing the economies of sub-Saharan Africa on the export of raw materials. This leaves countries vulnerable in the global economy, where prices for raw materials can vary widely year to year. While a few countries are diversifying and industrializing, most still sell raw materials and all rely on expensive imports of food, consumer products, and construction and transportation equipment.

Sub-Saharan Africa has had a centuries-long role in the global economy as a producer of human labor and raw materials, but the profits from turning these human resources and raw materials into higher-value manufactured and processed products have gone to wealthier countries. Raw materials that are traded (usually to other countries) for processing or manufacturing into more valuable goods are called **commodities**. Sub-Saharan commodities include cotton, cacao beans, coffee, timber, palm oil, unrefined oil, gas, precious stones, and metals. The profits of commodity production are usually too low to lift poor countries out of poverty. Because many other poor countries are producing the same commodities, competition between them on the world market often drives down profits.

commodities raw materials that are traded, often to other countries, for processing or manufacturing into more valuable goods

commodity dependence economic dependence on exports of raw materials

Commodity prices are also subject to wide fluctuations that create economic instability. This is because commodities are of more or less uniform quality and are traded as such on a global basis. For example, on the major commodities exchanges, a pound of copper may sell for U.S.$3 to $5, regardless of whether it is mined in South Africa, Chile, or Papua New Guinea. Because of the global scale of commodity trading, a change anywhere in the world that influences the supply or demand for a particular commodity can cause immediate instability in the pricing of that commodity. For instance, an earthquake in Chile could damage a major copper mine, restricting the supply of copper and sending up the price on world markets. In response, governments that make money from sales of raw copper may overestimate their future revenues and commit to expensive infrastructure projects. But the demand for copper can be decreased for many reasons; for example, innovations in the construction industry can reduce the use of copper in wiring and plumbing—resulting in sudden drops in the price of copper on world markets. Governments that depend on revenues from copper may suddenly need to halt infrastructure projects and cut essential services, such as electricity, road maintenance, or health care.

Since 2000 there has been a noticeable shift in many African countries, away from reliance on just one or two commodities. Nearly all have diversified to some extent, a trend that is discussed in the section "The Present Era of Diverse Globalization" on pages 301–303.

Successive Eras of Globalization

Successive eras of globalization have transformed Africa over the past several centuries. They include the colonial era, the transition to political independence but economic dependence (the era of structural adjustment), and the current era of diverse globalization. In this era, some African countries are moving away from **commodity dependence** on raw materials exports and toward new types of economic development and regional integration.

Most early European colonial administrations in Africa (Britain, France, Germany, Belgium) evolved directly out of private resource-extracting corporations, such as the German East Africa Company. The welfare of Africans and their future development was a low priority for most colonial administrators. Education and health care for Africans were generally neglected by colonial governments, which were guided by *mercantilism* (see page 125 in Chapter 3)—the idea that colonies existed to benefit their colonizers. Laborers and farmers were paid poorly and the colonial governments strongly discouraged any other economic activities that might compete with Europe. Thus, African economies were hindered from making a transition to the more profitable manufacturing-based industries that were transforming Europe and North America.

For years, commodity dependence, widespread poverty, and the lack of internal markets for local products and services characterized all sub-Saharan African economies, with the partial exception of South Africa.

South Africa is the only sub-Saharan African country with a manufacturing base. Early on, profits from its commodity exports (mainly minerals) were reinvested in the manufacturing of mining and railway equipment. In the late twentieth century, South Africa developed a service sector, with particular strengths in finance and communications, that in part support the mining and manufacturing industries. With only 6 percent of sub-Saharan Africa's population, South Africa today produces 30 percent of the region's economic output.

Unfortunately, the relics of colonialism and apartheid have left most sub-Saharan countries, including South Africa, with a **dual economy**—one part rich and industrialized, the other poor and primarily reliant on low-wage labor and small-scale informal enterprises. Even though the labor of black South Africans was essential to the country's prosperity, by the 1940s, 84 percent of black South Africans lived at a bare subsistence level. By 2009, fifty percent of black South Africans still lived below the poverty line (compared with just 7 percent of white South Africans), and only 22 percent of black South Africans had finished high school (compared with 70 percent of white South Africans).

dual economy an economy in which the population is divided by economic disparities into two groups, one prosperous and the other near or below the poverty level

The Era of Structural Adjustment By the 1980s, most African countries remained poor and dependent on their volatile and relatively low-value commodity exports. Attempts at investing in manufacturing industries failed (discussed below), and governments struggled to make payments on the large loans taken out for these projects. Infrastructure (roads, water, and utilities) was either not built or not maintained. A breaking point came in the early 1980s when an economic crisis swept through the region and much of the rest of the developing world, leaving most countries unable to repay their debts at all. In response, the IMF and the World Bank designed *structural adjustment programs* (SAPs; see pages 129–131 in Chapter 3) to enforce repayment of the loans. SAPs did have some useful results. They tightened bookkeeping procedures and thereby curtailed corruption and waste in bureaucracies. They closed some corrupt state-owned industrial and service monopolies, opened some sectors of the economy to medium- and small-scale business entrepreneurs, and made tax collection more efficient. But overall, SAPs had many unintended consequences, and surprisingly, they failed at their primary objective—reducing debt (Figure 7.16).

To facilitate loan repayment, SAPs required governments to sell off inefficient government-owned enterprises, often at bargain-basement prices. Also, government jobs in social services, education, health, and agricultural programs were slashed so that tax revenues could be devoted to loan repayment. If countries refused to implement SAP requirements, the international banks cut off any future lending for economic development.

As unemployment rose, so did political instability. Deteriorating infrastructure reduced the quality of remaining social services, transportation, and financial services, all of which scared away potential investors. SAPs also reduced food security because agricultural resources were shifted toward the production of cash crops for export. Between 1961 and 2009, per capita food production in sub-Saharan Africa actually decreased by 14 percent, making it the only region on earth where people were eating less well than in the past (see Figure 1.14 on page 26).

Informal Economies Africa's informal economies provided some relief from the hardships created by SAPs. Informal economies in Africa are ancient and wide-ranging, providing employment and useful services and products. People may grow and sell garden produce, prepare food, vend a wide array of products on the street, sell time cards (or credits) for mobile phones, make craft items and utensils, or do child and elder care. Others in the informal economy, however, earn a living doing less socially redeeming things such as sex work, distilling liquor, or smuggling scarce or illegal items including drugs, weapons, endangered animals, bushmeat, and ivory. Because most activities take place "under the radar," informal jobs may involve criminal acts, wildly unsafe procedures, and hazardous substances.

In most African cities, the role of the informal economy has grown from one-third of all employment to more than two-thirds. Such jobs are a godsend to the poor, but they create problems for governments because the informal economy typically goes untaxed, so less money is available to pay for government services or to repay debts. Moreover, as the informal sector grows, profits have declined as more people compete to sell goods and services to those with little disposable income. And although women typically dominate informal economies, when large numbers of men lose their jobs in factories or the civil service, they may crowd into the streets and bazaars as vendors, displacing the women and young people.

In response to the now widely recognized failures of SAPs, and the overemphasis on the power of markets to guide development, the IMF and the World Bank in 2000 replaced SAPs with *Poverty Reduction Strategy Programs*, or *PRSPs*, and *Sustainable Structural Transformation (SST)* strategies. These policies are similar to SAPs in that they push market-based solutions intended to reduce the role of government in the economy, but they differ in several ways. They focus on reducing poverty and diversifying economies into manufacturing, rather than on just "development" and debt repayment per se, and are generally promoting more democratic reforms. They also include the possibility that a country may have all or most of its debt "forgiven" (paid off by the IMF, the World Bank, or the African Development Bank) if the country follows the PRSP rules. Forty sub-Saharan countries had qualified for and been approved to receive debt relief by July 2010.

The Present Era of Diverse Globalization The current wave of globalization promises a different role for Africa, resulting in new sources of investment that are bringing jobs and infrastructure. As discussed at the beginning of the chapter, one significant trend is that young African professionals at home and abroad are supporting

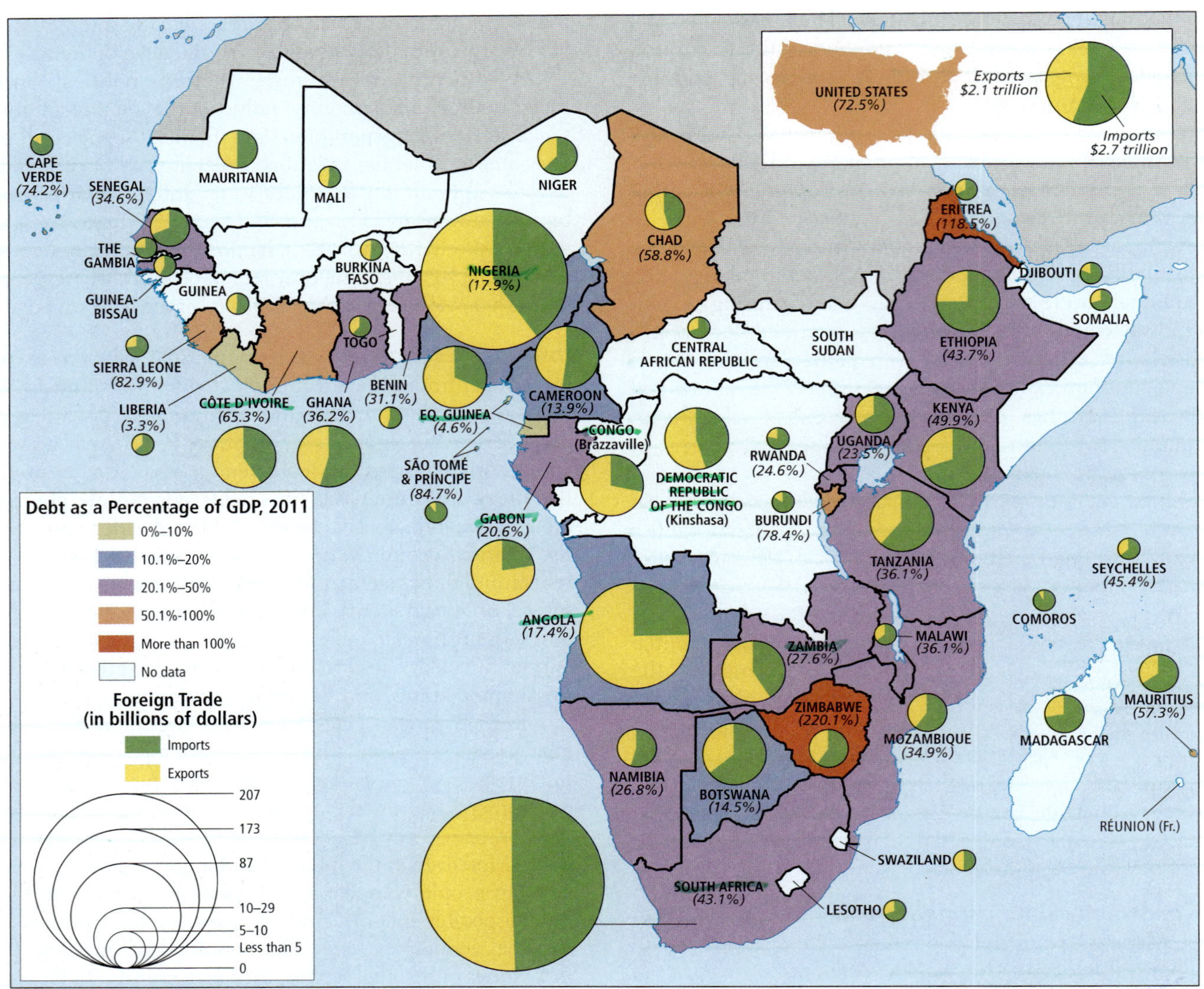

FIGURE 7.16 Economic issues: Public debt, imports, and exports. Public debt is increasing across Africa, partly as a result of borrowing to fund development projects; but the U.S. public debt (see inset) far exceeds that of most sub-Saharan countries. In all but 9 of the 48 countries in sub-Saharan Africa, imports exceed exports. These nine are Angola, Democratic Republic of the Congo (indicated in the text as Congo [Kinshasa]), Republic of Congo (indicated in the text as Congo [Brazzaville]), Equatorial Guinea, Gabon, Nigeria, South Africa, Zambia, and Côte d'Ivoire. Minerals (copper and cobalt) are the main export earners for Zambia; cocoa, coffee, and timber for Côte d'Ivoire; and the export earnings of the other 7 are from oil and petroleum products.

development in a multitude of ways. Another trend is foreign direct investment (FDI) at the corporate level. While Europe and the United States are still the largest sources of investment in sub-Saharan Africa, Asia's influence on African economies is increasing significantly, but there are potential problems involved in Asia's role.

In recent years, China and India have begun to view sub-Saharan Africa as a new frontier for their large and growing economies. These two countries now exert a powerful influence on the region through their demand for Africa's export commodities, direct investment in agribusiness, mining and industry, and the sale of their manufactured goods. Of the two, China's influence is the greater.

Together, China and India consume about 15 percent of Africa's exports, but their share is growing twice as fast as is that of any of Africa's other trading partners. However, the increasing Chinese and Indian demand for resources is a double-edged sword; it is not clear if Asian investments will ultimately prove beneficial or hurtful to sub-Saharan African economies. The improved infrastructure (roads, ports, utilities, and technology) could ultimately facilitate intra-African trade, linking countries that have never before been able to trade. China's and India's investment in African agriculture, undertaken to meet rising demands for better food in China and India, could result in more efficient production overall through *technology transfer* and thus higher earnings for Africa and greater food supplies for African internal urban markets. Overall, African food security could increase.

It is important, however, to keep an eye on the extent to which Asian investments in Africa actually benefit Africans rather than simply replicating the injustices of colonialism. For example, Chinese investments in mining and infrastructure development could have boosted local African economies by providing construction jobs for African laborers and professional design and management experience for mid-level educated Africans, but this benefit never materialized because African governments agreed to China's demand that all work be done by Chinese companies using Chinese workers (Figure 7.17). Most controversial has been the willingness of Chinese companies to deal with brutal and corrupt local leaders, such as Liberia's now convicted and imprisoned Charles Taylor (diamonds and timber extraction) or Zimbabwe's Robert Mugabe (mining), who bartered away their country's resources at bargain prices and used the profits to enrich themselves and to wage war against their own citizens.

Another example of private investment is from the Kuwaiti-based mobile company, Zain, which launched a borderless network that now covers 13 countries: Kenya, Tanzania, South Sudan, Uganda, Congo (Kinsasha), Congo (Brazzaville), Burkina Faso, Chad, Niger, Nigeria, Sierra Leone, Madagascar, and Zambia. Zain's customers can make calls across the network at local rates without incurring surcharges; recently, the company added Internet access, SMS, international roaming, and portal applications to the local rate.

The expansion of communication technology has inspired a wide range of innovations, such as imaginative ways to lay out networks in difficult terrain and ways to charge mobile phones in remote villages where there is no electricity, as the Malawian William Kamkwamba figured out how to do (for more on William Kamkwamba's story, see the vignette on page 305).

160. SOMALILAND EXPATRIATES RETURN HOME TO HELP NATIVE LAND DEVELOP

161. AFRICAN UNION APPEALS TO DIASPORA TO AID HOMELANDS

ON THE BRIGHT SIDE

Africans Investing in Africa

The role of African professionals in globalization is growing. As was noted in the opening vignette, perhaps the most dramatic sign of change for African economies is that educated Africans familiar with high tech and the globalized world are staying home and starting businesses that link their countries to the technological advantages that have spurred development elsewhere. Their incomes, along with *remittances* (money sent to family members from Africans working and living abroad, primarily in Europe and North America), buy mobile phones, start small businesses, fund education for children, build houses, and help the needy. Remittances are a more stable source of investment than *foreign direct investment*, in that they tend to come regularly from committed donors who will continue their support for years. They are also much more likely to reach poorer communities. But payments can stop if remitters lose their jobs in America or Europe.

Young, educated adults are profoundly changing the trajectory of Africa's development. They have been largely responsible for the rapid spread of mobile phones and have developed a wide array of new and important uses of cell phone technology. Figure 7.18 is a map of mobile phone subscriptions by country as of 2011. Between 1998 and 2010, mobile phone subscribers in sub-Saharan Africa increased by more than 100 times, from about 4 million to about 450 million. Google estimates that by 2016 there will be a billion mobile phones in sub-Saharan Africa, one for every person. An African innovation has been to make service available to the very poor by selling prepayment credits in very small units. In fact, many people now make a living selling these small credits. Private investment is improving connectivity by financing undersea cable connections.

FIGURE 7.17 China in Angola. The man on the right is one of the estimated 20,000 Chinese workers in Angola. China has given loans and aid to Angola in excess of U.S.$4 billion since 2004, and in return, China has been guaranteed a large portion of Angola's future oil production. In addition, 70 percent of Angola's development projects, such as building and upgrading railways, have been given to Chinese companies, most of which import workers from China.

Thinking Geographically

Why might China be particularly interested in investments in Africa's transportation infrastructure?

Regional and Local Economic Development

As they create alternatives to past development strategies, many African governments have been focusing on regional economic integration similar to that of the European Union. Hoping to unleash pent-up talent and consumer demand for African-made products, local agencies and public and private donors are pursuing grassroots development designed to foster very basic innovation at the local level that can then be marketed across the region.

The Potential of Regional Integration According to the World Trade Organization (WTO), less than 20 percent of the total trade of sub-Saharan Africa in 2011 was conducted between African countries. This is true partly because so many countries produce the same raw materials for export. And everywhere except South Africa, industrial capacity is so low that the raw materials cannot be absorbed within the continent, so African countries compete with each other and all other global producers to sell to the main

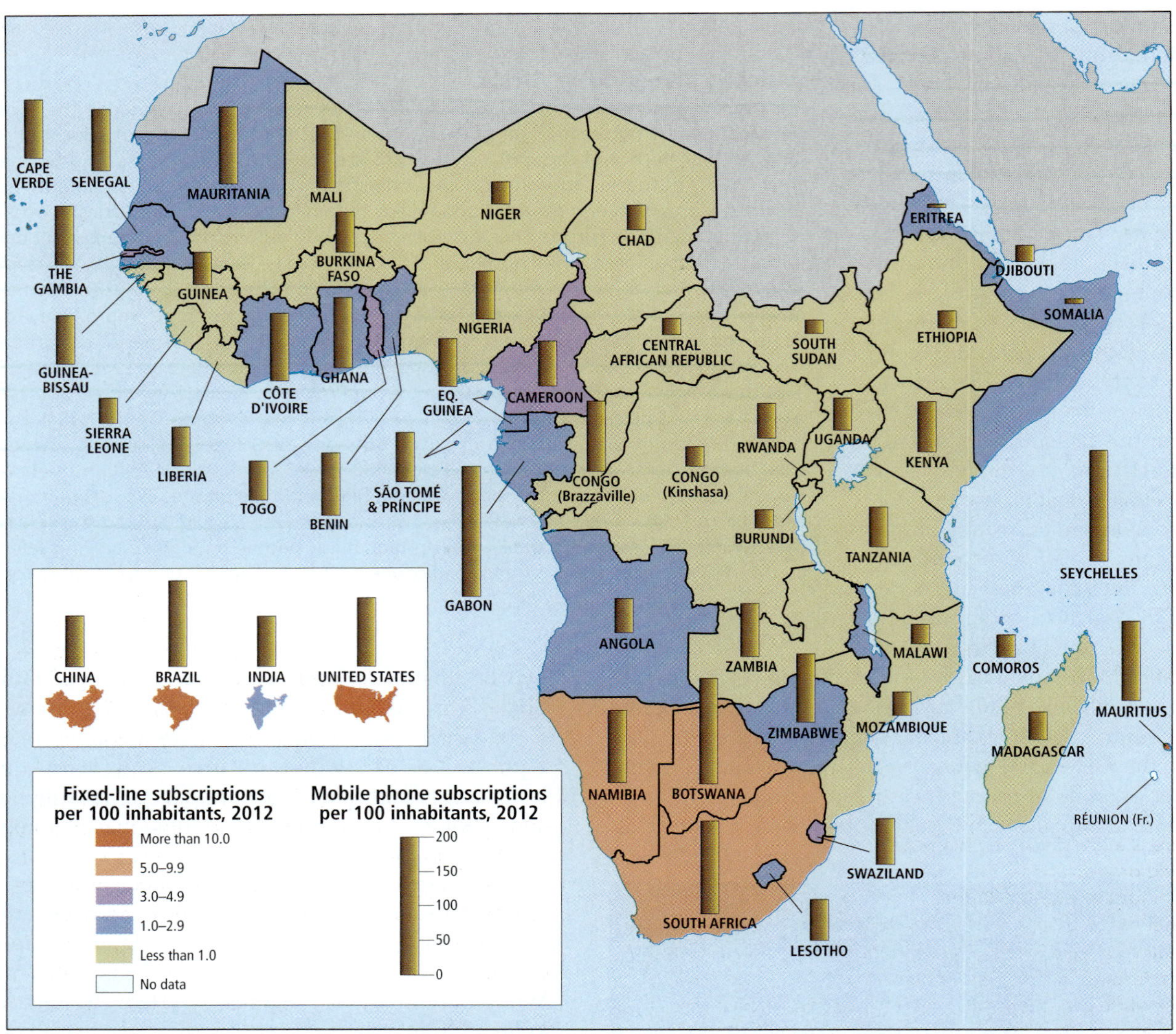

FIGURE 7.18 Mobile phone users in sub-Saharan Africa. In the past decade, mobile phone use in sub-Saharan Africa has boomed. By 2016, there will be more than 1 billion mobile phones, including in the hands of even very poor citizens, because of innovations that make it possible to buy service access in tiny amounts. One of the most popular uses of phones in Africa is to move money from one account to another.

customers, which currently are in Europe, North America, and industrialized Asia. This failure to trade with each other can be traced to divisive colonial policies, lack of transportation and communication grids, and arcane bureaucratic regulations.

African Regional Integration Over the last decades, regional trading blocs have been formed to encourage neighboring countries to trade with each other and cooperate in the production of manufactured (value-added) exports. The trade blocs are somewhat fluid; they form and then reorganize, as is depicted in Figure 7.19, a list and simplified map of the evolving African Regional Economic Communities (RECS). By combining the markets of several countries (as do NAFTA and the EU), regional trade blocs can create a market size sufficient to foster industrialization and entrepreneurialism. Africa's many different regional trade blocs share several goals: reducing tariffs between members, forming common currencies, reestablishing peace in war-torn areas, cooperating to upgrade transportation and communication infrastructure, and building regional industrial capacity.

These goals could be helped by the creation of *value chains*; the value chain approach analyzes all parts of a production chain of a final product—a tractor, a refrigerator, or chocolates—and the relationships among the parts. The goal is to facilitate production and marketing efficiencies and enable the flow of information, innovation, resources, and the products and profits so that the producers of raw materials are included and can live decent lives. Such value chains, which encourage efficient transport and specialization in products and labor skills, are second only to the fuel trade as the fastest-growing portions of global trade.

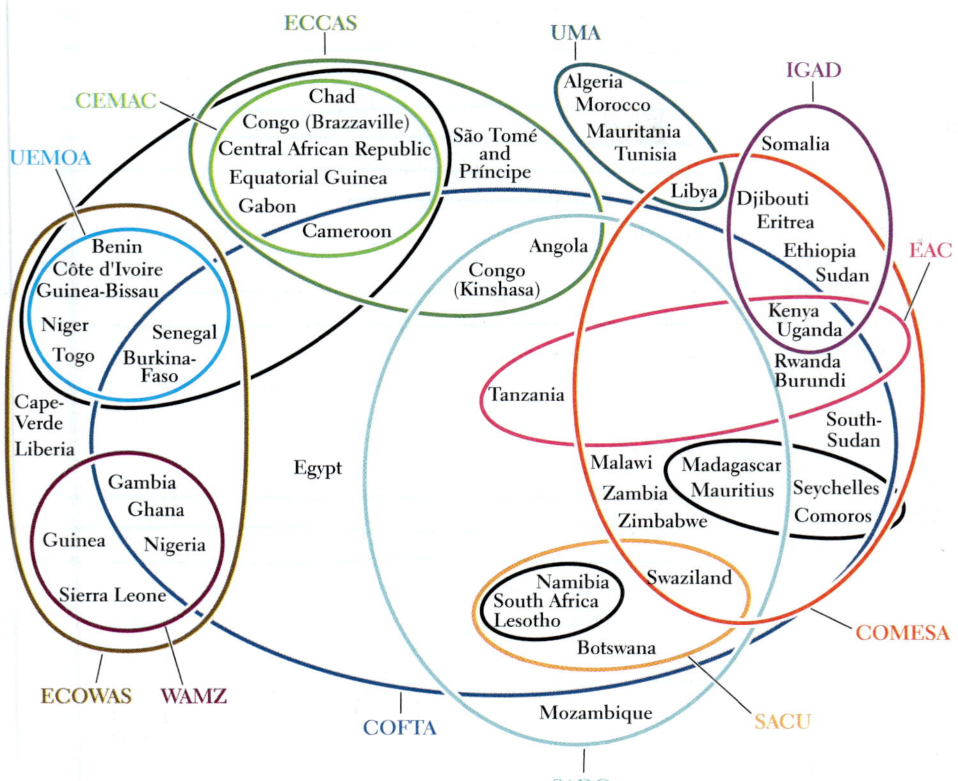

FIGURE 7.19 Principal trade organizations in sub-Saharan Africa. There is a lot of overlap between the countries and the regional trade organizations, but often one country is dominant in each group. South Africa and Nigeria, for example, are dominant in three regional trade organizations; but in fact, South Africa trades with nearly all countries in the region, according to well-established relationships. Because political stability is essential to trade, some of the strongest trade organizations, such as the Economic Community of West African States (ECOWAS), have become involved in the amelioration of conflicts.

Building a full-scale, continent-wide economic union along the lines of the European Union is a long-term goal. According to studies by the WTO, if sub-Saharan countries could increase trade with each other by just 1 percent, the region would show a total added income of $200 billion per year—a substantial internal contribution to poverty alleviation.

Local Development An increasingly common strategy for improving living standards in this region is **grassroots economic development**. Projects using this strategy are designed to provide sustainable livelihoods in rural and urban areas, and often use simple technology that requires minimal or no investment in imported materials. One such approach is **self-reliant development**, which consists of small-scale self-help projects that use local skills to create products or services for local consumption. One of the most important aspects of this type of development is that control of the projects remains in local hands, so that participants retain a sense of ownership and commitment in difficult economic times. One district in Kenya has more than 500 such self-reliant groups. Most members are women who terrace land so it can be readied for farming, building water tanks, and planting trees. They also build schools and form credit societies.

grassroots economic development economic development projects designed to help individuals and their families achieve sustainable livelihoods

self-reliant development small-scale development in rural areas that is focused on developing local skills, creating local jobs, producing products or services for local consumption, and maintaining local control so that participants retain a sense of ownership over the process

Transportation Needs The issue of improving rural transportation illustrates how a focus on local African needs can generate unique solutions. When non-Africans learn that transportation facilities in Africa are in need of development, they usually imagine building and repairing roads for cars and trucks. But a recent study that analyzed village transportation on a local level found that 87 percent of the goods moved are carried via narrow footpaths on the heads of women! Women "head up" (their term) firewood from the forests, crops from the fields, and water from wells. An average adult woman spends about 1.5 hours each day moving the equivalent of 44 pounds (20 kilograms) more than 1.25 miles (2 kilometers).

Unfortunately, the often-treacherous footpaths trod by Africa's load-bearing women have been virtually ignored by African governments and international development agencies, which tend to focus solely on roads for motorized vehicles (also badly needed). Grassroots-oriented nongovernmental organizations (NGOs) are now making much less expensive but equally necessary improvements to Africa's footpaths. Some women have been provided with bicycles, donkeys, and even motorcycles that can travel on the footpaths. This saves time and energy for women who can direct more of their efforts to becoming educated and generating income.

Energy Needs Africa exports oil, but its own energy needs, which are currently unmet even at the most basic level of home electricity, can also be addressed by local solutions, as the following vignette illustrates.

VIGNETTE In Malawi, 14-year-old William Kamkwamba was forced to drop out of school when a famine struck his country in 2001 and his family could no longer afford the $80 school fee. Depressed at the prospect of having no future, he went to a local library when he could. There he found a book in English called *Using Energy* that described an electricity-generating windmill. With an old bicycle frame, tractor fan blades, PVC pipes, and scraps of wood, he built a windmill that generated enough power (stored in a car battery) to light his home, run a radio, and charge neighborhood cell phones. More elaborate energy projects followed.

Now known as "the boy who harnessed the wind," in 2009 William appeared on Jon Stewart's *Daily Show* in the United States to explain how he plans to start his own windmill company and other ventures that will bring power to remote places across Africa. He graduated from the first pan-African prep school in South Africa and now attends Dartmouth College in the United States. William's web page is http://www.williamkamkwamba.com. ■

THINGS TO REMEMBER

Geographic Insight 2

- **Globalization and Development** Continuing reliance on the export of raw materials leaves countries vulnerable, especially because in the globalized economy, where all producers are in competition, prices can vary widely year to year. Most countries still primarily sell raw materials and must rely on expensive imports for daily necessities and manufactured equipment essential for further development.
- Prospective investors in sub-Saharan Africa have been discouraged by problems that structural adjustment programs (SAPs) either ignored or worsened. Recently, though, Africa has become a new frontier for Asia's large and growing economies, especially those of China and India, that both buy and sell in sub-Saharan Africa.
- A major and enduring source of investment funds in Africa is coming from members of the African diaspora, who send regular remittances to friends and family.
- Many African governments are focusing on regional economic integration along the lines of the European Union. Grassroots economic development is also being pursued.
- Africa's informal economies have provided some relief from the hardships created by SAPs. These economies are ancient, wide ranging, and agile; they provide employment and useful services and products.

Power and Politics

Geographic Insight 3

Power and Politics: Countries in the region are shifting from authoritarianism to elections and other democratic institutions, but the power structures in the region are resistant to democratization. While free and fair elections have brought about dramatic changes, distrusted elections have often been followed by surges of violence.

After years of rule by corrupt elites and the military, signs of a shift toward democracy and free elections are now visible across Africa. Yet progress is often blocked by conflict, and even when democratic reforms are enacted and free elections established, violence often accompanies these elections.

Ethnic Rivalry Africa's democratic and economic progress has been held back by frequent civil wars that are in many ways the legacy of colonial era policies of **divide and rule**. Divisions and conflicts between ethnic or religious groups were deliberately aggravated by European colonial powers. To make it hard for Africans to unite against colonial rule, the borders and administrative units of the African colonies were designed so that different and sometimes hostile groups were put together under the same jurisdiction (Figure 7.20 shows the situation in Nigeria). After independence, when Africans took over, governance was complicated because African officials inevitably belonged to one local ethnic group or another and so were not seen as impartial in their attempts to resolve conflicts. Moreover, during the colonial era, older indigenous traditions for ensuring ethical behavior, resolving ethnic conflict, and punishing greed by leaders had been eroded or erased. The result has been that in the absence of democratic conflict resolution, ethnic hostilities have devolved into civil wars (Figure 7.21C on page 308).

divide and rule the deliberate intensification of divisions and conflicts by potential rulers; in the case of sub-Saharan Africa, by European colonial powers

A Case Study: Conflict in Nigeria

Nigeria was and remains in part a creation of British divide-and-rule imperialism. Many disparate groups—speaking 395 indigenous languages—have been joined into one unusually diverse country.

The British reinforced a north–south dichotomy that mirrored the physical north (dry)–south (wet) patterns. Among the Hausa and Fulani ethnic groups in the north, the British ruled via local Muslim leaders who did not encourage public education. In the south, the Yoruba and Igbo ethnic groups, who were primarily a combination of animist and Christian, were ruled more directly by the British, who encouraged attendance at Christian missionary schools, open to the public. At independence, the south had more than 10 times as many primary and secondary school students as the north. It was more prosperous, and southerners also held most government civil service positions. Yet the northern Hausa dominated the top political posts, in part because of their long association with British colonial administrators. Over the years, bitter and often violent disputes erupted between the southern Yoruba–Igbo and northern Hausa–Fulani regarding the distribution of increasingly scarce clean water, development funds, jobs, and oil revenues, and the severe environmental damage done by oil extraction. **178. NEW NIGERIAN PRESIDENT INHERITS TURBULENT NIGER DELTA**

The politics of oil have complicated the troubles in Nigeria. Nigeria's oil is located on lands occupied by the Ogoni people, which lie in the south along the edges of the Niger River delta (see Figure 7.1D). The Ogani are a group of about 300,000 who are distinct from other ethnicities in Nigeria. Virtually none of the profits from oil production and export, and very little of the oil itself, goes to Ogoniland. While receiving few benefits from oil extraction, Ogoniland has suffered gravely from the resulting pollution (see the photo in Figure 1.4 on page 8). Oil pipelines crisscross Ogoniland, and spills and blowouts are frequent; between 1985 and 2009 there were hundreds of spills, many larger than that of the Exxon Valdez disaster in Alaska. Natural gas, a by-product of oil drilling, is burned off, even though it could be used to generate electricity—something many Ogoni lack. Royal Dutch Shell, a multinational oil company that is very active in the Niger River delta, acknowledges that while it has historically netted $200 million in profit yearly from Nigeria, in 40 years, it has paid a total of just $2 million to the Ogoni community.

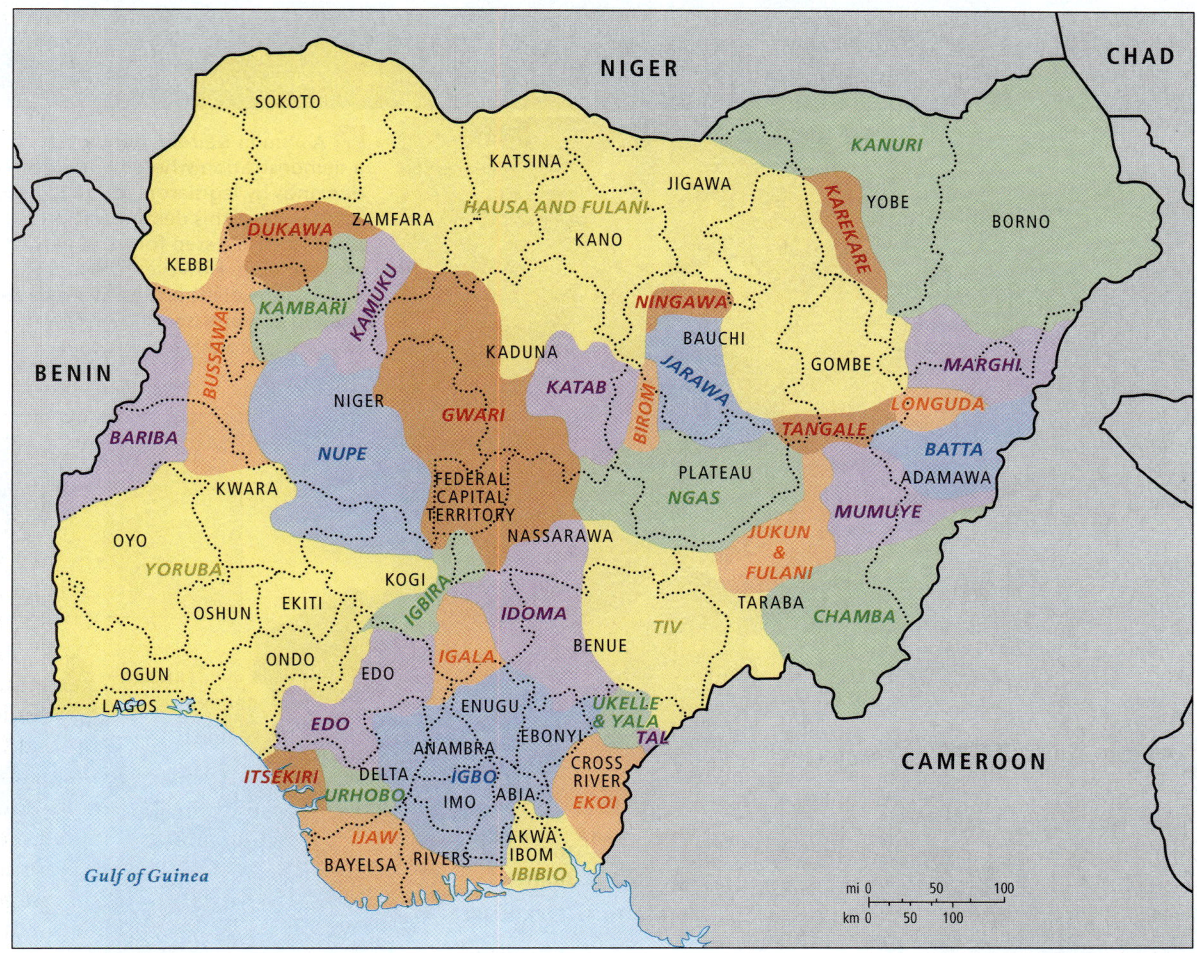

FIGURE 7.20 Nigeria's ethnic geography. There are thousands of ethnic groups distributed across sub-Saharan Africa. Nigeria alone has more than 400; but when Europeans set up political units, they either ignored ethnic boundaries or purposely divided large ethnic groups.

Geographic strategies have been used to reduce tensions in Nigeria. One approach has been to create more political states (Nigeria now has 30) and thereby reallocate power to smaller local units with fewer ethnic and religious divisions. However, when large, wealthy states were subdivided, reputedly to spread oil profits more evenly, the actual effect of the subdivision was to mute the voice and power of the public. ■

The Role of Geopolitics Cold War geopolitics between the United States and the former Soviet Union deepened and prolonged African conflicts that grew out of divide-and-rule policies. After independence, some sub-Saharan African governments turned to socialist models of economic development, often receiving economic and military aid from the Soviet Union and Yugoslavia. Other governments became allies of the United States, receiving equally generous aid (see Figure 5.13 on page 211). Both the United States and the USSR tried to undermine each other's African allies by arming and financing rebel groups.

In the 1970s and 1980s, southern sub-Saharan Africa became a major area of East–West tension. The United States aided South Africa's apartheid government in military interventions against Soviet-allied governments in Namibia, Angola, and Mozambique. Another area of Cold War tension was the Horn of Africa, where Ethiopia and Somalia fought intermittently throughout the 1960s, 1970s, and 1980s. At different times, the Soviets and Americans funded one side or the other. The failure by both sides of the Cold War to anchor the aid they dispensed to any requirements that it be used for sustainable development became a major stimulant for militarization and corruption. Once this source of cash was removed at the end of the Cold War, corrupt officials turned to selling off natural resources and commodities.

Al Qaeda in Mali In early 2012, Islamist geopolitics came to the north of Mali in the Sahara. The crisis began when the nomadic Tuareg ethnic group, accustomed for hundreds of years to using the whole of the Sahara without attention to country borders,

Figure 7.21 Photo Essay: Power and Politics in Sub-Saharan Africa

This region has long been plagued by political violence. A long-term shift toward democratization may help reduce bloodshed as governments become more responsive to their citizens. As in other regions, the most democratized countries have had fewer violent conflicts since 1945. Nevertheless, in any country, the transition to democracy can be tumultuous, and elections are often marked by violence.

A A man in Nairobi, Kenya, runs from police at a demonstration where he has been protesting demands by legislators for an increase in their salaries. That this demonstration could take place at all is a step forward, as is the presence of the media. However, the fact that participants in this nonviolent protest were beaten, tear-gassed, and arrested shows how limited the political freedoms are in Kenya.

B Ellen Johnson-Sirleaf of Liberia is the first woman to be elected head of state in this region. Her election in 2005 was considered free and fair by international observers and helped Liberia begin to resolve tensions that had produced two devastating civil wars.

Democratization and Conflict

Democratization index

- Full democracy
- Flawed democracy
- Hybrid regime
- Authoritarian regime
- No data

Armed conflicts and genocides with high death tolls since 1945

- Ongoing conflict
- 1000–10,000 deaths
- 10,000–60,000 deaths
- 60,000–180,000 deaths
- 180,000–500,000 deaths
- 500,000–1,000,000 deaths

CAPE VERDE, MAURITANIA, MALI, NIGER, CHAD, ERITREA, DJIBOUTI, SENEGAL, THE GAMBIA, GUINEA-BISSAU, GUINEA, SIERRA LEONE, LIBERIA, CÔTE D'IVOIRE, BURKINA FASO, GHANA, TOGO, BENIN, NIGERIA, CAMEROON, CENTRAL AFRICAN REPUBLIC, SOUTH SUDAN, ETHIOPIA, SOMALIA, UGANDA, KENYA, RWANDA, BURUNDI, TANZANIA, EQUATORIAL GUINEA, SÃO TOMÉ & PRÍNCIPE, GABON, CONGO (Brazzaville), CONGO (Kinshasa), SEYCHELLES, COMOROS, ANGOLA, ZAMBIA, MALAWI, MOZAMBIQUE, ZIMBABWE, MADAGASCAR, MAURITIUS, RÉUNION (Fr.), NAMIBIA, BOTSWANA, SOUTH AFRICA, SWAZILAND, LESOTHO

mi 0 500 1000
km 0 500 1000

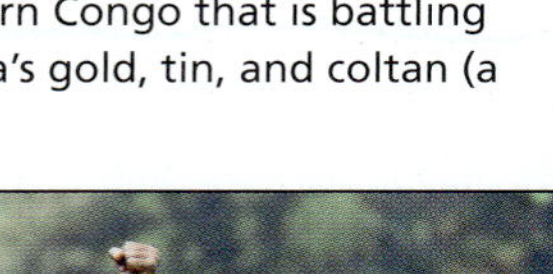

C Members of a rebel militia in eastern Congo that is battling the government for control of the area's gold, tin, and coltan (a mineral used in cell phones) resources.

D Refugees from Sudan relocate to South Sudan after a referendum in 2011 established the latter's independence from the former. South Sudan is plagued by political corruption and is involved in continuing hostilities with Sudan; both of these problems have constrained the export of its main resource, oil.

Thinking Geographically

After you have read about democratization and conflict in sub-Saharan Africa, you will be able to answer the following questions:

A In what ways does this picture show both the strengths and limits of political freedoms in Kenya?

B Why might it be significant that Ellen Johnson-Sirleaf was elected as Liberia's president without the help of quotas that reserve certain elected positions for women?

C What does this photo suggest about the strength of this rebel militia and the militia's methods of acquiring uniforms and arms?

D What does this photo convey about the general well-being of the people depicted?

increased their long-term efforts to have their right to move freely recognized. Mali's weak central government, located far to the south in Bamako, the capital, responded so ineptly that a faction of Tuareg rebel forces, buttressed by arms and fighting experience as mercenaries for Colonel Qaddafi in Libya (deposed in October 2011), seized the northern Mali town of Kidal and the legendary Muslim trading city of Tombouctou. Meanwhile, Al Qaeda members and other jihadists recruited across North Africa during the Arab Spring joined the Tuareg, but only temporarily. Calling themselves "Al Qaeda in Mali," this group apparently intended to take northern Mali from the Tuareg (some of whom are Sufi Muslims, while others are largely secular) and set up a regime based on Islamic Law (*shari'a*). Tuareg and Malian resistance to this grew over time.

The failed government response to both the Tuareg and Al Qaeda in Mali so irritated the Malian regular army that they themselves staged a coup in Bamako in the spring of 2012. However, under severe criticism from the African Union and the regional trade organization Economic Community of West African States (ECOWAS; see Figure 7.19), the Malian army eventually backed down. A new government for southern Mali was formed in August 2012, but Al Qaeda in Mali increased its hold on the north and feuded with the Tuareg over visions of what type of entity the new northern state would be. The disruption severely damaged Mali's tourist economy, which had been based on seasonal tourism to the Niger River basin and ancient sites in Djenné and Tombouctou. Tens of thousands of northern Malians, many of them Sufi Muslims (a version of Islam that believes in a benevolent, peaceful deity), fled south to Mauritania and told of harsh treatment by the Islamists.

By January 2013, neighboring states in North and sub-Saharan Africa worried that Al Qaeda's influence would destabilize the continent. On January 11, France, the former colonial power in Mali, took a controversial step when it began a bombing campaign and sent 2000 troops to assist the African Union in dislodging the militants. For further coverage of Islamist geopolitics and shifting alliances in the Sahara, see Chapter 6.

Conflict and the Problem of Refugees Conflicts create refugees, and refugees are commonplace across sub-Saharan Africa. Throughout the last decade of the twentieth century, refugees from Somalia, Ethiopia, Uganda, Liberia, Sierra Leone, Congo (Kinshasa), Congo (Brazzaville), Rwanda, Mozambique, Niger, and Mali poured back and forth across borders and were displaced within their own countries. Often they were trying to escape **genocide** (efforts to murder an entire ethnic, racial, religious, or political group).

With only 11 percent of the world's population, this region contains about 19 percent of the world's refugees, and if people displaced within their home countries are also counted, the region has about 28 percent of the world's refugee population (see Figure 7.21D). Women and children constitute 75 percent of Africa's refugees because many adult men who would be refugees are either combatants, jailed, or dead.

As difficult as life is for these refugees, the burden on the areas that host them is also severe. Even with help from international agencies, the host areas find their own development plans deferred by the arrival of so many distressed people, who must be fed, sheltered, and given health care. Large portions of economic aid to Africa have been diverted to deal with the emergency needs of refugees.

Successes and Failures of Democratization In sub-Saharan Africa, efforts to establish democratic institutions have produced mixed results. The number of elections held in the region has increased dramatically. In 1970, only 11 states had held elections since independence. By 2006, 25 out of the then 44 sub-Saharan African states had held open, multiparty, secret ballot elections, with universal suffrage. While the implementation of democratic elections has increased (see Figure 7.21B), the inauguration of other aspects of democracy, such as public participation in policy formation at the local and national levels, has been irregular and uneven. Too often, public frustration with election outcomes or with particular policies has resulted in violent demonstrations (see Figure 7.21A) or rebellion (see Figure 7.21C). When people have input at the policy level, governments tend to be more responsive to the needs of their people.

At times, flawed elections have brought about massive violence. Kenya, after several peaceful election cycles, had an election in 2008 so corrupt that deadly protest riots broke out. More than 1000 people died, and 600,000 were displaced by mobs of enraged voters. By 2010, a new Kenyan constitution gave some hope that political freedoms would be better protected. In 2008 in Nigeria, local elections sparked similar violence, and in the Congo (Kinshasa) in 2006, the first elections held there in 46 years resulted in violence that left more than 1 million Congolese refugees within their own country. In the 2012 parliamentary elections, the will of the voting public became more evident when the ruling party lost more than 40 percent of its legislative seats to opposition parties.

Zimbabwe may represent the worst case of the failure of democratization. In the 1970s, Robert Mugabe became a hero to many for his successful guerilla campaign in what was then called

genocide the deliberate destruction of an ethnic, racial, religious, or political group

Rhodesia (now Zimbabwe). The defeated white minority government had been allied with apartheid South Africa but was not formally recognized by any other country because of its extreme racist policies. Mugabe was elected president in 1980, following relatively free, multiparty elections. Over the years, however, his authoritarian policies impoverished and alienated more and more Zimbabweans and increased his personal wealth. In the 1990s, he implemented a highly controversial land-redistribution program that resulted in his supporters gaining control of the country's best farmland, much of which had been in the hands of white Zimbabweans. This move decimated agricultural production and contributed to a chronic food shortage and a massive economic crisis. The resulting political violence created 3 to 4 million refugees who fled mostly to neighboring South Africa and Botswana. Mugabe held on to his office through the rigged elections of 2002 and 2008, but was then forced to share power in 2009 with opposition leader Morgan Tsvangirai, who became prime minister. After some of Tsvangarai's policies were implemented, Zimbabwe began to have some modest economic growth. In 2012, though, Mugabe called for early elections and appeared to be financing his campaign with money from state-owned diamond mines.

159. ZIMBABWE'S ROBERT MUGABE—A PROFILE

In other places, however (most recently in South Sudan and Rwanda and some years ago in South Africa), elections have helped end civil wars and resettle refugees (see Figure 7.21D), as the possibility of becoming respected elected leaders induced former combatants to lay down their arms. In Sierra Leone and Liberia, public outrage against corrupt ruling elites resulted in elections that brought fortuitous changes of leadership (see Figure 7.21B).

Gender and Democratization

> **Geographic Insight 4**
> **Gender, Power, and Politics:** The education of African women and their economic and political empowerment are now recognized as essential to development. As girls' education levels rise, population growth rates fall; as the number of female political leaders increases, so does their influence on policies designed to reduce poverty.

One major characteristic of the democratization process has been the increase in the number of women across Africa who are assuming positions of political influence. This chapter opened with the story of Juliana Rotich, a businesswoman in Kenya, but her road to success was built on policy changes pushed by a number of female Kenyan political figures over the last decade. In Nigeria, Ngozi Okonjo-Iweala, the first female finance minister, now fights for the empowerment of women entrepreneurs as a director at the World Bank. In Liberia—the country devastated by the logging fraud, child-soldier recruitment, and general brutality of the dictator Charles Taylor (see the vignette on Silas Siakor on page 287)—President Ellen Johnson-Sirleaf (also a former World Bank economist) began to work on democratic and environmental reforms immediately after she took office (see Figure 7.21B). In Rwanda, racked by genocide and mass rapes in the 1990s, women now make up 51 percent of the national legislature, the highest percentage in the world. And Rwandan women are also leaders at the local level, where they make up 40 percent of the mayors. In Mozambique, 39.2 percent of the parliament is female; in Burundi, 36 percent; and in South Africa, 42.7 percent. All together, there are 17 African countries where the percentage of women in national legislatures is above the world average of 17.7 percent (the U.S. figure is 16.8 percent).

170. WOMEN HAVE STRONG VOICE IN RWANDAN PARLIAMENT

There has been a sea change in attitudes toward women in politics. The policies that establish quotas for the number of female members of parliament are usually a reflection of a larger, post-conflict commitment to the empowerment of women and to involving women in all aspects of political and civil society. Sometimes the quotas are written into a country's constitution, but it is important to note that many female leaders in Africa, including Ellen Johnson-Sirleaf and at least half of Rwanda's female parliamentarians, were elected without the aid of quotas.

THINGS TO REMEMBER

Geographic Insight 3 • **Power and Politics** After years of rule by corrupt elites and the military, signs of a shift toward democracy and free elections are now visible across Africa. Even when democratic reforms are enacted and free elections established, violence still accompanies some elections.

• Africa's democratic and economic progress has been held back by frequent civil wars that are in many ways the legacy of colonial-era divide-and-rule policies.

• Democratization is helping reduce the potential for civil conflict, and the countries with the fairest elections have had noticeably fewer violent conflicts.

• The trend toward democracy in sub-Saharan Africa tends to lead to governments that are more responsive to the needs of their people, but democratic reforms are not necessarily durable.

Geographic Insight 4 • **Gender, Power, and Politics** One major characteristic of the democratization process has been the increase in the number of women across Africa who are coming into positions of power, both political and economic.

Sociocultural Issues

To the casual observer, it may appear that a majority of sub-Saharan Africans live traditional lives in rural villages. It is true that with just 37 percent of its population in urban areas, Africa is the world's most rural region. A closer look, however, reveals that migration and urbanization are occurring so rapidly that the impact on African life is monumental.

Population Patterns: Growth, Density, and the Demographic Transition

> **Geographic Insight 5**
>
> **Population, Urbanization, Food, and Water:** In part because of lower infant mortality rates and the higher cost of rearing children in cities, urbanization has the positive effect of encouraging families to limit births. Nevertheless, urbanization often happens in an uncontrolled fashion, leaving many people to reside in impoverished slums without adequate services, sufficient food, and clean water.

In fewer than 50 years, sub-Saharan Africa's population has more than quadrupled, growing from about 200 million in 1960 to 828 million in 2009, to slightly under 1 billion in 2013. By 2050, the population of this region could reach just over 2 billion. Nonetheless, contrary to the perception outsiders often hold, Africa as a whole is much less densely populated than most of Europe and Asia—36 people per square kilometer, compared to the global average of 51 people per square kilometer. If fertility rates remain high, though, places that are relatively uncrowded now may change dramatically over the next few decades (Figure 7.22; see also Figure 1.8 on page 14). In some rural areas, the population density may far exceed the carrying capacity of the land, which will lead to the impoverishment of subsistence farmers. Increasingly, however, Africa's population is becoming concentrated in a few urban areas, such as the cities along the Gulf of Guinea coast in West Africa.

Right now, sub-Saharan Africa's annual rate of natural increase (2.6 percent) is the highest in the world; even so, overall rates are slowing in nearly every country, especially in urban areas. Nonetheless, people continue to have more children than would be necessary to simply maintain population numbers. Birth rates are as high as they are because many Africans view children as both an economic advantage and a spiritual link between the past and the future. Childlessness is considered a tragedy, as children ensure a family's genetic and spiritual survival.

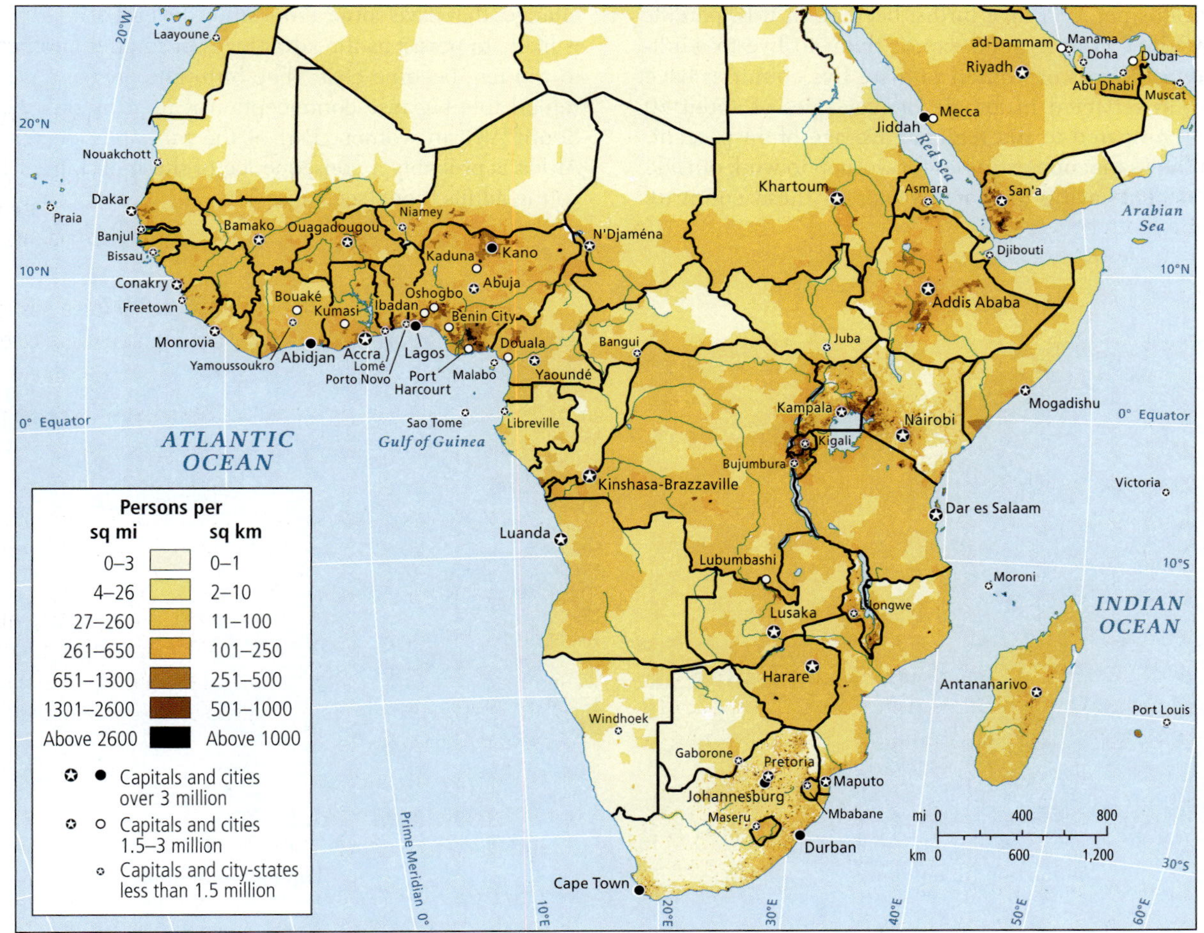

FIGURE 7.22 Population density in sub-Saharan Africa. Of all the world regions, populations are growing fastest in sub-Saharan Africa. The demographic transition is taking hold in a few of the more prosperous countries. In these countries, better health care and more economic and educational opportunities for women are enabling women to pursue careers and to choose to have fewer children.

Large families are also viewed as having economic value, as children and young adults still perform important work on family farms and in family-scale industries. Moreover, in this region of generally poor health care and the resultant high incidence of infant mortality, parents may have extra children in the hope of raising a few to maturity. In all but a few countries, the *demographic transition*—the sharp decline in births and deaths that usually accompanies economic development (see Figure 1.10 on page 16)—has barely begun, and more people now survive long enough to reproduce than did in the past. If, however, Africa is on the brink of a development surge, fertility rates could decline quite sharply as women choose roles other than motherhood.

At least five countries in the region have gone through the demographic transition. In South Africa, Botswana, Seychelles, Réunion, and Mauritius (the last three being small island countries off Africa's east coast), circumstances have changed sufficiently to make smaller families desirable and attainable. In all five countries, per capita incomes are five to ten times the sub-Saharan average of U.S.$2000. Advances in health care have cut the infant mortality rate to less than half the regional average of 88 infant deaths per 1000 live births; because of this, parents can have only a few children and expect them to live to adulthood. The circumstances of women in these five countries have also improved, as reflected in female literacy rates of about 80 to 90 percent, compared to the regional average of 54 percent. Research also shows that opportunities for women to work outside the home at decent-paying jobs are greater in these countries than in the rest of the region. Thus, many women are choosing to use contraception because they have life options beyond motherhood. Indeed, the percent of married women using contraception in these five countries is double or even triple the rate for sub-Saharan Africa as a whole, which is only 22 percent (the world average is 61 percent).

The population pyramids shown in Figure 7.23 demonstrate the contrast between countries that are growing rapidly, such as Nigeria, Africa's most populous country (with 170.1 million people and a natural increase rate of 2.6 percent) and countries that have already gone through the demographic transition, such as South Africa (with 51.1 million people and a natural increase rate of 0.6 percent). Nigeria's pyramid is very wide at the bottom because over half the population is under the age of 20. In just 15 years, this entire cohort will be of reproductive age. Only 15 percent of Nigerian women use any sort of birth control.

In contrast to Nigeria, South Africa's pyramid has contracted at the bottom because its birth rate has dropped from 35 per 1000 to 21 per 1000 over the last 20 years. This decrease is primarily an effect of economic and educational improvements as well as social changes that have come about since the end of apartheid (1994). It is likely to persist as the advantages of smaller families, especially to women, become clear. The birth rate decrease is all the more remarkable because contraception is used by only 60 percent of South African women. Part of the birth rate decrease in South Africa is probably a consequence of the high rate of HIV among young adults there (see pages 316–317).

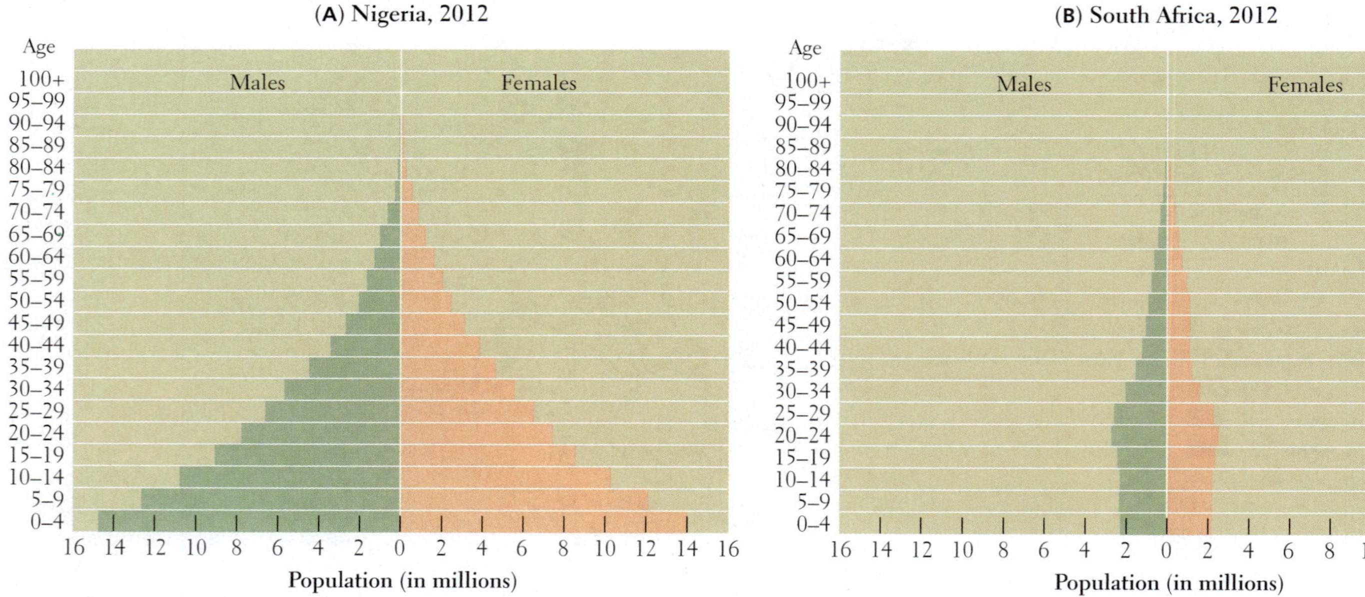

FIGURE 7.23 Population pyramids for Nigeria and South Africa. Note that the two pyramids are at the same scale. Nigeria had a population of 170.1 million people in 2012 and a growth rate of 2.6%, as indicated by its very wide pyramid base. It has the largest population on the continent, nearly twice that of Ethiopia, the second largest in terms of population. South Africa, on the other hand, has a population of 51.1 million and a growth rate of only 0.9%; its population is beginning to decline, as shown by the shrinking of the pyramid base. The decline is in part due to the AIDS epidemic.

ON THE BRIGHT SIDE

Urban Food Gardens

A 2012 UN Food and Agricultural Organization (FAO) report, "The Greening of African Cities," emphasizes that the growing of produce in and around cities, on even the tiniest scraps of land, has an advantage over both rural market gardening and imported food in supplying urban folks with safe, nutritious food. Fruits and vegetables can be highly perishable, so if urban consumers can produce their own food, they are likely to eat better, to create less waste, and to have lower food transport costs. Also, food waste can be immediately recycled as compost, reducing or eliminating the need for fertilizers; the greenbelts created by urban gardens can reduce urban air and water pollution.

Urbanization and Population Growth Sub-Saharan Africa is undergoing a massive shift in population from tiny rural settlements of a few houses to substandard residential facilities in dense urban areas. This has major implications for population growth. On average, rural sub-Saharan African women give birth to about six children, while urban African women give birth to four. Urban life strengthens all the factors that influence the demographic transition—increased education, economic opportunities, and better health care. While sub-Saharan Africa's urban fertility rate of four children per woman is still almost double the world average, the demographic transition has only just begun, and indications are that urban sub-Saharan Africa's birth rates will continue to decline.

Part of the reason that urban fertility rates are still as high as they are is that compared to cities in other regions, sub-Saharan Africa's cities have relatively low living standards, especially with regard to access to clean water and sanitation (**Figure 7.24A, C** on page 314). Although improved over rural conditions, poverty and disease are still relatively widespread in cities, and because of the low access to birth control and the low levels of education and economic opportunities for women (when compared to global levels), urban families continue to have more than two children.

Country-to-Country and Rural-to-Urban Migration The migration of sub-Saharan people looking for work in North Africa, Europe, Turkey, or the United States is now a familiar phenomenon, but the vast majority of the African migrants who are seeking work or escaping violence move from country to country within the continent. In 2005, the United Nations estimated that there were 17 million such internal migrants, most going to West Africa and Southern Africa, because jobs are more numerous in both areas. Country-to-country migrants often go to rural areas as agricultural laborers, but increasingly more educated migrants seek out cities. The five sub-Saharan African countries with the largest number of immigrants in 2005 were Côte d'Ivoire (2.4 million), Ghana (1.7 million), South Africa (1.1 million), Nigeria (1.0 million), and Tanzania (0.8 million). Together they have 40 percent of the migrants in Africa. Even though the migrants typically make few demands, the impact on receiving countries that are unable to provide adequately for their own citizens is substantial. Like migrants who leave the continent, internal migrants live frugally and send much of their earnings home to families.

Although urban birth rates remain relatively high, migration from rural areas to cities is actually the biggest factor behind sub-Saharan cities' growth rate of 5 percent per year, the highest rate in the world. In the 1960s, only 15 percent of sub-Saharan Africans lived in cities; now about 37 percent do (see the Figure 7.24 map). Estimates are that by 2030, half of all Africans will be urban dwellers. In 1960, just one sub-Saharan African city—Johannesburg, South Africa—had more than 1 million people; in 2009, fifty-two did. The largest sub-Saharan African city is Lagos, Nigeria, where various estimates put the population at between 11 and 13 million; by 2020, Lagos is projected to have 20 million people (see the Figure 7.24 map). Much of this growth is taking place in primate cities (see Chapter 3, page 145). For example, Kampala, Uganda, with 1.8 million people, is almost 10 times the size of Uganda's next largest city, Gulu.

Because governments and private investors have paid little attention to the need for affordable housing, most migrants have to construct their own dwellings using found materials. The vast, unplanned, one-story slums surrounding older urban centers house 72 percent of Africa's urban population. Transportation in these huge and shapeless settlements is a jumble of government buses and private vehicles, and in waterfront locales, boats (see Figure 7.24A). Parents often have to travel long hours through extremely congested traffic to reach distant jobs, getting much of their sleep while sitting on a crowded bus and leaving their children unsupervised and susceptible to the influence of gangs (see Figure 7.24B).

Food Security in Urban Areas Food supplies for urban residents can be brought in from the countryside, or be imported from distant lands, or, as is increasingly common, be grown in cities by urban residents. In some cities, residents have taken the initiative to produce their own food in the tiny spaces between houses (see Figure 7.24D) and in the wastelands that surround urban shantytowns.

If urban gardening is to become a viable food security solution, some regulation will be necessary. At present, urban farmers may be using *brownfields*—land where the soil is contaminated by past industrial activities and the wastewater used to irrigate may not be safe—but this can be corrected. Rainwater can be harvested or treated wastewater can be used to safely supply needed nutrients to the plants; raised cultivation beds can be sealed off from contaminated soil. Also, because few urban gardeners have title to the land they cultivate, urban garden land must not be subject to confiscation for development without properly compensating the gardeners and alloting them replacement land. Finally, the use, and especially overuse, of fertilizers must be controlled; this can be done through educating farmers about them. Adult education courses reminiscent of those taught to home gardeners in the United States during World War II could revolutionize food production in African cities.

Figure 7.24 Photo Essay: Urbanization in Sub-Saharan Africa

Urban populations are exploding because of migration from rural areas and relatively high birth rates in cities. By 2030, half of the people in the region will live in cities, most in slums plagued by violence and inadequate access to water, sanitation, and education. Largely ignored by most governments, people in slums survive by helping themselves.

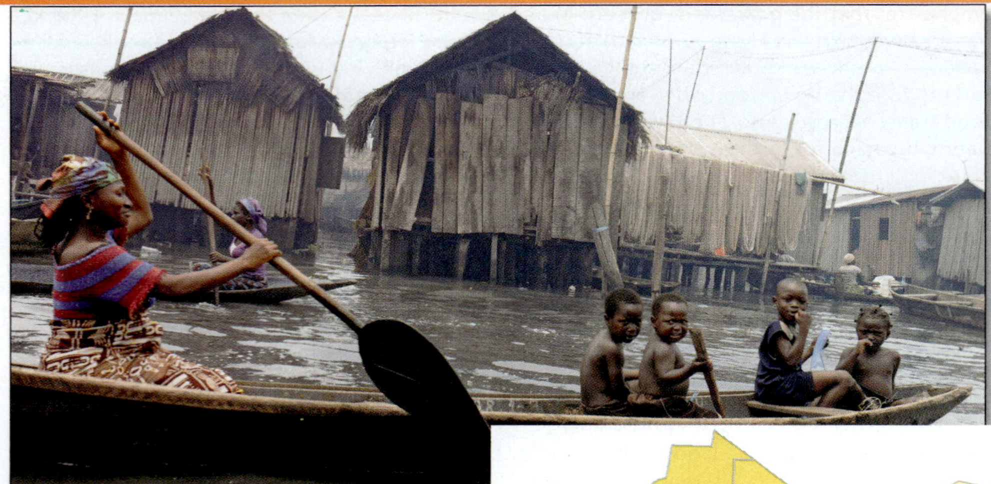

A A part of Makoko slum in Lagos, Nigeria, the largest city in this region and one of the fastest growing in the world. Makoko is home to over 100,000 people. Waterborne diseases such as malaria are common, as is flooding, which kills and displaces residents on a regular basis. Because of these hazards and the fact that the community sits on potentially valuable waterfront property, the municipal government of Lagos periodically evicts some of Makoko's residents and demolishes their homes.

B Opposing gangs battle for control of territory in Kinshasa, Congo. Gangs control most slum neighborhoods in this region. Violence is widespread and includes organized attacks and robberies of passersby, the police, and other gangs.

C A girl drinks from a public water tap in Soweto, South Africa. Slum houses rarely have indoor plumbing, and people obtain water from public spigots. Latrines provide the only sanitation.

CAPE VERDE
MAURITANIA
MALI
NIGER
CHAD
ERITREA
DJIBOUTI
Dakar
SENEGAL
THE GAMBIA
GUINEA-BISSAU
GUINEA
SIERRA LEONE
LIBERIA
Bamako
BURKINA FASO
BENIN
80 Kano
NIGERIA
Ibadan
Benin City
CÔTE D'IVOIRE
GHANA
43 Abidjan
86 Accra
28 Lagos
A
TOGO
CAMEROON
Yaoundé
EQUATORIA GUINEA
SÃO TOMÉ & PRÍNCIPE
GABON
CONGO
CENTRAL AFRICAN REPUBLIC
SOUTH SUDAN
Addis Ababa
123
ETHIOPIA
SOMALIA
KENYA
Nairobi
89 D
RWANDA
UGANDA
DEMOCRATIC REPUBLIC OF THE CONGO
BURUNDI
25 Kinshasa-Brazzaville
B
TANZANIA
119 Dar es Salaam
SEYCHELLES
71 Luanda
ANGOLA
ZAMBIA
MALAWI
COMOROS
121 Lusaka
Harare 118
ZIMBABWE
MOZAMBIQUE
MADAGASCAR
MAURITIUS
RÉUNION (Fr.)
NAMIBIA
BOTSWANA
C
Pretoria
Johannesburg 32
SWAZILAND
SOUTH AFRICA
120 Durban
LESOTHO
84 Cape Town
mi 0 400 800
km 0 600 1,200

Population Living in Urban Areas

83%–100%		29%–46%	
65%–82%		11%–28%	
47%–64%		No data	

Population of Metropolitan Areas 2013

20 million
10 million
5 million
3 million

Note: Symbols on map are sized proportionally to metro area population

① **Global rank** (population 2013)

D A "sack garden," one of many innovative self-help projects found in Kibera slum in Nairobi, Kenya. High food costs and high unemployment rates make gardening attractive to many urban Africans.

Thinking Geographically

After you have read about urbanization in sub-Saharan Africa, you will be able to answer the following questions about urbanization in sub-Saharan Africa:

A What do the types of houses, the water, and the boats suggest about the location of this slum?

B What are some factors that contribute to the prevalence of gangs in African slums?

C How might the lack of clean water and sanitation create an incentive for people to have large families?

D How does this project in Nairobi embody the idea of "self-reliant development"?

The Scarcity of Clean Water in Sub-Saharan Cities Public health is a major urban concern, as many water distribution systems are susceptible to being contaminated by harmful bacteria from untreated sewage (see Figure 7.24C). Only the largest sub-Saharan African cities have sewage treatment plants, and few of these extend to the slums that surround them. The result is frequent outbreaks of waterborne diseases such as cholera, dysentery, and typhoid. It is estimated that unsafe water and sanitation facilities result in an annual loss of $28.4 billion in Africa alone.

Although the safety of city water supplies is improving, there are frequent periodic water shortages in as many as 11 out of 14 sub-Saharan cities. To control usage during a crisis, city officials limit service to a few hours a day, causing people to store water in vessels that may become contaminated. Recent reassessments of the availability of groundwater resources may eventually alleviate these urban water shortages, but recall that the natural distribution of groundwater does not easily match up with the location of the big cities (see pages 290–292; see also Figure 7.9 on page 292).

Population and Public Health Infectious diseases, including HIV-AIDS, are by far the largest killers in sub-Saharan Africa, responsible for about 50 percent of all deaths. Some diseases are linked to particular ecological zones. For example, people living between the 15th parallels north and south of the equator are most likely to be exposed to sleeping sickness (trypanosomiasis), which is spread among people and cattle by the bites of tsetse flies. The disease attacks the central nervous system and, if untreated, results in death. Several hundred thousand Africans suffer from sleeping sickness, and most of them are not treated because they cannot afford the expensive drug therapy.

Africa's most common chronic tropical diseases, schistosomiasis and malaria, are linked to standing fresh water. Thus, their incidence has increased with the construction of dams and irrigation projects. Schistosomiasis is a debilitating, though rarely fatal, disease that affects about 170 million sub-Saharan Africans. It develops when a parasite carried by a particular freshwater snail enters the skin of a person standing in water. Malaria, spread by the anopheles mosquito (which lays its eggs in standing water), is more deadly. The disease kills at least 1 million sub-Saharan Africans annually, most of them children under the age of 5. Malaria also infects millions of adult Africans who are left feverish, lethargic, and unable to work efficiently because of the disease.

Until recently, relatively little funding was devoted to controlling the most common chronic tropical diseases. Now, however, major international donors are funding research in Africa and elsewhere. More than 60 research groups in Africa are

ON THE BRIGHT SIDE

The TivaWater Jug

The problem of unsafe water for households may have a solution that is simple, affordable, and may create local jobs. A for-profit group, working with grassroots organizations in Uganda but subsidized by American investors, has designed a simple plastic jug, called the TivaWater jug, with an interior sand and clay filtering device that purifies water (Figure 7.25). The jug is made to be affordable to urban poor people in Ugandan shantytowns. In line with modern self-sufficiency development theory for Africa, the jug, while initially made in the United States, is now manufactured in Uganda, is sold and distributed by small businesses there, and will eventually be exported by Ugandans to other African countries.

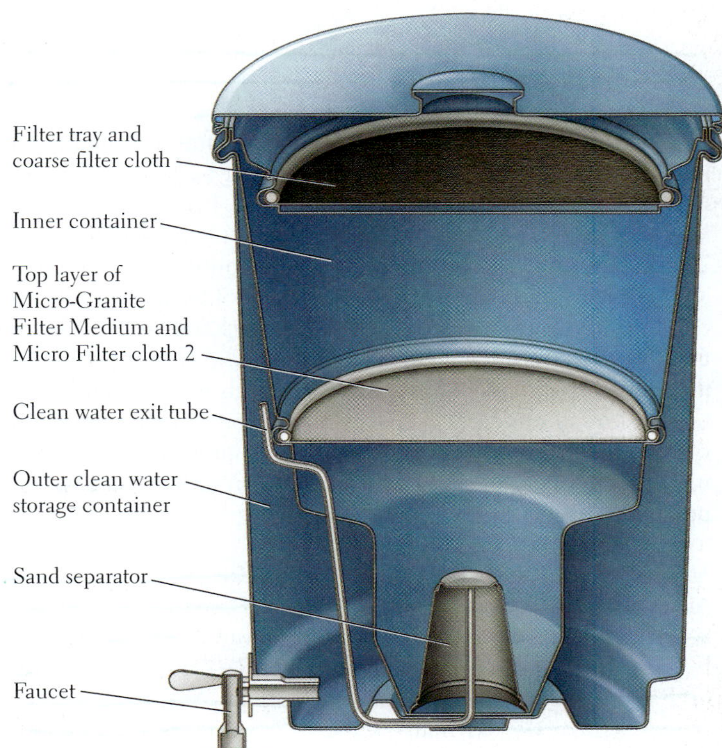

FIGURE 7.25 The TivaWater jug. This specially designed water-filtering jug, made of unbreakable food-grade plastic, removes dirt, bacteria, and viruses, using a new version of biosand technology. It is affordable through subsidies to the very poor, is easy to maintain, and has a faucet that dispenses the water.

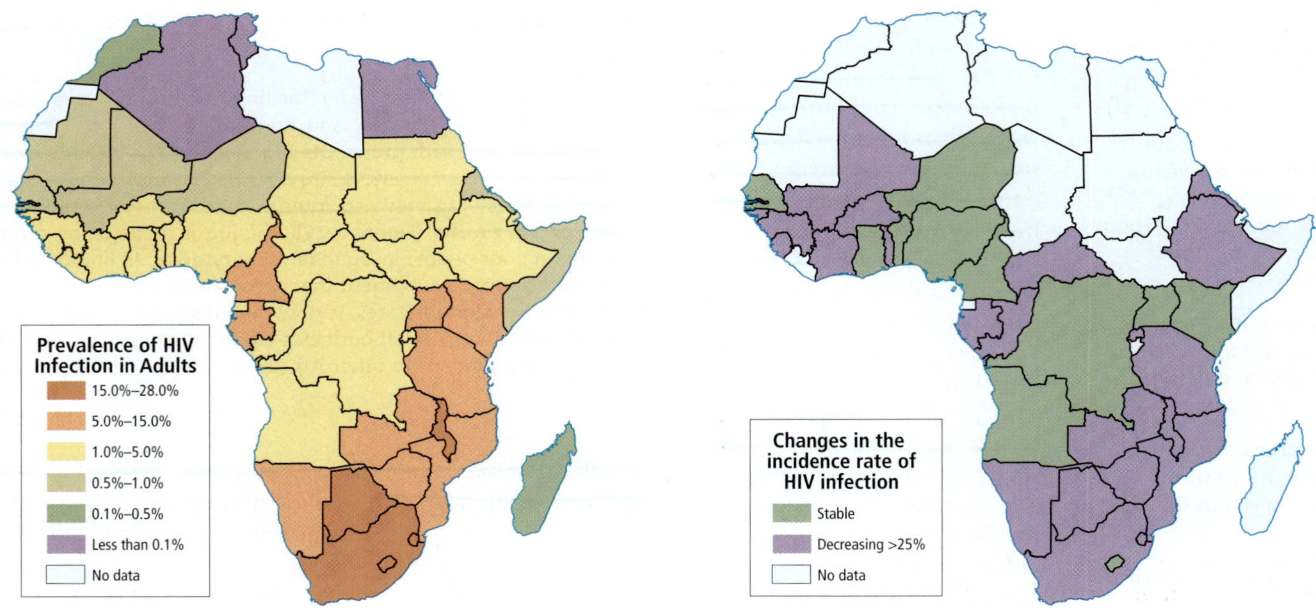

FIGURE 7.26 Global prevalence of HIV-AIDS (2009) and the incidence rate of infection (2001–2009). New HIV infections fell in sub-Saharan Africa from 2.2 million in 2001 to 1.8 million in 2009. In the same period, rates declined more than 25 percent in 22 of the countries. The rate has remained stable elsewhere in the region. The number of deaths from AIDS has declined in sub-Saharan Africa, in large measure because of the availability of antiretroviral medications.

working on a vaccine that will prevent malaria in most people. The distribution of simple, low-cost, and effective mosquito nets is also reducing the transmission rates of malaria.

Polio is not a tropical disease, but it remains a challenge in many sub-Saharan countries, despite having been controlled in Europe and the Americas some decades ago. Part of the problem is a lack of science education. For example, in Nigeria in the early 2000s, Muslim clerics in the north of the country preached that a polio vaccination project was a Western plot to sterilize African children. The incidence of the disease shot up and, in 2012, more than 90 cases were reported. Now, after public education efforts, the clerics have relented and instead preach that vaccination is required of all children. A successful polio eradication program must include public education, testing, improved sewage treatment, cleaning up of standing water, and an orderly vaccination program.

HIV-AIDS in Sub-Saharan Africa A leading cause of death in Africa, and the leading cause of death for women of reproductive age, is acquired immunodeficiency syndrome (AIDS), caused by the human immunodeficiency virus (HIV) (Figure 7.26). As of 2010, more than 16 million sub-Saharan Africans had died of the disease, and as many as 80 million more AIDS-related deaths are expected by 2025. In 2010, sub-Saharan Africa had two-thirds of the estimated worldwide total of 33 million people living with HIV. In Botswana, one of the richest countries in Southern Africa, nearly 29.2 percent of the adult female population is infected. The Southern Africa region alone accounted for 31 percent of global AIDS deaths in 2010. While the epidemic is subsiding a bit, it has significantly constrained economic development.

Worldwide, women account for half of all people living with HIV, but in Africa, HIV-AIDS affects women more than men: 59 percent of the region's HIV-infected adults are women. The reasons for this pattern are related to the social status of women in sub-Saharan Africa and elsewhere.

The rapid urbanization of Africa has hastened the spread of HIV, which is much more prevalent in urban areas. It is not uncommon for poor, new urban migrants, removed from their village support systems, to become involved in the sex industry in order to survive economically. In some cities, virtually all sex workers are infected. Meanwhile, many urban men, especially those with families back in rural villages, visit prostitutes on a regular basis. These men often bring HIV back to their rural homes. As transportation between cities and the countryside has improved, bus and truck drivers have also become major carriers of HIV to rural villages.

A number of myths and social taboos surround HIV-AIDS and exacerbate the problem of controlling the spread of the disease, making education a key component in combating the epidemic. For example, some men think that only sex with a mature woman can result in infection, so very young girls are increasingly sought as sex partners (sometimes referred to as "the virgin 'cure'"). Elsewhere, infection is considered such a disgrace that even those who are severely ill refuse to get tested, yet they remain sexually active.

The Social Costs of HIV-AIDS Across the continent, the consequences of the HIV-AIDS epidemic are enormous. Children contract HIV primarily in utero or from nursing; just over 400 children under the age of 15 die each day from AIDS-related causes. Millions of parents, teachers, skilled farmers, craftspeople, and trained professionals have been lost. More than 15 million children have been orphaned; many have no family left to care for them or to pass on vital knowledge and life skills. The disease has severely strained the health-care systems of most countries. Demand for treatment is exploding, drugs are prohibitively expensive, and many health-care workers themselves are infected. Because so many young people are dying of AIDS, decades of progress in improving the life expectancy of Africans have been erased. For example, in 1990, adult life expectancy in South Africa was 63 years, but in 2012 it was 54 years.

HIV-AIDS Education and Medication The most effective means of prevention are the massive education programs that lower rates of infection among those who can read and understand explanations about how HIV is spread. For example, Senegal started HIV-AIDS education in the 1980s, and levels of infection there have so far remained low. Major education campaigns in Uganda lowered the incidence of new HIV infections from 15 percent to just 4.1 percent between 1990 and 2004. By contrast, infection rates have soared in areas such as South Africa, where a few top politicians have put forth untenable theories about the causes of HIV infection or have denied that HIV-AIDS is a problem.

HIV-AIDS medications (known as antiretroviral therapy) are prohibitively expensive for families in sub-Saharan Africa. The United Nations and a number of private philanthropies cover some of the costs. The medications are not a cure and must be taken daily for the rest of one's life, but they do make it possible to live with the disease. Pharmaceutical firms in India and China and elsewhere have come out with generic antiretroviral versions that have reduced the cost to about $350 per year, about one-sixth of the average annual income in the region. About five out of every six recipients worldwide are in sub-Saharan Africa, and most of those are in Southern Africa. As yet, only half of those who need the medicines are receiving them; the goal is for all of those with HIV-AIDS to be covered by 2015. The cost to the world of keeping parents alive with antiretroviral therapy is far less than dealing with the social effects of millions of orphans.

THINGS TO REMEMBER

- The overall low population density figures in sub-Saharan Africa—36 people per square kilometer, compared to the global average of 51 people per square kilometer—can be misleading. Densities are extremely high in some places and very low in less habitable areas.
- Of all the world regions, populations are growing fastest in sub-Saharan Africa. The demographic transition is taking hold in a few of the more prosperous countries where better health care and lower levels of child mortality, along with more economic and educational opportunities for women, are enabling women to pursue careers and to choose to have fewer children.

Geographic Insight 5

- **Population, Urbanization, Food, and Water** While urbanization helps reduce the birth rate, too often it leaves many people to reside in slums in poverty and without proper food, water, and adequate services. Largely ignored by most governments, people in slums survive by providing for themselves, as exemplified by the emerging urban garden movement.
- Sub-Saharan Africa has higher rates of infectious diseases than does any other region in the world, with the world's worst epidemics of malaria, polio, schistosomiasis, sleeping sickness, and HIV-AIDS.

Gender Issues

Long-standing African traditions dictate a fairly strict division of labor and responsibilities between men and women. In general, women are responsible for domestic activities, including raising the children, attending to the sick and elderly, and maintaining the house. Women collect water and firewood and prepare nearly all the food. Men are usually responsible for preparing land for cultivation. In the fields intended to produce food for family use, women sow, weed, and tend the crops as well as process them for storage. In the fields where cash crops are grown, men perform most of the work and retain control of the money earned. When husbands in search of cash income migrate to work in mines or in urban jobs, women take over nearly all of the agricultural work, often using simple hand tools, not machines or even draft animals. When there are small agricultural surpluses or handcrafted items to trade, it is women who transport and sell them in the market. Throughout Africa, married couples often keep separate accounts and manage their earnings as individuals, but when a wife sells her husband's produce at the market, she usually gives the proceeds to him.

Ideas in Africa about how males and females should relate come from a variety of sources. Even in the ancient past, social controls tempered gender relationships. Most marriages were social alliances between families; therefore, husbands and wives spent most of their time doing their tasks with family members of their own sex rather than with each other. Then as today, women primarily influenced other women and men influenced men. It is important to note that having multiple wives—the practice of **polygyny** (Figure 7.27)—is more common in sub-Saharan Africa (where it has ancient pre-Muslim, pre-Christian roots) than in Muslim North Africa (as discussed in Chapter 6 on page 256). Economist James Fenske, in a 2012 study of the historical causes, distribution, and persistence of polygyny in Africa, produced the map in Figure 7.27. He noted that while polygyny is

polygyny the practice of having multiple wives

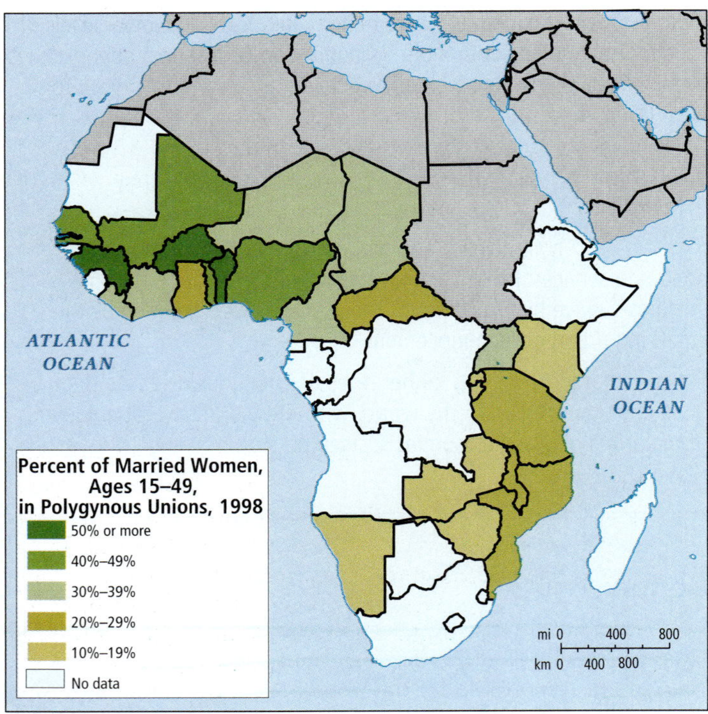

FIGURE 7.27 Polygyny in Africa. The map depicts the spatial distribution of polygyny in sub-Saharan Africa. Seventy-two percent of the women indicated they are the only wife but that their husbands have relations with other women; 19 percent are in two-wife marriages; 7 percent in three-wife marriages; and the remaining 2 percent are in marriages with four or more wives.

on a marked decline, the reasons for this are unclear. Current improvements in women's education don't seem to reduce its incidence, but economic development and declining infant mortality do.

Women face significant sexual and physical abuse in the home and also during civil unrest, when rape and beatings are more common. Many of the female politicians who have come to power in the last decade are now addressing these issues. Most have emphasized the need for more women to be in the position to influence and enforce policies to reduce violence against women. In Liberia, for instance, President Ellen Johnson-Sirleaf has defined women's rights as a national security issue. Liberia has recruited women to serve in both the police and the armed forces.

A practice known as **female genital mutilation (FGM)** (formerly called female circumcision) has been documented in 27 countries in the central portion of the African continent (Figure 7.28). It also is known in Yemen, India, Indonesia, Iraq, Israel, Malaysia, the United Arab Emirates and, in some cases, in the United States, Canada, Australia, and New Zealand. The practice predates Islam and Christianity, and today is performed among all social classes and in Christian, Muslim, and animist (see the definition below) religious traditions. In the procedure, which is usually performed without anesthesia, a young girl's entire clitoris and parts of her labia are removed. In the most extreme cases (called infibulation), her vulva is stitched nearly shut. This mutilation far exceeds that of male circumcision, eliminating any possibility of sexual stimulation for the female and making urination and menstruation difficult. Intercourse is painful and childbirth is particularly devastating because the flesh scarred by the mutilation is inelastic. A 2006 medical study conducted with the help of 28,000 women in six African countries showed that women who had undergone FGM were 50 percent more likely to die during childbirth, and their babies were at similarly high risk. The practice also leaves women exceptionally susceptible to infection, especially HIV infection. **171. FEMALE GENITAL MUTILATION STILL COMMON IN SOMALILAND**

female genital mutilation (FGM) removing the labia and the clitoris and sometimes stitching the vulva nearly shut

The practice is probably intended to ensure that a female is a virgin at marriage and that she thereafter has a low interest in intercourse other than for procreation. While in decline today, FGM is still widespread among some groups. Among the Kikuyu of Kenya, 40 years ago, nearly all females would have had FGM, but today only about 40 percent do. Kikuyu women's rights leaders have had some success in curbing FGM by replacing it with right-of-passage ceremonies that joyously mark a girl's transition to puberty.

Many African and world leaders have concluded that the practice is an extreme human rights abuse, and it is now against the law in 16 countries. The most successful eradication campaigns are those that emphasize the threat FGM poses

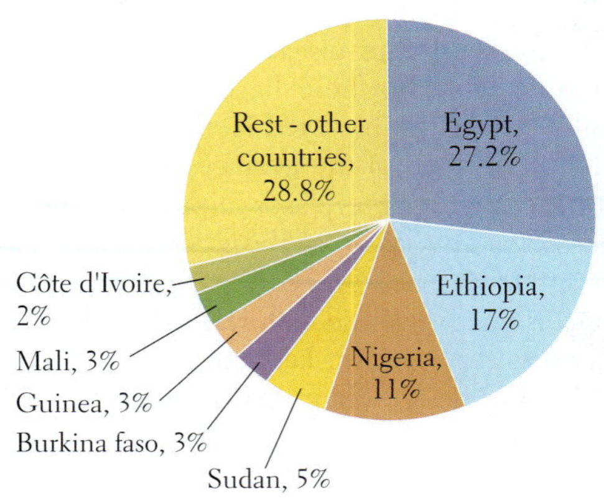

FIGURE 7.28 Where girls and women with female genital mutilation live in Africa. The World Health Organization estimates that there are 91.5 million girls and women living in Africa today who have undergone some form of FGM. Of those, 12.4 million are girls between 10 and 14 years of age. Close to half of them are in two countries: Egypt and Ethiopia. Incidences of FGM have been documented in 27 countries in Africa and in 8 other countries. There have also been some cases documented in Europe, North America, Australia, and New Zealand, which are probably due to migration.

to a woman's health, thus making it socially acceptable to not undergo this ritual. The World Health Organization in 2008 took a strong stand against FGM, saying:

Female genital mutilation has been recognized as discrimination based on sex because it is rooted in gender inequalities and power imbalances between men and women and inhibits women's full and equal enjoyment of their human rights. It is a form of violence against girls and women, with physical and psychological consequences.

WORLD HEALTH ORGANIZATION, *Eliminating Female Genital Mutilation—An Interagency Statement*
(GENEVA: WORLD HEALTH ORGANIZATION PRESS, 2008), P. 10.

Religion

Africa's rich and complex religious traditions derive from three main sources: indigenous African belief systems (many of them animist), Islam, and Christianity.

Indigenous Belief Systems Traditional African religions, found in every part of the continent, are probably the most ancient on Earth, since this is where human beings first evolved. Figure 7.29 highlights the countries in which traditional beliefs remain particularly strong. Traditional beliefs and rituals often seek to bring departed ancestors into contact with living people, who in turn are the connecting links in a timeless spiritual community that stretches into the future. The future is reached only if present family members procreate and perpetuate the family heritage through storytelling.

Most traditional African beliefs can be considered **animist** in that spirits, including those of the deceased, are thought to exist everywhere—in trees, streams, hills, and art, for example. In return for respect (expressed through ritual), these spirits offer protection from sickness, accidents, and the ill will of others. African religions tend to be fluid and adaptable to changing circumstances. For example, in West Africa, Osun, the god of water, traditionally credited with healing powers, is now also invoked for those suffering economic woes.

Religious beliefs in Africa, as elsewhere, continually evolve as new influences are encountered. If Africans convert to Islam or Christianity, they commonly retain aspects of their indigenous religious heritage. The three maps in Figure 7.29 show a spatial overlap of belief systems, but they do not convey the philosophical blending of two or more faiths, which is widespread. In the Americas, the African diaspora has influenced the creation of new belief systems developed from the fusion of Roman Catholicism and African beliefs. Voodoo in Haiti, Santería in Cuba, and Candomblé in Brazil are examples of this fusion.

Islam and Christianity Islam began to extend south of the Sahara in the first centuries after Muhammad's death in 632 C.E., and today, about one-third of sub-

animism a belief system in which spirits, including those of the deceased, are thought to exist everywhere and to offer protection to those who pay their respects

FIGURE 7.29 Religions in Africa. Notice that the various religions in Africa overlap in distribution; in many countries, one is dominant but others are present.

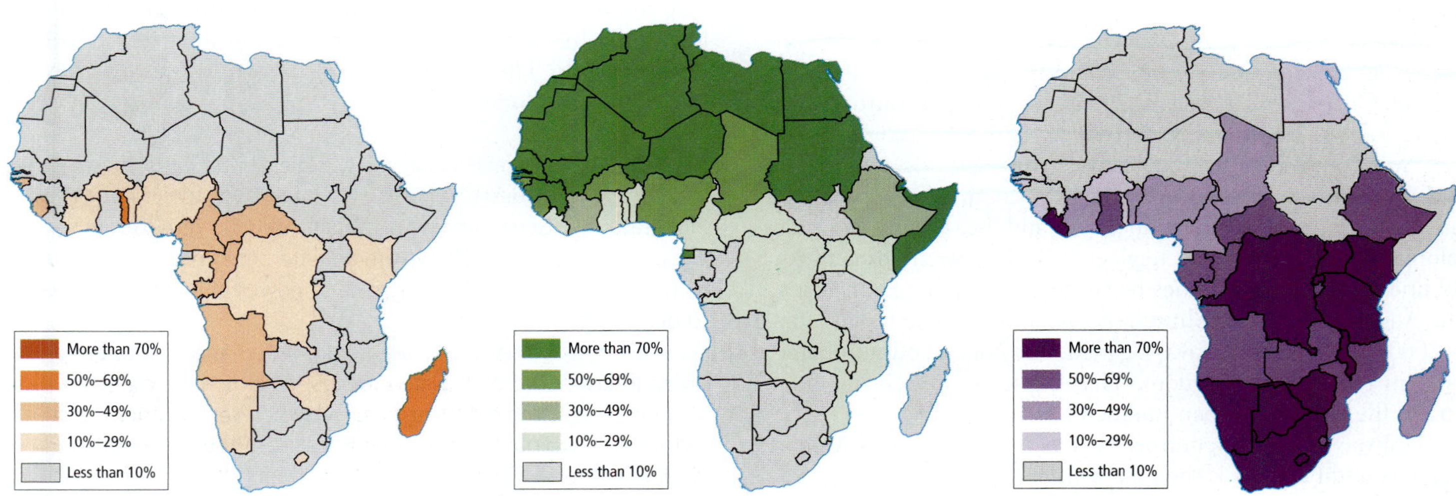

(A) Distribution of traditional religions.

(B) Distribution of Islam.

(C) Distribution of Christianity (Judaism is diminishing).

FIGURE 7.30 LOCAL LIVES FESTIVALS IN SUB-SAHARAN AFRICA

A During Timkat, a festival that marks the baptism of Jesus Christ, Ethiopian Orthodox Christian priests in Addis Ababa carry replicas of the Ten Commandments, wrapped in cloth, on their heads.

B South African Muslims spot the new moon that marks the end of Ramadan, a month of fasting during which observant Muslims do not eat, drink, quarrel, deliberately mislead people with language, or have sex between sunrise and sunset. Throughout Africa, many Muslims mark the end of Ramadan with prayers for peace and unity as well as drumming, singing, dancing, and general celebration.

C The 8-day festival of the Umhangla ("Reed Dance" ceremony) protects the virginity of young unmarried women who gather in the king of Swaziland's village and repair the surrounding fence with tall reeds. Later, in costumes, the women dance for the royal family, foreign dignitaries, tourists, and other spectators. Traditionally, the king also chooses a new wife. The women who wear red feathers in their hair are the daughters of the king.

Saharan Africans are Muslim (see Figure 7.29B). Islam is now the predominant religion throughout the Sahel and parts of East Africa, where Muslim traders from North Africa and Southwest Asia brought the religion. Powerful Islamic empires have arisen here since the ninth century. The latest of these empires challenged European domination of the region in the late nineteenth century.

Today about half of Africans are Christian. Christianity first came to the region via Egypt and Ethiopia shortly after the time of Christ, well before it spread throughout Europe (see Figure 7.29C). Christianity did not come to the rest of Africa until nineteenth-century missionaries from Europe and North America became active along the west coast. Many Christian missionaries provided the education and health services that colonial administrators had neglected. In the twentieth century, old-line established churches began to gain adherents in Africa. The Anglican Church (Church of England) grew so rapidly that by 2000, Anglicans in Kenya, Uganda, and Nigeria outnumbered those in the United Kingdom. The Anglican Church in Africa attracts the educated urban middle class.

Festivals across the continent celebrate the variety of African religious traditions. As is the case with festivals everywhere, such events offer an opportunity to represent ideal versions of culture, a chance to revel in joy and renewed religious dedication, or a chance to make important social contacts, reinforce identity, and perform community service (Figure 7.30).

Evangelical Christianity and the Gospel of Success In contrast to the Anglican Church, modern evangelical versions of Christianity appeal to the less-educated, more recent urban migrants—the most rapidly growing segment of the population. These independent, evangelical sects interpret Christianity as being aligned with traditional beliefs in the importance of sacrificial gifts to spiritual leaders and ancestors, the power of miracles, and the idea that devotion brings wealth. The result is one of the world's fastest-growing Christian movements.

VIGNETTE Preachers at the Miracle Center in Kinshasa describe the Gospel of Success forcefully: "The Bible says that God will materially aid those who give to Him. . . . We are not only a church, we are an enterprise. In our traditional culture you have to make a sacrifice to powerful forces if you want to get results. It is the same here."

Generous gifts to churches are promoted as a way to bring divine intervention to alleviate miseries, whether physical or spiritual. Practitioners donate food, television sets, clothing, and money. One woman gave 3 months' salary in the hope that God would find her a new husband.

Like all religious belief systems, the gospel of success is best understood within its cultural context. Many of the believers, new to the city, feel isolated and are looking for a supportive community to replace the one they left behind. In return for

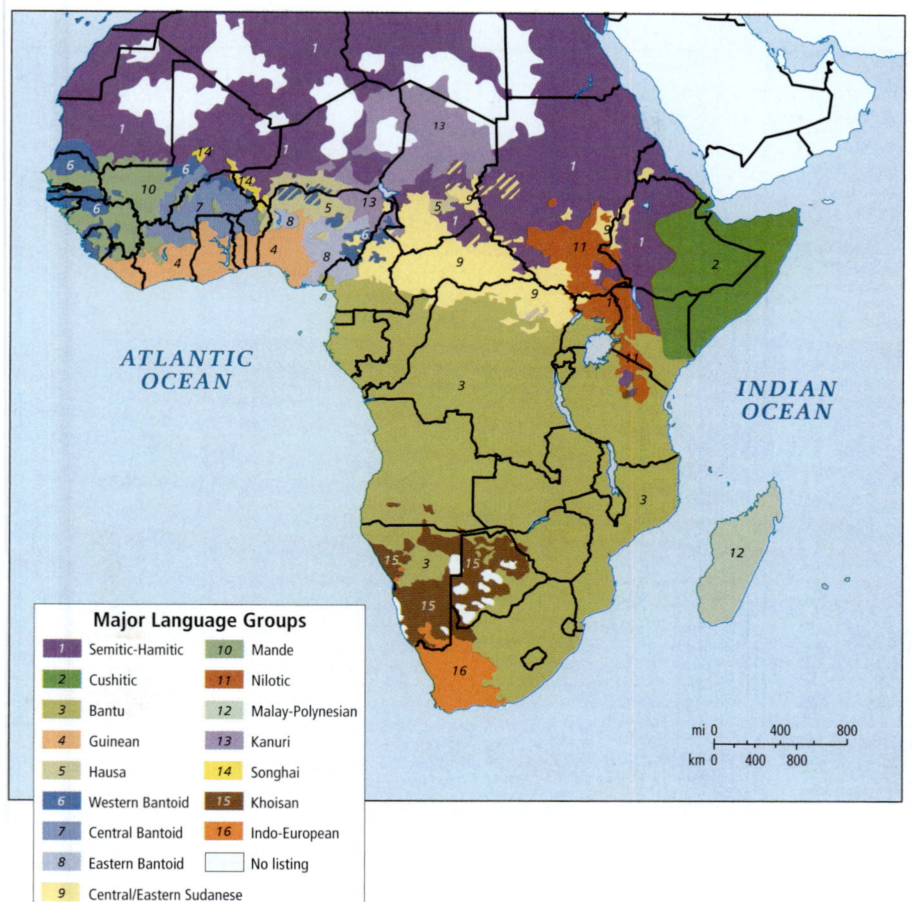

FIGURE 7.31 Major language groups of Africa.

material contributions to the church and volunteer services, members receive social acceptance and community assistance in times of need. ■

Ethnicity and Language *Ethnicity*, as we have seen throughout this book, refers to the shared language, cultural traditions, and political and economic institutions of a group. The map of Africa in Figure 7.31 shows a rich and complex mosaic of languages. Yet despite its complexity, this map does not adequately depict Africa's cultural diversity (compare it with Figure 7.20 on Nigeria on page 307).

Most ethnic groups have a core territory in which they have traditionally lived, but very rarely do groups occupy discrete and exclusive spaces. Often several groups share a space, practicing different but complementary ways of life and using different resources. For example, one ethnic group might be subsistence cultivators, another might herd animals on adjacent grasslands, and a third might be craft specialists working as weavers or metalsmiths.

People may also be very similar culturally and occupy overlapping spaces but identify themselves as being from different ethnic groups. Hutu farmers and Tutsi cattle raisers in Rwanda share similar occupations, languages, and ways of life. However, European colonial policies purposely exaggerated ethnic differences as a method of control, assigning a higher status to the Tutsis. Hutus and Tutsis now think of themselves as having very different ethnicities and abilities. In the 1990s and again in 2004, the Hutu-run government encouraged attacks on the minority Tutsi people. Hutus were also killed, but those killed tended to be people who opposed the genocide. The best estimates indicate that more than 1 million people, most of them Tutsis, died during the two genocide catastrophes.

Some African countries have only a few ethnic groups; others have hundreds. Cameroon, sometimes referred to as a microcosm of Africa because of its ethnic complexity, has 250 different ethnic groups. Different groups often have extremely different values and practices, making the development of cohesive national policies difficult. Nonetheless, the vast majority of African ethnic groups have peaceful and supportive relationships with one another.

To a large extent, language correlates with ethnicity. More than a thousand languages, falling into more than a hundred language groups, are spoken in Africa; Bantu is the largest such group (see Figure 7.31). Most Africans speak their native tongue and a **lingua franca** (language of trade). Some languages are spoken by only a few dozen people, while other languages, such as Hausa, are spoken by millions of people from Côte d'Ivoire to Cameroon. Lingua francas—such as Hausa, Arabic, and Swahili—are taking over because they better suit people's needs or have become politically dominant; some African languages are dying out. Former colonial languages such as English, French, and Portuguese (all classed as Indo-European languages in Figure 7.31) are also widely used in commerce, politics, education, and on the Internet.

lingua franca language of trade

THINGS TO REMEMBER

- Gender relationships in sub-Saharan Africa are complex and variable, but in this region as in others, women are generally subordinate and are often physically abused. Nonetheless, change is underway.
- Female genital mutilation has terrible consequences for women's health. It has been recognized as sexual discrimination because it is rooted in gender inequalities and power imbalances between men and women.
- Traditional African religions are among the most ancient on Earth and are found in every part of the continent. Today, however, about one-third of sub-Saharan Africans are Muslim and about half are Christian.

FIGURE 7.32 Maps of human well-being.

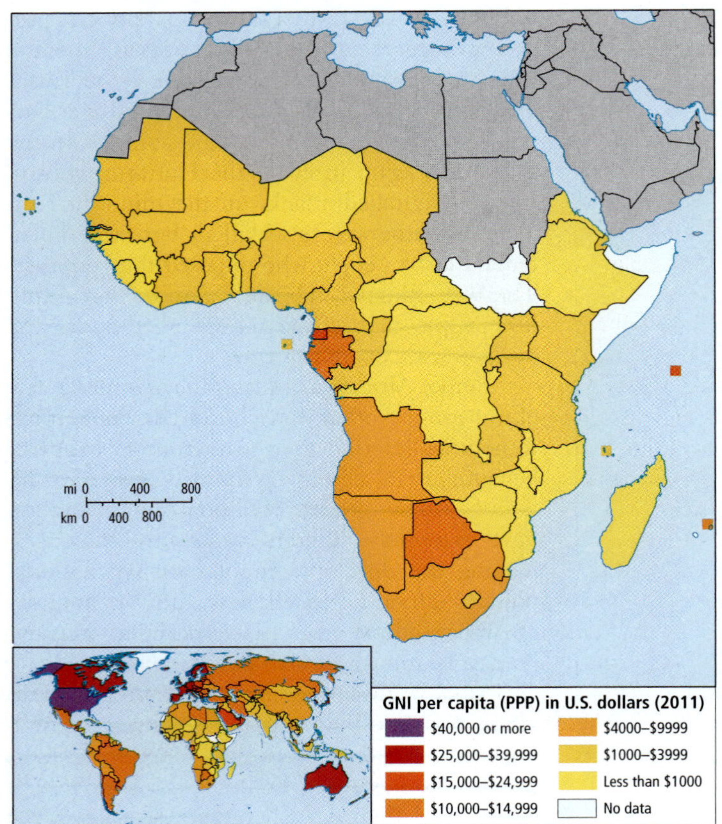

(A) Gross national income (GNI).

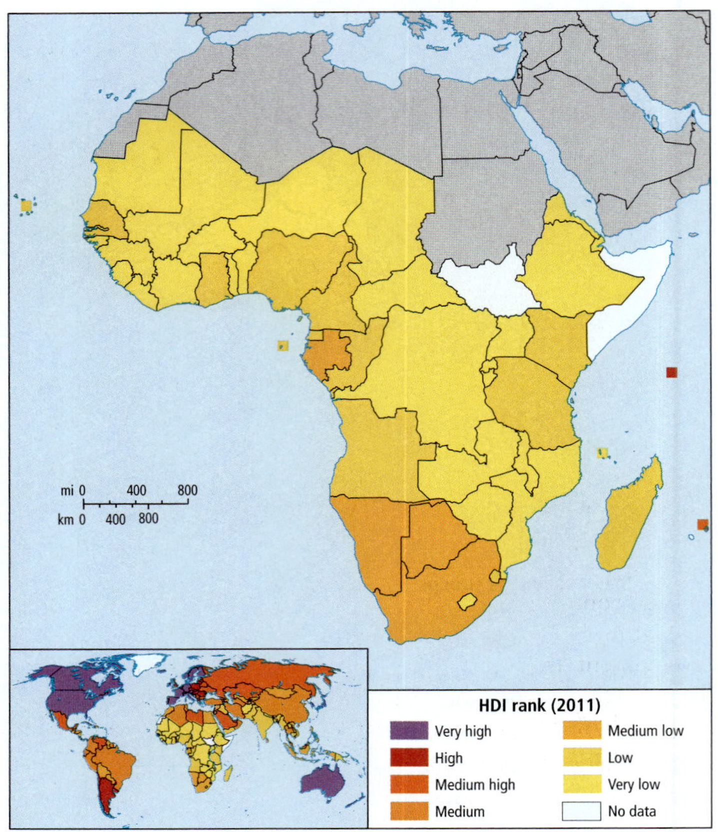

(B) Human Development Index (HDI).

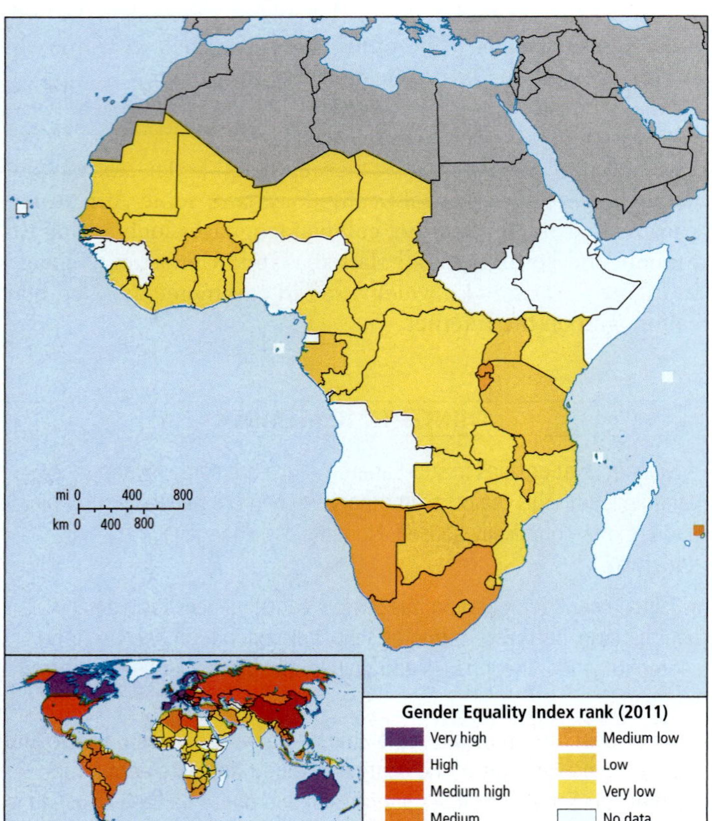

(C) Gender Equality Index (GEI).

Many in the evangelical movement, a subset of Christianity, promote the Gospel of Success as a way of life.

- Sub-Saharan Africa is a mosaic of more than 1000 languages that are roughly similar to the mosaic of ethnic groups. Hausa, Arabic, and Swahili are now the most commonly used languages of trade.

Geographic Patterns of Human Well-Being

As we have observed in previous chapters, gross national income (GNI) per capita can be used as a general measure of well-being because it shows broadly how much money people in a country have to spend on necessities. However, it is only an average figure and does not indicate how income is actually distributed across the population. Thus, it does not show whether income is equally apportioned or whether a few people are rich and most are poor. (As discussed in earlier chapters, the term GNI is often followed by *PPP*, which indicates that all figures have been adjusted for cost of living and so are comparable.)

The map in Figure 7.32A indicates three things regarding GNI in this region. First, most countries are yellow or light yellow in color, indicating that they are in the two lowest GNI ranks (incomes are under $4000). None are in the top two categories of GNI per capita rank (dark red or purple colors), and only two very small countries, Equatorial Guinea and the Seychelles, are in the medium-high rank (lighter red). A look at the world map inset for GNI shows that of all regions on Earth, sub-Saharan Africa has the lowest figures. A band of countries with somewhat higher GNIs runs along the west side of the continent from Equatorial Guinea through South Africa. In these countries, GNI per capita ranges from U.S.$4000 to U.S.$24,999, but only little Gabon, Botswana, and the tiny island countries of Mauritius and Seychelles, far to the east in the Indian Ocean, are in the medium category, averaging U.S.$10,000 to U.S.$14,999 per year. Gabon and Botswana depend on extractive industries—selling timber, oil, manganese, and diamonds—and in both, the income disparity is rather stark. South Africa, which is actually considered Africa's most developed country because of its industry, has a GNI that averages less than U.S.$9999. There is no data for South Sudan and Somalia; but as discussed elsewhere, both are dangerously troubled countries.

Figure 7.32B depicts countries' rank on the Human Development Index (HDI), which is a calculation (based on adjusted real income, life expectancy, and educational attainment) of how adequately a country provides for the well-being of its citizens. In this region, HDI ranks stretch from high (Seychelles) and medium-high (Mauritius) to very low, with only Botswana, Gabon, South Africa, and Namibia ranking medium-low. As noted in Chapter 6 on page 262, if on the HDI scale a country drops below its approximate position on the GNI scale, one might suspect that the lower HDI ranking is due to the poor distribution of wealth and social services: A critical mass of people are simply not getting a sufficient share of the national income to have good health care and an education. A host of countries along the west coast from Burkina Faso to South Africa rank lower on HDI than they do on GNI. Only the Seychelles and Mauritius have higher ranks on HDI (high and medium-high, respectively) than on GNI.

Figure 7.32C shows how well countries are ensuring gender equality in three categories: reproductive health, political and educational empowerment, and access to the labor market. First, as is true across the globe, nowhere in this region are women paid equally to men; but notice on the world inset map that gender disparity is less extreme in sub-Saharan Africa than in North Africa and Southwest Asia. Certainly, it is important to remember that GNI per capita figures (a surrogate for average income) are generally very low for this region, but the greater equality of income does mean that women are economically active; other evidence on gender (discussed previously) indicates that women in Africa are improving their social and economic status and are beginning to take leadership roles.

Geographic Insights: Sub-Saharan Africa Review and Self-Test

1. Climate Change, Food, and Water: Sub-Saharan Africa is particularly vulnerable to climate change because subsistence occupations are sensitive to even slight variations in temperature, rainfall, and water availability. In large part because of poverty, political instability, and having little access to cash, the region does not have much resilience to the effects of climate change.

- Why are many African food production systems particularly vulnerable to climate change?
- Why are African farmers and herders particularly sensitive to changes in temperature and water availability?
- How do poverty and having little access to cash influence the vulnerability of many Africans to climate change?

2. Globalization and Development: During the era of colonialism, Europeans set up the pattern that continues to this day of basing the economies of sub-Saharan Africa on the export of raw materials. This leaves countries vulnerable in the global economy, where prices for raw materials can vary widely year to year. While a few countries are diversifying and industrializing, most still sell raw materials and all rely on expensive imports of food, consumer products, and construction and transportation equipment.

- How did European colonization affect economic development and food production in Africa?
- Why does being a supplier of raw materials put a country at a disadvantage in the global economy?
- What are the current positive and negative trends in food production in sub-Saharan Africa?
- Are there any exceptions in the region to the general pattern of economic dependence on exports of cheap raw materials?
- Why is the development of manufacturing industries in sub-Saharan Africa considered such an important step in raising standards of living?

3. Power and Politics: Countries in the region are shifting from authoritarianism to elections and other democratic institutions, but the power structures in the region are resistant to democratization. While free and fair elections have brought about dramatic changes, distrusted elections have often been followed by surges of violence.

- What forces are working for and against democracy in sub-Saharan Africa?
- To what extent can the weakness of democracy in sub-Saharan Africa be traced to institutions put in place during the colonial era?
- What role have elections played in both diffusing and inciting violence?
- What are some examples of authoritarian political structures that have given way to more democratic systems of government?

4. Gender, Power, and Politics: The education of African women and their economic and political empowerment are now recognized as essential to development. As girls' education levels rise, population growth rates fall; as the number of female political leaders increases, so does their influence on policies designed to reduce poverty.

- How are issues of gender influencing trends toward democratization?
- Why have national crises led, in some cases, to an increase in women's political empowerment?
- What is the relationship between rights and education for women and the alleviation of poverty?

5. Population, Urbanization, Food, and Water: In part because of lower infant mortality rates and the higher cost of rearing children in cities, urbanization has the positive effect of encouraging families to limit births. Nevertheless, urbanization often happens in an uncontrolled fashion, leaving many people to reside in impoverished slums without adequate services, sufficient food, and clean water.

- On what grounds would the claim be made that sub-Saharan Africa is not densely populated?
- How is urbanization influencing population growth in sub-Saharan Africa? Relate these trends to the demographic transition.
- What do population pyramids tell us about a country's population?
- What special spiritual role do children play in family and community life in sub-Saharan Africa?
- Discuss the importance of urban gardening as an innovation in sub-Saharan Africa's modernization.
- What role do infectious diseases play in sub-Saharan African development, life expectancies, and use of resources?
- Why do women bear the brunt of the HIV-AIDS epidemic in sub-Saharan Africa?

Critical Thinking Questions

1. If you were to investigate the origins of lumber and lumber products sold where you live, where would you start? What would make you think that you are or are not participating in the African timber trade? How is this trade related to the concept of *neocolonialism*, the revival of colonial-like resource extraction in the modern period?

2. Describe the ways in which the environment is linked to the spread of the chronic communicable diseases of Africa. Why might some scientists argue that malaria is a worse threat to Africa's development than is HIV-AIDS?

3. Describe the trajectory of population growth in Africa? What are the factors that contribute to this pattern? How might the developed countries help the situation?

4. Reflect on the many passages throughout this chapter regarding gender roles in Africa, then develop a statement of the points that you think are the most influential in the economic development process.

5. Describe the important ways in which women are exerting an influence on African political processes.

6. Technology is changing life in Africa. What do you think are the most crucial ways in which technology will modify the ability of ordinary people to better their lives and to participate in civil society?

7. What are the various circumstances that make young heterosexual women so susceptible to HIV-AIDS in sub-Saharan Africa? How does the social and economic status of the average woman (compared to the average man) influence the spread of HIV among women?

8. Looking at the various African locations where food is in short supply, explain the main factors that account for the shortages. How would you respond to those who say that drought and climate are the principal causes of famine?

9. If you were asked to make a speech to the Rotary Club in your town about hopeful signs out of Africa, what would you include in your talk? Which pictures from this book (or elsewhere) would you show?

Chapter Key Terms

agroforestry 287
animism 319
apartheid 298
carbon sequestration 285
commercial agriculture 290
commodities 300
commodity dependence 300
divide and rule 306
dual economy 301
female genital mutilation (FGM) 318
genocide 309
grassroots economic development 305
groundwater 292
Horn of Africa 285
intertropical convergence zone (ITCZ) 285
lingua franca 321
mixed agriculture 289
pastoralism 292
polygyny 317
Sahel 284
self-reliant development 305
shifting cultivation 289
subsistence agriculture 289

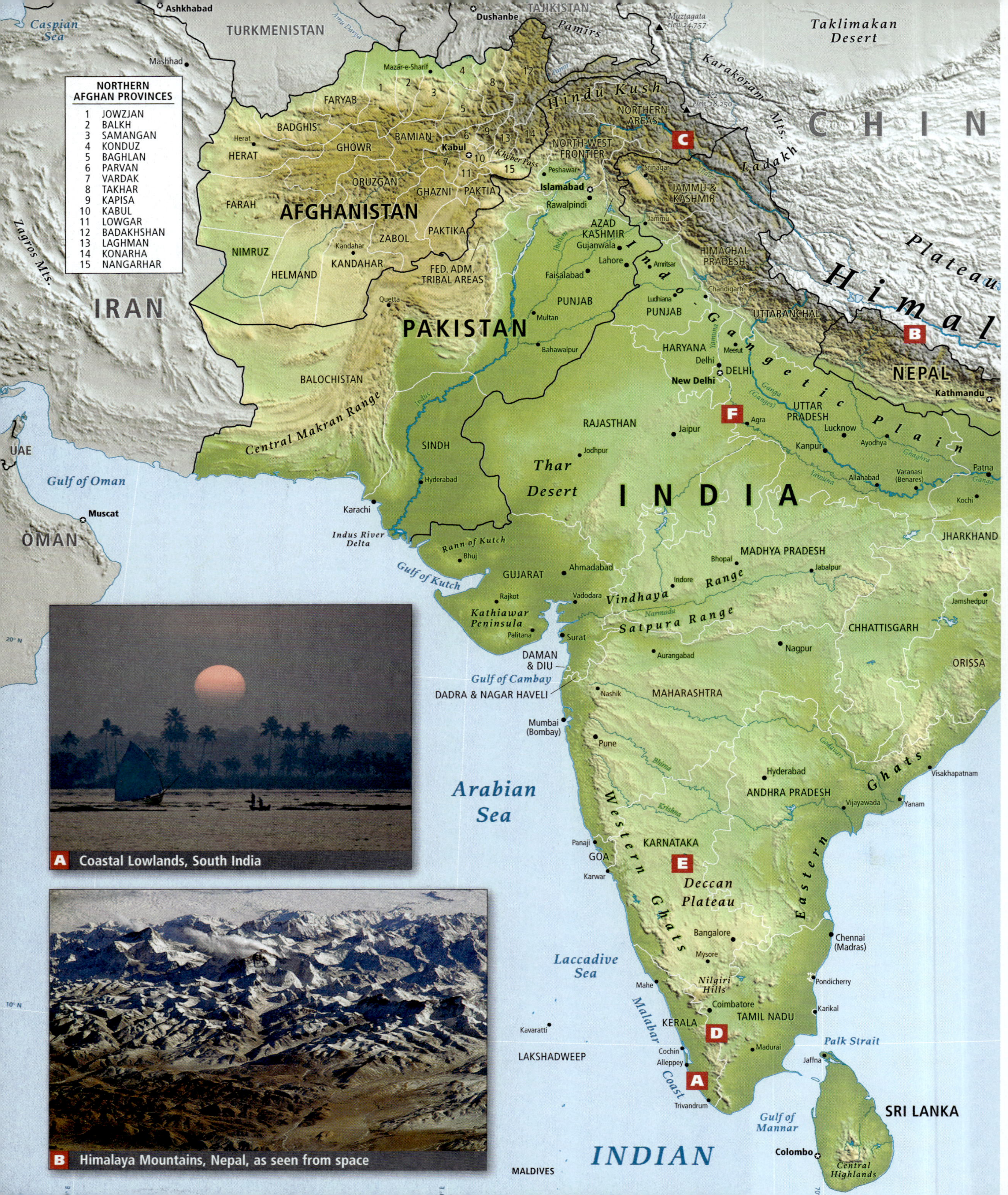

A Coastal Lowlands, South India
B Himalaya Mountains, Nepal, as seen from space

8 South Asia

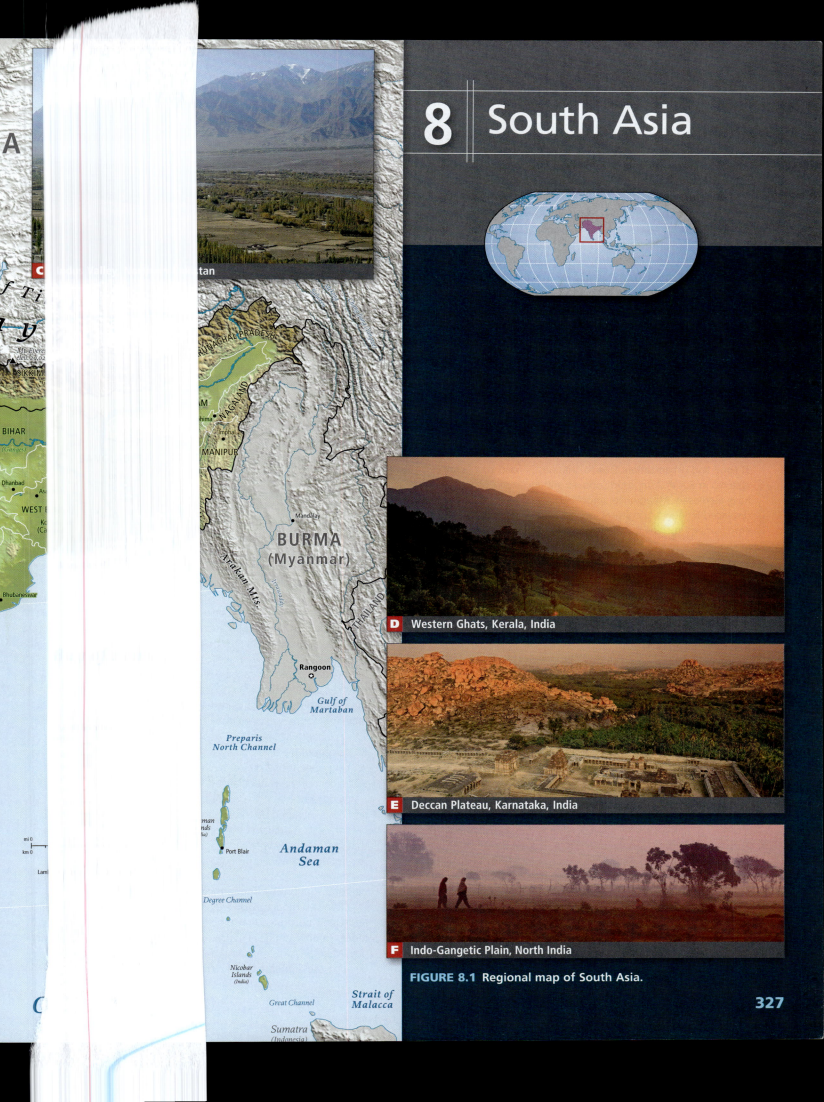

FIGURE 8.1 Regional map of South Asia.

GEOGRAPHIC INSIGHTS: SOUTH ASIA

After you read this chapter, you will be able to discuss the following geographic insights as they relate to the nine thematic concepts:

1. Climate Change and Water: Climate change puts more lives at risk in South Asia than in any other region in the world, primarily due to water-related impacts. Sea level rise, droughts, floods, and the increased severity of storms imperil many urban and agricultural areas over the short term. Over the longer term, glacial melting threatens rivers and aquifers.

2. Gender and Population: The assumptions that males are more useful to families than females and that females must be supplied with a dowry have resulted in a devaluation of females, leading to a gender imbalance in populations across South Asia. Those who can afford in utero gender-identifying technology may abort female fetuses, while those with fewer resources sometimes commit female infanticide. As a result, adult males significantly outnumber adult females. A second relationship between gender and population is that when females are given opportunities to work outside the home, they tend to choose to have fewer children.

3. Food and Urbanization: Due to changes in food production systems in South Asia, food supplies and incomes of wealthy farmers have increased. Meanwhile, impoverished agricultural workers, displaced by new agricultural technology, have been forced to seek employment in cities, where they often end up living in crowded, unsanitary conditions, and must purchase food.

4. Globalization and Development: Globalization benefits populations unevenly. Educated and skilled South Asian workers with jobs in export-connected and technology-based industries and services enjoy better lives. Low-skilled workers, however, in both urban and rural areas, are left with demanding but very low-paying jobs.

5. Power and Politics: India, South Asia's oldest, largest, and strongest democracy, has shown that democratic institutions can ameliorate conflict. Across the region, when ordinary people have been able to participate in policy-making decisions and governance—especially at the local level—intractable conflict has been diffused and combatants have been convinced to take part in peaceful political processes.

The South Asian Region

South Asia (see **Figure 8.1** on pages 326–327), like so many of today's world regions, began its modern history as the result of European colonization. During that time, economic, political, and social policies rarely put the needs of South Asian people first. Toward the end of the colonial era, major upheavals precipitated by departing colonists left the region with a legacy of distrust and difficult political borders that still lingers.

The nine thematic concepts highlighted in this book are explored as they arise in the discussion of regional issues, with interactions between two or more themes featured, as in the geographic insights above. Vignettes, like the one that follows, illustrate one or more of the themes as they are experienced in individual lives.

GLOBAL PATTERNS, LOCAL LIVES Narendra Modi, the chief minister of the state of Gujarat in western India (see Figure 8.1), is passionate about securing water for his state. In 2006, draped in garlands by well-wishers, he embarked on a flamboyant hunger strike to protest a decision by India's national government in New Delhi to limit the height of the Sardar Sarovar Dam on the Narmada River. Four Indian states lie in the drainage basin of the Narmada. Modi wanted to trap more water behind a higher dam so that as many as 50 million of his constituents in Gujarat would have greater access to water, primarily for irrigation. Gujarat produces much of the food needed by India's west coast cities.

Far away in New Delhi, at the same time as Modi's hunger strike, Medha Patkar, the leader of the "Save the Narmada River" movement, was in day 18 of her hunger strike protesting against the same dam. As water rose in the dam's reservoir, 320,000 farmers and fishers in Madhya Pradesh, the state in which the reservoir is located, were being forced to move. Although Indian law requires that these people be given land or cash to compensate them for what they had lost, so far only a fraction had received any compensation. Some of the farmers demonstrated their objection by forfeiting their right to compensation and refusing to move, even as the rising waters of the reservoir consumed their homes. Forcibly removed by Indian police, many have since relocated to crowded urban slums.

Environmental and human rights problems have been at the heart of the Sardar Sarovar Dam controversy since the project began in 1961. The natural cycles of the Narmada River have been severely disturbed. Once a placid, slow-moving river—and one of India's most sacred—the disruption of flows and the flooding upstream of the dam have caused massive die-offs of aquatic life, resulting in high unemployment levels among farmers and fishers. Many have also lost their homes to the rising reservoir waters (**Figure 8.2A**).

In addition to cost overruns that are more than three times original cost estimates, the ultimate benefits of the dam have also been called into question. A major justification for the dam was that, in addition to irrigation water, it would provide drinking water for 18 million people in the greater river basin as well as 1450 megawatts of electricity. But the demand for irrigation waters to serve drought-prone areas in the adjacent states of Gujarat and Rajasthan (see the Figure 8.2 map) is now so strong (and increasing due to climate warming) that there will not be sufficient water to generate even 10 percent of the 1450 planned megawatts on a sustained basis. Furthermore, 80 percent of the areas in Gujarat most vulnerable to drought are not yet connected to the project by the necessary canals, therefore, 90 percent of the water planned for irrigation still flows into the sea.

Concerned that the economic benefits would be small and easily negated by the environmental costs, the World Bank withdrew its funding of the dam some years ago. Ecologists say that far less costly water-management strategies, such as rainwater harvesting, intensified groundwater recharge (artificially assisted replenishment of the aquifer), and watershed management would be better options for the farmers of Gujarat and Rajasthan.

Minister Modi quickly ended his hunger strike when the Indian Supreme Court ruled that the Sardar Sarovar Dam could be raised higher (it now stands at 121.92 meters, or 400 feet). The following day, Medha Patkar ended her fast as well, because in the same decision the Supreme Court ruled that all people displaced by the

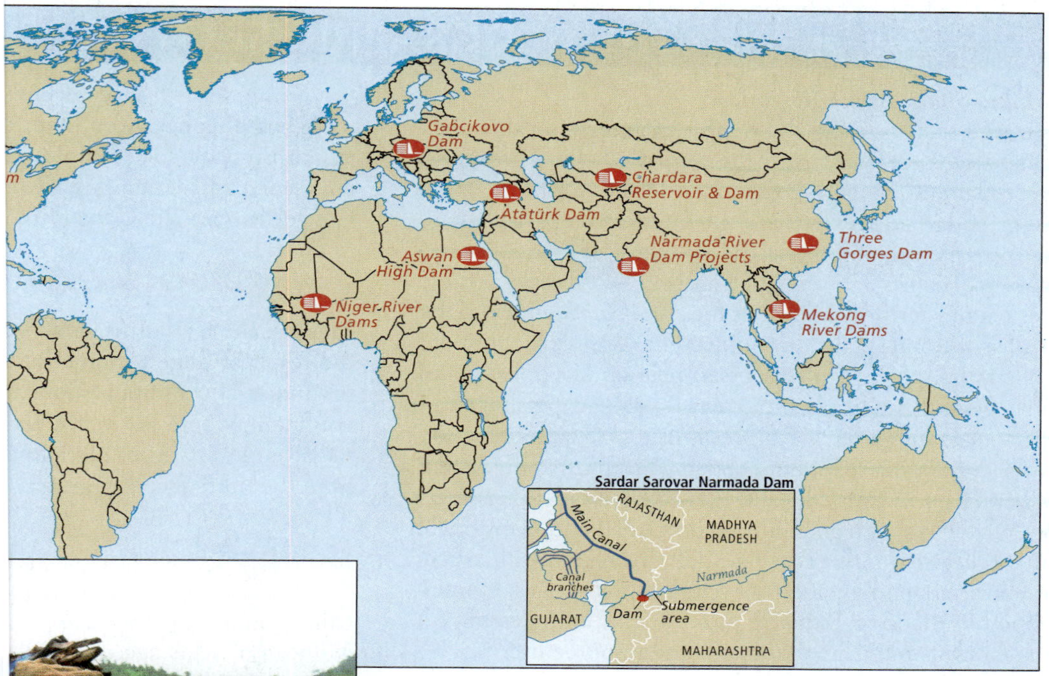

FIGURE 8.2 The Sardar Sarovar Dam on the Narmada and other major dams around the world. The Sardar Sarovar Dam on the Narmada River is only one of hundreds of thousands of dam projects throughout the world that together have displaced between 40 and 80 million people. The map shows some of the major dams around the world. **(A)** Before they were displaced by the reservoir of the Sardar Sarovar Dam, rural people stand outside homes that are now submerged.

Furthermore, the court decision
es are required for all dam
had been done for the Sardar
ember of 2012, the dam project
ad stalled again in court. Both
mote their positions. Meanwhile,
ominence and in 2013 began
ected chief minister of Gujarat.
nyurl.com/bhhanz3l; Friends of the
sues, Theories, and Policy; and the
nformation, see Text Credit pages.]

nagement in the Narmada River
ow facing South Asia and other
nies. Across the world, large, poor
asingly overtaxed environments,
ssures on fragile environments
hange. Improving the standard

of living for the poor nearly always requires increases in water and energy use. Misplaced efforts to meet these urgent needs often make neither economic nor environmental sense, but are driven to completion by political and social pressures. In the case of the Sardar Sarovar Dam, the wealthier, more numerous, and more politically influential farmers of Gujarat have tipped the scale in favor of a project that may be creating more problems than it is solving.

The role of water in South Asian life, including the ways water use intersects with other central issues, is a recurring topic in this chapter.

THINGS TO REMEMBER

- The Sardar Sarovar Dam and similar dams throughout the world are created primarily for irrigation and power generation, but their side effects often create more problems than they solve.
- The management of water—whether there is too much of it, too little of it, or where it is unfairly apportioned—has been a feature of South Asian life for a long time. Climate change now makes management more problematic.
- Improving the standard of living of poor populations in many parts of the world nearly always requires increases in water and energy use.

THE GEOGRAPHIC SETTING

What Makes South Asia a Region?

The countries that make up the South Asia region in this book are Afghanistan and Pakistan in the northwest; the Himalayan states of Nepal and Bhutan; Bangladesh in the northeast; India (including the Indian territories of the Lakshadweep, Andaman, and Nicobar Islands); and the island countries of Sri Lanka and the Maldives (Figure 8.3; see also Figure 8.1). Physically, these countries occupy territory known as the *Asian subcontinent*—the portion of a tectonic plate that joined the Eurasian continent nearly 60 million years ago (discussed below). Historically and culturally, these modern countries have a shared patchwork of religious, social, and political features linked to ancient conquerors from Central Asia, explorers and traders from the Arabian Peninsula and Southeast Asia, and more recent colonizers from Europe. A patchwork is a good image, because the long history of internal and external influences have left this region with a fragmented and overlapping pattern of religions, languages, ethnicities, economic theories, forms of government, attitudes toward gender and class, and ideas about land and resource use.

Terms in This Chapter

Because its clear physical boundaries set it apart from the rest of the Asian continent, the term **subcontinent** is often used to refer to the entire Indian peninsula, which includes Nepal, Bhutan, India, Pakistan, and Bangladesh (the term usually does not include Afghanistan).

subcontinent a term often used to refer to the entire Indian peninsula, including Nepal, Bhutan, India, Pakistan, and Bangladesh

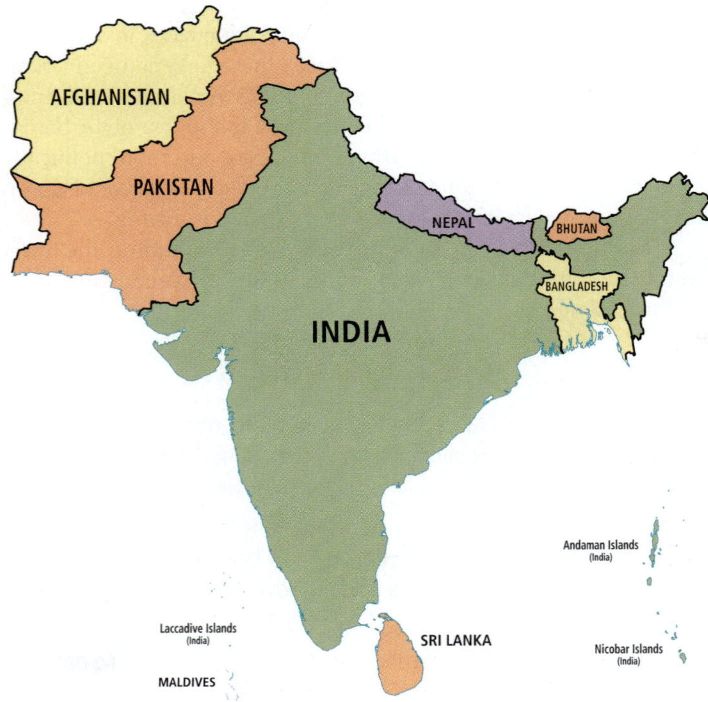

FIGURE 8.3 Political map of South Asia.

South Asians have recently adopted new place names to replace the names given them during British colonial rule. The city of Bombay, for example, is now officially *Mumbai*, Madras is *Chennai*, Calcutta is *Kolkata*, Benares is *Varanasi*, and the Ganges River is the *Ganga River*.

Physical Patterns

Many of the landforms, and even the climates, of South Asia are the result of huge tectonic forces. These forces have positioned the Indian subcontinent along the southern edge of the Eurasian continent, where the warm Indian Ocean surrounds it and the massive mountains of the Himalayas shield it from cold airflows from the north (see Figure 8.1).

Landforms

The Indian subcontinent and the territory surrounding it dramatically illustrate what can happen when two tectonic plates collide. Millions of years ago, the Indian-Australian Plate, which carries India, broke free from the eastern edge of the African continent and drifted to the northeast (see Figure 1.25 on page 46). When this plate collided with the Eurasian Plate about 60 million years ago, it left a giant peninsula jutting into the Indian Ocean. As the relentless pushing from the south continued, both the leading (northern) edge of what became South Asia and the southern edge of Eurasia crumpled and buckled. The result is the world's highest mountains—the Himalayas, which rise more than 29,000 feet (8800 meters)—as well as other mountain ranges that curve away from the central impact zone—in the west, the Hindu Kush and Karakoram Range and, in the east, the lesser-known ranges that border China and Burma (see Figure 8.1B and the figure map). The continuous compression also lifted up the Plateau of Tibet, which rose up behind the Himalayas to an elevation of more than 15,000 feet (4500 meters) in some places. The compression and mountain-building process continues into the present.

The three great rivers of South Asia—the Indus, the Ganga, and the Brahmaputra—all begin within 100 miles (160 kilometers) of one another in the Himalayan highlands near the Tibet, Nepal, and India borders; they are largely fed by glacial meltwater (see the Figure 8.1 map and Figure 8.5 map on page 333). The main river basins of the Indus and Ganga lie to the southwest and south of the Himalayas in what is called the Indo-Gangetic Plain (see Figure 8.1C, F). The Indus flows southwest through Pakistan and ultimately joins the Arabian Sea. The Ganga flows southeast to the Bay of Bengal. The Brahmaputra flows east through Tibet and then turns south through Arunachal Pradesh and Assam into Bangladesh. The mouths of the Ganga and Brahmaputra are blended together in a giant, ever-changing delta.

South of the Indo-Gangetic Plain lies the Deccan Plateau, an area of modest uplands (1000–2000 feet [300–600 meters] in elevation) interspersed with river valleys (see Figure 8.1E). This upland region is bounded on the east and west by two moderately high mountain ranges, the Eastern and Western Ghats (see Figure 8.1D). The Ghat mountain ranges descend to long but

y extensive river deltas and flood- iver valleys and coastal zones are ıre only slightly less so.

gh degree of tectonic activity and sia is prone to devastating earth- 7.7 quake that shook the state of , the 7.6 quake that hit the India– , and the 6.4 quake in Quetta ıt left 120,000 people homeless. to tidal waves, or *tsunamis*, that akes. In 2004, a massive tsunami east Asia wrecked much of coastal see Figure 8.1A), killing tens of any more in Southeast Asia.

RI LANKA, REGAINS SOME NORMALCY

aracterized by a seasonal rever- s (Figure 8.4). These monsoon winds are affected by the *intertropical convergence zone (ITCZ)*, which, as the name suggests, is a zone created when air masses moving south from the Northern Hemisphere and north from the Southern Hemisphere converge near the equator. As the warm air rises and cools, it drops copious amounts of precipitation. As described in Chapter 7 (see page 285), the ITCZ shifts north and south seasonally. The intense rains of South Asia's monsoon are likely caused by the ITCZ being sucked onto the land by a vacuum created when huge air masses over the Eurasian continent heat up and rise into the upper troposphere.

The **summer monsoon** (see Figure 8.4A) begins in early June, when the warm, moist ITCZ air first reaches the mountainous Western Ghats. The rising air mass cools as it moves over the mountains, releasing rain that nurtures patches of dense tropical rain forests and tropical crops on the upper slopes of the Western and Eastern Ghats and in the central uplands. Once on the other side of India, the monsoon gathers additional moisture and power in its northward sweep up the Bay of Bengal, sometimes turning into tropical cyclones.

summer monsoon rains that begin every June when the moist ITCZ air first reaches the mountainous western Ghats

nmer and winter monsoons in South Asia.

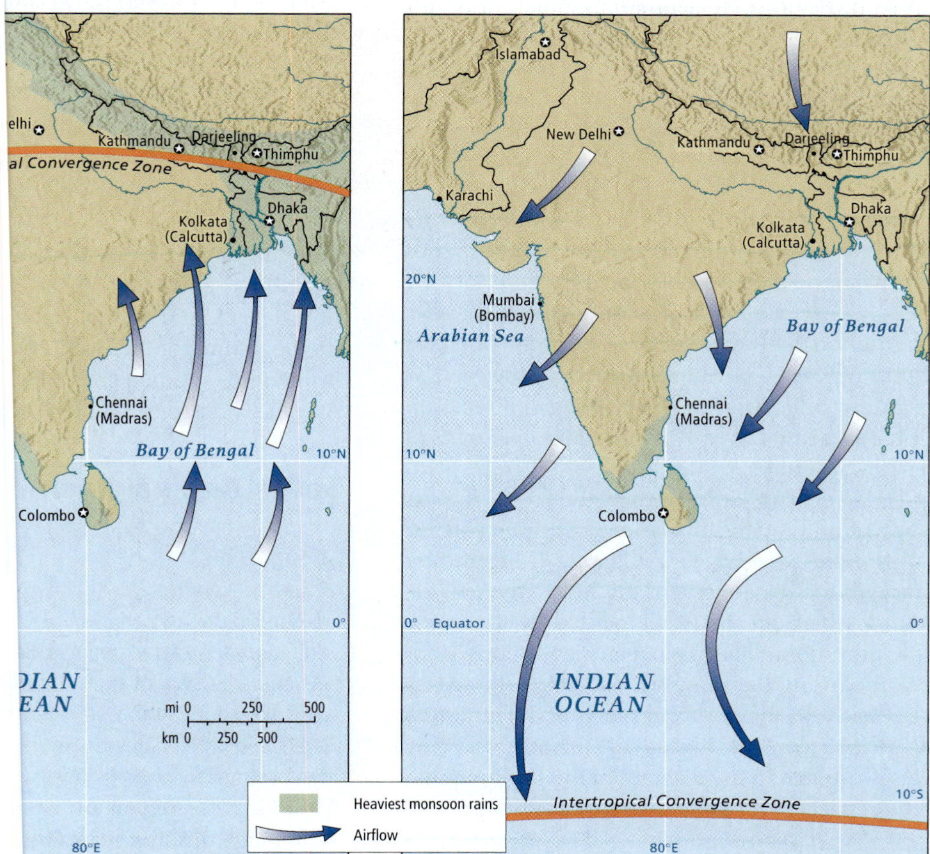

he ITCZ moves north across ge amounts of moisture from then deposited over India and

(B) In the winter, cool, dry air blows from the Eurasian continent south across India toward the ITCZ, which in winter lies far to the south.

As the monsoon system reaches the hot plains of Bangladesh and the Indian state of West Bengal in late June, columns of warm, rising air create massive, thunderous cumulonimbus clouds that drench the parched countryside. Monsoon rains sweep east to west, parallel to the Himalaya Mountains, reaching across northern India to Pakistan and finally petering out over the Kabul Valley in eastern Afghanistan by July. Rainfall is especially intense in the east, north of the Bay of Bengal, where the town of Darjeeling holds the world record for annual rainfall—about 35 feet, even though no rain falls for half the year. These patterns of rainfall are reflected in the varying climate zones (shown in Figure 8.5) and agricultural zones of South Asia (see Figure 8.28 on page 362). Although sufficient rain falls in central India to support forests, most land has long been cleared of forest (see the discussion on pages 339–342) and planted in crops. Patches of forest in the uplands and lowlands are now so fragmented that they no longer provide suitable habitat for India's wildlife. Forest vegetation is more common in the Himalayan highlands and foothills, but even in the mountains, human pressure has resulted in widespread deforestation, often leaving only patchy areas of formerly richly forested land.

Periodically, the monsoon seasonal pattern is interrupted and serious drought ensues. This happened in July and August of 2009, when the worst drought in 40 years struck much of South Asia (see the discussion below). Then in September of 2009, heavy rains came to central south India, causing crop-damaging floods that killed hundreds of people. Scientists are increasingly concluding that the extreme droughts and floods of recent years are not an anomaly but instead are part of the general global climate change.

The **winter monsoon** (see Figure 8.4B) is underway by November each year, when the cooling Eurasian landmass sends the cooler, drier, heavier air over South Asia. This cool, dry air from the north sinks down to the lower elevations and pushes the warm, wet air back south to the Indian Ocean. Very little rain falls in most of the region during this winter monsoon. However, as the ITCZ retreats southward across the Bay of Bengal, it picks up moisture, which is then released as early winter rains over parts of southeastern India and Sri Lanka.

The monsoon rains deposit large amounts of moisture over the Himalayas, much of it in the form of snow and ice that add to the existing mass of glaciers (see page 42 in Chapter 1). Meltwater from these glaciers feeds the headwaters of the Indus, the Ganga, and the Brahmaputra, which are then augmented by the many tributaries that flow into them. The rivers carry enormous loads of sediment, especially during the rainy season. When the rivers reach the lowlands, their velocity slows and much of the sediment settles out as silt. It is then repeatedly picked up and redeposited by successive floods. As illustrated in the diagram of the Brahmaputra River in Figure 8.6 on page 334, the movement of silt constantly rearranges the floodplain landscape, and this complicates human settlement and agricultural efforts. However, the seasonal deposit of silt nourishes much of the agricultural production in the densely occupied plains of Bangladesh. The same is true on the Ganga and Indus plains.

winter monsoon a weather pattern that begins by November, when the cooling Eurasian landmass sends the cooler, drier, heavier air over South Asia

THINGS TO REMEMBER

- Because of its high degree of tectonic activity and deep crustal fractures, all the countries of South Asia (Afghanistan, Pakistan, India, Nepal, Bhutan, Bangladesh, and Sri Lanka) are prone to devastating earthquakes.
- The three major rivers of this region originate from meltwater high in the northwest Himalaya Mountains and provide water to most of the region.
- The monsoons are a major influence on South Asia's climate. In June, the warm, moist air of the summer monsoon reaches India, where coastal mountains and rising hot air force the moist monsoon air up, cooling it and causing rain. Monsoon rains run in a variable band parallel to the Himalayas that reaches across northern India and Pakistan, finally petering out over the Kabul Valley in eastern Afghanistan by July.
- Periodic interruptions of the monsoon rainfall patterns can result in devastating droughts.
- Precipitation from monsoons is especially intense in the eastern foothills of the Himalayas.
- The three main rivers carry enormous loads of sediment and deposit it as silt, a process that is repeated by successive floods. The movement of silt constantly rearranges the floodplain landscape, complicating human settlement and agricultural efforts.

Environmental Issues

Humans have lived in South Asia for at least 50,000 years, but as recently as 1700 C.E. (just before British colonization), population density and human environmental impacts were light. By the beginning of the twenty-first century, population density and human impacts had grown exponentially; as a result, today South Asia has a range of serious environmental problems.

South Asia's Vulnerability to Climate Change

Geographic Insight 1

Climate Change and Water: Climate change puts more lives at risk in South Asia than in any other region in the world, primarily due to water-related impacts. Sea level rise, droughts, floods, and the increased severity of storms imperil many urban and agricultural areas over the short term. Over the longer term, glacial melting threatens rivers and aquifers.

All across the region of South Asia, people are dealing with a wide range of water issues that are made more serious by climate change. A discussion of eight of these issues follows.

1. Water Scarcity and Flooding Significant sections of South Asia have lived with water scarcity problems for millennia, due

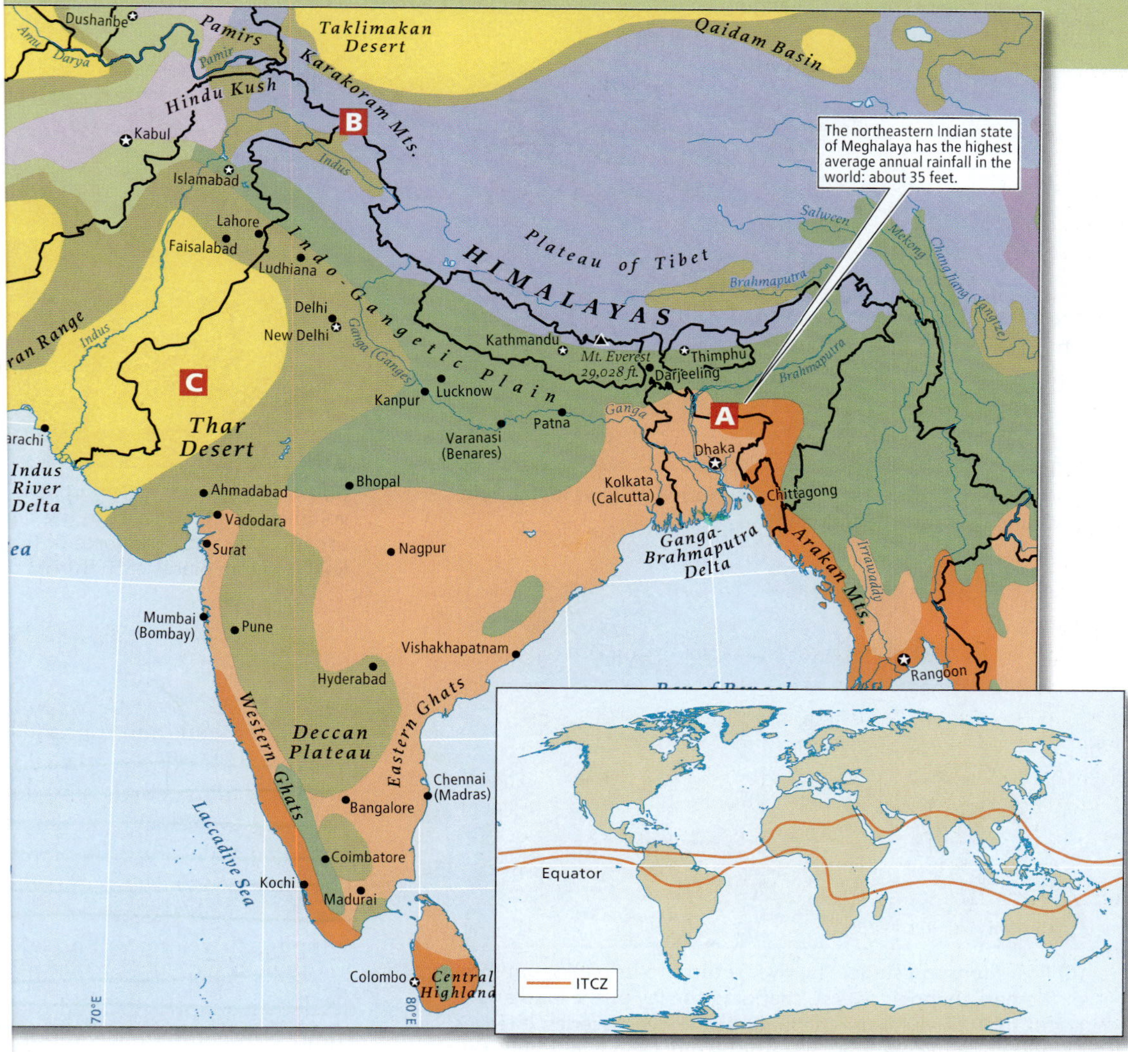

The northeastern Indian state of Meghalaya has the highest average annual rainfall in the world: about 35 feet.

Equator

ITCZ

A

B High altitude, Kashmir, India

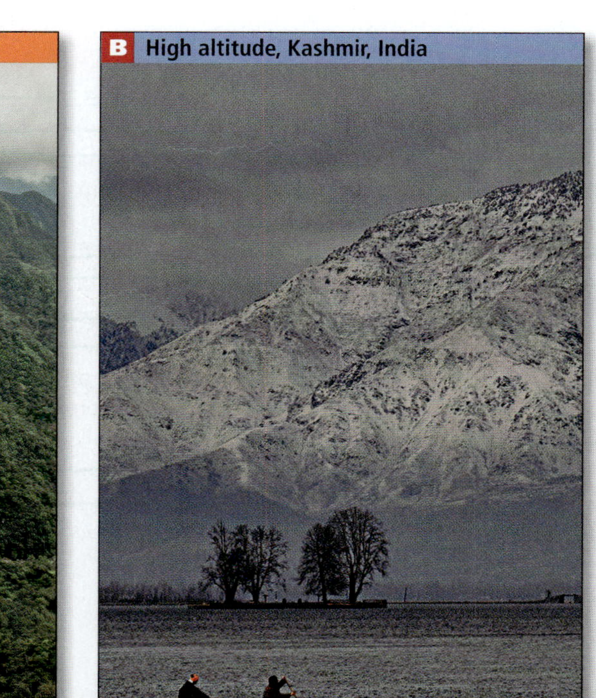

C Desert, Jaisalmer, India

FIGURE 8.6 The Brahmaputra River in Bangladesh at various seasonal stages. People who live along the river have learned to adapt their farms to a changing landscape, and along much of the river farmers are able to produce rice and vegetables nearly year-round.

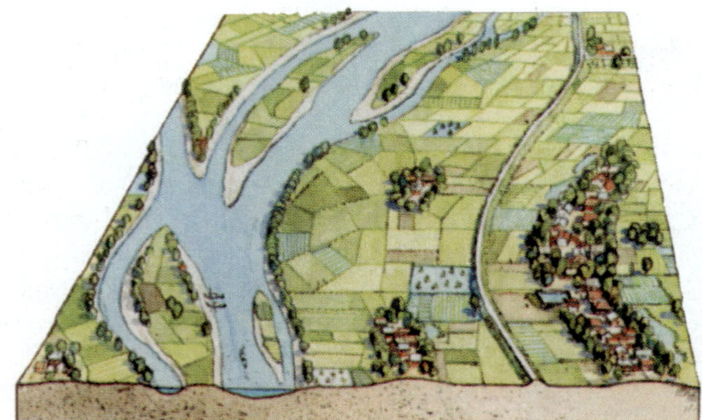

(A) Pre-monsoon stage: The river normally flows in multiple channels across the flat plain.

(B) Peak flood stage: During peak flood times, the great volume of water overflows the banks and spreads across fields, towns, and roads. The force of the water carves new channels, leaving some places cut off from the mainland.

(C) Post-monsoon stage: The river returns to its banks, but some of the new channels persist, changing the lay of the land and access for those who use the land. As the river recedes, it leaves behind silt and algae that nourish the soil. New ponds and lakes form and fill with fish.

primarily to low average rainfall, but also because of extreme seasonal variability. The Indus Valley of Pakistan, for example, has many ancient structures that captured water during wet seasons and floods and held it for use during dry times (see Figure 8.13A on page 344). Now, the shifting rainfall patterns and rising temperatures linked to climate change are creating drier conditions in much of South Asia, and also occasionally causing abnormal, heavy rainfall that can result in epic flooding, as was the case in Pakistan in 2010.

2. Glacial Melting Because South Asia's three largest rivers are fed by glaciers high in the Himalayas, the issue of glacial melting is of particular concern. (Smaller glacial-melt streams serve Afghanistan and Central Asia; see Figure 5.8A on page 206). As many as 703 million people, almost half of the region's population, depend on these glacially fed rivers for drinking water, domestic and industrial uses, and irrigation. (Irrigation takes more water than industrial and domestic uses combined.) While the immediate effect of glacial melting may be flooding, the real threat is that as the glaciers shrink, they will provide less water each year to recharge the ancient aquifers beneath the heavily populated Indo-Gangetic plains south of the Himalayan Range. The amount of water pumped on these plains for multiple uses already exceeds the rate of recharge; the long-term effect of the pumping, therefore, will be severe water shortages.

Both water conservation and increased water storage are needed for supplies to last through the dry winter monsoon. High Himalayan communities in Ladakh, on the India–China border, are attempting to retain autumn glacial melt in stone catchments, where the melt refreezes over the winter and is available for spring irrigation.

3. Sea Level Rise Tens of millions of poor farmers and fishers live near sea level in South Asia, most of them in Bangladesh, which has more people vulnerable to *sea level rise* (see page 42 in Chapter 1) than does any other country in the world (Figure 8.7A on page 336). With 162 million people already squeezed into a country the size of Iowa, as many as 17 million Bangladeshi people might have to find new homes if sea levels rise by a few feet. The biggest economic impacts of sea level rise would come from the submergence of parts of South Asia's largest cities, such as Dhaka in Bangladesh, Mumbai (Mumbai's vulnerability to sea level rise is discussed on pages 42–43 in Chapter 1; see also Figure 8.7C) and Kolkata in India, and Karachi in Pakistan.

Also threatened by continuing sea level rise are the Maldives Islands in the Indian Ocean, 80 percent of which lie 1 meter or less above the sea. Beach erosion is so severe that homes built only a few years ago in this richest of South Asia's countries are falling into the sea.

leveloping responses to climate available to do so are often very re 8.7D).

Vater As we learned in Chapter 1, 5 to 13 gallons (20 to 50 liters) : domestic needs: drinking, cook- capita domestic water consump- poor, so it would be reasonable ians consume 5 to 6 gallons per incomes rise, so not only do the ıme closer to 13 gallons per day, e than that. One flush of a toilet as more and more South Asians consume more water per capita, and it becomes easy to see that e across this region. We must also nsumption is only a fraction of a on. Much of South Asia's water is produce the manufactured goods outh Asians themselves consume ır export.

er The concept of *virtual water*, 8) is especially useful in assessing ces in the drier regions of South northwest India are all naturally e more so by thousands of years

ıe production of all agricultural regions (whether used locally or s that are so scarce that ordinary water than is considered healthy. mostly for local consumption, the nained in the region. Now, export and nuts—are all produced with ual water component (Table 8.1). and the virtual water is consumed ıce, these water-scarce regions are se of wealthy global consumers.

Component Table

	Virtual water content (in liters)
	70
	1300
	3400
	11,000
	16,600
	8000
ods	80

Water Footprint Network Web site, at /productgallery&product=wheat

The same is true for manufactured goods with high virtual water inputs, such as textiles, garments, hides, leather goods, and sporting goods. Furthermore, in these dry regions, water for crops and industry is frequently drawn down faster than it is naturally replenished, so none of this water use is sustainable. Most importantly, the costs of depleted water used in production are not sufficiently accounted for in the pricing of goods exported from these regions.

6. Inefficient Allocation and Use of Water There is wide spatial and temporal variation in the availability of water across the region, and this is only likely to increase with climate change. South Asia has more than 20 percent of the world's population, but only 4 percent of its freshwater, making access to water often difficult. Disputes are increasingly common, yet few citizens understand the gravity of the problem because political rivalries and bureaucratic inertia have delayed the development and implementation of national water policies. As a result, there are cases, like that described below, when scarce water is used to purposely create a flood and in so doing threatens the well-being of millions.

During the dry season, India occasionally diverts as much as 60 percent of the Ganga River flow to Kolkata to flush out channels where silt is accumulating and hampering river traffic (see the Ganga-Brahmaputra Delta on map in Figure 8.1 and the map in Figure 8.5). These diversions temporarily deprive Bangladesh of normal freshwater flow. Reductions in the freshwater levels in the Ganga-Brahmaputra Delta allows saltwater from the Bay of Bengal to penetrate inland, ruining agricultural fields. The diversion has also caused major alterations in Bangladesh's coastline, damaging its small-scale fishing industry. Thus, to serve the needs of Kolkata's 16 million people, the livelihoods of 40 million rural Bangladeshis have been put at risk, triggering protests in Bangladesh.

Similar water-use conflicts occur across the region. In Delhi (the ancient city and New Delhi are now commonly referred to simply as Delhi), just 17 five-star hotels, serving a few thousand guests, use about 210,000 gallons (800,000 liters) of water daily, which would be enough to serve the needs of 1.3 million people living in Delhi slums. When two or more states or countries share a water catchment, conflicts can become geopolitical in scale. In the late 1990s, India signed a treaty with Bangladesh promising a fairer distribution of water, but as of 2010, Bangladesh was still receiving a considerably reduced flow. India's draft water policy of 2012 merely mentions the need to resolve the issue.

182. COCA-COLA BLAMED FOR INDIA'S WATER PROBLEMS

7. Water Pollution First, it should be noted that most people do not drink water out of the tap in South Asia. Water must be boiled or otherwise purified before it is consumed.

Basic water safety is an issue across the region but is especially crucial in historic religious pilgrimage towns, such as Varanasi, where each year millions of Hindus come to die, be cremated, and have their ashes scattered over the Ganga River. The number of such final pilgrimages has increased with population growth and affluence, causing wood for cremation fires to become scarce. As a result, incompletely cremated bodies are being dumped into the river, where they pollute water used for drinking, cooking, and ceremonial bathing (Figure 8.8B on page 338). In an attempt to deal with this problem, the government recently

Figure 8.7 Photo Essay: Vulnerability to Climate Change in South Asia

Climate change is putting more lives at risk in South Asia than in any other region. However, many responses to climate change are being developed, which could increase resilience to climate-related hazards.

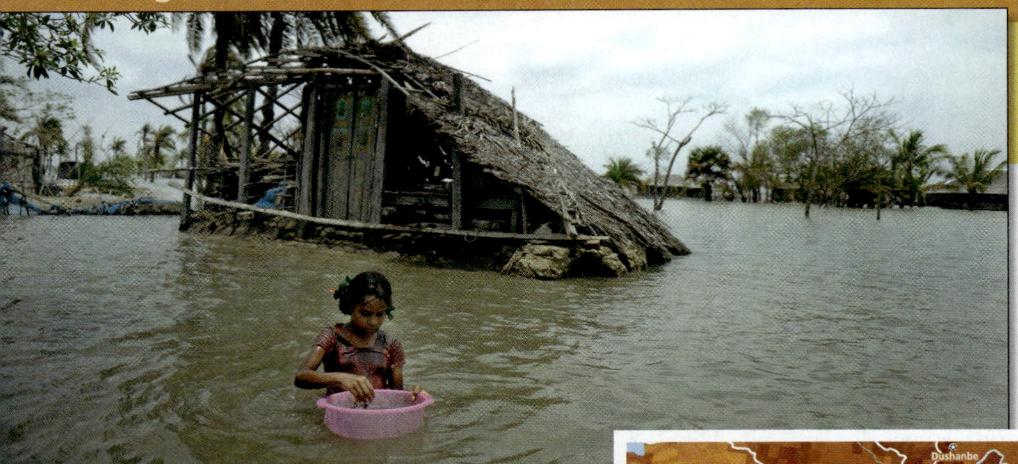

A A girl in coastal Bangladesh washes fish outside her home, which was flooded and partially ruined after a cyclone destroyed the river embankment that once protected her village from high tides. Bangladesh is highly exposed to sea level rise, increased storm intensity, and flooding related to glacial melting. Because it has a large population of poor people, Bangladesh is very sensitive to these climate hazards.

B In Afghanistan, a man improves irrigation infrastructure that may help reduce the sensitivity of nearby areas to water shortage. Widespread poverty and political instability contribute to Afghanistan's extremely high vulnerability to climate hazards.

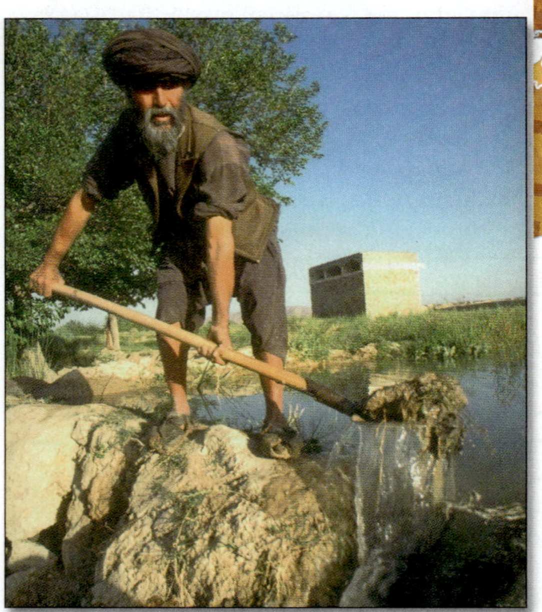

Mashhad
Dushanbe
Kabul
B
Islamabad
Rawalpindi
Lahore
Faisalabad
Delhi
New Delhi
Jaipur
Kathmandu
Thimphu
Lucknow
Kanpur
Patna
Varanasi
Muscat
Karachi
Hyderabad
Ahmadabad
Bhopal
Dhaka
A
Chittagong
Mandalay
Kolkata
Khulna
D
Nagpur
Surat
Nashick
Bhubaneshwar
Mumbai (Bombay)
C
Pune
Hyderabad
Rangoon
Bangalore
Chennai (Madras)
Colombo

Vulnerability to Climate Change
Extreme
High
Medium
Low

C A high tide during monsoon rains floods part of Mumbai's rail system. Low-lying Mumbai (formerly Bombay) is at high risk for sea level rise; while the city's wealth and well-developed emergency response systems increase its resilience to climate hazards, much of Mumbai remains quite sensitive.

D Workers add boulders to reinforce an embankment that guards against erosion south of Kolkata (formerly Calcutta) on an island that is highly exposed to sea level rise and storm surge. Such embankments can reduce sensitivity to water-level changes, but they are expensive to build and require maintenance.

installed an electric crematorium on the riverbank. It is attracting considerable business, as a cremation in this facility costs 30 times less than a traditional funeral pyre.

Of even greater concern now is the amount of industrial waste and sewage dumped into rivers and streams (Figure 8.8A). Most sewage enters these water bodies in raw form because city sewage systems (most built by the British early in the twentieth century) long ago exceeded their capacity. In Varanasi, pumps have been installed to move the sewage up to a new and expensive processing plant, but the plant is so overwhelmed by the volume of water during the rainy season that it can process only a small fraction of the city's sewage.

l Insecurity Insufficient water
ıral activities in South Asia: the
he production of food for domes-
use of water in South Asia is for
new ways of conserving water in
: 8.7B), the demand for irrigation
For most countries in the region,
se to 50 percent of the workforce
ce contribute 17 percent or more
measures will contribute to crop
eserving the possibility of irrigat-
; done to reduce CO_2 levels will
ınd drying trends.

ıdia and Pakistan have pioneered
at which water deposited dur-
olates through the soil and into
ın evaporating. This practice has
irrigation during the dry season.
ost water of any human activity
places), drip irrigation technol-
t efficient way to irrigate, would
though, the relatively high cost
ment has hampered widespread

ues Related ge

mate change is taken, Himalayan
ear, causing South Asia's largest
uring the winter, when cool, dry air flows off the Eurasian continent and rainfall is sparse (see Figure 8.4 and the Figure 8.5 map). South Asia is pioneering some innovative responses to the multiple threats posed by global climate change. India, by far the largest country in the region, has some experience in developing and implementing emergency plans. In 2012, after many years of bureaucratic stalling, India adopted a national water-use plan, which acknowledges the many water crises likely to be caused by climate change. Although implementation may be difficult, perhaps the most useful proposal in this plan is that water can no longer be considered a free resource. For health, safety, and conservation reasons, everyone must now be required to get their water from a public source and they must somehow be charged for the water they use.

ON THE BRIGHT SIDE

Technological Solutions

To address the issues of climate warming and drought, public and private entities are using alternative energy sources to reduce CO_2 emissions to the atmosphere. For example, India, which contributes the seventh-largest amount of greenhouse gases in the world, is also home to the world's largest producer of plug-in electric cars. The Mahindra Reva E20 sells in India for about U.S.$15,000, with a government subsidy. India's small but surging middle class now has the disposable income to afford these economy cars. Even factoring in emissions from the plants that generate the electricity used to charge the cars, electric cars contribute substantially lower levels of CO_2 emissions than do gasoline- or diesel-powered cars.

Solar and wind energy are the focus of public investment by several South Asian countries because they have the potential to decrease greenhouse gas emissions. In north and west South Asia, where cloud cover is less and where energy is in greatest demand by industries and high-tech firms, the development of solar power is being emphasized. Wind energy is most efficiently produced in the wind-prone state of Tamil Nadu in southern India. Nationally, the use of wind power increased 22 percent per year from 1992 to 2010; by 2011, it constituted 70 percent of India's renewable energy generation. By 2012, however, both state and private investment in wind power dropped and nuclear and thermal energy received more attention as it became clear that there is no easy way to transport wind energy long distances to the areas where energy is in great demand.

Modern technology is also being used to address water issues. In 2008, Veer Bhadra Mishra, who was a Brahmin priest and professor of hydraulic engineering at Banaras Hindu University in Varanasi, received approval from India's central government to build a series of processing ponds that will use India's heat and monsoon rains to clean Varanasi's sewage at half the cost of other methods. Mishra also preached a contemporary religious message to the thousands who visited his temple on the banks of the sacred, magnificent Ganga River: no longer is it valid to believe that the Ganga is a goddess who purifies all she touches while assuming that it is impossible to cause her damage. Rather, Mishra said, because the Ganga is their symbolic mother, it would be a travesty for Hindus to smear her with sewage and industrial waste.

Deforestation

Deforestation has been occurring in South Asia since the first agricultural civilizations developed between 5000 and 10,000 years

Figure 8.8 Photo Essay: Human Impacts on the Biosphere in South Asia

South Asia's huge population and growing industries have had major impacts on water and air quality, as well as on the extent and health of remaining ecosystems, such as forests.

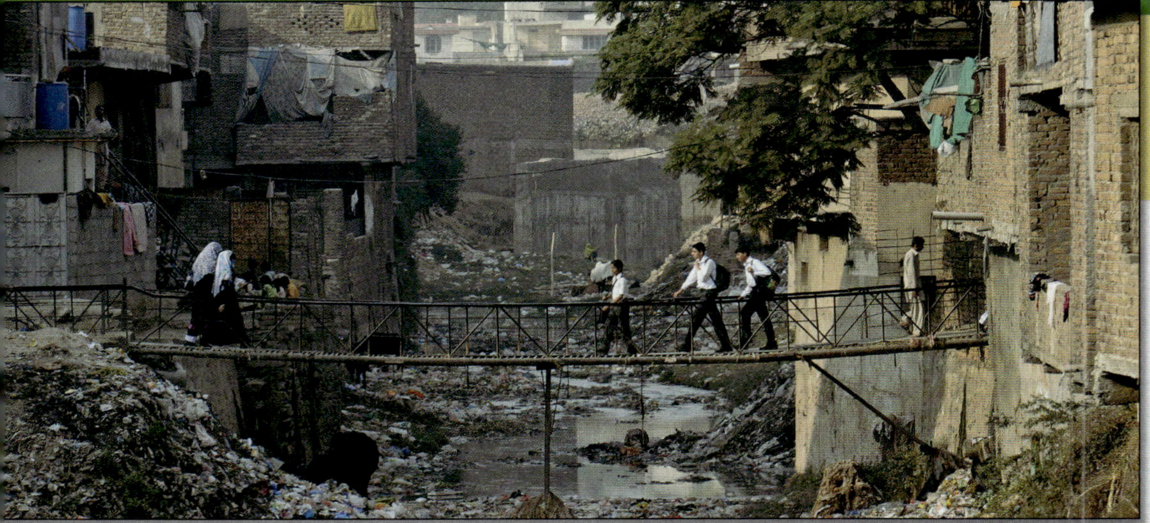

A Students walk on a bridge across a highly polluted and garbage-choked stream in a slum area in Rawalpindi, Pakistan. Nationally, only 8 percent of urban sewage is treated before it is discharged. Meanwhile, 40 percent of premature deaths in Pakistan are attributed to water-related diseases.

B A Hindu woman washes her baby in the highly polluted Ganga River in Varanasi, India. Many Hindus consider this a sacred act of ritual purification.

C A public bus billows smoke in Delhi, India, which now rivals Beijing, China, for having the worst air pollution in the world. Sixty-seven percent of air pollution in Delhi comes from vehicles.

D Women remove firewood from a highland forest in Arunachal Pradesh, India. Deforestation, driven by the needs of poor people as well as by large corporations, has resulted in habitat loss for many endangered species.

ago. Ecological historians have shown that, as the forests vanished, the northwestern regions of the subcontinent (from India to Afghanistan) became increasingly drier. The pace of deforestation has increased dramatically over the past 200 years. By the mid-nineteenth century, perhaps a million trees a year were felled for use in building railroads alone. Such radical deforestation jeopardized the well-being both of people and of animals in South Asia (Figure 8.9).

Causes of Deforestation and Resistance to It In the twenty-first century, South Asia's forests are still shrinking, due to commercial logging and expanding village populations that use wood for building and for fuel. Many of South Asia's remaining forests are in mountainous or hilly areas, where forest clearing dramatically increases erosion during the rainy season. In addition to the loss of CO_2-absorbing forests, one result of deforestation is massive landslides that can destroy villages and close roads. With fewer trees and less soil to retain the water, rivers and streams become clogged with runoff, mud, and debris. The effects can reach so far downstream that increased flooding in the plains of Bangladesh is linked to deforestation in the Himalayas.

Unlike China and many other nations facing similar problems, the countries of South Asia have a healthy and vibrant culture of environmental activism that has alerted the public to the consequences of deforestation. In 1973, for example, in the Himalayan district of Uttarakhand (then known as Uttaranchal), India, a sporting-goods manufacturer planned to cut down a grove of ash trees so that the factory, in the distant city of Allahabad, could use the wood to make tennis racquets. The trees were sacred to nearby villagers, however, and when their protests were ignored, a group of local women took dramatic action. When the loggers came, they found the women hugging the trees and refusing to let go until the manufacturer decided to find another grove.

The women's action grew into the *Chipko movement* (literally, "hugging"), which is also known as the *social forestry movement*. The movement has spread to other forest areas and has been responsible for slowing deforestation and increasing ecological awareness. Proponents of the movement argue that the management of forest resources should be turned over to local

LOCAL LIVES PEOPLE AND ANIMALS IN SOUTH ASIA

B A cow lounges in a fabric store in Varanasi, India. Cattle are allowed to roam free in some cities in India, due to the reverence Hindus widely hold for cows. The killing of cattle for any purpose is banned in 6 of India's 28 states and allowed without restriction in only 4. Scholars have found much evidence to suggest that cows have not always been sacred and were eaten by Hindus in the distant past.

C Asian elephants at work in Nagarhole National Park in Karnataka, India, carrying fallen branches and brush. Asian elephants (*Elephas maximus*) are a separate species distinct from African elephants (*Loxodonta africana*) and have, among several other unique characteristics, the largest volume of cerebral cortex of any land animal. This portion of the brain is used for memory, attention, perceptual awareness, thought, language, and consciousness. Asian elephants can be trained to follow instructions, a quality that has made them useful to humans for work and warfare for thousands of years. They are particularly prized for their ability to cross difficult terrain while carrying heavy objects such as logs.

communities. They say that people living at the edges of forests possess complex local knowledge of those ecosystems that has been gained over generations—knowledge about which plants are useful as building materials and for food, medicines, and fuel. Those who live in forested areas are more likely to manage the forests carefully because they want their descendants to benefit from forests for generations to come.

A Case Study: Nature Preserves in the Nilgiri Hills

The Mudumalai Wildlife Sanctuary and neighboring national parks in the Nilgiri Hills (part of the Western Ghats) harbor some of the last remaining forests in southern India. Here, in an area of about 600 square miles, live a few of India's last wild tigers and a dozen or more other rare species, such as sloth bears and barking deer (see Figure 8.9A). Even much smaller forest reserves play an important role in conservation. At 287 acres, Longwood Shola is a tiny remnant of the ancient tropical forests that once covered the Nilgiris.

Phillip Mulley, a naturalist, Christian minister, and leader of the Badaga ethnic group, points out that the indigenous peoples of the Nilgiris must now compete for space with a growing tourist industry (1.7 million visitors in 2005). In addition, huge tea plantations were cut out of forestlands by the Tamil Nadu state government to provide employment for Tamil refugees from the conflict in Sri Lanka. So while the forestry department and citizen naturalists are trying to preserve forestlands, the social welfare department, faced with a huge refugee population, is cutting them down. *[Source: Lydia and Alex Pulsipher and the government of Tamil Nadu. For detailed source information, see Text Credit pages.]* ■

180. TRIBAL PEOPLE IN INDIA WANT TO PROTECT INDIGENOUS WAYS OF LIFE

181. INDIA'S NATIONAL SYMBOL BECOMING MORE DIFFICULT TO SPOT

Industrial Air Pollution

In many parts of South Asia, the air as well as the water is endangered by industrial activity. Emissions from vehicles and coal-burning power plants are so bad that breathing Delhi's air is equivalent to smoking 20 cigarettes a day (see Figure 8.8C). The acid rain caused by industries up and down the Ganga River basin is destroying good farmland and such renowned heritage monuments as the Taj Mahal.

M.C. Mehta, a Delhi-based lawyer, became an environmental activist partly in response to the condition of the Taj Mahal, whose white marble was becoming pitted by acid rain. For more than 20 years, he has successfully promoted environmental legislation that has removed hundreds of the most polluting factories from India's river valleys. His efforts are also a response to a disastrous event that took place in central India in 1984. At that time, an explosion at a pesticide plant in Bhopal produced a gas cloud that killed at least 15,000 people and severely damaged the lungs of 50,000 more. The explosion was largely the result of negligence on the part of the U.S.-based Union Carbide Corporation (which owned the plant) and the local Indian employees who ran the plant. To help address the tragedy, the Indian government launched an ambitious campaign to clean up poorly regulated factories, a project that is far from complete.

Indus Valley civilization the first substantial settled agricultural communities, which appeared about 4500 years ago along the Indus River in modern-day Pakistan and northwest India

Harappa culture see *Indus Valley civilization*

THINGS TO REMEMBER

Geographic Insight 1

- **Climate Change and Water** Climate change—with the attendant sea level rise, water shortages, and crop failures—puts more lives at risk in South Asia than in any other region in the world. Low-lying, densely populated areas are particularly vulnerable to sea level rise. The melting of glaciers in the Himalayas that feed the region's major rivers and replenish aquifers during the dry season is especially worrisome.
- India has water-allocation problems that impact farmers, towns, and cities, especially along major rivers; water management in India also affects millions of people in Bangladesh.
- India now has a national water management plan; one of its chief features is the concept that water can no longer be a free resource. All must be charged for water use in some way.
- Pakistan, India, and Bangladesh have major air and water pollution problems related to industrial waste, lack of sewage treatment facilities, and end-of-life religious pilgrimages.
- The Chipko ("tree hugging") social forestry movement in Uttarakhand, India, has spread to other forest areas, slowing deforestation and increasing ecological awareness.

Human Patterns over Time

A variety of groups have migrated into South Asia, many of them as invaders who conquered peoples already there. Despite much blending over the millennia, the continued coexistence and interaction of many of these groups make South Asia both a richly diverse and an extremely contentious place.

The Indus Valley Civilization

There are indications of early humans in South Asia as far back as 200,000 years ago, but the first real evidence of modern humans in the region is about 38,000 years old. The first large agricultural communities, known as the **Indus Valley civilization** (or **Harappa culture**), appeared about 4500 years ago along the Indus River in what is modern-day Pakistan and northwest India. The architecture and urban design of this civilization were quite advanced for the time. Homes featured piped water and sewage disposal. Towns were well planned, with wide, tree-lined boulevards laid out in a grid. Evidence of a trade network that extended to Mesopotamia and eastern Africa has also been found.

Vestiges of the Indus Valley civilization's agricultural system survive to this day in parts of the valley, including techniques for storing monsoon rainfall to be used for irrigation in dry times (shown in Figure 8.13A on page 344). Remnants of language, and possibly superficial biological traits such as skin color,

dian peoples of southern India,
ıe Indus region beginning about

f the decline of the Indus Valley
scholars believe that complex
l (drier climate) changes brought about a gradual demise. Others argue that foreign invaders brought a swift collapse, instigating out-migration.

A Series of Invasions

The first recorded invaders to join the indigenous people of South Asia came from Central and Southwest Asia into the rich Indus Valley and Punjab about 3500 years ago. Many scholars believe that these people, referred to as Indo-European (a linguistic term), in conjunction with those of the Harappa and other indigenous cultures, instituted some of the early elements of classical Hinduism, the major religion of India today. One of those elements was the still-influential caste system (discussed on pages 350–351), which divides society into hereditary hierarchical categories.

Wave after wave of other invaders arrived, including the Persians, the armies of the Greek general Alexander the Great, and numerous Turkic and Mongolian peoples. Defensive structures against these invaders can still be found across northwest South Asia. Jews came to the Malabar Coast of Southwest India more than 2500 years ago, Christians shortly after the time of Jesus. Arab traders came by land and sea to India long before the emergence of Islam; and then starting about 1000 years ago, Arab traders and religious mystics introduced Islam to what are now Afghanistan, Pakistan, and northwest India. By sea, the Arabs brought Islam to the coasts of southwestern India and Sri Lanka. In 1526, the **Mughals**, a group of Turkic-Persian people from Central Asia, invaded from the north, intensifying the growth of Islam. The Mughals reached the height of their power and influence in the seventeenth century, controlling the north-central plains of South Asia. The last great Mughal ruler (Aurangzeb) died in 1707, but the cultural legacy of the Mughals remained, even as the power and range of the dynasty declined. One aspect of this legacy is the 520 million Muslims now living in South Asia. The Mughals also left a unique heritage of architecture, art, and literature that includes the Taj Mahal, miniature painting, and a rich tradition of lyric poetry (see Figure 8.13B on page 344). The Mughals contributed to the evolution of the Hindi language, which became the language of trade of the northern subcontinent and which is still used by more than 400 million people.

Mughals a dynasty of Central Asian origin that ruled India from the sixteenth century to the nineteenth century

As Mughal rule declined, a number of regional states and kingdoms rose and competed with one another (Figure 8.10). The absence of one strong power created an opening for yet

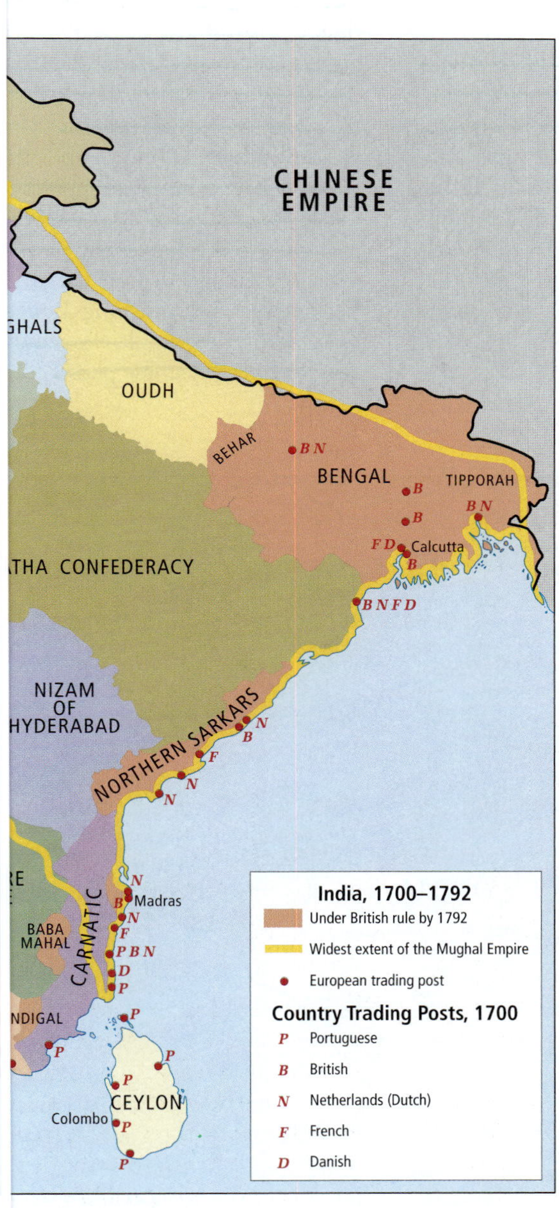

Asia. By 1700, several European nations had
coast of India and Ceylon (now Sri Lanka). After
ıngzeb in 1707, the Mughals' ability to assert strong
declined. A number of emergent regional states
ritory and power. Among the strongest was the
f a number of small states dominated by the Maratha
martial skills. Weakness of administrative control at
tween these South Asian states paved the way for
eighteenth century.

another invasion. By the late 1700s, several European trading companies were competing to gain a foothold in the region. Of these, Britain's East India Company was the most successful. By 1857, the East India Company, acting as an extension of the British government, put down a rebellion against European intrusion and became the dominant power in the region.

Globalization and the Legacies of British Colonial Rule

The British controlled most of South Asia from the 1830s through 1947 (Figure 8.11). By making the region part of the British Empire, the British accelerated the process of globalization in South Asia, transforming the region politically, socially, and economically. Even areas not directly ruled by the British felt the influence of their empire. Afghanistan repelled British attempts at military conquest, but the British continued to intervene there, trying to make Afghanistan a "buffer state" between British India and Russia's expanding empire. Nepal remained only nominally independent during the colonial period, and Bhutan became a protectorate of the British Indian government.

The Deindustrialization of South Asia As in their other colonies, the British used South Asia's resources primarily for their own benefit, often with disastrous results for South Asians. One example was the fate of the textile industry in Bengal (modern-day Bangladesh and the Indian state of West Bengal).

Bengali weavers, long known for their high-quality muslin cotton cloth, initially benefited from the increased access that traders gave them to overseas markets in Asia, the Americas, and Europe. By 1750, South Asia had an advanced manufacturing economy that produced 12 to 14 times more cotton cloth than Britain alone and more than all of Europe combined. However, during the eighteenth and nineteenth centuries, Britain's own highly mechanized textile industry—based on cotton grown in India, various other colonies, and the American South—developed cheaper cloth that then replaced Bengali muslin. This shift happened first in the British colonies in the Americas, then in Europe, and eventually throughout South Asia. The British textile industry was further aided by the British East India Company, which began severely punishing Bengalis who continued to run their own looms. As a result, the venerable South Asian textile industry survived in only a few places (see Figure 8.13C on page 345). As

FIGURE 8.11 The British Indian Empire, 1860–1920. After winning control of much of South Asia, Britain controlled lands from Baluchistan to Burma, including Ceylon and the islands between India and Burma.

ıl official put it, "While the mills s of Bengali weavers bleached on

ıhed out of their traditional liveli- were compelled to find work as h Asia already had an abundance grated to emerging urban centers. ed an already difficult situation, ›le starved to death. Throughout events forced millions of South of indentured laborers migrating Americas, Africa, Asia, and the are still to be found.

VITH INDIAN ACCENTS

ative Legacies Contemporary in institutions put in place by vast empire. These governments ıings of their colonial forebears, ·ocedures, resistance to change, ›f from the people they govern. ·oved functional over time, there il disturbances in virtually every nts were not instituted on a large empire, but since independence 'e been able to use the freedoms their concerns and to make many overnments. Still, the struggle to titutions continues.

The tremendous changes brought istance movements among South itant movements intent on push- h as the unsuccessful rebellion of y the British East India Company ı as the Indian National Congress cal means to agitate for greater he route to South Asian political ıilitant and political actions were ey movements gained worldwide ruggle, were successful.

rategy In the early twentieth cen- ıg lawyer from Gujarat, emerged South Asia's independence move- **bedience** to nonviolently defy laws eriminated against South Asians. ceful protesters, he would notify was about to break a discrimina- ıored the act, the demonstrators ıd the law would be weakened. If rce against the peaceful demon- t of the masses. Throughout the ıe was used to slowly but surely oss South Asia. The most famous f 1930, when Gandhi led tens of ı to the sea where they made salt

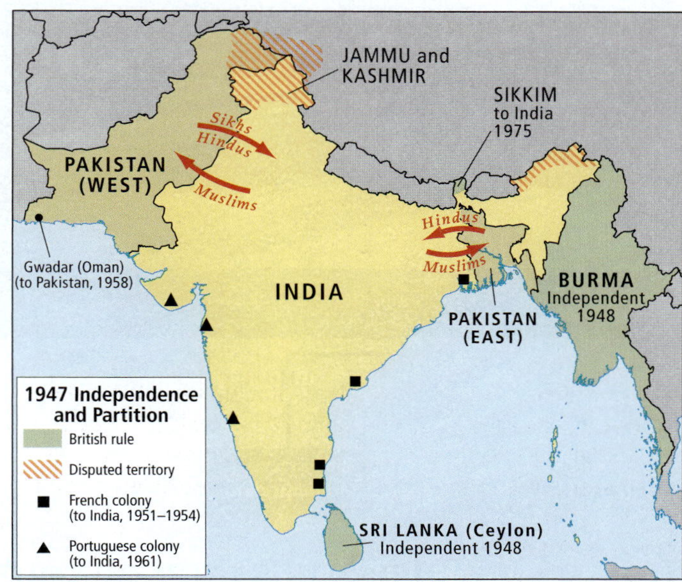

FIGURE 8.12 Independence and Partition. India became independent of Britain in 1947, and by 1948, the old territory of British India was partitioned into the independent states of India and East and West Pakistan. The Jammu and Kashmir region was contested space, and remains so today. Sikkim went to India, and both Burma and Sri Lanka became independent. Following additional civil strife, East Pakistan became the independent country of Bangladesh in 1971.

by evaporating seawater, thus breaking the law that made it illegal for South Asians to produce salt (see Figure 8.13D on page 345). The purpose of the law had been to facilitate British rule by controlling a vital human nutritional necessity. Breaking the law on such a massive scale created an international media frenzy that catapulted Gandhi to global notoriety; it moved millions of South Asians to support independence from Britain. The hunger strikes described in the vignette that opens this chapter about the Sardar Sarovar Dam on the Narmada are another strategy that reflect Gandhi's legacy of nonviolent political protest.

The Partition of India in 1947 When British India was granted independence by Britain in 1947, it was divided into two independent countries: India, which was predominantly Hindu; and Pakistan (East and West), which was predominantly Muslim (Figure 8.12). This division—called **Partition**—which Gandhi greatly lamented, was perhaps the most enduring and damaging outcome of colonial rule. (Afghanistan, Bhutan, and Nepal were never officially British colonies; Ceylon [now Sri Lanka] became independent in 1948.)

Muslim political leaders, concerned about the fate of a minority Muslim population in a united India with a Hindu majority, first suggested the idea of two nations. Though Partition was highly controversial, it became part of the independence agreement between the British and the

civil disobedience protesting of laws or policies by peaceful direct action

Partition the breakup following Indian independence that resulted in the establishment of Hindu India and Muslim Pakistan

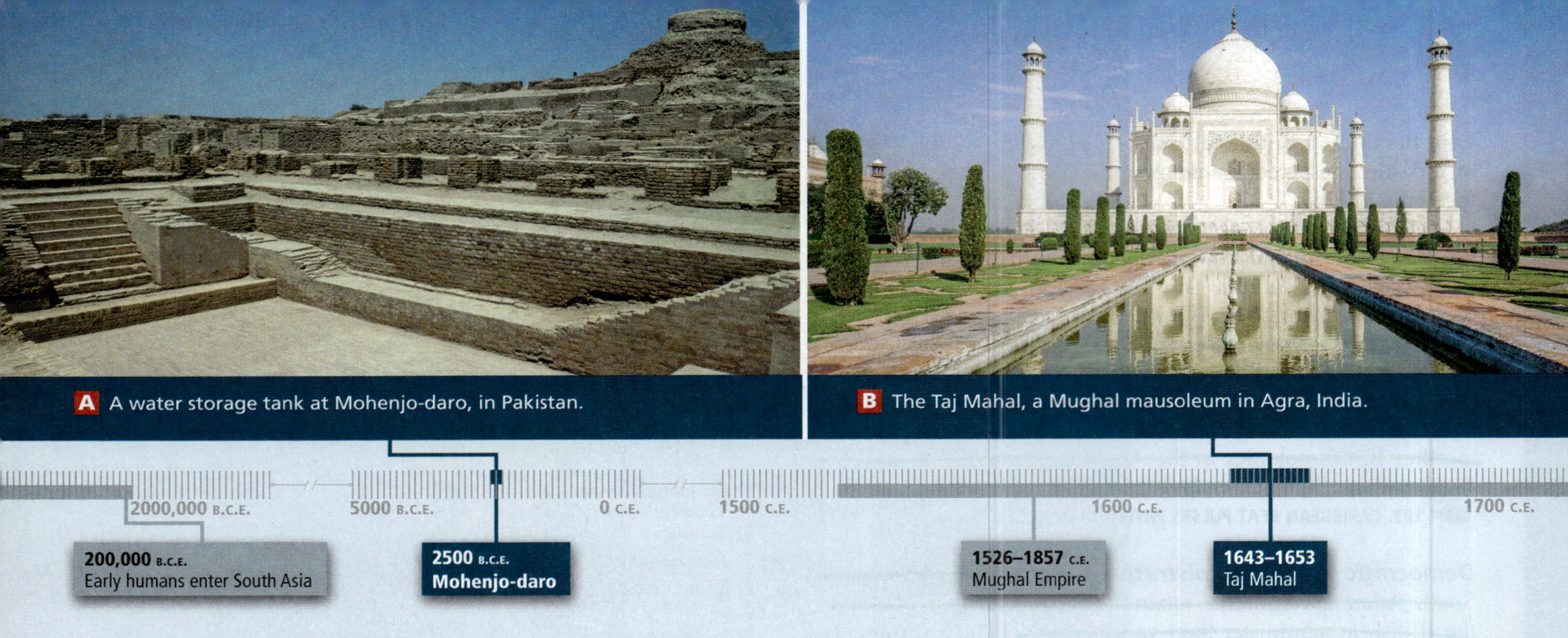

A A water storage tank at Mohenjo-daro, in Pakistan.

B The Taj Mahal, a Mughal mausoleum in Agra, India.

FIGURE 8.13 A VISUAL HISTORY OF SOUTH ASIA

Thinking Geographically

After you have read about the human history of South Asia, you will be able to answer the following questions:

A This structure was designed to store water from what source?

B Suggest some principle features of this structure and grounds.

Indian National Congress (India's principal nationalist party). Northwestern and northeastern India, two very different places historically and culturally, but where the populations were predominantly Muslim, became a single country consisting of two parts, known as West and East Pakistan, separated by northern India (see Figure 8.12). Although both India and Pakistan maintained secular constitutions, with no official religious affiliation, the general understanding was that Pakistan would have a Muslim majority and India a Hindu majority. Fearing that they would be persecuted if they did not move, more than 7 million Hindus and Sikhs migrated to India from their ancestral homes in what had become West or East Pakistan. A similar number of Muslims left their homes in India for one of the parts of Pakistan. In the process, civil society broke down. Families and communities were divided, looting and rape were widespread, and between 1 and 3.4 million people were killed in numerous local outbreaks of violence. In 1971, after bloody conflict bordering on civil war, Pakistan was divided into Bangladesh (formerly East Pakistan) and Pakistan (formerly West Pakistan). **184. 60TH ANNIVERSARY OF INDIA–PAKISTAN PARTITION ON AUGUST 15TH**

Partition was the tragic culmination of the divide-and-rule approach the British used throughout their empire (see page 306 in Chapter 7). The approach heightened tensions between South Asian Muslims and Hindus, thus creating a role for the British as seemingly indispensable and benevolent mediators. Instead of relieving tensions, the partition of India and Pakistan laid the groundwork for the wars and skirmishes, strained relations, and the ongoing arms race between India and Pakistan that persist to this day.

The Post-Independence Period In the more than 60 years since the departure of the British, South Asians have experienced both progress and setbacks. Democracy has expanded steadily, albeit somewhat slowly. India is now the world's most populous democracy. It is gradually dismantling age-old traditions that hold back poor, low-caste Hindus, women, and other disadvantaged groups, and a vibrant, if still small, middle class is emerging (see Figure 8.13E). Pakistan has had repeated elections and some of its governments have been modestly effective, but militaristic authoritarian and corrupt regimes have been more the norm, with Pakistan's and India's long border feud (discussed on page 368) sapping resources and national spirit. Bangladesh, despite the damage it sustained in its war of independence from Pakistan in 1971, has actually had a more stable, responsive, and less militaristic (though not trouble-free) government than Pakistan.

Under British rule, agricultural modernization lagged, and during World War II, the Bengal Famine took an estimated 4 million lives. After independence, progress in agricultural production was slow until the late 1960s, when the green revolution brought marked improvements (see pages 361–362). The move to modernized farming on large tracts of land with far fewer agricultural workers brought prosperity to some South Asians, relieved food scarcities, and made food exports possible; but it also forced millions to migrate to the cities. This intensified already vast economic disparities in urban South Asia (see Figures 8.27A, B on page 360).

Industrialization became a main goal after independence, partly in response to the dismantling of industry during the colonial period. The emphasis on technical training has produced

C T
Beng

y in

D Gandhi leading the Salt March to the sea.

E A shopping mall outside New Delhi, which is a center of employment for India's growing middle class.

C V
of Br
than

hat as of 1750, before the onset
ced 12 to 14 times more cotton
nbined?

D What was the immediate result of the Salt March?

E What does this photo suggest about India's modern economy.

lled engineers whose talents are
n most countries in the region,
vice economies now constitute
agriculture (which nonetheless
or more of the workforce). The
tor is growing especially rapidly

REMEMBER

on dating to 4500 years ago is
any innovations in water

y times, primarily from the north by
ongolia.

- Following some 300 years of Mughal rule, the British controlled most of South Asia from the 1830s through 1947, profoundly influencing the region politically, socially, and economically.
- British colonial rule in South Asia channeled resources to Europe, and in so doing depressed flourishing industries and inhibited development.
- In the early twentieth century, Mohandas Gandhi emerged as a central political leader of India's independence movement, using nonviolent civil disobedience that eventually led to independence.
- Instead of relieving tensions, Partition—the 1947 division of British India into two countries, India and Pakistan—laid the groundwork for the repeated wars and skirmishes, strained relations, and ongoing arms race between India and Pakistan.

CURRENT GEOGRAPHIC ISSUES

Asia can be overwhelming. This
is urbanizing rapidly and as this
aily life in villages and cities are
alities that touch all lives.

sian village or urban neighbor-
le cultural variety. Differences

based on caste, economic class, ethnic background, gender, religion, and even language are usually accommodated peacefully by long-standing customs, such as religious and ethnic festivals and foodways (**Figure 8.14**; see also Figure 8.19 on page 352) that guide cross-cultural interaction. Because of this cultural variety, everyone in South Asia is, in one way or another, a member of a minority. The widespread experience of being part of a minority group is to some extent an equalizing factor and has become part of the national dialogue about difference.

The Texture of Village Life

The vast majority—about 70 percent—of South Asians live in hundreds of thousands of villages. Even many of those now living in South Asia's giant cities were born in a village or occasionally visit an ancestral rural community, so for most people village life is a lived experience.

VIGNETTE The writer Richard Critchfield, who studied village life in more than a dozen countries, wrote that the village of Joypur (Bangladesh) in the Ganga-Brahmaputra Delta is set in "an unexpectedly beautiful land, with a soft languor and gentle rhythm of its own." In the heat of the day, the village is sleepy; naked children play in the dust, women meet to talk softly in the seclusion of courtyards, and chickens peck for seeds.

In the early evening, mist rises above the rice paddies and hangs there "like steam over a vat." It is then that the village comes to life, at least for the men. The men and boys return from the fields, and after a meal in their home courtyards, the men come "to settle in groups before one of the open pavilions in the village center and talk—rich, warm Bengali talk, argumentative and humorous, fervent and excited in gossip, protest, and indignation." They discuss their crops, an upcoming marriage, or national politics. *[Source: Richard Critchfield. For detailed source information, see Text Credit pages.]* ■

VIGNETTE The anthropologist Faith D'Aluisio and her colleague Peter Menzel give another peek into village life as night falls in Ahraura, a village in the state of Uttar Pradesh in north-central India. In the enclosed women's quarters of a walled compound, Mishri is finishing her day by the dying cooking fire as her 1-year-old son tunnels his way into her sari to nurse himself to sleep. Mishri, who is 27, lives in a tiny world bounded by the walls of the courtyard she shares with her husband, five children, and several of her husband's kin. Like many villages in northern India, her village observes the practice of *purdah*, in which women keep themselves apart from men (see the discussion on page 354).

That Mishri can observe purdah is a mark of status because it shows she need not help her husband in the fields. Within the compound, she works from sunup to sundown, chatting only momentarily with two women who cover their faces and scurry from their own courtyards to hers for the short visit. Mishri is devoted to her husband, who was chosen for her by her family when she was 10; out of respect, she never says his name aloud. *[Source: Adapted from Faith D'Aluisio and Peter Menzel, Women in the Material World, 1996.]* ■

The Texture of City Life

South Asian cities are often depicted as having split identities. On the one hand are sleek, modern skyscrapers bearing the logos of powerful global high-tech firms, and professional workshops filled with bright, motivated students (Figure 8.15C on page 348); and on the other hand are chaotic, crowded, and violent urban environments with overstressed infrastructures and menial jobs (see Figure 8.15B, D). Many South Asian cities exemplify this dichotomy; Mumbai is chief among them.

Bombay is the name by which most Westerners know South Asia's wealthiest city. It is now called Mumbai, after the Hindu goddess Mumbadevi. Mumbai has the largest deepwater harbor on India's west coast. Its metropolitan area, with over 21 million people, hosts India's largest stock exchange, has multiple IT firms, and is the home of the nation's central bank. It pays about

FIGURE 8.14 LOCAL LIVES **FESTIVALS IN SOUTH ASIA**

A A village festival in Pakistan features *kabaddi*, a popular South Asian sport in which teams take turns sending a "raider" across a field center line. That person must tag, or in some cases wrestle to the ground, members of the other team and then return to his or her own side without taking a breath. Kabaddi has been played at the Indian National Games since 1939 and at the Asian Games since 1991.

B Celebrants in Kolkata, India, during Holi—a festival celebrating the end of winter and beginning of spring. Holi evolved from temple worship practices involving the application of color to statues. In a riotous and celebratory atmosphere, people of different ages, genders, castes, and economic backgrounds temporarily disregard their differences and hurl the colors of the coming spring at each other.

C Pilgrims during the 2010 Kumbh Mela bathe in the Ganga River at Haridwar. During this event, which is held every 3 years, Hindus purify themselves by bathing in the sacred waters of the river. In 2013, the 45-day-long event attracted over 100 million participants.

the entire country and brings in
le revenue. Its annual per capita
's next wealthiest city, the capital

ends into the realm of culture
reative arts industries, including
ndustry. The industry is known as
es popular Hindi movies portray-
onflicts. (The term *Bollywood* is
Hollywood.) The stories, played
ied by popular music and dance,
r huge audiences from the physi-

e elegant high-rise condomini-
growing middle class, Mumbai's
treet level, the urban landscape
bers of people living on the side-
buildings, and in large, rambling
. The largest of the shantytowns
o a million people in less than
ntain 15,000 one-room factories
ucts that are sold in the global
om India's recycled plastic and
wnship in Karachi, Pakistan, for
ost populous slums, but it is also
he middle class, at least among

o urban life in South Asia. Were
s widely separated and culturally
Kathmandu, and Peshawar, one
main avenues and shantytowns,
ds of tightly compacted, reconsti-
of living are relatively high and
r, not anonymous as in Western

t fishing village that predates the
and is now squeezed between
es and the Bay of Mumbai
villagers still fish every day. Koli
packed homes ringed by fishing
buses is soon lost in quiet calm
ed passageway that winds through
le directions. At first Koli appears
ointed homes, some with marble
onto the dimly lit but pleasant
this is no warren of destitute
mmunity of educated bureaucrats,
nstitute South Asia's rising urban
notes of Lydia and Alex Pulsipher and

ercent of the region's popula-
Asia has several of the world's
2012, Mumbai had 22.9 mil-
n; Delhi, 19.5 million; Dhaka,

15.4 million; and Karachi, 11.1 million. All of these cities have grown quickly, and South Asia's current urban population of about 460 million people could expand to as many as 712 million by 2025.

Many middle-class South Asians move to cities for education, training, or business opportunities (see Figure 8.15C). Because people come to cities where schooling is more available, large cities have higher literacy rates than rural areas. Mumbai and Delhi both have literacy rates above 80 percent, approximately 25 percent above the average for the country; Dhaka, in Bangladesh, and Karachi, in Pakistan, both have 63 percent literacy, more than 15 percent above the rate for each country as a whole.

Like sub-Saharan Africa, the cell phone is bringing urban-style communication into the countryside. Mumbai-based reporter Anand Giridharadas found that over half the population of South Asia now has access to a cell phone. Even though this access is unevenly distributed, for many the cell phone has become a means of creating social privacy and sometimes circumventing social norms. With a cell phone, a person can conduct relationships, keep secrets, access information, and generally undermine the oppression of authority hierarchies. One measure of how rapidly cell phone technology has taken over is that many more South Asians now have a cell phone than have the use of a flush toilet. A result of this rapid change is that the expectations young South Asians have about what the future holds for them are rising briskly, with IT access viewed as essential.

Risks to Children in Urban Settings In rural areas, by age 5 a child can perform useful tasks, thereby contributing to family well-being. But what about children in urban areas? A report by the International Labor Organization (2008) found that 23 million children between the ages of 5 and 14 are working in South Asia, many of them in urban areas. They work primarily in the informal sector as domestic servants; in export-oriented factories like those small enterprises in urban slums; and as dump scavengers. Some are forced into the sex trade. Clearly, many of these occupations are entirely unsafe and inappropriate for children who should be protected from such exploitation.

But if we look more closely at one occupation that children have, that of carpet weaving, some interesting issues arise. Carpet weaving is an ancient artistic and economic enterprise in South Asia; traditionally, it has been a family-run enterprise, with women and children the weavers and men the merchants. For thousands of years, young children have learned weaving skills from their parents and have become proud members of the family's home-based production unit. But in today's rapidly modernizing society, the role of children as family workers must be balanced with the need to have children attend school. One of the chief benefits of urban life is education, where children learn the skills that will enable them to survive and prosper in a modern economy. All citizens need to read, write, and do math.

As global trade has increased the demand for fine, handwoven carpets, most of the profit has gone not to the weaving families but to South Asian middlemen and to foreign traders from Europe

Figure 8.15 Photo Essay: Urbanization in South Asia

While only 30 percent of South Asia's population is urban, the region is home to some of the world's largest cities, such as Mumbai, Delhi, Dhaka, Kolkata, and Karachi. Throughout the area, between 50 and 90 percent of the urban population live in slums plagued by shoddy construction and inadequate access to water and sanitation.

A High-rise luxury apartments overlooking Dhobi Ghat slum in Mumbai, India, illustrate the enormous disparities in wealth found in urban South Asia.

10,000 people work in Dhobi Ghat, washing clothes and linens from hotels and hospitals in the concrete enclosures shown here.

Dwellings of people who work as clothes washers.

AFGHANISTAN
PAKISTAN
NEPAL
BHUTAN
INDIA
BANGLADESH
SRI LANKA
Kabul 134
Rawalpindi 144
Faisalabad
Lahore 49
Delhi 10
Jaipur 125
Lucknow
Kanpur 111
Patna 135
Karachi 26
Ahmadabad 69
Surat 131
Nagpur
Nashick 140
Mumbai (Bombay) 4
Pune 59
Hyderabad 50
Kolkata 13
Dhaka 14
Chittagong 102
Bangalore 47
Chennai 41

Population Living in Urban Areas

83%–100%
65%–82%
47%–64%
29%–46%
11%–28%
No data

Population of Metropolitan Areas 2013

20 million
10 million
5 million
3 million

Note: Symbols on map are sized proportionally to metro area population

(1) **Global rank** (population 2013)

mi 0 400
km 0 600

B A bicycle rickshaw driver in New Delhi, India. Most poor, rural South Asians who move to the cities work in low-paying, physically demanding jobs.

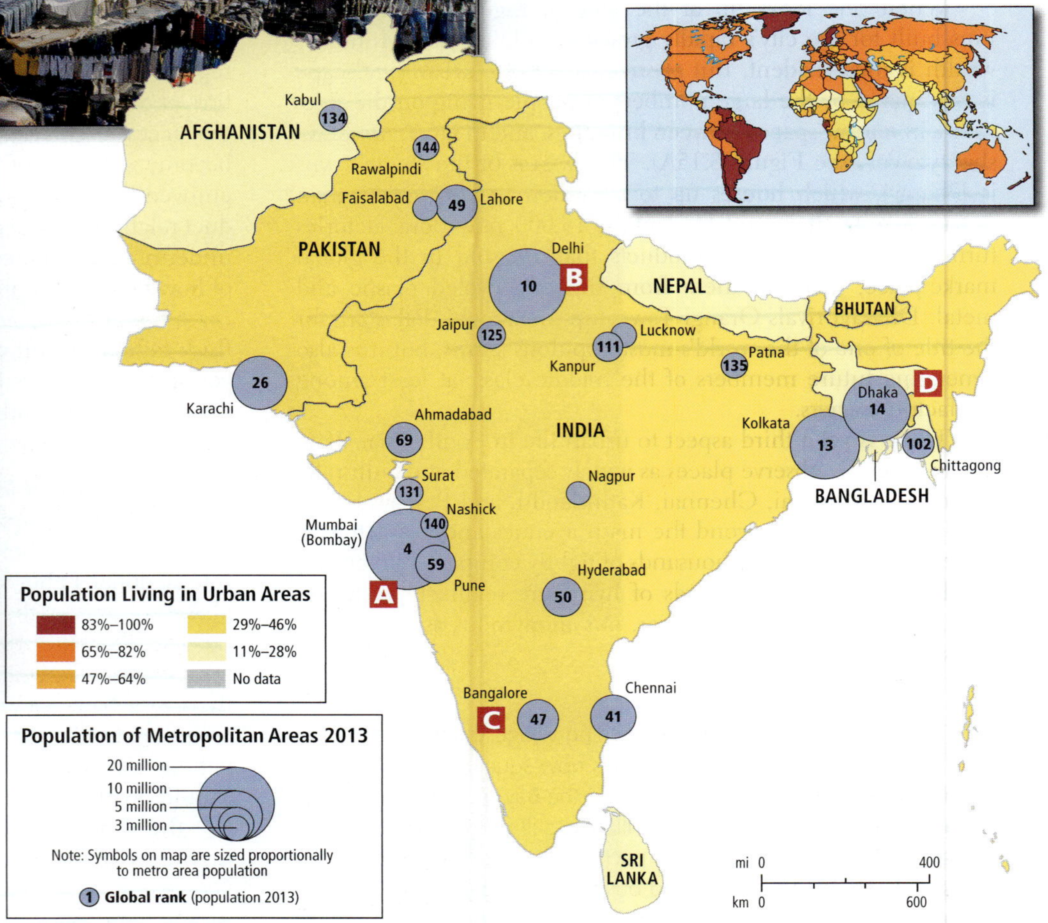

C A computer class in Mysore, India. Many middle class South Asians come to cities and stay there in order to take advantage of educational opportunities and the availability of jobs.

D A slum in Dhaka, Bangladesh, that grew up in a railroad right-of-way. Many slums develop in areas with clear risks to human habitation and few safeguards against potentially deadly hazards.

and America. The possibility of making large profits has led some unscrupulous carpet merchants to set up factories where kidnapped children are forced to produce carpets. Obviously, kidnapping and enslavement must be stopped, but the question remains: Is there room for cottage industry employment of children within the family circle?

ON THE BRIGHT SIDE

Can Child Labor Be "Fair Labor"?

The United Nations, South Asian governments, and NGOs like RugMark are now addressing this issue of child labor in the handwoven carpet industry. They have instituted an active program to curb child labor abuses while remaining open to the positive experience for a child of learning a skill and being part of a family production unit. India has established a national system to certify that exported carpets are made in shops where the children go to school, have an adequate midday meal, and receive basic health care. Such carpets bear the label "Kaleen" or "RugMark."

Risks to Women and Men in Urban Settings When rural adults move to cities, the loss of village constraints on behavior, the need to find a way to make a living, the crowded conditions, the lack
ew experiences can leave them
their own poor judgment.
purdah (seclusion) can leave
xperienced and undereducated
most often recruited or forced
l acts for a fee. The vast, urban, slum-based brothels in which sex workers work are legendary in the cities of South Asia and have been documented in, among other places, *Half the Sky*, a book and video by Nicholas Kristof and Sheryl WuDunn.

Brothels are notoriously exploitative of women wherever they are found; but the age of the mobile phone is changing the geography of sex work, at once allowing sex workers to move out of brothels, control those who will become their clients, and manage their incomes, which used to be seized by pimps and madams. However, being spatially autonomous also leaves them less likely to have access to condoms or to learning about how to avoid HIV exposure, for which sex workers are at high risk.

Rural men new to cities are also at risk; on-the-job accidents, exposure to HIV, extreme pressure to support families on tiny wages, and the loss of camaraderie with fellow villagers are but a few of the hardships felt by men.

sex work the provision of sexual acts for a fee

i. Now surrounded by the vibrant
ortuguese colonists arrived, Koli
beautiful bay. The village remains
li, but this exterior view of Koli
sight to strangers, dwellings have
esidents are educated and work
r of the city.

Language and Ethnicity

There are many distinct ethnic groups in South Asia, each with its own language or dialect. In India alone, 18 languages are officially recognized, but there are actually hundreds of distinct languages. This complexity results partly from the region's history of multiple invasions from outside. However, some groups were isolated for long periods of time, which also contributed to this complexity. As shown in Figure 8.17, the Dravidian language-culture group, represented by numbers 14 through 19, is an ancient group that predates the Indo-European invasions by a thousand years or more. Today, Dravidian languages are found mostly in southern India, but a small remnant of the extensive Dravidian past can still be found in the Indus Valley in south-central Pakistan.

By the time of British colonization, Hindi—an amalgam of Persian-based and Sanskrit-based Indo-European languages—was the dominant language throughout northern India and what is now Pakistan. Today, variants of Hindi serve as national languages for both India and Pakistan (there, called Urdu), though it is the first, or native, language of only a minority. English is a common second language throughout the region. As the language of the colonial bureaucracy, English remains a language used at work by professional people of all categories. Between 10 and 15 percent of South Asians speak, read, and write English. Many others use a version of spoken English.

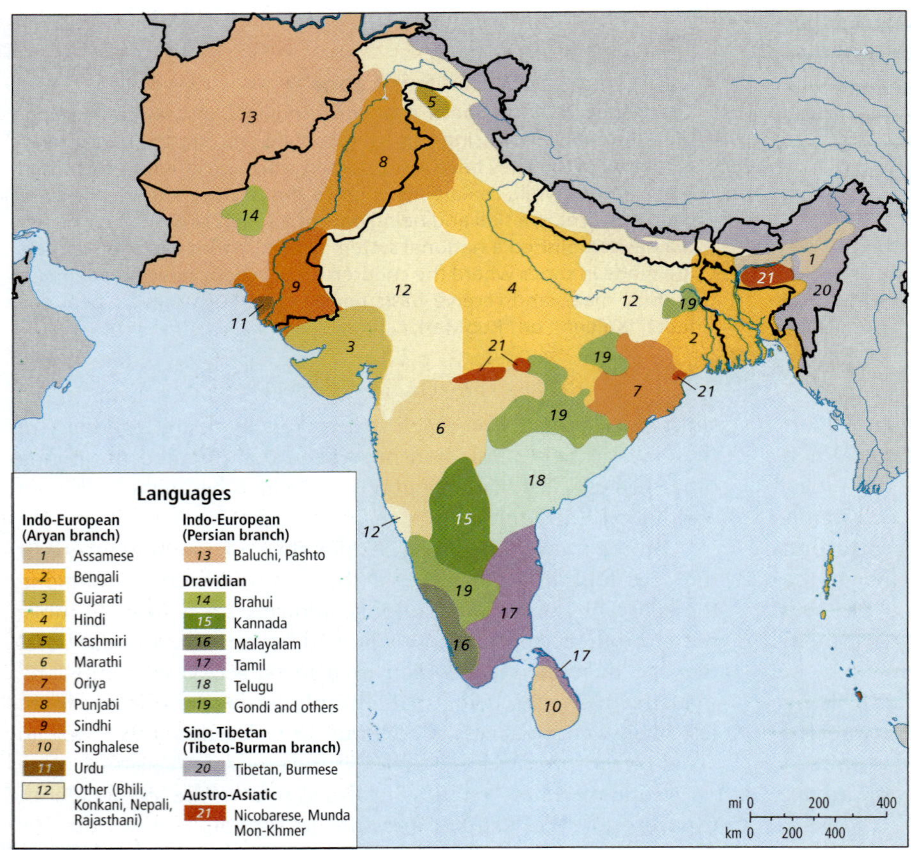

FIGURE 8.17 Major language groups of South Asia. The modern pattern of language distribution in South Asia is testimony to the fact that this region has long been a cultural crossroads.

Religion, Caste, and Conflict

The main religious traditions of South Asia are Hinduism, Buddhism, Sikhism, Jainism, Islam, and Christianity (Figure 8.18). (For a discussion of Islam, Christianity, and Judaism, see pages 248–250 in Chapter 6.)

Hinduism a major world religion practiced by approximately 900 million people, 800 million of whom live in India; a complex belief system, with roots both in ancient literary texts (known as the *Great Tradition*) and in highly localized folk traditions (known as the *Little Tradition*)

caste system a complex, ancient Hindu system for dividing society into hereditary hierarchical classes

jati in Hindu India, the subcaste into which a person is born, which traditionally defined the individual's experience for a lifetime

varna the four hierarchically ordered divisions of society in Hindu India underlying the caste system: *Brahmins* (priests), *Kshatriyas* (warriors/kings), *Vaishyas* (merchants/landowners), and *Sudras* (laborers/artisans)

Hinduism is a major world religion practiced by approximately 900 million people, 800 million of whom live in India. It is a complex belief system, with roots both in ancient literary texts (known as the *Great Tradition*) and in highly localized folk traditions (known as the *Little Tradition*).

The Great Tradition is based on 4000-year-old scriptures called the *Vedas*. Its major tenet is that all gods are merely illusory manifestations of the ultimate divinity, which is formless and infinite. Some devout Hindus worship no gods at all, and instead engage in meditation, yoga, and other spiritual practices designed to bring people to a state described as "infinite consciousness." The average person, however, is thought to need the help of personified divinities in the form of gods and goddesses (the Little Tradition). While all Hindus recognize some deities (such as Vishnu, Shiva, Ganesh, and Krishna), many deities are found only in one region, one village, or even one family.

Some beliefs are held in common by nearly all Hindus. One is the belief in reincarnation, the idea that any living thing that desires the illusory pleasures (and pains) of life will be reborn after it dies. A reverence for cows, which are seen as only slightly less spiritually advanced than humans, also binds all Hindus together (see Figure 8.9B). This attitude, along with the Hindu prohibition on eating beef, may stem from the fact that cattle have been tremendously valuable in rural economies as the primary source of transport, field labor, dairy products, fertilizer, and fuel (dried animal dung is often burned).

Caste, an Explanation Hinduism includes the **caste system**, a complex and ancient way of dividing society into hereditary hierarchical categories (see Figure 8.23 on page 356). One is born into a given subcaste, or community (called a ***jati***), that traditionally defined much of one's life experience—where one would live, where and what one could eat and drink, with whom one would associate, one's marriage partner, and often one's livelihood. The classical caste system has four main divisions or tiers, called ***varna***, within which are many hundreds of *jatis* and sub-*jatis*, which vary from place to place.

Brahmins, members of the priestly caste, are the most advantaged in caste hierarchy. Thus they must conform to those behaviors that are considered most ritually pure (for example, strict vegetarianism and abstention from alcohol). As is the case with many castes, Brahmins are found in many occupations outside their place as priests in the *varna* system (as barbers and hairdressers, for example). Below Brahmins, in descending rank are *Kshatriyas*, who are warriors and rulers; *Vaishyas*, who are landowning (small-plot) farmers and merchants; and *Sudras*, who are low-status laborers and artisans. A fifth group, the *Dalits*—"the oppressed," or untouchables—is actually considered to be so lowly as to have no caste. Dalits perform those tasks that caste Hindus consider the most despicable and ritually polluting: killing animals, tanning hides, cleaning, and disposing of refuse. A sixth group, also outside the caste system, is the *Adivasis*, who are thought to be descendants of the region's ancient aboriginal inhabitants.

Although *jatis* are associated with specific subcategories of occupations, in modern economies, this aspect of caste is more symbolic than real. Members of a particular *jati* do, however, follow the same social and cultural customs, dress in a similar manner, speak the same dialect, and tend to live in particular neighborhoods or villages. This spatial separation arises from

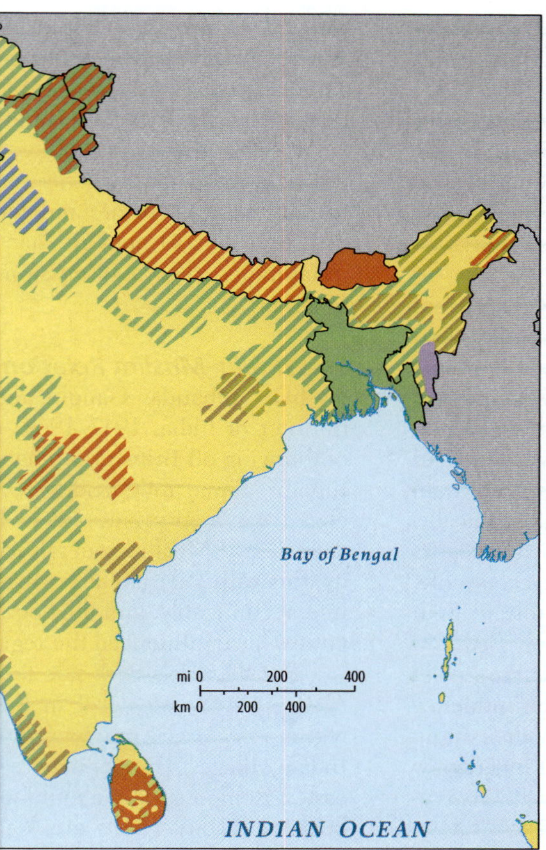

South Asia. Notice that while
are variable and overlapping
traditions in India, Nepal, and
an, Bangladesh, Pakistan, and
is diversity.

ears of ritual pollution, such as sharing of water or food with the familiar space of one's own able circle of family and friends y in times of trouble. The social lps explain the persistence and outsiders, seems to put a burden er ranks.

caste and class are not the same status, and there are class difcause of differences in wealth. Brahmins and Kshatriyas) owned and lower-caste groups (Sudras) class tended to coincide. There Today, as a result of legally manl economic opportunities, caste ted. Some Vaishyas and Sudras nd extraordinarily wealthy busin families struggle to achieve a By and large, however, Dalits

Caste, Politics, and Culture In the twentieth century, Mohandas Gandhi began an organized effort to eliminate discrimination against "untouchables." As a result, India's constitution bans caste discrimination. However, in recognition that caste is still hugely influential in society, upon independence from Britain, India began an affirmative action program. The program reserves a portion of government jobs, places in higher education, and parliamentary seats for Dalits and Adivasis. Together, the two groups now constitute approximately 23 percent of the Indian population and are guaranteed 22.5 percent of government jobs. Extended in 1990, this program now includes other socially and educationally "backward castes" (the term used in India), such as disadvantaged *jatis* of the Sudras caste, allotting them an additional 27 percent of government jobs. However, reserving nearly half of government jobs in this way has resulted in considerable controversy. In 2006, medical students successfully protested against quotas in elite higher-education institutions for lower-caste applicants. The Indian Supreme Court ruled in favor of the students.

At the local level, most political parties design their vote-getting strategies to appeal to subcaste loyalties. They often secure the votes of entire *jati* communities with such political favors as new roads, schools, or development projects. These arrangements fly in the face of the official ideologies of the major political parties and of Indian government policies, which actively work to eliminate discrimination on the basis of caste. Nonetheless, the role of caste in politics seems to be increasing at this time, as several new political parties that explicitly support the interests of low castes have emerged.

Among educated people in urban areas, the campaign to eradicate discrimination on the basis of caste may appear to have succeeded, but the reality is more complex. Throughout the country, there are now some Dalits serving in powerful government positions. Members of high and low castes ride city buses side by side, eat together in restaurants, and attend the same schools and universities. For some urban Indians—especially educated professionals who meet in the workplace—caste is deemphasized as the crucial factor in finding a marriage partner. However, less than 5 percent of marriages cross even *jati* lines, let alone the broader gulf of *varna*. Nearly everyone notices the tiny social clues that reveal an individual's caste, and in rural areas, where the majority of Indians still reside, the divisions of caste remain prevalent.

Geographic Patterns of Religious Beliefs The geographic pattern of religion in South Asia is complex and overlapping. As Figure 8.18 shows, there is a core Hindu area in central India, with other faiths more common on the fringes of the region.

The approximately 520 million Muslims in South Asia form the majority in Afghanistan, Pakistan, Bangladesh, and the Maldives; and Muslims are a large and important minority in

India, where at 140 million, they form about 12 percent of the population. They live mostly in the northwestern and central Ganga River plain.

Buddhism began about 2600 years ago as an effort to reform and reinterpret Hinduism. Its origins are in northern India, where it flourished early in its history before spreading eastward to East and Southeast Asia. About 10 million people—only 1 percent of South Asia's population—are Buddhists. They are a majority in Bhutan and Sri Lanka.

Jainism, like Buddhism, originated as a reformist movement within Hinduism more than 2000 years ago. Jains (about 6 million people, or 0.6 percent of the region's population) are found mainly in cities and in western India. They are known for their educational achievements, promotion of nonviolence, and strict vegetarianism.

Sikhism was founded in the fifteenth century as a challenge to both Hindu and Islamic systems. Sikhs believe in one god, hold high ethical standards, and practice meditation. Philosophically, Sikhism accepts the Hindu idea of reincarnation but rejects the idea of caste. (In everyday life, however, caste plays a role in Sikh identity.) The 21 million Sikhs in the region live mainly in Punjab, in northwestern India. Their influence in India is greater than their numbers because throughout India many Sikhs hold positions in the government, in military and security forces, and in the police.

Buddhism a religion of Asia that originated in India in the sixth century B.C.E. as a reinterpretation of Hinduism; it emphasizes modest living and peaceful self-reflection leading to enlightenment

Jainism a religion of Asia that originated as a reformist movement within Hinduism more than 2000 years ago; Jains are known for their educational achievements, nonviolence, and strict vegetarianism

Sikhism a religion of South Asia that combines beliefs of Islam and Hinduism

The first Christians in the region are thought to have arrived in the far southern Indian state of Kerala with St. Thomas, the Apostle of Jesus, in the first century C.E. Today, Christians and Jews are influential but tiny minorities along the west coast of India. A few Christians live on the Deccan Plateau and in northeastern India near Burma.

Animism, the most ancient religious tradition, is practiced throughout South Asia, especially in central and northeastern India, where there are indigenous people whose occupation of the area is so ancient that they are considered aboriginal inhabitants. (For a discussion of animism, see page 319 in Chapter 7.)

The Hindu–Muslim Relationship Indian independence leaders like Mohandas Gandhi and Jawaharlal Nehru (first prime minister of India, 1947–1964) emphasized the common cause of throwing off British rule that once united Muslim and Hindu Indians. Since independence, members of the Muslim upper class have been prominent in Indian national government and the military. Muslim generals have served India willingly, even in its wars with Pakistan after Partition. Hindus and Muslims often interact amicably, and they occasionally marry each other. Both groups have influenced the region's cuisine (Figure 8.19).

But there is a dark side to the Hindu–Muslim relationship. At the community level in South Asia, relations between the region's two largest religious groups are often quite tense. In some Indian villages, Hindus may regard Muslims as members of low castes. Religious rules about food are often the source of discord because dietary habits are a primary means of distinguishing caste. After Partition in 1947, some wealthy Hindu landowners remained in what was then East Pakistan (now Bangladesh). In some Bangladeshi villages today, while Muslims may be a majority, Hindus are often somewhat wealthier. Although the two

FIGURE 8.19 LOCAL LIVES FOODWAYS IN SOUTH ASIA

A Boys in Hampi, Karnataka, eat a south Indian *thali*, or feast, featuring rice, lentils, and various other vegetarian dishes served on a banana leaf. Most South Asian cuisine is eaten with the hands; in South India and Sri Lanka, banana leaves often serve as plates.

B *Chapatis* are cooked on a griddle at a wedding in Rajasthan in northwest India. Chapati is a kind of bread made of stone-ground wheat flour and cooked on a griddle or curved iron *tava* located above a fire or oven. The bread is unleavened—it is made of dough that doesn't rise because it doesn't contain yeast. Most popular in northern South Asia, where wheat is more widely grown than rice, chapatis are torn into pieces and used to scoop up vegetable or meat dishes.

C Street food in a crowded marketplace in Dhaka, Bangladesh, during Ramadan. On offer are whole chickens, lamb, and various vegetable dishes, including potatoes and lentils. Usually the food is taken home and shared with family members.

for many years, they view each
resulting from religious or eco-
ly called **communal conflict**—can
its, as described in the following

eth Roy, who studies communal
Asia, recounts an incident that
ı cows" in the village of Panipur
ure 8.20). The incident started
essly or provocatively allowed
til field of a Hindu. The Hindu
reacted complacently, the Hindu
fall, Hindus had allied themselves
with the owner of the cow. More
m the surrounding area, and

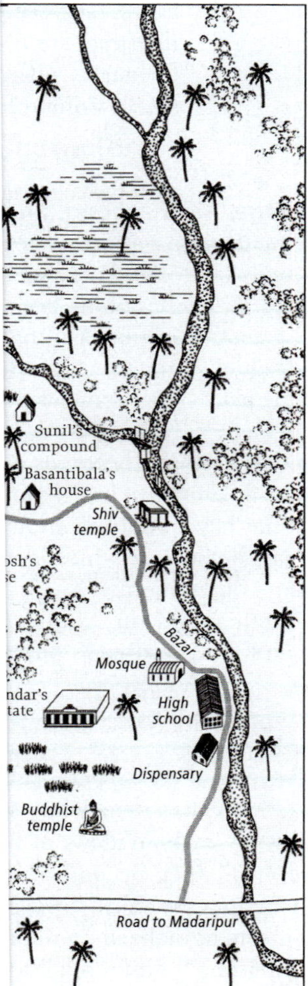

with cows. This map
ıdonym) illustrates how
and Hindu communities
s Muslim and Hindu areas,
ıce, and numerous other

soon there were thousands of potential combatants lined up facing each other. Fights broke out. The police were called. In the end, a few people died when the police fired into the crowd of rioters. *[Source: Beth Roy. For detailed source information, see Text Credit pages.]* ■

At the state and national level, Hindu–Muslim conflict has taken on strong political overtones, aspects of which are discussed below under the topic of religious nationalism in the section on political issues.

communal conflict a euphemism for religiously based violence in South Asia

THINGS TO REMEMBER

- The cultural, ethnic, religious, and rural/urban variety of life in South Asia means that most everyone is part of a minority group in one way or another, which is to some extent a protection against communal conflict.
- Mumbai is South Asia's wealthiest city and is a showcase of the economic disparity that characterizes South Asian cities.
- Village life is of course a major feature of rural South Asia, but city neighborhoods can share many of the intimate qualities of village life.
- There are many distinct ethnic groups in South Asia, each with its own language or dialect. Today, variants of Hindi are the principal languages of India and Pakistan, while Bengali is the official language of Bangladesh. English is a common second language throughout South Asia.
- Caste and class are not the same thing: "class" refers to economic status, and "caste" to hereditary hierarchical social categories. There are class differences within caste groups because of differences in wealth. Even though India's constitution bans caste discrimination, caste is still hugely influential in Indian society.
- There is a geographic pattern to Hinduism and Islam in South Asia, but it is not absolute and people of different religions often live in close proximity. Relations between Muslims and Hindus can be quite tense, occasionally resulting in violent confrontation. Other religious traditions also play prominent local roles in the life of South Asia.

Geographic and Social Patterns in the Status of Women

The overall status of women in South Asia is notably lower than the status of men. Even so, young women today have more educational and employment opportunities than did women of a generation ago, and a number of women hold and have held very high positions of power. However, on average, women's literacy rates, social status, earning power, and welfare are generally lower than those of men, especially in the belt that stretches from the northwest in Afghanistan across Pakistan, western India, Nepal, Bhutan, and the Ganga Plain into Bangladesh. Women fare better in eastern and central India and considerably better in southern India and in Sri Lanka. In these latter regions, where literacy rates are higher (Figure 8.21), different marriage, inheritance, and religious practices give women better access to education and resources. **187. SUFI ROCK SINGER FALU BLENDS OLD WITH NEW**

Urban women in South Asia generally have more individual freedom than rural women have, with many now

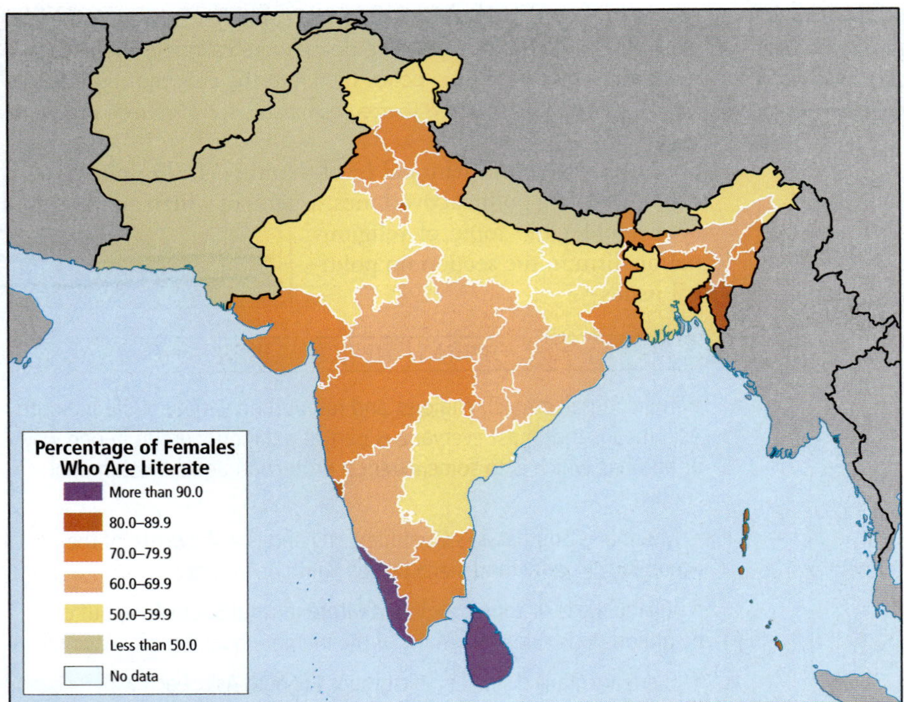

FIGURE 8.21 Female literacy in South Asia. Female literacy lags behind male literacy in all countries in South Asia, and is below 50 percent in four countries. Female literacy is crucial to improving the lives not only of women, but also of children; women who can read often seize opportunities to earn an income and nearly always use this income to help their children. In India, overall literacy has risen in the past decade, and in 2011, female literacy there reached 65 percent, up from 54.5 percent in 2005.

pursuing professional careers and some becoming involved in politics. Generally speaking, rural women are less free than urban women. In rural India, middle- and upper-caste Hindu women are often more restricted in their movements than are lower-caste women because they have a status to maintain. Meanwhile, lower-caste women who go into public spaces may have to contend with sexual harassment and exploitation from upper-caste men.

The socioeconomic status of Muslim women in South Asia is notably lower than that of their Hindu and Christian counterparts. In India, some of this relates to the generally lower incomes and standard of living for Muslims, which also usually means lower educational levels. Muslim women also work outside the home less than non-Muslim women in India. Low rates of education and workforce participation for women also prevail in Muslim-dominated countries such as Pakistan and Afghanistan, although this is less the case in Bangladesh. 191. REMEMBERING PAKISTAN'S FORMER PRIME MINISTER BENAZIR BHUTTO

purdah the practice of concealing women from the eyes of nonfamily men

dowry a price paid by the family of the bride to the groom (the opposite of *bride price*); formerly a custom practiced only by the rich

Purdah The practice of concealing women from the eyes of nonfamily men, especially during women's reproductive years, is known as **purdah**. It is observed in various ways across the region. The practice is strongest in Afghanistan and across the Indo-Gangetic Plain, where within both Muslim and Hindu communities, women are often secluded within structures (Figure 8.22) and wear veils or head coverings. Purdah is less strict in central and southern India, but even there, separation between unrelated men and women is maintained in public spaces. In general, low-caste Hindus do not observe this custom, but that is changing. In recent decades, as some low-status households have increased their incomes, they have adopted purdah as a sign of their rising wealth.

Purdah practices have influenced the architecture of South Asia. Homes are often in walled compounds that seclude kitchens and laundries as women's spaces. In grander homes, windows to the street are usually covered with lattice screens, known as *jalee*, that allow in air and light but shield women from the view of outsiders.

Marriage, Motherhood, and Widowhood Throughout South Asia, most marriages are arranged by the parents of the prospective bride and groom. Especially in wealthier, better-educated families, the wishes of the bride and groom are considered, but in some cases they are not (Figure 8.23 on page 356). This is especially true of child marriage, when a young girl (often as young as 12 and in some places even younger) is married off to a much older man.

Usually a bride goes to live in her husband's family compound, where she becomes a source of domestic labor for her mother-in-law. Most brides work at domestic tasks for many years until they have produced enough children to have their own crew of small helpers, at which point they gain some prestige and a measure of autonomy.

Motherhood in South Asia determines much about a woman's life. A woman's power and mobility increase when she has grown children and becomes a mother-in-law herself. On the other hand, in some communities, the death of a husband, regardless of cause, is a disgrace to a woman and can completely deprive her of all support and even of her home, children, and reputation. Widows may be ritually scorned and blamed for their husband's death. Widows of higher caste rarely remarry, and in some areas, they become bound to their in-laws as household labor or may be asked to leave the family home. Most simply become marginalized "aunties" in extended families and help with household duties of all sorts.

Dowry and Violence Against Females A **dowry** is a sum of money paid by the bride's family to the groom's family at the time of marriage. Dowries originated as an exchange of wealth between Muslim landowners or high-caste families that practiced purdah. With her ability to work reduced by purdah, a woman was considered a liability for the family that took her in.

ɔf purdah. A lattice screen
ılee, lattice screens are often
women are secluded. Like the
ıthwest Asia (see Figure 6.18 on
ıd let in light but shield women

r to be a cause of the growing mestic violence against females ɜh.

ɜ, only wealthy families gave the ubstantial sum that symbolized ɛ a daughter a measure of secu-

affluence and education have ıd made it much more common ame educated, their families felt ɛir worth as husbands and gave ɛr and larger dowries. Soon, the ste families wanting to upgrade must pay to get their daughters ı. A village proverb captures this dilemma: "When you raise a daughter, you are watering another man's plant."

TABLE 8.2 Percentages of South Asian Women in Parliament, 2013	
Country	**Percent in Parliament**
Sri Lanka	5.8
Bhutan	8.5
Pakistan	19.5
Nepal	33.2
Bangladesh	19.7
India	11.0
Afghanistan	27.7

Source: Women in National Parliaments, Inter-Parliamentary Union Web site at: http://www.ipu.org/wmn-e/classif.htm

Gender and Democratization

As countries in South Asia have moved toward more democracy, the status of women in the region has risen. India, Pakistan, Bangladesh, and Sri Lanka have all had female heads of state (prime ministers) in the past. However, it is important to note that all of these women were either wives or daughters of previous heads of state. Women have been notably less successful in local elections; and at the parliamentary level, in Bhutan, India, and Sri Lanka, women remain very poorly represented (Table 8.2).

The very low percentage of women in India's parliament (see Table 8.2) inspired a confederation of Muslim and Hindu women's groups to lobby for legislation that would temporarily (for a 15-year trial period) reserve for women one-third of the seats in the lower house of Parliament and in state assemblies. Such quotas are already in place in Pakistan (19.5 percent), and Bangladesh (19.7 percent), but are so far not quite met in Nepal (33.2 percent).

If recent voter turnout and political activism trends continue, there is likely to be a major improvement in female representation in the region's national parliaments and in local offices.

Women and the Taliban in Afghanistan Women in Afghanistan have frequently suffered brutal repression since a conservative Islamist movement, the **Taliban**, gained control of the government there in the mid-1990s. Prior to that time, rights for women in Afghanistan were slowly but steadily improving, and upper-class women had many freedoms; they could dress in Western styles and they had the right to attend gender-integrated university courses. The Taliban support strict and radical interpretations of Islamic law, forcing females, including urban professional women, to live in seclusion. In regions where the Taliban retain control, girls and women are not allowed to work outside the home or attend

Taliban an archconservative Islamist movement that gained control of the government of Afghanistan for a while in the mid-1990s

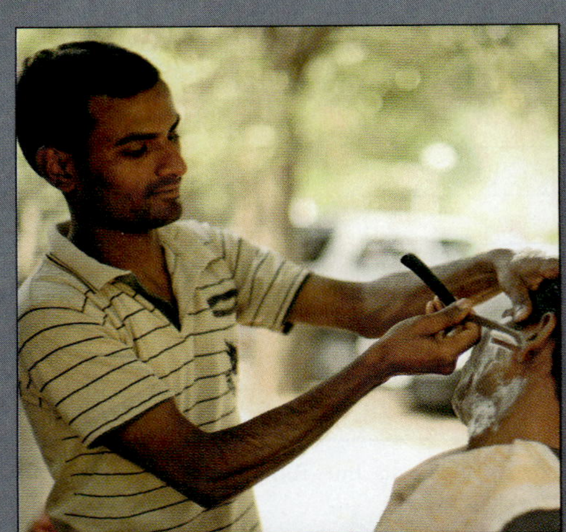

FIGURE 8.23 Caste and marriage. Caste remains a particularly powerful force with respect to marriage.

(A) A barber, often a member of a "barber caste" or *jati*, practices his trade in Chandigarh, India. In many areas, members of the barber caste act as go-betweens when families are arranging marriages for their children.

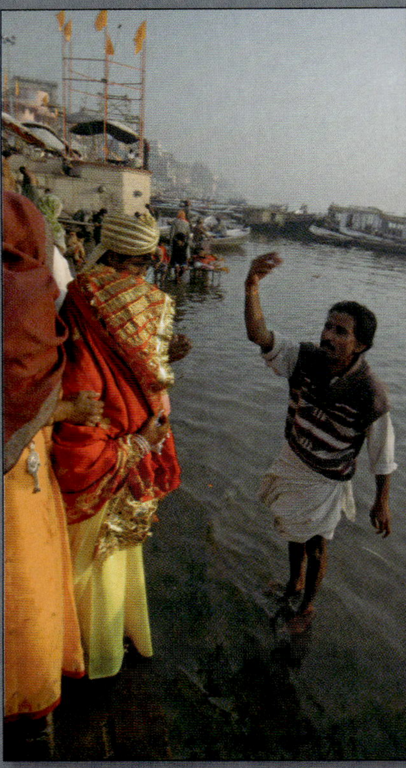

(B) A Brahmin priest officiates at a Hindu wedding ceremony in Varanasi, India. The specifics of the ceremony vary considerably, depending on the caste of the bride and groom.

(C) A *hijra* performs at a wedding ceremony in Rawalpindi, Pakistan. *Hijras* are men who leave their families, often at a young age, to join a "*hijra* family." Here they undergo a gradual transition toward femininity, sometimes culminating in castration. They also must learn the *hijra* trades, which include singing and dancing for weddings and birth ceremonies, sometimes fortune telling, and often sex work. Because of their unique identity, *hijras* are allowed to move between different castes with greater ease than most.

school. In virtually all parts of the country, despite the decline of Taliban control, women must wear a heavy, completely concealing garment, called a *burqa* (or burka), whenever they are out of the house. (Men also must follow a dress code, though a less restrictive one.) Although the Taliban were driven from official power in November of 2001, they maintain control of the rural southern provinces and mountainous zones near Pakistan, where cultural and religious conservatism continues to adversely affect Afghan women. Even efforts to provide women and girls with a basic education, such as that described in the following vignette about Radio Sahar, run into hostility, and oftentimes violence.

189. REPORT: DOMESTIC VIOLENCE WIDESPREAD IN AFGHANISTAN

190. *FRONTRUNNER* DOCUMENTARY TELLS STORY OF AFGHAN POLITICIAN WHO INSPIRES WOMEN

VIGNETTE From behind her microphone at Radio Sahar ("Dawn"), Nurbegum Sa'idi speaks to a female audience on a wide range of topics. Located in the city of Herat, Radio Sahar is one in a network of independent women's community radio stations that has sprung up in Afghanistan since early 2003. Radio Sahar provides 13 hours of daily programming consisting of educational items that address cultural, social, and humanitarian matters as well as music and entertainment. For example, one recent broadcast, aimed at informing women of their legal rights, followed the life of a young woman who was physically abused by her husband and his entire family. The woman took the brave step of asking for a divorce. As a result, she was forced into hiding, where she was counseled on the steps she might take next. Another program explored the various concepts of what democracy is and how it might work in Afghanistan. A reported 600,000 Afghan women and youth listen to Radio Sahar while they do their chores.

Girls on the Air, a film by Valentina Monti, reveals the diversity of the ideas and hopes for the future of the young journalists that founded Radio Sahar. A clip can be seen at http://www.youtube.com/watch?v=c6KxtDHbuuY. *[Source: Internews Afghanistan. For detailed source information, see Text Credit pages.]*

anistan ale indi- ears rec- causes. ı as that ; creativ- ı males, o closed d others, ess, despite the many expressions stan, Pakistan, and elsewhere in history of alternative expressions

REMEMBER

women in this region is within the

vomen in very high positions of average have more educational and nen of a generation ago. However, egion is notably lower than the e on policy remains low even as the nents rises above a tiny minority.

ne socioeconomic status of women, to move through the landscape, ers in elective office.

ı and Hindu households, but the ıtly lower than that of their Hindu pecially so in the north and west, kistan where tribal customs include ion and violence against women Generally speaking, women are less ctions of this region.

ssumptions that males are more that females must be supplied with ion of females, leading to a gender uth Asia. Those who can afford in / may abort female fetuses, while nes commit female infanticide. As utnumber adult females. A second opulation is that when females are the home, they tend to choose to

populated region in the world on page 14). The region already ıan China (1.35 billion), which

ON THE BRIGHT SIDE

Cell Phones and Literacy

In order to reach women held in deep seclusion, the Afghan government is now making available basic reading and writing lessons on special mobile phones distributed free to Afghan women. The reading/writing software was developed by an Afghan IT firm with USAID assistance. As literacy is achieved by a woman, she can add other subjects with free apps.

has almost twice the land area of South Asia. By 2025, India alone, with 1.46 billion people, will have overtaken China's 1.40 billion. By 2050, China's population (if the one-child policy persists) will be shrinking at the rate of about 20 million people per decade, while India's may still be growing. However, a notable trend is that current rates of growth have slowed to about 1.6 percent and fertility is declining significantly. Only in Afghanistan and Pakistan are fertility rates above 3 children. In Bangladesh and all other countries of the region, women now average just 2.5 children (Figure 8.25). Families are choosing to have fewer children for a variety of reasons, such as improved educational opportunities for women, better health care, and urbanization.

Population densities in this region are highest in the cities that lie just south of the Hindu Kush and Himalayas, and are at their peak in the Ganga-Brahamaputra delta (see Figure 8.24), much of which is a densely occupied, rural agricultural and fishing area.

Slowing Population Growth: Health Care, Urbanization, and Gender Efforts to slow population growth through better health care have been underway for more than 50 years. Today, urbanization and the rising status of women are helping to reduce the birthrate.

With improved health care, such as diarrhea control techniques (infant rehydration), far fewer babies are dying in infancy. In 1992, infant mortality rates were 91 per 1000 live births in India, 109 in Pakistan, and 120 in Bangladesh. By 2012, those rates were reduced to 47 in India, 68 in Pakistan, and 43 in Bangladesh. With better assurance that their babies will survive to adulthood and be able to care for their elderly parents, couples now choose more often to have just two children.

In addition to improved access to health care as an incentive to have smaller families, the shift in population from rural to urban areas reduces the economic incentive for large families. Whereas children in rural areas (starting at an early age) contribute their labor to farming, thus increasing the family income, the labor of children in urban areas is less likely to boost the family income. Children in urban areas are more likely to go to school, and thus represent a cost to the family for school fees, books, and uniforms. Improvements in the status of women also slow population growth. Research throughout South Asia and elsewhere has shown that women who have the opportunity to go to school and/or to get a paying job tend to delay childbearing and have fewer children.

A comparison of the population pyramids as well as some other statistics for Sri Lanka and Pakistan helps illustrate that health care and the status of women influence birth rates (see Figure 8.25 and

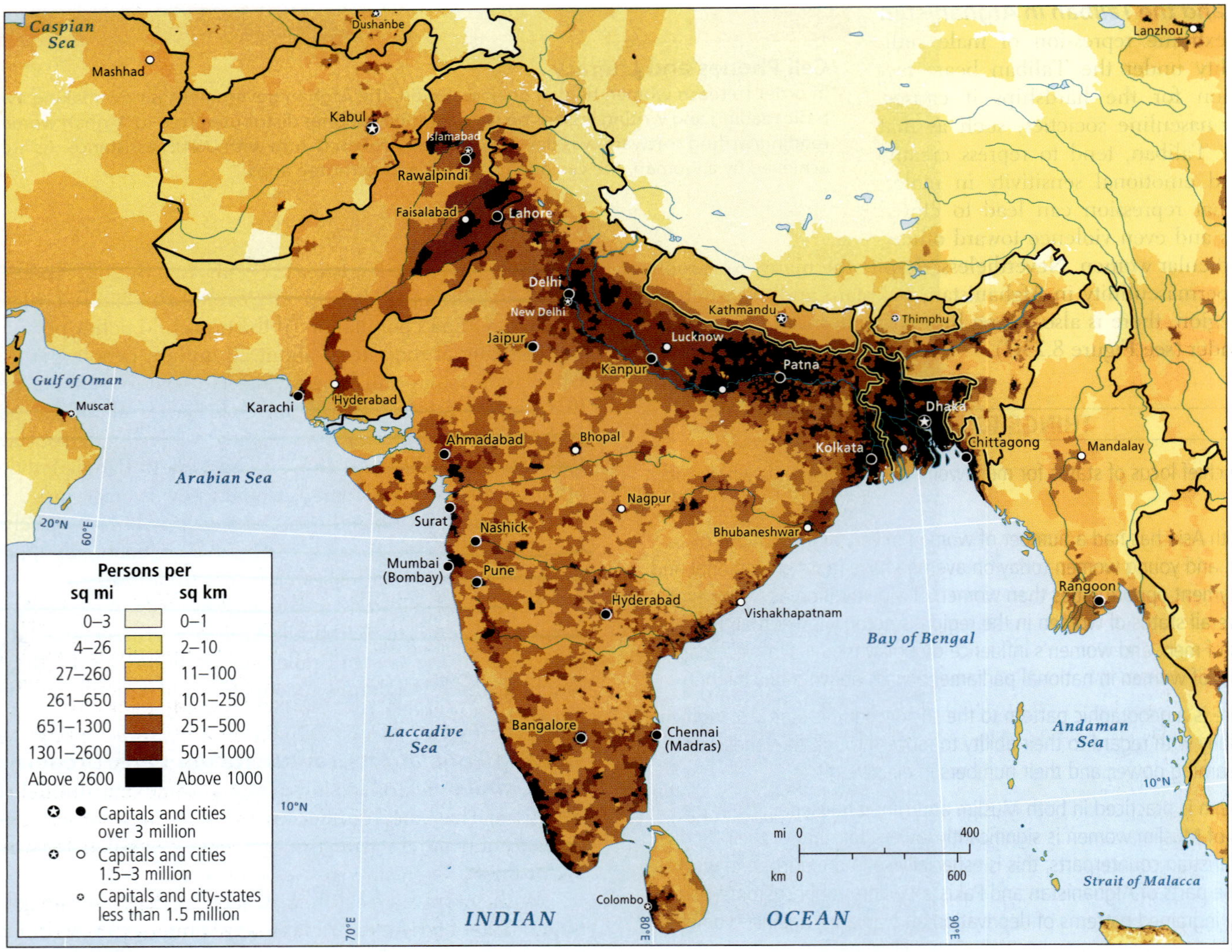

FIGURE 8.24 Population density of South Asia.

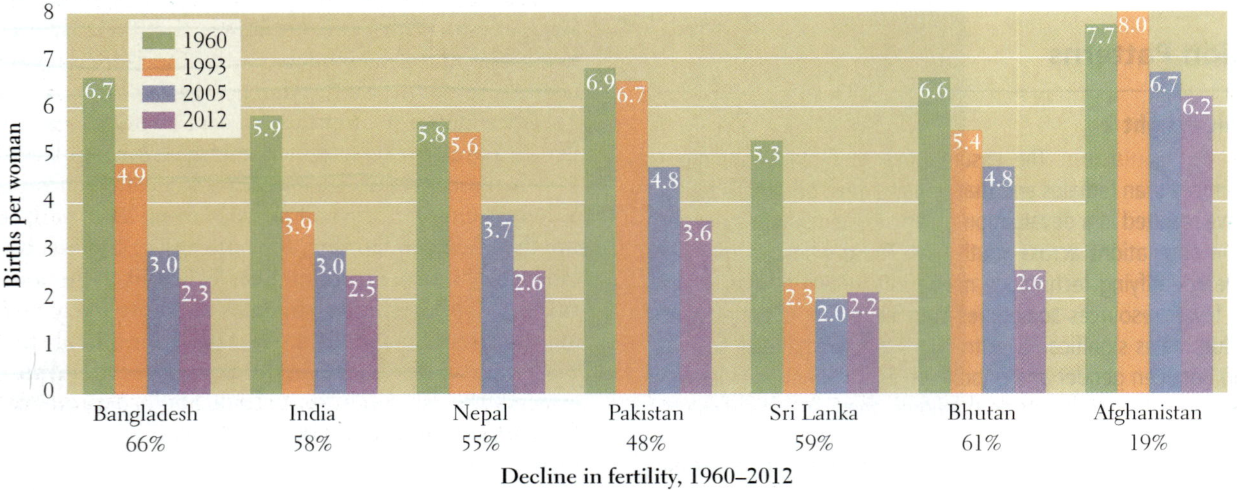

FIGURE 8.25 Total fertility rates in South Asia, 1960–2012. All South Asian countries except Afghanistan have had a substantial decline in fertility since 1960. Afghanistan's rate has declined less than 20 percent over the past 50 years, probably because of almost continuous conflicts, the overwhelmingly rural population (77 percent), and the low literacy rates for both men (43.1 percent) and women (12.6 percent).

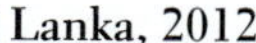

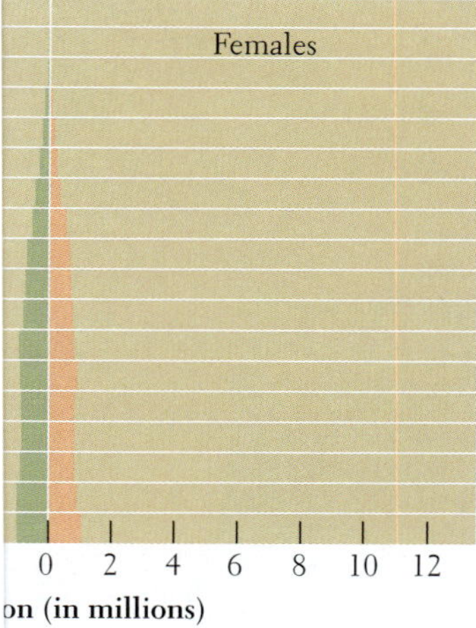

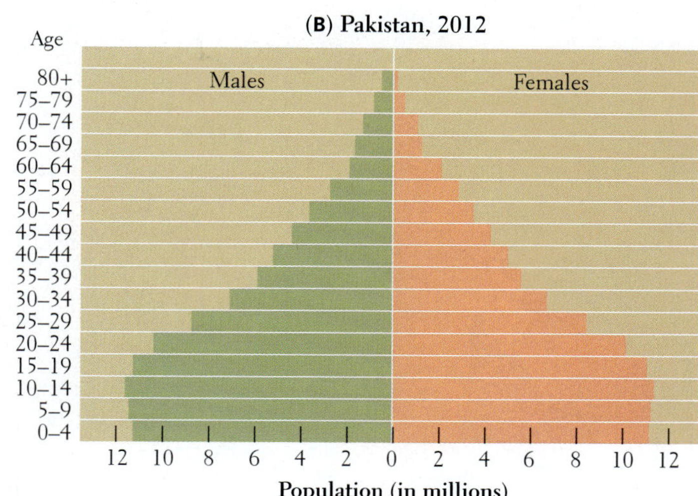

ıids for Sri Lanka and Pakistan.

at is far along but not completely on (see Figure 1.10 on page 16), NI per capita (U.S.$4943) than ɛ is generally better in Sri Lanka, infant mortality rate (in 2012,)00 live births versus Pakistan's ı worked harder to educate and n. Three key indicators of wom- : that are associated with reduced in Sri Lanka than in Pakistan: us 30 percent), the percentage ıigh school (89 percent versus of women who work outside the ɛnt). Surprisingly, Sri Lanka has ation rate of just 15 percent, far ɛrcent.

ınal statistics compared to those nid does indicate that birth rates that the bottom of the pyramid :ably). If this trend continues, n countries except Afghanistan, *nographic dividend*: a bulge of ɔrce and choosing to have fewer e time and energy to study, to conomy and civil society. Down ountries that have had a demo- ıl with an aging population for :egivers and taxpayers to support **'ERSISTS IN INDIA DESPITE NEW LAWS**

ı populations have a significant ıltural customs that make sons ontribute to a family's wealth. ; "May you be the mother of a s couples who wish to have sons ecialize in identifying the sex of a fetus. The intention is to abort female fetuses. This practice is now illegal, but it persists. Poorer South Asians may neglect the health of girl children, with some even committing female infanticide (the dowry implications are discussed on pages 354–355). **186. GIRLS PAY PRICE FOR INDIA'S PREFERENCE FOR BOYS**

After the 2001 census, India took a stronger stand against selecting for male children. A follow-up sex-ratio survey in five Indian states from 2004 to 2006 showed some improvement in all five states, with Tamil Nadu coming closest to a normal sex ratio, probably due to an especially aggressive campaign for women's health and against sex-selective abortions.

Gender imbalance of the magnitude India is facing could create serious problems, such as surges in crime and drug abuse related to the presence of so many young men with no prospect of having a family. In all cultures, the possibility of having a family is a stabilizing influence for young men. The efforts of the state of Kerala to overcome gender imbalance by putting more attention to women's development bear watching. Kerala's government has made women's development a priority, funding education for women well beyond basic levels. The results are reflected in its female literacy rate, which is the highest in India at 87.8 percent (the average for India as a whole is 54.5 percent; see Figure 8.21 for the state female literacy map), and in the number of women who work outside the home. In Kerala, daughters, far from being viewed as an economic liability to the family, are seen as an asset. Women can travel the public streets alone and in groups; they commonly work in public places, many in high positions in government, education, health care, IT industries, and other professions. **188. GLOBAL POPULATION BOOM PUTS 'MEGA' PRESSURE ON CITIES IN DEVELOPING WORLD**

Measures of Human Well-Being The maps of human well-being in South Asia (Figure 8.27) show that the overall state of human well-being in South Asia is marginal, with nearly the entire region ranking in the medium-low levels. The region as a whole has an average GNI per capita (PPP) below U.S.$4000

FIGURE 8.27 Maps of human well-being.

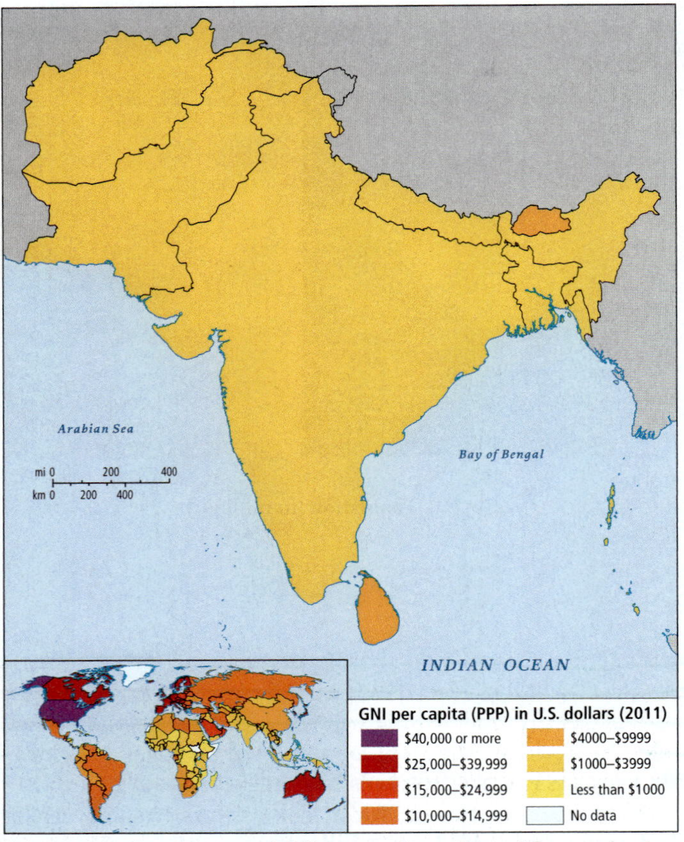

(A) Gross national income (GNI) per capita, adjusted for purchasing power parity (PPP).

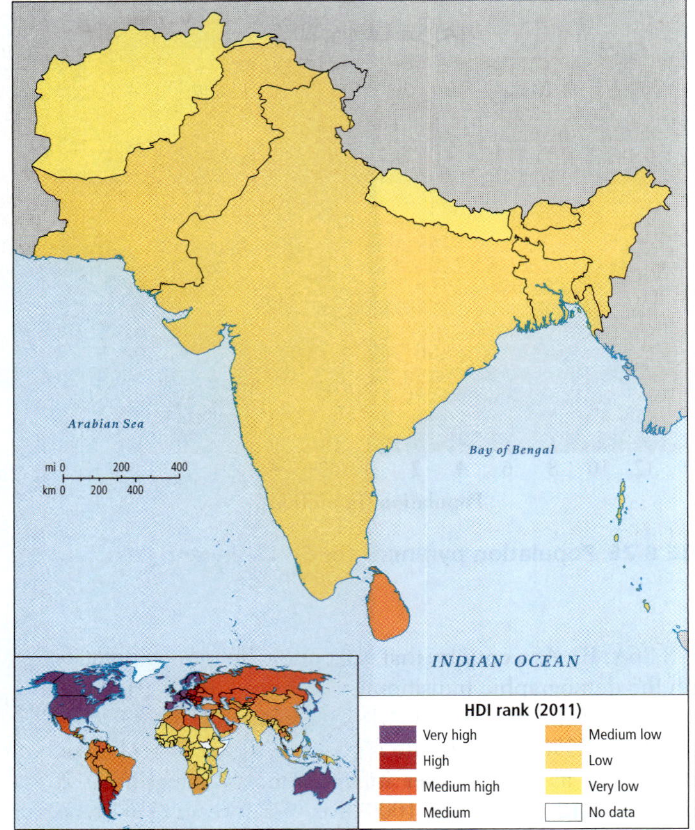

(B) Human Development Index (HDI).

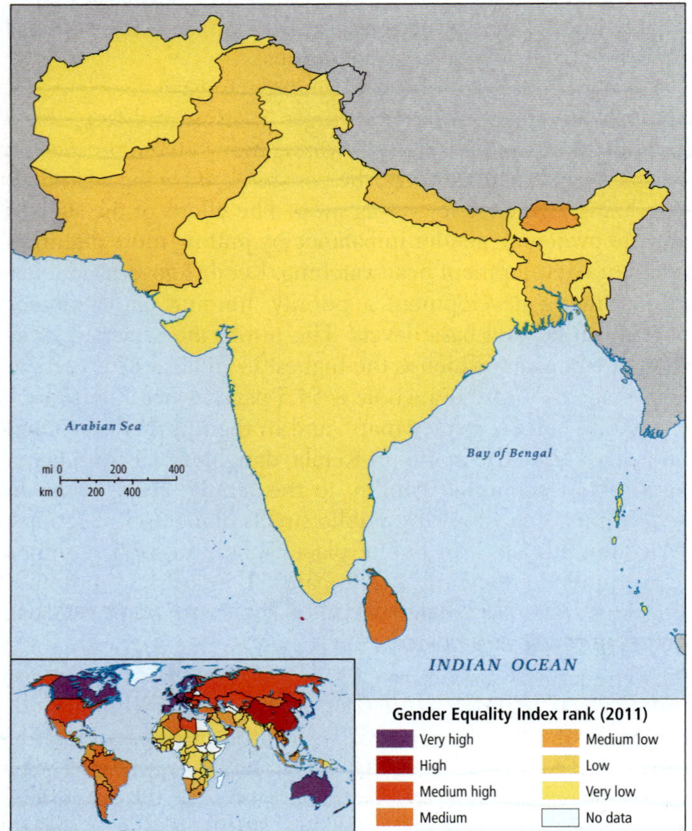

(C) Gender Equality Index (GEI).

(see Figure 8.27A). Of course, the huge country of India has many cities and a few states that rank above this level (Haryana, for example; see Figure 8.30 on page 365), but these variances are lost in the countrywide average. A bright spot in these data is that disparity of wealth (the gap between rich and poor) is quite narrow in South Asia. Some might say this is related to the broad-based democracies found in this region. Certainly the provision of services is better and more consistent here than in Africa, for example. In per capita GNI, however, Nepal ranks very low ($1160), Afghanistan does only slightly better ($1416), and Bangladesh is also only marginally better ($1529).

On the Human Development Index (HDI), there is even less variation (see Figure 8.27B). All of the South Asian countries rank either low or very low, with the slight exception of Sri Lanka (medium). These HDI levels indicate that the well-being of citizens in these countries is worse than the worldwide average in terms of adjusted real income, life expectancy, and educational attainment. In fact, India, Pakistan, and Bangladesh have all lost HDI rank in recent years, a fact that may be related to the global economic recession.

Figure 8.27C shows how well countries are ensuring gender equality (GEI) in three categories: reproductive health, political and educational empowerment, and access to the labor market. Again, South Asia ranks in the low and very low categories, with only Bhutan and Sri Lanka ranking in the medium and medium- low categories. Based on the global thumbnail map, it is clear that South Asia is in a league with

Asia, and Mexico, but virtually

in human development is not
icksliding a bit during the global
omic progress across the region,
chnological and theoretical inno-
ady contributed to development
iows.

REMEMBER

opulation There is a significant and
der imbalance in South Asia,
ıral customs that make sons more
lth than daughters. Too often this
iination and oppression that affects
Also, population growth tends to
ucated.

oss the region, but the momentum
ans that growth will continue for the

n well-being in South Asia is
on ranking at the medium-low

economic incongruities. India is
ndreds of millions of desperately
l leader in the computer software
ovation. It is celebrated for being
rld; yet its poor are often left out
sed or even harmed by economic
, overall, most countries in this
ı their way up.

ıan 50 percent of the workers in
r of agriculture to most national
raging less than 20 percent of
lustrial sector employs far fewer
ıe-quarter and one-third of GNI
Bhutan, and Afghanistan. The
e rapidly than either agriculture
more than half of the GNI in
ın and Bhutan. Rapid develop-
een the dream for all parts of the
nial era in the 1940s. Successful
nic modernization include an
ı-tech industries as well as the
xtile and leather manufacturing
rofinancing strategies pioneered
ut still promising, is the develop-
, which involves all three sectors
ces. Especially in India, but also

in Pakistan and Bangladesh, agribusiness has drastically increased the amount of food available.

Food Production and the Green Revolution

Geographic Insight 3

Food and Urbanization: Due to changes in food production systems in South Asia, food supplies and incomes of wealthy farmers have increased. Meanwhile, impoverished agricultural workers, displaced by new agricultural technology, have been forced to seek employment in cities, where they often end up living in crowded, unsanitary conditions and must purchase food.

Agricultural production per unit of land has increased dramatically over the past 50 years, especially in parts of India; nonetheless, agriculture remains the least efficient economic sector, meaning that it has the lowest return on investments of land, labor, and cash. Figure 8.28 shows the distribution of agricultural zones in South Asia.

Until the 1960s, agriculture across South Asia was based largely on traditional small-scale systems (the average farm holding in South Asia is still less than 2 acres) that managed to feed families in good years, but often left them hungry or even starving in years of drought or flooding. Moreover, these systems did not produce sufficient surpluses for the region's growing cities. Even now, cities must import food from outside South Asia. Much of South Asia's agricultural land is still cultivated by hand, and for decades, agricultural development was neglected in favor of industrial development, especially in India (discussed below). Nonetheless, by the 1970s, important gains in agricultural production had begun.

Beginning in the late 1960s, the *green revolution* (see page 26 in Chapter 1) boosted grain harvests dramatically through the use of new agricultural tools and techniques. Such innovations included seeds bred for high yield and for resistance to disease and wind damage; fertilizers; mechanized equipment; irrigation; pesticides; herbicides; and double-cropping (producing two crops consecutively per year). To a lesser extent, the increase in the amount of land under cultivation also contributed to a greater yield. Where the new techniques were used, yield per unit of farmland improved by more than 30 percent between 1947 and 1979, and both India and Pakistan became food exporters.

Despite the achievements of the green revolution, the shortcomings have been numerous. Some Indian states, such as Punjab and Haryana, which have extensive irrigation networks, have increased prosperity tremendously for some, but many poor farmers unable to afford the special seeds, fertilizers, pesticides, and new equipment could not compete and had to give up farming. Most migrated to the cities, where their skills matched only the lowest-paying jobs.

South Asia needs alternatives to standard green revolution strategies because it will have difficulty maintaining current levels of food production over the long term. The green revolution's dependency on chemical fertilizers, pesticides, and high levels of

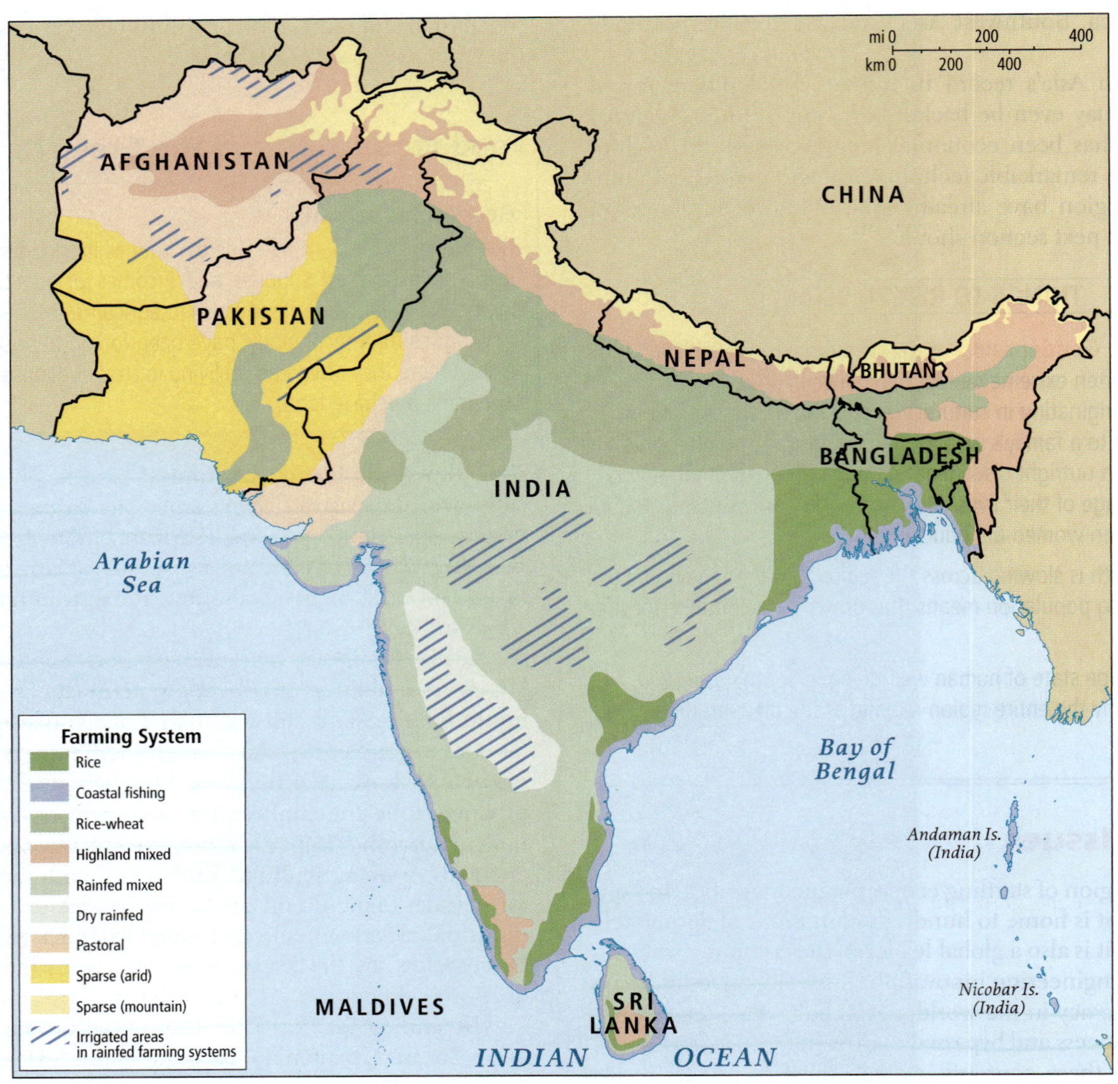

FIGURE 8.28 Major farming systems of South Asia.

irrigation all contribute to aquifer depletion, waterway pollution, increased erosion, and the loss of soil fertility through the buildup of salt in soils. *Soil salinization* (see page 243 in Chapter 6) is already reducing yields in many areas, such as the Pakistani Punjab, which is Pakistan's most productive—but highly irrigated—agricultural zone. **192. TECHNOLOGY KEY TO PRODUCING MORE FOOD**

The methods of agroecology are a potential remedy for some of the failings of green revolution agriculture. **Agroecology** often involves the revival and use of traditional methods, such as fertilizing crops with animal manure, intercropping (planting several species together) with legumes to add nitrogen and organic matter, water conservation, and using natural predators to control pests. Although the knowledge required is extensive, unlike green revolution techniques, the methods of agroecology are advantageous to poor farmers because the necessary resources are readily available in most rural areas and the knowledge can be handed down orally from generation to generation.

agroecology the practice of traditional, nonchemical methods of crop fertilization and the use of natural predators to control pests

One reason that the green revolution has managed to increase food supplies but not eliminate hunger and malnutrition is that food tends to go to those with money to spend. As noted previously, agricultural modernization usually pushes unskilled farm workers off the land and into precarious underemployment. Between 1970 and 2001, the amount of food produced per capita in South Asia increased 18 percent, and the proportion of undernourished people dropped from 33 percent of the population to 22 percent. Nonetheless, 22 percent of nearly 1.6 billion is 352 million people—roughly half of the world's total undernourished population. Because of corruption and social discrimination, government programs that provide food to the poor have generally failed to reach those most in need. For example, in India, despite massive resources devoted to improve child nutrition, about half of the children show signs of malnutrition.

sian
r

a highly le out of and mal- ies it has **licrocredit** ly under vould be uth Asia, as in most of the world, l in administering the small loans women, need. Instead, the poor enders, who often charge interest ore *per month*. In the late 1970s, cs professor in Bangladesh, started Bank," which makes small loans, es who wish to start businesses. **BANGLADESH'S "BANKER TO THE POOR"** n pay for the start-up costs of de such wide-ranging ventures chicken raising, small-scale egg pit toilets, and the distribution of . Potential borrowers (more than) are organized into small groups for repaying any loans to group o repay a loan, then everyone in the loan is repaid. This system, gs, creates incentives to repay te on the loans is extremely high, averaging around 98 percent—much higher than most banks achieve.

ON THE BRIGHT SIDE

Microcredit: A Bangladeshi Innovation

So far, the Grameen Bank has been an enormous success in Bangladesh, where it has loaned over U.S.$8 billion to more than 8 million borrowers. Similar microcredit projects have been established in India and Pakistan and throughout Africa, Middle and South America, North America, and Europe. In 2006, Dr. Yunus was awarded the Nobel Peace Prize for his work in microcredit. In 2009, he received the Presidential Medal of Freedom from President Obama.

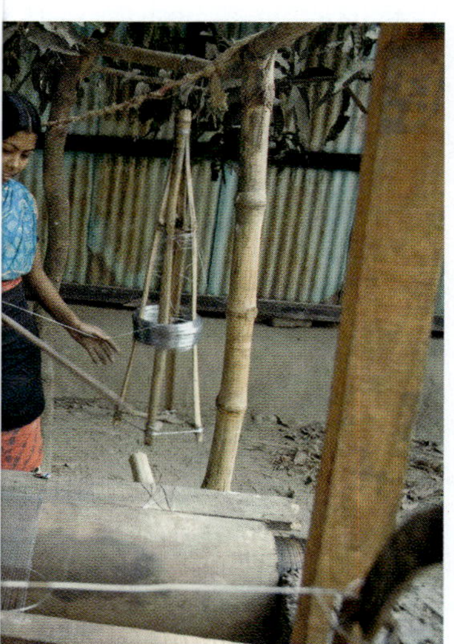

ıall-scale entrepreneurs. para, Bangladesh, have participated ıe Grameen Bank's microcredit ıing nets at the rate of about two ets. Kushum's first loan was for 3000 ıin a year, and her current yearly ving a model established in 1974 ad Yunus, Kushum's peer lending iness and pay loan installments.

Industry over Agriculture: A Vision of Self-Sufficiency

After independence from Britain in 1947, the new leaders in India, Pakistan, Bangladesh, and Sri Lanka tended to favor industrial development over agriculture. Influenced by socialist ideas and industrial success in the Soviet Union, they believed that government involvement in industrialization was necessary to ensure the levels of job creation that would cure poverty. They also wanted their respective countries to become self-sufficient and be independent of manufactured goods imported from the industrialized world (see pages 128–129 in Chapter 3 for a discussion of import substitution industrialization). To reach their goal of self-sufficiency, governments took over the industries they believed to be the linchpins of a strong economy: steel, coal, transportation, communications, and a wide range of manufacturing and processing industries.

South Asian industrial policies in the decades after independence generally failed to meet their goals. The emphasis on industrial self-sufficiency was not suited to countries that had such large agricultural populations. In India, for example, governments invested huge amounts of money in a relatively small industrial sector—even today industry employs only 14 percent of the population, compared with the 52 percent employed in agriculture. Since such a small portion of the population directly benefited from this investment, industrialization failed to significantly increase South Asia's overall prosperity.

Another problem was that the measures intended to boost employment often contributed to inefficiency. One policy encouraged industries to employ as many people as possible, even if they were not needed. So, for example, until recently, it took more than 30 Indian workers to produce the same amount of steel as 1 Japanese worker. Consequently, for years, Indian steel was not competitive in the world market. In addition, as in the former Soviet Union, decisions about which products should be produced were made by ill-informed government bureaucrats and were not driven by consumer demand. Until the 1980s, items that would improve daily life for the poor majority, such as cheap cooking pots or simple tools, were produced only in small quantities and were of inferior quality. At the same time, there was a relative abundance of large kitchen appliances and large cars that only a tiny minority could afford.

microcredit a program based on peer support that makes very small loans available to very low-income entrepreneurs

Economic Reform: Globalization and Competitiveness

Geographic Insight 4

Globalization and Development: Globalization benefits populations unevenly. Educated and skilled South Asian workers with jobs in export-connected and technology-based industries and services enjoy better lives. Low-skilled workers, however, in both urban and rural areas, are left with demanding but very low-paying jobs.

During the 1990s, much of South Asia began to undergo economic reforms aimed at increasing competitiveness and creating secure jobs in the private sector. In many other world regions, structural adjustment programs (SAPs; see pages 129–131 in Chapter 3) were mandated by the International Monetary Fund (IMF) and the World Bank. In India, by contrast, in response to an earlier financial crisis in the 1980s, the government itself initiated economic reforms. Although privatization of India's public sector industries and banks has proceeded slowly, it has been arguably more successful than SAPs and similar reforms imposed by the IMF and the World Bank in other countries.

India is now using the economic reforms to free its private companies from a maze of regulations, enabling both foreign and Indian companies to invest heavily in manufacturing and other industries (Figure 8.30). Some Indian companies are quite innovative, such as Reva, which builds economical electric cars for India's middle class. Foreign auto companies are also flocking to India, drawn by its large and cheap workforce, its excellent educational infrastructure (for the middle and upper classes), and especially by its large, pent-up domestic demand for manufactured goods of all sorts. Nearly every major global automobile company is currently establishing significant manufacturing facilities somewhere in India—producing everything from economical first cars for Indian families to luxury brands such as Mercedes-Benz for wealthier Indians. So many global manufacturers have flocked to India in recent years that the country is challenging China as an exporter of manufactured goods. The hope is that as the current recession subsides and manufacturing jobs increase, many of India's urban poor will see their incomes rise, and then they too will increase their consumption of Indian-made products, thus fueling further growth. By 2009 (the latest date for which comparable figures are available), GDP per capita (PPP) had risen in all Indian states (see Figure 8.30), and the average per capita income (PPP) in India in 2011 was U.S.$3468, nearly ten times higher than it was in 1992.

India's current manufacturing boom is benefiting from the previous boom in offshore outsourcing that started in the 1990s. In **offshore outsourcing**, a company contracts to have some of its business functions performed in a country other than the one where its products or services are actually developed, manufactured, and sold. Companies in North America and Europe have been outsourcing jobs to cities such as Bangalore, Mumbai, and Ahmadabad to take advantage of India's large, college-educated, and relatively low-cost workforce. Jobs outsourced to India include those in IT, data entry, Web design, engineering, telephone support, pharmaceutical research, and "back office" work. Many of the workers in these jobs are women. By holding outsourced jobs with North American firms, South Asian workers have gained experience that situates them perfectly for jobs now being created by South Asian–owned firms in South Asia.

offshore outsourcing the contracting of certain business functions or production functions to providers in areas where labor and other costs are lower

Major global finance firms are increasingly hiring the highly skilled workers on Mumbai's "Wall Street"—Dalal Street—to provide finance and accounting services. Stiff global competition makes cost cutting imperative for finance firms, and India's relatively low salaries provide a solution. While a junior analyst from an Ivy League school costs $150,000 a year in the United States, a graduate of a top Indian business school costs only $35,000 a year in India. That Indian employee's salary, however, buys a much higher standard of living than the U.S. employee would have. In fact, well-educated Indian migrants in the United States are now returning home to take these seemingly low salaries because they can still live well and join this exciting development phase in their home country. This trend of return-migration could eventually happen across the South Asian region.

To facilitate trade between countries and to lure extra-regional investors into this huge market, South Asia has attempted to create region-wide free trade agreements. Agreements were scheduled to be implemented in 2012, but progress has been slow and uneven. Because of its size, India is the biggest player in free trade discussions. India is a major supplier for Sri Lanka and Bangladesh but buys little from them in return. And although the potential for trade between India and Pakistan is great and could include cars, cotton, chemicals, and food imported from India, as yet the two countries have only a small trade relationship, with India buying very little from Pakistan. The ongoing political and border disputes inhibit progress.

Letting in Foreign Firms and Free Trade Until the 1990s, South Asia, and specifically India, avoided exposing its economy to foreign competition. Even today, while much of India's economy is now open to foreign competition, retail companies like IKEA and Walmart are not being allowed in for fear they will drive local companies, producers, and even farmers out of business. Resistance to further opening for foreign competition also comes from older generations, especially the many older government officials who have power over how regulations are developed and implemented. These people were raised and educated during the early post-colonial era when South Asian economic policies centered on import substitution (see pages 128–129 in Chapter 3). Many of the region's growing middle-class consumers have long supported opening economies further to foreign competition, especially in the retail sector. Since consumers are also voters, the older generation may eventually be swayed.

Even within India there are blocks to freer trade across state borders. The maze of regulations mentioned earlier extends down to the interstate level, and there is little reconciliation of the states' varying tax policies. Trade across state borders is so hampered that it has been suggested that India needs a free trade agreement with itself.

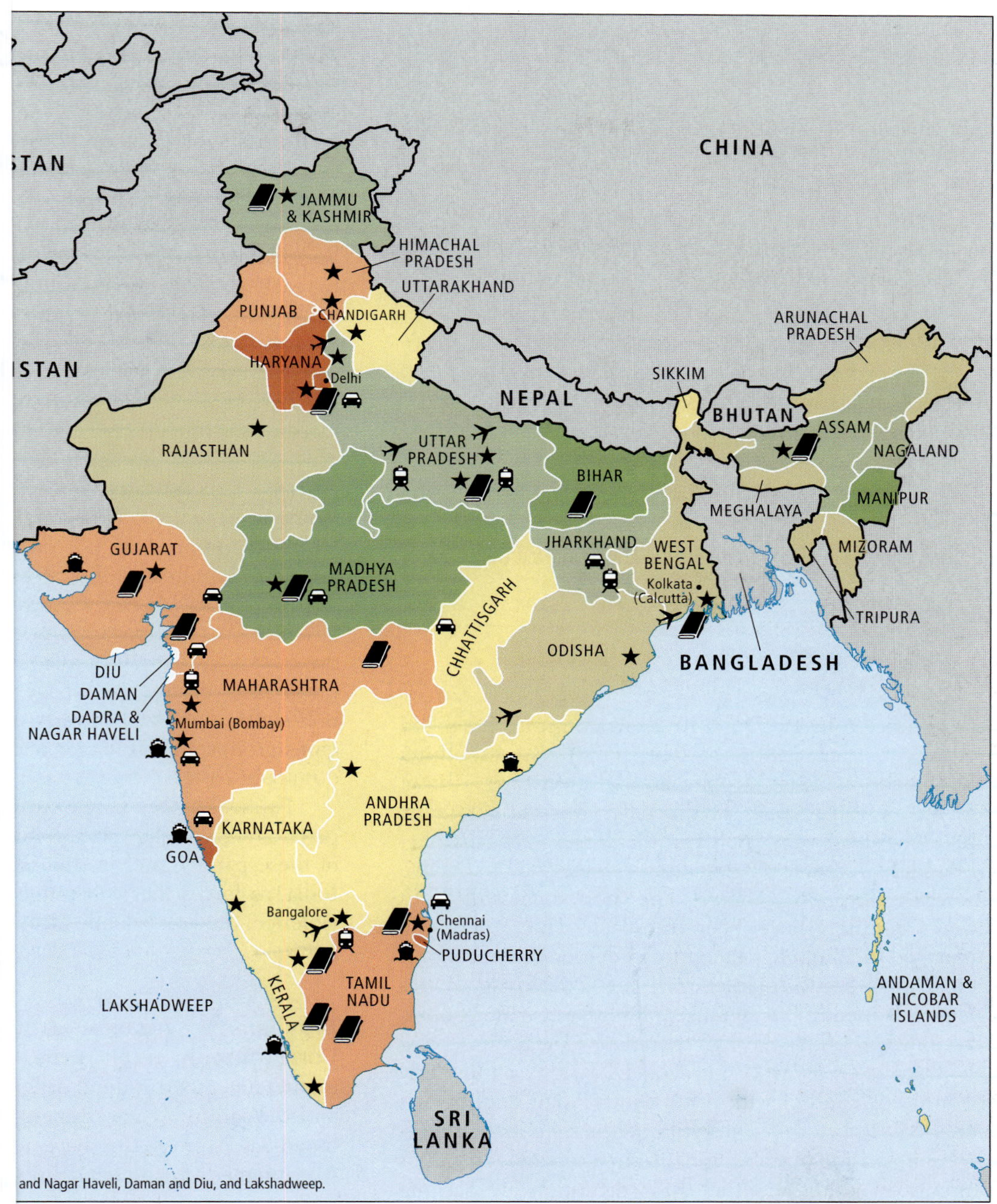

ıe per capita (PPP) and industrial and IT centers in India. The presence ed with higher incomes, and because of this, planners sometimes try to e places. In some areas, poverty may be so great that even fairly intensive ıle to raise average incomes above base levels only slowly. (The per capita ce 2009 for India's states was not available at time of writing.)

mplemented economic reforms ect of much debate. The highly e and upper classes have gained ıss now stands at 50 million—as ıillion—but is likely to grow draby 2030. However, the new ecoer urban/rural income disparities.

THINGS TO REMEMBER

Geographic Insight 3

• **Food and Urbanization** While food production has improved dramatically, many poor farmers in India, unable to afford special seeds, fertilizers, pesticides, and new equipment required for green revolution agriculture, can no longer compete and have gone hungry. They often give up farming;

some have moved to urban areas, hoping to find jobs so they can purchase food. Other farmers have successfully adopted agroecology methods.

- The Grameen Bank and its strategy of microcredit have been very successful in Bangladesh; this method of small-loan financing has spread around the world, including to the United States.
- The service sector is growing rapidly in all countries and dominates most economies in terms of the contributions made to GNI (or GDP), but this sector still employs a relatively small proportion of workers.

Geographic Insight 4

- **Globalization and Development** To take advantage of India's large, college-educated, low-cost workforces, foreign companies are outsourcing an increasing number of jobs to Indian cities like Bangalore, Mumbai, and Ahmadabad. India's current manufacturing boom is benefiting from the offshore outsourcing that started in the 1990s. However, low-skilled workers in rural India are not benefiting nearly as much as others from these changes.
- Foreign firms are now allowed to invest directly in India if they meet certain rules.

Political Issues

Since independence in 1947, South Asian countries have had much success in peacefully resolving conflicts, smoothing potentially bloody transfers of power, and nurturing vibrant public debate over the issues of the day. However, in many cases supporters of opposing political ideologies—for instance, religious freedom versus a religious state—have tried to resolve their differences through violence. The most successful paths out of these conflicts have been those in which opposing groups managed to compete peacefully in democratic elections. Nevertheless, every South Asian country and every foreign country that has intervened in the region's politics have missed opportunities to resolve conflicts through democratic means, resorting instead to the use of force (Figure 8.31A, B). A particularly virulent source of conflict within all South Asian countries is corruption, often linked to purposeful bureaucratic inefficiency, especially the soliciting of bribes to perform a service. A number of movements to address conflict and corruption are gaining acceptance and momentum, especially among the increasingly politically aware middle class, who, because their jobs are now more often in the private sector, have the courage to challenge government officials.

Religious Nationalism

The association of a particular religion with a particular territory or political unit—be it a neighborhood, a city, or an entire country—to the exclusion of other religions, is commonly called **religious nationalism**. The ultimate goal of such movements is often political control over a given territory.

religious nationalism the association of a particular religion with a particular territory or political unit

Although both India and Pakistan were formally created as secular states, religious nationalism has long been a reality in both countries, shaping relations between people and their governments. Rejecting the idea of multiculturalism, and referring back to the days of Partition, India is increasingly thought of as a Hindu state, while Pakistan calls itself an Islamic Republic, and Bangladesh a People's Republic. In each country, many people in the dominant religious group strongly associate their religion with their national identity.

In India, urban men from middle- and upper-caste groups are the predominant supporters of Hindu nationalism (sometimes called *Hindutva*). Hindutva proponents not only promote Hinduism, they fear the erosion of their castes' political influence and particularly resent the extension of the quota system for government jobs and admission to universities to lower-caste groups (see page 351). Conflict results because politically mobilized lower castes and members of other religious groups are no longer willing to follow the dictates of the dominant castes.

Political parties based on religious nationalism have gained popularity throughout South Asia. Although their members think of these parties, such as the Bharatiya Janata Parishad (BJP) in India, as forces that will purge their country of corruption and violence, they are usually no less corrupt or violent than secular parties.

Thinking Geographically

After you have read about power and politics in South Asia, you will be able to answer the following questions:

A Which army was deprived of the fuel carried by the convoy? Who attacked whom?

C What direction was the camera facing when this photo was taken?

D What circumstances led Maoists in Nepal to wage war against King Gyanendra?

Movements Against Government Inefficiency and Corruption In recent years, some, frustrated by government inefficiency, corruption, religious nationalism, caste politics, and the failure of governments to deliver on their promises of broad-based prosperity, have formed political blocs for reform. Bureaucrats who demand bribes have lately been the focus of such activism. Confronted by a bureaucrat who asked for nearly $200 to issue a legitimate income tax refund, one Indian couple in Bangalore launched the Web site I Paid a Bribe, aimed at collecting information about crooked officials. The idea quickly caught on and such sites are now in more than 17 countries. A more militant anticorruption crusader in India is Anna Hazare of Maharashtra, a former military man who advocates that those convicted of corruption lose a hand as punishment. Although most Indians quickly backed off such extreme measures, Hazare, despite his extremism, has attracted a following in the new middle class, especially among educated women, who are increasingly active in politics.

Two aspects of the high-tech revolution in South Asia are likely to make corruption more difficult for bureaucrats and elected officials. The first is the spread of cell phones and their use in banking and money transfers, which will now be more

Photo Essay: Power and Politics in South Asia

have been sparked by governmental authoritarianism that eroded political freedoms. Supporters of political franchisement, imprisonment, and sometimes execution. However, in some cases, growing respect for political aceful reconciliation between former combatants.

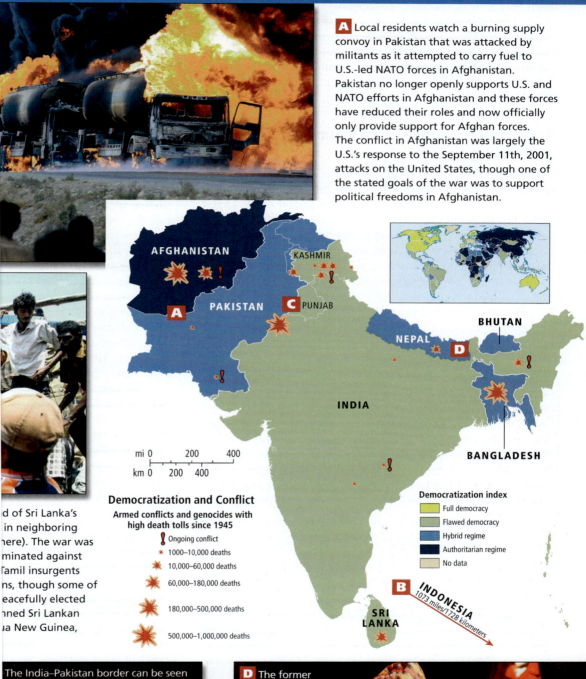

A Local residents watch a burning supply convoy in Pakistan that was attacked by militants as it attempted to carry fuel to U.S.-led NATO forces in Afghanistan. Pakistan no longer openly supports U.S. and NATO efforts in Afghanistan and these forces have reduced their roles and now officially only provide support for Afghan forces. The conflict in Afghanistan was largely the U.S.'s response to the September 11th, 2001, attacks on the United States, though one of the stated goals of the war was to support political freedoms in Afghanistan.

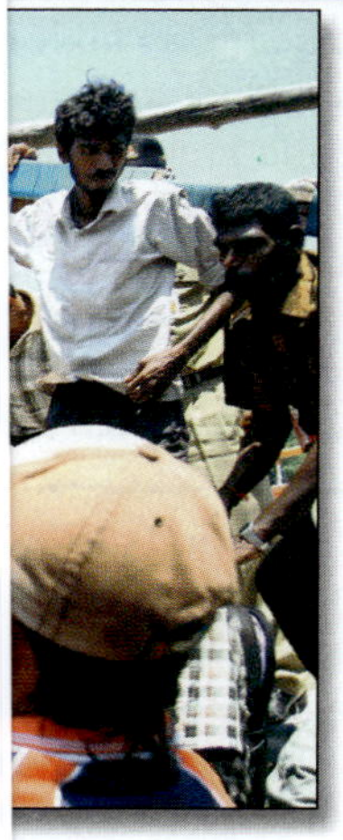

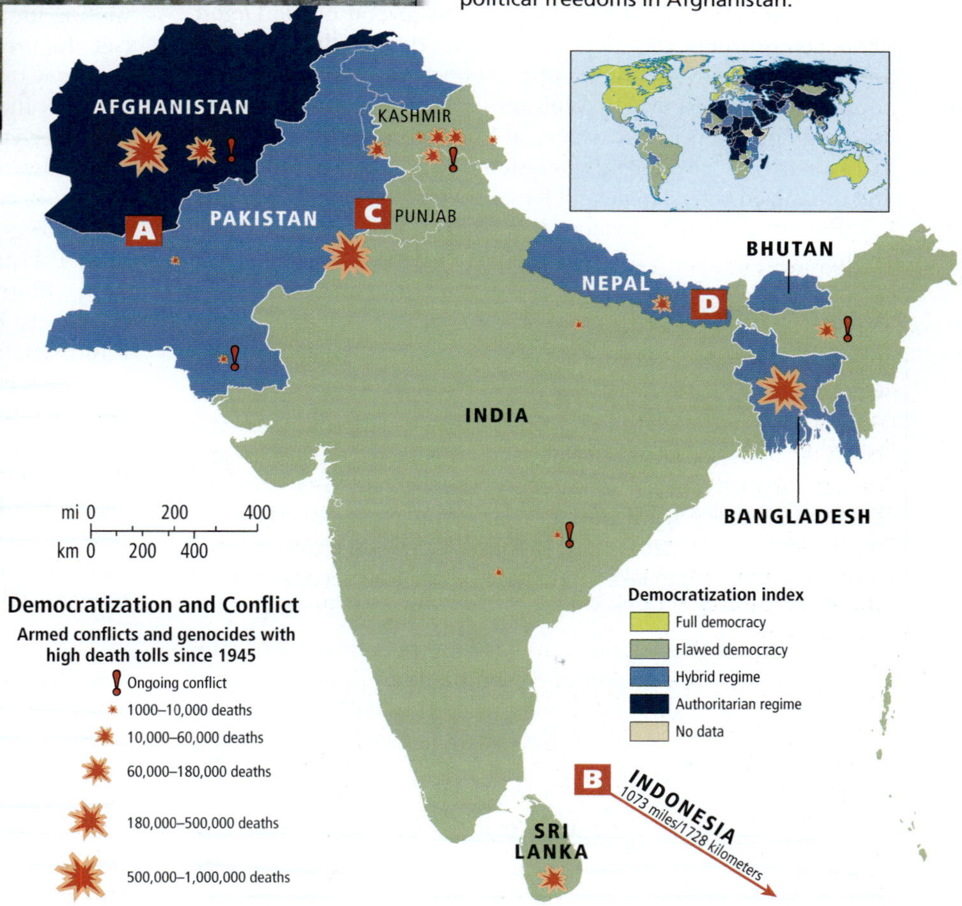

d of Sri Lanka's in neighboring here). The war was minated against Tamil insurgents ns, though some of eacefully elected nned Sri Lankan ua New Guinea,

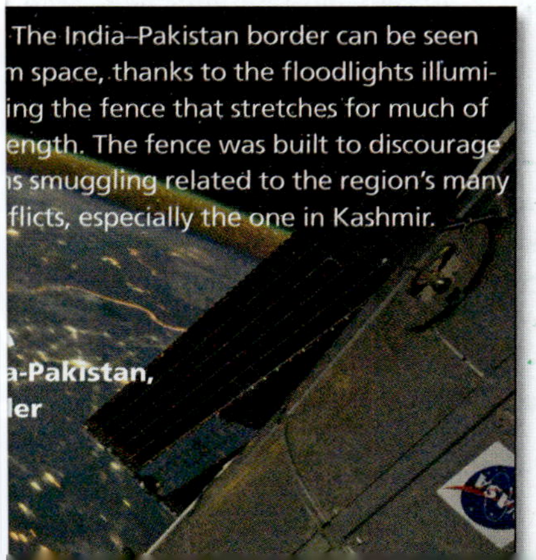

The India–Pakistan border can be seen m space, thanks to the floodlights illumiing the fence that stretches for much of ength. The fence was built to discourage s smuggling related to the region's many flicts, especially the one in Kashmir.

a-Pakistan, ler

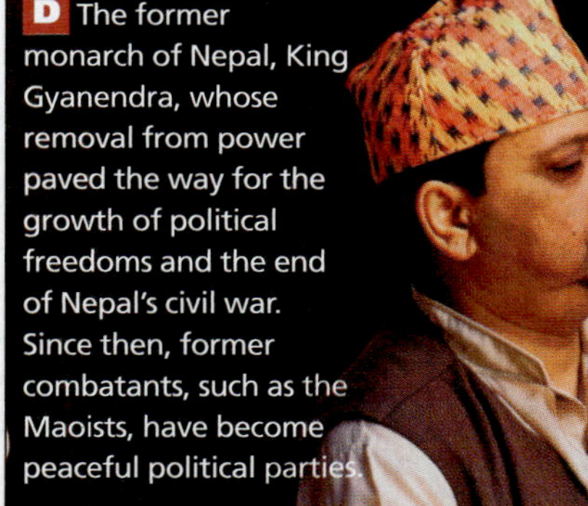

D The former monarch of Nepal, King Gyanendra, whose removal from power paved the way for the growth of political freedoms and the end of Nepal's civil war. Since then, former combatants, such as the Maoists, have become peaceful political parties.

traceable. The second is the *Aadhaar Project*, an ambitious effort to provide all 1.2 billion Indians with a unique photo/digitized ID card (UID). By October of 2012, a total of 208 million had been issued, with complete coverage of India expected before 2020. The UID project is designed to facilitate the distribution of government benefits of all types, many of which currently fail to reach the poorest people because there are few formal records that can be used to identify them. Currently, much of the money intended to reach these people ends up in the hands of corrupt officials.

The Growing Influence of Women and Young Voters

In the state elections of 2012, women and young voters were particularly active, coming to the polls in large numbers, with specific issues and candidates in mind. In several of the largest states, voter turnout was 50 percent higher than in the past. The anticorruption movements, mentioned on pages 366–368, widely covered in the press and by Web sites, apparently motivated voters. Large numbers of new voters, at least a third of whom are 18 to 19 years old, have recently registered to vote. These educated and urbanized voters, male and female, who have far more access to information than had voters in the past, could bring in a new political era.

Power and Politics

Geographic Insight 5

Power and Politics: India, South Asia's oldest, largest, and strongest democracy, has shown that democratic institutions can ameliorate conflict. Across the region, when ordinary people have been able to participate in policy-making decisions and governance—especially at the local level—intractable conflict has been diffused and combatants have been convinced to take part in peaceful political processes.

In some parts of this region, conflicts have been made worse by an unwillingness on the part of governments and warring parties to recognize the results of elections, or even to let issues come to a vote. In some cases conflicts have been defused, at least in the short run, by holding elections and letting former combatants run for office.

The most intense armed conflicts in South Asia today are **regional conflicts**, in which nations dispute territorial boundaries or a minority actively resists the authority of a national or state government. Most of these hostilities represent failed opportunities to resolve problems through the democratic processes.

regional conflict a conflict created by the resistance of a regional ethnic or religious minority to the authority of a national or state government; currently these are the most intense armed conflicts in South Asia

Conflict in Kashmir Since 1947 and the post-independence dividing of India and Pakistan, between 60,000 and 100,000 people have been killed in violence in Kashmir. At the root of the violence is a struggle for territory between India and Pakistan, neither of which are willing to let the people of Kashmir resolve the dispute democratically (see the Figure 8.31 map).

Kashmir has long been a Muslim-dominated area, and in 1947, some Kashmiris believed that for this reason it should be turned over to Pakistan. Although the Hindu maharaja (king) of Kashmir at the time wanted Kashmir to remain independent, the most popular Kashmiri political leader and significant portions of the populace favored joining India. They preferred India's stated ideals of having a nonreligiously based government to Pakistan's less robust safeguards for secularism. When Pakistan-sponsored raiders invaded western Kashmir in 1947, the maharaja quickly agreed to join India. A brief war between Pakistan and India resulted in a cease-fire line that became a tenuous boundary that is still being debated.

Pakistan attempted to invade Kashmir again in 1965 but was defeated by India. India and Pakistan are technically still waiting for a UN decision about the final location of the border (see Figure 8.31C). The Ladakh region of Kashmir (see the Figure 8.1 map) is the object of a more limited border dispute between India and China.

After years of military occupation, most Kashmiris now support independence from both India and Pakistan. However, neither country is willing to hold a vote on the matter. Anti-Indian Kashmiri guerrilla groups equipped with weapons and training from Pakistan have carried out many bombings and assassinations. Blunt counterattacks launched by the Indian government have killed large numbers of civilians and alienated many Kashmiris.

Another complication in the Kashmir dispute is the fact that both India and Pakistan—which came close to war against each other in 1999 and again in 2002—have nuclear weapons. Because of the nationalistic fervor of the protagonists, many see the conflict in Kashmir as more likely to result in the use of nuclear weapons than any other conflict in the world. Analysts agree that any use at all of nuclear weapons would have severe repercussions for all on Earth.

War and Reconstruction in Afghanistan In the 1970s, political debate in Afghanistan became polarized. On one side were several factions of urban elites, who favored modernization and varying types of democratic reforms. Opposing them were rural, conservative religious leaders, whose positions as landholders and ethnic leaders were threatened by the proposed reforms. Divisions intensified as successive governments, all of which came to power through military coups, became more and more authoritarian. Political opponents were imprisoned, tortured, and killed by the thousands, resulting in a growing insurgency outside the major cities.

In 1979, fearing that a civil war in Afghanistan would destabilize neighboring Soviet republics in Central Asia, the Soviet Union invaded Afghanistan. Rural conservative leaders (often erroneously called "warlords") and their followers formed an anti-Soviet resistance group, the *mujahedeen*. As Afghan resistance to Soviet domination increased, the mujahedeen became ever more strongly influenced by militant Islamist thought and by Persian Gulf Arab activists who provided funding and arms. At the time,

; for Cold War allies against the
stan in supporting the mujahe-
hans who favored democratic
his turbulent time, hoping to go
rned.
r heavy losses—14,000 Soviet
ollars wasted—gave up and left
for a time as mujahedeen fac-
; to the 1.5 million civilians and
n the Soviets.
religious, political, and military
merged from among the muja-
control corruption and crime
ecially those related to the role,
oduced in earlier decades by the
de more extreme by the Russian
l to strictly enforce *shari'a*, the
page 253 in Chapter 6). Efforts
society of non-Muslim influ-
g women (see pages 355–356),
slamic education, and publicly
n, to which many Afghan men
le privately promoting its sale to
, the Taliban controlled 95 per-
capital, Kabul. **198. TALIBAN**
TION
otember 11, 2001, the United
moving the Taliban, who were
aden and his international Al
Taliban were overpowered by
d heavily by the United States
eventually by NATO (see the

launched the war in Iraq that
ops, and financial resources
immediately the Taliban were
g the ability of Afghanistan's
rity and to meet the needs of
rural areas in both Pakistan
now aided by widespread dis-
, which is seen as corrupt (see
ost people in Afghanistan favor
Muslim principles, but func-
ng in 2004 have resulted in an
hing from 7.4 million in 2004
t presidential elections will be

was killed in a raid by U.S.
, Pakistan. Another raid killed
mander, and sporadic drone
and civilians. Because the
without the knowledge or con-
y suggest a weakening of the
o-called War on Terror. With
ngly diminished, calls within
withdrawal from Afghanistan
nderway and scheduled to be

completed in 2014 (see Chapter 2 on page 75), though the withdrawal is unlikely to be total.

Sri Lanka's Civil War The Singhalese have dominated Sri Lanka since their migration from Northern India several thousand years ago. Today they make up about 74 percent of Sri Lanka's population of 20.5 million. Most Singhalese are Buddhist. Tamils, a Hindu ethnic group from South India, make up about 18 percent of the total population of Sri Lanka. About half of these Tamils have been in Sri Lanka since the thirteenth century, when a Tamil Hindu kingdom was established in the northeastern part of the island. The other half were brought over by the British in the nineteenth century to work on tea, coffee, and rubber plantations. Some Tamils have done well, especially in urban areas, where they dominate the commercial sectors of the economy. However, many others have remained poor laborers isolated on rural plantations.

Upon its independence in 1948, Sri Lanka had a thriving economy, led by a vibrant agricultural sector and a government that made significant investments in health care and education. It was poised to become one of Asia's most developed economies. But, driven by nationalism, Singhalese was made the only official language and Tamil plantation workers were denied the right to vote. Efforts were also made to deport hundreds of thousands of Tamils to India. In the 1960s, the government shifted investment away from agricultural development and toward urban manufacturing and textile industries, which were dominated by Singhalese. By 1983, the Tamil minority, lacking political power and influence, chose guerilla warfare against the Singhalese, mounting an army known as the Tamil Tigers.

For more than 30 years, the entire island was subjected to repeated terrorist bombings and kidnappings. Peace agreements were attempted several times, but in the end it was an overwhelming military victory by the government, combined with an effective crackdown on international funding for the Tamils, that forced the Tamil surrender in May of 2009. However, dissatisfaction with the peace process has resulted in an ongoing flow of Tamil and other refugees from Sri Lanka (see Figure 8.31B).

Despite long years of violence that severely curtailed the tourist industry, economic growth in other sectors has been surprisingly robust in Sri Lanka. Driven by strong growth in food processing, textiles, and garment making, Sri Lanka is today one of the wealthiest nations in South Asia on a per capita GDP basis and provides for the human well-being of its citizens.

Nepal's Rebels After a civil war that ended generations of monarchy, Nepal has endured political turmoil. An elected legislature and multiparty democracy were introduced into Nepal in 1990, but until 2008 a royal family governed with little respect for the political freedoms of the Nepalese people.

In 1996, Maoist revolutionaries, inspired by the ideals of the late Chinese leader Mao Zedong (but with no apparent support from China), took advantage of public discontent and waged a "people's war" against the Nepalese monarchy. After a decade of civil war, during which 13,000 Nepalese died, the Maoists had both military control of much of the countryside and strong political support from most Nepalese. Persistent poverty and lack of the most basic development under the dictatorial rule of the latest monarch, King Gyanendra (see Figure 8.31D), led to massive protests that forced Gyanendra to step down in 2006. Soon thereafter, the Maoists declared a cease-fire with the government. In 2008, the Maoists won sweeping electoral victories that gave them a majority in parliament and made their former rebel leader prime minister. Then, in May of 2009, when his many conditions for reforming Nepalese society remained unmet, the Maoist prime minister, in a tactical parliamentary move, resigned and took his party into opposition against a new prime minister and his weak 22-party coalition. Although the Maoist opposition agreed to participate in writing a new constitution, divisive issues relating to power sharing and dividing the country into ethnic states have obstructed the creation of a broadly acceptable constitution. Observers concur, however, that if the various factions can see that peaceful democratic processes will give them a voice, Nepal will probably not return to civil war.

200. FUTURE OF NEPAL'S KING GYANENDRA IN QUESTION

Terrorist Attacks in Mumbai in 2008 A 3-day series of coordinated terrorist attacks in Mumbai in November 2008 left 160 dead in luxury hotels, a Jewish center, and a railroad station. The attacks continue to muddy Indian relations with Pakistan, where the plot was hatched. The attacks apparently are only distantly linked to religious nationalism as it has played out since 1947. They appear, on the one hand, to be related to Islamic militancy centered in Afghanistan and Pakistan. On the other hand, investigations have revealed that the terrorists were impoverished young men from Punjab recruited by Lashkar-e-Taiba, a Pakistan-based militant group, with origins based in the Kashmir dispute that is suspected of links with Al Qaeda. There is little doubt, however, that unchecked religious nationalism in the region helps create the antagonistic context for such violence.

THINGS TO REMEMBER

Geographic Insight 5

- **Power and Politics** Since independence in 1947, South Asian countries have had much success in peacefully resolving conflicts, smoothing potentially bloody transfers of power, and nurturing vibrant public debate over the issues of the day.
- Religious nationalist movements are increasingly attractive to people frustrated by government inefficiency, corruption, and caste politics, and by the failure of governments to deliver on their promises of broad-based prosperity.
- New populations of voters—women and educated youth—are changing the outcomes of elections across the region.

Geographic Insights: South Asia Review and Self-Test

1. Climate Change and Water: Climate change puts more lives at risk in South Asia than in any other region in the world, primarily due to water-related impacts. Sea level rise, droughts, floods, and the increased severity of storms imperil many urban and agricultural areas over the short term. Over the longer term, glacial melting threatens rivers and aquifers.

- Why is the issue of melting glaciers in the Himalayas so important in this region? Compare the short- and long-term effects of this change.
- What additional threats do coastal areas of South Asia face?

2. Gender and Population: The assumptions that males are more useful to families than females and that females must be supplied with a dowry have resulted in a devaluation of females, leading to a gender imbalance in populations across South Asia. Those who can afford in utero gender-identifying technology may abort female fetuses, while those with fewer resources sometimes commit female infanticide. As a result, adult males significantly outnumber adult females. A second relationship between gender and population is that when females are given opportunities to work outside the home, they tend to choose to have fewer children.

- What has created the strong preference for sons among many South Asian families? Where in South Asia is the preference for sons the weakest?
- What is the link between fertility and education for women of South Asia?

3. Food and Urbanization: Due to changes in food production systems in South Asia, food supplies and incomes of wealthy farmers have increased. Meanwhile, impoverished agricultural workers, displaced by new agricultural technology, have been forced to seek employment in cities, where they often end up living in crowded, unsanitary conditions and must purchase food.

- What has been the impact on food production of the introduction of new seeds, fertilizers, pesticides, and equipment in South Asia? How has this shift resulted in the growth of cities?
- What kind of jobs and housing do people in South Asia usually find in urban areas?

4. Globalization and Development: Globalization benefits populations unevenly. Educated and skilled South Asian workers with

ıology-based industries and services
.ers, however, in both urban and
ȝ but very low-paying jobs.

ɩia have gained from the recent
ıbout South Asian workers is
ors in high-tech industries?

to the global economy?

ɩia are benefiting the least from

th Asia's oldest, largest, and strongest
ıtic institutions can ameliorate

conflict. Across the region, when ordinary people have been able to participate in policy-making decisions and governance—especially at the local level—intractable conflict has been diffused and combatants have been convinced to take part in peaceful political processes.

• Which conflicts in South Asia have been made worse by an unwillingness on the part of governments and warring parties to allow free and fair elections?

• What democratic strategy has been used to diffuse some conflicts in the region, at least for the short term?

• What are some signs that political participation is catching on in South Asia?

stions

reate are sources of irrigation water
some of the factors that have
Asia, as well as elsewhere around the
ıater and electricity.

ıat India has been part of
ɩ. Tie this history to what is

ȝ features of the British colonial era
obalization reinforcing or erasing

e resulting in urbanization?

ıpact on extreme poverty in

6. Describe how South Asia is vulnerable to climate change. Identify some responses to the climate crisis that are emerging from this region.

7. Describe the main challenges to democracy in South Asia.

8. Identify the factors that have led to a recent boom in manufacturing industries in South Asia.

9. What factors have encouraged high population growth in South Asia in the past? What factors may encourage slower population growth in the future?

10. Describe the main factors that create South Asia's gender imbalance.

Jainism 352
jati 350
microcredit 363
Mughals 341
offshore outsourcing 364
Partition 343
purdah 354
regional conflict 368
religious nationalism 366

sex work 349
Sikhism 352
subcontinent 330
summer monsoon 331
Taliban 355
varna 350
winter monsoon 332

A Plateau of Tibet, China

9 East Asia

C Three Gorges Dam, Chang Jiang, Hubei Province, China

D Hong Kong Island and South China Sea

E Mount Fuji, Japan

FIGURE 9.1 Regional map of East Asia.

GEOGRAPHIC INSIGHTS: EAST ASIA

After you read this chapter, you will be able to discuss the following geographic insights as they relate to the nine thematic concepts:

1. Climate Change and Water: East Asia has long suffered from alternating droughts and devastating floods, and scientists fear that these hazards will only intensify with climate change. Meanwhile, the rivers of China, many of which originate in the Tibetan and neighboring plateaus, are dependent both on seasonal rainfall cycles and on glacial meltwater that are expected to diminish over the next few decades.

2. Food and Globalization: East Asian countries have enough economic development to be able to buy food on the global market, but they are less able to produce sufficient food domestically. Much agricultural land has been lost to urban and industrial expansion and to mismanagement; meanwhile the increase in the region's affluence has led to a demand for more imported food.

3. Urbanization and Development: China's move to join the global economy more than three decades ago has led to spectacular urban growth, a stronger emphasis on consumerism, and a massive rise in industrial and agricultural pollution of the air, water, and soil.

4. Power and Politics: As economic development has grown across East Asia, pressures for more political freedoms have also risen. Japan, South Korea, and Taiwan are now among the more politically free places in the world, and pressure for similar levels of freedom is intensifying in China and North Korea.

5. Population and Gender: China's "one-child policy" has created a shortage of females. Long-standing cultural preferences for male children mean that if their one child will be female, many parents choose abortion or put the baby up for adoption. The shortage of females is accentuated by the fact that many women are deciding to have a career rather than to marry and have children.

The East Asian Region

East Asia (see Figure 9.1 on pages 372–373) is the most populous world region, with 1.6 billion people. The region is dominated by China, which has a population of 1.35 billion. China's dominance is not due only to its population, but also to its physical size and its economic role in the global economy. Just as the astounding economic rise of Japan, South Korea, and Taiwan has been the model for developing countries over the past 50 years, the opening of China to the global economy will continue to transform East Asia and the world for 50 years more. The changes underway in this region today are immense. Incomes have risen and cities have boomed, but so have air and water pollution and greenhouse gas emissions. At the same time, there are signs that new, cleaner technologies may mitigate these pollution problems.

The nine thematic concepts in this book are explored as they arise in the discussion of regional issues, with interactions between two or more themes featured, as in the geographic insights above. Vignettes, like the one that follows about the plight of workers in the East Asian economy, illustrate one or more of the themes as they are experienced in individual lives.

GLOBAL PATTERNS, LOCAL LIVES In 2000, at age 18, Li Xia (Li is her family name) left her farming village in China's Sichuan Province (**Figure 9.2A**) for the city of Dongguan in the Guangdong Province of southern China. She was accompanied by two friends. A few months earlier, the government had taken their families' farmland for an urban real estate project, paying compensation of only U.S.$2000 per family. The three young women accepted an offer to work in a Dongguan toy factory so they could send money back to their families, who now have to pay cash for food and housing (see Figure 9.2B).

When the young women arrived in the city, they joined 5 million other recent internal migrants. Sixty percent, like Li Xia, had migrated illegally, without government-approved residency rights, and thus were dependent on their employers for housing. As did many others, Li Xia and her friends soon found that the labor recruiters had lied about their wages. Per month they would be paid not U.S.$45, as promised, but U.S.$30. In addition, they would work 12-hour shifts in 100°F heat, receive no overtime pay, and have only 1 day off a month. But there was no point in protesting. Their families in Sichuan needed the money they would send home, and the recruiters purposely brought in thousands of extra workers to replace any complainers.

Xia felt better when she saw that the toy factory was a clean, modern building known locally as the "Palace of Girls" (nearly all 3500 employees were female). Within a day, she had completed her training, signed a 3-year contract, and mastered her task of putting eyes on stuffed animals destined for toddlers in the United States and Europe. Her enthusiasm faded, however, when she learned that she and her friends would be spending much of their money on the expensive but low-quality food provided by the company, and would be sharing one small room and a tiny bath with eight other women and a rat or two.

In less than 6 months, Xia broke her contract and returned home to her village in Sichuan. There, using ideas and assertiveness she had gained from her time in Dongguan, she opened a snack stand. In just 1 month, she made ten times her investment of U.S.$12. But Xia yearned to return to the southern coast to try again for a well-paid factory job. A few months later, she returned to Dongguan with her sister, leaving the stand to her sister's husband (who also cared for the couple's baby).

Since her first arrival, the city had grown by 20 percent and now had 1400 foreign companies trying to hire thousands of workers. Through connections, Xia and her sister found jobs requiring midlevel skills in a Taiwanese-owned fiber optics factory at twice the wages Xia had earned at the toy factory. One year after her first trip to Dongguan, Xia was making a bit more than U.S.$100 a month, enough to live relatively comfortably with only three roommates. She had prospects for a raise, and she was sending money home and once again saving, this time to open a bar in her home village. *[Sources: Washington Post, National Public Radio, and Wall Street Journal. For detailed source information, see Text Credit pages.]* ■

The experiences of Li Xia and her sister illustrate first how the needs of rural areas are being subverted to the needs of China's burgeoning cities. Developers are increasingly targeting rural land, and farmers are rarely given a fair price for their land.

lopment, and urbanization in the new
nore than 145 million workers—over one-third
"migrant workers" who had moved from rural
thout the government's permission and hence
nship they would be entitled to had they stayed
another 243 million rural migrants will join them,
population of over 1 billion people. Geographer
the number of workers moving long distances
1990s, with the majority moving to Guangdong
the "world's factory," and into the lower Chang

(A) A village in rural Sichuan province.

(B) Workers in Dongguan, Guangdong Province, check stuffed toys for defects.

dividual Chinese farmer does not
ises it from the local government,
he land to high-bidding develop-
globalization, and changes in
ast Asia as millions of rural young
coastal cities to work in factories
There they produce goods for sale
ills, and gain new confidence.
rural-to-urban migration in China
household registration) **system**, an
he Maoist era that effectively ties
ir birth. Today the system is less
lly expected to obtain permission
to migrate. In a desperate search
ple like Li Xia are ignoring the
become part of what is called the
China to describe those who have
health care in the place to which

they have migrated. These migrants generally work in menial, low-wage jobs and make agonizing sacrifices to send money home to children and spouses living in rural areas. Between 2000 and 2010, the number of people migrating from rural to urban areas in China in search of work more than doubled (see Figure 9.17C on page 398). This amounts to approximately half of the working people in China's cities. The Chinese government has so far retained the hukou system because it keeps the cities from being overwhelmed by migrants, but the government enforces it only weakly because everybody realizes that urban industries are in need of a growing workforce. Hukou is therefore a deliberate policy that both enables mobility of labor and keeps that labor vulnerable and exploitable.

***hukou* system** the system in China that ties people to their place of birth; each person's permanent residence is registered and any person who wants to migrate must obtain permission from authorities to do so

floating population the Chinese term for people who live illegally in a place other than their required *(hukou)* household registration location; many are jobless or underemployed people who have left economically depressed rural areas for the cities

THINGS TO REMEMBER

- The migration of people from rural China to cities includes 160 million workers who have moved without legal papers and who now constitute half of China's urban labor force.
- Most rural workers who have moved to China's cities without official permission are in low-paying jobs with little or no access to social services or education.

THE GEOGRAPHIC SETTING

What Makes East Asia a Region?

China is part of the region of East Asia, home to nearly one-fourth of humanity. The vast East Asian territory stretches from the Taklimakan Desert in far western China to Japan's rainy Pacific coastline, and from the frigid mountains of Mongolia in the north to the tropical landscapes of Hainan Island, China's southernmost province (see Figure 9.1; see also Figure 9.4B on page 378). East Asia (Figure 9.3) is comprised of the countries of China, Mongolia, North Korea, South Korea, Japan, and Taiwan (the last has operated as an independent country since World War II but is claimed by China as a province). These countries are grouped together because of their cultural and historical roots, many of which are in China. Because of China's great size, historical influence, enormous population, and huge economy, this chapter gives it particular emphasis. Japan, whose large and prosperous economy makes it a major player on the world stage, is also emphasized.

Terms in This Chapter

East Asian place-names can be very confusing to those unfamiliar with East Asian languages. We give place-names in English transliterations of the appropriate Asian language, taking care to avoid redundancies. For example, *he* and *jiang* are both Chinese words for *river*. Thus the Yellow River is the Huang He; and the Long River, also called the Yangtze in English, is the Chang Jiang in China; it is redundant to add the term *river* to either name. The word *shan* appears in many place-names and usually means "mountain."

FIGURE 9.3 Political map of East Asia.

Pinyin (a spelling system based on Chinese sounds) versions of Chinese place-names are now commonplace. For example, the city once called Peking in English is now *Beijing*, and Canton is *Guangzhou*.

The region historically known as Manchuria is here referred to as *China's Far Northeast* to emphasize its geographical location. Although China refers to Tibet as *Xizang*, people around the world who support the idea of Tibetan self-government avoid using that name. This text uses *Tibet* for the region (with Xizang in parentheses), and *Tibetans* for the people who live there.

Physical Patterns

A quick look at the regional map of East Asia (see Figure 9.1) reveals that the topography here is perhaps the most rugged in the world. East Asia's varied climates result from the meeting of huge warm and cool air masses and the dynamic interaction between land and oceans. The region's large human population has affected the variety of ecosystems that have evolved there over the millennia and that still contain many important and unique habitats.

Landforms

The complex topography of East Asia is partially the result of the slow-motion collision of the Indian subcontinent with the southern edge of Eurasia over the past 60 million years. This tremendous force created the Himalayas and lifted up the Plateau of Tibet (see Figure 9.1A, also depicted in gray and gold in the Figure 9.1 map), which can be considered the highest of four descending steps that define the landforms of mainland East Asia, moving roughly west to east.

The second step down from the Himalayas is a broad arc of basins, plateaus, and low mountain ranges (depicted in yellowish tan in the Figure 9.1 map). These landforms include the broad, rolling highland grasslands and deep, dry basins and deserts of western China (such as the Taklimakan Desert) and Mongolia (see Figure 9.1B), as well as the Sichuan Basin and the rugged Yunnan–Guizhou Plateau to the south, which is dominated by a system of deeply folded mountains and valleys that bend south through the Southeast Asian peninsula.

The third step, directly east of this upland zone, consists mainly of broad coastal plains and the deltas of China's great rivers (shown in shades of green in the map in Figure 9.1). Starting from the south, this step is defined by three large lowland river basins: the Zhu Jiang (Pearl River) basin, the massive Chang Jiang basin (see Figure 9.1C), and the lowland basin of the Huang He on the North China Plain. Each of these rivers has a

:ing subject to periodic flooding,
;riculture; but now coastal cities
ling in wetlands. Each is now a
ıdustrialization (see Figure 9.17
nd hills (shown in light brown)
ı's Far Northeast and the Korean
rd step.
he continental shelf, covered by
East China Sea, and the South
including Hong Kong, Hainan,
ıis continental shelf; all are part
e 9.1D).
different geological origin: they
nental shelf. They rise out of the
c in the tectonically active zone
ıd Eurasian plates grind against
ın of the Pacific Ring of Fire (see
ire Japanese island chain is par-
volcanic eruptions, earthquakes,
. The volcanic Mount Fuji, the
a recognizable national symbol
ı 1707. However, the mountain
p internal rumblings have been
f 2011, the largest earthquake in
tering 9.2 on the Richter scale)
ın, near the city of Sendai. The
illed tens of thousands of people
reactors located on the coast,
clear accident ever in the world

s few flat portions, and most flat
d. Consequently, the large num-
:gion have had to be particularly
agriculture. They have cleared
ınges, until recently using only
9.5A on page 380). They have
ı melted snow, drained wetlands
ns, and applied their complex
nimal husbandry to help plants
conditions.

rasting climate zones, shown in
rior west and the wet (monsoon)
: the term *monsoon* refers to the
ds that flow from the Eurasian
:ans during winter and from the

heats up and cools off more rap-
s of large landmasses in the mid-
d in winter and extremely hot in
hly corresponding to the first two
e, is an extreme example of such
te because it is very dry. In fact,
ocean than any other place on

Earth's surface (also known as the "pole of inaccessibility"). With little vegetation or cloud cover to retain the warmth of the sun after nightfall, summer daytime and nighttime temperatures may vary by as much as 100°F (55°C).

Grasslands and deserts of several types cover most of the land in this dry region (see Figure 9.4C). Only scattered forests grow on the few relatively well-watered mountain slopes and in protected valleys supplied with water by snowmelt. In all of East Asia, humans and their impacts are least conspicuous in the large, uninhabited portions of the deserts of Tibet (Xizang), the Tarim Basin in Xinjiang, and the Mongolian Plateau.

The Monsoon East The monsoon climates of the east are influenced by the extremely cold conditions of the huge Eurasian landmass in the winter and the warm temperatures of the surrounding seas and oceans in the summer. During the dry winter monsoon, descending frigid air sweeps south and east through East Asia, producing long, bitter winters on the Mongolian Plateau, on the North China Plain, and in China's Far Northeast (see Figure 9.4D). While occasional freezes may reach as far as southern China, winters there are shorter and less severe. The cold air of the dry winter monsoon is partially deflected by the east-west mountain ranges of the Qin Ling, and the warm waters of the South China Sea moderate temperatures on land.

During the summer monsoon, as the continent warms, the air above it rises, pulling in wet, tropical air from the adjacent seas. The warm, wet air from the ocean deposits moisture on the land as seasonal rains. As the summer monsoon moves northwest, it must cross numerous mountain ranges and displace cooler air. Consequently, its effect is weakened toward the northwest. Thus, the Zhu Jiang basin in the far southeast is drenched with rain and has warm weather for most of the year (see Figure 9.4A), whereas the Chang Jiang basin, which lies in central China to the north of the Nan Ling range, receives about 5 months of summer monsoon weather. The North China Plain, north of the Qin Ling and Dabie Shan ranges, receives even less monsoon rain—about 3 months each year. Very little monsoon rain reaches the dry interior.

Korea and Japan have wet climates year-round, similar to those found along the Atlantic Coast of the United States, because of their proximity to the sea. They still have hot summers and cold winters because of their northerly location and exposure to the continental effects of the huge Eurasian landmass. Japan and Taiwan actually receive monsoon rains twice: once in spring, when the main monsoon moves toward the land, and again in autumn, as the winter monsoon forces warm air off the continent. This retreating warm air picks up moisture over the coastal seas, which is then deposited on the islands. Much of Japan's autumn precipitation falls as snow, in particular in northern latitudes and at higher altitudes.

Natural Hazards The entire coastal zone of East Asia is intermittently subject to **typhoons** (tropical cyclones or hurricanes). Japan's location along the northwestern

tsunami a large sea wave caused by an earthquake

typhoon a tropical cyclone or hurricane in the western Pacific Ocean

Climate Zones

Tropical humid climates (A)
- Tropical wet
- Tropical wet/dry

Arid and semiarid climates (B)
- Desert
- Steppe

Temperate climates (C)
- Midlatitude, moist all year
- Subtropical, winter dry
- Mediterranean, summer dry

Cool humid climates (D)
- Continental, winter dry
- Continental, moist all year

Coldest climates (E)
- High altitude

Winds

The summer monsoon pulls in warm, tropical air containing huge amounts of moisture that is then deposited on the land as seasonal rains.

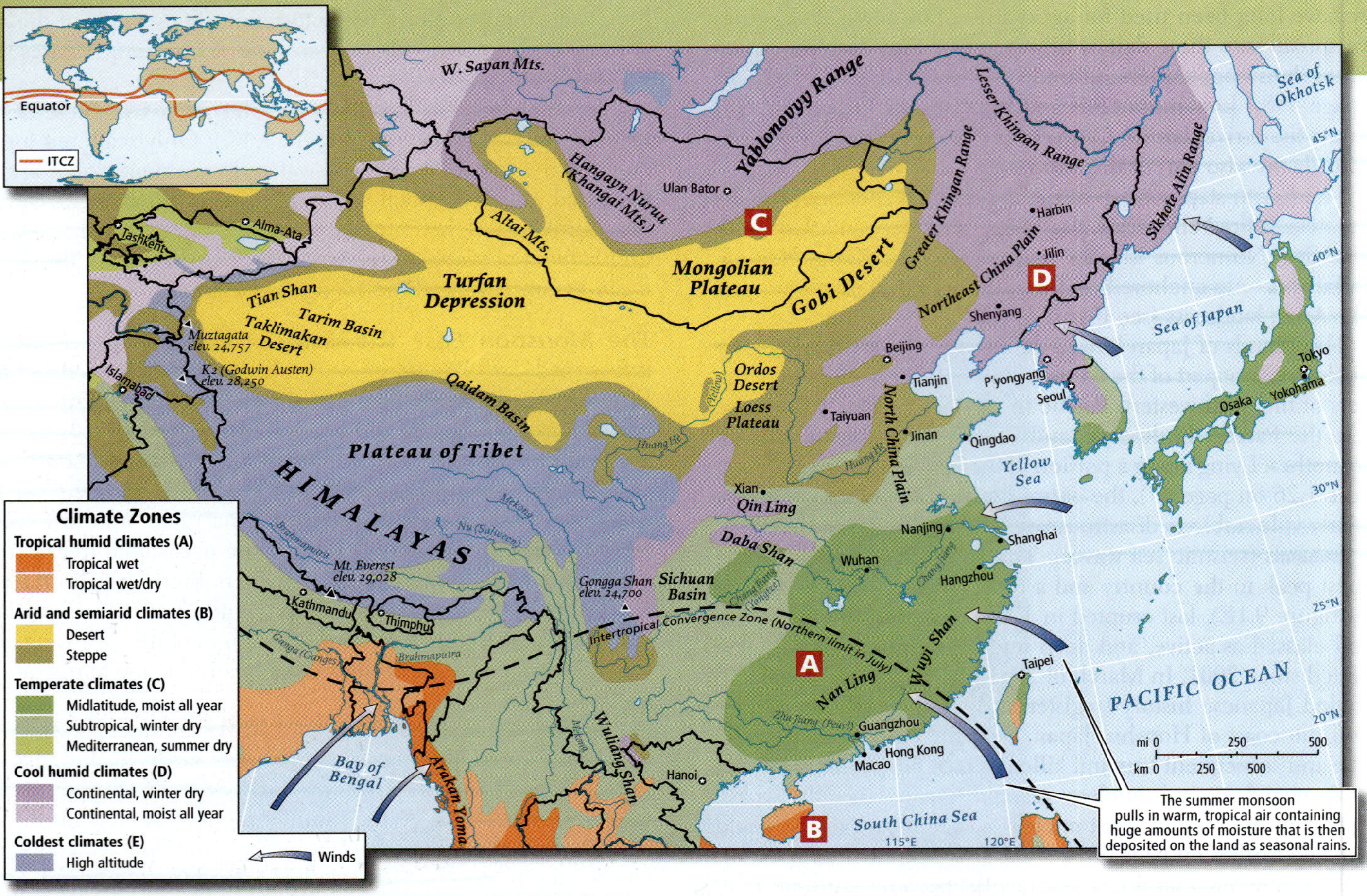

A Midlatitude, moist all year, Guilin, China

B Tropical wet, Hainan Island, China

C Steppe, Mongolia

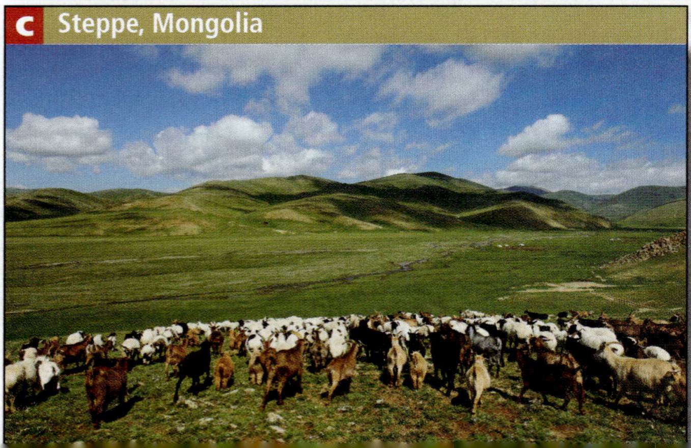

D Continental, winter dry, Jilin, China

re (see Figure 1.26 on page 47) arthquakes, and tsunamis. These it threat in Japan; the heavily uthwest through the Inland Sea n Honshu) is particularly endanerious natural hazard in Taiwan terior.

REMEMBER

l zones, or "steps," that form the

tivity along the Pacific Ring of Fire.

sting climates: the dry continental t.

atural hazards, including earth-, and tropical storms (typhoons).

es

onmental problems result from ied with rapid urbanization and nomic development (Figure 9.5).

ons and Vulnerability

East Asia has long suffered from ng floods, and scientists fear that th climate change. Meanwhile, the nate in the Tibetan and neighboring asonal rainfall cycles and on glacial inish over the next few decades.

l warming, is a growing concern United States as the world's largase gases (but still produces less n a per capita basis; see page 23 Japan is also a major emitter. g on limiting its emissions, peresponsibility as an increasingly wer. However, even with large se of new technologies, China's up almost a quarter of the world's by 2030 as its urban households appliances, and computers.

cial Melting Glaciers on the on moisture and store it as ice lowly released. Two of China's l the Chang Jiang, are partially now melting so rapidly that scilly disappear. One effect could be significantly lower flows in these two great rivers during the winter, when little rain falls. Both have already begun to run low during winter, and trade has been affected because riverboats and barges have become stranded on sandbars. Another effect could be a reduction in the amount of irrigation water available for dry-season farming.

Water Shortages Nearly every year, abnormally low rainfall or abnormally high temperatures create a drought somewhere in China. These droughts often cause more suffering and damage than any other natural hazard.

Droughts can be worsened by human activity on a local or regional scale. When people begin to live or farm in dry environments, as many millions have done in China and Mongolia during the twentieth century, *desertification* (see page 243 in Chapter 6) can result. People clear vegetation to grow crops, which require more water than the natural vegetation, so the crops must be irrigated with water pumped to the surface from underground aquifers (see Figure 9.5A). Many dry areas in China are subject to strong winds that can blow away topsoil once the vegetation is removed. Furthermore, irrigated crops are often less able than natural vegetation to hold the soil; this was, for example, part of what helped create the 1930s Dust Bowl in the central United States (see page 74 in Chapter 2). Huge dust storms have plagued China in recent years. High dunes of dirt and sand have appeared almost overnight in some areas that border the desert, threatening crops, roads, and homes. Dust storms can also move over long distances from western China and affect large cities, such as Beijing, near the coast. Particulate matter from these dust storms even circles the entire globe in the upper atmosphere.

VIGNETTE In the Ningxia Huizu Autonomous Region on the Loess Plateau in northwest China, Wang Youde squints out at what are now sand-colored low hills barren of vegetation. He explains how these vast tracts of former farmland have been transformed into deserts by a combination of human error and climate change.

The area was already prone to drought and wind (*loess* means "wind-deposited soil"), and then agricultural expansion into areas that have this fragile, wind-deposited soil led to the removal of thick, natural, deep-rooted grasses. One can still see the agricultural terraces on the arid slopes where now not even grass grows. Wang Youde's family and 30,000 others fled the area when Youde was 10 years old because one day a sand dune covered their village. From his early childhood, he remembers flowers, bird song, and occasional snowfalls; all have vanished. Now Youde is back, heading up a project to revegetate thousands of hectares with drought-resistant plants that will hold the soil. By hand, squares of braided straw or stones are laid down to keep water from running off the land and to protect planted seedlings. The seedling survival rate is only 20 to 30 percent. Yet, Youde says, "Every time we see an oasis that we have created we are very satisfied . . . [b]ecause we have poured sweat and blood into our work." His adult children are helping him, hoping to remain with the family and escape the hardships of migrating to find urban factory work. *[Source: Asia Pacific News. For detailed source information, see Text Credit pages.]* ■

Figure 9.5 Photo Essay: Vulnerability to Climate Change in East Asia

China is particularly vulnerable to drought, desertification, flooding, and other hazards that climate change may intensify.

A The terraced hillsides of China's eastern Gansu Province are in a dry upland zone, where agriculture depends on rainfall or irrigation water taken from rivers or underground aquifers. Rainfall is already unpredictable and could become more so with climate change. Melting glaciers on the Plateau of Tibet threaten to reduce river flows, and overuse of groundwater for irrigation is depleting aquifers. Large poor, rural populations are increasingly being forced to relocate to other parts of China.

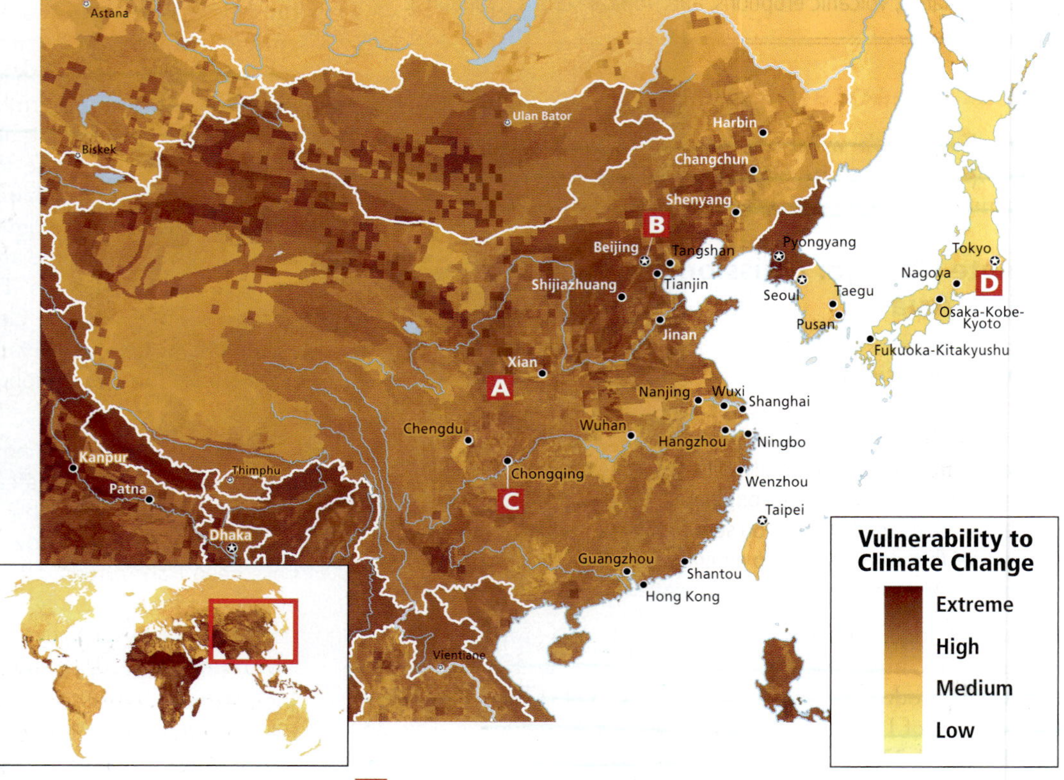

B A canal outside of Beijing, China, a city that faces severe water shortages and massive increases in population. China's national government is creating huge canals and other infrastructure to divert water from distant rivers in southern and western China to Beijing and surrounding areas.

C Flooding hits the Chang Jiang at Chongqing, China, carrying off boats and barges. Climate change could result in increased flooding. This is especially likely in eastern China, where large amounts of rain fall during the summer monsoon.

D A solar power station in Kawasaki, Japan. While it has made only small reductions in greenhouse gas emissions, Japan has the third-largest installed solar power–generating capacity in the world (behind Germany and Italy).

Water shortages are particularly intense in the North China Plain, which produces half of China's wheat and a third of its corn. Here the water table is falling more than 10 feet per year because of the increased use of groundwater for irrigation and urban needs. Meanwhile, withdrawals of water from the Huang He often make the river's lower sections run completely dry during the winter and spring. A gigantic development called the South–North Water Transfer Project is underway to rectify the problem (see Figure 9.5B). The goal is to divert water from the Chang Jiang basin in central China to the Huang He basin in the north. al access to water, diversion canals e locations, from the Plateau of ovinces in the east. The project is il 2050.

elated vulnerabilities, China is n to stretch existing supplies. ban water is recycled, and many centage. A major new national wastewater discharged by indusofficial support for these conser- eas, and Taiwan have monsoon n a generally wetter climate and ught than China and Mongolia.

e same shifting patterns of rain- an also worsen flooding. Under- nts of rain deposited on eastern oon periodically can cause cata- ivers. If global warming leads to tterns, flooding could be much ngineers have constructed elab- voirs, and artificial lakes to help stems failed in 1998, when heavy used some of the worst flooding of approximately 4000 people. uently on the Chiang Jiang and hern China (significant floods 012). Every year there are large ning and mudslides.

gy Vulnerability Until the f 2011, Japan, along with many ase the use of nuclear power as missions. However, the tsunami ear plant and destroyed its cool- tdown of several of the plant's nuclear reactors. The severity of the nuclear disaster that followed in the wake of the tsunami halted or curtailed many plans for expanding nuclear power in Japan and across the globe. The accident forced the evacuation of over 200,000 people; temporarily contaminated the water supply of Tokyo; contaminated the ground, food, and livestock for miles around the reactor; released thousands of gallons of water containing 10,000 times the normal level of iodine-131 (a radioactive substance) into the Pacific Ocean; and sent enough radiation into the atmosphere that scientists detected measurable radiation increases in the United States and Europe. The accident is expected to boost reliance both on renewable resources and on fossil fuel in the near future.

ON THE BRIGHT SIDE

Responses to the Climate Crisis

East Asia is increasingly responding to the warming aspect of global climate change. Japan has led such efforts for decades. In 1997, the Japanese city of Kyoto hosted the meeting in which countries first committed to reduce their greenhouse gas emissions. Japanese automakers such as Toyota were among the first to develop and sell hybrid gas-electric vehicles, and Japan is now third only to Germany and Italy in its installed solar power–generating capacity (see Figure 9.5D). However, Japan has made only small actual reductions in its greenhouse gas emissions.

Food Security and Sustainability in East Asia

Geographic Insight 2

Food and Globalization: East Asian countries have enough economic development to be able to buy food on the global market, but they are less able to produce sufficient food domestically. Much agricultural land has been lost to urban and industrial expansion and to mismanagement, but the increase in the region's affluence has led to a demand for more imported food.

East Asia's **food security**—the capacity of the people in a geographic area to consistently provide themselves with adequate food—is increasingly linked to the global economy. The wealthiest countries in the region are already very dependent on food imports from around the globe. About 75 percent of the food consumed in Japan, South Korea, and Taiwan is imported. In China, on the other hand, self-sufficiency in grain production is important to national identity because devastating famines were a recurring problem before and during the Communist Revolution. China is now nearly self-sufficient with regard to basic necessities, but it relies on imports of commodities to supply growing demands for grain and soybeans for animal feed and luxury food items (see below).

From one perspective, high levels of imported food are a sign of increasing food security in East Asia, because countries with competitive globalized economies can afford to bring in food from elsewhere. However, recent dramatic

food security when people consistently have access to sufficient amounts of food to maintain a healthy life

increases in global food prices illustrate the perils of dependence on food imports, especially for the poor. In 2007–2008, the world market price for grain and other basic food commodities shot up, and it did so again in 2010–2011. The reasons for the increases in food prices are multifold, and include high oil prices (because producing food is often energy intensive), the use of corn to produce ethanol fuel (ethanol is used as a replacement for gasoline in the Americas and elsewhere) rather than as a food source, and the growing demand for agricultural commodities in China and other developing economies.

217. FOOD PRICES SKYROCKETING IN CHINA

Food Production Just over half of East Asia's vast territory can support agriculture (Figure 9.6). In much of this area, food production has been pushed well beyond what can be sustained over the long term. As a result, East Asia's fertile zones are shrinking. In China, roughly one-fifth of the agricultural land has been lost since the Communist Revolution of 1949, largely because of urban and industrial expansion and agricultural mismanagement that created soil erosion and desertification (see the vignette on page 379).

As urban populations grow more affluent, they consume more meat and other animal products that require more land and resources than did the plant-based national diet of the past. Soybeans, which used to be important as an export for China, are one example of this change in diet. More than 70 percent of the soybeans used in China are now imported, primarily from the United States, Brazil, and Argentina. Soybeans are no longer used mainly for human food but for animal feed, for farmed fish food, and especially for high-grade cooking oil. Figure 9.7 shows, however, that vegetable dishes with an ancient history do remain popular in many parts of East Asia (see Figure 9.7A), and that meat and dairy dishes are still prepared in traditional ways in places where raising farm animals has been the norm (see Figure 9.7C).

Rice Cultivation The region's most important grain is rice, and over the millennia its cultivation has transformed landscapes throughout central and southern China, Japan, Korea, and Taiwan. In these areas, rainfall is sufficient to sustain **wet rice cultivation**, which can be highly productive.

wet rice cultivation a prolific type of rice production that requires the plant roots to be submerged in water for part of the growing season

FIGURE 9.6 China's agricultural zones. Part of the economic reforms instituted in recent years in China includes more regional specialization in agricultural products.

ires elaborate systems of water
he plants must be submerged in
ason. Centuries of painstaking
l rivers into intricate irrigation
ides have been transformed into
ly distribute the water. Writing
Sichuan Province, geographer
ow "[e]verywhere one can hear
t brings life and growth to the
iltivation techniques, combined
ining, have led to the loss of
it the most mountainous, dry,
suitable agricultural land left,
under wet rice cultivation is

Many East Asians depend heav-
otein (see Figure 9.7B). People
ong Kong eat more than twice
o Americans. The Japanese, in
pact on the seas of not only this
world. With less than 2 percent
n consumes 15 percent of the
are some 4000 coastal fishing
ens of thousands of small crafts
ly. Large Japanese fishing ves-
anneries and freezers, harvest
can do so because waters at a
d, typically 200 nautical miles
are considered international

waters and outside the jurisdiction of any individual country. In part because of Japan's global fishing practices, environmentalists have criticized the country for overfishing. For example, Japanese fishing off western Africa has reduced the catches of local fishers so much that many have been forced to migrate to Europe for work. However, there is little room for Japan to expand its fish imports. The global fish catch has been static since 1990 because much of the world's fishing grounds are either fully exploited or in decline.

People in China are also dependent on fish protein in their diets. China is, in fact, the world's largest fish producer, and much of the output is consumed domestically. However, most of China's fishing industry is based on aquaculture rather than on wild-caught fish.

People and Animals The use of animals in East Asia is not solely for the purpose of food production. For example, the role of East Asia's aquaculture is also to contribute to the aesthetic beauty of carefully manicured landscapes. Specifically, carp have been domesticated and bred in Japan to have a variety of colors and patterns (Figure 9.8C). These carp are commonly found in the ponds of Japanese gardens. As an adaptive species that does well under a wide variety of aquatic conditions, ornamental carp are now found around the world.

A practice that is less geographically widespread is traditional Chinese medicine, which, in part, involves the capturing of wild animals. Such animals are found in markets around China because they are believed to have healing properties (see Figure 9.8A). Unfortunately, some species are endangered due to this practice.

LOCAL LIVES FOODWAYS IN EAST ASIA

fermented
s—
a food
Specific
times
ppens
nal
ferments.

B A wealthy restaurant owner in Japan poses with a 448 lb (220 kg) tuna that he bought for U.S.$1.8 million at the Tsukiji fish market in Tokyo, Japan. All of this fish will be served raw as sushi, which is a combination of cooked, vinegared rice and other ingredients, usually raw seafood. With roughly 2 percent of the world's population, Japan consumes 15 percent of the global fish catch.

C A woman in rural Mongolia prepares *byaslag*, a cheese made in Mongolia of yak, cow, sheep, or goat milk. Unlike most cheeses originating in Europe, byaslag is not aged but rather is desiccated in the dry air of Mongolia. It is one of many unique products made with animal milk in Mongolia.

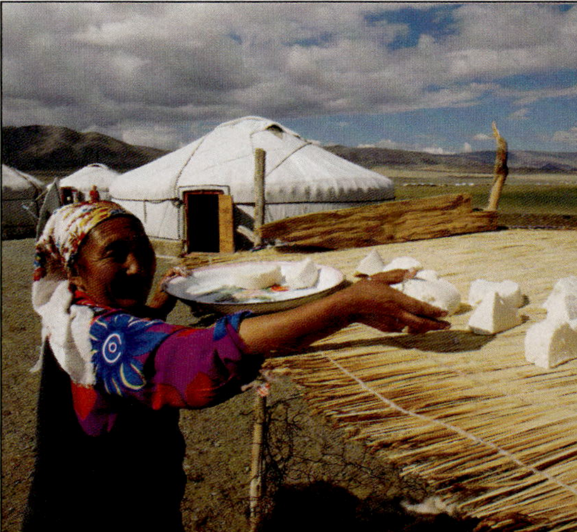

In Mongolia, many species of animals are fundamental to nomadic husbandry, a way of life that is common in areas of dry grasslands around the world (see Figure 9.8B). As Mongolia modernizes, such nomadic lives, there and elsewhere in East Asia, are becoming less common than they were in the past.

THINGS TO REMEMBER

Geographic Insight 1

- **Climate Change and Water** East Asia has long suffered from lengthy searing droughts and devastating floods. These disruptions are likely to increase with global climate change.
- China's main river systems are being affected by the melting of its glaciers, which ultimately leads to water shortages.
- The region's most important grain is rice, and over the millennia, rice cultivation has transformed landscapes throughout central and southern China, and Japan, the Koreas, and Taiwan.

Geographic Insight 2

- **Food and Globalization** About three-quarters of the food consumed in Japan, South Korea, and Taiwan is imported. So far, China is self-sufficient with regard to basic necessities such as rice, but it is highly dependent on imports for other foods.
- People in East Asia consume much fish and other seafood. Some of this food is harvested in waters far away from East Asia.

Three Gorges Dam: The Power of Water

The Three Gorges Dam (see Figure 9.9A on page 386) is at this time the largest dam in the world at 600 feet (183 meters) high and 1.4 miles (2.3 kilometers) wide. It was designed to improve navigation on the Chang Jiang and control flooding, but it is most lauded for its role in generating hydroelectricity for this energy-hungry country that is working to reduce its greenhouse gas emissions. But the dam also comes with a social cost. The Chinese environmental activist Dai Qing notes that China has 22,000 large dams, all of which have displaced people—perhaps as many as 60 million—without any attention to their rights as stakeholders in the projects.

Many experts involved with the Three Gorges project see serious design flaws. Of greatest concern is the dam's position above a seismic fault. The dam was built at the east end of the Three Gorges because the deep canyons provided a prodigious reservoir for water to power turbines, providing electricity for all of central China, from the sea to the Plateau of Tibet. Unfortunately, the enormous weight of the water and its percolation through geological fissures in the 370-mile-long (600-kilometer-long) reservoir behind the dam could trigger earthquakes. Already at issue are huge landslides along the gorges, lubricated by the rising reservoir water. Meanwhile, cracks in the dam raise doubts about its structural integrity. Similar defects led to the failure of China's much smaller Banqiao Dam during a 1975 typhoon, which was responsible

FIGURE 9.8 LOCAL LIVES PEOPLE AND ANIMALS IN EAST ASIA

A A stall in a night market in Beijing sells dried seahorses, lizards, and starfish, all of which are used in traditional Chinese medicine (TCM). Millions of animals are captured each year for use in TCM, and some are being pushed dangerously close to extinction.

B Sheep shearing in Mongolia, where livestock raising occupies 75 percent of the land. It was the backbone of the economy for thousands of years, until the 1970s, when mining and services took the lead. Sheep, goats, yaks (a long-haired Central Asian bovine), horses, and cattle are the most commonly raised animals. Many of these animals are still raised by nomadic and seminomadic herders. However, most of the people who were once herders are now moving to the city because of economic downturns and an increasingly variable climate.

C *Nishikigoi* (also known as koi), a variety of carp, in a garden in Himeji, Japan. Carp were domesticated in both ancient China and Rome as a food source. Nishikigoi, which means "brocaded carp," are an ornamental variety developed in Japan in the 1820s. Currently, the trade in ornamental nishikigoi is larger than the trade of carp raised for food.

n the flooding and an ensuing
;, its potential to generate power.
ne buildup of eroded silt behind

ld be a financial as well as human
osts are $25 billion, but the real
times this figure, due in part to
ental and social impacts and to
ot officials.
ncalculable costs associated with
ople who once lived where the
Thirteen major cities have been
arge towns, hundreds of small
2,000 acres (25,000 hectares) of
nmunities were rebuilt immedi-
ed location and people only had
others had to migrate to distant
voir has also destroyed important
some of China's most spectacu-
ignificant environmental costs as
xample, a fish that can weigh as
n and is as rare as China's giant
turgeon used to swim more than
p the Chang Jiang past the loca-
the sturgeon's reproductive pro-
errupted. **209. THREE GORGES**
ED

e Gorges Dam came just as UN
deciding that the benefits dams
y outweigh the many problems
national funding sources such as
r support for the Three Gorges
r the social and environmental
f the project. However, Chinese
rgy, construction companies that
the dam and its many ancillary
cials eager to impress the world
a continue not only to support
t also to look for other locations
can gain influence and profit by

on Success

nroughout East Asia, but the air
ticularly poor. After South Asia,
ks among the worst in the world.
lation grow—especially in urban
households, industry, and trans-
rning is the primary source of
largest consumer and producer
ent of all the coal burned on the
, D). Between 1980 and 2011,
w sixfold. By 2006, China was
r plant online every week. The
somewhat since then in favor of

other energy sources. Nevertheless, at any given point in time, there are still hundreds of new coal-fired plants in the planning stages in China.

Air Pollution in China The combustion of coal releases high levels of pollutants—suspended particulates, smog-causing ozone, and sulfur dioxide—all of which can cause respiratory ailments. In Chinese cities, these emissions can be many times higher than what the World Health Organization (WHO) defines as safe. Figure 9.10 on page 387 shows that almost no Chinese cities have air quality that is classified as "good" with regard to one type of pollutant—particulates. The worst pollution is often in cities, where homes are in close proximity to industries that depend heavily on coal for fuel. The use of coal to heat homes can also be detrimental, as the following vignette shows.

VIGNETTE Every winter, the elderly couple got through the biting Beijing winters by feeding 1200 one-kilogram coal bricks into a small iron stove. The ashes and coal dust blackened their belongings, and they worried about carbon monoxide poisoning. Then, in 2009 they were given a new electric space heater by the city government. In a move to clean up Beijing's air and reduce the city's contribution to China's greenhouse gas emissions, the city replaced nearly 100,000 old coal stoves with electric heaters and cut electric nighttime rates to just 3 cents per kilowatt-hour. The couple was astonished with the change the new cheap heater made in their lives. But of course, 100,000 more homes were now dependent on electrical power, most of it generated by coal-fired power plants. *[Source: New York Times/ International Herald Tribune Global Edition. For detailed source information, see Text Credit pages.]*

Sulfur dioxide from coal burning also contributes to acid rain, which is displaced to the northeast by prevailing winds, reaching the Koreas, Japan, Taiwan, and beyond. Japan and Taiwan are particularly afflicted (see Figure 9.9B, D, and map). Particulates from China's coal burning are transported globally by high-altitude, west-to-east flowing jet streams, thus affecting air quality in North America and Europe. Locally, toxic mercury from coal plants accumulates in aquatic life and, ultimately, in humans.

Air pollution from vehicles is also severe, even though the use of cars for personal transportation in China is just beginning. For years, China's vehicles have had very high rates of lead and carbon dioxide emissions. Quite recently, the government has begun to work on the problem. As of 2008, new Chinese cars had to meet EU standards for mileage and emissions—which are much stricter than those for U.S. cars. Even so, to provide decent air quality during the 2008 Summer Olympics, China removed half of Beijing's cars from the road. A number of polluting industries were temporarily shut down as well. The actions had the intended effect: Beijing's air was noticeably better during the Olympic summer. **203. WORLDWATCH INSTITUTE: 16 OF 20 OF WORLD'S MOST POLLUTED CITIES IN CHINA**

Figure 9.9 Photo Essay: Human Impacts on the Biosphere in East Asia

Economic development in East Asia has often proceeded without adequate environmental protection or safeguards for human health. Here we focus on the Three Gorges Dam, air pollution, and energy issues.

A China's Three Gorges Dam is the largest dam in the world. It is capable of supplying about 3 percent of China's electricity needs, resulting in significantly less air pollution than the several large coal-fired power plants that it replaces. However, the dam has also caused significant environmental damage related to its 370-mile-long (600-km-long) reservoir, which has altered the habitats of endangered species; submerged the dwellings of 1.2 million people; and become a cesspool of untreated sewage, industrial waste, and garbage that once flowed downstream.

B Pollution from vehicles chokes the motorcade of a political campaign in Taipei, Taiwan.

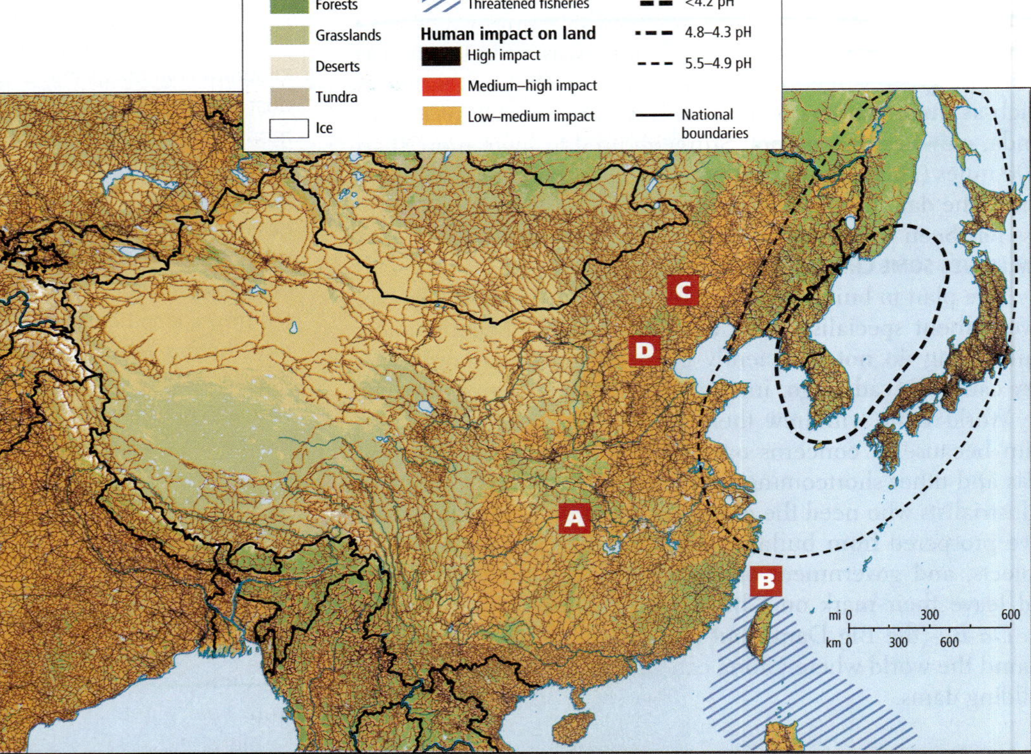

C An open-pit coal mine in China's province of Inner Mongolia. China is already the world's largest producer and consumer of coal, and its demand for coal is likely to increase in the future. China is already the largest contributor of greenhouse gases in the world.

D A toxic haze of air pollution engulfs Beijing and the North China Plain, often blowing eastward to reach Korea, Japan, and even the United States and Europe. China's coal-fired power plants, industries, and vehicles are emerging as major global environmental concerns.

Air Pollution Elsewhere in East Asia Public health risks related to air pollution are also serious in the largest cities of Japan (Tokyo and Osaka), Taiwan (Taipei), Mongolia (Ulan Bator), and South Korea (Seoul), and in adjacent industrial zones. Even with antipollution legislation and increased enforcement, high population densities and rising expectations for better living standards make it difficult to improve environmental quality. Taiwan is a case in point.

Taiwan has some of the dirtiest air in the world. Some of the main causes are the island's extreme population density of 250 people per square mile (646 per square kilometer), its high rate of industrialization,
ialized south China. There are
otorcycles) in Taiwan for every
lion exhaust-producing vehicles
there are nearly eight
uare mile (three per
iste gases. The govern-
iat the air is six times
ites or Europe.
depend on hydro- or
in factories and heat
ast Asia. North Korea
ises very few cars, so,
lable for North Korea,
be low for the region.
imerous industries and
hese are the sources of
r pollution. Mongolia
air; but in Ulan Bator,
ired some imaginative
ons of Ulan Bator are
lers, which supply hot
dividual dwellings.

EMEMBER

ods, and the
has built many dams,
s Dam, which itself

in East Asia, combined
as resulted in some of

Human Patterns over Time

East Asia is home to some of the most ancient civilizations on earth. Settled agricultural societies have flourished in China for more than 7000 years, which makes it the oldest continuous civilization in the world.

Chinese civilization evolved from several hearths, including the North China Plain, the Sichuan Basin, and the lands of interior Asia that were inhabited by nomadic pastoralists. On East Asia's eastern fringe, the Korean Peninsula and the islands of Japan and Taiwan were profoundly influenced by the culture of China, but they were isolated enough that each developed a distinctive culture and maintained political independence most of the time. In the early twentieth century, Japan industrialized rapidly by integrating European influences that China disdained. As Figure 9.11 shows, both ancient and contemporary traditions are celebrated in East Asia today.

Bureaucracy and Imperial China

Although humans have lived in East Asia for hundreds of thousands of years, the region's earliest complex civilizations appeared in various parts of what is now China about 4000 years ago. Written records exist only from the civilization that was located in north-central China. There, a small, militarized, feudal (see pages 166–168 in Chapter 4 for a discussion of feudalism) aristocracy controlled vast estates on which the majority

FIGURE 9.10 China's most polluted and least polluted cities in terms of air quality. Based on a 2011 report from the WHO, the map shows the 20 cities in China that have the most air pollution and the 10 cities with the least air pollution, as measured by the number of PM_{10} particles per cubic meter. PM_{10} particles are microscopic; by definition, they are less than 10 micrometers in diameter. These tiny particles can get into people's lungs and potentially cause serious health problems. The cities of northern and central China suffer from higher levels of air pollution than those in the south.

of the population lived and worked as semi-enslaved farmers and laborers. The landowners usually owed allegiance to one of the petty kingdoms that dotted northern China. These kingdoms were relatively self-sufficient and well defended with private armies.

An important move away from feudalism came with the Qin empire (beginning in 221 B.C.E.), which instituted a trained and salaried bureaucracy in combination with a strong military to extend the monarch's authority into the countryside. One part of the legacy of the Qin empire is shown in Figure 9.12B on page 390.

The Qin system proved more efficient than the old feudal allegiance system it replaced. The estates of the aristocracy were divided into small units and sold to the previously semi-enslaved farmers. The empire's agricultural output increased because the people worked harder to farm the land they now owned. In addition, the salaried bureaucrats were more responsible than the aristocrats they replaced, especially about building and maintaining levees, reservoirs, and other tax-supported public works that reduced the threat of flood, drought, and other natural disasters. Although the Qin empire was short-lived, subsequent empires maintained Qin bureaucratic ruling methods, which have proved essential in governing a united China.

Confucianism a Chinese philosophy that teaches that the best organizational model for the state and society is a hierarchy based on the patriarchal family

Confucianism Molds East Asia's Cultural Attitudes

Closely related to China's bureaucratic ruling tradition is the philosophy of **Confucianism**. Confucius, who lived from 551 to 479 B.C.E., was an idealist who was interested in reforming government and eliminating violence from society. He thought human relationships should involve a set of defined roles and mutual obligations. Confucian values include courtesy, knowledge, integrity, and respect for and loyalty to parents and government officials. These values diffused across the region and are still widely shared throughout East Asia (see Figure 9.12A).

The Confucian Bias Toward Males The model for Confucian philosophy was the patriarchal extended family. The oldest male

FIGURE 9.11 LOCAL LIVES FESTIVALS IN EAST ASIA

A A man dressed as a samurai, Japan's preindustrial-era warrior class, for a festival in Fukushima, Japan. Thousands of parades, processions, and street festivals are held each year in Japan, often celebrating events in local history or legend.

B Participants in the Boryeong Mud Festival, an annual South Korean event created in 1996 to celebrate the medicinal properties of the mud found near the town of Boryeoung. The 2-week-long festival draws about 2.2 million people each year and features a mud pool, mud slides, a mud prison, and cosmetics and treatments that are based on the medicinal qualities of the mud.

C A man participates in *naadam*, or "games," which form the basis of festivals held throughout Mongolia in early July to celebrate the country's nomadic history. There are three games practiced: archery, horseracing, and wrestling. Men and women compete in archery, though separately.

ʼas responsible for the well-being
ıer family members were aligned
to age and gender. Beyond the
ıal order held that the emperor
hina, charged with ensuring the
eaucrats were to do his bidding
e bureaucrats.

ıcian philosophy penetrated all
oncerning the ideal woman, for
ıs wrote: "A woman's duties are
: wine, look after her parents-in-
ll! When she is young, she must
marriage, she must submit to her
d, she must submit to her son."
oles for women affected society
ed that sons were the more valu-
, while daughters were primarily

ınts For thousands of years,
naintain the power and position
ıtic administrators at the expense
klore, merchants were character-
ınd disruptive to the social order.
f merchants were sorely needed.
e status of merchants waxed and
t merchants unable to invest in
ks. At other times, however, the
:ianism was less influential, and
:ished. Under communism, mer-
ve image; then when a market
social status of merchants once

ne, and Recovery Although
mes facilitated the expansion of
its resistance to change also led
ıxes were periodically levied on
revolts that weakened imperial
articularly invasions by nomadic
Mongolia and western China,
ve defenses, such as the Great
along China's northern border.
reaucracy always recovered from
;, the nomads were indistinguish-
ame time, Chinese culture and
kture of different influences and

esult in important links between
d. In the 1200s, the Mongolian
nd his descendants were able to
pushed west across Asia as far as
207 in Chapter 5). It was during
(also known as the Yuan empire)
Marco Polo made the first direct
rope. These connections proved
much more significant for Europe, which was dazzled by China's wealth and technologies, than for China, which saw Europe as backward and barbaric.

Indeed, from 1100 to 1600, China remained the world's most developed region, despite enduring several cycles of imperial expansion, decline, and recovery. It had the largest economy, the highest living standards, and the most magnificent cities. Improved strains of rice allowed dense farming populations to expand throughout southern China and supported large urban industrial populations. Nor was innovation lacking: Chinese inventions included paper making, printing, paper currency, gunpowder, and improved shipbuilding techniques.

Why Did China Not Colonize an Overseas Empire?

During the well-organized Ming dynasty, 1368–1644, Zheng He, a Chinese Muslim admiral in the emperor's navy, directed an expedition that could have led to China conquering a vast overseas empire. From 1405 to 1433, Zheng He sailed 250 ships—the biggest and most technologically advanced fleet that the world had ever seen. Zheng He took his fleet throughout Southeast Asia, across the Indian Ocean, and all the way to the east coast of Africa.

The lavish voyages of Zheng He were funded for almost 30 years, but they never resulted in an overseas empire like those established by European countries a century or two later. These newly explored regions simply lacked trade goods that China wanted. Moreover, back home the empire was continually threatened by the armies of nomads from Mongolia, so any surplus resources were needed for upgrading the Great Wall. Eventually the emperor decided that Zheng He's explorations were not worth the effort. In the years following Zheng He's voyages, China reduced its contacts with the rest of the world as the emperor focused on repelling invaders from Mongolia. As a result, the pace of technological change slowed, leaving China ill prepared to respond to growing challenges from Europe after 1600. China did, however, become a regional colonizing power by extending control to include territories in Central Asia and Southeast Asia (see Figure 9.13).

European and Japanese Imperialism

By the mid-1500s, during Europe's Age of Exploration, Spanish and Portuguese traders interested in acquiring China's silks, spices, and ceramics found their way to East Asian ports. To exchange, they brought a number of new food crops from the Americas, such as corn, peppers, peanuts, and potatoes. These new sources of nourishment contributed to a spurt of economic expansion and population growth during the Qing, or Manchurian, dynasty (1644–1912), and by the mid-1800s, China's population was more than 400 million, which can be compared to about 270 million in Europe at the same time.

By the nineteenth century, European merchants gained access to Chinese markets and European influence increased markedly. In exchange for Chinese silks and ceramics, British merchants supplied opium from India, which was one of the few

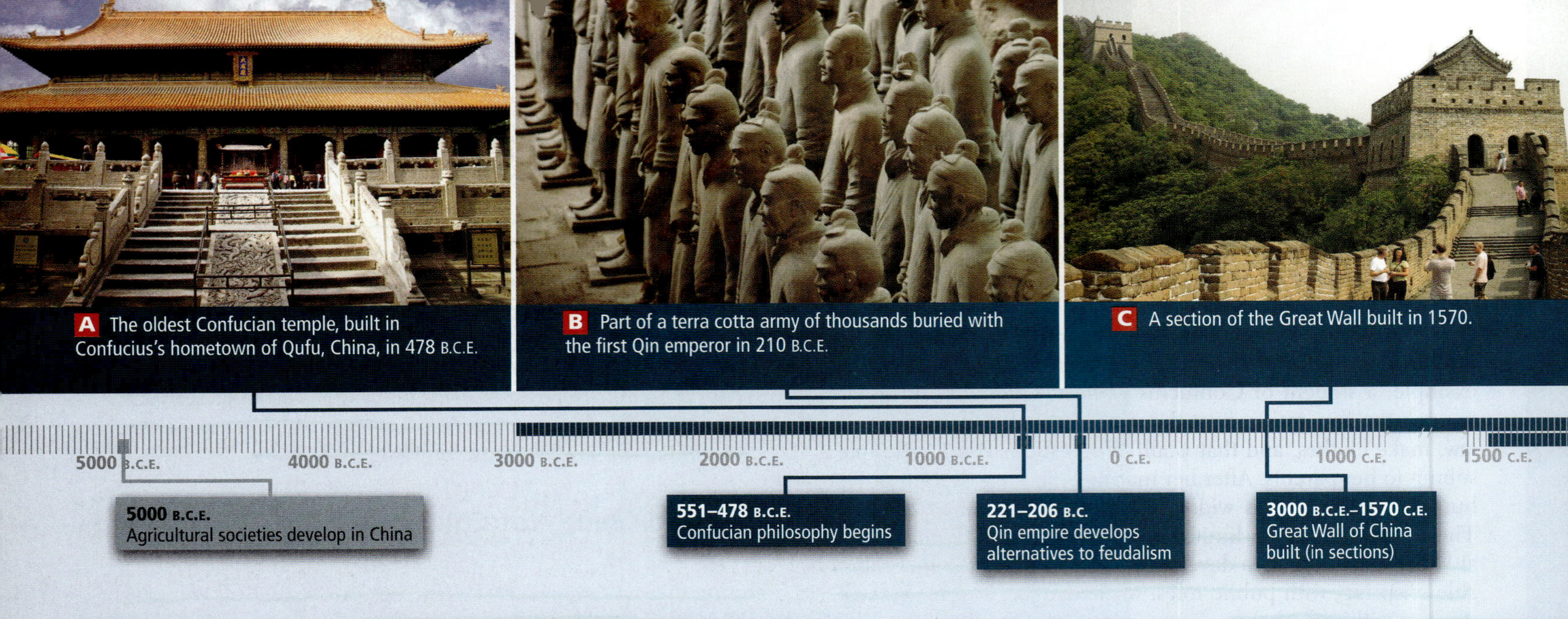

FIGURE 9.12 A VISUAL HISTORY OF EAST ASIA

Thinking Geographically

After you have read about the human history of East Asia, you will be able to answer the following questions:

A The model for Confucian philosophy is what?

B Although the Qin empire was short-lived, what did subsequent empires do to survive?

FIGURE 9.13 The extent of Chinese empires, 221 B.C.E.–1850 C.E. The Chinese state has expanded and contracted throughout its history.

things that Chinese merchants would trade for. The emperor attempted to crack down on this drug trade because of its debilitating effects on Chinese society. The result was the Opium Wars (1839–1860), in which Britain badly defeated China. Hong Kong became a British possession, and British trade, including opium, expanded throughout China, well into the twentieth century (see Figure 9.12D).

The final blow to China's long preeminence in East Asia came in 1895, when a rapidly modernizing Japan won a spectacular naval victory over China in the Sino-Japanese War (and to solidify its emerging regional dominance, Japan also defeated Russia). After this first defeat by the Japanese, the Qing dynasty made only halfhearted attempts at modernization, and in 1912 it was overthrown by an internal revolt and collapsed. Beginning with the decline

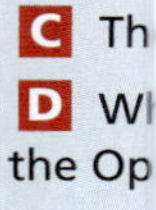

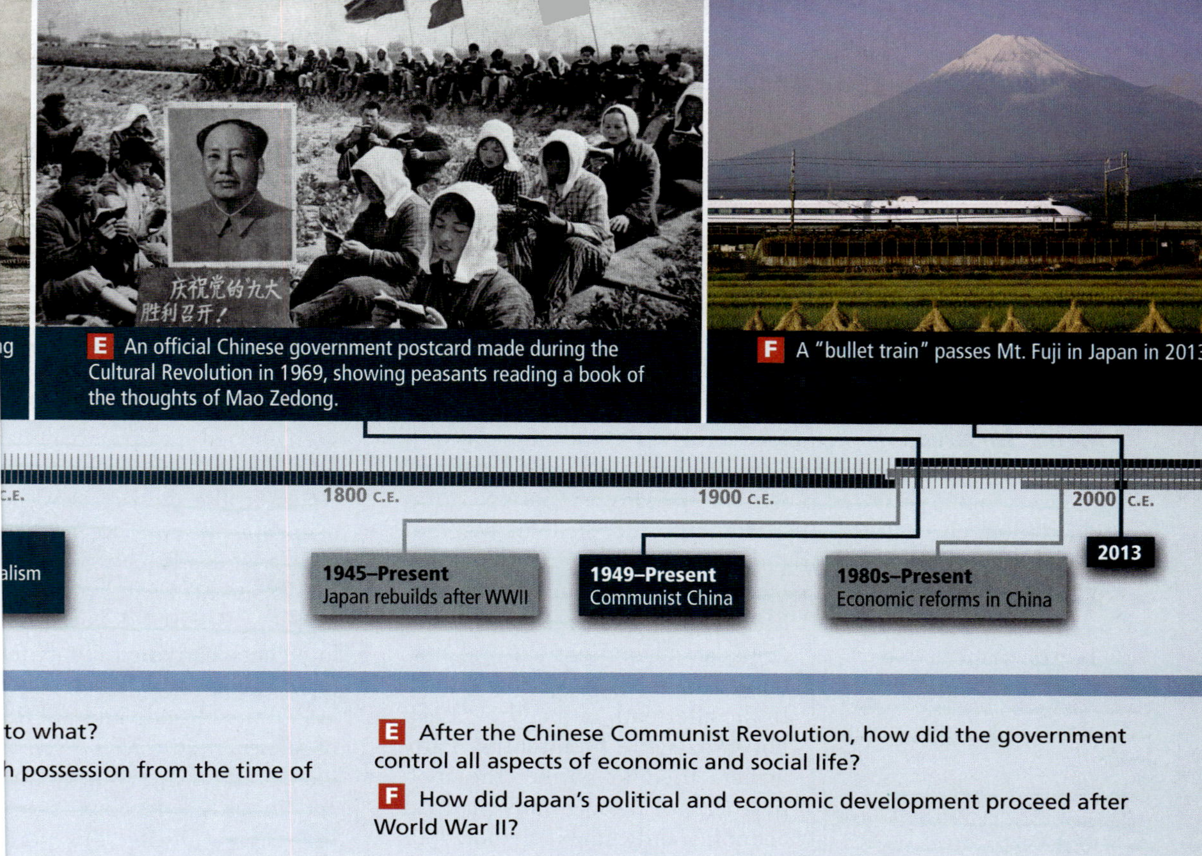

D Br the Fir

g

E An official Chinese government postcard made during the Cultural Revolution in 1969, showing peasants reading a book of the thoughts of Mao Zedong.

F A "bullet train" passes Mt. Fuji in Japan in 2013.

C Th

D Wl the Op

to what?

h possession from the time of

E After the Chinese Communist Revolution, how did the government control all aspects of economic and social life?

F How did Japan's political and economic development proceed after World War II?

China's Communist Party took
ntry was governed by provincial
ture of Chinese, Japanese, and
in the major cities.

eth Century

ntral state authority, two rival
n the early twentieth century.
the Kuomintang (KMT), was
ppealed to workers as well as
he Chinese Communist Party
its base in rural areas among
l the upper hand, uniting the
's invasion of China in 1931

of most major Chinese cit-
Japanese effectively and were
cities not in Japanese control.
stant guerilla war against the
. This resistance gained the
occupation caused 10 million
read killing of civilians in the
ape of Nanjing." Such events,
omfort women" (forced pros-
the occupied territories, still
een Japan and its East Asian
ithdrew in 1945, defeated at
ited States, Russia, and other
ar CCP pushed the KMT out

of the country and into exile in Taiwan. In 1949, the CCP, led by Mao Zedong, proclaimed the country the "People's Republic of China," with Mao as president.

Mao's Communist Revolution Mao Zedong's revolutionary government became the most powerful China ever had. It dominated all the outlying areas of China—the Far Northeast, Inner Mongolia, and western China (Xinjiang)—and launched a brutal occupation of Tibet (Xizang). The People's Republic of China was in many ways similar to past Chinese empires. The Chinese Communist Party replaced the Confucian bureaucracy and Mao Zedong became a sort of emperor with unquestioned authority. China received support from the Soviet Union, but the two communist states remained wary of each other for decades.

Among the early beneficiaries of the revolution were the masses of Chinese farmers and landless laborers. On the eve of the revolution, huge numbers lived in abject poverty. Famines were frequent, infant mortality was high, and life expectancy was low. The vast majority of women and girls held low social status and spent their lives in unrelenting servitude.

The revolution drastically changed this picture. All aspects of economic and social life became subject to central planning by the Communist Party. Land and wealth were reallocated, often resulting in an improved standard of living for those who needed it most. Heroic efforts were made to improve agricultural production and to reduce the severity of floods and droughts. The masses, regardless of age, class, or gender, were mobilized to construct almost entirely by hand huge public works projects—roads, dams, canals, whole mountains terraced into fields for rice and other crops. "Barefoot doctors" with rudimentary medical training dispensed basic medical care, midwife services, and nutritional advice to people in the remotest locations. Schools were built in the smallest of villages. Opportunities for women became available, and some of the worst abuses against them were stopped—such as the crippling binding of women's feet to make them small and childlike. Most Chinese people who are old enough to have witnessed these changes say that the revolution did a great deal to improve overall living standards for the majority.

Mao's Missteps Nonetheless, progress came at enormous human and environmental costs. During the **Great Leap Forward** (a government-sponsored program of massive economic reform initiated in the 1950s), 30 million people died, many from famine brought on by poorly planned development objectives; others because they were persecuted for opposing the reforms. Meanwhile, deforestation, soil degradation, and agricultural mismanagement became widespread. In the aftermath of the Great Leap Forward, some Communist Party leaders tried to correct the inefficiencies of the centrally planned economy only to be demoted or jailed as Mao Zedong remained in power.

Great Leap Forward an economic reform program under Mao Zedong intended to quickly raise China to the industrial level of Britain and the United States

Cultural Revolution a political movement launched in 1966 to force the entire population of China to support the continuing revolution

In 1966, partially in response to the failures of the Great Leap Forward, a political movement known as the **Cultural Revolution** was launched to enforce support for Mao and punished dissenters. Everyone was required to study the "Little Red Book" of Mao's sayings (see Figure 9.12E). Educated people and intellectuals were a main target of the Cultural Revolution because they were thought to instigate dangerously critical evaluations of Mao and Communist Party central planning. Tens of millions of Chinese scientists, scholars, and students were sent out of the cities to labor in mines and industries or to jail, where as many as 1 million died. Children were encouraged to turn in their parents. Petty traders were punished for being capitalists, as were those who adhered to any type of organized religion. The Cultural Revolution so disrupted Chinese society that by Mao's death in 1976, the Communists had been seriously discredited.

Changes After Mao Two years after Mao's death, a new leadership formed around Deng Xiaoping. Limited market reforms were instituted, but the Communist Party retained tight political control. In 2009, after more than 30 years of reform and remarkable levels of economic growth, China's economy became the third largest in the world, behind that of the European Union and the United States. It passed Japan (although Japan's per capital wealth is still five times that of China's) when China managed to weather the world recession beginning in 2008 better than expected. However, the disparity of wealth in China has been increasing for some years, human rights are still often abused, and political activity remains tightly controlled even as discontent boils over into open protests against government foibles.

Japan Becomes a World Leader

Although China's influence was dominant in East Asia for thousands of years, for much of the twentieth century, Japan, with only one-tenth the population and 5 percent of the land area of China, controlled East Asia economically and politically. Japan's rise as a modern global power resulted largely from its response to challenges from Europe and North America.

Beginning in the mid-sixteenth century, active trade with Portuguese colonists brought new ideas and technology that strengthened Japan's wealthier feudal lords (*shoguns*), allowing them to unify the country under a military bureaucracy. However, the shoguns monopolized contact with the European outsiders, allowing no Japanese people to leave the islands, on penalty of death.

A second period of radical change came with the arrival in Tokyo Bay in 1853 of a small fleet of U.S. naval vessels. The foreigners, carrying military technology far more advanced than Japan's, forced the Japanese government to open the economy to international trade and political relations. In response, a group of reformers (the Meiji) seized control of the Japanese government, setting the country on a crash course of modernization and industrial development that became known as the Meiji restoration. During this period, Japanese students were sent abroad and experts were recruited from around the world, especially from Western nations, to teach everything from foreign languages to modern military technology. Investments emphasized infrastructure development, especially in transportation and communication. The improvements that resulted enabled Japan's economy to grow rapidly, surpassing China's in size in the early twentieth century.

Between 1895 and 1945, Japan fueled its economy with resources from a vast colonial empire. Equipped with imported European and North American military technology, its armies occupied first Korea, then Taiwan, then coastal and eastern China, and eventually Indonesia and much of Southeast Asia, as shown in Figure 9.14. Many of the people of these areas still harbor resentment about the brutality they suffered at Japanese hands. Japan's imperial ambitions ended with its defeat in World War II and its subsequent occupation by U.S. military forces until 1952. **204. JAPANESE STILL RESOLVING FEELINGS ABOUT THE WAR**

In the immediate post-war era, the U.S. government imposed many social and economic reforms. Japan was required to create a democratic constitution and reduce the emperor to symbolic status. Its military was reduced dramatically, forcing it to rely on U.S. forces to protect it from attack. With U.S. support, Japan rebuilt rapidly after World War II, and it eventually became a giant in industry and global business, exporting automobiles, electronic goods, and many other products. Japan's economy is still among the world's largest, wealthiest, and most technologically advanced (see Figure 9.12F and Figure 9.17D on page 398).

Chinese and Japanese Influences on Korea, Taiwan, and Mongolia

The history of the remaining East Asia region is largely grounded in what transpired in China and Japan.

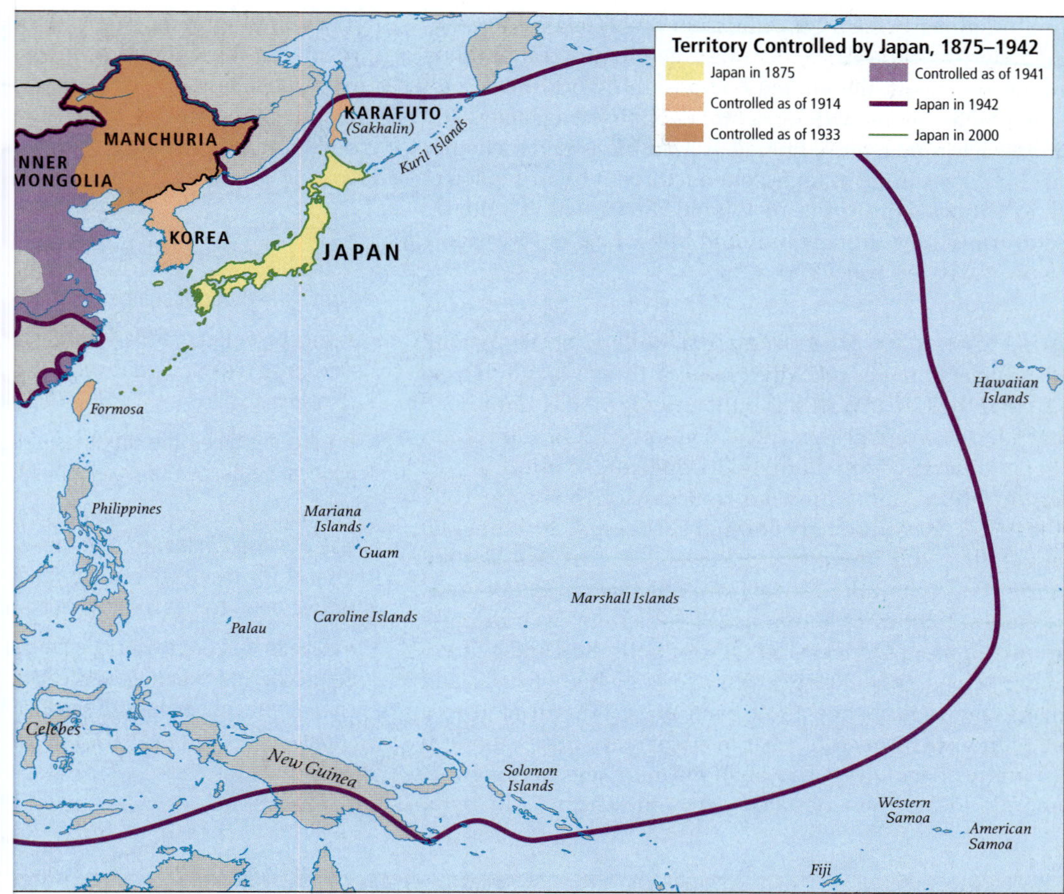

ansions, 1875–1942. Japan colonized Korea, Taiwan (then known as Formosa),)utheast Asia, and several Pacific islands to further its program of economic ff European imperialism in the early twentieth century.

ermath Korea was a unified but 45. At the end of World War II, iet Union, as victorious allies, nese colony. The Soviet Union nunist regime to the north, while the southern half of the Korean forms similar to those in Japan. w its troops in the late 1940s, rea. The United States returned ear war against North Korea and Communist China.

oth sides and devastation of the Korean War ended in 1953 in of a demilitarized zone (DMZ) s as the de facto border between s 2.5 mile (4 kilometers) wide because it is a buffer zone not the surrounding area is heavily f off from the rest of the world, d, impoverished, and defensive, nal attention by hinting at its nuclear potential (see Figure 9.21D on page 403). Meanwhile, South Korea has developed into a prosperous and technologically advanced market economy. Relations between the two countries remain tense, with occasional skirmishes breaking out along the border.

201. INTER-KOREAN COOPERATION GROWS

210. HIGH-TECH KOREA EYES 'UBIQUITOUS' FUTURE

Taiwan's Uncertain Status For thousands of years, Taiwan was a poor, agricultural island on the periphery of China; then between 1895 and 1945 it became part of Japan's regional empire. In 1949, when the Chinese nationalists (the Kuomintang) were pushed out of mainland China by the Chinese Communist Party, they set up an anti-communist government in Taiwan, naming it the Republic of China (ROC). For the next 50 years, with U.S. aid and encouragement, the ROC became a modern industrialized economy, which quickly dwarfed that of China and remained dominant until the 1990s, serving as a prosperous icon of capitalism right next door to massive Communist China.

Today, Taiwan remains an economic powerhouse. Taiwanese investors have been especially active in Shanghai and the cities of China's southeast coast. Yet mainland China has never relinquished claim to Taiwan. As China's economic and military power have increased, the United Nations, the World Bank, and the United States, along with most other countries, agencies, and institutions, have judiciously tiptoed around the issue of whether Taiwan should continue as an independent country or be downgraded to simply a province of China. Taiwan itself remains divided over just how strongly it should hold on to its sovereignty (see Figure 9.21C on page 403).

Mongolia Seeks Its Own Way For millennia, Mongolia's nomadic horsemen periodically posed a threat to China, so much so that the Great Wall was built and repeatedly reinforced and extended to combat them (see Figure 9.12C). China has long been obsessed with both deflecting and controlling its northern neighbor; and China did control Mongolia from 1691 until the 1920s. Revolutionary communism spread to Mongolia soon thereafter, and Mongolia continued as an independent communist country under Soviet, not Chinese, guidance until the breakup of the Soviet Union in 1989. Communism brought education and basic services. Literacy for both men and women rose above 95 percent. Deeply suspicious of both Russia and China, Mongolia has since 1989 been on a difficult road to a market economy. In need of cash to participate in the modern world, families have elected to abandon nomadic herding and permanently locate their portable *ger* homes near Ulan Bator and search for paid employment. While the economy has developed, the rapid and drastic change in lifestyle has also resulted in broken homes, increasing personal debt, and poverty, in a society that formerly took pride in an egalitarian if not prosperous standard of living.

THINGS TO REMEMBER

- The histories of East Asian countries are deeply intertwined, and Confucian thought still permeates the entire region.
- For six centuries, from 1100 to 1600, China was the world's most developed region, despite cycles of imperial expansion, decline, and recovery.
- In recent times, the countries of East Asia have followed very different political paths toward development.
- Japan has long-term cultural ties to the mainland of East Asia and has also exercised influence over the region, primarily through conquest. For most of the twentieth century, it has been the dominant economy in the region, only recently losing that status to China in 2010.
- Taiwan and South Korea emerged after World War II as rapidly industrializing countries, while Mongolia until recently retained communist connections to the Soviet Union. North Korea remains a communist country and is extremely defensive and cut off from the wider world.

CURRENT GEOGRAPHIC ISSUES

Although the countries of East Asia adopted new economic systems only after World War II, most of them are making progress toward creating a better life for their citizens. In fact, Japan and China's Hong Kong Special Administrative Region have among the highest standards of living in the world, and South Korea is not far behind.

Economic and Political Issues

After World War II, the countries of East Asia established two basic types of economic systems. The Communist regimes of China, Mongolia, and North Korea relied on central planning by the government to set production goals and to distribute goods among their citizens. In contrast, first Japan, and then Taiwan and South Korea, established **state-aided market economies** with the assistance and support of the United States and Europe. In this type of economic system, market forces, such as supply and demand and competition for customers, determine many economic decisions. However, the government intervenes strategically, especially in the financial sector, to make sure that certain economic sectors develop in a healthy fashion. Investment in the country by foreigners is also limited so that the government can retain more control over the direction of the economy and so that economic development benefits domestic interests. In the cases of Japan, South Korea, and Taiwan, government intervention was designed to enable **export-led growth**. This economic development strategy relies heavily on the production of manufactured goods destined for sale abroad, primarily to the large economies of North America and Europe, while limiting imports for local consumers.

state-aided market economy an economic system based on market principles such as private enterprise, profit incentives, and supply and demand, but with strong government guidance; in contrast to the free market (limited government) economic system of the United States and, to a lesser degree, Europe

export-led growth an economic development strategy that relies heavily on the production of manufactured goods destined for sale abroad

More recently, the differences among East Asian countries have diminished as China and Mongolia have set aside strict central planning and adopted reforms that rely more on market forces. China also relies heavily on exports of its manufactured goods to North America and Europe. Politically, however, contrasts remain stark. While Japan, South Korea, Taiwan, and Mongolia have become democracies, the North Korean dictatorship maintains total control over politics and the media. In China, there is some experimentation with democracy at the local level, but the central government generally acts as a force against widespread democratic participation.

The Japanese Miracle

Throughout the nineteenth century, the economies of Japan, Korea, and Taiwan were minuscule compared with China's. During the twentieth century, though, all three grew tremendously, in part because of ideas that originated in Japan.

pan's recovery after its crippling
II is one of the most remarkable
ot for Kyoto, which was spared
rchitectural significance, all of
yed by the United States. Most
mbed Hiroshima and Nagasaki
ed Tokyo with incendiary bombs.

UCLEAR BOMBING OF HIROSHIMA

very was its state-aided market
guided private investors in creates. The overall strategy was one
with the Japanese government
h the United States and Europe.
that large and wealthy foreign
import Japanese manufactured
ery based on a state-aided and
y was imitated in South Korea,
Asia, and eventually in China
veloping world. Also central to
ted abroad, were arrangements
corporations, and labor unions
ment by a single company for
ely modest pay.

d trade and labor arrangements
growth of 10 percent or more
ne 1970s. The leading sectors
e and electronics manufactur-
h as Sony, Panasonic, Nikon,
words in North America and
ese companies sold at much
been possible in the then rela-
y Asian economies. Although
h during the 1990s and again
war "economic miracle" con-
dwide impact as a model, and
r in the world economy. Japan
ts of the world for its industries
ses and investments in various
ate jobs for millions of people

apan Over the years, Japan has
anufacturing that have boosted
to other industrial economies,
ts of industries. The **just-in-time**
s that are part of the same pro-
eliver parts to each other when
or example, factories that make
ound the final assembly plant,
before they will be used. This
n more efficient and reducing

***kaizen* system** of continuous
and business management. In
constantly surveyed for errors,
defective parts are produced.
tly adjusted and improved to

FIGURE 9.15 Japan's just-in-time system. In this example of the just-in-time system, related industries are clustered together around a Toyota plant in Toyota City, Japan. The facilities supply each other with various automobile parts and other inputs when they are needed in the production process.

save time and energy. Both the just-in-time and the kaizen systems have been imitated by companies around the world and have been taken overseas by Japanese companies that invest abroad. For example, Toyota uses both systems in all of its U.S. plants (see page 88 in Chapter 2).

Mainland Economies: Communists in Command

After World War II, economic development on East Asia's mainland proceeded on a dramatically different course than it did in Japan. Communist economic systems transformed poverty-ridden China, Mongolia, and North Korea. Private property was abolished, and the state took full control of the economy, loosely following the example of the Soviet *command*, or *centrally planned*, economy (see Chapter 5 on page 210). These sweeping changes transformed life for the poor majority, but ultimately proved less resilient and successful than was hoped.

By design, most people in the communist economies could not consume more than the bare necessities. On the other hand, the policy—called the "iron rice bowl" in China—of guaranteeing nearly everyone a job for life, sufficient food, basic health care, and housing was better than what they had before. One drawback was that overall productivity remained low.

just-in-time system the system pioneered in Japanese manufacturing that clusters companies that are part of the same production system close together so that they can deliver parts to each other precisely when they are needed

***kaizen* system** a system of continuous manufacturing improvement, pioneered in Japan, in which production lines are constantly adjusted, improved, and surveyed for errors to save time and money and ensure that fewer defective parts are produced

The Commune System When the Communist Party first came to power in China in 1949, its top priority was to make monumental improvements in both agricultural and industrial production. The communist governments in North Korea and Mongolia held similar goals, though these countries had much smaller populations and resource bases to work with.

In the years following World War II, an aggressive agricultural reform program joined small landholders together into cooperatives so that they could pool their labor and resources to increase production. In time, the cooperatives became full-scale communes, with an average of 1600 households each. The communes, at least in theory, took care of all aspects of life. They provided health care and education and built rural industries to supply such items as simple clothing, fertilizers, small machinery, and eventually even tractors. The rural communes also had to fulfill the ambitious expectations that the leaders in Beijing had of better flood control, expanded irrigation systems, and especially, of increased food production.

The Chinese commune system met with several difficulties. Rural food shortages developed because farmers had too little time to farm. They were required to spend much of their time building roads, levees, terraces, and drainage ditches, or working in the new rural industries. Local Communist Party administrators often compounded the problem by overstating harvests in their communes to impress their superiors in Beijing. The leaders in Beijing responded by requiring larger food shipments to the cities, which caused even greater food shortages in the countryside.

Although the Chinese agricultural communes were inefficient and created devastating food scarcities during the Great Leap Forward, they did eventually result in a stable food supply that kept Chinese people well fed.

In North Korea, the Communists have pushed military strength at the expense of broader economic development. Meanwhile, agriculture has been so neglected that production is precarious, with food aid from its immediate neighbors or the United States a yearly necessity, and famine a constant threat. In recent years, North Korea has used rocket launchings, nuclear tests, and even threat of war to intimidate its neighbors, and is selling its military technology abroad to pay for food imports. The U.S. government intermittently suspends food aid to encourage more peaceful policies from the North Korean leadership, but with limited results.

In Mongolia, communist policy followed the Soviet model of collectivization but was tailored to Mongolia's economy, which at the time was largely one of herding and agriculture. There were minimal changes to the nomadic pastoralist way of life, but the roles of mining and industry were emphasized, and the contribution to GDP by herding and forestry declined to less than 20 percent.

regional self-sufficiency an economic policy in Communist China that encouraged each region to develop independently in the hope of evening out the wide disparities in the national distribution of resources production and income

responsibility system in the 1980s, a decentralization of economic decision making in China that returned agricultural decision making to the farm household level, subject to the approval of the commune

regional specialization the encouragement of specialization rather than self-sufficiency in order to take advantage of regional variations in climate, natural resources, and location

Focus on Heavy Industry The Communist leadership in China, North Korea, and Mongolia believed that investment in heavy industry would dramatically raise living standards. Massive investments were made in the mining of coal and other minerals and in the production of iron and steel. Especially in China, heavy machinery was produced to build roads, railways, dams, and other infrastructure improvements that leaders hoped would increase overall economic productivity. However, much as in India (see pages 344–345 in Chapter 8), the vast majority of the population remained impoverished agricultural laborers who received little benefit from industries that created jobs mainly in urban areas. Not enough attention was paid to producing consumer goods (such as cheap pots, pans, hand tools, and other household items) that would have driven modest internal economic growth and improved living standards for the rural poor. Even in the urban areas, growth remained sluggish because, as also was the case in the Soviet command economy, small miscalculations by bureaucrats resulted in massive shortages and production bottlenecks that constrained economic growth.

Struggles with Regional Disparity For centuries, China's interior west has been poorer and more rural than its coastal east. The interior west has been locked into agricultural and herding economies, while even in the most restricted times, the economies of the east have benefited from trade and industry. The first effort to address regional disparities, right after the revolution, was an economic policy that focused on a combination of investment from the central government and the development of **regional self-sufficiency**. Each region was encouraged to develop as an independent entity with both agricultural and industrial sectors that would create jobs and produce food and basic necessities.

Market Reforms in China

In the 1980s, China's leaders enacted market reforms that changed the country's economy in four ways. First, economic decision making was decentralized and given the name **responsibility system**. Under the responsibility system, farmers now could make household-level decisions, subject to the approval of the commune, about how to use land and which crops to grow. Second, market reforms allowed existing and newly established businesses to sell their produce and goods in competitive markets. Third, **regional specialization**, rather than regional self-sufficiency, was encouraged in order to take advantage of regional variations in climate, natural resources, and location, and thereby encourage national economic integration. In practice, this has meant that coastal urban areas focus on export-oriented manufacturing and advanced service production, while resource-based industries and production for domestic consumption grow in the interior provinces. Finally, the government allowed *foreign direct investment* in Chinese export-oriented enterprises and the

ıa. This four-part shift to a more tically improved the efficiency re produced and distributed. On ities have widened, and on the sparities are sometimes extreme ting population discussed at the rect consequence of this spatially

ve transformed not only China's es of wider East Asia and indeed is the world's largest producer ng consumers across the globe. countries, Mongolia has partici- derately, mainly as a raw material ıot participated at all.

Urbanization, Development, and Globalization in East Asia

> **Geographic Insight 3**
> **Urbanization and Development:** China's move to join the global economy more than three decades ago has led to spectacular urban growth, a stronger emphasis on consumerism, and a massive rise in industrial and agricultural pollution of the air, water, and soil.

Across East Asia, cities have grown rapidly over the last several decades as urban industrial development has accelerated, focused on production for the global market (see the Figure 9.17 map on page 398). China has undergone the most massive urbanization in the world's history. Since 1980, when market reforms were

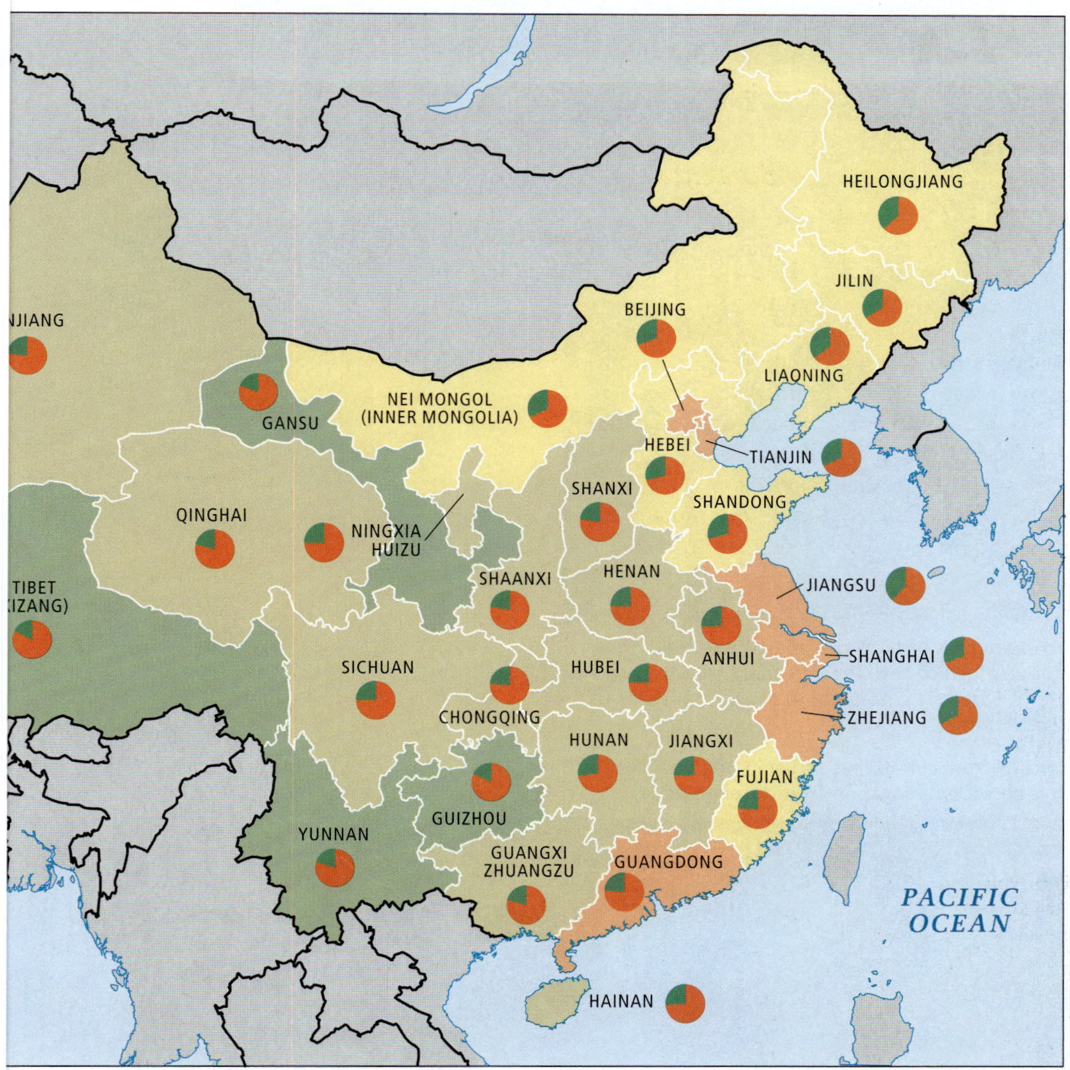

ional GDP and rural–urban income per capita disparities, 2008. Notice the cross China as indicated by the colors of the provinces as well as the rural–urban vince as represented by the pie diagrams.

Figure 9.17 Photo Essay: Urbanization in East Asia

Some of the largest and wealthiest urban areas in the world are in East Asia. China's cities are undergoing particularly rapid growth.

A A man searches for usable lumber in the remains of an old neighborhood in downtown Shanghai that is being demolished to make way for high-rise apartments for the city's growing number of wealthy residents. The inhabitants of this neighborhood were forcibly evicted by the city government and resettled in cheaper apartments on the urban fringe. In the name of modernization, few attempts at renovating historic neighborhoods, called *hutongs*, are being made in urban China.

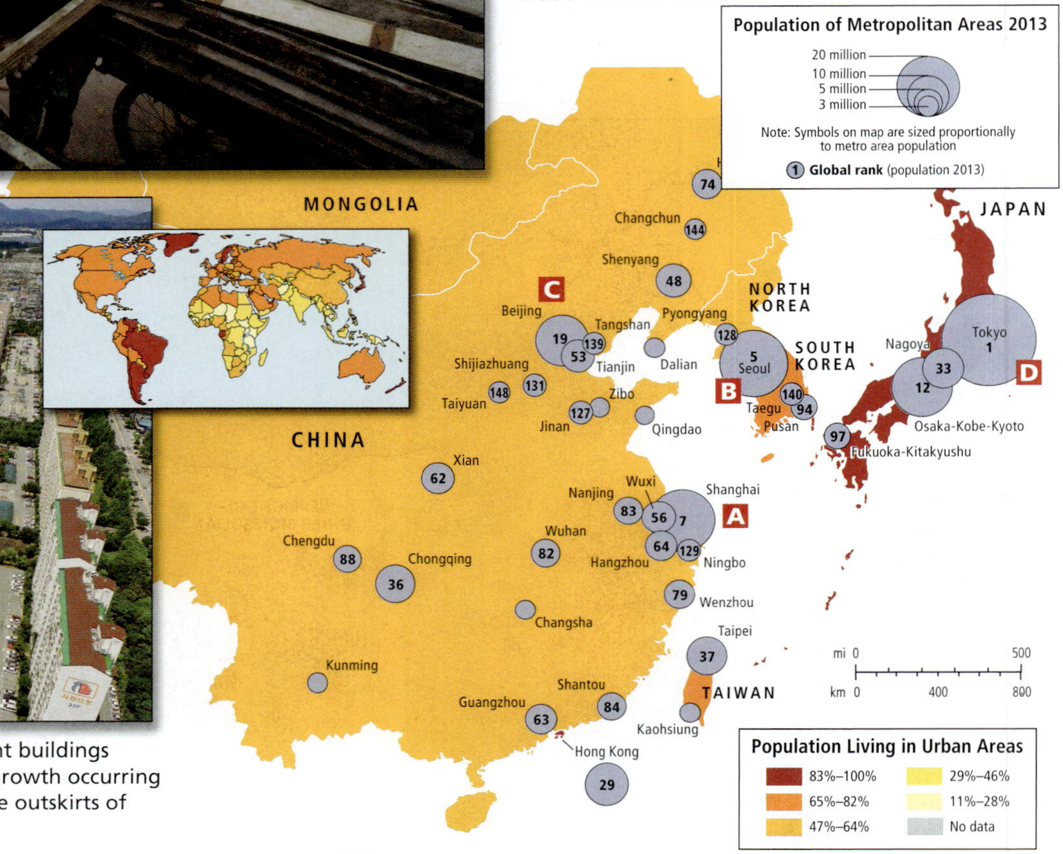

Population of Metropolitan Areas 2013

20 million
10 million
5 million
3 million

Note: Symbols on map are sized proportionally to metro area population

① Global rank (population 2013)

MONGOLIA
CHINA
NORTH KOREA
SOUTH KOREA
JAPAN
TAIWAN

Changchun 144, Shenyang 48, Beijing 19, Tangshan 139, Tianjin 53, Pyongyang, Dalian, Seoul 5, Shijiazhuang 131, Taiyuan 148, Zibo, Jinan 127, Qingdao, Taegu 140, Pusan 94, Nagoya 33, Tokyo 1, Osaka-Kobe-Kyoto 12, Fukuoka-Kitakyushu 97, Xian 62, Nanjing 83, Wuxi 56, Shanghai 7, Wuhan 82, Hangzhou 64, Ningbo 129, Chengdu 88, Chongqing 36, Wenzhou 79, Changsha, Taipei 37, Kunming, Shantou 84, Guangzhou 63, Kaohsiung, Hong Kong 29, 74, 128

mi 0 500
km 0 400 800

Population Living in Urban Areas

83%–100%
65%–82%
47%–64%
29%–46%
11%–28%
No data

B Large expanses of high-rise apartment buildings characterize much of the current urban growth occurring throughout East Asia. Shown here are the outskirts of Seoul, South Korea.

C Migrants from rural areas arrive in Beijing. Most of China's urban growth is based on migration from rural areas.

D People camp out to buy the latest iPhone in Ginza, Tokyo, the most lavish shopping district in the world's largest and wealthiest city.

initiated, the urban population has more than tripled, and is now nearly 700 million. This explosive economic development has been fueled largely by the highly globalized economies of China's big coastal cities, whose urban factories now supply consumers throughout the world. The urban economic growth has been so spectacular that in 2012, nineteen of the 20 fastest-growing cities in the world, as measured by increase in GDP per capita, were in China. Despite the rapid growth, urbanites still represent only 51 percent of China's total population, so more growth is likely, whereas in every other country in the region—even North Korea—the majority of the population has been urban for at least two decades. Most East Asian cities are extremely crowded, so future growth is a challenge for planners and residents (see Figure 9.17A, B).

rns of public and private invest-
on have led to rural areas lagging
iobs and income, education, and
: 9.16 depicts the problem. GDP
ı China's interior provinces than
within each province there is a
n places, shown in pie diagrams
Notice that rural–urban GDP
are less extreme in northeastern
interior and western provinces.
patterns are found in all other
North Korea.

FOR MODERN DEVELOPMENT
TE PROPERTY LAW WILL COST THEM

ecial Economic Zones One
egional disparities is to increase
ɔped areas by opening them up
Vary of the disruption that could
economy to international trade,
ities as sites where foreign tech-
ɔe imported to China and free
onomic zones (SEZs) (Figure 9.18)
her world regions (see Chapter 3
the program was expanded to
ı the interior. These new loca-
and technology development

ON THE BRIGHT SIDE

Transportation Improvements

Recently, ETDZs in the interior have had higher rates of growth than those on the coast. One reason is that China spends 9 percent of its GDP on transportation improvements (recently, the United States spent 2.4 percent), which has made central and western China more accessible. New highways and railroads may eventually enable the interior to catch up with the economic development of the coast. The world's longest high-speed rail now takes a traveler from Beijing to southern China in 10 rather than 22 hours, and it is routed through central China, not along the coast.

zones (ETDZs in Figure 9.18). Like SEZs, the ETDZs provide footholds for international investors and multinational companies eager to establish operations in the country.

This program was successful. Today the SEZs and ETDZs are China's greatest **growth poles**, meaning that their development, like a magnet, is drawing yet more investment and migration. The first coastal SEZs were spectacularly successful. In only 25 years, many coastal cities, including Dongguan (the city Li Xia migrated to in the vignette that opens this chapter), grew from medium-sized towns or even villages into some of the largest urban areas in the world. Foreign direct investment remains concentrated on the eastern coast, which accounts for 94 percent of China's exports, and as shown in Figure 9.16, GDP per capita (PPP) remains noticeably higher on the eastern coast than in China's interior.

While Hong Kong's democracy has been curtailed, political freedom is more evident there than elsewhere in China, and its role as China's link to the global economy has continued. Before 1997, some 60 percent of foreign investment in China was funneled through Hong Kong, and since then Hong Kong has remained the financial hub for China's booming southeastern coast (Figure 9.19 on page 401). Several of China's SEZs and ETDZs are located in close proximity to Hong Kong. Because so much of this investment comes from Japan, Korea, and Taiwan, Hong Kong is an important regional financial hub as well. **220. AS HONG KONG ENTERS SECOND DECADE UNDER CHINA, CITY PONDERS PLACE IN ECONOMIC GIANT**

Shanghai's Latest Transformation Shanghai has a long history as a trendsetter. Its opening to Western trade in the early nineteenth century spawned a period of phenomenal economic growth and cultural development that led it to be called the "Paris of the East." As a result of China's recent reentry into the global economy, Shanghai is undergoing another boom, which has enriched some people and dislocated others (see Figure 9.17A). On average, residents of Shanghai are now as affluent as people in Central European countries like Poland and Hungary, and if current growth trends continue, their wealth will soon surpass that of many other countries as well. Shanghai today has the world's busiest cargo port, and the region around Shanghai is now responsible for as much as a quarter of China's GDP.

special economic zones (SEZs) free trade zones within China, which are commonly called export processing zones (EPZs) elsewhere

growth poles zones of development whose success draws more investment and migration to a region

FIGURE 9.18 Foreign investment in East Asia. The map shows China's original special economic zones (SEZs) and its more recently designated economic and technology development zones (ETDZs). The colors on the map reflect levels of foreign investment in each of the countries of the region and in each of China's provinces.

In less than a decade, the city's urban landscape has been remade by the construction of more than a thousand business and residential skyscrapers; subway lines and stations; highway overpasses; bridges; and tunnels. Shanghai's shopping district on Nanjing Road is as imposing as any such district around the world. For hundreds of miles into the countryside, suburban development linked to Shanghai's economic boom is gobbling up farmland, and displaced farmers have rioted.

Pudong, the new financial center for the city, sits across the Huangpu River from the Bund—Shanghai's famous, elegant row of big brownstone buildings that served as the financial capital of China until half a century ago. Previously, Pudong was a maze of dirt paths and sprawling neighborhoods of low, tile-roofed houses, but its former residents were pushed out to make way for soaring high-rises. New construction is restricted in the historic section of the Bund, which is another reason why the centrality of Pudong has been attractive as a new business center. The Oriental Pearl Tower has become a visual symbol for the economic vitality of Pudong, the city of Shanghai, and all of China.

Shanghai's long-standing role as a window to the outside world has meant that it often is host to quirky behavior that is less common elsewhere in the country. For example, for years the people of Shanghai have relaxed in public in their pajamas—light, loose cotton tops and bottoms stamped with images of puppies or butterflies. The city government has tried to squelch the custom, but the citizens have proved recalcitrant. The police seem to understand that images of them arresting old and young alike for wearing pajamas would be ludicrous, so for now the issue is unresolved.

Urban Labor Surpluses and Shortages Spectacular urban growth in East Asia was based on hundreds of millions of new urban migrants who were willing to put up with adverse conditions to earn a little cash. However, so many factories, businesses, and shopping malls were built or were under construction that by 2005, experienced and skilled workers of all types were increasingly in short supply. Those with management skills were especially scarce.

To attract employees, some factory owners in China offered higher pay, better working conditions, and shorter workdays

[nvestment in China
;eign capital in billions of U.S. dollars)

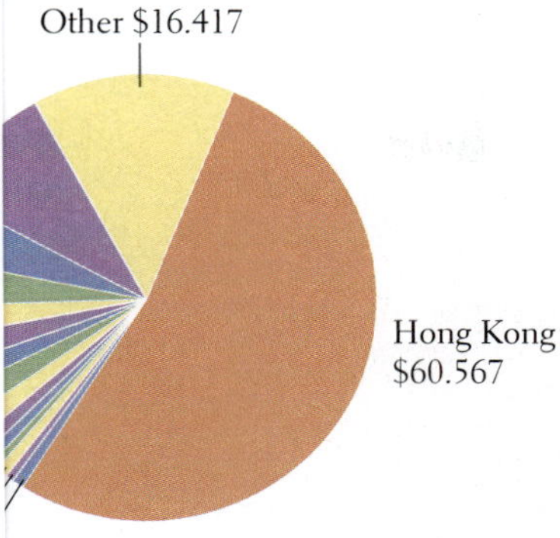

ent capital in China, a reached nearly $115 billion
›n in 2011. Note that investment
s part of China, is considered "for-
t actually originate in various coun-
l through Hong Kong banks, as was
3ritish territory. The same holds true
n Taiwan, the Virgin Islands, and

tra costs these changes imposed
;r the cheapest place to manufac-
moved to Vietnam, Bangladesh,
even cheaper labor. The more
tems, such as cell phones and
tively expensive components and
ι the lowest possible wages at the
higher wages for workers, at least
ecession, which was felt in Asia
: yet again. Demand for China's
; quickly laid off workers or shut
n workers returned to their farms
: being able to help their families
uture would hold for them. The
l robust after the recession, in no
rchasing power and consumption
ιished the dependency on shaky
eans that the outlook for most
These ups and downs in China's
to the story of Li Xia, whom we
ιis chapter.

ster returned home to rural
d time. With the money she had
ιgguan, Xia tried to open a bar in
the front room of her parents' house. She hoped to introduce the popular custom of karaoke singing she had enjoyed in the city, but people in her village could not stand the noise, and family tensions rose. Xia's sister, delighted to be home with her husband and baby, quickly found a job in a small new fruit-processing factory—a rural enterprise. Her husband began farming again to fill the new demand for organic vegetables among the middle class in the Sichuan city of Chongqing.

In early 2007, amid all this success in her family and the failure of her own bar, news that training was now available in Dongguan for skilled electronics assemblers convinced Xia to try again. Past experience with the bureaucracy and her knowledge of Dongguan helped Xia to sign up for the electronics training. The cost of tuition was to come out of her wages, which were only U.S.$350 a month rather than the U.S.$400 she had thought she would be paid. Also, the waiting list for an apartment was long. She would need several years to repay her tuition, so she went back to shared quarters with her friends.

In November of 2008, Xia heard rumors that a global recession was causing orders for electronics to be cut and that she might soon be laid off. But her factory limped along with a reduced staff. Then, miraculously, in June of 2009, orders picked up. Consumers in America had continued to buy electronics even as they downsized their homes and cars. Electronic inventories in America were down, so orders for products from Xia's factory went up significantly. And with so many laid-off workers returning home, the apartment wait list had shortened drastically. Now the apartment developers were only too happy to give her a firm move-in date. Meanwhile, the landscape of Dongguan has filled with shuttered export industries and empty shopping malls in the wake of the global downturn, so Xia's migration back and forth between the city and the countryside is likely to persist.

[Sources: USA Today, National Public Radio, Wall Street Journal, New York Times, China Daily, Plastics News. For detailed source information, see Text Credit pages.] ■

East Asia's Role as Mega-Financier As rampant consumers during the first decade of the twenty-first century, Americans borrowed both from themselves and others. By 2012, foreign banks owned 35 percent of U.S. public debt—and of that 35 percent, Chinese and Japanese banks owned 42 percent, or about U.S.$2.3 trillion (Figure 9.20). China and Japan both had a great deal of cash because of their successful export economies, and they lent the money partly to ensure that consumers in the United States would continue to buy their goods. On the one hand, many in the United States are afraid that China, especially, can wield a lot of economic power because it holds so much of the U.S. debt. On the other hand, both sides are equally dependent on each other for continued economic success, so Asian financing of U.S. government debt is likely to continue.

Elsewhere in the world, East Asians are able to use their reserves to finance development and thereby secure privileged trade deals. In recent years, China's lending to developing countries has outpaced even that of the World Bank. Often

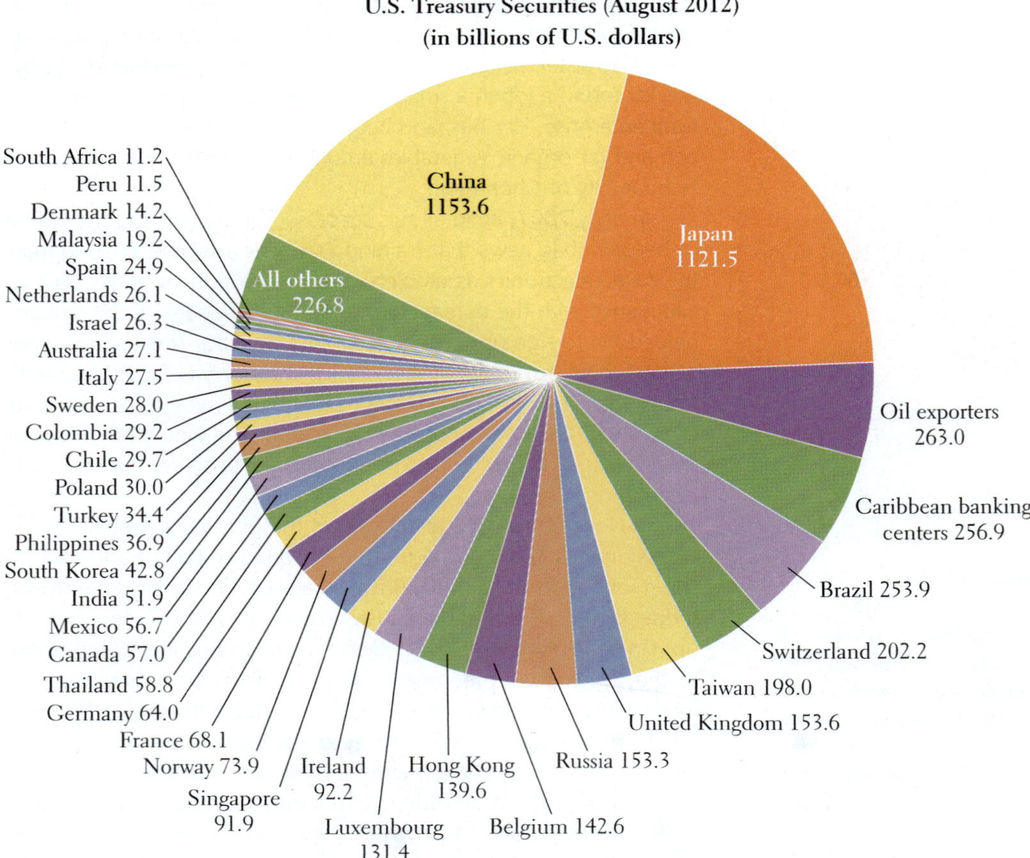

FIGURE 9.20 Foreign holders of U.S. Treasury securities, August 2012. The total dollar amount of U.S. securities held by foreign countries is over $5 trillion; note, though, that the United States holds over $6 trillion in securities from foreign countries.

these loans are made in return for guarantees that the recipient country will supply China with needed industrial materials, especially fossil fuels.

THINGS TO REMEMBER

- Key to Japan's rapid economic recovery after World War II was close cooperation between the government and private investors to create new, export-based industries, primarily those in the automotive and electronics sectors.
- Japanese management innovations known as the just-in-time and kaizen systems also were important to economic recovery; they were so successful that they diffused and created clusters of economic activity in different places around the world.
- In the 1980s, China's leaders enacted market reforms that changed the country's economy in significant ways, and subsequently those market reforms transformed the region and indeed the whole world.

Geographic Insight 3 • **Urbanization and Development** China's cities are growing quickly, with explosive economic development fueled by foreign direct investment and access to overseas markets.

Thinking Geographically

After you have read about power and politics in East Asia, you will be able to answer the following questions:

A What happened in Tiananmen Square in 1989?

C When was democracy in Taiwan established?

D What is North Korea purchasing with money earned from sales of its military technology abroad?

Power and Politics in East Asia

Geographic Insight 4
Power and Politics: As economic development has grown across East Asia, pressures for more political freedoms have also risen. Japan, South Korea, and Taiwan are now among the more politically free places in the world, and pressure for similar levels of freedom is intensifying in China and North Korea.

The demand for greater political freedom is growing throughout East Asia. Japan's current political structure was established after World War II, South Korea's in the late 1980s, and Taiwan's in the mid-1990s. All have steadily expanded political freedoms since their inception. Mongolia has had a dramatic expansion of political freedom since abandoning socialism in 1992. China, however, remains under the tight control of an authoritarian regime, and North Korea is even more tightly held. The map in Figure 9.21 shows each country's level of democracy. The red starbursts indicate places where civil unrest has broken out since 1945.

With China now a globalized economy, many wonder how much longer the Communist Party can remain in control without instituting democratic reforms throughout the entire

Political freedoms are relatively well protected in Japan, Taiwan, and South Korea, but China and North Korea have fewer outlets that citizens can use to directly influence the government. Nevertheless, pressure for more access to public political participation is growing.

A The annual vigil in Hong Kong commemorating those who died in the 1989 Tiananmen Square massacre in Beijing has become a focus for China's pro-democracy movement. As many as 150,000 people have shown up for the vigil in recent years. Hong Kong is the only part of China where the right to engage in political protests is protected by law.

ukuni
's war
nals
hina,
s
n.

ntry's flag on a model of
he East China Sea that is
. Many Taiwanese see this
m the country's continued
nsiders Taiwan a rebellious

D North Korea's longest-range ballistic missile is paraded in P'yongyang. In 2006, North Korea tested this missile and others, along with several nuclear weapons. China, Japan, South Korea, Russia, and the United States all expressed varying levels of concern.

country. The Communist Party officially claims that China is a democracy, and indeed some elections have long been held at the village level and within the Communist Party. However, representatives to the National People's Congress, the country's highest legislative body, are appointed by the Communist Party elite, who maintain tight control throughout all levels of government.

The Party selected a new leadership group (the *Politburo*) in 2012, but radical change is unlikely in the near future. Nevertheless, most experts on China agree that a steady shift toward greater political freedom is underway. This change might be inevitable as the population becomes more prosperous, educated, and widely traveled, thereby becoming exposed to places that have more freedom and more open government. Demands for political change built to a crescendo less than a decade after market reforms began, culminating in a series of pro-democracy protests that drew hundreds of thousands to Beijing's Tiananmen Square in 1989. These protests were brutally repressed, with thousands (the precise number is uncertain) of students and labor leaders massacred by the military. The Tiananmen Square event made clear that China's integration into the global economy will take place on political terms dictated by the state. Since then, pressure for change has continued to mount both within China (see Figure 9.21A) and internationally, but it seems that the link between economic development and democratization is only tenuous.

International Pressures Informed consumers and environmentalists in developed countries have long criticized China for its "no holds barred" pursuit of economic growth. Much of China's growth has been built on environmentally destructive activities and harsh conditions for workers, both of which effectively lower production costs so that the prices of goods can be globally competitive. The pressure for political change grew, though, once China was admitted to the World Trade Organization (WTO) in 2001 because its WTO membership meant that the Chinese government could no longer dictate economic policy to the same degree that it had in the past, even though the WTO does not enforce environmental and labor standards.

The 2008 Olympics in Beijing highlighted a number of issues that have also created an impetus for increasing democracy. Before the games, the international media shined its spotlight on human rights abuses of workers, protesters, prisoners, ethnic minorities such as Tibetans and Uygurs (see pages 411–413), and spiritual groups. The most prominent of such groups is Falun Gong, a Buddhist-inspired movement that has been perceived as a threat by the Chinese government because of its independence and capacity to mount a global media campaign for its cause. The extravagance of the 2008 Olympics also became a subject of criticism by the global media, with many journalists pointing out that no democracy would ever be able to devote so much tax revenue (estimated at $40 billion) to such an event, especially not in a country with as many pressing human and environmental problems as China has. Today, many of Beijing's Olympic venues are unused; at the same time, residents have benefited from the improvements to transportation and infrastructure that were part of the Olympic rebuilding of Beijing.

211. OLYMPIC RELAY CUT SHORT BY PARIS PROTESTS

213. BEIJING OLYMPICS: POLITICAL BATTLEGROUND?

221. REPORTS OF SALE OF EXECUTED FALUN GONG PRISONERS' ORGANS IN CHINA CALLED "SHOCKING"

Information Technology and Political Freedom The spread of information via the Internet has increased the push for political freedom from within China. Since the revolution in 1949, China's central government has controlled the news media in the country. By the late 1990s, though, the expanding use of electronic communication devices was loosening central control over information. By 2001, roughly 23 million people were connected to the Internet in China; by 2012, well over 500 million (or 40 percent of the population) were connected (Figure 9.22). However, Internet access is not evenly distributed. A huge digital divide has emerged in China; there is a "connected China," especially in large coastal cities where most people have Internet access, and a "disconnected China" in the rural interior. As development spreads, it is likely that the divide will diminish over time.

Not too long ago, telephones were very rare and people had to obtain permission to use them, but now millions of Chinese people have access to an international network of information. It is much more difficult for the government to give inaccurate explanations for problems caused by inefficiency and corruption. Journalists can now check the accuracy of government explanations by phoning witnesses or the principal actors directly and then posting their findings on the Internet. Analysts both inside and outside China see the availability of the Internet to ordinary Chinese citizens as a watershed event that supports democracy.

Nevertheless, the Internet in China is not the open forum that it is in most Western countries. Starting in 2013, Chinese Internet users are being required to register their actual names when using an online account. This is an attempt by the government to monitor speech and anonymous criticisms against party officials. And if people writing blogs in China use the words "democracy," "freedom," or "human rights," they may receive the following reminder: "The title must not contain prohibited language, such as profanity. Please type a different title." Such censorship, informally dubbed the "Great Firewall," has been aided by U.S. technology firms such as Yahoo and Microsoft, which allow the Chinese government to use software that blocks access to certain Web sites for users in China. Google, now in dispute with the Chinese government over state-sponsored censorship, invasion of human rights activists' email accounts, and cyber attacks, has moved operations to Hong Kong, where more freedom is allowed (see Figure 9.22B). The final chapter of Google's Internet ideological "war" with the Chinese government over freedom of Internet access and privacy has yet to be written. Meanwhile, the Chinese company Baidu has overtaken Google as the market-leading browser in China.

Urbanization, Protests, and Technology May Enhance Political Freedom Protests by workers for better pay and living

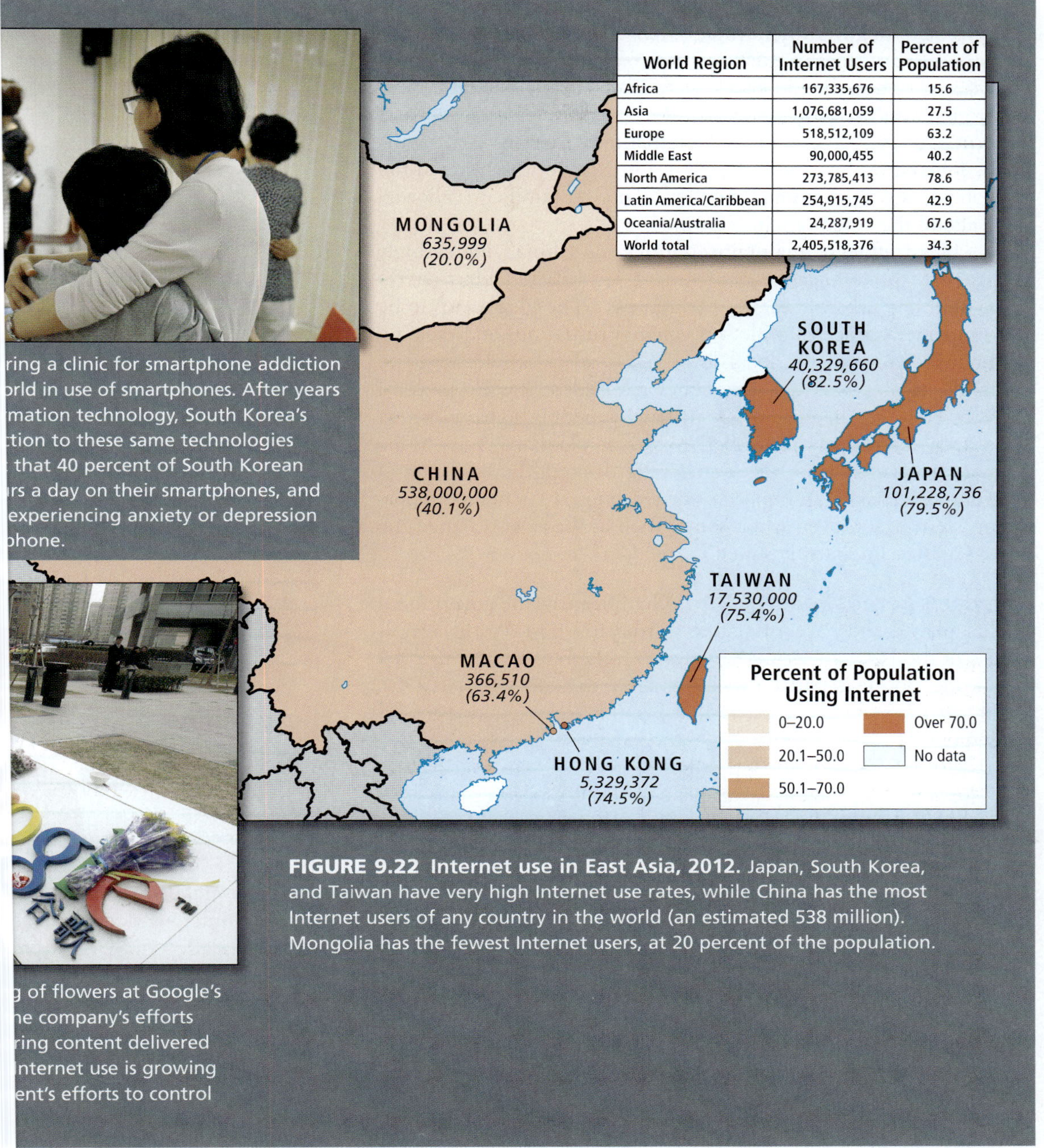

World Region	Number of Internet Users	Percent of Population
Africa	167,335,676	15.6
Asia	1,076,681,059	27.5
Europe	518,512,109	63.2
Middle East	90,000,455	40.2
North America	273,785,413	78.6
Latin America/Caribbean	254,915,745	42.9
Oceania/Australia	24,287,919	67.6
World total	2,405,518,376	34.3

ring a clinic for smartphone addiction
orld in use of smartphones. After years
mation technology, South Korea's
ction to these same technologies
that 40 percent of South Korean
rs a day on their smartphones, and
experiencing anxiety or depression
phone.

g of flowers at Google's
e company's efforts
ring content delivered
Internet use is growing
ent's efforts to control

FIGURE 9.22 Internet use in East Asia, 2012. Japan, South Korea, and Taiwan have very high Internet use rates, while China has the most Internet users of any country in the world (an estimated 538 million). Mongolia has the fewest Internet users, at 20 percent of the population.

and urban dwellers displaced
have exposed average citizens
t by their compatriots. China's
were 58,000 public protests in
er rose to 74,000 in 2004, and
government has stopped issu-
. The nonprofit group Human
roximately 100,000 to 200,000
year.
ng *interest groups* that pressure
n particular problems. One of
s was formed by parents whose

children died after schools collapsed during a massive earthquake (which registered 7.8 on the Richter scale) in Sichuan in May 2008. They demanded changes in policies that allowed schools to be poorly constructed and pushed for compensation for their lost children. Today, the Internet has emerged as yet another tool for collective action. For example, Renren, China's dominant social network site, has more than 150 million user accounts and has the capacity to be an instrument for community activism.

Groups that work toward gaining compensation for low pay or the loss of property are actually less threatening to the central government than those that work for political reform. In December 2008, people seeking more political freedoms signed a manifesto called the 08 *Charter*, calling for a decentralized federal system of government (that is, more power to the provinces), democratic elections, and the end of the Communist Party's political monopoly. Liu Xiaobo, coauthor of the 08 Charter, was arrested, and in December 2009 he was sentenced to 11 years in prison, even as prominent international human rights activists lobbied on his behalf. In 2010, Liu Xiaobo was awarded the Nobel Peace Prize, though neither he nor his family was allowed to attend the ceremony in Norway.

In an effort to maintain control over China's increasingly articulate protestors, the government has allowed elections to be held for "urban residents' committees." The idea seems to be to provide a peaceful outlet for voicing frustrations and creating limited change at the local level. Many of these elections are hardly democratic, with candidates selected by the Communist Party. However, in cities where unemployment is high and where protests have been particularly intense, elections tend to be more free, open, and truly democratic. It could be that interest group protests are an important first step toward actual participatory democracy, even when some, such as those who signed the 08 Charter, are harshly stifled.

Japan's Recent Political Shifts Significantly, the government that played such a central role during the post–World War II rise of the Japanese economy was controlled from 1955 to 2009 by one political party, the Liberal Democratic Party (LDP). This has long led to criticism that Japan is lacking a meaningful democracy. However, in 2009, the Japanese people elected a government led by the opposition party, the Democratic Party of Japan (DPJ), only to reverse course again and bring the LDP back to power in 2012. But no matter who is running the country, they have to deal with government debt and the prospect of cutting many social services that had supported Japan's high standard of living. The main difference may be that the more nationalistic LDP takes a hawkish stance against China and indeed all neighboring countries (see Figure 9.21B). Having closer relations with China may be advantageous for Japan, though, as China has the potential to be a major market for Japanese exports in the future.

THINGS TO REMEMBER

Geographic Insight 4

- **Power and Politics** While radical change is unlikely in the near future, most experts on China agree that a steady shift toward greater democracy is underway. This change might be inevitable as the population becomes more prosperous, educated, and widely traveled, thereby becoming exposed to places that have more freedoms and more open government.
- By 2012, China had 500 million people using the Internet, representing 40 percent of the country's population. China has strict controls on Internet access and use, but social networking is emerging as a tool to promote democracy.
- Chinese citizens who share a grievance against the government are increasingly cooperating with one another in their protests. In December 2008, a manifesto for change, the 08 Charter, was signed and submitted to the government by 350 protestors.
- In Japan, one political party has controlled the government for all but 3 years since 1955, leading to criticism that Japan lacks a vigorous democracy.

Sociocultural Issues

East Asia's economic progress has led to social and cultural change throughout the region. Population growth, long a dominant issue, has slowed, while other demographic issues are coming to the fore. Modernized economies that bear some features of capitalism are changing work patterns and family structures.

Population Patterns

Although East Asia remains the most populous world region, families there are having many fewer children than in the past. Of all world regions, only Europe has a lower rate of natural increase (0.0 percent increase per year, compared to East Asia's 0.4 percent). In China, this is partially due to government policies that harshly penalize families for having more than one child. But in China as well as elsewhere, urbanization and changing gender roles are also resulting in smaller families, regardless of official policy. Only in the two poorest countries—Mongolia (with 2.9 million people) and North Korea (with 24.6 million)—are women still averaging two or more children each, but even there family size is shrinking.

Responding to an Aging Population

Low birth rates mean that fewer young people are being added to the population, and improved living conditions mean that people are living longer across East Asia. The overall effect is that the average age of the populations is rising. Put another way, populations are aging. For Mongolia, South Korea, North Korea, and Taiwan, it will be several decades before the financial and social costs of supporting numerous elderly people will have to be addressed. China faces especially serious future problems with elder care because the one-child policy and urbanization have so drastically reduced family kin-groups. Japan, on the other hand, has already been dealing with the problem of having a large elderly population that requires support and a reduced number of young people to do the job.

Japan's Options Japan's population is growing slowly and aging rapidly, raising concerns about economic productivity and humane ways to care for dependent people. The demographic transition (see page 16 in Chapter 1) is well underway in this highly developed country, where 86 percent of the population live in cities. Japan has a negative rate of natural increase (−0.2), the lowest in East Asia and on par with that of Europe. If this trend continues, Japan's population is projected to plummet from the current 128 million to 95.5 million by 2050.

At the same time, the Japanese have the world's longest life expectancy, at 83 years (Figure 9.23). As a result, Japan also has the world's oldest population, 24 percent of which is over the age of 65. By 2055, this age group will account for approximately 40 percent of the population. By 2050, Japan's labor pool could be reduced by more than a third, but it would still need to produce enough to take care of more than twice as many retirees as it does

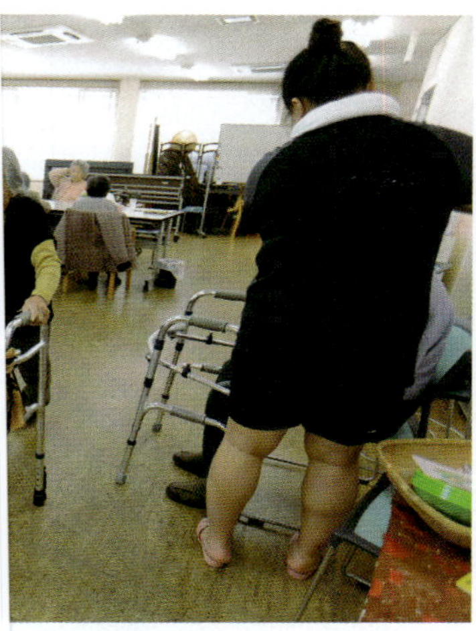

ulation. An elderly woman
Island, Japan. Japan has the
ears) and also the world's oldest
are over the age of 65.

changes will have a momentous
ne search for solutions is under-
mmigration to bring in younger
ntribute to the tax rolls, as the
ne. Another is to keep the elderly
d take care of themselves.
t workers from other countries
ne aging crisis. Many Japanese
foreigners, and the few small
connections to China or Korea
he children of foreigners born
p, and some who have been in
ll thought of as foreign. Today,
gistration card" at all times.
are dribbling into Japan in a
ys. Many are "guest workers"
he most dangerous and lowest-
ding that they will eventually
of Japanese people who once
and Peru). Regardless, Japan's
making up only 1.7 percent
opulation. A recent UN report
to admit more than 640,000
tain its present workforce and
its GDP.
's demographic changes, the
s sums of money in robotics.
already widespread in such
facturing, and their industrial
being developed to care for
gh hospitals, to look after chil-
2025, the government plans to
workforce with robots.

of China's population over 65
l change rapidly as conditions

improve and life expectancies increase by 5 to 10 years to become more like those in China's affluent East Asian neighbors. However, a crisis in elder care is already upon China for two other reasons: the high rate of rural-to-urban migration and the shrinkage of family support systems because of the one-child policy (discussed below).

When hundreds of millions of young Chinese were lured into cities to work, most thought rural areas would benefit from remittances, and this has happened. However, few anticipated that the one-child family would mean that for every migrant, two aging parents would be left to fend for themselves, often in rural, underdeveloped areas. China's parliament passed a law in 2013 that requires family members to visit and support their elderly relatives. While such a law may be unenforceable, it shows how worried the government is about the social consequences of the prospect of an aging society, including the increasing spatial mismatch between the elderly who need care and their younger relatives who often have moved elsewhere in search of employment. At the same time, individual Chinese people are being proactive: few outsiders realize that the throngs of elderly Chinese people who exercise in public parks are part of a movement to keep the elderly fit that is linked to far-sighted policies aimed at lowering the costs of supporting the aged.

China's One-Child Policy

> **Geographic Insight 5**
>
> **Population and Gender:** China's "one-child policy" has created a shortage of females. Long-standing cultural preferences for male children mean that if their one child will be female, many parents choose abortion or put the baby up for adoption. The shortage of females is accentuated by the fact that many women are deciding to have a career rather than to marry and have children.

In response to fears about overpopulation and environmental stress, China has had a one-child-per-family policy since 1979, when it also instituted broader economic reforms. The policy is enforced with rewards for complying and with large fines for not complying. As a result, China's rate of natural increase (0.5 percent) is the same as that of the United States, and less than half the world average (1.2 percent). If the one-child family pattern continues, China's population will start to shrink sometime between 2025 and 2050 (Figure 9.24B), creating many of the same economic and social problems that Japan is now facing. For example, China's working-age population will not expand much in the future. It is possible that further mechanization of Chinese agriculture will enable rural workers to shift to urban jobs, but with a limited supply of workers, employers may end up raising wages in order to compete for employees, which could diminish China's competitive advantage in global manufacturing.

The one-child policy has also transformed Chinese families and Chinese society at large. For example, an only child has no siblings, so within two generations, the kinship categories of brother, sister, cousin, aunt, and uncle have disappeared from most families, meaning that any individual has very few, if any, related age peers with whom to share family responsibilities. The effect on society of the one-child family is that most children are doted on by several adults, and children are not taught by siblings

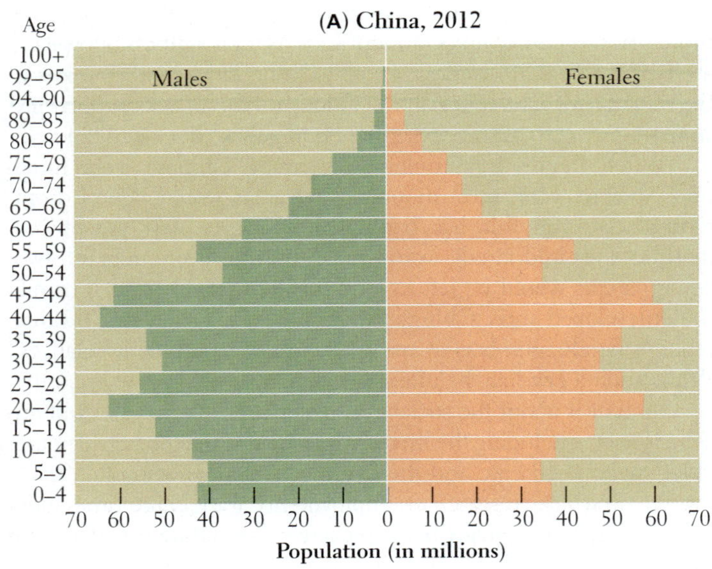

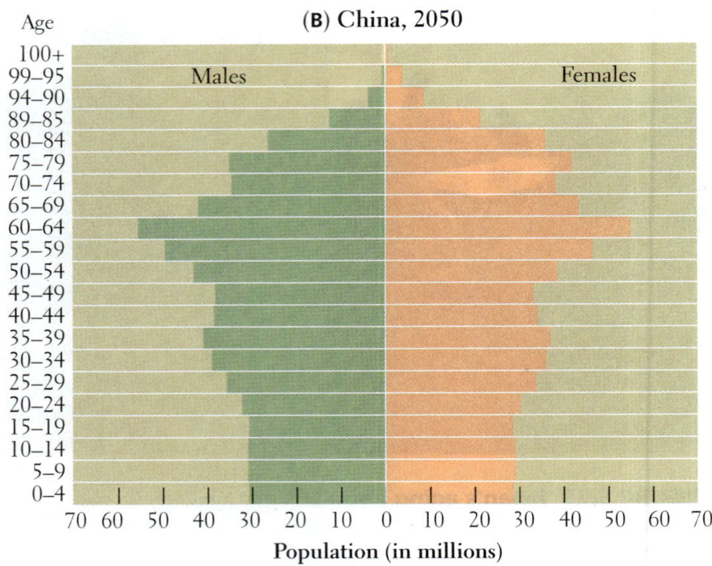

FIGURE 9.24 Population pyramids for China, 2012 and 2050 (projected).

to share. Conscious efforts must be made to instill self-sufficiency in only children. The one-child policy has sometimes been enforced brutally. In various times and places, the government has waged a campaign of forced sterilizations and forced abortions for mothers who already have one child.

Cultural Preference for Sons: Missing Females and Lonely Males The prospect of a couple's only child being a daughter, without the possibility of having a son in the future, is the aspect of the one-child policy that has caused families the most despair. For years, the makers of Chinese social policy have sought to eliminate the old Confucian-based preference for males by empowering women economically and socially. In many ways, the policy makers are succeeding in empowering young women. For example, Chinese women participate to a high degree in the workforce. Nevertheless, the preference for sons persists.

The population pyramid in Figure 9.24A illustrates the extent to which the preference for sons has created a gender imbalance in China's population. The normal sex ratio at birth is 105 boys to 100 girls, which evens out by the age of 5. In China, however, there are 113 boys for every 100 girls born. In fact, for nearly every category until age 70, males outnumber females. The census data show that there are already 50 million more men than women. What happened to the missing girls?

There are several possible answers. Given the preference for male children, the births of these girls may simply have gone unreported as families hoped to conceal their daughter and try again for a son. There are many anecdotes of girls being raised secretly or even disguised as boys. Also, adoption records indicate that girls are given up for international adoption much more often than boys are. Or the girls may have died in early infancy, either through neglect or infanticide. Finally, some parents have access to medical tests that can identify the sex of a fetus. There is evidence that in China, as elsewhere around the world, some of these parents choose to abort a female fetus.

The cultural preference for sons persists elsewhere in East Asia as well. A deficit of girls appears on the 2000 population pyramids for Japan, the Koreas, Mongolia, and Taiwan. Nonetheless, evidence shows that attitudes may be changing. In Japan, South Korea, and Mongolia, the percentage of women receiving secondary education equals or exceeds that of men.

A Shortage of Brides A major side effect of the preference for sons is that there is now a growing shortage of women of marriageable age throughout East Asia. China alone had an estimated deficit of 11 million women aged 20 to 35 in 2012. Females are also effectively "missing" from the marriage rolls because many educated young women are too busy with career success to meet eligible young men.

Research suggests that at least 10 percent of young Chinese men will fail to find a mate; and poor, rural, uneducated men will have the greatest difficulty. The shortage of women will lower the birth rate yet further, which will contribute to the expected shrinkage of the population over the next century. When single men age, without spouses, children, or even siblings, there will be no one to care for them. Furthermore, China's growing millions of single young men are emerging as a threat to civil order, as they may be more prone to drug abuse, violent crime, HIV infection, and sex crimes. Cases of kidnapping and forced prostitution of young girls and women are already increasing.

Population Distribution

In East Asia, people are not evenly distributed on the land (Figure 9.25). Mongolia is only lightly settled, with one modest urban area. China, with 1.35 billion people, has more than one-fifth of the world's population. However, 90 percent of these people are clustered on the approximately one-sixth of the total land area that is suitable for agriculture, and roughly half of these live in

centrated
ern third
ıa Plain,
to Hong
ı of the
outheast,
ddle and
sin.
e Korean
ettled, as
an. In Japan, settlement is con-
nes from the cities of Tokyo and
nshu, south through the coastal
islands of Shikoku and Kyushu.
f the most extensive and heavily
in the world, accommodating
lation. The rest of Japan is moun-
.

ON THE BRIGHT SIDE

Changes in the One-Child Policy?

Recently, some officials in China have supported abolishing the one-child policy, or adopting a two-child policy. Enforcement of the one-child policy is already lax in rural areas, and many ethnic minorities have long been exempt. Given China's growing urban populations, which face numerous pressures for small families, it is unlikely that abandoning the one-child policy would produce a population boom.

Human Well-Being

"The objective of development is to create an enabling environment for people to enjoy long, healthy and creative lives."

Mahbub ul Haq, Founder of the UN Human Development Report

The best ways to measure how well specific countries enable their citizens to enjoy healthy and rewarding lives is disputed, but it is

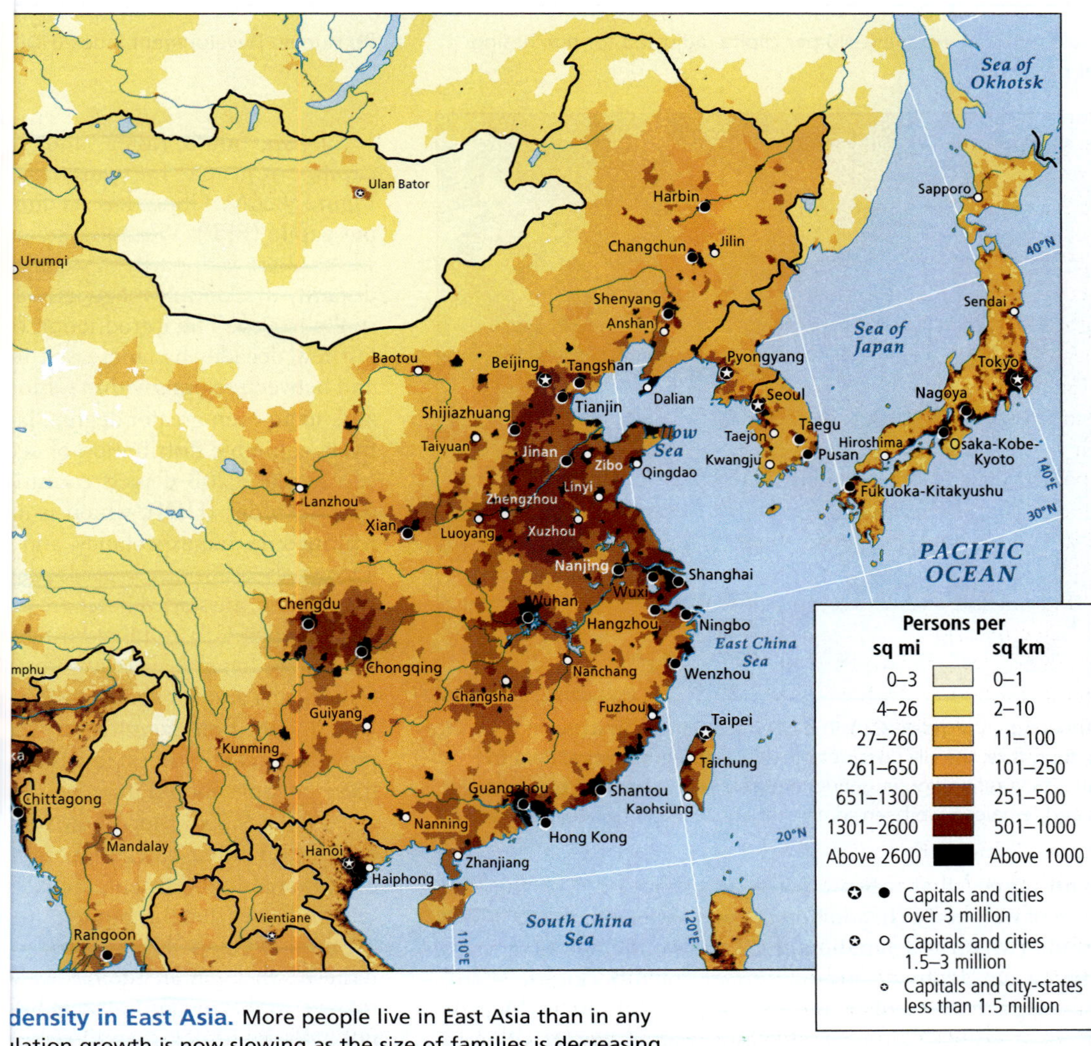

density in East Asia. More people live in East Asia than in any
ılation growth is now slowing as the size of families is decreasing.
; the challenge of caring for large elderly populations. Meanwhile,
ıexpected consequence of its one-child-per-family policy: a shortage

FIGURE 9.26 Maps of human well-being.

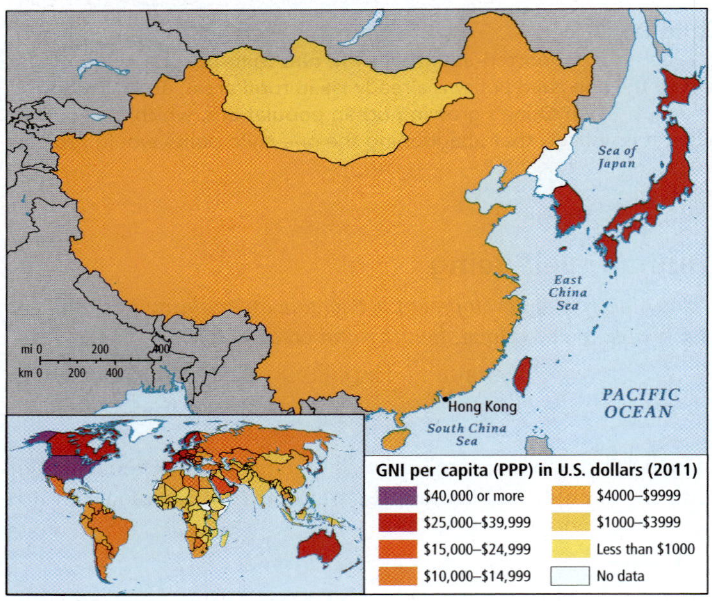

(A) Gross national income (GNI) per capita, adjusted for purchasing power parity (PPP).

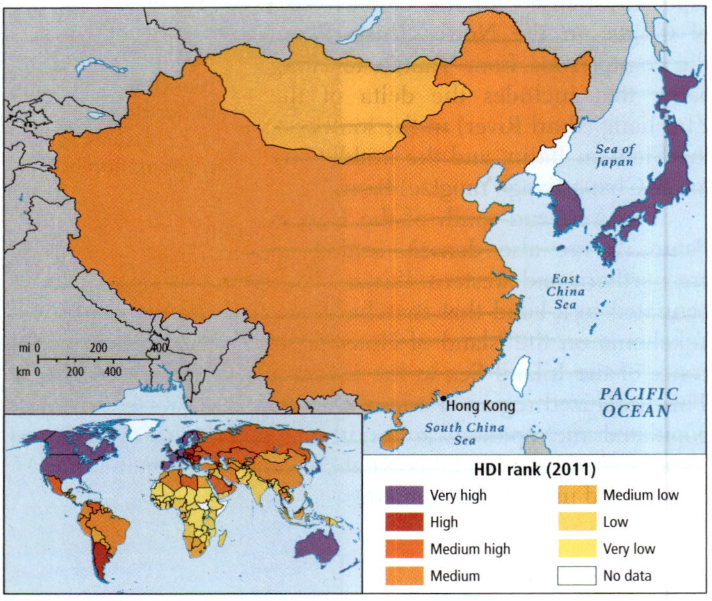

(B) Human Development Index (HDI).

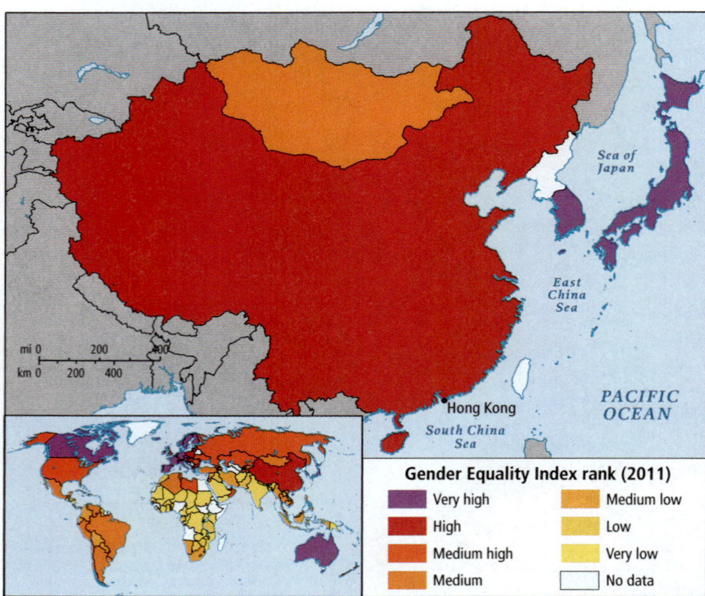

(C) Gender Equality Index (GEI). The UN treats Taiwan as a province of China; therefore, the UN does not provide separate statistical data for Taiwan. China ranks high in gender equality in part because pay scales for males and females have been somewhat equalized under Communism.

generally agreed that income per capita should not be the sole measure of well-being. In addition to gross national income (GNI) per capita (which is a variation of GDP), we include here maps on human development and on gender equality (Figure 9.26). On all maps, Japan and South Korea (plus Taiwan in Map A) stand out as quite different from China and Mongolia. (Taiwan is not covered in Maps B and C because in UN data, Taiwan is included with China). North Korea is not included on any of the maps because it does not submit data to the UN.

Japan, South Korea, and Taiwan all have high GNI per capita, adjusted for purchasing power parity (PPP) (see Figure 9.26A). China has medium-low, and Mongolia low, GNI per capita (PPP). What the income figures mask is the extent to which there is disparity of wealth in a country. Ironically, wealth disparity in Communist-governed China is larger than elsewhere in the region. The introduction of a market economy during the last few decades has increased inequalities between individuals and between regions within China. In contrast, both South Korea and, to an even greater degree, Japan are characterized by a relatively equitable distribution of economic resources.

Figure 9.26B depicts countries' rank on the Human Development Index (HDI), which is a calculation (based on adjusted real income, life expectancy, and educational attainment) of how adequately a country provides for the well-being of its citizens. Again, there are stark differences between the very high global ranking for Japan (15) and that for South Korea (12), and the medium to medium-low global rankings for China (101) and Mongolia (110). Still, China and Mongolia have made significant progress over the last 20 years. Both countries say that their specific goals now are to provide equitable access to basic services for all of their people, so these rankings should improve in the years to come.

Figure 9.26C shows how well countries are ensuring gender equality in three categories: reproductive health, political and education empowerment, and access to the labor market. A high rank indicates that the genders are tending toward equality. Again, the more developed Japan and South Korea rank very high, which indicates that gender equality in both countries is well developed relative to the rest of the world. While China and Mongolia do not rank as high, it is worth noting that they rank higher in measures of gender equality than they do in GNI and HDI. China had a high gender ranking in 2013 and Mongolia a medium ranking. This may be a result of the emphasis on gender equity under

ıf Mongolia, traditional cultural ısures of gender equality.

REMEMBER

ı in any world region, but population ; decreases. Most of the region now ȷe elderly populations.

ıgest life expectancy, at 83 years, ·24 percent of Japanese people are e group is projected to account for

d Gender China is grappling with ısequence of its one-child-per-family ɔf women. It is also struggling with nd the extended family.

ıtively high in East Asia, especially in l South Korea.

st Asia

: one dominant ethnic group, but cultural diversity. In China, for : citizens call themselves "people ack about 2000 years to the Han nly in the early twentieth century, ng to create a mass Chinese identi- s people who share a general way of The main language spoken by the ınly one of many Chinese dialects. s number about 117 million peo- thnic or culture groups scattered ɪside the Han heartland of eastern these areas have been designated orities theoretically manage their ·, the Communist Party in Beijing ıs, especially those considered to ıave resources of economic value. nic groups here.

Auslims of various ethnic origins ɔrities in China. All are originally re Turkic people who historically thers specialized in trading. They hina's northwest and to think of m mainstream China.

:E-gurs) and Kazakhs, who are ı in the autonomous region of t (see the Figure 9.1 map and ' were nomadic herders, and some n China's minorities is related to Asia. Contact among these groups 's market reforms began and the : 9.28). The Beijing government, ɪrea's oil, other mineral resources, ınd for national development.

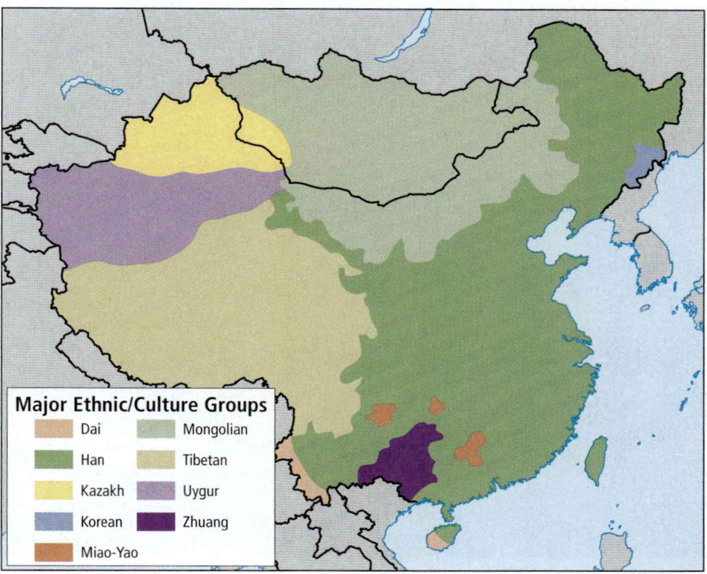

FIGURE 9.27 Major ethnic groups of China. This map shows the areas traditionally occupied by the Han and by ethnic minorities. It does not show the recent resettling of Han in Xinjiang and Tibet, nor does it show the Hui, people from many ethnic groups whose ancestors converted to Islam and who are found in disparate locations along the old Silk Road and in coastal southeastern China.

Accordingly, it has sent troops and by now many millions of Han settlers to Xinjiang through what is known as the "Go West" policy. The Han settlers fill most managerial jobs in the bureaucracy, in mineral extraction, the military, and power generation. An important secondary role of the Han is to dilute the power of Uygurs and Kazakhs within their own lands. In Xinjiang, there are now almost as many Han as Uygurs (8 versus 9 million), plus small numbers of other minorities.

The Beijing government has rushed to develop Xinjiang and its capital Urumqi with special development zones (ETDZs), but the Uygurs have been left out of most policy-making roles and indeed have been excluded from participating in the economic boom. For example, one young Uygur man in Urumqi writes of his discontent: "I am a strong man, and well-educated. But [Han] Chinese firms won't give me a job. Yet go down to the railroad station and you can see all the [Han] Chinese who've just arrived. They'll get jobs. It's a policy to swamp us."

Until recently, the Uygur people of Xinjiang expressed their resistance to Han dominance merely by reinvigorating their Islamic culture. Islamic prayers were increasingly heard publicly, more Muslim women were wearing Islamic dress, Uygur was spoken rather than Chinese, and Islamic architectural traditions were being revived. Then, more active resistance groups, formed by Uygur separatists, began carrying out attacks on Chinese targets. In 2009, violence erupted in the streets of Urumqi between Uygurs and the Han and about 200 people were killed. The Beijing government responded by harshly punishing the rioters and broadcasting the accusation that all Uygur separatists, even those committed to nonviolence, are Islamic fundamentalists bent on terrorism. In 2013, China's government began arresting nonviolent Uygur Internet-based activists.

FIGURE 9.28 The Silk Road market in Kashi. A Uygur man tests a horse before selling it at the Sunday market in Kashi, a city in China's Xinjiang Uygur Autonomous Region. Kashi has been an important trading center along the Silk Road for at least 2000 years.

Distinct from the Uygurs and Kazakhs are the Hui, who altogether number about 10 million. The original Hui people were descended from ancient Turkic Muslim traders who traveled the Silk Road from Europe across Central Asia to Kashgar (now Kashi) and on to Xian in the east (see the map of the Silk Road in Figure 5.9 on page 207; see also Figure 9.28). Subgroups of Hui live in the Ningxia Huizu Autonomous Region and throughout northern, western, and southwestern China. There they continue as traders, farmers, and artisans. The long tradition of commercial activity and adoption of the Chinese language among the Hui has facilitated their success in China. Many are active in the new free market economies of southeastern China as businesspeople, technicians, and financial managers, using their money not just to buy luxury goods, but also to revive religious instruction and to fund their mosques, which are now more obvious in the landscape.

The Tibetans In contrast to the prosperous and assimilated Hui are the Tibetans, an impoverished ethnic minority of nearly 5 million individuals scattered thinly over a huge, high, mountainous region in western China. The history of Tibet's political status vis-à-vis China is long and complex, characterized by both cordial relations and conflict. During China's imperial era (prior to the twentieth century), Tibet maintained its own government but faced the constant threat of invasion and of Chinese meddling in its affairs. In the early 1900s, Tibet declared itself separate and free from China and conducted its affairs as an independent country. It was able to maintain this status until 1949–1950, when the Chinese Communist army "liberated" Tibet, promising a "one-country, two-systems" structure, which suggested a great deal of regional self-governance. A Tibetan uprising in 1959 led China to abolish the Tibetan government and violently reorder Tibetan society. Since the 1950s, the Chinese government has referred to Tibet as the Xizang Autonomous Region. The Chinese government suppressed Tibetan Buddhism, which is an institution that Tibetans rally behind as a symbol of independence, by destroying thousands of temples and monasteries and massacring many thousands of monks and nuns. In 1959, the spiritual and political leader of Tibet, the Dalai Lama, was forced into exile in India along with thousands of his followers. **216. DALAI LAMA CALLS FOR AUTONOMY BUT NOT INDEPENDENCE**

By the 1990s, the Beijing government's strategy was to overwhelm the Tibetans with secular social and economic modernization and with Han Chinese settlers rather than outright military force (though China maintains a military presence in Tibet). To attract trade and quell foreign criticism of its treatment of Tibetans, China is spending hundreds of millions of dollars on housing and on roads, railroads, and a tourism infrastructure that capitalizes on European and American interest in Tibetan culture. China presents its actions in Tibet as part of its overall strategy to integrate the entire country economically and socially. Schools are being built and jobs opened up to young Tibetans. A new railway link connecting Tibet more conveniently to the rest of China was completed in 2006 (Figure 9.29). The Han in Tibet see the railway as a public service that will promote Tibetan development, but Tibetan activists see it as a conveyor belt for more Han dominance over the Tibetan economy and culture. **214. TIBETAN BUDDHIST NUNS' PROTEST SONGS BRING PUNISHMENT FROM CHINA**

Within Tibetan culture, women have held a somewhat higher position than in other cultures of East Asia. Buddhism did introduce many patriarchal attitudes to Tibet, but these did not curtail other more equitable traditions. Among nomadic herders, women could have more than one husband, just as men were free to have more than one wife; at marriage, a husband often joins the wife's family. By comparison, Han Chinese culture has typically regarded the women of Tibet and other western minorities as barbaric precisely because their roles were not circumscribed: they were not secluded, they rode horses, they worked alongside the men in herding and agriculture, and they were assertive.

Indigenous Diversity in Southern China In Yunnan Province in southern China, more than 20 groups of ancient native peoples live in remote areas of the deeply folded mountains that stretch into Southeast Asia. These groups speak many different languages, and many have cultural and language connections to the indigenous people of Tibet, Burma, Thailand, or Cambodia. Gender relations are different here than among the Han. A crucial difference may be that among several groups, most notably the Dai, a husband moves in with his wife's family at marriage and provides her family with labor or income, which is a pattern we also noted among the Tibetans. A husband inherits from his wife's family rather than from his birth family.

Taiwan's Many Minorities In Taiwan and the adjacent islands, the Han account for 95 percent of the population, but Taiwan

bet Railway. The western section of is the world's highest railway, with rs) of line that are at an altitude above rip from Beijing to Lhasa now takes less 1 less to transport goods and people via ne existing highway. However, many will become a conduit for more Han

(A) Han Chinese passengers aboard the train combat altitude sickness by sleeping and using supplemental oxygen supplied to every passenger.

(B) A view of the train as it passes through the mountains of the Tibetan Plateau.

minorities. Some have cultural ;, and agricultural and hunting connection to ancient cultures . The mountain dwellers among milation more than the plains 1 reservations set aside for indig- ɔut most are now being absorbed -influenced Taiwanese life. The ıous in Taiwan. They are divided cestral home in China, and the who fled to Taiwan around the People's Republic of China in ərs."

ɛ several indigenous minorities ɛd considerable discrimination. y group is the **Ainu**, character- ards, and thick, wavy hair. Now 50,000, the Ainu are a racially and culturally distinct group thought to have migrated many thousands of years ago from the northern Asian steppes. They once occupied Hokkaido and northern Honshu and lived by hunting, fishing, and some cultivation, but they are now being displaced by forestry and other development activities. Few full-blooded Ainu remain because, despite prejudice, they have been steadily assimilated into the mainstream Japanese population (Figure 9.30). In 2008, the Japanese parliament, somewhat belatedly, officially recognized the Ainu as an indigenous minority.

East Asia's Most Influential Cultural Export: The Overseas Chinese

China has had an impact on the rest of the world not only through its global trade, but also through the migration of its people to nearly all corners of the world. The first recorded emigration by the Chinese

Ainu an indigenous cultural minority group in Japan characterized by their light skin, heavy beards, and thick, wavy hair, who are thought to have migrated thousands of years ago from the northern Asian steppes

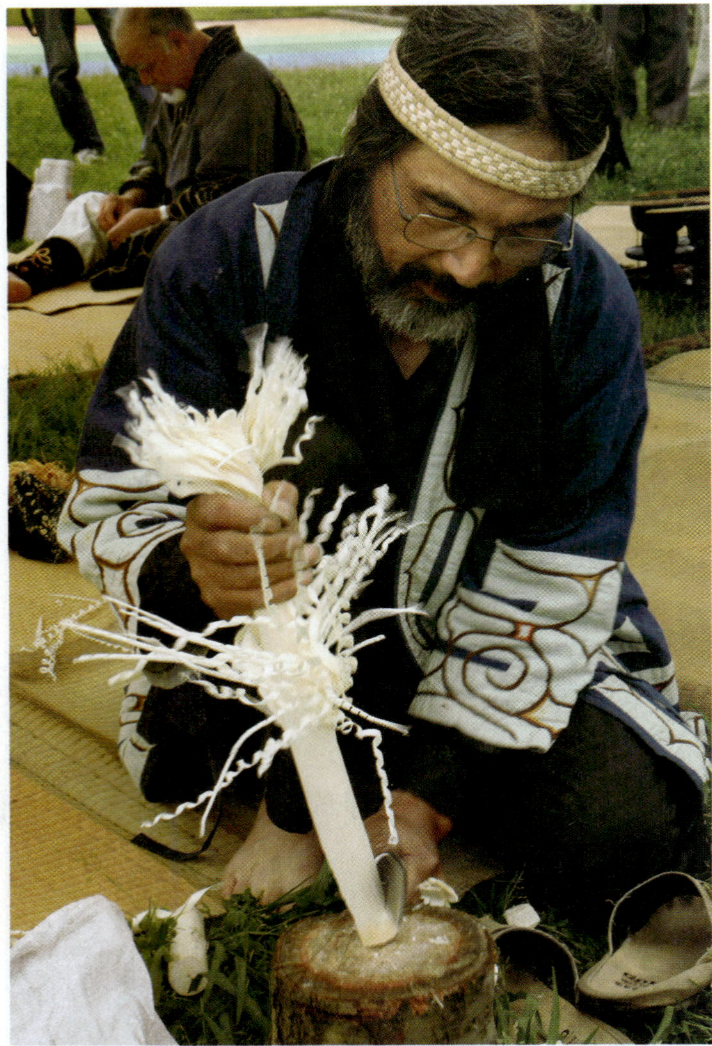

FIGURE 9.30 The Ainu of Japan. An Ainu man creates an item to be used in ancestor worship in Tokyo. In 2008, Japan's government formally recognized the Ainu as an indigenous people of Japan, more than 500 years after the Ainu began to be marginalized by mainstream Japanese culture.

took place more than 2200 years ago. Following that, China's contacts then spread eastward to Korea and Japan, westward into Central and Southwest Asia via the Silk Road, and by the fifteenth century, to Southeast Asia, coastal India, the Arabian Peninsula, and even Africa.

Trade was probably the first impetus for Chinese emigration. The early merchants, artisans, sailors, and laborers came mainly from China's southeastern coastal provinces. Taking their families with them, some settled permanently on the peninsulas and islands of what are now Indonesia, Thailand, Malaysia, and the Philippines. Today they form a prosperous urban commercial class known as the *Overseas Chinese*. The quintessential Overseas Chinese state in Southeast Asia is Singapore, where 77 percent of the population is ethnically Chinese.

In the nineteenth century, economic hardship in China and a growing international demand for labor spawned the migration of as many as 10 million Chinese people to countries all over the world. By the middle of the twentieth century, many others fleeing the repression of China's Communist Revolution joined those from earlier migrations despite thorough assimilation into adopted homelands. As a result, "Chinatowns" are found in most major world cities. Somewhat unfairly, the term *Overseas Chinese* has been extended to apply to Chinese emigrants and their descendants.

THINGS TO REMEMBER

- There are several distinct minority groups in East Asia: the Uygurs, Kazakhs, and Tibetans in western China; the Hui, scattered throughout central China; and smaller indigenous groups in southern China, Taiwan, and Japan.
- There is resentment and sometimes open resistance to Han Chinese domination in the various minority homelands.
- Throughout East Asia, minorities have experienced discrimination.
- Millions of "Overseas Chinese" live in cities and towns around the world, especially in Southeast Asia. Most have settled permanently in these new homelands.

Geographic Insights: East Asia Review and Self-Test

1. Climate Change and Water: East Asia has long suffered from alternating droughts and devastating floods, and scientists fear that these hazards will only intensify with climate change. Meanwhile, the rivers of China, many of which originate in the Tibetan and neighboring plateaus, are dependent both on seasonal rainfall cycles and on glacial meltwater that are expected to diminish over the next few decades.

- What are some physical and social conditions that have contributed to the destructive droughts and floods experienced by East Asians? What has been done to protect the population from such hazards?
- How might the melting of China's highest glaciers affect the country's rivers? Which areas might be more affected than others?
- In which ways is China more vulnerable to global warming than Japan, the Koreas, or Taiwan?

2. Food and Globalization: East Asian countries have enough economic development to be able to buy food on the global market, but they are less able to produce sufficient food domestically. Much agricultural land has been lost to urban and industrial expansion and to mismanagement, but the increase in the region's affluence has led to a demand for more imported food.

food on the global market but
me. Why might this situation
blems might this create both for

Japan's, role in the global
oceans?

ent: China's move to join the
cades ago has led to spectacular
on consumerism, and a massive
ollution of the air, water, and soil.

global economy and how might

d the most? What has determined
rural stagnation?

ost prevalent and how are they
nt in China?

mic development has grown across
cal freedoms have also risen.
e now among the more politically
ure for similar levels of freedom is
orea.

- How might increasing levels of economic development and pressures for greater political freedoms help bring about greater democratization in China?
- What is the background of the adversarial relationship between China and Taiwan?
- How can conflict on the Korean Peninsula affect global peace and security?
- What role did Japan play in East Asia during the colonial period?

5. Population and Gender: China's "one-child policy" has created a shortage of females. Long-standing cultural preferences for male children mean that if their one child will be female, many parents choose abortion or put the baby up for adoption. The shortage of females is accentuated by the fact that many women are deciding to have a career rather than to marry and have children.

- Why do most families want a male child? What problems have been created by the resulting shortage of women?
- What problems of an aging society are arising in Japan related to its rapidly aging population? What might be the solutions to such problems?
- Are there any social consequences to the disappearance of siblings, cousins, aunts, and uncles?

stions

tinental East Asia (western China
o extremes in temperature?

ulnerable to global warming than

fflicted with recurring famines.
ern, China's access to food has
ent. How?

c political system have resulted in
and design of the Three Gorges

e or criticize China's current political
mic and political change?

ion have taken a different course if
had inspired China to conquer a vast
t Britain?

7. Contrast Japan's pre–World War II policies in East Asia with its present role in the region. Describe the principal similarities or differences.

8. In what ways has the one-child-per-family policy helped maintain stability in China? How might it affect China in the future in terms of gender balance and economic growth?

9. How has East Asia's spatial pattern of urbanization been shaped by integration into the global economy?

10. What information technologies are now available to the Chinese, and what role can such technologies have in the democratization of the country?

growth poles 399
***hukou* system** 375
just-in-time system 395
***kaizen* system** 395
regional self-sufficiency 396
regional specialization 396
responsibility system 396
special economic zones (SEZs) 399
state-aided market economy 394
tsunami 377
typhoon 377
wet rice cultivation 382

A Gorges, Burma (Myanmar)

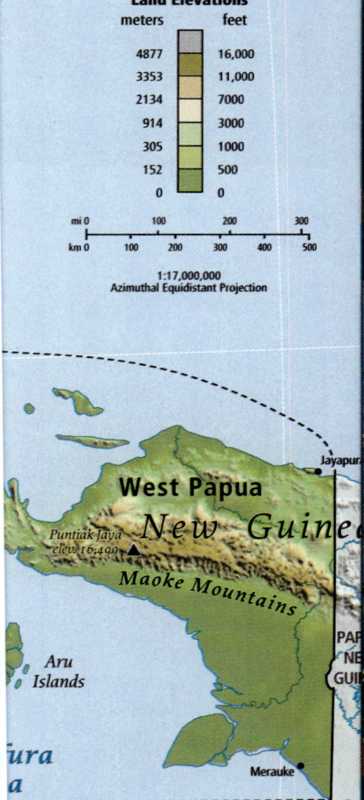

10 Southeast Asia

D Volcanoes, Java, Indonesia

E Post-Earthquake/Tsunami, Sumatra, Indonesia

FIGURE 10.1 Regional map of Southeast Asia.

GEOGRAPHIC INSIGHTS: SOUTHEAST ASIA

After you read this chapter, you will be able to discuss the following geographic insights as they relate to the nine thematic concepts:

1. Climate Change, Food, and Water: Conversion of forests to oil palm plantations, rice farms, subsistence plots, and settlements all add to greenhouse gas emissions, which intensify climate change and leave many people vulnerable to both flooding and occasional droughts.

2. Globalization and Development: Globalization has brought spectacular economic successes and occasional, dramatic economic declines. Urban incomes soar during periods of growth, but the poor are vulnerable during periods of global economic decline.

3. Power and Politics: Militarism, corruption, and favoritism are recurring issues in Southeast Asia. Although modernization and prosperity have brought significant increases in democratic participation, there have also been reversals, especially during periods of economic downturn. These sometimes result in violence, often against ethnic minorities.

4. Population and Gender: Economic change has brought better job opportunities and increased status for women, who are then choosing to have fewer children. The improving role and status of women has increased awareness that sex trafficking is an abuse of human rights.

5. Urbanization, Food, and Development: The development of export-oriented modernized agriculture and agribusiness have forced farmers who were once able to produce enough food for their families to migrate to cities, where jobs that fit their skills are scarce, where they must purchase their food, and where the only affordable housing is in slums that lack essential services. Surplus skilled and semiskilled workers, especially women, often migrate abroad for employment.

The Southeast Asian Region

The region of Southeast Asia (see **Figure 10.1** on pages 416–417) has at-tracted global attention for several decades because so many of its countries had emerged poverty-stricken from World War II only to embark on a rapid journey to relative prosperity. There have been ups and downs: difficulties in finding smooth paths to stable democratic processes and institutions; urbanization at too rapid a pace; and a tendency to rely on development strategies that carry grave environmental and social side effects. Still, the region has modeled many ideas now emulated by other developing regions: rapid industrialization, expansion of the middle class, education for the masses, empowerment for women, improvement of food security, public health care, and slower population growth.

The nine thematic concepts in this book are explored as they arise in the discussion of regional issues, with interactions between two or more themes featured, as in the geographic insights above. Vignettes, like the one that follows about the rights of a group of people indigenous to the state of Sarawak on the island of Borneo, illustrate one or more of the themes as they are experienced in individual lives.

GLOBAL PATTERNS, LOCAL LIVES In December 2005, a group of indigenous people in the Malaysian state of Sarawak, on the island of Borneo (see the Figure 10.1 map), attended a public meeting wearing orangutan masks. They carried signs informing onlookers that, although the government protects natural areas for orangutan, it ignores the basic right of indigenous people to live on their own ancestral lands and practice their forest skills (**Figure 10.2A**).

Over the years, Sarawak forest dwellers have tried many tactics to save their lands from deforestation by logging companies preparing to expand oil palm plantations (see Figure 10.2B). Palm oil is a major food and cosmetic export; it is sold primarily to Europe,

FIGURE 10.2 Indigenous people in Sarawak, Borneo.

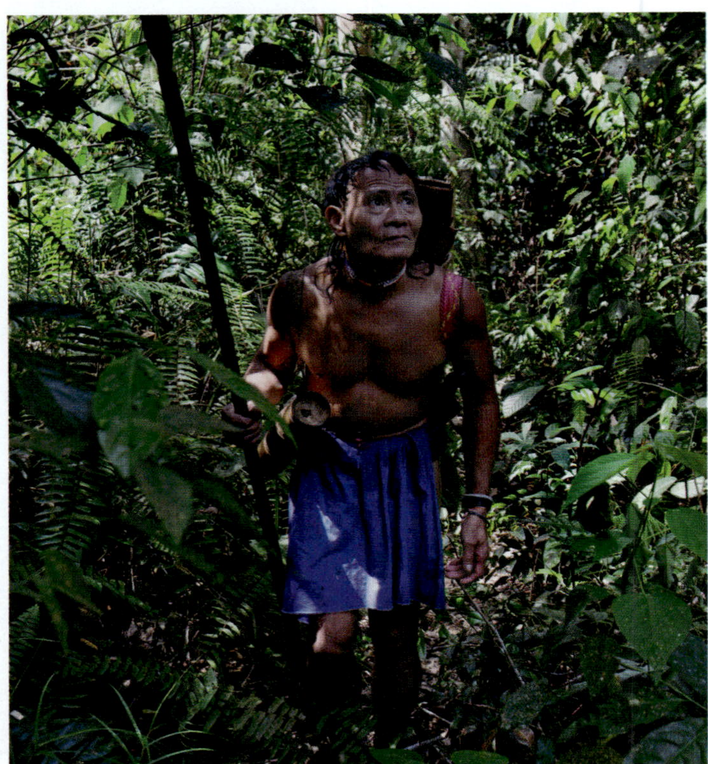

(A) An indigenous man in Sarawak hunts in the forests of Malaysian Borneo.

(B) Palm oil plantations in Sarawak.

In the mid-twentieth century, the
licensing logging companies to cut
indigenous peoples (**Figure 10.3**)
0s, 90 percent of Sarawak's
d and 30 percent had been clear-
replaced by oil palm plantations.
ple found that even on uncleared
nile, streams and rivers became
l by fertilizers and pesticides
, virtually none of the profits from
nunities affected by the conversion

ns in Berkeley, California, who
s of the deforestation in Sarawak,
offered to become a "sister
arawak, the Uma Bawang, giving
the help of the Borneo Project,
nity-based mapping project in
and-tape techniques, they began
iological content of their forest
home. Since then, indigenous people from across Sarawak have learned how to use global positioning systems (GPS), geographic information systems (GIS), and satellite imagery to make more sophisticated maps. The power of these maps was demonstrated in 2001, when they helped win a precedent-setting court case that protected indigenous lands from encroachment by oil palm plantation.

This favorable court ruling was challenged in 2005, when the government appealed, but in 2009, the Malaysian federal court ruled in favor of the Uma Bawang. The decision stated that native customary land rights had been protected since 1939, when British colonial officials had directed the district lands and survey departments to map the boundaries of native lands. Now indigenous people have the right to sue the government for past illegal leases to logging companies; as of 2011, some 203 such cases were in litigation. The outcome of this litigation is as yet unclear; in the Indonesian part of Borneo, deforestation for oil palm plantations is proceeding rapidly, as shown in the 2011 video, "Indigenous Community Witnesses End of Forest for Palm Oil," at http://tinyurl.com/d5pmr49. *[Source: The Borneo Wire. For detailed source information, see Text Credit pages.]* ■

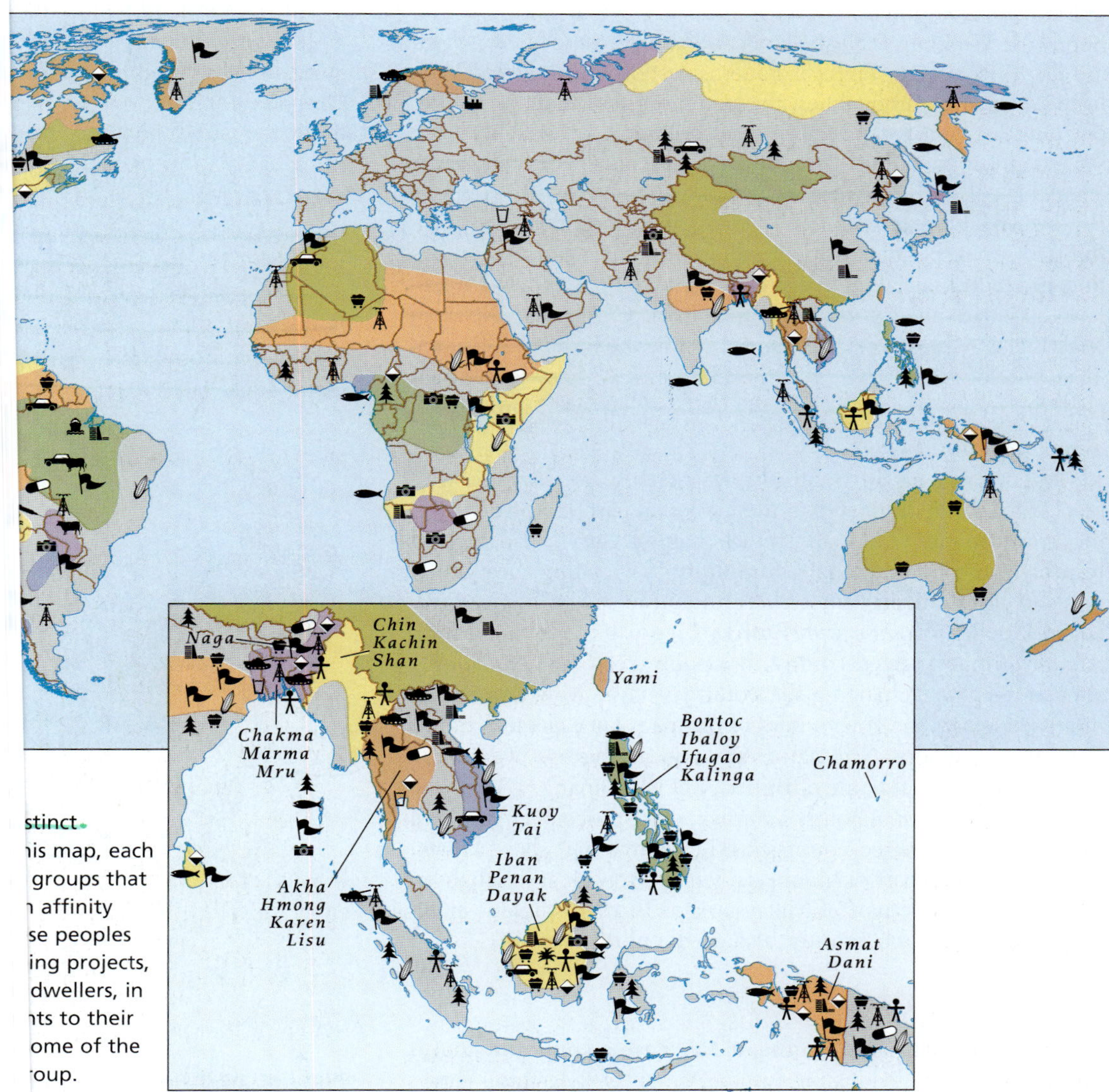

F
v
ir
c
a
t
h
si
o
tr
g

stinct-
is map, each
groups that
affinity
se peoples
ing projects,
dwellers, in
nts to their
ome of the
oup.

This account of the tactics that indigenous groups are using to secure their rights to control and manage their ancestral lands complements the documentation of similar struggles in Middle and South America (page 110) and highlights a number of themes in this book: globalization, power and politics, development, and access to food and water. As the interplay of these themes becomes better understood, issues that previously seemed to have only local significance are shown to be international or even global in nature. This connection of the local to the global presents both challenges and opportunities for the world's indigenous people (see Figure 10.3). On the one hand, the global market provides the demand for timber and palm oil; on the other, it is support from the global arena that enables efforts like the Borneo Project to help communities like that of the Uma Bawang to have their land claims validated. In fact, mapping as a tactic for helping indigenous groups secure their legal rights to ancestral lands has now become a global phenomenon. Hundreds of indigenous groups throughout the world are collaborating with mapping specialists, many of them geographers, often with the result that their land rights are for the first time formally recognized by governments. In 2007, the United Nations passed the Declaration on the Rights of Indigenous People; in 2010, President Obama signed the declaration for the United States, which previously had voted against it. The United Nations' *State of the World's Indigenous Peoples* report documents several of these efforts. It is available online at http://tinyurl.com/ybpfgpp.

THINGS TO REMEMBER

- Indigenous land claims are no longer just local issues. Indigenous groups have shifted their efforts to secure their rights to ancestral lands to the global scale, highlighting a trend in which many issues that were once local or national are now becoming global concerns.

THE GEOGRAPHIC SETTING

What Makes Southeast Asia a Region?

Southeast Asia is physically a manifestation of tectonic forces, which are described in the Physical Patterns section that follows. Aside from physical commonalities, the peninsula and island countries of Southeast Asia today share a deep cultural past, related to but separate from the cultures of southwestern China. They (all except Thailand) also share more recent experiences with European colonialism. During and after World War II, most countries went through severe political turmoil followed by the rapid modernization and industrialization that stretches into the present. The mainland countries are Burma, Thailand, Laos, Cambodia, Vietnam, Malaysia, and Singapore; the island countries are Indonesia, Brunei, Timor-Leste (East Timor), and the Philippines. Malaysia occupies part of the peninsula and parts of the island of Borneo (Figure 10.4).

Terms in This Chapter

Many governments in Southeast Asia choose to dispense with place-names that originated in their colonial past. However, when the governments that make these changes earn broad lack of respect in the international community by violating the human rights of their citizens, their chosen name may not be acknowledged. Such is the case with Burma (Figure 10.4), where a military government seized control in a coup d'état in 1990, shortly after an election, changing the country's name to Myanmar. As discussed below, the government of Burma is once again in transition; in press reports, both names are used. In this text, we use the country's traditional name of Burma, not Myanmar.

Another potential point of confusion is Borneo, a large island that is shared by three countries. The part of the island known as Kalimantan is part of Indonesia; while Sarawak and Sabah on the north coast are part of Malaysia; and Brunei is a very small, independent, oil-rich country, also on the north coast.

Physical Patterns

The physical patterns of Southeast Asia have a continuity that is not immediately obvious on a map. A map of the region shows a unified mainland region that is part of the Eurasian continent, and a vast and complex series of islands arranged in chains and groups. These landforms are actually related in tectonic origin. Climate is another source of continuity, with most of the region tropical or subtropical.

Landforms

Southeast Asia is a region of peninsulas and islands (see Figure 10.1). Although the region stretches over an area larger than the continental United States, most of that space is ocean; the area of all the region's landmass amounts to less than half that of the contiguous United States. Burma, Thailand, Laos, Cambodia, and Vietnam occupy the large Indochina peninsula

FIGURE 10.4 Political map of Southeast Asia.

arc
of is

t extends south of China. This
insula itself sprouts the long,
Malay Peninsula that is shared
nd Thailand, a main part of
Singapore, which is built on a
tip. The **archipelago** (a series of
out to the south and east of the
ountries of Indonesia, Malaysia,
nor), and the Philippines (see
has some 17,000 islands, and the

landforms of the Southeast
go are the result of the same
shed when India split off from
lly collided with Eurasia (see
result of this collision, which
inous folds of the Plateau of
t 20,000 feet (6100 meters).
bend out of the high plateau,
. There, they descend rapidly
e Indochina peninsula. Deep
at stretch toward the sea, each
0 feet (600 to 900 meters) and
e mountains of the Yunnan–
he Andaman Sea, the Gulf of
a Sea (see Figure 10.1A and
ivers of the Indochina penin-
Salween in Burma; the Chao
g, which flows through Laos,
Black and Red rivers of northern
have major delta formations—especially the Irrawaddy, Chao Phraya, and the Mekong—that are intensively cultivated and settled (see Figure 10.1B).

The curve formed by Sumatra, Java, the Lesser Sunda Islands (from Bali to Timor), and New Guinea conforms approximately to the shape of the Eurasian Plate's leading southern edge (see Figure 1.25 on page 46). Where the Indian-Australian Plate plunges beneath the Eurasian Plate along this curve, hundreds of earthquakes and volcanoes occur, especially on the islands of Sumatra and Java (see Figure 10.1D, E). There are also volcanoes and earthquakes in the Philippines, where the Philippine Plate pushes against the eastern edge of the Eurasian Plate and is in turn pushed by the Pacific Plate. The volcanoes of Southeast Asia are considered part of the Pacific Ring of Fire (see Figure 1.26 on page 47; see also Figure 10.1C, D).

Volcanic eruptions and the mudflows and landslides that follow eruptions complicate and endanger the lives of many Southeast Asians. Earthquakes are especially problematic because of the tsunamis they can set off (see Figure 10.1E). The tsunami of December 2004, triggered by a giant earthquake (9.1 in magnitude) just north of Sumatra, swept east and west across the Indian Ocean, taking the lives of 230,000 people (170,000 in Aceh Province, Sumatra) and injuring many more. It was one of the deadliest natural disasters in recorded history. There has been a series of strong earthquakes since, with several along coastal Sumatra in August and September of 2009, April and May of 2010, and April of 2012; but they did not generate notable tsunamis. Over the long run, the volcanic material creates new land and provides minerals that enrich the soil for farming.

The now-submerged shelf of the Eurasian continent, known as Sundaland (Figure 10.5) extends under the Southeast Asian peninsulas and islands. It was above sea level during the recurring

FIGURE 10.5 Sundaland 18,000 years ago, at the height of the last ice age. The now-submerged shelf of the Eurasian continent that extends under Southeast Asia's peninsulas and islands was exposed during the last ice age and remained above sea level until about 16,000 years ago, when that ice age was ending.

ice ages of the Pleistocene epoch, when much of the world's water was frozen in glaciers. The exposed shelf allowed ancient people and Asian land animals (such as elephants, tigers, rhinoceroses, and orangutans; Figure 10.6) to travel south to what, when sea levels rose, became the islands of Southeast Asia.

Climate and Vegetation

The largely tropical climate of Southeast Asia is distinguished by continuous warm temperatures in the lowlands—consistently above 65°F (18°C)—and heavy rain (Figure 10.7). The rainfall is the result of two major processes: the monsoons (seasonally shifting winds) and the movement of the intertropical convergence zone (ITCZ), an area centered roughly around the equator where surface winds converge from the northern and southern hemispheres and rise upward, resulting in rainfall. The wet summer season extends from May to October, when the warming of the Eurasian landmass sucks in moist air from the surrounding seas and pulls the ITCZ northward (see the Figure 10.7 map). Between November and April, there is a long dry season on the mainland, when the seasonal cooling of Eurasia causes dry, cool air from the interior of the continent to flow out toward the sea, pushing the ITCZ southward (see Figure 11.5 on page 464). On the many islands, however, the winter can also be wet because the air that flows from the continent warms and picks up moisture as it passes south and east over the seas. The air releases its moisture as rain after ascending high enough over elevated landforms to cool. With rains coming from both the monsoon and the ITCZ, the island part of Southeast Asia is one of the wettest areas of the world.

Irregularly every 2 to 7 years, the normal patterns of rainfall are interrupted, especially in the islands, by the El Niño phenomenon (see Figure 11.6 on page 465). In an El Niño event, the usual patterns of air and water circulation in the Pacific are reversed. Ocean temperatures are cooler than usual in the western Pacific near Southeast Asia. Instead of warm, wet air rising and condensing as rainfall, cool, dry air sits on the ocean surface. The result is severe drought, often with catastrophic results for farmers and for tinder-dry forests that regularly catch fire. The cool El Niño air can trap smoke and other pollutants at Earth's surface, creating unusually toxic smog and such low visibility that planes have sometimes crashed.

The soils in Southeast Asia are typical of the tropics. Although not particularly fertile, they will support dense and prolific vegetation when left undisturbed for long periods. The warm

FIGURE 10.6 LOCAL LIVES **PEOPLE AND ANIMALS IN SOUTHEAST ASIA**

A A cat at a temple in Chiang Mai, Thailand. Considered the guardians of statues in temples throughout Thailand, cats are written about in ancient poetry and celebrated by Thailand's royal family. Stray cats are given refuge at certain temples.

B Clownfish swim amid sea anemones that capture prey with paralyzing toxins. Immune to these toxins, the clownfish eat tiny animals that could harm the anemone; in return, they are provided with a safe home. Clownfish, whose natural habitat is this region's saltwater, have also been successfully bred for aquarium tanks.

C A pig is decorated and paraded around a village in Vietnam during Tet, the Vietnamese New Year. Pork is especially prized in Vietnam and is an essential part of holiday meals. This pig will be served as an offering to village gods.

Photo Essay: Climates of Southeast Asia

...pical air, combined with the movements of the intertropical convergence zone (ITCZ) across the ...ast Asia one of the wettest regions in the world.

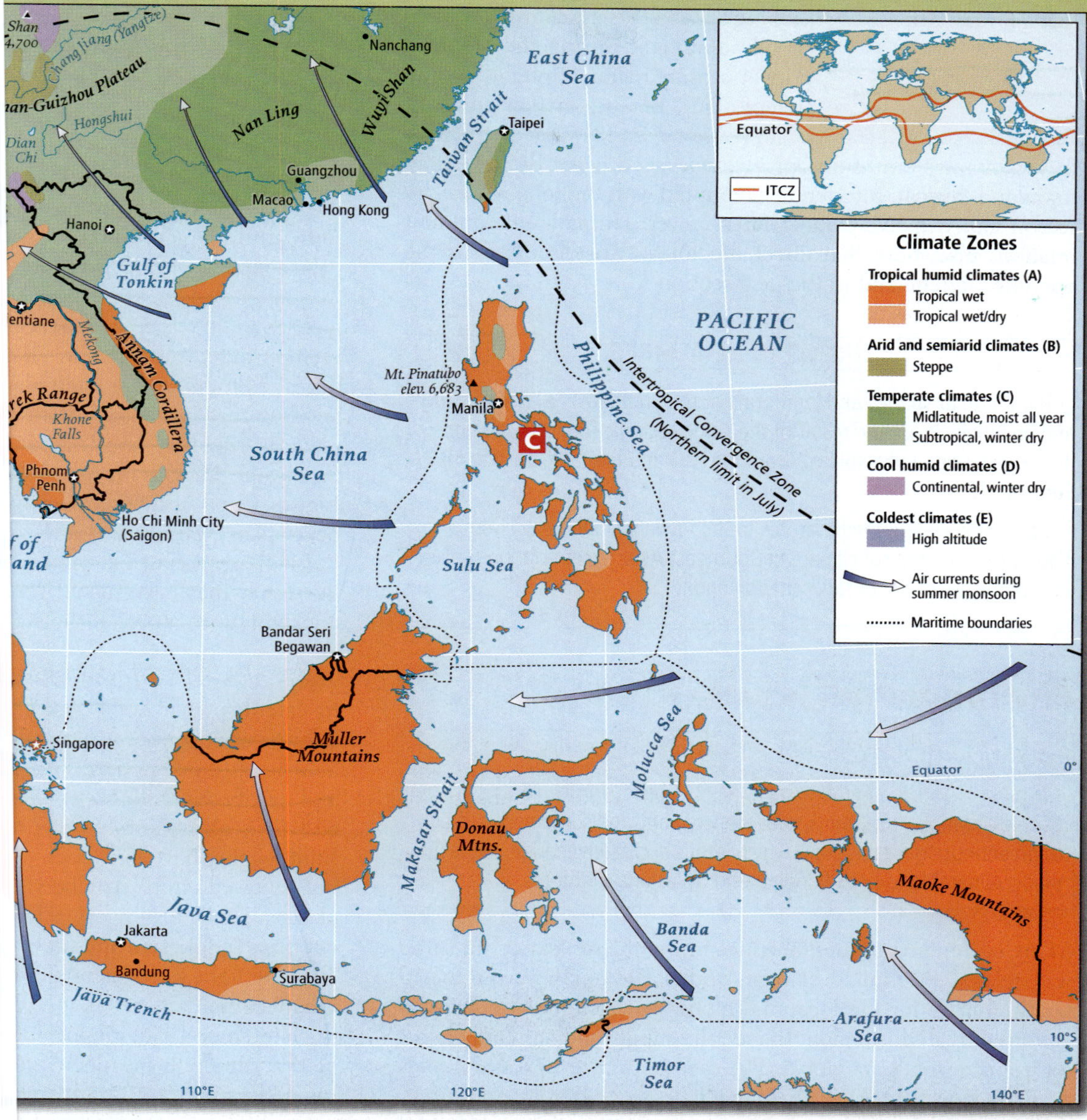

Climate Zones

Tropical humid climates (A)
- Tropical wet
- Tropical wet/dry

Arid and semiarid climates (B)
- Steppe

Temperate climates (C)
- Midlatitude, moist all year
- Subtropical, winter dry

Cool humid climates (D)
- Continental, winter dry

Coldest climates (E)
- High altitude

Air currents during summer monsoon

Maritime boundaries

Equator

ITCZ

Shan 4,700
Chang Jiang (Yangtze)
...an-Guizhou Plateau
Hongshui
Dian Chi
Nanchang
Nan Ling
Wuyi Shan
East China Sea
Taiwan Strait
Taipei
Guangzhou
Macao
Hong Kong
Hanoi
Gulf of Tonkin
...entiane
Mekong
Annam Cordillera
...rek Range
Khone Falls
Phnom Penh
Ho Chi Minh City (Saigon)
...f of ...and
South China Sea
Mt. Pinatubo elev. 6,683
Manila
Philippine Sea
PACIFIC OCEAN
Intertropical Convergence Zone (Northern limit in July)
Sulu Sea
Bandar Seri Begawan
Singapore
Muller Mountains
Molucca Sea
Makasar Strait
Donau Mtns.
Maoke Mountains
Java Sea
Jakarta
Bandung
Surabaya
Java Trench
Banda Sea
Arafura Sea
Timor Sea
110°E
120°E
140°E
0°
10°S

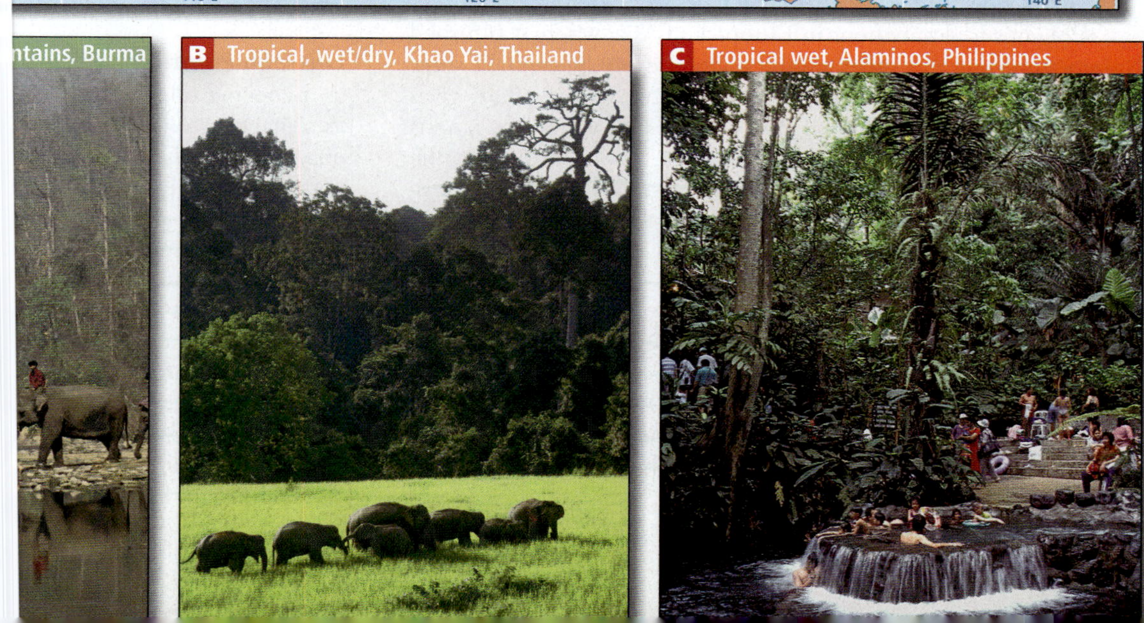

A ...ntains, Burma

B Tropical, wet/dry, Khao Yai, Thailand

C Tropical wet, Alaminos, Philippines

temperatures and damp conditions promote the rapid decay of **detritus** (dead organic material) and the quick release of useful minerals. These minerals are taken up directly by the roots of the living forest rather than enriching the soil. Because rainfall is usually abundant during the summer wet season (when drought is only episodic), this region has some of the world's most impressive forests in both tropical and subtropical zones (see Figure 10.7). On the mainland, human interference, coupled with the long winter dry season can negatively affect forest cover. Human–environment relations elsewhere that affect the forests are discussed in the opening vignette and in the next section.

detritus dead organic material (such as plants and insects) that collects on the ground

THINGS TO REMEMBER

- The irregular shapes and landforms of the Southeast Asian mainland and archipelago are the result of the same tectonic forces that were unleashed when India split off from the African Plate and crashed into Eurasia.
- Continuous warm temperatures in the lowlands and heavy rain distinguish the tropical climate of Southeast Asia, except during those irregular years of the El Niño phenomenon.

Environmental Issues

Geographic Insight 1

Climate Change, Food, and Water: Conversion of forests to oil palm plantations, rice farms, subsistence plots, and settlements all add to greenhouse gas emissions, which intensify climate change and leave many vulnerable to both flooding and occasional droughts.

Many of the environmental issues in Southeast Asia are in some way related to climate change, and the temptation may be to lay the blame for the local and global environmental effects on actions only in this region. In fact, the supply and demand chains for tropical woods are global. The map in Figure 10.8 and photos A and B show the supply chain that takes wood from various tropical regions to destinations in Europe and Asia, where it is turned into furniture and other products. Figures 10.8C and D show who is importing and exporting sawn wood; Figure 10.8E gives an approximation of the different activities for which the land is deforested legally. It is extremely difficult, however, to know the extent of illegal deforestation. The World Wildlife Fund estimated that in 2004, more than 80 percent of timber production in Indonesia alone was illegal.

The regional environmental impacts of deforestation and the agriculture that often follows it are covered in Figure 10.9 on page 426. One effect is that indigenous people of the forest lose their living space and resources when the land is given over for export agriculture. Another is the loss of habitat for orangutans, the Sumatran tiger, the Sumatran rhinoceros, and tens of thousands of other, less spectacular species that are lost when rain forests are converted to agricultural uses. Finally, the emissions of greenhouse gases that accompany deforestation processes foster climate change at a global scale. **222. INDONESIA'S POLLUTED ENVIRONMENT THREATENS HAWKSBILL TURTLE**

Climate Change and Deforestation

We know that deforestation is a major contributor to global climate change because enormous amounts of carbon dioxide (CO_2) are released when forests are removed and the underbrush is burned. Fewer trees also mean that less CO_2 is absorbed from the atmosphere. According to 2010 United Nations Food and Agriculture Organization (UNFAO) reports, although Southeast Asia contains only 5 percent of the world's forests, in the last 10 years it has accounted for nearly 25 percent of deforestation globally. The region has the world's second-highest rate of deforestation, after sub-Saharan Africa. Every day, 13 to 19 square miles (34 to 50 square kilometers) of Southeast Asia's rain forests are destroyed, much of it done illegally. A great deal of this takes place in Malaysia and Indonesia, where the logging of tropical hardwoods for sale on the global market is a major activity (see Figure 10.8). Additionally, when the forests are converted to oil palm plantations, especially when waste from tree removal is burned, significant amounts of CO_2 are released. For decades, these activities have made Indonesia among the world's biggest contributors to global climate change.

The Costs of Logging, Legal and Illegal It is now possible to calculate that deforestation costs Southeast Asian countries billions of dollars per year in lost forest services, such as renewable food, construction, and fuel resources; rainwater capture and purification; flood amelioration; biodiversity preservation; and CO_2 absorption. Efforts to use government regulations to limit logging have long been constrained by rampant corruption, but renewed efforts may be yielding better results. With a vision of sustainable logging as a long-term source of income for his country, Indonesian president Susilo Bambang Yudhoyono has lead the fight to reduce deforestation (see "On the Bright Side: Challenging Deforesters" on page 427). But the president is starting late; since 1950, tropical forest cover on the Indonesian island of Borneo alone has decreased by about 60 percent. Oil palm plantations now constitute the predominant use of the land there.

Oil Palm Plantations Palm oil, which is used around the world as a cooking oil, in food products, in soaps and cosmetics, and as a machine oil, is at the center of debates over global climate change in Southeast Asia. The oil palm originated in West Africa and was brought to Southeast Asia early in the colonial era. Today, Indonesia and Malaysia are the world's largest palm oil producers, and the expansion of their oil palm plantations has resulted in extensive deforestation. Much of the deforestation is in lowland peat swamps, which in a natural state are capable of storing large amounts of carbon in their soils. To create space for oil palm plantations, the swamps are drained and most of their vegetation is burned (see Figure 10.9A). Often the fires spread underground to the peat beneath the forests, smoldering for years after the surface fires have been extinguished. These subsurface fires release enormous amounts of carbon into the atmosphere.

timber. Most of the consumer demand for tropical timber is in North America, Europe, and … of tropical timber sold to China are made into furniture, plywood, and flooring, which are …loped world. The trade shown on the map is mostly legal, but much of the wood that fuels … is harvested illegally in Southeast Asia.

…orneo.

(B) Tropical hardwoods are made into toy train tracks in Dongguan, China.

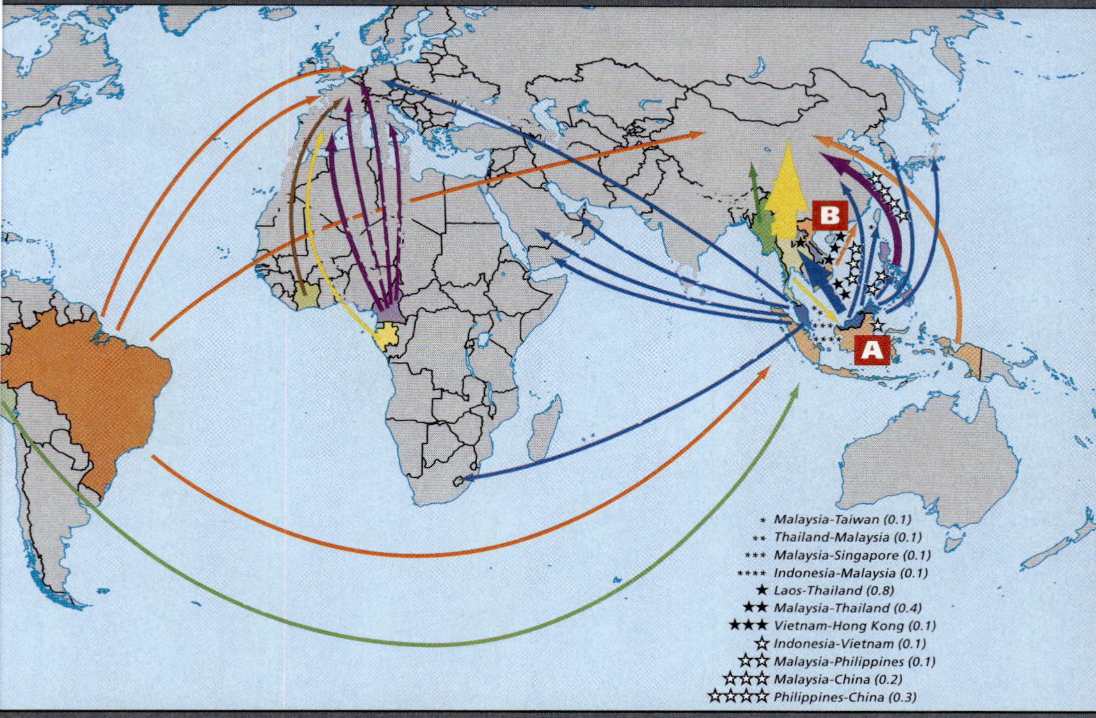

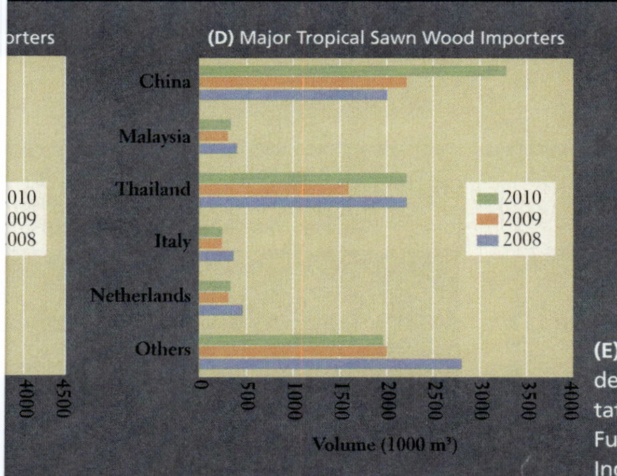

Sources of Legal Deforestation

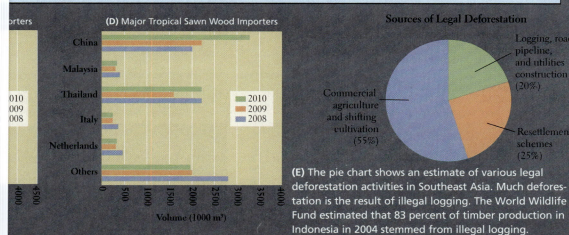

Logging, road, pipeline, and utilities construction (20%)

Commercial agriculture and shifting cultivation (55%)

Resettlement schemes (25%)

(E) The pie chart shows an estimate of various legal deforestation activities in Southeast Asia. Much deforestation is the result of illegal logging. The World Wildlife Fund estimated that 83 percent of timber production in Indonesia in 2004 stemmed from illegal logging.

…ly

…when there are fewer trees due to legal or illegal logging (see photo A)?

Figure 10.9 Photo Essay: Human Impacts on the Biosphere in Southeast Asia

Deforestation and the conversion of land to agricultural uses in Southeast Asia are having a profound impact both on regional ecosystems and on the entire biosphere.

A A worker sprays a palm oil plantation with pesticides in Puntianai, on the island of Sumatra in Indonesia. Deforestation in much of Malaysia and Indonesia is now driven by the expansion of oil palm plantations.

B A wild female orangutan and her baby eat flowers in Tanjung Puting National Park, Kalimantan, Indonesia. The deforestation and fragmentation of habitats caused by roads have made orangutans an endangered species.

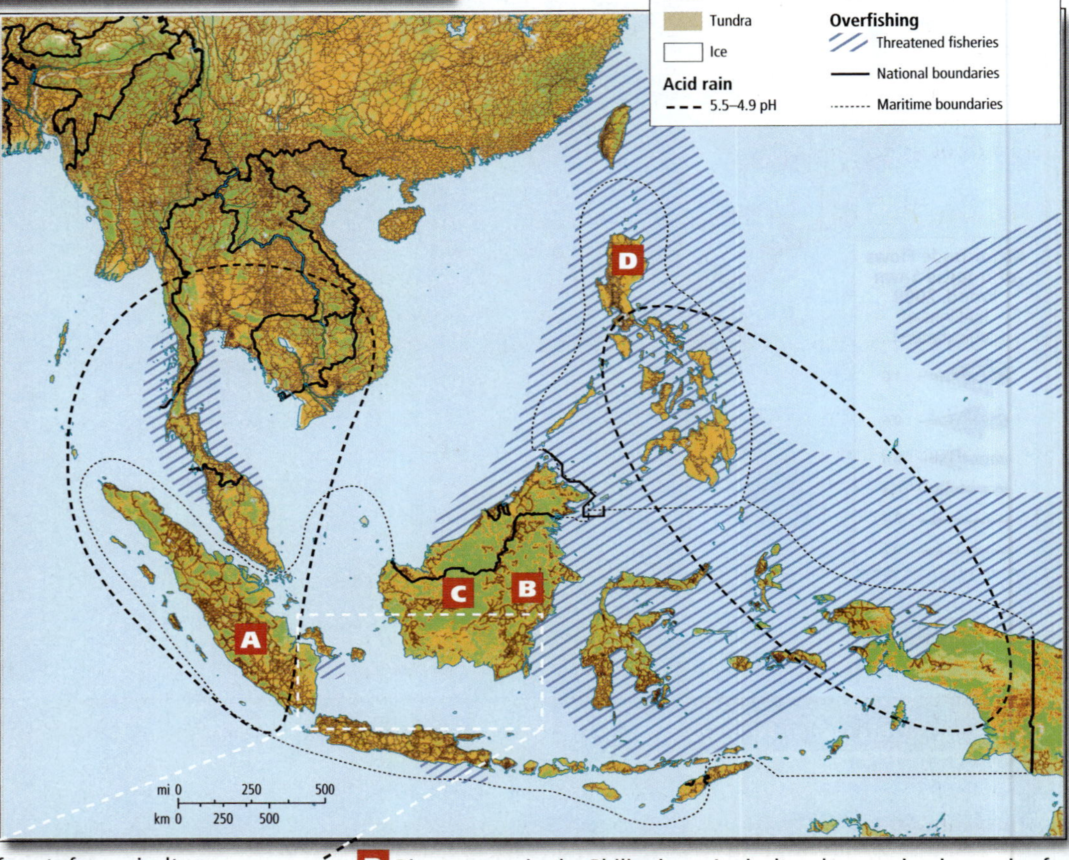

C A satellite image of fires used to clear forests for agriculture in Indonesia and Malaysia. The CO_2 produced contributes to climate change and degrades local air quality.

D Rice terraces in the Philippines. Agricultural expansion is a major force behind deforestation in the region, and wet rice agriculture, because it releases methane, is a significant contributor of greenhouse gases.

In recent years, smoke from burning forests and peat has periodically covered much of the region, dramatically reducing air quality, even forcing rural people to wear masks.

There have been efforts to promote palm oil as a potential solution to global climate change because it can be converted into fuel for automobiles. The claim is that because the palm trees absorb CO_2 from the atmosphere just as the original forest did, palm oil could be considered a "carbon neutral" fuel. This is a fallacy because palm trees do not absorb carbon at a rate equivalent to natural rain forests—and if forests are burned to clear land for plantations, it can take decades to offset the initial carbon emissions produced by deforestation. Not
on by growing oil palms could
rough the burning of subsurface
for oil palm planting.

od Production

lobal climate change in two ways.
deforestation, and second, some
uce significant greenhouse gases.

wn as *slash-and-burn* or *swidden*
as been practiced sustainably for
d uplands of mainland Southeast
ands (Figure 10.10B). To main-
et environments where nutrients
al farmers move their fields every
fallow for 15 years or more. The
d fields not only regenerates the
sorbs significant amounts of CO_2
if fallow periods are shortened or
ility can collapse, making future
cases, fertility can be temporarily
mical or organic fertilizers, but
ost farmers and chemical fertil-
ear or be too unhealthy for food
e of forest for too long eventually
nbaked clay called *laterite*.
are relatively low, subsistence
shifting cultivation indefinitely,
riods and still support human
larger areas than other types of
ty increases, farmers are usually

ON THE BRIGHT SIDE

Challenging Deforesters

Indigenous peoples, threatened by rapacious deforestation, are using new technologies to challenge corporations that are trying to appropriate their resources. Leaders like President Yudhoyono of Indonesia are attempting to enforce existing regulations against deforestation. Yudhoyono has shifted the focus from arresting low-level workers (such as laborers and truck drivers) to removing and fining provincial forestry officials and military personnel, who have been running illegal logging operations for years. This move has cost Yudhoyono politically, and although deforestation has decreased somewhat, it is difficult to measure the decrease accurately.

forced to shorten fallow periods, thus inhibiting forest regrowth and soil regeneration. Even though individual plots are small, because shifting cultivation requires clearing forest at each move, eventually the plots become close to one other. Shifting cultivation now accounts for a significant portion of the region's deforestation. Moreover, burning associated with shifting cultivation can result in wildfires. This is especially true during an El Niño period, when rainfall is low. These wildfires have increased in recent years, particularly in Indonesia, further contributing to deforestation and carbon emissions there.

Wet Rice Cultivation Another major contributor to global climate change is Southeast Asia's most productive form of agriculture. *Wet rice cultivation* (sometimes called *paddy rice*) entails planting rice seedlings by hand in flooded and often terraced fields that are first cultivated with hand-guided plows pulled by water buffalo or tractors (see Figure 10.9D). Wet rice cultivation has transformed landscapes throughout Southeast Asia. It is practiced throughout this generally well-watered region, especially on rich volcanic soils and in places where rivers and streams bring a yearly supply of silt.

The flooding of rice fields also results in the production of methane, a powerful greenhouse gas responsible for about 20 percent of global climate change. It is estimated that up to one-third of the world's methane is released from flooded rice fields where organic matter in soil undergoes fermentation as oxygen supplies are cut off. Wet rice has been cultivated for thousands of years, but the increase in human population in the last 25 years has driven a 17 percent expansion of the area devoted to wet rice. The map in Figure 10.10 shows patterns of field and forest crops across the region.

Commercial Agriculture During the recent decades of relative prosperity, small farms once operated by families have been combined into large commercial farms owned by local or multinational corporations. These farms produce cash crops for export, such as rubber, palm oil, bananas, pineapples, tea, and rice (see Figure 10.10A and the figure map). Commercial farming on large tracts of deforested land, using mechanization, irrigation, and chemicals for fertilizer and pest control, reduces the need for labor. The objectives of commercial farming are to generate big yields and quick profits, not to develop long-term, sustainable agriculture. Many commercial farmers have achieved dramatic

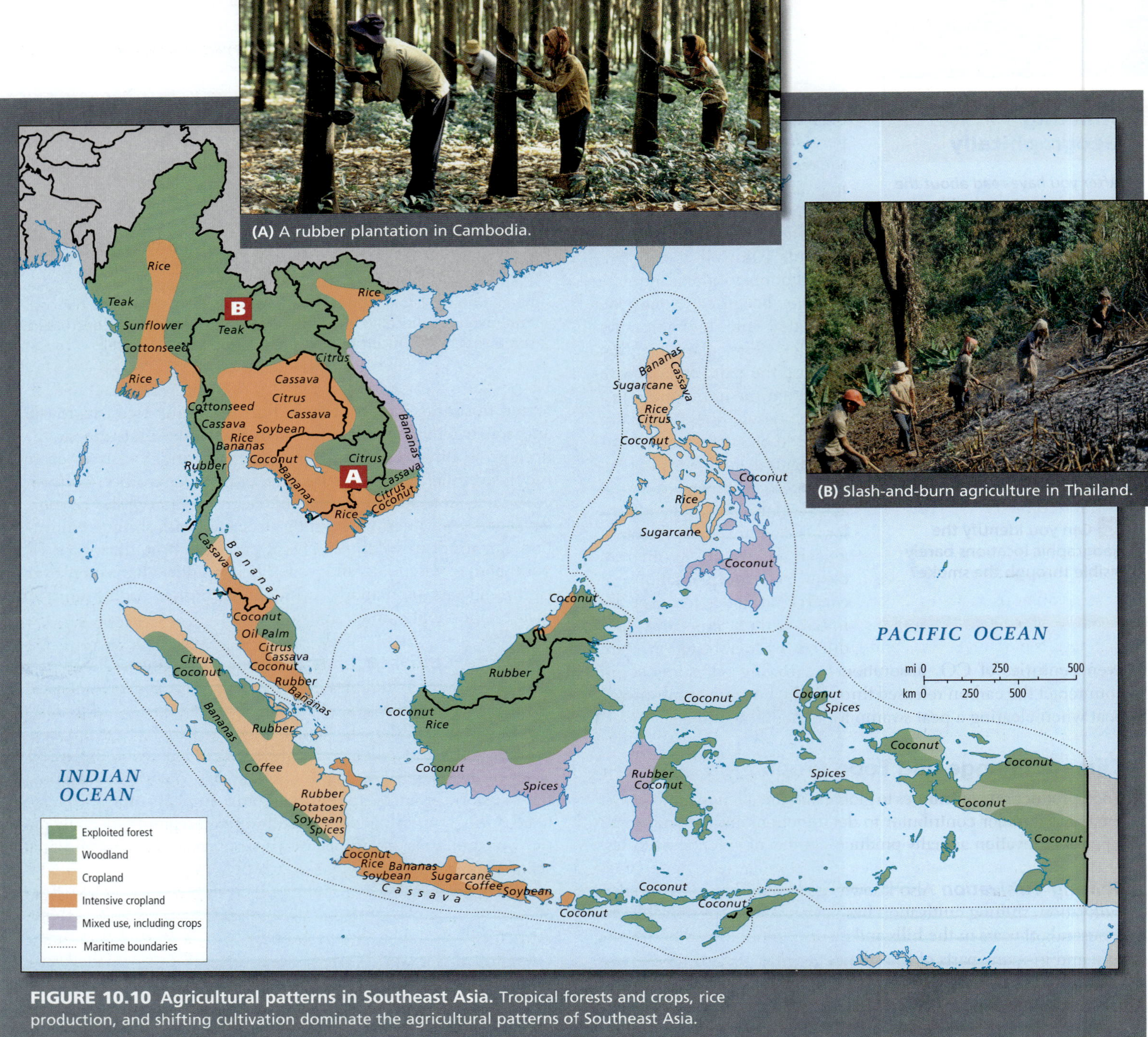

FIGURE 10.10 Agricultural patterns in Southeast Asia. Tropical forests and crops, rice production, and shifting cultivation dominate the agricultural patterns of Southeast Asia.

Thinking Geographically

On plantations of all types, native plants are replaced by marketable crop plants. What might be a way in which using slash-and-burn agriculture helps farmers avoid this particular impact (see photo B)?

boosts in harvests (especially of rice) by using high-yield crop varieties, the result of green revolution research that has been applied in many parts of the world (see Chapter 8 on pages 361–362).

As we have noted in relation to other world regions, large-scale commercial farming has significant negative environmental effects, including the loss of wildlife habitat and hence biodiversity (see Figure 10.9B); increases in soil erosion, flooding, and chemical pollution; and depletion of groundwater resources. In addition, poor and subsistence farmers, unable to afford the new technologies, find they cannot compete with large commercial farms and are forced to migrate to cities to look for work.

Climate Change and Water

Many of Southeast Asia's potential vulnerabilities to global climate change are related to water resources. Four areas of vulnerability may affect the region's economy and food supply: glacial melting, increased evaporation, coral reef bleaching, and storm surge flooding (Figure 10.11 on page 430).

ased Evaporation Like much ia's largest rivers (the Irrawaddy, ɛrs) are fed during the dry season glaciers high in the Himalayas. at a rate that could result in their

ıtes, the immediate risk is cata- areas have long been adapted to have occurred in this region for ɛas are unprepared for floods (see

reduced dry-season flows in the has been the primary source of cent of this region's rice harvest s of the major rivers. The loss of y farmers' incomes. In addition, rtages in cities. Coming on top n recent years, this would place of poor people throughout the

ssociated with current trends in oration rates will rise, resulting in ake levels, and lower fish catches or aquatic animals. Evaporation groundwater levels can also cause and freshwater aquifers.

climate change is expected to theast Asia, threatening the coral region's fishing and tourism. A e composed of the calcium-rich ng creatures called coral polyps are subject to **coral bleaching** (see nich results when photosynthetic xpelled by a variety of human or ing water temperature related to nic acidification resulting from the atmosphere, can both cause is thought to have increased 340 years, a rate of change that reatures to adapt. Bleaching also rom urban sources such as sew- ff, or industrial water pollution. leaching as well. Under normal d with just one of the bleaching s or months. However, severe e corals to die. Unprecedented ecting roughly half of the world's 92, 2004, and 2006, and regional year. **224. NEW SPECIES OF IA**

n Southeast Asia's seas depend urvival. The thousands of rural nmunities throughout coastal theast Asia that depend on fish food are thus also threatened by l bleaching. So far, however, the greatest observable impacts on people have been in the tourism industry. In the Philippines, the coral bleaching of 1998 brought a dramatic decline in tourists wanting to dive the country's usually spectacular reefs, resulting in a loss of about U.S.$30 million to the economy.

Storm Surges and Flooding Although the relationship is not yet entirely understood, violent tropical storms seem to be increasing as the climate warms. Normally the Philippines can expect six typhoons (known in the Atlantic as hurricanes) in a year, but in 2009, in October alone, four typhoons struck these islands. Climatologists studying data on tropical storms predict that the entire Southeast Asian region will have a somewhat higher number of typhoons over the next few years, and that, in particular, the duration of peak winds along coastal zones will increase. This is significant because many poor, urban migrants have crowded into precarious dwellings in low-lying coastal cities such as Manila, Bangkok, Rangoon, and Jakarta, where coping with high winds, flooding, and the aftermath of storms will be a common experience (see Figure 10.11A).

Responses to Climate Change

One goal of all governments in this region is to reduce the amount of fossil fuel consumption, and some are delivering on this goal despite the frequently high start-up costs of doing so. Both the Philippines and Indonesia have significant potential for generating electricity from *geothermal energy* (heat stored in Earth's crust). This energy is particularly accessible near active volcanoes, which both countries have in abundance. The Philippines already generates 27 percent of its electricity from geothermal energy; it is second only to the United States in the amount of geothermal power it generates. By some estimates, geothermal energy could eventually provide a majority of Indonesia's energy needs. Solar energy is another attractive source, given that the entire region lies near the equator, the part of the planet that receives the most solar energy. For most countries, however, wind is the most cost-effective option, especially in Laos and Vietnam, where many population centers are in high-wind areas.

THINGS TO REMEMBER

Geographic Insight 1

- **Climate Change, Food, and Water** Climate change is linked to deforestation, usually done to produce food for local and global consumers. Commercial farming of all types adds to greenhouse gas emissions, which intensify climate change. Many people in this region are vulnerable to both flooding and occasional droughts, the expected effects of climate change.
- The promise of palm oil as an alternative energy resource has proven false.
- Much of this region's rice harvest depends on the dry-season flows of the major rivers of mainland Southeast Asia, which are threatened by glacial melting.
- Climate change poses special challenges for people in low-lying cities or in typhoon zones.

cora wher the c

Figure 10.11 Photo Essay: Vulnerability to Climate Change in Southeast Asia

Much of Southeast Asia's vulnerability to climate change relates to water. Here we explore vulnerabilities related to tropical storms, flooding, and coral reefs.

A A father and his children on Mindanao, Philippines, sift through the wreckage of Typhoon Bopha in search of coconuts. Typhoons (known in the Atlantic as hurricanes) are projected to increase in frequency as a result of the warmer temperatures created by climate change.

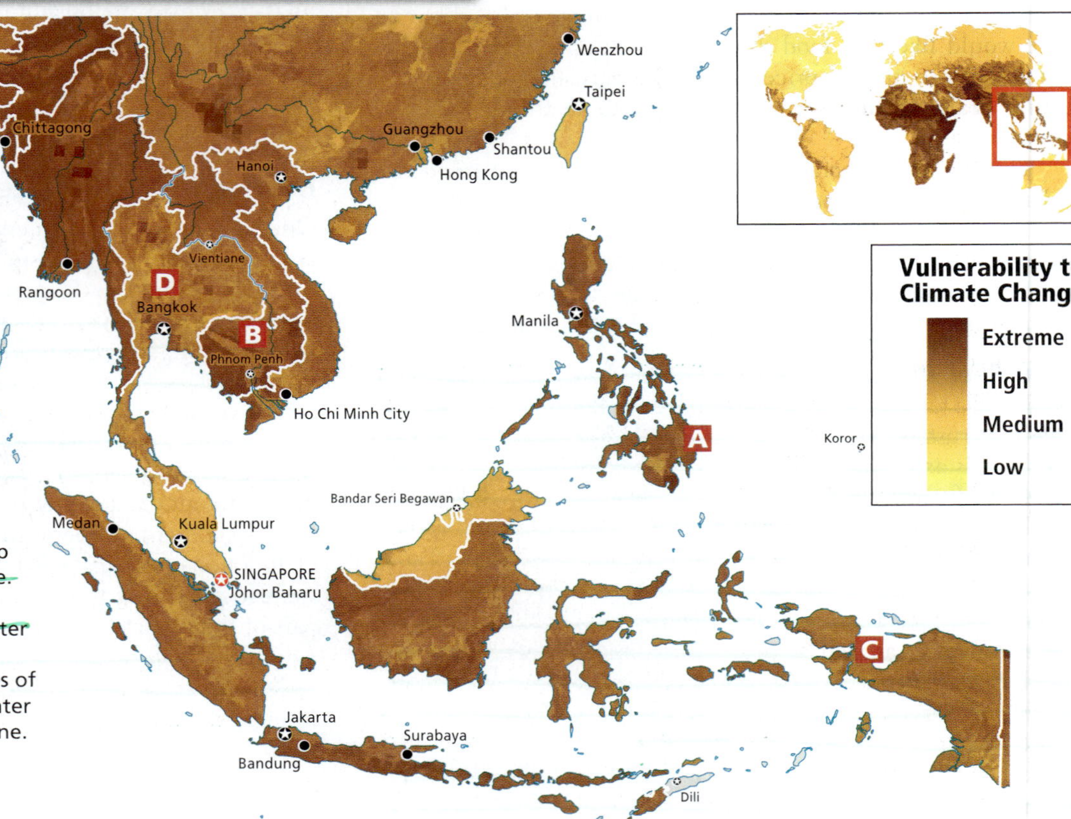

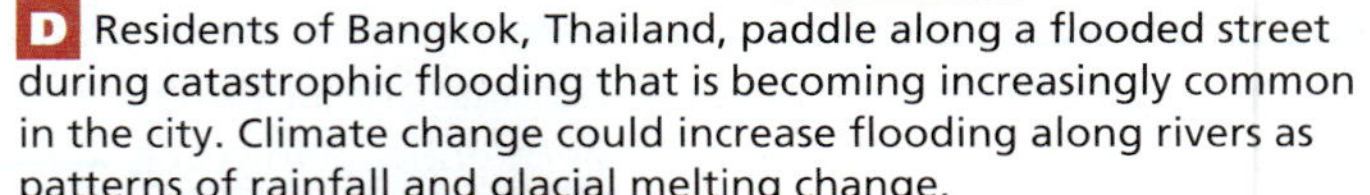

B A husband and wife harvest shrimp and fish on Cambodia's Tonlé Sap Lake. Sixty percent of Cambodia's protein intake comes from this inland freshwater lake. Higher temperatures related to climate change may be increasing rates of evaporation from the lake, causing water levels to drop and fish catches to decline.

C A partially bleached coral in Cenderawasih Bay, West Papua, Indonesia. Climate change is raising ocean temperatures, resulting in more coral bleaching events that severely degrade reefs, which are home to many aquatic life forms.

D Residents of Bangkok, Thailand, paddle along a flooded street during catastrophic flooding that is becoming increasingly common in the city. Climate change could increase flooding along rivers as patterns of rainfall and glacial melting change.

Human Patterns over Time

First settled in prehistory by migrants from the Eurasian continent, Southeast Asia was later influenced by Chinese, Indian, and Arab traders. Later still, it was colonized by Europe (from the 1500s to the early 1900s ; see Figure 10.12), and the Philippines were occupied by the United States from 1898 to 1946. During World War II, much of the region was occupied by Japan. By the late twentieth century, occupation and domination by outsiders had ended and the region was profiting from selling manufactured goods to its former colonizers.

ast Asia

ions of Southeast Asia arose from
in time. In the first migration,
, **Australo-Melanesians**, a group of
s now northern India and Burma,
f Sundaland and into present-day
l live in Indonesia's easternmost
ote pockets on other islands and
Australia, New Guinea and other
11.1 map on pages 458–459).
rom 6000 to 10,000 years ago,
n southern China began moving
ation gained momentum about
of skilled farmers and seafarers
nesians, migrated first to what is
nes, and then into the islands of
eninsula (see Figure 10.13A on
uresome sea travelers eventually
a and to Madagascar (off the east
nd eastward to the far reaches of
.13 on page 474).

ces

years, Southeast Asia has been
a steady circulation of cultural
xternal. Overland trade routes
t traders, religious teachers, and
from China and India, as well
sia. These newcomers brought
otton textiles), and food plants
) deep into the Indonesian and
roughout the mainland. The
m the west in the spring and

summer, facilitated access for merchant ships from South Asia and the Persian Gulf. The ships sailed home on winds blowing from the east in the autumn and winter, carrying spices, bananas, sugarcane, silks, and other East and Southeast Asian items, as well as people, to the wider world.

Religious Legacies Spatial patterns of religion in Southeast Asia reveal an island–mainland division that reflects the history of influences from India, China, Southwest Asia, and Europe. Both Hinduism and Buddhism arrived thousands of years ago via monks and traders who traveled by sea and along overland trade routes that connected India and China through Burma (see Figure 10.13B). Many early Southeast Asian kingdoms and empires switched back and forth between Hinduism and Buddhism as their official religion. Spectacular ruins of these Hindu–Buddhist empires are scattered across the region; the most famous is the city of Angkor in what is now Cambodia. At its zenith in the 1100s, Angkor was among the largest cities in the world, and its ruins are now a World Heritage Site (see Figure 10.16A on page 439). Today, Buddhism dominates mainland Southeast Asia, while Hinduism remains dominant only on the Indonesian islands of Bali and Lombok.

In Vietnam, people practice a mix of Buddhist, Confucian, and Taoist customs that reflect the thousand years (ending in 938 C.E.) when Vietnam was part of various Chinese empires. China's traders and laborers also brought cultural influences to scattered coastal zones throughout Southeast Asia.

Islam is now dominant in most of the islands of Southeast Asia. Islam came mainly through South Asia after India was conquered by Mughals in the fifteenth century. Muslim spiritual leaders and traders converted many formerly Hindu–Buddhist kingdoms in Indonesia, Malaysia, and parts of the southern Philippines, where Islam is still dominant. Roman Catholicism is the predominant religion in Timor-Leste and in most of the Philippines, colonized respectively by Portugal and Spain.

Australo-Melanesians groups of hunters and gatherers who moved from the present northern Indian and Burman parts of southern Eurasia to the exposed landmass of Sundaland about 60,000 to 40,000 years ago

Austronesians Mongoloid groups of skilled farmers and seafarers from southern China who migrated south to various parts of Southeast Asia between 10,000 and 5000 years ago

European Colonization

Over the last five centuries, several European countries established colonies or quasi-colonies in Southeast Asia (Figure 10.12). Drawn by the region's fabled spice trade, the Portuguese established the first permanent European settlement in Southeast Asia at the port of Malacca, Malaysia, in 1511. Although better ships and weapons gave the Portuguese an advantage, their anti-Islamic and pro-Catholic policies provoked strong resistance in Southeast Asia. Only on the small island of Timor-Leste did the Portuguese establish Catholicism as the dominant religion.

Arriving in 1521, by 1540 the Spanish had established trade links across the Pacific between the Philippines and their colonies in the Americas (see Figure 10.13C). Like the Portuguese, they practiced a style of colonial domination grounded in Catholicism, but they met less resistance because of their greater tolerance of

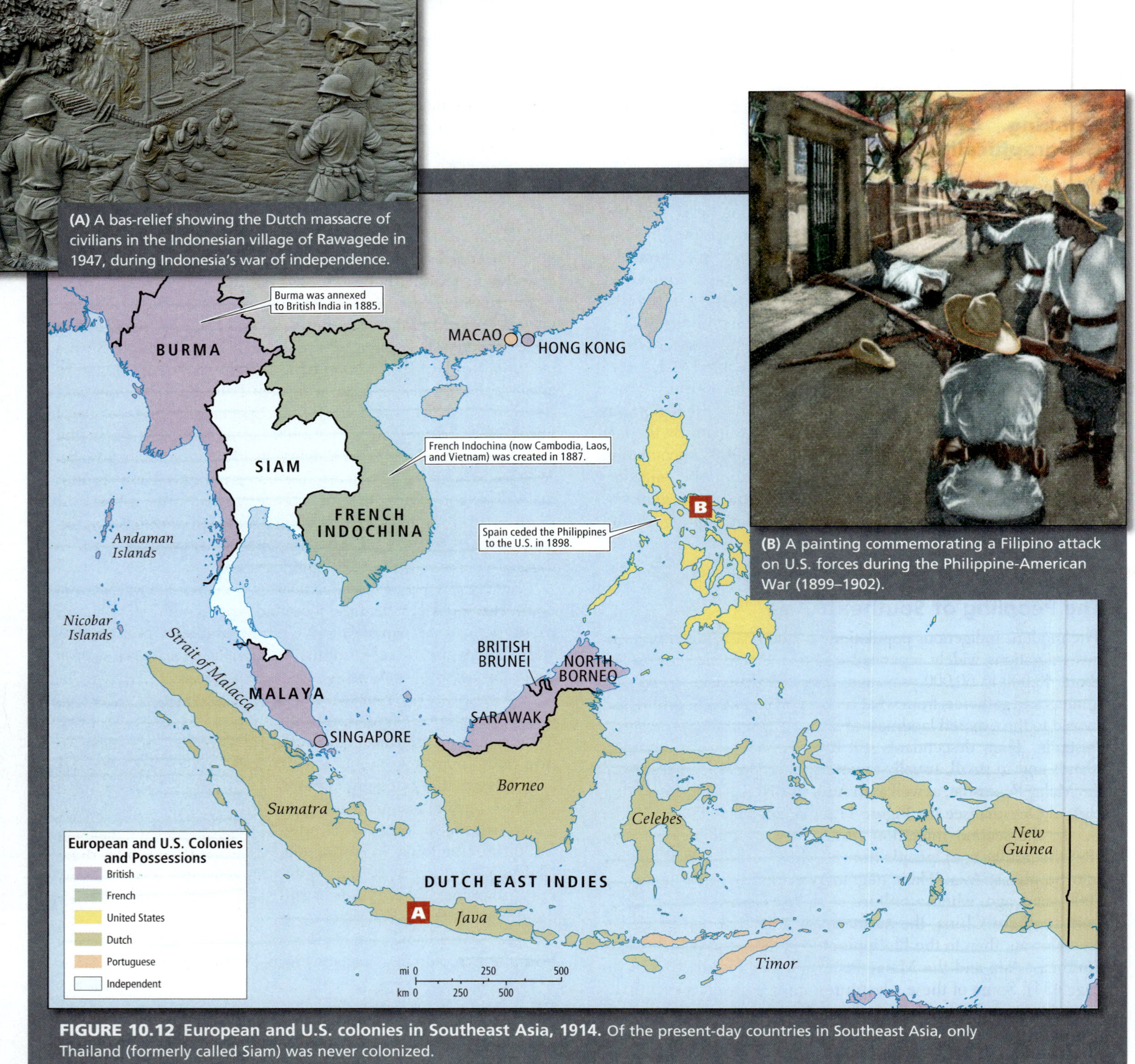

(A) A bas-relief showing the Dutch massacre of civilians in the Indonesian village of Rawagede in 1947, during Indonesia's war of independence.

(B) A painting commemorating a Filipino attack on U.S. forces during the Philippine-American War (1899–1902).

FIGURE 10.12 European and U.S. colonies in Southeast Asia, 1914. Of the present-day countries in Southeast Asia, only Thailand (formerly called Siam) was never colonized.

non-Christians. The Spanish ruled the Philippines for more than 350 years; as a result, except for the southernmost islands, the Philippines is the most deeply Westernized and certainly the most Catholic part of Southeast Asia.

The Dutch were the most economically successful of the European colonial powers in Southeast Asia. From the sixteenth to the nineteenth centuries, they extended their control of trade over most of what is today called Indonesia, known at the time as the Dutch East Indies. The Dutch became interested in growing cash crops for export. Between 1830 and 1870, they forced indigenous farmers to leave their own fields and work part time without pay on Dutch coffee, sugar, and indigo plantations. The resulting disruption of local food production systems caused severe famines and provoked resistance that often took the form of Islamic religious movements. Resistance movements hastened the spread of Islam throughout Indonesia, where the Dutch had made little effort to spread their Protestant version of Christianity.

Beginning in the late eighteenth century, the British established colonies at key ports on the Malay Peninsula. They held these ports both for their trade value and to protect the Strait

e for sea trade between Britain's
the nineteenth century, Britain
modern Malaysia in order to ben-
ıd plantations. Britain also added
ovided access to forest resources
outhwest China.

Southeast Asia as Catholic mis-
ıth century. They worked mostly
n the modern states of Vietnam,
e nineteenth century, spurred by
uropean powers for access to the
ench formally colonized the area,
h Indochina.

only country not to be colonized
known as Siam). Like Japan, it
h both diplomacy and a vigorous
odernization; Thailand retains a

dence

e began in the late nineteenth
first against Spain in 1896. They
ited States, which began in 1898
ar (see Figure 10.12B). However,
Southeast Asia did not win inde-
Var II (Figure 10.12A). By then,
ts colonies had been weakened by
ring which Japan had conquered
Asia that had been controlled by
(see Figure 10.13D). Japan held
ed by the United States at the end
the colonial powers had granted
region, and all of Southeast Asia

ost bitter battle for independence
a (the territories of Vietnam, Laos,
he nineteenth century). Although
independent in 1949, France
omic power over them. Various
tably Vietnam's Ho Chi Minh,
against continued French domi-
accepted military assistance from
Soviet Union, even though they
ists and despite ancient antipathies
millennia of domination. In this
ht to mainland Southeast Asia.

defeated the French at Dien Bien
hough the United States was against
dominance, it stepped in because it
vas increasingly worried about the
pread of international communism
hould the anticolonial resisters, now
upported by Communists, succeed.
The **domino theory**—the geographi-
cally based idea that if one country "fell" to communism, other nearby countries would follow—was a major influence in this decision, because both North Korea and China had recently become communist. The Vietnamese resistance, which controlled the northern half of the country, attempted to wrest control of the southern half of Vietnam from the U.S.-supported and quite corrupt South Vietnamese government. The pace of the war accelerated in the mid-1960s. After many years of brutal conflict, public opinion in the United States forced a U.S. withdrawal from the conflict in 1973. The civil war continued in Vietnam, finally ending in 1975, when the north defeated the south and established a new national government.

More than 4.5 million people died during the Vietnam War, including more than 58,000 U.S. soldiers. Another 4.5 million on both sides were wounded, and bombs, napalm, and chemical defoliants ruined much of the Vietnamese environment. Land mines continue to be a hazard to this day, and the effects of the highly toxic defoliant known as Agent Orange are still producing debilitating birth defects among many rural Vietnamese and Laotian people (see Figure 10.13E). The withdrawal from Vietnam in 1973 ranks as one of the most profound defeats in U.S. history. After the war, the United States crippled Vietnam's recovery by imposing severe economic sanctions that lasted until 1993. Since then, the United States and Vietnam have become significant trading partners.

The "Killing Fields" in Cambodia In Cambodia, where the Vietnam War had spilled over the border, a particularly violent revolutionary faction called the Khmer Rouge seized control of the government in 1975. Inspired by the vision of a rural communist society, they attempted to destroy virtually all traces of European influence. They targeted Western-educated urbanites in particular, forcing them into labor camps, where more than 2 million Cambodians—one-quarter of the population—starved or were executed in what became known as the "killing fields."

In 1979, Vietnam deposed the Khmer Rouge and ruled Cambodia through a puppet government until 1989. A 2-year civil war then ensued. Despite a major UN effort to establish a multiparty democracy in Cambodia throughout the 1990s, the country remained plagued by political tensions between rival factions and by government corruption. In March 2009, Kang Kek Iew, the first of the Khmer Rouge leaders to be tried for war crimes and genocide, was forced to listen to and watch lengthy accounts of the torture of men, women, and children that he supervised. He was convicted in 2010 and sentenced to 35 years in prison. Most operatives in the killing fields will never be prosecuted. **223. CAMBODIAN HIP-HOP ARTIST TELLS STORY THROUGH RAP**

THINGS TO REMEMBER

- Over the last five centuries, several European countries, and later the United States and Japan, established colonies or quasi-colonies that covered almost all of Southeast Asia.
- Colonial rule in Southeast Asia began in 1511 with the Portuguese and came to a violent end in Vietnam and Cambodia, where war took the lives of millions, including more than 58,000 U.S. soldiers.

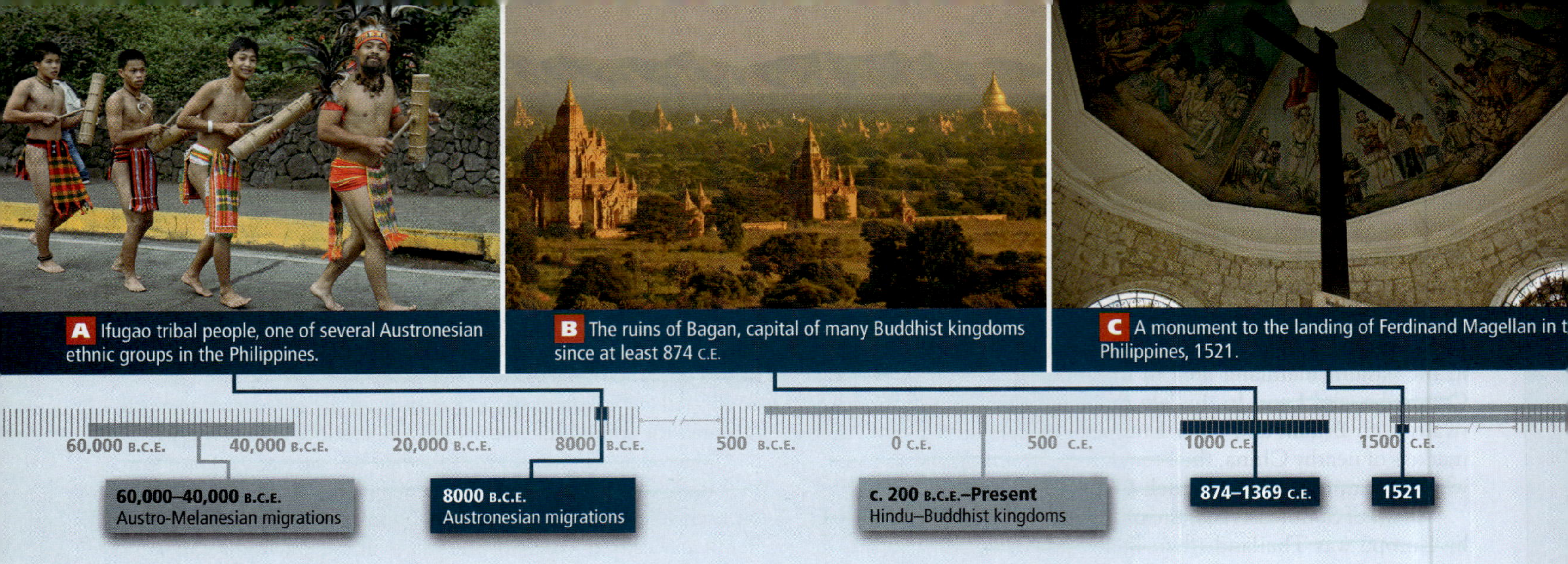

A Ifugao tribal people, one of several Austronesian ethnic groups in the Philippines.

B The ruins of Bagan, capital of many Buddhist kingdoms since at least 874 C.E.

C A monument to the landing of Ferdinand Magellan in t[he] Philippines, 1521.

FIGURE 10.13 A VISUAL HISTORY OF SOUTHEAST ASIA

Thinking Geographically

After you have read about the human history of Southeast Asia, you will be able to answer the following questions:

A From where did those who populated this region in prehistory come?

B How did Hinduism and Buddhism arrive in Southeast Asia?

CURRENT GEOGRAPHIC ISSUES

Like Middle and South America, Africa, and South Asia, Southeast Asia is expanding its links to the global economy. However, it has had more success than other areas in achieving widespread, if modest, prosperity. It has done so largely by following the example of some of East Asia's most successful countries, such as Japan, Korea, Taiwan, and more recently, China.

Economic and Political Issues

Initially, trade among countries within the region was inhibited by the fact that they all exported similar goods—primarily food and raw materials—and traditionally imposed tariffs against one another. They imported consumer products, industrial materials, machinery, and fossil fuels mostly from the developed world. Several decades ago, the economic and political situation in the region changed dramatically, when during the 1990s, Southeast Asian countries had some of the highest economic growth rates in the world. The growth was based on an economic strategy that emphasized the export of manufactured goods—initially clothing and then more sophisticated technical products. In the late 1990s, though, economic growth stagnated; as people's financial expectations were dashed, political instability increased. The high level of corruption revealed by the crisis resulted in stronger calls for democratization, which has expanded unevenly across the region.

Strategic Globalization: State Aid and Export-Led Economic Development

> **Geographic Insight 2**
>
> **Globalization and Development:** Globalization has brought spectacular economic successes and occasional dramatic economic declines. Urban incomes soar during periods of growth, but the poor are vulnerable during periods of global economic decline.

From the 1960s to the 1990s, some national governments in Southeast Asia created strong and sustained economic expansion by emulating two strategies for economic growth pioneered earlier by Japan, Taiwan, and South Korea (see pages 396–397 in Chapter 9). One was the formation of state-aided market economies. National governments in Indonesia, Malaysia, Thailand, and to some extent, the Philippines, intervened strategically in financial institutions to make sure that certain economic sectors developed; in addition, investment by foreigners was limited so that the governments could have more control over the direction of the economy. The other strategy was export-led economic development, which focused investment on industries that manufactured products for export, primarily to developed countries. These strategies amounted to a limited and selective embrace of globalization in that global markets for the region's products were sought but foreign sources of capital were not.

D J
durin

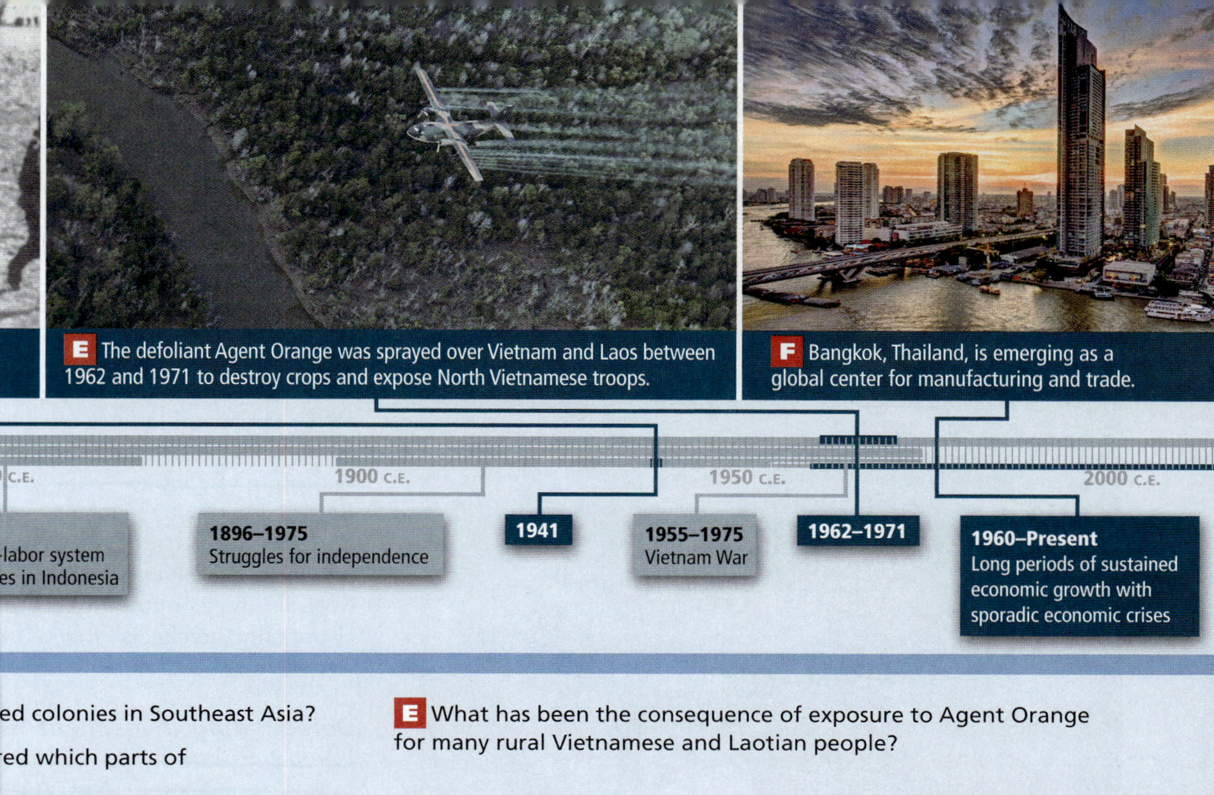

E The defoliant Agent Orange was sprayed over Vietnam and Laos between 1962 and 1971 to destroy crops and expose North Vietnamese troops.

F Bangkok, Thailand, is emerging as a global center for manufacturing and trade.

C V ed colonies in Southeast Asia?

D D ed which parts of
Sout

E What has been the consequence of exposure to Agent Orange for many rural Vietnamese and Laotian people?

dramatic departure from those
. In Middle and South America
l governments relied on import
roduced manufactured goods
ist, export-led industrial growth
rn much more money in the
leveloped world. Standards of
ecially in Malaysia, Singapore,
r important results of Southeast
ise in wealth disparities and
: lower population growth rates,
ortality rates, and longer life

the 1970s, some governments in
ditional strategy for encouraging
ne, they sought foreign sources of
rces could invest in were limited
e areas. Such Export Processing
on in Chapter 3) are places in
et up their facilities using inex-
o produce items only for export.
example of companies set up by
ated or greatly reduced so long
, not sold in-country. Since the
nomic development in Malaysia,
Philippines and are now used
d South America.

The Feminization of Labor Between 80 and 90 percent of the workers in the EPZs are women, not only in Southeast Asia but in other world regions as well (Figure 10.14). The **feminization of labor** has been a distinct characteristic of globalization over the past three decades (see page 375 in Chapter 9). Employers prefer to hire young, single women because they are perceived as the cheapest, least troublesome employees. Statistics do show that, generally, women across the globe will work for lower wages than men, will not complain about poor and unsafe working conditions, will accept being restricted to certain jobs on the basis of sex, and are not as likely as men to agitate for promotions. The reasons for this are complex and are discussed throughout this book; there are many exceptions, of course.

feminization of labor the increasing employment of women in both the formal and informal labor force, usually at lower wages than those of men

Working Conditions In general, the benefits of Southeast Asia's "economic miracle" have been unequally apportioned. In the region's new factories and other enterprises, it is not unusual for assembly-line employees to work 10 to 12 hours per day, 7 days per week, for less than the legal minimum wage, and without benefits. Labor unions that typically would address working conditions and wage grievances are frequently repressed by governments; international consumer pressure to improve working conditions at U.S. companies like Nike has been only partially effective. By 2010, for example, Nike had sidestepped the entire issue of customer complaints by outsourcing its manufacturing to non-U.S. contractors in the region. U.S. fair labor NGOs pursued

FIGURE 10.14 The feminization of labor. Women assemble circuit boards in a factory in Vietnam.

Thinking Geographically

Why do many employers prefer workers to be young, single women?

one Nike subcontractor in Jakarta, Indonesia, where workers were required to work an hour a day, off the books, in order to meet production quotas. A $1 million settlement will give each of the 4500 workers $222. There are more than 160,000 workers for Nike subcontractors in Indonesia alone, so this settlement is but a drop in the bucket.

A much more powerful force that is improving working conditions and driving up pay is the growth of the service sector throughout the region. In Singapore, the Philippines, and Thailand, the service sector already dominates the economy in employment and as a percentage of GDP. In Cambodia and Timor-Leste, the service sector contributes, respectively, 40 percent and 55 percent of the country's GDP, though agriculture still employs the majority of workers in both countries. In other countries in the region, the service sector is approaching parity with the industrial or agricultural sectors. Only Brunei is still dominated by its industrial sector, which is entirely based on oil production. Many service sector jobs require at least a high school education; competition for the smaller number of educated workers means that wages and working conditions are already better than in manufacturing and are likely to improve faster.

crony capitalism a type of corruption in which politicians, bankers, and entrepreneurs, sometimes members of the same family, have close personal as well as business relationships

Economic Crisis and Recovery: The Perils of Globalization

The economic crisis that swept through Southeast Asia in the late 1990s forced millions of people into poverty and changed the political order in some countries. A major cause of the crisis was the lifting of controls on Southeast Asia's once highly regulated financial sector. There were some geographic aspects to this crisis: the banks involved were located primarily in Singapore and the major cities of Thailand, Malaysia, and Indonesia. These were also the countries to feel the immediate effects of the crisis. Eventually the effects filtered into the hinterland, and workers in remote areas lost jobs and access to essentials.

Deregulating Investment As part of a general push by the International Monetary Fund (IMF) to open national economies to the free market, Southeast Asian governments relaxed controls on the financial sector in the 1990s. Soon, Southeast Asian banks were flooded with money from investors in the rich countries of the world who hoped to profit from the region's growing economies. Flush with cash and newfound freedoms, the banks often made reckless decisions. For example, bankers made risky loans to real estate developers, often for high-rise office building construction. As a result, many Southeast Asian cities soon had far too much office space, with millions of investment dollars tied up in projects that contributed little to economic development.

One of the forces that led bankers to make such bad decisions was a kind of corruption known as **crony capitalism**. In most Southeast Asian countries, as elsewhere, corruption is related to the close personal and family relationships between high-level politicians, bankers, and wealthy business owners. In Indonesia, for example, the most lucrative government contracts and business opportunities were reserved for the children of former president Suharto, who ruled the country from 1967 to 1997. His children became some of the wealthiest people in Southeast Asia. This crony capitalism expanded considerably with the new foreign investment money, much of which was diverted to bribery or unnecessary projects that brought prestige to political leaders.

The cumulative effect of crony capitalism and the lifting of controls on banks was that many ventures failed to produce any profits at all. In response, foreign investors panicked, withdrawing their money on a massive scale. In 1996, before the crisis, there was a net inflow of U.S.$94 billion to Southeast Asia's leading economies. In 1997, inflows had ceased and there was a net outflow of U.S.$12 billion.

The IMF Bailout and Its Aftermath The IMF made a major effort to keep the region from sliding deeper into recession by instituting reforms designed to make banks more responsible in their lending practices. The IMF also required structural adjustment policies (SAPs; see pages 129–131 in Chapter 3), which required countries to cut government spending (especially on

licies intended to protect domes-

nic chaos, and much debate over
l or hurt a majority of Southeast
ver. In the largest of the region's
1999; by 2006, the crisis, while
risks of globalization, had been

vth During the Southeast Asian
90s, Singapore, Malaysia, and
attracting new industries and
000s, China attracted more than
investment (FDI) as Southeast
also became an opportunity for
aysia, Indonesia, Thailand, and
on China's growth by winning
na's infrastructure in areas such
listribution, and shopping mall

sed its comparative advantages
nt. In poorer countries, such as
lower than in China. This has
ll manufacturing that had been
some of the region's wealthier
, and Thailand—have been posi-
that offer more highly skilled
structure than China does.

nning in 2007 The region was
lobal recession that became full
rowth slowed markedly because
cted disposable cash worldwide,
ountries (especially the United
hich are Southeast Asia's biggest
lebt that seriously curtailed their
and Southeast Asia appeared to
overy was based on the pent-up
vithin the domestic economies of
re consumers now had some dis-
bility to respond to this demand
pened up intraregional trade and
was carefully modulated by rela-
regulations aimed at controlling
as the establishment of EPZs,
ional multinational corporations

nts of 2009 did not last. By 2012,
nd slow recoveries in the United
emand for Southeast Asian prod-
y sophisticated economy that is
nd the transfer of goods through
usceptible to slowdowns in rich
ries like Indonesia, Malaysia, and
nere is demand for their products
bodia, Laos, and Burma are still
to places like China. Meanwhile,

China needed to restructure and reorganize investment, labor migration, and environmental policies in order to fend off civil unrest and loss of economic momentum. To do so, it began rebalancing its economy with the aim of encouraging internal sales of Chinese-produced products rather than those produced abroad. This further reduced the demand for raw materials and manufactured goods from Southeast Asia.

Regional Trade and ASEAN

During the 1980s and 1990s, Southeast Asian countries traded more with China and the rich countries of the world than they did with each other. This issue of insufficient reciprocal trade within Southeast Asia was the reason behind the creation of the **Association of Southeast Asian Nations (ASEAN)**, an increasingly strong organization of all ten Southeast Asian nations (see Figure 10.4 for locations).

Regional Integration ASEAN started in 1967 as an anti-Communist, anti-China association, but it now focuses on agreements that strengthen regional cooperation, including agreements with China. One example is the Southeast Asian Nuclear Weapons–Free Zone Treaty signed in December 1995. Another is the ASEAN Economic Community, or AEC, a trade bloc patterned after the North American Free Trade Agreement and the European Union. Now, ASEAN focuses on increasing trade between all ten members of the association.

In January of 2010, ASEAN established the ASEAN Free Trade Area (AFTA) to facilitate regional trade. By that same year, the total intraregional trade between ASEAN countries was more than trade with any one outside country or region, with ASEAN countries representing 25.8 percent of total imports and 25 percent of total exports (Figure 10.15). By 2012, a variety of other issues were being addressed, such as the free flow of services, skilled labor, investment and capital. Consumer protection, intellectual property rights, and leveling the playing field for competition were also receiving attention. ASEAN's overall goals are to develop a sustainable regional economic community and to decrease development gaps between and within ASEAN member states. Externally, ASEAN has ratified free trade agreements with Australia, New Zealand, China, India, Japan, and South Korea.

Tourism International tourism is an important and rapidly growing economic activity in most Southeast Asian countries (Figure 10.16 on page 439). Between 1991 and 2001, the number of international visitors to the region doubled to more than 40 million; in 2010, international visitors reached over 73.7 million. As in other trade matters, Southeast Asians are themselves increasingly touring neighboring countries; in 2010, for example, 34.8 million tourists to ASEAN countries were from within the region (Figure 10.17 on page 440). This is a positive trend because familiarity between neighbors lays the groundwork for various forms of regional cooperation, such as infrastructure improvements.

In response to its popularity with global and regional tourists, ASEAN members have been working to improve the region's

Association of Southeast Asian Nations (ASEAN) an organization of Southeast Asian governments that was established to further economic growth and political cooperation

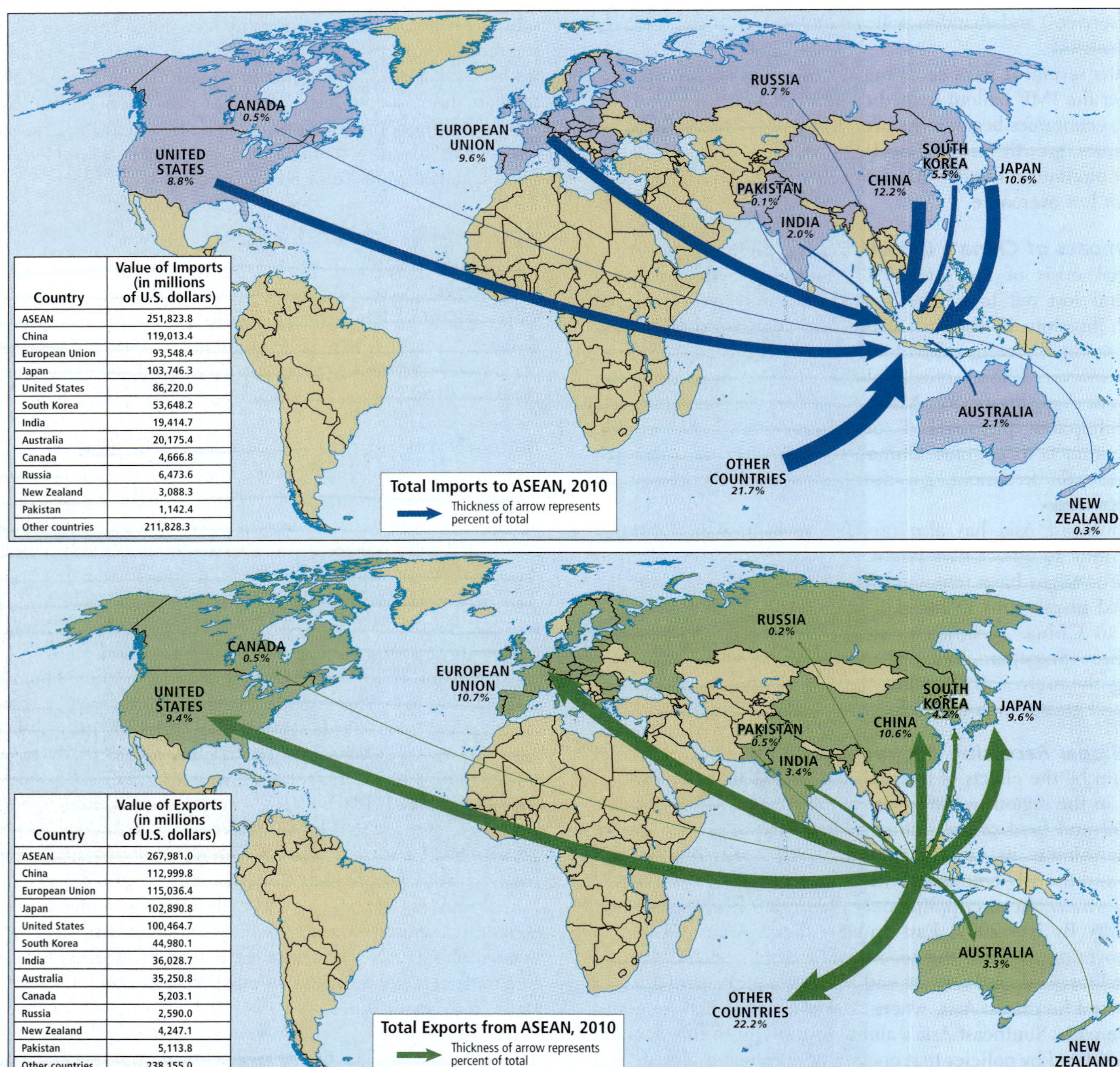

Country	Value of Imports (in millions of U.S. dollars)
ASEAN	251,823.8
China	119,013.4
European Union	93,548.4
Japan	103,746.3
United States	86,220.0
South Korea	53,648.2
India	19,414.7
Australia	20,175.4
Canada	4,666.8
Russia	6,473.6
New Zealand	3,088.3
Pakistan	1,142.4
Other countries	211,828.3

Country	Value of Exports (in millions of U.S. dollars)
ASEAN	267,981.0
China	112,999.8
European Union	115,036.4
Japan	102,890.8
United States	100,464.7
South Korea	44,980.1
India	36,028.7
Australia	35,250.8
Canada	5,203.1
Russia	2,590.0
New Zealand	4,247.1
Pakistan	5,113.8
Other countries	238,155.0

FIGURE 10.15 ASEAN imports and exports, 2010. The ASEAN countries are very active in world trade as importers and exporters. The largest share of trade, however, is among the ASEAN countries themselves, with imports at 25.8 percent and exports at 25 percent.

transportation infrastructure. One such project is the Asian Highway, a web of standardized roads that loop through the mainland and connect it with Malaysia, Singapore, and Indonesia (the latter via ferry) (see the Figure 10.16 map). Eventually, the Asian Highway will facilitate ground travel through 32 Eurasian countries from Russia to Indonesia and from Turkey to Japan.

The surge of tourism in Southeast Asia has also raised concerns about becoming too dependent on an industry that leaves economies vulnerable to events that precipitously stop the flow of visitors (natural disasters, political upheavals), or that leave local people vulnerable to the sometimes destructive demands of tourists (see the discussion of sex tourism on page 455). Examples of disasters are the tsunami of December 2004 that killed several thousand international tourists in Thailand and Indonesia, and various human-made disasters such as the terrorist bombings in Bali (2002, 2005) and in southern Thailand (2006, 2012). In addition, tourism can divert talent from contributing to national development, as the following vignette illustrates.

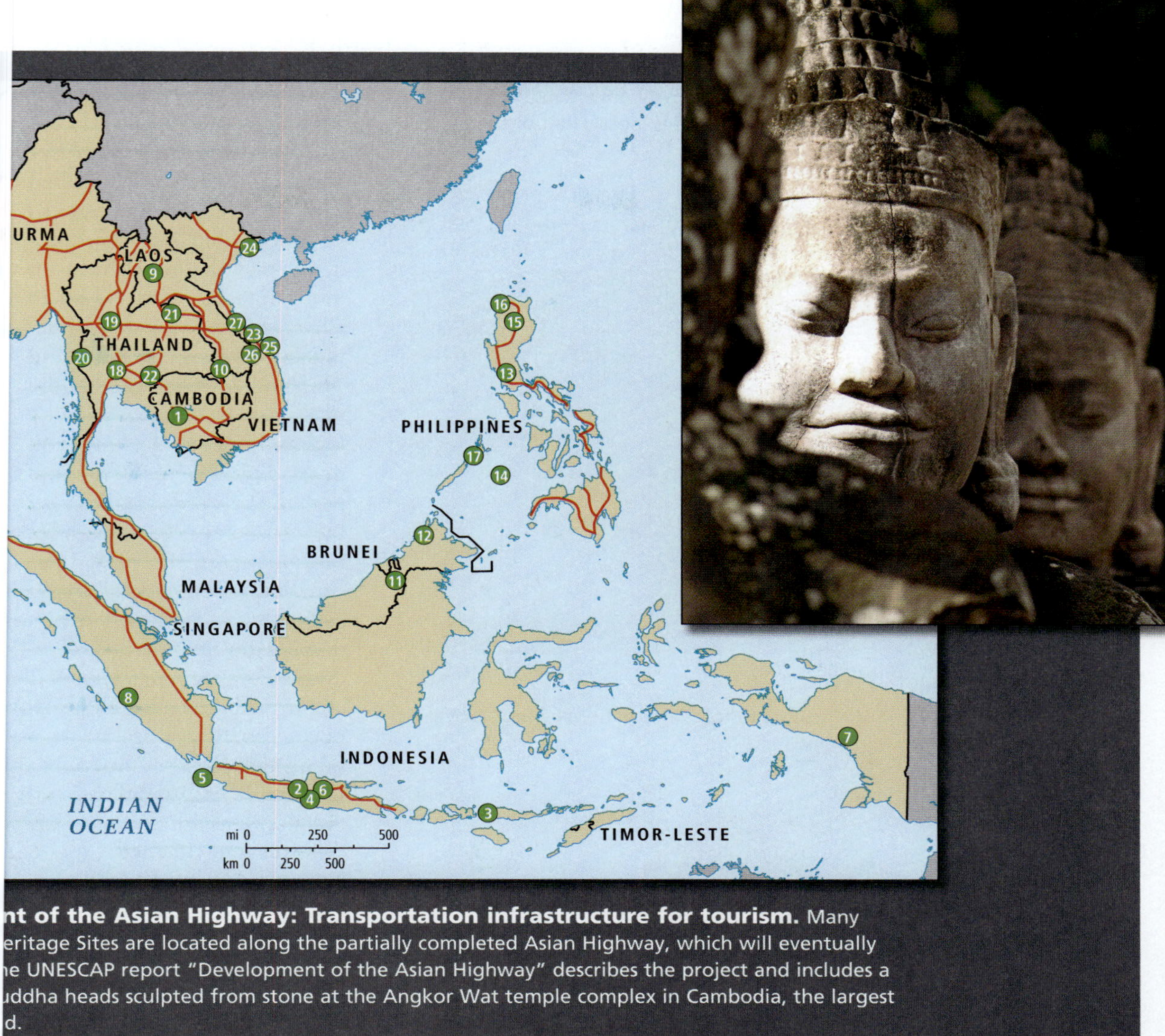

nt of the Asian Highway: Transportation infrastructure for tourism. Many eritage Sites are located along the partially completed Asian Highway, which will eventually e UNESCAP report "Development of the Asian Highway" describes the project and includes a uddha heads sculpted from stone at the Angkor Wat temple complex in Cambodia, the largest d.

patiently for the mechanic to rear tire on his Chinese-made ycle tour guide in Vietnam who along Highway 14 in the Central o Chi Minh Trail. Tan's current 3 days filling his digital camera tions covered in snowy blossoms, eir cocoons are unwound and genous minority children who met al curiosity. Tan finds his clients in els and negotiates a daily rate for and U.S.$75, not insignificant in a capita annual income of U.S.$2790. een subject to global forces. As sold produce to U.S. soldiers at a lost his brother, a soldier for South r spent 10 years in a Communist hardships, Tran obtained an education rate of 400 percent shrank teacher, Tran could not adequately er skilled Vietnamese people, Tran took advantage of Vietnam's transition to a market economy and established a business that caters to tourists. This means that he, along with many other educated Vietnamese people, no longer works in occupations crucial to Vietnam's future, like education. *[Source: The field work of Karl Russell Kirby. For detailed source information, see Text Credit pages.]* **230. BANKERS, ANALYSTS SEE RESURGENT ASIA 10 YEARS AFTER ECONOMIC CRISIS**

THINGS TO REMEMBER

Geographic Insight 2

- **Globalization and Development** Southeast Asia has had long periods of strong economic growth punctuated by brief but dramatic periods of decline (recession).
- Incomes have increased significantly, but many poor people remain highly vulnerable to periods of global economic decline, when jobs, food, and adequate living quarters are hard to come by.
- Regional integration, including the tourism infrastructure of the region, is expected to foster growth and encourage cooperation with China.

FIGURE 10.17 Visitor arrivals to ASEAN countries, 2010. Of the 73.7 million tourists to ASEAN countries, nearly half of them came from within the region.

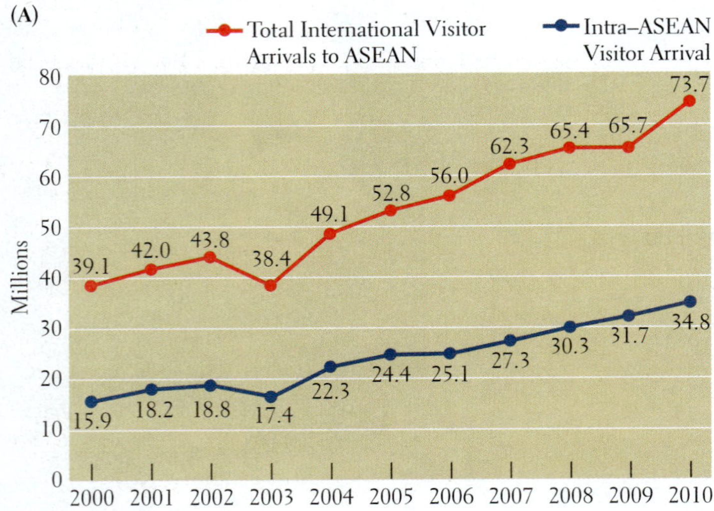

(A) Proportion of ASEAN visitors to total international visitor arrivals.

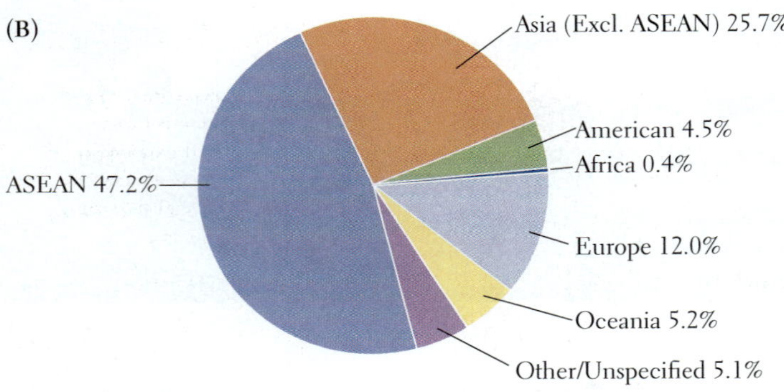

(B) Origin of international visitors to ASEAN countries, 2012.

Pressures For and Against Democracy

> **Geographic Insight 3**
>
> **Power and Politics:** Militarism, corruption, and favoritism are recurring issues in Southeast Asia. Although modernization and prosperity have brought significant increases in democratic participation, there have also been reversals, especially during periods of economic downturn. These sometimes result in violence, often against ethnic minorities.

Movements toward democracy have been uneven in Southeast Asia. Across the region, there are significant barriers to democratic participation (refer to the Figure 10.18 map), but the types of barriers vary. These are all countries with multicultural populations who share relatively few common characteristics. Efforts to smooth over these differences take many forms; Indonesia's solution, *Pancasila*, is discussed below on page 442.

Several countries are plagued with violent conflict: the repressive military regime in Burma, which has controlled with an iron fist and thwarted open elections and pro-democracy demonstrations for more than 20 years, began to allow small but possibly significant changes in 2012; and Thailand, long regarded as the most stable democracy in the region, now is dealing with deep divisions between supporters of a royalist military on the right and a large populist movement on the left. The situation is complicated by a rebellion and terrorism by Islamic militants in southern tourist zones. Indonesia has tried to use democratic reforms to reduce the tensions that in the past produced violence in many of its distinctive islands, but faces demoralizing setbacks in Aceh province in Sumatra. The Philippines, plagued with a long line of dictators, now has elected governments that have resolved many problems (except low wages and high rates of unemployment); however, militant Muslims in the southern islands continue an insurgency that periodically devolves into violence. **322. THAILAND'S PROTESTERS HIGHLIGHT RIFTS, POLITICAL PARTICIPATION**

> **Thinking Geographically**
>
> *After you have read about power and politics in Southeast Asia, you will be able to answer the following questions:*
>
> **A** What details in this picture are clues that these people are refugees?
>
> **B** Why would a government ban a flag?
>
> **C** What does this picture suggest about the MILF?

Can Democracy Work in Indonesia? An important yet still tentative shift toward democracy took place in Indonesia in the wake of the economic crisis of the late 1990s. After three decades of semidictatorial rule by President Suharto, the economic crisis spurred massive demonstrations that forced Suharto to resign. For more than a decade, democratic parliamentary and presidential elections brought a new political era to the country, but by 2012, corruption scandals had erupted among ministers in President Yudhoyono's government (Yudhoyono's campaigned in 2009 on a platform of reducing corruption), and Yudhoyono was proving to be an indecisive leader.

Indonesia is the largest country in Southeast Asia, and the most fragmented—physically, culturally, and politically. It comprises more than 17,000 islands (3000 of which are inhabited), stretching over 3000 miles (8000 kilometers) of ocean. It is also the most culturally diverse, with dozens of ethnic groups and multiple religions. Although Indonesia has the largest Muslim population in the world, there are also many Christians, Buddhists, Hindus, and adherents of various folk religions. With all these potentially divisive forces, many wonder whether this multi-island country of 238.2 million might be headed for disintegration.

Until the end of World War II, Indonesia was not a nation at all but rather a loose assemblage of distinct island cultures, which Dutch colonists managed to hold together as the Dutch East Indies. When Indonesia became an independent country in 1945, its first president, Sukarno, hoped to forge a new nation out of these many parts, founded on a fairly strong communist ideology. To that end, he articulated a national philosophy

Photo Essay: Power and Politics in Southeast Asia

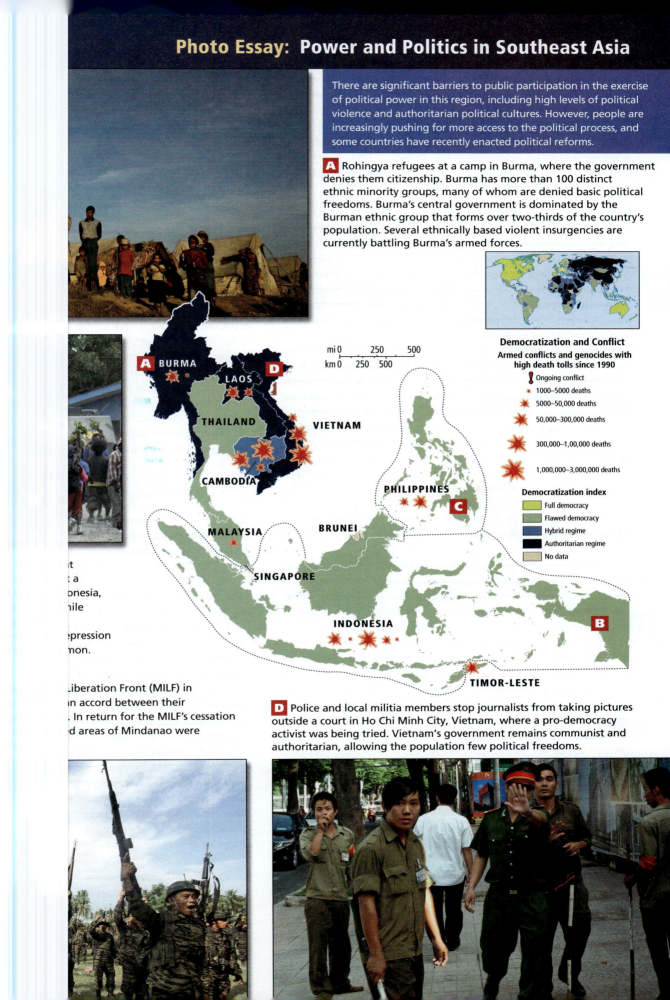

There are significant barriers to public participation in the exercise of political power in this region, including high levels of political violence and authoritarian political cultures. However, people are increasingly pushing for more access to the political process, and some countries have recently enacted political reforms.

A Rohingya refugees at a camp in Burma, where the government denies them citizenship. Burma has more than 100 distinct ethnic minority groups, many of whom are denied basic political freedoms. Burma's central government is dominated by the Burman ethnic group that forms over two-thirds of the country's population. Several ethnically based violent insurgencies are currently battling Burma's armed forces.

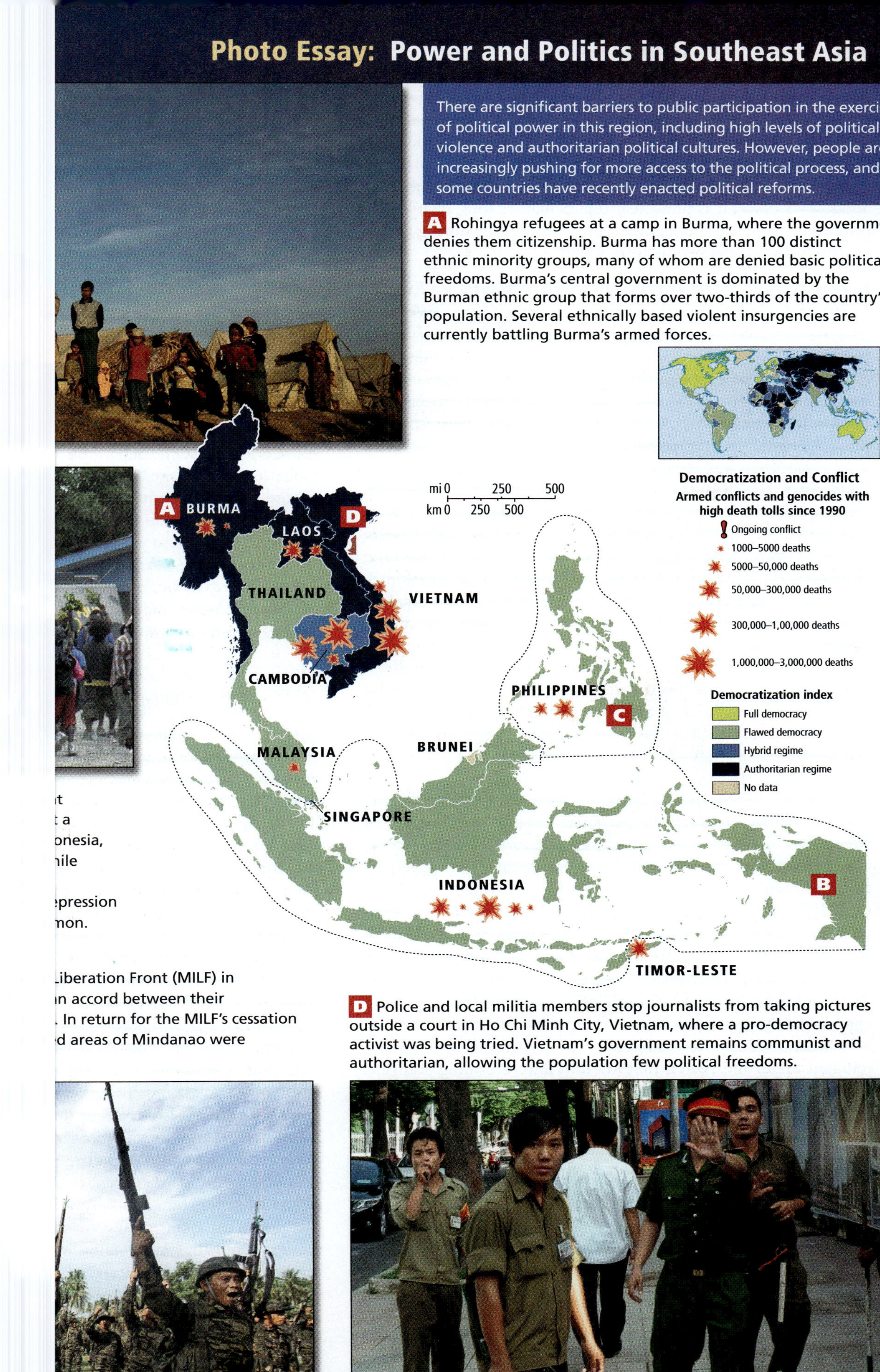

t
t a
onesia,
hile

epression
mon.

Liberation Front (MILF) in
n accord between their
. In return for the MILF's cessation
d areas of Mindanao were

D Police and local militia members stop journalists from taking pictures outside a court in Ho Chi Minh City, Vietnam, where a pro-democracy activist was being tried. Vietnam's government remains communist and authoritarian, allowing the population few political freedoms.

known as *Pancasila,* which aimed at holding the disparate nation together, primarily through nationalism and concepts of religious tolerance. In 1965, during the height of the Cold War, Suharto, a staunchly anti-Communist general in the Indonesian army, ousted Sukarno in a coup and ruled the country for another 33 years. It is now clear that Suharto's long regime was responsible for a purge of suspected Communists, during which as many as a million people were killed.

Encouraging Cohesiveness with Government Policies in Indonesia Despite his abrupt removal from office, Sukarno's unifying idea of *Pancasila* endures as a central theme of life in Indonesia (and less formally throughout the region). *Pancasila* embraces five precepts: *belief in God,* and the observance of *conformity, corporatism* (often defined as "organic social solidarity with the state"), *consensus,* and *harmony.* These last four precepts could be interpreted as discouraging dissent or even loyal opposition, and they seem to require a perpetual stance of boosterism. For many Indonesians, the strength of *Pancasila* is the emphasis on harmony and consensus that seems to ensure there will never be a religiously based government. Others note that conformity and corporatism counteract the extreme ethnic diversity and geographic dispersion of the country. But conformity and corporatism have also had a chilling effect on participatory democracy and on criticism of the government, the president, and the army. Indeed, the first orderly democratic change of government did not take place until national elections in 2004. Since then, there have been several peaceful elections, and the government uneasily allows citizens to publicly protest policies (see Figure 10.18B). But corruption remains a threat and there is a feeling of nostalgia for the Suharto model of authoritarian government.

Separatist movements that have sprouted in four distinct parts of Indonesia in recent years demonstrate the fragility of *Pancasila.* The only rebellion to succeed was in Timor-Leste, which became an independent country in 2002. However, its case is unique in that this area was under Portuguese control until 1975, when it was forcibly integrated into Indonesia. Two other separatist movements (in the Molucca Islands and in West Papua; see Figure 10.1) developed largely in response to Indonesia's forced **resettlement schemes**. Also known as *transmigration schemes,* between 1965 and 1995, these programs relocated approximately 8 million people from crowded islands such as Java to less densely settled islands. The policies were originally initiated under the Dutch in 1905 to relieve crowding and provide agricultural labor for plantations in thinly populated areas. After independence, Indonesia used resettlement schemes both to remove troublesome people and to bring outlying areas under closer control of the central government in Jakarta.

resettlement schemes government plans to move large numbers of people from one part of a country to another to relieve urban congestion, disperse political dissidents, or accomplish other social purposes; also called *transmigration*

The failure of the cohesiveness sought by *Pancasila* is particularly troubling in the far western province of Aceh, in the north of Sumatra. Conflicts there originally developed because most of the wealth yielded by Aceh's resources, especially oil, was going to the central government in Jakarta. The Acehnese people protested what they saw as the expropriation of oil without compensation. Jakarta sent the military to quell rebellion. Protestors, especially young people, viewed the military presence as adding insult to injury. Many were accused of resorting to violence and were hunted down, charged with terrorism, and jailed or worse. The conflict seemed unresolvable. Then in 2004, the earthquake and tsunami in Aceh, which killed more than 170,000 Acehnese, suddenly brought many outside disaster relief workers to Sumatra because the Indonesian government was incapable of rendering sufficient aid to the victims. Global press coverage of the relief effort mentioned the recent political violence; this created a powerful incentive for separatists and the government to cooperate in order to receive outside aid.

A resulting peace accord signed in 2005 brought many former combatants into the political process as democratically elected local leaders. Separatists laid down their arms. But when global press coverage stopped, the implementation of the peace accord ceased and another outside force crept into the situation. Islamic militants, who favor Shari'a, are gaining adherents among disaffected Acehnese youth. Shari'a (discussed more completely in Chapter 6, page 253) is usually subject to the laws of the land, but can be extreme, especially regarding the rights of women. Shari'a was never previously practiced in Aceh, and moderate leaders in Aceh are worried.

Southeast Asia's Authoritarian and Militaristic Tendencies Despite the ebb and flow of participatory democracy in Indonesia, Thailand, and most recently in Burma (March 2012), authoritarianism backed by a strong military is still a powerful force in Southeast Asia. Authoritarianism is rationalized primarily as a counter to destabilization caused by "too much" democracy. For example, in 2006, the military took over Thailand's government after a corrupt but charismatic prime minister marshaled masses of protestors (Red Shirts) who took to the streets, disrupting business and the lucrative tourism industry. Despite recent elections, the matter remains unresolved and the Red Shirts periodically take over the streets of Bangkok.

In the far south of Thailand, where Islamic militants have set off car bombs in European tourism areas, there are calls for military intervention. Undemocratic socialist regimes still control Laos and Vietnam (see Figure 10.18D); as of 2012, the military dictatorship running Burma has relaxed only slightly. Cambodia's democracy is precarious, and violence there is common. Brunei is an authoritarian sultanate. Even in the Philippines, the region's oldest democracy, powerful and corrupt leaders have subverted the democratic process repeatedly. The wealthier and usually stable countries of Malaysia and Singapore continue to use authoritarian versions of democracy.

Some Southeast Asian leaders, such as Singapore's former prime minister, Lee Kuan Yew, have argued that Asian values are not compatible with Western ideas of democracy. These leaders assert that Asian values are grounded in the Confucian view that individuals should be submissive to authority, so Asian countries should avoid the highly contentious public debate of open electoral politics. Nevertheless, when confronted with governments that abuse their power, people throughout Southeast ebelled (see Figures 10.18A, B). 1-educated son of Lee Kuan Yew, v prime minister, has expressed political freedoms than did his

vilian-led governments to always nilitary components of Southeast power. In virtually every country, o restore civil calm. Military rank ery top elected official has had a 1 Suu Kyi, the advocate for peace na, is the daughter of a military litary's role in politics is likely to

itarism, the countries of the region each other. However all are keep- nese navy. As a region surrounded usands of miles of coastline, beef- n obvious security measure. Also, a in 2004 showed that no military able to effectively manage such a countries recently have invested , primarily from Europe and the vealthy trade-centered Singapore, on weapons, spent one quarter of roduces armored troop carriers for

onomic Issues Like authoritari- med as a counterforce to political rrorist violence short-circuits the neart of democratic processes and for authoritarian militaristic mea- nt to recognize that terrorist move- text of both economic deprivation factors that are often interrelated. 1 Philippines (see Figure 10.18C), ort from people who feel shut out nic advancement. Increasingly, the 1 seems to be to listen to both the es of those who might be attracted

TRUGGLE IN INDONESIA
'S MUSLIM SOUTH INTENSIFIES

O REMEMBER

ritarianism, deep cultural divisions, and e equitable sharing of political power in

novements, prevalent in most countries mmon values on which to base

Politics Generally speaking, those the most developed economies have st increases in political freedoms, but there are both exceptions (Malaysia and Singapore) and countries that have undergone reversals (Thailand). Militarism, corruption, and favoritism are recurring issues in Southeast Asia.

- Violent conflict and even terrorism remain threats across the region. Nonetheless, the experience of recent years shows that movements toward more political freedom are strong.

Sociocultural Issues

Southeast Asia is home to 602 million people (almost double the U.S. population) who occupy a land area that is about one-half the size of the United States. Because of the region's long and complex history, the people in South east Asia have a great diversity of cultures and religious traditions.

Population Patterns

Geographic Insight 4
Population and Gender: Economic change has brought better job opportunities and increased status for women, who are then choosing to have fewer children. The improving role and status of women has increased awareness that sex trafficking is an abuse of human rights.

Southeast Asia's population is large and growing, but economic development, changing gender roles, and population-control efforts are slowing the growth rate. At current rates, Southeast Asia's population is projected to reach 796 million by 2050, by which time much of this population will live in cities (only 42 percent do now, though Singapore is 100 percent urban) (Figure 10.19). However, population projections could be inaccurate, both because rates of natural increase are continuing to slow markedly and because many Southeast Asians are migrating to find employment outside the region.

Population Dynamics Population dynamics vary considerably among the countries of this region. The variety is due in part to differences in economic development, government policy, prescribed gender roles, and broader religious and cultural practices. Most countries are nearing the last stage of the demographic transition, where births and deaths are low and growth is minuscule or slightly negative. In the last several decades, overall fertility rates in Southeast Asia have dropped rapidly (Figure 10.20 on page 445). Whereas women formerly had 5 to 7 children, they now have 1 to 3, with Laos (3.9) and Timor-Leste (5.7) being the major exceptions. However, because in most countries populations are still quite young (between one-quarter and one-third of the people are aged 15 years or younger), modest population growth is likely for several decades because so many are just coming into their reproductive years.

Brunei, Singapore, and Thailand have reduced their fertility rates so steeply—below replacement levels—that they will soon need to cope with aging and shrinking populations. Regionally, education levels are the highest in Singapore, where most educated women work outside the home at skilled jobs and professions. The Singapore government is now so concerned about the

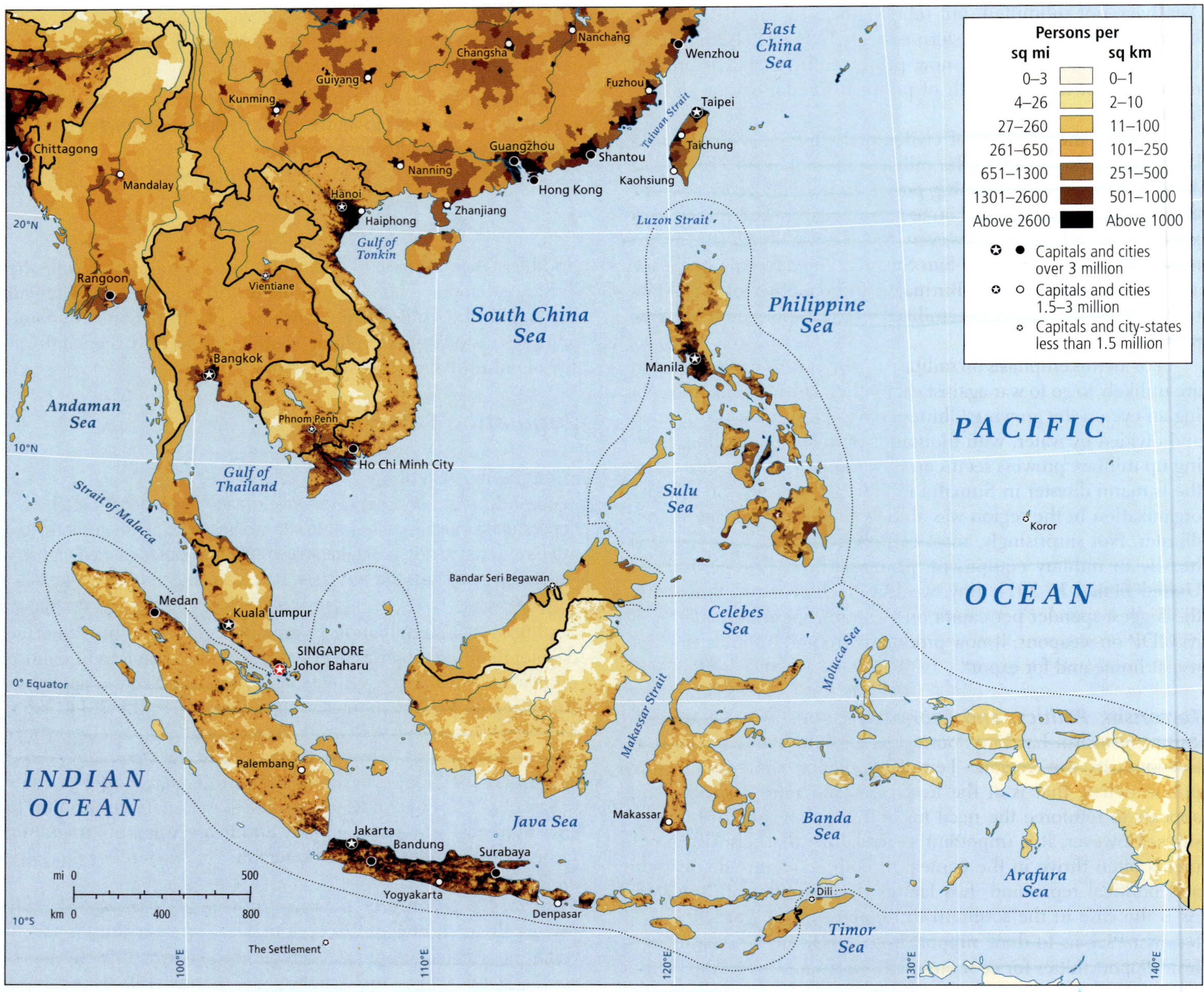

FIGURE 10.19 Population density in Southeast Asia. Population growth in Southeast Asia is slowing, largely related to economic development, urbanization, changing gender roles, and government policies. Fertility rates have declined sharply in all countries since the 1960s, but are still high in the poorest areas.

low fertility rate that it offers young couples various incentives for marrying and procreating.

An important source of population growth for Singapore is the steady stream of highly skilled immigrants that its vibrant economy attracts. Many of these immigrants come from elsewhere in the region, though the country also draws highly skilled workers from the United States and Europe. Singapore also attracts illegal immigrants from elsewhere in Southeast Asia, primarily in the building trades and low-wage services. Nearly all countries other than Singapore are losing population to emigration.

Government-sponsored birth control programs are unevenly distributed. Thailand's low fertility rate of 1.6 children per adult woman was achieved in part via a government-sponsored condom campaign, and in part by rapid economic development, which made many couples feel that smaller families would be best. Also, as women gained more opportunities to work and study outside the home, they decided to have fewer children. High literacy rates for both men and women, along with Buddhist attitudes that accept the use of contraception, have also been credited for the decline in Thailand's fertility rate.

The poorest and most rural countries in the region show the usual correlation between poverty, high fertility, and infant mortality. Timor-Leste, which suffered a violent, impoverishing civil disturbance before it gained independence from Indonesia,

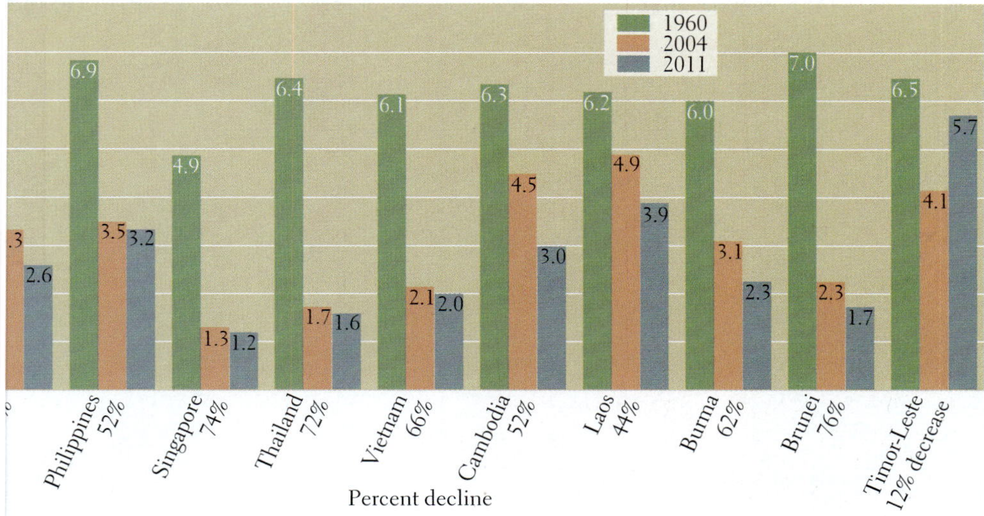

l fertility rates, 1960, 2004, and 2011. Total fertility rates declined for all Southeast Asian o 2011, but did increase in Timor-Leste between 2004 and 2011.

y, high fertility and high infant rths). However, its oil resources hey are used to help the welfare happens, both fertility and infant l. In Cambodia and Laos, fertility en, respectively. Infant mortality is for both Cambodia and Laos. n, where people are only slightly d, the fertility rate (2.0 children mortality rate (16 per 1000 births) are explained by the fact that as ides basic education and health to all, regardless of income. In re than 96 percent for men and as they are only 64 percent for Laos. In addition, Vietnam's rapling in foreign investment, which women, and careers are replacal focus of many women's lives. anges weren't happening rapidly ving population would jeopardize development, the government of somewhat elastic two-child family brought on gender imbalance, as rtion if the fetus is female.

s a higher per capita income than , or Vietnam, is an anomaly in ecause it is predominantly Roman Figure 10.26 map on page 452), a allow birth control. Fertility there gion (3.2 per adult female); infant ge (22 per 1000 live births); and high despite female literacy rates

VIGNETTE In Roman Catholic Philippines, Gina Judilla, who works outside the home, has had six children with her unemployed husband. They wanted only two, but because of the strong role of the Catholic Church and the political pressure it exerts, birth control was not available to them. Abortion is legal only to save the life of the mother, so with every succeeding pregnancy, she tried folk methods of inducing an abortion. None worked. Now she can afford to send only two of her six children to school.

A move by family planners to provide national reproductive health services and sex education is underway. A recent survey showed that 48 percent of all pregnancies in the Philippines in 2010 were unintended, which often led to "backstreet" abortions. Nonetheless, in 2011, only one-third of Philippine women had access to modern birth control methods; in 2005, because of pressure by conservative religious groups within the United States, the United States stopped offering birth control aid to the Philippines. *[Source: Likhaan Center for Women's Health, Inc. and the New York Times. For detailed source information, see Text Credit pages.]* ■

Population Pyramids The youth and gender features of Southeast Asian populations are best appreciated by looking at the population pyramids for Indonesia. Figure 10.21 shows the 2012 population and the projected population for 2050. The wide bottom of the 2012 pyramid indicates that most people are age 30 and under, but the projections to 2050 show that eventually, with declining birth rates, those 30 and under will be outnumbered by those 35 and older. Indonesia will accumulate ever-larger numbers in the upper age groups, and the pyramid will eventually be more box-shaped, as those for Europe are now. The gender disparities (more males than females) that have developed over the last 20 years can be seen by carefully examining the lengths of the bars on the male and female sides of the pyramid for those age 15–19 and under. The difference is slight but significant because there is a rather consistent deficit

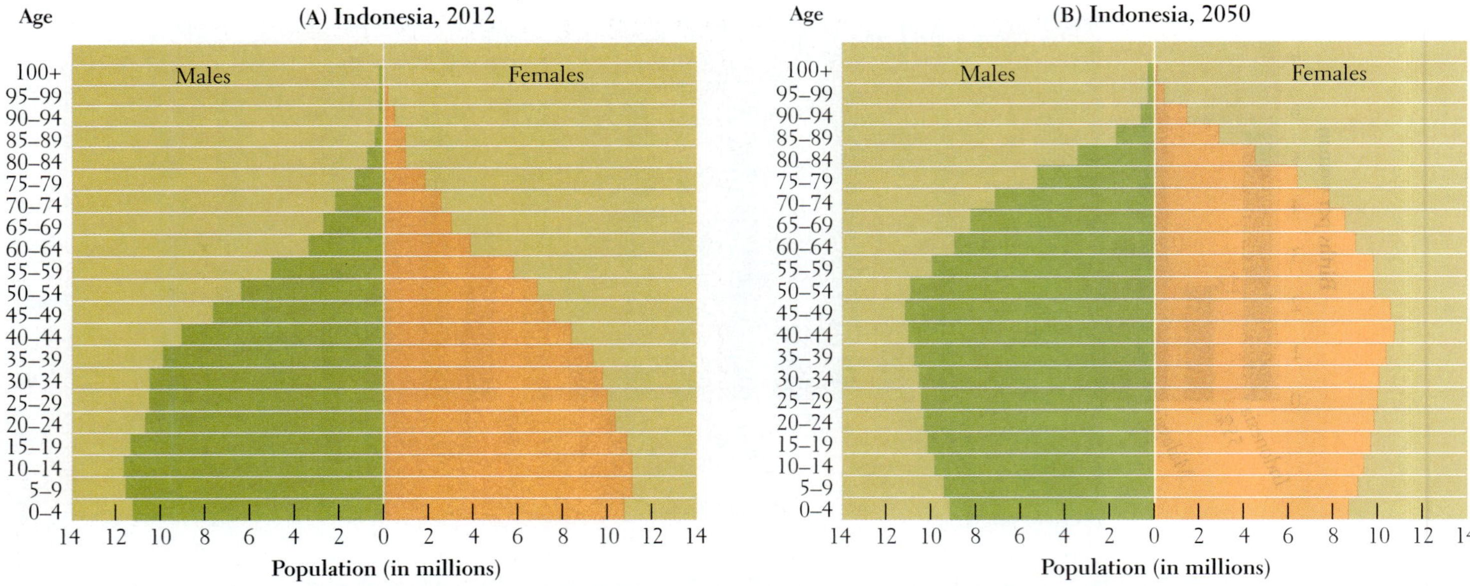

FIGURE 10.21 Population pyramids for Indonesia, 2012 and 2050. In 2012, Indonesia's population was 248.2 million and is projected to be 313 million in 2050.

of females; apparently they were selected out before birth, as is the case in Vietnam.

Southeast Asia's Encounter with HIV-AIDS As in sub-Saharan Africa (see Chapter 7), HIV-AIDS is a significant public health issue in Southeast Asia, although infection and death rates are now declining across the region. Cambodia, Thailand, and Burma currently have the highest infection rates. In Thailand in 2001, AIDS was the leading cause of death, overtaking stroke, heart disease, and cancer, but it is now just the third leading cause of death. Estimates are that about 500,000 Thais are infected, down from 1 million 12 years ago. Men between the ages of 20 and 40 have the highest rates of infection, but the rate of infection among women is rising. Thailand is one of the few countries that has been able to drastically reduce the incidence of HIV-AIDS; it did so through a well-funded program that increased the use of condoms, decreased STDs dramatically, and reduced visits to sex workers by half.

Elsewhere in the region, HIV rates are expected to increase rapidly in rural areas and in secondary cities where conservative religious leaders and faith-based international agencies (many from the United States) restrict sex education and AIDS-prevention programs, such as the promotion of condom use. Sex education is viewed as promoting promiscuity. At the same time, popular customs that support sexual experimentation (at least among men)—the high mobility of young adults, the reluctance of women to insist that their husbands and boyfriends use condoms, and intravenous drug use (primarily by men)—make aggressive prevention programs all the more essential. Also contributing to the spread of HIV among young women is sex tourism and sex-related human trafficking. **233. ACTIVISTS: AIDS LINKED TO WOMEN'S RIGHTS ABUSES**

Geographic Patterns of Human Well-Being If human well-being is defined as the ability to enjoy long, healthy, and creative lives, as suggested by Mahbub ul Haq, the founder of the annual United Nations Human Development Report (UNHDR), then personal income statistics are not be the best way to measure success. Nonetheless, income is relevant to well-being, as evidenced by this series of three maps of Southeast Asia. Figure 10.22A shows gross national income (GNI) per capita (PPP) by country; Figure 10.22B shows each country's rank on the Human Development Index (HDI); and Figure 10.22C depicts gender equality.

As shown in Figure 10.22A, two small countries, Singapore and Brunei, are the only places where annual per capita GNI (PPP) is in the highest category. For both countries, GNI (PPP) is about U.S.$50,000, which is a bit higher than in the United States and Australia and comparable to that in the richest countries in Europe (see the inset map of the world). Elsewhere in the region, per capita GNI (PPP) is considerably lower, with Malaysia the only country in the next highest range, at U.S.$13,710. Thailand is near the top of the medium range; and below it are Vietnam, Indonesia, the Philippines, Laos, and Cambodia, with per capita GNI (PPP) rates that vary from U.S.$1820 to U.S.$4730. In the Philippines, GNI (PPP) fell in recent years because the global recession cut the size of remittances and because of political troubles in the southern islands. In the lowest category are Timor-Leste and Burma.

Figure 10.22B shows each country's rank on the Human Development Index (HDI), which is a calculation (based on adjusted real income, life expectancy, and educational attainment) of how adequately a country provides for the well-being of its citizens. Singapore and Brunei again do well at providing for the overall well-being of their citizens, as they rank in the two highest categories. Malaysia ranks high; Thailand and the Philippines rank medium; Indonesia, Vietnam, Laos, and Cambodia all rank medium low, while Burma ranks low. Many have lost rank over the last 5 years because of the global recession.

well-being.

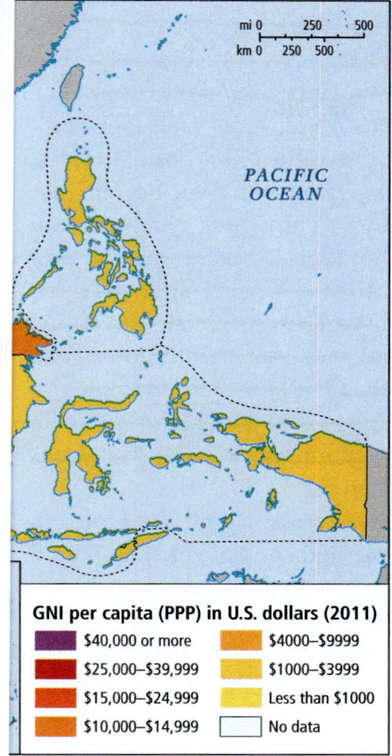

capita, adjusted for purchasing

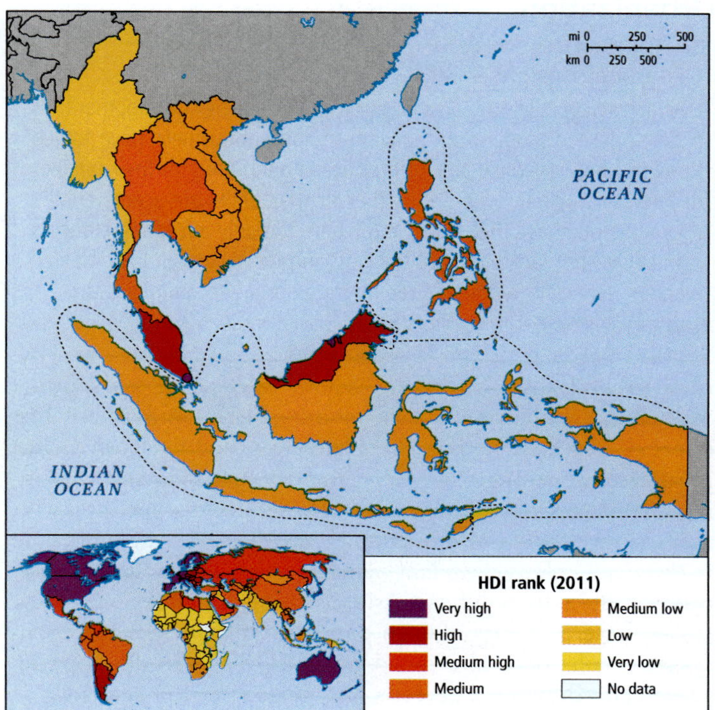

(B) Human Development Index (HDI).

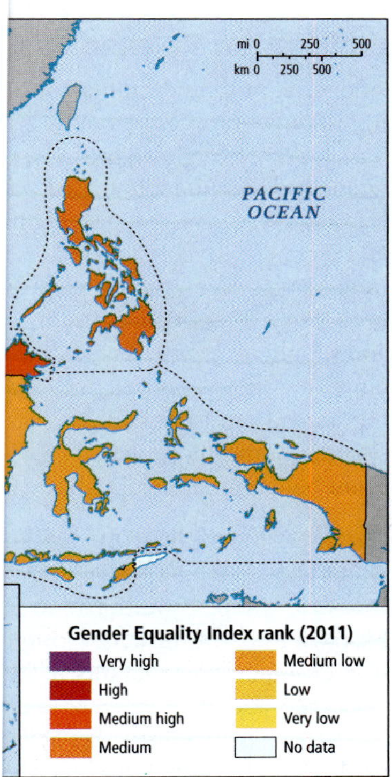

vell countries are ensuring gender
oductive health, political and edu-
ss to the labor market. The coun-
ave the highest degree of gender
ingapore ranks very high, at eighth

in the world, but there are no data for Brunei, which ranks high in income and well-being. Malaysia ranks medium high; and Vietnam, which ranks low on income and well-being, ranks medium high in this category, perhaps because of residual benefits of the general communist (socialist) philosophy regarding gender equality. The remaining countries rank medium low to low.

THINGS TO REMEMBER

- While population growth rates have slowed across the region, they are slowing the most in the developed economies of Singapore, Brunei, and Thailand.
- The variations in population growth are linked to relative prosperity, gender roles, and availability of birth control.
- By 2050, some if not all countries in this region will have to address the issue of having an aging population.

Geographic Insight 4

- **Population and Gender** Even though countries in the region are far from having gender equity, population patterns are changing as job and educational opportunities for women improve and birth control is more widely available. Empowerment of women is increasing awareness that sex trafficking is an abuse of their human rights.
- HIV-AIDS is a threat, to varying degrees, throughout the region. Preventative measures have proven quite successful in Thailand.
- Human well-being varies from very high (Singapore) to low (Timor-Leste, Burma, Laos, and Cambodia), with most countries ranking in the medium categories.

Urbanization

> **Geographic Insight 5**
>
> **Urbanization, Food, and Development:** The development of export-oriented modernized agriculture and agribusiness have forced farmers who were once able to produce enough food for their families to migrate to cities, where jobs that fit their skills are scarce, where they must purchase their food, and where the only affordable housing is in slums that lack essential services. Surplus skilled and semiskilled workers, especially women, often migrate abroad for employment.

Southeast Asia as a whole is only 42 percent urban, but the rural–urban balance is shifting steadily in response to declining agricultural employment and booming urban industries. The forces driving farmers into the cities are called the *push factors* in rural-to-urban migration. They include the rising cost of farming related to the use of new technologies and competition with agribusiness. *Pull factors,* in contrast, are those that attract people to the city, such as abundant manufacturing jobs and education opportunities. In Southeast Asia, as in all other regions, these factors have come together to create steadily increasing urbanization. Malaysia is already 64 percent urban; the Philippines, 63 percent; Brunei, 72 percent; and Singapore, 100 percent.

Throughout Southeast Asia, employment in agriculture has been declining since new production methods were introduced that increase the use of labor-saving equipment and reduce the need for human labor. Meanwhile, chemical pesticide and fertilizer use has also grown. While such additives can increase harvests dramatically and increase the supply of food to cities, they also drive the cost of production higher than what most farmers can afford. Many family farmers have sold their land to more prosperous local farmers or to agribusiness corporations and moved to the cities. These people, skilled at traditional farming but with little formal education, often end up in the most menial of urban jobs (**Figure 10.23C**) and live in circumstances that do not allow them to grow their own food (see Figure 10.23A).

Labor-intensive manufacturing industries (garment and shoe making, for instance) are expanding in the cities and towns of the poorer countries, such as Cambodia, Vietnam, and parts of Indonesia and Timor-Leste. In the urban and suburban areas of the wealthier countries—Singapore, Malaysia, Thailand, and parts of Indonesia and the northern Philippines—technologically sophisticated manufacturing industries are also growing. These include automobile assembly, chemical and petroleum refining, and computer and other electronic equipment assembly. Riding on this growth in manufacturing are innumerable construction projects that often provide employment to recent migrants (see Figure 10.23C, D).

Cities like Jakarta, Manila, and Bangkok, among the most rapidly growing metropolitan areas in the world, are *primate cities*—cities that, with their suburbs, are vastly larger than all others in a country. Bangkok is more than 20 times larger than Thailand's next-largest metropolitan area, Udon Thani; and Manila is more than 10 times larger than Davao, the second-largest city in the Philippines. Thanks to their strong industrial base, political power, and the massive immigration they attract, primate cities can dominate whole countries; and in the case of Singapore, the city constitutes the entire country.

foreign exchange foreign currency that countries need to purchase imports

Rarely can such cities provide sufficient housing, water, sanitation, or even decent jobs for all the new rural-to-urban migrants. Many millions of urban residents in this region live in squalor, often on floating raft-villages on rivers and estuaries. Of all the cities in Southeast Asia, only Singapore provides well for nearly all of its citizens (see Figure 10.23B). Even there, however, a significant undocumented, noncitizen population lives in poverty on islands surrounding the city. The experience of rural-to-urban migrants who go to Bangkok or Jakarta is more typical; migrants there often live in slums on the banks of polluted, trash-ridden waterbodies (see Figure 10.23A).

Thinking Geographically

After you have read about urbanization in Southeast Asia, you will be able to answer the following questions:

A Why might these people be living on the water in Jakarta?

B If government-built apartment buildings in Singapore are intended for the middle class, what are the common facilities that house most of the illegal, noncitizen population?

C What about this photo suggests that these men may be of rural origins?

D Why might this photo be viewed as a triumph for women's rights in Bangkok?

Emigration Related to Globalization

The same push and pull factors behind urbanization are also driving millions to migrate out of Southeast Asia. These migrants are a major force of globalization, as they supply much of the world's growing demand for low- and middle-wage workers who are willing to travel or live temporarily in foreign countries. For example, 40 percent of the foreign workers in Taiwan are Indonesian. These migrants are also a globalizing force within their home countries because their remittances (monies sent home) boost family incomes and supply governments with badly needed **foreign exchange** (foreign currency) that countries use to purchase imports. Filipinos working abroad are their country's largest source of foreign exchange, sending home more than U.S.$6 billion annually and increasing household annual income by an average of 40 percent.

The Merchant Marine Skilled male seamen from Southeast Asia make up a significant portion of the international merchant marines, where conditions are considerably better than they are for most migrant workers. Nonetheless, in the merchant marines, it is customary for workers to be paid according to their homeland's pay scales, which makes employing seamen from low-wage Southeast Asian countries attractive to shipowners. The seamen work aboard international freighters or on luxury cruise liners as deckhands, cooks, engine mechanics—and a few become officers. Generally, seamen work for 6 months at a time—with only a few hours a day for breaks—saving nearly every penny. At the end of a tour of duty, they return home to their families for another 6 months, where they often contribute financially to the well-being of an extended group of kin and friends and send their children to advanced education.

Photo Essay: Urbanization in Southeast Asia

g as manufacturing and service sector industries pull in people from rural areas and as changes in agriculture
ties are struggling to cope with rapid growth.

A A slum area in Jakarta, Indonesia, where 62 percent of the population lives in slums. Jakarta will grow by almost 50 percent by 2020.

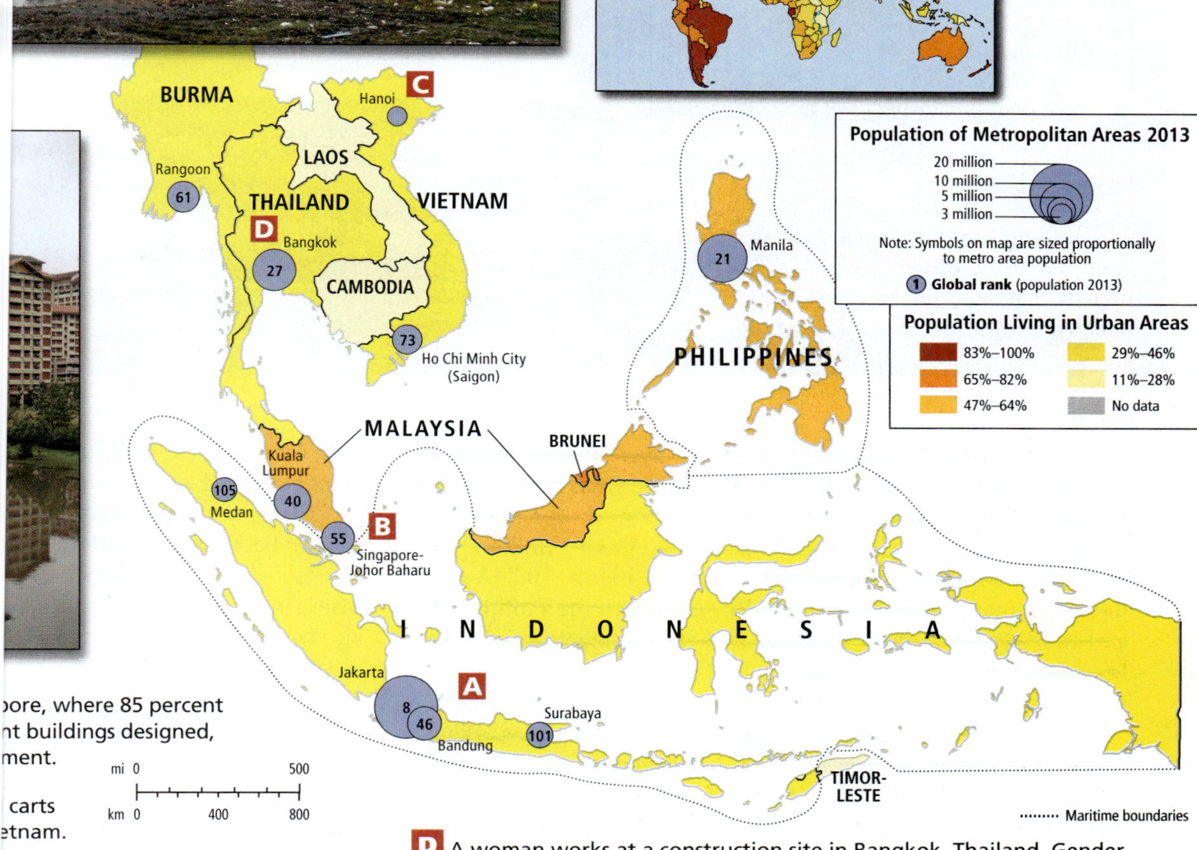

pore, where 85 percent
nt buildings designed,
ment.

carts
etnam.
age,

D A woman works at a construction site in Bangkok, Thailand. Gender roles are changing as more women move to the cities.

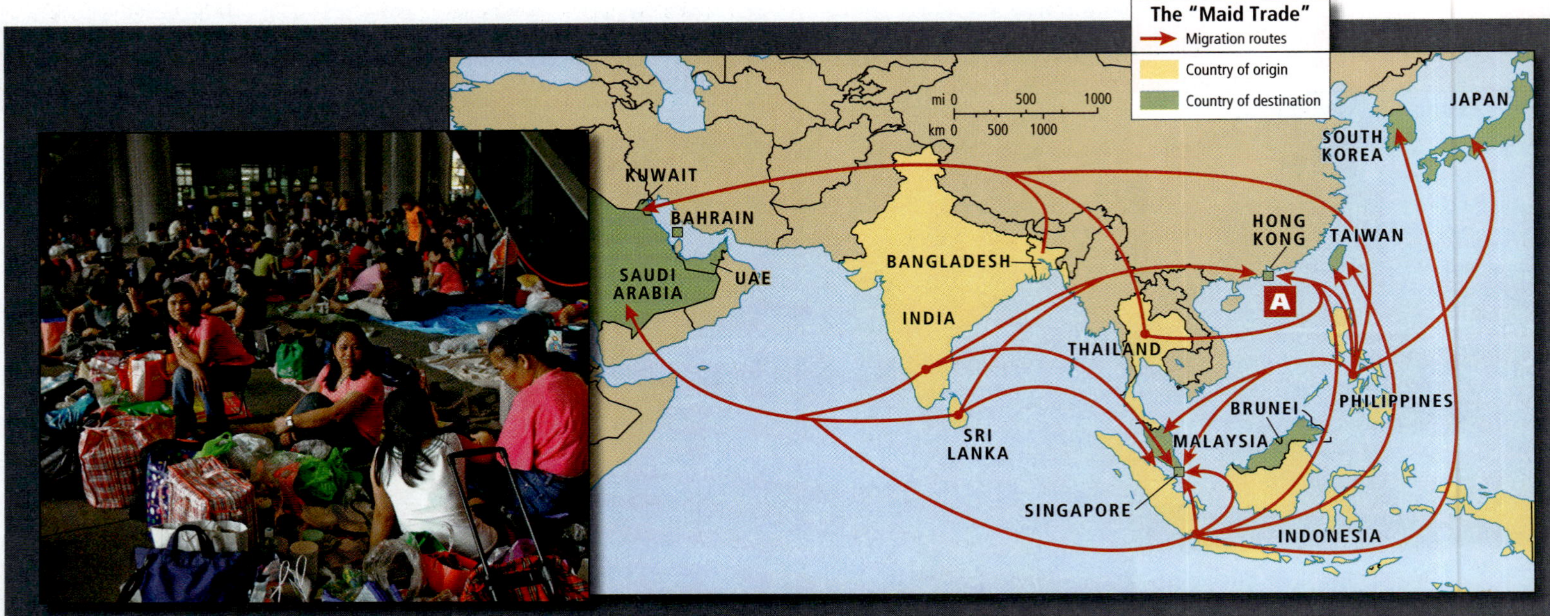

FIGURE 10.24 Globalization: The "maid trade." By 2009, more than 3 million women, primarily from the Philippines, Indonesia, and Sri Lanka, were working as domestic servants across Asia. **(A)** Filipina housemaids and nannies in Hong Kong relax on blankets on their day off. The Philippine government mandates that they must have Sundays off, so many public areas in central Hong Kong are occupied on Sundays by small groups of women talking, trading goods, playing games, and packing up things to send home. **(B)** Map of the "maid trade," showing the origins and destinations of the maids.

Thinking Geographically

How does this photo portray the inventiveness of women in the maid trade?

The Maid Trade Women constitute well over 50 percent of the more than 8 million emigrants from Southeast Asia. Many skilled nurses and technicians from the Philippines work in European, North American, and Southwest Asian cities. About 3 million participate in the global "maid trade" (shown in Figure 10.24A and the figure map). Most are educated women from the Philippines and Indonesia who work under 2- to 4-year contracts in wealthy homes throughout Asia. An estimated 1 to 3 million Indonesian maids now work outside of the country, mostly in the Persian Gulf.

The maid trade has become notorious for abusive working conditions and for having employers who often do not pay what they promise. In Saudi Arabia, the NGO Human Rights Watch is monitoring the cases of Muslim Indonesian women who were brutally abused—two were killed—by members of a privileged Saudi family. The Philippines went so far as to ban the maid trade in 1988, but reestablished it in 1995 after better pay and working conditions were negotiated with the countries that receive the workers.

VIGNETTE Every Sunday is amah (nanny) day in Hong Kong. Gloria Cebu and her fellow Filipina maids and nannies stake out temporary geographic territory on the sidewalks and public spaces of the central business district (see Figure 10.24A). Informally arranging themselves according to the different dialects of Tagalog (the official language of the Philippines) they speak, they create room-like enclosures of cardboard boxes and straw mats where they share food, play cards, give massages, and do each other's hair and nails. Gloria says it is the happiest time of her week, because for the other 6 days she works alone, caring for the children of two bankers.

Gloria, who is a trained law clerk, has a husband and two children back home in Manila. Because the economy of the Philippines has stagnated, she can earn more in Hong Kong as a nanny than in Manila in the legal profession. Every Sunday she sends most of her income (U.S.$125 a week) home to her family. *[Sources: Kirsty Vincin and the Economist. For detailed source information, see Text Credit pages.]*

THINGS TO REMEMBER

Geographic Insight 5

- **Urbanization, Food, and Development** Rural-to-urban migration instigated by agricultural modernization and mechanization has led to overcrowded slums that lack basic services, where farmers who were once able to produce sufficient food for their families now must purchase food.
- Southeast Asia as a whole is only 43 percent urban, but the rural–urban balance is shifting steadily.
- Often the focus of migration is the capital of a country, which may become a primate city—a city that, with its suburbs, is vastly larger than all others in a country.
- Thousands of people emigrate from the region each year to seek temporary or long-term employment in the Middle East, Europe, North America, Asian cities, or on the world ocean.

Pluralism

ıral pluralism in that it is inhabited
· different backgrounds. Over the
: come to the region from India,
ras, China, Southwest Asia, Japan,
ıese groups have remained distinct,
ated pockets separated by rugged
· religious practices and traditions
ural influences (Figure 10.25).
s Legacies section (page 431), the
utheast Asia include Hinduism,
sm, Islam, Christianity, and *ani-*
ı, a belief system common among
ral features such as trees, rivers,
ry spiritual meaning. These natu-
festivals and rituals to give thanks
ng of the seasons, and these ideas
religious traditions of the region.
ctice are complex; the patterns of
inland division. All but animism
were brought primarily by traders,
ıd colonists. Buddhism is dominant
rma, Thailand, and Cambodia. In
of Buddhism, Confucianism, and
Islam is dominant in Indonesia
untry), on the southern Malay Peninsula, and in Malaysia, and it is increasing in popularity in the southern Philippines. Roman Catholicism is the predominant religion in Timor-Leste and the Philippines, where it was introduced by Portuguese and Spanish colonists, respectively. Hinduism first arrived with Indian traders thousands of years ago and was once much more widespread; now it is found only in small patches, chiefly on the islands of Bali and Lombok, east of Java. Recent Indian immigrants who came as laborers in the twentieth century (during the latter part of the European colonial period) have reintroduced Hinduism to Burma, Malaysia, and Singapore, but only as minority communities.

cultural pluralism the cultural identity characteristic of a region where groups of people from many different backgrounds have lived together for a long time but have remained distinct

All of Southeast Asia's religions have changed as a result of exposure to one another. Many Muslims and Christians believe in spirits and practice rituals that have their roots in animism. Hindus and Christians in Indonesia, surrounded as they are by Muslims, have absorbed ideas from Islam, such as the seclusion of women. Muslims have absorbed ideas and customs from indigenous belief systems, especially ideas about kinship and marriage, as illustrated in the following vignette.

VIGNETTE Although arranged marriages have historically been the norm across Southeast Asia, in most urban and rural areas now, marriages are love matches. Such is the case for Harum and Adinda, who live in Tegal, on the island of Java

LOCAL LIVES FESTIVALS IN SOUTHEAST ASIA

elebrated in
and passersby
developed from
to wash statues
t water on the
, since the festival
of the year, a
r on anyone.

B A calligrapher at the Temple of Literature in Hanoi, Vietnam, elegantly writes down intentions for Tet (the New Year, celebrated in January or February). Exchanges of such calligraphy are hung in people's homes; the expressed intentions surround a particular idea or quality such as happiness, wealth, virtue, knowledge, talent, or long life.

C An elaborately costumed dancer in the Ati-Atihan Festival in Kalibo, Aklan, in the Philippines. Originally celebrating the migration to the Philippines of a group of indigenous people from the island of Borneo, Ati-Atihan includes non-Christian traditions, but is also celebrated by Christian Filipinos as a day honoring the infant Jesus.

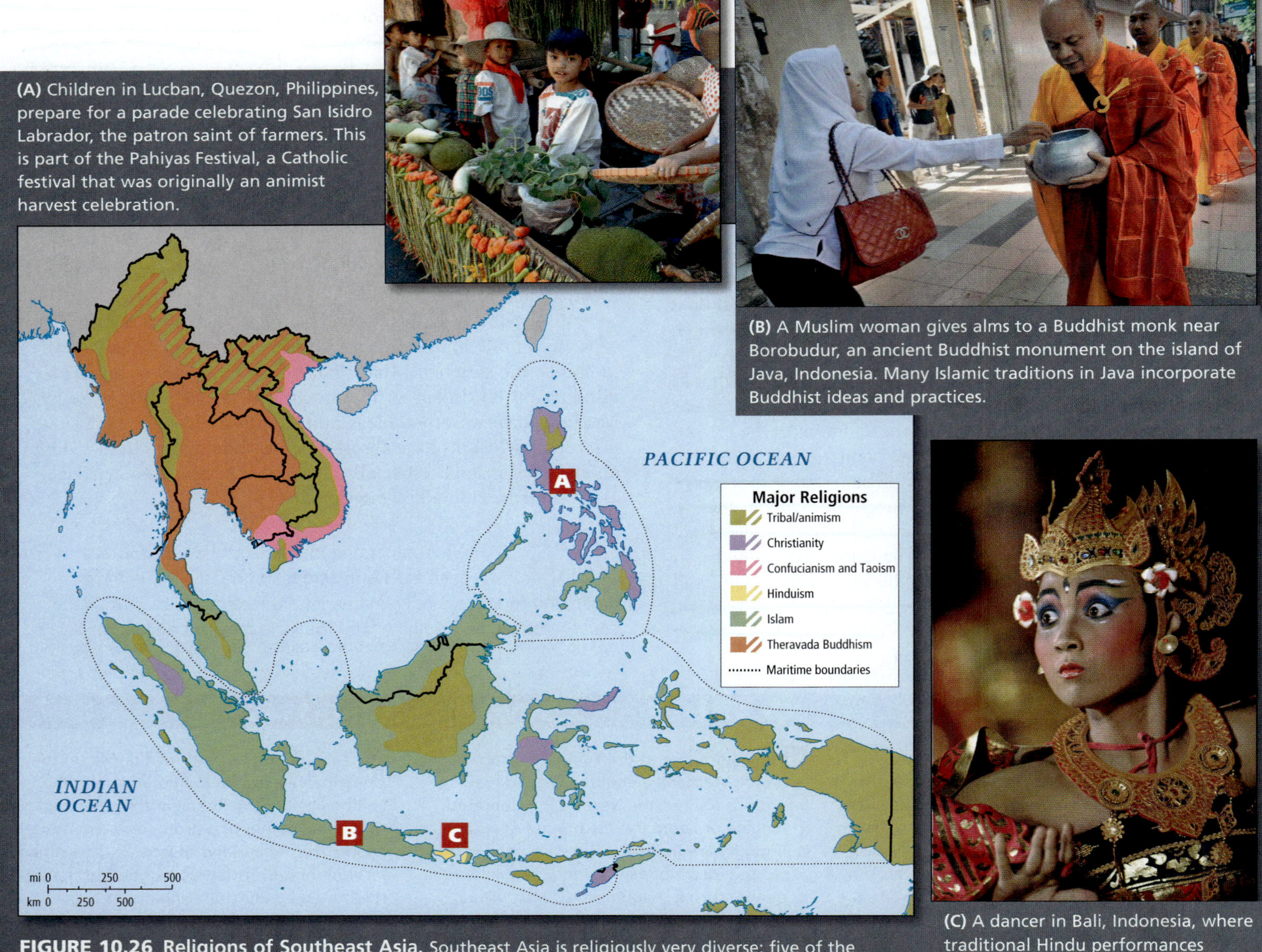

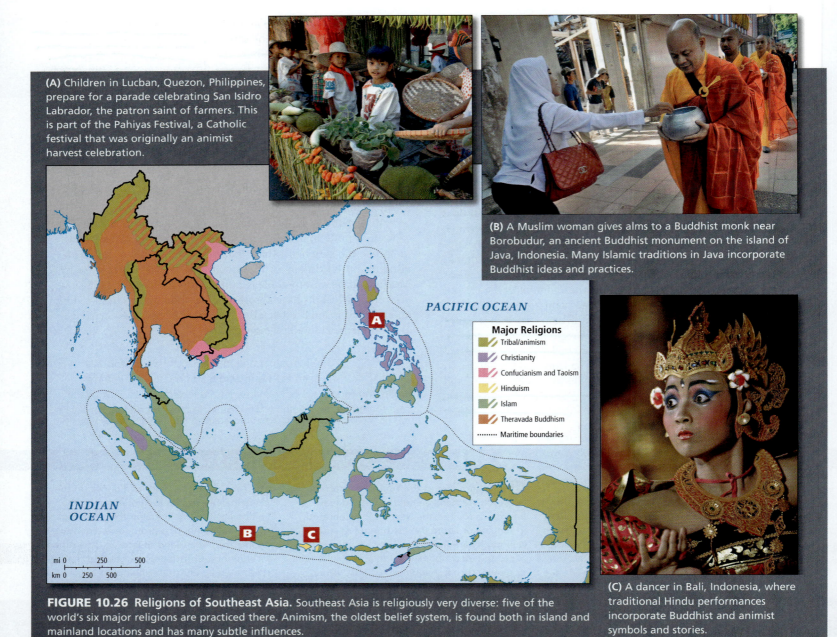

(A) Children in Lucban, Quezon, Philippines, prepare for a parade celebrating San Isidro Labrador, the patron saint of farmers. This is part of the Pahiyas Festival, a Catholic festival that was originally an animist harvest celebration.

(B) A Muslim woman gives alms to a Buddhist monk near Borobudur, an ancient Buddhist monument on the island of Java, Indonesia. Many Islamic traditions in Java incorporate Buddhist ideas and practices.

(C) A dancer in Bali, Indonesia, where traditional Hindu performances incorporate Buddhist and animist symbols and stories.

FIGURE 10.26 Religions of Southeast Asia. Southeast Asia is religiously very diverse: five of the world's six major religions are practiced there. Animism, the oldest belief system, is found both in island and mainland locations and has many subtle influences.

in Indonesia. They met in high school and some 10 years later, after saving a considerable sum of money, decided to formally ask both sets of parents if they could marry. Harum, an accountant, would normally be expected to pay the wedding costs, which could run to many thousands of dollars; however, Adinda was able to contribute from her salary as a teacher. Like nearly all Javanese people, both are Muslim. Because Islam does not have elaborate marriage ceremonies, colorful rituals from Christianity, Buddhism, and indigenous animism will enhance the elaborate and festive occasion.

In preparation, both bride and groom participate in unique Javanese rituals that remain from the days when marriages were arranged and the bride and groom did not know each other (**Figure 10.27**). A *pemaes,* a woman who prepares a bride for her wedding and whose role is to inject mystery and romance into the marriage relationship, bathes and perfumes the bride. She also puts on the bride's makeup and dresses her, all the while making offerings to the spirits of the bride's ancestors and counseling her about how to behave as a wife and how to avoid being dominated by her husband. The groom also takes part in ceremonies meant to prepare him for marriage. Both are counseled that their relationship is bound to change over the course of the decades as they mature and as their family grows older.

Despite the elaborate preparations for marriage, divorce in Indonesia (and also in Malaysia) is fairly common among Muslims, who often go through one or two marriages early in life before they settle into a stable relationship. Although the prevalence of divorce is lamented by society, it is not considered outrageous or disgraceful. Apparently, ancient indigenous customs predating Islam

dding. In the *dahar klimah* phase
the bridegroom makes three small
the bride; she then does the
hem that they should share joyfully

in life, and this attitude is still
. Nourse and Walter Williams. For
Credit pages.]

Diversity and Homogeneity

dinarily culturally diverse, a fact
customs (Figure 10.28). But in
es as the many groups continue
to global culture. For example,
nt ethnicities (Chinese, Tamil,

Malay, Bangladeshi) spend much of their spare time following the same European soccer teams, playing the same video games, visiting the same shopping centers, eating the same fast food, and talking to each other in English. While rural areas retain more traditional influences—of the world's 6000 or so actively spoken languages, 1000 can be found in rural Southeast Asia—in cities, one main language usually dominates trade and politics.

One group that has brought cultural diversity to Southeast Asia is the Overseas (or ethnic) Chinese (see Chapter 9, pages 413–414). Small groups of traders from southern and coastal China have been active in Southeast Asia for thousands of years; over the centuries, there has been a constant trickle of immigrants from China. The forbearers of most of today's Overseas Chinese, however, began to arrive in large numbers during the nineteenth century, when the European colonizers needed labor for their plantations and mines. Later, those who fled China's Communist Revolution after 1949 sought permanent homes in Southeast Asian trading centers. Today, more than 26 million Overseas Chinese live and work across Southeast Asia as shopkeepers and as small business owners. A few are wealthy financiers, and a significant number are still engaged in agricultural labor.

Chinese success at small-scale commercial and business activity throughout the region has reinforced the perception that the Chinese are diligent, clever, and extremely frugal, often working very long hours. Although few are actually wealthy, with their region-wide family and friendship connections and access to start-up money, some have been well positioned to take advantage of the new growth sectors in the globalizing economies of the region. Sometimes externally funded, new Chinese-owned enterprises have put out of business older, more traditional establishments depended upon by local people of modest incomes (both ethnic Malay and ethnic Chinese).

In recent years, when low- and middle-income Southeast Asians were hurt by the recurring financial crises, some tended to

LOCAL LIVES — FOODWAYS IN SOUTHEAST ASIA

heat
uce, is
Thailand.
ion
nd of
uslim
n South
ars
a food
est Asia.

B Throughout Southeast Asia, cold desserts feature shaved ice, coconut milk, evaporated or condensed milk, sweeteners, beans, noodles, corn, jelly, and other toppings. They are called *ais kacang* in Malaysia, *halo-halo* in the Philippines, *cendol* in Indonesia, and *ching bo leung* in Vietnam. Ice machines on European ships in the early twentieth century turned these desserts from beverages into ice cream–like sundaes.

C A *banh mi* sandwich, a street food in Vietnam and Laos, is a perfect example of the union of European (in this case, French) and Southeast Asian cuisine. European ingredients include French bread (a baguette), mayonnaise, and pâté. Vietnamese and Laotian ingredients include coriander, pickled vegetables, hot peppers, cucumber, sliced pork or headcheese, and fish sauce. This sandwich now has a U.S. audience and fans.

ON THE BRIGHT SIDE

Embracing Unity Within Diversity

For the most part, cultural diversity in Southeast Asia has led to amicable relations across ethnic and religious lines. Some countries actively promote national unity policies (e.g., *Pancasila* in Indonesia) as an antidote to diversity.

blame their problems on the Overseas Chinese. Waves of violence resulted. Chinese people were assaulted, their temples desecrated, and their homes and businesses destroyed. Conflicts involving the Overseas Chinese have taken place in Vietnam, Malaysia, and in many parts of Indonesia (Sumatra, Java, Kalimantan, and Sulawesi), as well. Some Overseas Chinese have attempted to diffuse tensions through public education about Chinese culture. Others have shown their civic awareness by financing economic and social aid projects to help their poorer neighbors, usually of local ethnic origins (Malay, Thai, Indonesian).

Gender Patterns in Southeast Asia

As observed in the Population Patterns section, gender roles are being transformed across Southeast Asia by urbanization and the changes it brings to family organization and employment. Here we look at some surprising traditional patterns of gender roles in extended families. However, moving to the city shifts people away from extended families and toward the nuclear family. The gains that women have made in political empowerment, educational achievement, and paid employment have not erased gender disparities.

Family Organization, Traditional and Modern Throughout the region, it has been common for a newly married couple to reside with, or close to, the wife's parents. Along with this custom is a range of behavioral rules that empower the woman in a marriage, despite some basic patriarchal attitudes. For example, a family is headed by the oldest living male, usually the wife's father. When he dies, he passes on his wealth and power to the husband of his oldest daughter, not to his own son. (A son goes to live with his wife's parents and inherits from them.) Hence, a husband may live for many years as a subordinate in his father-in-law's home. Instead of the wife being the outsider, subject to the demands of her mother-in-law—as is the case, for example, in South Asia—it is the husband who must show deference. The inevitable tension between the wife's father and the son-in-law is resolved by the custom of *ritual avoidance*—in daily life they simply arrange to not encounter each other much. The wife manages communication between the two men by passing messages and even money back and forth. Consequently, she has access to a wealth of information crucial to the family and has the opportunity to influence each of the two men.

Another traditional custom that empowers women is that women are often the family financial managers. Husbands turn over their pay to their wives, who then apportion the money to various household and family needs.

Urbanization and the shift to the nuclear family mean that young couples now frequently live apart from the extended family, an arrangement that takes the pressure to defer to his father-in-law off the husband. Because this nuclear family unit is often dependent entirely on itself for support, wives usually work for wages outside the home. Although married women lose the power they would have if they lived among their close kin, they are empowered by the opportunity to have a career and an income. The main drawback of this compact family structure, as many young families have discovered in Europe and the United States, is that there is no pool of relatives available to help working parents with child care and housework. Further, no one is left to help elderly parents maintain the rural family home.

Political and Economic Empowerment of Women Women have made some impressive gains in politics in Southeast Asia. Economically, they still earn less money than men and work less outside the home, but this will likely change if their level of education in relation to that of men continues to increase.

Southeast Asia has had several prominent female leaders over the years, most of whom have risen to power in times of crisis as the leaders of movements opposing corrupt or undemocratic regimes. In the Philippines, Corazon Aquino, a member of a large and powerful family, became president in 1986 after leading the opposition to Ferdinand Marcos, whose 21-year presidency was infamous for its corruption and authoritarianism. She is credited with helping reinvigorate political freedoms in the Philippines. Gloria Macapagal-Arroyo, from another powerful family, became president in 2001 after opposing a similarly corrupt president, and then she was accused of corruption herself. Still, her administration kept the economy sound throughout the global recession starting in 2007. In 2010, Benigno Aquino III, son of Corazon, became president.

In Indonesia, Megawati Sukarnoputri became president in 2001 after decades of leading the opposition to Suharto's notoriously corrupt 31-year reign. And in Burma, for more than two decades, a woman, Aung San Suu Kyi, has led opposition to the military dictatorship.

All of these women leaders were wives or daughters of powerful male political leaders, which raises some questions of nepotism. However, family favoritism cannot account for the several countries where the percentage of female national legislators is well above the world average of 18 percent: Timor-Leste (29 percent), Laos (25 percent), Vietnam (24 percent), Singapore (22 percent), the Philippines (22 percent), and Cambodia (21 percent).

Despite their increasing successes in politics and their acknowledged role in managing family money, women still lag well behind men in terms of economic empowerment (see Figure 10.22C). Throughout the region, men have a higher rate of employment outside the home than women and are paid more for doing the same work. But changes may be on the way. In Brunei, Malaysia, the Philippines, and Thailand, significantly more women than men are completing training beyond secondary school. If training qualifications were the sole consideration for employment, women would appear to have an advantage over men. This advantage may be significant if service sector economies, which generally require more education, become dominant in more countries. The service economy already dominates in Singapore, the Philippines, Thailand, and Indonesia.

318. THAILAND'S "THIRD SEX" WANTS ACCEPTANCE, LEGAL SUPPORT

der:

ne of sev- industry, ernational **ex tourism** he sexual erved for- n Asia during World War II, the War. Now, primarily civilian men ive out their fantasies during a few is found throughout the region land. In 2011, nineteen million up from 250,000 in 1965, and much as 70 percent were looking ry is officially illegal, some Thai publicly praised sex tourism for its her the economic crises, because t officials also favor sex tourism source of untaxed income from

of sex tourism is a high demand d has attracted organized crime. vorkers vary from 30,000 to more ne. Once girls and women have , gangs often coerce them into graphers estimate that 20,000 to nst their will—some as young as els. Their wages are too low to freedom. In the course of their k, they must service more than clients per day, and they are tinely exposed to physical abuse sexually transmitted diseases, ecially HIV.

sex indu trave fanta

ON THE BRIGHT SIDE

Women's Improving Status in Southeast Asia

Building on strong traditional roles at the family level, women now occupy prominent political roles in many countries, and their rising educational levels suggest that even in the most conservative countries, changes in gender roles will be significant. One result of the rising political status of women is that there is more focus on the abuse of women and girls in the sex industry.

old Watsanah K. (not her real name) every morning, attends after- al skills, and then goes to work at ok's red light district. There she will ica, Japan, Taiwan, Australia, Saudi to have sex with her. She leaves a while, and then goes to sleep.

Watsanah was born in northern Thailand to an ethnic minority group who are poor subsistence farmers. She married at 15 and had two children shortly thereafter. Several years later, her husband developed an opium addiction. She divorced him and left for Bangkok with her children. There, she found work at a factory that produced seat belts for a nearby automobile plant. In 1997, Watsanah lost her job as the result of the economic crisis that ripped through Southeast Asia. To feed her children, she became a sex worker.

Although the pay, between U.S.$400 and U.S.$800 a month, is much better than the U.S.$100 a month she earned in the factory, the work is dangerous and demeaning. Sex work, though widely practiced and generally accepted in Thailand, is illegal, and the women who do it are looked down on. As a result, Watsanah must live in constant fear of going to jail and losing her children. Moreover, she cannot always make her clients use condoms, which puts her at high risk of contracting AIDS and other sexually transmitted diseases. "I don't want my children to grow up and learn that their mother is a prostitute," says Watsanah. "That's why I am studying. Maybe by the time they are old enough to know, I will have a respectable job." *[Source: Alex Pulsipher and Debbi Hempel; Kaiser Family Foundation; BBC News. For detailed source information, see Text Credit pages.]* ■

THINGS TO REMEMBER

- The major religious traditions of Southeast Asia include Hinduism, Buddhism, Confucianism, Taoism, Islam, Christianity, and animism. All originated outside the region, with the exception of the animist belief systems.
- In some areas, women's economic and political empowerment builds on cultural traditions that give special responsibilities to women.
- Some countries have become centers for the global sex industry, which puts many women at risk of violence and disease.

Geographic Insights: Southeast Asia Review and Self-Test

1. Climate Change, Food, and Water: Conversion of forests to oil palm plantations, rice farms, subsistence plots, and settlements all add to greenhouse gas emissions, which intensify climate change and leave many people vulnerable to both flooding and occasional droughts.

- How does deforestation lead to more greenhouse gas emissions?
- How is food production for global markets related to deforestation?
- How is deforestation related to increased risks of floods and droughts?

2. Globalization and Development: Globalization has brought spectacular economic successes and occasional, dramatic economic declines. Urban incomes soar during periods of growth, but the poor are vulnerable during periods of global economic decline.

- How is globalization linked to the economic success of Southeast Asia?
- How is globalization linked to rapid urbanization and the growth of slums in this region?
- Why have global recessions caused such dramatic economic declines for the many workers in Southeast Asia?

3. Power and Politics: Militarism, corruption, and favoritism are recurring issues in Southeast Asia. Although modernization and prosperity have brought significant increases in democratic participation, there have also been reversals, especially during periods of economic downturn. These sometimes result in violence, often against ethnic minorities.

- What is the political history backdrop to the current dynamics of political power in Southeast Asia?
- How have modernization and prosperity helped to improve democratic participation in this region?
- What are some of the explanations for the violence against ethnic or religious minorities that tend to occur during hard economic times?
- What strategies have governments used to alleviate the ethnic tensions?
- What are the signs that ordinary citizens are trying to increase democratic participation?

4. Population and Gender: Economic change has brought better job opportunities and increased status for women, who are then choosing to have fewer children. The improving roles and status of women have increased awareness that sex trafficking is an abuse of human rights.

- Describe the circumstances that have led to better job opportunities for women.
- How are the improving roles and status of women changing attitudes toward female sex work?

5. Urbanization, Food, and Development: The development of export-oriented modernized agriculture and agribusiness have forced farmers who were once able to produce enough food for their families to migrate to cities, where jobs that fit their skills are scarce, where they must purchase their food, and where the only affordable housing is in slums that lack essential services. Surplus skilled and semiskilled workers, especially women, often migrate abroad for employment.

- How have development strategies, such as mechanized agriculture and agribusiness, changed the degree of food self-sufficiency in this region?
- How are development and urbanization linked to the decision of many to migrate abroad for employment?

Critical Thinking Questions

1. What do you think are the most serious threats to Southeast Asia posed by global climate change? What are the most promising responses to global climate change emerging from this region?

2. Which countries in the region have been most successful in controlling population growth? How do population issues differ among Southeast Asian countries? How do the population issues of the region as a whole compare with those of Europe or Africa?

3. Discuss the role of crony capitalism in the economic crisis of the 1990s in this region.

4. Did those responsible for resolving the economic crisis of the 1990s opt for reregulation or deregulation of financial institutions? Why did they make this choice?

5. In what ways did state aid to market economies and export-led growth amount to a strategic and limited embrace of globalization?

6. How did the economic crisis of the 1990s result in expanded democracy in some countries? Which countries in this region are the least democratic? Why do you think that is the case?

7. Describe the spatial distribution of major religious traditions of this region. How have these religious traditions influenced each other over time?

8. What factors have produced major flows of refugees in Southeast Asia?

9. Relationships and gender roles in traditional Southeast Asian families are not necessarily what an outsider might expect. Describe the characteristics of these relationships that interested you most. Why? Discuss some of the ways in which gender roles vary from those in South Asia or East Asia.

10. How has the role of the Overseas Chinese evolved over time? Why have some people in this region resented them? To what extent is this resentment understandable, if unfair? Where have the Overseas Chinese been most and least sensitive to public opinion?

tions

coral bleaching 429
crony capitalism 436
cultural pluralism 451
detritus 424
domino theory 433

feminization of labor 435
foreign exchange 448
resettlement schemes 442
sex tourism 455

A Australia's Eastern Highlands, Blue Mountains
B Australian Desert
CHINA
JAPAN
TAIWAN
PHILIPPINES
South China Sea
Philippine Sea
Sulu Sea
Celebes Sea
BRUNEI DARUSSALAM
MALAYSIA
Borneo
INDONESIA
Banda Sea
Arafura Sea
Timor Sea
TIMOR-LESTE
PALAU
Micronesia
Melanesia
FEDERATED STATES OF MICRONESIA
MARSHALL ISLANDS
NAURU
KIRIBATI (GILBERT ISLANDS)
Equator
PAPUA NEW GUINEA
SOLOMON ISLANDS
TUVALU (ELLICE ISLANDS)
VANUATU (NEW HEBRIDES)
FIJI
SAMOA (WESTERN SAMOA)
TONGA
International Date Line
Coral Sea
Tasman Sea
INDIAN OCEAN
AUSTRALIA
WESTERN AUSTRALIA
NORTHERN TERRITORY
QUEENSLAND
SOUTH AUSTRALIA
NEW SOUTH WALES
VICTORIA
TASMANIA
NEW ZEALAND
Great Sandy Desert
Gibson Desert
Great Victoria Desert
Nullarbor Plain
Great Australian Bight
Great Dividing Range
Great Barrier Reef
Gulf of Carpentaria
New Caledonia (France)
North Island
South Island
Chatham Islands (New Zealand)
Kermadec Islands (New Zealand)
Campbell Island (New Zealand)

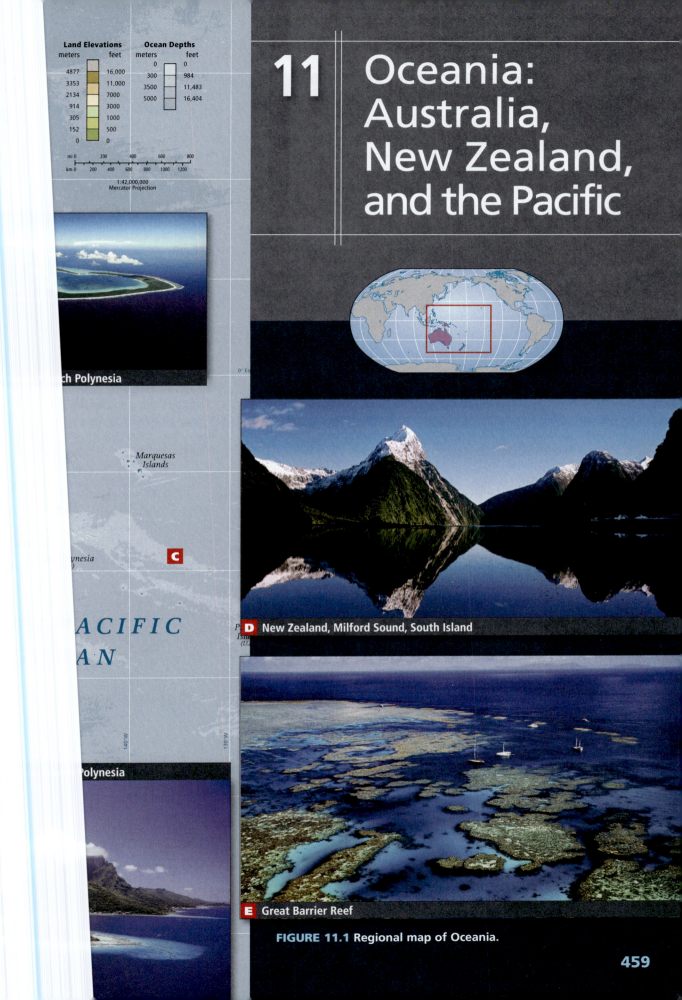

11 Oceania: Australia, New Zealand, and the Pacific

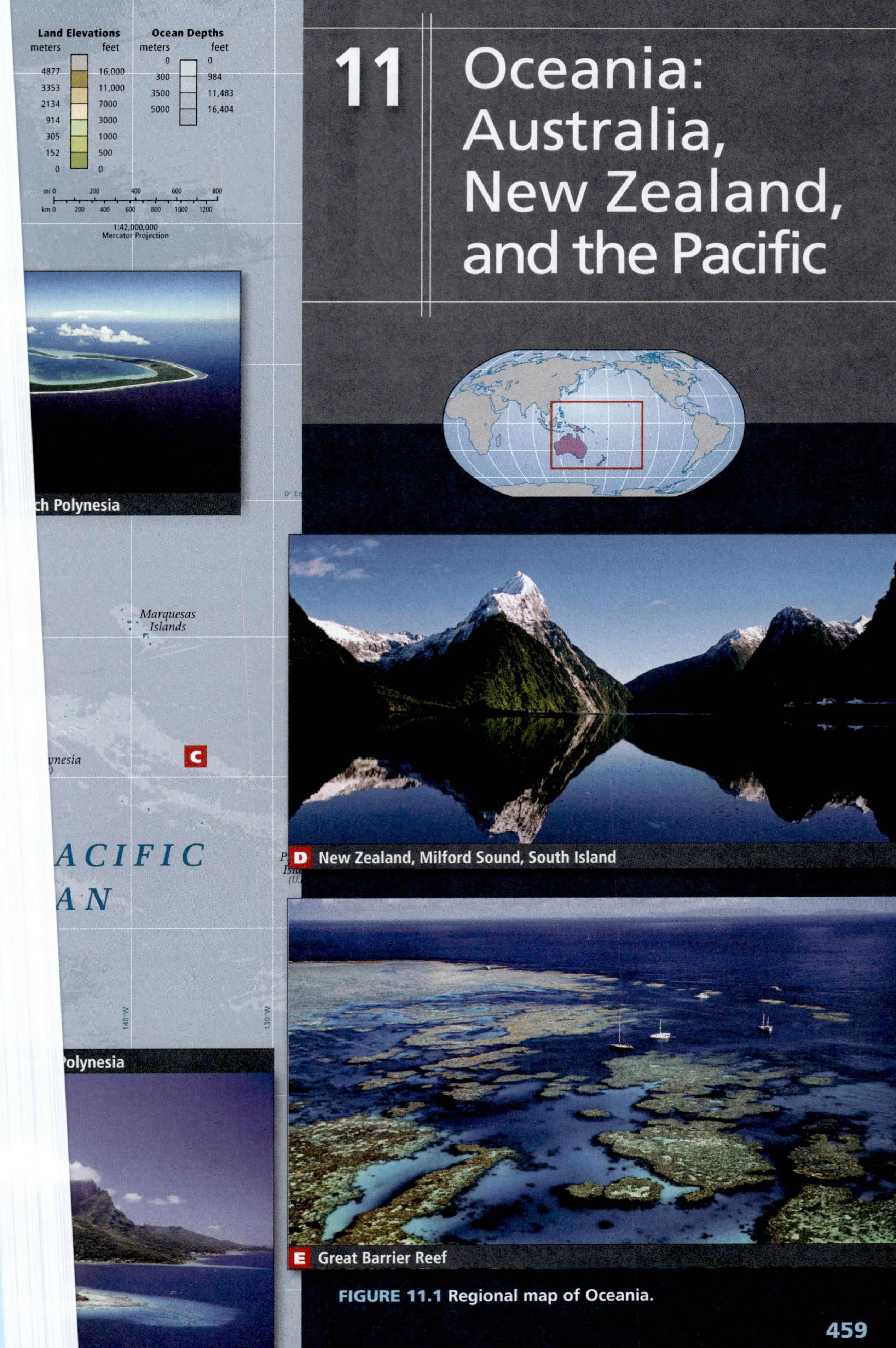

FIGURE 11.1 Regional map of Oceania.

GEOGRAPHIC INSIGHTS: OCEANIA

After you read this chapter, you will be able to discuss the following geographic insights as they relate to the nine thematic concepts:

1. **Climate Change and Water:** Oceania's sea level and freshwater resources are being impacted by climate change.

2. **Food and Development:** European systems for producing food and fiber introduced into Oceania have greatly changed environments.

3. **Globalization and Development:** Globalization, coupled with a greater focus on neighboring Asia (rather than on long-time connections with Europe), has transformed patterns of trade and economic development across Oceania.

4. **Population and Urbanization:** This largest region of the world is only lightly populated, but it is highly urbanized.

5. **Gender, Power, and Politics:** Changes in political empowerment have taken different forms across Oceania, but there is a general trend toward a more active role for women.

The Oceanian Region

Oceania (see Figure 11.1 on pages 458–459; see also Figure 11.3) is the largest but least populated world region. The idyllic, far-flung landscapes of the Pacific disguise the fact that the people of this region face particularly complex challenges. The nine thematic concepts in this book are explored as they arise in the discussion of regional issues, with interactions between two or more themes featured, as in the geographic insights on this page. Vignettes, like the one that follows about the Kiribati archipelago and the challenges its people face due to climate change and rising seas, illustrate one or more of the themes as they are experienced in individual lives.

GLOBAL PATTERNS, LOCAL LIVES Aurora (a pseudonym) is a nursing student in Brisbane, Australia, who is undergoing a wrenching personal adaptation due to climate change. Aurora is from Kiribati, a Pacific nation of 33 tiny islands barely 6½ feet above sea level. Rising seas are swamping the Kiribati archipelago, which straddles the equator just west of the international date line (180° longitude). Aurora pines for her island's blue lagoons. While she studies for a new future and prepares to become her household's primary income earner, she does so without the daily close company of her raucous extended family.

Several years ago, Aurora and her fellow citizens (100,000 people live in Kiribati) began to notice alarming environmental changes (**Figure 11.2**). Their drinking water was getting brackish (salty), several shoreline villages had sunk below sea level, and the climate became so dry that skilled gardeners could no longer raise their customary cabbages, taro crops, and cucumbers. Drought conditions were worsened by rising salty tides that saturated formerly fertile soil (see Figure 11.2A, B). Scientists predict that in a few decades, with rising seas and more frequent droughts and storms, much of Kiribati will be uninhabitable.

Aurora's nursing education, funded by a government program called AusAID, is one of several strategies developed by Kiribati president Anote Tong to encourage his people to gradually "migrate with dignity." President Tong sees climate change as a huge challenge for his country, and he hopes that other South Pacific islands, as well as Australia and New Zealand, will allow Kiribati people to resettle there permanently. Through the AusAID program, Aurora and other young people receive an education that will help them support their large extended families after they have found employment outside Kiribati and their family members have joined them.

FIGURE 11.2 Kiribati.

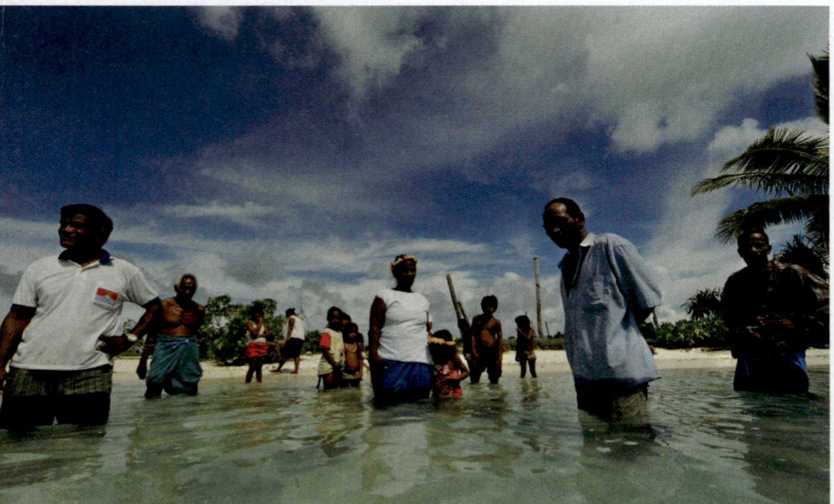

(A) Inhabitants of a now-relocated Kiribati village stand where their village used to be. Kiribati's coastline is eroding as sea level rises.

(B) A swamp taro pit that was rendered infertile by intruding salt water. Swamp taro is a major food source for inhabitants of low-lying islands. It is cultivated in a carefully managed pit that makes use of a layer of fresh (not salty) groundwater that exists along the coastline. As sea level rises, saltwater intrudes into this layer of fresh groundwater, killing the plants or stunting their growth.

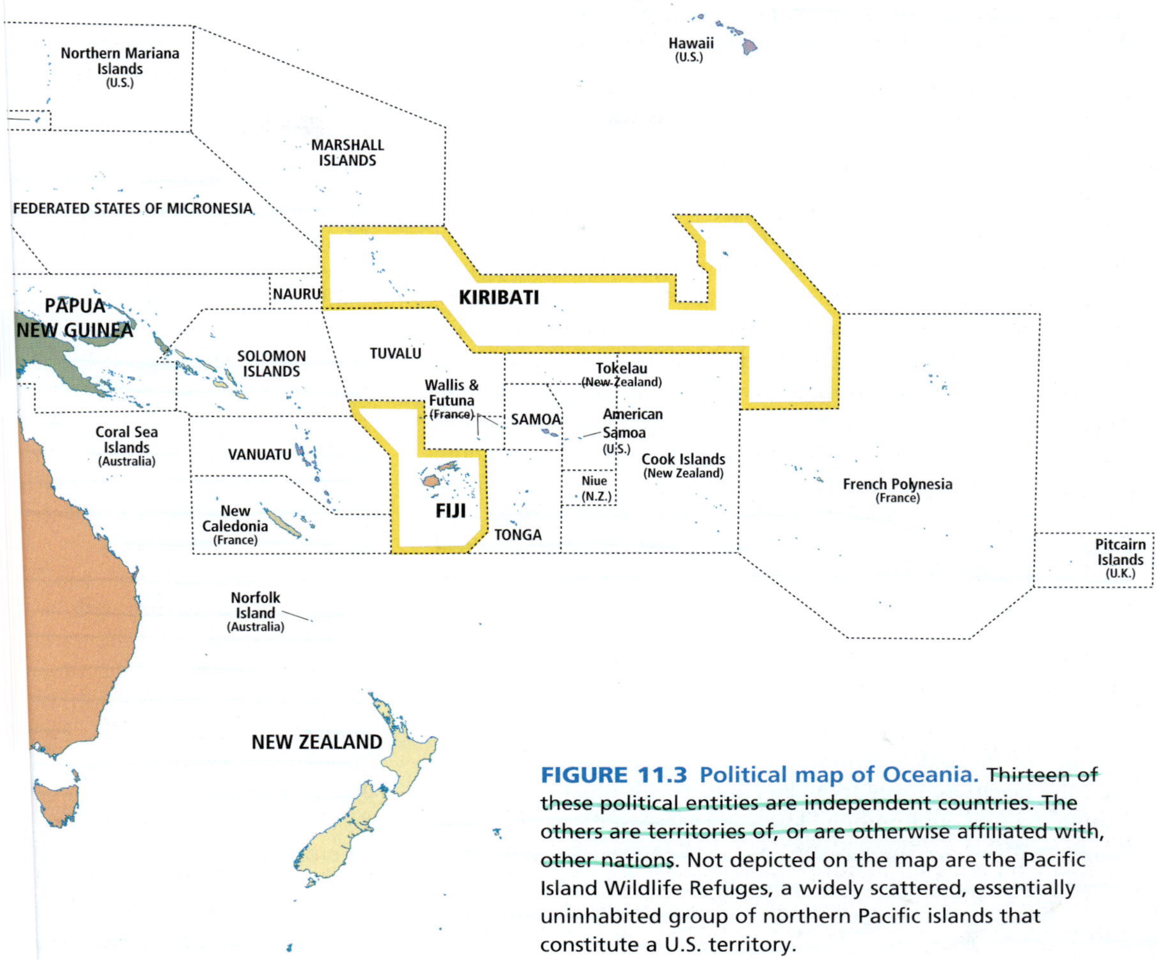

FIGURE 11.3 Political map of Oceania. Thirteen of these political entities are independent countries. The others are territories of, or are otherwise affiliated with, other nations. Not depicted on the map are the Pacific Island Wildlife Refuges, a widely scattered, essentially uninhabited group of northern Pacific islands that constitute a U.S. territory.

ɔther nursing student says
as his nation sinks into the
to leave Kiribati. They see the
: as Christians they believe that
to perish just like that."
ı Broadcasting Corporation)
ɔproached landowners on
to the south (**Figure 11.3**),
are twofold: to produce food
ijian soil to replace the soil
ɔroposition will surely trigger
xperiencing social and political
ces: National Public Radio; BBC
ɔn, see Text Credit pages.] ■

Perhaps nowhere else on Earth are the effects of climate change so measurably real as they are in the low-lying islands of the Pacific, where changes in rainfall patterns and sea level are not abstractions. Of special note is the fact that these islands have contributed little to the greenhouse gases that have triggered rapid climate change—not so much because of the low-emission lifestyles of the islands' inhabitants, but rather because the islands' total population is small. As we shall see, seemingly remote Pacific places are anything but remote from general global influences. Beyond climate change, people in this region are dealing with changes brought on by tourism, environmental degradation caused by modern mining techniques, freshwater scarcity, ocean pollution, and rapidly changing cultural and trading patterns.

THE GEOGRAPHIC SETTING

New Zealand, Papua New
s scattered across the Pacific
que world region in that it is
covers the largest portion of

Earth's surface of any region, yet it is home to only 38.7 million people, the vast majority of whom live in Australia (22.7 million), Papua New Guinea (6.9 million), and New Zealand (4.4 million). In this book, we also include the U.S. state of Hawaii (1.7 million) in Oceania. The thousands of other Pacific islands are home to only 3 million people. The Pacific Ocean is the link that unites Oceania as a region, profoundly influencing life even on dry land; but across the region, the ocean also acts as a biological and cultural barrier.

Terms in This Chapter
Some maps in this chapter show different place-names for the same locations. This reflects the political evolution of the region, where some islands previously were grouped under one name that remains in everyday use but are now grouped into country units with another name. For example, the Caroline Islands, located north of New Guinea, still go by that name on maps and charts, but they have been divided into two countries: Palau (a small group of islands at the western end of the Caroline Islands) and the Federated States of Micronesia, which extends over 2000 miles west to east, from Yap to Kosrae. In addition, there are three names commonly used to refer to large cultural groupings of islands: Micronesia, Melanesia, and Polynesia (see Figure 11.13 on page 474). These groupings are not political units; rather, they are based on ancient ethnic and cultural links.

Physical Patterns

The Pacific Ocean serves as both a link and a barrier in Oceania. Plants and animals have found their way from island to island by floating on the water, swimming, or flying; humans have long used the ocean as a way to make contact with other peoples for visiting, trading, and raiding. The movement of the water in the Pacific, with its varying temperatures, influences climates across all the landmasses of the Pacific Rim. Unfortunately, as we shall see, the ocean also serves as a conveyer of pollution, especially discarded plastics (see page 471).

The wide expanses of water also serve as a barrier, profoundly limiting the natural diffusion of plant and animal species and keeping Pacific Islanders spatially isolated from one another. The vast ocean has imposed solitude and fostered self-sufficiency and subsistence economies among its human occupants well into the modern era.

Gondwana the great landmass that formed the southern part of the ancient supercontinent Pangaea

Great Barrier Reef the longest coral reef in the world, located off the northeastern coast of Australia

Continent Formation

The largest landmass in Oceania is the ancient continent of Australia at the southwestern edge of the region (see the Figure 11.1 map). The Australian continent is partially composed of some of the oldest rock on Earth and has been relatively stable for more than 200 million years, with very little volcanic activity and only an occasional mild earthquake. Australia was once a part of the great landmass called **Gondwana** that formed the southern part of the ancient supercontinent Pangaea (see Figure 1.25 on page 46). What became present-day Australia broke free from Gondwana and drifted until it eventually collided with the part of the Eurasian Plate on which Southeast Asia sits. That impact created the mountainous island of New Guinea to the north of Australia.

Australia is shaped roughly like a dinner plate with an irregularly broken rim with two bites taken out of it: one in the north (the Gulf of Carpentaria) and one in the south (the Great Australian Bight). The center of the plate is the great lowland Australian desert, with only two hilly zones and rocky outcroppings (see Figure 11.1B). The eastern rim of Australia is composed of uplands; the highest and most complex of these are the long, curving Eastern Highlands (labeled "Great Dividing Range" in Figure 11.1; see also Figure 11.1A). Over millennia, the forces of erosion—both wind and water—have worn most of Australia's landforms into low, rounded formations. Some of these, like Uluru (Ayers Rock), are quite spectacular (Figure 11.4).

Off the northeastern coast of the continent lies the **Great Barrier Reef**, the largest coral reef in the world and a World Heritage Site since 1981 (see Figure 11.1E). It stretches in an irregular arc for more than 1250 miles (2000 kilometers) along the coast of Queensland, covering 135,000 square miles (350,000 square kilometers). The Great Barrier Reef is so large

FIGURE 11.4 Uluru (Ayer's Rock). This land formation, near the center of Australia, is a smooth remnant of ancient mountains. The site is held sacred by central Australian Aborigines. It is also one of Australia's most popular tourist destinations.

imate by interrupting the westward- ne mid–South Pacific circulation d to the south, where it warms the . Threats to the health of the Great page 467.

e (and are still being) created by a the movement of tectonic plates. rn reaches of Oceania—including and the main islands of Fiji—are andmass; they are large, moun- plex. Other islands in the region rn part of the Ring of Fire (see y in this latter group are situated tonic plates are either colliding the Mariana Islands east of the were formed when the Pacific lippine Plate. The two islands of n the eastern edge of the Indian- vard by its convergence with the

produced through another form th **hot spots**, places where particu- d from Earth's core breaches the past 80 million years, the Pacific t spots, creating a string of volca- kilometers) long. The youngest are active, are on or near the

ee forms: volcanic high islands, vhich are coral platforms raised islands are usually volcanoes ntainous, rocky formations that and rugged landscapes contain New Zealand, the Hawaiian nd are examples of high islands a low-lying island, or chain of nave built up on the circular or (see Figure 11.1C). These reefs oon that was once the volcano's ition, atoll islands tend to have ents and very limited supplies coral platforms sitting on top es. An example of this type of lies in French Polynesia in the

ches nearly from pole to pole, uated within the tropical and The tepid water temperatures imates year-round to nearly all (Figure 11.5). Seasonal varia- n the southernmost reaches of

Moisture and Rainfall With the exception of the vast arid interior of Australia, much of Oceania is warm and humid nearly all the time. New Zealand and the high islands of the Pacific receive copious rainfall; before human settlement, they supported dense forest vegetation. Much of that forest is gone after 1000 years of human impact (see Figure 11.5B; see also Figure 11.8D on page 468). Travelers approaching New Zealand, either by air or by sea, often notice a distinctive long white cloud that stretches above the north island. Seven hundred years ago, early **Maori** settlers (members of the Polynesian group) also noticed this phenomenon, and they named that place *Aotearoa*, "land of the long white cloud," a name that is now applied to all of New Zealand.

The legendary **Roaring Forties** (named for the 40th parallel south) are powerful air and ocean currents that speed around the far Southern Hemisphere virtually unimpeded by landmasses. These *westerly winds* (blowing west to east), which are responsible for Aotearoa's distinctive bodies of moist air, deposit a drenching 130 inches (330 centimeters) of rain per year in the New Zealand highlands and more than 30 inches (76 centimeters) per year on the coastal lowlands (see Figure 11.5B). At the southern tip of New Zealand's North Island, the wind averages more than 40 miles per hour (64 kilometers per hour) about 118 days per year. Farmers in the area stake their cabbages to the ground so the plants will not blow away.

By contrast, two-thirds of Australia is overwhelmingly dry (see Figure 11.5A). The dominant winds affecting Australia are the north and south *easterlies* (blowing east to west) that converge east of the continent. The Great Dividing Range blocks the movement of moist, westward-moving air so that rain does not reach the interior (an orographic pattern; see Figure 1.28 on page 49). As a result, a large portion of Australia receives less than 20 inches (50 centimeters) of rain per year, and humans have found rather limited uses for this interior territory. But the eastern (windward) slopes of the highlands receive more abundant moisture. This relatively moist eastern rim of Australia was favored as a habitat by both the indigenous people and the Europeans who displaced them after 1800. During the southern summer, the fringes of the monsoon that passes over Southeast Asia and Eurasia bring moisture across Australia's northern coast. There, annual rainfall varies from 20 to 80 inches (50 to 200 centimeters).

Overall, Australia is so arid that it has only one major river system, which is in the temperate southeast where most Australians live. There, the Darling and Murray rivers drain one-seventh of the continent, flowing west and south into the Indian Ocean near Adelaide. One measure of Australia's overall dryness is that the entire average *annual* flow of the Murray-Darling river system is equal to just *one day's* average flow of the Amazon in Brazil.

hot spots individual sites of upwelling material (magma) that originate deep in Earth's mantle and surface in a tall plume; hot spots tend to remain fixed relative to migrating tectonic plates

atoll a low-lying island or chain of islets, formed of coral reefs that have built up on the circular or oval rims of a submerged volcano

Maori Polynesian people indigenous to New Zealand

Roaring Forties powerful air and ocean currents at about 40° S latitude that speed around the far Southern Hemisphere virtually unimpeded by landmasses

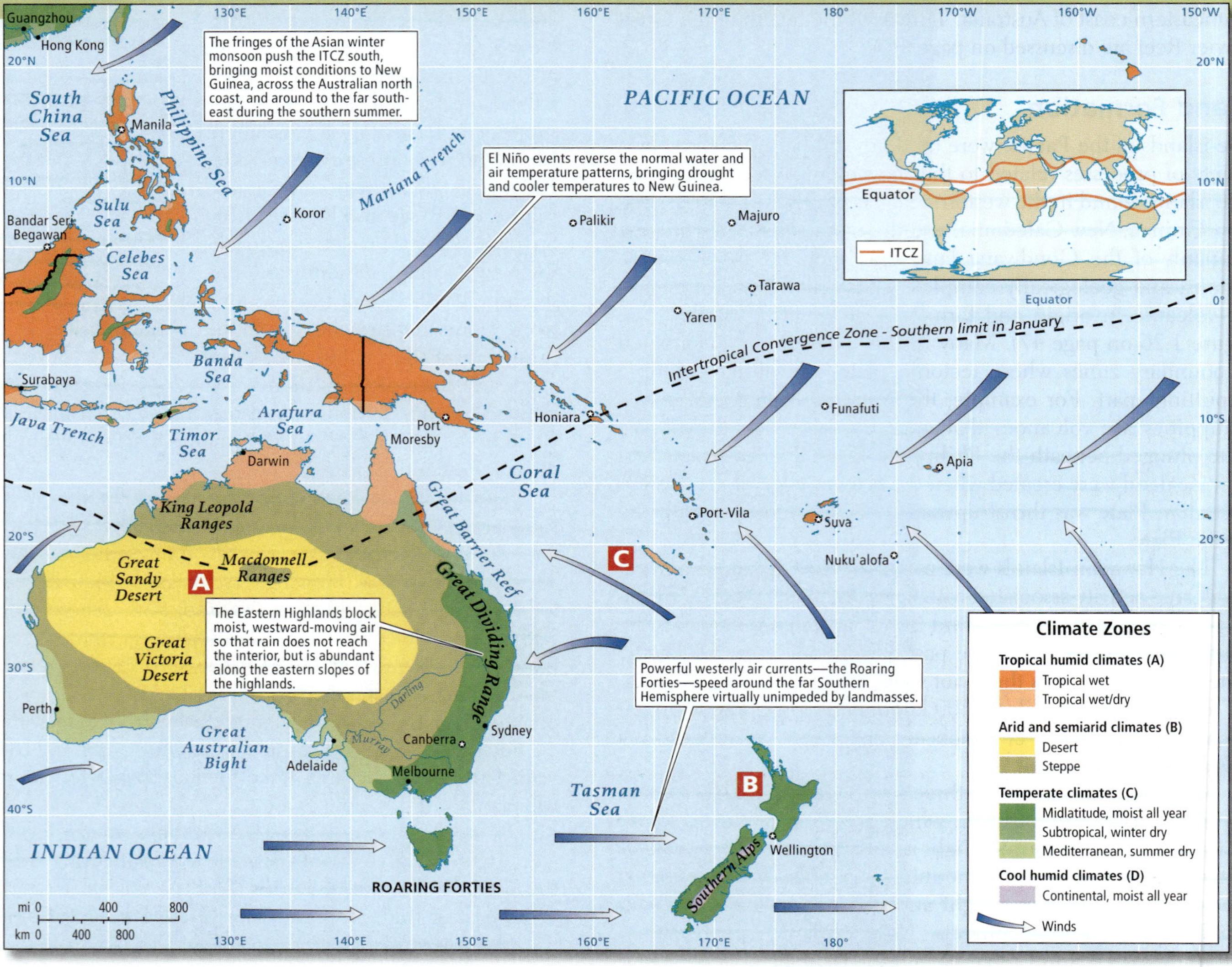

Figure 11.5 Photo Essay: Climates of Oceania

The fringes of the Asian winter monsoon push the ITCZ south, bringing moist conditions to New Guinea, across the Australian north coast, and around to the far south-east during the southern summer.

El Niño events reverse the normal water and air temperature patterns, bringing drought and cooler temperatures to New Guinea.

The Eastern Highlands block moist, westward-moving air so that rain does not reach the interior, but is abundant along the eastern slopes of the highlands.

Powerful westerly air currents—the Roaring Forties—speed around the far Southern Hemisphere virtually unimpeded by landmasses.

Intertropical Convergence Zone - Southern limit in January

ROARING FORTIES

PACIFIC OCEAN

INDIAN OCEAN

South China Sea

Philippine Sea

Sulu Sea

Celebes Sea

Banda Sea

Arafura Sea

Timor Sea

Java Trench

Mariana Trench

Coral Sea

Tasman Sea

Great Australian Bight

Great Barrier Reef

Great Dividing Range

King Leopold Ranges

Macdonnell Ranges

Great Sandy Desert

Great Victoria Desert

Southern Alps

Darling

Murray

Guangzhou, Hong Kong, Manila, Bandar Seri Begawan, Koror, Palikir, Majuro, Tarawa, Yaren, Surabaya, Port Moresby, Darwin, Honiara, Funafuti, Apia, Port-Vila, Suva, Nuku'alofa, Perth, Adelaide, Melbourne, Canberra, Sydney, Wellington

130°E, 140°E, 150°E, 160°E, 170°E, 180°, 170°W, 160°W, 150°W

20°N, 10°N, 0°, 10°S, 20°S, 30°S, 40°S

Equator

ITCZ

mi 0 400 800

km 0 400 800

Climate Zones

Tropical humid climates (A)
- Tropical wet
- Tropical wet/dry

Arid and semiarid climates (B)
- Desert
- Steppe

Temperate climates (C)
- Midlatitude, moist all year
- Subtropical, winter dry
- Mediterranean, summer dry

Cool humid climates (D)
- Continental, moist all year

Winds

A Desert, Alice Springs, Australia

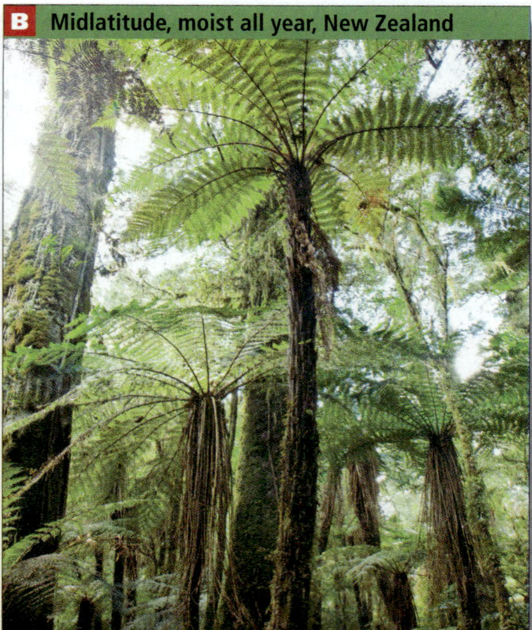

B Midlatitude, moist all year, New Zealand

C Tropical wet/dry, New Caledonia

ıntainous high islands also exhibit
vith a wet windward side and a
11.5C). Rainfall amounts on the
rably across the region. Some of
ɔath of trade winds, which deliver
rain per year, on average. These
ıriety of plants and animals. Other
hose on the equator, receive con-
lominated by grasslands that sup-

3 the *El Niño* phenomenon, a
on of air and water in the Pacific
to 7 years. Although these cycli-
ot yet well understood, scientists
how the oscillations may occur

-1998 illustrates the effects of this
1997, the island of New Guinea
ıre 11.1 map) had received very
Crops failed, springs and streams
tinder-dry forests. The cloudless
and away from elevations above
emperatures at high elevations
for stretches of a week or more.
unaccustomed to chilly weather
ıer end of the system, along the
ınd South America, the warmer-
sually strong storms, high ocean
d rainfall (see Figure 11.6C).
n, in which normal conditions
dentified and named *La Niña*.
ña patterns can bring unusually
uding tornadoes and blizzards)
to North America. La Niña is
the 2010–2011 major floods in
re especially damaging because
ted lengthy drought.

es an isolated continent and
animal life (*fauna*) and plant
endemic, meaning they exist in
se on Earth. This is especially
es of birds known in Australia,
nic), but many Pacific islands

ustralia The uniqueness of
ıe result of the continent's long
vely homogeneous landforms,
broke away from Gondwana
than 65 million years ago,
imal and plant species have
d in isolation. One spectac-
esult of this long isolation

endem
particul

FIGURE 11.6 A model of the El Niño phenomenon.

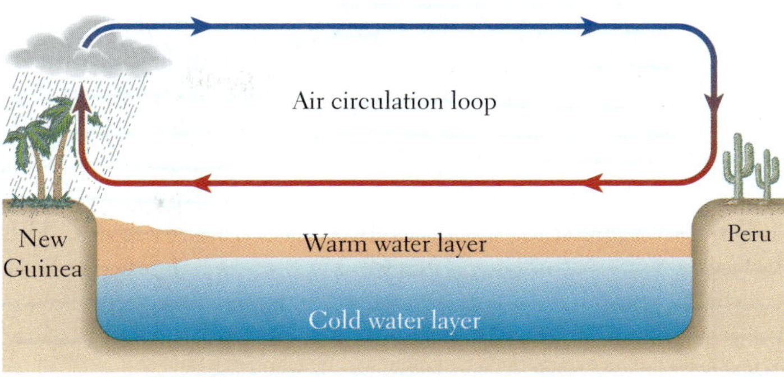

(A) Normal equatorial conditions. Water in the equatorial western Pacific (the New Guinea/Australia side) is warmer than water in the eastern Pacific (the Peru side). Due to prevailing wind patterns, the warm water piles up in the west. Warm air rises above the warm-water bulge in the western Pacific and forms rain clouds. The rising air cools and, once in the higher atmosphere, moves in an easterly direction. In the east, the dry, cool air descends, bringing little rainfall to Peru.

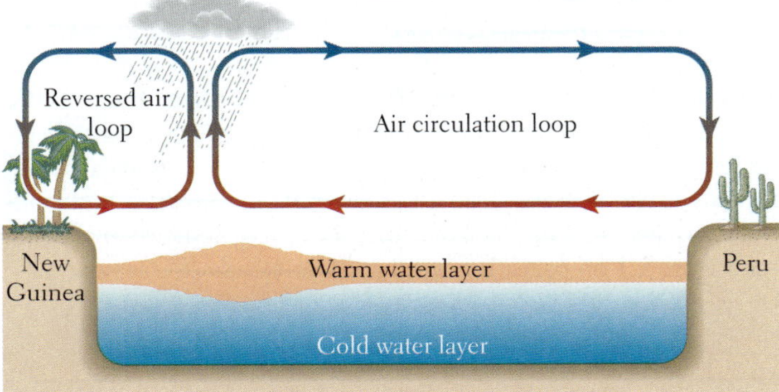

(B) Developing El Niño conditions. As an El Niño event develops, the ocean surfaces' warm-water bulge (orange) begins to move east. The air rising above it splits into two formations, one circling east to west in the upper atmosphere and one west to east.

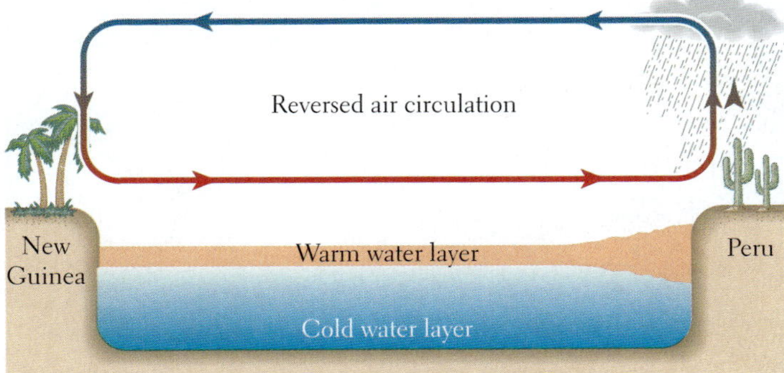

(C) Fully developed El Niño. Slowly, as the bulge of warm water at the surface of the ocean moves east, it forces the whole system into the fully developed El Niño, with air at the surface and in the upper atmosphere flowing in reverse of normal **(A)**. Instead of warm, wet air rising over the mountains of New Guinea and condensing as rainfall, cool, dry, cloudless air descends to sit at Earth's surface. Meanwhile, in the east, the normally dry, clear coast of Peru has clouds and rainfall.

marsupials mammals that give birth to their young at a very immature stage and nurture them in a pouch equipped with nipples

monotremes egg-laying mammals, such as the duck-billed platypus and the spiny anteater

is the presence of more than 144 living species of endemic marsupial animals. **Marsupials** are mammals that give birth to their young at a very immature stage and then nurture them in a pouch equipped with nipples. The best-known marsupials are kangaroos; other marsupials include wombats, koalas, and bandicoots. The **monotremes**, egg-laying mammals that include the duck-billed platypus and the spiny anteater, are endemic to Australia and New Guinea.

Most of Australia's endemic plant species are adapted to dry conditions. Many of the plants have deep taproots to draw moisture from groundwater and small, hard, pale green or shiny leaves to reflect heat and to hold moisture. Much of the continent is grassland and scrubland with bits of open woodland; there are only a few true forests, found in pockets along the Eastern Highlands, the southwestern tip, and in Tasmania (Figure 11.7). Two plant genera account for nearly all the forest and woodland plants: *Eucalyptus* (450 species, often called gum trees) and *Acacia* (900 species, often called wattles).

Plant and Animal Life in New Zealand and the Pacific Islands Naturalists and evolutionary biologists have long been interested in the species that inhabited the Pacific islands before humans arrived. Charles Darwin formulated many of his ideas about evolution after visiting the Galápagos Islands of the eastern Pacific (see Figure 3.1 on pages 108–109) and the islands of Oceania.

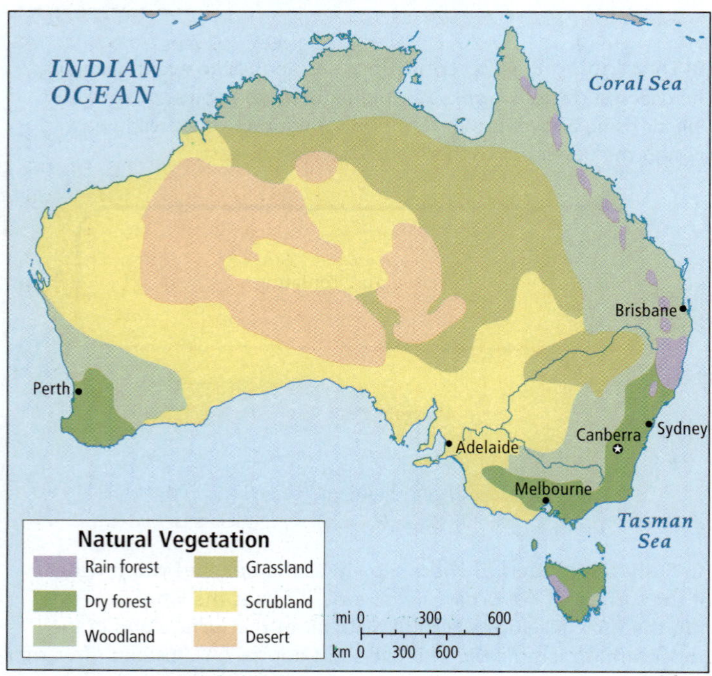

FIGURE 11.7 Australia's natural vegetation. Much of Australia is grassland and scrubland. A few forests can be found in the Eastern Highlands, in the far southwest, and in Tasmania.

Islands gain plant and animal populations from the sea and air around them as organisms are carried from larger islands and continents by birds, storms, or ocean currents. Once these organisms "colonize" their new home, they may evolve over time into new species that are unique to one island. High, wet islands generally contain more varied species because their more complex environments provide niches for a wider range of wayfarers and thus more opportunities for evolutionary change.

Once they arrive, human inhabitants modify the flora and fauna of islands. In prehistoric times, Asian explorers in oceangoing canoes brought plants, such as bananas and breadfruit, and animals, such as pigs, chickens, and dogs, to Oceania. European settlers later brought grains, vegetables, fruits, invasive grasses, cattle, sheep, goats, rabbits, housecats, and rats. Today, human activities from tourism to military exercises to urbanization continue to change the flora and fauna of Oceania. **240. BREADFRUIT ADVOCATES SAY IT COULD SOLVE HUNGER IN TROPICAL REGIONS**

Generally, the diversity of land animals and plants is richest in the western Pacific, near the larger landmasses. It thins out to the east, where the islands are smaller and farther apart. The natural rain forest flora is rich and abundant on New Zealand, New Guinea, and the high islands of the Pacific. However, the natural fauna is much more limited on these islands. While New Guinea has fauna comparable to Australia, to which it was once connected via Sundaland (see Figure 10.5 on page 421), New Zealand and the Pacific islands have no indigenous land mammals, almost no indigenous reptiles, and only a few indigenous species of frogs. New Zealand and the islands were never connected to Australia and New Guinea by a land bridge that land animals could cross. Two indigenous birds in New Zealand, the kiwi and the huge moa (a bird that grew up to 12 feet [3.7 meters] tall), were a major source of food for the Maori people. The moa was hunted to extinction before Europeans arrived. Today, New Zealand may well be the country with the most introduced species of mammals, fish, and fowl, nearly all brought in by European settlers.

THINGS TO REMEMBER

- This largest region of the world is primarily water and has a very small population of just 38.7 million.
- The largest land area in Oceania is the continent of Australia.
- The thousands of islands of the Pacific were created either as a result of plate tectonics or volcanic activity.
- The climate of most of the region is warm and humid nearly all the time.
- The fauna and flora of the region are unique in many distinctive ways that have informed our knowledge about evolution. Many species, such as marsupials and monotremes, are endemic to the region.

Environmental Issues

Geographic Insight 1

Climate Change and Water: Oceania's sea level and freshwater resources are being impacted by climate change.

ːonmental relatively ›lic aware- is keen. ːily warm- of threats ıe human e species. riculture, ı popula- ch of the

Australia, is a minor contributor has some of the world's highest per capita basis (see Figure 1.23 tates, Australia's emissions result ›d transportation systems to con- ːk of cities and towns. Also, as ›endence on coal for electricity ›ns. However, because Australia ı (22.7 million people in 2010), › than 1 percent of global emis- ıe negligible contributions of ands of the Pacific, they, as well › to the effects of global climate)).

g may raise sea levels by melt- Obviously, this issue is of great at already barely rise above the levels rise 4 inches (10 centi- by the International Panel on ›west-lying Pacific atolls, such water within 50 years. Other rowded coastal zones, will see ıore vulnerable to storm surges

rabilities Much of Oceania roughts and floods that could ›. Parts of Australia and some ready undergoing prolonged that are requiring changes in is that the severe droughts are ›ut may represent permanent t could also worsen wildfires. ›sue in Australia in February ıral firestorm near Melbourne

er across the region, *virtual* ecomes an issue. All export- ı consume or degrade fresh- duction processes (such as ːtion; meat, wool, and wheat

ON THE BRIGHT SIDE

Water Conservation

Oceania is a world leader in implementing alternative water technologies. Some are simple but effective age-old methods, such as harvesting rainwater from roofs and the ground for household use. Most buildings in rural Australia, New Zealand, and many Pacific islands get at least part of their water this way, relieving surface and groundwater resources. Australia and New Zealand are now stretching water resources further with extensive use of highly efficient drip irrigation technologies in agriculture (see Figure 11.9D on page 470) and new low-cost water-filtration techniques.

production; and tourism-related construction and maintenance) are essentially extracting virtual water from places that are already under water stress. If the true costs of this freshwater depletion were counted and added to the price of the products, these exports might no longer be competitive on the world market—at least not until all global producers saw virtual water accountability to be in their best interests and raised their prices accordingly.

Another concern is the warming of the oceans, which can result in stronger tropical storms. Warmer ocean temperatures threaten coral reefs and the fisheries that depend on them by causing *coral bleaching* (see Figure 11.9C), a phenomenon affecting all reefs in this region in recent years, but especially the Great Barrier Reef. Because so many fish depend on reefs, coral bleaching also threatens many fishing communities. This is especially true on some of the Pacific islands that have few other local food resources.

Responses to Potential Climate Change Crises Oceania is pursuing a number of alternative energy strategies to reduce greenhouse gas emissions. Australia remains dependent on fossil fuel sales (primarily coal, but also crude oil and natural gas) to Asia, but it is pursuing renewable energy strategies for use at home. These include not only increasing power production from geothermal, solar, and biomass sources, but also, because of water shortages, decreasing the emphasis on hydropower. New Zealand has set a goal of obtaining 95 percent of its energy from renewable sources by 2025. Much of this energy will come from wind power, which has excellent potential in this region, especially in areas near the Roaring Forties (see Figure 11.11 on page 473). On low Pacific islands, solar energy is now the most widely used alternative to costly and polluting imported fuel.

Invasive Species and Food Production

Geographic Insight 2

Food and Development: European systems for producing food and fiber introduced into Oceania have greatly changed environments.

Many of the unique endemic plants and animals of Oceania have been displaced by **invasive species**, organisms that spread into regions outside of their native range, adversely affecting economies or environments. Many *exotic*, or *alien*, plants and animals were brought to Oceania by Europeans to support

invasive species organisms that spread into regions outside of their native range, adversely affecting economies or environments

Figure 11.8 Photo Essay: Human Impacts on the Biosphere in Oceania

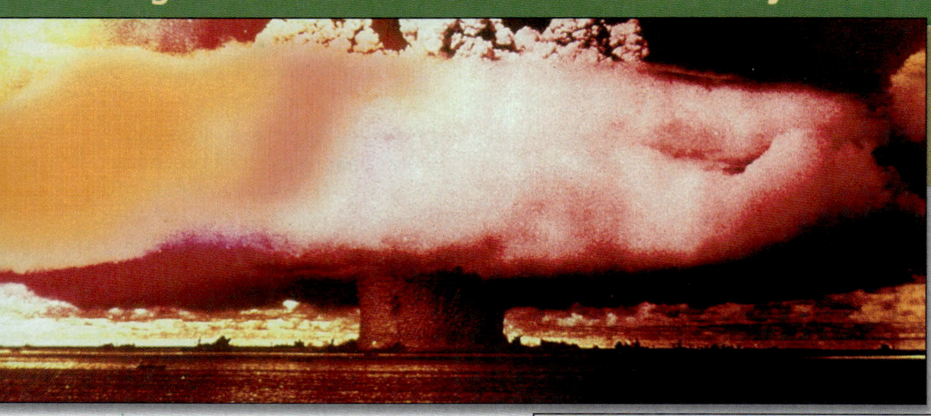

Despite its vast size and relatively small population, Oceania has been severely impacted by human activity. Much of damage has been done by people from distant countries, as well as by countries in Oceania that export resources outside the region.

A In 1946, on Bikini Atoll in the Marshall Islands (then a U.S. territory), the United States conducted one of the first underwater tests of a nuclear weapon and its effects on naval vessels. Since then, the United States and France have conducted over 300 nuclear tests in Oceania, some without sufficient attention to nuclear contamination.

B Once hunted to near extinction to protect introduced sheep herds, the Tasmanian devil is now threatened by low genetic diversity, which leaves it vulnerable to disease.

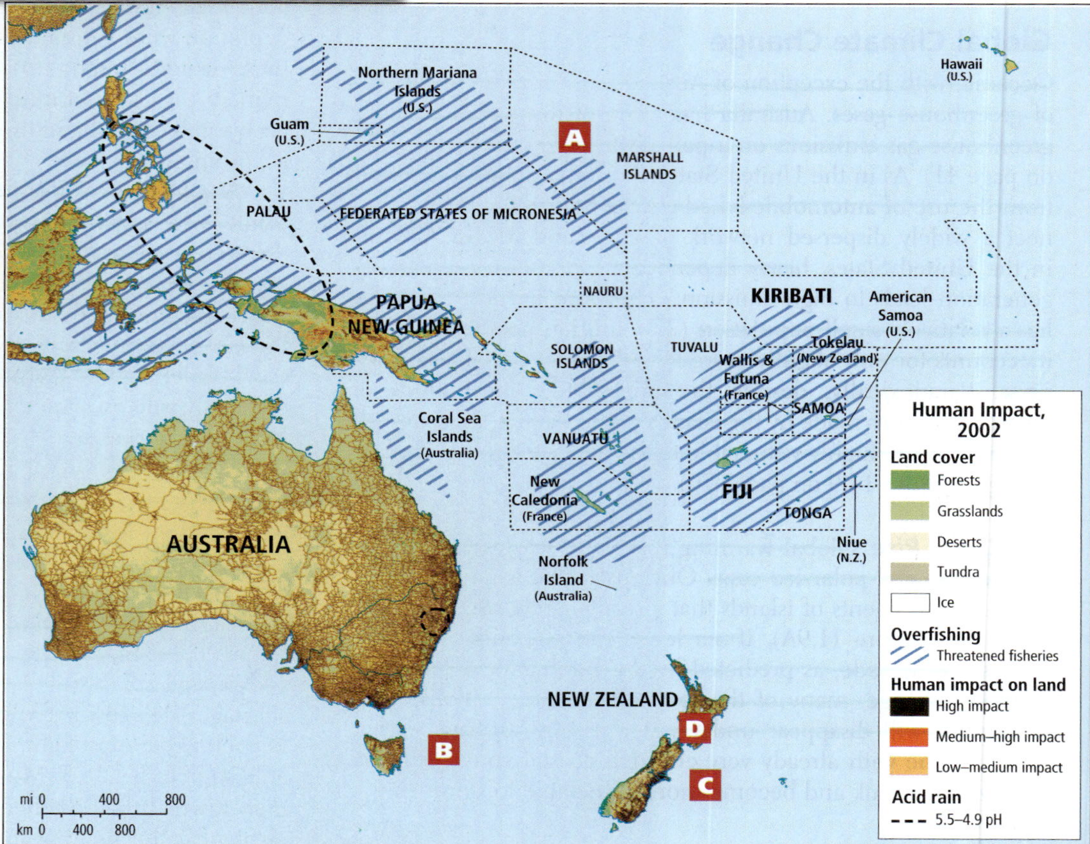

C An endangered great white shark tangled in a fish net off the coast of New Zealand. Fleets from around the world come to Oceania to take advantage of its fisheries, many of which are now overexploited. Nevertheless, many Pacific islands still sell fishing rights to foreign fleets because they need the money.

D A herd of sheep in New Zealand, where ranches, farms, roads, and urban areas cover 90 percent of the lowlands. Forests once covered 85 percent of New Zealand, but after two centuries of export-oriented agriculture and forestry, only 23 percent of the country remains forested. In recent decades, there have been increased efforts to conserve the remaining forests.

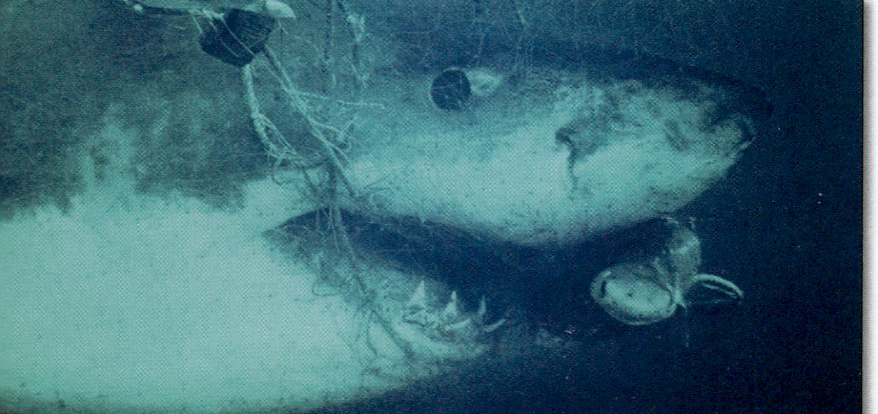

their food production systems. Ironically, many of these same species are now major threats to food production.

Australia When Europeans first settled the continent, they brought many new animals and plants with them, sometimes intentionally, sometimes unintentionally. European rabbits are among the most destructive of the introduced species. Early British settlers who enjoyed eating rabbits brought them to Australia. Many were released for hunting, but with no natural predators, the rabbits multiplied quickly, consuming so much of the native vegetation that many indigenous animal species starved. Moreover, rabbits became a major source of agricultural crop loss and reduced
s to support herds of introduced
control the rabbit population by
d cats backfired as these animals
themselves (Figure 11.10C on
driven several native Australian
ithout having much effect on the
introduced diseases have proven
e rabbit population, though rab-
resistance to them.
ge impact on Australian ecosys-
rid and soils in many areas are
t land use in Australia is the graz-
l animals—primarily sheep, but
t of the land has been given over
he world in exports of sheep and

wild dogs of Australia (see Fig-
uced sheep and young cattle. To
e herds, the Dingo Fence—the
extending 3488 miles (5614 kilo-
ately a major ecological barrier
ile, kangaroos (the natural prey
on the sheep side of the fence,
ned beyond sustainable levels.

s environment has been trans-
nd food production systems even
environment. No humans lived
) years ago, when the Polynesian
n they arrived, dense midlatitude
f the land. The Maori were cul-
d taro as well as other nonnative

ON THE BRIGHT SIDE

Environmental Awareness in New Zealand

New Zealand is a global environmental leader. It formed the world's first environmentally focused national political party in 1973 and spearheaded the world's first "nuclear-free" zone in 1984. New Zealand's government promotes a "clean and green" image internationally, but most New Zealanders acknowledge the severity of existing environmental problems, as well as the need for further action.

plants, rats, and birds. By the time of European contact (1642), forest clearing and overhunting by the Maori had already degraded many environments and driven several bird species to extinction.

European settlement in New Zealand dramatically intensified environmental degradation associated with food production and invasive species. Attempts to re-create European farming and herding systems in New Zealand resulted in environments that today are actually hostile to many native species, a growing number of which are becoming extinct. Only 23 percent of the country remains forested, with ranches, farms, roads, and urban areas claiming more than 90 percent of the lowland area (Figure 11.11 on page 473).

Most of the cleared land is used for export-oriented farming and ranching. Grazing has become so widespread that in New Zealand today there are 15 times as many sheep as people, and 3 times as many cattle. Both of these activities have severely degraded environments. Soils exposed by the clearing of forests proved infertile, forcing farmers and ranchers to augment them with agricultural chemicals. The chemicals, along with feces from sheep and cattle, have severely polluted many waterways, causing the extinction of some aquatic species.

Pacific Islands In the Pacific islands, many unique species of plants and animals have been driven to extinction as islands were deforested and mined or converted to commercial agriculture. This was the case in Hawaii, which is home to more threatened or endangered species than any other U.S. state, despite having less than 1 percent of the U.S. landmass. There, extensive conversion of tropical forests to land used for the export crops of sugar cane and pineapples caused the extinction of numerous plant, bird, and land species. **239. HAWAII CONSIDERED AMERICA'S ENDANGERED SPECIES CAPITAL**

Globalization and the Environment in the Pacific Islands

As the Pacific islands have become more connected to the global economy over the years, flows of resources and pollutants have increased dramatically. Mining, nuclear pollution, and tourism are all examples of how globalization has transformed environments in the Pacific islands.

Mining in Papua New Guinea and Nauru Foreign-owned mining companies that took advantage of poorly enforced or nonexistent environmental laws are responsible for major environmental damage in Oceania. In the Ok Tedi Mine on Papua New Guinea, huge amounts of mine waste devastated river systems

Figure 11.9 Photo Essay: Vulnerability to Climate Change in Oceania

Oceania is vulnerable to a wide variety of hazards related to climate change, including sea level rise, increased tropical storm intensity, and less certain water availability. Fortunately, several countries in this region are already implementing solutions that are increasing resilience to climate hazards.

A A low-lying atoll in Tuvalu. With its highest point only 14.7 feet (4.5 meters) above sea level, Tuvalu is highly vulnerable to sea level rise. Combined with increased flooding during tropical storms, higher sea levels could make many low islands in this region uninhabitable. Tuvalu's government is already negotiating the future resettlement of parts of its population to nearby nations such as New Zealand.

B A wildfire outside of Brisbane, Australia. Higher temperatures are bringing stronger and more extensive wildfires to this region.

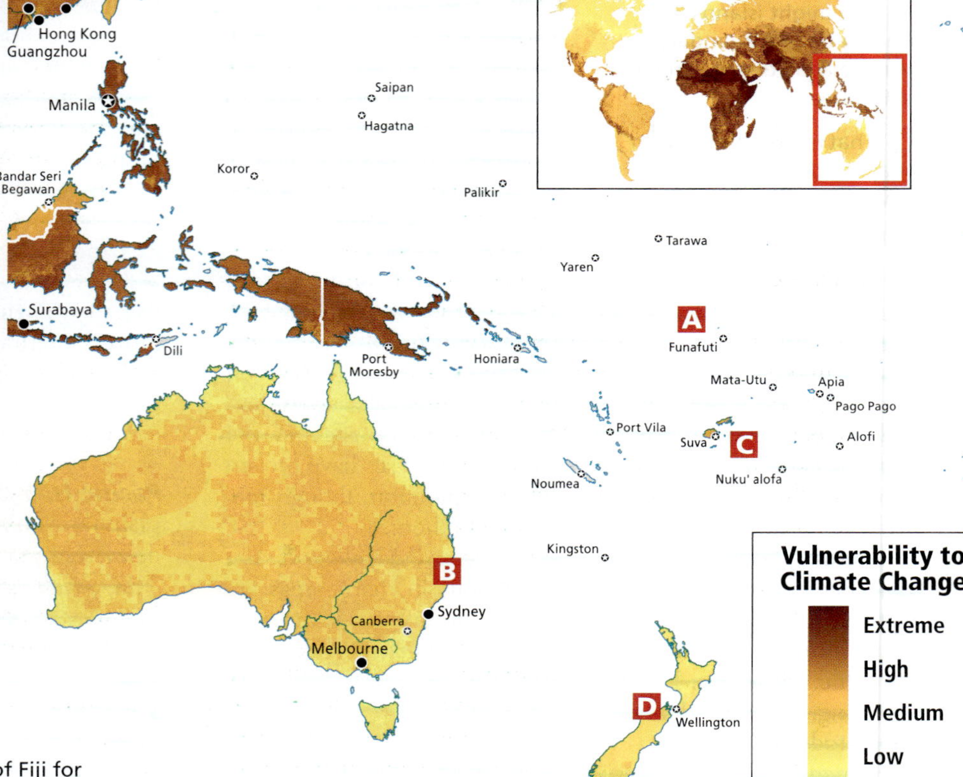

C A marine biologist monitors the reefs off the coast of Fiji for signs of coral bleaching. Caused by warmer temperatures, coral bleaching threatens many fish species that live on coral reefs. In turn, human communities on Pacific islands are threatened by the loss of fisheries.

D A vineyard fitted with a drip irrigation system, visible at the base of the vines, in Marlborough, New Zealand. These systems provide resilience in the face of drought and use substantially less water than other irrigation methods.

(**Figure 11.12** on page 473). The environmental degradation forced tens of thousands of indigenous subsistence cultivators into new mining market towns, where their horticulture skills were of little use and where they needed cash to buy food and pay rent.

In 2007, thirty thousand of these people sued the Australian parent mining company, which was then BHP Billiton, for U.S.$4 billion. Two villagers, Rex Dagi and Alex Maun, traveled to Europe and the United States to explain their cause and meet with international environmental groups. They and their supporters convinced U.S. and German partners in the Ok Tedi Mine to divest their shares.

The most extreme case of environmental disaster due to mining took place on the once
nd of Nauru, one-third the size
of the Solomon Islands (see the
th was based on proceeds from
hosphates derived from eons of
used to manufacture fertilizer.
ned first by Germany, then by
nd for a time in the early 1970s
income in the world. Today, the
the proceeds ill spent, and the
nining equipment sits on miles
est once stood.

litical aspects of globalization
l. Nuclear weapons testing by
m the 1940s to the 1960s (dur-
e acceptance of nuclear waste
e become major environmen-
(see Figure 11.8A). In New
secret service blew up the ship
tinuclear environmental group
86 Treaty of Rarotonga estab-
Free Zone. Most independent
eaty, which bans nuclear weap-
nping on their lands. Because
nd the United States, however,
es such as the Marshall Islands

ntil recently was considered a
onmental problems. Foreign-
en accelerated the loss of wet-
by clearing coastal vegetation
for hotel construction, golf courses, and waterfront-related entertainment.

Tourism has also strained island water resources because of showering, laundering, and other services that consume fresh water. Furthermore, inadequate methods of disposing of sewage and trash from resorts have polluted many once-pristine areas. Ecotourism (see Chapter 3, pages 119–120), aimed at reducing these impacts, is now a common element of development throughout the Pacific, but environmental impacts from ecotourism are still generally high.

The Great Pacific Garbage Island When oceanographer Charles Moore found himself surrounded by a massive floating island of plastic garbage in the north Pacific in 1997, he thought it was an anomaly. But his investigations revealed that the throwaway concept of modern living was responsible. The island included plastic beverage bottles and caps along with Lego blocks, trash bags, toothbrushes, footballs, and kayaks—indeed, virtually every consumer product made. By 2008, several such garbage masses were floating in the Pacific and Atlantic oceans. Because modern plastics do not degrade, millions of seabirds and sea mammals die each year after ingesting bits of this trash. Whatever goes into the ocean eventually ends up on someone's dinner plate. The following video tells more of the story: http://www.youtube.com/watch?v=FrAShtolieg.

The UN Encourages Pacific Resource Protection Implementing the 1994 United Nations Convention on the Law of the Sea (UNCLOS) involves acknowledging that the globalization of Pacific island economies has undermined the environment. Based on the idea that all the problems of the world's oceans are interrelated and need to be addressed as a whole, UNCLOS established rules governing all uses of the world's oceans and seas and has been ratified by 157 countries (although not the United States).

The treaty allows islands to claim rights to ocean resources 200 miles (320 kilometers) out from their shores. Island countries can now make money by licensing privately owned fleets from Japan, South Korea, Russia, the United States, and elsewhere to fish within these offshore limits. As of yet, there is no overarching enforcement agency, and protecting the fisheries from overfishing by these rich and powerful licensees has turned out to be an enforcement nightmare for tiny island governments with few resources. Similarly, it has proven difficult to monitor and control the exploitation of seafloor mineral deposits by foreign mining companies.

241. ENDANGERED HAWAIIAN MONK SEAL POPULATION CONTINUES TO DECLINE

242. NEW SPECIES OF UNDERSEA LIFE FOUND NEAR INDONESIA

243. SCIENTISTS WARN OF DEPLETION OF OCEAN FISH IN 40 YEARS

THINGS TO REMEMBER

Geographic Insight 1

- **Climate Change and Water** Sea level rise threatens to submerge some Pacific islands, and water supplies throughout the region may be strained as rainfall patterns change. If sea levels rise the predicted 4 inches per decade, many of the lowest-lying atolls will disappear under water, leaving a multitude of refugees. Warming oceans threaten coral reefs that are important to fisheries and local livelihoods.

- In an effort to combat global warming, renewable energy alternatives to fossil fuels are being pursued across the region. New Zealand has set a goal of obtaining 95 percent of its energy from renewable sources by 2025.

Geographic Insight 2

- **Food and Development** Throughout Oceania, the introduction of food production systems from elsewhere has resulted in the spread of ecologically and economically damaging invasive plant and animal species.

- Globalization and the patterns of consumption by people who live far from the Pacific are seriously affecting human and animal life in this region.

Human Patterns over Time

VIGNETTE *"With courage, you can travel anywhere in the world and never be lost. Because I have faith in the words of my ancestors, I'm a navigator."*

MAU PIAILUG

In 1976, Mau Piailug made history by sailing a traditional Pacific island voyaging canoe across the 2400 miles (3860 kilometers) of deep ocean between Hawaii and Tahiti. He did so without a compass, charts, or other modern instruments, using only methods passed down through his family. To find his way, he relied mainly on observations of the stars, the sun, and the moon. When clouds covered the sky, he used the patterns of ocean waves and swells, as well as the presence of seabirds, to tell him of distant islands over the horizon.

Piailug reached Tahiti 33 days after leaving Hawaii and made the return trip in 22 days. His voyage resolved a major scholarly debate over how people settled the many remote islands of the Pacific without navigational instruments, thousands of years before the arrival of Europeans. Some thought that navigation without instruments was impossible and argued that would-be settlers simply drifted about on their canoes at the mercy of the winds, most of them starving to death on the seas, with a few happening upon new islands by chance. It was hard to refute this argument because local navigational methods had died out almost everywhere. However, in isolated Micronesia, where Piailug lives, indigenous navigational traditions survive.

After the successful 1976 voyage, Piailug trained several students in traditional navigational techniques. His efforts have become a symbol of cultural rebirth and a source of pride throughout the Pacific. In 2007, the protégés of Mau Piailug sailed from Hawaii through the Marshall Islands to Yokohama, Japan, to celebrate peace and the human need to stay connected with nature. *[Source: Facts on File, 2007. For detailed source information, see Text Credit pages.]*

FIGURE 11.10 LOCAL LIVES **PEOPLE AND PETS IN OCEANIA**

A A rainbow lorikeet, a small parrot that is native in much of Oceania. Rainbow lorikeets are popular pets because of their plumage and curious disposition. However, in parts of Australia and New Zealand, they are considered pests because they consume fruit in orchards, are noisy, and leave behind large droppings.

B An Australian dingo, a type of wild dog that lives mainly in the Australian outback. Dingoes were likely brought to Australia by Aborigines who used them as guard dogs and possibly as a food source. Shepherds consider dingoes to be pests, and in the 1880s they built a 3488-mile (5614-km) fence to keep dingoes out of southeastern Australia.

C A cat in New Zealand. Introduced throughout Oceania by Europeans who brought them in as pets and to control rodent populations, cats quickly became feral. This had disastrous impacts on native species of marsupials and birds. Cats have been completely removed from several of New Zealand's smaller islands, including those designated as native bird sanctuaries.

ıatural resources of New
ıettlement and the clearing of land
v Zealand remains forested.

a

nts of Oceania are Australia's
ıe Australoids) migrated from
0 years ago (Figure 11.13; see
), at a time when sea level was
ome memory of this ancient
boriginal oral traditions, which
aphic features that are now sub-
e same time that the Aborigines
nania, related groups were set-

Melanesians, so named for their
ively dark skin tones, a result
igh levels of the protective pig-
t *melanin*, migrated through-
New Guinea and other nearby
ds, giving this area its name,
esia. Archaeological evidence
ates that they first arrived more
50,000 to 60,000 years ago
Sundaland (see Figure 10.5
page 421), a now-submerged
exposed during the Pleistocene
h. They lived in isolated pock-
hich resulted in the evolution
ındreds of distinct yet related
ages. Like the Aborigines, the

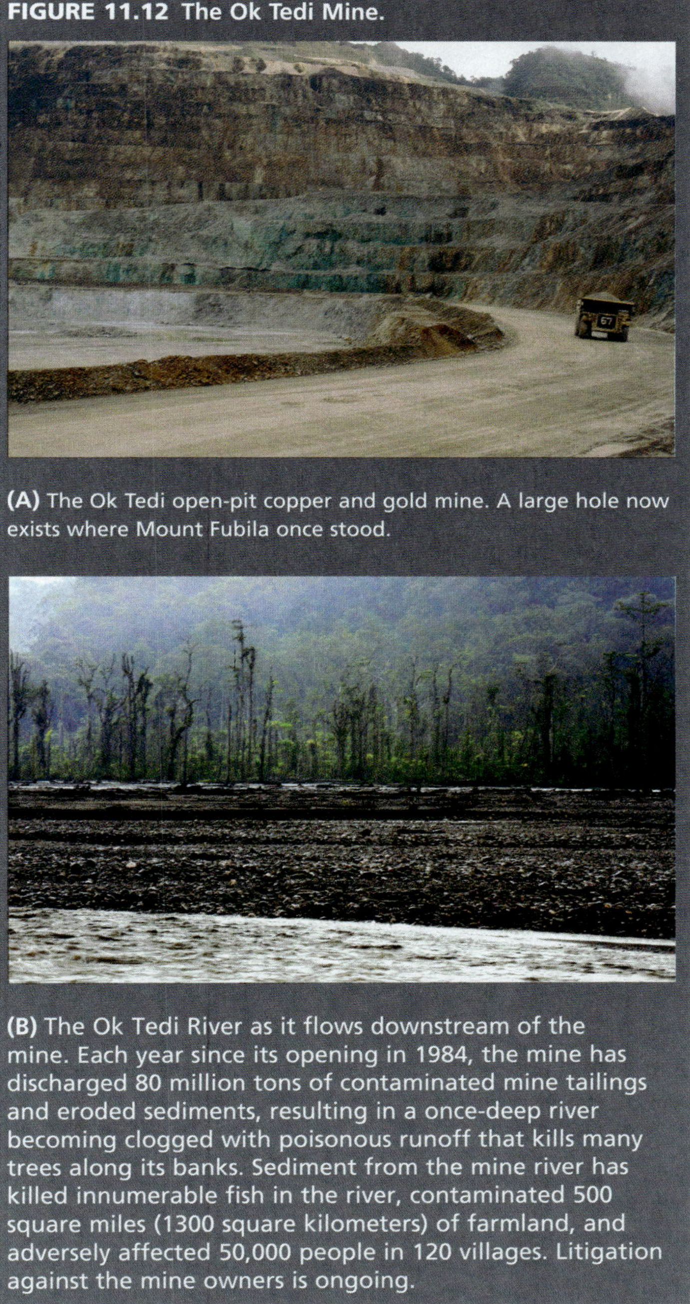

FIGURE 11.12 The Ok Tedi Mine.

(A) The Ok Tedi open-pit copper and gold mine. A large hole now exists where Mount Fubila once stood.

(B) The Ok Tedi River as it flows downstream of the mine. Each year since its opening in 1984, the mine has discharged 80 million tons of contaminated mine tailings and eroded sediments, resulting in a once-deep river becoming clogged with poisonous runoff that kills many trees along its banks. Sediment from the mine river has killed innumerable fish in the river, contaminated 500 square miles (1300 square kilometers) of farmland, and adversely affected 50,000 people in 120 villages. Litigation against the mine owners is ongoing.

Thinking Geographically

(A) How many indigenous subsistence cultivators were forced off their land and into new mining market towns as a result of sediment and chemical pollution from the Ok Tedi Mine in Papua New Guinea? **(B)** What evidence shows that this part of the polluted river flows through lowlands where flooding has taken place?

early Melanesians survived mostly by hunting, gathering, and fishing, although some groups—especially those inhabiting the New Guinea highlands—eventually practiced agriculture.

Much later, between 5000 to 6000 years ago and as recently as 1000 years ago, linguistically related *Austronesians* (a group of skilled farmers and seafarers from southern China who migrated

Abor
inhab
the Au
Asia 5
the Su
expos

Mela
namec
tones,
protec
throug
nearby

Melar
islands
west o
New C

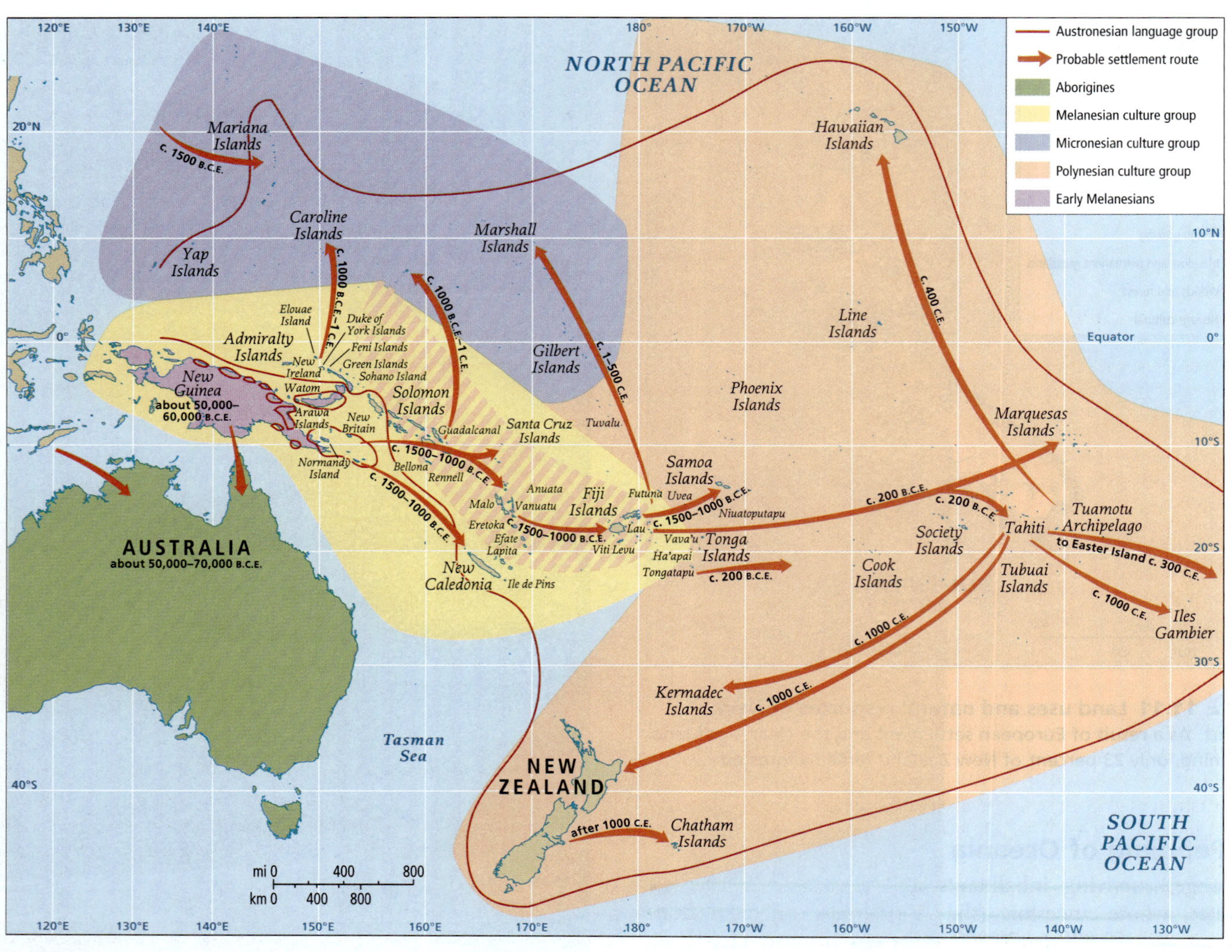

FIGURE 11.13 Primary indigenous culture groups of Oceania. By 50,000 to 70,000 years ago, humans had come to New Guinea and Australia. About 25,000 years ago, people began moving across the ocean to nearby Pacific islands. Movement into the more distant islands began with the arrival of Austronesians, who went on to inhabit the farthest reaches of Oceania.

to Southeast Asia; see Chapter 10, page 431) settled **Micronesia** and **Polynesia**, sometimes mixing with the Melanesian peoples they encountered. Micronesia consists of the small islands that lie east of the Philippines and north of the equator (see the Figure 11.1 map). Polynesia is made up of numerous islands situated inside a large irregular triangle formed by New Zealand, Hawaii, and Easter Island. (Easter Island, also called Rapa Nui, is a tiny speck of land in the far eastern Pacific, at 109° W 27° S, not shown in the figures in this chapter.) Recent experiments run by Polynesians (see the voyaging vignette on page 472) have provided evidence that ancient sailors could navigate over vast distances, using seasonal winds, astronomic calculations, bird and aquatic life, and wave patterns to reach the most far-flung islands of the Pacific. The Polynesians were fishers, hunter-gatherers, and cultivators who developed complex cultures and maintained trading relationships among their widely spaced islands.

Micronesia the small islands that lie east of the Philippines and north of the equator

Polynesia the numerous islands situated inside an irregular triangle formed by New Zealand, Hawaii, and Easter Island

In the millennia that have passed since first settlement, humans have continued to circulate throughout Oceania. Some apparently set out because their own space was crowded and full of conflict, or food reserves were declining. It is also likely that Pacific peoples were enticed to new locales by the same lures that later attracted some of the more romantic explorers from Europe and elsewhere: sparkling beaches, magnificent blue skies, scented breezes, and captivating landscapes.

Arrival of the Europeans

The earliest recorded contact between Pacific peoples and Europeans took place in 1521, when the Portuguese navigator Ferdinand Magellan (exploring for Spain) landed on the island of

ounter ended badly. The islanders,
tried to take a small skiff. For this
kill the offenders and burn their
onths later, Magellan was himself
came the Philippines, which he
heless, by the 1560s, the Spanish
ade route between Manila in the
e west coast of Mexico. Explorers
wed, first taking an interest mainly
The British and French explored
teenth century (Figure 11.14C).
lly divided among the colonial
ntury. By that time, the United
d joined France and Britain in
d groups. As in other regions,
nia emphasized extractive agri-
tive people were often displaced
xotic diseases to which they had
leclined sharply.

stralia

under European or U.S. rule at
parts of the region are Australia
tion of these two countries by
parallels with North America.
was a major impetus for "set-
North American colonies were
somewhere else to send their
ury Britain, a relatively minor
night be punished with 7 years
ire 11.14D).
nd Irish convicts arrived in
convicts chose to stay in the
served. They are given credit
l egalitarian spirit. They were
oluntary immigrants from the
the availability of inexpensive
its arrived until World War II.
at later than Australia, in the
also derives primarily from
as never a penal colony.
ustralia, New Zealand, and
t of indigenous peoples by
a and New Zealand, native
ilated by infectious diseases,
The few who lived on terri-
rable were able to maintain
ver, the vast majority of the
ing poverty, either in urban
nches. Today, native peoples
naladies such as alcoholism
ogress is being made toward
In 2008, the newly elected
udd, officially apologized to
hey have received since the

Closely related to attitudes toward indigenous people were attitudes toward immigrants of any color other than white. By 1901, a whites-only policy governed Australian immigration, with favored migrants coming from the British Isles and (after World War II) southern Europe. This discrimination persisted until the mid-1970s, when the White Australia policy on immigration was ended. In New Zealand, where similar racist attitudes prevailed, there was never an official whites-only policy, and by the 1970s, students and immigrants were arriving from Asia and the Pacific islands. Controversy over immigration in Australia and New Zealand has recently centered on the arrival of refugees by boat from various parts of Asia (see Figure 11.23D on page 488).

Oceania's Shifting Global Relationships

During the twentieth century, Oceania's relationship with the rest of the world went through three phases: from a predominantly European focus to identification with the United States and Canada to the currently emerging linkage with Asia.

Until roughly World War II, the colonial system gave the region a European orientation. In most places, the economy depended largely on the export of raw materials to Europe (see Figure 11.14E). Thus, even when a colony gained independence from Britain, as Australia did in 1901 and New Zealand in 1907, people remained strongly tied to their mother countries. Even today, the Queen of England remains the titular head of state in both countries. During World War II, however, the European powers provided only token resistance to Japan's invasion of much of the Pacific and its bombing of northern Australia. This European impotence began a change in the region's political and economic orientation. **244. VETERANS REMEMBER TRAGEDY OF WAR IN PACIFIC**

After the war, the United States, which already had a strong foothold in the Philippines, became the dominant power in the Pacific, and U.S. investment became increasingly important to the economies of Oceania. Australia and New Zealand joined the United States in a Cold War military alliance, and both fought alongside the United States in Korea and Vietnam, suffering considerable casualties and experiencing significant antiwar activity at home. U.S. cultural influences were strong, too, as North American products, technologies, movies, and pop music penetrated much of Oceania.

By the 1970s, another shift was taking place as many of the island groups were granted self-rule by their European colonizers, and Oceania became steadily drawn into the growing economies of Asia. Since the 1960s, Australia's thriving mineral export sector has become increasingly geared toward supplying raw materials to Asian manufacturing industries (first Japan in the 1960s, and increasingly China since the 1990s). Similarly, since the 1970s, New Zealand's wool and dairy exports have gone mostly to Asian markets. Despite occasional backlashes against "Asianization," Australia, New Zealand, and the rest of Oceania are becoming increasingly transformed by Asian influences. Many Pacific islands have significant Chinese, Japanese, Filipino, and Indian minorities, and the small Asian minorities of Australia and New Zealand are increasing (see Figure 11.14F). On some Pacific islands, such as Hawaii, Asians now constitute the largest portion (42 percent in Hawaii) of the population.

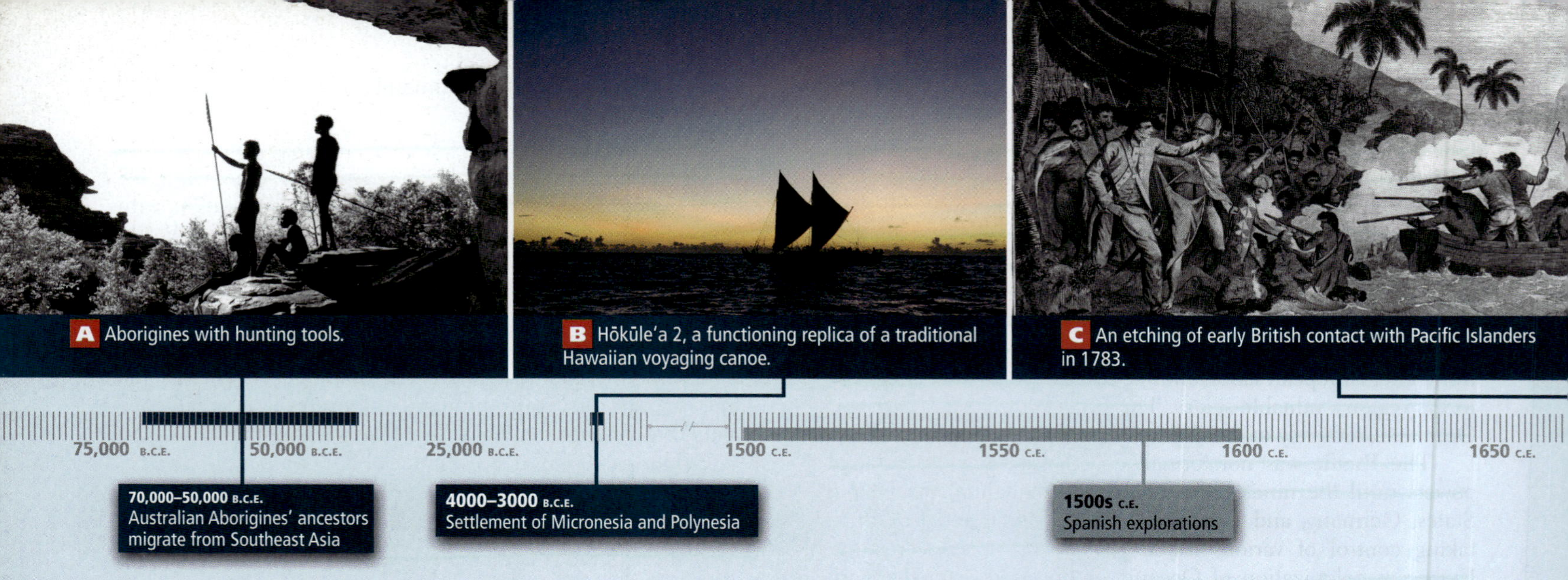

A Aborigines with hunting tools.

B Hōkūle'a 2, a functioning replica of a traditional Hawaiian voyaging canoe.

C An etching of early British contact with Pacific Islanders in 1783.

FIGURE 11.14 A VISUAL HISTORY OF OCEANIA

Thinking Geographically

After you have read about the human history of Oceania, you will be able to answer the following questions:

A From where did the ancestors of Aborigines migrate to Oceania?

B How did Polynesians use and navigate boats like the one in this photo?

THINGS TO REMEMBER

- Oceania can be divided into four distinct indigenous cultural regions: Australia and Tasmania, settled originally by Aborigines; Melanesia, settled by Melanesians, so identified because of their skin tone derived from melanin; and Micronesia and Polynesia, settled by a variety of Austronesian peoples.
- Through colonization, Europeans were active in Oceania from the early sixteenth century until the end of World War II. During the 50 years after the war, the United States, Australia, and New Zealand were the principal powers in the Pacific.
- Beginning in the 1970s, Asian countries have had increasing influence throughout Oceania.

CURRENT GEOGRAPHIC ISSUES

Oceania's old relationships were built on historical factors, such as the European settlement of Australia and New Zealand and the depth of ancient Pacific cultural affiliations. By contrast, its new relationships are influenced by economic and geographic considerations, particularly its physical proximity to Asia.

Economic and Political Issues

> **Geographic Insight 3**
>
> **Globalization and Development:** Globalization, coupled with a greater focus on neighboring Asia (rather than on long-time connections with Europe), has transformed patterns of trade and economic development across Oceania.

The forces of globalization, driven largely by Asia's growing affluence and enormous demand for resources, are shifting trade, migration, and tourism patterns within Oceania.

Globalization, Development, and Oceania's New Asian Orientation

One could say that globalization in Oceania began when the first European explorers came into the region, beginning the trend of influence by outsiders (primarily Europeans) on settlement, culture, and economics. More recently, the United States has exerted a powerful influence on trade and politics in the region. For the past several decades, however, globalization has reoriented this region toward Asia, which buys more than 76 percent of Australia's exports (mainly coal, iron ore, and other minerals). In 2011, China and India each purchased major shares of Australia's coal deposits, not just the output of mines. Both countries use coal to generate energy and are trying to secure future access to more coal. Asia also buys nearly 35 percent of New Zealand's exports (mainly meat, wool, and dairy products), as well as many other products and services from islands across Oceania (Figure 11.15 on page 478).

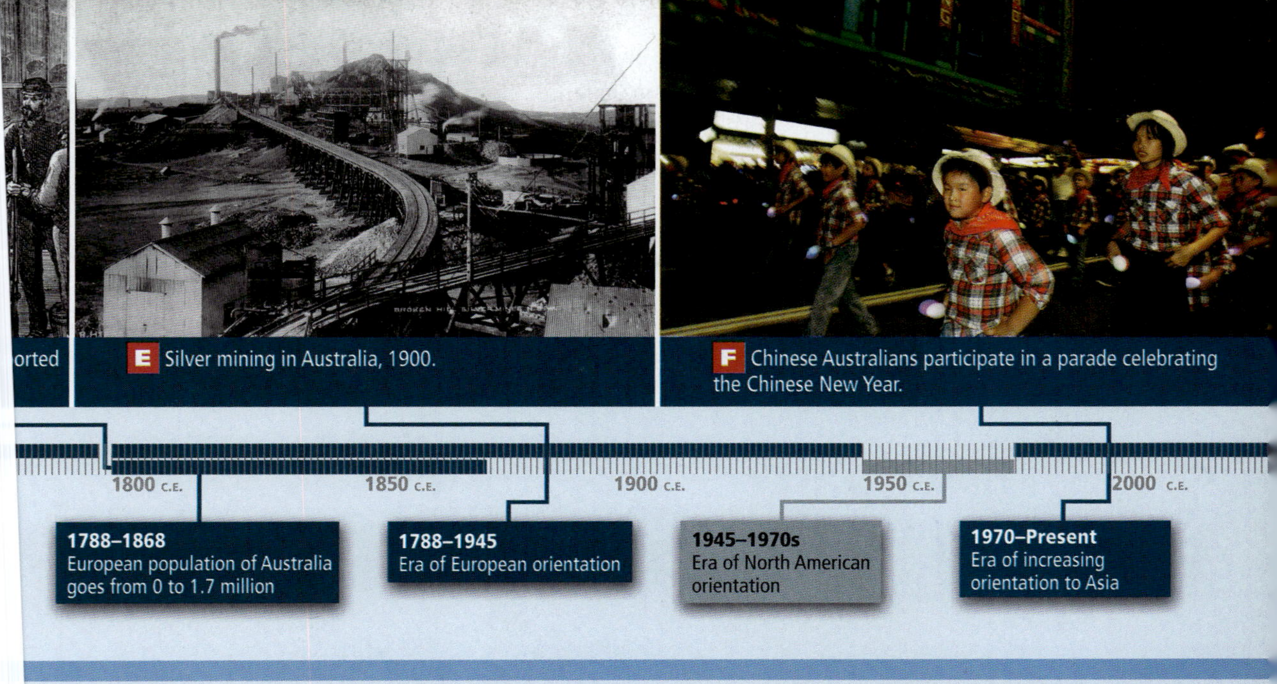

D to A ...orted

E Silver mining in Australia, 1900.

F Chinese Australians participate in a parade celebrating the Chinese New Year.

C V ...erested in when they came to Ocea... and eighteenth centuries?

D T... England and Ireland ended in 1868. Wha... ›nvicts chose to remain in Australia?

E Until roughly World War II, on what did the economy in most parts of Oceania depend?

F Give an example of a Pacific Island in which Asians constitute more than one-third of the population.

source of the region's imports. ...uring in Oceania, most manu-...m China, Japan, South Korea, ...nia's leading trading partners. ...d have free trade agreements ...us negotiation with Asia's two ...an.

...er along in their reorientation ...d New Zealand. Not only are ...s from the Pacific islands sold ...panies increasingly own these ... Fishing fleets from Asia regu-...ific island nations. Asians also ...t trade, both as tourists and as ... And increasing numbers of ... the Pacific islands, exerting ...nfluence.

...nomic Development ...d New Zealand

...sia's global economic rise has ...lso increased competition with ...s. Throughout Oceania, local ...preferential trade with Europe. ...use EU regulations stemming ... World Trade Organization ...nger protected in their trade ...e stiff competition from larger ...much cheaper labor.

Australia and New Zealand are somewhat unusual in having achieved broad prosperity largely on the basis of exporting raw materials over a long period of time and to a wide range of customers (see Figures 11.14E and 11.15). Preferential trade with Europe allowed higher profits for many export industries. Very important is the fact that strong labor movements in Australia and New Zealand meant that these industries' profits were more equitably distributed throughout society than the profits in other raw materials–based economies in Middle and South America and sub-Saharan Africa. Australian coal miners' unions successfully agitated not just for good wages, but also for the world's first 35-hour workweek. Other labor unions won a minimum wage, pensions, and aid to families with children long before such programs were enacted in many other industrialized countries. For decades, these arrangements were highly successful. Both Australia and New Zealand enjoyed living standards comparable to those in North America but with a more egalitarian distribution of income. However, since the 1970s, competition from Asian companies has meant that increasing numbers of workers in Australia and New Zealand have lost jobs and seen their hard-won benefits scaled back or eliminated.

Competition from Asian companies also led to lower corporate profits in Oceania. As corporate profits fell, so did government tax revenues, which necessitated cuts to previously high rates of social spending on welfare, health care, and education. The loss of social support, especially for those who have lost jobs, has contributed to rising poverty in recent years. Australia now has the second-highest poverty rate in the industrialized world. (The United States has the highest.)

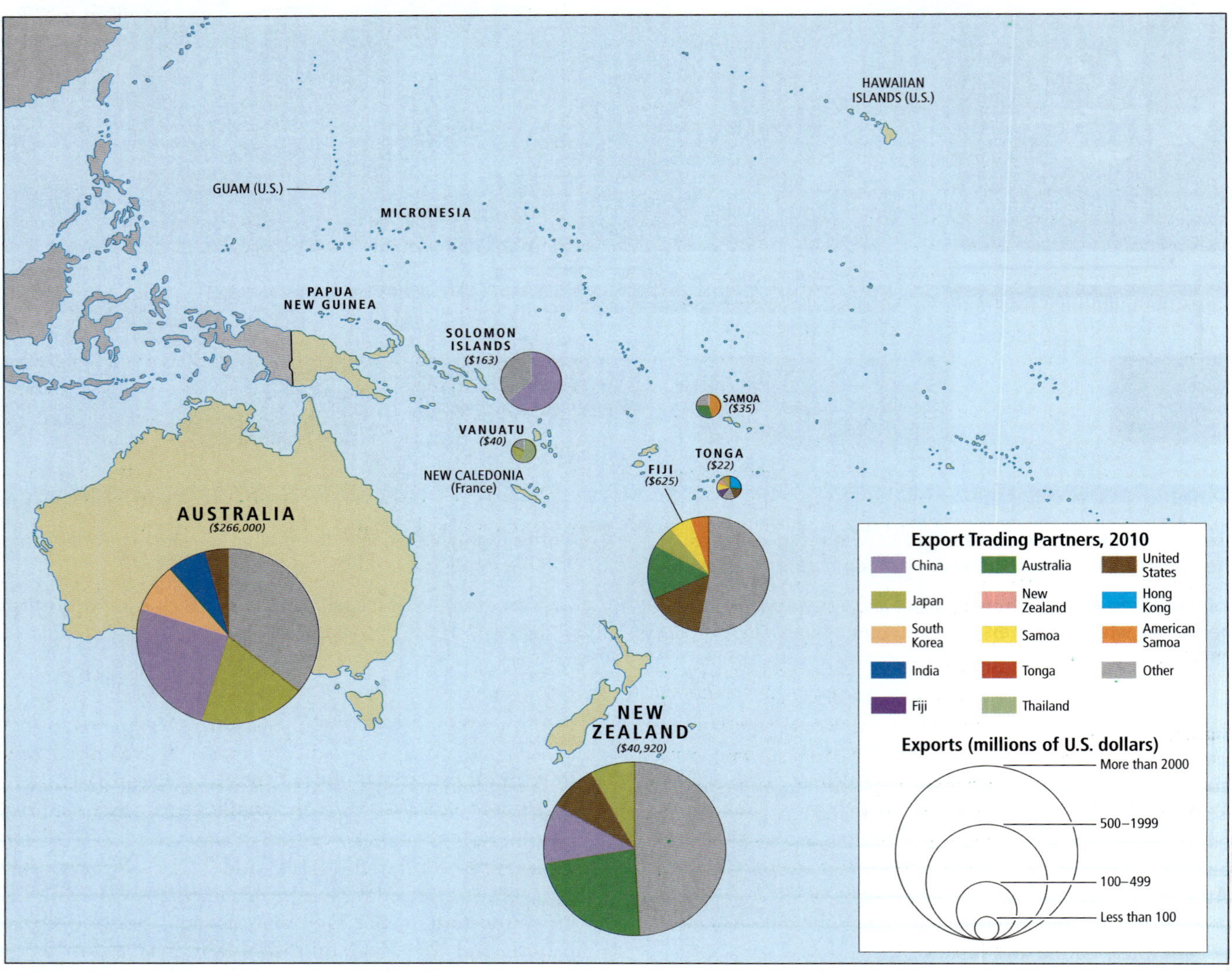

FIGURE 11.15 Exports from Oceania. The colors of each pie chart indicate a country's export trading partners. The "other" sections can include trade with Canada, Mexico, the Caribbean, non-EU Europe, sub-Saharan Africa, and other locales, some of them new trading partners. (Figures for Hawaii do not include exports to other parts of the United States.)

Maintaining Raw Materials Exports as Service Economies Develop The shift toward trading more with Asia than with Europe or North America has had little effect on Oceania's dependence on exporting unprocessed raw materials, because most Asian economies have a much greater need to import raw materials (rather than manufactured goods). Nevertheless, although their dollar contribution to national economies remains high, industries that export raw materials are of decreasing prominence in the economies of Australia and New Zealand, in that they now employ fewer people because of mechanization. This shift to lower labor requirements has been essential for these industries to stay globally competitive with other countries that use much cheaper labor.

Today, the economies of both Australia and New Zealand are dominated by diverse and growing service sectors, which have links to the region's export sectors. Extracting minerals and managing herds and cropland have become technologically sophisticated enterprises that depend on many supporting services and an educated workforce. Australia is now a world leader in providing technical and other services to mining companies, sheep farms, and winemakers. Meanwhile, New Zealand's well-educated workforce and well-developed marketing infrastructure have helped it break into luxury markets for dairy products, meats, and fruits. Perhaps the most visible success has been New Zealand's global marketing of the indigenous kiwifruit. (In fact, *kiwi* is now a slang term for anyone from New Zealand.)

Economic Change in the Pacific Islands In general, the Pacific islands are also shifting away from extractive industries, such as mining and fishing, and toward service industries such

On many resources ss of eco- holds still d rely on ation for he islands subsistence agriculture engages pulation, although it accounts for conomy. Remittances sent home landers working abroad are essen- omies and account for more than nds. However, remittances rarely

d undereducated nations (the parts of Papua New Guinea, for has been termed a **MIRAB econo-** mittances, aid, and bureaucracy. esent colonial powers supports supply employment for the edu- economy has little potential for vth. Nevertheless, Islanders who be self-sufficient in food and ter while saving extra cash for l and occasional purchases of ufactured goods are sometimes to have achieved **subsistence ence.** Where there is poverty, it ten related to geographic isola- which means a lack of access formation and economic oppor- y. Although computers and the rks are not yet widely available the potential to significantly

MIR on m burea

subs where regar some travel manu

ON THE BRIGHT SIDE

Subsistence Affluence Practices Could Go Global

Many of the qualities of subsistence affluence practiced by Pacific Islanders, including local self-sufficiency and resource conservation strategies, have the potential for being elaborated upon and adapted elsewhere in the region and across the world.

esses of Tourism

economy throughout Oceania, pan, Korea, Taiwan, Southeast igure 11.16). In 2008 (the lat- es are available), 17.7 million the tourists visiting Oceania, ercent from Japan alone; just wn from 17 percent in recent th America; 19 percent were nt came from other locations. s, the number of tourists far uam, for example, annually alent to five times its popula- iana Islands annually receive tions. Such large numbers of and graciously accommodat- al inhabitants. And although onomies, these visitors create extra burdens on water and d of living that may be far out of reach for local people. Perhaps nowhere in the region are the issues raised by tourism clearer than in Hawaii.

A Case Study: Conflict over Tourism in Hawaii

Since the 1950s, travel and tourism have been the largest industries in Hawaii, producing nearly 18 percent of the gross state product in 2008. Tourism is related in one way or another to nearly 75 percent of all jobs in the state. (By comparison, travel and tourism account for 9 percent of GDP worldwide.) In 2008, tourism employed one out of every six Hawaiians and accounted for 25 percent of state tax revenues.

Dramatic fluctuations in tourist visits, often driven by forces far removed from Oceania, can wreak havoc on local economies. Decreases in tourism affect not just tourist facilities but supporting industries, too. For example, construction thrives by building condominiums, hotels, resorts, and retirement facilities. The Asian recession of the late 1990s, the terrorist attacks of September 11, 2001, and the global recession of 2008–2009 all affected Hawaii's economy by creating dramatic slumps in tourist visits. In 2011, however, Hawaii's tourism industry rebounded: more than 7 million visitors arrived that year, up 11 percent over 2009.

FIGURE 11.16 Tourism in Oceania. Tourism plays a major role in the economies of Oceania. In 2008, more than 17.7 million tourists visited the region, with just under one-third of these going to Australia. The origins of the tourists reflect changing trade patterns in the region, with more and more coming from Asia. Here you see hotels on Honolulu's Waikiki Beach, which provide lodging for a majority of Hawaii's 6 million yearly visitors.

ON THE BRIGHT SIDE

APEC, a Regional Economic Association with a Broader View

APEC, which stands for "Asia-Pacific Economic Cooperation," is a forum for 21 "Pacific Rim" countries to discuss economic issues. However, in recent years it has extended its focus to address issues such as securing the Asia–Pacific food supply, reducing vulnerability to global climate change, finding energy-efficient means of transportation, developing financial institutions for small businesses, and training teams to handle pandemics. While APEC has no means of compelling its member countries to act, it is raising the profile of many crucial long-term concerns for Oceania.

Sometimes mass tourism can seem like an invading force to ordinary citizens. For example, an important segment of the Honolulu tourist infrastructure—hotels, golf courses, specialty shopping centers, import shops, and nightclubs—is geared to visitors from Japan, and many such facilities are owned by Japanese investors. Hawaiian citizens and other non-Japanese shoppers and vacationers can feel out of place.

Another example of the impositions of mass tourism is the demand by tourists for golf courses on many of the Hawaiian Islands, which resulted in what Native (indigenous) Hawaiians view as desecration of sacred sites. Land that in precolonial times was communally owned, cultivated, and used for sacred rituals was confiscated by the colonial government and more recently sold to Asian golf course developers. Now the only people with access to the sacred sites are fee-paying tourist golfers. As of 2012 there were more than 90 golf courses in Hawaii, and the golf industry alone contributed $1.6 billion to the state's economy—more than twice the amount from agriculture. Golf's total impact is $2.5 billion, which represents about 12.5 percent of the state's tourism sector income. Relocation to Hawaii by retired Americans looking for a sunny spot—often called *residential tourism*—has also had an effect on property values and the use of sacred lands by local citizens. *[Source: Hawaii Tourism Authority; and a field report from Conrad M. Goodwin and Lydia Pulsipher. For detailed source information, see Text Credit pages.]* ■

Sustainable Tourism Some Pacific islands have attempted to deal with the pressures of tourism by adopting the principle of *sustainable tourism*, which aims to decrease tourism's imprint and minimize disparities between hosts and visitors. Samoa, for example, has created the Samoan Tourism Authority in conjunction with the South Pacific Regional Environment Programme. (*Samoa* refers to the independent country that was formerly known as Western Samoa; that country is politically distinct from American Samoa, a U.S. territory.) With financial aid from New Zealand, the Authority develops and monitors sustainable tourism components (beaches, wetlands, and forested island environments) and provides *knowledge-based tourism experiences* for visitors (information-rich explanations of political, social, and environmental issues).

The Future: Diverse Global Orientations?

Despite the powerful forces pushing Oceania toward Asia, important factors still favor strong ties with Europe and North America. In spite of increasing trade links and recent efforts by China to expand diplomatic and cultural relations with Australia, both Australia and New Zealand remain staunch military allies of the United States. Over the years, both have participated in U.S.-led wars in Korea, Vietnam, Afghanistan, and Iraq. In 2012, in an apparent effort to check the growing influence of the Chinese military in the South China Sea and the Indian Ocean, the Australian government gave the U.S. Marine Corps access to a large tract of land near Darwin (located in Australia's Northern Territory). The United States and Australia also opened discussions regarding the use of the Cocos Islands (Australian possessions in the Indian Ocean) for reconnaissance purposes.

In some of the Pacific islands, strong links to Europe and North America are also upheld by continuing administrative control. In Micronesia, the United States governs Guam and the Northern Mariana Islands; in Polynesia, American Samoa is a U.S. territory. Just as the Hawaiian Islands are a U.S. state, the 120 islands of French Polynesia—including Tahiti and the rest of the Society Islands, the Marquesas Islands, and the Tuamotu Archipelago—are Overseas Lands of France. Any desire people in these possessions have for independence has not been sufficient to override the financial benefits of aid, subsidies, and investment money provided by France and the United States.

In 1989, the Asia Pacific Economic Cooperative (APEC), composed of 21 members (Australia, Brunei, Canada, Chile, China, Hong Kong, Indonesia, Japan, South Korea, Malaysia, Mexico, New Zealand, Papua New Guinea, Peru, the Philippines, Russia, Singapore, Taiwan, Thailand, the United States, and Vietnam) was organized to enhance economic prosperity and strengthen the Asia–Pacific community. APEC is the only intergovernmental group in the world that operates on the basis of nonbinding commitments and open dialogue among all participants. Unlike the WTO or NAFTA or even the European Union (after which it is partially patterned), APEC does not oblige its members to participate in any treaties. Decisions made within APEC are reached by consensus (meaning discussions continue and agreements are adjusted until all consent) and commitments are undertaken on a voluntary basis. APEC's member economies account for approximately 40 percent of the world's population, just over 50 percent of global production, and more than 40 percent of global trade.

THINGS TO REMEMBER

Geographic Insight 3

- **Globalization and Development** Globalization is reorienting Oceania (especially Australia and New Zealand) toward Asia as a major destination for exports and an increasing source of Oceania's imports and tourists.
- Service industries are becoming the dominant income source for most of the region's economies, although extractive industries remain important; subsistence is a crucial economic supplement.
- Tourism is a significant and growing part of the economies in Oceania, but it can produce stresses.

e attempt (APEC) to forge interna-
/, climate change, and energy-efficient

ns

ıuge portion of the planet, its total
ı people, close to that of the state
re 11.17). The people of Oceania
larger than the contiguous United
pieces across an ocean larger than
acific islands, including Hawaii's

1.74 million, have nearly 4.75 million people; Australia has 22.7 million; Papua New Guinea, 6.9 million; and New Zealand, 4.4 million.

Disparate Population Patterns in Oceania

> **Geographic Insight 4**
> **Population and Urbanization:** This largest region of the world is only lightly populated, but it is highly urbanized.

There are varying population patterns in Oceania. Like many wealthy countries, Australia, New Zealand, and Hawaii have

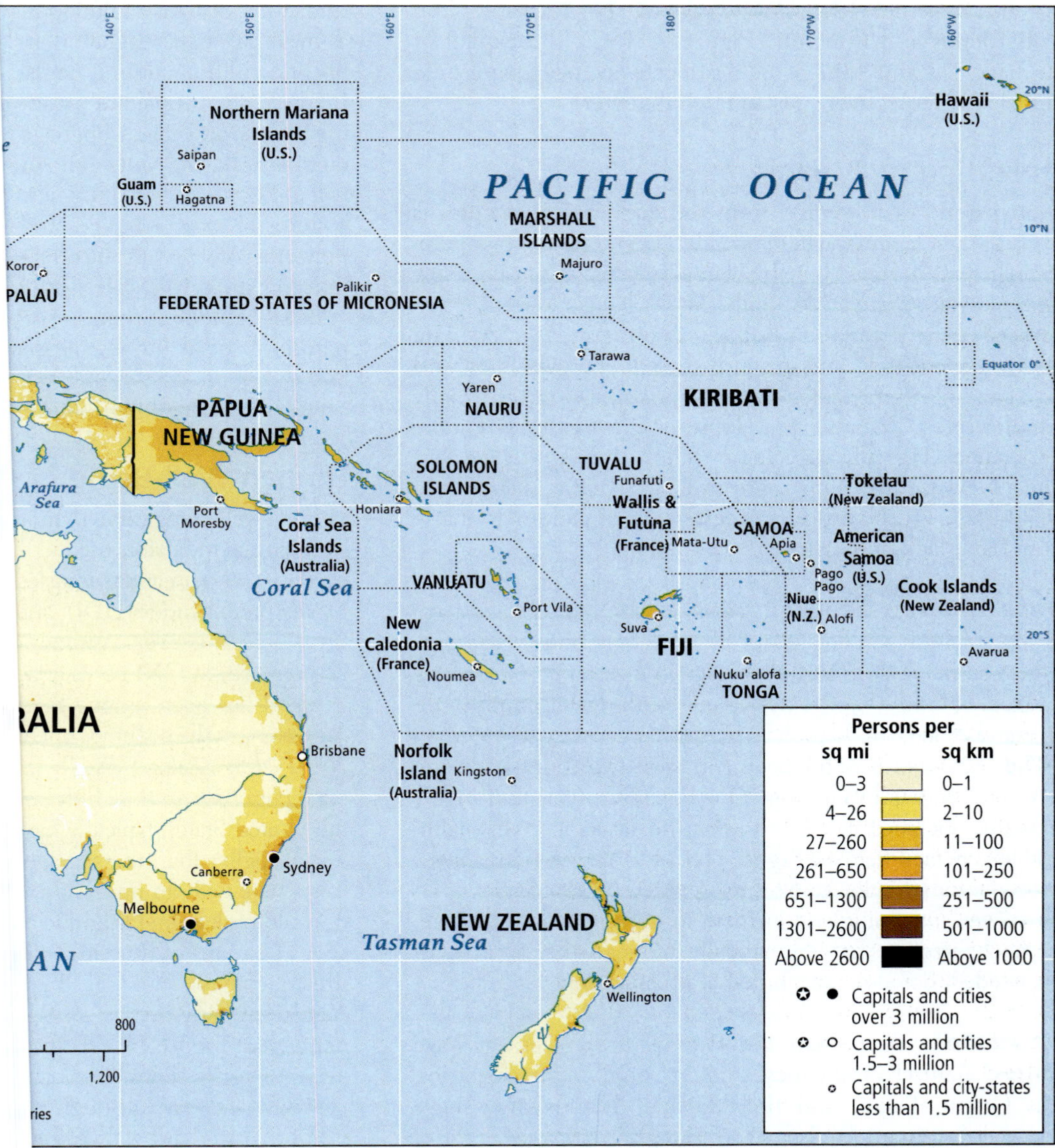

ensity in Oceania. Population growth is slowing throughout Oceania
es, where health care is better and women are more likely to pursue
families less likely. Australia, New Zealand, and Hawaii are furthest along
banized, with smaller families and increasingly old populations.

highly urbanized, relatively older, and more slowly growing populations, with life expectancies of close to 80 years. The Pacific islands and Papua New Guinea, like many developing countries, have much more rural, younger, and rapidly growing populations, with life expectancies in the 60s and low 70s. The overall trend throughout the region, however, is toward smaller families, aging populations, and urbanization.

Population densities remain low in Australia, at 7.8 people per square mile (3 per square kilometer) for the country as a whole and in Australia's settled southeast about 50 per square mile (20 per square kilometer). New Zealand's density is also low, at 41.4 per square mile (16 per square kilometer). In the Pacific islands, densities vary widely. Some are sparsely settled or uninhabited, while others—including some of the smallest, poorest, and lowest in elevation (and thus exposed to rising sea levels)—are extremely densely populated. For example, the Marshall Islands and Funafuti (Figure 11.18C), the capital of Tuvalu, have 772 and 4847 people, respectively, per square mile (298 and 1871, respectively, per square kilometer).

Urbanization in Oceania

The global trend of migration from the countryside to cities is highly visible in Oceania, where overall 66 percent of the population lives in urban areas, but in selected places (Australia, New Zealand, Hawaii, Guam, Marshall Islands, Nauru, Palau) 80 to 100 percent of the population is urban (see Figure 11.18A). The shift from agricultural and resource-based economies toward service economies is a major driver of urbanization, especially in Australia, New Zealand, and some of the wealthier Pacific islands such as Hawaii and Guam. These trends are weakest in Papua New Guinea and many smaller Pacific islands (see Figure 11.18C, D). Nauru is a special case of extremely dense settlement and lack of balanced development.

Australia and New Zealand Australia and New Zealand have among the highest percentages of city dwellers outside Europe. More than 82 percent of Australians live in a string of cities along the country's relatively well-watered and fertile eastern and southeastern coasts. Similarly, 86 percent of New Zealanders live in urban areas. The vast majority of the people in these two countries live in modern comfort, work in a range of occupations typical of highly industrialized societies, and have access to tax-supported healthcare and leisure facilities (see Figure 11.18A). Vibrant, urban-based service economies employ about three-quarters of the population in both countries. Declining employment in mining and agriculture, where mechanization has dramatically reduced the number of workers needed, has also contributed to urbanization.

Pacific Islands Throughout the Pacific, urban centers have transformed natural landscapes, and in some small countries, such as Guam, Palau, and the Marshall Islands, they have become the dominant landscape. Although cities are places of opportunity, they can also be sites of cultural change, conflict, and environmental hazards (see Figure 11.18C).

The great majority of Pacific island towns and all the capital cities are located in ecologically fragile coastal settings. Many of these waterfront towns were established during the colonial era as ports or docking facilities and were situated in places suitable for only limited numbers of people. Consequently, little land is available for development and access to housing is inadequate. Squatter settlements have been a visible feature of these urban areas for decades.

Multiculturalism has been enhanced by urbanization across the region. Many urban residents are letting go of the rural ways of their childhoods, as well as their ethnic identity and cultural commitments. Increasingly, people are marrying in town and across language divisions, having only one or two children, and creating new patterns of social alliances and networks. Along with the adoption of urban lifestyles, such cultural blending results in new social tensions, changing the very nature of social life in the island Pacific. Urban unemployment and unrest are on the rise, and low economic growth restricts the revenue available to governments to manage urban development.

Thinking Geographically

After you have read about urbanization in Oceania, you will be able to answer the following questions:

A Approximately what percent of Australians live in cities?

C Notice the primary features of this photo. In what environmental zone is this settlement likely located?

D Given the contents of this photo, the caption, and what you have read about population trends, what would be your rationale for assuming this girl likely is not the only child in her family?

Human Well-Being in Oceania

Human well-being varies dramatically across this region as measured by the usual indicators used in this book—gross national income per capita (GNI per capita, adjusted for PPP), rank on the UN Human Development Index (HDI), and the Gender Equality Index (GEI). Over the last decade, Australia typically has ranked among the top 20 countries in GNI per capita PPP (in 2011, it ranked 18th). New Zealand has been among the top 25, but in 2011 it fell to 35th. Australia usually occupies one of the top 5 slots on the HDI (2nd in 2011). New Zealand usually falls at the lower end of highly developed countries on the HDI, but rose to 5th in 2011. Hawaii, part of the United States, typically has a GNI per capita PPP that is about $2000 higher than the average for the United States, and because it is a state with an unusually well-developed social welfare system, it would rank higher than the overall U.S. level of 4th on the HDI.

The maps of human well-being (Figure 11.19 on page 484) illustrate these rankings. Figure 11.19A shows that, except for Australia, New Zealand, and Hawaii, Oceania has low levels of GNI per capita. Figure 11.19B shows medium to low HDI rankings—again with the exceptions of Australia (2), New Zealand (5), and probably Hawaii. Although the Pacific islands have very low levels of income and well-being as measured in official statistics, it should be remembered that *subsistence affluence* (discussed on page 479) and strong communitarian values (see the discussion on page 487 about the *Pacific Way*) can result in higher-than-expected actual well-being.

Photo Essay: Urbanization in Oceania

There are two general patterns of urbanization in Oceania. While Australia, New Zealand, and Hawaii are already highly urbanized places with high standards of living, Papua New Guinea and many Pacific islands are much more rural and have lower standards of living.

A Tourists climb the Harbor Bridge in Sydney, Australia, the largest city in Oceania and one that is consistently ranked among the most livable cities in the world, along with Melbourne and Perth, Australia, and Auckland, New Zealand.

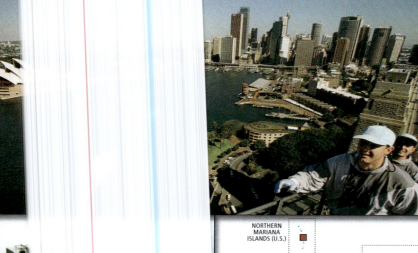

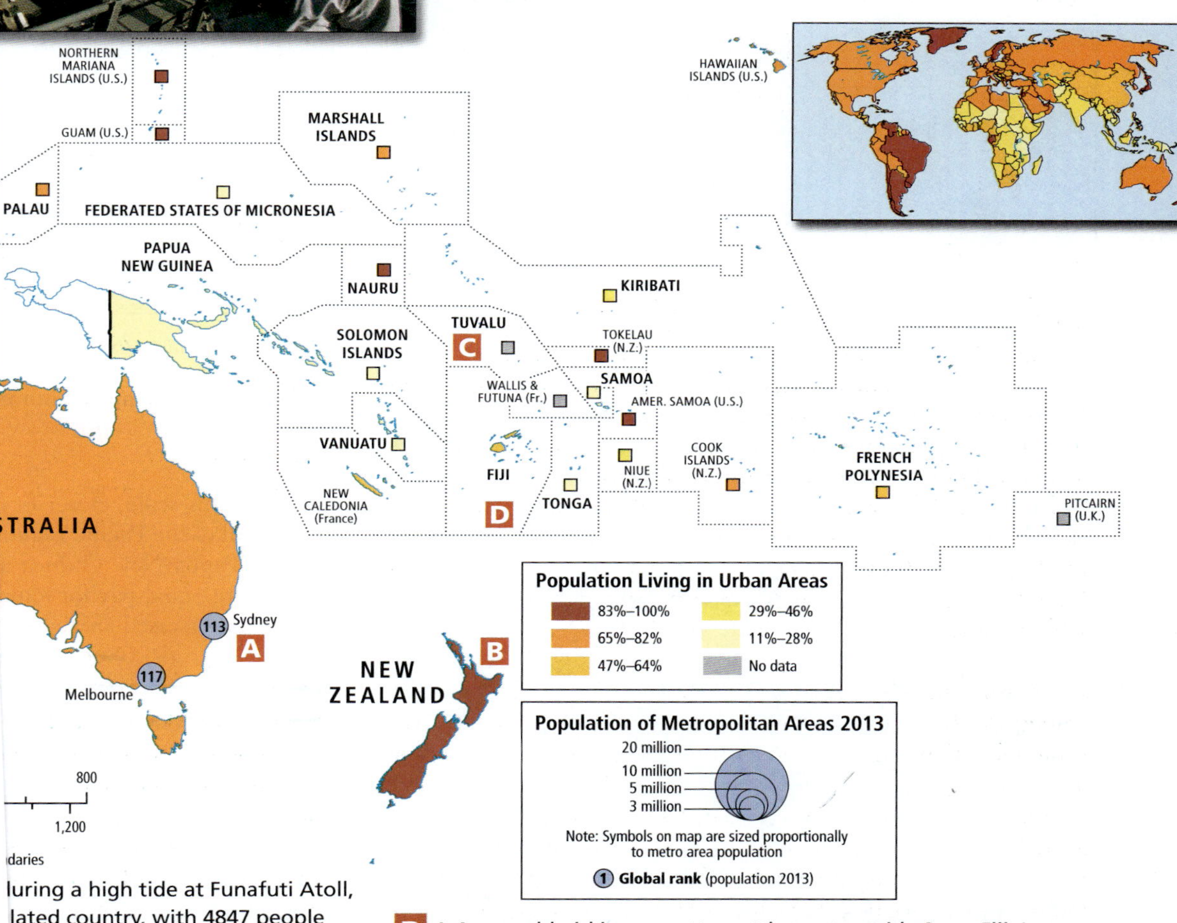

B A fr…
Christma…
park in A…
New Zea…

…during a high tide at Funafuti Atoll,
…lated country, with 4847 people
…meter), Tuvalu is one of the poorest
…r capita of about U.S.$1600.

D A 4-year-old girl in a squatter settlement outside Suva, Fiji. Incomes are relatively low here, as are access to health care and education.

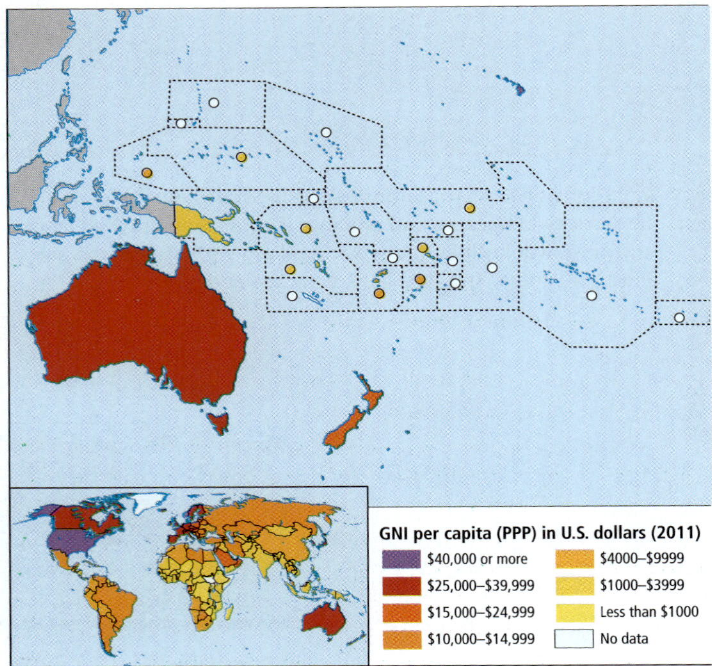

(A) Gross national income (GNI) per capita, adjusted for purchasing power parity (PPP).

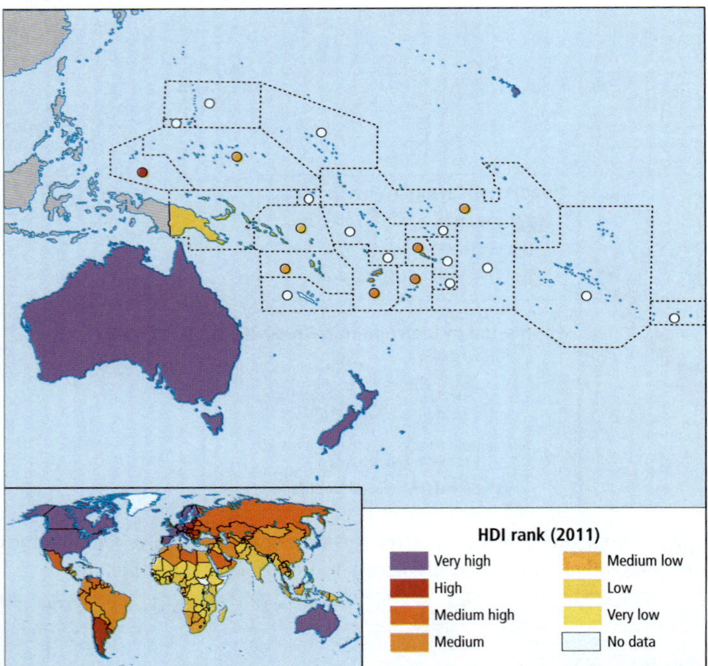

(B) Human Development Index (HDI).

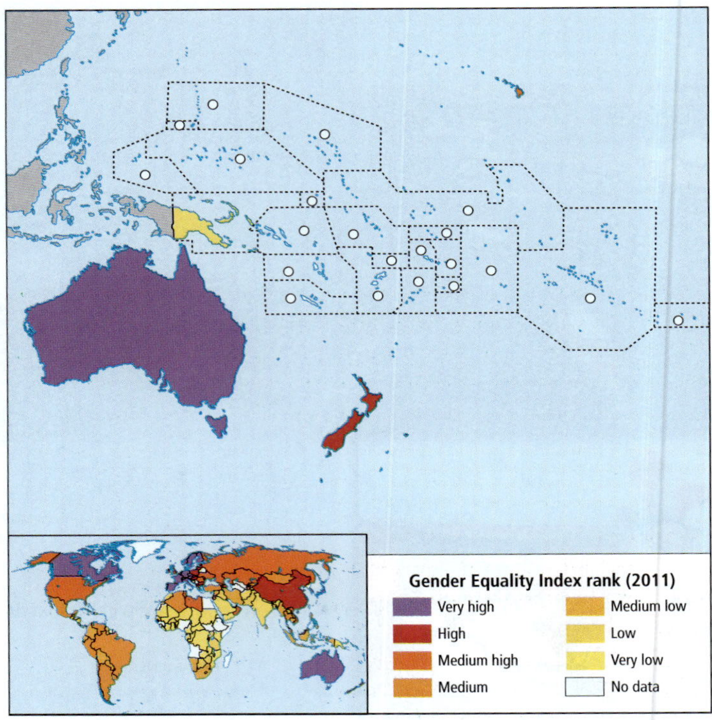

(C) Gender Equality Index (GEI).

FIGURE 11.19 Maps of human well-being.

Figure 11.19C shows how well countries are ensuring gender equality (GEI) in three categories: reproductive health, political and education empowerment, and access to the labor market. With a high rank indicating that genders are tending toward equality, Australia (18) and New Zealand (32) had the best records in the region in 2011. Hawaii, however, ranked lower on this scale, closer to the level of the United States as a whole. All of the Pacific islands ranked lower yet, with the exceptions of Papua New Guinea and Vanuatu. However, the apparently more equal pay for females and males in these two places should be assessed in light of the generally very low incomes on those islands. Overall across the region, few countries have reported sufficient statistics to determine the gap in pay between men and women.

THINGS TO REMEMBER

Geographic Insight 4

- **Population and Urbanization** Like many wealthy countries, Australia, New Zealand, and Hawaii have highly urbanized, relatively older, and more slowly growing populations, with life expectancies of about 80 years. The Pacific islands and Papua New Guinea, like many developing countries, have populations that tend to be more rural, young, and rapidly growing, with life expectancies in the 60s and low 70s.
- Despite having overall very light population density, urbanization is the dominant settlement pattern in Oceania, associated with rapid and deep cultural change.
- The great majority of Pacific island towns and all of the capital cities are located in ecologically fragile coastal settings.
- Human well-being varies dramatically across the region, but subsistence affluence and communitarian values help alleviate some apparently poor circumstances.

:s

Oceania away from Europe and
has been accompanied by new
. In addition, a growing sense of
th Asia has heightened awareness
ıre.

ıed

: of European descent in Australia
hemselves as Europeans in exile.
complete until they had made a
r the European continent. In her
:don (1902), Louise Mack wrote:
ıway there at the other end of the
great in art and music, but born
and hear and come close to these
ir home[land]s."
were accompanied by racist atti-
:oples and Asians. Most histories
twentieth century failed to even
ıter writings described them as
ere numerous projects to take
ɔarents and acculturate them to
; schools. From the 1920s to the
ɔolicies barred Asians, Africans,

and Pacific Islanders from migrating to Australia and discouraged them from entering New Zealand. As we have seen, trading patterns in that era further reinforced connections to Europe.

Weakening of the European Connection When migration from the British Isles slowed after World War II, both Australia and New Zealand began to lure immigrants from southern and eastern Europe, many of whom had been displaced by the war. Hundreds of thousands came from Greece, Italy, and what was then Yugoslavia. The arrival of these non-English-speaking people began a shift toward a more multicultural society. The whites-only immigration policy was abandoned and people began to arrive from many places. There was an influx of Vietnamese refugees in the early 1970s, during the frantic exodus that followed the United States' withdrawal from Vietnam. More recently, skilled workers from India and elsewhere in Asia have met the growing demand for information technology (IT) specialists throughout the service sector.

As of 2010, more than one-fourth of the Australian population was foreign born. The fastest-growing group was from India. Nonetheless, while new immigration policies are increasing the numbers of immigrants from China, Vietnam, and India, people of Asian birth or ancestry remain a minor percentage of the total population in both Australia and New Zealand. In 2006, the latest year for which complete statistics are available, of Australia's foreign-born residents, 42 percent were from Europe; 15 percent were from Asia (Figure 11.20). New Zealand has similar

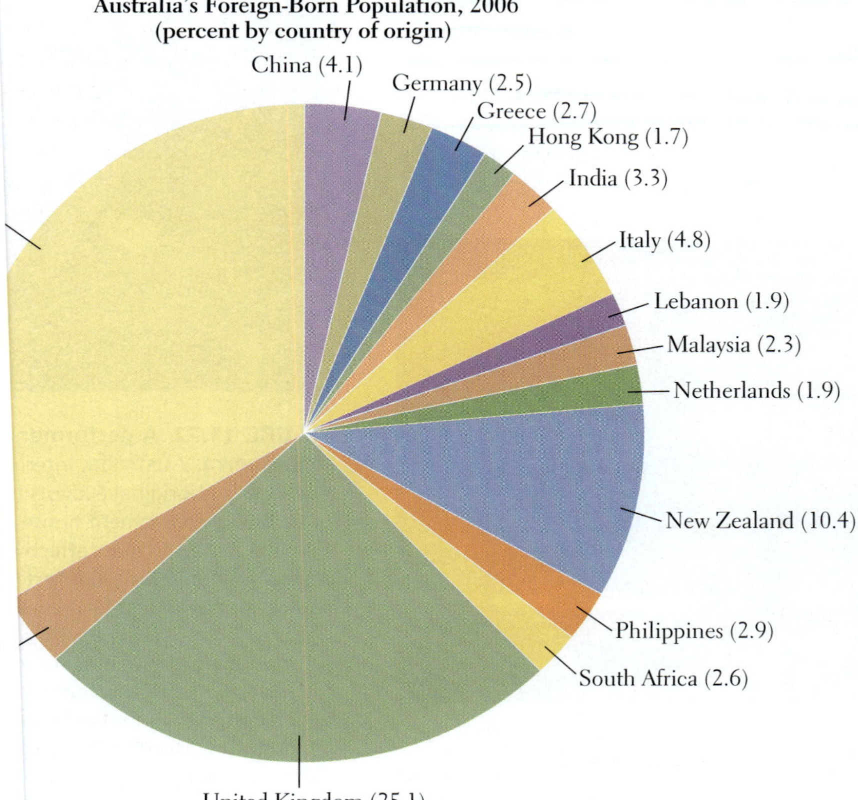

Australia's cultural diversity in 2006. More than 24 percent (4.95 million) of Australia's
ɿn in other places, making Australia one of the world's most ethnically diverse nations.

proportions among its foreign-born residents, and most immigrants continue to come from Europe. Although (due to low birth rates and high immigration rates) Europeans are decreasing as a percentage of the population in both Australia and New Zealand, they will still constitute two-thirds or more of both countries' populations by 2021 (see http://tinyurl.com/7ejcmp3).

The Social Repositioning of Indigenous Peoples in Australia and New Zealand Perhaps the most interesting population change in Australia and New Zealand is one of identity. For the first time in 200 years, the number of people in both countries who claim indigenous origins is increasing. Between 1991 and 1996, the number of Australians claiming Aboriginal origins rose by 33 percent. By 2009, the Aboriginal population was estimated at 528,600. In New Zealand, the number claiming Maori background rose by 20 percent between 1991 and 2009 to 652,900.

These increases are due to changing identities, not to a population boom. More positive attitudes toward indigenous peoples have encouraged open acknowledgment of Aboriginal or Maori ancestry. Also, marriages between European and indigenous peoples are now more common. As a result, the number of people with a recognized mixed heritage is increasing.

As society has acknowledged that discrimination has been the main reason for the low social standing and impoverished state of indigenous peoples, respect for Aboriginal and Maori culture has also increased. The Australian Aborigines base their way of life on the idea that the spiritual and physical worlds are intricately related (Figure 11.21). The dead are everywhere present in spirit, and they guide the living in how to relate to the physical environment. Much Aboriginal spirituality refers to the *Dreamtime*, the time of creation when the human spiritual connections to rocks, rivers, deserts, plants, and animals were made clear. However, very few Aboriginal people continue to practice their own cultural traditions or live close to ancient homelands. Instead, many live in impoverished urban conditions. In New Zealand, where the Maori constitute about 15 percent of the country's population and Auckland now has the largest Polynesian population (including Native Maori) of any city in the world, there are now many efforts to bring Maori culture more into the mainstream of national life.

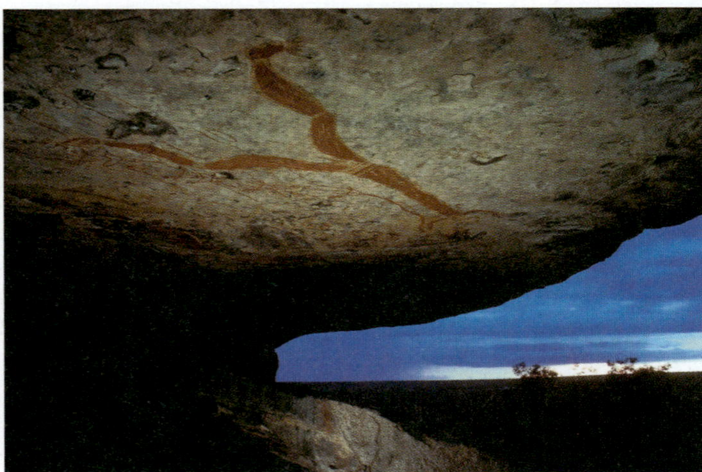

FIGURE 11.21 Aboriginal rock art. A Mimi spirit painted on a rock at Kakadu National Park, Australia. To the Aborigines, Mimi spirits are teachers who pass between this world and another dimension via crevices in rocks. They are responsible for many teachings on hunting, food preparation, the use of fire, dance, and sexuality.

Aboriginal Land Claims In 1988, during a bicentennial celebration of the founding of white Australia, a contingent of some 15,000 Aborigines protested that they had little reason to celebrate. During the same 200 years, they were assumed to have no prior claim to any land in Australia, had lost basic civil rights, and had effectively been erased from the Australian national consciousness. Into the 1960s, it was even illegal for Aborigines to drink alcohol.

British documents indicate that during colonial settlement, all Australian lands were deemed to be available for British use. The Aborigines were thought to be too primitive to have concepts of land ownership because their nomadic cultures had "no fixed abodes, fields or flocks, nor any internal hierarchical differentiation." The Australian High Court Mabo vs. Queensland decision, regarding native land rights, declared this position void in 1993. After that, Aboriginal groups began to win some land claims, mostly for land in the arid interior previously controlled by the Australian government. Figure 11.22 shows the Aboriginal Tent Embassy in 2012, versions of which have stood on the grounds of Parliament in Canberra for more than 40 years. The Aboriginal

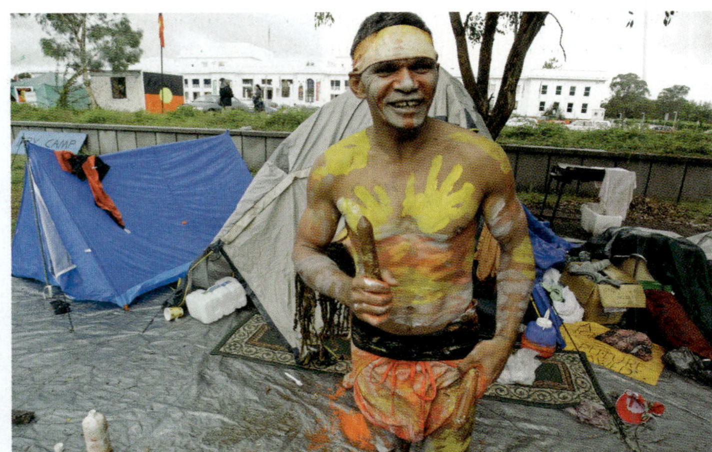

FIGURE 11.22 A performer at the Aboriginal Tent Embassy in Canberra, Australia. Intermittently since 1972, and continuously since 1992, Aboriginal activists have camped out on the grounds of Australia's parliament house in Canberra, Australia. Considered by many to be the most effective political action ever taken by Australian Aborigines, the first tent embassy was a response to the Australian government's denial of land ownership and other land rights to Aborigines in territories they had continuously occupied for thousands of years. As Aboriginal land rights have gained recognition, the tent embassy has championed other causes, including opposition to mining that threatens Aboriginal communities and cultural sites, as well as the plight of the Aboriginal urban poor, such as the community of Redfern in Sydney, Australia. The tent embassy remains controversial and has been targeted by arsonists. The Australian government plans a more permanent structure but will then ban camping at the embassy.

tal in rais-
stices and
Aboriginal
her efforts
and lands

Zealand,
European-
ndigenous
omewhat more amicably than in
gned the Waitangi Treaty with the
anting only rights of land usage,
4B on page 490). The Maori did
nmodity, but rather as an asset of
l larger kin groups to fulfill their
wson writes: "To the Maori the
eatures of land and water bodies
ual meaning and the Maori cre-
d that the treaty had given them
d with British migrants and to
mining, and forestry.
ll but 6.6 percent of their former
ne government. Maori numbers
in the early 1800s to 42,000 in
upy the lowest and most impov-
iety. In the 1990s, however, the
ure, and they established a tribu-
nterests and land claims through
lf a million acres of land and
ansferred back to Maori control.
have notably higher unemploy-
d poorer health than the New

rent

ave emerged in Oceania over
em of government that domi-
ia—versus the **Pacific Way**, the
that influences governments

n Australia and New Zealand
universal suffrage, debate, and
ased on traditional notions of
ers to a way of settling issues
any Pacific Islanders. It favors
ng over open confrontation,
(especially the usually patri-
llages) over free speech and
he Pacific Way can embody
rights and corruption from
.23B, C).
al and cultural philosophy
f its independence from the
uently gained popularity in

ON THE BRIGHT SIDE

The Value of the Pacific Way to Grassroots Sustainability

Regardless of its global political status, the Pacific Way is likely to endure, especially as a concept that upholds regional identity and traditional culture. Further, some organizations now use the Pacific Way as the basis of an integrated approach to economic development and environmental issues. For example, the South Pacific Regional Environmental Programme builds on traditional Pacific island economic activities—such as fishing and local traditions of environmental knowledge and awareness—to promote grassroots economic development and environmental sustainability.

many Pacific islands, most of which gained independence in the 1970s and 1980s. The Pacific Way carries a flavor of resistance to Europeanization and has often been invoked to uphold the notion of a regional identity shared by Pacific islands that grows out of their unique history and social experience. It was particularly influential among educators given the task of writing new textbooks to replace those used by the former colonial masters. The new texts focused students' attention away from Britain, France, or the United States and toward their own cultures. Appeals to the Pacific Way have also been used to uphold attempts by Pacific island governments to control their own economic development and solve their own political and social problems.

In politics, the Pacific Way has occasionally been invoked as a philosophical basis for overriding democratic elections that challenge the traditional power of indigenous Pacific Islanders. In 1987, 2000, and 2006, indigenous Fijians used the Pacific Way to justify coups d'état against legally elected governments (see Figure 11.23B). All three of the overthrown governments were dominated by Indian Fijians, the descendants of people from India whom the British brought to Fiji more than a century ago to work on sugar plantations.

Fiji's population is now about evenly divided between indigenous Fijians and Indian Fijians. Indigenous Fijians are generally less prosperous and tend to live in rural areas where community affairs are still governed by traditional chiefs. By contrast, Indian Fijians hold significant economic and political power, especially in the urban centers and in areas of tourism and sugar cultivation. In response to the coups, many Indian Fijians left the islands, resulting in a loss of badly needed skilled workers, which has slowed economic development.

Political responses around Oceania to the Fiji coups have been divided. Australia, New Zealand, and the United States (via APEC and the state government of Hawaii) have demanded that the election results stand and the Indian Fijians be returned to office. But much of Oceania has referenced the Pacific Way in arguments supporting the coup leaders. As in Fiji, those who govern many of the Pacific islands are leaders of indigenous descent who have not always had the strongest respect for political freedoms, especially when their hold on power is threatened (see Figure 11.23C).

On the global stage, Fiji has been suspended from the Commonwealth of Nations (a union of former British colonies) for subverting majority-rule democracy. As a result, it is ineligible for Commonwealth aid and is not allowed to participate in Commonwealth sports events. Because sports play a central role in

Pacific Way the regional identity and way of handling conflicts based on notions of power and problem solving originating in the traditional culture of many Pacific Islanders

Figure 11.23 Photo Essay: Power and Politics in Oceania

Politics in this region vary significantly between the more Europeanized areas of Australia, New Zealand, and Hawaii, and the more indigenous and traditional political cultures of New Guinea and many Pacific islands.

A A funeral at an immigration detention facility on Christmas Island, Australia, for refugees from Iraq and Iran who drowned while trying to reach Australia by boat. Immigration has long been a political hot-button issue in Australia. Controversy surrounds Australia's policy of sending refugees who arrive by boat to detention facilities on Pacific islands, some not governed by Australia. Here they may wait for years in prison-like conditions for their claims to be judged. Some Australians argue that because relatively few refugees come to their country, they should be treated better and accepted more quickly. Others see refugees as a security risk or an economic drain and argue that the long detentions discourage refugees from coming to Australia.

!! WARNING !!
MILITARY
CHECK POINT
SLOW DOWN

B A Fijian soldier during the military coup of 2006, which overturned a fair election. Military takeovers also occurred in Fiji in 1987 and 2000.

Democratization and Conflict Armed conflicts and genocides with high death tolls since 1990

13,000 deaths

Democratization index

- Full democracy
- Flawed democracy
- Hybrid regime
- Authoritarian regime
- No data

mi 0 250 500
km 0 250 500

HAWAIIAN ISLANDS (U.S.)
NORTHERN MARIANA ISLANDS (U.S.)
GUAM (U.S.)
MARSHALL ISLANDS
PALAU
FEDERATED STATES OF MICRONESIA
NAURU
PAPUA NEW GUINEA
KIRIBATI
CHRISTMAS ISLAND (Australia)
SOLOMON ISLANDS
TUVALU
TOKELAU (N.Z.)
SAMOA
AMER. SAMOA (U.S.)
COOK ISLANDS (N.Z.)
WALLIS AND FUTUNA (Fr.)
VANUATU
FIJI
TONGA
NIUE (N.Z.)
FRENCH POLYNESIA
NEW CALEDONIA (France)
PITCAIRN (U.K.)
AUSTRALIA
NEW ZEALAND

C Members of a violent criminal gang guard the entrance to their headquarters in Port Moresby, the capital of Papua New Guinea. Decades of rampant corruption have led to a breakdown of law and order in the capital and many other areas of Papua New Guinea.

D Corrie Bodney, an elder of the Ballaruk Aboriginal tribe, stages a sit-in at Perth International Airport, which is located on land the Ballaruk have occupied for thousands of years. In 2013, the government of Western Australia offered several Aboriginal tribes over U.S.$1 billion to settle a larger claim, which covered the entire city of Perth.

Pacific identity (see "Sports as a Unifying Force" below), this latter sanction carries significant weight.

Forging Unity in Oceania

Although wide ocean spaces and the great diversity of languages in the region sometimes make communication difficult, travel, sports, and festivals (Figure 11.24) are three forces that help bring the people of Oceania closer together.

Languages in Oceania The Pacific islands—most notably Melanesia—have a rich variety of languages. In some cases, the islands in a single chain have several different languages. A case in point is Vanuatu,
c islands to the east of northern
re spoken by a population of just
e for every 1600 people.
tant part of a community's cul-
cross-cultural understanding.
he Pacific, the need for com-
is served by a number of **pidgin**
ilar to be mutually intelligible.
wed from several languages by
ships. Over time, pidgins can
es, capable of fine nuances of
in is in such common use that
, then it can literally be called
Guinea, a version of pidgin

which unity is manifested in
, people travel in small planes
ch as Fiji, where jumbo jets
ourne, and Honolulu. Cook
canoes of the modern age,"
ancers from across the region
risbane; businesspeople from
to take a short course at the
ook Islands teacher can take
ts fans can visit multiple loca-

as a Unifying Force
d games are a major feature
life throughout Oceania.
ion has shared sports tradi-
, and borrowed them from,

pidgin a
made up o
several lan
trading rela

cultures around the world. Surfing evolved in Hawaii and, like outrigger sailing and canoeing, derives from ancient navigational customs that matched human wits against the power of the ocean. On hundreds of Pacific islands and in Australia and New Zealand, rugby, volleyball, soccer, and cricket are important community-building activities. Baseball is a favorite in the parts of Micronesia that were U.S. trust territories. Women compete in the popular sport of netball (similar to basketball but without a backboard).

Pan-Oceania sports competitions are the single most common and resilient link among the countries of the region. Attendance at regional sports events is so desirable that low-income islanders will hold yard sales and raffles to amass the cash necessary to make the trip. The centrality of such competitions in daily life encourages regional identity and provides opportunities for ordinary citizens to travel extensively around the region and to other parts of the world.

The *haka* (Figure 11.25 on page 491) is an example of how, in the postcolonial modern era, indigenous culture in Oceania is being revived, celebrated, and appropriated in new places by those who wish to project a multicultural image. The haka is a highly emotional and physical dance traditionally performed by the Maori to motivate fellow warriors and intimidate opponents before entering battle. Dances like this have long been a part of many cultures in the islands of Oceania, but the haka has now become an integral part of rugby, the region's most popular sport (Figure 11.26 on page 491). Before almost every international match for the past century, the All Blacks (the New Zealand men's rugby team) have performed the haka: chanting, screaming, jumping, stomping their feet, poking out their tongues, widening their eyes to show the whites, and beating their thighs, arms, and chests.

Outside of Oceania, those who perform the haka include the rugby teams at Jefferson High in Portland, Oregon, and Middlebury College in Vermont, and the football teams at Brigham Young University and the University of Hawaii. All of these teams have players who are of Polynesian heritage. Most practitioners speak of the haka as filling them with the necessary exuberance, aggression, and spirituality to play a vigorous and successful game. To see videos of a haka, go to http://www.YouTube.com and type in *haka*.

THINGS TO REMEMBER

- Oceania's long-standing cultural and economic links to Europe are being challenged by reinvigorated native traditions and identities and by economic globalization, which is increasing the region's links to Asia.
- The number of Asian immigrants into Australia and New Zealand has been increasing over the past two decades, while the number of Europeans has been declining. Asians, however, still comprise only a small minority of the populations of both Australia and New Zealand.
- Governance in Oceania is for the most part based on democratic principles with regular elections; however; traditional power holders have staged coups that have negated or threatened political freedoms.
- Sports and festivals are unifying forces for the region, inspiring fundraisers that allow ordinary citizens to travel to games, thus reinforcing regional and ethnic identity.

FIGURE 11.24 LOCAL LIVES **FESTIVALS IN OCEANIA**

A A young Aboriginal dancer at Garma Festival, which is held to encourage the practice of traditional dance, singing, visual art, and ceremony of the Yolngu people. The festival is held every year in Arnhem Land, which overlooks the Gulf of Carpentaria in Australia's Northern Territory.

B Waitangi Day in New Zealand, a national holiday that commemorates the signing of a treaty between the indigenous Maori of New Zealand and the British. The long boats shown here are Maori canoes, known as *waka*, and are part of a reenactment of the signing of the treaty.

C The aerial theater and comedy troupe named Dislocate performs at the Sydney Festival, a 3-week-long international arts festival that is held every January in Sydney, Australia.

Gender Roles in Oceania

> **Geographic Insight 5**
> **Gender, Power, and Politics:** Changes in political empowerment have taken different forms across Oceania, but there is a general trend toward a more active role for women.

Perceptions of Oceania are colored by many myths about how men and women are and should be. As always, the realities are more complex.

Gender Myths and Realities Because of Oceania's cultural diversity, there are many different roles for men and women. In the Pacific islands, men traditionally were cultivators, deepwater fishers, and masters of seafaring. In Polynesia, men also were responsible for many aspects of food preparation, including cooking. In the modern world, men fill many positions, but idealized male images continue to be associated with vigorous activities.

In Australia and New Zealand, the hypermasculine, white, working-class settler has long had prominence in the national mythologies. In New Zealand, he was a farmer and herdsman.

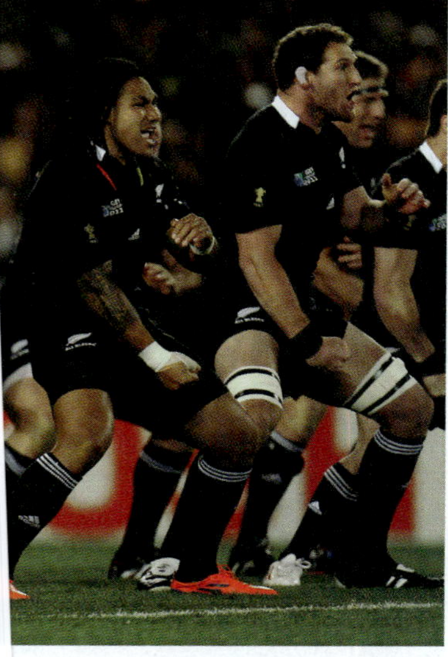

ori tradition. A haka performed
eam, the All Blacks, before a match

ten a many-skilled laborer—a
cutter, or digger (miner)—who
nse of humor. Labeled a "swag-
went from station (large farm)
ing hard but sporadically, gam-
king again until he had enough
it in the city (**Figure 11.27**).

There, he often felt ill at ease and chafed to return to the wilds. Now immortalized in songs ("Waltzing Matilda," for example), novels, and films, these men are portrayed as a rough and nomadic tribe whose social life is dominated by male camaraderie and frequent brawls. No small part of this characterization of males derived from the fact that many of Australia's first immigrants were convicts.

Today, as part of larger efforts to recognize the diversity of Australian society, new ways of life for men are emerging and are breaking down the national image of the tough male loner. Nonetheless, the old model persists and remains prominent in the public images of Australian businessmen, politicians, and movie stars.

Perhaps the most enduring myth Europeans created regarding Oceania was their characterization of the women of the Pacific islands as gentle, simple, compliant love objects. (Tourist brochures still promote this notion.) There is ample evidence to suggest that Pacific Islanders did have more sexual partners in a lifetime than Europeans did. However, the reports of unrestrained sexuality related by European sailors were no doubt influenced by the exaggerated fantasies one might expect from all-male crews living at sea for months at a time. The notes of Captain James Cook are typical: "No women I ever met were less reserved. Indeed, it appeared to me, that they visited us with no other view, than to make a surrender of their persons." Over the years, such notions about Pacific island women have

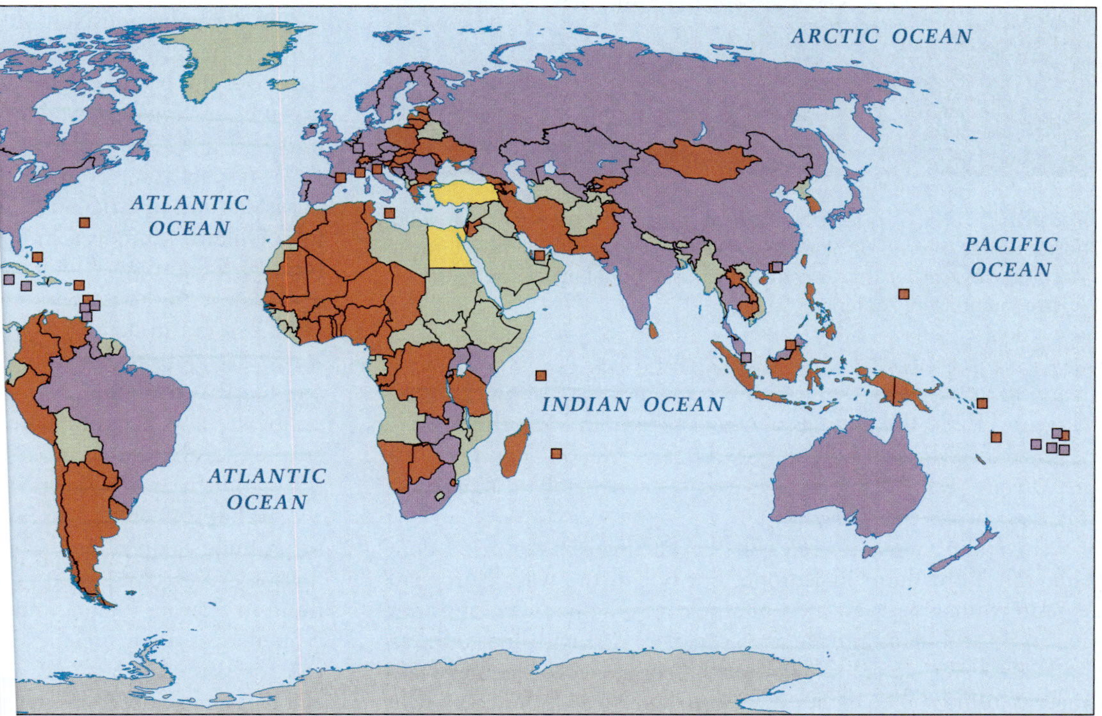

FI **world.** In more than 136 countries, women, men, boys, and girls play rugby. All
of rugby teams, and 58 of these countries have women's national teams. The men's
Wc 1987 and the women's World Cup competition began in 1991. In April 2010,
Ne n men and women, but the rankings of the men's teams can change weekly.

FIGURE 11.27 Gender and national mythology. Australian wild horse hunter George Girdler epitomizes the hypermasculine, white, working-class settler that is central to the national mythologies of Australia and New Zealand.

FIGURE 11.28 *Arearea* ("*Amusement*") by Paul Gauguin. In this 1892 painting, Tahitian women are rendered in a European Romantic pastoral style that emphasizes their gentle, compliant demeanor.

been encouraged by the paintings and prints of Paul Gauguin (Figure 11.28), the writings of novelist Herman Melville (*Typee*), and the studies of anthropologist Margaret Mead (*Coming of Age in Samoa*), as well as by movies and musicals such as *Mutiny on the Bounty* and *South Pacific*.

In reality, women's roles in the Pacific islands varied considerably from those in Europe, but not in the ways European explorers imagined. Women often exercised a good bit of power in family and clan, and their power increased with motherhood and advancing age. In Polynesia, a woman could achieve the rank of ruling chief in her own right, not just as the consort of a male chief. Women were primarily craftspeople, but they also contributed to subsistence by gathering fruits and nuts and by fishing. And in some places—Micronesia, for example—lineage was established through women, not men.

Gender, Democracy, and Economic Empowerment Today there is a trend toward equality across gender lines throughout Oceania, but there is persistent inequality as well. A striking disparity is emerging between Australia and New Zealand (where women are gaining political and economic empowerment) and Papua New Guinea and the Pacific islands (where change is much slower).

In Australia and New Zealand, women's access to jobs and policy-making positions in government has improved, particularly over the last few decades. New Zealand and the Australian province of South Australia were among the first places in the world to grant European women full voting rights (1893 and 1895, respectively). New Zealand has elected two female prime ministers, and in 2010 Australia elected a woman prime minister, Julia Gillard. Moreover, in both countries, according to the 2011 Inter-Parliamentary Union report on women in Lower Houses of Parliament, the proportion of females in national legislatures (32.2 percent in New Zealand, 24.7 percent in Australia) is well above the global average of 18 percent. In Papua New Guinea and the Pacific islands, women are generally far less empowered politically and economically. No woman has yet been elected to a top-level national office, and women are a tiny minority in national legislatures, if present at all.

In both Australia and New Zealand, young women are pursuing higher education and professional careers and postponing marriage and childbearing until their thirties. (This is also a trend in Hawaii, Guam, and the islands with French affiliation.) Nonetheless, both societies continue to reinforce the housewife role for women in a variety of ways. For example, the expectation is that women, not men, will interrupt their careers to stay home to care for young or elderly family members. Women in Australia receive, on average, only about 70 percent of the pay that men receive for equivalent work. This is, however, a smaller gender pay gap than in many other developed countries.

Guinea and the Pacific islands, ›s vary greatly over the course of ıphasis on community, male and ibute to family assets through the ›s. Traditionally, men are the boat nd house builders. And they are ıough women often supply some r gathering and cultivating efforts n marketplaces are women, and, sell are also made and transport- men today fulfill traditional roles ıctice a wide range of domestic ısketry. In middle age, however, ı to school and take up careers. ıolarships, some Pacific Islander n or job training that takes them ‹ raised their children. Thus the chapter's opening vignette) will study far from home and then support her peers and elders is in line with evolving gender roles in Pacific ways of life. Aurora can expect that with accumulating age and experience, she will be boosted into a position of considerable power in her community.

THINGS TO REMEMBER

Geographic Insight 5

- **Gender, Power, and Politics** Gender roles and relationships vary considerably in Oceania. In Australia and New Zealand, compared with elsewhere in the region, women generally are better off in terms of political power and pay equity with men.
- In some traditional societies, especially those on Pacific islands with more than basic education, women can inherit or personally accrue considerable power in their own communities over the course of a lifetime.

LOCAL LIVES FOODWAYS IN OCEANIA

›er Valley ›ry. ›r of wine ›pain.

B Taro is harvested in Hawaii. Grown in flooded fields, taro was originally brought to Oceania from Southeast Asia. It is now a major part of diets throughout the Pacific islands. While the leaves are also eaten, the root is particularly prized as a source of calories.

C Food is removed from a Maori earth oven, or *hangi,* in New Zealand. First a pit is dug, and then a fire is made to heat stones placed in the pit. Baskets of food are placed over the hot stones (which are covered with cloth and then earth) for several hours until the food is cooked.

Geographic Insights: Oceania Review and Self-Test

1. Climate Change and Water: Oceania's sea level and freshwater resources are being impacted by climate change.

• Which parts of Oceania are most vulnerable to changes brought about by climate change, and what changes are taking place? Which parts of the region contribute the most and least to greenhouse gas emissions?

• Why are coral reefs important ecologically, and how are they threatened by climate change?

2. Food and Development: European systems for producing food and fiber introduced into Oceania have greatly changed environments.

• How have European farm animals and crops introduced into Australia and New Zealand affected the economies of those countries? What is the purpose of the Dingo Fence? Has it worked?

• How do foreign patterns of consumption affect this region, and what are the threats to Oceania's food security?

3. Globalization and Development: Globalization, coupled with a greater focus on neighboring Asia (rather than on longtime European connections), has transformed patterns of trade and economic development across Oceania.

• What does Asia buy from the Pacific islands, and what are Asia's trade connections to Australia and New Zealand?

• How does the loss of preferential trading ties with Europe affect the region's workers and tax revenues?

• Explain the effect of raw materials exports on Oceania's GNIs.

• Explain why many Pacific Islanders can be said to enjoy subsistence affluence despite having rather low monetary incomes.

4. Population and Urbanization: This largest region of the world is only lightly populated, but it is highly urbanized.

• Why, despite its low average population densities, is the region highly urbanized?

• Contrast the population growth rates in Australia and New Zealand with those in the rest of Oceania. In which parts of this region is the aging of the population of most concern?

• Which parts of Oceania are both densely populated and threatened by climate change?

5. Gender, Power, and Politics: Changes in political empowerment have taken different forms across Oceania, but there is a general trend toward a more active role for women.

• What are the interesting new trends in ethnic identities and the positioning of indigenous peoples in Oceania?

• How do ideas about the proper exercise of political power in the Pacific islands differ from those found in Australia and New Zealand and some Europeanized islands? To what extent is the political philosophy known as the Pacific Way a counter to European ideas of democracy?

• How do sports events and festivals help to forge a common identity and unity in this far-flung region?

• Describe the possible changing gender roles of Pacific island women over the course of their lifetimes.

• Describe how male roles have evolved in Australia and New Zealand, from colonial times to the present. Which domestic roles of Polynesian men contradict the image asserted by the haka?

Critical Thinking Questions

1. As Australia and New Zealand move away from intense cultural and economic involvement with Europe, new policies and attitudes have evolved to facilitate increased involvement with Asia. If you were a college student in Australia or New Zealand, how might you experience these changes? Think about fellow students, career choices, language learning, and travel choices.

2. Discuss the emerging cultural identity of the Pacific islands, taking note of the extent to which Australia and New Zealand share or do not share in this identity. What factors are helping to forge a sense of unity across Oceania and beyond? (First, review the spatial extent of Oceania.)

3. Discuss the many ways in which Asia has historic, and now increasingly economic, ties to Oceania. In your discussion, include patterns of population distribution, mineral exports and imports, technological interactions, and tourism.

4. To what extent can the countries of Oceania exercise control over their future as the climate changes?

5. Australia and New Zealand differ from each other physically. Compare and contrast the two countries in relation to water, vegetation, and prehistoric and modern animal populations.

6. Indigenous peoples worldwide are beginning to speak out on their own behalf. Discuss how the indigenous peoples of Australia, New Zealand, and the Pacific islands are serving as leaders in this movement and what measures they are taking to reconstitute a sense of cultural heritage.

7. How is tourism both boosting economies and straining environments and societies throughout the Pacific islands? Describe the solutions that are being proposed to reduce the negative impacts of tourism.

8. Compare how women have or have not been empowered, politically and economically, in Australia, New Zealand, Papua New Guinea, and the Pacific islands.

9. Compared with other regions, Australia and New Zealand are somewhat unusual in having become broadly prosperous on the basis of raw materials exports. How would you explain this achievement?

Maori 463
marsupials 466
Melanesia 473
Melanesians 473
Micronesia 474
MIRAB economy 479
monotremes 466
Pacific Way 487
pidgin 489
Polynesia 474
Roaring Forties 463
subsistence affluence 479

Photo Essay: Antarctica

Despite its remote location and few human inhabitants, Antarctica is undergoing environmental change at an accelerating pace as temperatures rise and surrounding fisheries are exploited.

A Scientists measure sea level at Antarctica's Ross Sea. While most sea level rise predicted for the near future is caused by the thermal expansion of oceans, sea levels could rise 200 feet (60 meters) if Antarctica's massive glaciers were all to melt.

B A fishing vessel in the Southern Ocean, one of thousands that are placing increasing pressure on Antarctica's fisheries. Species like the Patagonian toothfish (marketed as Chilean sea bass) are severely overfished. Tiny shrimp known as *krill*, which form the basis of many oceanic ecosystems, are also increasingly being over fished.

C An emperor penguin dives beneath a hole in sea ice. Emperor penguins are found only in Antarctica and are proving extremely sensitive to climate change. In warmer years, declines in sea ice, the penguins' ideal hunting habitat, have resulted in widespread starvation among the penguins. Meanwhile, in colder years, fewer penguin chicks hatch.

D Amundsen–Scott South Pole Station, the dome-shaped structure, is a U.S. research station. Outside fly the flags of the first countries to sign the Antarctic Treaty of 1959, which bans all military and resource extraction–related activity, making the continent a scientific and nature reserve.

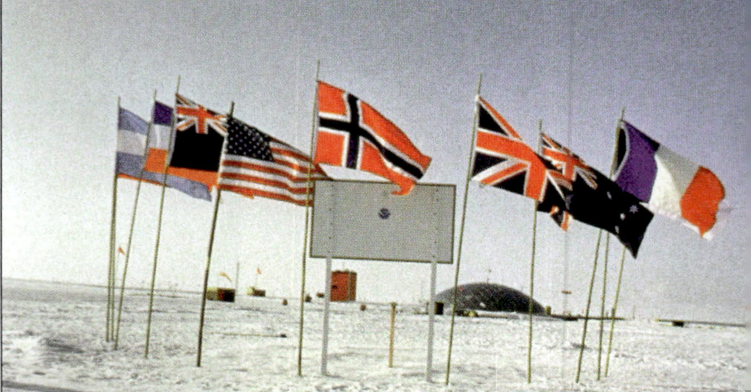

GLOSSARY

and are to be understood in the context in which they appear in the text.

viving inhabitants of Oceania, whose
ıted from Southeast Asia 50,000 to
ıland landmass that was exposed

ɔf a minority culture to the host
ctively and be self-supporting
that has formed through the interaction
with sulfur dioxide and nitrogen oxides
il fuels, making it acidic
farming conducted by large-scale
ce, finance, package, and distribute

roducing food through animal
ıals and the cultivation of plants
ɔf traditional, nonchemical
d the use of natural predators to

mically useful crops of trees on
s and crops, to reduce dependence
and to provide income to the farmer
l minority group in Japan
heavy beards, and thick, wavy hair
d thousands of years ago from the

which spirits, including those of the
rywhere and to offer protection to

ɔch during which humans have had
Earth's biosphere
ıandating racial segregation. South
ı 1948 until 1994.
rground reservoirs of water
chain, of islands
f old ways of life and the adoption
or country
ons (ASEAN) (p. 437) an
ɔvernments that was established to
tical cooperation
ı of islets, formed of coral reefs that
rims of a submerged volcano
s of hunters and gatherers who
Indian and Burman parts of
ındmass of Sundaland about

oups of skilled farmers and
o migrated south to various parts of
d 5000 years ago
stem that subordinates individual
ɔr of elite regional and local

ıigh-central Mexico noted for
he Spanish conquest
forms to be found in a given area
arth's integrated physical spheres,
cluded as part of nature
per 1000 people in a given
ly per year)

Bolsheviks (p. 209) a faction of Communists who came to power during the Russian Revolution

brain drain (p. 145) the migration of educated and ambitious young adults to cities or foreign countries, depriving the communities from which the young people come of talented youth in whom they have invested years of nurturing and education

brownfields (p. 93) old industrial sites whose degraded conditions pose obstacles to redevelopment

Buddhism (p. 352) a religion of Asia that originated in India in the sixth century B.C.E. as a reinterpretation of Hinduism; it emphasizes modest living and peaceful self-reflection leading to enlightenment

capitalism (pp. 36, 171) an economic system based on the private ownership of the means of production and distribution of goods, driven by the profit motive and characterized by a competitive marketplace

capitalists (p. 209) usually a wealthy minority that owns the majority of factories, farms, businesses, and other means of production

carbon sequestration (p. 285) the removal and storage of carbon taken from the atmosphere

carrying capacity (p. 26) the maximum number of people that a given territory can support sustainably with food, water, and other essential resources

cartel (p. 264) a group of producers strong enough to control production and set prices for products

cartographer (p. 4) geographers who specialize in depicting geographic information on maps

cash economy (p. 17) an economic system that tends to be urban but may be rural, in which skilled workers, well-trained specialists, and even farm laborers are paid in money

caste system (p. 350) a complex, ancient Hindu system for dividing society into hereditary hierarchical classes

Caucasia (p. 221) the mountainous region between the Black Sea and the Caspian Sea

central planning (p. 171) a communist economic model in which a central bureaucracy dictates prices and output with the stated aim of allocating goods equitably across society according to need

centrally planned, or **socialist, economy** (p. 210) an economic system in which the state owns all land and means of production, while government officials direct all economic activity, including the locating of factories, residences, and transportation infrastructure

Christianity (p. 248) a monotheistic religion based on the belief in the teachings of Jesus of Nazareth, a Jew, who described God's relationship to humans as primarily one of love and support, as exemplified by the Ten Commandments

civil disobedience (p. 343) protesting of laws or policies by peaceful direct action

civil society (p. 36) the social groups and traditions that function independently of the state and its institutions to foster a sense of unity and informed common purpose among the general population

clear-cutting (p. 65) the cutting down of all trees on a given plot of land, regardless of age, health, or species

climate (p. 47) the long-term balance of temperature and precipitation that characteristically prevails in a particular region

climate change (p. 40) a slow shifting of climate patterns due to the general cooling or warming of the atmosphere

Cold War (pp. 171, 210) a period of conflict, tension, and competition between the United States and the Soviet Union that lasted from 1945 to 1991

commercial agriculture (p. 290) farming in which crops are grown deliberately for cash rather than solely as food for the farm family

commodities (p. 300) raw materials that are traded, often to other countries, for processing or manufacturing into more valuable goods

commodity dependence (p. 300) economic dependence on exports of raw materials

Common Agricultural Program (CAP) (p. 178) an EU program, meant to guarantee secure and safe food supplies at affordable prices, that places tariffs on imported agricultural goods and gives subsidies to EU farmers

communal conflict (p. 353) a euphemism for religiously based violence in South Asia

communism (pp. 36, 171, 209) an ideology, based largely on the writings of the German revolutionary Karl Marx, that calls on workers to unite to overthrow capitalism and establish an egalitarian society in which workers share what they produce; as practiced, communism was actually a socialized system of public services and a centralized government and economy in which citizens participated only indirectly through Communist Party representatives

Communist Party (p. 210) the political organization that ruled the USSR from 1917 to 1991; other communist countries, such as China, Mongolia, North Korea, and Cuba, also have Communist parties

Confucianism (p. 388) a Chinese philosophy that teaches that the best organizational model for the state and society is a hierarchy based on the patriarchal family

contested space (p. 133) any area that several groups claim or want to use in different and often conflicting ways, such as the Amazon or Palestine

coral bleaching (p. 429) color loss that results when photosynthetic algae that live in the corals are expelled

coup d'état (p. 135) a military- or civilian-led forceful takeover of a government

Creoles (p. 125) people mostly of European descent born in the Caribbean

crony capitalism (p. 436) a type of corruption in which politicians, bankers, and entrepreneurs, sometimes members of the same family, have close personal as well as business relationships

cultural homogenization (p. 155) the tendency toward uniformity of ideas, values, technologies, and institutions among associated culture groups

cultural pluralism (p. 451) the cultural identity characteristic of a region where groups of people from many different backgrounds have lived together for a long time but have remained distinct

Cultural Revolution (p. 392) a political movement launched in 1966 to force the entire population of China to support the continuing revolution

culture (p. 52) all the ideas, materials, and institutions that people have invented to use to live on Earth that are not directly part of our biological inheritance

czar (p. 208) title of the ruler of the Russian empire

death rate (p. 13) the ratio of total deaths to total population in a specified community, usually expressed in numbers per 1000 or in percentages

delta (p. 47) the triangular-shaped plain of sediment that forms where a river meets the sea

democratization (p. 33) the transition toward political systems guided by competitive elections

demographic transition (p. 16) the change from high birth and death rates to low birth and death rates that usually accompanies a cluster of other changes, such as change from a subsistence to a cash economy, increased education rates, and urbanization

desertification (p. 243) a set of ecological changes that converts arid lands into deserts

detritus (p. 424) dead organic material (such as plants and insects) that collects on the ground

development (p. 19) a term usually used to describe economic changes such as the greater productivity of agriculture and industry that lead to better standards of living or simply to increased mass consumption

diaspora (p. 248) the dispersion of Jews around the globe after they were expelled from the eastern Mediterranean by the Roman Empire, beginning in 73 C.E.; the term can now refer to other dispersed culture groups

dictator (p. 135) a ruler who claims absolute authority, governing with little respect for the law or the rights of citizens

digital divide (p. 86) the discrepancy in access to information technology between small, rural, and poor areas and large, wealthy cities that contain major government research laboratories and universities

divide and rule (p. 306) the deliberate intensification of divisions and conflicts by potential rulers; in the case of sub-Saharan Africa, by European colonial powers

domestication (p. 23) the process of developing plants and animals through selective breeding to live with and be of use to humans

domino theory (p. 433) a foreign policy theory that used the idea of the domino effect to suggest that if one country "fell" to communism, others in the neighboring region were also likely to fall

double day (p. 187) the longer workday of women with jobs outside the home who also work as caretakers, housekeepers, and/or cooks for their families

dowry (p. 354) a price paid by the family of the bride to the groom (the opposite of *bride price*); formerly a custom practiced only by the rich

dual economy (p. 301) an economy in which the population is divided by economic disparities into two groups, one prosperous and the other near or below the poverty level

early extractive phase (p. 128) a phase in Central and South American history, beginning with the Spanish conquest and lasting until the early twentieth century, characterized by a dependence on the export of raw materials

ecological footprint (p. 22) the amount of biologically productive land and sea area needed to sustain a person at the current average standard of living for a given population

economic core (p. 71) the dominant economic region within a larger region

economic diversification (p. 265) the expansion of an economy to include a wider array of activities

economies of scale (p. 175) reductions in the unit cost of production that occur when goods or services are efficiently mass produced, resulting in increased profits per unit

ecotourism (p. 119) nature-oriented vacations, often taken in endangered and remote landscapes, usually by travelers from affluent nations

El Niño (p. 114) periodic climate-altering changes, especially in the circulation of the Pacific Ocean, now understood to operate on a global scale

emigration (p. 13) out-migration (see also *migration*)

endemic (p. 465) belonging or restricted to a particular place

erosion (p. 47) the process by which fragmented rock and soil are moved over a distance, primarily by wind and water

ethnic cleansing (p. 37) the deliberate removal of an ethnic group from a particular area by forced migration

ethnic group (p. 52) a group of people who share a common ancestry and sense of common history, a set of beliefs, a way of life, a technology, and usually a common geographic location of origin

ethnicity (p. 96) the quality of belonging to a particular culture group

required) currency of the European

in the European Union that use the

supranational organization that
West, South, North, and Central

a Christian movement that focuses
werment of the individual through
mation; some practitioners preach to
—that a life dedicated to Christ will
ver
p. 131) specially created legal spaces
try where, to attract foreign-owned
t charged
omic development strategy that
of manufactured goods destined for

at consists of related individuals
ents and children
nd agriculture
uity throughout the international
alternative to free trade
ns and shantytowns built by the
barriadas in other countries
p. 318) removing the labia and the
he vulva nearly shut
ement that women stay out of

creasing employment of women
abor force, usually at lower wages

, fertile land formed by the uplands
stems and the Zagros Mountains,
earliest known agricultural

nese term for people who live
r required *(hukou)* household
less or underemployed people
ssed rural areas for the cities
d a river where sediment is

of a state to consistently supply a
the entire population
31) investment funds that come
ountry
rency that countries need to

the economy that take place in

formed from the remains of dead

rnational exchange of goods,

owned energy company; it is the

cial group defines the differences

ed roles for males and females
agriculture, the practice of
dely divergent species to achieve

lestruction of an ethnic, racial,

gentrification (p. 93) the renovation of old urban districts by affluent investors, a process that often displaces poorer residents

Geographic Information Science (GISc) (p. 7) the body of science that underwrites multiple spatial analysis technologies and keeps them at the cutting edge

geopolitics (p. 36) the strategies that countries use to ensure that their own interests are served in relations with other countries

glasnost (p. 213) literally, "openness"; the policies instituted in the late 1980s under Mikhail Gorbachev that encouraged greater transparency and openness in the workings of all levels of Soviet governments

global economy (p. 30) the worldwide system in which goods, services, and labor are exchanged

globalization (p. 30) the growth of interregional and worldwide linkages and the changes these linkages are bringing about

global scale (p. 12) the level of geography that encompasses the entire world as a single unified area

global warming (p. 40) the warming of the Earth's climate as atmospheric levels of greenhouse gases increase

Gondwana (p. 462) the great landmass that formed the southern part of the ancient supercontinent Pangaea

grassroots economic development (p. 305) economic development projects designed to help individuals and their families achieve sustainable livelihoods

Great Barrier Reef (p. 462) the longest coral reef in the world, located off the northeastern coast of Australia

Great Leap Forward (p. 392) an economic reform program under Mao Zedong intended to quickly raise China to the industrial level of Britain and the United States

Green (p. 164) environmentally conscious

greenhouse gases (p. 40) gases, such as carbon dioxide and methane, released into the atmosphere by human activities, which become harmful when released in excessive amounts

green revolution (p. 26) increases in food production brought about through the use of new seeds, fertilizers, mechanized equipment, irrigation, pesticides, and herbicides

gross domestic product (GDP) per capita (p. 19) the total market value of all goods and services produced within a particular country's borders and within a given year, divided by the number of people in the country

gross national income (GNI) per capita (p. 15) the total production of goods and services in a country in a given year divided by the mid-year population

groundwater (p. 292) water naturally stored in aquifers as long as 5000 years ago during wetter climate conditions

Group of Eight (G8) (p. 212) an organization of eight countries with large economies: France, the United States, Britain, Germany, Japan, Italy, Canada, and Russia

growth poles (p. 399) zones of development whose success draws more investment and migration to a region

guest workers (p. 184) legal workers from outside a country who help fulfill the need for temporary workers but who are expected to return home when they are no longer needed

Gulf states (p. 254) Saudi Arabia, Kuwait, Bahrain, Oman, Qatar, and the United Arab Emirates

hacienda (p. 128) historically, a large agricultural estate in Middle or South America, not specialized by crop and not focused on market production

hajj (p. 253) the pilgrimage to the city of Makkah (Mecca) that all Muslims are encouraged to undertake at least once in a lifetime

Harappa culture (p. 340) see *Indus Valley civilization*

Hinduism (p. 350) a major world religion practiced by approximately 900 million people, 800 million of whom live in India; a complex belief system, with roots both in ancient literary texts (known as the

Great Tradition) and in highly localized folk traditions (known as the *Little Tradition*)

Hispanic (Latino) (p. 61) a term used to refer to all Spanish-speaking people from Middle and South America, although their ancestors may have been black, white, Asian, or Native American

Holocaust (p. 171) during World War II, a massive execution by the Nazis of 6 million Jews and 5 million gentiles (non-Jews), including ethnic Poles and other Slavs, Roma (Gypsies), disabled and mentally ill people, gays, lesbians, transgendered people, and political dissidents

Horn of Africa (p. 285) the triangular peninsula that juts out from northeastern Africa below the Red Sea and wraps around the Arabian Peninsula

hot spots (p. 463) individual sites of upwelling material (magma) that originate deep in Earth's mantle and surface in a tall plume; hot spots tend to remain fixed relative to migrating tectonic plates

***hukou* system** (p. 375) the system in China that ties people to their place of birth; each person's permanent residence is registered and any person who wants to migrate must obtain permission from authorities to do so

human geography (p. 3) the study of patterns and processes that have shaped human understanding, use, and alteration of the Earth's surface

humanism (p. 167) a philosophy and value system that emphasizes the dignity and worth of the individual, regardless of wealth or social status

human well-being (p. 20) various measures of the extent to which people are able to obtain a healthy and socially rewarding standard of living in an environment that is safe and sustainable

humid continental climate (p. 157) a midlatitude climate pattern in which summers are fairly hot and moist, and winters become longer and colder the deeper into the interior of the continent one goes

immigration (p. 13) in-migration (see also *migration*)

import substitution industrialization (ISI) (p. 129) policies that encourage local production of machinery and other items that previously had been imported at great expense from abroad

Incas (p. 122) indigenous people who ruled the largest pre-Columbian state in the Americas, with a domain stretching from southern Colombia to northern Chile and Argentina

income disparity (p. 127) the gap in income between rich and poor

indigenous (p. 112) native to a particular place or region

Indus Valley civilization (p. 340) the first substantial settled agricultural communities, which appeared about 4500 years ago along the Indus River in modern-day Pakistan and northwest India

industrial production (p. 19) processing, manufacturing, and construction

Industrial Revolution (p. 23) a series of innovations and ideas that occurred broadly between 1750 and 1850, which changed the way goods were manufactured

informal economy (p. 20) all aspects of the economy that take place outside official channels

infrastructure (p. 71) road, rail, and communication networks and other facilities necessary for economic activity

interregional linkages (p. 30) economic, political, or social connections between regions, whether contiguous or widely separated

intertropical convergence zone (ITCZ) (p. 285) a band of atmospheric currents that circle the globe roughly around the equator; warm winds from both north and south converge at the ITCZ, pushing air upward and causing copious rainfall

intifada (p. 275) a prolonged Palestinian uprising against Israel

invasive species (p. 467) organisms that spread into regions outside of their native range, adversely affecting economies or environments

Iron Curtain (p. 171) a long, fortified border zone that separated western Europe from (then) eastern Europe during the Cold War

Islam (p. 236) a monotheistic religion that emerged in the seventh century C.E. when, according to tradition, the archangel Gabriel revealed the tenets of the religion to the Prophet Muhammad

Islamism (p. 237) a grassroots religious revival in Islam that seeks political power to curb what are seen as dangerous secular influences; also seeks to replace secular governments and civil laws with governments and laws guided by Islamic principles

isthmus (p. 112) a narrow strip of land that joins two larger land areas

Jainism (p. 352) a religion of Asia that originated as a reformist movement within Hinduism more than 2000 years ago; Jains are known for their educational achievements, nonviolence, and strict vegetarianism

jati (p. 350) in Hindu India, the subcaste into which a person is born, which traditionally defined the individual's experience for a lifetime

jihadists (p. 254) especially militant Islamists

Judaism (p. 248) a monotheistic religion characterized by the belief in one god (Yahweh), a strong ethical code summarized in the Ten Commandments, and an enduring ethnic identity

just-in-time system (p. 395) the system pioneered in Japanese manufacturing that clusters companies that are part of the same production system close together so that they can deliver parts to each other precisely when they are needed

***kaizen* system** (p. 395) a system of continuous manufacturing improvement, pioneered in Japan, in which production lines are constantly adjusted, improved, and surveyed for errors to save time and money and ensure that fewer defective parts are produced

Kyoto Protocol (p. 43) an amendment to a United Nations treaty on global warming, the Protocol is an international agreement, adopted in 1997 and in force in 2005, that sets binding targets for industrialized countries for the reduction of emissions of greenhouse gases

landforms (p. 44) physical features of the Earth's surface, such as mountain ranges, river valleys, basins, and cliffs

latitude (p. 4) the distance in degrees north or south of the equator; lines of latitude run parallel to the equator, and are also called parallels

legend (p. 4) a small box somewhere on a map that provides basic information about how to read the map, such as the meaning of the symbols and colors used

liberation theology (p. 148) a movement within the Roman Catholic Church that uses the teachings of Jesus to encourage the poor to organize to change their own lives and to encourage the rich to promote social and economic equity

lingua franca (p. 321) language of trade

living wages (p. 33) minimum wages high enough to support a healthy life

local scale (p. 12) the level of geography that describes the space where an individual lives or works; a city, town, or rural area

longitude (p. 4) the distance in degrees east and west of Greenwich, England; lines of longitude, also called meridians, run from pole to pole (the line of longitude at Greenwich is 0° and is known as the prime meridian)

machismo (p. 144) a set of values that defines manliness in Middle and South America

Maori (p. 463) Polynesian people indigenous to New Zealand

map projections (p. 7) the various ways of showing the spherical Earth on a flat surface

maquiladoras (p. 131) foreign-owned, tax-exempt factories, often located in Mexican towns just across the U.S. border from U.S. towns, that hire workers at low wages to assemble manufactured goods which are then exported for sale

marianismo (p. 143) a set of values based on the life of the Virgin Mary, the mother of Jesus, that defines the proper social roles for women in Middle and South America

pment of a free market economy in

t give birth to their young at a very
em in a pouch equipped with nipples
climate pattern of warm, dry
rs
d when several cities expand so that

nd the islands south of the equator
n Islands, New Caledonia, Fiji, and

traloids named for their relatively dark
of the protective pigment melanin;
nea and other nearby islands
licy by which European rulers
d wealth of their realms by managing
ort, and commerce in their colonies
created in 1991 that links the
Uruguay, and Paraguay to create a

uropean, African, and indigenous

50,000 or more and their

d on peer support that makes very
ncome entrepreneurs
s that lie east of the Philippines and

k, a region that includes Mexico,
of the Caribbean
eople from one place or country to
nic reasons
ny based on migration, remittance,

hat involves raising a variety of
n, often to take advantage of several

on of nomadic pastoral people
who by the thirteenth century
pire that stretched from Europe

here is only one god
mals, such as the duck-billed

ich in summer months, warm,
gs copious rainfall, and in winter,
ental interior toward the ocean
Asian origin that ruled India
n century
in which many culture groups

isiness organization that operates
bution facilities in multiple

terests or culture of a particular
he idea that a group of people
ing cultural traits should be
hey are loyal and obedient
perty and place under
ompensation
hose way of life and economy
nals that are moved seasonally to

nongovernmental organizations (NGOs) (p. 37) associations outside the formal institutions of government in which individuals, often from widely differing backgrounds and locations, share views and activism on political, social, economic, or environmental issues

nonpoint sources of pollution (p. 201) diffuse sources of environmental contamination, such as untreated automobile exhaust, raw sewage, and agricultural chemicals, that drain from fields into water supplies

North American Free Trade Agreement (NAFTA) (pp. 85, 132) a free trade agreement made in 1994 that added Mexico to the 1989 economic arrangement between the United States and Canada

North Atlantic Drift (p. 157) the easternmost end of the Gulf Stream, a broad warm-water current that brings large amounts of warm water to the coasts of Europe

North Atlantic Treaty Organization (NATO) (p. 177) a military alliance between European and North American countries that was developed during the Cold War to counter the influence of the Soviet Union; since the breakup of the Soviet Union, NATO has expanded membership to include much of Eastern Europe and Turkey, and is now focused mainly on providing the international security and cooperation needed to expand the European Union

nuclear family (p. 99) a family consisting of married or common-law parents and their children

occupied Palestinian Territories (oPT) (p. 237) Palestinian lands occupied by Israel since 1967

offshore outsourcing (p. 364) the contracting of certain business functions or production functions to providers in areas where labor and other costs are lower

oligarchs (p. 212) in Russia, those who acquired great wealth during the privatization of Russia's resources and who use that wealth to exercise power

OPEC (Organization of the Petroleum Exporting Countries) (p. 264) a cartel of oil-producing countries—including Algeria, Angola, Iran, Iraq, Kuwait, Libya, Nigeria, Qatar, Saudi Arabia, the United Arab Emirates, Ecuador, and Venezuela—that was established to regulate the production and price of oil and natural gas

organically grown (p. 84) products produced without chemical fertilizers and pesticides

orographic rainfall (p. 48) rainfall produced when a moving moist air mass encounters a mountain range, rises, cools, and releases condensed moisture that falls as rain

Ottoman Empire (p. 249) the most influential Islamic empire the world has ever known; begun in the 1200s when nomadic Turkic herders from Central Asia converged in western Anatolia (Turkey)

Pacific Rim (p. 102) a term that refers to all the countries that border the Pacific Ocean

Pacific Way (p. 487) the regional identity and way of handling conflicts based on notions of power and problem solving originating in the traditional culture of many Pacific Islanders

Partition (p. 343) the breakup following Indian independence that resulted in the establishment of Hindu India and Muslim Pakistan

pastoralism (p. 292) a way of life based on herding; practiced primarily on savannas, on desert margins, or in the mixture of grass and shrubs called *open bush*

patriarchal (p. 254) relating to a social organization in which the father is supreme in the clan or family

perestroika (p. 213) literally, "restructuring"; the restructuring of the Soviet economic system that was done in the late 1980s in an attempt to revitalize the economy

permafrost (p. 199) permanently frozen soil that lies just a few feet beneath the surface

physical geography (p. 3) the study of the Earth's physical processes: how they work and interact, how they affect humans, and how they are affected by humans

pidgin (p. 489) a language used for trading; made up of words borrowed from the several languages of people involved in trading relationships
plantation (p. 128) a large factory farm that grows and partially processes a single cash crop
plate tectonics (p. 44) the scientific theory that the Earth's surface is composed of large plates that float on top of an underlying layer of molten rock; the movement and interaction of the plates create many of the large features of the Earth's surface, particularly mountains
political ecologists (p. 22) geographers who study the interactions among development, politics, human well-being, and the environment
polygyny (p. 317) the practice of having multiple wives
Polynesia (p. 474) the numerous islands situated inside an irregular triangle formed by New Zealand, Hawaii, and Easter Island
population pyramid (p. 14) a graph that depicts the age and gender structures of a political unit, usually a country
populist movements (p. 148) popularly based efforts, often seeking relief for the poor
precipitation (p. 48) dew, rain, sleet, and snow
primary sector (p. 19) an economic sector of the economy that is based on extraction (see also *extraction*)
primate city (p. 145) a city, plus its suburbs, that is vastly larger than all others in a country and in which economic and political activity is centered
privatization (pp. 129, 213) the selling of formerly government-owned industries and firms to private companies or individuals
purchasing power parity (PPP) (p. 20) the amount that the local currency equivalent of U.S.$1 will purchase in a given country
purdah (p. 354) the practice of concealing women from the eyes of nonfamily men
push factors (p. 93) factors that get people to consider the drastic move of leaving family, friends, and a familiar place to strike out into the unknown, with what are usually unknown resources
push/pull phenomenon of urbanization (p. 27) conditions, such as political instability or economic changes, that encourage (push) people to leave rural areas, and urban factors, such as job opportunities, that encourage (pull) people to move to urban areas
quaternary sector (p. 19) a sector of the economy that is based on intellectual pursuits such as education, research, and IT (information technology) development
Qur'an (or **Koran**) (p. 240) the holy book of Islam, believed by Muslims to contain the words Allah revealed to Muhammad through the archangel Gabriel
race (p. 54) a social or political construct that is based on apparent characteristics such as skin color, hair texture, and face and body shape, but that is of no biological significance
rate of natural increase (RNI) (p. 13) the rate of population growth measured as the excess of births over deaths per 1000 individuals per year without regard for the effects of migration
region (p. 9) a unit of the Earth's surface that contains distinct patterns of physical features and/or distinct patterns of human development
regional conflict (p. 368) a conflict created by the resistance of a regional ethnic or religious minority to the authority of a national or state government; currently these are the most intense armed conflicts in South Asia
regional self-sufficiency (p. 396) an economic policy in Communist China that encouraged each region to develop independently in the hope of evening out the wide disparities in the national distribution of resources production and income
regional specialization (p. 396) the encouragement of specialization rather than self-sufficiency in order to take advantage of regional variations in climate, natural resources, and location
religious nationalism (p. 366) the association of a particular religion with a particular territory or political unit
resettlement schemes (p. 442) government plans to move large numbers of people from one part of a country to another to relieve urban congestion, disperse political dissidents, or accomplish other social purposes; also called *transmigration*
responsibility system (p. 396) in the 1980s, a decentralization of economic decision making in China that returned agricultural decision making to the farm household level, subject to the approval of the commune
Ring of Fire (p. 47) the tectonic plate junctures around the edges of the Pacific Ocean; characterized by volcanoes and earthquakes
Roaring Forties (p. 463) powerful air and ocean currents at about 40° S latitude that speed around the far Southern Hemisphere virtually unimpeded by landmasses
Roma (p. 171) the now-preferred term in Europe for Gypsies
Russian Federation (p. 198) Russia and its political subunits, which include 21 internal republics
Russification (p. 217) the czarist and Soviet policy of encouraging ethnic Russians to settle in non-Russian areas as a way to assert political control
Sahel (p. 284) a band of arid grassland, where steppe and savanna grasses grow, that runs east-west along the southern edge of the Sahara
Salafism (p. 268) an extreme, purist Qur'an-based version of Islam that has little room for adaptation to modern times
salinization (p. 240) a process that occurs when large quantities of water are used to irrigate areas where evaporation rates are high, leaving behind salts and other minerals
scale (of a map) (p. 4) the proportion that relates the dimensions of the map to the dimensions of the area it represents; also, variable-sized units of geographical analysis from the local scale to the regional scale to the global scale
Schengen Accord (p. 184) an agreement signed in the 1990s by the European Union and many of its neighbors that allows for free movement across common borders
seawater desalination (p. 243) the removal of salt from seawater—usually accomplished through the use of expensive and energy-intensive technologies—to make the water suitable for drinking or irrigating
secondary sector (p. 19) an economic sector of the economy that is based on industrial production (see also *industrial production*)
secular states (p. 268) countries that have no official state religion and in which religion has no direct influence on affairs of state or civil law
self-reliant development (p. 305) small-scale development in rural areas that is focused on developing local skills, creating local jobs, producing products or services for local consumption, and maintaining local control so that participants retain a sense of ownership over the process
services (p. 19) sales, entertainment, and financial services
sex (p. 17) the biological category of male or female; does not indicate how males or females may behave or identify themselves
sex tourism (p. 455) the sexual entertainment industry that services primarily men who travel for the purpose of living out their fantasies during a few weeks of vacation
sex work (p. 349) the provision of sexual acts for a fee
shari'a (p. 253) literally, "the correct path"; Islamic religious law that guides daily life according to the interpretations of the Qur'an
Shi'ite (or **Shi'a**) (p. 254) the smaller of two major groups of Muslims who have different interpretations of shari'a; Shi'ites are found primarily in Iran and southern Iraq
shifting cultivation (pp. 122, 289) a productive system of agriculture in which small plots are cleared in forestlands, the dried brush

and the clearings are planted with
sed for only 2 or 3 years and then
egrowth
that is located east of the Ural

th Asia that combines beliefs of Islam

ho originated between the Dnieper
ay Poland, Ukraine, and Belarus
rea characterized by crowding,
ate access to food, clean water,

ustrial emissions, car exhaust, and
rs as a yellow-brown haze over many
problems
provided by the government—such
fits, and health care—that prevent
poverty
ion, social protection) (p. 189) in
at provide citizens with benefits
igher education and housing,

nt south of Central America
Soviet Socialist Republics
gement of a phenomenon across the

goods, people, services, or
ong places
399) free trade zones within
ed export processing zones (EPZs)

) an economic system based on
enterprise, profit incentives, and
ng government guidance; in
d government) economic system of
degree, Europe
ed plains
s) (p. 129) policies that require
ess government involvement in
rvices; sometimes imposed by the
Monetary Fund as conditions for

sed to refer to the entire
, Bhutan, India, Pakistan, and

re one tectonic plate slides under

granted by a government to an
activity, such as farming, that is
est
le whereby people are
ecessities and have some
and occasional purchases of

ng that provides food for only the
on small farms
my in which families produce
d shelter
he peripheries of cities
egin every June when the moist
ous western Ghats
groups of Muslims who have

sustainable agriculture (p. 27) farming that meets human needs without poisoning the environment or using up water and soil resources

sustainable development (p. 22) the effort to improve present standards of living in ways that will not jeopardize those of future generations

taiga (p. 199) subarctic coniferous forests

Taliban (p. 355) an archconservative Islamist movement that gained control of the government of Afghanistan for a while in the mid-1990s

temperate midlatitude climate (p. 157) as in south-central North America, China, and Europe, a climate that is moist all year with relatively mild winters and long, mild to hot summers

temperature-altitude zones (p. 114) regions of the same latitude that vary in climate according to altitude

tertiary sector (p. 19) an economic sector of the economy that is based on services (see also *services*)

theocratic states (p. 268) countries that require all government leaders to subscribe to a state religion and all citizens to follow rules decreed by that religion

total fertility rate (TFR) (p. 13) the average number of children that women in a country are likely to have at the present rate of natural increase

trade deficit (p. 87) the extent to which the money earned by exports is exceeded by the money spent on imports

trade winds (p. 114) winds that blow from the northeast and the southeast toward the equator

trafficking (p. 228) the recruiting, transporting, and harboring of people through coercion for the purpose of exploiting them

tsunami (p. 377) a large sea wave caused by an earthquake

tundra (pp. 199) a region of winters so long and cold that the ground is permanently frozen several feet below the surface

typhoon (p. 377) a tropical cyclone or hurricane in the western Pacific Ocean

UNASUR (p. 132) a union of South American nations that was organized in May of 2008; it supersedes Mercosur and the Andean Community of Nations, two previous customs unions

underemployment (p. 214) the condition in which people are working too few hours to make a decent living or are highly trained but working at menial jobs

Union of Soviet Socialist Republics (USSR) (p. 196) the multinational union formed from the Russian empire in 1922 and dissolved in 1991; commonly known as the *Soviet Union*

United Nations (UN) (p. 37) an assembly of 193 member states that sponsors programs and agencies that focus on economic development, security, general health and well-being, democratization, peacekeeping assistance in "hot spots" around the world, humanitarian aid, and scientific research

United Nations Human Development Index (HDI) (p. 20) index that calculates a country's level of well-being, based on a formula of factors that considers income adjusted to PPP, data on life expectancy at birth, and data on educational attainment

United Nations Gender Equality Index (GEI) (p. 20) a composite measure reflecting the degree to which there is inequality in achievements between women and men in three dimensions: reproductive health, political and educational empowerment, and labor-force participation. A high rank indicates that the genders are tending toward equality.

urban growth poles (p. 147) locations within cities that are attractive to investment, innovative immigrants, and trade, and thus attract economic development like a magnet

urban sprawl (p. 65) the encroachment of suburbs on agricultural land

urbanization (p. 27) the process whereby cities, towns, and suburbs grow as populations shift from rural to urban livelihoods

varna (p. 350) the four hierarchically ordered divisions of society in Hindu India underlying the caste system: *Brahmins* (priests), *Kshatriyas* (warriors/kings), *Vaishyas* (merchants/landowners), and *Sudras* (laborers/artisans)

veil (p. 255) the custom of covering the body with a loose dress and/or of covering the head—and in some places the face—with a scarf

virtual water (p. 38) the volume of water used to produce all that a person consumes in a year

water footprint (p. 38) the water used to meet a person's basic needs for a year, added to the person's annual virtual water consumption

weather (p. 47) the short-term and spatially limited expression of climate that can change in a matter of minutes

weathering (p. 47) the physical or chemical decomposition of rocks by sun, rain, snow, ice, and the effects of life-forms

welfare state (p. 171) a government that accepts responsibility for the well-being of its people, guaranteeing basic necessities such as education, affordable food, employment, and health care for all citizens

West Bank barrier (p. 275) a 25-foot-high concrete wall in some places and a fence in others that now surrounds much of the West Bank and encompasses many of the Jewish settlements there

wet rice cultivation (p. 382) a prolific type of rice production that requires the plant roots to be submerged in water for part of the growing season

winter monsoon (p. 332) a weather pattern that begins by November, when the cooling Eurasian landmass sends the cooler, drier, heavier air over South Asia

world region (p. 12) a part of the globe delineated according to criteria selected to facilitate the study of patterns particular to the area

world regional scale (p. 12) the regional scale of analysis that encompasses all the regions of the world

World Trade Organization (WTO) (p. 32) a global institution made up of member countries whose stated mission is the lowering of trade barriers and the establishment of ground rules for international trade

Zionists (p. 273) those who have worked, and continue to work, to create a Jewish homeland (Zion) in Palestine

PHOTO CREDITS

1.29d2 Nikki Kahn/The Washington Post/Getty Images
1.29e1 Paul J. Richards/AFP/Getty Images
1.29e2 P. Morris/Hulton Archive/Getty Images
1.30a Yehuda Raizner/AFP/Getty Images
1.30b Sonu Mehta/Hindustan Times/Getty Images
1.30c Romeo Gacad/AFP/Getty Images
1.30d Luis Acosta/AFP/Getty Images
1.30e Peter Delarue/AFP/Getty Images
1.30f TPG/Getty Images

CHAPTER 2
2.1a Karen Bleier/AFP/Getty Images
2.1b Jeffrey Phelps/Getty Images
2.1c Streeter Lecka/Getty Images
2.1d George Rose/Getty Images
2.1e Chris Graythen/Getty Images
2.2 Paul Harris/Getty Images
2.4a Lowell Georgia/National Geographic/Getty Images
2.4b Linda Davidson/The Washington Post via Getty Images
2.4c DEA/F. Barbagallo/De Agostini/Getty Images
2.5a James P. Blair/National Geographic/Getty Images
2.5b Wild Horizons/UIG via Getty Images
2.5c NASA
2.5e Daniel Acker/Bloomberg via Getty Images
2.5f Rick Eglinton/Toronto Star via Getty Images
2.5g Mandel Ngan/AFP/Getty Images
2.6a Gabriel Bouys/AFP/Getty Images
2.6b Josh Ritchie/Getty Images
2.6c Education Images/UIG via Getty Images
2.6d Craig Warga/NY Daily News Archive/Getty Images
2.9a Cahokia Mounds State Historic Site, painting by Michael Hampshire
2.9b Photo by MPI/Getty Images
2.9c Library of Congress
2.9d Library of Congress
2.9e Library of Congress
2.9f Library of Congress
2.12a Tara Walton/Toronto Star via Getty Images
2.12b Courtesy USAID
2.12c Yoshikazu Tsuno/AFP/Getty Images
2.12d Till Muellenmeister/AFP/Getty Images
2.12e Thony Belizaire/AFP/Getty Images
2.18 Mac Goodwin
2.21a Bob Martin/Sports Illustrated/Getty Images
2.21b Mario Tama/Getty Images
2.21c Carol M. Highsmith/Buyenlarge/Getty Images
2.21d Barry Williams/Getty Images
2.22a Emory Kristof/National Geographic/Getty Images
2.23a Education Images/UIG via Getty Images
2.23b Kasia Wandycz/Paris Match via Getty Images
2.23c Yuri Yuriev/AFP/Getty Images
2.24 Preston C. Mack/Getty Images
2.26a Scott Olson/Getty Images
2.26b Christinne Muschi/Toronto Star via Getty Images
2.26c Andrew McKinney/Dorling Kindersley/Getty Images
2.28a Mario Tama/Getty Images
2.28b David McNew/Newsmakers/Getty Images
2.28c Photo by Todd Korol/Sports Illustrated/Getty Images

CHAPTER 3
3.1a Michel Setboun/Gamma-Rapho via Getty Images
3.1b Hoberman Collection/UIG via Getty Images
3.1c Visions of America/UIG via Getty Images
3.1d Alain Buu/Gamma-Rapho via Getty Images
3.1e NASA
3.1f Antonio Scorza/AFP/Getty Images
3.1g Shaun Botterill - FIFA/FIFA via Getty Images
3.1h DEA/G. Dagli Orti/De Agostini/Getty Images
3.1i James P. Blair/National Geographic/Getty Images
3.2 Alex Pulsipher
3.5a Holger Leue/Lonely Planet Images/Getty Images
3.5b Veronique Durruty/Gamma-Rapho via Getty Images
3.5c DEA/A. Garozzo/De Agostini/Getty Images
3.7a Raphael Gaillarde/Gamma-Rapho via Getty Images
3.7b Education Images/UIG via Getty Images
3.7c Michele Burgess/Photolibrary/Getty Images
3.8a Wesley Bocxe/Photo Researchers/Getty Images
3.8b Michael Fairchild/Peter Arnold/Getty Images
3.8c Glowimages/Getty Images
3.8d Michael Nichols/National Geographic/Getty Images
3.8e JC Patricio/Flickr Open/Getty Images
3.9a Michael Schwab/Flickr/Getty Images
3.9b Mark Carwardine/Peter Arnold/Getty Images
3.9c Inga Spence/Photolibrary/Getty Images
3.10a John Moore/Getty Images
3.10b Yuri Cortez/AFP/Getty Images
3.10c Thony Belizaire/AFP/Getty Images
3.10d John Coletti/AWL Images/Getty Images
3.11a Bjorn Holland/The Image Bank/Getty Images
3.11b Ilufs Vinagre/Flickr/Getty Images
3.11c Danita Delimont/Gallo Images/Getty Images
3.11d Spanish School/The Bridgeman Art Library/Getty Images
3.11e M.H. Zahner/Library of Congress
3.14 Wysocki Pawel/Getty Images
3.17a Greg Elms/Lonely Planet Images/Getty Images
3.17b Virgina Sherwood/Bravo/NBCU Photo Bank via Getty Images
3.17c Juanmonino/E+/Getty Images
3.18a Shaul Schwarz/Getty Images
3.18b Luis Ramirez/AFP/Getty Images
3.18c Luis Gonzalo Vinagre Solans/FlickrVision/Getty Images
3.24a Viviane Ponti/Lonely Planet Images/Getty Images
3.24b Danita Delimont/Gallo Images/Getty Images
3.24c Claudio Cruz/LatinContent/Getty Images
3.25a Jeremy Edwards/Jumper/Photodisc/Getty Images
3.25b Mathias T Oppersdorff/Photo Researchers/Getty Images
3.25c Ernesto Benavides/AFP/Getty Images
3.27 Mario Tama/Getty Images
Table 3.1 (top) Alain Jocard/AFP/Getty Images
Table 3.1 (below top) Francois Ancellet/Gamma-Rapho via Getty Images
Table 3.1 (center) Louise Heusinkveld/Getty Images
Table 3.1 (below center) Travis Dove/National Geographic/Getty Images
Table 3.1 (bottom) Neil Fletcher and Matthew Ward/Dorling Kindersley/Getty Images

CHAPTER 4
4.1a Ingmar Wesemann/E+/Getty Images
4.1b Beate Zoellner/F1online/Getty Images
4.1c Karl Johaentges/LOOK/Getty Images
4.1d Ingolf Pompe/LOOK/Getty Images

4.1e Ultraforma/E+/Getty Images
4.2a Jeroen Peys/E+/Getty Images
4.2b Andy Sacks/Stone/Getty Images
4.4a David Henderson/OJO Images/Getty Images
4.4b Ellen Rooney/Robert Harding World Imagery/Getty Images
4.4c Raimund Linke/Radius Images/Getty Images
4.5a Hans-Guenther Oed/STOCK4B/Getty Images
4.5b Ross McRoss/Flickr/Getty Images
4.5c Quim Llenas/Cover/Getty Images
4.5d NASA
4.7a Jasper Juinen/Getty Images
4.7b Asahi Shimbun via Getty Images
4.7c Ermal Meta/AFP/Getty Images
4.9a Ethel Davies/Robert Harding World Imagery/Getty Images
4.9b Izzet Keribar/Lonely Planet Images/Getty Images
4.9c De Agostini Picture Library/Getty Images
4.9d Werner Forman/Universal Images Group/Getty Images
4.9e Prisma/Universal Images Group/Getty Images
4.9f German School/The Bridgeman Art Library/Getty Images
4.10 Paul Biris Photography/Flickr/Getty Images
4.11a Rudi Von Briel/Peter Arnold/Getty Images
4.11b Manfred Pfefferle/Oxford Scientific/Getty Images
4.11c davespilbrow/Getty Images
4.13a Tom Stoddart/Getty Images
4.13b Christophe Simon/AFP/Getty Images
4.13c Oli Scarff/Getty Images
4.13d Weber/Getty Images
4.16a Jonathan Gelber/Getty Images
4.16b Dorling Kindersley/Getty Images
4.16c Image Source/Getty Images
4.17 Lydia Pulsipher
4.18a Denis Doyle/Getty Images
4.18b Mike Hewitt/Allsport/Getty Images
4.18c Peter Still/Redferns/Getty Images
4.20a Slow Images/Photographer's Choice/Getty Images
4.20b Odd Andersen/AFP/Getty Images
4.20c Samuel Aranda/Getty Images
4.20d Sonnet Sylvain/hemis.fr/Getty Images

CHAPTER 5

5.1a Viktor Drachev/AFP/Getty Images
5.1b Amar Grover/AWL Images/Getty Images
5.1c Walter Bibikow/The Image Bank/Getty Images
5.1d John Cancalosi/Getty Images
5.1e Norberto Cuenca/Flickr/Getty Images
5.1f Pal Hermansen/Stone/Getty Images
5.1g Mark Newman/Getty Images
5.2 Sasha Mordovets/Getty Images
5.5a Russian Look/UIG via Getty Images
5.5b Fotosearch/Getty Images
5.5c Daisy Gilardini/Getty Images
5.6a Randy Olson/National Geographic/Getty Images
5.6b NASA image created by Jesse Allen, Earth Observatory, using data obtained from the University of Maryland's Global Land Cover Facility. Image interpretation provided by Dr. Gareth Rees and Dr. Olga Tutubalina, Scott Polar Research Institute.
5.6c WIN-Initiative/Getty Images
5.6d Alexander Zemlianichenko Jr./Bloomberg via Getty Images
5.7b NASA Earth Observatory
5.7c NASA Earth Observatory
5.8a Martin Moos/Lonely Planet Images/Getty Images
5.8b Vyacheslav Oseledko/AFP/Getty Images
5.8c Andrew Peacock/Lonely Planet Images/Getty Images
5.10a placchic/Flickr/Getty Images
5.10b Nick Laing/AWL Images/Getty Images
5.10c Martin Child/Photodisc/Getty Images
5.10d Ilya Efimovich Repin/The Bridgeman Art Library/Getty Images
5.10e Galerie Bilderwelt/Getty Images
5.10f Bill Ingalls/NASA/Getty Images
5.11 Elena Ermakova/Flickr/Getty Images
5.17a Noorullah Shirzada/AFP/Getty Images
5.17b Sergei Supinsky/AFP/Getty Images
5.17c STR/AFP/Getty Images
5.19a Vyacheslav Oseledko/AFP/Getty Images
5.19b Oleg Nikishin/Getty Images
5.19c Viktor Drachev/AFP/Getty Images
5.21a Vyacheslav Oseledko/AFP/Getty Images
5.21b Dmitry Kostyukov/AFP/Getty Images
5.21c WIN-Initiative/The Image Bank/Getty Images
5.22a Dmitry Mordvintsev/E+/Getty Images
5.22b Ekaterina Nosenko/Flickr/Getty Images
5.22c Michael Runkel/Robert Harding World Imagery/Getty Images
5.25a Alexandra Grablewski/Lifesize/Getty Images
5.25b Lara Hata/Photodisc/Getty Images
5.25c Victor Drachev/AFP/Getty Images

CHAPTER 6

6.1a Frans Lemmens/The Image Bank/Getty Images
6.1b Belahensene's Photography/Flickr/Getty Images
6.1c Sylvester Adams/Photodisc/Getty Images
6.1d Malcolm MacGregor/AWL Images/Getty Images
6.1e Franz Aberham/Getty Images
6.1f Radius Images/Getty Images
6.1g Chris Caldicott/Axiom Photographic Agency/Getty Images
6.3 Mohammed Abed/AFP/Getty Images
6.5a Saleh Al-Rashaid/Flickr/Getty Images
6.5b J. Boyer/StockImage/Getty Images
6.5c Frederic Soreau/Photononstop/Getty Images
6.5d Ozcan Malkocer/Flickr/Getty Images
6.7a Nicolas Thibaut/Photononstop/Getty Images
6.7b Gilles Bassignac/Gamma-Rapho/Getty Images
6.7c Jason Larkin/arabianEye/Getty Images
6.7d Marwan Ibrahim/AFP/Getty Images
6.8 Adapted from A. M. MacDonald et al., *Environ. Res. Lett.* 7 (2012), Fig. 2. © IOP Publishing Ltd; CP13/082 British Geological Survey © NERC 2011. All rights reserved. Boundaries of surficial geology of Africa, courtesy of the US Geological Survey; Country boundaries sourced from ArcWorld © 1995–2011 ESRI. All rights reserved.
6.9a Gary Yeowell/The Image Bank/Getty Images
6.9b Abid Katib/Getty Images
6.9c Sabah Arar/AFP/Getty Images
6.10a James Baigrie/Photodisc/Getty Images
6.10b Scott Nelson/Getty Images
6.10c Christophe Boisvieux/hemis.fr/Getty Images
6.13a Salah Malkawi/Getty Images
6.13b Clive Streeter/Dorling Streeter/Getty Images
6.13c Reza/Getty Images
6.15a Wu Swee Ong/Flickr/Getty Images
6.15b Rabi Karim Photography/Flickr/Getty Images
6.15c Ayhan Altun/Flickr/Getty Images
6.15d DEA/V. Pirozzi/De Agostini Picture Library/Getty Images
6.15e AFP/Getty Images
6.15f Chip Somodevilla/Getty Images
6.17a Abdelhak Senna/AFP/Getty Images
6.17b Mohammed Abed/AFP/Getty Images
6.17c Behrouz Mehri/AFP/Getty Images
6.18a Part of G. Eric and Edith Matson Photograph Collection/Library of Congress
6.18b Eric Lafforgue/Gamma-Rapho/Getty Images
6.18c Werner Forman/Universal Images Group/Getty Images
6.19a Ozan Kose/AFP/Getty Images
6.19b John Borthwick/Lonely Planet Images/Getty Images
6.19c Fethi Belaid/AFP/Getty Images
6.23a Motivate Publishing/Gallo Images/Getty Images
6.23b Abdelhak Senna/AFP/Getty Images
6.23c Phillip Hayson/Photolibrary/Getty Images
6.28a Purestock/Getty Images
6.28b Khaled Desouki/AFP/Getty Images
6.28c Dimitar Dilkoff/AFP/Getty Images
6.28d Adem Altan/AFP/Getty Images
6.28e Mahmud Turkia/AFP/Getty Images
6.30a Warrick Page/Getty Images
6.30b David H Wells/arabianEye/Getty Images

CHAPTER 7

7.1a Martin Harvey/Workbook Stock/Getty Images
7.1b oversnap/Vetta/Getty Images
7.1c Nigel Pavitt/AWL Images/Getty Images
7.1d Pius Utomi Ekpei/AFP/Getty Images
7.1e Botafogo/Flickr/Getty Images
7.2 Simon Maina/AFP/Getty Images
7.5a Jacques Jangoux/Photo Researchers/Getty Images
7.5b Nigel Hicks/Dorling Kindersley/Getty Images
7.5c David Yarrow Photography/The Image Bank/Getty Images
7.5d Image courtesy of the Image Science & Analysis Laboratory, NASA Johnson Space Center
7.6a Raphael Gaillarde/Gamma-Rapho/Getty Images
7.6b Georges Merillon/Gamma-Rapho/Getty Images
7.6c Wade Davis/Archive Photos/Getty Images
7.6d Matt Fletcher/Lonely Planet Images/Getty Images
7.7a Photostock Israel/Oxford Scientific/Getty Images
7.7b Lynn Johnson/National Geographic/Getty Images
7.7c Graeme Williams/Gallo Images/Getty Images
7.8a Tony Karumba/AFP/Getty Images
7.8b Nigel Pavitt/AWL Images/Getty Images
7.8c Ivan Vdovin/AGE Fotostock/Getty Images
7.8d Hoberman Collection/Universal Images Group/Getty Images
7.9 Adapted from A. M. MacDonald et al., *Environ. Res. Lett.* 7 (2012), Fig. 2. © IOP Publishing Ltd; CP13/082 British Geological Survey

8.7a Munir uz Zaman/AFP/Getty Images
8.7b Robert Nickelsberg/Time Life Pictures/Getty Images
8.7c Punit Paranjpe/AFP/Getty Images
8.7d Deshakalyan Chowdhury/AFP/Getty Images
8.8a Farooq Naeem/AFP/Getty Images
8.8b Tim Graham/Getty Images
8.8c Prashanth Vishwanathan/Bloomberg/Getty Images
8.8d Thierry Falise/LightRocket/Getty Images
8.9a Tom Brakefield/Stockbyte/Getty Images
8.9b Buena Vista Images/The Images Bank/Getty Images
8.9c Dibyangshu Sarkar/AFP/Getty Images
8.13a Luca Tettoni/Robert Harding World Imagery/Getty Images
8.13b Mlenny Photography/E+/Getty Images
8.13c Science & Society Picture Library/Getty Images
8.13d Mansell/Time & Life Pictures/Getty Images
8.13e Hemant Chawla/The India Today Group/Getty Images
8.14a Amir Mukthar/FlickrVision/Getty Images
8.14b Sanjay Kanojia/AFP/Getty Images
8.14c Pedro Ugarte/AFP/Getty Images
8.15a Mark Williamson Stock Photography/The Image Bank/Getty Images
8.15b Amit Bhargava/Bloomberg/Getty Images
8.15c Namas Bhojani/Bloomberg/Getty Images
8.15d Godong/Universal Images Group/Getty Images
8.16 Alex Pulsipher
8.19a Huw Jones/Lonely Planet Images/Getty Images
8.19b Dario Mitidieri/Photonica World/Getty Images
8.19c AFP/Getty Images
8.22 Niko Guido/E+/Getty Images
8.23a Mark Kolbe/Getty Images
8.23b Religious Images/UIG/Getty Images
8.23c Behrouz Mehri/AFP/Getty Images
8.29 Majority World/UIG/Getty Images
8.31a Banaras Khan/AFP/Getty Images
8.31b AFP/Getty Images
8.31c NASA
8.31d R. Fox Photography/Lonely Planet Images/Getty Images

CHAPTER 9

9.1a Keren Su/China Span/Getty Images
9.1b Stuart Keasley/Robert Harding World Imagery/Getty Images
9.1c ChinaFotoPress/Getty Images
9.1d Bruce Dale/National Geographic/Getty Images
9.1e Koichi Kamoshida/Getty Images
9.2a Keren Su/China Span/Getty Images
9.2b Feng Li/Getty Images
9.4a Haibo Bi/E+/Getty Images
9.4b Travelasia/Asia Images/Getty Images
9.4c Bruno Morandi/The Image Bank/Getty Images
9.4d Fotosearch/Getty Images
9.5a Simon Yu/Flickr/Getty Images
9.5b Goh Chai Hin/AFP/Getty Images
9.5c ChinaFotoPress/Getty Images
9.5d ngkaki/E+/Getty Images
9.7a Martin Moos/Lonely Planet Images/Getty Images
9.7b Yoshikazu Tsuno/AFP/Getty Images
9.7c Bruno Morandi/The Image Bank/Getty Images
9.8a Blue Jean Images/Getty Images
9.8b Jacques Marais/Gallo Images/Getty Images
9.8c Alexander Safonov/Flickr/Getty Images
9.9a STR/AFP/Getty Images
9.9b Sam Yeh/AFP/Getty Images
9.9c China Photos/Getty Images
9.9d Provided by the SeaWiFS Project, NASA/Goddard Space Flight Center, and ORBIMAGE
9.11a Toru Yamanaka/AFP/Getty Images
9.11b Jung Yeon-Je/AFP/Getty Images
9.11c Bruno Morandi/Robert Harding World Imagery/Getty Images
9.12a Imagemore Co., Ltd./Getty Images
9.12b Robert G. Brown/Design Pics/Perspectives/Getty Images
9.12c Melanie Stetson Freeman/The Christian Science Monitor/Getty Images
9.12d Hulton Archive/Getty Images
9.12e AFP/Getty Images
9.12f DAJ/Amana Images/Getty Images
9.15 Kurita Kaku/Gamma-Rapho/Getty Images
9.17a Hong Wu/Getty Images
9.17b Multi-Bits/The Image Bank/Getty Images
9.17c Peter Parks/AFP/Getty Images
9.17d Tomohiro Ohsumi/Bloomberg/Getty Images
9.21a Andrew Ross/AFP/Getty Images
9.21b Kazuhiro Nogi/AFP/Getty Images
9.21c STR/AFP/Getty Images
9.21d Pedro Ugarte/AFP/Getty Images
9.22a Jung Yeon-Je/AFP/Getty Images
9.22b Doug Kanter/Bloomberg/Getty Images
9.23 Yuriko Nakao/Bloomberg/Getty Images
9.28 Robert Van Der Hilst/The Image Bank/Getty Images
9.29a Palani Mohan/Getty Images
9.29b View Stock/Getty Images
9.30 Toshiyuki Aizawa/Bloomberg/Getty Images

CHAPTER 10

10.1a Steve Winter/National Geographic/Getty Images
10.1b Wilbur E. Garrett/National Geographic/Getty Images
10.1c Per-Andre Hoffmann/LOOK/Getty Images
10.1d Øystein Lund Andersen/E+/Getty Images
10.1e Kazuhiro Nogi/AFP Photo/Getty Images
10.2a Yvan Cohen/LightRocket/Getty Images
10.2b Mattias Klum/National Geographic/Getty Images
10.6a John S. Lander/LightRocket/Getty Images
10.6b Per-Andre Hoffmann/LOOK/Getty Images
10.6c Hoang Dinh Nam/AFP/Getty Images
10.7a Frans Lemmens/Photographer's Choice/Getty Images
10.7b Feargus Cooney/Lonely Planet Images/Getty Images
10.7c Danita Delimont/Gallo Images/Getty Images
10.8a Auscape/UIG/Getty Images
10.8b Chicago Tribune/McClatchy-Tribune/Getty Images
10.9a Universal Images Group/Getty Images
10.9b Paula Bronstein/Getty Images
10.9c Jesse Allen/Earth Observatory/MODIS/NASA
10.9d Per-Andre Hoffmann/LOOK/Getty Images
10.10a Thierry Falise/Gamma-Rapho/Getty Images
10.10b Eco Images/Universal Images Group/Getty Images
10.11a STR/AFP/Getty Images
10.11b Lemaire Stephane/hemis.fr/Getty Images
10.11c Reinhard Dirscherl/WaterFrame/Getty Images
10.11d Roland Neveu/LightRocket/Getty Images
10.12a Romeo Gacad/AFP/Getty Images
10.12b Prisma/Universal Images Group/Getty Images
10.13a John S. Lander/LightRocket/Getty Images
10.13b Joe & Clair Carnegie/Libyan Soup/Getty Images
10.13c John S. Lander/LightRocket/Getty Images
10.13d Keystone/Getty Images
10.13e Dick Swanson/Time Life Pictures/Getty Images
10.13f Igor Prahin/Flickr/Getty Images
10.14 Steve Raymer/Asia Images/Getty Images
10.16 Joakim Leroy/E+/Getty Images
10.18a Paula Bronstein/Getty Images

10.18b Tjahjono Eranius/AFP/Getty Images
10.18c Karlos Manlupig/AFP/Getty Images
10.18d Ian Timberlake/AFP/Getty Images
10.23a Bay Ismoyo/AFP/Getty Images
10.23b C. P. Cheah/Flickr/Getty Images
10.23c Hoang Dinh Nam/AFP/Getty Images
10.23d Pairoj/AFP/Getty Images
10.24 Ted Aljibe/AFP/Getty Images
10.25a STR/AFP/Getty Images
10.25b Hoang Dinh Nam/AFP/Getty Images
10.25c Jeremy Jones Villasis/FlickrVision/Getty Images
10.26a Jay Directo/AFP/Getty Images
10.26b Clara Prima/AFP/Getty Images
10.26c Paul Kennedy/Lonely Planet Images/Getty Images
10.27 Adek Berry/AFP/Getty Images
10.28a John Elk/Lonely Planet Images/Getty Images
10.28b Seet Ying Lai Photography/Flickr/Getty Images
10.28c Rebecca Skinner/Lonely Planet Images/Getty Images

CHAPTER 11
11.1a Ignacio Palacios/Lonely Planet Images/Getty Images
11.1b Michael Runkel/Robert Harding World Imagery/Getty Images
11.1c Jean-Pierre Pieuchot/Stone/Getty Images
11.1d Fotosearch/Getty Images
11.1e Danita Delimont/Gallo Images/Getty Images
11.1f Gonzalo Azumendi/age fotostock/Getty Images
11.2a Justin Mcmanus/The AGE/Fairfax Media/Getty Images
11.2b Justin Mcmanus/The AGE/Fairfax Media/Getty Images
11.4 Michael Dunning/Photographer's Choice/Getty Images
11.5a John White Photos/Flickr/Getty Images
11.5b ML Harris/The Image Bank/Getty Images
11.5c Guiziou Franck/hemis.fr/Getty Images
11.8a Scott Camazine/Photo Researchers/Getty Images
11.8b Danita Delimont/Gallo Images/Getty Images
11.8c Kim Westerskov/Stone/Getty Images
11.8d Raimund Linke/Getty Images
11.9a Torsten Blackwood/AFP/Getty Images
11.9b Images by Ni-ree/Flickr/Getty Images
11.9c Mark Conlin/Oxford Scientific/Getty Images
11.9d David Wall Photo/Lonely Planet Images/Getty Images
11.10a Vanessa Mylett/Flickr/Getty Images
11.10b Robin Smith/Photolibrary/Getty Images
11.10c Kirsten Gamby/Flickr/Getty Images
11.12a Andrew Peacock/Lonely Planet Images/Getty Images
11.12b The Asahi Shimbun/Getty Images
11.14a Pictorial Parade/Archive Photos/Getty Images
11.14b Stephen Alvarez/National Geographic/Getty Images
11.14c Hulton Archive/Getty Images
11.14d Hulton Archive/Getty Images
11.14e Popperfoto/Getty Images
11.14f Lisa Maree Williams/Getty Images
11.16 Gerard Sioen/Gamma-Rapho/Getty Images
11.18a Torsten Blackwood/AFP/Getty Images
11.18b Sandra Mu/Getty Images
11.18c Torsten Blackwood/AFP/Getty Images
11.18d Alex Ellinghausen/The Sydney Morning Herald/Fairfax Media/Getty Images
11.21 Auscape/UIG/Getty Images
11.22 Anoek De Groot/AFP/Getty Images
11.23a Mark Kolbe/Getty Images
11.23b William West/AFP/Getty Images
11.23c Torsten Blackwood/AFP/Getty Images
11.23d The West Australian/AFP/Getty Images
11.24a Glenn Campbell/The Sydney Morning Herald/Fairfax Media/Getty Images
11.24b Kenny Rodger/Getty Images
11.24c Wendell Teodoro/WireImage/Getty Images
11.25 Cameron Spencer/Getty Images
11.27 George Silk/Time Life Pictures/Getty Images
11.28 Imagno/Hulton Fine Art Collection/Getty Images
11.29a Peter Stoop/Fairfax Media/Getty Images
11.29b Robert Madden/National Geographic/Getty Images
11.29c Chris Jackson/Getty Images
Epilogue-A Paul Nicklen/National Geographic/Getty Images
Epilogue-B Paul Sutherland/National Geographic/Getty Images
Epilogue-C Paul Nicklen/National Geographic/Getty Images
Epilogue-D Hulton Archive/Getty Images

times change or are deleted during the publication process and may not be currently available. The authors realize this and will .com/pulsipher6e as new data and information become known.

Population Reference Bureau, at http://www.prb.org/pdf11/2011population-data-sheet_eng.pdf and *World Gazetteer*, at http://world-gazetteer.com/wg.php?x=&men=gcis&lng=en&des=wg&srt=npan&col=abcdefghinoq&msz=1500&pt=a&va=&srt=pnan.

Fig 1.20: *United Nations World Water Development Report 2: Water—A Shared Responsibility*. Published jointly in 2006 by the UN Educational, Scientific and Cultural Organization (UNESCO), Paris; and Berghahn Books, New York, pp. 391–392.

Fig 1.23: Adapted from the UN Department of Economic and Social Affairs, Statistics Division, *Environmental Indicators: Greenhouse Gas Emissions*, 2009, at http://unstats.un.org/unsd/environment/air_greenhouse_emissions.htm.

Fig 1.25: Adapted from Frank Press, Raymond Siever, John Grotzinger, and Thomas H. Jordan, *Understanding Earth*, 4th ed. (New York: W. H. Freeman, 2004), pp. 42–43.

Fig 1.26: Adapted from United States Geological Survey, *Active Volcanoes and Plate Tectonics, "Hot Spots" and the "Ring of Fire,"* at http://vulcan.wr.usgs.gov/Glossary/PlateTectonics/Maps/map_plate_tectonics_world.html; and Frank Press, Raymond Siever, John Grotzinger, and Thomas H. Jordan, *Understanding Earth*, 4th ed. (New York: W. H. Freeman, 2004), p. 27.

Fig 1.28 (a): Adapted from Frank Press, Raymond Siever, John Grotzinger, and Thomas H. Jordan, *Understanding Earth*, 4th ed. (New York: W. H. Freeman, 2004), p. 281.

Fig 1.32: Map courtesy of UNEP/GRID-Arendal at http://www.grida.no/graphicslib/detail/skin-colour-map-indigenous-people_8b88, Emmanuelle Bournay, cartographer. Data source: G. Chaplin, "Geographic Distribution of Environmental Factors Influencing Human Skin Coloration," *American Journal of Physical Anthropology*, *125*, 292–302, 2004; map updated in 2007.

CHAPTER 2

Fig 2.3: Map: USGS/National Wetlands Research Center.

Fig 2.8: Graphic adapted from *National Geographic* (March 1993): 84–85, with supplemental information from High Plains Underground Water Conservation District 1, Lubbock, Texas, at http://www.hpwd.com; and Erin O'Brian, Biological and Agricultural Engineering, National Science Foundation Research Experience for Undergraduates, Kansas State University, 2001.

Fig 2.10: Data adapted from "Black Population from Year 1790–2010," BlackDemographics.com, at http://www.blackdemographics.com/population.html.

Fig 2.11: Adapted from James A. Henretta, W. Elliot Brownlee, David Brody, and Susan Ware, *America's History*, 2nd ed. (New York: Worth, 1993), pp. 400–401; and James L. Roark, Michael P. Johnson, Patricia Cline Cohen, Sarah Stage, Alan Lawson, and Suan M. Hartmann, *The American Promise: A History of the United States*, 3rd ed. (Boston: Bedford/St. Martin's, 2005), p. 601.

Fig 2.13 (a): Adapted from "U.S. Imports by Country of Origin," U.S. Energy Information Administration, at http://www.eia.gov/dnav/pet/pet_move_impcus_a2_nus_epc0_im0_mbblpd_m.htm, and "Supply and Disposition," U.S. Energy Information Administration, at http://www.eia.gov/dnav/pet/pet_sum_snd_d_nus_mbblpd_m_cur.htm.

Fig 2.13 (b): Adapted from *Statistical Handbook*, Canadian Association of Petroleum Producers, at http://www.capp.ca/library/statistics/handbook/Pages/default.aspx; and *Statistical Handbook for Canada's Upstream Petroleum Industry*, Canadian Association of Petroleum Producers, November 2011, at http://www.capp.ca/GetDoc.aspx?DocId=184463&DT=NTV.

Fig 2.15: Adapted from *National Geographic*, February 1990: 106–107, and augmented with data from: "Table 1: International Trips to Canada," Statistics Canada, at http://www.statcan.gc.ca/pub/66-001-p/2011012/t001-eng.htm; *International Visitation to the United States: A Statistical Summary of U.S. Visitation (2011)*, U.S. Department of Commerce, Office of Travel and Tourism Industries, at http://tinet.ita.doc.gov/outreachpages/download_data_table/2011_Visitation_Report.pdf; "Foreign Trade: Trade in Goods with Canada," U.S. Census Bureau, at http://www.census.gov/foreign-trade/balance/c1220.html; "U.S. Relations with Canada," U.S. Department of State, August 23, 2013, at http://www.state.gov/r/pa/ei/bgn/2089.htm; Data source: Statistics Canada, "Table 1: International Trips to Canada," at http://www.statcan.gc.ca/pub/66-001-p/2011012/t001-eng.htm and http://tinet.ita.doc.gov/outreachpages/download_data_table/2011_Visitation_Report.pdf; and "Immigration Overview: Permanent and Temporary Residents: Canada—Permanent Residents by Category and Source Area," Citizenship and Immigration Canada, at http://www.cic.gc.ca/english/resources/statistics/facts2010/permanent/08.asp.

Fig 2.16: Revised December 26, 2012 according to the results of 2012 U.S. election results, at "Record Number of Women Will Serve in Congress; New Hampshire Elects Women to All Top Posts," CAWP, November 7, 2012, at http://www.cawp.rutgers.edu/press_room/news/documents/PressRelease_11-07-12.pdf; and for the world, "World Classification," Women in International Parliaments, Inter-Parliamentary Union, at http://www.ipu.org/wmn-e/classif.htm.

Fig 2.17: Adapted from Arthur Getis and Judith Getis, eds., *The United States and Canada: The Land and the People* (Dubuque, IA: William C. Brown, 1995), p. 165.

Fig 2.19: Data source: "Wal-Mart Stores, Inc. Data Sheet—Worldwide Unit Details January 2012," Walmart, at http://walmartstores.com/pressroom/news/10821.aspx.

Fig 2.20: Adapted from "Women's Earnings and Employment by Industry, 2009," U.S. Department of Labor, Bureau of Labor Statistics, February 16, 2011, at http://www.bls.gov/opub/ted/2011/ted_20110216.htm.

Fig 2.21: Population data adapted from *2011 World Population Data Sheet*, Population Reference Bureau, at http://www.prb.org/pdf11/2011population-data-sheet_eng.pdf and *World Gazetteer*, at http://world-gazetteer.com/wg.php?x=&men=gcis&lng=en&des=wg&srt=npan&col=abcdefghinoq&msz=1500&pt=a&va=&srt=pnan.

Fig 2.25: Data source: "Percent of People Who Are Foreign Born—United States—Places by State; and for Puerto Rico Universe: Total Population, 2010 American Community Survey 1-Year Estimates," *American FactFinder*, U.S.

Census Bureau, at http://factfinder2.census.gov/faces/tableservices/jsf/pages/productview.xhtml?pid=ACS_10_1YR_GCT0501.US13PR&prodType=table.
Fig 2.27: Adapted from Christine Gambino and Thomas Gryn, "The Foreign Born With Science and Engineering Degrees: 2010," *American Community Survey Briefs*, Figure 2, p. 3, U.S. Census Bureau, November 2011, at http://www.census.gov/prod/2011pubs/acsbr10-06.pdf.
Fig 2.29: Adapted from Jorge del Pinal and Audrey Singer, "Generations of Diversity: Latinos in the United States," *Population Bulletin* 52 (October 1997): 14; U.S. Census Bureau, "Race by Sex, for the United States, Urban and Rural, 1950, and for the United States, 1850 to 1940," Census of Population: 1950, Volume 2, Part 1, United States Summary, Table 36, 1953, at http://www2.census.gov/prod2/decennial/documents/21983999v2p1ch3.pdf; http://2010.census.gov/news/releases/operations/cb11-cn125.html; and http://2010.census.gov/news/releases/operations/cb11-cn125.html and http://www12.statcan.gc.ca/census-recensement/2006/dp-pd/tbt/Rp-eng.cfm?LANG=E&APATH=3&DETAIL=0&DIM=0&FL=A&FREE=0&GC=0&GID=0&GK=0&GRP=1&PID=92342&PRID=0&PTYPE=88971,97154&S=0&SHOWALL=0&SUB=0&Temporal=2006&THEME=80&VID=0&VNAMEE=&VNAMEF=.
Fig 2.30: Adapted from "What Happened to Incomes in 2010? Analysis of Census Income Estimates for 2010," Diagram 1, National Urban League Policy Institute, September 15, 2011, at http://www.nul.org/sites/default/files/Income_2010_CPS.pdf.
Fig 2.31: Adapted from Jerome Fellmann, Arthur Getis, and Judith Getis, *Human Geography* (Dubuque, IA: Brown & Benchmark, 1997), p. 164.
Fig 2.32: Data from "Selected Population Profile in the United States: 2007–2009 American Community Survey 3-Year Estimates," *American FactFinder*, U.S. Census Bureau, at http://factfinder2.census.gov/faces/tableservices/jsf/pages/productview.xhtml?pid=ACS_09_3YR_S0201&prodType=table.
Fig 2.33: Data sources: (A) U.S. Census Bureau, "Money Income of Households—Distribution by Income Level and Selected Characteristics: 2007," Income, Expenditures, Poverty, & Wealth: Household Income, Table 676, 2010, at http://www.census.gov/compendia/statab/cats/income_expenditures_poverty_wealth/household_income.html; (B) U.S. Census Bureau, "Mean Earnings by Highest Degree Earned: 2007," Education: Educational Attainment, Table 227, 2010, at http://www.census.gov/compendia/statab/cats/education/educational_attainment.html.
Fig 2.35: Data from *International Data Base*, U.S. Census Bureau, at http://www.census.gov/population/international/data/idb/informationGateway.php.
Fig 2.37: Maps created from data in the *Human Development Report 2011 Statistical Annex*, Tables 1 and 4, United Nations Development Programme, at http://www.undp.org/content/dam/undp/library/corporate/HDR/2011%20Global%20HDR/English/HDR_2011_EN_Tables.pdf.

CHAPTER 3

Fig 3.6: Illustration by Tomo Narashima, based on fieldwork and a drawing by Lydia Pulsipher.
Fig 3.12: Adapted from *Hammond Times Concise Atlas of World History* (Maplewood, NJ: Hammond, 1994), pp. 66–67.
Fig 3.13: Adapted from *Hammond Times Concise Atlas of World History* (Maplewood, NJ: Hammond, 1994), p. 69.
Fig 3.15: Source: "Foreign Direct Investment in Latin America and the Caribbean, 2011," Figure 1.8, p. 38, Economic Commission for Latin America and the Caribbean, United Nations, April 16, 2012, at http://www.cepal.org/publicaciones/xml/2/46572/2012-182-LIEI-WEB.pdf.
Fig 3.16: Adapted from *Goode's World Atlas*, 21st ed. (Chicago: Rand McNally, 2005), p. 137, Minerals and Economic map.
Fig 3.19: Source: *World Drug Report 2010*, United Nations Office on Drugs and Crime, at http://www.unodc.org/unodc/en/drug-trafficking/index.html.
Fig 3.21: Adapted from *2011 World Population Data Sheet*, Population Reference Bureau, at http://www.prb.org/pdf11/2011population-data-sheet_eng.pdf.
Fig 3.22: Maps created from data in the *Human Development Report 2011 Statistical Annex*, Tables 1 and 4, United Nations Development Programme, 2010, at http://www.undp.org/content/dam/undp/library/corporate/HDR/2011%20Global%20HDR/English/HDR_2011_EN_Tables.pdf.
Fig 3.23: Data from Internet World Stats: "Internet Users in the World: Distribution by World Region—Q2 2012," at http://www.internetworldstats.com/stats.htm; "Latin American Internet Usage Statistics," at http://www.internetworldstats.com/stats10.htm; "Internet Usage and Population in the Caribbean," at http://www.internetworldstats.com/stats11.htm.
Fig 3.25: Population data adapted from *2011 World Population Data Sheet*, Population Reference Bureau, at http://www.prb.org/pdf11/2011population-data-sheet_eng.pdf; *World Gazetteer*, at http://world-gazetteer.com/wg.php?x=&men=gcis&lng=en&des=wg&srt=npan&col=abcdefghinoq&msz=1500&pt=a&va=&srt=pnan.
Fig 3.26: From *Yearbook of the Association of Pacific Coast Geographers* 57(28): 1995; printed with permission.

CHAPTER 4

Fig 4.6: Source: European Environment Agency, at http://www.eea.europa.eu/data-and-maps/figures/map-of-summer-chlorophyll-a-concentrations-observed-in-1.
Fig 4.8: Data adapted from European Commission Mobility and Transport, at http://ec.europa.eu/transport/themes/infrastructure/revision-t_en.htm.
Fig 4.12: Adapted from Alan Thomas, *Third World Atlas* (Washington, DC: Taylor & Francis, 1994), p. 29.
Fig 4.13: Source: Democratization Index adapted from "The Democracy Index 2011: Democracy Under Stress," *Economist* Intelligence Unit, at http://www.eiu.com/public/topical_report.aspx?campaignid=DemocracyIndex2011.
Fig 4.14: Source: *United Nations Human Development Report 2009*, Table H, United Nations Development Programme, at http://hdr.undp.org/en/reports/global/hdr2009/; *Human Development Report 2011*, Sustainability and Equity: A Better Future for All, United Nations Development Programme, at http://hdr.undp.org/en/reports/global/hdr2011/.
Fig 4.15: Source: *Europe in Figures: Eurostat Yearbook 2009* (Eurostat Statistical Books, Luxembourg: Office for Official Publications of the European Commission, 2009), pp. 388 and 392, Figures 10.5, 10.6, 10.9, and 10.10, at http://epp.eurostat.ec.europa.eu/cache/ITY_OFFPUB/KS-CD-09-001/EN/KS-CD-09-001-EN.PDF. Source: "International Trade in Goods," Figures 2, 3, 6, and 7, Eurostat, at http://epp.eurostat.ec.europa.eu/statistics_explained/index.php/International_trade_in_goods#Analysis_of_main_trading_partners.
Fig 4.20: Population data adapted from *2011 World Population Data Sheet*, Population Reference Bureau, at http://www.prb.org/pdf11/2011population-data-sheet_eng.pdf; *World Gazetteer*, at http://world-gazetteer.com/wg.php?x=&men=gcis&lng=en&des=wg&srt=npan&col=abcdefghinoq&msz=1500&pt=a&va=&srt=pnan.
Fig 4.21: Adapted from "Population Pyramid of Germany," "Population Pyramid of Sweden," and "Population Pyramid of the European Union," *International Data Base*, U.S. Census Bureau, 2012, at http://www.census.gov/population/international/data/idb/informationGateway.php.
Fig 4.22: Sources: *Human Development Report 2009*, United Nations Development Programme, at http://hdrstats.undp.org/en/indicators/6.html; data in table is from *Migrants in Europe: A Statistical Portrait of the First and Second Generation*, Eurostat Statistical Books, 2011, Table 1, at http://epp.eurostat.ec.europa.eu/cache/ITY_OFFPUB/KS-31-10-539/EN/KS-31-10-539-EN.PDF.
Fig 4.23: Source: "The Future of the Global Muslim Population," Pew Research Center Forum on Religion and Public Life, January 27, 2011, pp. 121–122 and 161–162, at http://www.pewforum.org/2011/01/27/the-future-of-the-global-muslim-population/.
Fig 4.24: Adapted from: *Europe in Figures: Eurostat Yearbook 2011*, Figure 5.3, Eurostat, 2011, at http://epp.eurostat.ec.europa.eu/cache/ITY_OFFPUB/KS-CD-11-001/EN/KS-CD-11-001-EN.PDF.
Fig 4.25: Source: *Inter-Parliamentary Union*, Women in National Parliaments, as of February 2013 http://www.ipu.org/wmn-e/classif.htm.
Fig 4.26: Adapted from: *Europe in Figures: Eurostat Yearbook 2011*, ch. 6 (especially Table 6.4), Eurostat, 2011, at http://epp.eurostat.ec.europa.eu/cache/ITY_OFFPUB/CH_06_2011/EN/CH_06_2011-EN.PDF.
Fig 4.27: Maps created from data in the *Human Development Report 2011 Statistical Annex*, Tables 1 and 4, United Nations Development Programme, at http://www.undp.org/content/dam/undp/library/corporate/HDR/2011%20Global%20HDR/English/HDR_2011_EN_Tables.pdf.

CHAPTER 5

Fig 5.7: Adapted from *National Geographic*, February 1990, pp. 72, 80–81; NASA Earth Observatory, at http://earthobservatory.nasa.gov/Features/WorldOfChange/aral_sea.php.
Fig 5.9: Adapted from "Maps of the Silk Road," the Silk Road Project and the Stanford Program on International and Cross-Cultural Education, at http://www.silkroadproject.org/tabid/177/default.aspx.
Fig 5.12: Adapted from Robin Milner-Gulland with Nikolai Dejevsky, *Cultural Atlas of Russia and the Former Soviet Union*, rev. ed. (New York: Checkmark Books, 1998), pp. 56, 74, 128–129, 177.
Fig 5.13: Adapted from Clevelander, at http://en.wikipedia.org/wiki/Image:New_Cold_War_Map_1980.png.
Fig 5.14: Map adapted from U.S. Department of Energy, Energy

Fig 6.6: Adapted from United Nations Environmental Programme, *Vital Water Graphics: An Overview of the State of the World's Fresh and Marine Waters*, 2nd ed., 2008, at http://www.unep.org/dewa/vitalwater/article69.html.
Fig 6.8: Source: A. M. MacDonald, H. C. Bonsor, B. E. O. Dochartaigh, and R. G. Taylor, 2012, "Quantitative Maps of Groundwater Resources in Africa," *Environmental Research Letters* 7. doi:10.1088/1748-9326/7/2/024009.
Fig 6.11: Adapted from United Nations Environment Programme, *Vital Water Graphics: Problems Related to Freshwater Resources*, "Turning the Tides" map, at http://www.unep.org/dewa/assessments/ecosystems/water/vitalwater/22.htm.
Fig 6.12: Map adapted from Bruce Smith, *The Emergence of Agriculture* (New York: Scientific American Library, 1995), p. 50.
Fig 6.14: Adapted from Richard Overy, ed., *The Times History of the World* (London: Times Books, 1999), pp. 98–99.
Fig 6.16 (a): Adapted from *Hammond Times Concise Atlas of World History* (Maplewood, NJ: Hammond, 1994), pp. 100–101. **(b)** Adapted from *Rand McNally Historical Atlas of the World* (Chicago: Rand McNally, 1965), pp. 36–37; *Cultural Atlas of Africa* (New York: Checkmark Books, 1988), p. 59.
Fig 6.20: Adapted from the *Human Development Report 2011*, Sustainability and Equity: A Better Future for All, United Nations Development Programme, at http://hdr.undp.org/en/reports/global/hdr2011/.
Fig 6.22: Adapted from "Population Pyramids for Iran," *International Data Base*, U.S. Census Bureau, 2012, at http://www.census.gov/population/international/data/idb/informationGateway.php.
Fig 6.23: Population data adapted from *2011 World Population Data Sheet*, Population Reference Bureau, at http://www.prb.org/pdf11/2011population-data-sheet_eng.pdf; *World Gazetteer*, at http://world-gazetteer.com/wg.php?x=&men=gcis&lng=en&des=wg&srt=npan&col=abcdefghinoq&msz=1500&pt=a&va=&srt=pnan.
Fig 6.24: Maps created from data in the *Human Development Report 2011 Statistical Annex*, Tables 1 and 4, United Nations Development Programme, at http://www.undp.org/content/dam/undp/library/corporate/HDR/2011%20Global%20HDR/English/HDR_2011_EN_Tables.pdf.
Fig 6.25: Adapted from U.S. Department of Energy, Country Analysis Briefs, at http://www.eia.doe.gov/emeu/cabs/Region_me.html; U.S. Department of Energy, "Selected Oil and Gas Pipeline Infrastructure in the Middle East," at http://www.eia.doe.gov/cabs/Saudi_Arabia/images/Oil%20and%20Gas%20Infrastructue%20Persian%20Gulf%20(large)%20(2).gif; "Who Has the Oil? A Map of the World Oil Reserves," at http://gcaptain.com/ who-has-the-oil-a-map-of-world-oil-reserves/; adapted from "Countries: Proved Reserves," U.S. Department of Energy, Energy Information Administration, at http://www.eia.gov/countries/index.cfm?view=reserves.
Fig 6.26: Adapted from Organization of the Petroleum Exporting Countries, *Annual Statistical Bulletin 2008*, *2009*; adapted from *Annual Statistical Bulletin 2012* (Vienna: OPEC, 2012), pp. 47–48 and 53–54, at http://www.opec.org/opec_web/static_files_project/media/downloads/publications/ASB2012.pdf.
Fig 6.27: Adapted from *Human Development Report 2010*, Table 15, and *Human Development Report 2011*, Table 10 (New York: United Nations Development Programme, 2010).
Fig 6.29: Adapted from Colbert C. Held, *Middle East Patterns—Places, Peoples, and Politics* (Boulder, CO: Westview Press, 1994), p. 184; *The Israeli Settlements in the Occupied Territories*, 2002, Foundation for Middle East Peace, at http://www.firstpr.com.au/nations; "Map: Golan Heights," at http://www.fmep.org/reports/archive/vol.-22/no.-6/map-golan-heights; and Geoffrey Aronson, "The Occupation Returns to Center Stage," *Foundation for Middle East Peace Settlement Report* 22 (6), at http://www.fmep.org/reports/archive/vol.-22/no.-6/the-occupation-returns-to-center-stage; http://thinkprogress.org/security/2012/05/14/483860/eu-settlements-threaten-two-states/?mobile=nc; http://thinkprogress.org/wp-content/uploads/2012/05/APN-Settlement-Map1.png.

CHAPTER 7

Fig 7.9: Adapted from A. M. MacDonald, H. C. Bonsor, B. E. O. Dochartaigh, and R. G. Taylor, 2012, "Quantitative Maps of Groundwater Resources in Africa," *Environmental Research Letters* 7. doi:10.1088/1748-9326/7/2/024009; boundaries of surficial geology of Africa courtesy of the U.S. Geological Survey; country boundaries sourced from ArcWorld © 1995–2011 ESRI. All rights reserved.
Fig 7.12: Adapted from the work of Joseph H. Harris, in Monica B. Visona et al., *A History of Art in Africa* (New York: Harry N. Abrams, 2001), pp. 502–503.
Fig 7.14: Adapted from Alan Thomas, *Third World Atlas* (Washington, DC: Taylor & Francis, 1994), p. 43.
Fig 7.16: Data from *The World Fact Book 2012*: *Africa: South Africa*, Central Intelligence Agency, at https://www.cia.gov/library/publications/the-world-factbook/geos/sf.html.
Fig 7.19: Adapted from UN Conference on Trade and Development, *Economic Development in Africa Report 2009: Strengthening Regional Economic Integration for Africa's Development* (New York: United Nations, 2009), p. 12, Figure 1, at http://www.unctad.org/en/docs/aldcafrica2009_en.pdf.
Fig 7.20: Adapted from *Ethnic Map of Nigeria*, Online Nigeria: Community Portal of Nigeria at www.onlinenigeria.com/mapethnic.asp#.
Fig 7.22: Data courtesy of Deborah Balk, Gregory Yetman, et al., Center for International Earth Science Information Network, Columbia University, at http://www.ciesin.columbia.edu.
Fig 7.23: Adapted from *International Data Base*, U.S. Census Bureau, at http://www.census.gov/population/international/data/idb/informationGateway.php.
Fig 7.24: Population data adapted from *2011 World Population Data Sheet*, Population Reference Bureau, at http://www.prb.org/pdf11/2011population-data-sheet_eng.pdf; *World Gazetteer*, at http://world-gazetteer.com/wg.php?x=&men=gcis&lng=en&des=wg&srt=npan&col=abcdefghinoq&msz=1500&pt=a&va=&srt=pnan.
Fig 7.26: Adapted from "Global Report 2010 Fact Sheet," UNAIDS, at http://www.unaids.org/documents/20101123_FS_SSA_em_en.pdf; *Global Report: UNAIDS Report on the Global AIDS Epidemic/2010*, UNAIDS, at http://www.unaids.org/globalreport/documents/20101123_GlobalReport_full_en.pdf.
Fig 7.27: Adapted from James Fenske, "African Polygamy: Past and Present," *Editorial Express*, February 15, 2012, at https://editorialexpress.com/cgi-bin/conference/download.cgi?db_name=CSAE2012&paper_id=115.
Fig 7.28: Adapted from "An Update on WHO's Work on Female Genital Mutilation (FGM), Progress Report," World Health Organization, 2011, at http://www.who.int/reproductivehealth/publications/fgm/rhr_11_18/en/index.html.
Fig 7.29: Maps adapted from Matthew White, "Religion in Africa," in *Historical Atlas of the Twentieth Century* (October 1998), at http://users.erols.com/mwhite28/afrorelg.htm; revised with new data from the CIA, *The World Factbook*, 2009, at https://www.cia.gov/cia/publications/factbook/index.html.
Fig 7.31: Adapted from Edward F. Bergman and William H. Renwick, *Introduction to Geography—People, Places, and Environment* (Englewood Cliffs, NJ: Prentice-Hall, 1999), p. 256;

Jost Gippert, *TITUS Didactica*, at http://titus.uni-frankfurt.de/didact/karten/afr/afrikam.htm.
Fig 7.32: Maps created from data in the *Human Development Report 2011 Statistical Annex*, Tables 1 and 4, United Nations Development Programme, at http://www.undp.org/content/dam/undp/library/corporate/HDR/2011%20Global%20HDR/English/HDR_2011_EN_Tables.pdf.

CHAPTER 8

Fig 8.6: Adapted from *National Geographic*, June 1993, p. 125.
Fig 8.10: Adapted from William R. Shepherd, *The Historical Atlas* (New York: Henry Holt, 1923–1926), p. 137; Gordon Johnson, *Cultural Atlas of India* (New York: Facts on File, 1996), p. 111.
Fig 8.11: Adapted from Gordon Johnson, *Cultural Atlas of India* (New York: Facts on File, 1996), p. 158.
Fig 8.12: Adapted from *National Geographic*, May 1997, p. 18.
Fig 8.15: Population data adapted from *2011 World Population Data Sheet*, Population Reference Bureau, at http://www.prb.org/pdf11/2011population-data-sheet_eng.pdf; *World Gazetteer*, at http://world-gazetteer.com/wg.php?x=&men=gcis&lng=en&des=wg&srt=npan&col=abcdefghinoq&msz=1500&pt=a&va=&srt=pnan.
Fig 8.17: Adapted from Alisdair Rogers, ed., *Peoples and Cultures* (New York: Oxford University Press, 1992), p. 204.
Fig 8.18: Adapted from Gordon Johnson, *Cultural Atlas of India* (New York: Facts on File, 1996), p. 56.
Fig 8.20: Courtesy of the University of California Press.
Fig 8.21: Adapted from "District Wise Female Literacy Rate of India," Maps of India, at http://www.mapsofindia.com/census2001/femaleliteracydistrictwise.htm; *United Nations Human Development Report 2009* (New York: United Nations Development Programme), Table J, "Gender-Related Development Index and Its Components," at http://hdr.undp.org/en/reports/global/hdr2009/; adapted from "Map of Literacy Rate in India," Maps of India, at http://www.mapsofindia.com/census2011/literacy-rate.html; "Literacy," *The World Fact Book*, Central Intelligence Ageny, at https://www.cia.gov/library/publications/the-world-factbook/fields/2103.html#af.
Fig 8.25: Sources: *2012 World Population Data Sheet*, Population Reference Bureau, at http://www.prb.org/pdf12/2012-population-data-sheet_eng.pdf; "Bhutan—Fertility Rate: Fertility Rate, Total (births per woman)," Index Mundi, at http://www.indexmundi.com/facts/bhutan/fertility-rate; "Afghanistan—Fertility Rate: Fertility Rate, Total (births per woman)," Index Mundi, at http://www.indexmundi.com/facts/afghanistan/fertility-rate; "Afghanistan," *The World Fact Book 2012*, Central Intelligence Agency, at https://www.cia.gov/library/publications/the-world-factbook/geos/af.html.
Fig 8.26: Adapted from *International Data Base*, U.S. Census Bureau, at http://www.census.gov/population/international/data/idb/informationGateway.php.
Fig 8.27: Maps created from data in the *Human Development Report 2011 Statistical Annex*, Tables 1 and 4, United Nations Development Programme, at http://www.undp.org/content/dam/undp/library/corporate/HDR/2011%20Global%20HDR/English/HDR_2011_EN_Tables.pdf.
Fig 8.28: Adapted from John Dixon and Aidan Gulliver with David Gibbon, *Farming Systems and Poverty: Improving Farmers' Livelihoods in a Changing World* (Rome and Washington, DC: FAO and World Bank, 2001), at http://www.fao.org/farmingsystems/FarmingMaps/SAS/01/FS/index.html.
Fig 8.30: Data from the Directorate of Economics & Statistics of each of the respective state governments, 2006; GDP data from the Reserve Bank of India, Table 8, "Per Capita Net State Domestic Product at Factor Cost—State-Wise (at Current Prices)," at http://rbidocs.rbi.org.in/rdocs/Publications/PDFs/008T_BST130913.pdf; data from "Comparing Indian States and Territories with Countries: An Indian Summary," *Economist*, at http://www.economist.com/content/indian-summary.

CHAPTER 9

Fig 9.2: Map adapted from Kam Wing Chan, "Internal Migration in China: Trends, Geography and Policies," *Population Distribution, Urbanization, Internal Migration and Development: An International Perspective* (New York: United Nations Department of Economic and Social Affairs, Population Division, 2011), pp. 81–109.
Fig 9.6: Adapted from "World Agriculture," *National Geographic Atlas of the World*, 8th ed. (Washington, DC: National Geographic Society, 2005), p. 19; "China: Economic, Minerals" map, *Goode's World Atlas*, 21st ed. (New York: Rand McNally, 2005), pp. 39 and 207.
Fig 9.10: Adapted from "China's Most Polluted Cities—WHO Index," *China Briefing*, September 28, 2011, at http://www.china-briefing.com/news/2011/09/28/chinas-most-polluted-cities-who-index.html; "Particle Pollution and Your Health," *AirNow*, at http://airnow.gov/index.cfm?action=particle_health.page1#1.
Fig 9.13: Adapted from *Hammond Times Concise Atlas of World History* (Maplewood, NJ: Hammond, 1994).
Fig 9.14: Adapted from *Hammond Times Concise Atlas of World History* (Maplewood, NJ: Hammond, 1994).
Fig 9.16: Adapted from Invest in China, "Per Capita Cash Income of Rural Households by Region (Third Quarter, 2009)," at http://www.stats.gov.cn/english/statisticaldata/Quarterlydata/t20091102_402598053.htm; "Income of Urban Households by Region (Third Quarter, 2009)," at http://www.fdi.gov.cn/pub/FDI_EN/Economy/Investment%20Environment/Macro-economic%20Indices/Population%20&%20GDP/t20091120_114779.htm.
Fig 9.17: Population data adapted from *2011 World Population Data Sheet*, Population Reference Bureau, at http://www.prb.org/pdf11/2011population-data-sheet_eng.pdf; *World Gazetteer*, at http://world-gazetteer.com/wg.php?x=&men=gcis&lng=en&des=wg&srt=npan&col=abcdefghinoq&msz=1500&pt=a&va=&srt=pnan.
Fig 9.18: FDI data from Invest in China, at http://www.fdi.gov.cn/common/info.jsp?id5ABC00000000000022787. Specific Web site no longer available without registering at http://www.fdi.gov.cn/pub/FDI_EN/Statistics/default.htm.
Fig 9.19: Adapted from Invest in China, February 7, 2012, at http://www.fdi.gov.cn/1800000121_10000041_8.html.
Fig 9.20: Adapted from "Major Foreign Holders of Treasury Securities (in Billions of Dollars)," U.S. Treasury, at http://www.treasury.gov/resource-center/data-chart-center/tic/Documents/mfh.txt.
Fig 9.22: Data from "Internet Users in Asia, 2012 Q2," Internet World Stats, at http://www.internetworldstats.com/stats3.htm.
Fig 9.24: Adapted from *International Data Base*, U.S. Census Bureau, 2012, at http://www.census.gov/population/international/data/idb/informationGateway.php.
Fig 9.26: Maps created from data in the *Human Development Report 2011 Statistical Annex*, Tables 1 and 4, United Nations Development Programme, at http://www.undp.org/content/dam/undp/library/corporate/HDR/2011%20Global%20HDR/English/HDR_2011_EN_Tables.pdf.
Fig 9.27: Adapted from Chiao-min Hsieh and Jean Kan Hsieh, *China: A Provincial Atlas* (New York: Macmillan, 1995), p. 12. The Web site http://www.index-china.com/minority/minority-english.htm includes a comprehensive survey of minorities in China.
Fig 9.29: Adapted from "Travel and Tours," China Tibet Train, at http://www.chinatibettrain.com/index.html.

CHAPTER 10

Fig 10.3: Adapted from "Struggling Cultures," *National Geographic Atlas of the World*, 8th ed. (Washington, DC: National Geographic Society, 2005), p. 15; "Globalization: Effects on Indigenous Peoples" map, International Forum on Globalization, 2007, at http://www.ifg.org/programs/indig/IFGmap.pdf.
Fig 10.8: Map adapted from "Annual Review and Assessment of the World Timber Situation," by the International Tropical Timber Organization, Figure 7, p. 13, at http://www.itto.int/en/annual_review/; graphs from *Annual Review and Assessment of the World Timber Situation 2010* (Yokohama, Japan: International Tropical Timber Organization, 2010), pp. 14–15, at http://www.itto.int/annual_review/.
Fig 10.8 (e): Data from "Forestry Issues—Deforestation: Tropical Forests in Decline," at http://www.canadian-forests.com/Deforestation_Tropical_Forests_in_Decline.pdf; http://www.panda.org/about_our_earth/about_forests/deforestation/forestdegradation/forest_illegal_logging/.
Fig 10.9: Map adapted from *United Nations Environment Programme, 2002, 2003, 2004, 2005, 2006* (New York: United Nations Development Programme), at http://maps.grida.no/go/collection/globio-geo-3.
Fig 10.10: Adapted from *Hammond Citation World Atlas* (Maplewood, NJ: Hammond, 1996), pp. 74, 83, 84.
Fig 10.12: Adapted from *Hammond Times Concise Atlas of World History* (Maplewood, NJ: Hammond, 1994), p. 101.
Fig 10.15: Data source: "ASEAN External Trade Statistics," Table 19, ASEAN, February 15, 2012, at http://www.aseansec.org/external-trade-statistics/.
Fig 10.16: Adapted from "World Heritage List," UN World Heritage Convention, at http://whc.unesco.org/en/list/; "Tourism Attractions Along the Asian Highway," UN Economic and Social Commission for Asia and the Pacific, 2004, at http://www.unescap.org/ttdw/common/tis/ah/tourism%20attractions.asp; *Asia Times Online*, at http://www.atimes.com/atimes/Asian_Economy/images/highways.html.
Fig 10.17: Source: "ASEAN Tourism Marketing Strategy (ATMS) 2012-2015," Figures 2-2 and 2-3 (Jakarta: ASEAN Secretariat, March 2012), p. 13, at http://www.aseansec.org/tourism-statistics/.
Fig 10.20: Data from Sasha Loffredo, *A Demographic Portrait of South and Southeast Asia* (Washington, DC:

men=gcis&lng=en&des=wg&srt=npan &col=abcdefghinoq&msz=1500&pt=a &va=&srt=pnan.

Fig 10.24: Map adapted from Joni Seager, *The Penguin Atlas of Women in the World* (New York: Penguin Books, 2003), p. 73; with updated information from the Migration Policy Institute, at http://www.migrationinformation.org/Profiles/display.cfm?ID5364.

Fig 10.26: Map adapted from *Oxford Atlas of the World* (New York: Oxford University Press, 1996), p. 27.

CHAPTER 11

Fig 11.6: Adapted from Environmental Dynamics Research, Inc., 1998; Ivan Cheung, George Washington University, Geography 137, Lecture 16, October 29, 2001.

Fig 11.7: Adapted from Tom L. McKnight, *Oceania* (Englewood Cliffs, NJ: Prentice Hall, 1995), p. 28.

Fig 11.11: Adapted from Richard Nile and Christian Clerk, *Cultural Atlas of Australia, New Zealand, and the South Pacific* (New York: Facts on File, 1996), p. 194.

Fig 11.13: Adapted from Richard Nile and Christian Clerk, *Cultural Atlas of Australia, New Zealand, and the South Pacific* (New York: Facts on File, 1996), pp. 58–59.

Fig 11.15: Data from *The World Factbook*, Central Intelligence Agency, at https://www.cia.gov/library/publications/the-world-factbook/geos/xx.html; U.S. Department of State's background notes on Kiribati and other countries, at http://www.state.gov/p/eap/ci/index.htm.

Fig 11.18: Population data adapted from *2011 World Population Data Sheet*, Population Reference Bureau, at http://www.prb.org/pdf11/2011population-data-sheet_eng.pdf; *World Gazetteer*, at http://world-gazetteer.com/wg.php?x=&men=gcis&lng=en&des=wg&srt=npan&col=abcdefghinoq&msz=1500&pt=a&va=&srt=pnan.

Fig 11.19: Maps created from data in the *Human Development Report 2011 Statistical Annex*, Tables 1 and 4, United Nations Development Programme, at http://www.undp.org/content/dam/undp/library/corporate/HDR/2011%20Global%20HDR/English/HDR_2011_EN_Tables.pdf.

Fig 11.20: Data from Australian Bureau of Statistics, "Main Countries of Birth," *Year Book Australia*, 2008, Table 7.39, at http://www.ausstats.abs.gov.au/ausstats/subscriber.nsf/0/8D6ED0E197FE38A6CA2573E7000EC2AD/$File/13010_2008.pdf.

Fig 11.26: For additional information, see http://www.irb.com/aboutirb/organisation/index.html

TEXT SOURCES AND CREDITS

78 World Development Indicators Database, "Gross National Income 2010, Atlas Method," World Bank, July 1, 2011, at http://siteresources.worldbank.org/DATASTATISTICS/Resources/GNI.pdf.

81 Trading Economics, "Unemployment Rate—Countries—List," at http://www.tradingeconomics.com/country-list/unemployment-rate; "Snapshots: Health Care Spending in the United States & Selected OECD Countries," The Henry J. Kaiser Family Foundation, April 12, 2011, at http://www.kff.org/insurance/snapshot/OECD042111.cfm.

83 Adapted from NPR Staff, *All Things Considered*, "What Recession? It's Boom Time for Nebraska Farms," February 25, 2011, at http://tinyurl.com/6jyljoa.

84 Dan Charles, "Genetically Modified Corn Helps Common Kind, Too," *All Things Considered*, National Public Radio, October 7, 2010, at http://www.npr.org/templates/story/story.php?storyId=130405227.

86 "Top 20 Internet Countries by Users—2012 Q2," Internet World Stats, at http://www.internetworldstats.com/top20.htm.

88 Jean Boivin, "The 'Great' Recession in Canada: Perception vs. Reality," Bank of Canada, March 28, 2011, at http://www.bankofcanada.ca/2011/03/publications/speeches/great-recession-canada-perception-reality; Tom Murse, "How Much U.S. Debt Does China Really Own?" About.com, at http://usgovinfo.about.com/od/moneymatters/ss/How-Much-US-Debt-Does-China-Own.htm.

92 Adapted from Nate Berg, "As Canada's First Nations Start Developing Their Land, Is Sprawl Inevitable?" *The Atlantic Cities*, February 9, 2012, at http://www.theatlanticcities.com/jobs-and-economy/2012/02/canadas-first-nations-start-developing-their-land-sprawl-inevitable/1182/.

93 Helena Norberg-Hodge, "A Tale of Two Cities: Beijing and Detroit," *Yes!*, January 26, 2012, at http://www.yesmagazine.org/happiness/a-tale-of-two-cities-beijing-and-detroit; Jeannie Kever, "Houston Region Is Now the Most Diverse in the U.S.," *Houston Chronicle*, March 5, 2012, at http://www.chron.com/news/houston-texas/article/Houston-region-is-now-the-most-diverse-in-the-U-S-3382354.php.

96 Adapted from Lydia Pulsipher's field notes in April 2012 from personal correspondence with Elizabeth Kennedy and Stuart Aitken, geographers who participate in the ISYS Unaccompanied Minors project.

99 *Statistics Canada*, "Canada," at http://www12.statcan.ca/english/census01/products/analytic/companion/rel/canada.cfm; Religion and Public Life Project, Pew Research, at http://www.pewforum.org/.

100 http://www.childstats.gov/americaschildren/eco1.asp. "Child Poverty," Conference Board of Canada, January 2013, at http://conferenceboard.ca/HCP/Details/society/child-poverty.aspx.

CHAPTER 3

110 Alex Pulsipher's field notes; Amazon Watch, 2006; Oxfam America, 2005; Juan Forero, "Rain Forest Residents, Texaco Face Off In Ecuador," *Morning Edition*, National Public Radio, April 30, 2009, at http://www.npr.org/templates/story/story.php?storyId=103233560; Gonzalo Solano, "Damages in Chevron Ecuador Suit Jump Billions," *San Francisco Chronicle*, September 18, 2010, at http://www.sfgate.com/cgi-bin/article.cgi?f=/c/a/2010/09/17/BUKH1FFIOR.DTL; Patrick Radden Keefe, "Reversal of Fortune," *New Yorker*, January 9, 2012, pp. 38–49; Reuters, "Ecuador: Villagers Try Again to Hold Chevron to $18bn Ruling," June 29, 2012, at http://www.independent.co.uk/news/world/americas/ecuador-villagers-try-again-to-hold-chevron-to-18bn-ruling-7897095.html.

120 Alex Pulsipher's field notes in Ecuador.

125 Table 3.1 adapted from Biran R. Kermath, Bradley C. Bennett, and Lydia M. Pulsipher, "Food Plants in the Americas: A Comprehensive Survey," November 2009.

131 NPR reports by John Idste, August 14, 2001, and August 16 and 25, 2003, and by Gary Hadden, August 27, 2003; David Bacon, "Anti-China Campaign Hides Maquiladora Wage Cuts," ZNet, February 2, 2003, at http://www.zcommunications.org/anti-china-campaign-hides-maquiladora-wage-cuts-by-david-bacon; Maquila Portal, April 2006, at http://www.maquilaportal.com/cgi-bin/public/index.pl; "How Rising Wages Are Changing the Game in China," *Bloomberg Businessweek*, March 27, 2006, at http://www.businessweek.com/magazine/content/06_13/b3977049.htm; Pete Engardio, "So Much for the Cheap 'China Price,'" *Bloomberg Businessweek*, June 4, 2009, at http://www.businessweek.com/magazine/content/09_24/b4135054963557.htm; "Avalanche of Chinese Investment in Mexico," Maquila Portal, December 1, 2011, at http://www.maquilaportal.com/index.php/blog/show/Avalanche-of-Chinese-investment-in-Mexico.html; Gustavo de Lima Palhares and Higor Uzzun Sales, "An Economic Analysis of the Role of Foreign Investments in Brazil," April 18, 2012, Council on Hemispheric Affairs, at http://www.coha.org/an-economic-analysis-of-the-role-of-foreign-investments-in-brazil.

132 "Foreign Direct Investment in Latin America and the Caribbean, 2011," UN Economic Commission for Latin America and the Caribbean, April 16, 2012, p. 37, at http://www.cepal.org/publicaciones/xml/2/46572/2012-182-LIEI-WEB.pdf; Alexei Barrioneuvo, "China's Interest in Farmland Makes Brazil Uneasy," *New York Times*, May 26, 2001, at http://www.nytimes.com/2011/05/27/world/americas/27brazil.html?pagewanted=all; Juan Forero, "Latin America: Once A Risky Bet, Now EU's Hero?" *Morning Edition*, National Public Radio, December 5, 2011, at http://www.npr.org/2011/12/05/143131926/once-a-risky-bet-latin-america-tapped-to-aid-eurozone.

134 Adapted from Andrew Wheat, "Toxic Bananas," *Multinational Monitor* 17: 6–7, September 1996 (updated 2007); David Magney, "Costa Rican Bananas," David Magney Environmental Consulting, December 16, 2005 (updated July 9, 2007), at http://www.magney.org/photofiles/CostaRica-Bananas1.htm.

145 Lydia Pulsipher's field notes on Brazil, updated with the help of John Mueller, Fortaleza, Brazil, 2006.

CHAPTER 4

154 Conversations with geographer Margareta Lelea, a Romanian specialist, post-doctoral researcher, Departments of Entomology and Human and Community Development, University of California, Davis; Doreen Carvajal and Stephen Castle, "A U.S. Hog Giant Transforms Eastern Europe,"

New York Times, May 5, 2009, at http://www.nytimes.com/2009/05/06/business/global/06smithfield.html?ref=europe.

159 "Russia, the EU, and Energy," *FTI Consulting*, May 11, 2012, at http://www.fticonsulting.com/global2/critical-thinking/articles/russia-the-eu-and-energy.aspx; "Net Energy Import Dependency (ENER 012)," European Environment Agency, April 30, 2012, at http://www.eea.europa.eu/data-and-maps/indicators/net-energy-import-dependency/net-energy-import-dependency-assessment-2; Thomas Gerke, "22.24 GW of PV-Solar Output in Germany—New Record!" Clean Technica, at http://cleantechnica.com/2012/05/25/2224-gw-pvsolar-output-germany-new-record/.

160 Steven Graning, "Taking It to the Streets," *Green Horizon*, August 18, 2008, at http://www.greenhorizon-online.com/index.php/Insight/taking-it-to-the-streets.html.

164 Adapted from Rob Gifford, "Gardeners Brighten London Under Cover of Dark," *Morning Edition*, National Public Radio, May 15, 2006, at http://www.npr.org/templates/story/story.php?storyId=5404229.

176 "Data: United States," World Bank, at http://data.worldbank.org/country/united-states.

179 Lydia Pulsipher's conversations with Vera Kuzmic and Dusan Kramberger, 1993–2012.

CHAPTER 5

196 Ellen Barry and Andrew E. Kramer, "In Biting Cold, Protesters Pack the Center of Moscow," *New York Times*, February 4, 2012, at http://www.nytimes.com/2012/02/05/world/europe/tens-of-thousands-protest-putin-in-moscow-russia.html?pagewanted=all; Simon Saradzhyan and Nabi Abdullaev, "Putin Election Victory Doesn't Pave an Easy Path Through His Third Presidential Term," *Christian Science Monitor*, March 5, 2012, at http://www.csmonitor.com/Commentary/Opinion/2012/0305/Putin-election-victory-doesn-t-pave-an-easy-path-through-his-third-presidential-term; Ellen Barry and Michael Schwirtz, "After Election, Putin Faces Challenges to Legitimacy," *New York Times*, March 5, 2012, at http://www.nytimes.com/2012/03/06/world/europe/observers-detail-flaws-in-russian-election.html?pagewanted=all; Michael Schwirtz, "Fear of Return to '90s Hardship Fuels Support for Putin," *New York Times*, March 3, 2012, at http://www.nytimes.com/2012/03/04/world/europe/in-russia-vote-fear-of-hardship-fuels-putin-support.html?pagewanted=all; Amanda Walker, "Last Anti-Putin Rally Before Russian Election," *Sky News*, February 26, 2012, at http://news.sky.com/home/world-news/article/16177315.

201 Blacksmith Institute, http://www.blacksmithinstitute.org/wwpp2007/finalReport2007.pdf; "City of Norilsk Still Tops Pollution List," *Moscow Times*, June 24, 2011, at http://www.themoscowtimes.com/news/article/city-of-norilsk-still-tops-pollution-list/439413.html; United Nations Scientific Committee on the Effects of Atomic Radiation, "The Chernobyl Accident: UNSCEAR's Assessments of the Radiation Effects," at http://www.unscear.org/unscear/en/chernobyl.html.

204 "Shrinking Aral Sea," NASA Earth Observatory, August 25, 2000, at http://earthobservatory.nasa.gov/Features/WorldOfChange/aral_sea.php; "Central Asia: Uzbekistan," *World Factbook*, Central Intelligence Agency, at https://www.cia.gov/library/publications/the-world-factbook/geos/uz.html; "Cotton Fact Sheet: Uzbekistan," ICAC, at http://www.icac.org/econ_stats/country_fact_sheets/fact_sheet_uzbekistan_2011.pdf.

205 "CO_2 Emissions per Capita," World Bank, at http://www.google.com/publicdata/explore?ds=d5bncppjof8f9_&met_y=en_atm_co2e_pc&idim=country:RUS&dl=en&hl=en&q=russia+carbon+emissions#!ctype=l&strail=false&bcs=d&nselm=h&met_y=en_atm_co2e_pc&scale_y=lin&ind_y=false&rdim=region&idim=country:RUS&ifdim=region&hl=en_US&dl=en&ind=false.

212 "Business Day Topics: Gazprom," *New York Times*, at http://topics.nytimes.com/top/news/business/companies/gazprom/index.html.

214 "Central Asia: Russia," *World Factbook*, Central Intelligence Agency, at https://www.cia.gov/library/publications/the-world-factbook/geos/rs.html; "Corruption Perceptions Index 2011," *Transparency International*, at http://cpi.transparency.org/cpi2011/results/; Kenneth Rapoza, "Russia Plagued by Corruption Perception," *Forbes*, June 17, 2011, at http://www.forbes.com/sites/kenrapoza/2011/06/17/russia-plagued-by-corruption-perception/; Kenneth Rapoza, "Russian Organized Crime Strategic Threat to US–NSC," *Forbes*, July 25, 2011, at http://www.forbes.com/sites/kenrapoza/2011/07/25/russian-organized-crime-strategic-threat-to-us-nsc/; Friedrich Schneider, Andreas Buehn, and Claudio E. Montenegro, "Shadow Economies All Over the World: New Estimates for 162 Countries from 1999 to 2007," Policy Research Working Paper 5356, World Bank, at http://www-wds.worldbank.org/external/default/WDSContentServer/IW3P/IB/2010/10/14/000158349_20101014160704/Rendered/PDF/WPS5356.pdf.

216 William Liefert and Olga Liefert, "Russian Agriculture During Transition: Performance, Global Impact, and Outlook," *Applied Economic Perspectives and Policy*, Oxford Journals, December 11, 2011, at http:/aepp.oxfordjournals.org/content/34/1/37.full; "ISTC Partner to Improve Russian Crop Production," International Science and Technology Center, at http://www.istc.ru/ISTC/ISTC.nsf/va_webpages/CropsEng.

220 "Deaths of Journalists in Russia," *Journalists in Russia*, at http://journalists-in-russia.org/; John Lancaster, "Tomorrowland," *National Geographic*, February 2012, at http://ngm.nationalgeographic.com/2012/02/astana/lancaster-text.

221 "2013 Cost of Living Rankings: African, European, and Asian Cities Dominate the Top 10 Most Expensive Locations for Expatriates," *Mercer*, at http://www.mercer.com/costoflivingpr#City_rankings.

223 www.esa.un.org/unpd/wpp/index.htm.

225 "2011 World Population Data Sheet," Population Reference Bureau, at http://www.prb.org/pdf11/2011population-data-sheet_eng.pdf; Timothy Heleniak, "Russia's Demographic Decline Continues," Population Reference Bureau, June 2002, at http://www.prb.org/Articles/2002/RussiasDemographicDeclineContinues.aspx; "Half of All Premature Deaths of Russian Adults Down to Alcohol," University of Oxford, June 26, 2009, at http://www.ox.ac.uk/media/news_stories/2009/090626.html.

227 "Overcoming Barriers: Human Mobility and Development," Human Development Report 2009, United Nations Development Program, at http://hdr.undp.org/en/reports/global/hdr2009/.

228 "Trafficking in Persons to Europe for Sexual Exploitation," *The Globalization of Crime—A Transnational Crime Threat Assessment*, United Nations Office on Drugs and Crime, at http://www.unodc.org/documents/publications/TiP_Europe_EN_LORES.pdf; "Human Trafficking," United Nations Office on Drugs and Crime, at http://www.unodc.org/unodc/en/human-trafficking/what-is-human-trafficking.html.

229 "Level of Religiosity in Russia Is Lower Than in the World—Sociological Data," *Interfax*, November 16, 2005, at http://wwrn.org/articles/19525/.

CHAPTER 6

234 Mona Eltahawy, "Why Do They Hate Us?" *Foreign Policy*, May/June, 2012, http://www.foreignpolicy.com/articles/2012/04/23/why_do_they_hate_us; "Debating the War on Women," *Foreign Policy*, April 24, 2012, http://www.foreignpolicy.com/articles/2012/04/24/debating_the_war_on_women; "Talk to Al Jazeera: Why Arab Women Still 'Have No Voice'," http://www.aljazeera.com/indepth/opinion/2012/05/201255134117758375.html.

242 A. M. MacDonald, H. C. Bonsor, B. E. O. Dochartaigh, and R. G. Taylor, "Quantitative Maps of Groundwater Resources in Africa," *Environmental Research Letters*, April 19, 2012, http://iopscience.iop.org/1748-9326/7/2/024009/pdf/1748-9326_7_2_024009.pdf.

254 Toni Johnson and Lauren Vriens, "Islam: Governing Under Sharia," Council on Foreign Relations, January 9, 2013, at http://www.cfr.org/religion/islam-governing-under-sharia/p8034.

268 Fareed Zakaria, "Arab Spring's Hits and Misses," *Washington Post*, January 30, 2013, at http://www.washingtonpost.com/opinions/fareed-zakaria-arab-springs-hits-and-misses/2013/01/30/fc72dcc2-6b15-11e2-af53-7b2b2a7510a8_story.html.

273 "Israelis and Palestinians Killed in the Current Violence," *If Americans Knew*, at http://www.ifamericansknew.org/stats/deaths.html; Daniel Levy, "Seven Lean Years of Peacemaking," *New York Times*, September 11, 2012, at http://www.nytimes.com/2012/09/11/opinion/seven-lean-years-of-peacemaking.html; Robert Worth, "Earth Is Parched Where Syrian Farms Thrived," *New York Times*, October 13, 2010, at http://www.nytimes.com/2010/10/14/world/middleeast/14syria.html?_r=0.

CHAPTER 7

282–283 Adapted from "Juliana Rotich—This Is What I See: Crowdsourced Crisis Mapping in Real Time," *99 Faces*, at http://99faces.tv/julianarotich/; Ushahidi Downloads, at http://download.ushahidi.com/; "Africa: The Next Chapter," *TED Radio Hour*, National Public Radio, at http://www.npr.org/2012/06/29/155904209/africa-the-next-chapter.

283 https://africaknowledgelab.worldbank.org/akl/sites/africaknowledgelab.worldbank.org/files/report/PER%20Congo%20English%20Post%20GoC.pdf.

287 Adapted from "Silas Kpanan'Ayoung Siakor: A Voice for the Forest and Its People,"

Sarovar Project: Winners and Losers," *Environmental Justice, Issues, Theories and Policy,* March 2012, at http://environmentalgeographies.wordpress.com/2012/03/20/the-sardar-sarovar-project-winners-and-losers/; Megha Bahree, "Does India Manage Its Water Like a 'Banana Republic?'" *India Real Time,* Wall Street Journal, April 9, 2012, at http://blogs.wsj.com/indiarealtime/2012/04/09/does-india-manage-its-water-like-a-banana-republic/.

340 Lydia and Alex Pulsipher's field notes, Nilgiri Hills, June 2000; Government of Tamil Nadu, *Tamil Nadu Human Development Report* (Delhi: Social Sciences Press, 2003), at http://data.undp.org.in/shdr/tn/TN%20HDR%20final.pdf.

347 "India Sees Explosion of Mobile Technology," *Tell Me More,* National Public Radio, July 20, 2010, at http://www.npr.org/2010/07/20/128645261/india-see-explosion-of-mobile-technology.

353 Beth Roy, *Some Trouble with Cows—Making Sense of Social Conflict* (Berkeley: University of California Press, 1994), pp. 18–19.

356 Internews Afghanistan, March 2008, at http://www.internews.org/bulletin/afghanistan/Afghan_200803.html; "Fearless Women Use Radio to Make Ripples of Change in Herat, Afghanistan," September 7, 2011, at http://cima.ned.org/fearless-women-use-radio-make-ripples-change-herat-afghanistan.

366 Jim Yardley, "Protests Awaken a Goliath in India," *India's Way, New York Times,* October 29, 2011, at http://www.nytimes.com/2011/10/30/world/asia/indias-middle-class-appears-to-shed-political-apathy.html.

CHAPTER 9

374 Adapted from Peter S. Goodman, "In China's Cities, a Turn from Factories," *Washington Post,* September 25, 2004, at http://www.washingtonpost.com/wp-dyn/articles/A48818-2004Sep24.html; Louisa Lim, "The End of Agriculture in China," *Reporter's Notebook,* National Public Radio, May 19, 2006; with background information from Kathy Chen, "Boom-Town Bound," *Wall Street Journal,* October 29, 1996, p. A6; "Life Lessons," *Wall Street Journal,* July 9, 1997.

375 "China's 'Floating Population' Exceeds 221 mln," *China.org.cn,* March 1, 2011, at http://www.china.org.cn/china/2011-03/01/content_22025827.htm.

379 Adapted from Maria Siow, "Desertification: One of the Challenges Faced by China," *Asia Pacific News,* October 29, 2009, at http://www.channelnewsasia.com/stories/eastasia/view/1014326/1/.html.

381 Associated Press, "China Flooding Has Killed 701 in 2010; Worst Toll Since 1998," *USA Today,* July 21, 2010, at http://usatoday30.usatoday.com/news/world/2010-07-21-china-flooding_N.htm; "China Faces Worst Flooding in 12 Years on Yangtze," Reuters, July 14, 2010, at http://www.reuters.com/article/2010/07/15/us-asia-weather-china-idUSTRE66E0ES20100715.

383 http://www.st.nmfs.noaa.gov/st1/fus/fus10/08_perita2010.pdf; "FAO State of World Fisheries, Aquaculture Report—Fish Consumption," *The Fish Site,* September 10, 2012, at http://www.thefishsite.com/articles/1447/fao-state-of-world-fisheries-aquaculture-report-fish-consumption; Mike Stones, "Fish Consumption Hits All Time High: FAO Report," *Food Navigator,* February 2, 2011, at http://www.foodnavigator.com/Financial-Industry/Fish-consumption-hits-all-time-high-FAO-report.

385 "Typhoon Nina—Banqiao Dam Failure," *Encyclopædia Britannica,* at http://www.britannica.com/EBchecked/topic/1503368/Typhoon-Nina-Banqiao-dam-failure; "2012 EPI and Pilot Trend Results: Air Pollution (Effects on Human Health)," Environmental Performance Index, Data Explorer: Indicator Profiles, Yale University, at http://epi.yale.edu/dataexplorer/indicatorprofiles?ind=eh.air; Emily Alpert, "The Worst Air Pollution in the World," *World Now: Asia, Los Angeles Times,* February 2, 2012, at http://latimesblogs.latimes.com/world_now/2012/02/worst-air-pollution-in-the-world.html; Bryan Walsh, "The 10 Most Air-Polluted Cities in the World," *Ecocentric, Time,* September 27, 2011, at http://science.time.com/2011/09/27/the-10-most-air-polluted-cities-in-the-world/. "Total Coal Consumption (Thousand Short Tons)," International Energy Statistics, U.S. Energy Information Administration, at http://www.eia.gov/cfapps/ipdbproject/iedindex3.cfm?tid=1&pid=1&aid=2&cid=CH,&syid=1980&eyid=2011&unit=TST; Timothy Hurst, "China's Massive Coal-Fired Power Plant Boom Visualized," *Ecopolitology,* August 26, 2010, at http://ecopolitology.org/2010/08/26/chinas-massive-coal-fired-power-plant-boom-visualized/; Damian Carrington, "More Than 1,000 New Coal Plants Planned Worldwide, Figures Show," *Guardian,* November 19, 2012, at http://www.guardian.co.uk/environment/2012/nov/20/coal-plants-world-resources-institute.

396 Mark Manyin and Mary Beth Nikitin, "Foreign Assistance to North Korea," Congressional Research Service 7-5700, June 11, 2013, at http://www.fas.org/sgp/crs/row/R40095.pdf; Michael Wines, "Beijing's Air Is Cleaner, but Far from Clean," *New York Times,* October 16, 2009, at http://www.nytimes.com/2009/10/17/world/asia/17beijing.html?partner=rss&emc=rss&_r=0.

399 "America's Transport Infrastructure: Life in the Slow Lane," *Economist,* April 28, 2011, at http://www.economist.com/node/18620944; Yukon Huang, "Reinterpreting China's Success Through the New Economic Geography," Carnegie Endowment for International Peace, November 10, 2010, at http://carnegieendowment.org/2010/11/10/reinterpreting-china-s-success-through-new-economic-geography/1lyq; Matthew Stadlen, "China Opens World's Longest High-Speed Rail Route," *News China,* BBC, December 25, 2012, at http://www.bbc.co.uk/news/world-asia-china-20842836.

401 Paul Wiseman, "Chinese Factories Struggle to Hire," *USA Today,* April 11, 2005, at http://www.usatoday.com/money/world/2005-04-11-china-labor_x.htm; Louisa Lim, "The End of Agriculture in China," *Reporter's Notebook,* National Public Radio, May 19, 2006, at http://www.npr.org/templates/story/story.php?storyId=5411325; Mei Fong, "A Chinese Puzzle: Surprising Shortage of Workers Forces Factories to Add Perks; Pressures on Pay—and Prices," *Wall Street Journal,* August 16, 2004, p. B1; David Barboza, "Labor Shortage in China May Lead to Trade Shift," *New York Times,* April 3, 2006, Business Section, p. 1, at http://www.nytimes.com/2006/04/03/business/03labor.html?pagewanted=all; Qiu Quanlin, "Labor Shortage Hinders Guangdong Factories," *China Daily,* August 25, 2009, at http://www.chinadaily.com.cn/china/2009-08/25/content_8612599.htm; Nina Ying Sun, "Labor Shortage Returns to China," *Plastics News,* August 10, 2009, at http://www.pnchina.com/en/Detail.aspx?id=3831&cat=0; "Major Foreign Holders of Treasury Securities (in Billions of Dollars) Holdings 1/ At End of Period," Data and Charts Center, U.S. Department of the Treasury, at http://www.treasury.gov/resource-center/data-chart-center/tic/Documents/mfh.txt; "The Biggest Holders of US Government Debt," *CNBC Explains,* at http://www.cnbc.com/id/29880401/The_Biggest_Holders_of_US_Government_Debt.

404 Mark Byrnes, "Beijing's Olympic Ruins," *Atlantic Cities,* July 6, 2012,

at http://www.theatlanticcities.com/jobs-and-economy/2012/07/beijings-olympic-ruins/2499/; http://www.businesstoday.org/magazine/temporarily-cancelled-running-bull/post-olympics-beijing; "Internet Users in Asia, 2012 Q2," *Internet World Stats,* June 30, 2012, at http://www.internetworldstats.com/stats3.htm; "China Approves Tighter Rules on Internet Access," *News China,* BBC, December 28, 2012, at http://www.bbc.co.uk/news/world-asia-20857480.
405 "World Report 2012: China," Human Rights Watch, at http://www.hrw.org/world-report-2012/world-report-2012-china; "Chinese Social Network Renren Sees Losses Increase," *News Business,* BBC, May 14, 2012, at http://www.bbc.co.uk/news/business-18068128.
408 "Gender Statistics," World Bank, at http://data.worldbank.org/data-catalog/gender-statistics; "Field Listing: Sex Ratio," *World Factbook,* Central Intelligence Agency, at https://www.cia.gov/library/publications/the-world-factbook/fields/2018.html; Rob Brooks, "China's Biggest Problem? Too Many Men," *CNN Opinion,* March 4, 2013, at http://www.cnn.com/2012/11/14/opinion/china-challenges-one-child-brooks/index.html; http://www.census.gov/population/international/data/idb/region.php.
412 Preeti Bhattacharji,"Uighurs and China's Xinjiang Region," *Backgrounder,* Council on Foreign Relations, May 29, 2012, at http://www.cfr.org/china/uighurs-chinas-xinjiang-region/p16870.
416 "World Directory of Minorities and Indigenous Peoples—Japan: Ainu," *RefWorld,* UNHCR, June 2008, at http://www.unhcr.org/refworld/topic,463af2212,49747c6f2,49749cfe23,0,,,.html; "East & Southeast Asia: Singapore," *World Factbook,* Central Intelligence Agency, at https://www.cia.gov/library/publications/the-world-factbook/geos/sn.html.

CHAPTER 10

418–419 Adapted from the *Borneo Wire,* Jessica Lawrence, ed., Spring 2006, at http://borneoproject.org/article.php?id=623; "Community Stops Illegal Logging and Bulldozing Towards Protected Rainforests," Jessica Lawrence, ed., *Borneo Wire,* Fall 2006, at http://borneoproject.org/article.php?list=type&type=39; Mark Bujang, "A Community Initiative: Mapping Dayak's Customary Lands in Sarawak," presented at the Regional Community Mapping Network Workshop, November 8–10, 2004, Diliman, Quezon City, Philippines; Julia Zappei, "Malaysia Highest Court Affirms Tribes' Land Rights," Associated Press, May 10, 2009, at http://borneoproject.org/article.php?id=762; "US Announces Its Support of the UN Declaration of the Rights of Indigenous Peoples (UNDRIP), December 16, 2010," International Forum on Globalization, at http://ifg.org/programs/indig/USUNDRIP.html; video of deforestation in Indonesian Borneo, at http://news.mongabay.com/2012/0326-muaratae-video.html.
437 Coco Masters, "Japan to Immigrants: Thanks, but You Can Go Home Now," *Time,* April 20, 2009, at http://www.time.com/time/world/article/0,8599,1892469,00.html.
437 "ASEAN Economic Community Scorecard: Charting Progress Toward Regional Economic Integration," Jakarta: ASEAN Secretariat, March 2012, at http://www.aseansec.org/documents/scorecard_final.pdf.
439 The fieldwork of Karl Russell Kirby, Department of Geography, University of Tennessee, 2010.
445 Likhaan Center for Women's Health, 2010, at http://www.likhaan.org/; Carlos H. Conde, "Bill to Increase Access to Contraception Is Dividing Filipinos," *New York Times,* October 25, 2009, at http://www.nytimes.com/2009/10/26/world/asia/26iht-phils.html.
450 Adapted from notes by Kirsty Vincin, an Australian teacher in Hong Kong, November 17, 2009, and "The Filipina Sisterhood: An Anthropology of Happiness," *Economist,* December 20, 2001, at http://www.economist.com/world/asia/displaystory.cfm?story_id=E1_RRPJDJ.
451–453 Adapted from personal communications with anthropologist Jennifer W. Nourse, University of Richmond, a specialist in Southeast Asia, 2010; Walter Williams, *Javanese Lives* (Piscataway, NJ: Rutgers University Press, 1991), pp. 128–134.
455 Adapted from the field notes of Alex Pulsipher and Debbi Hempel, 2000; coverage of the HIV-AIDS conference in Thailand, July 11–16, 2004, by the Kaiser Family Foundation; "Life as a Thai Sex Worker," *BBC News,* February 22, 2007, at http://news.bbc.co.uk/2/hi/world/asia-pacific/6360603.stm.

CHAPTER 11

460–461 Adapted from audio segments and articles written by Brian Reed for National Public Radio, "Preparing for Sea Level Rise, Islanders Leave Home," at http://www.npr.org/2011/02/17/133681251/preparing-for-sea-level-rise-islanders-leave-home, and "Could People from Kiribati Be 'Climate Change Refugees?'" February 17, 2011, at http://www.npr.org/templates/story/story.php?storyId=131216964, and "Kiribati Mulls Fiji Land Purchase in Battle Against the Sea," *BBC News Asia,* March 8, 2012, at http://www.bbc.co.uk/news/world-asia-17295862.
480 Richard Nile and Christian Clerk, *Cultural Atlas of Australia, New Zealand, and the South Pacific* (New York: Facts on File, 1996), pp. 63–65; *Voyage Weblog,* Polynesian Voyaging Society, 2007, at http://pvs.kcc.hawaii.edu/2007voyage/index.html.
485 Mary Jones, "Australian Immigration Reveals Indians Biggest Minority Community in Australia," *Canada Updates,* May 31, 2010, at http://www.canadaupdates.com/content/australian-immigration-reveals-indians-biggest-minority-community-australia.
489 Adapted from articles by Phil Wilkins, "Tonga Can Only Match the Kiwis in the Haka," Brisbane, Australia, October 25, 2003; Mark Falcous, "The Decolonizing National Imaginary: Promotional Media Constructions During the 2005 Lions Tour of Aotearoa," *New Zealand Journal of Sport and Social Issues,* November 2007, 31(4): 374–393.

INDEX

dicates a figure.

mixed, **289**, 291, 294
shifting cultivation, *24*, *25*, **122**, 126, *133*, **289**, 427, *428*
slash and burn, 427, *428*
subsidies, **178**–179
subsistence, 23, **289**, 290, 479
wet rice cultivation, **382**–383, *426*, 427
zones, *133*, 332, 382
Agroecology, **362**, 366
Agroforestry, 119, **287**, 294
Aguan River, *44*
Aguarico River, 110
Aguilar, Javier, 60
AIDS. *See* HIV/AIDS
Aila, Cyclone, *45*
Ainu, **413**, *414*
Air pollution
Bulgaria, 161
Canada, 62, 65, 66, *68*, *80*, 91, 92
Central Europe, 159, 161
China, 385–387, *387*
East Asia, 385–387
Europe, 159–161
Hungary, 161
industrial, 159, 170, 201–203, 205
Middle and South America, 120
smog, **66**, 68, 385, 422
South Asia, 340
urban, 201, 205, 339
U.S., 62, 65, 66, *68*, *80*, 91, 92
Air pressure, 48, 49
Alaminos, *423*
Alaska
arctic climate zone, *63*
Bering land bridge, 70, 72, 122, *122*
continental climate, *51*
Exxon Valdez disaster, 306
pipeline, *64*, 65
Alaskan Malamutes, 92
Albania, 156, *163*, *188*
Alberta, 58, 97, 98
Alcohol abuse, 60, 110, 112, 196, 225, 226, 231, 350, 475, 486
Alexander the Great, 341, 370
Alexandria, 241, 244, *244*
Algeria, *166*, *239*
Alhambra, *166*
Alice Springs, 458, *464*
Allah, 240, 248, 251, 268
Allende, Salvador, 138
Alps, *153*, 157
Altruism, 54
Amazon Basin, *109*, **113**–120, 150
Amazon River, 113, 114, *119*
American Samoa, 480
Amu Darya, 203
Amundsen-Scott South Pole Station, *496*
Anatolia Project, 245
Ancient Greece/Rome, 165–166
Andes
Chile, *108*
domesticated plants, *125*
formation, 112
glacial melting, 120, *121*
guinea pigs, *117*
high-altitude climate zone, *50*
mining, 128
rain shadows, 114, *115*
Angkor Wat temple, *439*
Angola, *172*, *303*
Animal husbandry, 23, 165, *166*
Animals and plants. *See also* Local lives; People; Vegetation; *specific animals*; *specific countries*
endemic species, 462, **465**–466, 467, **472**, 473, 476, 485, 486, *486*, 489
habitat loss, 62, 65, 91, 338
introduced species, 466, 468, 469
invasive species, 287, 466, **467**–468, 469, 472
nonnative species, 65, 287, 467, 469
Animism, 306, **318**, 352, 451, *452*
Antarctic Plate, 46, 47
Antarctic Treaty, *496*
Antarctica, photo essay, *496*
Anteaters, 466
Anthropocene, **22**, 23
Aotearoa, 463
Apartheid, 269, 295, 297, **298**–299, 300, 301, 307, 309, 312, 323
APEC (Asia Pacific Economic Cooperative), 480, 481
Appalachian Mountains, 59, 61, 62, 157, 199
Apple Corporation, 31, 33
Aqaba, Gulf of, *232*
Aquifers, **61**, 68, 69, 70, 83
Aquino, Benigno, III, 454
Aquino, Corazon, 374, 454
Arab Spring movements
commencement of, 36, 268
described, 234–236, 267–269
Egypt and, 283
Internet and, 86
issues of, 235
Libya and, 265, 283
Maghreb and, 277
NATO and, 178
North Africa and Southwest Asia, 267–269, 275–276
Syria and, 35, *271*, 275–276
Turkey and, 174
Ushahidi and, 283
Uzbekistan and, 229
women's status and, 234–235, 235, 255, 279
Arab world, 237
Arabian horses, *247*
Arabian Plate, 46, 239, 285
Arab–Israeli conflict, 272–273
Aral Sea, 203–204, *204*, *206*
Ararat, Mt., *232*
Archipelagos, *417*, **421**, 424, 431, 460, 474, 480
Arctic climate, *50*, *51*, *63*, *200*
Arctic Ocean, 161, 199, 230
Arctic Sea, *67*, 198
Argentina
asado, *135*
ISI policies, 128–**129**, 133, 151
pampas, *109*
UNASUR, **132**, 133, 151
women in parliament, 82
Arid/semiarid climates, *50*
Arizona–Mexico border, 96
Armenia, 229, 229
Arranged marriages, 451–453
Arunachal Pradesh, 330, 338
asado, *135*
ASEAN (Association of Southeast Asian Nations), 32, **437**, *438*, *440*
Asia. *See* Central Asia; East Asia; North Africa and Southwest Asia; South Asia; Southeast Asia
Asia Pacific Economic Cooperative. *See* APEC
Asian economy, New Zealand/Australia and, 477–479
Asian Highway, 438, 439
Asian recession, 479
Asian snakehead fish, 65
Asian-Americans, 96, 97
Asia-Pacific region, **414**, 480
Assam, 330
Assimilation, **142**, 182, **185**–186, *186*, 191, 412, 413, 414
Association of Southeast Asian Nations. *See* ASEAN
Astana, Kazakhstan, 221–222, *224*, 225
Aswan dams, 245
Atacama Desert, 114, *115*
Atheism, 98, 229, 230
Ati-Atihan Festival, *451*
Atlanta, Georgia, *90*, 91
Atlantic Coast, Hurricane Sandy, 42, 43, *45*, *121*
Atlas Mountains, 232, 237, 238, 277
Atolls, **463**, 467, *468*, *470*, 471, 472, *483*
Auckland, 483, 486
Aung San Suu Ki, 443, 454
AusAID program, 460–461
Australia
Aborigines, 462, **473**, *473*, 476, 485, 486, *486*, 489
animals and plants, 465–466
Asian economy and, 477–479
Blue Mountains, 458
climate change vulnerability, 460, 467, *470*, 471, 494

Australia *(Continued)*
colonization of, 475
continent formation, 462–463
GEI, 481–482, *484*
gender roles, 460, 490–493, 494
GNI per capita PPP, 481–482, *484*
HDI, 481–482, *484*
invasive species, 466, **467**–468, 469, 472
women in parliament, 82
Australian Desert, 462
An Australian Girl in London (Mack), 485
Australo-Melanesians, **431**
Austria, 14–16, *15*, 18, *188*
Austronesians, 431, 473, *474*
Authoritarianism
defined, **33**
democratic systems compared to, 2, 33–34, 36, 37, 56
European imperialism and, *172*
Automobiles
electric, 337, 364, 381
U.S. production, 88, 89
Autonomous regions
internal republics, 198, 217, 219, 231
Ningxia Huizu Autonomous Region, 379, 412
Xinjiang Uygur Autonomous Region, 412
Average population density, 13, *14*
Ayers Rock, 458, 462
Azerbaijan, 174, 220, 229, *229*
Aztecs, 117, **122**, 123, 144

Baby boomers, 102, 103, 182
Badaga ethnic group, 340
Bagan, 434
Baghdad, 249, *255*, 272
Bago Mountains, *423*
Bahrain, *41*
Baikonur Cosmodrome, *209*
Bali, 421, 431, 452
Balkans, 156
Baltic Sea, pollution, *161*
Baltic states. *See* Estonia; Latvia; Lithuania
Banana haciendas, Costa Rica, 133–134
Bangalore, 366
Bangkok, 430, 435, 448, *449*
Bangladesh
Joypur, 346
microcredit, 230, **363**, 363, 366, 370
sea level rise, 336
subcontinent and, **330**
Banqiao Dam, 384
Baptists, 98, 99, 229
Barbados, *18*, 20, 141
Barbeques, 94, *135*
Barcelona, 182, 183
Barrios, 28, 145
Basel, Rhine River, *153*
Bateria, *144*
Bauxite, 128, 133
Bazaars, 215, 301
Bechtel, 31, 38, 120
Bees, 92
Beijing, 385, 402, *403*, 404
Beijing–Tibet Railway, *413*
Belarus, 194, *221*, 223, 225, *229*
Belém, tropical wet climate zone, *115*
Belgian town festival, *166*
Belgium, *167*, *187*, *188*
Belief systems, 53, 94, 145, 248, 320
Belt-tightening programs, 32, 129
Benares. *See* Varanasi
Bengal Famine, 344
Bengal tiger, 339
Bengali language, 353
Benguela Current, 285
Bering land bridge, 70, 72, 122, *122*
Berlin Conference, 297
Bhopal gas tragedy, 340
Bhutan, 330, 339
Bhutto, Benazir, 354
Bible Belt, 98
Bicycle rickshaw driver, *348*
Bikini Atoll, 468
Bin Laden, Osama, 75, 369
Biodiversity, **113**–114. *See also* Amazon Basin; Human impact on biosphere
Biopiracy, *419*
Biosphere, **22**–23
Birth control, 13, 258, 260, 312, 313, 444, 445, 447
Birth rates, **13**
Black Death, 12
Black market, 203, 214
Black Sea, pollution, *161*
Bleaching, coral reef, 428, **429**, *430*, 431, 467, *470*
Bloodhound, *168*
Blue Mountains, Australia, *458*
Boer War, 298
Boers, 297–298
Bolivia, 38, 120, *121*, *127*, 131, 132, 133, *137*, 137–138, 151
Bollywood, 347
Bolsheviks, **209**–210, 211
Bombay. *See* Mumbai
Bombings. *See* Terrorism
Bopha, Typhoon, *430*
Borneo, 418, 420
Borneo Project, 419, 420
Borscht, *228*
Bosnia civil war, *173*
Bosnia and Herzegovina, 156, *188*, 190, 192
Boston, 71
Botswana, *18*, 82, 280, 310
Bottled water, 39
BP. *See* British Petroleum
Brahmin priests, 337, 350, 351, 356, 451
Brain drain, 145
Brazil
asado, *135*
Belém, *115*
BRIC countries and, 131–132, 212
Candomblé in, 148, *150*, 319
deforestation in, *118*
elite landscapes, 147
favelas, 28, **145**, 147, 148, 150
FDI in, 131–132
Fortaleza, 145
greenhouse gas emissions, 117
income disparities, *127*
landless movement, 134
Minas Gerais soy fields, *118*
MST movement, 134, 142
Rondonia cattle, *118*
skin color and, 142
UNASUR, **132**, 133, 151
Brazilian Amazon, *24*
Brazzaville. *See* Congo
Breadbasket, 72
BRIC, 131–132, 212
Bride price, 354
Britain. *See* United Kingdom
British East India Company, 342, 343, 345
British Indian Empire, 342, *342*
British Isles, 31, 475, 485
British Petroleum (BP), 31, 62, 65, 70, 78
Brownfields, **93**, 313
Brundtland, Gro Harlem, 188
Brunei, 420, 421, 436, 442, 446, 447, 448
Buddhism, 53, 99, 350, 351, **352**, 353, 369
Buffer state, 342
Bukhara, 208
Bulgaria, 13, 174, *192*
Burlaks, *209*
Burma (Myanmar), 25, 55, 82
Burning Man, 97
Bush, George H. W., 272
Bush, George W., 75, 272
Bushmeat, 293, 301
Business, social networking and, 86

Cacao market, 297
Cahokia, 70, 72
Cairo, 234, 235, *261*
Calcutta. *See* Kolkata
Calgary Stampede, 97
California, *49*, 60, *60*, *73*
Call centers, 171
Cambodia, 82, *417*, 433
Camel race, *247*
Cameroon, 297, 320
Campo de Dalía, greenhouses, *160*
Canada
abroad, 75–76
agribusiness, 68, *69*, **82**–83
Arctic tundra, *51*
children, poverty, 100
civil society and, 75, 78
climate change vulnerability, *45*, 60, 66, 67, 70
democratic system, 79–80
ethnicity in, 96–98
families, 99–101
First Nations people, 74, 92, 96
food production, 60, 66, 82–84
G8 and, 212
GEI, *104*, 104–105
global economic downturn and, 88–89
GNI per capita PPP, *104*, 104–105
greenhouse gas emissions, 60, 66, 70
HDI, *104*, 104–105
health-care system, 80–81, *81*
immigration and, *80*, 93–96
income disparities, *18*
infant mortality rates, *81*, 96
mobility in, 102
oil dependency, 76, 78, *78*
oil imports, *78*
population patterns, 60, 101–105
religions, 98–99, *99*
terms usage, 61
trade, 60, *80*, 86–88
urbanization, 70, 89–93, *90*
U.S.–Canada relationships, 78–82
water pollution, 60, *68*, 69, 70
water scarcity, 60, 66, 68–69
women in parliament, 82, *188*
women's earnings, 89, *89*
Canadian Rockies, 58
Candomblé, 148, *150*, 319
Canton. *See* Guangzhou
CAP (Common Agricultural Program), **178**–179
Cape Town, 284, 323
Capitalism
communism compared to, 171, 209
crony, **436**, 456
defined, **36**, **171**, **209**
Carbon dioxide, 3, 40, 43, 110, 117, 120, 150, 385, 424
Carbon footprint, 23
Carbon sequestration, **285**
Caribbean
Caribbean Plate, 46, 112, 113
FDI to, 129–132, *130*
Internet usage, *143*
Carnaval, Rio de Janeiro, *144*
Caroline Islands, 462
Carpathian Mountains, 157, 199
Carrying capacity, **26**–27, 311
Cartels, 137, **264**
Cartography, **4**
Casablanca, *261*, 277
Cash crops, 71, 134, 323, 427, 432
Cash economies, **17**
Caspian Sea, 199, 203, 215
Cassava root, 289
Caste system, 341, **350**–351
Castles, 166, *167*
Castro, Fidel, 137, 138
Castro, Raúl, 138
Çatalhöyük, 248
Catalonia, *163*
Catholic Church. *See* Roman Catholicism
Cats, 92, *247*, *422*, *472*
Cattle, in Rondonia, *118*
Caucasia
defined, **221**
former Soviet Union and, 197
informal economy, 214
political instability, *218*
Caucasus Mountains, *194*, 199, 201, 215, 219
Cell phones. *See* Mobile phones
Central Africa, 323
Central African Republic, 288
Central America

floating population, **375**, 397
foreign investment in, 401, *401*
GDP per capita, 397
glacial melting, 379, *380*, 384, 414
Great Leap Forward, **392**, 396
greenhouse gas emissions, *41*, 117
hukou system, **377**, 399
imperialism, 394, 412
income disparities, *127*
indigenous groups, 413
Internet and, 404–405
Japanese miracle, 394–396
Li Xia (rural-to-urban migration), 374–375, 399, 401
market reforms in, 396–397
Mexico–China relations, 131
Muslims, 411–412
one-child policy, 357, 374, 406, 407–409, 415
population growth, 407
responsibility system, **396**
SEZs and, **399**, 400, 414, 435
Three Gorges Dam, 329, *373*, 385, *386*, 387, 415
twentieth century and, 391–392
urbanization, *375*
U.S.–China economic relations, 87–88
women in parliament, 82
workers, in Angola, *303*
yurts (gers), 205
China's Far Northeast (Manchuria), 376, 377, 391
Chinese Communist Party, 391, 393, 412
Chipko (tree hugging), 339, 340
Chivu, Grigore, 154–155
Cholera, 40, 313
Christ, 248
Christianity. *See also* Roman Catholicism
Baptists, 98, 99, 229
defined, **248**
evangelical, 98, 99, 229, 319, 321
Evangelical Protestantism, **149**
North America, 98–99, *99*
Orthodox Church, 207, 208, 221, 229, 320
Protestant Reformation, *166*, 167
Protestantism, 53, 149
sub-Saharan Africa and, 319–320
in world, *53*
Chukchi people, *223*
Cienfuegos, Camillo, *136*
Cisco, 31
Cities
livability, 90, 91, 92, 92–93
megalopolis areas, **91**
primate, **145**, 146, 221, 224, 313, 448, 450
suburbs and, **91**
world, 169, 414
Civil disobedience, **343**, 345
Civil rights, 80, 167, 298, 486, 487
Civil society
Canada and, 75, 78
defined, **36**
NGOs and, 36, 37
Pakistan and, 344, 359
PRSFs and, 33
sub-Saharan Africa and, 299, 300, 310, 324
Civil War, U.S., 71, *73*
Civil wars
Bolsheviks and, 209–210, 211
Bosnia, *173*
Congo (Kinshasa/Democratic Republic of Congo), *34*
Syria, *35*
Clam chowder, *94*
Clear-cutting, *64*, **65**, 105, 287
Climate, **47**, **49**
Climate change
Antarctica, *496*
coral reef bleaching and, 428, **429**, *430*, 431, 467, *470*
defined, 40
desertification, **243**, 288, 292–293, 294, 379, *380*, 382
drivers of, 42
hurricanes and, *44*
impacts, 42
Kyoto Protocol, 43, 205
as thematic concept, 40–43
UN Framework Convention on Climate Change, 43
water and, 42–43
Climate change vulnerability
Australia, 460, 467, *470*, 471, 494
Canada, *45*, 60, 66, *67*, 70
Europe, 154, 162, 163, *163*, 192
exposure and, 42, 43, 44, 45, 56, 120
India, *45*
Middle and South America, 110, 117–119, 120, *121*, 150
Morocco, *45*
Mumbai, 42–43
New Zealand, 460, 467, *470*, 471, 494
North Africa and Southwest Asia, 242–244, *244*, 278
North America, 60, 66, 67, 70, 106
Oceania, 460, 467, *470*, 471, 494
photo essay, *44–45*
resilience and, 42, 43, 44, 45, 56, 120
Russia and post-Soviet states, 204–205, *206*
sensitivity and, 42, 43, 44, 45, 56, 120
South Asia, 328, 332–340, *336*, 370
Southeast Asia, 424–429, *430*
Spain, *45*
sub-Saharan Africa, 285–293, *291*
Sudan, *45*
Uganda, *45*
U.S., *45*, 60, 66, *67*, 70
U.S.–Mexico border, *67*
Climate zones (climate regions), **49–50**
Europe, *158*
global map, *50–51*
Middle and South America, *115*
North Africa and Southwest Asia, 237–239, *238*
North America, *63*
Oceania, *464*
Russia and post-Soviet states, *200*
South Asia, 331–332, *333*
Southeast Asia, 422–424, *423*
sub-Saharan Africa, 285, *286*
Clinton, Bill, 81
Clinton, Hillary, 82
Clouds, 48, 49, 465
Clownfish, *422*
Coal ash spill, 65
Coal mining, *25*, *64*, 65, *160*, *386*, 477
Coastal lowlands, *59*, *62*, 140, 157, 284, 326, 463
Coca-Cola, 31, 276, 335
Cocaine, 137
Cochabamba water utility, 38, 120
Cocos Plate, 46, 47, 112
Coffee, 33
Cold War
alliances, 9
capitalism and, 36
Cuba and, 138
defined, **171**, **210**
Europe, 171
global map, *211*
Russia and post-Soviet states, 210–211, *211*
U.S. military presence in Europe and, *77*
Coldest climates, *50*
Colombia
conflict in, 34, *136*
democratization and, 34
drug trade, *137*, 137–138
Incas and, **122–123**, *123*
income disparities, *127*
UNASUR, **132**, 133, 151
Colonial regimes, in North Africa and Southwest Asia, 252
Colonialism, European, 168–169
decolonization and, 171–172
divide-and-rule tactics, 300, **306**, 307, 310, 344
mercantilism, **125**, 128, **168**–169, *169*, 300
Middle and South America, 110, 112, *126*
New Zealand, 475
North Africa and Southwest Asia, 252
South Africa, 297–299, *298*
South Asia, 342–343
Southeast Asia, 431–433, *432*
sub-Saharan Africa, 297–299, *298*
U.S., 70–71, 72, *169*
wealth transfers, colonies to Europe, *169*
Colonias, 145
Color revolutions, 217, 218, 231
Colorado River, 68
Columbus, Christopher, 111, 121, 123, 124
Command economy, 210, 212–213, 396

Commercial agriculture
  defined, **290**
  Industrial Revolution and, 23
  Pacific Islands, 469
  Southeast Asia, 427–428
  sub-Saharan Africa, 290, 297
Commodities, **300**–301, 302, 307
Commodity dependence, **300–301**
Common Agricultural Program (CAP), **178**–179
Communal conflict, **353**
Commune system, 396
Communism
  capitalism compared to, 171, 209
  central planning, **171**, 210, 211, 212, 217, 219, 230, 231, 391, 392, 394, 395
  Chinese Communist Party, 391, 393, 412
  defined, **36**, **171**, **209**
  domino theory, **433**
  East Asia and, 395–396
  eastern Europe and, 156, 171, 177, 485
  Marx and, 36, 170, 171, 209
Communist Party, **210**
Communist Revolution, 208–210
The Communist Manifesto (Marx), 170
Comoros, 305
Condom use, 140, 349, 444, 446
Condominiums, 270, 347, 479
Confit, of duck legs, *178*
Confucianism, 53, **388**–389, *390*, 391, 394, 408, 431, 442
Congo (Brazzaville/Republic of Congo), 283, 284, 302, 303, 304, *305*, 309
Congo (Kinshasa/Democratic Republic of Congo), 35, 309, *314*, 319–320
Conservative welfare systems, 190
Container ship industry, *164*
Contested space, **133**–135, *343*
Continent formation, Australia/Oceania, 462–463
Continental climates, *50*, *51*, **157**
Continental plates, 44, *112*
Contraception. *See* Birth control
Convict heritage, Australia, 475
Cook, James, 491
Cool humid climates, *50*
Coon cat, Maine, 92
Copenhagen Accord, 43, 205
Copper mines, *129*, 300
Coptic Christians, Egyptian, *242*
Coral reefs
  atolls, **463**, 467, *468*, *470*, 471, 472, *483*
  bleaching, 428, **429**, *430*, 431, 467, 470
Corn production, 24, 83–84, *135*
Corporate agriculture, 179
Corruption, Russia and post-Soviet states, 214, 220
Corsica, summer dry climate zone, *158*
Costa Rica, 82, 133–134, *143*
Côte d'Ivoire, 297, 302, 313, 318, 320
Council of the European Union, 175
Coup d'état, **135**, 137, 138, 268, 272, 276, 309, 420, 442, 487, *488*
Cows, Hinduism and, 339
Creoles, **125**
Croatia, 155, 156, 174, *188*, 189
Crony capitalism, **436**, 456
Crowdsourcing, 270, 282, 282–283, *283*
Crowley's model of urban land use, 147, *147*
Cuba, 82, 137, 138, 148, 319
Cultivation, shifting, *24*, *25*, **122**, 126, *133*, **289**, 427, *428*
Cultural diversity
  East Asia, 411–414
  Middle and South America, 142, 149
  Russia and post-Soviet states, 217–220, *219*
  values and, 52–54
Cultural geography, 3, 52–56
Cultural homogenization, **155**, 453
Cultural pluralism, **451**
Cultural preference, for boys, 15
Cultural Revolution, *391*, **392**
Culture
  acculturation, 74, **142**, 485
  ethnicity compared to, 52
  material, 8, *8*
Culture groups, **52**
Culture of poverty, 97–98
Cumulonimbus clouds, 332
Curitiba, 147, 148
cuy, *117*
Cyclones, *45*, 331, *336*, 377–378, 467
Cyprus, 155, 174, 176, *188*
Czars, **208**, 217, 222
Czech Republic, 182, *183*, *188*

Dadaab Refugee Camp, *291*
Dai Qing, 384
Dalai Lama, 412
Dalal Street, 364
Dalits, 350, 351
Dalmatians, *168*
Dams
  Aswan Dams, 245
  Banqiao Dam, 384
  Haditha Dam, *241*
  at Manantan, *288*
  Niger River and, 292
  Sardar Sarovar Dam, 328–329, *343*
  Southeastern Anatolia Project, 245
  Three Gorges Dam, 329, 373, 385, 386, 387, 415
  Turkey and, 245, *245*
Danube River, *153*, 157, 162, 192
Darfur region, *244*
Darjeeling, 332
Dark Continent, 294
Darwin, Charles, 466, 480
De Soto expeditions, 70, 71
Dead zones, 69
Death rates, **13**
Debt crises, 131, 147, 176–178
Deccan Plateau, 330, 352
Decorated pig, *422*
Deepwater Horizon oil spill, 62, 65, 78
Deforestation. *See also* Logging
  agroforestry and, 119, **287**, 294
  Amazon Basin, 110, 117–120, 150
  Brazil, *118*
  Brazilian Amazon, *24*
  Burma, *25*
  carbon dioxide emissions, 3, 40, 43, 110, 117, 120, 150, 385, 424
  ecotourism and, *119*, 119–120
  Middle and South America, 110, 117–120, 150
  oil palm trees and, 22
  in Sarawak, 418–419
  shifting cultivation, *24*, *25*, **122**, 126, *133*, **289**, 427, *428*
  South Asia, 337–340
  Southeast Asia, 424, *425*, 427
  sub-Saharan Africa, 285–289
  tree hugging and, 339, 340
Deindustrialization, 177, 342–343
Del Monte, 133
Delta Works, *163*
Deltas, 47, 62
Democratic Republic of Congo. *See* Congo
Democratic systems, U.S.–Canada, 79–80
Democratization
  authoritarianism compared to, 2, 33–34, 36, 37, 56
  Canada and, 75–76
  defined, **33**, **37**
  expansion of democracy and, 34–36
  factors for, 36, 37
  geopolitics and, 36–37
  globalization, 37
  international cooperation and, 37
  New Zealand, 35
  North America and, 60, 76–77, 106
  Pacific Way and, 482, **487**, 494
Democratization/conflict issues
  Europe, 154, 169–172, 192–193
  Middle and South America, 135–138, 150–151
  North Africa and Southwest Asia, *270–271*, 270–272
  Oceania, 460, 487–489, *488*, 494
  Russia and post-Soviet states, 217–220, *218*
  South Asia, 328, 370
  Southeast Asia, 440–443, *441*
  sub-Saharan Africa, 306–310, *308*
  world, *34–35*
Demographic transition, *16*, **16**–17, 311–312
Deng Xiaoping, 392
Denmark, 112, 159, 174, *188*
Derry, Northern Ireland, *173*
Desalination, 160, 243
Desert, Namibia, *51*
Desertification, **243**, 288, 292–293, 294, 379, *380*, 382
Detritus, 289, **424**
Development. *See also* GEI; Globalization; GNI per capita PPP; HDI; Human impact on biosphere; Sustainability
  China, *375*
  defined, **19**, **23**
  East Asia, 397–402
  Europe, 154, 192
  export-led growth, **394**, 456
  grassroots economic development, 303, **305**, 306, 324, 487
  indigenous peoples and, *111*
  measures of, 19–20, 23
  Middle and South America, 110, 119–120, 124–126, 150–151
  North Africa and Southwest Asia, 263–265
  North America and, 60, 106
  Oceania, 460, 494
  political ecology and, 22
  self-reliant, **305**, 315
  South Asia and, 328, 370
  as thematic concept, 19–23
Dhaka, 29, 334, *348*, 352
Dia de los Muertos, *144*
Diamond mining, 297–298, 303, 310
Diaspora, **248**, 303, 306, 319
Dictators, **135**
Diffusion, 462
Digital divide, **86**, 404
Dingoes, 469, 472, *472*
Diseases, 71, 75, 315. *See also* HIV/AIDS
Dislocate aerial theater, *490*
Disobedience, civil, **343**, 345
Divide-and-rule tactics, 300, **306**, 307, 310, 344
Djenné, 295, 296, 309
Doctors Without Borders, 37
Dogrib people, 74
Dogs, 92, *117*, *168*, *223*, *293*, *472*
Dolphins, Amazon River, *119*
Domesticated plants, Middle and South America, *125*
Domestication, **23**
Dominican Republic, *26*, 123, 128
Domino theory, **433**
Donetsk, 230
Double day, 18, **187**, 227
Douz, Tunisia, *241*
Dowry, 328, **354**–355, 357, 359, 370
Drip irrigation, 240, 337, 467, 470
Drug trade, *136*, *137*, 137–138
Dual economies, **294**
Dubai, 260, 261, 264, 265, 266
Duck legs, *178*
Duck-billed platypus, 466
Dumping, 68, 120, 179, 471
Durban, 323
Dust Bowl era, 73, 74, 379
Dysentery, 313
Dzerzhinsk, 201

Eagle, golden, *223*
Early extractive phase, Middle and South America, **128**
Early human species, 23–24, *294*, 295

Ecology
agroecology, **362**, 366
political, **22**
Economic and technology development zones (ETDZs), 399, 400, 412
Economic Community of West African States (ECOWAS), 305, 309
Economic core, **71**–72
Economic development. *See* Development
Economic diversification
California, *73*
defined, **265**
North Africa and Southwest Asia, 243, 265–266
Uzbekistan, 204
Economic issues. *See also* Global economic downturn; Income disparities
East Asia, 394–406
Europe, 174–180
Middle and South America, 126–133
North Africa and Southwest Asia, 262–272
North America, 82–89
Oceania, 476–480
Russia and post-Soviet states, 211–216
South Asia, 361–366
Southeast Asia, 434–439
sub-Saharan Africa, 300–310, *302*
Economic reforms, Russia and post-Soviet states, 212–215
Economies. *See also* Informal economies; Service economies
cash, **17**
command, 210, 212–213, 396
dual, **294**
formal, **20**
free market, 129, 412
knowledge, 85–86
of scale, **175**
sectors of, 19
state-aided market economies, **394**, 395, 434
subsistence, **16**
Ecotopia, 74
Ecotourism, *119*, 119–120, 293–294, 471
ECOWAS. *See* Economic Community of West African States
Ecuador
Amazon Basin and, 110, 111, *119*
ecotourism, 119
oil development in, 100, 110, *111*
Secoya people, 110, 112
UNASUR, **132**, 133, 151
Education expenditures, North Africa and Southwest Asia, 267
Egypt
Alexandria, 241, 244, *244*
Aswan dams, 245
Coptic Christians, *242*
FGM and, 256
Islamist movements and, 53
Mubarek and, 234, 235, 269
El Niño phenomenon, **114**, *115*, 422, 424, 427, *464*, 465, *465*
El Salvador, *146*
Elderly, living arrangements in U.S., *103*
Electric cars, 337, 364, 381
Elephants, *293*, *339*
Elite urban areas, 28, 147
Eltahawy, Mona, 234–235
Emigration, **13**, 448, 450
Emir, 255
Emperor penguin, *496*
Empires
British Indian Empire, 342, *342*
Great Zimbabwe Empire, 294, 295
Mali, *294*, 296
Mughal Empire, 249, 341, 344
Ottoman Empire, 166, **249**–250, 251, 279
Russian Empire, 196, 205, 207–208, *209*, 230
Endemic species, **465**–466, 467
Environmental issues. *See also* Air pollution; Climate change vulnerability; Deforestation; Globalization; Human impact on biosphere; Oil spills; Urbanization; Water pollution
East Asia, 379–387
Europe, 159–164
indigenous peoples and, *111*
Middle and South America, 117–121
North Africa and Southwest Asia, 239–246
North America, 62–70
Oceania, 466–472
Russia and post-Soviet states, 201–205
South Asia, 332–340
Southeast Asia, 424–434
sub-Saharan Africa, 285–294
EPZs. *See* Export Processing Zones
Equatorial Guinea, 321
Erosion, **47**
Escarpments, *281*, 284
Estonia, 171, 176, *188*, 190, 230
ETDZs. *See* Economic and technology development zones
Ethanol, 24, 83–84, 117, 287, 382
Ethiopia, 53, 297
Ethnic cleansing, 37, 192
Ethnic groups, **52**
Badaga, 340
Nigeria, 307
Russia and post-Soviet states, *219*
South Asia, 349–350
Ethnic xenophobia, 172
Ethnicity, **96**
Canada and, 96–98
culture compared to, 52
sub-Saharan Africa and, 320–321
U.S. and, 96–98, 97
EU. *See* European Union
EU-28, 156, 159, *189*
Euphrates River, 165, 239, 245, 246, 247, 278
Eurasian Plate, 46, 47, 157, 199, 239, 330, 331, 377, 421, 462
Euro (€), **176**–178
Euro zone, **20**
Europe (world region), 152–193. *See also* Colonialism
climate, 157
climate change vulnerability, 154, 162, 163, *163*, 192
climate zones, *158*
Cold War and, 171
components of, 155, *156*
conflicts and genocides, *173*
country alliances and relationships, 9, 9–10, *10*
deindustrialization and, 177
democratization/conflict issues, 154, 169–172, 192–193
development, 154, 192
eastern, 156, 171, 177, 485
economic issues, 174–180
energy resources, 159
environmental issues, 159–164
feudalism, 166, 167
food security, 154, 178, 180, 193
GEI, 190–191, *191*
gender roles, 154, 187–188, 193
geographic issues, 174–191
geographic setting, 155–173
globalization and, 154, 168–169, 192
GNI per capita PPP, 190–191, *191*
green policies, 162, **164**
HDI, 190–191, *191*
human impact on biosphere, 159–162, *160*
human patterns over time, 164–172
immigration and, 184–187
Industrial Revolution, **23**, 30, 111, 167, **170**, 193, 250, 296
landforms, 156–157
medieval period, 166, 172, 182, 249
migration and, 184–187, *185*
multimodal transport, 164
Muslims in, 185–187, *186*
physical patterns, 156–158
political issues, 174–180
political map, *156*
population patterns, 154, 181–184, 193
power and politics, 154, *172*–*173*, 192–193
regional map, *152*–*153*
Russia (and post-Soviet states) interaction with, *198*
seawater pollution, *161*
service economies, 179–180
social welfare/protection systems, *189*, 189–190
sociocultural issues, 180–191
southeastern, 37, 156, 174, 190
subregions, 156, *156*
terms usage, 156
tourism, 179–180
transportation, 164, *165*

Europe *(Continued)*
urbanization, 154, 166–172, *183*, 192–193
vegetation, 157
virtual water and, 154, 161, 162, 164, 192
visual timeline, *166–167*
water pollution, 154, *161*, 161–162, 192
wealth transfers, colonies to Europe, *169*
western, 156
women at work, 187, *187*
women in parliaments, *188*
world region characteristics, 154–156
World Wars and, 171–172
European conquest, of Middle and South America, 123–124
European Economic Community, 174, *175*
European Parliament, 175
European Russia, 199
European settlements, North America, 70–74
European slave trade, 295, *296*, 296–297
European Union (EU)
agricultural subsidies and, **178–179**
CAP and, **178–179**
creation of, 154–155, 174–176
cultural homogenization and, **155**
defined, **154–155**
economies of scale, **175**
EU-28, 156, 159, *189*
euro and, 176–178
Gazprom and, 212
governing institutions, 174–175
greenhouse gas emissions, 154, 159, 162
members of, *175*
NAFTA compared to, *175*
NATO and, 177–178
Nobel Peace Prize to, 176
trading partners, 177, *177*
transport networks, *165*
U.S. economy compared to, 176
Evangelical Christianity, 98, 99, 229, 319, 321
Evangelical Protestantism, **149**
Evolutionary science, 56, 466
Exclave, 230
Expansions, Japan, 393, 393
Export Processing Zones (EPZs), **131**, 399, 400, 414, 435
Export trading partners, Oceania, *478*
Export-led growth, **394**, 456
Exposure, 42, 43, 44, 45, 56, 120. *See also* Climate change vulnerability
Extended families
East Asia, 388–389, 411
Middle and South America, 110, **142**–143, 144, 145, 149, 151
South Asia, 354
External processes, 44. *See also* Landforms
Extraction, **19**
Exxon Valdez disaster, 306

Facebook, 86, 228, 270, 276
Fair trade, **33**, 106
Falafels, *242*
Families. *See also* Extended families
Canada, 60, 99–101
Middle and South America, 110, **142**–143, 144, 145, 149, 151
North Africa and Southwest Asia, 254
North America, 60, 99–101
nuclear, **99–100**, 106, 142, 454
U.S., 60, 99–101
Family farms, 82–83
Famines, 24, 36, 37, 344
East Asia, 381, 385, 391, 392, 396, 415
sub-Saharan Africa, 290, 305
FAO (UN Food and Agricultural Organization), 27, 313
Far East, Russian, 230
Farmland preservation, North America, 91–92
Fauna and Flora. *See* Animals and plants
Favelas, 28, **145**, 147, 148, 150
FDI. *See* Foreign direct investment
Federal Security Service, 217
Federated States of Micronesia, 462
Female genital mutilation (FGM), 256, **318**
Female infanticide, 328, 357, 359, 370, 408
Female seclusion, **254**, 255
Feminization of labor, **435**, *436*
Fertile Crescent, *246*, **246–247**, *247*, *250*, 251
Fertility rates, TFR, **13**
Festivals. *See* Local lives
Feudalism, 166, 167, 387, 388, 390
FGM. *See* Female genital mutilation
Finland, *158*, *188*
First Nations people, 74, 92, 96
Fishbone pattern, *24*
Fisheries, threatened, *64*, *496*
Five Pillars of Islamic Practice, 251, 253
Floating population, **375**, 397
Flooding, Southeast Asia, 332, 334
Floodplains, **47**, 278, 331, 332
Flora and fauna. *See* Animals and plants
Florida, *67*, *93*
Food and Agricultural Organization. *See* FAO
Food production. *See also* Agriculture
Canada, 60, 66, 70, 82–84
food security and, 24–27
GM and, 27, 84
Middle and South America, 110, 133–135, 151
North Africa and Southwest Asia, 240–242
North America, 60, 66, 70, 82–84, 106
Oceania, 460, 467, 469, 494
Russia and post-Soviet states, 215–216
South Asia, 328, 361–365, 370
Southeast Asia, 427–428
as thematic concept, 23–27
U.S., 60, 66, 70, 82–84
Food security
Africa, 303
agriculture deemphasis, 279
defined, **24**, **381**
Europe, 154, 178, 180, 193
food production and, 24–27
Mali, 288
SAP and, 301
urban areas, 313
Food self-sufficiency, 178, 456
Foodways. *See* Local lives
Footprints
carbon, 23
ecological, **22**–23
water, 38, 39, 162, 335*t*
Foreign direct investment (FDI)
Canada–U.S. interaction, *80*
China, 401, *401*
defined, **131**
East Asia, 400, *400*
to Middle and South America, 129–132, *130*
sub-Saharan Africa, 303
Foreign exchange, **448**
Foreign holders, of Treasury securities, 402, *402*
Foreign political involvement, in Middle and South America, 138
Forests. *See* Deforestation; Logging; Rain forests; Tropical forests
Formal economies, **20**
Fortaleza, Brazil, 145
Fossil fuel, 212, **237**, 262, 263–266
Fossil water, 68, 240, *242*, 243, 292
France, *82*, *127*, *188*
Free market economies, 129, 412
Free trade, **32–33**
Free Trade Area of the Americas. *See* FTAA
French Guiana, *113*
French Revolution, *167*, 170
Freshwater availability, *239*
Frontal precipitation, 48–49
FTAA (Free Trade Area of the Americas), 87, 132
Fuji, Mount, 373, 377
Fujimori, Alberto, 142
Fukushima nuclear crisis, 159, 203, 381

G8 (Group of Eight), **212**
Galápagos Islands, 466
Gandhi, Mohandas, 343, 345, 351, 352
Ganga River (Ganges River), 330, 332, 335, 337, 338, 339, 340
Ganga–Brahmaputra delta, 335, 346
Ganges River. *See* Ganga River
Garma festival, *490*
Garreau, Joel, 105
Garzweiler open pit coal mine, *160*
Gauguin, Paul, 492
Gaza, 237. *See also* occupied Palestinian Territories
Gazprom, **212**
GDP (gross domestic product) per capita, **19–20**
GDP per capita PPP
China, 397
defined, 20
global map, *21*
India, *365*
GEI (Gender Equality Index)
Australia, 481–482, *484*
Canada, *104*, 104–105
defined, **20**
East Asia, *410*, 410–411
Europe, 190–191, *191*
global map, *21*
Middle and South America, 140–142, *141*
Moldova, 226
New Zealand, 481–482, *484*
North Africa and Southwest Asia, 262, *263*
Oceania, 481–482, *484*
Russia and post-Soviet states, 225–227, *227*
South Asia, 359–361, *360*
Southeast Asia, 446, *447*
U.S., *104*, 104–105
Gender, **17–19**. *See also* Women
sex compared to, 17, 19
as thematic concept, 17–19
U.S. politics and, 81–82
Gender equality, shift to, 2, 17, 19
Gender Equality Index. *See* GEI
Gender inequalities
cultural preference for boys, 15
North Africa and Southwest Asia, 254–257
Russia and post-Soviet states, 226–229
South Asia, 328, 370
sub-Saharan Africa, *317*, 317–318, *318*
Gender roles, **17–19**
Australia, 460, 490–493, 494
Europe, 154, 187–188, 193
Middle and South America, 110, 142–144, 151
New Zealand, 460, 490–493, 494
North Africa and Southwest Asia, 247–248, 254–255, 258–260
North America, 60, 106
Oceania, 460, 490–493, 494
Gendered spaces, 254–255
General Motors, 88
Genetic modification (GM), **27**, **84**
Genetically modified organisms (GMOs), **84**
Genocide
defined, **37**, **309**
ethnic cleansing and, 37, 192
Hutus–Tutsis conflict, 320
killing fields, 433
Rwanda, 310
Gentrification, **93**
Geodesy, 7

Gers (yurts), 205
Ghettos, 28
Giant panda, 385
Gift of Givers, 36, 37
Girls. *See* Women
Girls on the Air, 356
Glacial melting
  Andes, 120, *121*
  Antarctica, *496*
  Bering land bridge and, 70, 72, 122, *122*
  China, 379, *380*, 384, 414
  Great Lakes and, 61
  ice ages and, 47, 61
  impact of, 42
  Middle and South America, 119, *121*
  Oceania, 467
  Russia and post-Soviet states, 203, 205, *206*
  South Asia, 328, 330, 332, 334, 337, 340, 370
  Southeast Asia, 428, 429, *430*
Glasnost, **213**
Glastonbury Festival of Contemporary Performing Arts, *180*
Global economic downturn
  California produce and, 60
  Canada and, 88–89
  Chinese–Mexican relations and, 131
  corn production, 83–84
  euro and, 176–178
  food security and, 24, 26
  Southeast Asia and, 437
  U.S. and, 88–89
Global economy. *See also* Multinational corporations
  defined, **30**
  photo interpretation and, 8, *8*
  workers in, 31–32, *32*
Global Footprint Network's calculator, 23
Global maps
  climate zones, *50–51*
  Cold War, *211*
  democratization/conflict, *34–35*
  domesticated plants, Middle and South America, *125*
  freshwater availability, *239*
  GDP per capita PPP, *21*
  GEI, *21*
  greenhouse gas emissions, *41*
  HDI rank maps, *21*
  HIV/AIDS, *316*
  income disparities, *127*
  indigenous groups, *111*, *419*
  languages, *55*
  malnutrition, 26, *26*
  metropolitan areas, *28*
  national water footprints, *39*
  oil reserves, *265*
  Pangaea breakup, *46*
  population density, *14*
  religions, *53*
  skin color, *55*
  Walmart, *87*
Global patterns, local lives
  AusAID program, 460–461
  Central Valley unemployment, 60, *60*
  Li Xia (rural-to-urban migration), 374–375, 399, 401
  pig farms, Romania, 154–155, *155*
  Putin protesters, 196
  Sarawak forest dwellers, 418–419
  Sardar Sarovar Dam, 328–329
  Secoya people, 110, 112
  Ushahidi platform, 270, 282, 282–283, *283*
  women's status, Arab Spring, 234–235, *235*
Global scale, **12**
Global warming
  Antarctica, *496*
  defined, 40
  polar ice caps and, 42, 70, 199, 467
Globalization
  accelerated, 31, 168
  defined, **30**
  democratization and, 37
  East Asia and, 383, 397–402
  EU and, 176–177
  Europe and, 154, 168–169, 192
  fossil fuel exports and, 263–266
  free trade compared to, 32–33
  interregional linkages and, 30, 33
  maid trade and, 450, *450*
  Middle and South America, 110, 117, 119, 126, 150–151
  NGOs and, 37
  North Africa and Southwest Asia and, 259–262, *261*, 263–265
  North America and, 60, 86–88, 106
  of nuclear pollution, 201
  Oceania and, 460, 469–472, 494
  outsourcing and, 88, 89, **364**, 366, 435
  SAPs and, 129–130
  South Asia and, 328, 364–365, 370
  Southeast Asia and, 448, 450
  sub-Saharan Africa and, 300–303
  as thematic concept, 30–33
GM. *See* Genetic modification
GMOs. *See* Genetically modified organisms
GNI (gross national income) per capita PPP, **15–16**, 20
Golan Heights, 274, 275
Gold Rush, 73, 74
Golden Comanche, Chief, 97
Golden eagle, 223
Gondwana, 46, **462**, 463, 465
Google, 31, 303, 404, *405*
Gorbachev, Mikhail, 210, 213, 227
Gorges, 384, 421
Gorillas, 293
Gospel of Success, 159, 319, 320, 321
Grameen Bank, microcredit, 230, **363**, 363, 366, 370
Grand Coulee Dam, 329
Grandmother hypothesis, 18
Grasslands. *See* Steppes
Grassroots economic development, 303, **305**, 306, 324, 487
Gravlax, *178*
Great Barrier Reef, **462–463**, 467
Great Basin, 62, 74
Great Lakes, 61, 66–68, 72, 73
Great Lakes Agreement, 66, 68
Great Leap Forward, **392**, 396
Great Man-Made River, 243
Great Plains, 68, *69*, 72, 74, 83
Great Rift Valley, *281*
Great Zimbabwe Empire, 294, 295. *See also* Zimbabwe
Greece, 165–166, 176, *188*
Green living, 70
Green policies, Europe, 162, **164**
Green revolution agriculture
  described, **26–27**
  Middle and South America, 134–135, 148
  North America, 83, 84
  Slovenia, 179
  South Asia, 361–365
Greene, Mark, 92
Greenhouse gas emissions
  Brazil, 117
  Canada, 60, 66, 70
  China, *41*, 117
  EU, 154, 159, 162
  Europe, 159
  Indonesia, 117
  Kyoto Protocol, 43, 205
  North America, 60, 66, 70
  Russia and post-Soviet states, *41*, 204–205
  urban sprawl, 28, **65**, 79, 89–93, 99, 101, 106
  U.S., *41*, 60, 66, 70, 117
  world, *41*
Greenhouse gases, *40*, **40–41**
Greenhouses, Campo de Dalía, *160*
Grey parrots, African, 293
Groundwater, 290–291, **292**, 292, 294, 313
Groundwater pumping, 243
Groundwater recharge, 328
Group of Eight (G8), **212**
Growth poles, **147**
Grozny, Chechnya, *218*, 220
Guangdong Province, 374, 375
Guangzhou (Canton), 376
Guatemala, *125*, *127*, *144*
Guerrilla Gardeners, 164
Guest workers, **184–185**, 259, *259*, 260, 277, 407
Guevara, Che, *136*
Guianas, 142
  French Guiana, *113*
  Guyana, 132
  Suriname, *118*
Guilin, 378
Guinea pigs, Andes, *117*
Gujarat, 328, 329, 331, 343
Gulf of Aqaba, 232
Gulf of Carpentaria, 462, *490*
Gulf of Mexico
  BP oil spill, 31, 62, 65, 70, 78
  Chicago River and, 68
  dead zone in, 69
  hurricanes and, 66
  Mississippi River delta and, 62

Gulf of Thailand, 421
Gulf states. *See also* Persian Gulf; Saudi Arabia
  Bahrain, *41*
  defined, **254**
  female seclusion, **254**, 255
  Kuwait, *11*, *18*, 256, 261, 262, 263, 264, 265, 278
  Oman, 234, 254, 262, 266, 268, 270
  OPEC and, 76, 78
  Qatar, 22, 234, 235, 254, 255, 257, 258, 259, 262, 263, 265, 266, 269
  remittances, 260, 265
  UAE, *11*, 234, 254, 255, 260, 318
Gulf War, *241*, 263, 272, 278
Guyana, 132
Gyanendra, King, 366, 367, 370
Gypsies (Roma), **171**, 186

Habitat loss, 62, 65, 91, 338
Haciendas, **128**, 133, 134
Haditha Dam, *241*
Haiti
  Hispaniola island and, 123, 128
  Hurricane Sandy, 42, 43, *45*, *121*
  malnutrition, *26*
  refugees, *76*
  voodoo in, 148, 319
hajj, **253**
Half the Sky, 349
hangi, *493*
Hanoi, 449, 451
Harappa culture. *See* Indus Valley civilization
Hawaii
  Honolulu, 52, *479*, 480, 489
  Polynesia and, 462, **474**, 476, 480, 490, 492, 494
  tourism and, 479–480
  tropical wet climate, *50*
Hazardous wastes, 62, 301
Hazare, Anna, 366
HDI (United Nations Human Development Index), **20**
  Australia, 481–482, *484*
  Canada, *104*, 104–105
  East Asia, *410*, 410–411
  Europe, 190–191, *191*
  Middle and South America, 140–142, *141*
  New Zealand, 481–482, *484*
  North Africa and Southwest Asia, 262, *263*
  Oceania, 481–482, *484*
  Russia and post-Soviet states, 225–227, *227*
  South Asia, 359–361, *360*
  Southeast Asia, 446, *447*
  sub-Saharan Africa, 321–322, *322*
  U.S., *104*, 104–105
HDI rank maps, *21*
Health expenditures, North Africa and Southwest Asia, 267
Health-care systems, Canada–U.S., 80–81, *81*
Hemispheres, 6, *6*
Herat, 370
Herding. *See* Pastoralism
Herzegovina. *See* Bosnia and Herzegovina
Hezbollah, 276
High-altitude climates, *50*, *51*
High-impact areas, *24*
Highlands, Middle and South America, 112–113
Highway System, Interstate, 85, 88, 89
Hill rice, *25*
Hindu Kush, 357, 370
Hinduism
  Buddhism and, **352**
  caste system and, 341, **350**–351
  cows and, 339
  Hindu–Muslim relationship, 352–353
  Jainism and, 350, 351, **352**
  purdah practice, 346, 349, **354**, 355, 357
  Sikhism and, 53, 344, 350, *351*, **352**
  in world, *53*
Hindutva, 366
Hiroshima, 395
Hispanics
  Central Valley unemployment, 60, *60*
  culture of poverty, 97–98
  defined, **61**
  income disparities, 96–97, *98*
  Latinos and, **61**, 93, 97, 102
  living arrangements, *103*
Hispaniola island, 123, 128
History. *See* Human patterns over time
HIV/AIDS (human immunodeficiency virus/ acquired immunodeficiency syndrome)
  FGM and, 256, 318
  global map, *316*
  Middle and South America, 140
  population patterns and, 13
  Southeast Asia, 446, *447*
  sub-Saharan Africa, 13, 312, 313, 315, *316*
Ho Chi Minh, 433
Ho Chi Minh Trail, 439
Hodder, Ian, 247
Holi festival, *346*
Holocaust, **171**, 250, 263
Homo erectus, 295
Homo sapiens, 295, 299
Homo sapiens sapiens, 54
Homogenization, cultural, **155**, 453
Honduras, 96, *121*
Hong Kong, *177*, 383, 390, 394
Honolulu, 52, *479*, 480, 489
Horizontal zones, West Africa, 323
Horn of Africa, **285**, 291, 307
Horses, 97, *180*, *221*, *247*
Hot spots, 37, 47, **463**
Households, U.S., 100, *100*
Huang He (Yellow River), 376, 379, 381
Hub-and-spoke network, 85
hukou system, **377**, 399
Humans
  early, 23–24, *294*, 295
  well-being, **20–22**
Human Development Index. *See* HDI
Human geography. *See* Geography
Human immunodeficiency virus. *See* HIV/AIDS
Human impact on biosphere, 22–23
  Anthropocene and, 22, 23
  East Asia, 386, 387
  Europe, 159–162, *160*
  impact areas, *24*
  Middle and South America, *118*
  North Africa and Southwest Asia, *241*
  North America, *64*
  Oceania, *468*
  photo essay, *24–25*
  Russia and post-Soviet states, *202*
  South Asia, 337–340, *338*
  Southeast Asia, *426*
  sub-Saharan Africa, 285–289, *288*
Human patterns over time (history). *See also* Peopling; Visual timeline
  China, 387–394
  East Asia, 387–394
  Europe, 164–172
  Middle and South America, 122–126
  North Africa and Southwest Asia, 246–251
  North America, 70–75
  Oceania, 472–476
  Russia and post-Soviet states, 205–211
  South Asia, 340–345
  Southeast Asia, 431–433
  sub-Saharan Africa, 294–300
Human Rights Watch, 229, 405, 450
Human well-being, **20–22**, *21*. *See also* GEI; GNI per capita PPP; HDI
Humanism, **167**
Humid continental climates, *50*, **157**
Hungary, 9, 13, *188*
Hunnicutt case, corn production, 83–84
Hurricanes
  climate change and, *44*
  cyclones, *45*, 331, *336*, 377–378, 467
  defined, 116
  Florida and, *67*
  frontal precipitation and, 49
  Gulf of Mexico, *66*
  Katrina, *67*
  Louisiana wetlands and, 58, 61, *61*
  Middle and South America, *44*, 116, 120, *121*
  Mitch, *44*, *121*
  Sandy, 42, 43, *45*, *121*
  typhoons, **377**–378, 379, 384, 429, *430*, 431
Husbandry, animal, 23, 165, *166*
Hussein, Saddam, 272, 278
Hutus–Tutsis conflict, 320
Hydroelectric industries. *See* Dams
Iberian Peninsula, 44, 123, 155, 248, 249
IBM, 31
Ice ages, 47, 61, *421*, 422, 431, 473
Ice caps, polar, 42, 70, 199, 467
Iceland, 82, 174, 184, *188*
IKEA, 364
"Il Palio" horse race, *180*
IMF. *See* International Monetary Fund
Immigration, **13**. *See also* Migration; U.S.–Mexico border
  Canada and, *80*, 93–96
  Europe and, 184–187
  European settlements, North America, 70–74
  international immigrants, 184
  push/pull phenomenon and, 27, 93, 448
  U.S. and, *80*, 93–96, *94*
Imperialism. *See also* Colonialism
  China, 394, 412
  European, *172*, 306, 393
  Japanese, 389–390, *391*
  Russian, expansion, 198, 209
Import quotas, 32
Import substitution industrialization (ISI), 128–**129**, 133, 151, 266, 267, 363, 364, 435
In the Land of God and Man: Confronting Our Sexual Culture (Paternostro), 140
Incas, **122**–123, *123*
Income disparities (wealth disparities), *18*, 18–19, 96–97, 98, **127**, *127*. *See also* Poverty; *specific countries*
India
  BRIC countries and, 131–132, 212
  climate change vulnerability, *45*
  GDP income per capita PPP, 365
  nuclear weapons, 368
  Pakistan–India Partition, 343, **343**–344, 345, 352, 366
  sea level rise, *45*
  subcontinent and, **330**
  Taj Mahal, 340, 341, *344*
Indian Ocean tsunami (2004), 37, 331, 421, 438, 442, 443
Indian peninsula, 330
Indian–Australian Plate, 46, 47, 330, 421, 463
Indigenous groups. *See also* Middle and South America; Native Americans
  Aborigines, 462, **473**, *473*, 476, 485, 486, *486*, 489
  Ainu, **413**, *414*
  Alaska Natives, 71
  Aztecs, 117, **122**, 123, 144
  China and, 413
  defined, *111*, **112**
  environmental/development issues and, *111*
  First Nations people, 74, 92, 96
  global map, *111*, *419*
  Incas, **122**–123, *123*
  Occidental Petroleum and, 110
  Oceania, *474*

- Internal processes, 44. *See also* Landforms; Pangaea; Plate tectonics
- Internal republics, 198, 217, 219, 231
- International cooperation, democratization and, 37
- International immigrants, 184
- International Monetary Fund (IMF)
  - bailout, 436–437
  - EU and, 132, 176
  - Middle and South America help, 132
  - PRSPs, 33, 131, 301
  - SAPs, 32–33, 129
- International Space Station, *209*
- Internet. *See also* Social networking
  - Arab Spring and, 86
  - China and, 404–405, *405*
  - crowdsourcing, 270, 282, 282–283, 283
  - digital divide, **86**, 404
  - East Asia, 404–405, *405*
  - knowledge economy and, 86
  - Middle and South America, 142, *143*
  - North Africa and Southwest Asia and, 269–270
  - *Ushahidi* platform, 270, 282, 282–283, 283
- Interregional linkages, **30**, 33
- Interstate Highway System, 85, 88, 89
- Intertropical convergence zone (ITCZ), **285**, 286, 294, 331, 332, 422, *423*
- Intifada, **275**
- Introduced species, 466, 468, 469
- Inuit, 92
- Invasions, into South Asia, 341–342
- Invasive species, 287, 466, **467**–468, 469, 472
- Ipswich, Massachusetts, 66
- Iran, 10–11, *11*, 82, 240
- Iranian Spring, 269
- Iran–Iraq War (1980–1988), 263, 278
- Iraq, 10–11, *11*, 240, 272
- Iraq War, 75, 76, 278
- Iraq–Iran War (1980–1988), 263, 278
- Ireland, Northern, 54, 173
- Iron Curtain, **171**, 210
- Iron works, German, *167*
- Irrigation, drip, 240, 337, 467, 470
- Islam. *See also* Muslims
  - Allah and, 240, 248, 251, 268
  - defined, **236**
  - Muhammad and, 236, 240, 248, 249 314, 251, 253, 255, 319
  - North Africa and Southwest Asia, **236**–237, 248–250, *249*
  - Ottoman Empire, 166, **249**–250, 251, 279
  - Pillars of Islamic Practice, 251, 253
  - Qur'an, **240**, 355
  - Ramadan, 253, 276, 320, 352
  - shari'a and, **253**, 254
  - spread of, 248–250, *249*
  - theocratic states, 210, **268**–269, 278
  - women's rights and, 255–257
  - in world, *53*
- Islamic extremists
  - Hezbollah, 276
  - Islamism, **237**, 268, 269
  - jihadists, **254**, 262, 268, 309
  - mujahedeen movement, 211, 368–369
  - Al Qaeda, 75, 77, 307, 309, 369
  - Salafism, **268**, 276, 277
  - Taliban, 229, **355**–357, 369
- Islamic fundamentalism, 229, 370, 412
- Islamism, **237**, 268, 269
- Island formation, Oceania, 463
- Israel
  - Arab–Israeli conflict, 272–273
  - creation of, 250, *251*
  - drip irrigation, 240
  - infant mortality rates, *121*
  - Israeli–Palestinian conflict, 272–275, *273*, *274*
- Isthmus, Central America, **112**, 122, 142
- IT industries. *See* Information technology industries
- Italy, *50*, *188*, 212
- ITCZ. *See* Intertropical convergence zone

- J curve, 13, *13*
- Jablonski, Nina, 56
- Jainism, 350, 351, **352**
- Jamaica, *41*, *82*
- Jamestown, 71
- Jammu and Kashmir region, *343*
- Japan
  - Ainu people, **413**, *414*
  - automobile production, 88
  - expansions, 393, *393*
  - export-led growth, **394**, 456
  - Fukushima nuclear crisis, 159, 203, 381
  - G8 and, 212
  - greenhouse gas emissions, *41*
  - imperialism, 389–390, *391*, 392
  - income disparities, *18*
  - Japanese miracle, 394–396
  - just-in-time systems, **395**, 402
  - political shifts, 406
  - population aging, 406–407
  - state-aided market economies, **394**, 395, 434
  - U.S. military base, *77*
  - women in parliament, *188*
- *jatis*, **350**–351, *356*
- Al Jazeera, 235, 269–270, 272
- Jesus of Nazareth
  - Ati-Atihan Festival, *451*
  - Christianity and, 248
  - Islam and, 236
  - in liberation theology, **148**–149
  - *Semana Santa* procession, 144
  - St. Thomas and, 352
  - Timkat festival, *320*
- Jews. *See also* Israel
  - Holocaust and, **171**, 250, 263
  - Judaism, 53, **248**
  - yarmulkes, 187
- Jihadists, **254**, 262, 268, 309
- Jobs. *See* Unemployment; Women at work; Work
- Johnson-Sirleaf, Ellen, 287, 308, 309, 310, 318
- Jordan, 13, 16, *18*, *127*, *247*
- Joypur, 346
- Judaism, 53, **248**. *See also* Jews
- Just-in-time system, **395**, 402

- *kabaddi*, *346*
- Kabul, *25*, 369
- Kaizen system, **395**, 402
- Kakadu National Park, *486*
- Kalahari Desert, 285
- Kaliningrad, 230
- Kamkwamba, William, 305–306
- Kandahar, 370
- Karachi, 334, 347, 348, 370
- Karaoke, 401
- Kasanka National Park, 293–294
- Kashmir, *343*, 368
- Kassem, Hisham, 269
- Katrina, Hurricane, *67*
- Katz, Cindi, 256
- Kayford Mountain mine site, *64*
- Kazakhstan
  - Aral Sea issues, 203–204, *204*, *206*
  - Astana, 221–222, *224*, 225
  - Central Asia world region, 10–11, *11*
  - Central Asian states and, 230
  - golden eagle, *223*
  - nuclear testing in, 203
  - steppes, *200*
- KaZantip, *221*
- Kentucky Fried Chicken, 87, 88
- Kenya
  - Maasai Mara game reserve, 293
  - mobile phones, 282–283
  - presidential election issues, 77, 309
  - *Ushahidi* platform and, 270, 282, 282–283, 283
  - women in parliament, 82
- KGB, 217, 220
- Khan, Genghis, 207, 389
- Khao Yai, *423*
- Khartoum, 6, 7, 11, 284
- Khmer Rouge, 433
- Khubz, *242*
- Kiev, 207, *224*
- Kilimanjaro, Mount, 280, 284, 285
- Killing fields, 433
- Kinshasa. *See* Congo
- Kiribati, 460–461
- Kiribati Archipelago, 460
- Knowledge economy, 85–86
- Koli village, 347
- Kolkata (Calcutta), 334, 336, *346*, 347, *348*
- Koran (Qur'an), **240**, 355
- Korean Peninsula, 377, 387, 393, 409, 415
- Kosovars, 192
- Kosovo Liberation Army, *173*, 192
- Krill, *496*
- Kristof, Nicholas, 349

Kryshtanovskaya, Olga, 217
Kuala Lumpur, 147
Kühtai, Alps in, *153*
Kumbh Mela, *346*
Kurds, 52, 174, 271, 272
Kuwait, *11*, *18*, 256, 261, 262, 263, 264, 265, 278
Kuzmic, Vera, 179
Kyoto Protocol, 43, 205
Kyrgyzstan
- Central Asia world region, 10–11, *11*
- Central Asian states and, 230
- Nauryz festival, *221*
- Osh, *230*
- Tulip Revolution, 217
- women in parliament, 82, 229

La Paz, *121*
La Rambla, *183*
La Tomatina, *180*
Labor force. *See* Women at work; Work
Labrador retrievers, 92
Lagarde, Christine, 132
Lahore, 370
Lake Chad, 293, 323
Land bridge, Bering, 70, 72, 122, *122*
Landforms, **44–49**. *See also* Plate tectonics
- East Asia, 376–377
- Europe, 156–157
- Middle and South America, 112–114
- North Africa and Southwest Asia, 237–239
- North America, 61–62
- Russia and post-Soviet states, 199
- South Asia, 330–331
- Southeast Asia, 420–422
- sub-Saharan Africa, 284–285

Land-for-peace formula, 275
Landless movement, 134
Languages
- Bengali, 353
- lingua franca, **320**–321
- Oceania, 489
- Romance, 165–166
- South Asia, 349–350, *350*
- sub-Saharan Africa, 320–321, *321*
- Tibetic–Burmic, *55*
- in world, *55*

Laos, *417*, 433, 435
Las Vegas, *90*, 93, 102
Latin America, 112, 129–132, *130*. *See also* Middle and South America
Latinos, **61**, 93, 97, 102. *See also* Hispanics
Latitude, **4**
Latitude lines (parallels), **4**, 6, 7
Latvia, 156, 171, *188*, 190, 230
Legends (maps), **4**, *5*
Legislatures. *See* Women in parliaments
Lenin, Vladimir, 209, 210
Lettuce harvest, Central Valley, 60, *60*
Li Xia (rural-to-urban migration), 374–375, 399, 401

Liberation theology, **148**–149
Liberia, 287, 308, 309, 310, 318
Libya, Arab Spring and, 265, *270*
Liechtenstein, *188*
Lima, squatter settlement, *146*
Lines of latitude (parallels), **4**, 6, 7
Lines of longitude (meridians), **4**, 6, 7
Lingua franca, **320**–321
Linkages, interregional, **30**, 33
Lithuania, 156, 171, 188, 230
Liu Xiaobo, 406
Livability, 90, 91, 92, 92–93
Living, green, 70
Living wages, **33**, 57
Llamas, 114, *117*
Llanos, 113
Local lives. *See also* Global patterns, local lives; Vignettes
- animals, 92, *117*, *168*, *223*, *247*, *293*, *339*, *384*, *422*, *472*
- festivals, *97*, *144*, *180*, *221*, *253*, *320*, *346*, *388*, *451*, *490*
- foodways, *94*, *135*, *178*, *228*, *242*, *289*, *352*, *383*, *453*, *493*

Local scale, **12**
Location (photo interpretation), 8, 8
Lodz, industrial pollution, *160*
Loess Plateau, 372, 379
Loess soil, 379
Logging. *See also* Deforestation
- Brazilian Amazon, *24*
- clear-cutting, *64*, **65**, 105, 287
- fraud, in Liberia, 287
- New England and Canadian Atlantic Provinces, 105
- North America, *64*, 65
- Olympic National Park, *64*
- Pacific Northwest, 65, 73
- Sarawak forest dwellers and, 418–419

Lomo Saltado, *135*
London, 164, 169
Longitude, **4**
Longitude lines (meridians), **4**, 6, 7
Louisiana wetlands, 58, 61, *61*
Lowlands
- central, 58, 59, 61, 62
- coastal, 59, 62, 140, 157, 284, 326, 463
- Middle and South America, 113–114

Low-to-medium impact areas, *24*
Luang Prabang, 439
Luther, Martin, 167
Luxembourg, *188*

Maasai Mara game reserve, 293
Maathai, Wangari, 287
Macedonia, 156, *188*, 190
Machismo, **144**, 149
Machu Picchu, 122
Mack, Louise, 485
Al Madinah (Medina), 248, 253
Madras. *See* Chennai
Mafia, Russian, 220, 228
Magellan, Ferdinand, 434, 474, 475
Maharashtra, 366
Maid trade, 450, *450*

Maine coon cat, 92
Makkah (Mecca), 248, 250, 253, 296
Malacca, Malaysia, 31, 32, 431, 433
Malamutes, Alaskan, 92
Malaria, 40, 292, 314, 315, 317, 324
Malawi, 305–306
Malaysia
- foreign workers and, 31, 32
- Malacca, 31, 32, 431, 433
- Overseas Chinese in, 414
- women in parliament, 82

Mali, al Qaeda in, 307, 309
Mali Empire, *294*, 296
Malnutrition, global map, 26, *26*
Malta, 155, 164, *188*, 190
Manantan, dam at, 288
Manchuria. *See* China's Far Northeast
Mandela, Nelson, 298, 299
Mao Zedong, 391
Maori people, **463**
Maps, 4–7. *See also* Global maps; Political maps; Regional maps
Maquiladoras, 120, **131**, 133, 435
Marcos, Ferdinand, 454
Mardi Gras, 97, 142, *144*, *221*
Marianismo, **143**–144, 149
Market reforms, in China, 396–397
Marketization, water, 38–40, 69–70, 110, 120, 150
Marshall Islands, 468, 471, 482
Marshall Plan, 171
Marsupials, **466**, *472*
Marx, Karl, 36, 170, 171, 209
Maryland, temperate midlatitude climate, *63*
mashrabiyas, *255*
Maslenitsa festival, *221*
Material culture, 8, *8*
Mato Grosso, 150
Mau Piailug sailing trip, 472
Mauritania, 82, 296, 305, 309, 323
Mauritius, 311, 321, 322
Mawazine festival, *253*
Mayan temple, *109*
Mayans, 117, 144
Mead, Margaret, 492
Media, 220, 269–270. *See also* Internet; Social networking
Medieval period, 166, 172, 182, 249
Medina (Al Madinah), 248, 253
Mediterranean climates, *50*, **157**
Mediterranean Sea, pollution, *161*, 161–162
Medium-to-high impact areas, *24*
Mega-financier, East Asia as, 401–402
Megalopolis, **91**
Megawati Sukarnoputri, 454
Meghalaya, 333
Mekong River, *417*
Melanesia, 462, **473**, 476, 489
Melanesians, 431, **473**, 475
Melville, Herman, 492
Mercantilism, **125**, 128, **168**–169, *169*, 300. *See also* Colonialism
Mercator projection, 6, 7
Mercedes-Benz, 364
Mercer Quality of Living Survey, 92

Mercosur (Southern Common Market), 32, **132**, 151
Meridians. *See* Lines of longitude
Messiah, 248
Mestizos, 112, **125**, 137, 142
Metropolitan areas, 28, *90*, 91, **91**, 93, 101
Mexicali workers, Thompson Electronics, 131
Mexican hairless dog, *117*
Mexico
- China–Mexico relations, 131
- drug trade, *136*
- income disparities, *127*
- indigenous religion, 53
- informal economy, 132, 147, 148, 150
- ISI policies, 128–**129**, 133, 151
- maquiladoras, 120, **131**, 133, 435
- Sierra Madre, *108*, 112, 113
- Yucatán, *50*
- Zapatistas and, 134, 142

Mexico City, 29, 114, 120
Microcredit, 230, **363**, 363, 366, 370
Micronesia, 462, 472, 474, 475, 476, 489
Mid-Atlantic economic core, 71–72
Middle America, **112**, *143*
Middle and South America (world region), 108–151. *See also* Amazon Basin; Hispanics; Indigenous groups
- animals and plants, 113–114, *115*, *117*, 124, *125*
- biodiversity, 113–114
- Catholicism and, 112, 123, 140, 143, 148–149
- climate, 114–116
- climate change vulnerability, 110, 117–119, 120, *121*, 150
- climate zones, *115*
- colonialism and, 110, 112, *126*
- components of, 112, *113*
- conflicts and genocides, 34, *136*
- Crowley's model of urban land use, 147, *147*
- cultural diversity, 142, 149
- deforestation, 110, 117–120, 150
- democratization/conflict issues, 135–138, 150–151
- development, 110, 119–120, 124–126, 150–151
- domesticated plants, *115*, *125*
- drug trade, *137*, 137–138
- economic development phases, 128–131
- economic issues, 126–133
- ecotourism, *119*, 119–120
- environmental issues, 117–121
- European conquest of, 123–124
- extended families, 110, **142**–143, 144, 145, 149, 151
- FDI to, 129–132, *130*
- festivals, *144*
- food production, 110, 133–135, 151
- foreign political involvement in, 138
- GEI, 140–142, *141*

assimilation, **142**, 182, **185**–186, *186*, 191, 412, 413, 414
Astana, 222, *224*, 225
brain drain, 145
defined, **13**
diaspora and, **248**, 303, 306, 319
Europe and, 184–187, *185*
maid trade and, 450, *450*
Middle and South America, 145–148
North Africa and Southwest Asia and, 259–262, *261*
to primate cities, **145**, 146, 221, 224, 313, 448, 450
resettlement schemes, *425*, **442**
Milford Sound, *459*
Military expenditures, North Africa and Southwest Asia, 267
Milosevic, Slobodan, 192
Minas Gerais, *118*
Mindanao, *430*, *441*
Mineral zones, Middle and South America, *133*
Ming dynasty, 389, *390*
Mining. *See specific mines*
Minnesota, central lowlands, 59, 61
MIRAB economy, **479**
Mishra, Veer Bhadra, 337
Mississippi River, 61, *61*, 68, 69, 72–74
Mississippi River delta, 62
Mitch, Hurricane, *44*, *121*
Mixed agriculture, **289**, *291*, 294
Moa bird, 466
Mobile phones
Afghanistan, 357
Arab Spring and, 270
Europe, 180
GISc and, 7
Kenya, 282–283
South Asia, 347, 366
sub-Saharan Africa, 303, *304*
Mobility, in North America, 102
Modest welfare systems, 190
Modi, Narendra, 328, 329
Mohenjo-daro, *344*
Moldova
EU membership and, 174
GEI, 226
GNI per capita PPP, 225
population decline, 223
women in parliament, *229*
Monaco, *188*
Mongolia
gers (yurts), 205
history, 394
pirozhki, 228
steppes in, *51*
Mongols, **207**, 208, 229, 249
Monotheism, 236, **248**, 251
Monotremes, **466**
Monroe Doctrine, 138
Monsoons
defined, **48**
South Asia, 331–332, *334*, *336*, 423, 340
summer, 48, **331**, *331*, 332, 337, 377, 378, *380*, 381
winter, 331, **332**, 334, 377, *464*
Montenegro, 156, *188*, 192
Monti, Valentina, 356
Montserrat, Soufrière Hills Volcano, *109*, 113
Morales, Evo, 131
Mormon religion, *53*
Moro Islamic Liberation Front, 441
Morocco
Casablanca, 261, 277
climate change vulnerability, *45*
Mawazine festival, *253*
steppe, *239*
tagines, *242*
Wadi El Mellah, *233*
Moscow
described, 230
housing shortages, 223
informal economy, Natasha, 215
as primate city, *224*
St. Basil's Cathedral, *208*
Moses, 248
Mosques, 54, 93, *250*, 253, 269, 272
Mountaintop removal, 65
MST movement (Movement of Landless Rural Workers), 134, 142
Mubarak, Hosni, 234, 235, 269
Mudumali Wildlife Sanctuary, 340
Mugabe, Robert, 303, 309, 310
Mughal Empire, 249, 341, 344
Mughals, **341**–342, 344, 345, 370
Muhammad, Prophet, 236, 240, 248, 249, 250, 251, 253, 255, 319
Mujahedeen, 211, 368–369
Mullahs, 254
Mulley, Phillip, 340
Multiculturalism, **52**, 100, 192, 250, 253, 366, 440, 482, 485, 489
Multimodal transport, Europe, 164
Multinational corporations
Apple, 31, 33
Bechtel, 31, 38, 120
BP, 31, 62, 65, 70, 78
Chevron, 31, 110
Cisco, 31
Coca-Cola, 31, 276, 335
defined, **31**
Google, 31, 303, 404, *405*
IBM, 31
living wages and, **33**, 57
Shell, 31, 306
Toyota, 31, 381, 395, 395
Walmart, 31, 33, 85, 87, 87, 364
Multiplier effect, 71
Mumbai (Bombay)
climate change vulnerability, 42–43
monsoons, *336*
sea level rise, 43, *336*
slums, 39
Mumbai attacks, 370
Mummified cat, *247*
Muslim Brotherhood, 36, 234, 268, 269
Muslims. *See also* Islam; Islamic extremists
in China, 411–412
defined, **248**
in Europe, 185–187, *186*
Hindu–Muslim relationship, 352–353
purdah practice, 346, 349, **354**, *355*, 357
Shi'ite (Shi'a), 248, **254**, 272, 276, 278
Sunnis, 219, **253**, 254, 276, 278
Myanmar. *See* Burma
MySpace, 86

NAFTA (North American Free Trade Agreement)
defined, **85**
EU compared to, *175*
Mexico and, 86–87, 93–94, 106, 120, **132**, 133, 134, 151
North America and, **85**, 86–87, 93–94, 106
as regional trade bloc, 32
Nagasaki, 395
Namibia, *51*, 307, 321
Narmada River, Sardar Sarovar Dam, 328–329, 343
Natasha, informal economy and, 215
National water footprints, 39
Nationalism, **170**, 353, **366**–368
Nationalize, **131**
Native Americans
barbeques, *94*
culture of poverty, 97–98
disease/technology and, 71, 75
European arrival/settlements and, 71, 74
Mardi Gras, 97
reservations, 74, 174, 413
NATO (North Atlantic Treaty Organization), *173*, **177**–178, 192
Natural gas resources. *See* Oil/natural gas resources
Nauryz festival, *221*
Nazarbayev, Nursultan, 222
Nazca Plate, 46, 47, 112
Nazis, 171, 210, 273
Nefedova, Tatyana, 223
Nehru, Jawaharlal, 352
Nenets people, *223*
Nepal, 330, 369–370
Netherlands, *82*, 157, *163*, *188*
New Caledonia, *464*
New Delhi, 328, 335, 338, 339, 345, *348*
New England, 71, 72, 74, *94*
New Urbanism projects, 93, *93*
New York City, 20, 72, *73*, *90*
New Zealand
animals and plants, 466
Asian economy and, 477–479
climate change vulnerability, 460, 467, *470*, 471, 494
colonization of, 475
democratization, 35
GEI, 481–482, *484*
gender roles, 460, 490–493, 494
GNI per capita PPP, 481–482, *484*
HDI, 481–482, *484*
invasive species, 466, **467**–468, 469, 472
Polynesia and, 462, **474**, 476, 480, 490, 492, 494

Newroz holiday, 253
NGOs (nongovernmental organizations)
civil society and, 36, 37
defined, 37
Gift of Givers, 36, 37
Human Rights Watch, 229, 405, 450
Oxfam, 37
Somalia, 36
World Wildlife Fund, 37, 424, *425*
Nicaragua, 82
Nicholas II, Czar, 208
Niger, slavery in, 297
Niger River, 288, 292, 296, 306, 309, 323
Nigeria, 306–307, *307*
Nile Delta, 242, 244
Nile River, 233
9/11 attacks. *See* September 11th attacks
The Nine Nations of North America (Garreau), 105
Ningxia Huizu Autonomous Region, 379, 412
Nitrogen dioxide, 68
Nobel Prizes
Aung San Suu Ki, 443, 454
EU, 176
Liu Xiaobo, 406
Maathai, Wangari, 287
Mandela, Nelson, 299
Sen, Amartya, 20
Yunus, Muhammad, 363
Nobility, serfs and, 166
Nogales, Mexico, *121*
Nomadic pastoralists, **205**–206, 207, **292**, 387, 396
Non bread, *228*
Nongovernmental organizations. *See* NGOs
Nonpoint sources of pollution, **201**
Nonviolence, civil disobedience, 343, 345
Norilsk Nickel, *202*, 203
Norms, 52–54
North Africa and Southwest Asia (world region), 232–279. *See also* Arab Spring movements
Arab Spring and, 267–269, 275–276
children and, 256–257
climate, 237
climate change vulnerability, 242–244, *244*, 278
climate zones, 237–239, *238*
colonial regimes in, 252
components of, 236, *236*
conflicts and genocides, *270*, *271*
democratization/conflict issues, *270–271*, 270–272
development, 263–265
economic diversification and, 265–266
economic issues, 262–272
education expenditures, *267*
environmental issues, 239–246
family values, 254
fertility rates, 258
food production, 240–242
GEI, 262, *263*
gender inequalities, 254–257
gender roles, 247–248, 254–255, 258–260
gendered spaces, 254–255
geographic issues, 251–277
geographic setting, 236–251
geopolitical issues, 272–276
globalization and, 259–262, *261*, 263–265
GNI per capita PPP, 262, *263*
HDI, 262, *263*
human impact on biosphere, *241*
human patterns over time, 246–251
Internet and, 269–270
Islam, **236**–237, 248–250, *249*
landforms, 237–239
media and, 269–270
Middle East compared to, 234, 244
migration and, 259–262, *261*
oil/natural gas resources, 263–265, *264*
patriarchal attitudes, **254**, 255, 257, 275
physical patterns, 237–239
political issues, 262–272
political map, *235*
population patterns, 257–262
power and politics, 267–272, *270–271*
press freedom, 269–270
regional map, *232–233*
religions, 248–250, *249*, 251, 253–254
Russia (and post-Soviet states) interaction with, *198*
sociocultural issues, 251–262
terms usage, 237
urbanization, 259–262, *261*
vegetation, 237–239
visual timeline, *250–251*
water issues, 240–242
women's status, *236*, 255–257, 258–260, 270–272
world region characteristics, 234–236
North America (world region), 58–107
animals and plants, 62, 65, 91, 92
climate, 62
climate change vulnerability, 60, 66, *67*, 70, 106
climate zones, *63*
coal mining, *64*, 65
components of, 61
culture of poverty in, 97–98
democratization/conflict issues, 60, *76–77*, 106
development and, 60, 106
economic issues, 82–89
environmental issues, 62–70
ethnicity in, 96–98
European arrival, 70–71, *72*
European settlements, 70–74
families, 60, 99–101
farmland preservation, 91–92
food production systems, 60, 66, 70, 82–84, 106
gender roles, 60, 106
geographic issues, 75–105
geographic setting, 61–75
globalization and, 60, 86–88, 106
green living, 70
green revolution agriculture, 83, 84
greenhouse gas emissions, 60, 66, 70
habitat loss, 62, 65, 91
human impact on biosphere, *64*
human patterns over time, 70–75
landforms, 61–62
logging, *64*, 65
metropolitan areas, *90*, 91, 93, 101
Middle and South America compared to, 110–112
mobility in, 102
mountain ranges, 62
NAFTA and, 86–87, 93–94, 106
oil drilling, 62, 65
peopling of, 70–71
physical patterns, 61–62
political issues, 75–82
political map, *79*
population patterns, 60, 101–105, 106
power and politics, 60, *76–77*, 106
race in, 96–98
regional composition, 74
regional map, *58–59*
religions, 98–99, *99*
service sector and, 85–86
sociocultural issues, 89–101
technology and, 85–86
terms usage, 61
trade, 60, 86–88
transportation networks, *73*, 84–85
urban sprawl, **65**, 79, 89–93, 99, 101, 106
urbanization, 70, 89–93, *90*, 106
visual timeline, *72–73*
water pollution, 60, *68*, *69*, 70, 106
water scarcity, 60, 66, 68–69, 106
West Coast, 70, 74
women, in government-business, 81–82
world region characteristics, 60
North American Free Trade Agreement. *See* NAFTA
North American Plate, 46, 47, 61, 112
North Atlantic Drift, **157**, 158, 159
North Atlantic Treaty Organization. *See* NATO
North Europe, 156, *156*, *188*
North European Plain, *153*, 157, 158, 159, 167, 194, 199, 230
North Korea, *403*
Northern Hemisphere, 6, *6*
Northern Ireland, 54, 173
Northern settlements, colonial U.S., 71
Norway, 82, 174, 188, *188*
Novosibirsk, 221
Nuclear family, **99**–100, 106, 142, 454
Nuclear pollution, 201–203, 471
Nuclear power
Chernobyl disaster, 159, 201, *202*
Europe, 159
Fukushima disaster, 159, 203, 381
North America, 66
Nuclear weapons
on Hiroshima/Nagasaki, 395
India, 368
Kashmir conflict and, 368
Kazakhstan, 203
North Korea, *403*
Pakistan, 368
Syria, 276
terrorists and, 203
testing, Oceania, *468*, 471
Nunavut territory, 74, 81

Obama, Barack, 75, 81, 82, 100, 363, 420
Obama administration, 79
Obamacare, 81
Occidental Petroleum, 110
occupied Palestinian Territories (oPT), 234, 236, **237**, 248, 257, *274*, *275*
Occupy Wall Street, 33, 86
Ocean currents
Benguela Current, 285
global warming and, 42
Peru Current, 114, 115, 285
Roaring Forties, 463, *464*, 467
Oceania (world region), 458–495. *See also* Australia; New Zealand; Pacific Islands; Papua New Guinea
animals and plants, 465–466
APEC and, 480, 481
climate, 463–465
climate change vulnerability, 460, 467, *470*, 471, 494
climate zones, *464*
components of, 461, *461*
conflicts and genocides, *488*
continent formation, 462–463
democratization/conflict issues, 460, 487–489, *488*, 494
development, 460, 494
economic issues, 476–480
El Niño and, **114**, *115*, 422, 424, 427, *464*, 465, *465*
environmental issues, 466–472
export trading partners, *478*
food production, 460, 467, 469, 494
future of, 480
GEI, 481–482, *484*
gender roles, 460, 490–493, 494
geographic issues, 476–493
geographic setting, 461–476
globalization and, 460, 469–472, 494
GNI per capita PPP, 481–482, *484*
HDI, 481–482, *484*
human impact on biosphere, *468*
human patterns over time, 472–476
indigenous groups, *474*
invasive species, 466, **467**–468, 469, 472

- Orange Free State, 297, 298
- Orange Revolution, 217, 218
- Orangutans, 418, 422, 424, 426, 427
- Organic agriculture, 27, 56, 84, 154, 178, 179, 180
- Organically grown, **84**
- Organization of Petroleum Exporting Countries. *See* OPEC
- Organized crime, Russia and post-Soviet states, 214, 220
- Orographic rainfall, **48**, *49*, 237, 257, 463, 465
- Orthodox Christianity, 207, *208*, 221, 229, *320*
- Osh, Kyrgyzstan, *230*
- Ottoman Empire, 166, **249**–250, 251, 279
- Outsourcing, 88, 89, **364**, 366, 435
- Overseas Chinese, 414
- Owens Valley, *49*
- Ownership, of water, 38–40, 69–70, 110, 120, 150
- Oxfam International, 37

- Pacific Islands
  - animals and plants, 466
  - Austronesians and, 431, 473, *474*
  - economic change in, 478–479
  - El Niño and, **114**, *115*, 422, 424, 427, *464*, 465, *465*
  - endemic species, **465**–466, 467
  - Great Barrier Reef and, **462**–463, 467
  - informal economy, 493
  - subsistence affluence and, **479**, 482, 484, 494
- Pacific Mountain Zone, 195, 199
- Pacific Northwest, 65, 73, 74
- Pacific Ocean, Oceania and, 462
- Pacific Plate, 46, 47, 61, 199, 421, 461, 463
- Pacific Rim, **102**, 414
- Pacific Ring of Fire, 377, 379, 421
- Pacific Way, 482, **487**, 494
- Paddy rice, 427
- Pahiyas Festival, 452
- Pakistan
  - Central Asia world region, 10–11, *11*
  - civil society and, 344, 359
  - India–Pakistan Partition, *343*, **343**–344, 345, 352, 366
  - Indus Valley civilization, 327, **340**–341, 345
  - nuclear weapons, 368
  - subcontinent and, **330**
- Palace, of Peter the Great, *208*
- Palau, 462, 479, 482
- Palestinian Territories, 237. *See also* occupied Palestinian Territories
- Palm Jumeirah, 260, *261*, 264
- Pamirs, 205, *206*, 370
- Pampas, Argentina, *109*
- Panama, 138
- Panama Canal, 108, 128
- *Pancasila*, 440, 442, 443, 454
- Panda, giant, 385
- Pangaea
  - Africa and, 284–285
  - breakup of, 46, *46*, 61, 284–285
  - Gondwana and, 46, 462, 463, 465
  - hypothesis, 44
- Papua New Guinea, 469, 471, *473*
- Parachuting, *28*
- Paraguay, *41*, *135*
- Parakeets, *168*
- Parallels. *See* Lines of latitude
- Paramilitaries, 34
- *parilla*, *135*
- Paris, 169, *183*
- Parliaments. *See* Women in parliaments
- Partition, 273
  - India–Pakistan, *343*, **343**–344, 345, 352, 366
- Pastoralism (herding), **205**–206, 207, **292**, 387, 396
- Paternostro, Silvana, 140
- Patkar, Medha, 328
- Patriarchal attitudes
  - Abraham, 248
  - Confucianism, 388–389
  - North Africa and Southwest Asia, **254**, 255, 257, 275
  - Pacific Way, 487
  - Southeast Asia, 454
  - Tibet, 413
- Pay equity. *See* Income disparities
- Penguin, Emperor, *496*
- People
  - animals and, *92*, *117*, *168*, *223*, *247*, *293*, *339*, *384*, *422*, *472*
  - photo interpretation and, 8, 8
- People's Republic of China, 391, 393, 414
- Peopling. *See also* Human patterns over time
  - of Middle and South America, 122–123
  - of North America, 70–71
  - of Oceania, 473–475
  - of Southeast Asia, 431
  - of sub-Saharan Africa, 295–297
- Perestroika, **213**
- Permafrost, **199**, 201
- Persian Gulf. *See also* Gulf states
  - fossil fuels, 237
  - maid trade, 450
  - seawater desalination, 243
- Peru
  - drug trade, *137*, 137–138
  - income disparities, *127*
  - Lima, squatter settlement, *146*
  - *Lomo Saltado*, *135*
  - Peruvian Amazon, *118*
  - UNASUR, **132**, 133, 151
- Peru Current, 114, 115, 285
- Peter the Great, Czar, *208*, 222
- Pharmaceuticals, water pollution and, 69
- Philippine Archipelago, *417*, 431
- Philippine Plate, 46, 47, 421, 463
- Philippines
  - income disparities, *127*
  - Magellan and, 434, 474, 475
  - Mindanao, *430*, *441*
- Phoenix, urban sprawl, *91*
- Phones. *See* Mobile phones
- Photo essays
  - Antarctica, *496*
  - climate change vulnerability, *44–45*
  - climate zones, *50–51*
  - human impact on biosphere, *24–25*
  - North America climate zones, *63*
  - North America human impact on biosphere, *64*
  - power and politics, *34–35*
  - religions, *53*
  - urbanization, *28–29*
- Photogrammetry, 7
- Photograph interpretation, 8, 8
- Physical geography. *See* Geography
- Physical patterns
  - East Asia, 376–379
  - Europe, 156–158
  - Middle and South America, 112–116
  - North Africa and Southwest Asia, 237–239
  - North America, 61–62
  - Oceania, 462–466
  - Russia and post-Soviet states, 198–201
  - South Asia, 330–332
  - Southeast Asia, 420–424
  - sub-Saharan Africa, 284–285
- Piailug, Mau, 472
- Pidgin, **489**
- Pig, decorated, *422*
- Pig farms, Romania, 154–155, *155*
- Pillars of Islamic Practice, 251, 253
- Pinochet, Augusto, 138
- Piracy, 178, 277, *419*
- Pirozhki, 228
- Pizarro, Francisco, *123*
- Place of the Heart, *84*
- Plains, Botswana, *280*
- Plantations
  - defined, **128**
  - haciendas, **128**, 133, 134
  - Middle and South America, 128
  - multiplier effect and, 71
  - oil palm, 22, 31, 117, *118*, 418, 424, 426, 427
  - Southern colonies, U.S., 71, 72
- Plants and animals. *See* Animals and plants
- Plate tectonics
  - African Plate, 46, 61, 157, 239, 421, 424
  - Antarctic Plate, 46, 47
  - Arabian Plate, 46, 239, 285
  - Caribbean Plate, 46, 112, 113
  - Cocos Plate, 46, 47, 112
  - continental plates, 44, 112
  - defined, **44**
  - earthquakes and, 46–47
  - Eurasian Plate, 46, 47, 157, 199, 239, 330, 331, 377, 421, 462

Plate tectonics *(Continued)*
Indian–Australian Plate, 46, 47, 330, 421, 463
Nazca Plate, 46, 47, 112
North American Plate, 46, 47, 61, 112
oceanic plates, 44, 112
Pacific Plate, 46, 47, 61, 199, 421, 461, 463
Pangaea breakup and, 46, *46*, 61, 284–285
Philippine Plate, 46, 47, 421, 463
Ring of Fire and, 47, *47*, 377, 379, 421, 463
South American Plate, 46, 47, 112
subduction zones, **112**
volcanoes and, 46–47
Pluralism, cultural, **451**
Poaching, 293–294
Podrabinek, Alexandr, *197*
Poland
EU membership and, 9, 159
income disparities, *18*
industrial pollution, 159, *160*
women in parliament, 82, *188*
Polar ice caps, 42, 70, 199, 467
Political ecology, **22**
Political issues. *See also* Geopolitics; Power and politics
East Asia, 394–406
Europe, 174–180
Middle and South America, 126–133
North Africa and Southwest Asia, 262–272
North America, 75–82
Oceania, 476–480
Russia and post-Soviet states, 211–216
South Asia, 366–370
Southeast Asia, 440–443
sub-Saharan Africa, 300–310
Political maps
East Asia, *376*
Europe, *156*
Middle and South America, *113*
North Africa and Southwest Asia, *235*
North America, *79*
Oceania, *461*
Russia and post-Soviet states, *197*
South Asia, *330*
Southeast Asia, *420*
sub-Saharan Africa, *284*
Politics. *See* Power and politics
Pollution
industrial, 159, 170, 201–203, 205
nonpoint sources of, **201**
nuclear, 201–203, 471
urban, 201, 205, 339
Polygyny, 256, **317**, *317*
Polynesia, 462, **474**, 476, 480, 490, 492, 494
Population. *See also specific countries*
African Americans, in states, *72*
aging, 13, 17
decline, 13, 223, 225
densities, 13–14, *14*
floating, **375**, 397
global patterns, 12–13, *13*
HIV/AIDS and, 13
J curve, 13, *13*
RNI and, **13**
as thematic concept, 12–17
urban areas, 28–29
wealth and, 15–17
Population patterns
Canada, 60, 101–105
East Asia, 406–411
Europe, 154, 181–184, 193
HIV/AIDS and, 13
Middle and South America, 110, 138–142
North Africa and Southwest Asia, 257–262
North America, 60, 101–105, 106
Oceania, 460, 481–484, 494
Russia and post-Soviet states, 221–227
South Asia, 357–361
Southeast Asia, 443–447
sub-Saharan Africa, 310–317
United States, 60, 97, 101–105
Population pyramids, **14–15**
Belarus, *226*
Canada, *103*
China, *408*
EU, *184*
Germany, *184*
Indonesia, *446*
Iran, *259*
Israel, *259*
Jordan, 14–15, *15*
Kazakhstan, *226*
Kyrgyzstan, *226*
Nigeria, *312*
Pakistan, *359*
Russia, *226*
South Africa, *312*
Sweden, *184*
U.S., *103*
Populist movements, 137, **148**, 440
Portugal
Angola and, *172*
Portuguese–Spanish trade routes (circa 1600), *124*
wars of independence from Spain and Portugal, *123*, 124–125, *126*
women in parliament, *188*
Post-9/11. *See* September 11th attacks
Post-independence period, South Asia, 344–345
Post-Soviet states. *See* Russia and post-Soviet states
potjiekos, *289*
Poutine, *94*
Poverty. *See also* Income disparities; Slums
children, U.S. and Canada, 100
culture of, 97–98
microcredit and, 230, **363**, 363, 366, 370
populist movements, 137, **148**, 440
Poverty Reduction Strategy Papers. *See* PRSPs
Power and politics. *See also* Arab Spring movements; Democratization/conflict issues
East Asia, *403*
Europe, 154, *172–173*, 192–193
Middle and South America, 110, 126, 135–138
North Africa and Southwest Asia, 267–272, *270–271*
North America, 60, *76–77*, 106
Oceania, 460, 487–489, *488*, 494
photo essay, *34–35*
Russia and post-Soviet states, 217–220, *218*
South Asia, 328, 366–370, *367*, 370
Southeast Asia, 440–443, *441*
sub-Saharan Africa, 306–310, *308*
as thematic concept, 33–37
PPP (purchasing power parity), **20**. *See also* GDP per capita PPP
Prague, Czech Republic, 182, *183*
Precipitation
acid rain, **66**, *68*, *80*, *118*, 159, *160*, *202*, *241*, *288*, *338*, 340, 385, 386, *426*, *468*
defined, **48**, **49**
frontal, 48–49
Middle and South America, 115
orographic rainfall, **48**, *49*, 237, 257, 463, 465
rain shadows, 48, 49, 114, *115*
Precolonial South Asia, *341*, 341–342
Predjama Castle, *167*
Press freedom, North Africa and Southwest Asia, 269–270
Primary sector, **19**
Primate cities, **145**, 146, 221, 224, 313, 448, 450
Pripyat, 201, 202, 203
Privatization. *See also* SAPs
defined, **129**, **213**
in Russia and post-Soviet states, 213–214
water, 38–40, 69–70, 110, 120, 150
Prostitution, 140, 228, 230, 297, 316, 391, 409. *See also* Sex workers
Protestantism. *See* Christianity
PRSPs (Poverty Reduction Strategy Papers), 33, 131, 301
Puerto Misahualli, 119
Puerto Rico, *123*
Pull/push phenomenon, **27**, 93, 448
Punjab, 341, 352, 361
Purchasing power parity. *See* PPP
Purdah, 346, 349, **354**, 355, 357
Push/pull phenomenon, **27**, 93, 448
Pussy Riot, 196
Putin, Vladimir, 196, *197*, 217, 220

Qaddafi, Muammar, 243, 309
Al Qaeda, 75, 77, 307, 309, 369
Qanats, 240, 276
Qatar, 22, 234, 235, 254, 255, 257, 258, 259, 262, 263, 265, 266, 269. *See also* Gulf states
Quaternary sector, **19**
Québec, 79, 98
Qur'an (Koran), **240**, 355

Race
cultural geography and, 3, 52–56
defined, **54**
Middle and South America, 142
North America, 96–98
skin color, 54–56, *55*, 96, 98, 142
Racism, 56
Radio Sahar, 356
Railroads
Trans-Siberian Railroad, 214, *215*, 221
U.S., nineteenth-century, 72, *73*
Rain
acid, **66**, *68*, *80*, *118*, 159, *160*, *202*, *241*, *288*, *338*, 340, 385, 386, *426*, *468*
shadows, 48, 49, 114, *115*
Rain forests. *See also* Amazon Basin
Middle and South America, 113, 114, 116, 117, 119, *119*, 128
New Zealand, 466, 469
South Asia, 331
Southeast Asia, 424, 427
sub-Saharan Africa, 285, 323
Rainbow lorikeet, *472*
Rainwater harvesting, 120, 328
Rajasthan, 328, 352
Ramadan, 253, 276, 320, 352
Ramstein Air Base, *77*
Rapa Nui (Easter Island), *108*, 463, 474
Rate of natural increase (RNI), **13**
Recession. *See* Global economic downturn
Red Crescent, 37
Red Cross, 37
Red Sea, 232, 234, 239, 248
Reed Dance, *320*
Refugees
in Australia, 475, 485, *488*
Chechnya, 219–220
Darfur, *244*
Haitian, *76*
Kenya, *291*
in New Zealand, 475, 485
North Africa and Southwest Asia, 260, 262
in Pacific Islands, 472
Palestinian, 260, 262, 273, 275
Rohingya, *441*
from Sri Lanka, 340, 367, 369
in sub-Saharan Africa, 309
Sudan, 45, 308
Syrian, 35
in Taiwan, 414
in Uganda, *45*
U.S. immigrants, 95
in Yemen, *244*, 245
from Zimbabwe, 310
Regional conflicts, **368**
Regional development schemes, Soviet, 214
Regional Economic Communities, African, 304
Regional geography. *See* Geography

Rocky Mountains, 58, 62, 74
Rohingya refugees, *441*
Roma (Gypsies), **171**, 186
Roman Catholicism
European conquest and, 123
liberation theology, **148–149**
marianismo, **143**–144, 149
Middle and South America, 112, 123, 140, 143, 148–149
in world, *53*
Romance languages, 165–166
Romania, 13, 154–155, *155*, 174
Rome, ancient, 165–166
Rondonia, *118*
Rose Revolution, 217
Rose-ringed parakeet, *168*
Ross Sea, *496*
Rotary International, 36, 37
Rotich, Juliana, 282, 282–283, 310
Rub'al Khali dunes, 232
Rudimentary welfare systems, 190
Rugby, 489, *491*
RugMark, 349
Rural-to-urban migration
brain drain, 145
demographic transition, *16*, **16**–17, 311–312
Li Xia and, 374–375, 399, 401
North Africa and Southwest Asia, 260
sub-Saharan Africa, 313
Rus, 207
Russia (Russian Federation)
BRIC countries and, 131–132, 212
continental climate, *51*
defined, **198**
ethnic groups, *219*
G8 and, 212
greenhouse gas emissions, *41*, 204–205
income disparities, *18*
population decline, 13
Russian Empire, 196, 205, 207–208, *209*, 230
women in parliament, 82
Russia and post-Soviet states (world region), 194–231
agriculture in, 215–216, *216*
Central Asian states, 230
climate change vulnerability, 204–205, *206*
climate zones, *200*
climates, 199–201
Cold War and, 210–211, *211*
components of, 198, *198*
conflicts and genocides, *218*
corruption, 214, 220
cultural diversity, 217–220, *219*
democratization/conflict issues, 217–220, *218*
East Asia and, *198*
economic issues, 211–216
economic reforms, 212–215
environmental issues, 201–205
ethnic groups, *219*
Europe and, *198*
European Russia, 199
food production, 215–216
GEI, 225–227, *227*
gender inequalities, 226–229
geographic issues, 211–230
geographic setting, 198–211
GNI per capita PPP, 225–227, *227*
HDI, 225–227, *227*
human impact on biosphere, *202*
human patterns over time, 205–211
industrial areas and transport routes, *215*
informal economy, 214–215
landforms, 199
natural gas resource areas, 212, *213*
North Africa and Southwest Asia and, *198*
nuclear pollution, 201–203
oil resource areas, 212, *213*
organized crime, 214, 220
physical patterns, 198–201
political issues, 211–216
political map, *197*
population patterns, 221–227
power and politics, 217–220, *218*
privatization in, 213–214
regional map, *194–195*
religious revival, 229
resource extraction, 203
Russian Far East, 230
Sochi, 205
sociocultural issues, 220–230
South Asia and, *198*
Soviet Union and, 196–197
terms usage, 198
transport routes and industrial areas, *215*
unemployment, 214
vegetation, 199–201
visual timeline, *208–209*
water issues, 203–204
women legislators, *229*
world region characteristics, 196–197, 198
world regions' interaction with, *198*
World War II and, 210–211
Russian arctic, 202
Russian Empire, 196, 205, 207–208, *209*, 230
Russian Far East, 230
Russian Federation, **198**. *See also* Russia
Russian imperial expansion, 198, *209*
Russian Mafia, 220, 228
Russification, 217, 219, 222

Saarschleife, uplands, *153*
Sack garden, *314*
Sahara Desert, 232, 234, 235, 237, 239, 284
Sahel region, **284**, 285, 287, 292, 293, *294*, 295, 297, 319, 323
Sa'id, Nurbegum, 356
Salafism, **268**, 276, 277
Salinization, **240**, 243, 245, 362
San Marino, *188*
Sandy, Hurricane, 42, 43, *45*, *121*
Sankore Mosque, *294*
Santería, 148, 319
Santiago, Chile, *146*, 147
Santos, Juan Manuel, *34*
SAPs (structural adjustment policies), **32–33**, 129–130, 131, 301
Sarajevo, *173*
Sarawak forest dwellers, 418–419
Sardar Sarovar Dam, 328–329, 343
Sarkozy, Nicolas, 188
Saudi Arabia
desert climate zone, *239*
Gulf states and, 254
income disparities, *18*
Makkah (Mecca), 248, 250, 253, 296
Medina (Al Madinah), 248, 253
Scales, **4**, 5, **12**, **175**
Schengen Accord, **184**
Schleswig-Holstein, North European Plain, *153*
Schurz, William, 150
Scotland, moist midlatitude climate zone, *158*
Scott-Amundsen South Pole Station, *496*
Sea level rise
Antarctica and, *496*
Bangladesh, 336
India, *45*
Mumbai, 43, 336
Netherlands, 157, *163*
North Africa and Southwest Asia, 242, *244*
North America, 66, *67*
Oceania, 460, 467, *470*, 471
South Asia, 328, 332, 334–335, 336, 340, 370
Seasons, 49
Seawater desalination, **243**
Seclusion, female, **254**, 255
Secondary sector, **19**
Secoya people, 110, 112
Sectors, of economy, 19
Secular states, **268–269**, 276, 278, 366
Segupova, Marina, 196
Self-reliant development, **305**, 315
Self-sufficiency
food, Southeast Asia, 456
industrial, South Asia, 363
regional, **396**
Semana Santa procession, 144
Semiarid/arid climates, *50*
Semiautonomous regions, 198, 217, 219, 231
Sen, Amartya, 20
Sensitivity, 42, 43, 44, 45, 56, 120. *See also* Climate change vulnerability
September 11th attacks, 37, 75, 77, 307, 309, 369
Sequestration, carbon, **285**
Serbia, *173*
Serfs, 166
Service and Food Workers Union, 35
Service economies, 85–86, 179–180
Services, as economic sector, **19**
Sex, **17**, **19**. *See also* Gender
Sex tourism, 438, 446

Sex trafficking, 96, 228, 418, 443, 446, 447, 456
Sex workers, 228, 301, 316, **349**, 356, 446, 456
Seychelles, 311, 321, 322
SEZs. *See* Special economic zones
Shadows, rain, 48, 49, 114, *115*
Shakira, *253*
Shamans, 229
Shanghai, 28, *28*, 398, 399–400
Shantytowns, 28, 145, 150, 313, 315, 347
Shari'a, **253**, 254
Shark, white, *468*
Sheep, New Zealand, *468*
Sheikh Hamad bin Khalifa, 255, 269
Sheikha Moza, 255
Shell Oil Company, 31, 306
Shifting cultivation, *24*, *25*, **122**, 126, *133*, **289**, 427, *428*
Shi'ite Islam, 53
Shi'ite (Shi'a) Muslims, 248, **254**, 272, 276, 278
Shintoism, 53
Shiraz grapes, *493*
Shishmaref, *67*
Siakor, Silas, 287, 310
Siberia
  Central Siberian Plateau, *195*, 199
  defined, **199**
  West Siberian Plain, *195*, 199, 221
  western, 207, *208*, 223
Siberian huskies, 223
Sichuan Province, 374, 375, 383, 401
Sierra Leone, 287, 310
Sierra Madre, *108*, 112, 113
Sierra Nevada Mountains, 49
Sikhism, 53, 344, 350, *351*, **352**
Silk Road, 23, 30, 207, *207*, *208*, 211, 230
Silt, **114**
Singapore, 414, 421, 437
Singhalese nationalism, 369
Skin color, 54–56, *55*, 96, 98, 142
Skydiving, 28
Slash and burn agriculture, 427, *428*
Slavery, 71, 73, 287, 295, 296, 296–297, 310
Slavs, 171, **207**
Slovakia, 157, 176, *188*
Slovenia, 82, *167*, 179, *188*
Slums, **27**–28
  favelas, 28, **145**, 147, 148, 150
  ghettos, 28
  Mumbai, 39
  Nogales, Mexico, *121*
  shantytowns, 28, 145, 150, 313, 315, 347
  squatter settlements, 28, 145, 146, 147, 482, 483
  sub-Saharan Africa, *314*
  unplanned urbanization, 28, 43, 110, 120, 145, 150, 313
  wastewater and, 39
  water pollution and, 39–40
Smart growth, 92–93, 106
Smog, **66**, 68, 385, 422
Snakehead fish, Asian, 65
Soccer, 295, 299, 453
Sochi, Olympics in, 205
Social democratic welfare systems, 189–190
Social forestry movement, 339, 340
Social networking. *See also* Internet
  business and, 86
  Facebook, 86, 228, 270, 276
  MySpace, 86
  Twitter, 83, 86, 270
  YouTube, 86, 164, 255, 356, 471, 489
Social protection. *See* Social welfare/protection systems
Social safety nets, **80**–81, 82, 88, 93, 190, 191, 220, 225
Social Security Trust Fund, 88
Social welfare/protection systems, *189*, 189–190
Socialist economy. *See* Central planning
Sociocultural issues
  East Asia, 406–414
  Europe, 180–191
  Middle and South America, 138–149
  North Africa and Southwest Asia, 251–262
  North America, 89–101
  Oceania, 485–487
  Russia and post-Soviet states, 220–230
  South Asia, 345–361
  sub-Saharan Africa, 310–322
Soil salinization, **240**, 243, 245, 362
Solidarity movement, 174
Solomon Islands, 471, 473, 479
Somalia, *36*
Songkran, *451*
Soufrière, 31, 32
Soufrière Hills Volcano, *109*, 113
South Africa. *See also* North Africa and Southwest Asia; Sub-Saharan Africa
  European colonialism, 297–299, 298
  income disparities, *127*
  Mandela and, 298, 299
  Olympics and, 299
  subtropical climates, *50*
South America. *See also* Middle and South America
  defined, **112**
  haciendas, **128**, 133, 134
  Internet usage, *143*
South American Plate, 46, 47, 112
South Asia (world region), 326–371
  Buddhism, 350, 351, **352**, 353, 369
  caste system, 341, **350**–351
  children in, 347, 349
  city life, 346–347
  climate, 331–332, *333*
  climate change vulnerability, 328, 332–340, *336*, 370
  climate zones, 331–332, *333*
  colonialism and, 342–343
  components of, 330, *330*
  conflicts and genocides, *367*
  deforestation, 337–340
  deindustrialization, 342–343
  democratization/conflict issues, 328, 370
  development, 328, 370
  economic issues, 361–366
  environmental issues, 332–340
  ethnic groups, 349–350
  farming systems, *362*
  fertility rates, 357, 358
  food production, 328, 361–365, 370
  GEI, 359–361, *360*
  gender imbalance, 328, 370
  geographic issues, 345–370
  geographic setting, 330–345
  glacial melting, 328, 330, 332, 334, 337, 340, 370
  globalization and, 328, 370
  GND per capita PPP, 359–361, *360*
  green revolution agriculture, 361–365
  HDI, 359–361, *360*
  human impact on biosphere, 337–340, *338*
  human patterns over time, 340–345
  Indian Ocean tsunami (2004), 37, 331, 421, 438, 442, 443
  Indian peninsula and, 330
  Indus Valley civilization, 327, **340**–341, 345
  industrial self-sufficiency, 363
  invasions into, 341–342
  landforms, 330–331
  languages, 349–350, *350*
  microcredit and, **363**, 363, 366, 370
  monsoons, 331–332, *334*, *336* 423, 340
  Mughals, **341**–342, 344, 345, 370
  physical patterns, 330–332
  political issues, 366–370
  political map, 330
  population patterns, 357–361
  post-independence period, 344–345
  power and politics, 328, 366–370, *367*, 370
  precolonial, *341*, 341–342
  regional map, *326–327*
  religions, 350–352, *351*
  religious nationalism, 353, **366**–368
  Russia (and post-Soviet states) interaction with, *198*
  sociocultural issues, 345–361
  subcontinent and, **330**
  terms usage, 330
  thematic concepts overview, 328, 370
  urbanization, 328, 347–349, *348*, 370
  vegetation, 331–332
  village life, 346
  visual timeline, *344–345*
  water issues, 328, 332–337, 370
  women's status, 353–357
  world region characteristics, 328–330
South Carolina, Charleston, 31, 32, 59, 93
South China Sea, 377, 421
South Europe, 156, *156*, *188*
South Korea, 88, 394, 395, 434, 456
South Ossetia, 220
South Pole Station, Amundsen-Scott, *496*
Southeast Asia (world region)
  agricultural patterns, *428*
  ASEAN and, 32, **437**, 438, *440*
  climate, 422–424, *423*
  climate change vulnerability, 424–429, *430*
  climate zones, 422–424, *423*
  colonization, 431–433, *432*
  components of, 420, *420*
  conflicts and genocides, *441*
  cultural pluralism in, **451**
  deforestation, 424, *425*, 427
  democratization/conflict issues, 440–443, *441*
  economic issues, 434–439
  emigration, 448, 450
  environmental issues, 424–434
  fertility rates, 443, *444*, *445*
  food production, 427–428
  food self-sufficiency, 456
  GEI, 446, *447*
  geographic issues, 434–454
  geographic setting, 420–434
  globalization and, 448, 450
  GNI per capita PPP, 446, *447*
  HDI, 446, *447*
  HIV/AIDS in, 446, 447
  human impact on biosphere, *426*
  human patterns over time, 431–433
  IMF bailout, 436–437
  landforms, 420–422
  maid trade and, 450, *450*
  Overseas Chinese in, 414
  peopling of, 431
  physical patterns, 420–424
  political issues, 440–443
  political map, *420*
  population patterns, 443–447
  power and politics, 440–443, *441*
  regional map, *416–417*
  religions, 451, *452*
  sex tourism, 438, 446
  Sundaland and, 421, *421*, 431, 466
  terms usage, 420
  terrorism and, 443
  thematic concepts overview, 418
  tourism, 437–438
  urbanization, 448–450, *449*
  vegetation, 422, 424
  visual timeline, *434–435*
  water issues, 428–429
  world region characteristics, 418–420
Southeastern Anatolia Project, 245
Southeastern Europe, 37, 156, 174, 190
Southern Common Market. *See* Mercosur
Southern Hemisphere, 6, *6*

agriculture, early history, 295–297
apartheid and, 269, *295*, 297, **298**–299, 300, 301, 307, 309, 312, 323
civil society and, 299, 300, 310, 324
climate, 285
climate change vulnerability, 285–293, *291*
climate zones, 285, *286*
colonialism, *298*
commodities and, **300**–301, 302, 307
components of, 284, *284*
conflicts and genocides, *308*
debt issues, *302*
deforestation, 285–289
democratization/conflict issues, 306–310, *308*
desertification, 288, 292–293, 294
economic issues, 300–310, *302*
environmental issues, 285–294
ethnicity and, 320–321
European colonialism, 297–299, *298*
European slave trade, 295, *296*, 296–297
famines, 290, 305
fertility rates, 258, 311, 312
gender issues, *317*, 317–318, *318*
geographic issues, 300–322
geographic setting, 284–300
globalization and, 300–303
GNI per capita PPP, 321–322, *322*
groundwater, 290–291, **292**, *292*, 294, 313
HDI, 321–322, *322*
HIV/AIDS in, 13, 312, 313, 315, *316*
human impact on biosphere, 285–289, *288*
human patterns over time, 294–300
independence, from colonialism, 299
infant mortality rates, 282, 310, 311, 312, 317, 324
informal economies, 301, 306
landforms, 284–285
language groups, 320–321, *321*
mixed agriculture, **289**, *291*, 294
mobile phone usage, *304*
peopling of, 295–297
physical patterns, 284–285
political issues, 300–310
political map of, *284*
population patterns, 310–317
power and politics, 306–310, *308*
Regional Economic Communities, 304
regional map, *280–281*
religions, 318–320, *319*
Sahel region, **284**, 285, 287, 292, 293, *294*, 295, 297, 319, 323
sociocultural issues, 310–322
terms usage, 284
thematic concepts overview, 282, 323–324
trade, 295–297, *296*, *305*
trade organizations, *305*
urbanization, 312–317, *314*
vegetation, 285
visual timeline, *294–295*
water issues, 313
wildlife issues, 293–294
women in legislatures, 310
world region characteristics, 282–284
Subsidence, 62, 70, 243
Subsidies, **178**–179
Subsistence affluence, **479**, 482, 484, 494
Subsistence agriculture, 23, **289**, 290, 479
Subsistence economies, **16**
Subtropical climates, *50*
Suburbanization, 91
Suburbs, **91**
Success, Gospel of, 159, 319, 320, 321
Sudan, *45*, *244*
Sudras, 350, 351
Suharto, 436, 440, 442, 454
Suicide bombings, 273, 275
Sukarno, 440, 442
Sukarnoputri, Megawati, 454
Summer monsoons, 48, **331**, *331*, 332, 337, 377, 378, *380*, 381
Sundaland, 421, *421*, 431, 466, 473
Sunni Islam, *53*
Sunni Muslims, 219, **253**, 254, 276, 278
Suriname, *118*
Sustainability, **22**
East Asia, 381–384
organic agriculture, 27, 56, **84**, 154, 178, 179, 180
sustainable agriculture, **27**, 120, 126, 193
Swamp taro, 460, 469, 493
Sweden, *18*, *82*, *188*
Switzerland, *82*, 174, 184, *188*
Syr Darya, 203
Syria
Arab Spring and, *35*, *271*, 275–276
civil war, 35
nuclear weapons, 276
water resources, 240

Tagines, *242*
Tahrir Square, 234, 235
Taiga, **199**
Tailings, 64, 65
Taiwan
export-led growth, **394**, 456
minorities in, 413–414
state-aided market economies, **394**, 395, 434
uncertain status of, 393–394
Taj Mahal, 340, 341, *344*
Tajikistan
Central Asian states and, 230
glacial melting, *206*
GNI per capita, 225
women in parliament, *229*
Taliban, 229, **355**–357, 369
Tamil Nadu, 337, 340, 359
Tanzania, *291*
Tapirs, 110
Tariffs
ASEAN and, 32, **437**, *438*, *440*
CAP and, **178**–179
defined, **32**
Taro, 460, 469, 493
Tasmanian Devil, 468
Tatarstan, 198, 219, 229
Taylor, Charles, 287, 303, 310
Tea crops, for South Africa, *291*
Technology. *See also* Internet; Mobile phones
Native Americans and, 71, 75
North America and, 85–86
Tectonic plates. *See* Plate tectonics
Temperate climates, *50*
Temperate midlatitude climate, *63*, **157**, 159
Temperature, **48**–49
Temperature-altitude zones, *50*, **114**, 116, *116*
Temporary workers, 184, 265
Tenochtitlan, 122
Teotihuacan, *122*
Territorial claims, Antarctica, *496*
Terrorism. *See also* Islamic extremists; September 11th attacks
bin Laden, 75, 369
Chechnya and, 219–220
Mumbai attacks, 370
nuclear weapons and, 203
Al Qaeda, 75, 77, 307, 309, 369
Southeast Asia and, 443
suicide bombings, 273, 275
War on Terror, 75, 268, 369
Tertiary sector, **19**
Tet celebration, *451*
Texaco, in Ecuador, 110, *111*
TFR. *See* Total fertility rate
Thailand, *127*, *422*, *451*, 453
Thali, *352*
Thematic concepts, **12**–43. *See also* Climate change; Development; Food production; Gender; Globalization; Population; Power and politics; Urbanization; Water; *specific countries*
Theocratic states, 210, **268**–269, 278
Thomas, Saint, 352
Thompson Electronics, 131
Three Gorges Dam, 329, 373, 385, *386*, 387, 415
Tiananmen Square, 402, *403*, 404
Tianjin, 409
Tibet (Xizang), *51*, *53*, 412–413, *413*
Tibetic–Burmic languages, *55*
Tierra caliente, 114, *116*, 140
Tierra del Fuego, *108*, *115*
Tierra fria, 114, *116*, 140
Tierra helada, 114, *116*, 140
Tierra templada, 114, *116*, 140
Tigers, *339*
Tigris River, 165, 239, 245, 246, 247, 278
Tijuana, *144*
Timbuktu (Tombouctou), 277, 296

Timgad, *166*
Timkat festival, 320
Tito, Josip Broz, 192
TivaWater jug, *315*
Tombouctou (Timbuktu), 277, 296
Tong, Anote, 460–461
Tornadoes, 42, 62, 465
Tortellini, *178*
Torture, 137, 229, 368, 433
Total fertility rate (TFR), **13**
Tourism
  Antarctica, *496*
  Asian Highway, 438, 439
  Caribbean, 149
  ecotourism, *119*, 119–120, 293–294, 471
  Europe, 179–180
  Hawaii and, 479–480
  Maghreb, 277
  Oceania, 471, 479–480
  Pacific Northwest, 65
  sex, 438, 446
  Southeast Asia, 437–438
  U.S.–Canada, *80*
Toyota, 31, 381, 395, *395*
Trade
  ASEAN and, 32, **437**, *438*, *440*
  Asia-Pacific region, **414**, 480
  BRIC and, 131–132, 212
  Canada, 60, *80*, 86–88
  Central Asian states, 230
  commodities, **300**–301, 302, 307
  EU and, 177, *177*
  fair, **33**, 106
  free, **32**–33
  maid trade and, 450, *450*
  Middle and South America, *124*
  North America, 60, 86–88
  regional, 32, 132–135, 154, 304, 305, 309, 437–439
  SEZs and, **399**, 400, 414, 435
  Silk Road, 23, 30, 207, *207*, *208*, 211, 230
  Spanish–Portuguese trade routes (circa 1600), *124*
  sub-Saharan Africa, 295–297, *296*, 305
  U.S., 60, *80*, 86–88
  WTO, **32**, 37, 212, 303, 305, 404, 477, 480
Trade blocs
  FTAA, 87, 132
  in Middle and South America, 132–133
  UNASUR, **132**, 133, 151
Trade deficit, **87**, 176
Trade winds, **114**, 115, 465
Trafficking
  defined, **228**
  drug, 96, 137
  sex, 96, 228, 418, 443, 446, 447, 456
Trail of Tears, 73, 74
Tram, *183*
Trans-Alaska pipeline, *64*, 65
Trans-Amazon Highway, 117
Trans-European transport network, *165*
Transit system, Paris, *183*
Transmigration schemes, *425*, **442**
Transportation
  EU, *165*
  Europe, 164, *165*
  Middle and South America, 148
  North America, 84–85
  Russia and post-Soviet states, *215*
  U.S. (nineteenth century), *72*, *73*
Trans-Siberian Railroad, 214, *215*, 221
Transvaal, 297, 298
Treasury securities, foreign holders of, 402, *402*
Tree cropping, 119
Tree hugging (*Chipko*), 339, 340
Tripoli, 270, 277
Tropical cyclones, 331, 377–378
Tropical forests, 22, 119, 287, 340, 419, 424, 469
Tropical wet/dry climates, *50*
Trudeau, Pierre, 78
Tsawwassen First Nation, 92
Tsunamis, 37, 331, **377**, 379, 381, *417*, 421, 442, 443
Tuamotu Archipelago, 474, 480
Tulip Revolution, 217
Tundra, **199**
Tunisia, 240, *241*, 250, 256, 268
Tupai Island, *459*
Turkey
  Arab Spring and, 174
  Central Asia world region, 10–11, *11*
  continental climate zone, 239
  dam construction, 245, *245*
  income disparities, *127*
  Kurds and, 52, 174, 271, 272
  Ottoman Empire, 166, **249**–250, 251, 279
  Southeastern Anatolia Project, 245
  water resources, 240
  women in parliament, 82
Turkmenistan
  Central Asia world region, 10–11, *11*
  Central Asian states and, 230
  steppes, 201
  unemployment, 214
  women in parliament, 229
Tutsis–Hutus conflict, 320
Tuvalu, *470*, 471, 479, *483*
Twitter, 83, 86, 270
Two child policy, 409, 445
Two-state solution, 273, 275
Tymoshenko, Yulia, 217
Typhoid, 313
Typhoons, **377**–378, 379, 384, 429, *430*, 431

UAE (United Arab Emirates), *11*, *41*, 234, 254, 255, 260, 318
Udon Thani, 448
Uganda, *45*, *291*
UK. *See* United Kingdom
Ukraine
  borscht, 228
  Chernobyl nuclear power disaster, 159, 201, *202*
  Kiev, 207, *224*
  Orange Revolution, 217, 218
  population decline, 13, 223
  Pripyat, 201, *202*, 203
  Tymoshenko and, 217
  women in parliament, *229*
Ulan Bator, 387, 394
Uluru (Ayers Rock), 458, 462
Umbanda, 148, 149
Umhangla, *320*
UN. *See* United Nations
UN Food and Agricultural Organization. *See* FAO
UN Framework Convention on Climate Change, 43
UNASUR, **132**, 133, 151
UNCLOS (United Nations Convention on the Law of the Sea), 471
Underdevelopment, Middle and South America, 124–126
Underemployment, **214**, 260, 362, 375
Undernourishment. *See* Malnutrition
Unemployment
  Central Valley, 60, *60*
  deindustrialization and, 177, 342–343
  EPZs and, **131**, 399, 400, 414, 435
  Russia and post-Soviet states, 214
  Sardar Sarovar controversy and, 328–329
  underemployment, **214**, 260, 362, 375
UNHDI. *See* HDI
Union Carbide Corporation, 340
United Arab Emirates. *See* UAE
United Kingdom (U.K.)
  British Isles, 31, 475, 485
  G8 and, 212
  income disparities, *18*
  midlatitude climate, *51*
  women in parliament, *188*
United Nations (UN), **37**
United Nations Convention on the Law of the Sea (UNCLOS), 471
United Nations Human Development Index. *See* HDI
United States (U.S.)
  abroad, 75–78
  acid rain and, 66
  Afghanistan and, 75, *77*
  agribusiness, 68, *69*, **82**–83
  agriculture in, 83
  Canada–U.S. relationships, 78–82
  children, poverty, 100
  China–U.S. economic relations, 87–88
  climate change vulnerability, *45*, 60, 66, *67*, 70
  colonial period, 70–71, *72*, *169*
  democratic system, 79–80
  education, 60, 100, *100*
  ethnicity in, 96–98, 97
  EU economy compared to, 176
  families, 99–101
  food production, 60, 66, 70, 82–84
  G8 and, 212
  GEI, *104*, 104–105
  geopolitics and, 37, 75
  global economic downturn and, 88–89
  GNI per capita PPP, *104*, 104–105
  greenhouse gas emissions, *41*, 60, 66, 70, 117
  HDI, *104*, 104–105
  health-care system, 80–81, *81*
  households, 100, *100*
  immigration and, *80*, 93–96, *94*
  income disparities, *18*, *127*
  infant mortality rates, *81*, 96, 189
  Iraq War, 75, 76, 278
  Iraq–U.S. troubles (1963–2013), 272
  military base, in Japan, *77*
  mobility in, 102
  oil dependency, 76, 78, *78*
  oil imports, *78*
  politics, gender and, 81–82
  population patterns, 60, 97, 101–105
  presence, in Middle and South America, 137–138
  religions, 98–99, 99
  terms usage, 61
  trade, 60, *80*, 86–88
  transportation networks, 73, 84–85
  urbanization, 89–93, *90*
  virtual water and, 68
  water pollution, 60, 68, 69, 70
  water scarcity, 60, 66, 68–69
  women in parliament, 82, *188*, *229*
  women's earnings, 89, 89
Unplanned urbanization, 28, 43, 110, 120, 145, 150, 313. *See also* Slums
Untouchables, 350, 351
Uplands, Saarschleife, *153*
Ural Mountains, 157, *194*, 199, 207, 230
Urban growth poles, **147**
Urban pollution, 201, 205, 339
Urban sprawl, 28, **65**, 79, 89–93, 99, 101, 106
Urban squatters, 28, 145, 146, 147, 482, 483
Urbanization
  Canada, 89–93, *90*
  China, *375*
  defined, **27**
  East Asia, 397–402
  Europe, 154, 166–172, *183*, 192–193
  habitat loss and, 62, 65, 91, 338
  Middle and South America, 110, 145–148, *146*, 151
  mobility and, 102
  North Africa and Southwest Asia, 259–262, *261*
  North America, 60, 89–93, *90*, 106
  Oceania, 460, 482, 483, 494
  photo essay, *28–29*
  population distribution, 28–29
  push/pull phenomenon of, 27, 93, 448
  South Asia, 328, 347–349, *348*, 370
  Southeast Asia, 448–450, *449*

banana hacienda, Costa Rica, 133–134
Central Valley unemployment, 60, *60*
corn production, Hunnicutt case, 83–84
ecotourism, Ecuador, 119
electric heaters (Beijing), 385
favelas, in Fortaleza, 145
green food production, Slovenia, 179
informal economy, Moscow, 215
Koli village, 347
Mexicali workers, Thompson Electronics, 131
norms, Honolulu, 52
Piailug sailing trip, 472
Secoya people, 110, 112
Soufrière, 31, 32
Tsawwassen First Nation, 92
windmill, Malawi, 305–306
Village Bank. *See* Grameen Bank
Village life, South Asia, 346
Virginia, *59*, 71
Virtual water, **38**, **40**, 68, 154, 161, 162, 164, 192
Visual timeline
East Asia, *390–391*
Europe, *166–167*
Middle and South America, *122–123*
North Africa and Southwest Asia, *250–251*
North America, *72–73*
Oceania, *476–477*
Russia and post-Soviet states, *208–209*
South Asia, *344–345*
Southeast Asia, *434–435*
sub-Saharan Africa, *294–295*
Vladivostok, 214
Volcanoes
Ring of Fire and, 47, *47*, 377, 379, 421, 463
Soufrière Hills Volcano, *109*, 113
tectonic plates and, 46–47
Volga River, *194*, 199, 207, 209, 230
Volovada region, *200*
Voodoo, 148, 319

Wadi El Mellah, *233*
Wadi Mur, *244*
Wages, living, **33**, 57
Waitangi Day, *490*
Wall Street, Occupy, 33, 86
Walmart, 31, 33, 85, 87, *87*, 364
War on Drugs, 137–138
War on Terror, 75, 268, 369
Wars. *See specific wars*
Water. *See also* Aquifers; Dams; Glacial melting; Irrigation; Virtual water
access to, 38–40
bottled water, 39
climate change and, 42–43
fossil, 68, 240, 242, 243, 292
groundwater, 290–291, **292**, 292, 294, 313
marketization, 69–70, 110, 120, 150
ownership, 38–40, 69–70, 110, 120, 150
qanats, 240, 276
quality, 38–39
as thematic concept, 37–40
Water footprints, 38, 39, 162, 335*t*
Water issues/scarcity
Canada, 60, 66, 68–69
East Asia, 379, 381
global map of, *239*
Middle and South America, 110, 120, *121*, 150
North Africa and Southwest Asia, 240–242
North America, 60, 66, 68–69, 106
Oceania, 460, 494
Russia and post-Soviet states, 203–204
South Asia, 328, 332–337, 370
Southeast Asia, 428–429
sub-Saharan Africa, 313
U.S., 60, 66, 68–69
Water pollution, 37–40
Canada, 60, *68*, 69, 70
Europe, 154, *161*, 161–162, 192
industrial, 159, 170, 201–203, 205
Mississippi River, *68*
North America, 60, *68*, 69, 70, 106
Russia and post-Soviet states, 203–204
slums and, 39–40
South Asia, 335, 337
urban, 201, 205, 339
urbanization and, 39–40
U.S., 60, *68*, 69, 70
Wealth. *See also* Income disparities
mercantilism and, **125**, 128, **168**–169, *169*, 300
oligarchs and, 196, **212**, 220, 231
population growth rates and, 15–17
transfers, colonies to Europe, *169*
Weather, **47**. *See also* Climates
Weathering, **47**, **49**
Wegener, Alfred, 44
Welfare states, **171**
Welfare systems. *See* Social welfare/protection systems
West Bank, 237. *See also* occupied Palestinian Territories
West Bank barrier, **275**, *276*
West Coast, North America, 70, 74
West Europe, 156, *156*, *188*, 190
West Germany, 171, 174
West Papua, 53, *430*, *441*, 442
West Siberian Plain, *195*, 199, 221
West Virginia, Kayford Mountain, *64*
Western Europe, countries, 156
Western Hemisphere, 6, *6*
Western Siberia, 207, *208*, *223*
Wet rice cultivation, **382**–383, *426*, 427
Wetland loss, Louisiana, *58*, 61, *61*
Where/why questions, geography, 2–3
White shark, *468*
"Why Do They Hate Us?" (Eltahawy), 234–235
W...
W...
Wi...
Win...
Wom...
cul...
dow...
femal... ...cide, 328, 357, 359, 370, 408
female seclusion and, 254, 255
FGM and, 256, **318**
grandmother hypothesis, 18
maid trade and, 450, *450*
maquiladoras and, 120, **131**, 133, 435
marianismo and, 143–144, 149
purdah practice, 346, 349, **354**, 355, 357
sex trafficking, 96, 228, 418, 443, 446, 447, 456
Taliban and, 229, **355**–357, 369
veiling practice, 255, 256, 354
Women at work
Canada, 89, *89*
double day, 18, **187**, 227
Europe, 187, *187*
feminization of labor, **435**, *436*
North Africa and Southwest Asia, *257*, 269
prostitution, 140, 228, 230, 297, 316, 391, 409
Russia and post-Soviet states, 229
sex workers, 228, 301, 316, **349**, 356, 446, 456
U.S., 89, *89*
Women in parliaments, 81–82, *82*, *188*, *229*, 310
Women's status, 17–19, *19*
Arab Spring and, 234–235, *235*, 255, 279
Islam and, 255–257
North Africa and Southwest Asia, *236*, 255–257, 258–260, 270–272
Russia and post-Soviet states, 228–229
South Asia, 353–357
Work. *See also* Unemployment; Women at work
in China, *375*
in countries, *187*
in global economy, 31–32, *32*
guest workers, **184**–185, 259, *259*, 260, 277, 407
living wages, **33**, 57
outsourcing, 88, 89, **364**, 366, 435
temporary workers, 184, 265
underemployment, **214**, 260, 362, 375
World Bank, 32–33, 129, 131, 301
World cities, 169, 414
World Cup, *295*, 299, *491*
World maps. *See* Global maps
World regional scale, **12**

. *See also* East
...ope; Middle and
...n America; North Africa
...nd Southwest Asia; North
America; Oceania; Russia and
post-Soviet states; South Asia;
Southeast Asia; Sub-Saharan
Africa
World Trade Organization (WTO), **32**, 37, 212, 303, 305, 404, 477, 480
World War I, 171–172
World War II
  Europe and, 171–172
  Holocaust and, **171**, 250, 263
  Nazis, 171, 210, 273
  North Africa and Southwest Asia and, 250
  nuclear families after, 99–100
  Russia and post-Soviet states and, 210–211
  Soviet Union and, 210
  U.S. presence in Europe after, *77*
World Wildlife Fund, 37, 424, *425*
WTO. *See* World Trade Organization
WuDunn, Sheryl, 349

Xian, 18, 412
Xinjiang Uygur Autonomous Region, 412
Xizang. *See* Tibet
xolo, Mexican hairless dog, *117*
Yangtze River. *See* Chang Jiang
Yanukovych, Viktor, 217
Yarmulkes, 187
Yellow River. *See* Huang He
Yemen, *11*, *244*, 245, 318
Yokohama, 399, 410, 472
Yom Kippur War, 264, 275
YouTube, 86, 164, 255, 356, 471, 489
Yucatán, Mexico, *50*
Yucatan Lowlands, *109*
Yudhoyono, Susilo Bambang, 424
Yugoslavia breakup, 172, *173*, 177
Yunnan Province, 413
Yunnan-Guizhou Plateau, 376, 421
Yunus, Muhammad, 363
Yurts (gers), 205
zakat, 253
Zambia, 294
Zapata, Emiliano, 134
Zapatistas, 134, 142
Zebras, 293
Zharkeshov, Yernar, 221–222
Zheng He, 389, 415
Zimbabwe, 294, 295, 303, 309, 310
Zionists, **273**
Zones. *See specific zones*
Zoroastrians, *221*, 250, 253
Zuma, Jacob, 323

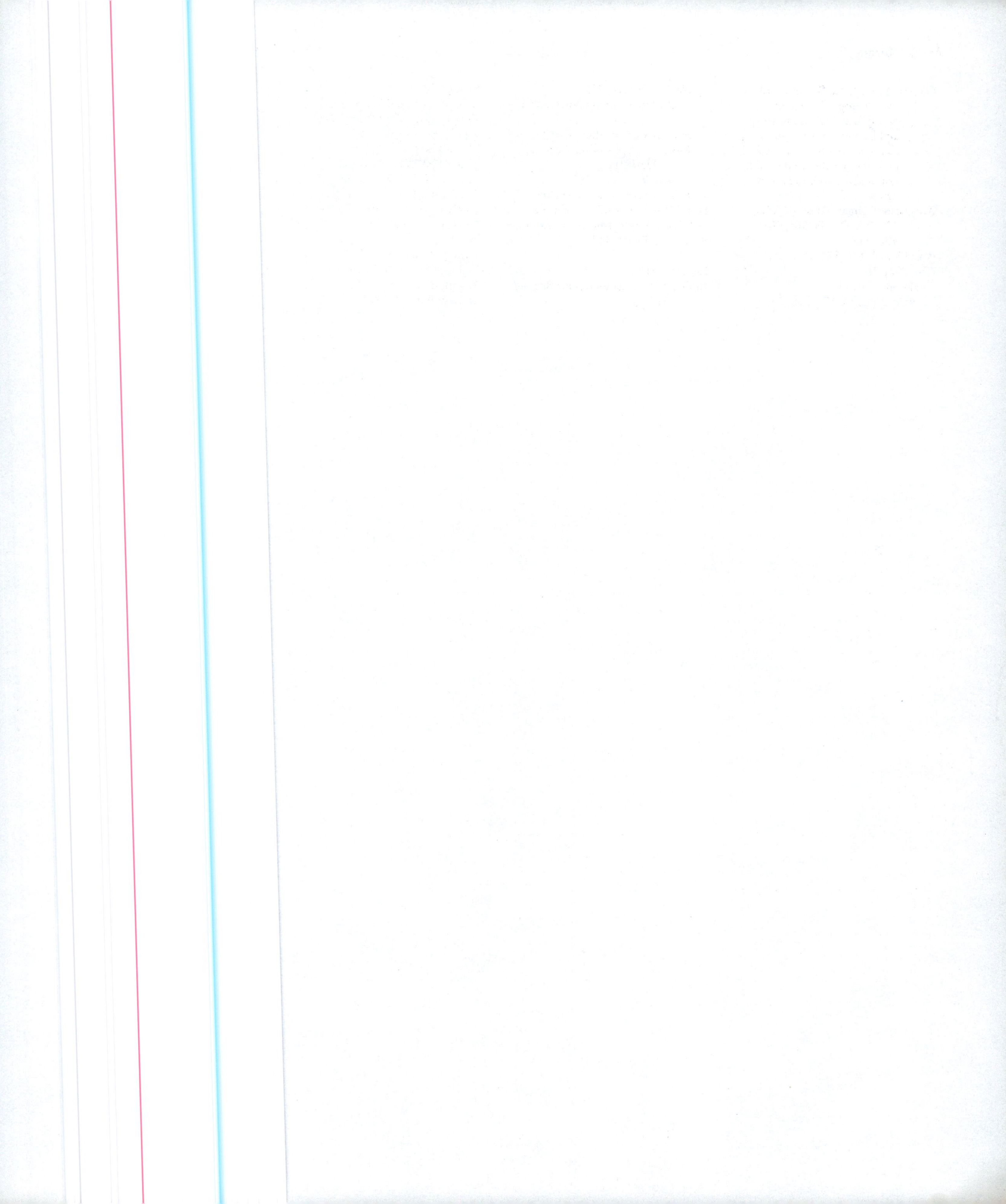